26th International Conference on Adaptive Structures and Technologies (ICAST 2015)

Kobe, Japan
14 - 16 October 2015

ISBN: 978-1-5108-2189-7

Printed from e-media with permission by:

Curran Associates, Inc.
57 Morehouse Lane
Red Hook, NY 12571

Some format issues inherent in the e-media version may also appear in this print version.

Printed by Curran Associates, Inc. (2016)

ICAST 2015
Optimum System Design Lab
School of System Informatics
Kobe University
1-1, Rokkodai, Nada
Kobe 657-8501
Japan

Phone: +81-78-803-6249
Fax: +81-78-803-6391

info@icast2015.com

Additional copies of this publication are available from:

Curran Associates, Inc.
57 Morehouse Lane
Red Hook, NY 12571 USA
Phone: 845-758-0400
Fax: 845-758-2633
Email: curran@proceedings.com
Web: www.proceedings.com

26th International Conference on Adaptive Structures and Technologies (ICAST 2015)

Kobe, Japan
14 - 16 October 2015

Program

14th October 2015

Session 1 / Morphing (Room: CON 1)
Chairs: G. A. Lesieutre and T. Iwasa

Keynote 1 (Room: CON 1)
Chair: H. Okubo

Poster introduction (Room: CON 1)
Chairs: K. Ishimura and N. Kogiso

13:00 **All posters are introduced in this session.**

Session2 / Applications (Room: CON 1)
Chairs: W. H. Liao and H. Furuya

15th October 2015

Session9 / Vibration (Room: CON 1)
Chairs: H. Tzou and K. Senda

18:30 Dinner (KOBE SHU-SHIN-KAN BREWERIES)

16th October 2015

OCTOBER 14-16, 2015
KOBE, JAPAN

The 26th International Conference on Adaptive Structures and Technologies

Co-organized by

- Kobe University
  - -Graduate School of Engineering
  - -Graduate School of System Informatics
- The Japan Society of Mechanical Engineers
- Japan Society for Aeronautical and Space Sciences

Supported by

- Kobe Convention & Visitors Association
- Tsutomu Nakauchi Foundation

Message from the 26th Conference Chair

WELCOME

Welcome to the 26th International Conference on Adaptive Structures and Technologies (ICAST 2015) held in Kobe, Japan. This is the fifth time that ICAST has been held in Japan since the conferences at Nagoya (1991, 2000), Sendai (1994), and Wakayama (1997), and the first time for it to be held in Kobe.

The topics being dealt with in ICAST are wide-ranging: materials, structures, actuators, sensors, mechanisms, systems, kinematics, dynamics, controls, designs, and so on, from basics to applications; the list goes on and on. Many research fields are relevant to the conference, since the study of adaptive structure has an inherently interdisciplinary nature. In ICAST 2015, we can discover, learn, and discuss the depth and the width of the research into and application of adaptive structures through more than eighty presentations and two keynote lectures.

The city of Kobe is located west of Osaka and is well known for its international port. Its multicultural nature has been attracting many visitors from all over the world. Kobe is also famous for its food and drink, such as Kobe Beef and Japanese Sake. From the beginning of the series of conferences, the intimate atmosphere of ICAST has fostered international collaboration and friendships among its participants. We anticipate that the food, the drink, and the cultural nature of Kobe will contribute to the fruitful mutual interaction of the conference participants.

We would like to express our sincere appreciation to the Graduate School of Engineering and Graduate School of System Informatics, Kobe University, for their co-organization of the event. We would like to thank the Kobe Convention & Visitors Association and the Tsutomu Nakauchi Foundation for their tangible and intangible support. We would like to acknowledge various forms of support from the Japan Society for Aeronautical and Space Sciences (JSASS) and the Japan Society of Mechanical Engineers (JSME).

We wish you a fruitful stay in Kobe, Japan.

Kazuyuki Hanahara, Conference Chair

Committes

Local organizing committee

Conference Chair

- Kazuyuki Hanahara, Kobe University

Vice Chair

- Hiroshi Furuya, Tokyo Institute of Technology
- Tadashige Ikeda, Nagoya University

Secretary General

- Hiroaki Tanaka, National Defense Academy

Secretary

- Hiraku Sakamoto, Tokyo Institute of Technology
- Yohsuke Nambu, Osaka Prefecture University

Advisory Board

- M. C. Natori, JAXA/ISAS
- Hiroshi Okubo, Kanagawa Institute of Technology

Member

- Kazuhiko Adachi, Chubu University
- Takeshi Akita, Chiba Institute of Technology
- Hiroshi Asanuma, Chiba University
- Ken Higuchi, Muroran Institute of Technology
- Kosei Ishimura, JAXA/ISAS
- Takashi Iwasa, Tottori University
- Atsushi Kanda, JAXA/IAT
- Nobuhisa Katsumata, Muroran Institute of Technology
- Naoko Kishimoto, Setsunan University
- Nozomu Kogiso, Osaka Prefecture University
- Kanjuro Makihara, Tohoku University
- Yasuyuki Miyazaki, Nihon University
- Yoji Okabe, University of Tokyo
- Yoshiro Ogi, Oxford Space Systems
- Nobukatsu Okuizumi, JAXA/ISAS
- Kazuya Saito, University of Tokyo
- Takahide Sakagami, Kobe University
- Kei Senda, Kyoto University
- Atsuhiko Senba, Meijo University
- Masato Tamayama, JAXA/IAT
- Ayako Torisaka Kayaba, Tokyo Metropolitan University
- Tomohiro Yokozeki, University of Tokyo

International organizing committee

- H. Abramovich, Technion Israel Institute of Technology, Israel
- A. Bergamini, EMPA/ETH Zürich, Switzerland
- D. Brei, University of Michigan, USA
- J. Cagnol, Ecole Centrale Paris, France
- M. Collet, CNRS, Lyon, France
- A. Concilio, CIRA, Naples, Italy
- L. F. Campanile, EMPA, Switzerland
- Y. Chen, National Research Council, Canada
- I. Chopra, University of Maryland, USA
- R. De Breuker, Delft University of Technology, The Netherlands
- P. Ermanni, ETH Zurich, Switzerland
- A. Ertürk, Georgia Tech, USA
- A. Flatau, University of Maryland, USA
- M. I. Friswell, Swansea University, UK
- F. Gandhi, Rensselaer Politechincal University, USA
- P. Gaudenzi, University of Rome, Italy
- Z. Gürdal, Skolkovo Institute of Science and Technology, Moscow, Russian Federation
- D. Guyomar, INSA Lyon, France
- K. Hanahara, Kobe University, Japan
- H. Hanselka, Fraunhofer Institute, Darmstadt, Germany
- D. J. Inman, University of Michigan,USA
- X. X. Jiang, Canadian Space Agency, Canada
- S. J. Kim, Seoul National University, Korea
- D. C. Lagoudas, Texas A&M University, USA
- C. K. Lee, National Taiwan University, Taiwan
- I. Lee, Korea Advanced Institute of Science and Technology, Korea
- J. S. Leng, Harbin Institute of Technology, China
- G. A. Lesieutre, The Pennsylvania State University, USA
- W. H. Liao, Chinese University of Hong Kong, China
- M. Martinez, Delft University of Technology, The Netherlands
- H. P. Monner, German Aerospace Center (DLR), Braunschweig, Germany
- M. C. Natori,JAXA/ISAS, Japan
- F. Nitzsche, Carleton University, Canada
- R. Ohayon, Conservatoire National des Arts et Métiers, France
- H. Okubo, Osaka Prefecture University, Japan
- J. Qiu, Nanjing University of Aeronautics and Astronautics, China
- D. A. Saravanos, University of Patras, Greece
- W. J. Staszewski, AGH University of Science & Technology, Poland
- N. Takeda, University of Tokyo, Japan
- M. Trindade, University of Sao Paula, Brasil
- H. S. Tzou, Zhejiang University, China
- D. H. Wang, Chongqing University, China
- K. W. Wang, University of Michigan, USA
- N. M. Wereley, University of Maryland, USA
- A. Wickenheiser, George Washington University, USA
- W. J. Wu, National Taiwan University, Taiwan

ICAST2015 #040

Variable Camber Morphing Wing Using Corrugated Composites

Tomohiro Yokozeki[1*], Hiroki Takahashi[1], and Yoshiyasu Hirano[2]

[1] Department of Aeronautics and Astronautics, The University of Tokyo, Tokyo, Japan
[2] Aeronautical Technology Directorate, Japan Aerospace Exploration Agency, Tokyo, Japan

Abstract

This paper describes the development of variable camber morphing wing, which is mainly composed of corrugated structures. The morphing wing with both leading edge and trailing edge morphing sections is proposed and the prototype model is designed by consideration of finite element structural analysis with actuation mechanisms and aerodynamic analysis. Through wind tunnel experiment with the manufactured prototype model, smooth actuation without harmful deformation under 20m/s air-flow is demonstrated. The observed deformation shape is well correlated with simulated shape by analysis. Thereby the feasibility of the present morphing wing mechanism and design process are verified.

1. INTRODUCTION

Recently, the cutback of airplane fuel consumption is strongly required because of the soaring fuel cost and increasing awareness of environment issues. Morphing wing concept is one of the possible solutions to response this requirement and has been flourishingly investigated all over the world. As the ideal morphing wing can continuously change its shape like the wing of a bird, it has a lot of advantages such as improvement of aerodynamic performance by adapting the shape corresponding to flight profiles and structural weight reduction by eliminating complex mechanisms of ailerons and flaps. In addition, the noise reduction and the expansion of flight envelope are also expected.

Many types of morphing wings are suggested and studied from both aerodynamic and structural points of view focused on shape optimization to achieve superior aerodynamic performance. Owing to the recent innovative development of smart materials and adaptive structures, both novel mechanisms and structural concepts (bi-stable composite structure, shape memory alloy, piezoelectric actuators, compliant structure, etc.) have been suggested as morphing wing structures These researches include conceptual studies, numerical simulations by computational fluid dynamics (CFD) or finite element method (FEM), elementary experiments using partial scale-down prototype models, demonstration in the wind tunnel, and flight tests of aircraft with morphing wings. Development of morphing wings would result from complex consideration of aerodynamic, structural, and control dynamics. Therefore, drive demonstration under air-flow and aerodynamic performance evaluation by actual wing size model are essential to validate the feasibility of the morphing wing.

Although morphing wing has a lot of advantages as mentioned above, there are few practical applications. The major problem of the morphing wing structure is the conflicting requirement that the structure is enough stiff to bear large aerodynamic loadings but, at the same time, flexible to deform easily.

* info@icast2015.com

Corrugated structure, which is composed of a wavy plate (see Figure 1), was proposed as a candidate structure for variable camber morphing wings[1]. When corrugated structure is applied to wing structure in the manner that corrugation is arranged in the span direction, it can withstand large load such as lift force in the span direction. In contrast, flexible deformation is possible in the chord owing to the large anisotropic stiffness characteristics. This is why corrugated structure is well suited to morphing wing structure.

Numerous studies on corrugated structure have been carried out to investigate the mechanical behavior of corrugated structure using analytical modelings, finite element analyses and experimental approaches. Kress and Winkler[2] analyzed equivalent stiffness of a corrugation using unit cell analysis based on curved shell theory. Xia et al.[3] developed an analytical homogenization model including geometric parameters. Dayyani et al.[4] experimentally and numerically investigated mechanical behavior of trapezoidal corrugated core. In case of large deformation behavior of corrugated structures, Thill et al.[5] conducted a large tensile experiment, and Schmitz and Horst[6] investigated large pure bending deformation. Thill et al.[7] also discussed the applicability of corrugated structure to morphing skin. Yokozeki et al.[8] manufactured a variable camber morphing wing prototype, which was composed of corrugated structure in trailing edge (TE) flap section, and demonstrated the actuation of morphing mechanism by the wind tunnel test and excellent aerodynamic performance of morphing wing compared to the conventional wing with a hinged flap. Corrugated structure enables us to make the wing more flexible without discontinuity of the structure owing to the large anisotropic mechanical properties.

The above-mentioned achievements of corrugated structures motivated the authors to develop a more flexible and high-performance morphing wing. It is known that leading edge (LE) morphing as well as trailing edge (TE) morphing is efficient to gain large lift coefficient. In this study, a morphing wing model with large morphing section, including both LE and TE sections as shown in Figure 2, is developed. A large sized morphing wing prototype model using corrugated structure is designed based on both structural and aerodynamic considerations. Finally, we demonstrated the feasibility of the present morphing wing including actuation mechanism and design methodology through the wind tunnel experiment.

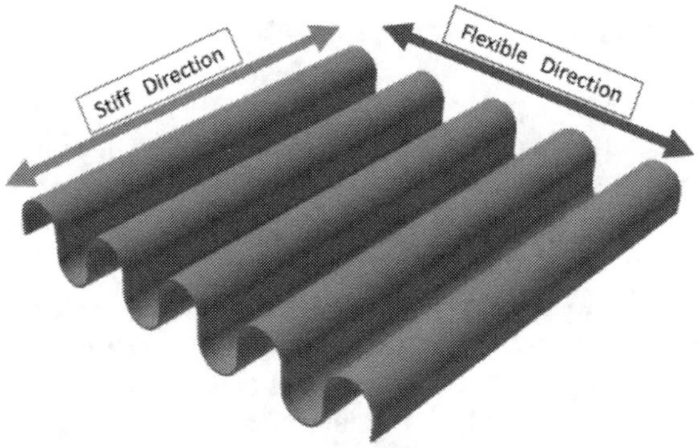

Figure 1. Corrugated structure with large anisotropy

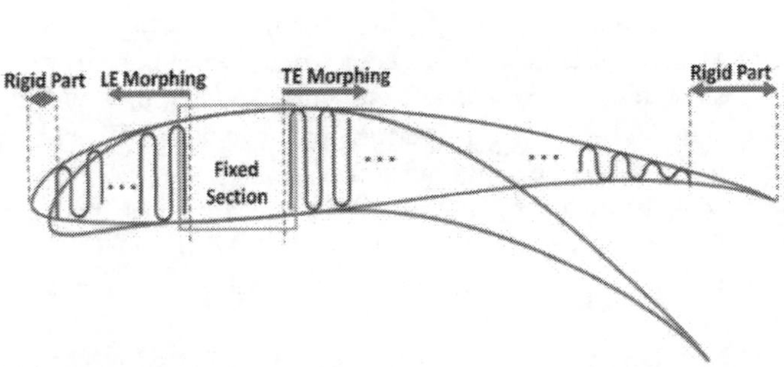

Figure 2. Morphing wing composed of corrugated structure: both LE and TE sections can be morphed independently

2. CONCEPT OF MORPHING WING

Wortmann FX63-137 airfoil (maximum thickness ratio: 13.7%, camber ratio: 5.97%), which is often adopted to motor gliders, is selected as the airfoil profile. The present morphing wing model is two dimensional wing with neither taper nor dihedral angle. It is assumed that a main stiff spar constitutes the fixed section from 20% to 35% of the chord length, and LE and TE morphing sections are placed before and after the fixed section respectively, as shown in Figure 2. This means that main load bearing structures are priori given and not designed in this study, and only flexible sections are to be analyzed in the present study. An actuation mechanism for morphing is incorporated in the fixed section.

The core of morphing section is made of corrugated structure of which the overall envelope coincides with the shape of FX63-137 airfoil. The upper skin is glued to corrugated structures and the lower surface is covered with a flexible sheet. The sheet is assumed to maintain smooth surface and not to prevent the morphing section from deforming. The most front part in LE morphing (from 0% to 5% chord length) and the most rear part in TE morphing (from 90% to 100% chord length) are not morphed in the present wing.

Wire tension is utilized to drive the morphing sections. An edge of the wire is attached to the corrugated structures, and the wire is placed through the corrugated structure near the bottom wing surface. The other edge of the wire is connected to the servomotor mounted inside the fixed section. Roll-up motion of the wire by the motor induces the tension to the wire, resulting in downward deformation of the morphing sections. This drive mechanism was demonstrated in our previous research[8], and is applied to the present morphing wing.

Deflection angle of the morphing wing (called morphing angle) is explained herein. Consider a plain-flap airfoil which has the same flap length as the morphing section length as shown in Figure 3. Flap angle can be defined as the rotating angle of the line connecting between the hinge position and the TE position. Similarly, hinge position is hypothetically considered in morphing wing as shown in the lower figure in Figure 4, and the morphing angle is defined as the rotating angle of the line connecting between the hinge position and the TE position of morphing wing. In the case of LE morphing, the same definition is adopted as the TE morphing angle. In the following sections, LE morphing angle is called θ_1, and TE morphing θ_2 (positive downwards).

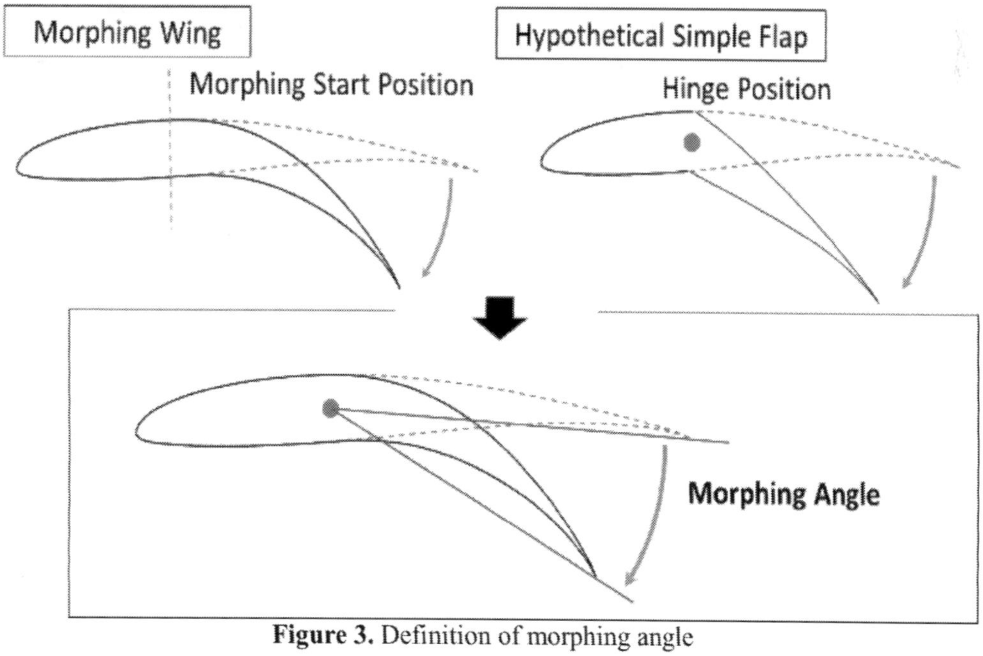

Figure 3. Definition of morphing angle

The objective of this study is to design and manufacture the morphing wing model for drive demonstration in the wind tunnel. The model dimension is fixed as 600mm in chord direction and 1200mm in span direction considering the sizes of the wind tunnel and the drive mechanisms installed in the wing. The model has the aluminum box structure as the fixed section (from 20% to 35% chord length) to carry drive mechanisms and morphing sections using corrugated structure. Two sets of servomotor and drive mechanisms (e.g. drive shaft, gearbox, pulleys) are mounted in the aluminum box part, and they enable the model to actuate both morphing sections independently. Control wirings are taken out from the bottom surface of the box part, and connected to the controller outside the wind tunnel. Two main struts, which are connected to the box part, are utilized to support the prototype on the balance in the wind tunnel test facility. An additional rear strut is also connected to the wing spanwise tips to support the prototype as shown in Figure 4, in order not to prevent the rear strut from interfering action of rear morphing. Sufficiently large acrylics elliptical wing tip boards (800mm×700mm×5mm thickness) are attached to both wing spanwise tips to make the two-dimensional aerodynamic flow.

The corrugated structures and upper surface are manufactured using carbon fiber reinforced plastic (CFRP) laminates, as easy fabrication of structure with complex shapes and variable thickness is possible. It is also expected that light and stiff CFRP enables structural weight reduction more than metal structure. Morphing sections are designed based on the following analytical methods.

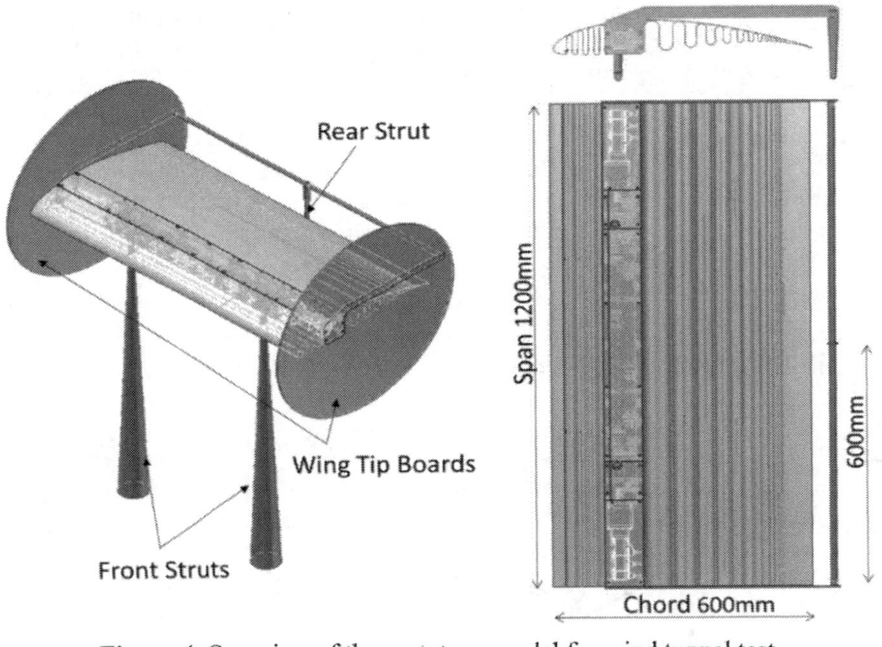

Figure 4. Overview of the prototype model for wind tunnel test

3. STRUCTURAL AND AERODYNAMIC ANALYSIS

3.1 Numerical Method

Stress and deformation of the present morphing wing is numerically calculated using the commercial finite element method (FEM) software (MSC Marc). The stress/deformation states in the morphing sections and the necessary tension of the actuation wires can be estimated with consideration of geometric nonlinearity (i.e. large deformation) and contacts between the actuation wire and the corrugated structure. As the present wing model is uniform in the span direction, two-dimensional numerical model is constructed using shear deformable beam elements (Beam elements in straight sections have about 2mm length, and the curved and contact sections consist of finer meshes). The material is assumed to behave linear elastic, and the aerodynamic force is also considered as described later. Assuming both LE and TE corrugated structures and upper skin are made of CFRP plain woven fabrics in the manner that weft fibers are aligned to the corrugation direction, material properties are set as Young's modulus E=52.5 [GPa], Poisson's ratio ν=0.3 and density ρ=1600 [kg/m3]. In the FEM model, fixed condition was applied to the aluminum box part which is enough stiff. The actuation wire is also modeled as beam elements. One end of the wire is tied to the node in the corrugated structure, and wire movement is simulated by increasing the displacement of the node at the other end of the wire. The wire-guides are modeled as two beam elements attached on the corrugated structure. The distance between the two beam elements is set to be 0.5mm. The frictionless contact option is set between the wire and the wire-guides.

In the aerodynamic analysis, wing profile analysis software XFLR5 is used to obtain pressure distribution over the airfoil. The deformed shapes of the morphing wing obtained by FEM without aerodynamic forces are input, and two-dimensional flow analysis is conducted by two-dimensional panel method. Reynolds number (Re) is set to 8.0×105 (corresponding air speed is 20 m/s), and the compressibility of flow is neglected. The pressure distribution over the deformed airfoil obtained from XFLR5 is, then, fed back into the FEM model. This method enables us to simply estimate the deformation of the prototype model suffering from aerodynamic force. The flexible sheet (i.e. lower surface) of the model is not taken into account in FEM, because the flexible sheet is not a structural component. It is assumed that the aerodynamic pressure on the lower surface is superimposed on the aerodynamic pressure of the upper surface in FEM.

This study focused on two design parameters: corrugation radius and thickness along chord direction. TE corrugation wave number was fixed, and the radius distribution was varied. In addition, thickness of the corrugated structure was changed at the middle point of corrugated structure to make the TE section of the wing deform more flexibly. Although it is possible to consider many radius combinations of the corrugated structure, the present investigation decided that the radius changed in the constant ratio along the chord direction to simplify the analysis. A lot of corrugation geometry and thickness distribution has been investigated using the present methodology by trial and error. As a result, the model utilized decreasing radius in the ratio of 0.9 along the chord direction and changing thickness at the second corrugation wave (see Figure.6) to obtain larger lift coefficient. The five corrugation radii located in the most TE part are 3.75mm by the restriction of manufacturing. The constant corrugation radius (6.4 mm) and thickness for the LE corrugated structure was adopted, because the effect of varied corrugation parameters on the deformed shape is not significant compared to TE morphing section.

In addition, the preliminary analysis suggested that, if only one pair of wire (wire 1) is connected to TE corrugation, aerodynamic load compels the TE morphing section to deform drastically. The TE morphing section is successfully prevented from the unexpected deformation by adding another pair of wire (wire2) connected to the middle point of the TE corrugation structure as shown in Figure 6. In order to drive two pairs of wires by one servomotor, wire1 and wire2 are pulled by a constant ratio using two pulleys which have different diameters. A single similar wire actuating system is applied to the LE morphing section. Details of FEM model with actuation mechanism are presented in Figure 5 with a magnified description which shows the modeling of wires including boundary conditions.

3.2 Numerical results

The deformed shapes of corrugated structure and upper skin obtained by FE analysis (including aerodynamic loads) are compared in Figure 6 for the cases of (θ_1, θ_2) = (0°,0°), (5°, 10°), and (10°, 20°). As can be seen in these figures, the wing surface is smoothly curved. The final deformations are input to XFLR5, and pressure distributions (Pressure coefficient, C_p, is defined as a dimensionless number given by dividing the wing loading by dynamic pressure.) are obtained at each morphing angle, as shown in Figure 7, which suggests that the smooth variable camber deformation is of benefit to obtain large lift coefficient. As a result of structural analysis considering aerodynamic pressure, the stresses/strains in the morphing wing structures and wires are lower than the allowables of the candidate materials. It is confirmed that morphing wing can be driven under the condition of 20m/s air flow without harmful static aeroelastic deformation.

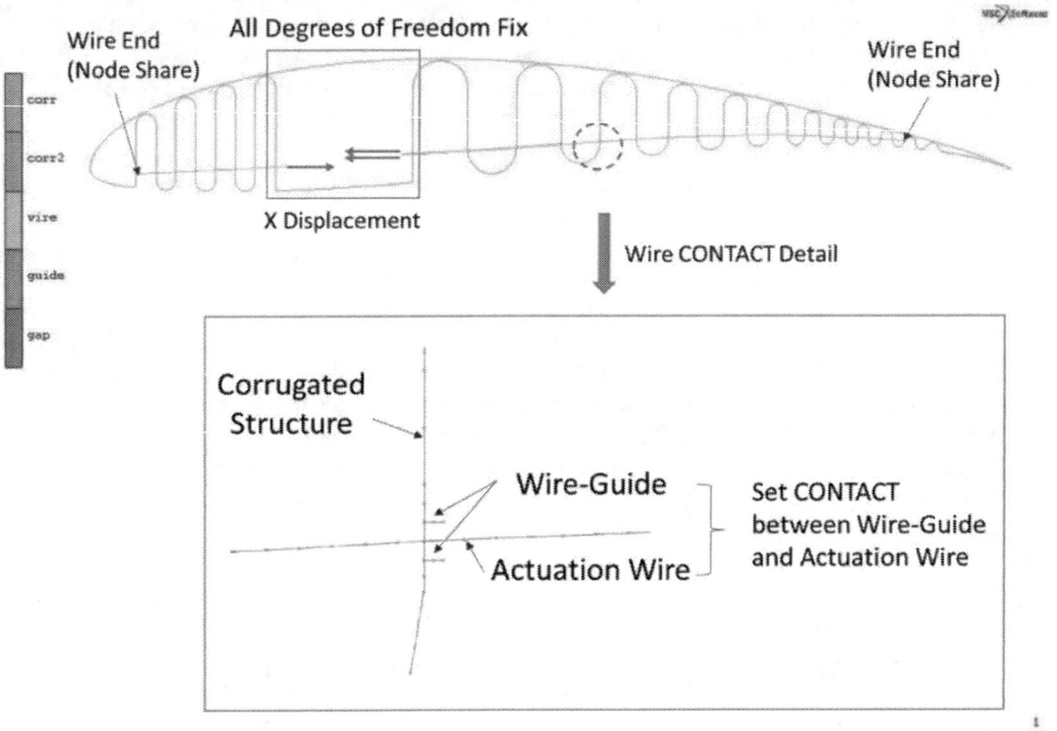

Figure 5. FEM model of morphing wing: boundary conditions and contact settings

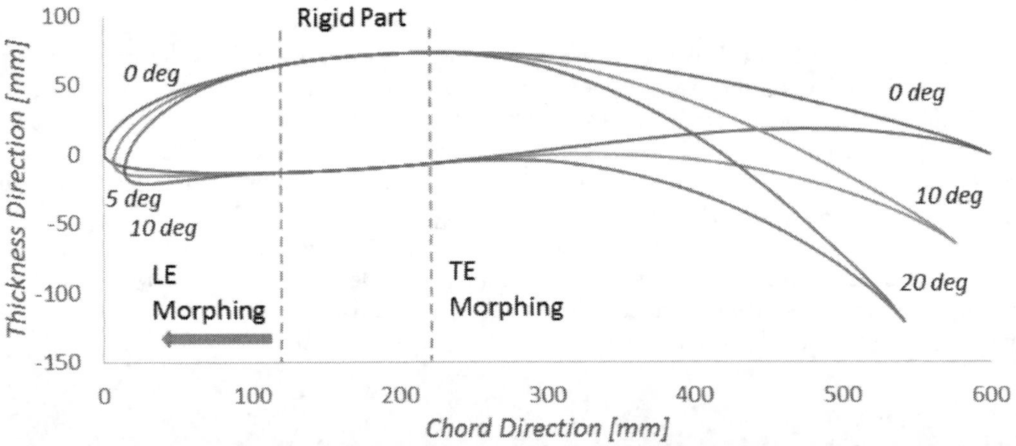

Figure 6. Predicted deformed shapes of morphing wing considering aerodynamic loads (20m/s)

In addition, the necessary actuation forces to obtain the morphing angles of (θ_1, θ_2) = (0°,0°), (5°, 10°), and (10°, 20°) in 20 m/s and 30 m/s airflow are also simulated and used for the design of actuation mechanism. The LE morphing actuation forces are shown in the left figure of Figure 8 as a function of LE morphing angle θ_1. The actuation forces to morph from θ_1=0° to 10° in 20m/s and in 30m/s flow are about 40% larger and 90% larger than in windless condition, respectively. The TE morphing actuation forces are shown in right figure of Figure 8 for wire 1 and wire 2. It should be mentioned that the driving force for wire 1 decreases with increasing TE morphing angle (see at θ_2=20° in right figure of Figure 10). It is due to the fact that air separation in the TE region takes place when θ_2 increases (i.e. 20°) and reduces the pressure at TE section, resulting in decrease of driving force for wire 1 which is connected to the most TE region. Because the actuation forces become much larger in airflow in general, the design approach considering aerodynamic loads is essential to estimate necessary driving forces.

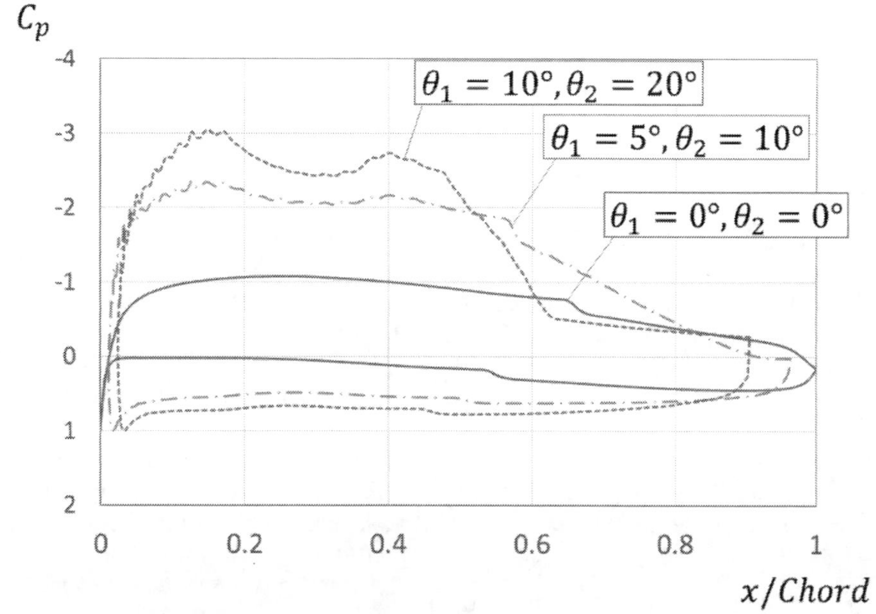

Figure 7. Pressure distribution of morphing wing (20m/s)

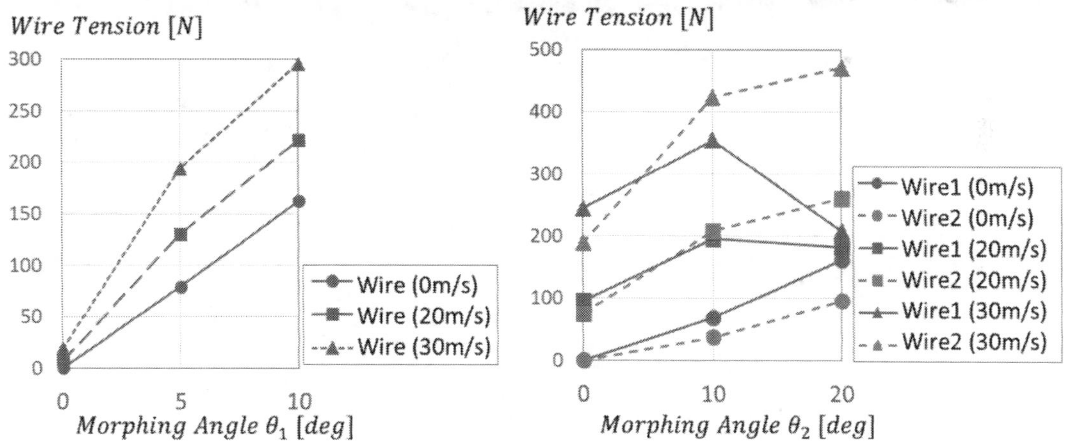

Figure 8. Necessary actuation forces: (left) LE morphing actuation, (right) TE morphing actuation

4. WIND TUNNEL TEST

4.1 Manufacturing of wing model

As mentioned previously, upper skin, LE and TE corrugated structures were made of CFRP laminate. The molds for the corrugations and the upper skin were prepared by cutting plaster boards using a numerical control machine. CFRP components were made by the autoclave using the prepared molds. First, plain woven fabric prepreg (T300-3K/2500, Toray) was layered over the mold and put into vacuum bag. The upper surface skin and the LE corrugated structure were made of CFRP with three layers of plain woven fabrics, and TE corrugated structure was made of CFRP with three layers of plain woven fabrics and partially with four layers (thickness change as described in the analytical section). Then, the mold with stacked CFRPs was covered by vacuum bag. It was put into the autoclave and kept at 135°C temperature and 0.3MPa pressure for two hours with vacuuming. After the curing process, the product was removed from the mold and trimmed in necessary dimension. All CFRP components were bonded and assembled to the aluminum box part. After 90% chord position, foam core was inserted between the upper skin and the bottom skin. The lower surface of the model was covered with a polypropylene sheet (0.2mm thickness). We drilled holes in the corrugated structures and glued the wire guide on the holes. The actuation wires were connected to the LE section, the most TE section (i.e. wire 1), and the middle point of the TE corrugated structure (i.e. wire 2). Finally, other ends of the wires were attached to the pulleys and prototype model was completed (see Figure 9). Note that two sets of servomotor (NX65MA-PS25, Oriental Motor Co.) were installed inside the box for actuation of LE and TE morphing sections.

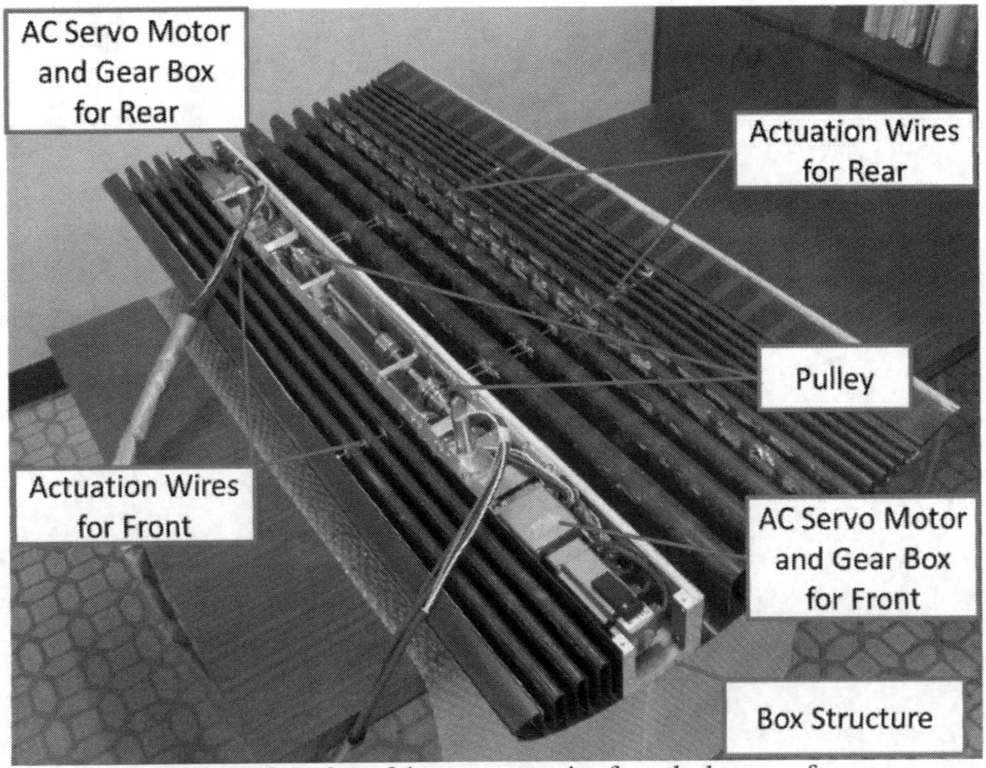

Figure 9. Overview of the prototype wing from the lower surface

4.2 Demonstration of morphing wing

The morphing motion under aerodynamic forces with the manufactured prototype model was demonstrated by the wind tunnel test. The wind tunnel test was performed at low-speed-wind-tunnel facility in JAXA. The test section is 2m×2m and the model was supported by three struts, two struts in front and one strut in rear as explained in section 2.2. The aerodynamic forces were measured by the balance scales connected to the struts and deformation shape of the wing was observed by digital video camera which was set in front of the test section (10mm×10mm square grids were grooved on the elliptical wing tip board). The wind tunnel experiment is overviewed in Figure 10.

The actuation system inside the model was controlled by the operation board which was placed outside the test section. By setting the amount of wire winding on the operation board, LE and TE morphing can be driven independently. The morphing angle scales are written in the wing tip board and we controlled morphing angle by adjusting the wing trailing and leading edge to the angle scale.

In order to verify the feasibility of the present morphing wing mechanism, drive demonstration in airflow was performed. The flow velocities were 0m/s, 5m/s, 10m/s, 15m/s and 20m/s, and the angle of attack, α, was fixed to 0°. When air flow was 20m/s, Reynolds number was about 8.0×105, which was used in the design analysis. The L.E. and rear morphing angles were swept from 0° to 10°and from 0° to 20° respectively. The deformed prototype model under 20m/s flow is shown in Figure 11 for (θ_1, θ_2) = (0°,0°), (5°,10°), (10°, 20°). During actuation in 20m/s flow, neither harmful deformation nor vibration was observed and morphing sections smoothly deformed to the target angle. According to this demonstration, the feasibility of the present morphing wing was confirmed. It should be noted that the downward morphing induced some bucklings (wrinklings) of the lower surface sheet. The solution to the flexible skins has to be further investigated.

Figure 10. Overview of wind tunnel test

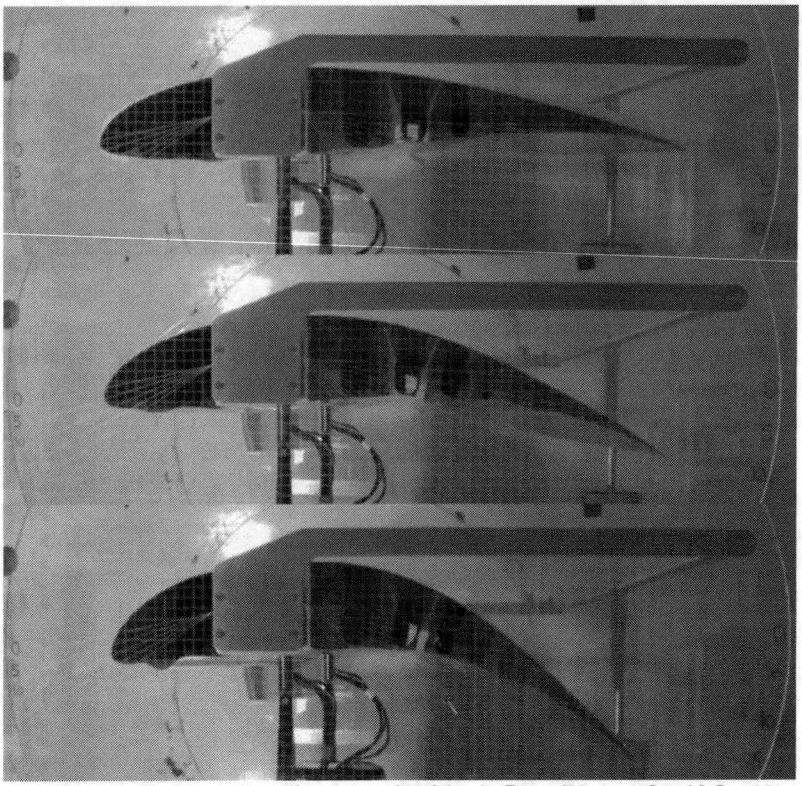

Figure 11. Deformation Shapes under 20m/s flow (Upper: θ_1=0°,θ_2=0°, Middle: θ_1=5°,θ_2=10°, Lower: θ_1=10°,θ==20°)

5. CONCLUSIONS

In this study, variable camber wing with both LE and TE morphing sections with use of corrugated structures was proposed and investigated analytically and experimentally. Based on structural analysis considering aerodynamic loads and actuation mechanisms, the wind tunnel test prototype model was designed and manufactured. The wind tunnel test using the manufactured morphing wing prototype model was performed, and successful actuation in 20m/s airflow was demonstrated. The feasibility of the present morphing wing mechanism was confirmed.

ACKNOWLEDGMENTS

This study was conducted under the financial support of Grant-in-Aid for Scientific Research (No. 24560970, No.15K06598) by Japan Society for the Promotion of Science.

REFERENCES

1. Yokozeki, T., Takeda, S., Ogasawara, T., Ishikawa, T., "Mechanical Properties of Corrugated Composites for Candidate Materials of Flexible Wing Structures," *Composites Part A*, Vol.37, No.10, 2006, pp. 1578-1586.

2. Kress, G., Winkler, M., "Corrugated Laminate Homogenization Model," *Composite Structures*, Vol.92, No.3, 2010, pp.795-810.
3. Xia, Y., Friswell M. I., Saavedra Flores, E., I., "Equivalent Models of Corrugated Panels," *International Journal of Solids and Structures*, Vol.49, No.13, 2012, pp.1453-1462
4. Dayyani, I., Ziaei-Rad, S., Salehi, H., "Numerical and Experimental Investigations on Mechanical Behavior of Composite Corrugated Core," *Applied Composite Materials*, Vol.19, No. 3-4, 2012, pp.705-721
5. Thill, C., Etches, J. A, Bond, I. P, Potter, K D., Weaver, P. M., Wisnom, M. R., "Investigation of Trapezoidal Corrugated Aramid/Epoxy Laminates under Large Tensile Displacements Transverse to the Corrugation Direction," *Composites Part A*, Vol.41 No.1, 2010, pp.168-176.
6. Schmitz, A., Horst, P., "Bending Deformation Limits of Corrugated Unidirectionally Reinforced Composites," *Composite Structures*, Vol.107, 2014, pp.103-111.
7. Thill, C., Etches, J. A., Bond, I. P., Potter, K. D., Weaver, P. M., "Composite Corrugated Structures for Morphing Wing Skin Applications," *Smart Materials and Structures*, Vol.19, No.12, 2010, 124009.
8. Yokozeki, T., Sugiura, A., Hirano, Y., "Development of Variable Camber Morphing Airfoil Using Corrugated Structure," *Journal of Aircraft*, Vol.51, No.3, 2014, pp.1023-1029.

ICAST2015: 26th International Conference on Adaptive Structures and Technologies
October 14-16th, 2015, Kobe, Japan

ICAST2015 #83

Development and Testing of an Unconventional Morphing Wing Concept with Variable Chord and Camber

D. Keidel[1,2], J. Sodja[2], N. Werter[2], R. De Breuker[2*] and P. Ermanni[1]

[1] Laboratory of Composite Materials and Adaptive Structures, ETH Zurich, Switzerland
[2] Faculty of Aerospace Engineering, Delft University of Technology, The Netherlands

Abstract

Driven by the need to improve the performance and energy-efficiency of aircraft, current research in the field of morphing wings is growing in significance. The most recently developed concepts typically adjust only one characteristic of the wing. Within this paper a new concept for morphing wings is developed and tested, enabling large changes of the chord length and camber simultaneously. By changing two characteristics of the wing instead of one, the range of flight missions can be extended more effectively. To achieve these large shape changes, highly adaptable leading and trailing sections are mounted onto a rigid wingbox. A thin polymer film is encompassing these three sections. By adjusting the length of this film, the outline of the wing can be changed significantly. A prototype has been designed and manufactured for wind tunnel tests. The leading and trailing sections are made of polyurethane foam, which can be compressed to 10 percent of its original volume. Different polyurethane foams are tested for optimal stiffness to withstand the aerodynamic loads acting on the wing, while being soft enough to accommodate the large deformations. The film encompassing the sections is retracted into the wingbox to achieve the desired shapes of the airfoil. On the one hand, this film needs to be very thin to fit into the wingbox, while on the other hand it needs to be stiff enough to withstand the pressure exerted by the foam. The manufactured prototype enables changes of the chord length by 30 percent and the camber by 10 percent of the chord length. These deformations were achieved without any significant kinking or buckling. Three different asymmetric airfoil shapes and two symmetric shapes with different chord lengths were tested in a series of wind tunnel tests. All five airfoil shapes deformed very little under wind loads. The lift and drag results were compared to generate values and matched very closely. The prototype fulfils all predefined requirements and performs very well over a wide range of airfoil shapes and wind speeds.

1. INTRODUCTION

Ever since the first human heavier-than-air flight by the Wright brothers, shape morphing wings were used to control the flight dynamics of aircraft in various ways. As these early aircraft were built from wood and canvas, morphing of the control surfaces could be easily achieved by the pilot. As the size and, hence wing loading of aircraft increased, materials and construction changed to much more rigid designs. Therefore hinged control surfaces were introduced to guarantee full flight control. This enabled

* R.DeBreuker@tudelft.nl

larger and heavier aircraft to be build. However, the performance of these aircraft was less than optimal for a range of different flight conditions.

In more recent years new technologies and materials enabled the design and actuation of adaptive load carrying structures such as wings, as outlined by Sofla et al. [1]. In the case of full size aircraft, due to strict regulations, only small incremental changes take place. However, for small-scale UAV research is done using various techniques and materials. Using UAV for testing allows cheaper and faster concept validation. Different concepts aim to change different characteristics of the wing, or even the entire aircraft. These concepts can be split roughly into two categories. Slow deformation for different flight missions, and fast deformations for active flight and loads control. The reviewed literature, summarized in Section 1.1, encloses mainly concepts from the first category, as the focus of this project is to design and manufacture a wing which enables slow, but very large deformations.

The objective of this project is to evaluate the possibilities of large deformation with simultaneous chord and camber morphing. Therefore a new morphing wing concept is designed and developed. The idea is to mount flexible polymer foam sections onto a stiff wingbox. A very thin skin encompasses the foam sections and the wingbox. By changing the length of the skin, the shape and size of the wing is changed. The detailed concept is explained in Chapter 2. A fully functional small-scale wing is manufactured and tested in a wind tunnel. The novelty of the concept is the combination of materials used and the larger than usual morphing capabilities. In order to achieve the aimed for deformations, materials such as foam and thin polymer film are used.

1.1 Background

The two main morphing concepts related to this project are morphing of the chord length and camber morphing. For both morphing techniques various approaches have been outline by Sofla et al. [1] and Barbarino et al. [2], but there has been very little research on the combination of both.

Changing the chord length impacts different characteristics of a wing. Most notable it affects the aspect ratio. If the wing span is kept constant and the chord length is increased, payload increases and take-off and landing speeds decrease. The LIG-7 built by Bakshaev in 1937, is a full size aircraft which can extend or retract six sections along each wing. By extending the sections, the landing speeds decrease by 25%, while the take-off distance decreases by 45%. More recent research is done on UAVs, such as the NextGen MFX-1, by Flanagan et al. [3]. The MFX-1 achieves a range of different wing shapes by changing the sweep angle, wing span and chord length. Changing the chord length, without influencing any other parameters increases the weight of the wing drastically, while only marginally altering the flight characteristics.

Changing the camber of the wing has a much higher impact on the wings performance, than changing the chord length. While keeping all other parameters constant, changing the camber affects the generated lift and drag of the wing. Morphing of the camber is used for active flight control, as well as adapting to different flight missions. The Active Aeroelastic Wing (AAW) program by NASA uses camber morphing on a full-scale wing, as described by Pendleton et al. [4] The wing of the AAW program has a leading edge with optimized stiffness distribution in order to achieve shape changes. Small UAV wings, which can perform camber changes are described by Barrett et al. [5] and Good [6]. The shape changes are limited by the inner components and the stiffness of the skin. Another disadvantage of these projects is, that only the rearmost part of the wing is actively deformed, while keeping the remaining wing unchanged.

The most important materials to achieve both chord and camber morphing are the polymer foam and the thin skin material. The polymer foam needs to match a variety of criteria, in order to result in a successful test wing. The foam needs to be stiff enough to withstand the wind and bending loads acting on the wing, while simultaneously being flexible enough to yield under the pressure of the skin. Furthermore, the stiffness of the foam needs to higher at the trailing edge to prevent the trailing edge of folding. According to Liu and Subhash [7], the compression behaviour of polymer foams can be generally split into three regions: (i) a linear elastic region; (ii) a plasticity-like stress plateau; and (iii) a densification region. The aim is to pre-compress the foam to the plasticity-like stress plateau, where it exerts roughly equal force on the skin during the entire compression. According to the stiffness of the foam, the skin material needs to be chosen. Thill et al. [8] summarizes different morphing skins and materials which themselves morphing. However, for this project the material needs to be very stiff with little elongation, as the material is rolled up, thereby morphing the shape of the wing. The skin needs to be strong enough to compress the foam, while being thin and flexible enough to be rolled up. In Section 2.3 different polymer foam and skin materials are evaluated and selected.

2. WING DESIGN AND MANUFACTURING

In order to go from an initial idea to a fully functional prototype, the wing concept is divided into sub-problems, which can be tackled separately. The prototype is manufactured using available facilities and materials and tested in a small sub-sonic low turbulence wind tunnel. The used materials are chosen based on certain criteria, such as durability, light-weight, cost, availability, and performance. Especially the foam and skin materials are tested in order to find an optimal combination of both, as there is little research on the combination of both in context with aircraft or wing design. In the following sections the design, shape and material evaluation are described in detail.

2.1 Concept Overview

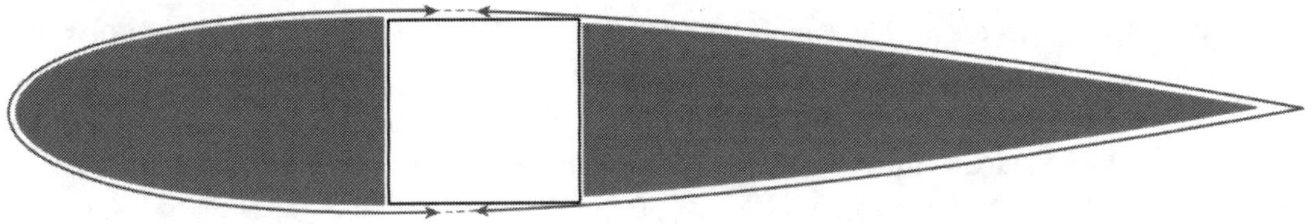

Figure 1. First basic concept idea (Grey: Foam encompassed by thin skin)

The underlying idea of this project is to develop a new morphing wing enabling large changes of the chord and camber. The idea is to split the airfoil into three main sections, as shown in Figure 1. The leading and trailing sections are made from very compliant foam and are mounted onto a rigid wingbox in the centre. All three sections are encompassed by a thin skin. The wingbox has an open cross section, which allows the skin to be retracted into the wingbox. By retracting the skin into the wingbox, the foam is compressed and the outline of the airfoil changes. As the skin can be retracted at each end individually, the airfoil can be morphed into different symmetric and asymmetric shapes with variable chord and camber. The shapes which are aimed for are described in Section 2.2. The skin is reeled onto a thin axle located in the wingbox, which in turn are actuated by stepper motors and controlled using an arduino micro controller. During the idea generation certain features are developed, which play a crucial role in the design of a functioning prototype. These features are outlined for clarity in the following list:

Foam Compartment The foam compartments are positioned between the wingbox and the leading and trailing foam sections. Each consists of a thin plate which pushes out the foam sections while the foam is extended. When the skin is retracted to fully compress the foam, the thin plates are retracted as well, thereby decreasing the overall cross section and the chord length of the airfoil in its compressed state.

Compartment Ledges In order to ensure a smooth outline if the airfoil, the hard edges of the compartment plates need to be smoothed out. In order to do so flexible ledges are introduced, which bend with the skin, but keep a smooth outline across the foam compartment. They span from the wingbox over the compartment plates.

Wing Mount The wing mount connects the wing to the balance during testing. Furthermore it keeps the motors in place. At the base of the wing mount two plates are mounted above each other and can be rotated to adjust the angle of attack of the wing.

End-Plates The end-plates are mounted at the root and the tip of the wing. They connect the wing to the wing mount. They further hold the guides of the compartment plates and some of the actuation in position. There are two sets of end-plates; one is the size of the wing, which simulates a finite wing, and the other extends outward from the wing, to simulate an infinite wing.

In order to find the best solution for this morphing wing, it is divided into sub-problems. These include: skin material, skin mounting, foam material, wingbox shape and design, and actuation. For each of these problems a variety of solutions is come up with. They are collected in a morphological box. From this, four different concepts are developed, incorporating different solutions. For these concepts, the functionality and interaction of the different components is assessed. The concepts are shown in Figure 2. The main difference between the concepts is the wingbox layout. Concept 1 has a stiff wingbox, with a fixed length, while the other concepts have flexible foam compartments, which decreases the overall length of the collapsed wing. The four concepts are evaluated using a list of criteria, with include: deformation from structural and aerodynamic loads, airfoil shape and deformation capabilities, and collapsibility. Based on this evaluation, concept 1 performs worst, while concepts 2 to 4 perform similarly well, due to the wingbox design and the materials used. For the final concept, the best aspects of each concept are combined. The wingbox is designed as small as possible, enlarging the foam compartments. The leading section foam is made of one part, while the trailing section includes a stiffer part made of styrofoam at the trailing edge, to ensure a sharp edge which maintains the aerodynamic performance. The skin is made of a very thin polymer film which takes up as little as possible volume within the wingbox, while being strong enough to prevent elongation. It is split into two parts, one for the leading section and one for the trailing section. The skin is reeled up onto four axles inside the wingbox, which are actuated separately by stepper motors. The compartment plates are actuated by strings, which are reeled up using servo motors for discrete positioning.

2.2 Airfoil Shape and Size

In order to define the exact size of the components, the airfoil shape and size needs to be defined. The undeformed and deformed airfoil shapes are chosen from the 4-digit NACA series. For this series of airfoil profile theoretical models for validation are available, which are used for comparison with the wind tunnel test results.

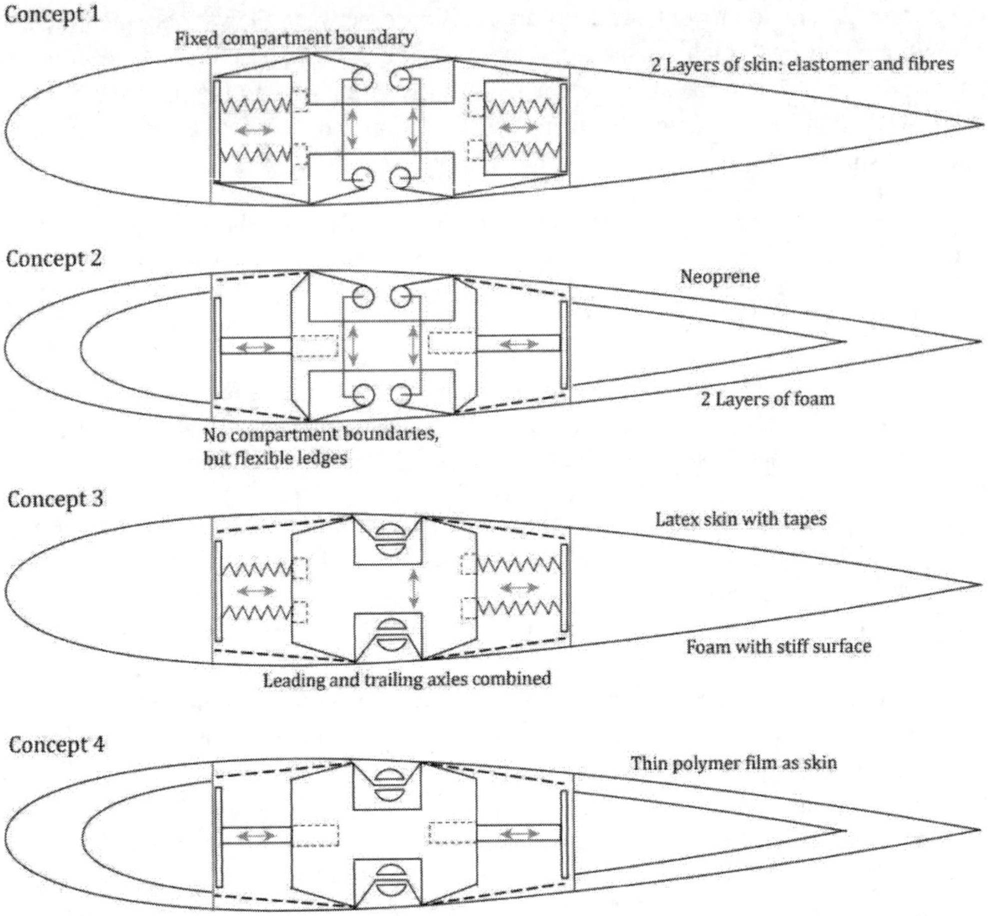

Figure 2. Four different concept ideas sketched out; dashed: flexible compartment ledges

As the used wind tunnel has a cross section of $400 \times 400mm$, the wing span is defined as $400mm$. The undeformed chord length is defined to be $300mm$ to maintain a large enough wingbox to contain the actuation mechanism.

The initial shape of the airfoil is chosen to be the NACA 0015 profile. This shape is chosen as a trade-off between induced drag and wingbox thickness. Compared to similar profiles, it has a very low induced drag over a range of lift coefficients, as shown in Figure 4. The maximum deformed shapes, which are tested are the NACA 0018 for symmetric shapes, and NACA 9315 for the asymmetric shapes. The asymmetric shape and the location of the wingbox are chosen, to minimize the concave curves at the underside of the wing, which the skin needs to achieve, as shown in Figure 3.

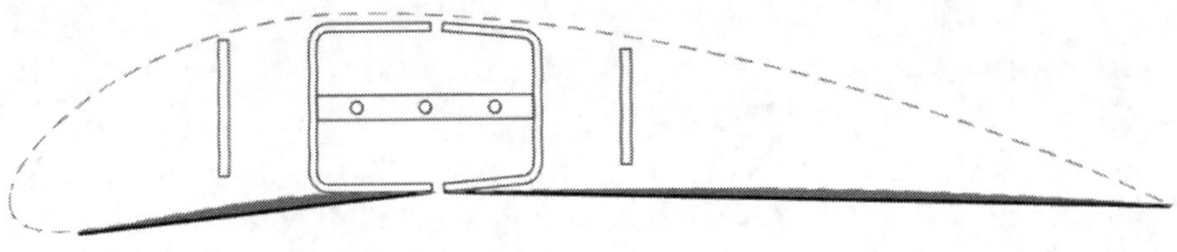

Figure 3. Concave curves of the skin on the underside of the wing are minimized

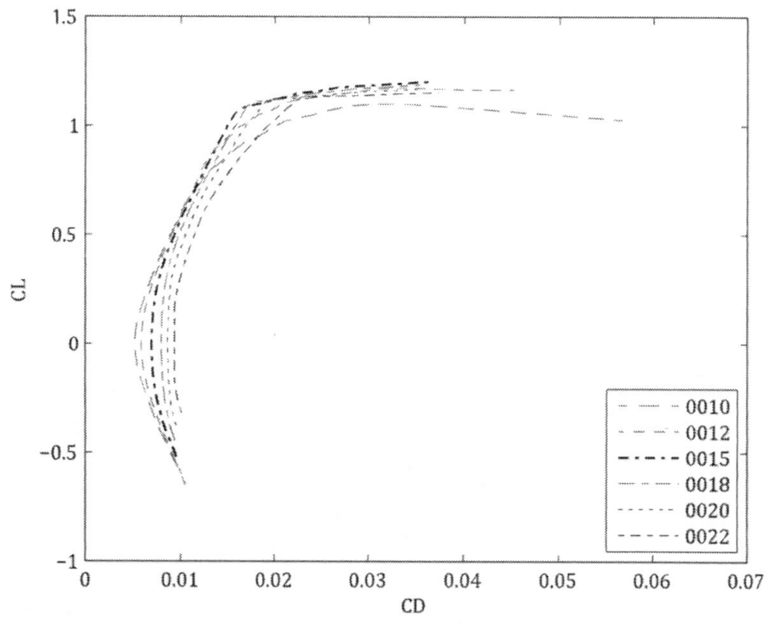

Figure 4. Theoretical lift-to-drag curves for different symmetric profiles

2.3 Material Evaluation

As a core component for successfully manufacturing and operating the wing, the foam material for the leading and trailing section is critical. Therefore, prior to testing the wing, different foams with different compression characteristics are evaluated. According to Liu and Subhash [7] polymer foam compression is split into three distinct regions: a linear elastic region; a plasticity-like stress plateau; and a densification region. After an initial linear elastic compression, the cell walls of the foam start to collapse, and the stress plateau is reached, where the stress stays roughly constant. After all cell walls have collapsed, the foam material starts to densify. For this project, the used foam shall only operate within the second region, the stress plateau, where the force exerted on the skin stays equal. If the foam is compressed to far, therefore reaching the densification regime, it will have a permanent effect on the consistency of the stress plateau. Different polymer foams with different characteristics are tested and compared, in order to find the best suited foam to use for the final wing. The test samples are $50 * 50 * 40mm$ and compressed multiple times to gain data on repetitive compression behaviour. From the different tested foams three materials are selected for further testing, which all have different stress plateau levels. The stress-strain curves for the first two compressions of the three chosen materials are shown in Figure 5. The required stress needed to compress the stiffest foam to the beginning of the densification phase is approximately $13 \times 10^{-3} N/mm^2$.

The skin of the wing has two crucial requirements to fulfil: (1) withstand the force acting from the foam, and (2) have a smooth and surface to minimize drag, while being flexible enough to be rolled up within the wingbox. The tested materials include different ethylene films (PFA, ET), and Kapton film. Each of the materials was tested for their tensile strength. The test results are shown in Figure 6. These tests show, that the Kapton film is most suited for the wing. The maximum yield stress, which the film can take is 45 N/mm^2.

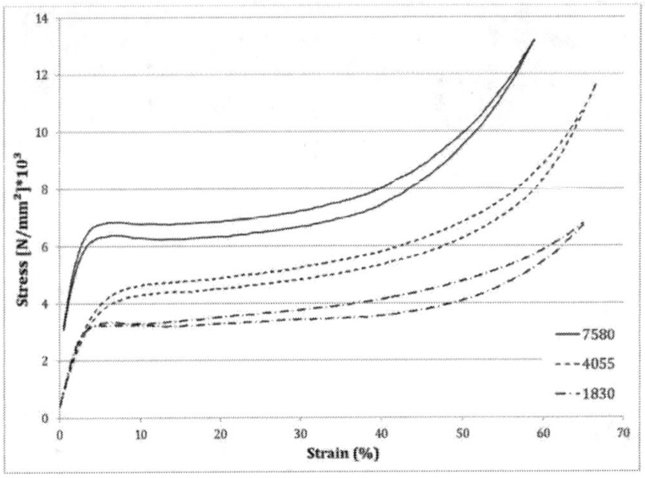

Figure 5. Stress-strain behaviour of the selected foams

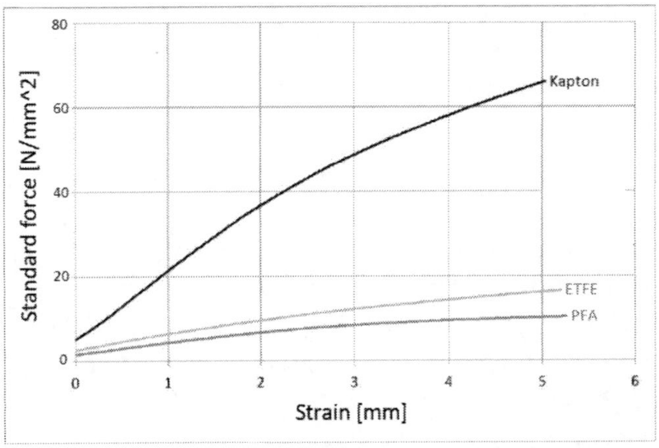

Figure 6. Stress-strain curves of the tested skin materials

The force required to compress the foam is equal to $F_{comp} = \sigma \times Area$ which is equal to $F_{comp} = 13 \times 10^{-3} \times 400 \times 30 = 156N$. The force that can be exerted by the skin without any permanent elongation is $F_k = \sigma \times Area$ which results in $F_k = 457.2N$. However, the skin is not pulled parallel to the film, but at an angle. Therefore the force, which can be exerted on the skin needs to be corrected. It is assumed, that when the foam is fully compressed, the angle between the skin and the x-axis is 45°. The corrected force is calculated using a safety factor of two:

$$F_{k,corr} = \frac{F_k \times cos(45)}{f_S}$$
$$F_{k,corr} = \frac{457.2 \times cos(45)}{2} = 161.6N$$

Therefore, the force which can be exerted on the kapton film is greater than the force required to compress the foam, as $161.6N > 156N$.

3. WIND TUNNEL EXPERIMENTS

According to the decisions taken in Chapter 2, the wing is manufactured and assembled. The finished, fully functional prototype is shown in Figure 7. The trailing foam section is slightly modified, as the trailing edge keeps collapsing. The polymer foam is to thin to exert enough force on the skin to keep the desired shape. Therefore the foam section is split into two parts. The front part remains polymer foam, while the rear part is replaced with a small styrofoam wedge. If the trailing edge has a rounded or blunt shape, it has a very negative effect on the lift and drag characteristics., as it creates turbulences. The wing is tested in a wind tunnel in order to compare the prototypes performance to the predicted lift and drag characteristics, generated using xfoil, an aerodynamic analysis code. The test setup and testing is described in the following sections.

Figure 7. Fully-functional test wing mounted in the wind tunnel with the infinite end-plates

3.1 Test Setup

A subsonic, low turbulence open wind tunnel with a cross section of $400 \times 400mm$ is used for the wind tunnel tests. The wind speed is measured using a venturi nozzle which is placed in the wind tunnel ahead of the test wing. As it is an open wind tunnel some correction factors need to be incorporated to compare the tests with the xfoil predictions, which are explained in Section 3.2.

The wing is tested in five different wingshape configurations: NACA0015, NACA0018, NACA3315, NACA6315, NACA9315. The angle of attack is varied between $0°$ and $18°$, in intervals of $3°$. The wing is tested at wind speeds between 8 and 18 m/s. Therefore the Mach number varies from 0.0233 to 0.0525, and the Reynolds number varies from 180,000 to 360,000. Each shape is tested over the range of angles of attack and wind speeds to attain more significant data. Each of the experiments over a range of wind speeds and angles of attack is repeated, in order to obtain more representative data.

In order to fully understand the behaviour of the test wing, different end-plates are tested in the wind tunnel. The first set of end-plates is roughly the outline of the wing, to simulate a finite wing, therefore named finite end-plates. The second set of end-plates extends the complete width of the wind tunnel and twice along the chord length. These simulate an infinite wing which theoretically generates more lift, as less lift is lost at the wing tips and are called infinite end-plates. The end-plates are connected to the wing and the balance and therefore increase the measured drag of the wing. The used balance measures forces and moments in x,y,directions. The moments are used to validate the measured forces. Because the entire wing, including the end-plates and wing mount are connected to the balance, the forces acting on everything are measured. Therefore the drag of the end-plates needs to be taken into account as well.

3.2 Theoretical Predictions

In order to attain validated data, which the wind tunnel test results can be compared to, the analysing tool xfoil is used. As the NACA profiles are predefined in the xfoil program the generated data is assumed to be valid and accurate. The theoretically predicted lift and drag curves only differ slighty when varying the Mach and Reynolds number. Therefore, the average wind speed of 13 m/s is used to generate the lift and drag curves, which the wind tunnel tests are compared to. For angles of 12° and above certain lift characteristics change but these are disregarded as these values tend to be very inaccurate due to the fact that the used wind tunnel has an open test section.

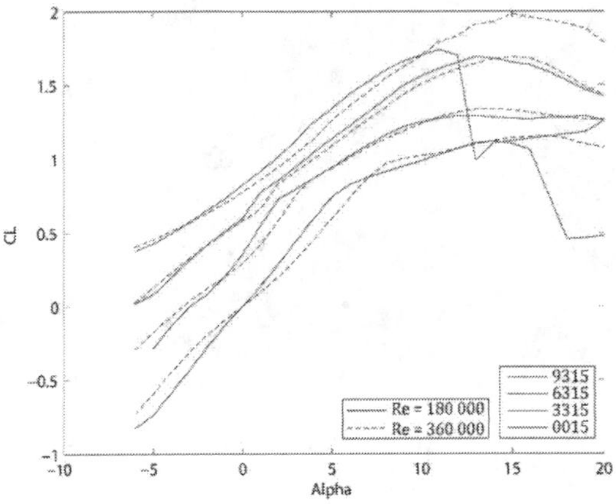

Figure 8. Change in lift coefficient depending on the Reynolds number

The open test section of the wind tunnel needs to be accounted for in the theoretical model, as the shape and size of the wing significantly influence the air flow of the wind tunnel. There are two factors that need to be accounted for to compare the wind tunnel tests to the theoretical predictions. Firstly, the difference between the geometric angle of attack α and effective angle of attack α_{eff} affects lift and drag measurements. The local flow direction near the wing changes depending on various factors, such as aspect ratio and lift. The geometric angle of attack is defined as $\alpha = \alpha_i + \alpha_{eff}$. Secondly the downward deflection of the free jet caused by the wing needs to be taken into account. This effect is dependant on the ratio between the height of the wind tunnel and the chord length of the wing, and described by Glauert [9]. The ultimate downward velocity w_0 is defined as

$$w_0 = v\frac{c}{h}c_l$$

where h is the height of the wind tunnel, v the wind speed, c the chord length and c_l the measured lift coefficient. The adjusted lift coefficient can be calculated as

$$c_l = \pi(a - a_0 - \frac{1}{2}\frac{c}{h}c_l)$$
$$c_l = c_l(f) - \frac{\pi}{2}\frac{c}{h}c_l$$

where a_0 is the angle of zero lift, a is the geometric angle of attack, and $c_l(f)$ is the lift coefficient predicted by xfoil. Therefore the new lift force is calculated to be

$$\frac{L_0}{C_L} = 1 + \frac{\pi}{2}\frac{c}{h}$$

where L_0 is the lift in the unlimited stream, and C_L is the lift in the free stream of the open wind tunnel. The combination of the above equations results in the corrected lift and drag

$$\frac{L_0}{C_L} = 1 + \frac{\pi}{2}\frac{c}{h} + \frac{\pi^2}{12}(\frac{c}{h})^2 = 2.64$$
$$C_D = c_d + \frac{1}{2}\frac{c}{h}C_L^2$$

Therefore the measured lift force needs to be multiplied by 2.64. The drag needs to be adjusted individually depending on the measured lift force and the predicted drag coefficient.

3.3 Test Results and Evaluation

The measured data is adjusted using the correction factors described in Section 3.2. In Figure 9 the corrected lift curves of the finite and infinite end-plates are compared to the predicted xfoil curves. Due to the loss of lift at the wing tips, the finite end-plate setup generates less lift compared to the infinite end-plate setup. The match between the measured curves and the predicted values is better at lower angles up to 12°. At high angles, above 15°, the results spread due to air flow separation. For the NACA9315, the air flow separates earlier at low wind speeds, as shown in Figure 8.

The measured drag curves and the adjusted xfoil predictions for the five tested profiles are shown in Figure 10. The drag of the infinite end-plates is generally higher, as the end-plates are connected to the balance and add to the measured drag force. Furthermore, the airflow changes its angle over the length of the chord and is not parallel to the balance. Therefore some of the measured drag is a result of the generated lift. The xfoil predicted drag and the measured drag have a worse match than for the lift curves, as the measured forces are smaller and less accurate. Furthermore factors, such as the induced drag and the end-plates add to the uncertainty of the measurements.

The achieved shape of the wing slightly differs from the shapes which were used for the xfoil predictions. Figure 11 shows the difference between the NACA9315 predicted outline and the real outline. This is due to the rigid wingbox, and the fact, that the compartment plates expand in a straight line. However, for the lift and drag curves it is shown, that the wing generates approximately as much lift and drag as predicted with xfoil. The generated lift forces change according to the change in airfoil shape. Furthermore the overall range of lift coefficient is much greater than for conventional wings.

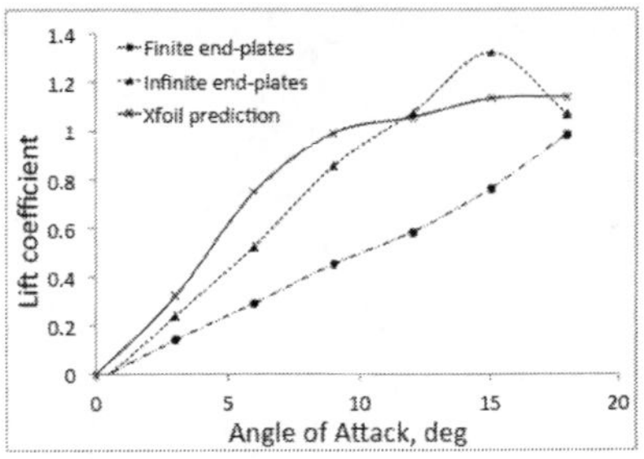

(a) Lift curve of NACA 0015

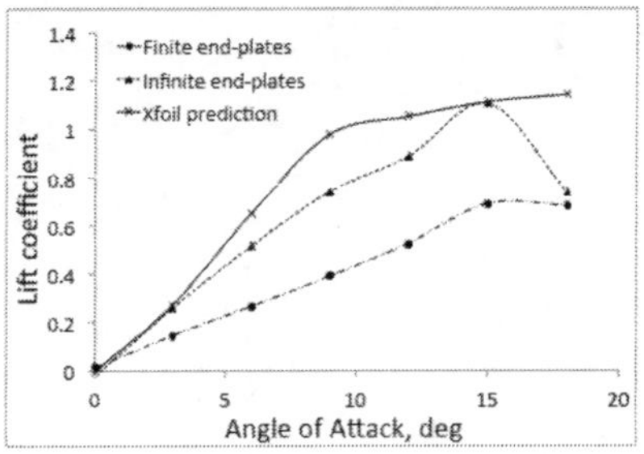

(b) Lift curve of NACA 0018

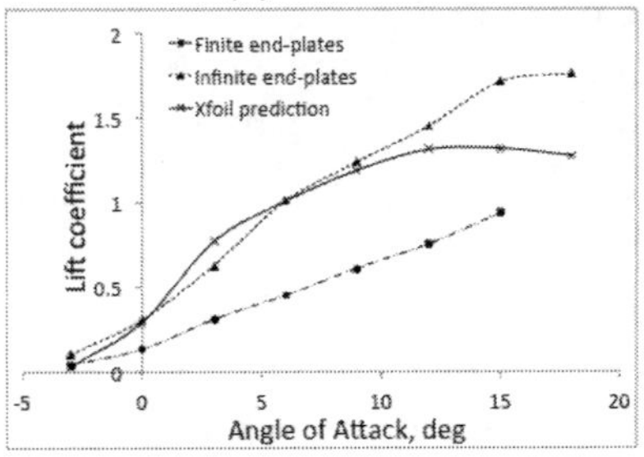

(c) Lift curve of NACA 3315

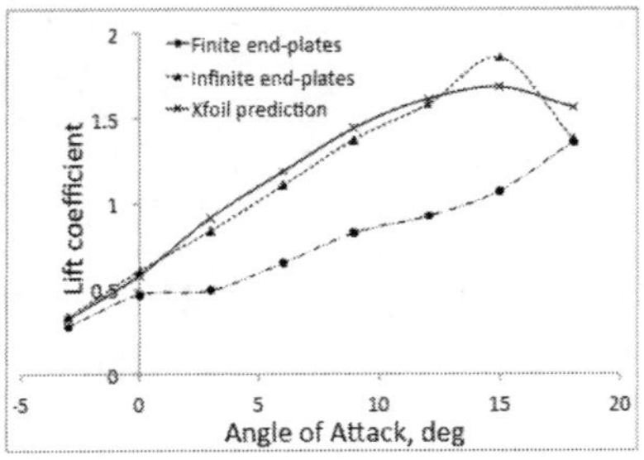

(d) Lift curve of NACA 6315

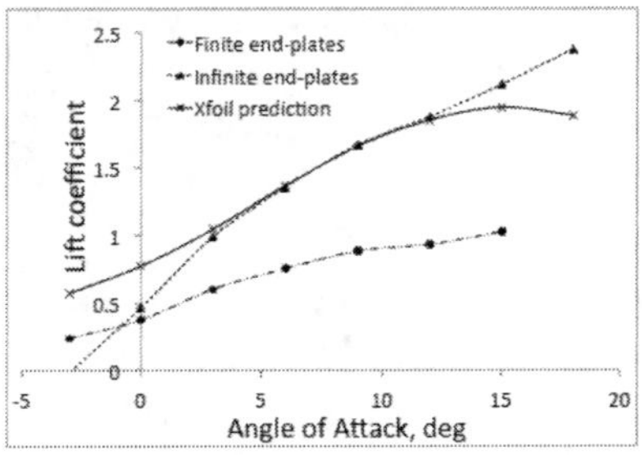

(e) Lift curve of NACA 9315

Figure 9. Lift curve comparison of finite and infinite end-plate setups and the xfoil predictions

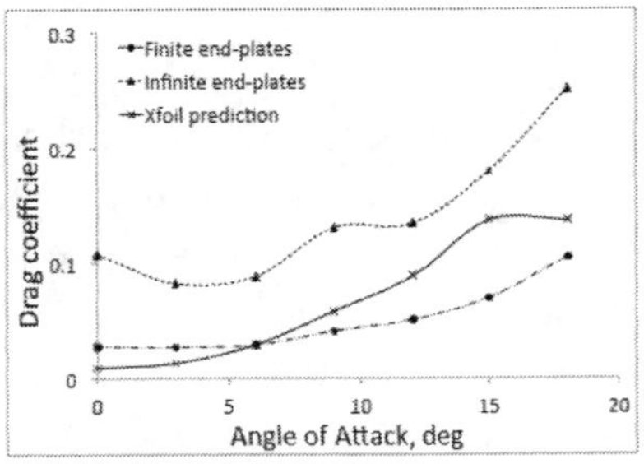

(a) Drag curve of NACA 0015

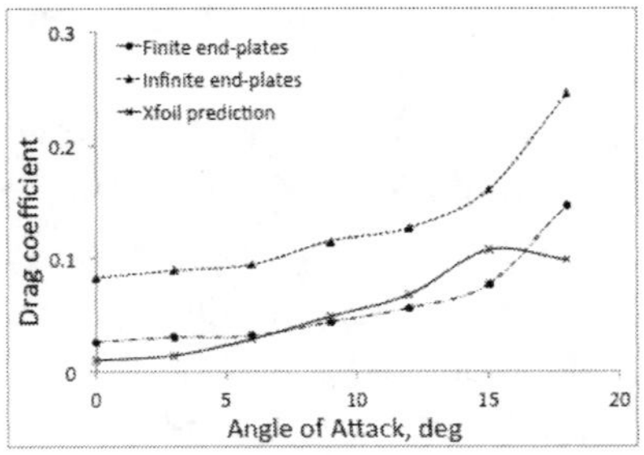

(b) Drag curve of NACA 0018

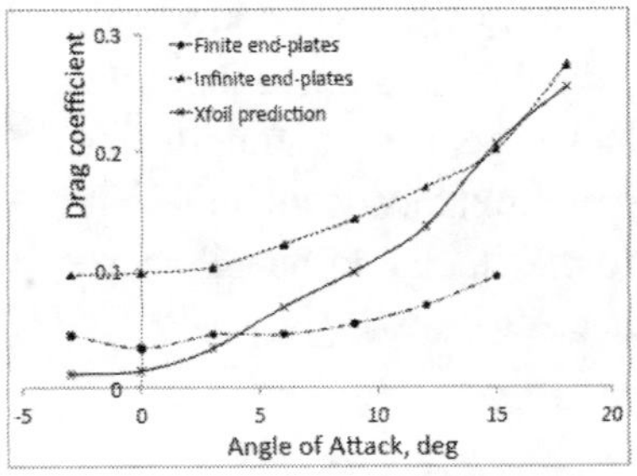

(c) Drag curve of NACA 3315

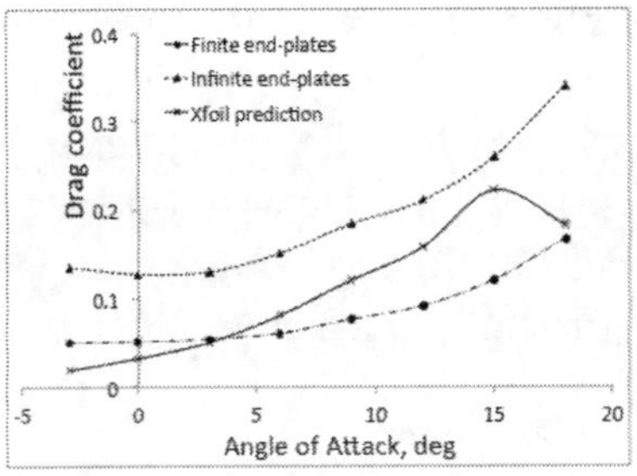

(d) Drag curve of NACA 6315

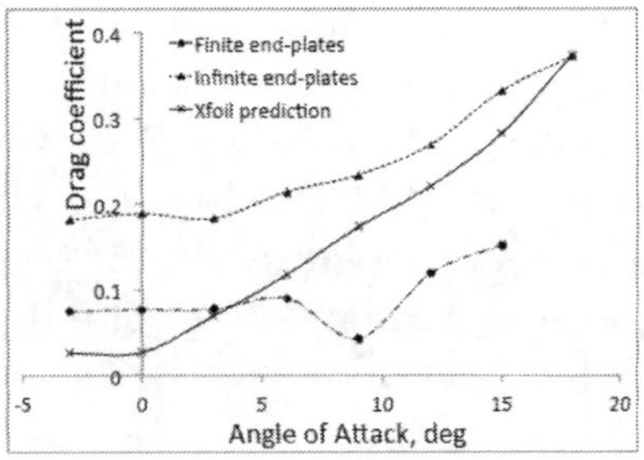

(e) Drag curve of NACA 9315

Figure 10. Drag curve comparison of finite and infinite end-plate setups and the corrected xfoil predictions

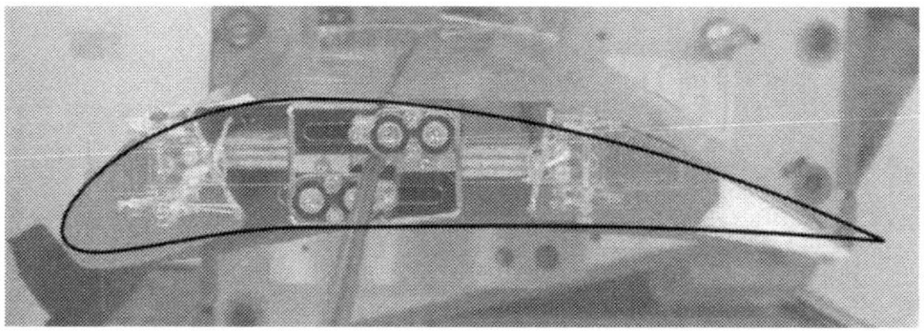

Figure 11. Difference between predicted outline and test wing outline for NACA9315 profile

4. CONCLUSION

During the project, a new concept for morphing wings was developed and tested. The concept allows for controlled changes of the chord length by 25% and camber changes of 9% of the original chord length. The deflection of the foam sections due to aerodynamic loads was minimal, while the measured lift coefficients are close to the predicted values. However, the achieved shapes had certain deviation of the predicted values due to the internal structure. Furthermore the measured drag had certain inaccuracies due to the test setup.

REFERENCES

1. A.Y.N. Sofla and S.A. Meguid and K.T. Tan and W.K. Yeo, "Shape morphing of aircraft wing: Status and challenges", Materials and Design, Vol. 31, No. 3, pp.1284-1292, 2009
2. Silvestro Barbarino and Onur Bilgen and Rafic M. Ajaj and Michael I. Friswell and Daniel J. Inman, "A Review of Morphing Aircraft, Journal of Intelligent Material Systems and Structures, Vol. 22, No.9 823-877, 2011
3. John S. Flanagan and Rolf C. Strutzenberg and Robert B. Myers and Jeffrey E. Rodrian, "Development and Flight Testing of a Morphing Aircraft, the NextGen MFX-1", 48th AIAA/ASME/ASCE/AHS/ASC Structures, Structural Dynamics, and Materials Conference, 2007
4. Edmund W. Pendleton and Denis Bessette and Peter B. Field and Gerald D. Miller and Kenneth E. Griffin, "Active Aeroelastic Wing Flight Research Program: Technical Program and Model Analytical Development", Journal of Aircraft, Vol. 37, No.4 pp. 554-561, 2000
5. R.M. Barrett and R.S. Gross and F.T. Brozoski, "Design and Testing of a Subsonic All-Moving Adaptive Flight Control Surface", AIAA Journal Vol.35, No.7, 1997
6. Matthew G. Good, "Development of a Variable Camber Compliant Aircraft Tail using Structural Optimization", Virginia Polytechnic Institute and State University, July 2003
7. Qunli Liu and Ghatu Subhash, "A Phenomenological Constitutive Model for Foams Under Large Deformations", Polymer Engineering and Science Vol.44, No.3, pp.463-473, 2004
8. C. Thill and J. Etches and I. Bond and K. Potter and P. Weaver, "Morphing skins", The Aeronautical Journal, Vol. 112, No. 1129, 2008
9. H. Glauert, "Wind Tunnel Interference in Wings, Bodies and Airscrews", Aeronautical Research Committee No 1566, 1933

ICAST2015: 26th International Conference on Adaptive Structures and Technologies
October 14-16th, 2015, Kobe, Japan

ICAST2015 #002

Generation of a Shock Control Bump by Pressurized Chambers

Markus Kintscher[1*], Nuno Alves de Sousa[2*], Hans Peter Monner[1], Martin Wiedemann,[1]

[1] Department of Adaptronics, DLR Institute of Composite Structures and Adaptronics, Braunschweig, Germany.
[2] Instituto Superior Técnico, Universidade de Lisboa, Portugal.

Abstract

In the design of a new generation of eco-efficient aircrafts, fuel reduction is the number one objective. To increase the efficiency of the next generation aircraft, disruptive technologies are under investigation to achieve a significant benefit and advantage in the market. Such technologies often mentioned as game changer technologies like for example laminar flows are under investigation since decades and have not yet reached a technology readiness level suitable for serial production of large passenger aircraft. Besides laminar flow and other game-changer technologies, there are technologies like for example variable camber and shock control, which offer less improvement in efficiency compared for example to laminar flow, but which are easier to get developed to the needed technology readiness level since only small changes in system architecture and/or the structural layout are necessary. Therefore in this paper a structural concept for an adaptive spoiler shock control bump is presented. The objective of the shock control bump is to spread a single transonic shock which causes a significant drag increase into several smaller shocks with less drag penalty. Since the shock bump performance depends on various parameters and flow conditions like height, density, temperature, flow velocity, etc. the optimal shock position and height is changing during flight. Therefore, a shock control bump concept which is able to adapt the bump to the varying flight conditions is presented. A simplified analytical model for the structure is used in the framework of an optimization module to investigate the interdependencies of structural parameters and actuation parameters for the accurate approximation of given aerodynamic target shapes. An outlook is given on the design issues for a 3D design of the pressure actuated adaptive bump spoiler concept.

Keywords: Shock control bump; wave drag minimization; morphing spoiler; pressure actuated adaptive structure; structural optimization.

* markus.kintscher@dlr.de
nunoalvesdesousa@me.com

1. INTRODUCTION

The reduction of DOC's and the improvement of mission flexibility by incremental improvements is a strategic key issue in the development of the next generation aircraft. For the enhancement of the performance in off-design conditions the adaptation of the wings geometry to changing flight conditions has large potential. There are several performance-enhancing technologies as identified by Schrauf et al.[1] which are under investigation (fig.1). Besides large scale morphing technologies, with the objective to adapt the complete wing as described by Smith[3] on an experimental F-111 fighter, it seems more feasible to adapt the airfoil geometry by focusing on individual movables like the leading edge, flap or the spoiler. Adaptive wing devices as for example a smart droop nose [2] or a smart flap trailing edge can contribute to a drag reduction of lift induced drag and viscous drag as well.

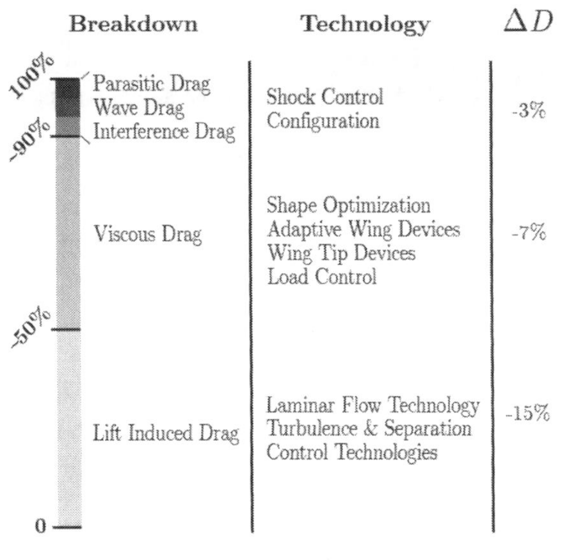

Figure 1: Drag breakdown of transport aircraft in cruise including technologies for drag reduction by Schrauf et al.[1] and estimated potential drag reduction (ΔD)

The most effective approach for the reduction of wave drag is the application of SCBs, according to the EUROSHOCK-program [4] as proposed by Ashill et. al [5] 1992. The control of the transonic shock especially for variable camber (VC) application is a promising concept for wave drag reduction. In transonic flight supersonic regions and shock waves are generated when exceeding the critical Mach number. These increase the entropy while the total pressure decreases. The effect is wave drag which can be reduced by the application of shock control bumps to significantly reduce the shock strength.

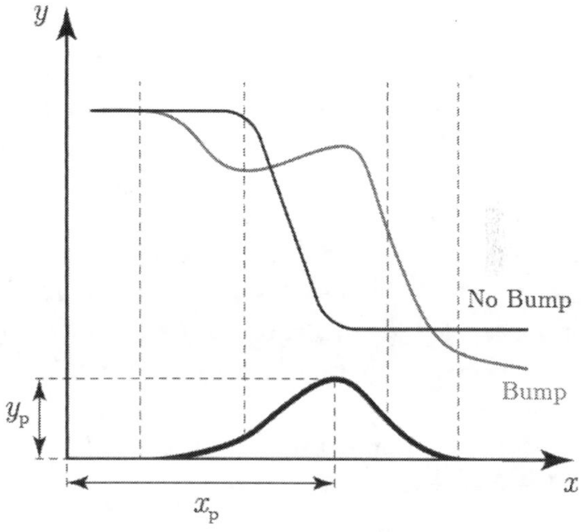

Figure 2: Definition of geometrical key parameters of a shock control bump (SCB)

To do so, a bump must be formed in the region of the shock as depicted in Figure 2. The shock position is located on the rising slope of the bump to induce isentropic compression waves and a pre-shock which reduces the Mach number in front of the main shock. From parametric investigations of Knauer [6] and Dargel [7-8] we know that the potential for drag reduction by the application of shock control bumps (SCBs) depends strongly on the correct bump height y_p and the position x_p of the bump peak. The bump height depends on the actual flight conditions and the airfoil geometry but can be estimated to be within about 0.5 percent of local chord length. Compared to the bump height and position, the detailed bump shape like ramps, polynomials and sinusoidal curves does not have a very strong effect on the bump effectiveness.

2. STRUCTURAL CONCEPT, MODEL AND OPTIMIZATION

For the exploitation of the full potential of the SCB/VC technology, an active system must be developed to be able to adapt the bump position and height to changing flight conditions. While there are numerous publications dealing with the aerodynamic design of SCBs, the structural realization of such a structure-system combination is still a key challenge. Since the shock control bump is located at the chord position of the wings trailing edge and spoilers, the available design space is extremely limited. The main drivers for a concept selection are therefore a compact design, a lightweight structure and a simple and low complexity mechanism for the deployment of the bump. Concepts for structural realization have been compared by Pritschow [9] while concepts using SMA actuated deformation have been realized and tested by Wadehn [10] and Bein [11].

2.1 Concept

According to the above mentioned design drivers a concept with pressurized chambers is selected as pioneering technology combining the key requirements. Using pressure for actuation was also proposed by Bein in [11] as an alternative actuation concept to the application of shape memory alloys. The deployment of a SCB by pressure and its adaptation to changing flight conditions is realized by differential pressurization of chambers in the spoiler body (fig.3).

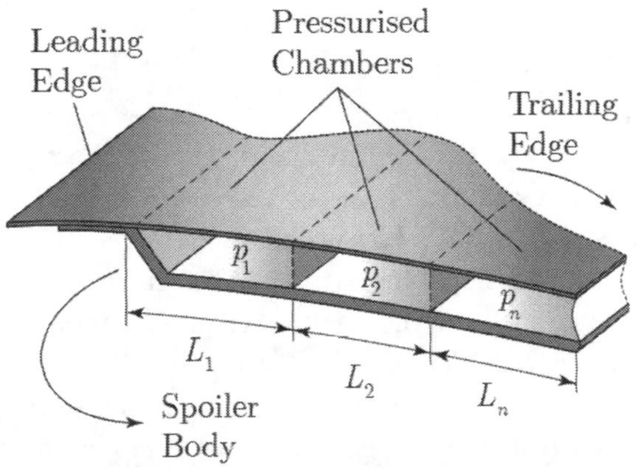

Figure 3: Concept of pressurized chambers for an adaptive spoiler shock control bump

Within the concept the overall structural stiffness is given by a conventional "rigid" spoiler lower body, while the flexibility for realization of a SCB is given by thinner upper spoiler cover. Pressurized chambers are formed by separating walls which can be realized for example by stringers with a cross-sectional shape allowing for deformation in direction of the chamber height which at the same time can serve as standard longitudinal stiffeners in span direction.

2.2 Analytical Model & Optimization

A simplified spoiler model is proposed to expedite the pre-design stage and to cope with the necessity of defining a set of design variables which offers enough degrees of freedom to generate the required bump shapes.

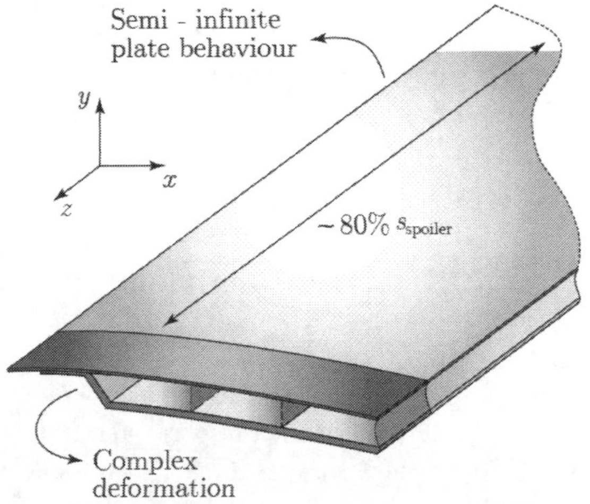

Figure 4: Different deformation behaviours

The upper spoiler skin is modelled as a transversely loaded plate while the main lower body is assumed to be comparably much stiffer. If the spoiler edges are free at both ends (*i.e.* the upper skin tips), it is then possible to identify two regions that exhibit different deformation behaviours. Figure 4 shows that near both spoiler tips (depicted in grey) there is a relatively small region where the deformation along the z-axis is complex.

The inner blue region, where the tip deformation has little influence, deforms almost uniformly along z. This can be called the "effective bump region" (EBR), and the plate is idealized as infinite along the span direction.

For transversely loaded Kirchhoff plates and an isotropic and homogeneous material, the deformation along y is given by

$$\frac{\partial^4 v}{\partial x^4} + 2\frac{\partial^4 v}{\partial x^2 \partial z^2} + \frac{\partial^4 v}{\partial z^4} = \frac{p}{D} \tag{1}$$

where p is the transverse loading. D is defined as follows

$$D = \frac{EH^3}{12(1-\nu^2)} \tag{2}$$

where E is the Young's Modulus, ν is the Poission coefficient and H is the half-plate thickness $H = h/2$.

Additionally, for the blue region $\partial/\partial z = 0$ and eq.1 becomes

$$E^* I \frac{\mathrm{d}v^4}{\mathrm{d}x^4} = p \tag{3}$$

where

$$E^* = \frac{E}{1-\nu^2} \quad \text{and} \quad I := \frac{H^3}{12} \tag{4}$$

Showing that for the blue region it is possible to implement a beam-like solution process where the upper skin of each chamber is modelled as a beam segment per unit span with a scaled Young's Modulus E^*.

Figure 5 shows the upper surface of the wing idealized as a series of beams which are actuated by the pressurized chambers inside the spoiler. A spring is used to model the interface between different bays whose pressure is kept constant.

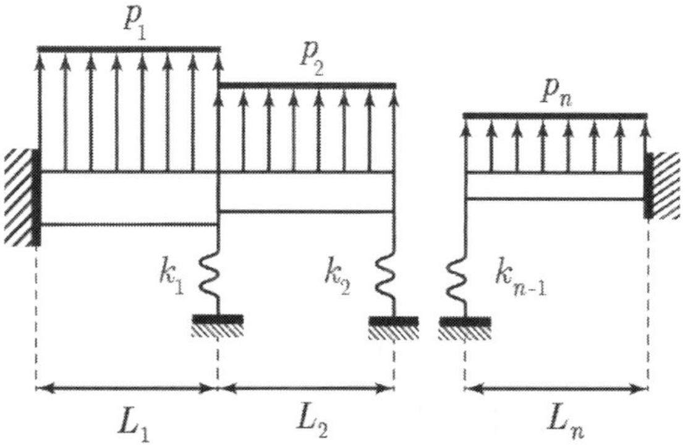

Figure 5: Simplified 2D model

The model is then used in an optimization environment where the desired bump shapes are fed to the optimization algorithm through an objective function as target shapes in order to find the optimized design parameters that minimize the difference between the target shape (*i.e.* the desired bump) and the optimized shape. The design parameters include:

- α_i - the relative position of each spring as a percentage of the total spoiler chord. This is then used to compute the length L_i of each chamber;
- t_i - the thickness of the upper skin of each chamber;
- k_i - the spring constant;
- p_i - the pressure inside each chamber.

It is also possible to choose the number of beam segments (*i.e.*: the number of pressurized chambers) which is directly related to the overall design complexity.

2.2.1 Structural Module

A structural model is developed based on the analytic element method (AEM) approach as described by Policarpo et al.[12]. This allows for a structural solver which is both accurate and efficient.

Since the pressure inside each chamber is assumed to be uniform

$$v_i(x) = \frac{p_i}{24H_i}x^2 + C_1^i x^3 + C_2^i x^2 + C_3^i + xC_4^i \quad (5)$$

where $H_i = E_i^* I_i$ is the flexural rigidity of beam i and C^i are the set of beam constants to be determined using the boundary conditions. The loads on the left (I), and right (J) nodes are respectively given in terms of shear force (V) and bending moment (M) by

$$\begin{Bmatrix} F_i^I \\ M_i^I \\ F_i^J \\ M_i^J \end{Bmatrix} = \begin{Bmatrix} -V(x=0) \\ -M(x=0) \\ V(x=L_i) \\ M(x=L_i) \end{Bmatrix} \quad (6)$$

Similarly, it is possible to write the nodal degrees of freedom $\{p\}$ (*i.e.* the nodal deflections and rotations) as a function of the C^i analytic constants. Noting that $\theta(x) = \mathrm{d}v/\mathrm{d}x$,

$$\begin{Bmatrix} v_i^I \\ \theta_i^I \\ v_i^J \\ \theta_i^J \end{Bmatrix} = \begin{Bmatrix} v(x=0) \\ v'(x=0) \\ v(x=L_i) \\ v'(x=L_i) \end{Bmatrix} \quad (7)$$

Using the shear force and bending moment definitions and eq.5 it is possible to rewrite eq.6 and eq.7 as follows,

$$\{F\} = [B]\{c\} + \{q\} \quad (8)$$

$$\{p\} = [C]\{c\} + \{g\} \quad (9)$$

where matrices $[B]$ and $[C]$ as well as vectors $\{q\}$ and $\{g\}$ are only functions of the design parameters. Solving equations 8 and 9 simultaneously holds a familiar FEM representation

$$[K]\{p\} = \{F\} + \{f\} \quad (10)$$

Where $[K]$ is the stiffness matrix, $\{F\}$ and $\{f\}$ are the nodal and distributed loads, respectively. The relation with the original matrices is given by

$$\begin{cases} [K] &= [B][C]^{-1}\{p\} \\ \{f\} &= [B][C]^{-1}\{g\} - \{g\} \end{cases} \quad (11)$$

Therefore, the solver implementation is very similar to the FEM approach for both assembly and solving algorithms.

The AEM maximizes accuracy because it retrieves the exact solution, while the FEM would retrieve the exact nodal displacements but underestimate the deflection between the nodes for a given distributed loading, leading to an underestimation of the required pressure. The AEM also maximizes speed because it minimizes the number of elements required to describe each pressurized chamber.

Reducing the overall number of elements also contributes to the efficiency of the optimizer algorithm.

2.2.2 Optimization Module

The optimization module uses the simplex search method of Lagarias et al.[13] and includes variable transformation to address bound constraints for the optimization parameters. This can be used to define both design criteria constraints (*e.g.*: minimum beam thickness and length) as well as constraints imposed by the pressure limits inside each chamber.

The initial set-up requires the user to input a base shape for the "unbumped" spoiler, a target shape, the number of pressurized chambers (*i.e.*: number of beams). A set of constraints for the optimization parameters may be provided as well.

Eq.12 shows the implementation of the objective function.

$$r_j = \max\left(\frac{\left|\overrightarrow{QP_k^j}\cdot \mathbf{n}\right|}{\|\mathbf{n}\|}\right) \quad (12)$$

The minimization objective is r_j and it represents the maximum difference between the current j-iteration deformed shape and the target shape. The deviation is computed by sampling a series of $P_k^j(x_k, y_k)$ points along the shape of the current iteration ($k = 1, 2, \ldots$ n.º of sample points) and then finding the minimum distance from each of these points to a generic $Q_(x, y)$ point along the target shape.

Fig.6 shows how the structural and optimization modules are integrated into the design algorithm.

The base and target shapes are first loaded and the user inputs the number of pressurized chambers. Each chamber is associated with a series of new optimization parameters. Adding chambers has therefore a very significant impact on the problem complexity. Moreover, increasing the number of optimization parameters has a disproportional effect on the number of calculations the optimizer must perform and consequently leads to a rapid growth of the algorithm runtime.

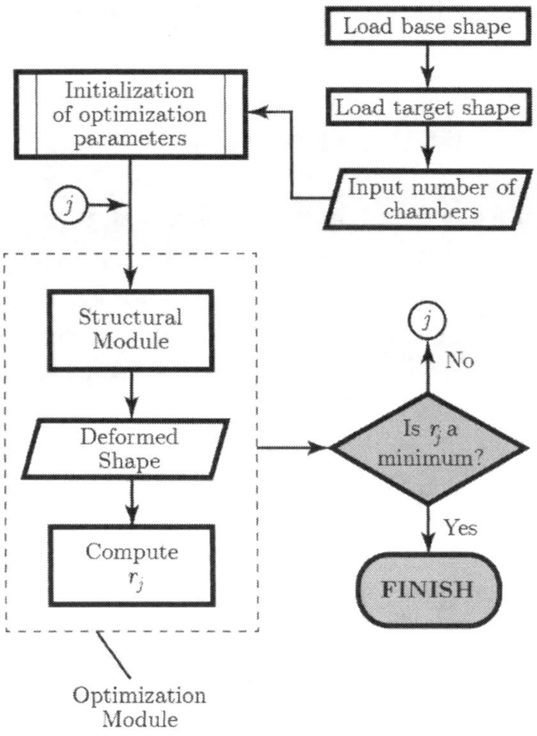

Figure 6: Flow diagram for the design algorithm

2.2.3 Target Shapes

As shown by Sommerer et all.[14] the position of the bump peak plays an important role for the drag reduction but the exact bump shape doesn't. This means that there is freedom to generate the target shapes in a way that is more efficient (*i.e.* requires less pressure to achieved the desired peak position) while still respecting mandatory shape constrains imposed by the nature of the structure (*i.e.* clamped leading and trailing edges). Therefore, if care is not taken while generating those shapes, a beam model with physical background might not able to approximate the given target shape, or it might do so using an excessive actuation pressure.

The target shapes are generated using cubic splines that interpolate the desired peak position and ensure compatibility with the clamped leading and trailing edges.

The next figure shows the typical "overshoot" that happens on the optimized shape when the target shapes are generated using a cubic spline containing only the peak position P and both leading (O) and trailing (L) edges. The target shape is very asymmetric around the peak position and the asymmetry increases as the peak moves towards the leading or trailing edges. Therefore, the peak of the optimized shape tends to deviate from point P so that it is possible to minimize the maximum difference between the curves.

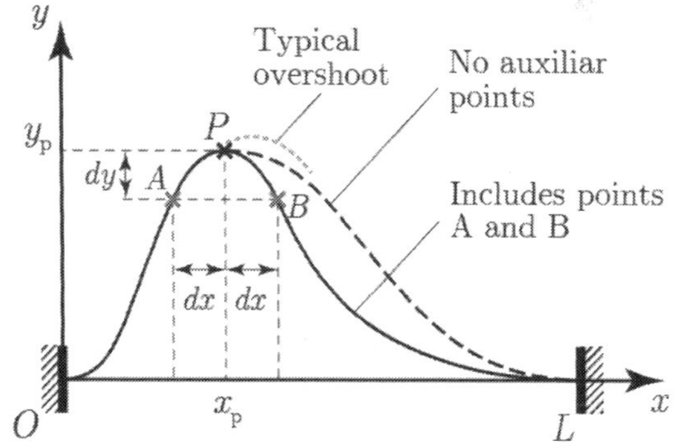

Figure 7: Bump shape geometry

The solution is enforcing some degree of local symmetry around point P using the auxiliary points $A(x_p - dx, y_p - dy)$ and $B(x_p + dx, y_p - dy)$. Values of dx and dy that better approximate the actual physical behaviour will then have to be determined.

When interpolating point B, the slope is averaged using x_p

$$\left.\frac{dy}{dx}\right|_B = \frac{x_p\left(-\left.\frac{dy}{dx}\right|_A\right) + \left(\frac{L_x}{2} - x_p\right) m_{\overline{BL}}}{\frac{L_x}{2}} \tag{13}$$

so that when $x_p = L/2$ the curve is perfectly symmetric (*i.e.* $dy/dx|_A = -dy/dx|_B$). The case $x_p = 0$ corresponds to the limit when B would be exactly vertical to O and $dy/dx|_B = m_{\overline{BL}}$, where $m_{\overline{BL}}$ is the slope of the straight line connecting those two points.

After generating the bump shape as in fig.7, the actual target shape is computed by adding the bump shape to the base spoiler profile in the direction perpendicular to the upper spoiler skin.

3. RESULTS

3.1 Target Shapes

Figure 8 illustrates the impact that the target shape has on the optimized peak position of a simple two chamber spoiler design with 0,8 m of chord length (c_{sp}) and a fibreglass skin. Different target shapes that include the same peak position $x_p = 0.35c_{sp}$, $y_p = 0.05c_{sp}$ are compared. The curves have been vertically spaced for visualization purposes only.

Target shape A uses no auxiliary points and the peak "overshoot" is the highest in both x and y directions. The other curves include auxiliary points with fixed $dy = 0.10y_p$ and varying values of dx.

Target B and C represent two extreme cases. Curve B has a large value of dx and the target peak radius is too large. The peak of the optimized shape has too much vertical displacement. Curve C has a small dx which means the target peak radius is too small and the optimized shape has too little vertical displacement. However, it is clear that adding the auxiliary points improves greatly the horizontal positioning of the peak in both cases because of the resultant local symmetry enforcement.

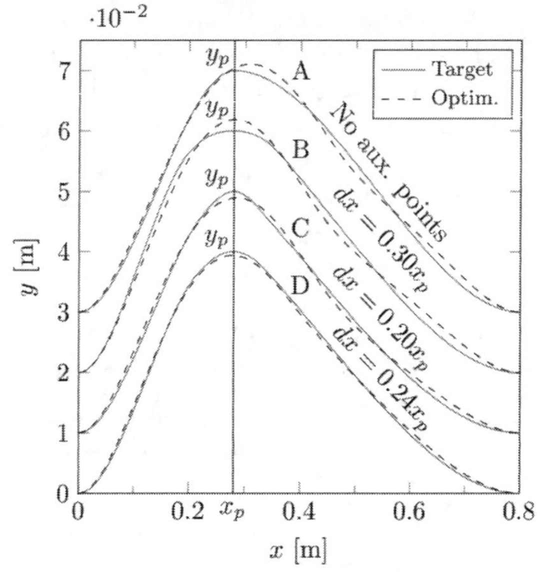

Figure 8: Effect of parameters dx and dy

Curve D shows that after proper adjustment of the values dx and dy it is possible to generate target shapes that not only meet the design constraints, but are also more easily realizable by the physical model.

3.2 Deformation Envelope

Different aircraft have different aerodynamic requirements regarding the bump shapes. Moreover, for a given aircraft, the optimal bump shape also varies under different flight conditions. This means that there is the need for an envelope of spoiler bump deformations. In this section, a simple flat spoiler is considered and a range of bump shapes are generated by varying the peak position along the chord of the spoiler.

Figure 9 shows the deformation envelope for a group of peak positions ranging from 20% to 45% of the total chord length while the peak height is kept constant and equal to 5% of the chord length. The dashed vertical lines represent the optimal position of the two chamber interface for each target.

The deformation envelope can be completed with an actuation envelope. The actuation envelope is computed by plotting the maximum chamber pressure against the peak position. Figure 10 shows the actuation envelope that results from the deformation envelope presented in fig.9 for the two chamber design.

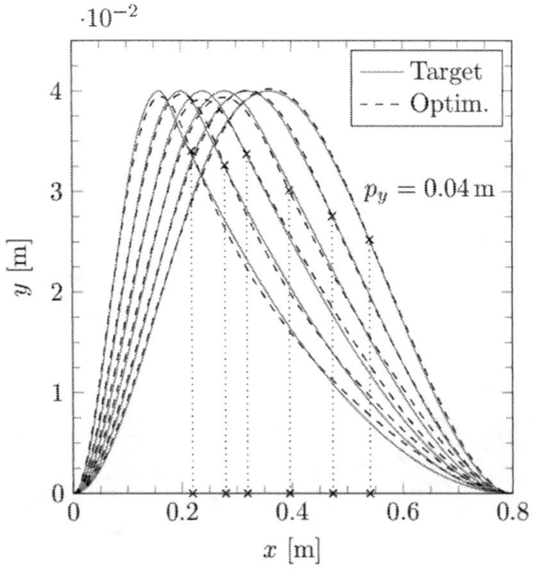

Figure 9: Typical deformation envelope

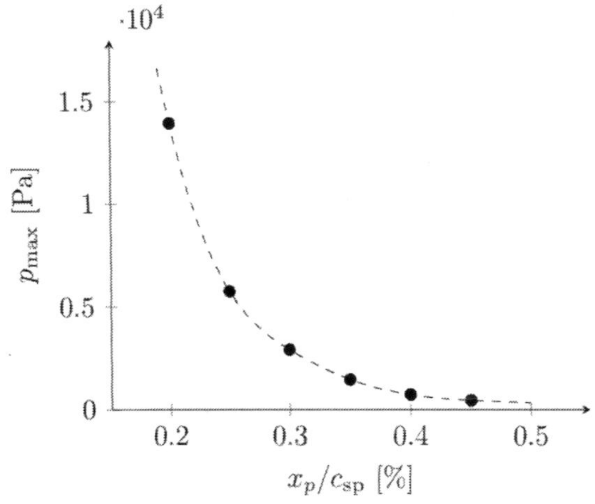

Figure 10: Typical actuation envelope

As a simple flat spoiler was considered for the "clean configuration", the minimum actuation pressure happens when the peak position is located symmetrically in the middle of the plate. As the peak moves towards the leading or trailing edges, the pressure starts increasing disproportionately as a consequence of the greater deformation level required for such configurations. The actuation envelope is therefore symmetric to $x_p/c_{\mathrm{sp}} = 0.5$.

It is clear that the actuation pressure can vary considerably as the optimal peak position changes due to aerodynamic reasons (*e.g.* change in flight conditions and/or aircraft weight). Which draws attention to the fact that, once a chamber configuration is chosen, generating bumps in off-design conditions can become extremely inefficient. This means that a detailed aerodynamic study must be carried out for each specific aircraft in order to ensure that the required deformation envelope (determined by the aerodynamics) is located in the most favourable region of the actuation envelope (to increase structural efficiency).

For more complex "clean configurations" (*i.e.* spoilers with curved base shapes) and complex aerodynamic requirements that leads to challenging deformation requirements and it might be necessary to use a tapered upper spoiler skin in order to cope with the fact that the chamber configuration is fixed in the final design.

3.3 Number of Chambers

The number of pressurized chambers affects directly the weight and complexity of the spoiler structure.

Therefore in this section, the influence of the number of chambers on the spoiler efficiency is investigated. The same spoiler is used to generate a deformation envelope identical to the one in fig.9 using now three chambers.

Figure 11 shows the actuation envelope for the three chamber configuration. The maximum actuation pressure is no longer symmetric in respect to the middle of the spoiler. Nevertheless, and as previously, as the peak position moves towards both the leading and trailing edges, the required maximum pressure increases significantly.

To better assess the effect of an additional chamber, figure 12 shows the relative difference between maximum chamber pressures for the two configurations in study.

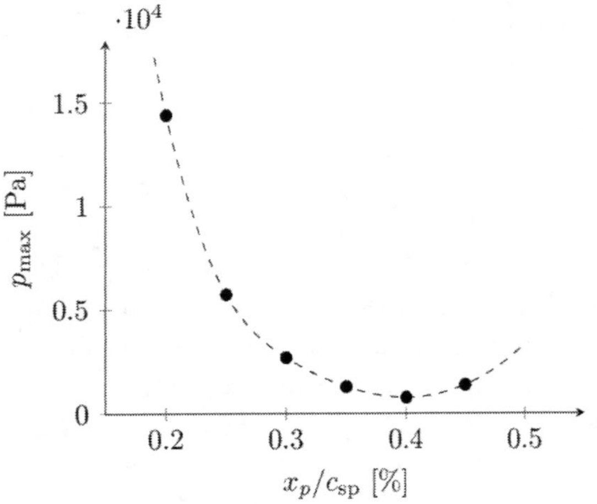

Figure 11: Actuation envelope for the three chamber spoiler.

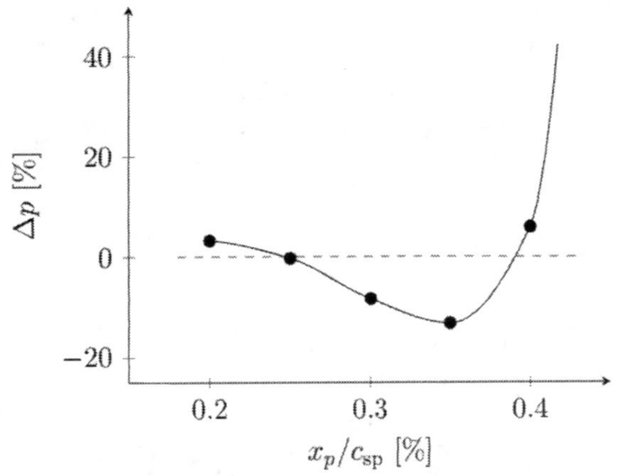

Figure 12: Pressure variation

It is possible to verify that, even though there is a loss of structural efficiency for the outermost regions of the spoiler, the inner region actually benefits from a lowering of up to 13% of the maximum pressure compared to the two chamber configuration. For the current spoiler, adding extra chambers wouldn't contribute to a substantial decrease of the actuation pressure.

Figure 13 shows the pressure history during the optimisation process for point $x_p/c_{\text{sp}} = 0.35$ of fig. 12 where the 3 chamber design lowers most effectively the actuation pressure. The spoiler with two chambers is depicted in black while the three chamber configuration is shown in red. The chamber number increases from leading to trailing edge.

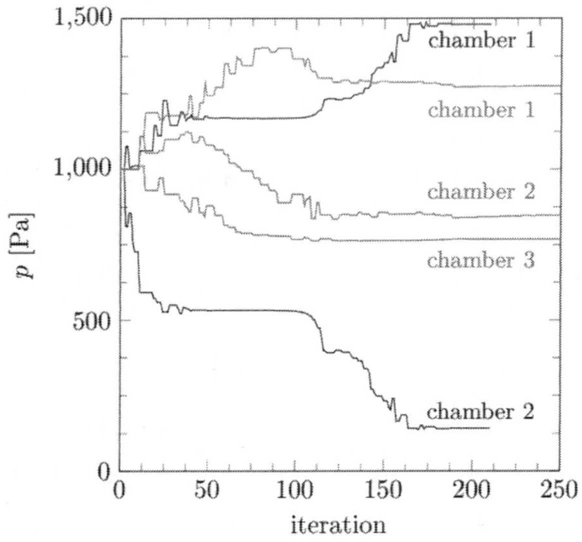

Figure 13: Pressure during the optimization for a peak position of $x_p/c_{\text{sp}} = 0.35$

The region of the actuation envelope where the three chamber configuration allows for a reduction of the maximum actuation pressure behaves as fig.13 and fig.14. Adding another chamber within the innermost region of the spoiler allows the reduction of the maximum actuation pressure (i.e. pressure in chamber 1 is lower for the three chamber design). The pressure in the remaining chambers increases.

Figure 14 shows that when a chamber is added within the innermost region of the spoiler, its end moves to the trailing edge. This facilitates generating the peak shape at lower pressure levels.

However, as a trade-off, because of the added flexural rigidity introduced by an additional interface between new chambers and the greater length of the first chamber, the local slope at the intersection of chambers 1 and 2 is higher. Chambers 2 and

3 are therefore used to recover the trailing edge geometry at the cost of an increased actuation pressure in those chambers because their interface contributes to the upper skin flexural rigidity and can be interpreted as "a cost of adding extra chambers".

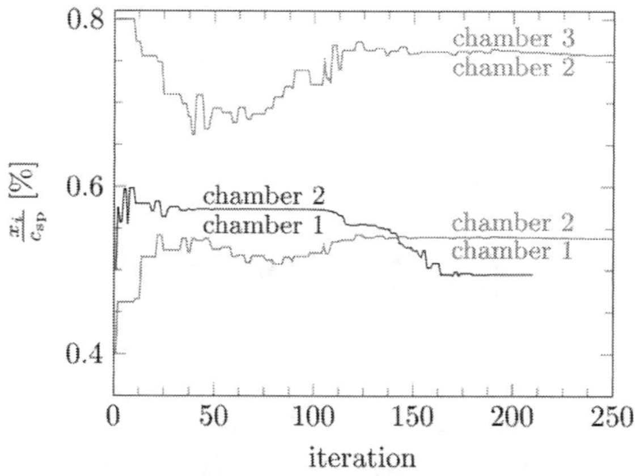

Figure 14: Chamber positioning

For more complex base shapes, and depending on the specific aerodynamic requirements, an assessment of the increased structural complexity that arises from additional chambers should always be carried out first. Specially if the deformation envelope needs to include some peak positions in the outermost regions.

3.4 Spoiler Ends

The design of the spoiler ends has a great impact on the structure performance and its capacity not only to conform to the desired peak position, but also to maintain an aerodynamically acceptable shape. In this section, three different designs for the spoiler ends are considered: a spoiler with open ends (and thus free tips); a spoiler with fixed ends and a spoiler whose ends have been closed using a 1 mm thin fibreglass plate like the upper spoiler skin.

The results are computed using a commercial FEM code and a complete 3D spoiler model. Fig.15 shows the deformation for each design: the dashed curves represent the deformation of the spoiler tips (*i.e.* upper skin edge at both spoiler ends) and the solid lines show the deformation in the middle of the spoiler.

For the spoiler configuration with free ends, the bump profiles on the spoiler tips and midplane present little variation. Moreover, the midplane bump is exactly coincident with the bump shape predicted by the structural solver that was developed. The full 3D model shows that the outermost region of the spoiler experiences some additional deformation, but for the innermost region, there is good concordance between the full 3D model and the simplified semi-infinite plate model.

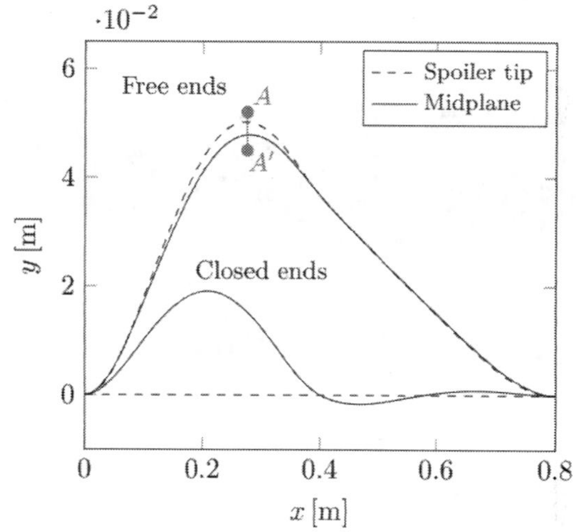

Figure 15: Pressure during the optimization

The designs with closed and fixed ends are much stiffer. Both designs present identical deformations and for that reason only the closed end variant is depicted in fig.15. The additional stiffness introduced by the closing plates has a dramatic effect on the deformation of the spoiler tip bump which now behaves as if it were fixed. The inner region is also greatly affected: not only is the deformation much lower than desired, but the peak position on the x axis is considerably different. The bump profile is also completely altered and a "wavy" region near the spoiler trailing edge appears.

Figure 16 presents the span-wise evolution of

the peak deflection obtained by inspecting the upper skin deformation along a plane AA' coincident with the peak position as shown in figure 15. The same was done along the fixed and closed end designs. The latter have almost the same span-wise deformation except for a region near the ends where the spoiler with the closing plates presents a slightly larger deformation. In any case, such configurations lead to completely different solutions, showing that the structural model developed can only be used for accurate predictions if the spoiler has free ends. Again, the red line in figure 15 shows that there is indeed a considerably larger region ("the effective bump region") where the current model provides accurate predictions and can be used as a useful design tool.

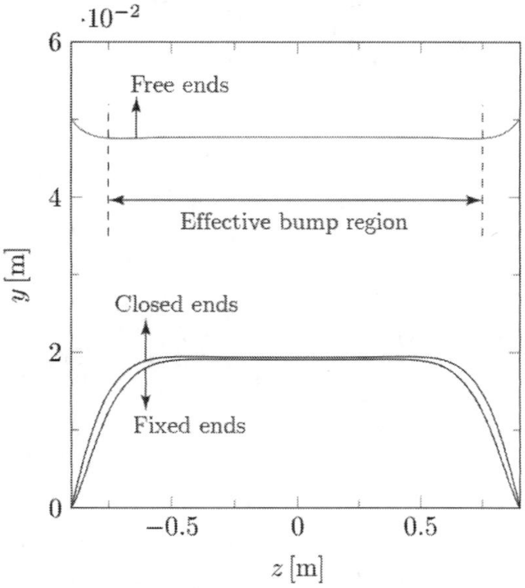

Figure 16: Span-wise peak deformation

On a final note, it should be mentioned that closing or fixing the spoiler ends increases the bump deformation complexity considerably and leads inevitably to a completely 3-dimensional problem where simplified models can hardly be implemented. Furthermore, even if the actuation pressure is increased in order to reach the desired peak deflection, the deformation would, as mentioned, lead to a "wavy" region near the trailing edge that presumably has a very negative effect on the aerodynamics. Therefore, the free end concept seems to be the most promising alternative as it provides a large region where the bump shape is maintained uniform using lowest actuation pressure possible and avoiding complex and aerodynamically unfavorable span-wise deformations.

4. CONCLUDING REMARKS

A structural concept of pressurized chambers is presented in this study for the realization of an adaptive spoiler shock control bump. Using a simplified 2D parametric model the influence of selected design parameters is investigated. Depending on the geometry of pressurized chambers the effect on the actuation pressure is studied. Furthermore, the effect of an increasing number of chambers on the actuation pressure and the chamber design is investigated. The deployment of the bump for various designs of the spoiler ends closing is presented.

Since the peak position (chord-wise location and deflection of the bump peak) is the driving aerodynamic parameter, a method for generating "easily realizable" bump shapes (from a structural standpoint) is proposed.

The concept suggested has enough freedom to conform to a wide range of bump shapes even for a simple 2 chamber design. It has been shown that increasing the number of chambers can lead to a reduction of the actuation pressure, but since each chamber interface contributes to an added flexural rigidity, this effect can only be used within a limited region of peak positions and up to a maximum number of chambers that varies according to the different spoiler base shapes.

The type of closing of the spoiler ends has been shown to have both a strong impact on the ability of the structure to conform to the desired peak position and also on the quality of the entire bump target shape.

ACKNOWLEDGEMENTS

The work is funded by the 5th Federal Aeronautical Research Program (LUFO V) by the Federal Ministry of Economic Affairs and Energy (BMWi) in the project LDAinOP, FkZ: 20A1302B.

REFERENCES

1. SCHRAUF G., GÖLLING, B., WOOD, N. *Key Aerodynamic Technologies for Aircraft Performance Improvement.* "KATnet- 5th Community Aeronautical Days 2006" Vienna, Austria, 19-21st June 2006
2. KINTSCHER, M. and WIEDEMANN, M. *Design of a smart leading edge device.* In: Adaptive, Tolerant and Efficient Composite Structures Springer. Pages 381-390. ISBN 978 3 642 29189 0.
3. SMITH, W. J., LOCK, W.W. *Variable Camber System Integration and Operational Performance of the AFTI/F-111 Mission Adaptive Wing.* "NASA Technical Memorandum 4370", 1992
4. STANEWSKI, E. *A Euroshock I and II: A survey.* "Proceedings IUTAM Symposium on Mechanics of Passive and Active Flow Control", 7–11 September, 1998, Göttingen, Germany, pp 35– 42. Kluwer Academic Publishers, 1998.
5. ASHILL, W. J., LOCK, W.W. *A novel technique for controlling shock strength of laminar-flow aerofoil sections.* "Proceedings 1st European Forum on Laminar Flow Technology March 16-18th, 1992", pp 175–183. DGLR, AAAF, RAeS, Hamburg, Germany, 1992 *Acta Astronautica,* Vol.12, No.7/8, 1985, pp. 599-607.
6. KNAUER, A, *Die Leistungsverbesserung transsonischer Profile durch Konturmodifikationen im Stoßbereich.* PhD thesis, DLR, 1998
7. DARGEL G. *Widerstandsreduktion durch Stoßkontrolle mit adaptiver Hautstruktur: Shock Control Bump.* "Proceedings DGLR" Jahrestagung, 1998
8. DARGEL G., ARCHAMBAUDJ. P. *Assessment of the capability of drag reduction of the shock control device sc bump on airfoil flows and application aspects on wing.* "Proceedings IUTAM Symposium on Mechanics of Passive and Active Flow Control 7–11th September, 1998, Göttingen, Germany" pp 57–62. Kluwer Academic Publishers, 1998
9. PRITSCHOW G., WADEHN W. and KEHL G. *Shape adaptation of fixed wing aircraft by shape memory alloys Proceedings of the 7th International Conference on New Actuators.* "Proceedings of the 7th International Conference on New Actuators", Actuator 2000 1012 June, 2000, Bremen, Germany, 2000.
10. WADEHN W., SOMMERER A., LUTZ T., FOKIN D., PRITSCHOW G., WAGNER S. *Structural Concepts and aerodynamic design of shock control bumps.* "Proceedings of the International Council of the Aeronautical Sciences", 23rd Congress of Internation Council of the Aeronautical Sciencies, 8-13 September 2002, Toronto Canada, 2002
11. BEIN T., HANSELKA H. and BREITBACH E. *An adaptive spoiler to control the transonic shock.* "Journal of Smart Mater", Struct. 9, 141148, 2000.
12. POLICARPO, H., NEVES, M.M *Using Symbolic Computation for Teaching Structural Mechanics I: Frames.* "CSEI2012 – Conferência Nacional sobre Computação Simbólica no Ensino e na Investigação", Lisboa, 2-3 Abril de 2012
13. LAGARIAS, J.C., REEDS J. A., WRIGHT M. H., WRIGHT P. E., *Convergence Properties of the Nelder-Mead Simplex Method in Low Dimensions.* "SIAM Journal of Optimization", Vol. 9 Number 1, pp. 112-147, 1998
14. SOMMERER, A., LUTZ, T., WAGNER, S. *Numerical Optimisation of Adaptive Transonic Airfoils with Variable Camber.* "ICAS 2000 Congress", Number 2111, pp. 1-10, 2000

ICAST2015 # 067

Stall Recovery of the Span-wise Morphing Trailing Edge Concept via an Optimized Nonlinear Model

Lawren L. Gamble[1*], Alexander M. Pankonien[2], and Daniel J. Inman[3]

[1] Ph.D. Candidate, Department of Aerospace Engineering, University of Michigan, Ann Arbor, USA
[2] Postdoctoral Researcher, Department of Aerospace Engineering, University of Michigan, Ann Arbor, USA
[3] Collegiate Professor and Chair, Department of Aerospace Engineering, University of Michigan, Ann Arbor, USA

Abstract

Inspired by the dynamic capability of a bird to continuously change its wing shape to suit an arbitrary flight condition, the current work uses model-informed experiments to investigate the effects of smooth, span-wise variation in camber of a morphing wing on aerodynamic stall via the Span-wise Morphing Trailing Edge (SMTE) concept. While previous research has shown that smooth morphing wings have a distinct advantage over conventional discrete control surfaces, in part by reducing drag by eliminating vortex formations that occur at the aileron's geometric discontinuities, the complex multidisciplinary challenge associated with assessing these advantages near stall motivates this study. The physical model used in this body of work employs macro-fiber composites (MFCs), a piezoelectric smart material, as the active sections of the camber-morphing wing with alternating elastomeric honeycomb sections. The conformal bending and large bandwidth properties of the MFC allow for a quick and precise trailing edge camber response while the elastomeric honeycomb eliminates geometric discontinuities. By using MFCs, which act as both the aileron structure and actuator, the system's mechanical complexity was reduced providing an advantage to other morphing concepts which rely on conventional actuators. A variation of nonlinear lifting-line theory was implemented to include span-wise morphing, and was optimized in order to predict the desired trailing edge deflections that minimize adapted drag. The wing performance during static stall was assessed for a 0.3 meter chord, aspect ratio 6, finite wing of half span in a wind tunnel. Additionally, flow visualization was implemented for verification. The resultant forces were compared to those of a conventional, articulated wing of equal dimensions with servo-driven ailerons.

1. INTRODUCTION

Nature frequently provides tremendous inspiration for engineers and within aerospace this inspiration is typically drawn from natural flyers like birds, which are comparable to the UAVs represented in this study. Birds have a unique ability to locally morph their wings in order to respond to a change in aerodynamic conditions and adapt to unfavorable disturbances. For example wing sweep in swifts, a high speed flyer, was shown to effectively broaden the aerodynamic performance over a large range allowing the swift to fine-tune its aerodynamic profile to achieve a specific flight condition. Most notably, their glide performance at high speeds is optimized at large sweep by reducing drag and maintaining adequate lift.[1] Alternately, the steppe eagle, a passive soaring flyer, can deliberately stall a section of the inside wing during banked turns to increase the rate of

* llgamble@umich.edu

maneuvering in roll and yaw.[2] This advanced manipulation of aerodynamics inspired some of the first aircraft morphing by engineers.

One of the first instances of morphing aircraft was the Wright Brothers Flyer that was developed in 1903 and utilized twisting wings to employ flight control. While rigid aircraft quickly dominated the field during World War 1 due to their robust design, major advances within morphing have been made since then. More recently, research within morphing has focused on cruise performance of large aircraft, notably the Variable Camber Continuous Trailing Edge Flap by NASA and Boeing [3], and the FlexSys Mission Adaptive Compliant Wing [4]. But while a slight reduction in drag can contribute to significant savings in fuel consumption at transonic cruising speeds, these aircraft remain designed for a specific cruise condition meaning that any deviation from this optimized case incurs additional aerodynamic and financial costs. An alternative approach to morphing would be to design the wing with the intention of adaption in order to accommodate for changes in flight conditions or mission requirements. Additionally, since small UAVs are capable of exotic flight maneuvers, adaptation to recover from the post-stall regime is highly significant. Yet, while the capabilities of a smooth morphing MALE/LALE wing have been proven to favorably adapt to off-design flight conditions in the linear regime by Pankonien *et al.* [5]-[6],[7], it has yet to be proven effective at large angles of attack past stall.

The current work aims to analyze the capabilities of a morphing wing with span-wise variation in trailing edge camber to adapt to an off-design flight condition and recover from stall, and compare its performance to a conventional articulated flap wing. The scenario of interest, as seen in Figure 1, assumes an on-design flight condition $C_{L,Design}$ (Point #1), and an off-design condition $\alpha_{current}$ (Point #2) defined as some change in geometric angle of attack past the stall angle. The objective is to adapt the wing from the off-design condition back to the target lift while maintaining the off-design angle of attack (Point #3). However, while the aircraft has adapted to the new flight condition it now operates at a higher drag, resulting in a drag penalty. This is shown in Figure 2 which demonstrates how a larger decrease in the drag profile from the unadapted condition results in a smaller drag penalty. A model-informed experimental approach was used to determine which configurations to test. The model was then optimized to achieve a target lift at a given angle of attack by varying the aileron deflections and shows that the adapted SMTE reduces the drag in the post-stall region nearly twice as much as the articulated wing while requiring substantially less actuation of the ailerons.

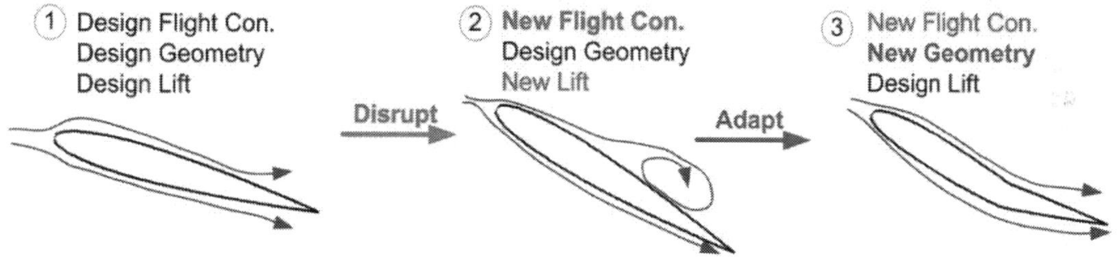

Figure 1. Graphic summary of the design scenario depicting the theoretical aileron response

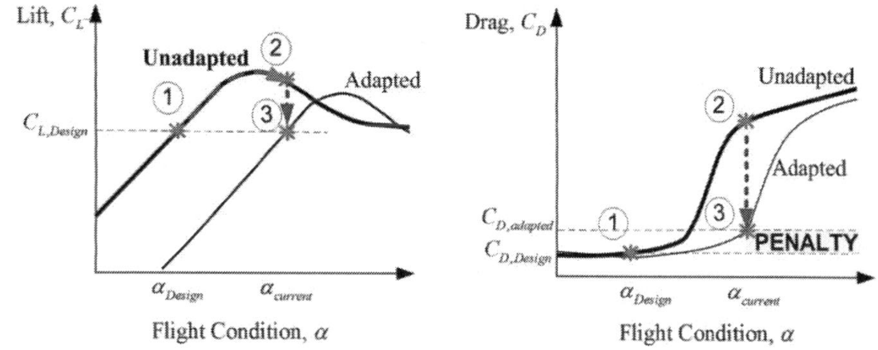

Figure 2. Graphic summary of the design scenario in terms of forces, depicting the adapted aerodynamic curves and associated drag penalty

2. NONLINEAR LIFTING LINE MODEL

While linear aerodynamic models have been thoroughly investigated, nonlinear models have arguably been less successfully implemented and are more rarely used to analyze morphing aircraft. Perhaps the first nonlinear method was developed in 1934 by Tani[8] and relied on assuming an initial circulation distribution, calculating the induced followed by the effective angle of attack, and using this to lookup the section lift coefficient from nonlinear experimental data. With this, a new circulation distribution is calculated and the process is iterated until the solution converges. This method was popularized by Sievells and Neely in a NACA report[9] but criticized for its inability to handle instances where sections of the wing are stalled, implicitly having a negative lift-curve slope[10]. Further analysis by Piszkin and Levinsky[11] indicated that in fact, the post-stall solution is non-unique and frequently converges with saw-tooth oscillations that evidently aren't a true aerodynamic phenomenon.

More recently, a method developed by Chattot[12] resolves many of the issues discussed previously while maintaining a similar iterative format. Chattot's method employs the use of an artificial viscosity term that other models fail to consider. Not only does Chattot's model show good agreement with experiments post-stall where the lift-curve slope is negative, but the circulation distribution also remains smooth and does not show saw-tooth oscillations. A step-by-step summary of this procedure is described below in dimensionless form and will be used in the following optimization.

1. Discretize the wing into jx span-wise sections with a cosine distribution for higher resolution at the tips near the trailing vortices

$$y_j = -\cos\left(\frac{j-1}{jx-1}\right) \qquad j = 1,2,\dots jx \tag{1}$$

2. Define the points of integration between each of the sections described above to avoid singularities when calculating the induced downwash

$$\eta_k = -\cos\left(\frac{k-1/2}{jx-1}\pi\right) \qquad k = 1,2,\dots jx-1 \tag{2}$$

3. Assume an initial circulation distribution Γ (elliptical is adequate)
4. Calculate the induced downwash w_{ind} at each section caused by the presence of the tip vortices, where integration has been replaced with a finite sum

$$w_{ind,j} = -\frac{1}{4\pi}\sum_{k=1}^{jx-1}\frac{\Gamma_{k+1}^n - \Gamma_k^n}{y_j - \eta_k} \qquad j = 2,3,\dots jx-1 \tag{3}$$

5. Linearize the integro-differential governing equation developed by Prandtl[13] using Newton's Method where c_j is the section chord, $C_{l,j}$ is the section lift coefficient, $\frac{dC_{l,j}}{d\alpha}$ is the section lift curve slope, $\alpha_{eff,j}$ is the section effective angle of attack, and noting that $a_j < 0$

$$\Gamma_j + \Delta\Gamma_j = \frac{1}{2}c_j\left(C_{l,j} + \frac{dC_{l,j}}{d\alpha}\Delta\alpha_{eff,j}\right) \tag{4}$$

where $\Delta\alpha_{eff,j} = \frac{\Delta w_{ind,j}}{1+w_{ind,j}^2}$ and $\Delta w_{ind,j} = -\frac{1}{4\pi}\left(\frac{1}{y_j-\eta_{j-1}} - \frac{1}{y_j-\eta_j}\right)\Delta\Gamma_j = a_j\Delta\Gamma_j$

The equation is then written in iterative form where n represents the iteration number, ω represents a damping factor, and the artificial viscosity coefficient μ is added in such a manner that the equation remains unchanged and the span-wise stations are now coupled

$$\left(1-\frac{1}{2}c_j\frac{dC_{l,j}}{d\alpha}\frac{a_j}{1+w_{ind,j}^2}+2\mu\right)\frac{\Delta\Gamma_j^n}{\omega}=\frac{1}{2}c_jC_{l,j}-\Gamma_j^n+\mu\left(\Gamma_{j+1}^n-2\Gamma_j^n+\Gamma_{j-1}^{n+1}\right) \quad (5)$$

$$\text{where}\ \begin{cases}\mu\geq\max\left(\frac{1}{4}c_j\frac{dC_{l,j}}{d\alpha}\frac{a_j}{1+w_{ind,j}^2};0\right)\\ \omega<1\ (underrelaxed)\end{cases}$$

6. Solve for the next circulation distribution $\Gamma_j^{n+1}=\Gamma_j^n+\Delta\Gamma_j^n$ by rewriting as $\Gamma_{j-1}^{n+1}=\Gamma_{j-1}^n+\Delta\Gamma_{j-1}$ and substitute into step 5. Thus, the only unknown in the linearized integro-differential equation is now $\Delta\Gamma_j^n$ and the new circulation distribution can be calculated
7. Steps 4-6 are repeated until the solution converges to a desired tolerance
8. Calculate the total lift, induced drag, and total drag for the finite wing using the converged circulation at each angle of attack where AR represents the wing's aspect ratio

$$C_L=\frac{AR}{2}\int_{-1}^{1}\Gamma(y)dy \quad (6)$$

$$C_{D,Total}=C_{D,profile}+\overbrace{\frac{AR}{-2}\int_{-1}^{1}\Gamma(y)w(y)dy}^{C_{Di}} \quad (7)$$

This method has mainly been used to analyze wing/winglet combinations[14], turbine blades[15], and swept wings[16], yet it is modeled in such a way that analyzing a morphing wing is relatively straightforward since the section-based method allows for span-wise variation in the airfoil characteristics i.e. $C_{l,j}$, $\frac{dC_{l,j}}{d\alpha}$, etc. The primary gap between the existing model and the morphing model developed in this paper is the infinite domain of airfoil characteristics required to characterize an arbitrarily morphed wing. By simulating the airfoil characteristics at the maximum, minimum, and neutral aileron deflections, a three-dimensional surface can be linearly interpolated that relates the airfoil lift, drag, and pitching moment to both the geometric angle of attack α and the aileron deflection η as shown in Figure 3. Thus, given the effective angle of attack and a shape-function describing the span-wise tip deflection, the section lift coefficients can be determined.

As proof of concept, this model was compared to previously collected data for a NACA0012 finite wing of aspect ratio 6 at a Reynolds number of $1.5\text{x}10^5$. The model was informed with section lift data simulated using OVERTURNS, a RANS solver[17] which was then used to calculate the section lift derivatives required for solving the integro-differential equation. It is important to note that the artificial viscosity has some degree of leniency since it is defined as a lower bound; thus, a brief analysis of its effect on the solution post-stall was conducted in order to tune the variable before proceeding to optimize the model. This is important because the artificial viscosity determines the shape of the lift curve post-stall. As shown in Figure 3, the experimental data closely matches the nonlinear model with a root mean squared error (RMSE) of 0.027 when the additional contribution to the artificial viscosity term is 0.5. This value will be used throughout the subsequent optimization. The success of these preliminary results validates the model and serves as the foundation for the optimizer.

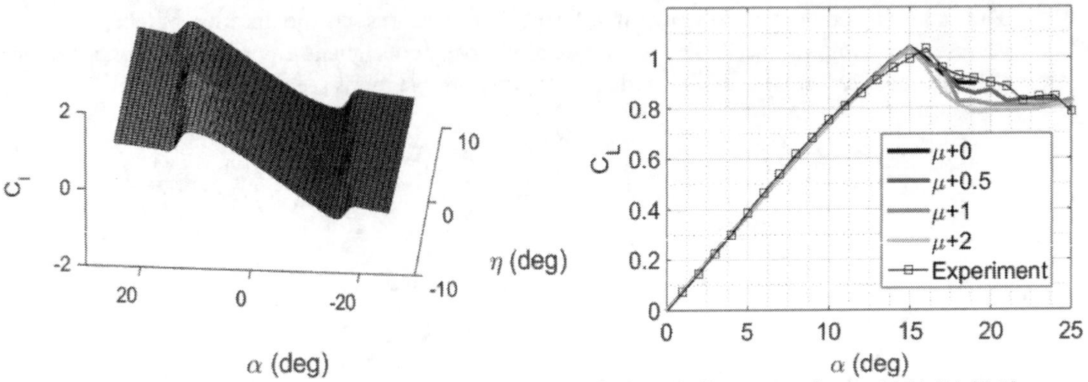

Figure 3. Three-dimensional surface of section lift data (left), comparison of NACA0012 experimental data to the nonlinear model with varying artificial viscosity (right)

3. OPTIMIZATION

Upon development and validation of the nonlinear model, an optimizer was implemented to inform the experiments. Using *fmincon*, a constrained minimization algorithm in Matlab, the optimal aileron configurations were predicted by minimizing the total drag at the off-design case. Two key parameters determine the results of the optimization. First the on-design lift, which constrains the model, was defined as some percentage of the maximum on-design lift $C_{L,max}$. Secondly, the off-design case was specified at some angle past stall. The particular case tested in this study was arbitrarily chosen to occur at 86% of $C_{L,max}$ and an $\alpha_{current}$ of 16°, one degree past stall.

Additionally, further modifications were made to the optimized model to account for the arbitrary variation in trailing edge displacement. While it is common practice to assume an elliptical distribution as the initial circulation guess, this is not an adequate assumption for an arbitrary configuration of aileron deflections. Thus, the objective function was seeded with circulation distributions at each geometric angle of attack that had been converged using the nonlinear model under the assumption of no trailing edge deflections. Lastly, the solution was bounded to 8.85° aileron deflection based on the experimental model discussed in the following section. A summary of the optimizer can be found in Table 1.

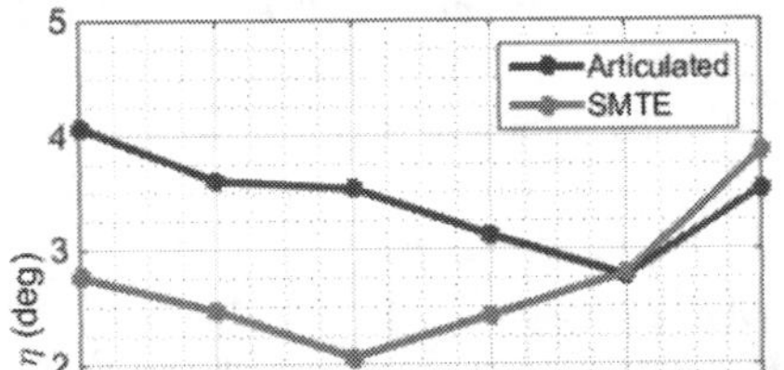

Figure 4. Optimal actuations at $86\%\ C_{L,Design}$ and $\alpha = 16°$

Table 1 Constrained optimization of nonlinear model

Minimize: $C_{D,total}(\eta)$
Given: $\alpha_{off\ design}$
Constrained by: $C_{L,Design}$
Subject to: $0 < \eta < 8.85°$

The data shown in Figure 4 represents the optimized actuations for the articulated wing and the SMTE. It is important to note that the trends noted by Pankonien *et al.*, namely that the SMTE requires less actuation to achieve the same lift distribution, are clearly still valid in the nonlinear range. Since both cases are optimized to achieve the same lift, this indicates that the SMTE is more effective at controlling the forces than the articulated flap wing.

4. EXPERIMENT

Upon development of the nonlinear model and optimizer, experiments were conducted in order to verify the results. The following sections summarize the experimental model and setup. Section 4.1 outlines the construction of both the articulated flap wing and the Span-wise Morphing Trailing Edge (SMTE) concept in addition to detailing the control mechanisms used to ensure the actual aileron displacements matched the prescribed displacements. Finally, Section 4.2 summarizes the procedure implemented for taking and storing the desired data.

4.1 Experimental Model

Two half-span finite wings were used in the following experiments. The first was a conventional articulated flap wing, driven by internalized mechanical servos. The construction of this wing can be seen in more detail in Figure 5. The second wing is the SMTE developed by Pankonien *et al.* which is known for improving control of aerodynamic forces and reducing the drag penalty associated with adaptation. This is due to the elimination of geometric discontinuities where chordwise and span-wise losses occur due to vortex formation. The SMTE, shown in Figure 6, was manufactured with repeating MFC (active) and honeycomb (inactive) sections, which produced smooth conformal actuation of the trailing edge. Both wings used closed loop control to regulate the actuator tip displacements. While the articulated flap wing relied on the internal controller of the servos, the SMTE utilized internalized Flex Sensors shown in Figure 6d which translated the tip deflection to a change in resistance which was read by a data acquisition system and used as the input to a PI controller.

While the internal mechanisms of both wings were unique, they were constructed for comparison purposes and thus share the same undeformed geometry. The wings were built with a symmetric NACA0012 airfoil of aspect ratio 6 with a 0.305m chord and 6 active sections each. The width of the articulated flaps spans the same width as the repeating active-inactive section in the SMTE. Additional information on the construction of the models can be found in Pankonien *et al.* 2014.

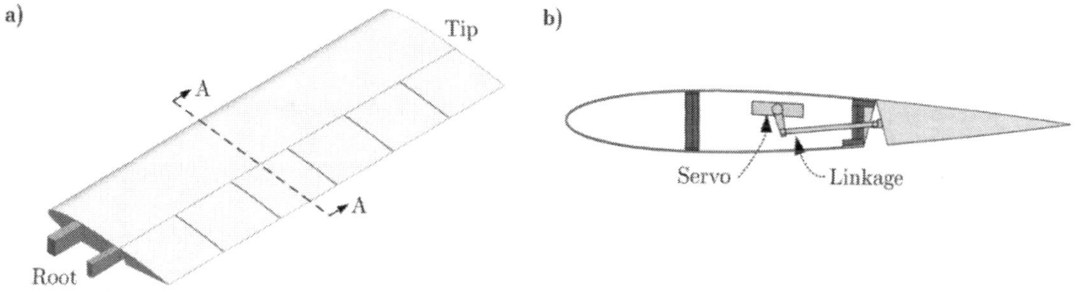

Figure 5. a) Isometric view and **b)** cross-sectional view of the articulated flap wing

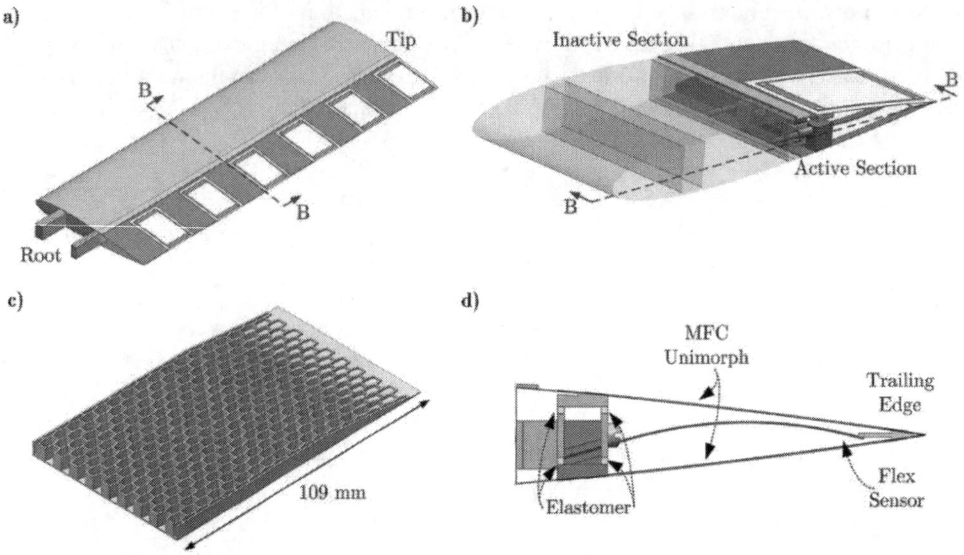

Figure 6. a) SMTE full isometric view **b)** repeated section isometric view, **c)** inactive honeycomb **d)** cross-sectional view

4.2 Experimental Setup

Wind tunnel tests were conducted in the 1.5m x 2.1m tunnel at the University of Michigan depicted in Figure 7a. The wings were vertically mounted to the 6-axis ATI Delta force balance above a ground plate that raised the reflecting plane of the half-span wing by 30 cm, outside of the boundary layer of the wind tunnel. The mounting rig was shielded by attaching a fixed cowling in the shape of a NACA0012 airfoil that inherently extended the wing to the full aspect ratio and reduced the gap between the finite wing and the ground plate to 0.5 cm. Below the ground plate, a shroud shielded the force balance from the airflow.

Alpha sweeps were conducted in order to obtain the C_L-α curves for the on-design and off-design configurations by holding the aileron configurations constant and recording data from 0 to 25 degrees. The angle of attack was set using two high-precision Keyence LK-G402 laser displacement sensors configured to measure the rotation of the mount. This was achieved by directing the fixed lasers to a cross bar on either side of the mount, perpendicular to the wing's orientation. As the mount rotated, the lasers effectively translated the change in out-of-plane displacement to a change in rotation via basic geometry. Lastly, the closed-loop control allowed for angular accuracy to up to 1/40th of a degree.

Over the course of the test, forces, dynamic pressure, temperature, and MFC tip displacement readings were time averaged and recorded. The 6-axis force balance collected all forces and moments although only lift and drag are reported in this study. The dynamic pressure was recorded using an Omega PX2650 differential pressure transducer from a pitot tube 2 meters upstream of the wing. Finally, the tunnel temperature was recorded with a thermocouple mounted upstream at the base of the tunnel.

The experiment was orchestrated in MATLAB using both a dSPACE MicroLabBox and cDAQ system. This allowed for precise control of the ailerons at a much higher rate than the sampling rate of the sensors. The cDAQ collected all external sensor data and controlled the angular position of the wing while the dSPACE controlled the aileron deflections and collected the Flex sensor data from the SMTE. As mentioned in Section 4.1, the servos are manufactured with their own internal controller; however, the dSPACE was used to execute a PI controller written in Simulink for the MFCs to prescribe the accurate tip displacements. A flow diagram of the experiment is depicted in Figure 7c.

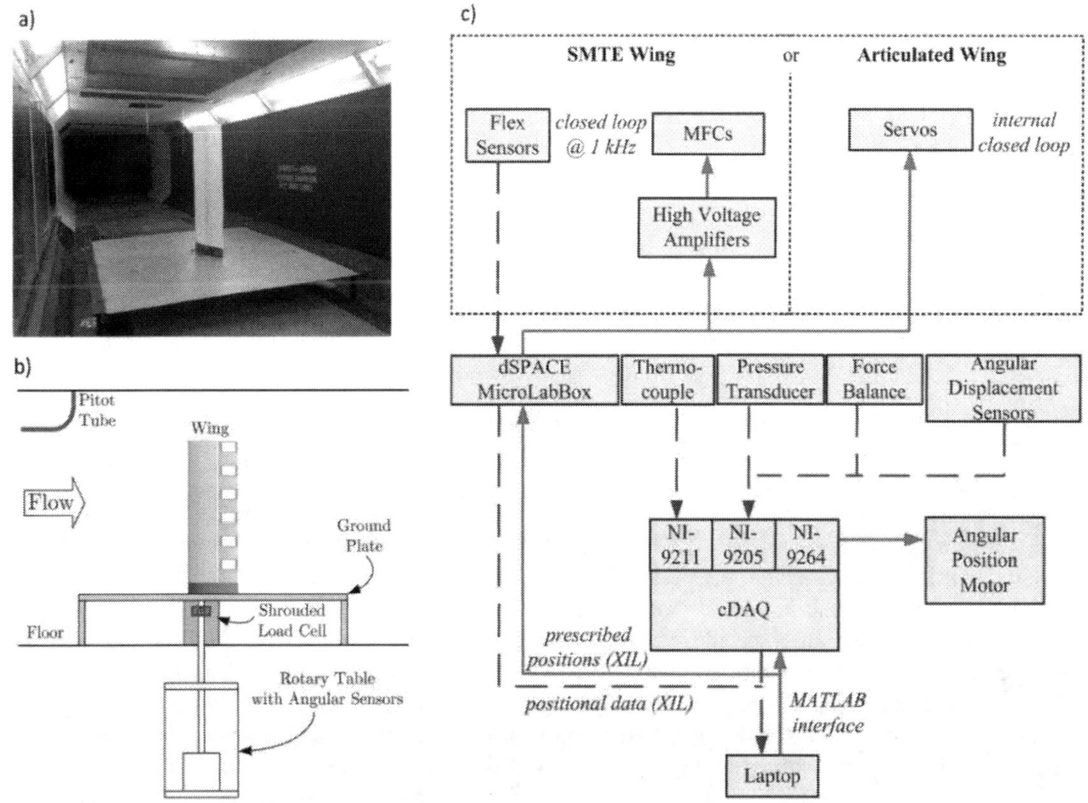

Figure 7. a) Mounted wing **b)** experimental setup **c)** experimental data flow

5. RESULTS & DISCUSSION

In this section, the aerodynamic forces of the SMTE and articulated wing are analyzed using the method described in Section 4. The goal was to prove that the optimized nonlinear model, when applied to span-wise variations in tip deflections, could predict the aileron deflections required to recover from stall. While a preliminary analysis presented in Figure 3 indicated that the nonlinear model could accurately predict the aerodynamic behavior of a uniform finite wing with an RMSE of 0.027 by tuning the artificial viscosity, the experiments will show whether or not this same model can be used to predict span-wise varying cases.

The optimized model was set to reach a design lift condition of 0.8816 (86% of $C_{L,max}$) at an off-design angle of attack at 16°. In both models, the tolerance of the lift constraint was relaxed to $1 \cdot 10^{-2}$ in order to meet the strict circulation convergence criteria of 0.01%. The resulting error between the target lift and the optimized lift at the off-design condition for the articulated and SMTE wings was 4.2% and 7.1% respectively as demonstrated in Figure 8. The adapted lift-curves have been generated by modeling the wing at the optimized configuration and provide insight as to whether the wing was able to recover from stall. Notice how the modeled stall angle for the adapted case is lower than the off-design angle of 16° for both the SMTE and articulated wings, indicating that while they have returned to the desired lift the off-design angle occurs post-stall on the new lift curves, thus only partial stall recovery was achieved. This is thought to explain the error between the target lift and optimized lift. While the optimizer was designed to minimize the drag, ideally avoiding the post-stall regime where drag dramatically increases, this is not always feasible and is dependent upon the shape of the 2D lift and drag curves, and the design inputs $C_{L,target}$ and $\alpha_{current}$. The particular configuration analyzed in this study happened to correspond to a case where the closest optimized solution occurs at one degree past stall.

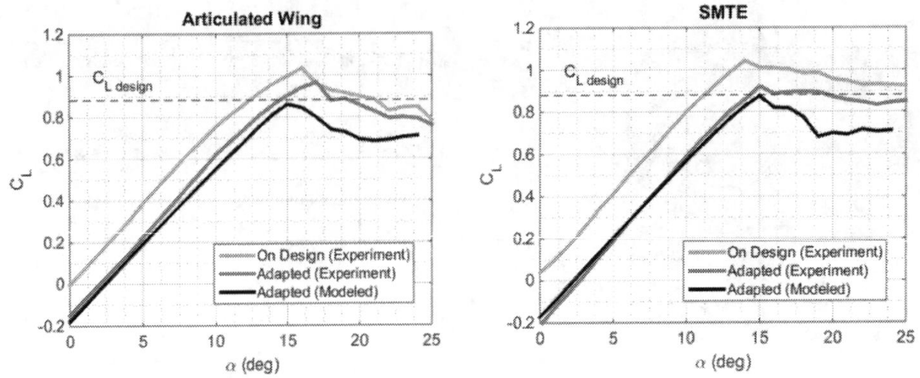

Figure 8. Lift curves at the unadapted (on design) and adapted configurations for the articulated and SMTE wings

Next, a brief comparison of the articulated and SMTE wings was conducted to determine if both wings were constructed identically. While ideally the wings are identical, there is some discrepancy between the articulated wing and the SMTE. This is most evident in the measurements of drag. Due to the inherent sensitivity in drag measurements, the drag data presented hereafter will be reported as the difference between the actual and minimum drag. The results presented in Figure 9 indicate that the articulated wing in fact stalls after the SMTE wing. This is important to note, as this trend will be carried through in the rest of the data. Thus, the following analysis will compare each wing to itself in the unadapted vs. adapted configurations to determine the increase in performance, eliminating any discrepancies between the wing designs.

The ability of the optimized model to predict the experimental data was then analyzed. In both the articulated and SMTE wings, the model predicts the pre-stall region (0 to 16°) relatively well with a RMSE of 0.047 and 0.0060 respectively. This discrepancy in error between the two wings is not unexpected, as the articulated wing's servo connections were comprised of linkages which may have provided some give when applied to an aerodynamic load. This may also explain the shift in the stall angle in the articulated wing. However, post-stall the experiments deviate significantly from the models. Although the curvature of the articulated wing bears resemblance to the modeled lift curve post-stall, the difference in stall angles results in a large post-stall error between the model and experiment. Conversely, the SMTE matches the model's stall angle accurately but post-stall the lift plateaus, a trend which was not apparent in the SMTE 2D lift curve and thus would not be represented in the optimized nonlinear model. This is thought to be caused by the compliance of the silicone skin on the SMTE's trailing edge which, if over-relaxed due to fatigue or age, could act as a sail at larger angles of attack. This further supports the decision to compare the wings based on their increase in performance from the unadapted configurations.

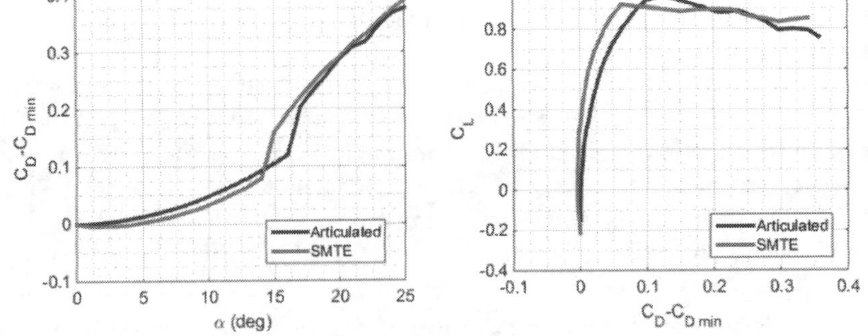

Figure 9. Drag comparison (left) and the lift-drag curves (right) for the unadapted articulated and SMTE wings

Lastly, it is important to note the differences between the articulated and SMTE wings with respect to their drag reduction. Figure 10 presents the drag of each wing at the adapted and unadapted configurations in addition to the change in drag between the unadapted and the adapted configurations. Both cases demonstrate that adaptation may delay the onset of stall as is evident by the shift in the drag curves between unadapted and adapted configurations, thus presenting a sharp decrease in the change in drag at stall in Figure 10 c and d. Furthermore, these results show that the SMTE reduces post-stall drag on average about twice the amount as the articulated wing in the post-stall region. This correlates to a maximum reduction of over 16% at 18° from the unadapted condition. This result is dramatic considering that the SMTE wing's aileron deflections are on average 30% smaller than the articulated wing's deflections. Even in the linear regime where the aerodynamic measurements match closely with the modeled results, the morphing wing substantially outperforms the articulated wing.

While the model appears to have partially succeeded in predicting optimal flap configurations to recover from stall, flow visualization shown in Figure 11 was used to verify whether or not flow separation was indeed reduced if not eliminated. A smoke wand was inserted into the flow upstream of the wing and slow motion video was taken to capture images of the smoke stream at a high frame rate. The aileron deflections presented in Figure 4 are used here. Both wings show similar trends with respect to flow separation. At the on-design flight condition, the flow clearly remains attached across the entire span of the wings as expected. Then, when moved to the off design flight condition the ¼ span and midspan show strong vortex formations indicating that the wings were indeed stalled. Most notable is the adapted condition at the same angle of attack as the off design case, which shows almost complete elimination of the separation in both wings. This data serves to prove that the nonlinear optimization indeed predicts aileron deflections to adequately reduce if not completely eliminate separation.

The results presented here should be viewed as just the first steps in analyzing the nonlinear aerodynamic performance of morphing finite wings. Further examination should include a variety of other design conditions to analyze the limitations of span-wise varying wings to adapt to stall. While the results presented here showed that the particular configuration tested was only partially successful at recovering from stall, this indicates that dual morphing, such as aileron deflection in addition to reflex camber, may be able to achieve stall recovery over a broader range of design criteria. Additionally, higher ordered modeling, such as 3D RANS, could be implemented for higher fidelity and further fluid analysis using advanced flow visualization techniques such as PIV could also be implemented.

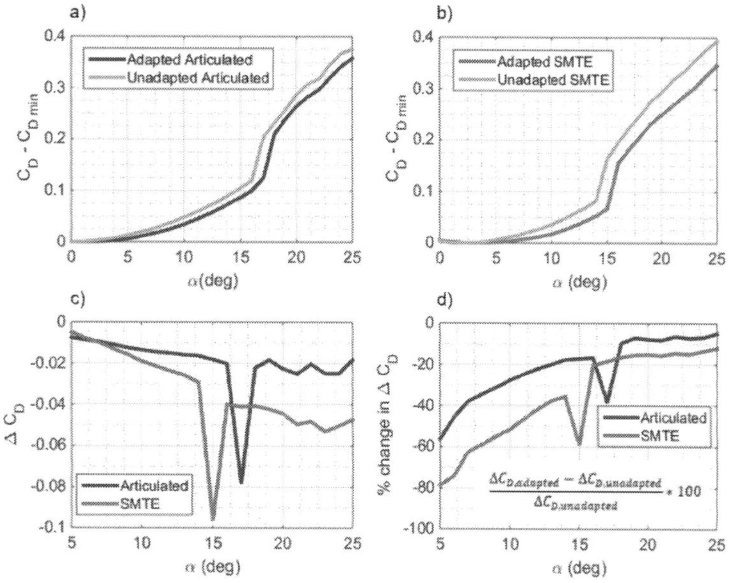

Figure 10. Experimental adapted and unadapted drag curves for **a)** articulated and **b)** SMTE wings. **c)** difference in drag between the adapted and unadapted cases shown in a) and b)

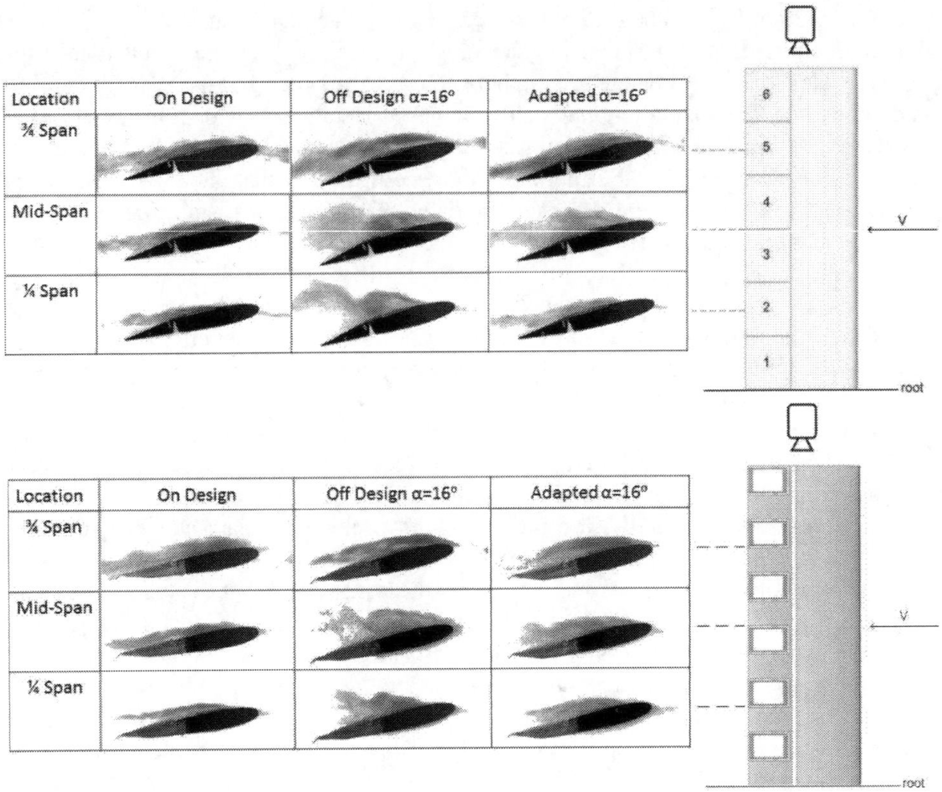

Figure 11. Flow visualization of the articulated (top) and SMTE (bottom) wings at the on design and off design flight conditions

6. CONCLUSION

This research indicated that the nonlinear aerodynamic model formulated in this study has great potential for predicting the optimal aileron deflections of a span-wise morphing wing to adapt and recover from stall. Using simulated 2D polars, a nonlinear lifting line model was successfully modified to include arbitrary span-wise variation to predict the aerodynamic behavior of a finite wing. The model was optimized to predict the adequate aileron deflections required to meet the design constraints by minimizing the total drag, ideally recovering from stall. Experimental tests were conducted to validate the nonlinear model and flow visualization was implemented to analyze whether the method of adaptation presented in this research eliminated vortex formations thus recovering from stall.

The model was able to accurately adapt to meet the target lift constraint to within 7%, yet the optimized point was still in the post-stall region of the lift curve and thus was not capable of completely recovering from stall due to the assumed design conditions and the limits of the 2D lift and drag data. When uniform in span-wise camber, the nonlinear model predicted the lift distribution with an RMSE of 0.027. Incongruities within the physical models hindered a favorable comparison with the optimized span-wise varying results past stall, yet showed good agreement in the linear regime. Most notably, the SMTE showed a 16.5% reduction in drag post-stall from the unactuated case versus a 9.6% reduction in the articulated wing, with significantly smaller actuations. The results presented here motivate future research into the limitations of this optimized model, specifically how the magnitude of the on-design lift and off-design angle of attack limits the ability of a morphing wing to recover from stall.

ACKNOWLEDGMENTS

The author would like to thank the Clarence L. "Kelley" Johnson collegiate professorship for the funding to conduct this research. This work is also supported in part by the US Air Force Office of Scientific Research under the grant "Avian Inspired Multifunctional Morphing Vehicles'' monitored by Dr. BL Lee.

REFERENCES

[1] D. Lentink, U. K. Muller, E. J. Stamhuis, R. de Kat, W. van Gestel, L. L. M. Veldhuis, P. Henningsson, A. Hedenstrom, J. J. Videler, and J. L. van Leeuwen, "How swifts control their glide performance with morphing wings," *Nature*, vol. 446, no. 7139, pp. 1082–1085, Apr. 2007.

[2] J. A. Gillies, A. L. R. Thomas, and G. K. Taylor, "Soaring and manoeuvring flight of a steppe eagle Aquila nipalensis," *J. Avian Biol.*, vol. 42, no. 5, pp. 377–386, Sep. 2011.

[3] J. Urnes, N. T. Nguyen, and J. Dykman, "Development of Variable Camber Continuous Trailing Edge Flap System," 13-Mar-2012.

[4] S. Kota, J. Hetrick, R. Osborn, D. Paul, E. Pendleton, P. Flick, and C. Tilmann, "Design and application of compliant mechanisms for morphing aircraft structures," in *Smart Structures and Materials 2003: Industrial and Commercial Applications of Smart Structures Technologies*, vol. 5054, E. H. Anderson, Ed. Bellingham: Spie-Int Soc Optical Engineering, 2003, pp. 24–33.

[5] A. M. Pankonien and D. J. Inman, "Span-wise Morphing Trailing Edge on a Finite Wing," in *Active and Passive Smart Structures and Integrated Systems 2015*, Bellingham, 2015, vol. 9431.

[6] A. Pankonien and D. J. Inman, "Experimental testing of span-wise morphing trailing edge concept," in *Active and Passive Smart Structures and Integrated Systems 2013*, Bellingham, 2013, vol. 8688.

[7] A. Pankonien and D. J. Inman,, "Aerodynamic Performance of a Span-wise Morphing Trailing Edge Concept," in *25th International Conference on Adaptive Structures and Technologies*.

[8] I. Tani, "A simple method of calculating the induced velocity of a monoplane wing," Rep 111 (1934).

[9] J. C. Sivells and R. H. Neely, "Method for calculating wing characteristics by lifting-line theory using nonlinear section lift data," Jan. 1947.

[10] A. Gopalarathnam and H. Segawa, "Use of Lift Superposition for Improved Computational Efficiency of Wing Post-Stall Prediction," in *Applied Aerodynamics Conference*, 2008, vol. 7049.

[11] S.T. Piszkin, and E. S. Levinsky, "Nonlinear Lifting Line Theory for Predicting Stalling Instabilities on Wings of Moderate Aspect Ratio," ADA027645, Jun. 1976.

[12] J.-J. Chattot, "Analysis and Design of Wings and Wing/Winglet Combinations at Low Speeds," in *42nd AIAA Aerospace Sciences Meeting and Exhibit*, American Institute of Aeronautics and Astronautics.

[13]O.G. Tietjens, and Prandtl, L., *Fundamentals of hydro-and aeromechanics*, vol. 1. Courier Corporation, 1957.

[14]J. J. Chattot, "Low speed design and analysis of wing/winglet combinations including viscous effects," *J. Aircr.*, vol. 43, no. 2, pp. 386–389, Apr. 2006.

[15]J. J. Chattot, "Optimization of Wind Turbines Using Helicoidal Vortex Model," *J. Sol. Energy Eng.*, vol. 125, no. 4, pp. 418–424, Nov. 2003.

[16]S. Gallay, S. Ghasemi, and E. Laurendeau, "Sweep effects on non-linear Lifting Line Theory near Stall," in *52nd Aerospace Sciences Meeting*, American Institute of Aeronautics and Astronautics.

[17]K. Duraisamy, "Studies in Tip Vortex Formation, Evolution and Control," University of Maryland, 2005.

ICAST2015 #016

Structural Analysis of Smart Fins Operated by Piezoelectric Actuators for Performance Improvement

Chul-Woo Park[1*], Chan Hoon Chung[1], YooJin Kang[1], SangJoon Shin[1], Bum Soo Yoon[2], Ji Hwan Park[2], Kwan Joon Yoon[2]

[1] Department of Aerospace Engineering, Seoul National University, Seoul, Korea
[2] Department of Aerospace Information Engineering, Konkuk University, Seoul, Korea

Abstract

In order to develop a smart fin operated by piezoelectric actuator, structural analysis of smart fins embedded with the unimorph or bimorph actuator was attempted in this paper. Structural behavior of two smart fins operated by piezoelectric actuators was analyzed regarding the resulting actuation of trailing edge under specific voltage input by using commercial software ANSYS. Through the analysis of structural behavior, modified design of the smart fin was suggested for improving actuation of trailing edge. The analysis will be extended to fluid-structure interaction (FSI) and flutter analysis to develop optimal control algorithm under the possible aerodynamic disturbance.

1. INTRODUCTION

Development of composite material has progressed because of its application for the field of aerospace engineering. Those advancements are called smart structures. Smart structures consist of sensors and actuators which are attached or embedded to existing structures. Analysis of smart structures has been conducted by several researchers. The artificial caudal fin was conducted for the performace of biomimeetic fish robot by Seok heo[1].The optimal positioning of piezoelectric actuators on a smart fin have been conducted by Mehrabian [2]. Chinnasamy conducted analysis using computational fluid dynamics (CFD) to evaluate the aerodynamic performance of smart material airfoil [3]. In this paper, analysis of smart fin embedded with piezoelectric actuator was conducted. Unimorph piezoelectric actuator was analyzed using ANSYS structural analysis to verify the behavior and performance of its tip deflection compared with experimental results. After the analysis of unimorph piezoelectric actuator, structural analysis of the smart fin embedded with unimorph actuator without external loads was conducted using ANSYS structural analysis to examine the structural behavior, and derive the resulting tip deflection under Mach number 0.1 flight.

2. THEORETICAL BACKGROUND

2.1 The Stress Form of the Piezoelectric Matrix

For structural analysis using ANSYS, stress form of the piezoelectric matrix was needed. Given the

* ssjoon@snu.ac.kr

strain form of the piezoelectric equations, it can be converted to the stress form and give expression for the three independent e tensor parameters. In matrix form it can be expressed as follows[4].

$$\varepsilon = d^T E + c^{-1}\sigma \tag{1}$$
$$P = \epsilon_0 \chi E + d\sigma \tag{2}$$

where ε is strain tensor, P is the electric polarization vector, E is the electric field vector, σ is the stress tensor, d are the components of the piezoelectric tensor, c are the components of the inverse elastic tensor, χ are the components of the electric susceptibility tensor, and ϵ_0 is the permittivity of free space.
Multiplying Eqn. (1) by the elasticity matrix c,

$$c\varepsilon = cd^T E + \sigma \tag{3}$$

so that it can be solved for the stress matrix σ.

$$\sigma = c\varepsilon - eE \tag{4}$$

where $e = cd^T$. And from Eqn. (2)

$$\begin{aligned} D &= P + \epsilon_0 E = \epsilon_0 \chi E + d\sigma + \epsilon_0 E \\ &= \epsilon_0(\chi + I)E + d\sigma \\ &= \epsilon^\sigma E + d\sigma \end{aligned} \tag{5}$$

where D is the electric displacement vector and ϵ^σ is the permittivity at constant stress.
Substituting the expression for the stress tensor σ which is solved from above

$$\begin{aligned} D &= \epsilon^\sigma E + d(c\varepsilon - eE) \\ &= (\epsilon^\sigma - de)E + dc\varepsilon \\ &= \epsilon^\varepsilon E + e^T \varepsilon \end{aligned} \tag{6}$$

where $\epsilon^\varepsilon = \epsilon^\sigma - de$ is the permittivity at constant strain. The elastic matrix c has 5 independent parameters and d has three independent parameters as follows[5].

$$c = \begin{bmatrix} c_{11} & c_{12} & c_{13} & 0 & 0 & 0 \\ c_{12} & c_{22} & c_{13} & 0 & 0 & 0 \\ c_{13} & c_{13} & c_{33} & 0 & 0 & 0 \\ 0 & 0 & 0 & c_{44} & 0 & 0 \\ 0 & 0 & 0 & 0 & c_{44} & 0 \\ 0 & 0 & 0 & 0 & 0 & 2(c_{11} - c_{12}) \end{bmatrix} \tag{7}$$

and

$$d = \begin{bmatrix} 0 & 0 & 0 & 0 & d_{15} & 0 \\ 0 & 0 & 0 & d_{15} & 0 & 0 \\ d_{31} & d_{31} & d_{33} & 0 & 0 & 0 \end{bmatrix} \tag{8}$$

Then the piezoelectric stress matrix is described as follows[6].

$$e = cd^T = \begin{bmatrix} c_{11} & c_{12} & c_{13} & 0 & 0 & 0 \\ c_{12} & c_{22} & c_{13} & 0 & 0 & 0 \\ c_{13} & c_{13} & c_{33} & 0 & 0 & 0 \\ 0 & 0 & 0 & c_{44} & 0 & 0 \\ 0 & 0 & 0 & 0 & c_{44} & 0 \\ 0 & 0 & 0 & 0 & 0 & 2(c_{11}-c_{12}) \end{bmatrix} \begin{bmatrix} 0 & 0 & d_{31} \\ 0 & 0 & d_{31} \\ 0 & 0 & d_{33} \\ 0 & d_{15} & 0 \\ d_{15} & 0 & 0 \\ 0 & 0 & 0 \end{bmatrix}$$

$$= \begin{bmatrix} 0 & 0 & c_{11}d_{31}+c_{12}d_{31}+c_{13}d_{33} \\ 0 & 0 & c_{12}d_{31}+c_{11}d_{31}+c_{13}d_{33} \\ 0 & 0 & c_{13}d_{31}+c_{13}d_{31}+c_{33}d_{33} \\ 0 & c_{44}d_{15} & 0 \\ c_{44}d_{15} & 0 & 0 \\ 0 & 0 & 0 \end{bmatrix} \tag{9}$$

3. DESCRIPTION OF THE DESIGN CANDIDATES FOR THE SMART FIN STRUCTURES

3.1 Design Candidate 1

The concept of Design Candidate 1 was suggested by Barret[7]. Instead of slanted actuator suggested by Barret, the straight unimorph and bimorph actuator were embedded along the chordwise direction in the hollow inner space of airfoil. It has a hinge on 1/4 of chord line which is constrained except the axial rotation. The actuator was separated from the smart fin in the initial state. When the voltage is applied on the actuator, the actuator will generate piezoelectric bending loads. Therefore the actuator will be in contact with projecting spot on the smart fin during actuation and the fin will be able to rotate on a hinge. It also adopts bearing to rotate smoothly. Figs. 1 and 2 illustrate the geometry of Design Candidates 1 and Table 1 summarizes the details of material properties used in the smart fin.

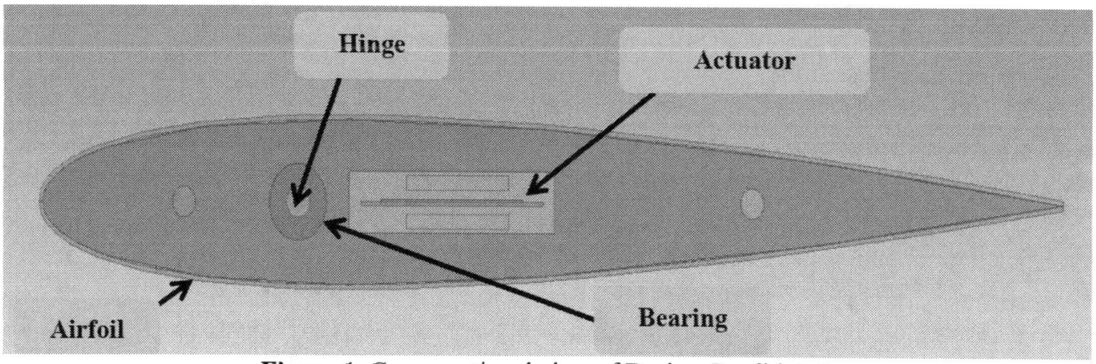

Figure 1. Cross-sectional view of Design Candidate 1

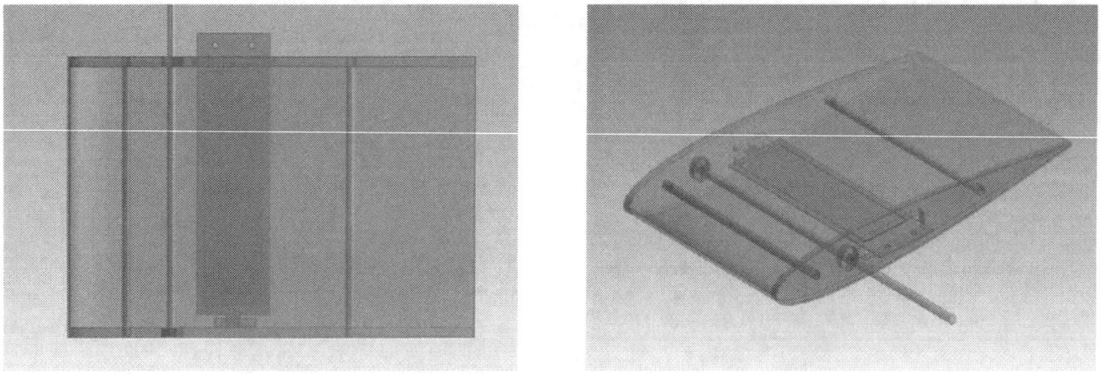

Figure 2. Upper and isometric view of Design Candidate 1

Table 1 Material properties used in the Design Candidate 1

	Parts	Material / Lay-up	Material Property
1	Airfoil	Aluminum-alloy	E = 71Gpa, υ = 0.33
2	Skin	Glass/Epoxy	E_1 = 21.7GPa, υ = 0.3
3	Hinge	Carbon/Epoxy	E_1 = 217.2GPa, υ = 0.29
4	Front/Back spar	Carbon/Epoxy	E_1 = 217.2GPa, υ = 0.29
5	Bearing	Aluminum-alloy	E = 71Gpa, υ = 0.33

Fig 3. shows the composition of the unimorph and bimorph actuator. The piezoelectric material used in the actuator was PMN-29PT.

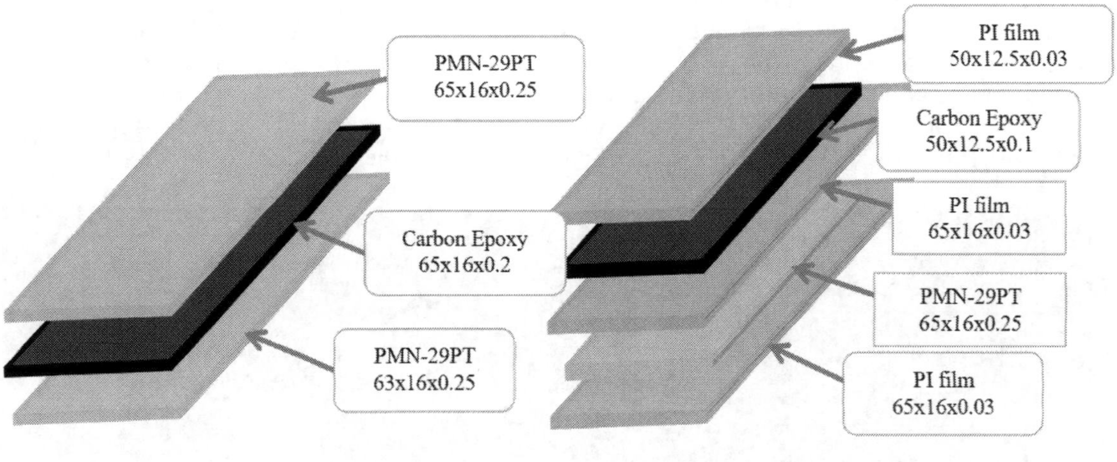

(a) Bimorph actuator (b) Unimorph actuator

Figure 3. Lay-up and dimensions of the bimorph and unimorph actuators in Design Candidate 1

3.2 Design Candid ate 2

Similar to Design Candidate 1, the straight unimorph actuator is embedded along the chordwise direction in the hollow inner space of the airfoil. However, as shown in Fig. 4, the additional geometric component called the support component is added between the actuator and airfoil to reduce the loss of piezoelectric bending actuation caused by the geometry of Design Candidate 1. Because the actuator is not bonded at the airfoil, it does not occur the torsional effect. It also adopts bearing to rotate smoothly. The actuator used in Design Candidate 2 is similar to the unimorph actuator used in Design Candidate 1. However, piezoelectric material called 3203HD was used in Design Candidate 2 instead of PMN-29PT. Fig. 5 shows the geometry of Design Candidate 2 and Table 2 summarizes the details of material properties used in the smart fin.

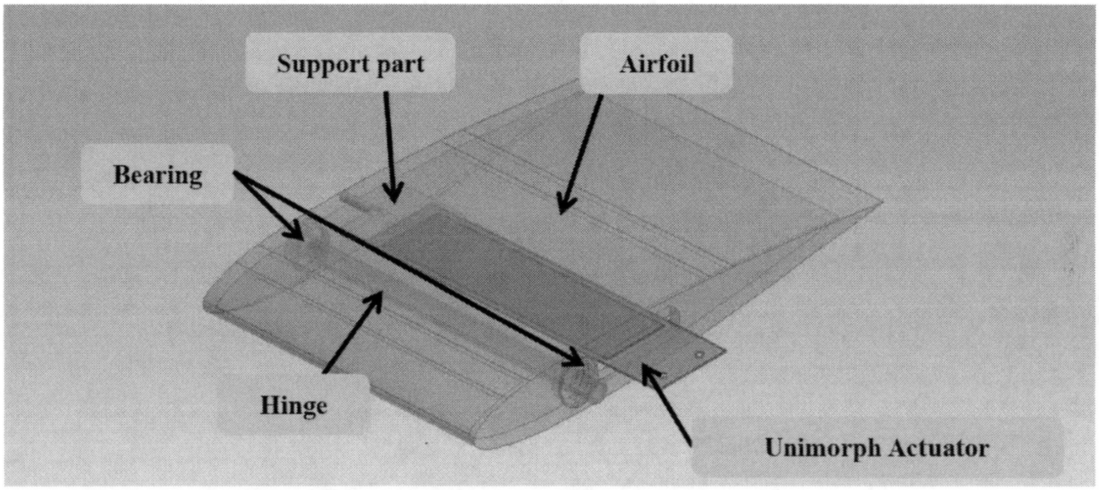

Figure 4. Isometric view of Design Candidate 2

Table 2 Material properties used in the Design Candidate 2

	Parts	Material / Lay-up	Material Property
1	Airfoil / Skin	ULTEM	E = 3.6Gpa, υ = 0.36
2	Hinge	Carbon/Epoxy	E_1 = 217.2GPa, υ = 0.29
3	Bearing	Aluminum-alloy	E = 71Gpa, υ = 0.33
4	Support part	Carbon/Epoxy	E_1 = 217.2GPa, υ = 0.29

Additional actuators are added for performance improvement as shown in Fig. 5. Four unimorph actuators are applied in parallel connected with the support component. It is expected that additional actuators may withstand more aerodynamic loads and improve the performance of actuation at trailing edge than a single actuator is applied in the smart fin.

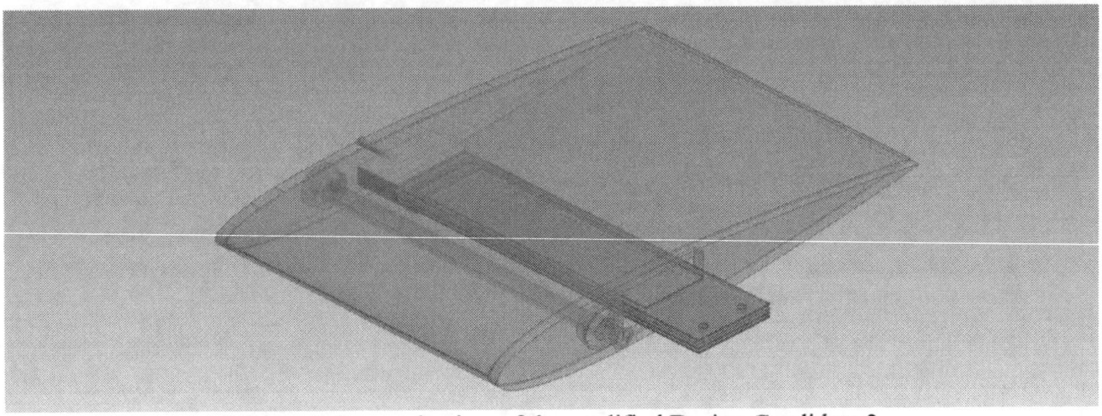

Figure 5. Isometric view of the modified Design Candidate 2

4. RESULTS

Structural analysis for the smart fin embedded with piezoelectric material is conducted using commercial finite element method analysis software, ANSYS. Solutions from ANSYS provide the simulation about every structural aspect of a product, including linear static analyses that simply provide stresses and deformations, modal analysis that determines vibration characteristics, to advanced transient nonlinear phenomena involving dynamic effects and complex behaviors. Analysis of the design candidate was conducted by using Piezo and MEMS module supported by ANSYS[8].

4.1 Structural analysis for Design Candidate 1

The actuator is located at 0.4 of chord line and 400 volt is applied on piezoelectric material. Actuator is separated from the smart fin in the initial configuration. It will be in contact with projecting spot on the smart fin during actuation and pin will rotate on a hinge. To realize frictional contact with projecting spot and rotation of hinge, the structural behavior was analyzed using the relevant nonlinear formulation provided in ANSYS. Augmented Lagrange formulation was used to prevent the penetration in the contact region. Also, the function of the pinball region provided in ANSYS was used to detect the penetration and solve several iterations until penetration did not occur[9].

With the linear increase of voltage input of 400V over 1 sec without external loads, the resulting actuation is found to be 0.83mm at the trailing edge and the angle of rotation is 0.55° when using the unimorph actuator. The angle of rotation is predicted by arctangent based on the hinge. Fig. 6 illustrates the resulting actuation of the trailing edge.

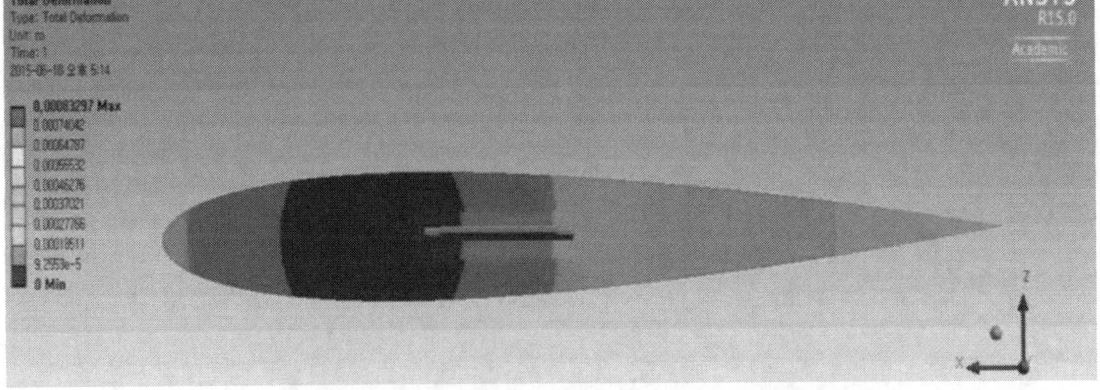

Figure 6. Resulting actuation of unimorph actuator

When using the bimorph actuator, the resulting actuation is 7.43mm at the trailing edge and the angle of rotation is 6.31°. 400 volt is applied on the upper component of piezoelectric actuator and 0 V on the lower component. Fig. 7 illustrates the resulting actuation of the trailing edge. However, as shown in Fig. 8, contact occurs between the actuator and smart fin during actuation when using the bimorph actuator. Also, the bimorph actuator has problems during actuation in experiment because of voltage coupling. Therefore, to solve the geometry limit of Design Candidate 1 and loss of the actuator bending force during actuation, Design Candidate 2 is suggested.

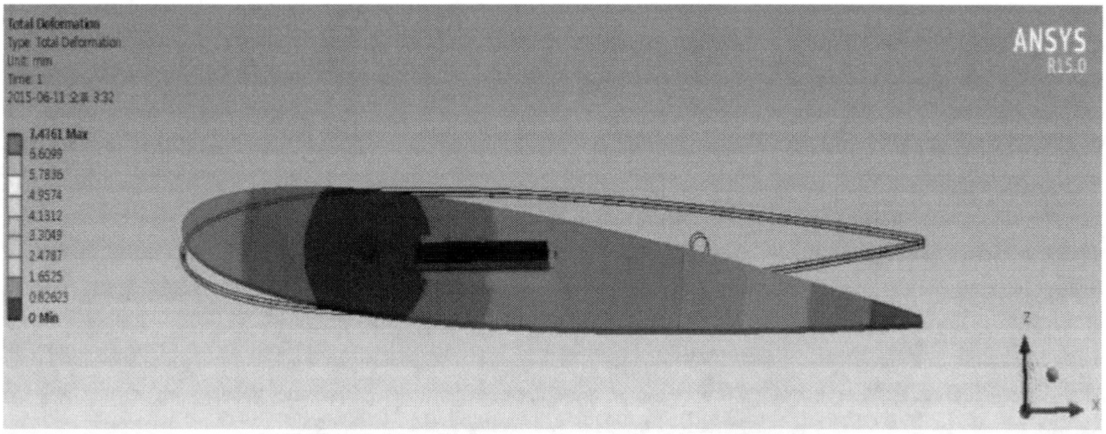

Figure 7. Resulting actuation of the bimorph actuator

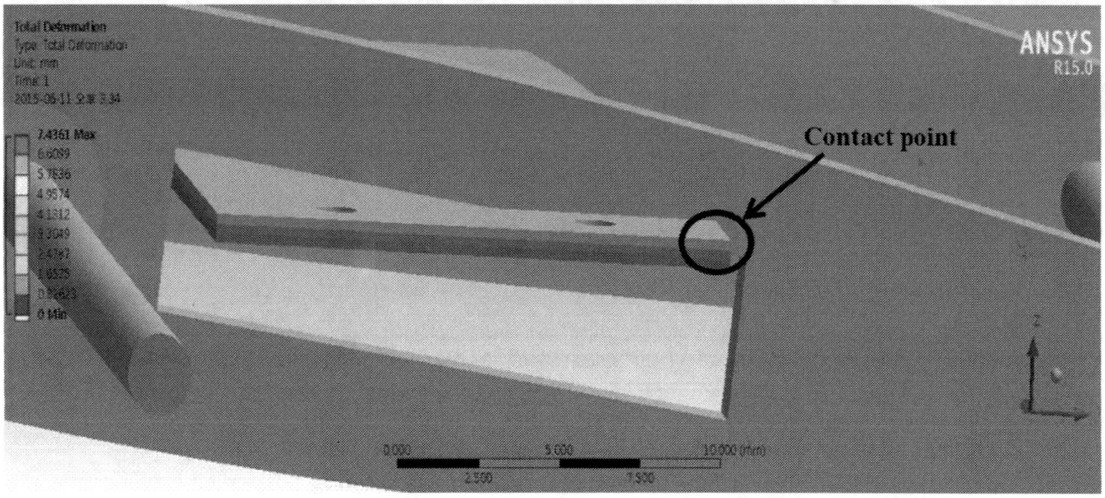

Figure 8. Geometry limit of Design Candidate 1

4.2 Structural analysis for Design Candidate 2

To solve loss of the actuator bending force during actuation, the additional component called the support component is now added between the airfoil and actuator. Also the airfoil is re-designed as a taper shape to solve the geometry limit of Design Candidate 1 by applying NACA0015 at the root airfoil and NACA0012 at the tip airfoil. Fig. 9 shows the modified geometry of the airfoil used in Design Candidate 2. The unimorph actuator is located at 0.4 of chord line and 400 V is applied on the piezoelectric material. With the linear increase of voltage input of 400V over 1 sec without external loads, the resulting actuation is 4.37mm at the trailing edge and the angle of rotation is 3.71°. Fig. 10 illustrates the resulting actuation of the trailing edge.

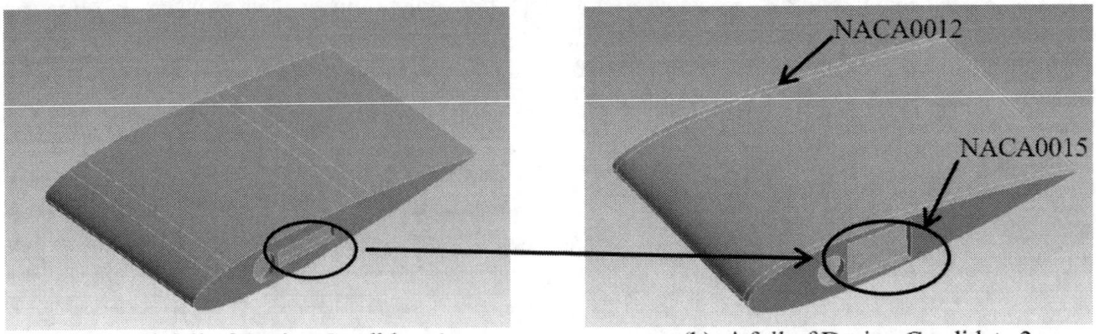

(a) airfoil of Design Candidate 1 (b) airfoil of Design Candidate 2

Figure 9. Taper shape of airfoil used in Design Candidate 2

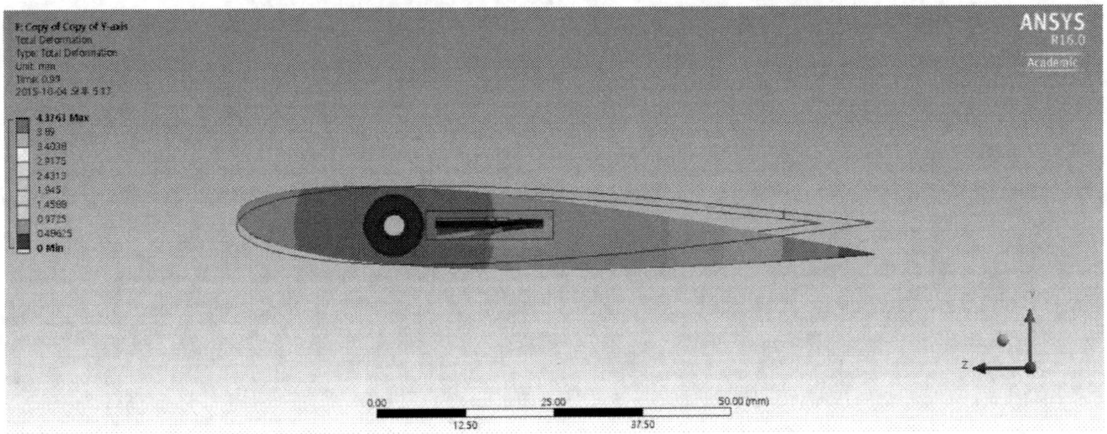

Figure 10. Resulting actuation of Design Candidate 2

The modified structures by adding four actuators are shown in Fig. 5. It is also conducted when 450 V is applied on piezoelectric material without external loads. The resulting actuation is found to be 6.67mm at the trailing edge and the angle of rotation is 5.66°, shown in Fig. 11. The results confirms that the performance of actuation at the trailing edge is improved by adding additional actuators.

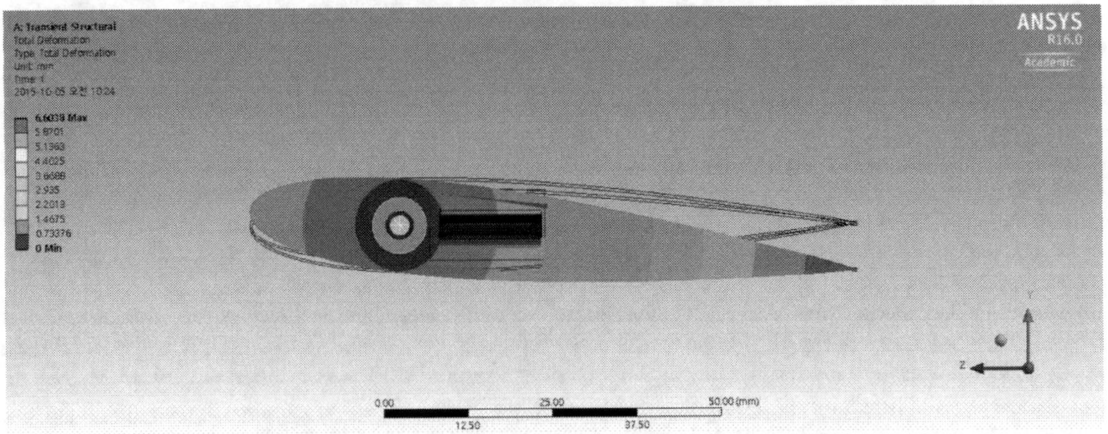

Figure 11. The resulting actuation of Design Candidate 2

4.3 Modal analysis of Design Candidate 2

Since the present smart fin is planned to be applied for the flight vehicles, further aeroelastic analysis such as flutter characteristics will be needed. In order to find the flutter characteristics of a structure, their mode shape would be needed. Modal analysis of Design Candidate 2 was conducted using commercial software ANSYS and are shown in Fig. 12.

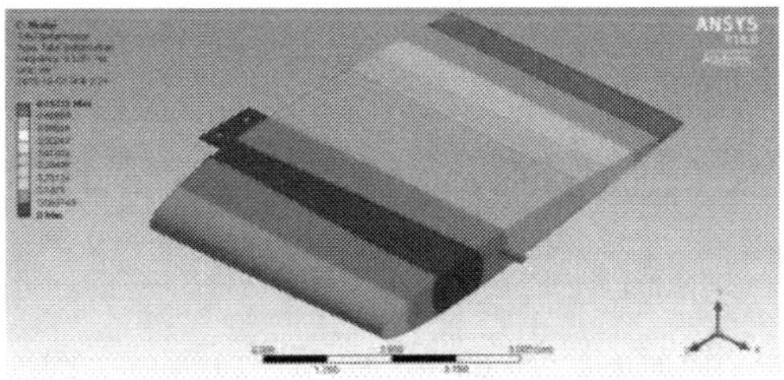

Mode 1 Freq. = 6.295Hz

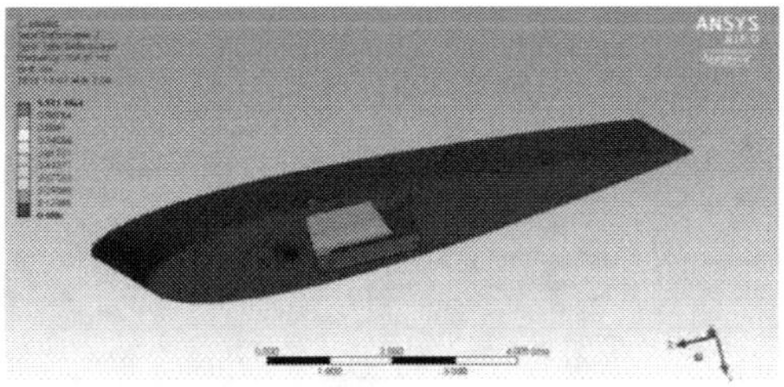

Mode 2 Freq. = 153.34Hz

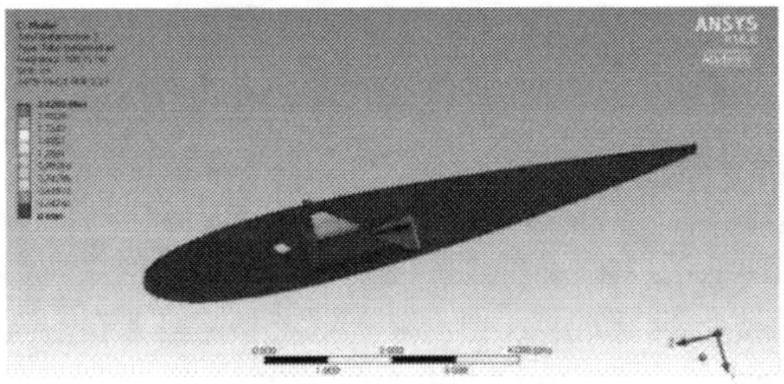

Mode 3 Freq. = 297.43Hz

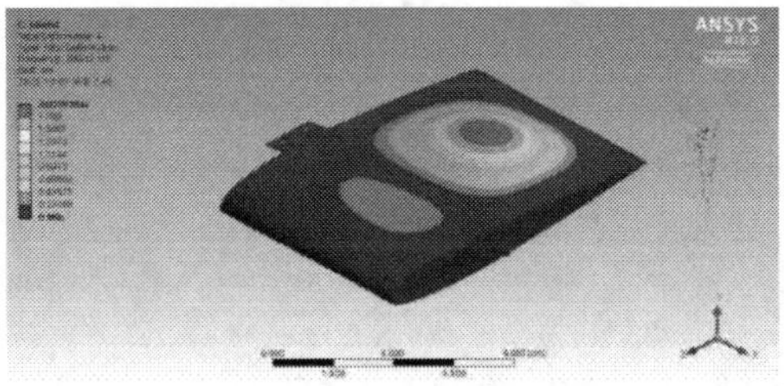

Mode 4 Freq. = 380.17Hz

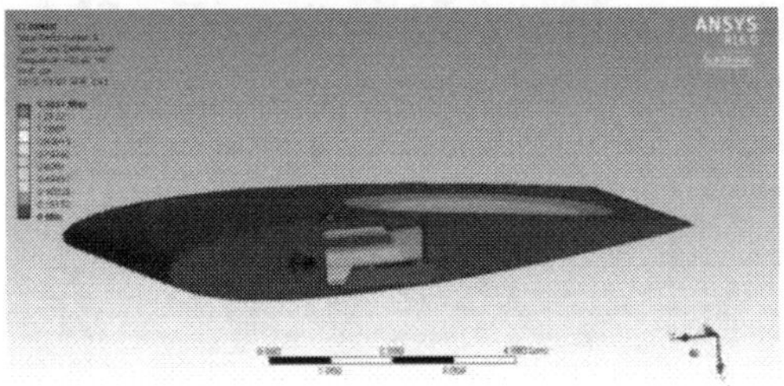

Mode 5 Freq. = 398.31Hz

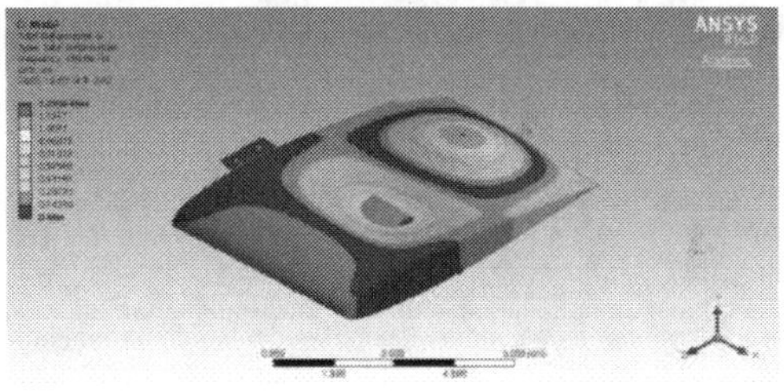

Mode 6 Freq. = 456.46Hz

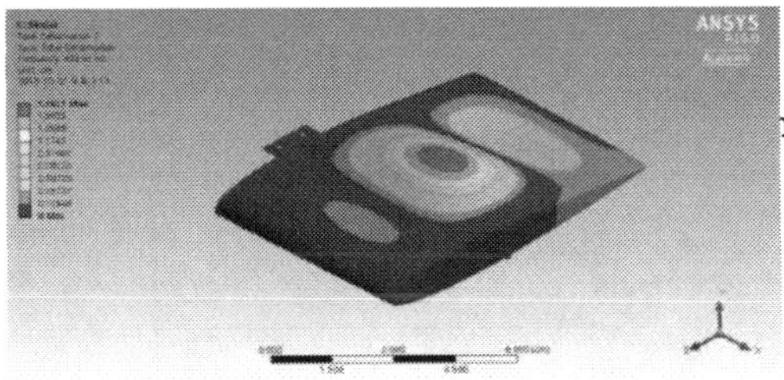

Mode 7 Freq. = 492.61Hz

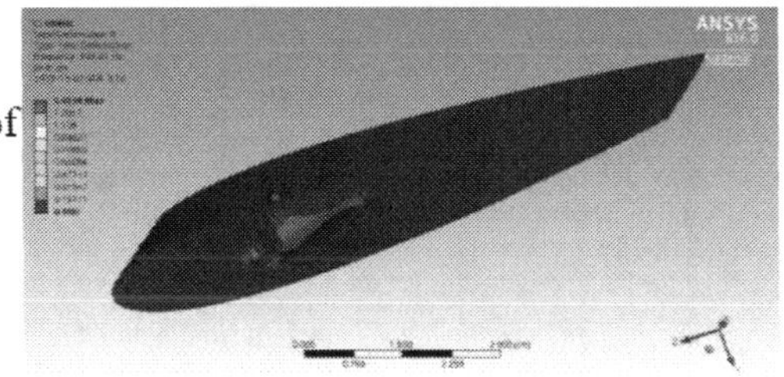

Mode 8 Freq. = 503.04Hz

Figure 12. Mode shapes of Design Candidate 2

According to Fig. 12, the first mode shape was caused by the hollow fin structure. Its results showed that the natural frequency of the hollow fin is lower than that of the actuator. Also, to examine the effect of the frictional coefficient between the contact region, three results of the modal analysis were compared while changing the frictional coefficient between support component and airfoil. Table 3 summarizes the three results of the present modal analysis, and the results show that it has little effect on the frictional coefficient.

Table 3 Three results of the modal analysis when changing the frictional coefficient

Frictional coefficient	Mode frequency (Hz)					
	1	2	3	4	5	6
0.2	6.243	151.02	293.52	379.36	393.36	455.97
0.5	6.305	153.83	297.91	380.29	399.44	456.58
0.7	6.325	154.91	298.75	380.52	402.02	456.86

Future work will include the structural behavior of modified Design Candidate 2 with aerodynamic loads. It will be analyzed using fluid-structure interaction (FSI) capability supported from ANSYS. Also, the comprehensive structural behavior of the smart fin will be analyzed regarding flutter characteristics of Design Candidate 2. Based on the analysis of flutter, optimal control law will be developed by considering the aerodynamic disturbance given to the smart fin. Closed-loop control will be required for feedback about tip deflection by adjusting the voltage to develop control algorithm for reducing nonlinear effect. Fig. 13 illustrates closed-loop control of the smart fin considering the aerodynamic disturbances. The analysis of aircraft structure with aerodynamics and control forces in Fig. 13 generally belongs to the field of aeroservoelasticity. The transfer function of PZT actuator will be obtained either by using ANSYS and fast Fourier transform (FFT) or by wind-tunnel experiment. The transfer function reflects characteristics of only the PZT actuator so the smart fin of the rightmost block in Fig. 13 may be just considered as a hollow fin with aerodynamics. The calculations of the closed-loop system will be fulfilled by ZAERO, which is a commercial computing program. ZAERO offers flutter margins or boundaries of the smart fin with control system using several methods such as rational function approach (RFA), doublet lattice method (DLM), p-k method or g-method for flutter analysis. There are also some modules and bulk data cards like ASE, GLOADS, ACTU, SISOTF which can be used for the aeroservoelastic analysis.

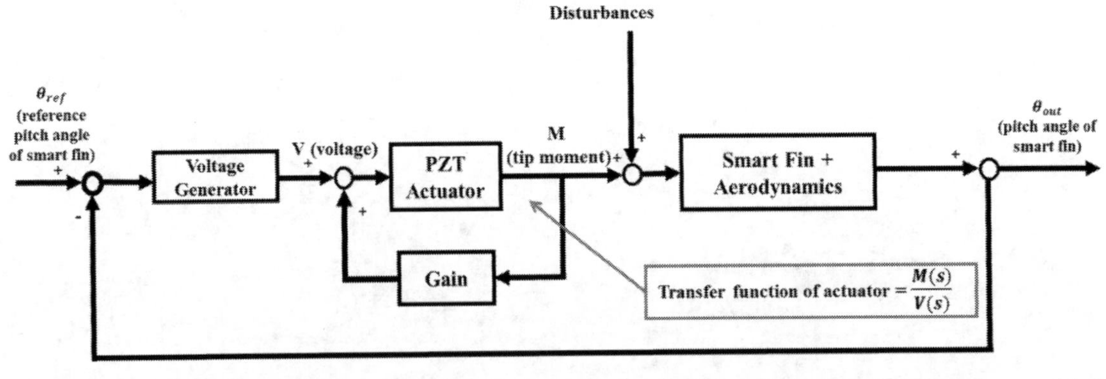

Figure13. Block diagram of aeroservoelastic control system of smart fin with PZT actuator considering aerodynamic disturbances

5. CONCLUSIONS

In this paper, composite smart fins with piezoelectric material embedded were analyzed in order to confirm their structural performance without external loads using commercial software ANSYS. Two design candidates were suggested for the present smart fin, and both have hinge on 1/4 of chord line which is constrained except axial rotation. Unimorph actuator was embedded in both design candidates. Actuation of the piezoelectric materials was analyzed under 400V or 450V input using ANSYS static structural analysis.

Design Candidate 1 contained the unimorph and bimorph actuator. The result showed rotation angle of 0.55° without external aerodynamic load when unimorph actuator was used. Its result showed too small angle of rotation, and thus the bimorph actuator was used to improve the performance of actuation on trailing edge. The rotation angle of smart fin was 6.31° without external loads. However, contact problem occurred between the airfoil and actuator during actuation. Also the bimorph actuator has problems in experiment, and loss of actuator bending force occurred during actuation. So, modified geometry of Design Candidate 2 was suggested.

Design Candidate 2 used the unimorph actuator. Compared with Design Candidate 1, additional component was used to connect between the actuator and airfoil. The result showed rotation angle of 3.71° without external loads. To improve the performance of actuation, four unimorph actuators were applied in parallel connecting with one support component. The result showed rotation angle of 5.66° without external loads.

Future work will include fluid-structure interaction (FSI) analysis of modified Design Candidate 2 in order to confirm structural behavior of modified Design Candidate 2 with external aerodynamic load. Also, flutter analysis of smart fin will be conducted based on modal analysis of smart fin, and optimal control algorithm will be developed by considering the disturbance given to the smart fin using flutter characteristics of the smart fin.

REFERENCES

1. Seok, H., Tedy, W., Park, H. C., Goo, N. S., "Effect on Artificial Caudal Fin on the Performance of a Biomimetic Fish Robot Propelled by Piezoelectric Actuators", *Journal of Bionic Engineering,* Vol. 4, 2007, pp. 151-158
2. Mehrabianm, A. R., Koma, A.Y., "Optimal positioning of piezoelectric actuators on a smart fin using bio-inspired algorithms," *Aerospace Science and Technology*, Vol. 11, 2007, pp. 174-182.
3. Chinnasamy, P., Chen, Y., "Application of Computational Fluid Dynamics on Smart Wing Design," *43rd AIAA Aerospace Sciences Meeting and Exhibit*, 10-13 Jan 2005, Reno, Nevada, USA
4. Emery, J., "Piezoelectricity", STEM Society, 3 April 1997, pp.9-11
5. Bent, A.A, Hagood, N.W., Rodgers, J.P., "Anisotropic Actuation with Piezoelectric Fiber Composites. *Journal of Intelligent Material Systems and Structures,* Vol. 5, 6 May 1995, pp.338-349
6. ANSYS INC. "Piezoelectric& MEMS exposure in ANSYS Workbench using ACT", March 2015, pp.21-37
7. Barret, R., Gross, R. S., and Brozoski, F., "Missile flight control using active flex spar actuators," *Smart Materials and Structures,* Vol.5, No.2, 1996, pp. 121~128
8. ANSYS INC. "Workbench - Mechanical Introduction 14.0", Ch.1-5, May 2012
9. ANSYS INC. "WB Mechanical Nonlinearities", Ch.3, June 2013

ICAST2015 #032

Bending Moment Reduction of a High Aspect Ratio Wing

Masato Tamayama[1*], Kanata Fujii[2], Hitoshi Arizono[1] and Tomohiro Yokozeki[2]

[1] Aeronautical Technology Directorate, Japan Aerospace Exploration Agency, Tokyo, Japan.
[2] Department of Aeronautics and Astronautics, University of Tokyo, Tokyo, Japan.

Abstract

In order to achieve the compromise between contradictory demands of reducing wing weight and still sustaining a good amount of lift for a high aspect ratio wing, e.g. that of High Altitude Long Endurance Aerial Vehicle, the concept to reduce wing root bending moment by optimizing lift distribution along wing span-wise direction is studied. The analytical wing model is a straight plate having tapered wing thickness and chord length along its span-wise direction. The wing has 4 flaps at each of its leading and trailing edges as the device to control wing lift distribution. Under some variation of flap chord-wise length, the set of flap deflection angle is optimized. The resultant lift distributions show the characteristics of increased lift at the wing inboard sections and reduced lift at the outboard sections. The wing bending moment is successfully reduced by about 50% comparing to the original configuration of flaps having 0° deflection angle.

1. INTRODUCTION

It is inevitable in aircraft design to include considerations for wing flexibility, and it becomes more important if aircraft become lighter. The more flexible the wing is, the worse its aerodynamic performance becomes. Nguyen, et al. showed the influence of remaining fuel amount on the aerodynamic performance for an elastic wing of transport type airplane[1)]. Reducing wing weight causes another problem: the margin of wing strength gets thinner. This is very serious: it directly influences the aircraft safety. In order to deal with these problems, the technique to control wing load distribution can be used[2, 3)]. For example, if the bending strength at the wing root is critical to sustain gust load, it can be relieved by reducing aerodynamic force at the wing outboard sections and, simultaneously, increasing it at the wing inboard sections to retain necessary gross lift. The resultant wing root bending moment will be reduced from that generated with original aerodynamic force distributions. This technique dose not bring only a benefit to us. For the case of reducing wing root bending moment by the above mentioned control, the lift distribution along the wing span deviates from the designed one, and the aerodynamic drag becomes worse. This control technique should not be applied to transport aircraft. The reason is that transport aircraft fly through the altitude range in which strong gust might attack themselves anytime. But, if a High Altitude Long Endurance Aerial Vehicle (HALE) is considered, the above mentioned control might be a solution to bring us a benefit. A sample of mission profile for the Global Hawk Unmanned Aerial Vehicle is presented in Ref.4. Its flight altitude is above 50,000 feet for over 90% of the flight time. The turbulence encountering rate at the altitude above 50,000 feet is much less than that at the altitude range of 30,000 ~ 40,000 feet[5)]. The HALE, therefore, encounters strong air turbulence just during ascend and descend flights of its long

* masato@chofu.jaxa.jp

endurance mission profile. If the above mentioned bending moment control technique is applied just to these ascend and descend paths, the aerodynamic disadvantage might be limited to minimum influence. Following this idea, an optimization to acquire the best performance for reducing wing bending moment is conducted in this study. As to the device to change the distribution of aerodynamic force along wing span-wise direction, 4 flaps at each of wing leading and trailing edges are selected. The resultant lift distributions show the characteristics of increased lift at the wing inboard sections and reduced lift at the outboard sections. The wing bending moment is successfully reduced by about 50% comparing to the original configuration of flaps: each flap has 0° deflection angle. In this paper, the specification of wing model, the optimization procedure and the optimized results are presented.

2. WING MODEL

The wing model is designed referring to the NASA Altair shown in Fig.1. The two-view drawing of wing model is shown in Fig.2. The root and tip chord lengths are 1.65m and 0.73m, respectively. The semi-span length is 13.1m. The aspect ratio can be calculated as 22 for the case of ignoring fuselage. The wing is straight, and the chord length and wing thickness vary linearly along the wing span-wise direction. A set of wing thickness, t_r and t_t at the wing root and tip sections, respectively, is taken as a parameter in the analysis to see the influence of wing flexibility. There are 4 flaps at each of wing leading and trailing edges as a device to change lift distributions along the wing span-wise direction. Figure 3 (a) shows the arrangement of flaps and those identifying names: the leading and trailing edge flaps are tagged as "LEF" and "TEF," respectively. Each flap name is followed by a

Figure 1. Reference HALE: NASA Altair
Source: http://commons.wikimedia.org/wiki/File:Altair_UAV.jpg

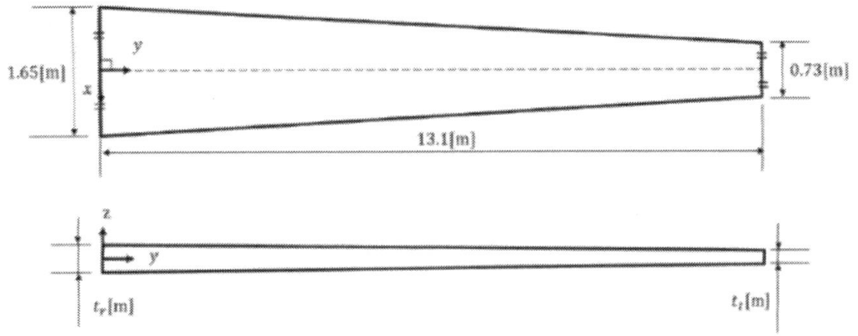

Figure 2. Two-view Drawing of Wing Model

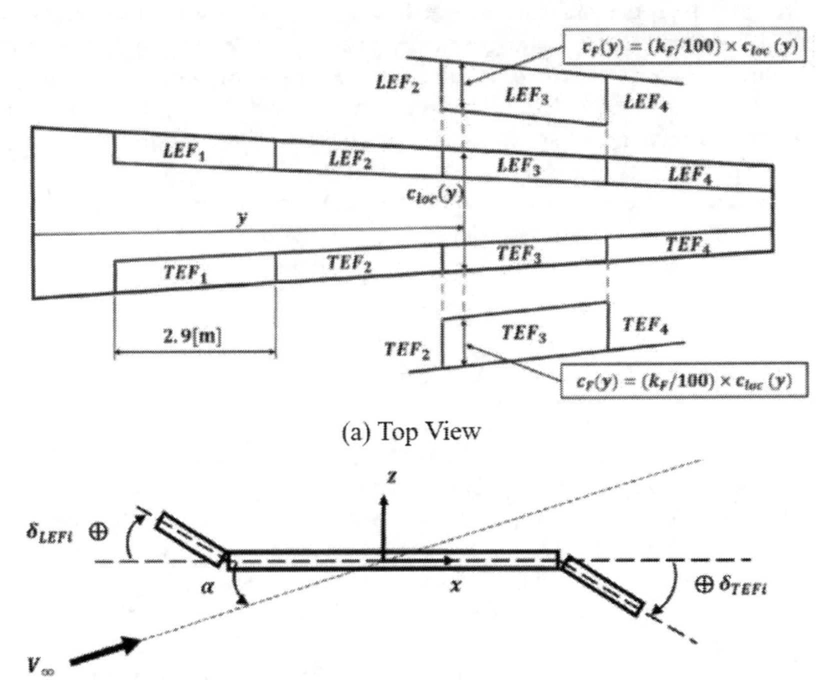

(a) Top View

(b) Definition of flap angle sign (δ_{LEFi} and δ_{TEFi}: i=1~4 are flap deflection angles)

Figure 3. Arrangement of Flaps

number counting from the most inboard flap. The chord-wise length is set as same value, $c_F(y)$, for both of the LEFs and TEFs at the same span location, y, which axis is originated from wing root and parallel to the wing center axis: this coincides with the wing elastic axis. In Fig.3, $c_{loc}(y)$ is the local chord length at the location y, and k_F is the percentage to determine flap chord-wise length to be used to calculate $c_F(y)$. The description for structural analysis model is presented in the following section. Figure 3 (b) shows the sign definition of flap angle. For the LEFs, nose-up is positive: for the TEFs, nose-down is positive.

3. ANALYSIS

As an optimizing process to determine a set of flap deflection angles for realizing maximum reduction of wing root bending moment, the Response Surface Method is used in this study. For the acquisition of original data to be fitted in the Response Surface Method, the static Finite Element Method (FEM) analysis under linear aerodynamic force is conducted. The MSC-NASTRAN with Aeroelasticity module is used for this purpose. The wing FEM model is shown in Fig.4 together with the modeling of flap connection and the wing root constraint. The FEM nodes are generated by equally dividing the wing into 20 in chord-wise direction and 45 in span-wise direction. The FEM element is a shell model with 4 nodes. The flaps are attached to the main wing with rigid bar elements placed at the 20% and 80% span-wise locations for each flap. The node located at the middle of wing root is applied with the constraint having 6 degrees of freedom. All of the nodes at the wing root are applied with rigid connection. Figure 5 shows the mesh used in aerodynamic calculation in the MSC-NASTRAN Aeroelasticity module. The node locations of aerodynamic mesh are same as those of structural FEM model. The orange-colored area shows flaps. For this area, the deflection can be set for individual flaps in aerodynamic analysis.

The assumed flight condition is shown in Tab.1. The flight speed refers to that of the NASA Altair[6)]. The altitude is set to almost the half of Altair's highest limit. The wing model is a plate without camber, and the angle of attack 3.0° is applied to the wing to generate lift.

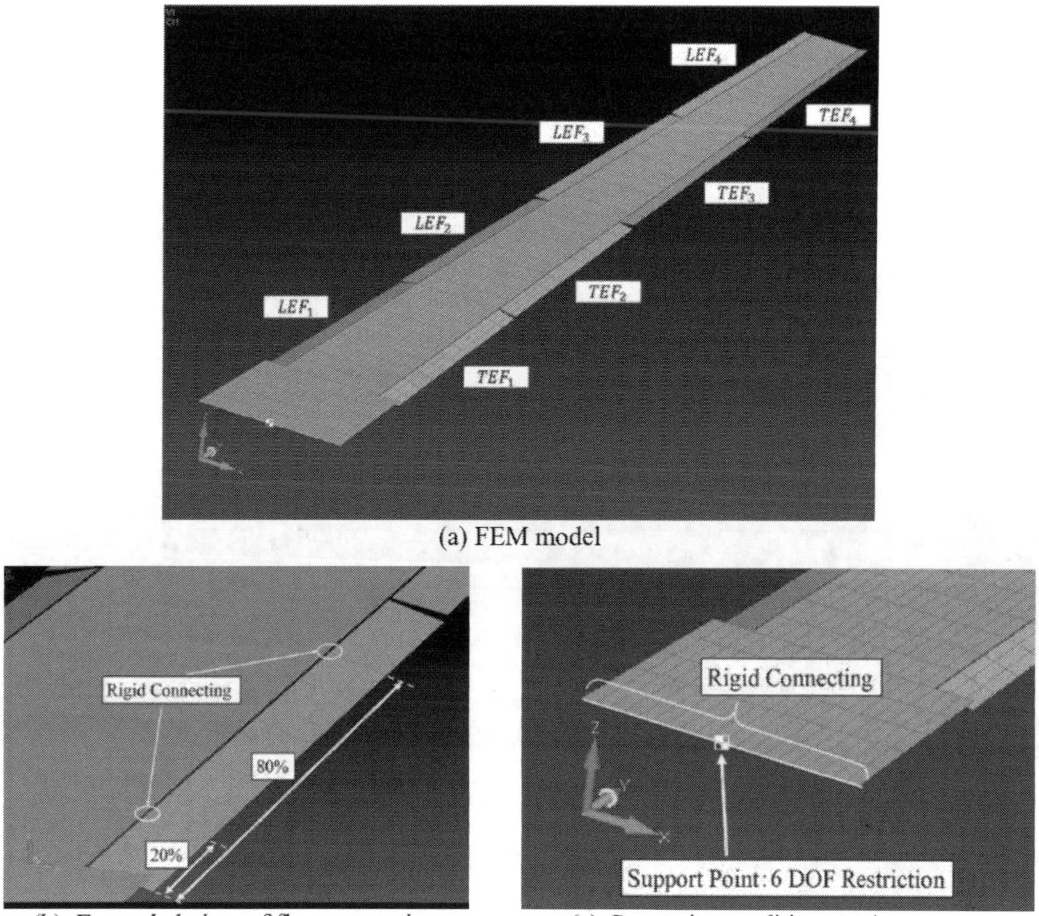

(a) FEM model

(b) Expanded view of flap connection (c) Constraint condition at wing root

Figure 4. Wing FEM model

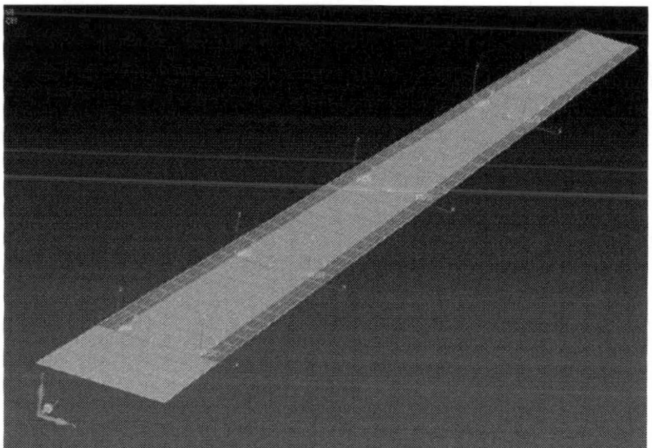

Figure 5. Aerodynamic model

Table 1. Flight condition

Altitude	7,600 m
Speed	108 m/s
Angle of attack	3.0°

Table 2. Analysis cases

Case Name	t_r [m]	t_t [m]	k_F [% chord length]
C1-10	0.10	0.050	10
C1-15	0.10	0.050	15
C1-20	0.10	0.050	20
C2-10	0.075	0.050	10
C2-15	0.075	0.050	15
C2-20	0.075	0.050	20
C3-10	0.075	0.025	10
C3-15	0.075	0.025	15
C3-20	0.075	0.025	20

The variables in optimization are the flap deflection angles: δ_{LEFi} and δ_{TEFi} (i=1~4). In the optimization, two parameters are varied to see their influence on the optimized set of flap deflections: the flap chord-wise size, k_F, explained in Fig.3 (a); the set of root and tip wing thickness, (t_r, t_t). The list of analysis cases are shown in Tab.2.

The constraints in optimization are set as follows:

$$\begin{cases} -15^{o} \le \delta_{LEFi} \le 15^{o}\ , -15^{o} \le \delta_{TEFi} \le 15^{o} \quad i = 1 \sim 4 \\ (\text{Variation of gross lift}) \le 1\% \end{cases} \tag{1}$$

The objective function is the wing root bending moment itself, and the optimization tries to find a set of flap deflection angles to minimize the objective function.

From the considerations to the optimized results under Eq.(1), it is determined to introduce the additional constraints as follows:

$$\begin{cases} (\text{the location of maximum wing bending moment}) = (\text{wing root}) \\ (\text{the minimum wing bending moment}) \ge 0 \end{cases} \tag{2}$$

In the next section, the optimized results under the constraints (1) are presented at first. And then, the results optimized under both the constraints (1) and (2) are presented.

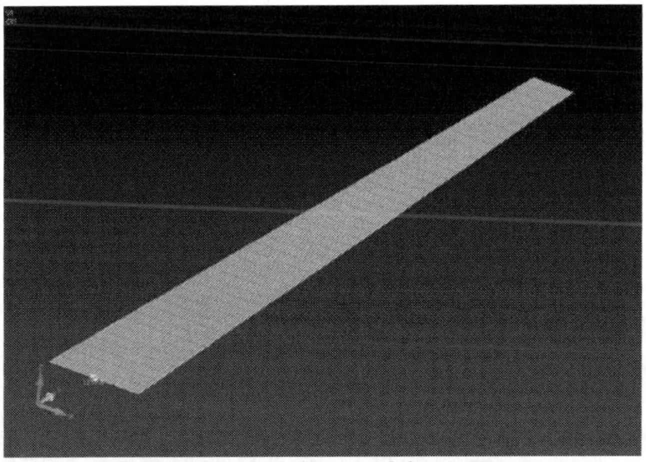

Figure 6. Baseline model: CASE0

Table 3. Gross lift, wing root bending moment and maximum von Mises stress of CASE0

Gross lift [N]	3.30×10^4
Bending moment [Nm]	9.15×10^4
Maximum von Mises stress [Pa]	3.44×10^7

Table 4. Set of optimized flap deflections under constraints of Eq.(1) [unit: °]

Case	δ_{LEF1}	δ_{LEF2}	δ_{LEF3}	δ_{LEF4}	δ_{TEF1}	δ_{TEF2}	δ_{TEF3}	δ_{TEF4}
C1-10	15.0	11.0	-15.0	-15.0	15.0	4.9	-15.0	-15.0
C1-15	15.0	11.2	-15.0	-15.0	15.0	4.8	-15.0	-15.0
C1-20	15.0	10.3	-15.0	-15.0	15.0	4.8	-15.0	-15.0

4. RESULTS

First of all, the analysis was conducted for the wing without flap deflections: Figure 6 shows the bird's-eye view of this model. This result is taken as the baseline in this study, and this case is called as "CASE0." The wing thickness parameter set is (t_r, t_t)=(0.10m, 0.050m): the wing thickness distribution along wing span is same as analyzing cases of C1-10, C1-15 and C1-20. The gross lift, wing root bending moment and maximum von Mises stress of CASE0 are presented in Tab.3. The gross lift is slightly bigger than the necessary force to sustain the Altair's gross weight of 3,175kgf (3.11×10^4 N). Although the maximum von Mises stress is shown around the wing root, the value is enough smaller than the yield stress of aluminum alloy, 5.05×10^8 Pa.

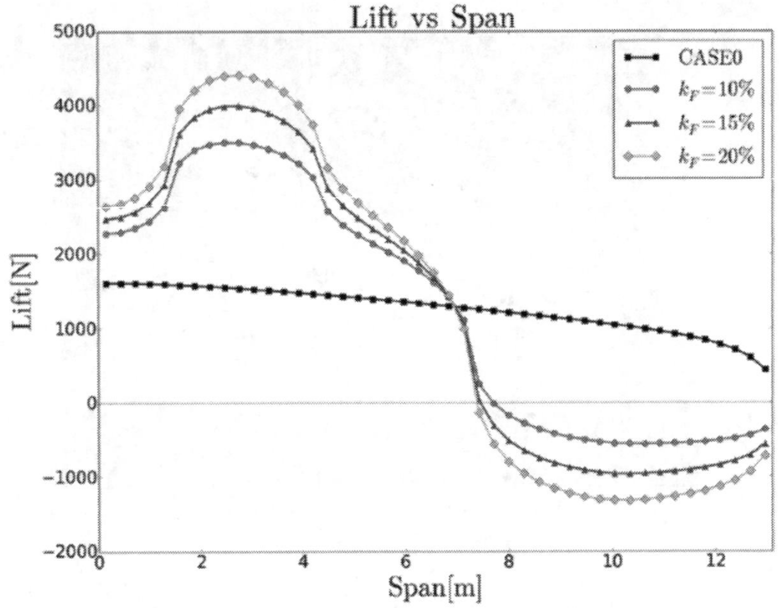

(a) Lift distribution

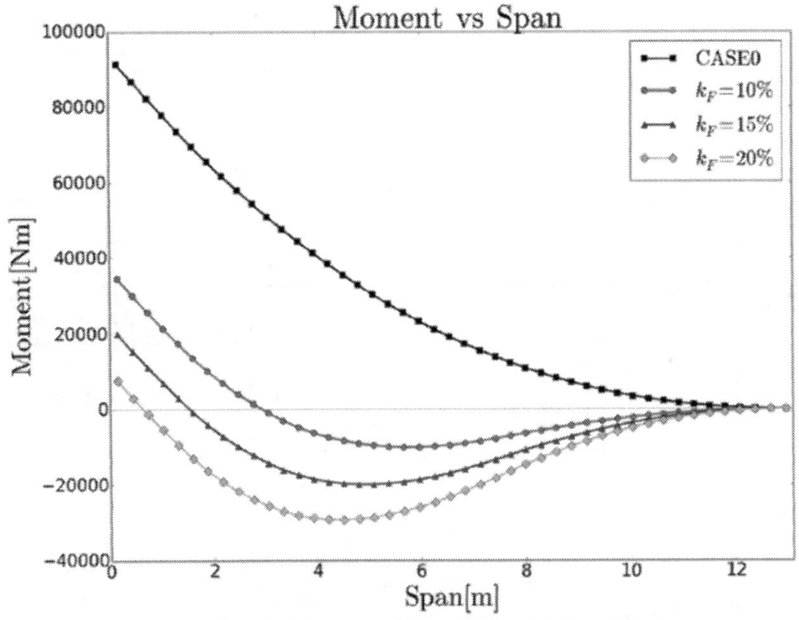

(b) Bending moment distribution

Figure 7. Results of optimization under constraints determined by Eq.(1)
(Black: CASE0, Red: C1-10, Blue: C1-15, Yellow: C1-20)

Next, the optimization under the constraints presented by Eq. (1) was conducted for the cases C1-10, C1-15 and C1-20. The set of optimized flap deflections is shown in Tab.4. Figures 7 (a) and (b) are the lift and bending moment distributions along wing span. The lift distribution at inboard locations increases and, simultaneously, that at outboard locations decreases: it seems to give a favorable lift distribution at first sight. A problem appears in the bending moment distribution. Although the wing root bending moment is reduced comparing to that of CASE0, the absolute value of bending moment in the middle span locations exceeds that of root bending moment. In order to resolve this problem and

Table 5. Set of optimized flap deflections under constraints of Eq.(1) and (2) [unit: °]

Case	δ_{LEF1}	δ_{LEF2}	δ_{LEF3}	δ_{LEF4}	δ_{TEF1}	δ_{TEF2}	δ_{TEF3}	δ_{TEF4}	Wing root bending moment [10^4Nm]	Percentage to CASE0 [%]
C1-10	15.0	14.9	-1.5	15.0	15.0	-6.7	-8.8	-7.8	5.21	57
C1-15	15.0	14.3	8.9	14.4	15.0	-11.6	-5.3	-6.9	5.00	55
C1-20	14.8	13.0	14.0	7.0	15.0	-13.5	-4.5	-6.3	4.71	51
C2-10	15.0	14.3	-1.3	15.0	15.0	-8.6	-9.6	-7.8	5.21	57
C2-15	15.0	8.3	15.0	14.9	15.0	-11.5	-7.9	-7.7	4.61	50
C2-20	12.4	12.9	15.0	15.0	14.9	-15.0	-6.2	-6.8	4.49	49
C3-10	14.9	11.1	15.0	15.0	15.0	-9.6	-12.8	-7.7	5.34	58
C3-15	15.0	14.0	13.4	15.0	15.0	-12.9	-9.3	-8.9	4.77	52
C3-20	14.8	14.5	15.0	15.0	15.0	-14.8	-9.5	-8.7	4.19	46

to acquire more practical result, the constraints described in Eq. (2) were introduced to supplement Eq. (1).

The results optimized under both the constraints Eq. (1) and (2) are shown in Tab. 5 and Fig. 8. Table 5 shows the set of optimized flap deflections for all of the cases listed in Tab. 2, and Fig. 8 show only the cases of C1-10, C1-15 and C1-20. In Fig. 8, the value of lift distribution at outboard locations increases comparing to that of Fig.7 (a). And, the bending moment distribution shown in Fig.8 (b) shows more practical one: it varies monotonously along wing span. The reduction percentage of wing root bending moment against that of CASE0 is also listed in Tab.5. The minimum value shown in Tab.5 is 46%, and the technique to reduce wing root bending moment by controlling wing lift distribution along wing span-wise direction appears to be valid.

In Ref.7, Yagihashi describes that the LEFs influence little on enhancing the reduction of wing root bending moment than the TEFs. Following this insight, it is better, at first, to see Tab.5 just focusing on the TEF deflections. For each case, δ_{TEF1} shows almost maximum value of 15°. It is considered that the most inboard trailing edge flap, TEF1, has a role to generate positive lift as much as possible. Other flap deflections, δ_{TEF2}, δ_{TEF3} and δ_{TEF4}, are negative values for all cases: they have a role to generate negative lift.

Before entering into the consideration for the LEF effect, Fig.9 are prepared: they show the flap deflections schematically at the mid span section of each flap segment for the cases of C1-10, C1-15 and C1-20. And consider the wing comprises 4 areas as shown in Fig.10. It is easy to image that the lift generated on the Area 1 should be large as much as possible under the constraints of Eq. (1) and (2). In view of wing flexibility, the pitch angle at the Area 1 should increase for the purpose of acquiring larger lift. On the other hand, δ_{TEF1} of 15° gives the wing a negative effect against increasing the pitching angle at the Area 1: the positive δ_{TEF1} generates nose-down pitching moment. Looking at LEF1, the positive δ_{LEF1} is effective to increase nose-up pitching moment at the Area 1, and it also generates positive lift itself. At the Areas 2 through 4, the negative angle of TEFs and the positive angle of LEFs have an effect to lead to the positive pitching angle at the Area 1, though the latter generates positive lift at the Areas 2 through 4: this causes the effect contrary to the fact that the lift at these Areas should be less than that of CASE0.

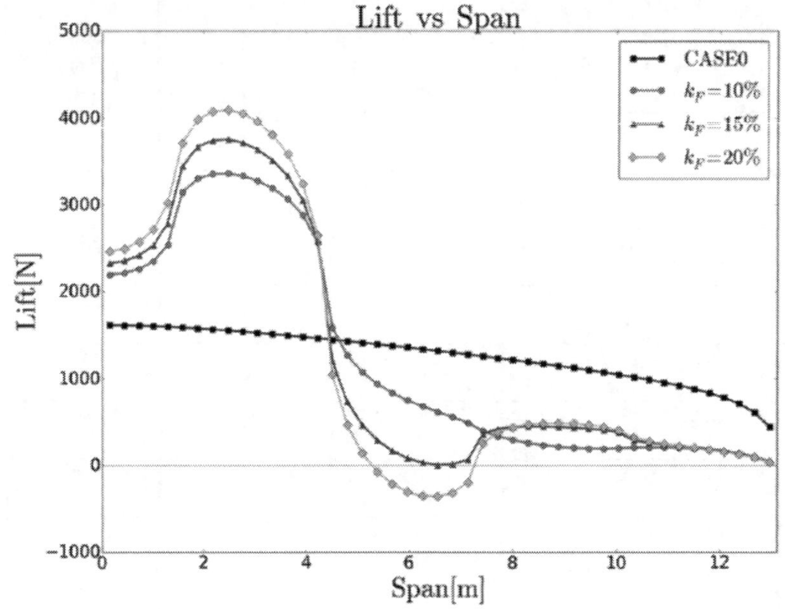

(a) Lift distribution

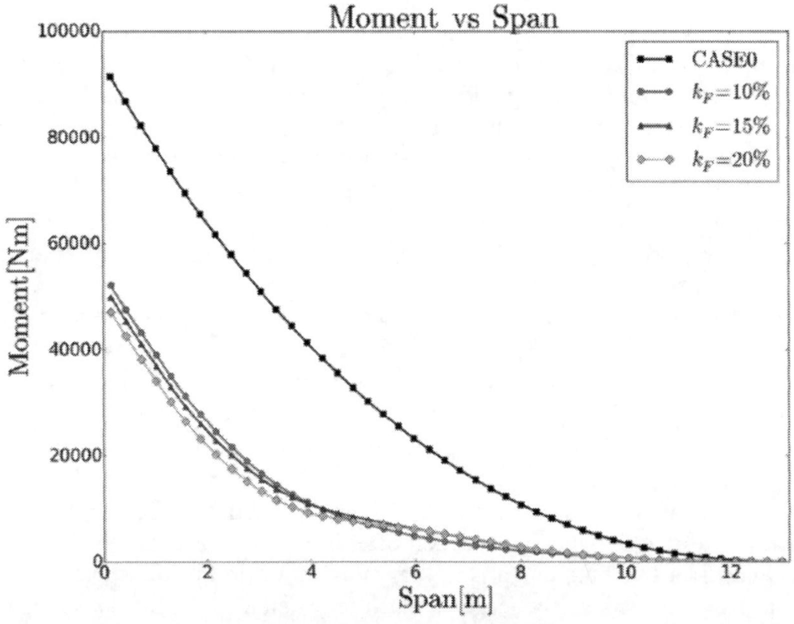

(b) Bending moment distribution

Figure 8. Results of optimization under constraints determined by Eq.(1) and (2)
(Black: CASE0, Red: C1-10, Blue: C1-15, Yellow: C1-20)

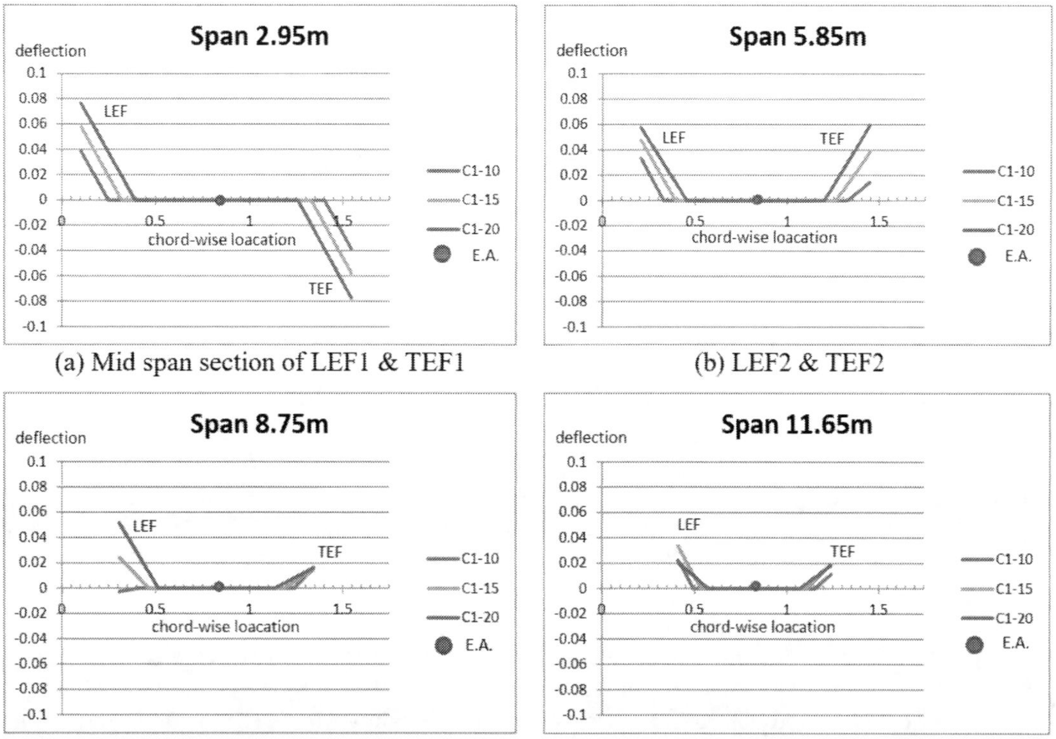

(a) Mid span section of LEF1 & TEF1 (b) LEF2 & TEF2

(c) LEF3 & TEF3 (d) LEF4 & TEF4

Figure 9. Flap deflections at the mid span section of each flap segment (Unit: m)

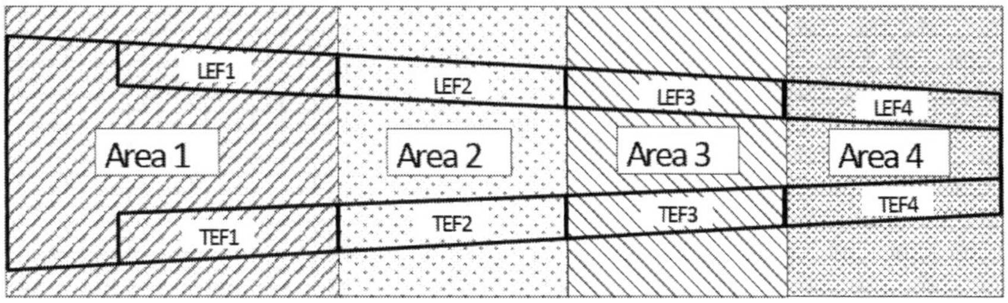

Figure 10. Definition of wing area

According to the considerations above, it is really insisted that the aeroelasticity approach must be applied to the optimization analysis to realize the most effective configuration of flap deflections.

5. CONCLUDINGS

In order to reduce the wing root bending moment of HALE airplane, the flap deflections have been optimized in this study. The wing model was selected as a straight plate having tapered wing thickness and chord length along its span-wise direction. The wing has 4 flaps at each of wing leading and trailing edges. The resultant lift distributions showed the characteristics of increased lift at the wing inboard sections and, on the contrary, reduced lift at the outboard sections. In the optimization, to generate the maximum lift at the wing inboard sections seemed to be the most important viewpoint. From the consideration for the optimized results, it was cleared that the aeroelasticity approach must be needed along with optimization technique to acquire the most effective result. According to the presented approach, the wing root bending moment was successfully reduced by about 50% comparing to the original configuration without flap deflections. As for the future works, it is needed to make the wing model more practical one from the present simplest model of tapered plate.

REFERENCES

1. Nguyen, N., "NASA Innovation Fund 2010 Project Elastically Shaped Future Air Vehicle Concept," *ARC-E-DAA-TN3743*, 2010.
2. Xu, J. and Kroo, I., "Aircraft Design with Maneuver and Gust Load Alleviation," *AIAA Paper 2011-3180*, 2011.
3. Regan, C.D. and Jutte, C.V., "Survey of Applications of Active Control Technology for Gust Alleviation and New Challenges for Lighter-weight Aircraft," *NASA/TM-2012- 216008*, 2012.
4. United States Air Force Air Combat Command, "CHAPTER 1: PURPOSE AND NEED FOR THE PROPOSED ACTION," ENVIRONMENTAL ASSESSMENT for Global Hawk Main Operating Base Beddown, 2001.
5. Steiner, R., "A Review of NASA High-Altitude Clear Air Turbulence Sampling Programs," *AIAA Paper 65-13*, 1965.
6. Gibbs, Y., "NASA Armstrong Fact Sheet: Altair," http://www.nasa.gov/centers/armstrong/news/FactSheets/FS-073-DFRC.html, 2015.
7. Yagihashi, M., "Study on Reduction of Wing Root Bending Moment by Controlling Trailing edge Flaps for a High Aspect Ratio Wing (there are no English title: translated by the present author)," *Master Thesis written in Japanese, the Waseda University*, 2015.

ICAST2015: 26th International Conference on Adaptive Structures and Technologies
October 14-16th, 2015, Kobe, Japan

ICAST2015 #058

An active back-flow flap for dynamic stall control - from concept to wind tunnel

Steffen Opitz[1*], Anthony D. Gardner[2], Christoph B. Merz[2] and Christian C. Wolf[2]

[1] Department of Multifunctional Materials, Institute of Composite Structures and Adaptive Systems, DLR (German Aerospace Center), Braunschweig, Germany.
[2] Department of Helicopters, Institute of Aerodynamics and Flow Technology, DLR (German Aerospace Center), Göttingen, Germany.

Abstract

The paper describes the concept, design and realization of a wind tunnel model featuring an active back-flow flap for dynamic stall control. In order to investigate influence of a back-flow flap on the stall behavior of an airfoil with oscillating pitch angle, an existing wind tunnel model is retrofitted with a spoiler-type flap on the upper surface. The presented work starts with the basic idea, describes the aerodynamic and structural design and ends with the first results from the wind tunnel testing. Within the paper special emphasis is given to the magnetic actuation system of the flap. The flap deployment under wind tunnel conditions and the corresponding deformation of the flap are discussed in detail. Finally the influence of the timing of the flap actuation on the flap motion and the aerodynamic performance of the flap will be elucidated.

1. INTRODUCTION

Dynamic stall is a well-known effect for helicopter airfoils occurring when a pitching airfoil stalls, forming separated flow in a dynamic stall vortex. A lift peak and a negative spike in pitching moment form and then a rapid drop in lift appears as the stall vortex moves downstream. The torsional impulse from the pitching moment peak is often a load-limiting case for the pitch links, and high drag is experienced compared to attached flow. Dynamic stall can be controlled by using passive devices, including vortex generators and changes in the leading edge contour [6, 7], but these have disadvantages at high Mach number and are limited in their control of deep stall. In contrast, active devices include actively retracting vortex generators [5], which avoid shocks at high Mach number, or air jets [11, 12], which add energy to the flow, and thus improve the flow control under deep stall conditions.
The leading-edge stall associated with many helicopter airfoils including the OA209 [1] airfoil used for these investigations, is characterized by strong backward flow along the suction side of the airfoil [9]. To control dynamic stall of this type, the single dynamic stall vortex should be broken into several smaller vortexes and the passage of the vortexes should be delayed [2].

2. STALL CONTROL WITH A BACK-FLOW FLAP

As part of the DLR project STELAR an alternative method of influencing dynamic stall using an active back-flow flap is presently being investigated [4]. In the past back-flow flaps have been tested on gliders as a passive method of reducing stall [8], as freely-hinged spoiler-like flaps on the suction side of the airfoil near the trailing edge. When trailing edge stall occurs, the back-flow lifts the flap and the region of stall can be significantly delayed. Even though the effectiveness of back-flow flaps has been demonstrated in flight experiments with sailplanes, it still has to be investigated whether they are also suited to improve the dynamic stall behavior of helicopter rotor blades [3].

* Steffen.Opitz@dlr.de

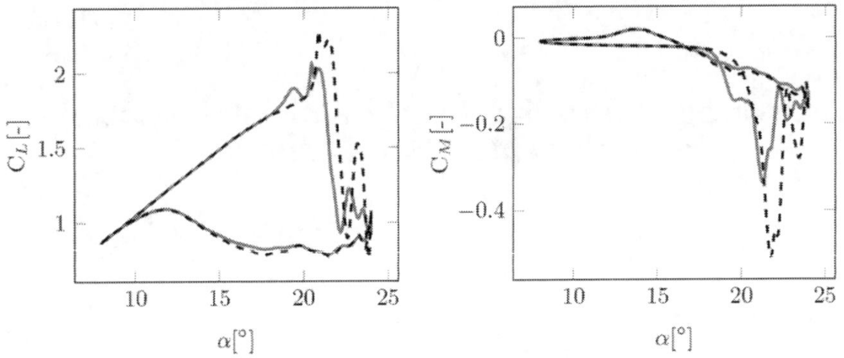

Figure 1. Comparison of lift (left) and pitching moment coefficient (right) for the OA209 airfoil at M=0.14, Re=920000, $\omega^* = 2\,\pi\,f\,c/v_\infty$=0.1, α=16±8°. The black lines are without flap actuation, and the red lines are with flap actuation (image from [4]).

Initial numerical investigations with an actively actuated flap showed promising results. The computation was performed in 2D using the unstructured DLR-TAU solver and the Spalart-Allmaras turbulence model using the settings of Richter et al. [10]. The flap was dynamically actuated to only be deployed during the stalled flow, by using the overset grid (chimera) method with automatic hole cutting. The back-flow flap breaks the main stall vortex into smaller structures, and this results in a reduction in the pitching moment peak of 34 % compared to the reference case without actuation, while maintaining the lift during the attached flow (see Figure 1).
To gather first experiences, simple wind tunnel experiments were performed with an oscillating 2D airfoil without instrumentation. Analyzing the results with passive flaps , it was noted that an existing airfoil could generally be retrofitted with a flow control flap. However, to have reproducible conditions during the wind tunnel experiment the back-flow flap approach requires an actuator to hold the flap closed when not needed (e.g. reference cases), and to accelerate the opening of the flap with a pulse of force to guarantee the immediate flap actuation. Further, an angle restriction is needed to limit the flap opening angle, so that the highly dynamic opening of the flap does not cause the flap to flip over, lying upside down upstream of the hinge line.

Summing up the results of the initial numerical investigations and the first wind tunnel experiment a list of requirements for the structure and the actuation mechanism was established. Based upon this list, a concept was developed and presented in detail in [13] and [14]. To prevent any steps or gaps the concept comprises a solid state hinge that connects the flap to a Glassfiber Reinforced Polymer (GFRP) shell that forms the outer model surface. Fiber reinforced elastomers that are highly bendable but which do not elongate significantly under tensile loading are intended to limit the flap opening angle. To provide the opening impulse for the flap the actuation concept uses flap integrated magnets that can be attracted or pushed away by a second set of magnets that moves underneath the GFRP shell. With this approach the flap can be held closed or swiftly opened.
Even though this concept could be shown to be generally feasible there were still a lot of open questions to be answered before the final wind tunnel testing of the back-flow flap. Major points were the detailed aerodynamic design (size and position of the flap), the detailed design of actuation mechanism and the instrumentation of the wind tunnel model. These issues will be addressed within this paper, presenting a detailed concept for the retrofit of a wind tunnel model. The paper ends with initial wind tunnel results from the testing of the active back-flow flap.

3. DESIGN OF WIND TUNNEL MODEL

3.1 Aerodynamic design

Computations using the method of Kaufman et al. [4] were performed to estimate the correct size and position of the flap. Therefore the flap size was varied in 3 steps from 5 % chord to 15 % chord. The position of the flap was altered in steps of 10 % chord between 10 % and 70 % of chord. The results of this study are depicted in Figure

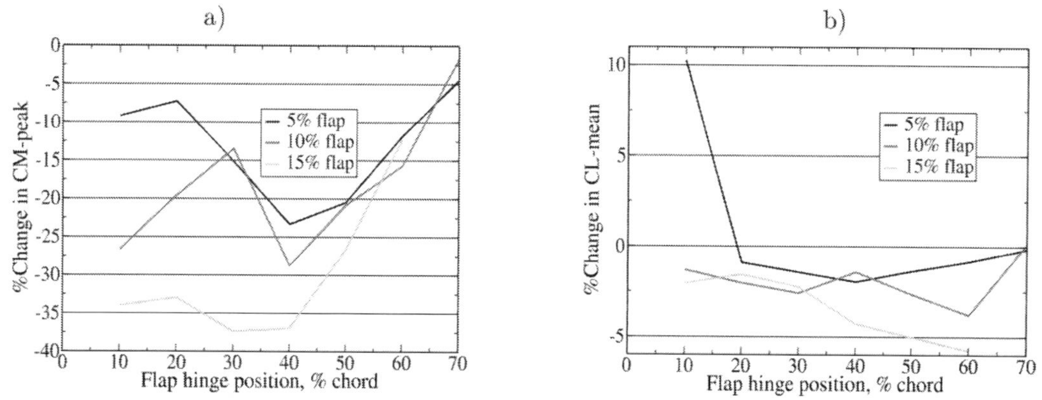

Figure 2. Numerical investigation of the flap size and position, compared to the reference case without a flap, a): Change in pitching moment coefficient peak (Dynamic stall load), b): Change in mean lift coefficient (Hysteresis).

2. These computations represent only a rough approximation of the real physics, since they could not model the body-on-body parts of the opening, and always had a gap between the flap and the airfoil. Further, the flap motion which was prescribed in the computations would certainly differ in reality.
Nevertheless some trends can be deduced. In most cases the reduction of the pitching moment peak becomes more effective with increasing flap length. At the same time a larger flap tends to cause a larger reduction of the mean lift coefficient. In contrast to the flaps used to improve the static stall behavior of gliders, positions near the trailing edge do not seem to be effective to reduce the pitching moment peak caused by dynamic stall. For all three investigated flap lengths the best position of the flap appears to be in the range between 30 % and 50 % of chord and tends to be closer to the leading edge for larger flaps. For the two larger flap configurations the mean lift tends to decrease when the flap is shifted towards the trailing edge.
Setting the flap to a length of 15 % chord with hinge position of 30 % chord had the greatest reduction in pitching moment, while not excessively reducing lift. The demonstrator for the basic concept of the solid state hinge and the magnetic actuation concept (see [13]) had an effective flap length of approx. 35 mm which corresponds to approx. 9.5 % chord of the later wind tunnel model. Laboratory tests have shown that in this configuration the actuation mechanism is able to deploy the flap to 30 ° in approx. 12.5 ms. This opening speed was sufficient to match the prescribed motion that was used in the numerical investigation when the airfoil oscillates with a frequency of 6 Hz. As the inertia of the flap rises with increasing flap size and consequently reduces the deployment speed, it was decided to realize a flap that extends over 12.5 % of the chord instead of the 15 % that gave the best results in the numerical prediction. Following the best simulated configuration the hinge line of the solid state hinge was placed at 30 % chord. Further on, the decision was made to limit the flap opening angle to a maximum of 30 °, to avoid over-opening during the back-flow of the fully separated flow. On the basis of the computations a reduction of the pitching moment peak by 30 % and a loss for $C_{L_{mean}}$ of 2.5 % is expected for this setting.

3.2 Structural design

During the first wind tunnel experiment with passive flaps it was posited that it is possible to equip an existing wind tunnel model with a kind of back-flow flap glove. Such a retrofit solution has multiple advantages. If an adequate concept is found the structure and the instrumentation of an existing model can be used to save costs and to reduce the development time for each tested configuration. Additionally the potential to retrofit existing rotor blades with a back-flow flap can be demonstrated to a certain extent. The first challenge related to a retrofit solution is to find a good compromise between available installation space and size of the model. Increasing the chord length leads to a larger blockage of the wind tunnel and also to a larger flap that might be difficult to deploy in a short time due to its inertia. For this reason the developed glove should be as short and as thin as possible. To maximize the mounting space available at the upper surface where the flap is installed, it was decided to partially use the lower side of the underlying model. The existing wind tunnel model that is intended to be equipped with the back-flow flap has a chord length of 300 mm. Already knowing the chordwise position and the size of the flap within the retrofit-glove from the aerodynamic design, a small optimization routine was used to place the original airfoil in the enlarged contour of the model with the back-flow flap. This placement was done in a way that a maximum of mounting space is available at the position of the actuation mechanism. The optimization was repeated for increasing chord length of the glove until the distance between new and old airfoil guaranteed a thickness of 5 mm at the position of the flap (see Figure 3). This requirement for a minimum thickness is driven by the size of

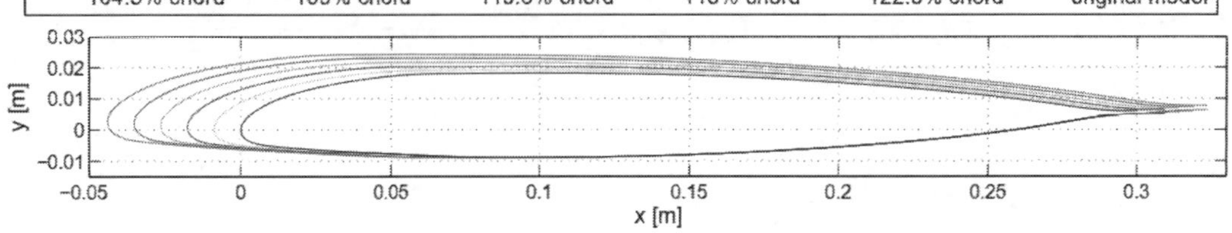

Figure 3. Positioning of the original airfoil of the wind tunnel model within the glove of the back-flow flap for different chord lengths of the modified model.

the neodymium magnets that will be used to actuate the flap. The magnets installed in the flap and the magnets sliding underneath the GFRP-shell have a thickness of 2 mm each. Consequently a minimum of 4 mm thickness is already needed without any structure that incorporates the magnets or friction layers that enable the sliding motion and protect the underlying wind tunnel model. Finally the described optimization procedure resulted in a scale of 1.225 for the glove which is equivalent to a chord length of 367.5 mm (red line in Figure 3).

3.3 Actuation

The left part of Figure 4 shows the general operation principle for the actuation mechanism of an active back-flow flap presented in [13] and [14]. The mechanism consists of two magnets. In previous publications it was

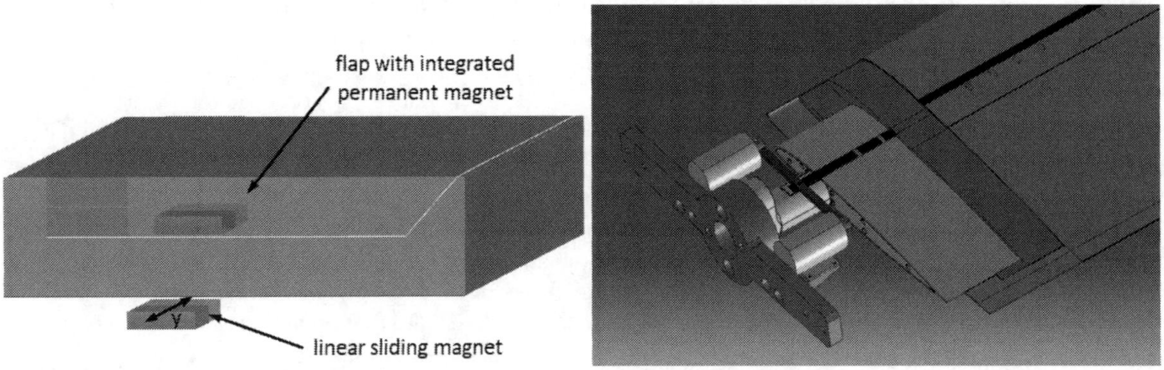

Figure 4. left: Actuation principle (image from [13]) right: Retrofit-concept (green: original model, orange: GFRP-shell, red: resin, blue: model attachment, gray: electromechanical solenoids, , black: push rod).

demonstrated how a magnet could be integrated into the GFRP-flap. The second, moving magnet and the adjacent actuation mechanism that generates the linear motion were not in the focus of interest so far. In [14] it is shown that the magnetic actuation system can lead to a snap opening of the flap when the bending stiffness of the solid state hinge is small. As the actuation concept allows to keep the flap closed and to deploy it, the bending stiffness of the hinge is preferred to be as small as possible, while still allowing a robust operation of the flap. This enables large opening angles since the restoring moment that has to be overcome by the magnetic forces stays small. The small bending stiffness and the nonlinear magnet forces used to deploy the flap allow the final wind tunnel model to have the same impulsive opening behavior described in [14]. This circumstance already prevents a linear control of the flap angle. Since the flap operation is digital, there is no need to have a linear control of the position of the moving magnets. This makes simple electromechanical solenoids located on both sides of the model an adequate

option (see Figure 4, right). To comply with the minimum of available mounting space the sliding magnets were integrated into a lightweight fiber composite push rod with a thickness of only 2.2 mm. To ensure a simultaneous motion of all magnet pairs the spacing between the magnets was adjusted to the actual spanwise location of the magnets in the flap after the manufacturing process. As the push rod exits the model in the region of the root attachment (see Figure 4, right), two electromechanical solenoids were used on both sides of the model.
The back-flow flap of the wind tunnel model is deployed by 24 pairs of magnets distributed over the span of the model. Numerical simulations of the magnetic field between the neodymium magnets in the flap and underneath the GFRP-shell have been used to estimate the repellent and attracting forces for the flap deployment (dashed black line in Figure 5) and at the same time the forces that are needed to move the magnets underneath their counterparts in the flap (solid black line in Figure 5). To switch between attraction and rejection of the flap magnets the push rod

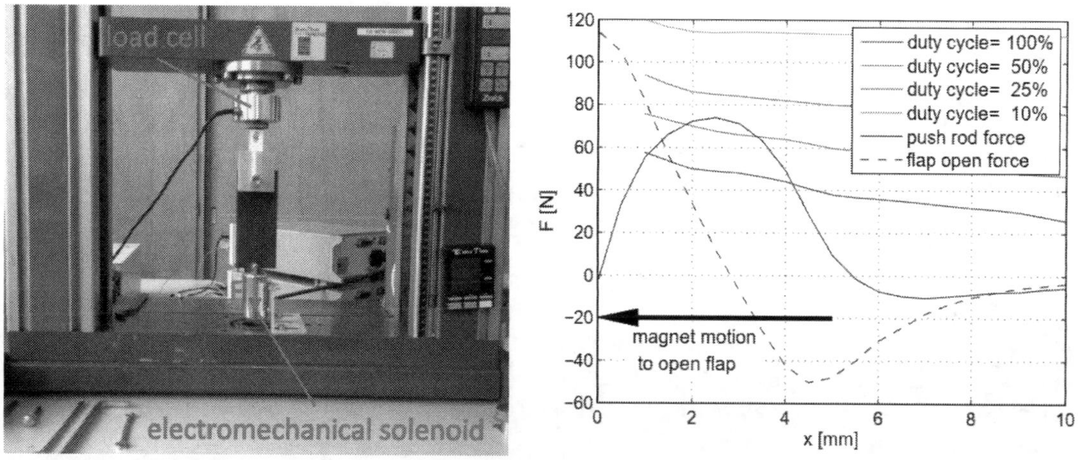

Figure 5. Characterization of electromechanical solenoids (left: test setup, right: maximum solenoid forces for different duty cycles).

has to be moved by approx. 5mm (arrow in Figure 5). According to the numerical prediction a maximum force of approx. 75 N is needed for this movement in the worst case of a blocked flap that does not deploy. To assure a reliable deployment of the flap the force generated by the electromechanical solenoids has to be larger than the push rod force over the whole duration of the push rod motion. Especially for large operating displacements the force of electromagnetic actuators tends to decrease. An adequate geometry of the solenoids armature is needed to overcome this issue. Further on the selection of suited electromechanical solenoids is a trade off between performance and weight. Designing the solenoids to be able to generate a continuous force of 75 N would results in a bulky and heavy actuation system. As these actuators are mounted on an oscillating wind tunnel model the additional mass constitutes an undesirable loading for the motor that drives the oscillation of the model. This additional load can cause restrictions in the maximum pitching frequency or amplitude as well as damages to the model, the drive train or the balance. Form the load characteristic it can be seen that after overcoming the maximum force the push rod force drops to zero and a smaller solenoid would be sufficient to hold the magnets in this position. For this reason it was decided to use smaller and lighter solenoids with a lower continuous force but to operate them at higher power in an intermittent duty. The diagram on the right side of Figure 5 shows the result of the characterization of the forces that can be generated by the electromechanical solenoid for different duty cycles over push rod position. Operating the solenoids at a 25 % duty cycle allows overcoming the push rod force to open the flap. Switching to a 10 % duty cycle an increase in the voltage and current to reach forces of up to 110 N is possible. This leaves enough margin to compensate inertia forces or friction.
The electromechanical solenoids are controlled by a micro-controller that analyzes a sinusoidal signal that represents the pitching angle of the model to allow the deployment of the flap synchronized to the model motion. The phase angle for the opening of the flap can be set by a potentiometer to tune the timing of the opening to the dynamic stall condition of each measurement point.

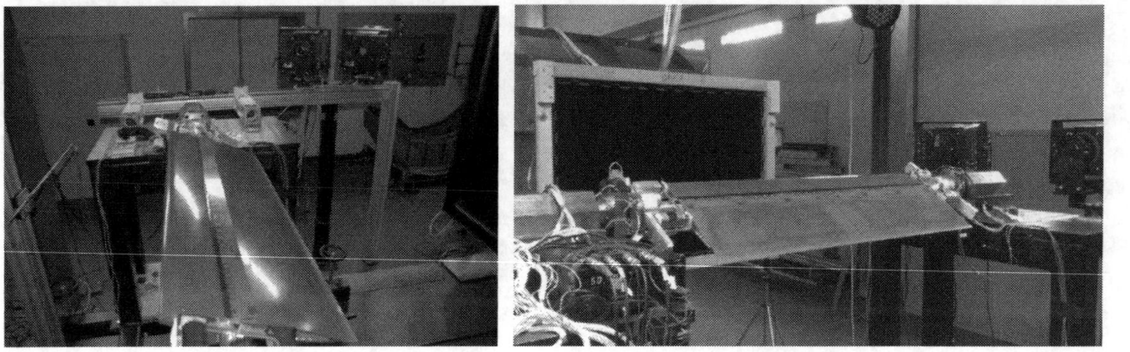

Figure 6. Model in the open test section of the wind tunnel.

4. WIND TUNNEL TESTING

The wind tunnel model with the installed back-flow flap glove is depicted in Figure 6. The model was mounted horizontally in the 1 m wide and 0.75 m high open test section of the 1MG low speed wind tunnel in Göttingen. The airfoil is supported on both sides by bearings, and driven from one side using an electromotor. In the left picture the actuation mechanism is holding the flap open. The forces on the airfoil were measured using the integration of pressures from the airfoil midline. Additionally a piezoelectric force balance was used as a check of the forces. The pressure integration on the airfoil includes sensors above and below the flap. The flap angle is measured by the Hall-effect sensors and is factored into the force computation. The pressure under the flap is taken to be a constant from the single pressure sensor present there, as an approximately constant pressure was indicated by the CFD computations of [4]. The pressure sensors are differential Kulites, type XCQ-093, which are referenced to the static pressure outside the free jet. To compute C_P, the pressure sensors are referenced to the dynamic pressure as measured by a Prandtl-probe mounted at the nozzle exit. The Prandtl probe is also used to compute the flow speed. The position of the push rod is quantified in the same way as the flap angle using another combination of Hall-effect sensor and magnet. Therefore the permanent magnet was attached to one end of the push rod while the Hall sensor was glued to the model root attachment. The angle of attack of the airfoil is measured using two laser rangefinders, which measure to the flat plate at the bottom of the force balance. The average of the calibrated angles of attack is used. For each measurement point all data was recorded for 40 s with a sampling frequency of 20kHz. The results presented in this paper are averaged over the cycles performed within the measurement period. The flap and its actuation mechanism have already been tested in the laboratory without any aerodynamic forces. Adding wind can have a strong effect on the opening behavior of the flap since the magnetic actuation mechanism exhibits no mechanical connection between push rod and flap. Further on, the whole concept relies on the utilization of the forces generated by the back-flow to fully deploy the flap. Figure 7 shows the flap opening angle measured with a hall sensor located near the mid plane of the model with and without wind (upper diagram) as well as the corresponding motions of the push rod (lower diagram). The shaded area underlaid to each curve depicts the standard deviation as a measure for the cycle to cycle differences. While these deviations are practically invisible for the push rod motion and the flap motion without wind, the amount of arbitrary motion caused by the wind becomes obvious. Looking on the flap motion without wind it can be seen that the dynamic opening causes a strong oscillation of the flap angle. This is due to the characteristic of the magnetic forces. When the flap opens the repellent force between the two magnets decreases with the square of their mean distance. In the equilibrium state the moment generated by the magnetic forces equals the elastic moment that is needed to deform the solid state hinge. When the flap is dynamically opened the kinetic energy stored in the flap motion leads to an overshoot in flap angle and the flap starts oscillating around this equilibrium.

As expected there is no influence from the wind on the push rod motion when the flap opens. Even though the push rod motion indicates a comparable excitation of the flap, operating the flap in the flow of the wind tunnel significantly alters the flap motion (red lines in Figure 7). It can be seen that the flap opening speed is slightly increased by the aerodynamic forces. The oscillation of the flap angle is still visible but less pronounced. After the first maximum of the flap angle oscillation the back-flow developed during dynamic stall supports the deployment of the flap which lead to a significantly higher mean value of the flap opening angle. Comparing the phase shift between the movement of the push rod in the "closed" position and the reaction of the flap, reveals that the authority

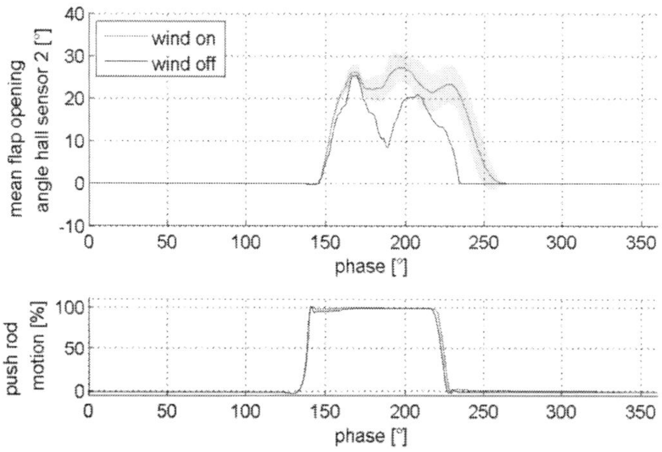

Figure 7. Influence of wind on the flap motion (α=22±8°, f=2.5 Hz, black lines: 0 m/s, red lines: 50 m/s).

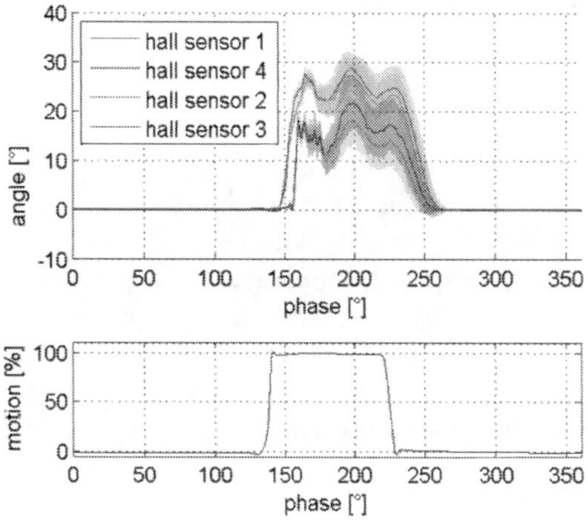

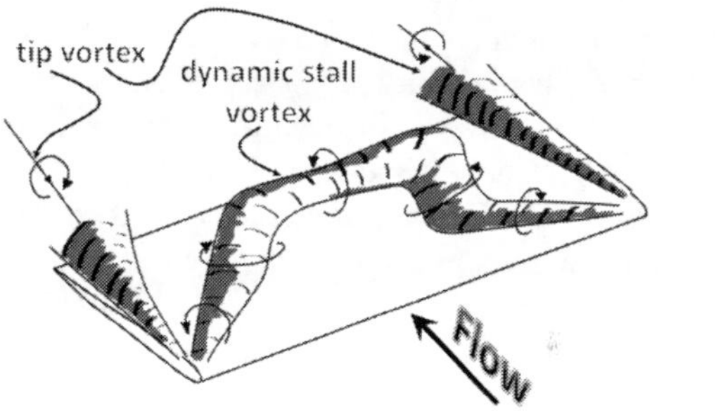

Figure 8. left: Flap opening angle for different span-wise locations (α=22±8°, f=2.5 Hz, wind speed 50 m/s); right: Sketch of vortices on an oscillating finite wing (Figure adapted from [15]).

of the actuation mechanism to close the flap is limited. This holds especially for large opening angles where the magnetic forces are small. Aerodynamic forces can further delay the closing motion. Principally the flap closes when the flow reattaches and the magnets keep the flap closed.

Since the model is tested in an open test section the edge effects at the sides of the model also influence the flap opening. This becomes obvious when the flap angles from the 4 different spanwise positions are compared (see Figure 8, left). The similarity of the signals from Hall-effect sensors 1 and 4 as well as from sensors 2 and 3 give a good impression of the symmetry of the flow conditions on both ends of the model. To ease the interpretation of the results a sketch of the typical structure of a dynamic stall vortex that develops on a finite rectangular oscillating wing (adapted from [15]) is depicted in the right part of Figure 8. The illustration shows the dynamic stall vortex with its characteristic "Omega"-shape in the middle of the model and two tip vortexes rolling up at both sides of the model. While the direction of rotation of the Omega vortex supports the flap deployment the tip vortexes push the edges of the flap onto the model surface and cause smaller mean flap opening angle towards both sides of the model. Further on the tip vortexes cause a delay in the flap opening close to the ends of the model. This can be directly seen in the phase shift of the flap angle from the sensors 1 and 4 compared to the sensors 2 and 3 that are located closer to the airfoil midline. The spanwise ends of the flap start to open when the mid region is already almost fully deployed. This difference in flap angle is an indicator for the flexibility of a GFRP-flap with a thickness of 1 mm, a chord length of approx. 41 mm and a spanwise extension of 1 m. A closer look at the signal sequences reveals that in the middle of the model the flap starts oscillating immediately after the snap

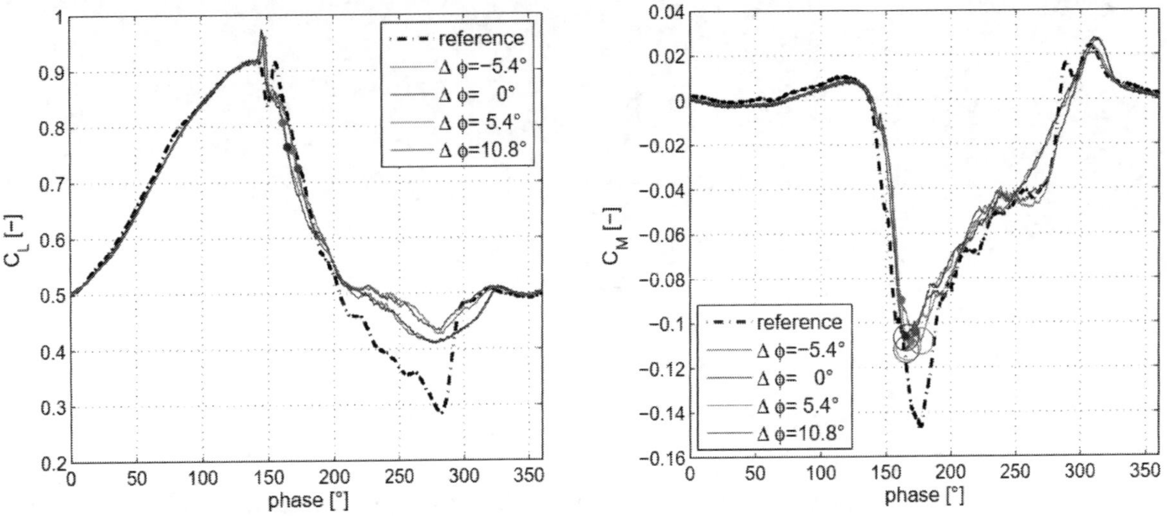

Figure 9. Lift and pitching moment in case of right flap deployment (α=22±8°, f=2.5 Hz, wind speed 50 m/s).

opening. In contrast to that the Hall effect sensors close to the ends of the model indicate that the flap performs higher frequency motions before entering in an equivalent oscillation state. These higher frequency motions are probably caused by the variety of vortexes that are shed from the edges of the model and the edges of the flap itself. The aerodynamic effect of the flap can be evaluated by either using the forces and moments captured by the piezoelectric balance that connects the model to the drive train of the motor or by analyzing the data measured with the integrated pressure sensors. The data from the piezoelectric balance comprises a mixture of the effects from the central stall vortex and the both edge vortexes. As these effects can not be separated in the post-processing the comparability to the initial two-dimensional CFD-computations is very limited. Further on the signal sequences of the forces and moments measured with the balance are polluted by impulsive shock forces. These are generated by the electromechanical solenoids that move into their hard stop when they alter the position of the push rod to trigger the flap opening or closing [16]. Consequently this paper presents the results gathered by pressure instrumentation.

This paper is not intended to give a detailed insight in the behavior of the flap for the whole variety of flow conditions covered by the test matrix of the wind tunnel experiment. As a first result the influence of the timing for the flap opening on the flap motion as well as on the lift and moment coefficient will be investigated. Therefore the model with the flap was tested at a wind speed of 50 m/s. The airfoil oscillated with an amplitude of 8° around a mean angle of 22° with a frequency of 2.5 Hz causing a pronounced dynamic stall. To maintain the readability of the diagrams the standard deviation is not plotted. Nevertheless these results represent the averaged data of approx. 100 cycles. The dotted lines in Figure 9 show the reference for the lift and pitching moment coefficient for a closed flap. The peak in pitching moment caused by the dynamic stall can be clearly recognized in the right diagram. An online estimation of the pitching moment from a weighted sum of the pressure sensor signals enabled to set the timing of the flap opening. In order to get a high reduction in pitching moment peak without disturbing the aerodynamics in the regime of attached flow the flap opening was manually tuned to match the onset of the pitching moment peak. Afterwards the timing was intentionally detuned to investigate the effect of an too early or too late actuation of the flap. The deployment was shifted in 8 steps of Δt= 6 ms in each direction within the oscillation cycle of the model. For a pitching frequency of 2.5 Hz a time step of Δt= 6 ms corresponds to a phase shift of $\Delta\phi$= 5.4 °. To ease the readability of the diagrams the results are presented in three groups. The first one shows the regime of optimal timing of the flap deployment. The second group presents test cases with an early opening of the flap while the third one summarizes the results of a delayed deployment. For each group separate diagrams will illustrate the flap motion and the resulting effect on lift and pitching moment.
Figure 9 shows that for the given measurement point the active back-flow flap can reduce the peak in pitching moment by approx. 28%. Lift and pitching moment remain almost unchanged until the onset of the dynamic stall. Even though the reduction in pitching moment peak is highest for the initial setting of the flap deployment timing the peak size remains in the same range when the timing of the deployment is altered by a small amount.

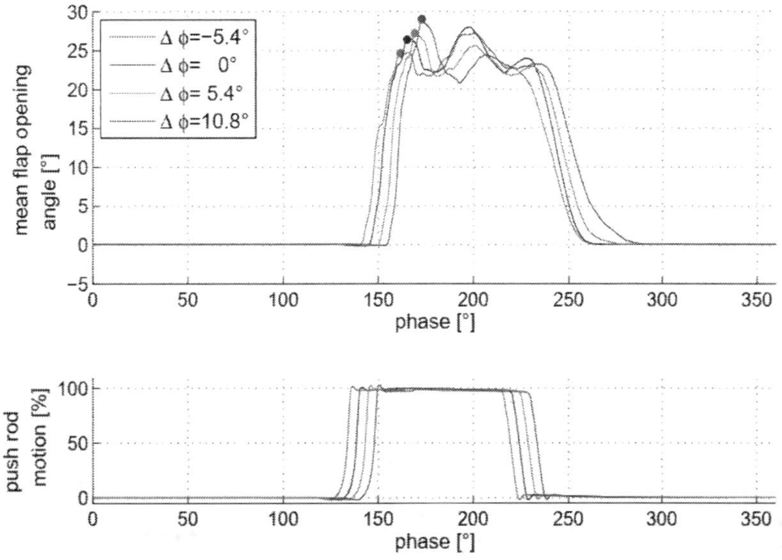

Figure 10. Flap angle in case of right flap deployment (α=22±8°, f=2.5 Hz, wind speed 50 m/s).

This means that to a certain extent the efficiency of the flap is robust and insensitive to any minor jitter in the timing. Comparing the lift with and without flap actuation it can be observed that the flap deployment leads to an increase of the lift in the region of stalled flow. This effect is quite interesting since it was expected to be much less pronounced (see Figure 1). Interpreting the increase in mean lift as another advantage of the back-flow flap it seems to be beneficial to delay the flap deployment by 5 ° to 10 °. Figure 10 depicts the averaged flap motion measured by the two hall sensors close to the airfoil midline. Having a look at the corresponding flap motions reveals that the phase region of increased lift clearly extends the opening period of the flap. The physical effect behind this observation has to be further investigated (e.g. by analyzing the PIV data). For all phase settings the flap angle swiftly increases. The point were the opening angle starts to decrease again is marked in Figure 10 as well as in Figure 9. It seems that the back-flow that supports the deployment is already stronger when the flap is opened at a later instant. The consequence is an increase of the first peak in the flap angle oscillation. For an early opening of the flap the rising edge of the flap opening angle develops a small bump. This effect is much clearer when the second group of measurement points with early opening is analyzed.

Figure 11 depicts the averaged flap opening angle measured by the Hall effect sensors close to the model midline in case of early opening. It can be seen that the stagnation point within the opening process moves to smaller flap angles. This means that the first oscillation starts before the back-flow fully deploys the flap. Until the first peak in the flap angle oscillation the flap always opens with approx. the same speed. In the region of small flap opening angles the driving forces are the repellent forces between the magnets. They determine the opening speed of the flap. The aerodynamic forces are of minor importance for this first part of the opening motion. After the point of first stagnation of the flap angle the opening speed of is significantly reduced. Beyond this point the aerodynamic forces are dominating the flap motion. As stall occurs at approx. 150 ° there is no back-flow that can rapidly deploy the flap. Only the pressure gradient on the upper airfoil surface supports the flap deployment. The pressure gradient from the trailing to the leading edge rises with the angle of attack. This is also supposed to be the reason for the slight increase in angle of the first peak in the flap opening when the flap is activated at a later instant in the oscillation period. The aerodynamic effect of the flap that opens too early is depicted in Figure 12. The point in the opening motion where the aerodynamic forces begin to dominate the flap motion can be recognized in the lift and in the pitching moment signal. In the lift signal the point coincides with the start of the formation of the first bump where the lift curves exceed the lift of the reference case. This bump is also visible in the CFD simulations with prescribed flap motion (see Figure 1) which indicates that the deployment in the simulation is also a little bit to early. Directly after the first oscillation in the flap opening angle the lift drops and stays below the lift of the reference case. At the beginning of the second half of the cycle lift signal is pretty comparable to the measurement points close to the optimal timing. Once again the lift in the region of detached flow is higher than for the reference

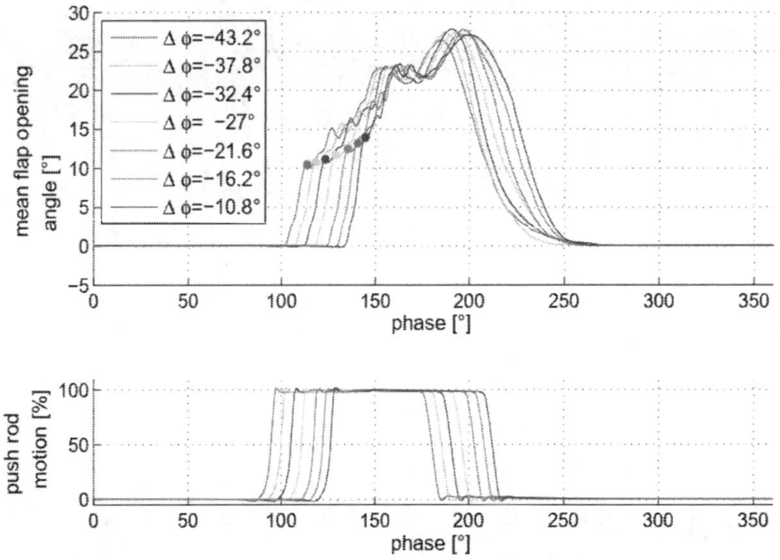

Figure 11. Flap angle in case of early flap deployment (α=22±8°, f=2.5 Hz, wind speed 50 m/s).

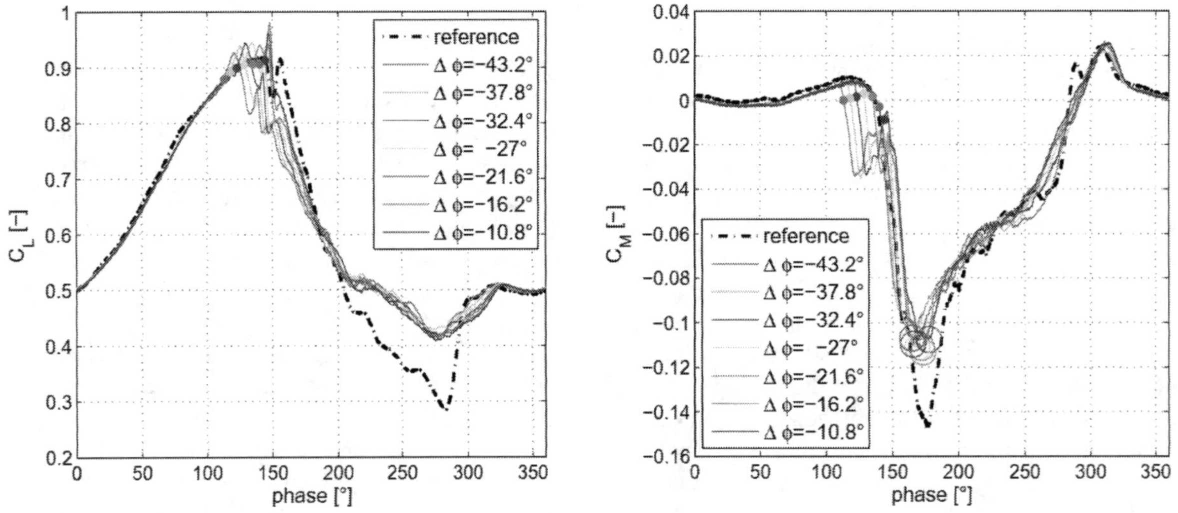

Figure 12. Lift and pitching moment in case of early flap deployment (α=22±8°, f=2.5 Hz, wind speed 50 m/s).

case without flap. Analyzing the pitching moment signal the points marked in the flap motion diagram can be correlated to the first drop in the pitching moment. Even though the early opening of the flap causes this first drop in pitching moment the effect of the back-flow flap in the second half of the cycle and therefore also on the main peak is comparable to the test cases close to the best timing of the flap. Consequently opening the flap to early is not crucial for the reduction of the pitching moment peak but will reduce the mean lift that can be generated over the oscillation cycle.

This picture changes when the flap is deployed too late (see Figure 13). Despite the rapid opening of the flap depicted in Figure 14 both the effect on the pitching moment and the effect on the lift become significantly smaller when the deployment is delayed too much. The activation of the flap gets simply too late to influence the pitching moment peak. Consequently the peak magnitude approaches to that of the reference condition without flap. As already noticed analyzing the test cases with optimal timing, the effect that increases the lift in the region between 200 ° and 300 ° seems to have its origin earlier in the oscillation cycle. This is confirmed by the fact that even

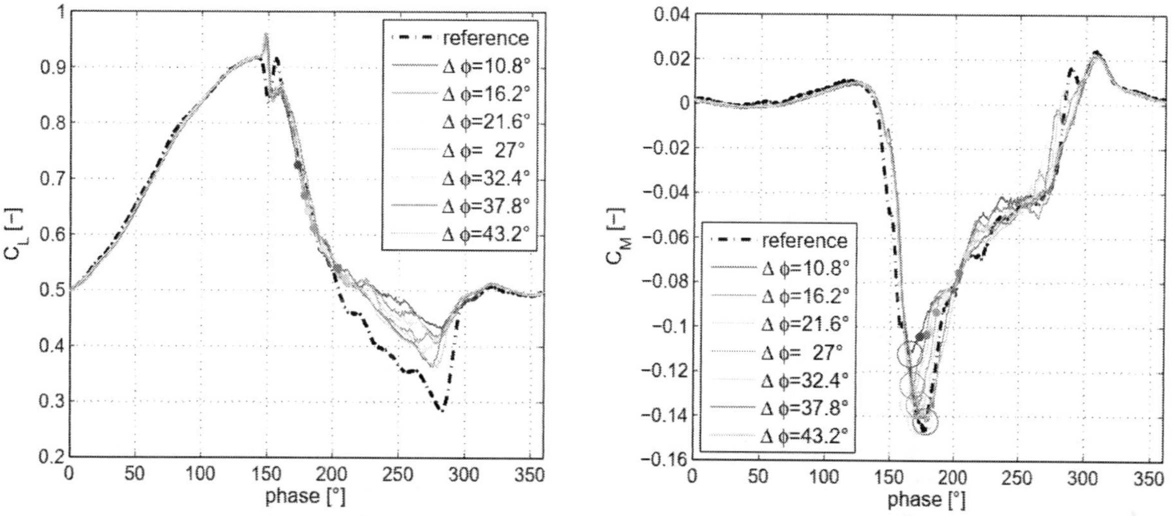

Figure 13. Lift and pitching moment in case of late flap deployment (α=22±8°, f=2.5 Hz, wind speed 50 m/s).

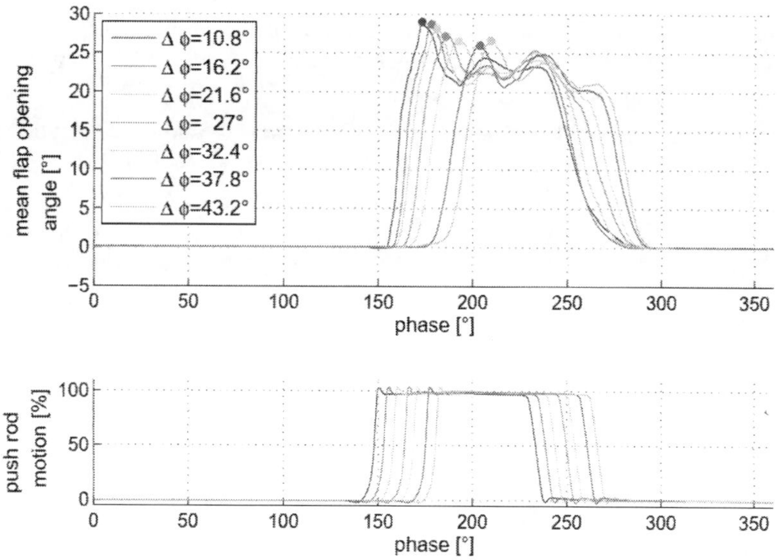

Figure 14. Flap angle in case of late flap deployment (α=22±8°, f=2.5 Hz, wind speed 50 m/s).

though the flap activation with the biggest delay keeps the flap opened for almost the complete phase region with increased lift the effect is much less pronounced than for smaller phase shifts of the opening impulse.

5. CONCLUSIONS

The paper presents the way from the basic principle of an active back-flow flap to the successful testing in a wind tunnel. The size and position of the flap were determined using numerical simulations. For the structural design of the wind tunnel model a solution that requires a very small mounting space for structure and actuation system has been presented. It was demonstrated that electromechanical solenoids can be operated in an intermittent mode to deliver the forces needed to drive the actuation mechanism while keeping the weight of the actuation system small. The effect of the wind on the flap motion was presented as well as the effects of the different vortexes that are caused by the operation of the model in the open test section of the wind tunnel. Finally, first wind tunnel results were able to show that the predicted reduction in the pitching moment peak of approx. 30 % can be reached, validating also the results of the numerical simulation. The expected reduction of the mean lift by approx. 2.5 % could not be confirmed by the experimental data. The wind tunnel results show an increase in mean lift which is beneficial for the back-flow flap concept but clearly shows the deficits of either the numerical prediction or the wind tunnel setup and needs further investigation. The effect of the timing of the flap actuation was studied in detail. In can be stated that the procedure to determine the optimal phase setting for the flap deployment in the wind tunnel experiment worked reliably. The effects are robust to small changes in the timing. A too early flap actuation mainly causes a decrease in mean lift but has only small effect on the reduction in the pitching moment peak. Deploying the flap too late diminishes the beneficial effects on both lift and pitching moment.

REFERENCES

[1] Gallot, J., Vingut, G., De Paul, M. V., Thibert, J. "Blade profile for rotary wing of an aircraft", United States Patent 4325675,(20.4.1982).

[2] Gardner, A.D., Richter, K., Rosemann, H., "Numerical investigation of air jets for dynamic stall control on the OA209 airfoil", CEAS Aeronautical Journal, Volume 1, Issue 1, Page 69-82, 2011. DOI 10.1007/s13272-011-0002-z

[3] Höfinger, Marc, "Rotorblatt mit integrierter passiver Oberflächenklappe" Deutsches Patent DE 10 2010 041 111 A1, 22.03.2012.

[4] Kaufmann,K., Gardner, A.D., Richter, K., "Numerical Investigations of a Back-Flow Flap for Dynamic Stall Control", New Results in Numerical and Experimental Fluid Mechanics IX, Notes on Numerical Fluid Mechanics and Multidisciplinary Design, Volume 124, 2014, pp 255-262. DOI: 10.1007/978-3-319-03158-3_26.

[5] LePape, A., Costes, M., Joubert, G., David, F., Deluc, J.-M., "Experimental Study of Dynamic Stall Control Using Deployable Leading-Edge Vortex Generators", AIAA Journal, Vol. 50, No. 10 (2012), pp. 2135-2145.

[6] Mai, H., Dietz, G., Geissler, W., Richter, K., Bosbach, J., Richard, H., de Groot, K., "Dynamic stall control by leading edge vortex generators", Journal of the American Helicopter Society, Volume 53, Number 1, pp26-36 (2008)

[7] Martin, P., Wilson, J., Berry, J., Wong, T., Moulton, M., McVeigh, M., "Passive Control of Compressible Dynamic Stall" AIAA Paper 2008-7506, 2008.

[8] Meyer, R.K.J., "Experimentelle Untersuchungen von Rückströmklappen auf Tragflügeln zur Beeinflussung von Strömungsablösungen. Dissertation Technische Universität Berlin, Mensch-und-Buch-Verlag, 2000.

[9] Mulleners, K., Raffel, M., "The onset of dynamic stall revisited", Experiments in Fluids, Vol. 52, (3), pp 779-793, 2012. DOI 10.1007/s00348-011-1118-y

[10] Richter, K., Le Pape, A., Knopp, T., Costes, M., Gleize, V., Gardner, A.D., "Improved Two-Dimensional Dynamic Stall Prediction with Structured and Hybrid Numerical Methods", Journal of the American Helicopter Society, Volume 56, Issue 4, 2011. doi:10.4050/JAHS.56.042007

[11] Weaver, D., McAlister, K.W., Tso, J., "Control of VR7 Dynamic stall by strong steady blowing", Journal of Aircraft, Vol. 41, No. 6, 2004.

[12] Gardner, A.D., Richter, K., Mai, H., Neuhaus, D., "Experimental Investigation of Air Jets for the Control of Compressible Dynamic Stall", Journal of the American Helicopter Society, Volume 58, Number 4, 2013. DOI: 10.4050/JAHS.58.042001.

[13] Opitz, S., Gardner, A.D., Kaufmann, K., "Aerodynamic and structural investigation of an active back-flow flap for dynamic stall control", CEAS Aeronautical Journal, Vol. 5, No. 3, 2014. pp. 279-291. DOI 10.1007/s13272-014-0106-3

[14] Opitz, S., Gardner, A.D., Kaufmann, K., "An active back-flow flap for a helicopter rotor blade", Advances in Aircraft and Spacecraft Science, Vol. 1, No. 1, 2014, pp. 69-91. DOI: 10.12989/aas.2014.1.1.069.

[15] Coton, F.N., Galbraith, R.A.McD., "An examination of dynamic stall on an oscillating rectangular wing", AIAA Paper 2003-3675, 2003.

[16] Opitz, S., Gardner, A.D., Merz, C.B., Wolf C.C., "Structural demonstration of an active back-flow flap under wind tunnel conditions", 41st European Rotorcraft Forum, Munich, Germany, 1-4 Sept, 2015

ICAST2015: 26th International Conference on Adaptive Structures and Technologies
October 14-16th, 2015, Kobe, Japan

ICAST2015 #070

Shape Control Experiment of Space Reflector for Extremely High Frequency Radio Astronomy

Hiraku Sakamoto[1*], Hiroaki Tanaka[2], Kosei Ishimura[3], Akihiro Doi[3], and Yusuke Kono[4]

[1] Tokyo Institute of Technology, Tokyo, Japan.
[2] National Defence Academy of Japan, Kanagawa, Japan.
[3] Institue of Space and Astronautical Science, Kanagawa, Japan.
[4] National Astronomical Observatory of Japan, Tokyo, Japan.

Abstract

The present paper reports the results of shape-control experiments using the smart secondary reflector prototype for satellite-borne radio-astronomy. The shape of the secondary reflector is modified by six special piezoelectric actuators with displacement magnification mechanisms. The experiments successfully demonstrated that the reception power of the antenna system has been improved by the shape control.

1. INTRODUCTION

The concept of smart space reflectors has been discussed [1–5], but experiments using actual antenna system have been rarely implemented thus far. For radio astronomy satellite, for example, high accuracy of reflector shape is required, especially to obverse frequencies as high as 100GHz. Since the high shape accuracy is difficult to achieve, "smart" reflector concept is attractive. Smart reflectors change the shape of the reflector on orbit to cancel the wavefront error due to the shape imperfection. In order to develop such innovative system, experimental studies are important. Experiments can demonstrate the technical feasibility and clarify the problems to realize new space systems.

The present authors have developed a smart reflector, which is a prototype of smart secondary reflector for Cassegrain radio astronomy antenna systems in space [6]. Figure 1 illustrates the space-borne radio astronomy observatory system where the secondary reflector corrects the wavefront error due to the shape errors of a large primary reflector. Figure 2 shows the prototype of the reconfigurable hyperbolic secondary reflector.

To this end, this study aims at evaluating the technical feasibility of the shape control concept for space antenna system, and clarifying the problems to design future space systems. In order to achieve this objectives, this study demonstrates a shape control of smart reflector using this space secondary reflector prototype. Specifically, this study proposes and evaluates the shape control algorithms to improve the reception power of the antenna system, with which smart secondary reflector prototype is integrated.

2. EXPERIMENTAL SETUP

* hsakamoto@mech.titech.ac.jp

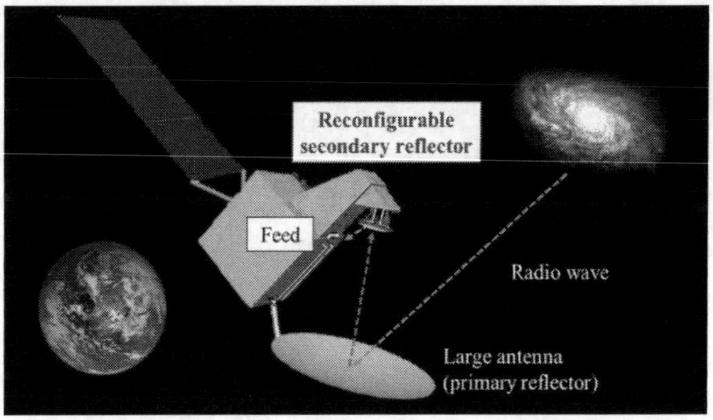

Figure 1: Highly precise space antenna system with smart secondary reflector.

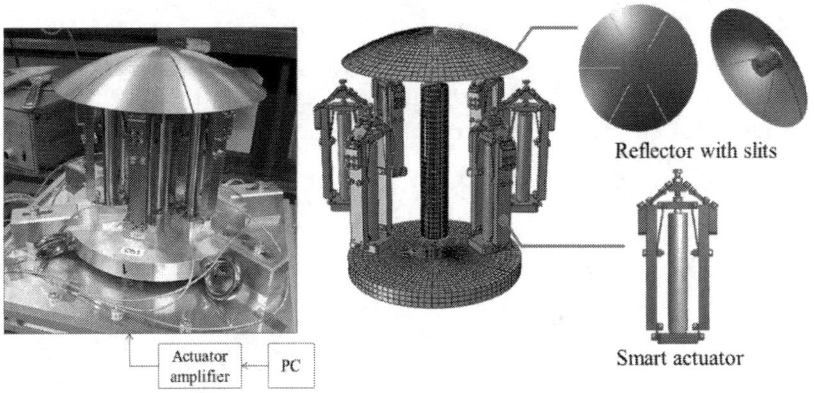

Figure 2: Prototype of smart hyperbolic secondary reflector.

The present study carries out an experiment using a ground-based antenna system, with the smart secondary reflector. The shape of the smart secondary reflector is controlled to improve the reception power of the antenna system. This section explains the experimental setup.

2.1 Prototype of space smart secondary reflector

This section reviews the prototype of the smart space reflector system for EHF (Extremely High Frequency) band that is used for the antenna experiment. Figure 2 shows a photo of the prototype, and Table 1 describes some of the specifications.

This reconfigurable reflector, developed for a preliminary investigation, has six "smart actuators". Six actuators are not a sufficient number to deal with multiple deformation modes, but are sufficient to evaluate basic performances of the reconfigurable reflector system. Based on the evaluation using this preliminary model, an appropriate number and locations for upgraded actuators will be determined in

Table 1: Specifications of smart reconfigurable reflector.

Category	*Specification*
Solid reflector	ϕ200mm, convex mirror made of aluminium
Smart actuator	Piezoelectric actuator (PSt 150/10/100 VS15) + Displacement magnification mechanism
Amplifier	Input: 0-10V, Output: 0-150V

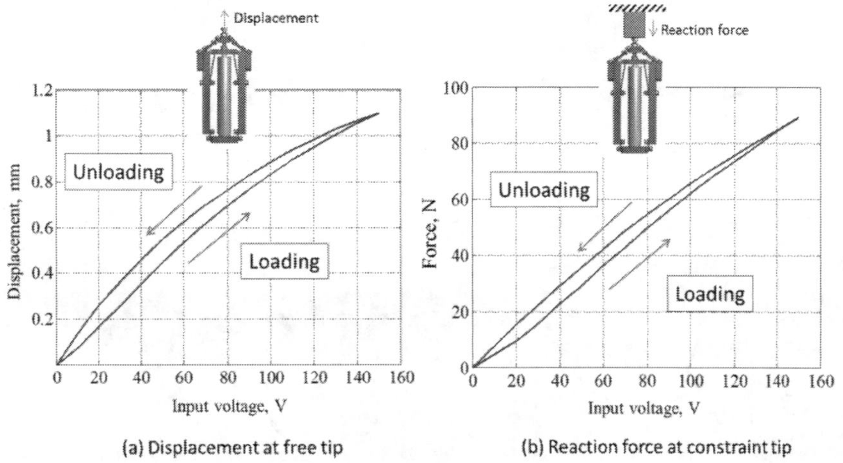

Figure 3: Properties of smart actuator in prototype reflector.

subsequent studies. Fig. 2 (top right) shows a hyperbolic aluminum reflector with slits. Six slits are made in order to avoid the buckling of the reflector when the actuators apply forces on the reflector in the out-of-plane direction. The reflector's diameter is 200mm and thickness is 0.5mm.

The prototype of the smart actuator is shown in Fig. 2 (bottom right). It consists of a stack piezoelectric actuator made by Piezomechanik GmbH, and the displacement magnification mechanism. The stack piezoelectric actuator generates the displacement at the tip of 0.1mm driven by the voltage of 150V, when no external load is applied at the tip. The small displacement of the stack piezoelectric actuator is magnified by the displacement magnification mechanism approximately 10 times.

The performance of the smart actuator is experimentally tested in a room temperature in advance. Two cases are tested: (i) the actuator tip is free, and (ii) the displacement of the actuator tip is completely constrained. For the test case (i), the relation between the applied voltage and the vertical displacement of the interface block on the top of the displacement magnification mechanism is shown in (a) of Fig. 3. The figure depicts the result of one cycle. The maximum displacement is 1.18mm. The loading-unloading path is quite repeatable.

For the test case (ii), when the tip of the interface block is fixed, the relation between the applied voltage and the generated force are obtained as shown in (b) of Fig. 3. Both results show that the

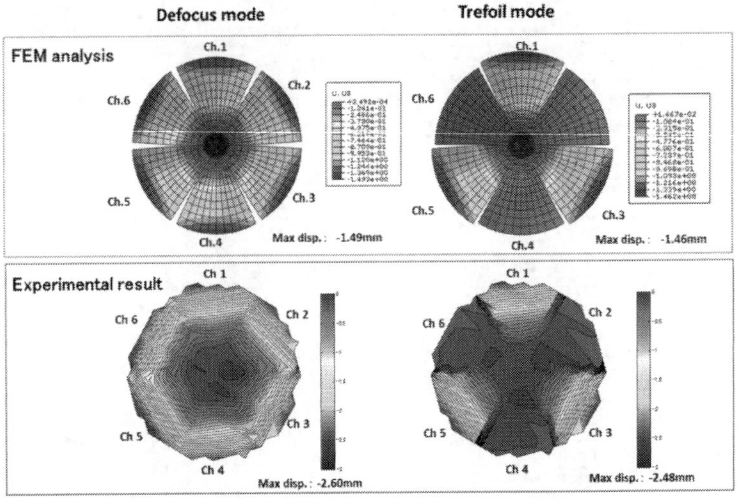

Figure 4: Deformation of smart reflector due to six smart actuators.

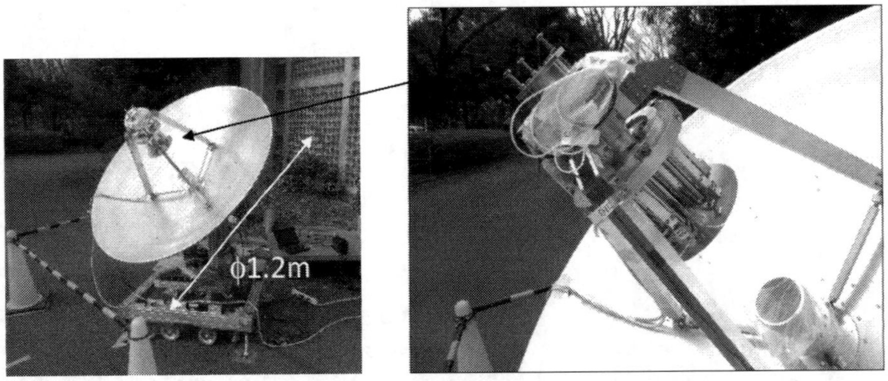

Figure 5: Smart secondary reflector integrated with primary reflector on ground.

existence of hysteresis of the actuators as well as the nonlinearity of the response curves. The hysteresis and nonlinearity should be considered in shape-control laws, such as implementing a feedback control.

Figure 4 shows the reflector's deformation due to the actuation of six smart actuators. Top figures show the result of finite-element method (FEM) analyses, and the bottom figures show the measurement in a ground laboratory. The results show that sufficient stroke of the reflector deformation is achieved for the EHF radio astronomy application.

2.2 Ground-based antenna system

The smart secondary reflector is integrated in the ground-based antenna system with the primary reflector of 1.2m diameter, as shown in Fig. 5. To measure the signal reception power, the experimental

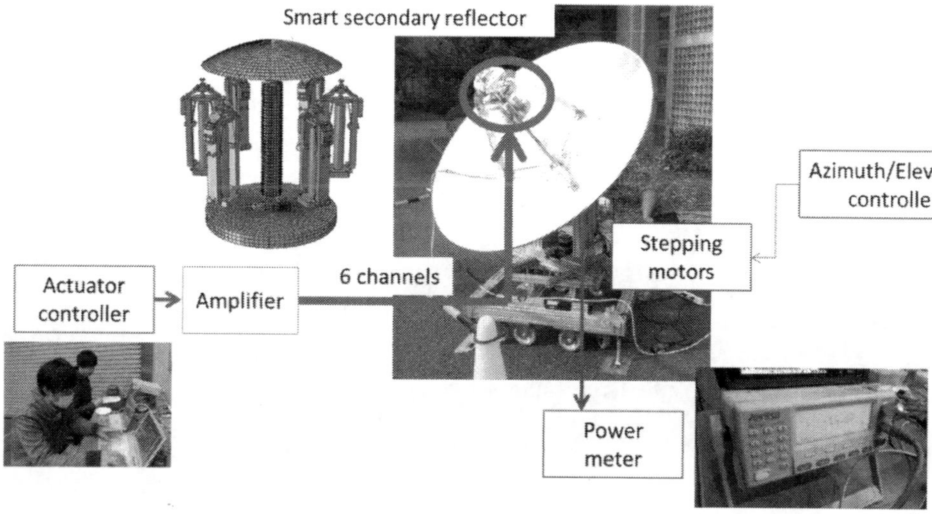

Figure 6: System diagram for shape-control experiment.

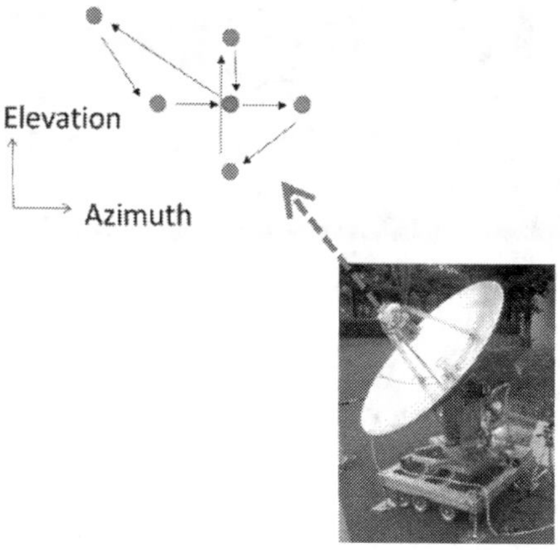

Figure 7: Five-point method to estimate peak power and azimuth/elevation offset angles.

system shown in Fig. 6 is constructed. In order to estimate the peak reception power at the current reflector shape, "five-point method" is used for each power measurement. Figure 7 shows the concept of the method. By fitting a Gaussian curve to a set of measurements with perturbed azimuth and elevation, the peak power and offset angle from the peak direction can be estimated, as explained in Fig. 8.

As a signal source, a communication satellite IPSTAR is used in this experiment. The frequency is approximately 20GHz. The shape control is conducted by using the steepest descent algorithm, combined with the golden-section line search algorithm.

3. RESULTS AND DISCUSSION

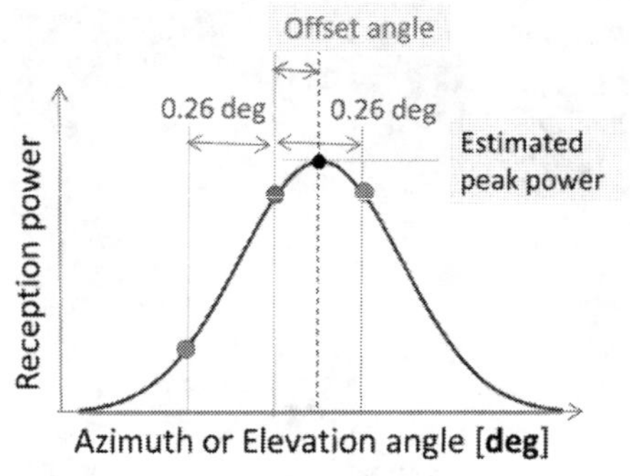

Figure 8: Rationale of five-point method.

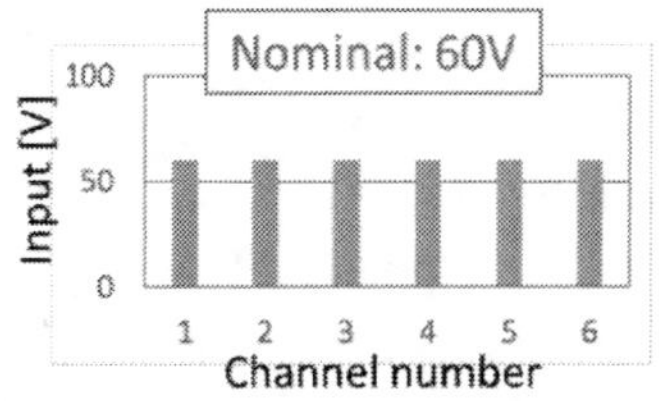

Figure 9: Nominal input voltages applied to six actuators.

In a nominal shape, 60V is uniformly applied to each of the six actuators, as shown in Fig. 9. First, the voltage applied to one of the six actuator is reduced to 30V to obtain a gradient (sensitivity) vector for the steepest descent algorithm, as shown in Fig. 10. Figure 11 shows the change of reception power when 30V is applied in one of the six channels in the horizontal axis. The red line shows the reception power for the nominal case, where 60V is applied for all six channels. Note that these reception powers for each reflector shape are obtained through the five-point method.

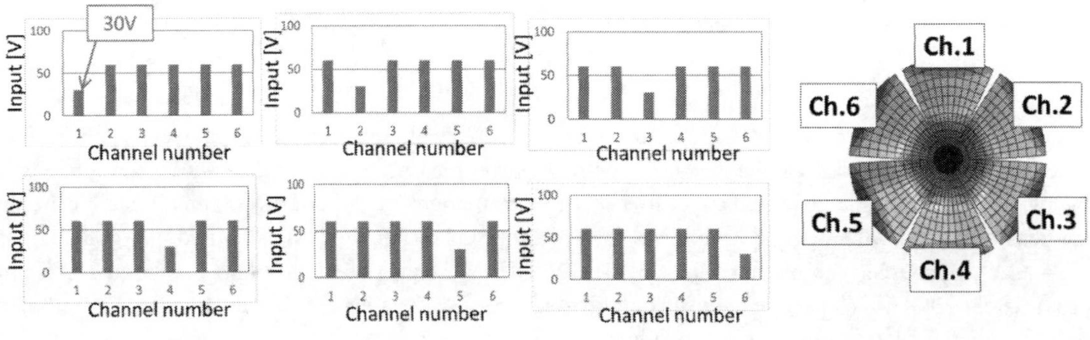

Figure 10: Six cases of input voltages to obtain gradient vector for shape control.

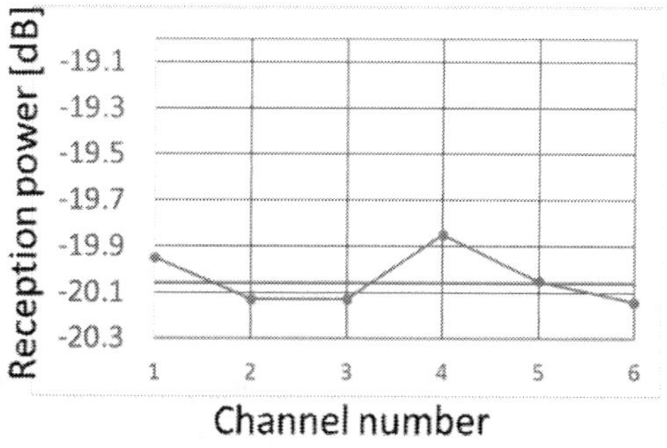

Figure 11: Results of six test cases to obtain gradient vector for shape control.

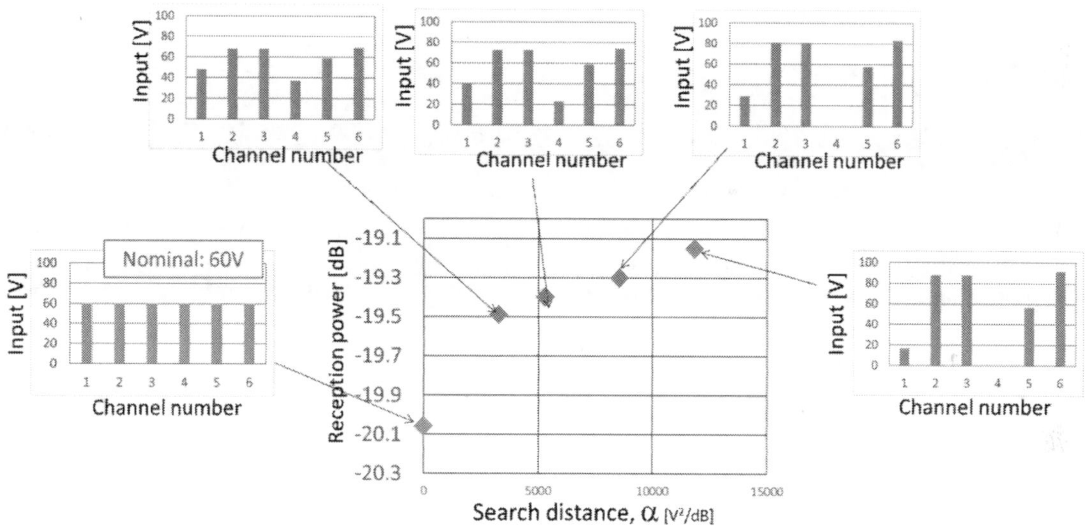

Figure 12: Shape control result using steepest-descent method with golden-section line search algorithm.

Second, various voltages are applied to the six channels calculated by the golden-section line search algorithm. Figure 12 shows the result. In the forth reflector shape, which is shown in green in the figure, the reception peak power is 0.86dB larger than the nominal value. This result shows the successful reception power improvement using the smart secondary reflector prototype and the shape-control algorithms used in this study.

4. CONCLUSION

The present study successfully demonstrated the shape control of smart secondary reflector to improve the reception power using preliminary prototype of future space astronomy antenna.

ACKNOWLEDGEMENTS

The authors thank those who supported the experiments in this study: Naoko Matsumoto and Kazuyoshi Yamashita at National Astronomical Observatory of Japan; Yasutaka Satou at Institue of Space and Astronautical Science; Akiya Inagaki, Kazuaki Hayasaki, and Kouta Gotou at Tokyo Institute of Technology. This study is supported by the grant from Institue of Space and Astronautical Science.

REFERENCES

5. * References

[1] Datashvili, L., Baier, H., Wei, B., Hoffman, J., Wehrle, E., Schreider, L., Mangenot, C., Prowald, J. S., Scolamiero, L., and Angevain, J. C., "Mechanical Investigations of in-Space-Reconfigurable Reflecting Surfaces," *32nd ESA Antenna Workshop on Antennas for Space Applications*, 2010, pp. 1–8.

[2] Hill, J. R., Wang, K. W., Fang, H., and Quijano, U., "Actuator Grouping Optimization on Flexible Space Reflectors," *Proc. SPIE 7977, Active and Passive Smart Structures and Integrated Systems*, 2011.

[3] Fang, H., Quijano, U., Bach, V., Hill, J., and Wang, K. W., "Experimental Study of a Membrane Antenna Surface Adaptive Control System," AIAA 2011-1828, 2011.

[4] Datashvili, L. S., Baier, H., Wei, B., Endler, S., and Schreider, L., "Design of a Morphing Skin Using Flexible Fiber Composites for Space-Reconfigurable Reflectors," AIAA Paper 2013-1522, 2013.

[5] Bradford, S., Wilkie, W. K., Agnes, G. S., and Peterson, L., "Controlling Wavefront in Lightweight Reflector Systems using Piezocomposite Actuator Arrays," AIAA Paper 2013-1525, 2013.

[6] Tanaka, H., Sakamoto, H., Inagaki, A., Ishimura, K., Doi, A., Kono, Y., Oyama, T., Watanabe, K., Oikawa, Y., and Kuratomi, T., "Development of Smart Reconfigurable Reflector Prototype for an Extremely High Frequency Antenna," *Proc. 25nd International Conference on Adaptive Structures and Technologies*, No. 030, Oct. 2014.

ICAST2015: 26th International Conference on Adaptive Structures and Technologies
October 14-16th, 2015, Kobe, Japan

ICAST2015 #061

Multiple-Piezoelectric Vibration Absorber System for Suppression of Vibration of Truss Structure under Random Force Excitation

Yohsuke Nambu[1*], Toshihide Takashima[1], and Masakatsu Chiba[1]

[1] Graduate School of Engineering, Osaka Prefecture University, Osaka, Japan.

Abstract

Vibration damping is important for the realization of high-precision space structures, because the vibration of the base of observation systems negatively effects accuracy of observation in direct. In previous study, we revealed that a multiple-piezoelectric vibration absorber (PVA) system improved the robustness of the PVA against the variation of characteristics of the piezoelectric transducers for vibration damping of a clamped beam. The paper aims to reveal the optimization method and effectiveness of the muliple-PVA system for more practical case, namely vibration damping of a truss structure under random excitation. Numerical simulations showed that the H_2 norms of the transfer functions of the truss with the non-optimized, single-, and double-PVA systems are 8.85×10^{-5}, 4.70×10^{-5}, and 4.52×10^{-5} respectively. Therefore, the H_2 norms of the transfer function of the truss with optimized PVAs is dramatically lower than one with non-optimized PVA. The performance of double-PVA system was slightly better than one of single-PVA system.

1. INTRODUCTION

In recent years, the need for high-precision and high-functionality space structural systems for realizing advanced space missions such as high-resolution observation and high-capacity communication has emerged. Vibration damping is important for the realization of high-precision space structures, because the vibration of the base of observation systems negatively effects accuracy of observation in direct. This research proposes multiple-piezoelectric vibration absorber system for suppression of vibration of truss structure under random force excitation.

Dynamic vibration absorbers (DVAs) were invented by H. Frahm in 1909. In 1928, Ormondroy and Den Hartog showed that a new DVA with a damper is more efficient than Frahm's DVA under broadband frequency excitation. [1] They also proposed an optimal tuning method that is known as the "Fixed Point Theorem."

Since the advent of this kind of DVA, many researchers and engineers have studied passive, active, and semi-active DVAs. [2, 3, 4] Today, the DVA is the most important vibration control device. Finding the optimal tuning point is a fundamental issue in DVA research. The many ways of determining the optimal tuning point that have been proposed fall into three categories, depending on estimated indicators. 1) Minimization of the H_∞ norm of the transfer function. 2) Minimization of the H_2 norm

* nambu@aero.osakafu-u.ac.jp

of the transfer function. 3) Maximization of the minimum modal damping ratio. The first indicator is used for damping the periodic responses of the main system. The second indicator is used for damping of random responses. The third indicator is used for damping of impulse responses. Asami *et al* [5] introduced analytical solutions for these optimization problems for a single DVA combined with a single-degree-of-freedom (SDOF) system.

Multiple-DVA systems contain several DVAs with natural frequencies that are set close to the natural frequency of the main structure. The use of multiple-DVA systems improves the effectiveness of the DVA without significantly increasing the mass. Zuo *et al* [6, 7] proposed an optimal tuning method based on a numerical calculation for multiple DVAs in a multiple-degree-of-freedom (MDOF) system. Their methods are based on calculation techniques used in optimal feedback control theorems such as the H_2 optimal control theorem and the H_∞ optimal control theorem. These methods calculate the stiffnesses and damping coefficients of several SDOF or MDOF DVAs attached to an MDOF system.

DVAs are powerful devices, but they are not robust. For example, aging degradation and temperature-dependency negatively affect the DVA performance. The robustness can be improved by using an active DVA. Ohnuki *et al* [8] and Alujevic *et al* [9] verified that the robustness of the DVA could be improved by employing active feedback control. Nambu *et al* [10] realized a smart DVA to suppress the vibration of tension-stabilized structures; it is a self-sensing active DVA using a small cantilever with a piezoelectric transducer.

Piezoelectric materials are often used in adaptive structures to suppress vibration. A piezoelectric vibration absorber (PVA) is a damping device composed of a piezoelectric transducer connected to an inductor (L) and a resistor (R). [11, 12]. PVAs have the advantages of small size and ease of tuning for both frequency and damping ratio. Moheimani[13] reviewed various research regarding vibration damping and control using a piezoelectric transducer shunted to several kinds of circuits. In the last two decades, the semi-active method has received great attention. Energy-recycling semi-active vibration control methods using a resonant circuit with a synchronized switch [14, 15, 16, 17, 18, 19, 20] are more effective than the conventional passive method. Semi-active vibration control methods are superior to passive methods in terms of stability and effectiveness.

Although the semi-active method has many advantages, the necessity of including the external sensor reduces the advantages of piezoelectric shunt damping. This research evaluates the effectiveness of passive damping systems. In previous study, we proposed a multiple-PVA system that is composed of several shunt circuits to improve the robustness of the PVA against the variation of characteristics of the piezoelectric transducers[21]. We also revealed that the multiple-PVA system for the vibration damping of the clamped beam was more robust than the single PVA in the variable temperature environment by numerical simulations and experiments. In this sturdy, we apply the multiplex technique to more practical case, namely vibration control of the truss structure and also propose optimization method of the multiple-PVA system for the random response. The paper aims to reveal the optimization method and effectiveness of the muliple-PVA system.

2. ANALYTICAL MODEL

A five bays truss structure is analyzed in this paper. One unit of this truss consists of eight nodes and 12 rods shown as Fig. 1. The rod is an aluminum hollow rod of whith wall thickness is 1 mm. The length of the virtical and horizontal rod is $L_k^{(1)} = 255$ mm, and one of the diagonal rod is $L_k^{(2)} = 380$ mm. The mass of the node is 69.4 kg. The joint between the rod and the node is the pin-joint where the moment transmitted through the joint is negligible. Every member of the truss is in pure compression or pure tension.

The bottom two nodes are fixed on the base. The other two nodes are connected with piezoelectric vibration absorbers like Fig. 2.

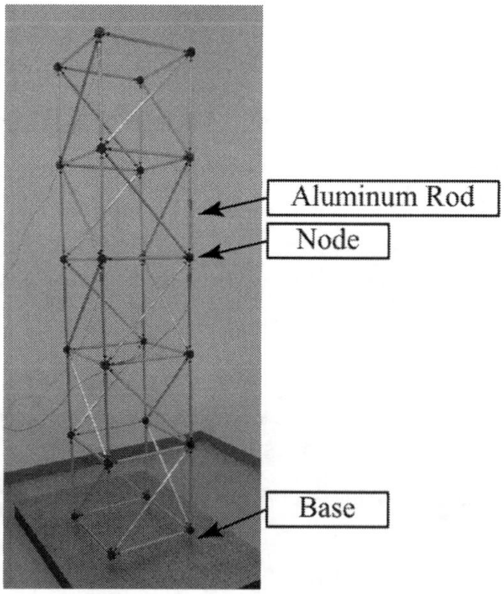

Figure 1: Photo of Truss Structure.

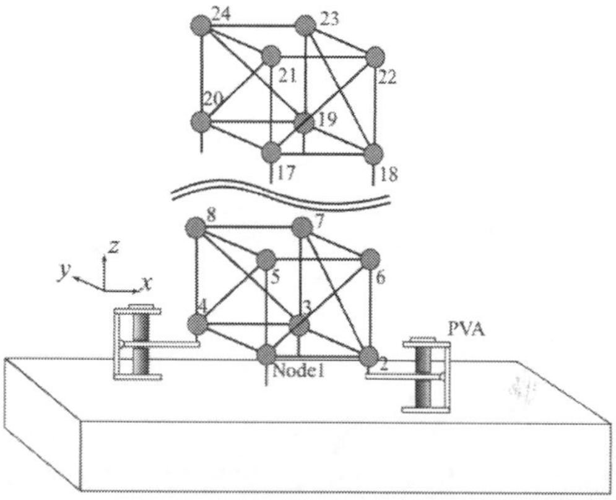

Figure 2: Model of Truss Structure with PVAs.

2.1 Stiffness Matrix and Mass Matrix

Assume that the kth rod is connected with ith node and jth node like Fig. 3. The rods are modeled as elastic bars and the node are modeled as mass-points.

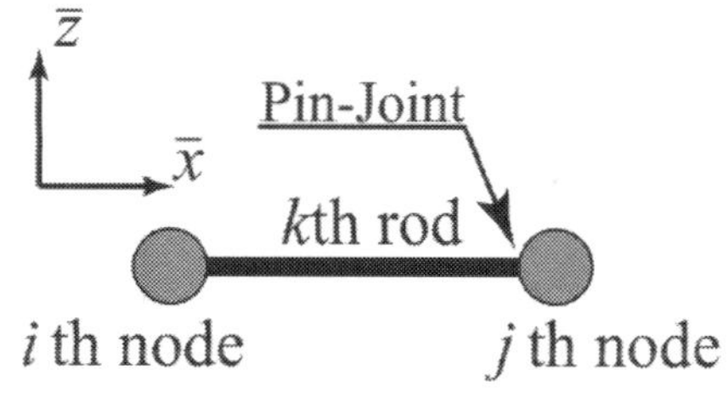

Figure 3: Model of Rod.

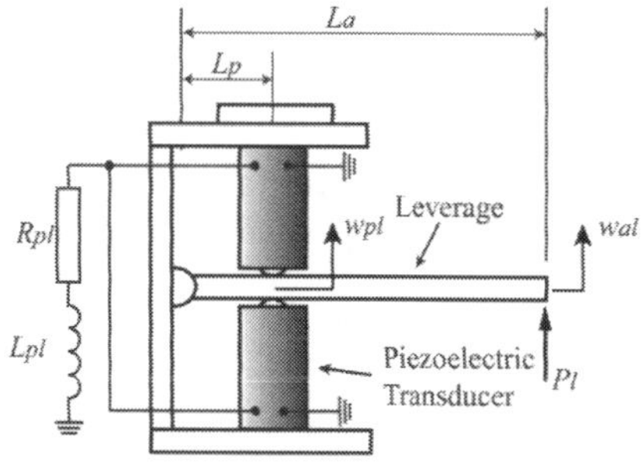

Figure 4: Model of PVA.

The constitutive equation of the rod is expressed as

$$\bar{f}^k_{xi} = -\bar{f}^k_{xj} = \frac{E_k A_k}{L_k}(\bar{u}_{xi} - \bar{u}_{jx}), \tag{1}$$

where $\bar{f}^k_{xi}, \bar{f}^k_{xj}$ are forces applied on each node i, j by the kth rode, and $\bar{u}_{xi}$, $\bar{u}_{jx}$ are displacement of each node along the rod. The constitutive equation regarding the matrix described above are as follows:

$$\begin{pmatrix} \bar{\boldsymbol{f}}^k_i \\ \bar{\boldsymbol{f}}^k_j \end{pmatrix} = \begin{pmatrix} \bar{\boldsymbol{K}}^k & -\bar{\boldsymbol{K}}^k \\ -\bar{\boldsymbol{K}}^k & \bar{\boldsymbol{K}}^k \end{pmatrix} \begin{pmatrix} \bar{\boldsymbol{u}}_i \\ \bar{\boldsymbol{u}}_j \end{pmatrix} \tag{2}$$

$$\bar{\boldsymbol{f}}^k_\alpha = {}^t\left(\bar{f}^k_{x\alpha}\ \bar{f}^k_{y\alpha}\ \bar{f}^k_{z\alpha}\right), \quad (\alpha = i, j) \tag{3}$$

$$\bar{\boldsymbol{u}}_\alpha = {}^t\left(\bar{u}_{x\alpha}\ \bar{u}_{y\alpha}\ \bar{u}_{z\alpha}\right), \quad (\alpha = i, j) \tag{4}$$

$$\bar{\boldsymbol{K}}^k = \begin{pmatrix} E_k A_k / L_k & 0 & 0 \\ 0 & 0 & 0 \\ 0 & 0 & 0 \end{pmatrix} \tag{5}$$

Assume that the coordinate transformation matrix is expressed as $\boldsymbol{P}_k$. Then

$$\begin{pmatrix} \boldsymbol{f}^k_i \\ \boldsymbol{f}^k_j \end{pmatrix} = \begin{pmatrix} \boldsymbol{P}_k & \boldsymbol{O}_{3\times3} \\ \boldsymbol{O}_{3\times3} & \boldsymbol{P}_k \end{pmatrix} \begin{pmatrix} \bar{\boldsymbol{f}}^k_i \\ \bar{\boldsymbol{f}}^k_j \end{pmatrix} \tag{6}$$

$$\begin{pmatrix} \boldsymbol{u}_i \\ \boldsymbol{u}_j \end{pmatrix} = \begin{pmatrix} \boldsymbol{P}_k & \boldsymbol{O}_{3\times3} \\ \boldsymbol{O}_{3\times3} & \boldsymbol{P}_k \end{pmatrix} \begin{pmatrix} \bar{\boldsymbol{u}}_i \\ \bar{\boldsymbol{u}}_j \end{pmatrix} \tag{7}$$

and we obtain constitutive equation in the global coordinate system.

$$\begin{pmatrix} \boldsymbol{f}^k_i \\ \boldsymbol{f}^k_j \end{pmatrix} = \begin{pmatrix} {}^t\boldsymbol{P}_k & \boldsymbol{O}_{3\times3} \\ \boldsymbol{O}_{3\times3} & {}^t\boldsymbol{P}_k \end{pmatrix} \begin{pmatrix} \bar{\boldsymbol{K}}^k & -\bar{\boldsymbol{K}}^k \\ -\bar{\boldsymbol{K}}^k & \bar{\boldsymbol{K}}^k \end{pmatrix} \begin{pmatrix} \boldsymbol{P}_k & \boldsymbol{O}_{3\times3} \\ \boldsymbol{O}_{3\times3} & \boldsymbol{P}_k \end{pmatrix} \begin{pmatrix} \boldsymbol{u}_i \\ \boldsymbol{u}_j \end{pmatrix} \tag{8}$$

where $\boldsymbol{P}_k{}^{-1} = {}^t\boldsymbol{P}_k$. We obtain

$$\boldsymbol{K}^k = {}^t\boldsymbol{P}_k \bar{\boldsymbol{K}}^k \boldsymbol{P}_k = \frac{E_k A_k}{L_k} \begin{pmatrix} P^2_{k11} & P_{k11}P_{k12} & P_{k11}P_{k13} \\ P_{k11}P_{k12} & P^2_{k12} & P_{k12}P_{k13} \\ P_{k11}P_{k13} & P_{k12}P_{k13} & P^2_{k13} \end{pmatrix} \tag{9}$$

where E_k is Young's modulas of the rod, A_k is the cross section area, and L_k is the legth. The elements of $\boldsymbol{P}_k$ are direction cosine of the rod in the global coordinate system.

$$P_{k11} = \frac{X_i - X_j}{L_k}, \quad P_{k12} = \frac{Y_i - Y_j}{L_k}, \quad P_{k13} = \frac{Z_i - Z_j}{L_k} \tag{10}$$

where (X_i, Y_i, Z_i) represents a placement of ith node.

The extended matrix of $\boldsymbol{f}^k_\alpha, \boldsymbol{u}_\alpha$ are expressed as

$$\tilde{\boldsymbol{f}}^k = {}^t\left(\boldsymbol{O}_{1\times3(i-1)}\ \boldsymbol{f}^k_i\ \boldsymbol{O}_{1\times3(j-i-1)}\ \boldsymbol{f}^k_j\ \boldsymbol{O}_{1\times3(N-2)}\right) \tag{11}$$

$$\boldsymbol{u} = {}^t\left(\boldsymbol{u}_1 \ldots \boldsymbol{u}_N\right) \tag{12}$$

$$\tilde{\boldsymbol{K}}^k = \begin{pmatrix} \boldsymbol{O} & & \cdots & & \boldsymbol{O} \\ & \boldsymbol{K}^k & \boldsymbol{O} & -\boldsymbol{K}^k & \\ & & \boldsymbol{O} & \boldsymbol{O} & \vdots \\ & \text{sym.} & & \boldsymbol{K}^k & \\ & & & & \boldsymbol{O} \end{pmatrix} \in \boldsymbol{R}^{3N\times3N} \tag{13}$$

Then the element stiffness equations in global coordinate system are expressed as

$$\tilde{\boldsymbol{f}}^k = \tilde{\boldsymbol{K}}^k \boldsymbol{u} \tag{14}$$

And the elastic energy of the rod is expressed as

$$\mathcal{U}_r^k = \frac{1}{2} {}^t\boldsymbol{u} \tilde{\boldsymbol{K}}^k \boldsymbol{u} \tag{15}$$

We assume that the mass of the rod is concentrated to the node.

$$\tilde{\boldsymbol{M}}_{rod}^k = \begin{pmatrix} \boldsymbol{O} & & \cdots & & \boldsymbol{O} \\ & \boldsymbol{M}^k & & & \\ & & \boldsymbol{O} & & \vdots \\ & \text{sym.} & & \boldsymbol{M}^k & \\ & & & & \boldsymbol{O} \end{pmatrix} \in \boldsymbol{R}^{3N \times 3N} \tag{16}$$

$$\boldsymbol{M}_{rod}^k = \text{diag}\left[\left(\tfrac{\rho_k A_k L_k}{2} \; \tfrac{\rho_k A_k L_k}{2} \; \tfrac{\rho_k A_k L_k}{2}\right)\right] \tag{17}$$

The effective mass of the nodes are expressed as

$$\boldsymbol{M} = \sum_{k=1}^{N_r} \boldsymbol{M}_{rod}^k + \boldsymbol{M}_{node} \tag{18}$$

where

$$\boldsymbol{M}_{node} = \text{diag}\left[\left(m_1 \; m_1 \; m_1 \; m_2 \; m_2 \; m_2 \cdots m_N \; m_N \; m_N\right)\right] \tag{19}$$

Then the kinetic energy of the truss is expressed as follows:

$$\mathcal{T} = \frac{1}{2} {}^t\dot{\boldsymbol{u}} \boldsymbol{M} \dot{\boldsymbol{u}} \tag{20}$$

2.2 Piezoelectric Transducer

A piezoelectric transducer system for the lth PVA consists of a base, an elastic leverage, and a piezoelectric transducer (Piezomechanik GmbH: PSt 150/4/20 VS9) like Fig. 4. The constitutive equations of piezoelectric transducers are expressed as

$$f_{pl} = k_p u_{pl} - b_{pl} q_l \tag{21}$$

$$V_{pl} = -b_p u_{pl} + \frac{q_l}{C_p} \tag{22}$$

The displacement of the edge of the leverage is expressed

$$u_{al} = \frac{P_l}{3E_a I_a} L_a (L_a - L_p)^2 + \gamma u_{pl} \tag{23}$$

$$f_{pl} = -\gamma P_l \tag{24}$$

where $\gamma = L_a/L_p$. Substituting Eq. (23) and Eq. (24) to Eq. (21) and Eq. (22), we obtain

$$P_l = k_a u_{al} - b_{a1} q_l \tag{25}$$
$$V_{pl} = -b_{a2} u_{al} + C_a^{-1} q_l \tag{26}$$

where

$$k_a = \frac{k_p}{\nu\gamma^2} \tag{27}$$
$$b_{a1} = \frac{b_p}{\nu\gamma} \tag{28}$$
$$C_a = \left\{1 + (\nu - 1)\gamma \frac{b_p^2 C_p}{k_p}\right\}^{-1} C_p \tag{29}$$
$$b_{a2} = \frac{1 - (\nu - 1)\gamma^3}{\gamma} b_p \tag{30}$$
$$\nu = 1 - \frac{k_p L_a^3}{3E_a I_a} \frac{(\gamma - 1)^2}{\gamma^4} \tag{31}$$

In this model, PVAs is connected to node 2 and node 4. Therefore,

$$u_{a1} = u_{z2}, \quad u_{a2} = u_{z4} \tag{32}$$

The vertual work of piezoelectric transducers is

$$\begin{aligned} \delta\mathcal{W}_p &= -\sum_{l=1}^{2} (\delta u_{al} P_l + \delta q_l V_{pl}) \\ &= -{}^t\delta \boldsymbol{u} \tilde{\boldsymbol{K}}_{\boldsymbol{p}} \boldsymbol{u} - {}^t\delta \boldsymbol{q} \boldsymbol{C}_{\boldsymbol{p}}{}^{-1} \boldsymbol{q} + {}^t\delta \boldsymbol{u} \tilde{\boldsymbol{B}}_{\boldsymbol{p1}} \boldsymbol{q} + {}^t\delta \boldsymbol{q} \tilde{\boldsymbol{B}}_{\boldsymbol{p2}} \boldsymbol{u} \end{aligned} \tag{33}$$

where

$$\tilde{\boldsymbol{K}}_{\boldsymbol{p}}(i,j) = (\delta_{i6}\delta_{6j} + \delta_{i12}\delta_{12j}) k_a \quad \in \boldsymbol{R}^{\boldsymbol{3N \times 3N}} \tag{34}$$
$$\tilde{\boldsymbol{B}}_{\boldsymbol{p1}}(i,j) = (\delta_{i6}\delta_{1j} + \delta_{i12}\delta_{2j}) b_{a1} \quad \in \boldsymbol{R}^{\boldsymbol{3N \times 2}} \tag{35}$$
$$\tilde{\boldsymbol{B}}_{\boldsymbol{p2}}(i,j) = (\delta_{i1}\delta_{6j} + \delta_{i2}\delta_{12j}) b_{a2} \quad \in \boldsymbol{R}^{\boldsymbol{2 \times 3N}} \tag{36}$$
$$\tilde{\boldsymbol{C}}_{\boldsymbol{p}} = \begin{pmatrix} C_a & 0 \\ 0 & C_a \end{pmatrix} \tag{37}$$
$$\boldsymbol{q} = {}^t(q_1 \; q_2) \tag{38}$$

2.3 Governing Equations

The Lagrangean of the truss structure is expressed as

$$\mathcal{L} = \mathcal{T} - \sum_{k=1}^{N_{rod}} \mathcal{U}_r^k \tag{39}$$

The vertual work by the external forces $\boldsymbol{F}$ applied on nodes is expressed as

$$\delta\mathcal{W} = {}^t\delta\boldsymbol{u}\,\boldsymbol{F} \tag{40}$$

and the vertual work by the external voltage $\boldsymbol{V_e}$ applied on the piezoelectric transducers is epressed as

$$\delta\mathcal{W}_e = {}^t\delta\boldsymbol{q}\,\boldsymbol{V_e} \tag{41}$$

Then the following equation is obtained by the Hamilton's principle.

$$\int_{t_1}^{t_2} (\delta\mathcal{L} + \delta\mathcal{W}_p + \delta\mathcal{W} + \delta\mathcal{W}_e)\,dt = 0 \tag{42}$$

$\delta\boldsymbol{u}$ and $\delta\boldsymbol{q}$ are respectively the variation of $\boldsymbol{u}$ and $\boldsymbol{q}$ that satisfy $\delta\boldsymbol{u}(t_1) = \delta\boldsymbol{u}(t_2) = \delta\boldsymbol{q}(t_1) = \delta\boldsymbol{q}(t_2) = 0$. Substituting Eqs. (15), (20), (33), (40), (41) to Eq. (42), we obtain

$$\int_{t_1}^{t_2} \Big[{}^t\delta\boldsymbol{u}\Big\{-\boldsymbol{M}\ddot{\boldsymbol{u}} - (\boldsymbol{K}+\boldsymbol{K_p})\boldsymbol{u} + \tilde{\boldsymbol{B}}_{\boldsymbol{p1}}\boldsymbol{q} + \boldsymbol{F}\Big\} + {}^t\delta\boldsymbol{q}\Big\{\tilde{\boldsymbol{B}}_{\boldsymbol{p2}}\boldsymbol{u} - \tilde{\boldsymbol{C}}_{\boldsymbol{p}}^{-1}\boldsymbol{q} + \boldsymbol{V_e}\Big\}\Big]\,dt = 0 \tag{43}$$

Hamilton's principle requires that Eq. (43) holds for all possible perturbations $\delta\boldsymbol{u}$ and $\delta\boldsymbol{q}$. Thus, we obtain

$$\boldsymbol{M}\ddot{\boldsymbol{u}} + (\boldsymbol{K}+\boldsymbol{K_p})\boldsymbol{u} - \tilde{\boldsymbol{B}}_{\boldsymbol{p1}}\boldsymbol{q} = \boldsymbol{F} \tag{44}$$

$$\tilde{\boldsymbol{B}}_{\boldsymbol{p2}}\boldsymbol{u} - \tilde{\boldsymbol{C}}_{\boldsymbol{p}}^{-1}\boldsymbol{q} + \boldsymbol{V_e} = 0 \tag{45}$$

In this paper, we assume that node $1, 3$ are fixed and node $2, 4$ can move only vertically.

$$u_{x1} = u_{y1} = u_{z1} = u_{x2} = u_{y2} = u_{x3} = u_{y3} = u_{z3} = u_{x4} = u_{y4} = 0 \tag{46}$$

Sustituting Eq. (46) to Eqs. (44), (45), we obtain the degenerated equation concerning $\hat{\boldsymbol{u}} \in \boldsymbol{R^n}$ $(n = 3N - 10)$ and $\boldsymbol{q} \in \boldsymbol{R^2}$.

$$\hat{\boldsymbol{M}}\ddot{\hat{u}} + (\hat{\boldsymbol{K}} + \hat{\boldsymbol{K}}_{\boldsymbol{p}})\hat{\boldsymbol{u}} - \hat{\boldsymbol{B}}_{\boldsymbol{p1}}\boldsymbol{q} = \hat{\boldsymbol{F}} \tag{47}$$

$$\hat{\boldsymbol{B}}_{\boldsymbol{p2}}\hat{\boldsymbol{u}} - \tilde{\boldsymbol{C}}_{\boldsymbol{p}}^{-1}\boldsymbol{q} + \boldsymbol{V_e} = 0 \tag{48}$$

The piezoelectric transducer is conneceted to a registor and a inductor.

$$\boldsymbol{V_e} = -\boldsymbol{L_p}\ddot{\boldsymbol{q}} - \boldsymbol{R_p}\dot{\boldsymbol{q}} \tag{49}$$

where

$$\boldsymbol{L_p} = \begin{pmatrix} L_{p1} & 0 \\ 0 & L_{p2} \end{pmatrix} \tag{50}$$

$$\boldsymbol{R_p} = \begin{pmatrix} R_{p1} & 0 \\ 0 & R_{p2} \end{pmatrix} \tag{51}$$

Substituting Eq. (49) to Eq. (48), we obtain

$$\boldsymbol{L_p}\ddot{\boldsymbol{q}} + \boldsymbol{R_p}\dot{\boldsymbol{q}} + \tilde{\boldsymbol{C}}_{\boldsymbol{p}}^{-1}\boldsymbol{q} - \hat{\boldsymbol{B}}_{\boldsymbol{p2}}\hat{\boldsymbol{u}} = 0 \tag{52}$$

2.4 Modal Equation and State Equation

The natural anguler velocities $\omega_1, \ldots, \omega_n$ and modal shapes $\phi_1, \ldots, \phi_n$ of truss structure is obtained by solving the following eigen value problem.

$$\left\{-\omega^2 \hat{\boldsymbol{M}} + (\hat{\boldsymbol{K}} + \hat{\boldsymbol{K}}_p)\right\} \phi = \mathbf{0} \tag{53}$$

Asume that the displacement $\hat{\boldsymbol{u}}$ is expressed by the modal displacement $\boldsymbol{\xi} \in \boldsymbol{R}^n$ as

$$\hat{\boldsymbol{u}} = \Phi \boldsymbol{\xi} \tag{54}$$

where

$$\boldsymbol{\Phi} = \begin{pmatrix} \phi_1 \ldots \phi_n \end{pmatrix} \tag{55}$$

Substituting Eq. (54) to Eqs. (47) and (52), we obtain

$$\ddot{\boldsymbol{\xi}} + \boldsymbol{\Omega}\boldsymbol{\xi} - \boldsymbol{B}'_{p1}\boldsymbol{q} = {}^t\boldsymbol{\Phi}\boldsymbol{F} \tag{56}$$

$$\ddot{\boldsymbol{q}} + \boldsymbol{\Xi}_p\dot{\boldsymbol{q}} + \boldsymbol{\Omega}_p\boldsymbol{q} - \hat{\boldsymbol{B}}'_{p2}\boldsymbol{\xi} = 0 \tag{57}$$

where

$$\boldsymbol{\Omega} = {}^t\boldsymbol{\Phi}(\hat{\boldsymbol{K}} + \hat{\boldsymbol{K}}_p)\boldsymbol{\Phi}, \quad \boldsymbol{B}'_{p1} = {}^t\boldsymbol{\Phi}\boldsymbol{B}_{p1}, \quad \boldsymbol{B}'_{p2} = \boldsymbol{L}^{-1}\boldsymbol{B}_{p2}\boldsymbol{\Phi},$$

$$\boldsymbol{\Xi}_p = \mathrm{diag}(2\zeta_{p1}\omega_{p1},\ 2\zeta_{p2}\omega_{p2}), \quad \boldsymbol{\Omega}_p = \mathrm{diag}(\omega_{p1}^2,\ \omega_{p2}^2)$$

$$\omega_{pl} = \sqrt{1/c_{pl}L_{pl}}, \quad \zeta_{pl} = R_{pl}/(2L_{pl}\omega_{pl}) \quad (l = 1, 2)$$

The state variables of the system of the truss and PVAs are

$$\boldsymbol{x}_s = {}^t(\boldsymbol{\xi}\ \boldsymbol{q}\ \dot{\boldsymbol{\xi}}\ \dot{\boldsymbol{q}}) \tag{58}$$

The state equations can be obtained from Eqs. (56), (57) :

$$\dot{\boldsymbol{x}}_s = \boldsymbol{A}_s\boldsymbol{x}_s + \boldsymbol{B}_s\boldsymbol{F}, \tag{59}$$

where

$$A_s = \begin{pmatrix} \boldsymbol{O} & \boldsymbol{O} & \boldsymbol{I} & \boldsymbol{O} \\ \boldsymbol{O} & \boldsymbol{O} & \boldsymbol{O} & \boldsymbol{I} \\ -\boldsymbol{\Omega} & \boldsymbol{B}'_{p1} & -\boldsymbol{\Xi} & \boldsymbol{O} \\ \boldsymbol{B}'_{p2} & -\boldsymbol{\Omega}_p & \boldsymbol{O} & -\boldsymbol{\Xi}_p \end{pmatrix} \tag{60}$$

$$B_s = \begin{pmatrix} \boldsymbol{O} \\ {}^t\boldsymbol{\Phi} \end{pmatrix} \tag{61}$$

Here we introduced $\boldsymbol{\Xi}$ as modal damping of the truss structure.

$$\boldsymbol{\Xi} = \mathrm{diag}\left[(2\zeta_1\omega_1 \ldots 2\zeta_n\omega_n)\right] \tag{62}$$

3. OPTIMIZATION METHOD

First, we have to define the estimated output $\boldsymbol{z}$:

$$\boldsymbol{z} = \boldsymbol{C_z x_s} \tag{63}$$

The closed loop transfer function matrix from $\boldsymbol{F}$ to $\boldsymbol{z}$ is accordingly expressed as

$$\boldsymbol{T_{fz}} = \boldsymbol{C_z}(s\boldsymbol{I} - \boldsymbol{A_s})\boldsymbol{B_s} \tag{64}$$

where $\boldsymbol{I}$ is a unit matrix.

If the system input $\boldsymbol{F}$ is unit white noise, which means $E[\boldsymbol{F}] = \boldsymbol{0}$ and $E[{}^t\boldsymbol{F}\boldsymbol{F}(t+\tau)] = \boldsymbol{\delta}(\tau)$, we obtain

$$\lim_{\mathrm{T}\to\infty} E\left[\frac{1}{\mathrm{T}}\int_0^{\mathrm{T}} ({}^t\boldsymbol{z}\boldsymbol{z})dt\right] = \frac{1}{2\pi}\int_{-\infty}^{\infty} \mathrm{trace}\left({}^t\boldsymbol{T_{fz}}(\omega)\boldsymbol{T_{fz}}(\omega)\right) d\omega = \|\boldsymbol{T_{fz}}\|_2^2, \tag{65}$$

where $E[\boldsymbol{F}(t)]$ is the root mean square (RMS) of $\boldsymbol{F}(t)$. Therefore, the indicator for random excitation should be the H_2 norm of $\boldsymbol{T_{fz}}$.

By the way, H_2 norm of $\boldsymbol{T_{fz}}$ is expressed as

$$\|\boldsymbol{T_{fz}}\|_2^2 = \mathrm{trace}\left({}^t\boldsymbol{C_z}\boldsymbol{R}\boldsymbol{C_z}\right), \tag{66}$$

where $\boldsymbol{R}$ is solution of the following Lyapunov equation.

$$\boldsymbol{A_s R} + {}^t\boldsymbol{R}\boldsymbol{A_s} + \boldsymbol{B_s}{}^t\boldsymbol{B_s} = 0 \tag{67}$$

Then the optimization problem is formulated as

$$\text{Objective Function}: \ \mathrm{trace}\left({}^t\boldsymbol{C_z}\boldsymbol{R}\boldsymbol{C_z}\right) \tag{68}$$
$$\text{Design Parameters}: \ \omega_{pl},\ \zeta_{pl} \quad (l = 1, 2) \tag{69}$$
$$\text{Constraints}: \ \omega_{el} > 0,\ \zeta_{el} > 0 \tag{70}$$

In the case of optimization of single-PVA system, the dynamic parameters of two PVAs are same, namely $\omega_{p1} = \omega_{p2}$ and $\zeta_{p1} = \zeta_{p2}$.

4. NUMERICAL SIMULATIONS

Here we demonstrate the optimization of single- and double-PVA system to suppress the random response of the truss structure. Parameter values used in the optimization are shown in Table 1. In this case, the modal frequencies and modal shapes of the truss were obtained as Fig. 5 shows.

Assume that the target mode is the second mode. Then the estimated output matrix $\boldsymbol{C_z}$ becomes

$$\boldsymbol{C_z} = (0\ 1\ 0 \ldots 0) \quad \in \boldsymbol{R}^{n+4} \tag{71}$$

The external force $\boldsymbol{F}$ is applied only on the 22th node along x-axis.

We used the " fmincon " function of MATLAB R2015b to solve this optimization problem. Note that $\boldsymbol{A_s}$ must be stabilized at all steps of the optimization, or else a unique solution of Lyapunov equation Eq. (67) will not exist. We calculated about several initial values that were given randomly.

The results are showed in Tab. 2. Fig. 6 shows the transfer functions of the truss with the non-optimezed, H_2 optimized single-, and double-PVA systems. They are characterized by increased local

Table 1: Parameter values used in the optimization.

Quantity	Value	Unit	Quantity	Value	Unit
ρ_k	2700	kg/m^3	k_p	2.500×10^8	N/m
E_k	71.00	GPa	b_p	6.482×10^6	N/C
A_k	2.573×10^{-5}	m^2	c_p	7.200	μF
$L_k^{(1)}$	0.2550	m	$L_k^{(2)}$	0.3800	m
E_a	200.0	GPa	I_a	1.667×10^{-9}	m^4
L_a	0.115	m	L_p	0.0015	m

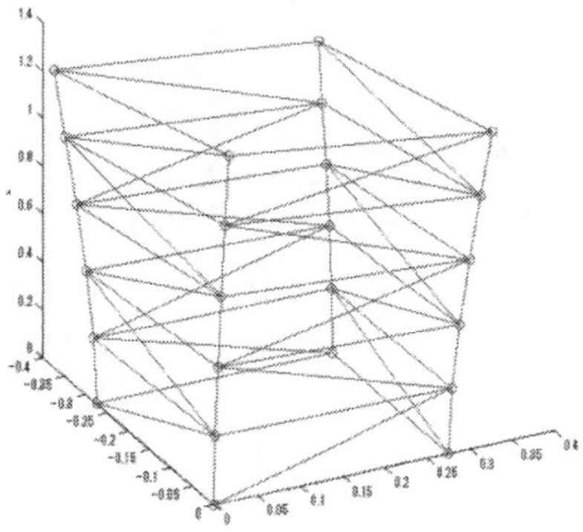

(a) 1st mode (35.3Hz).

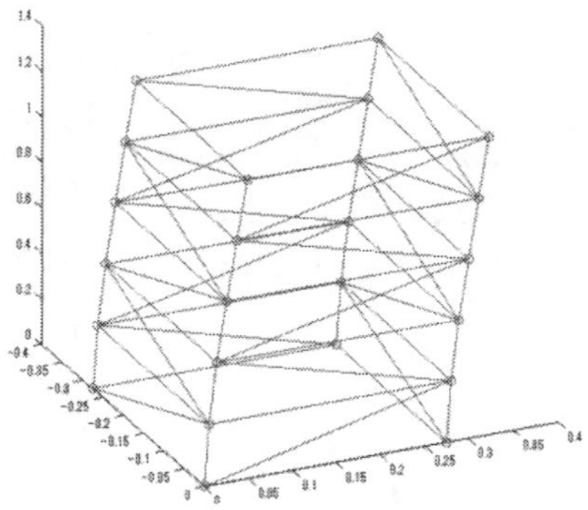

(b) 2nd mode (45.8Hz)

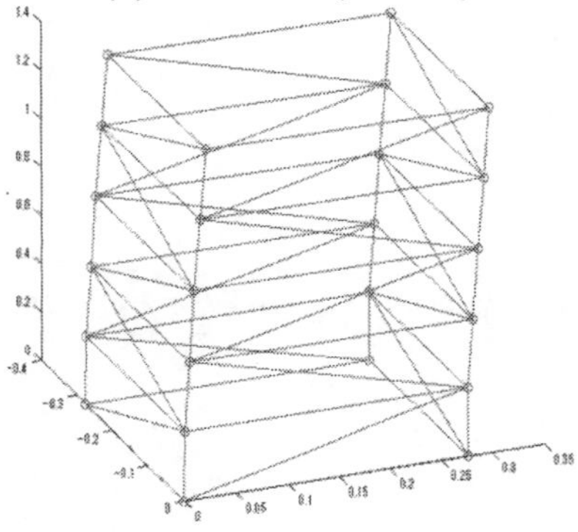

(c) 3rd mode (47.7Hz).

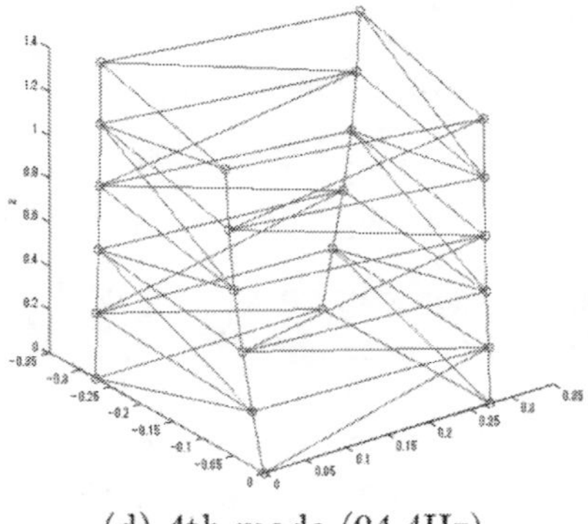

(d) 4th mode (94.4Hz).

Figure 5: Modal frequencies and modal shapes of the truss structure.

peaks with the number of PVAs. The H_2 norms of the transfer functions of the truss with the non-optimized, single-, and double-PVA systems are 8.85×10^{-5}, 4.70×10^{-5}, and 4.52×10^{-5} respectively. Therefore, the H_2 norms of the transfer function of the truss with optimized PVAs is dramatically lower than one with non-optimized PVA. The performance of double-PVA system was slightly better than one of single-PVA system. The optimal frequencies became distributed more widely, and the optimal damping ratios became lower, with an increased number of PVAs. This tendency is similar to that see in multiple-DVA systems.[6, 7]

For comparison, we also conducted H_∞ optimization using the method mentioned in our previous work[21]. As the result, it was revealed that the performance of H_2 optimized PVA system is better, however not so remarkable. Fig. 7 shows the transfer functions of the truss with the non-optimezed, H_2 optimized single-, and H_∞ optimized double-PVA systems.

Table 2: Optimal Parameters of PVA System

	ζ_{p1}	ζ_{p2}	ω_{p1}	ω_{p2}	$\|\boldsymbol{T_{fz}}\|_2^2$
H_2 Opt. Single	0.05185		308.2		4.703×10^{-5}
H_2 Opt. Double	0.03112	0.3432	319.6	294.8	4.530×10^{-5}
H_∞ Opt. Single	0.06382		308.9		4.748×10^{-5}

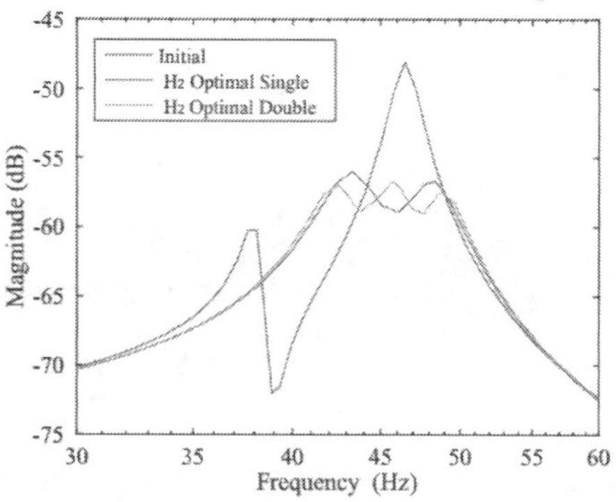

Figure 6: Comparison of Transfer Function $|\boldsymbol{T_{fz}}|$ about H_2 Optimized Single- and Double-PVA system.

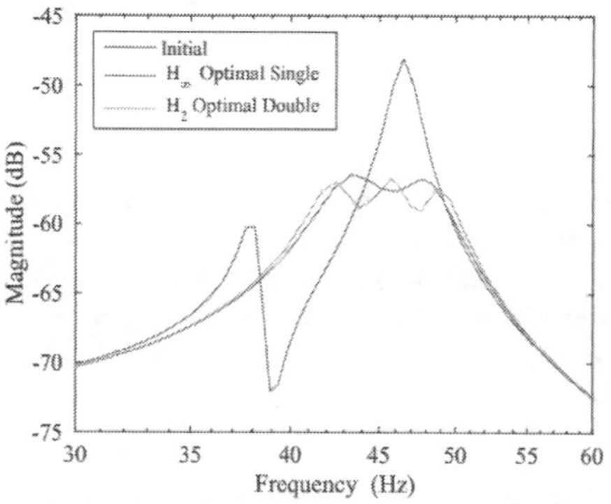

Figure 7: Comparison of Transfer Function $|\boldsymbol{T_{fz}}|$ about H_∞ Optimized Single-PVA System and H_2 Optimized Double-PVA system.

5. CONCLUSION

The optimization methods of multiple PVA system for vibration suppression of truss structure were proposed. Numerical Simulation revealed that the H_2 norms of the transfer function of the truss with optimized PVAs is dramatically lower than one with non-optimized PVA. The performance of double-PVA system was slightly better than one of single-PVA system. Experiment to reveal the effectiveness of multiple-PVA system and usefulness of this optimization method is under going.

ACKNOWLEDGEMENTS

This work was supported by MEXT/JSPS KAKENHI grant number 25820412.

REFERENCES

[1] Den Hartog, J. P., "Mechanical Vibrations", McGraw-Hill Inc.,US, 1956.

[2] Williams, K. A., Chiu, G. T. C., and Bernhard, R. J., "Dynamic Modeling of a Shape Memory Alloy Adaptive Tuned Vibration Absorber", Journal of Sound and Vibration, Vol. 280, No. 1-2, 2005, pp. 211-234.

[3] Nishimura, I., Yamada, T., Sakamoto, M., and Kobori, T., "Control Performance of Active-Passive Composite Tuned Mass Damper", Smart Materials and Structures, Vol. 7, No. 5, 1998, pp. 637-653.

[4] Pinkaew, T., and Fujino, Y., Effectiveness of Semi-Active Tuned Mass Dampers under Harmonic Excitation, Engineering Structures, Vol. 23, No. 7, 2001, pp. 850-856.

[5] Asami, T., Nishihara, O., and Baz, A. M., "Analytical Solutions to H_∞ and H_2 Optimization of Dynamic Vibration Absorbers Attached to Damped Linear Systems", AIAA Journal, Vol. 124, 2002, pp. 284-295.

[6] Zuo, L. and Nayfeh, S. A., "Minimax Optimization of Multi-Degree-of-Freedom Tuned-Mass Dampers", Journal of Sound and Vibration, Vol. 272, 2004, pp. 893-908.

[7] Zuo, L. and Nayfeh, S. A., "Optimization of the Individual Stiffness and Damping Parameters in Multiple-Tuned-Mass-Damper Systems", Journal of Vibration and Acoustics, Vol. 127, 2005, pp. 77-83.

[8] Ohnuki, O. , Nonami, K. , Nishimura H., and Ariga, Y., "Sliding mode control of multi-degree-of-freedom structure by means of snsorless active dynamic vibration absorber", Proceeding of Third International Conference on Motion and Vibration Control, 1996, pp. 58-63.

[9] Alujevic, N. , Tomac, I. , Gardonio, P., "Tuenable vibration absorber using acceleration and displacement feedback", Journal of Sound and Vibration, Vol. 331, No. 12, 2012, pp. 2731-2728.

[10] Nambu, Y., Yamamoto, S., and Chiba, M.., "A Smart Dynamic Vibration Absorber for Suppressing the Vibration of a String Supported by Flexible Beams", Smart Materials and Structures, Vol. 23, No. 2, 2014, pp. 25-32.

[11] Hagood, N. W. and Von Floton, A., "Damping of Structure Vibration with Piezoelectric Materials and Passive Piezoelectric Materials and Passive Electrical Networks", Journal of Sound and Vibration, Vol. 146, No. 2, 1991, pp. 243-268.

[12] Yamada, K., Matsuhisa, H., Utsuno, H., and Sawada K., "Optimum Tuning of Series and Parallel LR Circuits for Passive Vibration Suppression using Piezoelectric Elements", Journal of Sound and Vibration, Vol. 329, No. 24, 2010, pp. 5036-5057.

[13] Moheimani, S. O. R., "A Survey of Recent Innovations in Vibration Damping and Control using Shunted Piezoelectric Transducers", Control Systems Technology, IEEE Transactions on, Vol. 11, no. 4, 2003, pp. 482-494.

[14] Petit, L., Lefeuvre, E., Richard, C., and Guyomar, D., "A Broadband Semi passive piezoelectric technique for structural damping", Journal of Intelligent Material Systems and Structures, Vol. 18, No. 3, 2007, pp. 275-287.

[15] Richard, C., Guyomar, D., Audigier, D. and Bassaler, H., "Enhanced Semi Passive Damping using Continuous Switching of a Piezoelectric Device on an Inductor", Proc. SPIE Smart Structures and Materials Damping and Isolation, No. 3989, 2000, pp. 288-299.

[16] Clark, W. W., "Vibration Control with State-Switched Piezoelectric Materials", Journal of Intelligent Material Systems and Structures, No. 3672, 1999, pp. 263-271.

[17] Richard, C., Guyomar, D., Audigier, D. and Ching, G., "Semi-passive Damping using Continuous Switching of a Piezoelectric Device", Proc. SPIE Smart Structures Materials Passive Damping Isolation, Vol. 11, No. 4, 2000, pp. 104-111.

[18] Corr, L. R. and Clark, W. W., "Comparison of Low-frequency Piezoceramic Shunt Techniques for Structural Damping", Proc. SPIE Smart Structures and Materials Damping and Isolation, No. 4331, 2001, pp. 262-272.

[19] Onoda, J., Makihara, K. and Minesugi, K., "Energy-recycling Semi-active Method for Vibration Suppression with Piezoelectric Transducers", AIAA Journal, Vol. 41, No. 4, 2002, pp. 711-719.

[20] Makihara K., Onoda J. and Minesugi K., "A Self-sensing Method for Switching Vibration Suppression with a Piezoelectric Actuator", Smart Material and Structure, Vol. 16, 2007, pp. 455-461.

[21] Nambu Y., Takashima T., and Inagaki A.: "Robust Design Method and Thermostatic Experiment for Multiple Piezoelectric Vibration Absorber System", Smart Material and Structure, Vol. 24, No. 12, 2015, pp. 125016.

ICAST2015 #008

Beam Damage Uncertainty Quantification Using Guided Lamb Wave Responses

Gang Wang[1]

[1] Department of Mechanical and Aerospace Engineering, University of Alabama in Huntsville, Huntsville, Alabama, USA

Abstract

In this paper, a Bayesian inference approach is developed to conduct uncertainty quantification on the notch damage in a beam structure using guided Lamb wave responses. The proposed methodology not only determines the notch damage characteristics (i.e., location and extent) but also quantifies associated uncertainties of these inferred values. Thus, such information is essential for decision-making in the structural health monitoring (SHM) applications. First, a spectral finite element model is used to characterize Lamb wave propagation responses in a beam under PZT actuation and sensing. Very few elements are required to accurately capture the wave propagation. The PZT sensor can pick up the reflected wave responses from both boundaries and damages. Total 18 simulation cases were generated by varying notch damage extent, damage location, and noise level. Secondly, the Markov Chain Monte Carlo (MCMC) techniques are employed to estimate the notch damage location and extent from guided Lamb wave responses, in which the random walk Metropolis algorithm is used. Finally, both crack size/location and associated uncertainties are characterized. In summary, the proposed probabilistic damage detection is successfully demonstrated in beam structures using the guided wave responses, which can be extended to other SHM applications.

1. INTRODUCTION

Lamb waves, like other types of acoustic waves can travel long distances in plate-like structures due to their high wave speeds [1-6]. Lamb waves are highly sensitive to changes in densities. An incident wave on an interface with dissimilar densities reflects or refracts based on the density values, allowing it to detect transverse cracks and delamination. Lamb waves are particularly attractive for damage detection in structures because it is very sensitive to small damage amounts and the location and approximate extent of damage can be determined using less information. Embedded smart actuators/sensors, such as piezoelectric types, i.e., Lead-Zirconate-Titanate (PZT), have been used to by conducting Lamb wave transmission and reception, pulse-echo, pitch-catch, and phased array functions in order to achieve in-situ nondestructive evaluation for different structures [7-11]. Typically, this methodology does not require analytical structural models and interrogation algorithm is carefully designed with little wave propagation knowledge of the structure. By comparing to baseline signatures, the damage location, amount, and type can be determined accordingly.

In order to fundamentally understand the wave propagation in a beam under PZT actuation and sensing, one needs to characterize a piezoelectric beam system. Park et al. developed a coupled PZT beam model, in which the base beam is modeled using Timoshenko beam theory to account for the rotational inertia and shear deformation and the PZT layers are assumed to follow Euler-Bernoulli beam theory [12]. Lin et al. proposed Mindlin plate theory to model the piezoelectric beam system, which consists of disk shaped actuators [13]. Conventional finite element method (CFEM) has been used to predict the dynamic response of a piezoelectric beam system [14-15]. In order to obtain accurate wave propagation prediction, a large number of elements must be used. Typically, the size of an element is 10-20 times smaller than the wavelength. As the actuating frequency increases, the wavelength becomes smaller and so does the element size. This implies a large number of elements are required, which substantially increases the computational time and cost.

The spectral finite element method (SFEM) is an alternative approach to conduct dynamic analysis of structures [16-19]. It follows a similar solution procedure as in CFEM. The difference is that SFEM is formulated in the frequency domain and its interpolation functions are duplicated from exact solutions of the governing equations. Partial differential equations are transformed to the frequency domain using Discrete Fourier Transform (DFT) [17]. The transformation yields a set of ordinary differential equations (ODEs). Then, analytical solutions can be determined, which are frequency dependent. Similar to the CFEM, the nodal force and nodal displacement vector is introduced. A dynamic stiffness matrix is formulated to relate these two vectors. After assembling elements and applying appropriate boundary conditions, we can determine the structural responses at each frequency. The time domain solutions can then be constructed by conducting inverse FFT. Note that the SFEM has been successfully used in various structures such as rotating beam [20], sandwich beam analysis [21], and vibration based piezoelectric beam energy harvester [22]. The least possible elements are required to obtain accurate predictions. In a recent paper [23], a SFEM was applied to characterize wave propagations in a beam structure under piezoelectric material (i.e., PZT) actuation/sensing. Mathematical models were developed to account for both unimorph and bimorph configurations, in which PZT layers are modeled as either an actuator or a sensor. The Timoshenko beam theory was adopted for base beam and PZT layers to accommodate high-frequency acoustic wave propagations, i.e., 20-200 kHz. These electro-mechanical coupled equations were solved in the frequency domain in order to formulate the SFEM. It was shown that the PZT sensor can pick up the reflected waves from beam boundaries and damages. Very few spectral finite elements were required to accurately capture the wave propagation and the SFEM predictions were validated analytically and experimentally. It was also noticed that the contributions from PZTs have less significant effects on the wave propagation characteristics when using thin PZT layers in the beam structures.

As aforementioned, Lamb wave based damage detection techniques typically consist of comparing signals from the defective structure with a benchmark and do not require analytical structural models. Subsequently, the damage location and size can be calculated via the time of flight (TOF) estimation from the wave propagation and reflected wave responses. However, it has been recognized that a considerable amount of uncertainties exist in the damage detection process such as measurement errors and variability of material properties. Therefore, such deterministic damage detection method cannot provide comprehensive information for decision-making in the SHM applications.

One needs a novel probabilistic damage detection approach that accounts for the underlying uncertainties. Beck et al. proposed a Bayesian statistical framework to update a structural model and its associated uncertainties using dynamic response data [24]. Lam et al. applied the same Bayesian model updating method to characterize multiple cracks in a cantilever beam using transient vibration data [25]. The Euler-Bernoulli beam theory was adopted in order to determine analytical transient response, in which continuity conditions were applied at each crack location. A two-stage approach was used to determine the number of cracks, each crack depth and location. Experiments were conducted to verify this probabilistic crack detection approach. Other crack detection studies based on the same Bayesian probabilistic framework can be found in literature [26-28]. Ng et al proposed a probabilistic optimization approach to the characterization of damage in beams using guided waves. A SFEM was applied to a rod structure with notch damages in order to capture the longitudinal guided wave propagation. Subsequently, an optimization scheme was used to evaluate the posterior probability density functions (PDFs) of the uncertainty damage parameters by applying simulated guide wave data.

The objective of this paper is to extend the Bayesian approach to characterize notch damage in a beam structure using guided Lamb wave responses. First, a simplified SFEM is employed to model a piezoelectric beam based on the previous work [23]. The electro-mechanical coupling is neglected but the stiffness and mass contributions from PZT materials are included. Very few elements are required to provide accurate Lamb wave predictions under PZT actuation and sensing in a beam structure. Secondly, the Bayesian approach is employed for the damage parameter estimations. Instead of using probabilistic optimization method [25,28], the Markov Chain Monte Carlo (MCMC) techniques [29] are applied along with the Random Walk Metropolis algorithm [29-31] to conduct notch damage uncertainty quantification. Finally, the proposed probabilistic notch damage detection scheme is demonstrated to infer crack size/location from guided Lamb wave responses.

This article proceeds as follows. First, the SFEM and solution approach are discussed. Secondly, the Bayesian damage detection scheme is presented. The crack inference is demonstrated in the subsequent section. Finally, summary and some concluding remarks are given.

2. SFEM FORMULATION

2.1 Spectral Finite Element Method

As outlined in Figure 1, the basic steps in the development of SFEM can be described as follows [17-19]. The governing wave equations are first transformed from the time domain to the frequency domain using a discrete Fourier transform (DFT). For computational implementation, a FFT algorithm is adopted. The spatial variation is semi-explicitly obtained by solving the characteristic equation in k-space. This results in a complex shape function matrix representing the linear superposition of all the wave modes. Also, the complex dynamic strain–displacement matrix and the dynamically consistent force vector can both be derived. Following the CFEM, the complex dynamic stiffness matrix is then formed, which is exact. This makes the proposed SFEM an efficient model suitable for use within the framework of an automated FEM rather than a trade-off for broadband wave propagation analysis. But the fundamental difference from the CFEM is that all the spectral amplitudes that correspond to elemental nodal variables are evaluated at each frequency step (FFT sampling points) instead of pseudo-static variables evaluated at each time step or at each eigenfrequency. The global system is solved for unit spectral amplitude of the applied load history at each FFT sampling frequency. Therefore, the time domain response is obtained using the inverse FFT. Certainly the solution is different at each position, but once it is evaluated at a particular position then its inverse immediately gives the time history of the solution at that point.

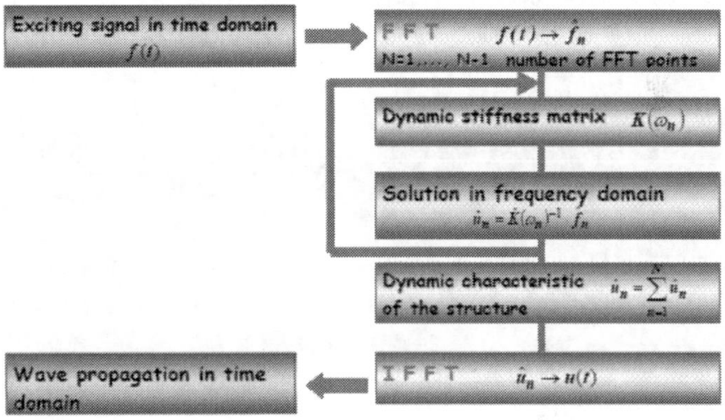

Figure 1. Solution procedure in SFEM

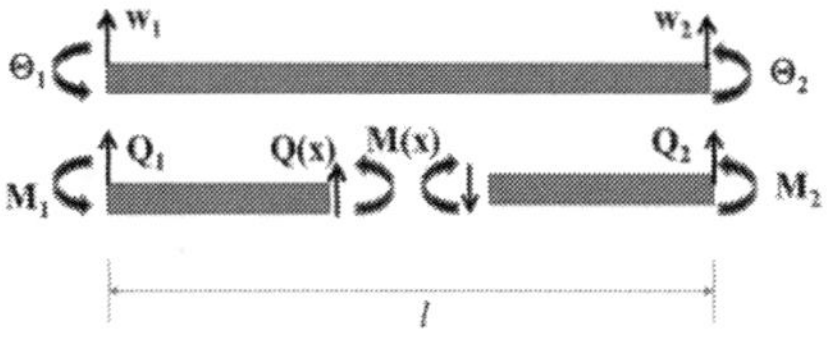

Figure 2. Nodal displacement and force vector

2.2 Piezoelectric Beam

A SFEM for an isotropic Timoshenko beam is discussed below. The governing equation is given by [19]

$$\rho A\frac{\partial^2 w}{\partial t^2} + GA\kappa\left(\frac{\partial \theta}{\partial x} - \frac{\partial^2 w}{\partial x^2}\right) = 0 \tag{1}$$

$$\rho I\frac{\partial^2\theta}{\partial t^2}-EI\frac{\partial^2\theta}{\partial x^2}+GA\kappa\left(\theta-\frac{\partial w}{\partial x}\right)=0 \tag{2}$$

Here, w(x,t) and θ(x,t) are transverse and angular displacements, respectively. Assume harmonic oscillation, and the solutions are given by

$$w(x,t)=W(x)e^{j\omega t} \tag{3}$$

$$\theta(x,t)=\Theta(x)e^{j\omega t} \tag{4}$$

Substituting Eq. (3)-(4) into Eq. (1)-(2), we obtain

$$GA\kappa\frac{d^2W}{dx^2}-GA\kappa\frac{d\Theta}{dx}+\rho A\omega^2W=0 \tag{5}$$

$$EI\frac{d^2\Theta}{dx^2}+GA\kappa\left(\frac{dW}{dx}-\Theta\right)+\rho I\omega^2\Theta=0 \tag{6}$$

Assume the general solutions to Eq. (5)-(6) are

$$W(x)=\alpha e^{kx} \tag{7}$$

$$\Theta(x)=\beta e^{kx} \tag{8}$$

And substituting Eq. (7)-(8) into (5)-(6), we obtain the following polynomial, which can be used to determine the wave number k

$$(EIGA\kappa)k^4+\left(GA\kappa\rho I\omega^2+EI\rho A\omega^2\right)k^2+\rho A\omega^2\left(\rho I\omega^2-GA\kappa\right)=0 \tag{1}$$

The four roots of the Eq. (1) are

$$k_1=\sqrt{\frac{-(GA\kappa I+EIA)\rho\omega^2+\sqrt{(GA\kappa I+EIA)^2\rho^2\omega^4-4(EIGA\kappa)(\rho A\omega^2)(\rho I\omega^2-GA\kappa)}}{2EIGA\kappa}} \tag{10}$$

$$k_2=-k_1 \tag{11}$$

$$k_3=\sqrt{\frac{-(GA\kappa I+EIA)\rho\omega^2-\sqrt{(GA\kappa I+EIA)^2\rho^2\omega^4-4(EIGA\kappa)(\rho A\omega^2)(\rho I\omega^2-GA\kappa)}}{2EIGA\kappa}} \tag{12}$$

$$k_4=-k_3 \tag{13}$$

Finally, the transverse displacement W and angular displacement Θ are given by

$$W(x,t)=\alpha_1e^{k_1x}+\alpha_2e^{k_2x}+\alpha_3e^{k_3x}+\alpha_4e^{k_4x} \tag{14}$$

$$\Theta(x,t)=\beta_1e^{k_1x}+\beta_2e^{k_2x}+\beta_3e^{k_3x}+\beta_4e^{k_4x} \tag{15}$$

Where

$$\alpha_i=c_i\beta_i=\frac{k_i}{k_i^2+\dfrac{\rho A\omega^2}{GA\kappa}}\beta_i \tag{16}$$

Similar to the conventional beam finite element, two nodal displacements per node (one transverse and angular displacement per node as shown in Figure 2) are assumed in order to replace four unknown wave coefficients.

$$\begin{bmatrix}W_1\\\Theta_1\\W_2\\\Theta_2\end{bmatrix}=P\begin{bmatrix}\beta_1\\\beta_2\\\beta_3\\\beta_4\end{bmatrix}=\begin{bmatrix}c_1&c_2&c_3&c_4\\1&1&1&1\\c_1e^{k_1l}&c_2e^{k_2l}&c_3e^{k_3l}&c_4e^{k_4l}\\e^{k_1l}&e^{k_2l}&e^{k_3l}&e^{k_4l}\end{bmatrix}\begin{bmatrix}\beta_1\\\beta_2\\\beta_3\\\beta_4\end{bmatrix} \tag{17}$$

where l is the length of the element. Based on the Timoshenko beam theory, the shear force and bending moment are given by

$$Q=GA\kappa\left(\frac{\partial w}{dx}-\theta\right) \tag{18}$$

$$M=EI\frac{\partial\theta}{\partial x} \tag{19}$$

Then one can define the elemental dynamic stiffness matrix K_e to relate the nodal displacements to nodal forces as shown below

$$\begin{bmatrix}Q_1\\M_1\\Q_2\\M_2\end{bmatrix}=[K_e]\begin{bmatrix}W_1\\\Theta_1\\W_2\\\Theta_2\end{bmatrix} \tag{20}$$

where

$$[K_e]=\begin{bmatrix} -GA\kappa(c_1k_1-1) & -GA\kappa(c_2k_2-1) & -GA\kappa(c_3k_3-1) & -GA\kappa(c_4k_4-1) \\ -EIk_1 & -EIk_2 & -EIk_3 & -EIk_4 \\ GA\kappa(c_1k_1-1)e^{k_1l} & GA\kappa(c_2k_2-1)e^{k_2l} & GA\kappa(c_3k_3-1)e^{k_3l} & GA\kappa(c_4k_4-1)e^{k_4l} \\ EIk_1e^{k_1l} & EIk_2e^{k_2l} & EIk_3e^{k_3l} & EIk_4e^{k_4l} \end{bmatrix}[P]^{-1} \tag{21}$$

Figure 3 shows a beam with surfaced bonded PZT bimorph actuators and a PZT sensor. Only total seven elements are needed to capture guided Lamb wave propagation in this cracked beam. The stiffness and mass contributions from PZT materials are included. The bending moment (M) introduced by a pair of PZT actuators under a given actuation voltage V_a and the collected PZT sensor voltage (V_s) can be determined as shown below, in which a closed circuit is assumed [32].

$$M = E_c b d_{31}(h+h_c)V_a \tag{22}$$

$$V_s = \frac{d_{31}E_c b_c}{C_p}\int_{l_c}\varepsilon dx \tag{23}$$

C_p is the capacitance of the PZT sensor. Associate beam material and geometry constants are listed below in Table 1. The length of each element is shown in Table 2.

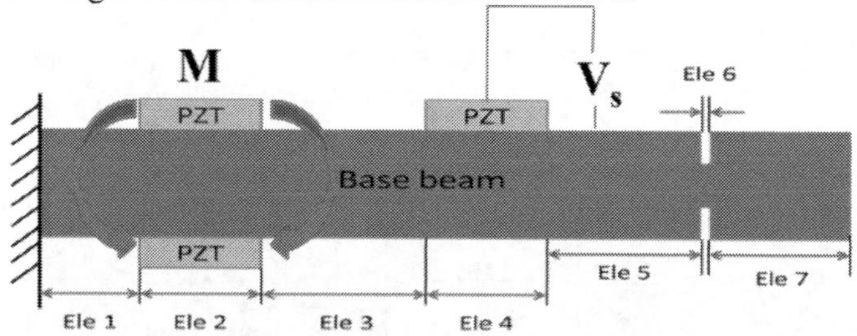

Figure 3. Piezoelectric beam configuration

Table 1 Piezoelectric Beam constants

	Base beam	PZT
Length [mm]	1000	10
Width [mm]	10	10
Height [mm]	5	0.5
Young's modulus [Gpa]	70	60.6
Poisson's ratio	0.33	0.31
Density [kg/m^3]	2700	7800
Compliance	N/A	76.42
Piezoelectric constant, d_{31} [pm/v]	N/A	-274
Piezoelectric constant, e_{33}^T [nF/m]	N/A	30.1

Table 2 Spectral finite elements of the cracked piezoelectric beam; Lc=damage location

Element Number	Element Type	Element Length (*mm*)
1	Base beam	10
2	Bimorph (Actuator)	10
3	Base beam	20
4	Unimorph (Sensor)	10
5	Base beam	Lc-50
6	Base beam (Crack)	2
7	Base beam	998-Lc

2.3 Guided Lamb Wave Response

Figure 4 shows a 100 kHz pure sine tone burst and a modulated tone burst using a Hanning window. It is clear that the Hanning window reduces the side lobes by a considerable amount [9,11,23]. Figure 5 shows the PZT sensor response by applying a 100kHz Hanning widowed 5-cycled sine signal. Clearly, the PZT sensor is able to pick up the excitation pulse, reflection from notch damage, and reflection from boundary conditions. In this case, the notch damage is assumed at 30% from beam root with 50% thickness loss. Comprehensive simulations were conducted in order to determine the optimal actuation frequency of an antisymmetric Lamb wave mode. Figure 6 shows the maximum PZT sensor voltage frequency responses. The 100 kHz actuation frequency was selected to generate all Lamb wave data. In all simulations, a 1.5% damping was assumed.

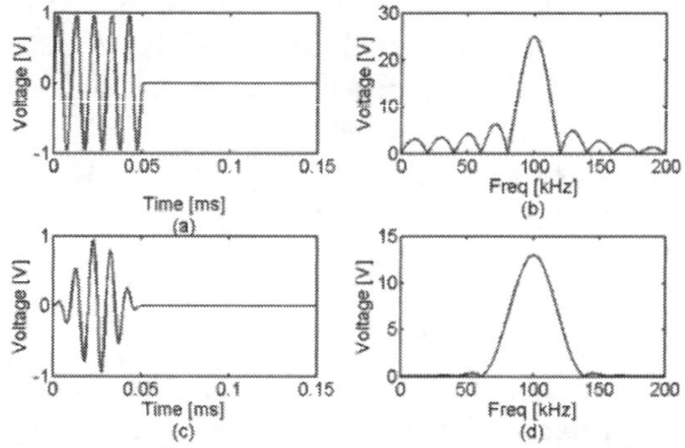

Figure 4. Time domain and frequency domain plots for a 100 kHz actuating frequency: (a-b) pure sine; (c-d) Hanning windowed sine

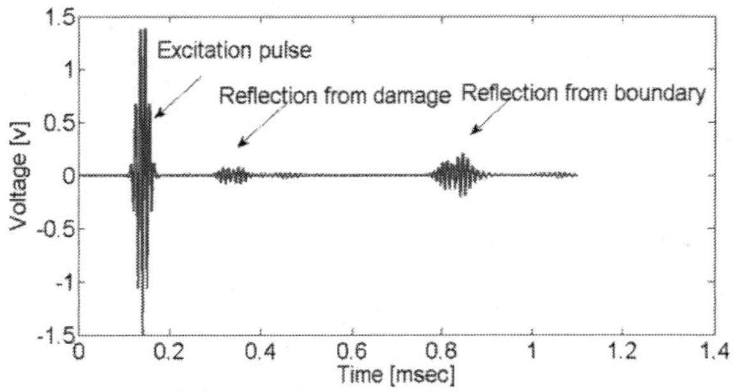

Figure 5. Simulated antisymmetric Lamb wave response in a damaged piezoelectric beam

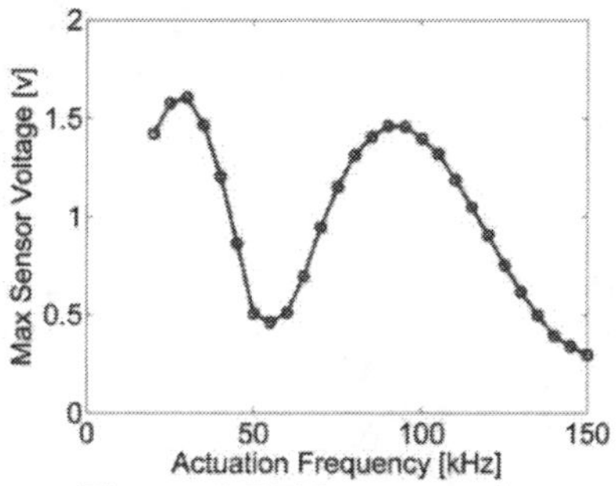

Figure 6. PZT sensor voltage (maximum value) frequency response

3. BAYESIAN INFERENCE

3.1 Bayesian Approach

In this section, a Bayesian approach is employed to infer crack depth and crack location from guided Lamb wave responses. A random variable vector (q) containing both crack depth and crack location is treated as with prior probability density function (PDF) (π_0). Based on the Bayes' theorem of inverse problem [31], the posterior probability density function (π) given the measurement data (v_{obs}) is defined by

$$\pi(q|v_{obs}) = \frac{\pi(v_{obs}|q)\pi_0(q)}{\int \pi(v_{obs}|q)\pi_0(q)dq} \tag{24}$$

The likelihood function, $\pi(v_{obs}|q)$, depends on the assumptions made on the distribution errors [31]. Assuming that the errors are independent and identically distributed, i.e., i.i.d. with constant variance σ, the likelihood function is given by

$$\pi(v_{obs}|q) = \frac{1}{(2\pi\sigma^2)^{n/2}} e^{-SS_q/2\sigma^2} \tag{25}$$

where

$$SS_q = \sum_{i=1}^{n} (v_i - u_i)^2$$
$$v = v_{obs}$$

v_i is the measurement at the ith time step. u_i is the model prediction at the ith time step. n is the total number of data. Note that the posterior PDF provides the distribution characteristics of random variable vector q based on the observation v. Different algorithms have been developed to evaluate this posterior PDF as discussed in Refs. [24-28]. The Markov Chain Monte Carlo techniques along with the Random Walk Metropolis algorithm [29-31] are adopted in this study.

3.2 Random Walk Metropolis Algorithm

Ref. [31] provides a detailed discussion on the Random Walk Metropolis algorithm. A slight modification was made for the notch crack case as shown below.

- Step a: Determine the chain number and parameter (crack depth and crack location)
- Step b: Determine initial parameter value, q_0, (i.e., find minimum error from the plot between model predictions and measurement data)
- Step c: Compute SS_{q_0} and initial variance estimation, $s_0 = \frac{SS_{q_0}}{n-1}$
- Step d: Construct covariance matrix, $V = s_0^2 \left[\chi(q_0)^T \chi(q_0)\right]^{-1}$ and R=*chol* (V). Note that χ is the Jacobian matrix which is evaluated numerically.
- Step e: Conduct iteration for each chain, k
  1. Sample $z_k = N(0,1)$
  2. Update $q_k = q_{k-1} + Rz_k$
  3. Sample $u_\alpha = U(0,1)$
  4. Compute $SS_{q^*} = \sum_{i=1}^{n} (v_i - u_i(q_k))^2$
  5. Compute $\alpha\left(q^* \middle| q_{k-1}\right) = \min\left(1, e^{-\left[SS_{q^*} - SS_{q_{k-1}}\right]/2s_{k-1}^2}\right)$
  6. If $u_\alpha < \alpha$, set $q_k = q^*$ $SS_{q_k} = SS_{q^*}$

     else

     set $q_k = q_{k-1}$ $SS_{q_k} = SS_{q_{k-1}}$
  7. Update $s_k = 1/\text{gamrnd}\ (a_{val}, 1/b_{val})$ and $a_{val} = 0.5(0.001 + n)$ $b_{val} = 0.5(0.001 s_0^2 + SS_{q_k})$

4. RESULTS

Simulations were conducted to determine guided Lamb wave responses by varying crack depth, crack location, and noise level. Three crack sizes were selected by varying the ratio between crack depth and beam height at 15%, 30%, and 50%. The crack location (i.e., Lc value) was set at either 30% or 50% away from the beam root. Three noise levels, i.e., 5%, 10%, and 15%, were assumed to account for measurement errors. Finally, total 18 simulations were conducted to serve as input data in order to demonstrate the proposed probabilistic damage detection approach.

Figure 7 shows the inferred crack size and location results, in which actual 15% crack depth, 30% location, and 5% added noise are assumed. By applying the Bayesian inference scheme as outlined in Section 3, both crack size and crack location can be inferred and the corresponding marginal paths are plotted in Figure 7a and Figure 7b, respectively. Corresponding histogram plots are shown in Figure 7c and Figure 7d as well. Figure 8-9 shows similar results for two other cases.

Clearly, both crack size and location can be inferred simultaneously and associated uncertainties can also be quantified.

Table 3 summarizes the inferred crack size mean, standard derivation (STD), and coefficient of variation (COV) values, in which the actual crack location is assumed at 30% from beam root. The COV is used to quantify uncertainty, which is a non-dimensional value of the dispersion of a

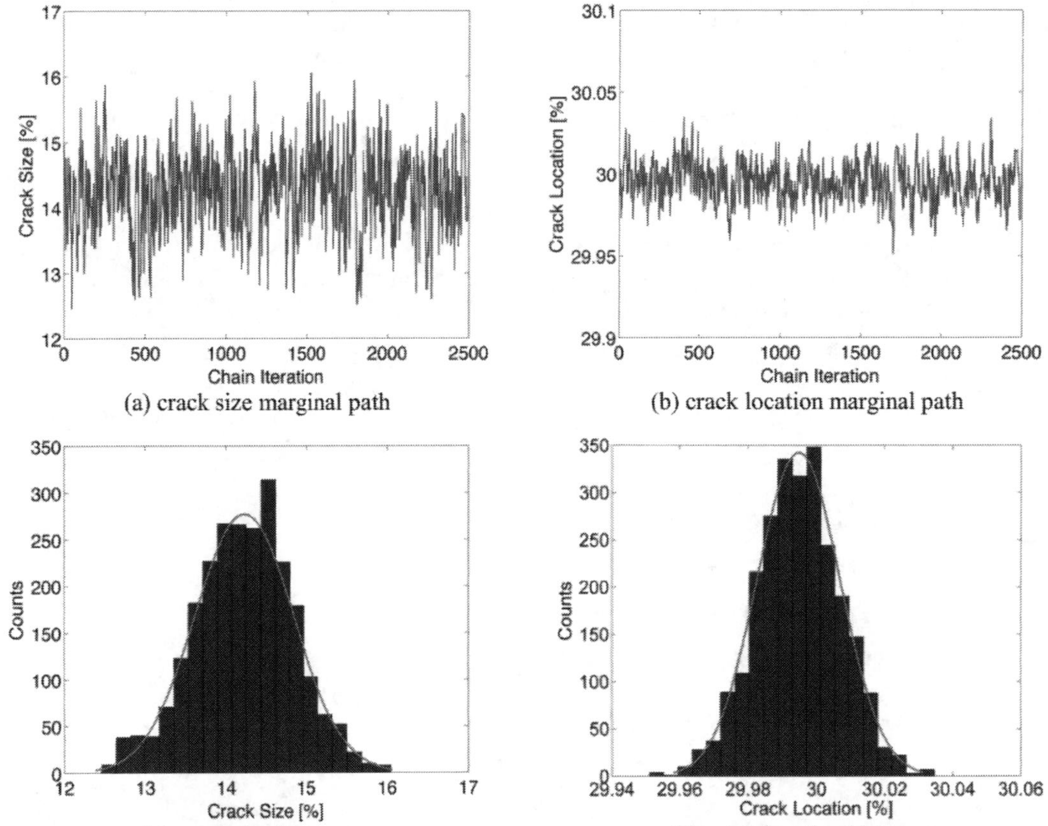

(a) crack size marginal path (b) crack location marginal path

(c) crack size histogram (b) crack location histogram

Figure 7. Results of the case with actual 15% crack size, 30% crack location, and 5% added noise

probability distribution with respect to the mean value in all cases. The COV values increase as the noise level increases in all crack size cases. However, the COV values reduce as the crack size increases under the same noise level. This suggests that larger crack size case exhibits less uncertainty.

In order to further demonstrate the uncertainty due to noise level and crack size, the marginal cumulative distribution functions (CDFs) in above cases are plotted as shown in Figures 10 and 11. The crack size was normalized by the value at which the marginal CDF corresponds to 0.5 so that one is able to quantitatively compare all cases. Figure 10 shows the marginal cumulative distribution of each identified crack size under 15% noises. The slope of each curve can be used to measure the uncertainty associated with each identified value. The steeper curve indicates the size of the crack is more concentrated around the median size of the crack (i.e., corresponding to the value at which the CDF equals to 0.5), which implies less uncertainty in the model. Current plots are consistent with the findings from the COV values as shown in Table 3. Figure 11 shows the marginal cumulative distribution of the normalized 50% crack under different noise levels. Clearly, more added noise yields higher uncertainty.

Table 4 summarizes the crack location results, in which the actual crack size was set at 30%. Similar to the crack size case as shown in Table 3, the COV values increase as the noise level increases in all crack location cases. However, the COV values reduce as the crack location moves toward the beam center under the same noise level. Less uncertainty is shown in the 50% location case compared to 30% location case. Also, the noise effects on the uncertainty are similar to the crack size case. Less noise indicates less uncertainty. Compared to the COV values associated with crack size listed In Table 3, the COV values associated with crack location are smaller. Guided Lamb wave signals usually present accurate information on the damage location compared to the damage extent. This indicates that less uncertainty appears in the identified crack location results.

In summary, it is clearly demonstrated that the proposed Bayesian approach can be used to infer both crack size and location from guided Lamb wave responses. In addition, associated uncertainties are quantified, which yields a probabilistic based structural health monitoring method for the cracked beam structures.

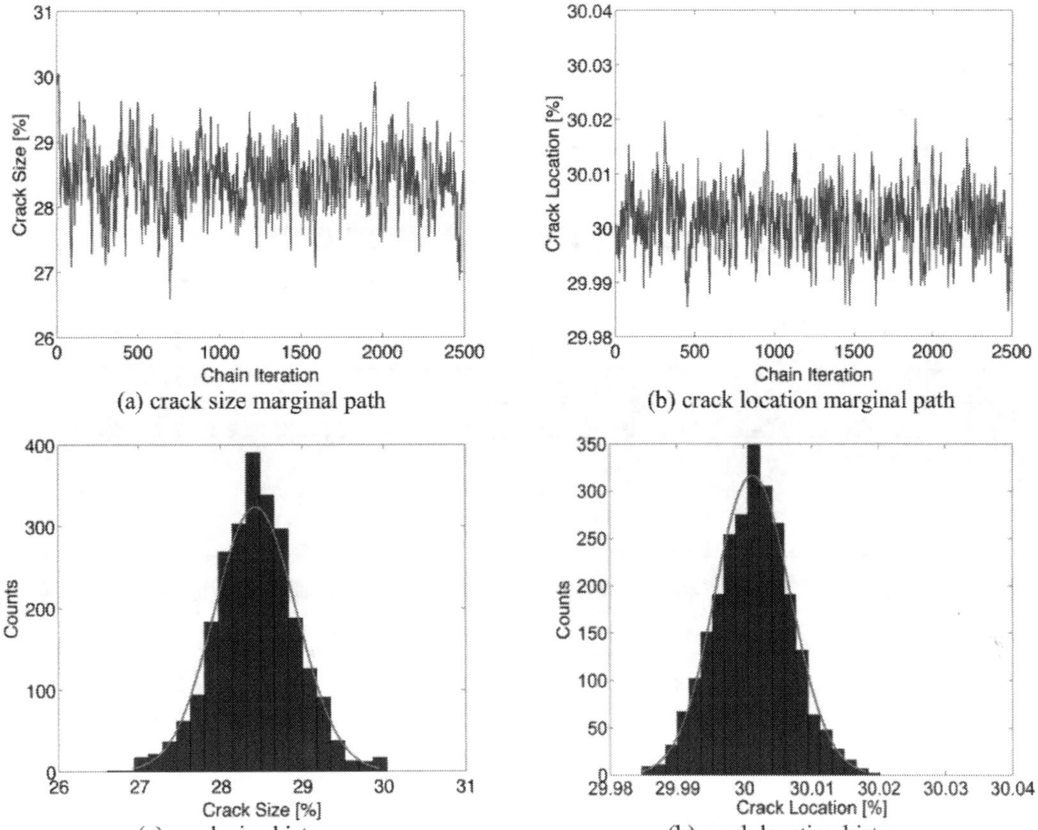

(a) crack size marginal path (b) crack location marginal path

(c) crack size histogram (b) crack location histogram

Figure 8. Results of the case with actual 30% crack size, 30% crack location, and 15% added noise

5. CONCLUSIONS

In this paper, a Bayesian approach was developed to conduct uncertainty quantifications on crack size and location based on guided Lamb wave responses. A SFEM was used to characterize Lamb wave propagation in a cracked beam under PZT actuation and sensing. Simulated voltage data collected from the PZT sensor with added noises were used to represent measurement signals. The MCMC techniques were employed to quantify crack size and location based on the Random Walk Metropolis algorithm. The proposed Bayesian approach not only infers crack size/location concurrently but also quantifies associated uncertainties in those identified values. Both inferred values and associated uncertainties can be used for decision-making in the beam SHM applications.

ACKNOWLEDGMENTS

This research is partially supported by the start-up package provided by the College of Engineering (COE) at the University of Alabama in Huntsville.

Table 3. **Identified crack depth data; Crack location Lc=0.3L**

Case	Actual Crack Size [c/h]	Mean Value	STD	COV [%]
5% response noise				
Case	Actual Crack Size [c/h]	Mean Value	STD	COV [%]
1	0.15	0.14225	0.006139	4.315641
2	0.30	0.28247	0.004390	1.554147
3	0.50	0.47519	0.003131	0.658894
10% response noise				
Case	Actual Crack Size [c/h]	Mean Value	STD	COV [%]
1	0.15	0.14293	0.007391	5.171063
2	0.30	0.28459	0.004329	1.521136
3	0.50	0.47616	0.003564	0.748488
15% response noise				
Case	Actual Crack Size [c/h]	Mean Value	STD	COV [%]
1	0.15	0.14399	0.007889	5.478853
2	0.30	0.28426	0.005042	1.773728
3	0.50	0.47671	0.004051	0.849783

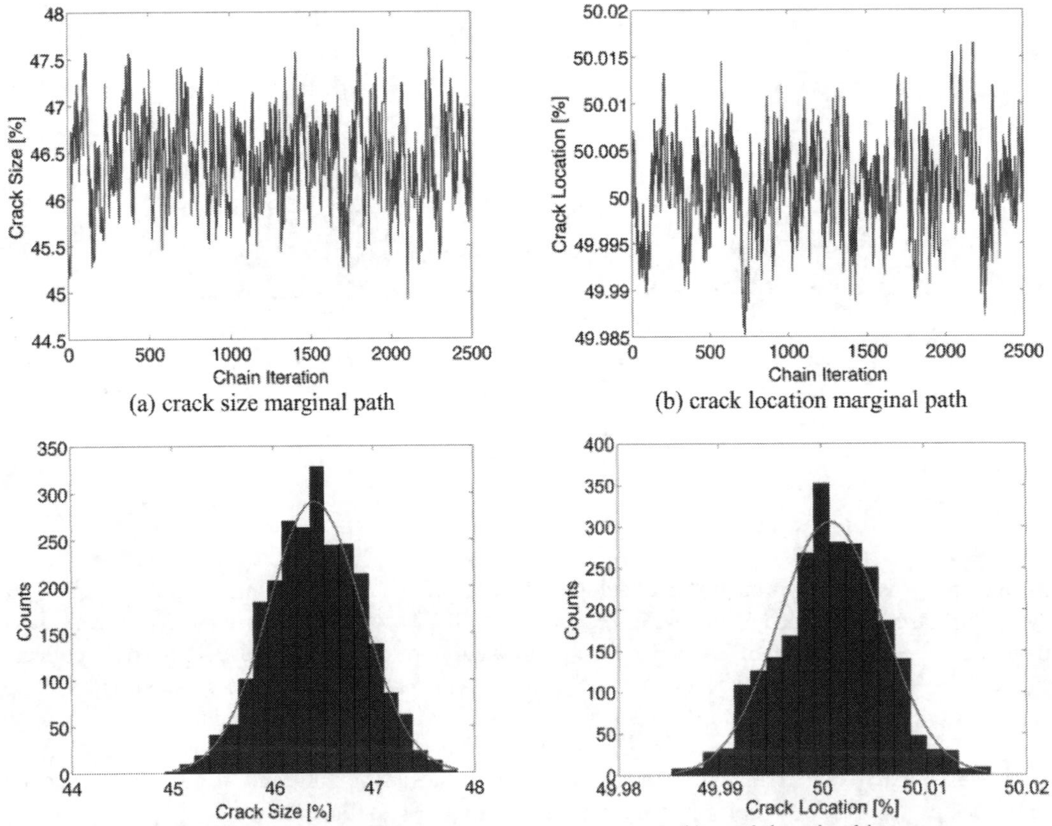

(a) crack size marginal path (b) crack location marginal path

(c) crack size histogram (b) crack location histogram

Figure 9. Results of the case with actual 50% crack size, 50% crack location, and 15% added noise

Table 4. Identified crack location data; Crack size=30%

5% response noise				
Case	Actual Crack Location [Lc/L]	Mean Value	STD	COV [%]
1	0.3	0.3	4.67E-05	0.01557
2	0.5	0.50022	5.96E-05	0.01191
10% response noise				
Case	Actual Crack Location [Lc/L]	Mean Value	STD	COV [%]
1	0.3	0.29998	4.86E-05	0.01620
2	0.5	0.50026	6.57E-05	0.01313
15% response noise				
Case	Actual Crack Location [Lc/L]	Mean Value	STD	COV [%]
1	0.3	0.30001	5.57E-05	0.01857
2	0.5	0.50026	6.88E-05	0.01375

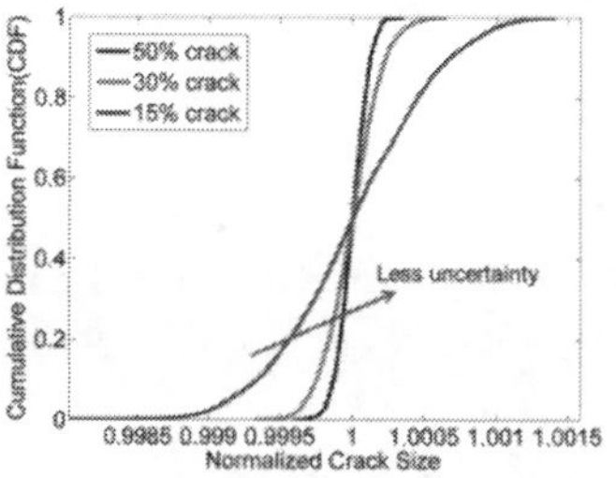

Figure 10. Uncertainty due to crack size (15% added noise; Lc=0.3 L)

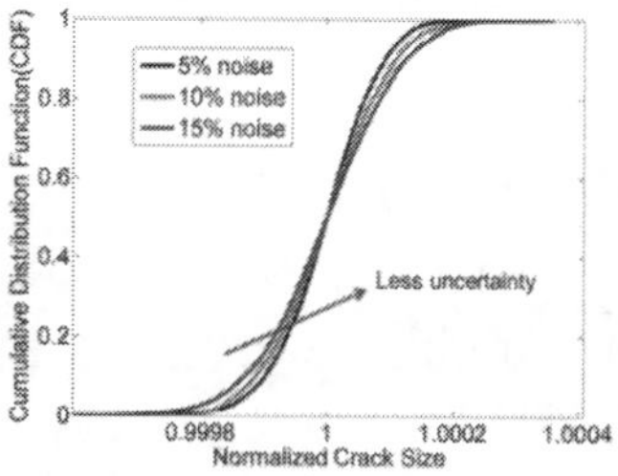

Figure 11. Uncertainty due to added noise (crack size=50%; Lc=0.3 L)

REFERENCES

1. Lamb, H., "On waves in an elastic plate," in Proceedings of the Royal Society of London. Series A, vol. 93, pp. 114–128, 1917.
2. Viktorov, I. A., *Rayleigh and Lamb Waves—Physical Theory and Applications*, Plenum Press, New York, NY, 1967.
3. Graff, K..F., *Wave Motion in Elastic Solids*, Dover Publications, Inc., 1975.
4. Achenbach, J. D., *Wave Propagation in Elastic Solids*, North-Holland Publishing Co., New York, NY,1984.
5. Miklowitz, J., *The Theory of Elastic Waves and Waveguides*, North Holland Publishing Co., New York, NY, pp. 409–430; 1984, North-Holland Series in Applied Mathematics and Mechanics, eds., H. A. Lauwerier and W. T. Koiter, 1978.
6. Auld, B. A, *Acoustic Fields and Waves in Solids*, Vol. 1 and 2, Second edition.; Kreiger Publishing Co., FL,1990.
7. Worlton, D.C., "Ultrasonic testing with Lamb waves," Nondestructive Testing 15, pp. 218–222, 1957.
8. Chang, F.K., eds, *Structural Health Monitoring: Current Status and Perspectives*, CRC Press, April 1998.
9. Giurgiutiu, V., *Structural Health Monitoring with Piezoelectric Wafer Active Sensors*; Academic Press: Burlington, MA, USA, 2007.
10. Su, Z.Q, Ye, L., and Lu, Y., "Guided Lamb wave for identification of damage in composite structures: A Review," *Journal of Sound and Vibration*, 295, pp.753-780, 2006.

11. Su, Z. and Ye, L., *Identification of Damage Using Lamb Waves: From Fundamentals to Applications*, Springer Verlag, London, UK, 2009.
12. Park, H. W., Kim, E. J., Lim, K. L., and Sohn, H., "Spectral element formulation for dynamic analysis of a coupled piezoelectric wafer and beam system," *Computers & Structures,* vol. 88, no. 9–10, pp. 567-580, 2010.
13. Lin, X., and Yuan, F. G., "Diagnostic Lamb waves in an integrated piezoelectric sensor/actuator plate: analytical and experimental studies," *Smart Materials and Structures,* vol. 10, pp. 907-913, 2001.
14. Benjeddou, A., "Advances in piezoelectric finite element modeling of adaptive structural elements: a survey," *Computers and Structures,* vol. 76, pp. 347-363, 2000.
15. Yang, Y. W., Lim, Y. Y., and Soh, C. K., "Practical issues related to the application of the electromechanical impedance technique in the structural health monitoring of civil structures: II. Numerical verification," *Smart Materials and Structures,* vol. 17, no. 3, pp. 1-12, 2008.
16. Beskos D. and Narayanan, G., "Dynamic response of frameworks by numerical Laplace transform," *Computer Methods in Applied Mechanics and Engineering*, 37, 1983, pp.289-307.
17. Doyle, J. F., *Wave Propagation in Structures: Spectral Analysis Using Fast Discrete Fourier Transforms*, Springer Verlag, 2012.
18. Leung, A. Y. T., *Dynamic stiffness and substructures*: Springer-Verlag, 1993.
19. Lee, U., *Spectral element method in structural dynamics*, Singapore ; Hoboken, NJ: J. Wiley & Sons Asia, 2009.
20. Wang, G. and Wereley, N. M., "Free vibration analysis of rotating blades with uniform tapers," *AIAA Journal,* vol. 42, no. 12, pp. 2429-2437, 2004.
21. Wang, G., and Wereley, N. W., "Spectral Finite Element Analysis of Sandwich Beams With Passive Constrained Layer Damping," *Journal of Vibration and Acoustics,* vol. 124, no. 3, pp. 376-386, 2002.
22. Wang, G., "Analysis of bimorph piezoelectric beam energy harvesters using Timoshenko and Euler–Bernoulli beam theory," *Journal of Intelligent Material Systems & Structures,* vol. 24, no. 2, pp. 226-239, 2013.
23. Venugopal, V.P. and Wang, G., "Modeling and Analysis of Lamb Wave Propagation in a Beam Under PZT Actuation and Sensing," *Journal of Intelligent Material Systems and Structures*, vol. 26, no. 13, pp. 1679-1698, 2015.
24. Beck JL. And Katafygiotis LS., "Updating Models and Their Uncertainties I: Bayesian Statistical Framework," *ASCE Journal of Engineering Mechanics*, Vol.124, No. 4, pp., 455-461, 1998.
25. Lam, HF., Ng, CT., and Veidt, M., "Experimental Characterization of Multiple Cracks in a Cantilever Beam Utilizing Transient Vibration Data Following a Probabilistic Approach," *Journal of Sound and Vibration,* Vol. 305, No. 1-2, pp. 34-49, 2007.
26. Lam, HF. and Ng, CT., "A Probabilistic Method for the Detection of Obstructed Cracks of Beam-type Structures Using Spatial Wavelet Transform," *Probabilistic Engineering Mechanics*, Vol. 23, No. 2-3, pp. 237-245, 2008.
27. Lam, HF., Lee, YY., Sun, HY., Cheng, GF., and Guo, X., "Application of the Spatial Wavelet Transform and Bayesian Approach to the Crack Detection of a Partially Obstructed Beam, *Thin-Walled Structures*," Vol. 43, No.1, pp.1-21, 2005.
28. Ng, CT., Veidt, M., and Lam, HF., "Guided Wave Damage Characterization in Beams Utilizing Probabilistic Optimization," *Engineering Structures*, Vol. 31, No. 12, pp.2842-2850, 2009.
29. Metropolis, N., Rosenbluth, A.W., Rosenbluth, M.N., Teller, A.H., Teller, E, "*Equations of State Calculations by Fast Computing Machine," Journal of Chemical Physics* **21** (6): 1087–1092, 1953.
30. Hastings, W. K., "Monte Carlo Sampling Methods Using Markov Chains and Their Applications," Biometrika, Vol. 57, No. 1, pp. 97-109, 1970.
31. Smith, Ralph C., *Uncertainty Quantification Theory, Implementation, and Applications,* Computational Science and Engineering Series, Society for Industrial and Applied Mathematics, Philadelphia, PA, 2014.
32. Chopra, I. and Sirohi, J., *Smart Structures Theory*, Cambridge Aerospace, Cambridge University Press, New York, 2013.

ICAST2015: 26th International Conference on Adaptive Structures and Technologies
October 14-16th, 2015, Kobe, Japan

ICAST2015 #019

Temperature Effects on an Acoustic Emission based SHM system - Applied to Composite Materials -

A. Vargalui[1], M.Martinez[2,1], D.S. Zarouchas[1*] and S. Pant[3]

[1] Group of Structural Integrity & Composites, Faculty of Aerospace Engineering, Delft University of Technology, Kluyverweg 1, 2629 HS Delft, the Netherlands
[2] Department of Mechanical and Aeronautical Engineering, Wallace H. Coulter School of Engineering, Clarkson University, 8 Clarkson Avenue Potsdam, New York, 13699-5725, USA.
[3] Structural Integrity, National Research Council Canada, 1200 Montreal Road, Ottawa, Ontario, K1A 0R6, Canada.

Abstract

This study focuses on understanding the effect of temperature variations and the position of the piezoelectric sensors with respect to fiber orientation angle, as it relates to acoustic emission wave velocity in composite structures. A hybrid panel consisting of Unidirectional Carbon Fiber (UDCF) combined with Unidirectional Glass Fiber (UDGF) along with foam-core sandwich panels with hybrid face-sheets were used for this case study. The group velocities of ultrasonic Lamb waves for three different temperatures (-40 °C, +25 °C and +70 °C) at three different positions (0°, 45° and 90° with respect to the fiber direction) were measured using an Acoustic Emission (AE) system. In order to perform the experiments, a special set-up was built consisting of an environmental chamber and a temperature cycling unit. In parallel, a Finite Element Model (FEM) was developed to simulate the wave propagation of several signals in the three aforementioned temperatures. An analytical algorithm found in the literature was used to verify the experimental results. The main findings of this work were that the Lamb wave group velocities are affected by the fiber orientation more than by temperature variations imposed on the system. In addition, it was found that the signals propagate 15 to 20 % faster in the hybrid laminate than the sandwich panel.

1. INTRODUCTION

Aircraft structures are subjected to high loads, fatigue cycles and extreme temperature variations. In order to accommodate these operating conditions while maintaining aircraft structural airworthiness, three main design philosophies - safe-life, fail-safe and damage-tolerant have been employed [1]. With the damage-tolerant approach, structures are designed to withstand damage up to a certain size while undergoing regular inspections. Determination of the inspection interval is done by setting a minimum detectable crack length and determining the number of load cycles (ΔN) it takes for the crack to grow to a critical length. Using this approach, the inspection interval is then set to $\Delta N/2$ providing enough chances to find the damage and repair it [2]. Most of the structures on an aircraft have to go through this process leading to high maintenance costs and significant downtime. This process is mostly applicable to commercial aircraft and it is based on the regulations imposed by the authorities such as the Federal Aviation Administration (FAA) and the European Aviation Safety Agency (EASA).

Acoustic Emission (AE) based Structural Health Monitoring (SHM) systems have the potential to assess the structural integrity of metallic / composite structures and to detect damage in real-time as it grows [3]. This assessment on the health of the structure is performed by monitoring the propagation characteristics of stress waves that occur due to the release of energy at various Environmental and Operational Conditions (EOC). EOC have a significant effect on the wave propagation phenomenon, potentially influencing the ability of the SHM system to detect and localize damage. Recent work performed at TU Delft, demonstrated the influence of the combined effects of load and temperature on the propagation of ultrasonic Lamb waves in an aluminum plate [4]. The results showed that even though the propagation velocity was affected, the location error of the system only increased by 1 cm in the most extreme case.

This study sought to extend that research to understand the effects of temperature on the propagation of ultrasonic Lamb waves in composite materials using analytical, numerical and experimental tools. The test specimens were a 300 by 300 mm^2, 0.8 mm thick hybrid laminate constructed from Unidirectional Carbon Fiber (UDCF) and Unidirectional Glass Fiber (UDGF) layers. The second specimen consisted on a 600 by 600 mm^2, 3.5 mm thick sandwich structure consisting of hybrid face-sheets and a 2 mm thick core of Rohacell 110A high performance structural foam. The waves were recorded using an AMSY-6 AE system from Vallen Systeme and four non-permanent broadband piezoelectric transducers. The experimental setup also consisted of a temperature chamber used to expose the test specimens to three temperature steps of: -40 °C, 25 °C and 70 °C.

An analytical MATLAB code was used to generate the dispersion curves for the hybrid material at the previously mentioned temperature steps. Next, the experimental setups were modeled in an FEM ABAQUS CAE environment. A 5-cycle Hanning window, with frequencies of (150, 300 and 400) kHz were used to excite a piezoelectric actuator. The response of the generated signals were then detected using three piezoelectric sensors located at angles of 0°, 45° and 90° with respect to the actuator. Finally, the experimental tests were performed and the results were compared with those of the FEM model and the dispersion curves.

2. MANUFACTURING PROCESS & MATERIAL CHARACTERIZATION

2.1 Manufacturing process

Three different types of panels were manufactured. The first type was UDCF and UDGF, the second type was a hybrid laminate with the following lay-up [$\pm45°_{GF}$, $0°_{CF}$, $\pm45°_{GF}$] while the third test specimen consisted of sandwich panel consisting of two face-sheets with the aforementioned hybrid lay-up and a 2 mm foam-core. The panels were produced using the following materials:

- UDO® CST 300/300 UDCF produced by SGL group [5]
- HexForce® 7581 glass fibre weave from Hexcel [6]
- Hexion® Epikote™ 04908 epoxy resin and Epicure™ 04908 curing agent [7]
- Rohacell 110A closed cell foam. [8]

The panels used in this work were based on the preliminary design of a solar powered Unmanned Aerial Vehicle (UAV) made by students at TU Delft. The UAV features front and back wings made of composite materials with the following layup: [$(\pm45^o)_{GF}$ / 90^o_{CF} / $(\pm45^o)_{GF}$] face-sheets and a 20 mm thick foam-core. In the beginning, it was intended to perform the tests on this structure, however, due to limitations regarding the frequency range of the piezoelectric transducers that were available, it was decided to use a 2 mm foam thickness instead.

Vacuum infusion process was used to manufacturing both the hybrid and the sandwich composite panels. The epoxy resin and the curing agent were selected due to the long pot life of the mixture (300 minutes). The long pot life was important especially for the infusion of the UDCF laminates which had an average infusion time of 2 hours. An infused laminate ready for the post-curing process can be seen in Figure 1.

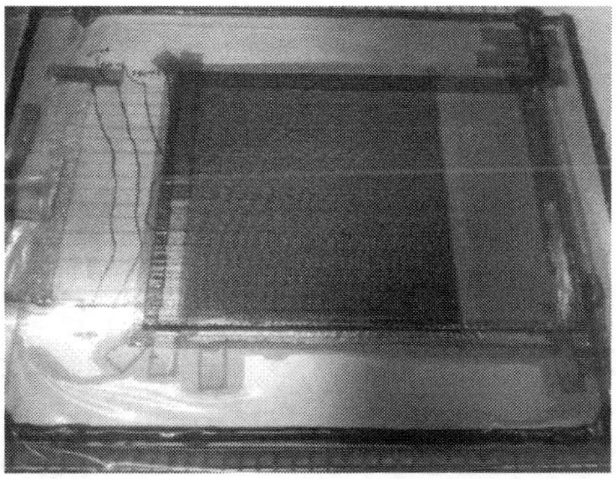

Figure1. Infusion process of the UDCF laminate.

2.2 Material Characterization

To determine E_1, E_2, G_{12} and ν_{12}, following test samples were manufactured: $[0_5]_{CF}$, $[90_7]_{CF}$, $[(0/90)_9]_{GF}$, and $[(\pm 45)_8]_{GF}$. All laminates were 300 mm by 300 mm. The number of layers was selected in order to have a laminate thickness consistent with the ASTM standard used [9].

The panels were cut into small rectangular pieces, whose dimensions were provided by the ASTM standard. The samples were weighed and measured and the relevant information was catalogued. Paper tabs were applied to the edges of the samples using fast curing acrylic glue in order to provide a good load introduction mechanism. Finally, the samples were instrumented with strain-gauges.

The tests were performed using a Zwick 250 kN tensile and compression bench. The wires of the strain-gauges were connected to a four-channel Peekel Strain machine that was used to amplify the signal from the strain-gauges. The Peekel Strain machine was then connected to a computer equipped with software capable of controlling the Zwick machine and monitoring the results of the measurements. The samples were then placed in the hydraulic grips of the testing machine. The grips were equipped with a mechanism that facilitates easy alignment of the sample. Next, the load was applied and the strains of the sample were recorded. The tests were stopped once the required strain levels to obtain the elastic properties were achieved. Figure 2 depicts the experimental set-up. Table 1 summarises the measured elastic properties of the UDCF and UDGF.

Figure 2. CF sample instrumented with strain-gauge and the Zwick 250 kN bench.

Table 1. The elastic properties of the UDCF and UDGF

Material	E_1 [GPa]	E_2 [GPa]	G_{12} [GPa]	G_{13} [GPa]	G_{23} [GPa]	v_{12}
UDCF	119.05	8.342	4.5	2.24	2.51	0.34
UDGF	24	23	3.52	2.24	2.51	0.14

3. NUMERICAL ANALYSIS AND EXPERIMENTAL SET-UP

As previously mentioned, the work presented in this study was performed using a mix of analytical modelling in MATLAB, FEA simulations in ABAQUS/CAE and experiments performed in the laboratory. The analytical model is not presented in this paper but the readers can refer to Pant et al. [10]. This section describes the setup of the numerical simulations and the experiments.

3.1 Finite element modelling

Two models were constructed using the Abaqus/CAE environment:

- A full model of the hybrid plate
- A quarter model for the sandwich panel. The sandwich panel was modelled using symmetry boundary conditions due to the large amount of computational resources needed to run a full sandwich model.

Figure 3 presents the hybrid and the sandwich panels together with the array of sensors. The plies of the laminates are orientated with respect to the x-direction for both structures.

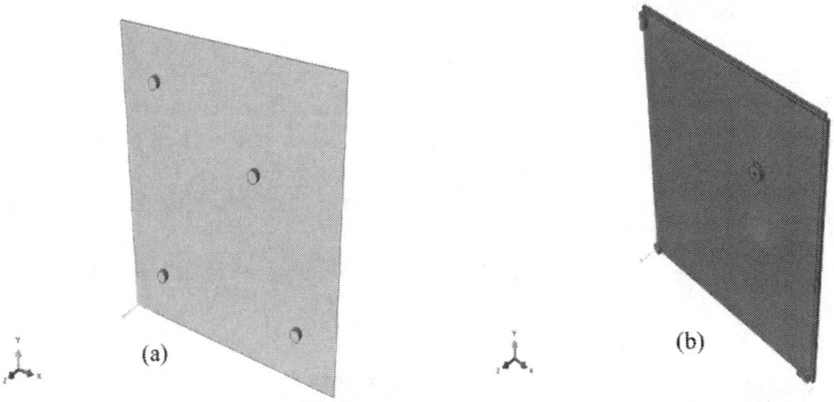

Figure 3. (a) Hybrid and (b) sandwich FEA models.

All of the parts were meshed using rectangular elements of different types depending on their geometry and material properties. The hybrid laminate and the face-sheets were modelled using the 4 node S4R shell elements with reduced integration (5 integration points) and hourglass control. The foam was modelled using the 8 node C3D8R solid elements with reduced integration and hourglass control. Finally, the piezoelectric transducers were modelled using the 8 node C3D8E elements and the infinite edges using CIN3D8 elements.

Two types of sensors were used in this work, both manufactured by Vallen Systeme. They were modelled in Abaqus using only the piezoelectric part with the following dimensions:

- VS 150-M sensors had a cylindrical shape with a diameter of 12.7 mm and a height of 6.35 mm
- VS 900-M sensors had a ring shape with an outer diameter of 12.7 mm, inner diameter of 2.6 mm and a height of 3.8mm.

A meshed model of a VS 150-M sensor can be seen in Figure 4. The poling direction of the material was orientated along the z axis shown in the figure above (Figure 3).

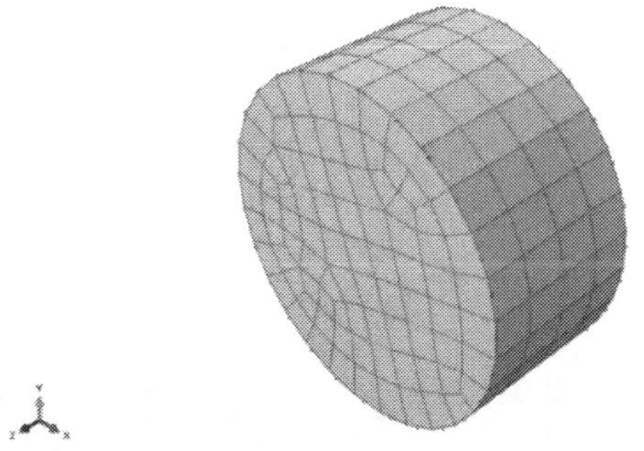

Figure 4. Meshed model of a 150-M piezoelectric element.

Several boundary conditions were necessary in order to model the wave propagation. Most of these are common between the hybrid and the sandwich model. Simply supported conditions were applied to the two edges on which the specimens were standing on. Furthermore, in order to measure or apply a voltage difference to the piezoelectric elements, a zero electric potential boundary condition was applied on the bottom surface of each sensor. The sensors were connected to the upper surface of the plates using the Tie constraint within Abaqus, which assumes perfect contact between the two surfaces.

In order to ensure the stability of the analysis, certain restrictions had to be applied to the sizes of the elements and the time step of the analysis. Chen et al. stated that 10 nodes per wavelength were sufficient to achieve a good spatial resolution and also found that a minimum of 20 points per cycle at the highest frequency must be used to ensure time stability [11].

Using this information, the maximum sizes of the elements were calculated for each of the frequencies and the results and the final values used in the FE models are shown in Table 2.

Table 2. The element sizes and the time steps for the three different signals

Frequency [kHz]	Max element size calculated [mm]	Used element size [mm]	Max time step calculated [μs]	Used time step [μs]
150	1.6	1	0.33	0.1
300	0.7	0.6	0.16	0.1
400	0.6	0.6	0.125	0.1

3.2 Experimental set-up

The AMSY-6 from Vallen Systeme was used to generate and gather Lamb waves. The system is a multi-channel AE measurement unit that consists of 8 parallel measurement channels and front-end software for post-processing. The AMSY-6 is capable of working with different sensors, designed for a wide range of applications. In the case of the work presented here, it was important to use sensors that could operate within the previously mentioned temperature range of -40 °C to 70 °C. Thus, the VS-150 M and VS-900 M sensors were selected. Both sensors have an operating temperature between -50 °C to 100 °C.

The environmental system consisted of a chamber, constructed out of URSA XPS, extruded polystyrene and a temperature cycling unit manufactured by CTS. The unit was connected to the chamber using two attachable tubes to ensure good airflow. The temperature on the plates was allowed to stabilize once the desired temperature had been reached. In order to minimize the noise detected by the sensor the environmental chamber was powered down.

Before bonding, the sensor positions were drawn as accurately as possible in order to ensure that the distance between them is close to 20 cm (standard distance used in these tests). Four transducers were used - one was used to actuate the signal, whereas other three were used to measure the propagated signal (one for each of the 0, 45 and 90° directions). Thermocouples were attached using two layers of tape due to the fact that electrically conductive materials interfere with the measurements. Figure 5 presents the experimental set-up, the environmental chamber and the hybrid plate with the attached sensors.

Figure 5. (a) Overall experimental set-up, (b) the environmental chamber and (c) the hybrid plate with the sensors attached.

4. RESULTS

4.1 Hybrid composite structure

In this section the results of the tests performed on the hybrid laminate are presented. The wave speeds from the experiment, FEA and analytical methods are plotted on the same graph for the 150 kHz signal at the three directions of interest. The results for the 300 kHz and 400 kHz signals have a similar trend as the 150 kHz signal and therefore they are not documented in this study.

Figure 6 shows the influence of temperature on the group velocity in the range of -40 °C to 70 °C in the 0°, 45° and 90° direction of the hybrid laminate.

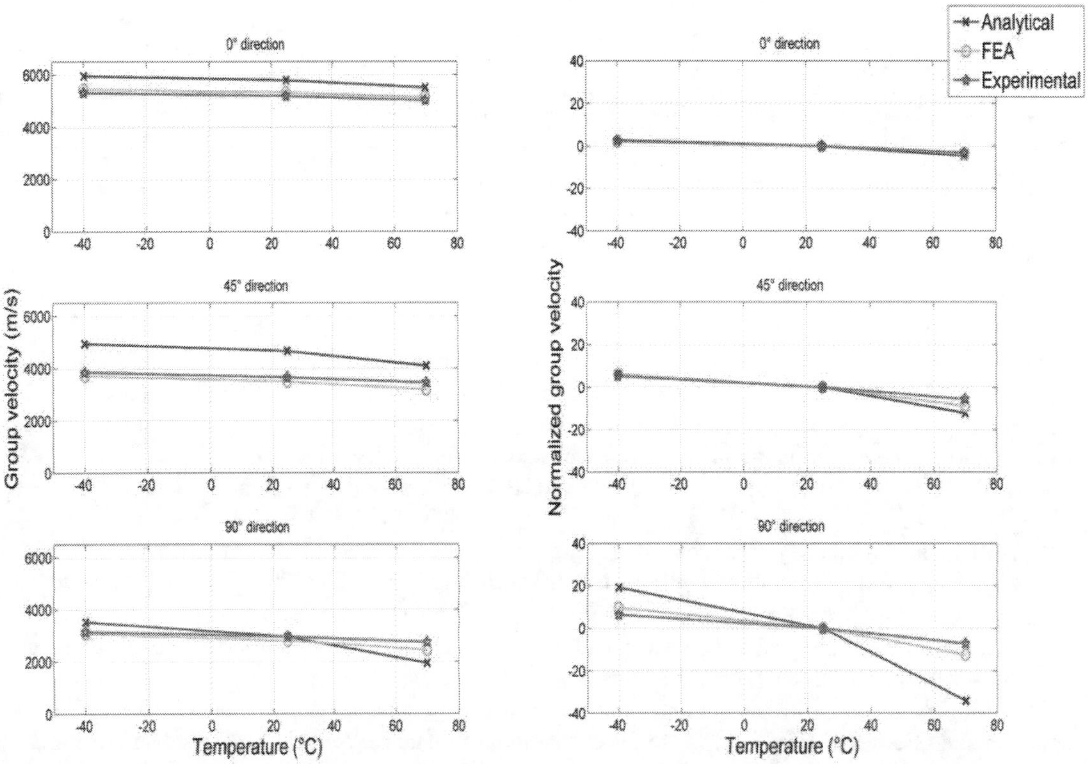

Figure 6. Group velocities for the range of -40 °C to +70 °C for the three main fibre directions.

The first aspect that can be observed in Figure 6 is that there is a constant difference of around 10% between the results of the analytical method and the results of the experiments. This difference can be attributed to some of the assumptions made when developing the analytical method such as: anisotropic layers behaving as linear elastic material perfectly bonded inter-laminar interface so that the displacements and stresses between the layers are continuous. Also the group velocity was calculated without taking into account the slowness curve of the phase velocity. The hybrid laminate used in these tests is highly anisotropic, as there is a significant difference between the stiffness of the 0°, 45° and 90° directions.

The results of the FEA were much closer to those of the experiments with a difference of around 3% at 25 °C. This difference is well within the expected error of the measurements due to uncertainties such as: material properties and accuracy of the sensor positions.

The second observation that can be made is that the group velocity was linearly dependent on the temperature of the medium in which the waves were propagating. Figure 7 illustrates an example of how this effect can be observed. The plot shows the transient response of the piezoelectric sensor corresponding to the 0° direction at temperatures of 25 °C (blue) and 70 °C (red) as predicted by the FEM. The difference between the time-of-arrival of the two signals is around 1.4 μs.

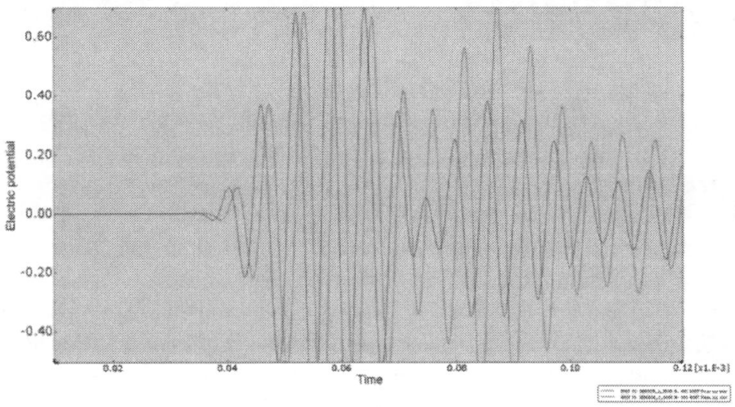

Figure 7. Hybrid laminate FEM waveforms at 25°C (blue) and 70°C (red)

The results indicate that both modelling techniques are slightly more accurate at lower temperatures, such as for example, -40 °C. This can be observed especially in the graphs for the 45° and 90° directions. The fact that this effect is present in both models suggests that the predicted values of the material properties for high temperatures are not as accurate as for the lower temperatures.

In order to put into perspective the results of the temperature dependency tests, the variation of the group velocity with respect to the material orientation is shown in Table 3. Only the results for the 25°C case are shown.

Table 3. Measured wave group velocities @ 25 °C.

	Speed [m/s]		
Frequency [kHz]	0°	45°	90°
150	5184	3657	2963
300	5277	3673	3084
400	5238	3629	3062

The experimental results show that there is a 42% drop in speed in the 45° direction and a 75% drop in speed in the 90° direction compared to the 0° direction of the laminate. This difference can be observed in Figure 8, which shows the in plane displacements generated by a 150 kHz signal in the hybrid panel which correspond to the S_0 mode. The elliptical shape of the wavefront highlights the difference in speed between the different directions. Similarly, the A_0 mode can be visualized by looking at the out-of-plane displacements, see Figure 8. The wave front has an almost circular shape, which means that the group velocity of this mode remains fairly constant at different orientations. This is consistent with results found in literature, such as the ones from Arredondo & Fritzen [12]. It should be noted that the variations due to the position of the sensors are significantly larger than the ones due to temperature changes.

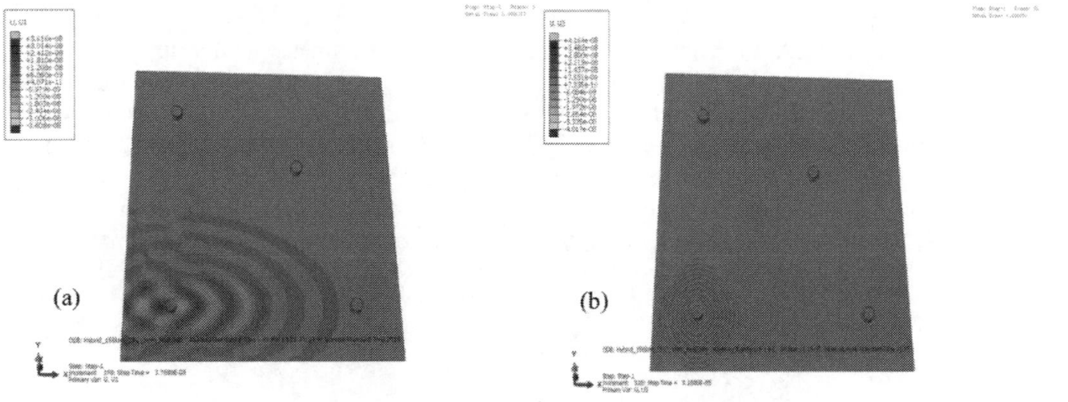

Figure 8. (a) A_o and (b) S_o Lamb wave modes.

4.1 Sandwich structure

In this section the results of the tests, performed on the sandwich panel are presented which generally follow the same trend that was observed in the laminate - the velocity decreases with increasing temperature. It should be noted that the experimental data are compared only with the FEM results. Figure 9 presents the group velocities with respect to the temperature for the three different orientations of the sensors.

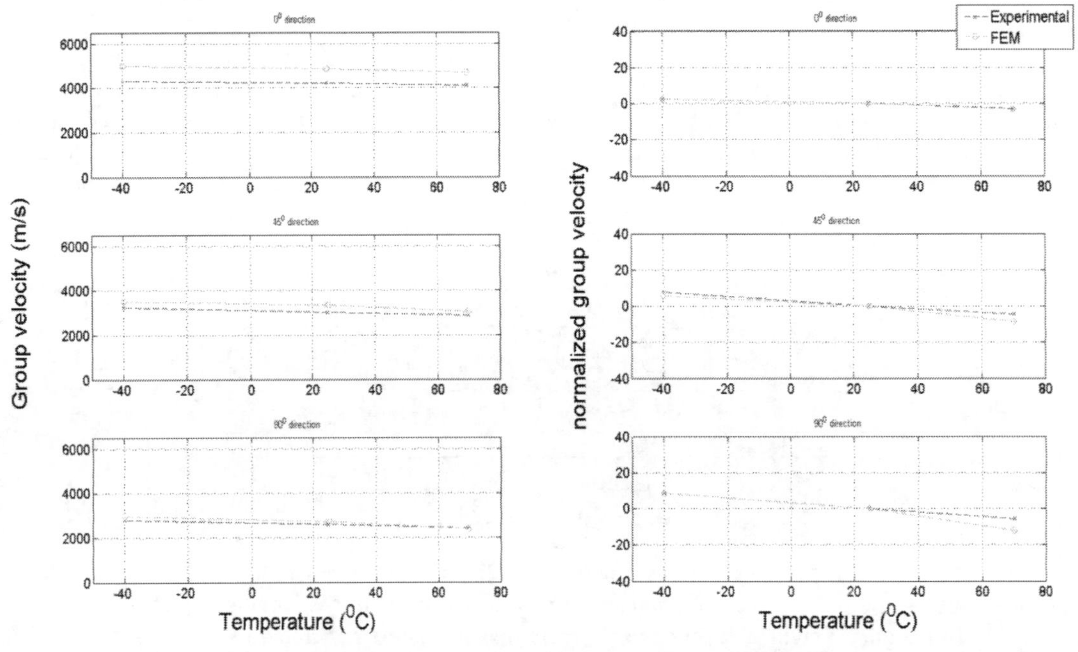

Figure 9. Group velocities with respect to the temperature for the three main fibre orientations.

The main observation form Figure 9 is that there is 15 to 20% reduction in the group velocity due to the presence of the core. This was expected considering the lower mechanical properties of the core. Moreover, a systematic error of around 15% can be observed which can be attributed to the fact that the foam-core was modelled as a homogeneous isotropic material. In reality, the foam has a closed cell structure and exhibits viscoelastic effects that may affect the wave propagation phenomena.

Furthermore, the variation in speed is once again higher in the 45° and 90° directions due to the larger influence of the resin. The difference between the experimental and FE results is also higher for the 70 °C case most likely due to the same factors mentioned for the hybrid laminate.

The directional dependence of the velocity for the sandwich panel is similar to the case of the hybrid laminate. The experimental wave speeds for the 25 °C case are shown in Table 4. The group velocity is highest in the direction of the CF layer, while in the 45° and 90° directions the speed is slower by 39% and 63% respectively.

Table 4. Measured wave group velocities at 25 °C

	Speed [m/s]		
Frequency [kHz]	0°	45°	90°
150	4214	3022	2580
300	4256	3016	2578
400	4267	2992	2576

Figure 10 presents the wavefronts of the S_o and A_o. A full model of the sandwich panel was built to understand the computational load requirement. Although the results of the full model were not used in the comparison, the simulation was helpful to demonstrate the wavefront of the signal.

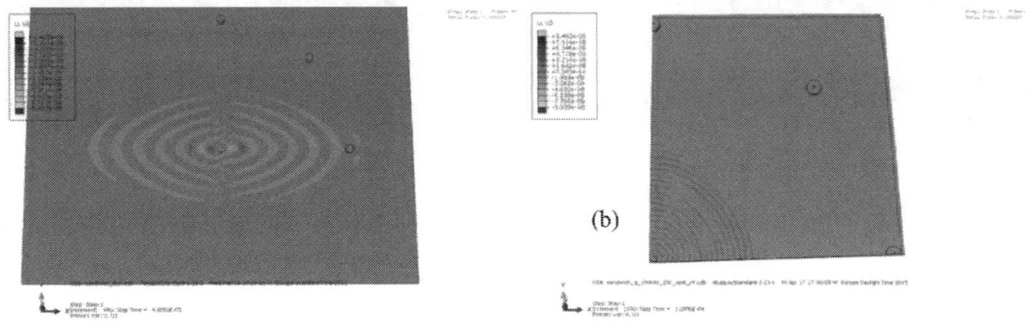
(a) (b)

Figure 10. (a) A_o (b) S_o modes of the waveform

5. CONCLUSIONS

Based on the results the following conclusions can be drawn:

- It is possible to develop a thorough understanding of the physics governing the propagation of Lamb waves in anisotropic materials influenced by the effects of temperature either experimentally or numerically.
- Analytical and FEM simulations of Lamb waves propagating in composite materials are able to predict with a good degree of accuracy the changes in wave speed for directions in which fibres are present. Considering that composite materials in most real applications have quasi-isotropic layups, this suggests that it is possible to model the thermal effects for most composite structures.
- The variations in wave velocities due to the position of the sensors with respect to the fibre orientation are significantly higher than the ones due to temperature changes.

REFERENCES

1. Balageas D., "Structural Health Monitoring", *Willey,*2010.
2. Boller C., Meyendorf N., "State-of-the-art in Structural Health Monitoring for Aeronautics", *International Symposium on NDT in Aerospace*, Furth/Bavaria, 2008.
3. Moore P.O. "Nondestructive testing handbook: Acoustic Emission testing", *American Society for Nondestructive testing*, Volume 6, third edition, 2005.
4. Boon M.J.G.N., Zarouchas D.S., Martinez M., Gagar D., Benedictus R., Foote P., "Temperature and load effects on acoustic emission signals for structural health monitoring applications", *7th European Workshop of Structural Health Monitoring*, Nantes, France, 2014.
5. http://www.sglgroup.com/cms/international/home/index.html?__locale=en.
6. http://hexply.com/hexforce/database/web/front/main/
7. http://www.swiss-composite.ch/pdf/t-Hexion-Harz-EPR04908.pdf
8. http://www.rohacell.com/sites/lists/PP-HP/Documents/ROHACELL-A-mechanical-properties-EN.pdf
9. ASTM International, Standard Test Method for Tensile Properties of Polymer Matrix Composite Materials, ASTM International, 2008.
10. Pant S., Laliberte J. Martinez M., Rocha B. "Derivation and experimental validation of Lamb wave equations for an n-layered anisotropic composite laminate", *Composite Structures,* vol.111, 2014, pp.566-579.
11. Chen L., Dong Y., Meng Q., Liang W. "FEM Simulation for Lamb wave evaluate the defects of plate", in *2012 International workshop on MMWCST*, Chengdu, 2012.
12. Arrendondo M.T., Fritzen, C.P. "Ultrasonic Guided Wave Dispersive Characteristics in Composite Structures Under Variable Temperature and Operational Conditions", *6th European Workshop of Structural Health Monitoring,* Dresden, Germany, 2012.

ICAST2015: 26th International Conference on Adaptive Structures and Technologies
October 14-16th, 2015, Kobe, Japan

ICAST2015 050

Evaluation of Viscoelastic Parameters of Active Structural Elements Using Shape Memory Polymers with a Painted Heater

Atsuhiko Senba[1*], Makoto Ii[2], and Masahiro Arai[2]

[1] Department of Vehicle and Mechanical Engineering, Meijo University, Nagoya, Japan
[2] Department of Aerospace Engineering, Nagoya University, Nagoya, Japan

Abstract

This study proposes a method for heating a shape memory polymer (SMP) specimen above its glass transition temperature by using a heater painted on its surface. To use such a heater in space applications, its effects on the shape recovery and shape fixation characteristics must be studied if the SMP is to be applied to deployable elements for realizing lightweight space structures such as solar sails. This study evaluates the viscoelastic characteristics of active structural elements using SMPs with a painted heater. A simple viscoelastic model is used to compare viscoelastic parameters such as the stress-relaxation ratio for an SMP with and without a heater layer that is painted on the surface of the SMP specimen. Tensile and stress relaxation tests were performed at five different temperatures to study the temperature dependence of the viscoelastic behavior. As a result, we found that the painted heater led to a clear change in the viscoelastic parameters. Furthermore, the viscoelastic parameters defined by a simple model were identified for the specimen with and without the heater layer. A comparison of the identified parameters indicated that the painted heater layer increased the stiffness and decreased the stress relaxation ratio in a specific temperature range, and the opposite relationship was obtained in other temperature ranges.

1. INTRODUCTION

Shape memory materials such as shape memory alloys and polymers have attracted considerable attention in space applications because they could be applied as active structural elements in future space structures.[1] This study focuses on temperature-induced shape memory polymers (SMPs) that have the ability to recover from a deformed state to their original state with a change in temperature.[2] When their temperature exceeds the glass transition temperature (T_g), the elastic modulus reduces drastically, and therefore, they can be easily deformed by external loads. The material deformation is then fixed, and the elastic modulus recovers when the temperature decreases below T_g. Thus, SMPs' shape recovery and fixation abilities are considered suitable for deployable space structures that can assume a stored configuration before launch and deploy in space.

Although many applications use SMPs as active structural elements in deployable space structures, few studies have focused on feasible heating devices for the SMP. Senba et al. investigated the use of a nichrome wire that is fixed on the SMP specimen. [9] However, such a wire cannot uniformly heat the surface of the SMP specimen. A commercial paint-type heater (©Carbo e-Therm) that can be used to

* senba@meijo-u.ac.jp

easily heat any surface of an SMP structural element has also been developed. Therefore, this study investigates SMP specimens coated with a paint heater on their surface. In particular, elastic characteristics including the viscoelastic parameters are evaluated by conducting tensile and stress relaxation tests.

This study aims to evaluate the viscoelastic parameters of SMP specimens with a painted heater. Toward this end, first, we conduct tensile and stress relaxation tests of SMP specimens with and without a painted heater layer. Then, the viscoelastic parameters for a linear standard model [3] are identified using the experimental data. Finally, the differences between the viscoelastic parameters in each identification result are compared, and the effect of the painted heater layer is discussed.

2. SHAPE MEMORY POLYMER WITH PAINTED HEATER

This section explains how to prepare the SMP specimen with a painted heater.

First, we obtain a thin SMP sheet with $T_g = 55\,^\circ C$ and thickness of approximately 1 mm. We cut the sheet to obtain a specimen for the tensile test. Fig. 1 shows the specimen size. Note that the effective length of the specimen is 60 mm, and each end of the specimen has 30 mm for fixing the end in the tensile tests. As shown in Fig. 2, the size of another specimen with the painted heater is the same; however, only the specimen thickness is slightly larger because of the approximately 0.1-mm thickness of the heater layer.

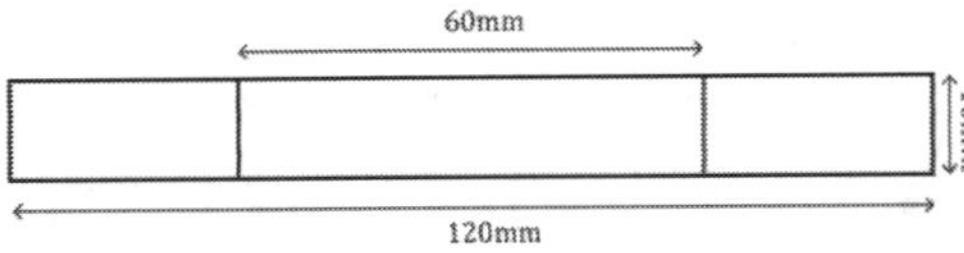

Figure 1: SMP specimen

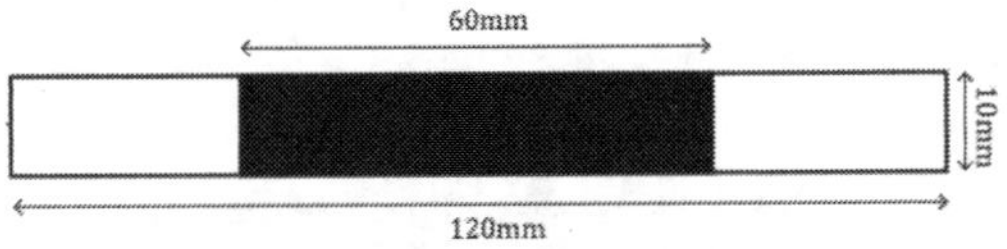

Figure 2: SMP specimen with painted heater

3. MODELING USING VISCOELASTIC MODEL

This section briefly explains the linear viscoelastic model and how to identify the model parameters using the data obtained from the tensile and stress-relaxation tests.

For simplicity, a linear model with two springs and a dash pot, as shown in Fig. 3, is used in this study; it should be noted that more advanced models have been proposed in various studies.[4-7] This model is used to evaluate the change in viscoelastic behavior after a heater is painted on the surface of the SMP specimen, as shown in Figs. 1 and 2.

3.1 Viscoelastic Model

The stress-strain relationship of the viscoelastic model used in this study is expressed as follows.

$$\dot{\varepsilon} = \frac{\dot{\sigma}}{E_1 + E_2} + \frac{E_1}{(E_1 + E_2)\mu}\sigma - \frac{E_1 E_2}{(E_1 + E_2)\mu}\varepsilon \tag{1}$$

By identifying the viscoelastic parameters in Eq. 1, we evaluate the effect of the paint heater layer on the surface of the SMP specimen.

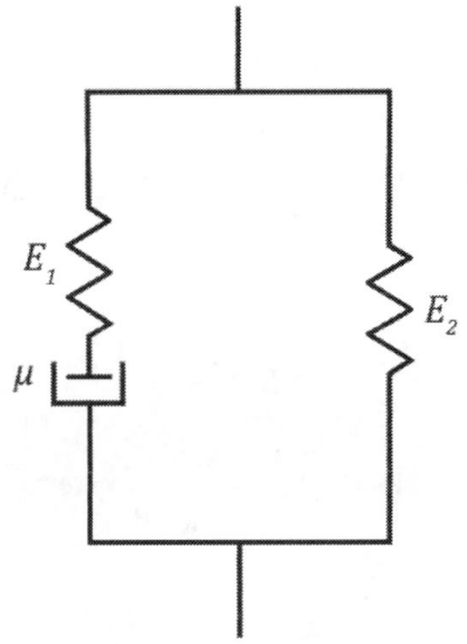

Figure 3: Schematic of a linear standard model

3.2 Identification of Viscoelastic Parameters

As described in the next section, we conduct tensile and stress relaxation tests for the SMP specimens. Then, test data are substituted into the following equation to estimate the viscoelastic parameters, E_1, E_2, and μ, in Eq. (1). With multiple sets of stress and strain data for the SMP specimens, multiple equations like Eq. (1) can be obtained. Then, these equations can be expressed in a matrix form as follows:

$$\begin{Bmatrix} \dot{\varepsilon}_1 \\ \dot{\varepsilon}_2 \\ \vdots \\ \dot{\varepsilon}_N \end{Bmatrix} = \begin{bmatrix} \dot{\sigma}_1 & \sigma_1 & \varepsilon_1 \\ \dot{\sigma}_2 & \sigma_2 & \varepsilon_2 \\ \vdots & \vdots & \vdots \\ \dot{\sigma}_N & \sigma_N & \varepsilon_N \end{bmatrix} \begin{Bmatrix} \frac{1}{E_1+E_2} \\ \frac{E_1}{(E_1+E_2)\mu} \\ \frac{E_1 E_2}{(E_1+E_2)\mu} \end{Bmatrix} \tag{2}$$

Then, a matrix $\mathbf{B}$ is defined as

$$\mathbf{B} = \begin{bmatrix} \dot{\sigma}_1 & \sigma_1 & \varepsilon_1 \\ \dot{\sigma}_2 & \sigma_2 & \varepsilon_2 \\ \vdots & \vdots & \vdots \\ \dot{\sigma}_N & \sigma_N & \varepsilon_N \end{bmatrix} \tag{3}$$

Next, the generalized inverse matrix for the matrix $\mathbf{B}$ is given as

$$\mathbf{B}^{+} = (\mathbf{B}^{\mathbf{T}}\mathbf{B})^{-1}\mathbf{B}^{\mathbf{T}} \tag{4}$$

where $\mathbf{B}^{\mathbf{T}}$ is the transpose of matrix $\mathbf{B}$. Then, a generalized inverse matrix $\mathbf{B}^{+}$ can be calculated to identify the viscoelastic parameters as follows:

$$\mathbf{x} = \mathbf{B}^{+}\mathbf{a} \tag{5}$$

where $\mathbf{a}$ and $\mathbf{x}$ are respectively given as follows:

$$\mathbf{a} = \begin{Bmatrix} \dot{\varepsilon}_1 \\ \dot{\varepsilon}_2 \\ \vdots \\ \dot{\varepsilon}_N \end{Bmatrix} \tag{6}$$

$$\mathbf{x} = \begin{Bmatrix} \frac{1}{E_1+E_2} \\ \frac{E_1}{(E_1+E_2)\mu} \\ \frac{E_1 E_2}{(E_1+E_2)\mu} \end{Bmatrix} \tag{7}$$

Finally, after the vector in the right-hand side in Eq. (7) is obtained, the parameters E_1, E_2, and μ are identified.

4. EXPERIMENT

This section shows the experimental results and identified viscoelastic parameters for the SMP specimen with and without a painted heater.

4.1 Experimental setup

Figure 4 shows the experimental setup for the tensile and stress relaxation tests. By using a thermostatic bath, we measure the stress of the SMP specimen for both loading and constant strain at five different temperatures: 35, 45, 55, 65, and 75 °C.

For the test using the SMP specimen in its rubbery state, tensile loading is applied up to a maximum strain of $\epsilon = 0.2$ at 75 °C, where the strain rate is $1\%/\text{min}$. Then, the maximum strain is kept constant, and we measure the stress for a sufficient duration to confirm stress relaxation. On the other hand, for the test using the SMP specimen in its glassy state, tensile loading is applied up to the maximum strain $\epsilon = 0.017$ at 75 °C, where the strain rate is $0.1\%/\text{min}$). Then, the maximum strain is kept constant, and we measure the stress for a sufficient duration to confirm stress relaxation.

4.2 Experimental Results

This section show the experimental results of the tensile and stress relaxation tests for the SMP specimens. Figures 5 and 6 show the observed stress data for the SMP specimen at two different temperatures, which are the lowest and highest temperatures in this experiment. Table 1 shows the observed maximum

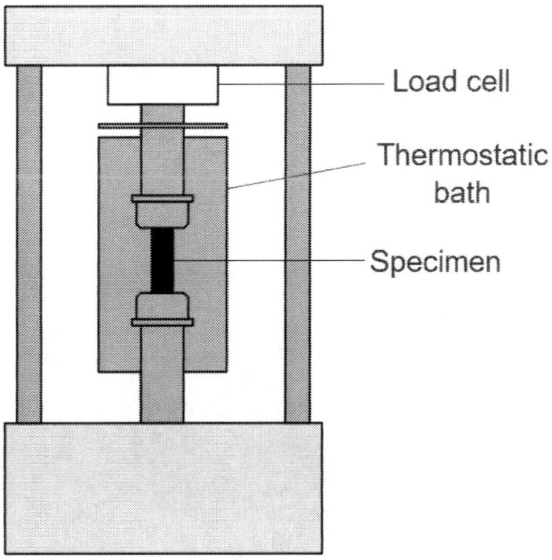

Figure 4: Experimental setup for tensile tests

and minimum stress in the tensile and stress relaxation tests for SMP specimens with and without the painted heater. The stress relaxation ratio, defined as $\sigma_{min}/\sigma_{max}$, is shown in Table 1.

As shown in Figs. 5 and 6, the effect of the painted heater can be captured in the maximum stress observed at each temperature, and the minimum stress after a particular time also showed some differences. These results suggested that the addition of the heater layer afforded higher stiffness to the SMP specimen in both the rubbery and the glassy states. Furthermore, the stress relaxation ratio at each temperature showed that the ratio for the specimen with the painted heater became less than that without the heater at each temperature. The decrease in the stress relaxation ratio was more obvious for the specimen in its rubbery state at 35 °C. Therefore, the paint heater has a different viscoelastic property compared to the SMP specimen. In other words, the experimental results suggest that the heater as well as the heater layer can be used as a material for controlling the stiffness and viscoelastic properties. It is interesting that necessary device for changing the SMP specimen in its glassy state to that in the rubbery state can increase the stiffness of the specimen.

Table 1: Comparison of maximum stress and stress relaxation ratio for SMP specimen with and without painted heater

Temp. [°C]	state	thickness of heater [μm]	max. stress [MPa]	min. stress [MPa]	relaxation ratio [%]
35	glassy	0	2.39	1.87	21.6
35	glassy	82	2.61	2.39	8.5
75	rubbery	0	0.16	0.05	70.5
75	rubbery	82	0.22	0.09	59.7

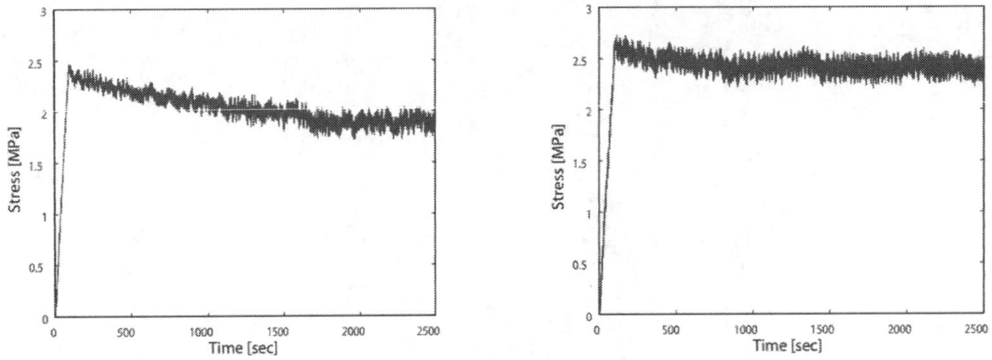

(a) (b)

Figure 5: Stress data for SMP specimen at 35 °C: (a) without painted heater and (b) with painted heater.

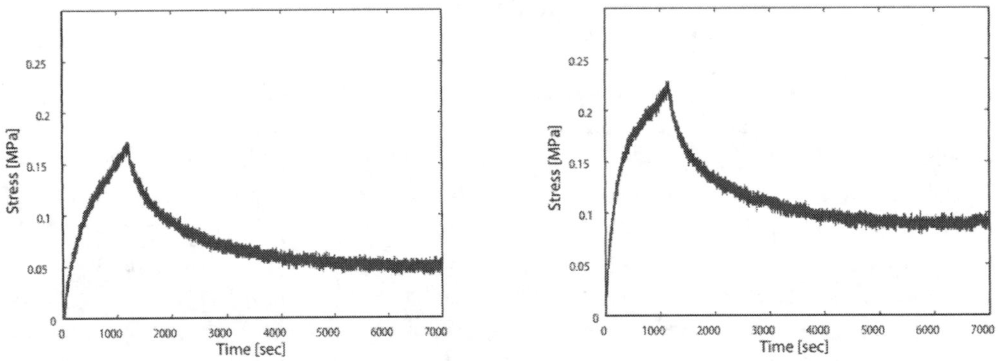

(a) (b)

Figure 6: Stress data for SMP specimen at 75 °C: (a) without painted heater and (b) with painted heater.

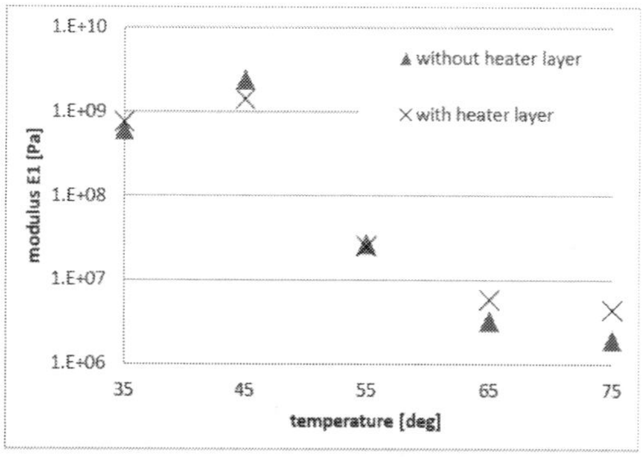

(a)

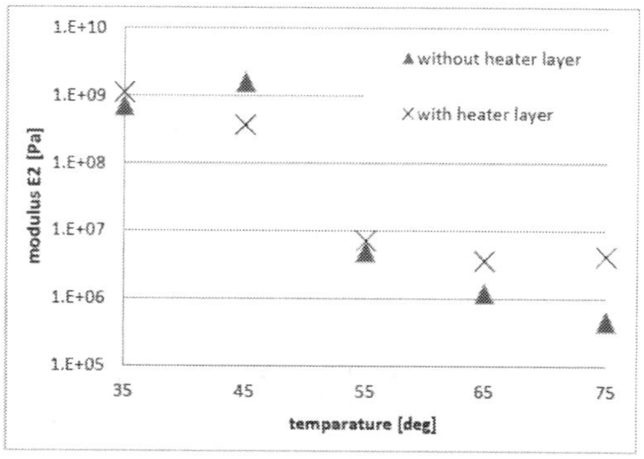

(b)

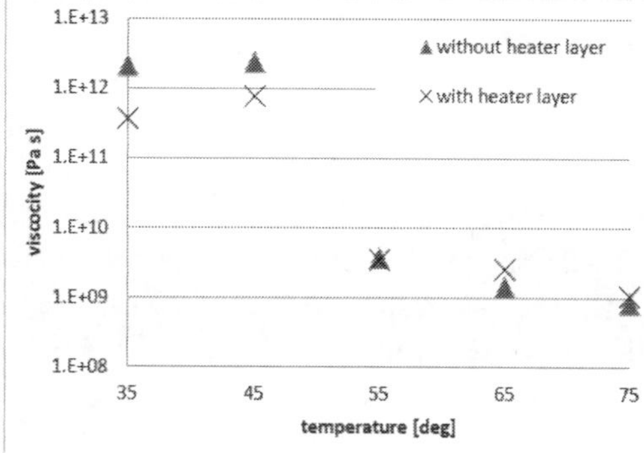

(c)

Figure 7: Identified viscoelastic parameters: (a) E_1, (b) E_2, and (c) μ.

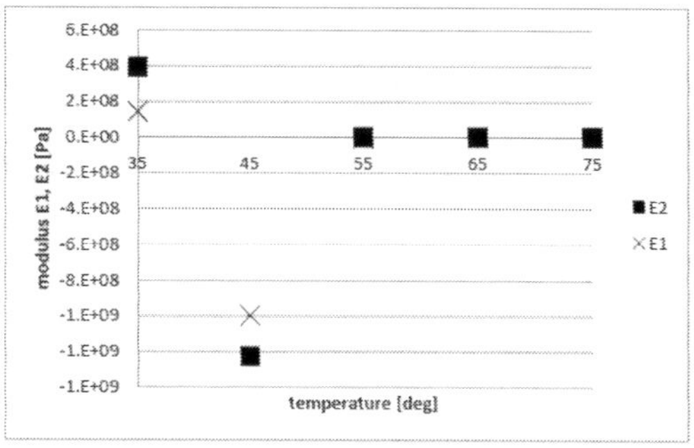

(a)

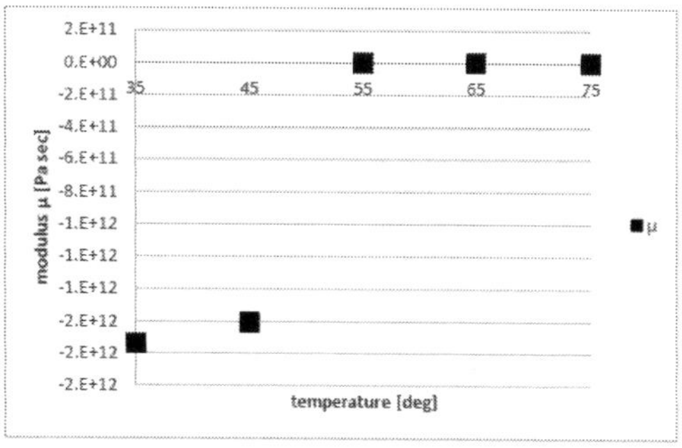

(b)

Figure 8: Difference between identified viscoelastic parameters with and without the paint heater: (a) difference in E_1 and E_2, and (b) difference in μ.

4.3 Discussion

This section discusses the effect of the painted heater layer of the SMP specimens on their identified viscoelastic parameters. Figure 7 shows the identified viscoelastic parameters for the SMP specimen at five different temperatures. Each result was an averaged value of the identified parameters using at least two sets of parameters identified by using multiple test data.

Obviously, the results in Figs. 7 (a) and (b) show that the specimen stiffness increased with the addition of the painted heater at higher temperatures of 65 and 75 °C, which are the higher than $\mathrm{T_g}(=55^\circ\mathrm{C})$ of the SMP specimen. On the contrary, the stiffness at lower temperatures of 35 and 45 °C showed more complicated characteristics. Figures 7 (a) and (b) show that the specimen stiffness with the paint heater at 45 °C was less than that without the paint heater. However, the stiffness at 35 °C showed the opposite result. In addition, Fig. 8 shows the difference in each viscoelastic parameter. As shown in Fig. 8 (c), the identified viscosity μ for the specimen with the paint heater was less than that for the specimen without the paint heater at lower temperatures of 35 and 45 °C.

Therefore, the experimental data at lower temperatures of 35 and 45 °C might contain errors due to signal noise; therefore, the identified parameters may contain a certain amount of error. Additional data must be obtained to further discuss this discrepancy at lower temperatures.

Finally, the accuracy of the viscoelastic model identified in the previous sections was discussed by reconstructing the experimental data observed in the tests. Figures 9 and 10 show the calculated and experimental data for tensile tests. The calculated data were reconstructed based on the solutions given below for the differential equation in Eq. (1) by using the identified viscoelastic parameters. The solution of the tensile loading process is given as

$$\sigma = \alpha\left\{E_2 t + \mu(1 - e^{-\frac{E_1}{\mu}t})\right\} \tag{8}$$

On the other hand, the solution of the stress relaxation process is given as

$$\sigma = \alpha\left\{E_2 t + \mu(1 - e^{-\frac{E_1}{\mu}t_1})\right\} e^{-\frac{E_1}{\mu}t} \tag{9}$$

Thus, each solution can be calculated by using the identified parameters.

As shown in Figs. 9 and 10, the calculated and experimental data were in good agreement at 35 °C, whereas these data did not show good agreement at 75 °C. These results indicated that the simple viscoelastic model used in this study cannot adequately express the viscoelastic behavior of the SMP specimen in its rubbery state. Further, the difference between the calculated and the experimental data were most obvious for the SMP specimen with the painted heater, as shown in Fig. 10 (b).

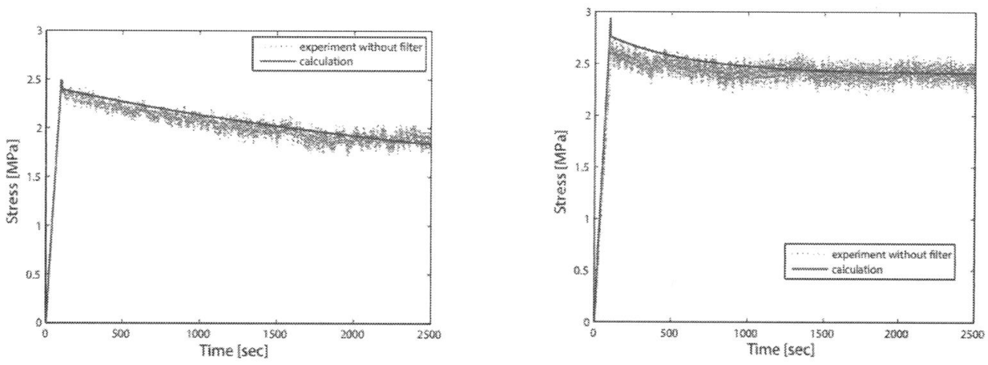

(a) (b)

Figure 9: Comparison between simulated and experimental data for SMP specimen at 35 °C: (a) without painted heater and (b) with painted heater.

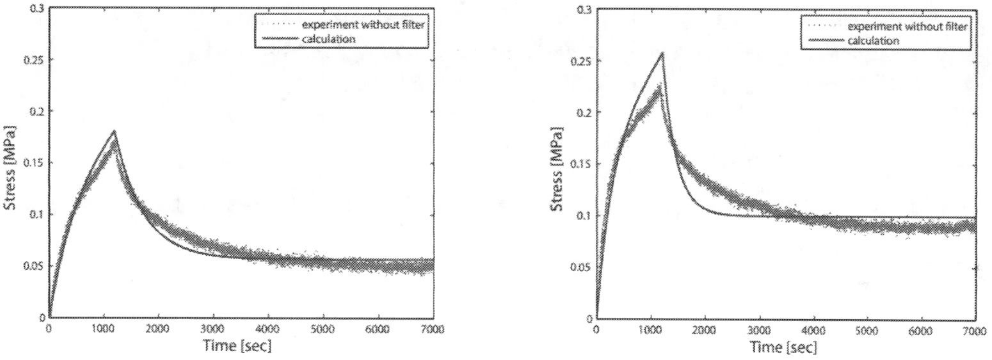

(a) (b)

Figure 10: Comparison between simulated and experimental data for SMP specimen at 75 °C: (a) without painted heater and (b) with painted heater.

5. CONCLUSIONS

This study investigated the viscoelastic characteristics of SMP specimens with a painted heater. A simple viscoelastic model was used to compare viscoelastic parameters such as the elastic modulus and viscosity for SMP specimens with and without a heater layer. As a result, all test data showed that the painted heater led to a clear change in the viscoelastic parameters. In particular, the stress relaxation ratio for the specimen with a heater layer became less than that for a specimen without the heater layer. In addition, the viscoelastic parameters defined by a simple model were identified for the specimen with and without the heater layer. The identified parameters suggested that the elastic modulus of the painted heater layer was greater than that of the SMP layer above the glass transition temperature (T_g) of the

SMP specimen. On the other hand, the viscosity of the specimen with the painted heater layer was lesser than that of the specimen without the painted layer below T_g.

ACKNOWLEDGEMENTS

This study was partially supported by JSPS KAKENHI (Nos. 24760663 and 26249131).

REFERENCES

1. Sokolowski, W., Tanb, S., Willisa, P., and Pryorc, M., "Shape memory self-deployable structures for solar sails", *Smart Materials*, Vol. 7267, 2008, pp. 1-12.
2. Liu, Y., Du, H., Liu, L., and Leng, J., "Shape memory polymers and their composites in aerospace applications", *Smart Materials and Structures*, Vol. 23, No. 1, 2014, pp. 2-16.
3. Lockett, F. J., "Nonlinear Viscoelastic Solids", Academic Press, 1972, pp. 33-36.
4. Tobushi, H., Hashimoto, T., Hayashi, S., and Yamada, E., "Thermomechanical constitutive modeling in shape memory polymer of polyurethane series", *Journal of Intelligent Material Systems and Structures*, Vol. 8, No. 8, pp. 711-718.
5. Kim, J. H., Kang, T. J., and Yu, W. R., "Thermo-mechanical constitutive modeling of shape memory polyurethanes using a phenomenological approach", *International Journal of Plasticity*, Vol. 26, No. 2, 2010, pp. 204-218.
6. Baghani, M., Naghdabadi, R., Arghavani, J., and Sohrabpour, S., "A constitutive model for shape memory polymers with application to torsion of prismatic bars", *Journal of Intelligent Material Systems and Structures*, Vol. 23, No. 2, 2011, pp. 107-116.
7. Baghani, M., Naghdabadi, R., and Arghavani, J., "A large deformation framework for shape memory polymers: Constitutive modeling and finite element implementation", *Journal of Intelligent Material Systems and Structures*, Vol. 24, No. 1, 2012, pp. 21-32.
8. Gu, J., Sun, H., and Fang, C. "A phenomenological constitutive model for shape memory polyurethanes", *Journal of Intelligent Material Systems and Structures*, Vol. 26, No. 5, 2015, pp. 517-526.
9. Senba, A. and Ogi, Y., "Development of Deployable Composite Thin-film Structures Using Shape Memory Polymer", *52nd AIAA/ASME/ASCE/AHS/ASC Structures, Structural Dynamics and Materials Conference 19th AIAA/ASME/AHS Adaptive Structures Conference*, AIAA 2011-2104, pp. 1-7.

ICAST2015: 26th International Conference on Adaptive Structures and Technologies
October 14-16th, 2015, Kobe, Japan

ICAST2015 #018

Efficiency analysis of energy harvesting using an electromagnetic transducer and a pulse width modulated step-up chopper circuit

Toru Ikegame[1*], Kentaro Takagi[1], and Ichiro Jikuya[2]

[1] Department of Mechanical Science and Engineering, Nagoya University, Aichi, Japan.
[2] Department of Aerospace Engineering, Nagoya University, Aichi, Japan.

Abstract

This paper analyzes a circuit for highly efficient energy harvesting using an electromagnetic transducer. Electromagnetic transducer considered in this paper is a voice coil motor, which comprises of a permanent magnet and a coil. Electromagnetic shunt damping is one of the self-sensing damping techniques using an electromagnetic transducer. In this technique the damping effect appears by dissipating thermal energy through the shunt circuit after the transducer converts kinetic energy into electrical energy. If the dissipated energy could be used to charge a battery or a capacitor, it would be possible to achieve both damping and power generation. In this way, a technique charging energy after converting an external energy such as kinetic energy into electrical energy is referred to as energy harvesting. To this end, a step-up converter, which is a kind of DC-DC converter, has been considered in the previous works. This circuit is composed of a switch and an inductor and has a strong nonlinearity because two different governing equations are switched at high speed. Exact analysis of this circuit is very difficult, and approximate analysis is only considered in most of the conventional studies. This paper considers the system that consists of a step-up chopper circuit and an electromagnetic transducer attached to a one-degree-of-freedom vibrating system. Firstly, this paper derives the analytical solution of the current flowing to a battery when the circuit switches at a constant duty ratio of pulse width modulation. Then, this paper derives the optimal duty ratio by maximizing the averaged current flowing to a battery. In order to verify the analytical solution, simulations are carried out using LTspice IV.

1. INTRODUCTION

A voice coil motor which is composed of coil and permanent magnet is one of electromagnetic transducer. Electromagnetic transducer has the following properties: electromotive force is generated by the movement in the magnetic field, and Lorenz force is generated by the current in the magnetic field. These properties are used as sensor and actuator, respectively. Therefore, the electromagnetic transducer is called a self-sensing actuator. One of sensor-less damping technique which is used for used the electromagnetic transducer is electromagnetic shunt damping [1]. This technique can suppress the vibration by attaching a shunt circuit to the coil of the electromagnetic transducer. The transducer converts vibration energy into electric energy, and then the resistance element of the shunt circuit converts electric energy into thermal energy and dissipates thermal energy. If the dissipated thermal energy can be stored as

* t_ikegame@nuem.nagoya-u.ac.jp

electric energy, it is possible to achieve both damping and power generation. In this way, storing kinetic energy as electric energy is called energy harvesting [2] [3].

The generated electromotive force is AC voltage since the electromotive force is generated by the vibration. And, since it is necessary to apply DC voltage to charge a battery, energy harvesting is possible by using full-wave rectifier [3]. But the conversion efficiency is reduced by the effect of the voltage drop in the full-wave rectifier and power storage can not be achieved when the electromotive force is small. For this reason, a step-up chopper circuit composed an inductor and a switch is utilized instead of a full-wave rectifier [2]. This circuit stores the electric energy as the magnetic energy, and charges the magnetic energy in the battery by switching. The advantage of this circuit is that it is possible to charge even if electromotive force is small. In this circuit, ON/OFF ratio of the switch (Duty ratio) is very important since this circuit repeats stopping or charging the battery. The design of the duty ratio is necessary to improve the efficiency since charging efficiency of this circuit depends on the duty ratio.

This paper analyzes the step-up chopper circuit when the circuit and the battery are attached to the electromagnetic transducer. In the circuit analysis, the duty ratio in CCM and the optimal duty ratio are derived. The optimal duty ratio is the duty ratio which maximizes energy amount. In order to verify the analytical solution, simulations are carried out using SPICE for the parameters in the previous literature [2].

2. MODELING

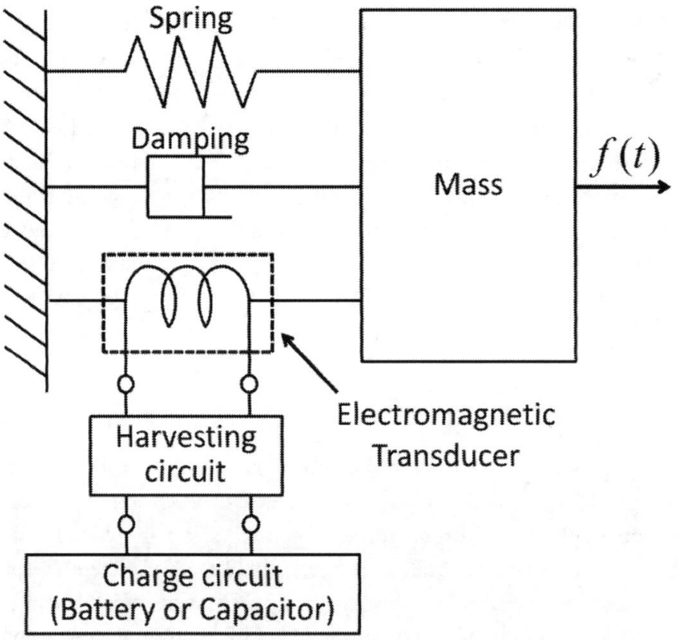

Figure 1: Vibration absorption system

Figure 1 shows a mechanical model attached to the electromagnetic transducer, the harvesting circuit

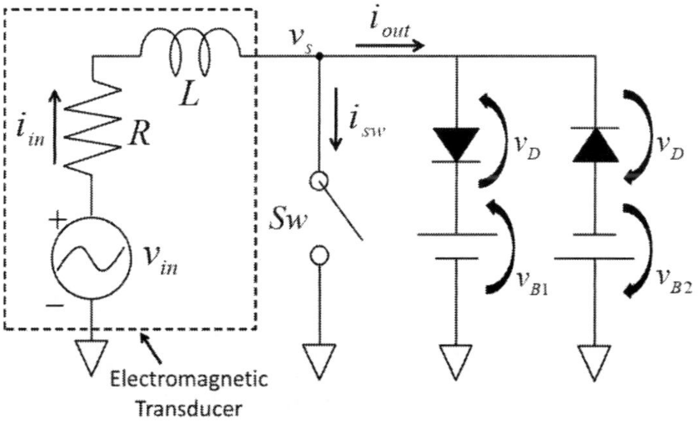

Figure 2: Harvesting circuit and charged circuit

and the charge circuit. In the electromagnetic transducer, electromotive force v_{in}[V] is generated by the movement of the mass. This paper focuses on the electrical model of electromagnetic transducer in Figure 2. v_{B1}, v_{B2}[V] are battery voltages for charging, v_D[V] is a voltage drop of a diode, v_s[V] is a terminal voltage of the electromagnetic transducer. i_{in}[A] is current generated by the electromotive force, i_{out}[A] is current flowing into the battery and the diode, i_{sw}[A] is current flowing into the switch. L[H] and R[Ω] are the self-inductance and the internal resistance of the electromagnetic transducer. We notice that an inductor is installed in the step-up chopper circuit. In this paper, the self-inductance in the electromagnetic transducer is utilized to construct the step-up chopper instead of introducing the other inductor. The circuit equation is given by following equation regardless of the switch state

$$v_{in}(t) - v_s(t) = L\frac{di_{in}(t)}{dt} + Ri_{in}(t). \tag{1}$$

By solving this differential equation, the following equation is derived by

$$i_{in}(t) = \mathrm{e}^{-\frac{R}{L}(t-t_0)} i_{in}(t_0) + \frac{1}{L}\int_{t_0}^{t} \mathrm{e}^{-\frac{R}{L}(t-\tau)}\{v_{in}(\tau) - v_s(\tau)\}d\tau, \tag{2}$$

where t_0[sec] is the time when the switch is turned ON. The switch is operated by pulse width modulation (PWM). Because the PWM frequency is higher than the frequency of the electromotive force, $v_{in}(t)$ can be regarded as quasi-static. The duty ratio, β, is defined by the ratio of ON period, T_{on}, over the whole switching period, T:namely, $\beta = T_{\mathrm{on}}/T$. t_0^n defines time to switch ON in the n-th, t_1^n defines time to switch OFF, t_2^n defines time to switch ON again. t_0^n, t_1^n and t_2^n are given by

$$t_0^n = \frac{n-1}{f_s},\ t_1^n = \frac{n-(1-\beta)}{f_s},\ t_2^n = \frac{n}{f_s}, \tag{3}$$

where $f_s = 1/T$ [Hz]

Because $v_s(t)$ is led to GND in Figure 2 when the switch is ON, $v_s(t) = 0$. Therefore, the current at ON, $i_{in}^{on}(t)$, is derived by

$$i_{in}^{on}(t) = \mathrm{e}^{-\frac{R}{L}(t-t_0^n)} i_{in}^{on}(t_0^n) + \frac{v_{in}}{L}\int_{t_0^n}^{t} \mathrm{e}^{-\frac{R}{L}(t-\tau)}d\tau, \tag{4}$$

$$= \left(i_{in}^{on}(t_0^n) - \frac{v_{in}}{R}\right) e^{-\frac{R}{L}(t-t_0^n)} + \frac{v_{in}}{R}. \tag{5}$$

$v_s(t)$ is expressed by the sum of the battery voltage and the voltage drop of the diode because the current flows to the battery and the diode when the switch is OFF. The current flows to the left side of the diode in Figure 2 if $i_{in}(t)$ is positive, and to the right side of the diode in Figure 2 if $i_{in}(t)$ is negative. In other words, $v_s(t) = v_{B1} + v_D$ if $i_{in}(t)$ is positive, $v_s(t) = -(v_{B2} + v_D)$ if $i_{in}(t)$ is negative. Here, we consider the case where $i_{in}(t)$ is positive and define $v_{BD} := v_{B1} + v_D$. The current at OFF, $i_{in}^{off}(t)$, is derived by

$$i_{in}^{off}(t) = e^{-\frac{R}{L}(t-t_1^n)} i_{in}^{off}(t_1^n) + \frac{v_{in} - v_{BD}}{L} \int_{t_1^n}^{t} e^{-\frac{R}{L}(t-\tau)} d\tau, \tag{6}$$

$$= \left(i_{in}^{off}(t_1^n) - \frac{v_{in} - v_{BD}}{R}\right) e^{-\frac{R}{L}(t-t_1^n)} + \frac{v_{in} - v_{BD}}{R}. \tag{7}$$

In addition, since the current is continuous in the switching, the following equation is considered by

$$i_{in}^{on}(t_1^n) = i_{in}^{off}(t_1^n). \tag{8}$$

3. CIRCUIT ANALYSIS

This section analyzes the duty ratio which is very important in the energy harvesting.

3.1 Duty ratio in CCM

The current $i_{in}(t)$ increases during ON and it decreases during OFF. If the current falls to zero during OFF, it is unsuitable for energy harvesting since large current does not flow. The above state is called DCM (Discontinuous Conduction Mode). Conversely, the current does not fall to zero during OFF, this state is called CCM (Continuous Conduction Mode). In this section, we obtain the condition when the step-up chopper operates in the CCM mode.

If the current falls to zero in the 1-st switching, it also falls to zero in the n-th switching. Therefore, if $i_{in}^{on}(t_0^1 = 0) = 0$, the current at t_1^1 given by follow equation by Equation (5).

$$i_{in}^{on}(t_1^1) = \frac{v_{in}}{R}\left(1 - e^{-\frac{R}{L}t_1^1}\right) \tag{9}$$

t_{12} $(t_1^1 < t_{12} < t_2^1)$ witch satisfies $i_{in}^{off}(t_{12}) = 0$ is derived by

$$i_{in}^{off}(t_{12}) = \left(i_{in}^{on}(t_1^1) - \frac{v_{in} - v_{BD}}{R}\right) e^{-\frac{R}{L}(t_{12}-t_1^1)} + \frac{v_{in} - v_{BD}}{R}, \tag{10}$$

$$= 0, \tag{11}$$

$$t_{12} = -\frac{L}{R} \ln \frac{v_{in} - v_{BD}}{v_{in} e^{-\frac{R}{L}t_1^1} - v_{BD}} + t_1^1. \tag{12}$$

If $t_{12} > t_2^1$, the current does not fall to zero during OFF. Therefore, by using $\gamma = R/(Lf_s)$ and Equation (3), β which satisfies $t_{12} > t_2^1$ derived by

$$\beta > 1 - \frac{1}{\gamma} \ln \frac{v_{BD}}{v_{BD} - v_{in}\left(1 - e^{-\gamma}\right)}. \tag{13}$$

Equation (13) is the condition under which the current is CCM. The following approximate equation is obtained by truncating the Maclaurin series with respect to $v_{in}\,(1-\mathrm{e}^{-\gamma})$.

$$\beta > 1 - \frac{1-\mathrm{e}^{-\gamma}}{\gamma v_{BD}} v_{in} \tag{14}$$

Figure 3 shows the boundary line of Equations (13) and (14). The parameter values used Figure 3 are shown in below Table 1.

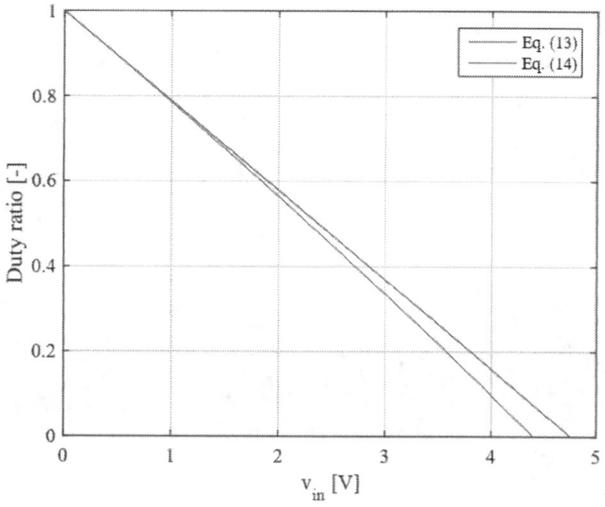

Figure 3: Boundary condition of CCM/DCM

3.2 Optimal duty ratio

In this subsection, we consider the case where $i_{in}(t)$ converges to the steady oscillation state. $i_{in}(t)$ increases while oscillating at frequency f_s, and it becomes steady oscillation state after the time passes. At this time, the initial value at ON and the final value at OFF match and this is expressed by

$$i_{in}^{on}(t_0^n) = i_{in}^{off}(t_2^n). \tag{15}$$

By utilizing Equation (15), $i_{in}^{on}(t_0^n)$ and $i_{in}^{on}(t_1^n)$ are obtained by

$$i_{in}^{on}(t_0^n) = i_{in}^{off}(t_2^n) = \frac{v_{in}}{R} - \frac{v_{BD}}{R}\frac{1-\mathrm{e}^{-\frac{R}{L}(t_2^n - t_1^n)}}{1-\mathrm{e}^{-\frac{R}{L}(t_2^n - t_0^n)}}, \tag{16}$$

$$i_{in}^{on}(t_1^n) = i_{in}^{off}(t_1^n) = \frac{v_{in}}{R} - \frac{v_{BD}}{R}\frac{\mathrm{e}^{-\frac{R}{L}(t_1^n - t_2^n)}-1}{\mathrm{e}^{-\frac{R}{L}(t_0^n - t_2^n)}-1}. \tag{17}$$

By rearranging Equation (16) and (17), the following equations are obtained by

$$i_{in}^{on}(t_0^n) = \frac{v_{in}}{R} - \frac{v_{BD}}{R}\frac{1-\mathrm{e}^{-(1-\beta)\gamma}}{1-\mathrm{e}^{-\gamma}}, \tag{18}$$

$$i_{in}^{on}(t_1^n) = \frac{v_{in}}{R} - \frac{v_{BD}}{R}\frac{\mathrm{e}^{(1-\beta)\gamma}-1}{\mathrm{e}^{\gamma}-1}. \tag{19}$$

The current can flow to the battery when the switch is OFF. By the charge during OFF, Q^{off}[C], the average current of one cycle of the switch during OFF, I^{off}_{mean}[A], is derived by

$$\begin{aligned}
Q^{off} &= \int_{t_1^n}^{t_2^n} i_{in}^{off}(t)dt, && (20)\\
&= \frac{L}{R^2}\left((1-\beta)\,\gamma(v_{in}-v_{BD}) + 2v_{BD}\mathrm{csch}\left(\frac{\gamma}{2}\right)\sinh\left(\frac{\beta\gamma}{2}\right)\sinh\left(\frac{(1-\beta)\,\gamma}{2}\right)\right), && (21)\\
I^{off}_{mean} &= Q^{off} f_s, && (22)\\
&= \frac{1}{R\gamma}\left((1-\beta)\,\gamma(v_{in}-v_{BD}) + 2v_{BD}\mathrm{csch}\left(\frac{\gamma}{2}\right)\sinh\left(\frac{\beta\gamma}{2}\right)\sinh\left(\frac{(1-\beta)\,\gamma}{2}\right)\right). && (23)
\end{aligned}$$

In the same way, the average current of one cycle of the switch during ON, I^{off}_{mean}[A], and the averaged current of one cycle of the switch, I_{mean}[A], are derived by

$$\begin{aligned}
I^{on}_{mean} &= \frac{1}{R\gamma}\left(\beta\gamma v_{in} - v_{BD}\left((1-\cosh(\beta\gamma))\coth\left(\frac{\gamma}{2}\right) + \sinh(\beta\gamma)\right)\right), && (24)\\
I_{mean} &= \frac{1}{R}\left(v_{in} + v_{BD}(\beta - 1)\right). && (25)
\end{aligned}$$

Figure 4 shows Equations (23), (24) and (25) when $v_{in} = 5$.

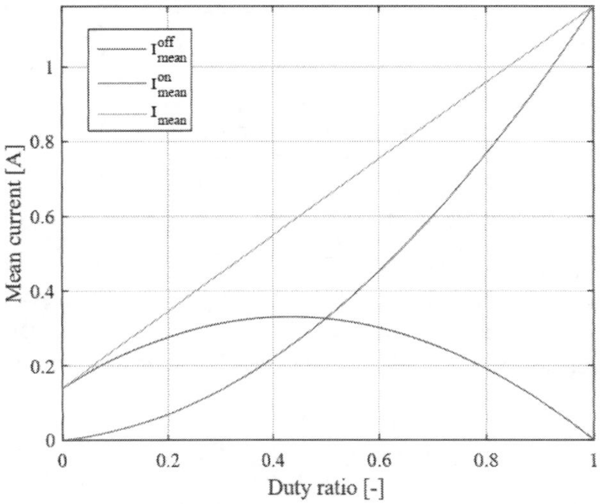

Figure 4: Mean currents ($v_{in} = 5$)

A performance of energy harvesting is evaluated by an electrical power which is obtained by multiplying the current flowing to the battery and the voltage of the battery. Here, the electrical power can be identified with the current by assuming that the battery voltage is constant. Maximizing the electrical power and maximizing the current flowing the battery have the same meaning. By truncating the Maclaurin series with respect to $\beta\gamma$ of I^{off}_{mean}, the following equation is obtained by

$$I^{off}_{mean} = -\frac{1}{R}\left(\frac{\gamma v_{BD}}{2}\coth\left(\frac{\gamma}{2}\right)\beta^2 + (v_{in} - 2v_{BD})\,\beta - (v_{in} - v_{BD})\right). \quad (26)$$

Table 1: Simulation parameters

L[H]	$R[\Omega]$	v_D[V]	v_B[V]	f_s[Hz]
0.02732	4.3	0.8	3.6	1000

The duty ratio maximizing the average current is obtained by the following equation.

$$\beta_{opt} = \frac{2}{\gamma}\tanh\left(\frac{\gamma}{2}\right)\left(1 - \frac{v_{in}}{2v_{BD}}\right) \tag{27}$$

We notice that β_{opt} depends on v_{in} and v_{BD}; in other words, β_{opt} is in the form of feedback control. By assuming that γ is small, $\frac{2}{\gamma}\tanh\left(\frac{\gamma}{2}\right) \approx 1$ and Equation (27) is approximated by

$$\beta_{opt} \approx 1 - \frac{v_{in}}{2v_{BD}}. \tag{28}$$

If $i_{in}(t)$ is negative, in a similar way to $i_{in}(t) > 0$, the optimal duty ratio is obtained by

$$\beta_{opt} = \frac{2}{\gamma}\tanh\left(\frac{\gamma}{2}\right)\left(1 + \frac{v_{in}}{2(v_{B2} + v_D)}\right), \tag{29}$$

$$\approx \left(1 + \frac{v_{in}}{2(v_{B2} + v_D)}\right). \tag{30}$$

4. SIMULATION RESULTS

A circuit simulation is conducted by using the simulation soft of LTspice IV [4]. Figure 5 shows the simulation results of $i_{in}[A]$ when Equation (28) or constant duty ratio (β=0.7.0.8,0.9) is used. Table 1 shows the parameter values used in the simulations. If $v_{in} = 1$ and $v_{BD} = 4.4$ are substituted to Equation (28) and Equation (14), $\beta_{opt} = 0.886$ and $\beta > 0.790$ are obtained. It is confirmed that the average current is large when the duty ratio increases as Equation (25).

Figure 6 shows the simulation results of the current flowing the battery when Equation (28) or constant duty ratio (β=0.7.0.8,0.9) is used. It is important to evaluate the area of the current in Figure 6 in order to evaluate the amount of current flowing to the battery.

The simulation in the case of AC voltage such as $v_{in}(t) = \sin(20\pi t)$ is conducted since the voltage generated by vibration is AC voltage. The terminal voltage of the capacitor is evaluated by replacing the battery with the capacitor of 10μF as shown in Figure 7 in order to visualize the harvesting energy. Figure 8 shows the time history of the terminal voltage on the left side of the capacitor in Figure 7 when Equation (28) or constant duty ratio (β=0.7.0.8,0.9) is applied. In Figure 8, it is confirmed that the large energy can be harvested by Equation (28), compared with constant duty ratio.

Figure 9 shows the time history of i_{in}. It is confirmed that the circuit is switched by the fluctuation near the peak of the current.

Figure 10 shows the time history of β_{opt}. It is confirmed that the variation of the duty ratio is large in the first cycle because the initial voltage of v_{B1} and v_{B2} and small. It is also confirmed that the variation of the optimal duty ratio decreases as the voltage of the capacitor increases. This phenomena is consistent with Equation (28).

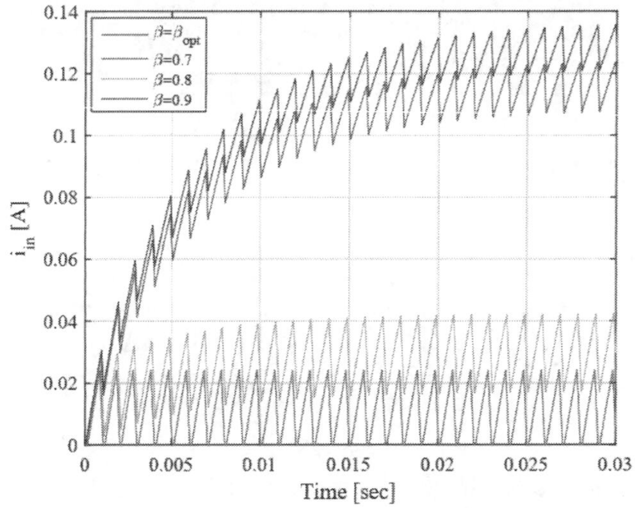

Figure 5: The current generated by electromagnetic transducer

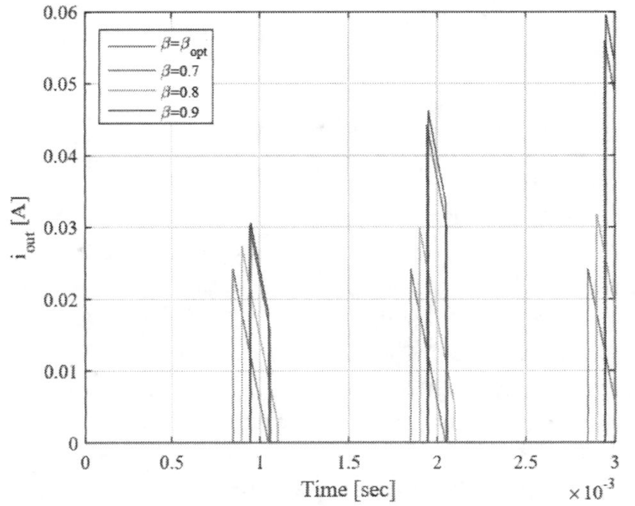

Figure 6: The currents flowing into the battery

5. CONCLUSIONS

The time responses of the currents in the step-up chopper are computed to investigate the relation between the duty ratio and the performance of energy harvesting. Then, the conditional equation that the current is CCM and the duty ratio for maximizing the current flowing to the battery are derived. In order to verify the effectiveness of the optimal duty ratio, the simulation is conducted by using the circuit simulation soft of LTspice IV. And the validity of analytical solution is verified in the case of

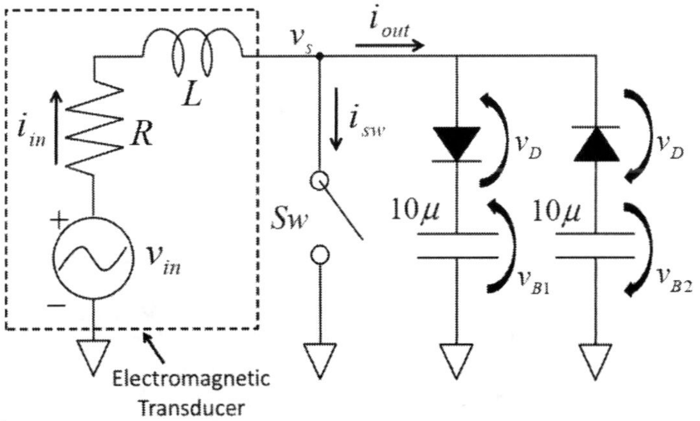

Figure 7: Regenerative circuit, where two capacitors are used to charge positive i_{in} or negative i_{in}

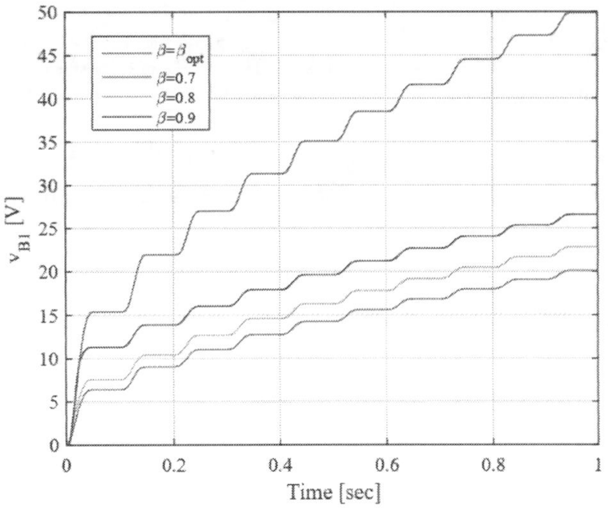

Figure 8: The voltage v_{B1} of the capacitor

DC generated voltage. In addition, the simulations in the case of AC generated voltage are conducted, it is confirmed that the optimal duty ratio contributes to harvest larger energy than the constant duty ratio. And, it is confirmed that the variation of the optimal duty ratio decreases when the voltage of the capacitor increases.

ACKNOWLEDGEMENTS

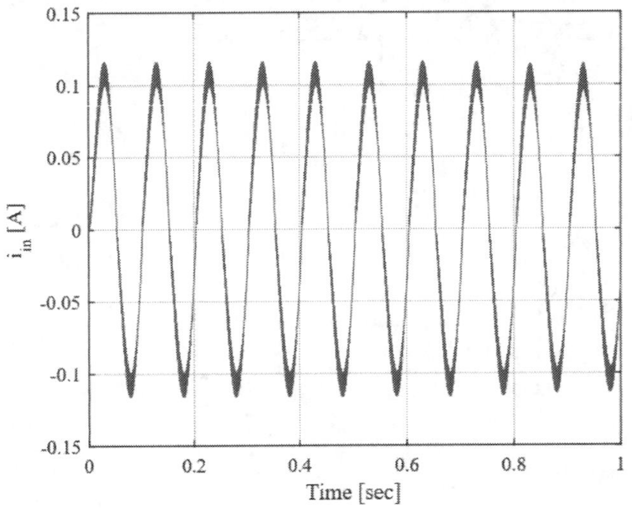

Figure 9: Current generated by electromagnetic transducer in β_{opt}

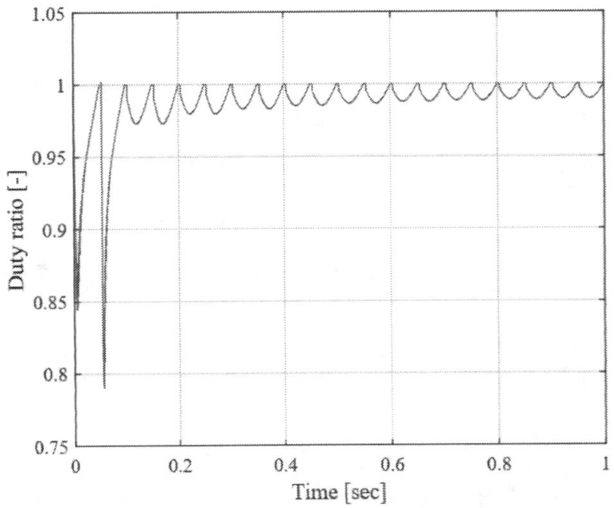

Figure 10: Optimal duty ratio β_{opt}

This work is supported by JSPS KAKENHI Grant Number 26420174.

REFERENCES

1. Behrens, S., Fleming, A. W. and Moheimani, R. O. R., "Passive Vibration Control via Electromagnetic Shunt Damping", *IEEE/ASME Transactions on Mechatronics*, Vol. 10, No. 1, 2005, pp. 118-122.
2. Kim, S. S. and Okada, Y., "Variable Resistive Type Energy Regenerative Damper Using Pulse Width Modulated Step-up Chopper", *Journal of Vibration and Acoustics*, Vol. 124, No. 1, 2002, pp. 110-115.
3. Zhu, S., Shen, W. and Xu, Y., "Linear electromagnetic devices for vibration damping and energy harvesting: Modeling and testing", *Engineering Structures*, Vol. 34, 2012, pp. 198-212.
4. Linear technology, LTspice IV, available from <http://www.linear.com/designtools/software/>, (accessed on 8 October, 2015).

ICAST2015 #30

Design and Performance Evaluation of an Adaptive Energy Dissipation System

Ruifu Zhang[1], Yinying Xia [2*], Junwei Li[1], and Hao Luo[1]

[1] Research Institute of Structural Engineering and Disaster Reduction, Tongji University, China

[2] Department of Engineering Design and Mathematics, University of the West of England, Bristol, UK

Abstract

An adaptive energy dissipation system, which can dissipate the energy from earthquake ensure the safety, is developed and discussed in this study. Nowadays, two major types of energy dissipation systems, viscous damper and metal damper, are widely used in structures. However, both of them have limitations. The viscous damper is working fine but has the limitations for working at high velocity situation such as major earthquake, which may further leads to structure damage and failure of the damper. Meanwhile, the durability of viscous damper is lower than primary structure. The mental damper can dissipate seismic energy significantly under the major earthquake, but has limited performance under small vibration and post-earthquake due to the residual displacement. On the other side, the mental energy dissipation damper may have residual deformation after earthquakes or due to temperature, which may lead to structural repair difficulty and high cost of repair. Therefore, the aim of this research is to design an adaptive energy dissipation system, which has the performance under the wide range of vibration and higher reliability and durability.

The proposed adaptive energy dissipation system is designed to be made of friction element for the purpose of reducing the seismic response of the isolated structure. An adaptive friction damper (AFD) offers a good solution for keep the performance in a wide range of vibrations. The friction force and stiffness of AFD can change obviously to offer adaptive energy dissipation under seismic load of different levels. During the minor earthquake vibrations, the AFD dissipate less seismic energy and limit wind vibration, while in the moderate and major earthquakes, seismic energy will be a large number of consumption. On the other side, AFD has the flexibility to adjust the shape to keep the original shape of AFD, even after the moderate or major earthquake, which will not cause significant influence of additional stress on structures due to the of residual deformation of damper.

A serious tests are performed to assess mechanical model of AFD. The performance of the design is further evaluated by numerical analysis, with comparison with two typical energy dissipation devices, one is viscous damper and the other is AFD.

1. INTRODUCTION

* yuying.xia@uwe.ac.uk

Isolation system is usually installed at the bottom of building in order to protect the structure from the damage of the earthquake. The isolator is the most important element in isolation system. The rubber bearing is the most common isolation bearing. And other isolators include high damping rubber bearing, friction pendulum bearing and so on [1-3]. The isolator is used to change the period of primary structure to reduce the seismic response. However, the isolation system can produce large displacement of isolation layer, which can lead to the difficulties of pipeline connection and building usage. The dampers are used to dissipate seismic energy and control the displacement of the isolation layer. These dampers can be categorized as rate-dependent dampers and rate-independent dampers [4-6]. The common rate-dependent damper is viscous damper and the rate-independent damper is the mental damper and friction damper. Although both rate dependent damper and rate independent damper can reduce the displacement of the isolation layer. However, both of them have limitations. The viscous damper is working fine but has the limitations for working at high velocity situation such as major earthquake, which may further leads to structure damage and failure of the damper. The metal energy dissipation damper can play a good performance under moderate and major earthquakes, but has limited energy dissipation performance in minor earthquake. In order to overcome excessive dissipation of isolation layer of structure under the strong earthquake. Many kinds of friction dampers are proposed to dissipation seismic energy [7-12]. However, the friction force of most of friction dampers is nearly same under earthquake of different levels, which means that the capacity of energy dissipation is constant. Sometimes the seismic requirements of structure are not satisfied by these passive dampers under the different earthquake levels.

The active control system is a solution for structure under the different earthquake levels [13, 14]. But the reliability of active control system is not as good as a passive control system. The variable friction damper and active control system has similar mechanical mechanism, but variable friction damper do not need servo mechanism [11, 12, 15, 16]. A proposed adaptive energy dissipation system is designed to be made of friction element for the purpose of reducing the seismic response of the isolated structure. An adaptive friction damper (AFD) offers a good solution for keep the performance in a wide range of vibrations. The friction force and stiffness of AFD can change obviously to offer adaptive energy dissipation under seismic load of different levels. During the minor earthquake vibrations, the AFD dissipate less seismic energy and limit wind vibration, while in the moderate and major earthquakes, seismic energy can be a large number of consumption. The test is performed to assess mechanical model of AFD. The mechanical performance of AFD is adaptive, so optimum design is needed to achieve optimal performance of isolated structure. An optimum design method is proposed in this study using endurance time history (ETH) [17-19] and genetic algorithm (GA) [20-22].

An SDOF model is used to show the performance of AFD and illustrate this optimal design method. After analytical results is obtained by ETH, 7 seismic waves is used to evaluate this kind of optimal design method. The performance of the isolated structure with AFD is further evaluated, with comparison with two typical energy dissipation devices, one is viscous damper and the other is AFD.

2. PRINCIPLE AND TEST OF AFD

2.1 Principle of AFD

The AFD proposed in the present study mainly consists of covering board, sliding board, bolt and butterfly spring, the schematic of which is shown in Figure 1. This damper is able to produce a friction force varying with the displacement by self-adapting the friction coefficient of the metal friction surface. The friction surfaces in the core of this damper is specially handled so that the friction coefficient μ_1 on this area is lower than the friction coefficient μ_2 on the both sides between the covering boards and the sliding board. Length a is the effective length of friction area of the sliding board, which is larger than the length of the friction area on the middle, denoted by b2 , and smaller than the length of the friction area on the covering boards, denoted by b1. The bolt is fastened to deform the butterfly spring, and a stable pressure is produced on the two covering boards, therefore the initial sliding force of this damper can be controlled by adapting the tightening of the bolt. The energy is able to be dissipated through the relative moving between the sliding board and the covering boards, and the adaptive stiffness of this friction damper can be realized by adaptive friction coefficient.

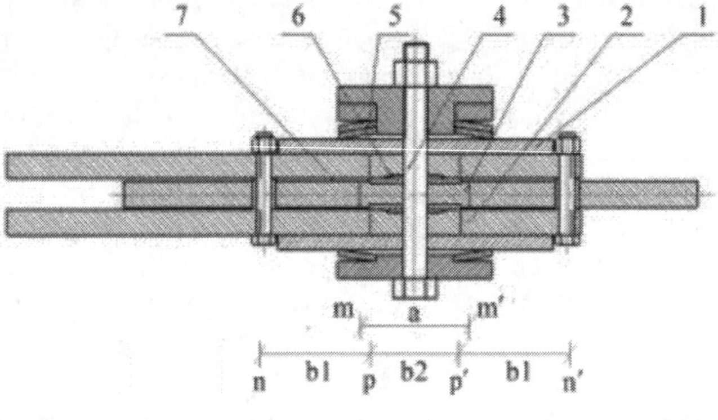

(1.upper covering board, 2. lower covering board, 3. sliding board, 4.preloaded bolt, 5. butterfly spring, 6.middle handling friction area, 7.side friction area)

Figure 1 Schematic of AFD

For example, while the damper moves to the right side under the external loads, and the left of the effective friction area on the sliding board, denoted by m point, is between the endpoints of the friction area on the covering boards, denoted respectively by n and p, the friction area in the energy dissipation system with different friction coefficients remains the constant, thus the damping force is constant according to the Coulomb's law. However, while m is in the core friction area on the covering boards, with the displacement increasing, the area with the friction coefficient μ_2 increases, whereas the area with a lower friction coefficient μ_1 decreases, and the combined effect is the increasing of the total friction coefficient, so that the provided damping force also increases. While the effective friction area on the sliding board is all located in the side friction area, the damping force provided by this system reaches the peak value.

2.2 Mechanical model of AFD

The hysteresis curve, damping force-displacement curve, is regarded as an important indicator to evaluate the effect of energy dissipation. Coulomb's friction model is often used to model the mechanical performance of the friction energy dissipation system, which is able to be expressed as follows,

$$F_{\mu} = \mu N Sgn(v) \tag{1}$$

where μ is the sliding friction coefficient, N is the pressure, and v is the relative sliding velocity on the contact surfaces.

For the proposed adaptive friction system, the pressure N remains a constant, while the sliding friction coefficient is expressed as the function of displacement, and the working friction force provided by the AFD can be obtained as follows,

$$F_{\mu} = \mu(x) N Sgn(v) \tag{2}$$

where x is the displacement of the AFD, and $\mu(x)$ represents the friction coefficient varying with the displacement.

Based on the working theory of the proposed adaptive friction energy dissipation system, the theoretical hysteresis curve of this system can be obtained, shown as Figure 2. The mechanism of this system is illustrated according to the working processes as follows:

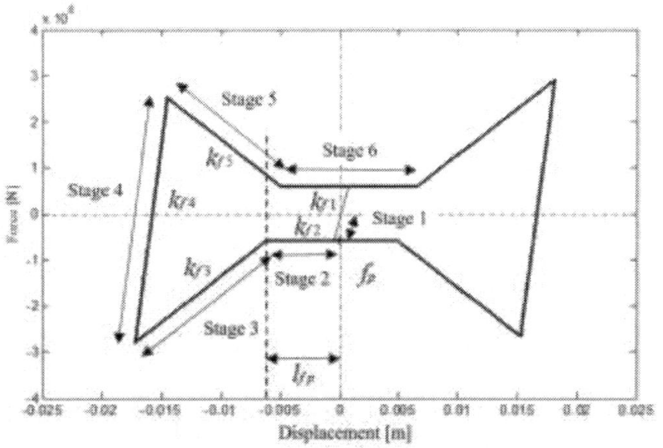

Figure 2 Hysteresis curve of AFD

Stage 1: initial sliding stage

When the axial force that the AFD bears is too small to move the two steel boards, a certain axial deformation is produced, and the stiffness of the AFD in this stage is denoted by k_{f1}. While the axial force is large enough, the two steel boards slide mutually, and the axial force at this moment is named the initial sliding force, expressed as f_p.

Stage 2: plain sliding stage

After initial sliding stage, the AFD start to slide plainly, and the force provided by this system remains a constant during this stage. The theoretical sliding friction force is slightly smaller than the initial sliding force. In order to be theoretically calculated, the sliding friction force is assumed as initial sliding force. In this stage, the force provided by the AFD is constant as f_p, and the stiffness and the plain sliding length are respectively represented by k_{f2} and l_{fp}.

Stage 3: positive stiffness slope sliding stage

After the sliding distant of the AFD exceeds the plain sliding length, because of the characteristic of friction coefficient varying with the increasing of displacement, the friction force between the upper and lower steel boards increases. In this stage, the axial force provided by the AFD increases with the displacement, which implies the property of the positive stiffness, denoted by k_{f3}.

Stage 4: transition stage

After the axial force reach the peak, the AFD bears a reverse load, and go into the transition stage. Because the gap between the upper and lower steel boards doesn't change, the friction coefficient remains a constant, therefore the maximum friction force produced on the friction surface also remains a constant. In this stage, the damper doesn't produce axial sliding, but elements produce inelastic deformation. The stiffness of the damper during this stage is assumed to be equal to the initial stiffness, represented by k_{f4}.

Stage 5: negative stiffness unloading stage

When the reserve load is larger than the initial sliding force, the AFD can slide reservedly and goes in the unloading stage. To analyze simply, the stiffness of the damper during this stage is assumed to be equal to the negative value of the stiffness during Stage 3, represented by k_{f5}.

Stage 6: plain sliding stage

When the sliding board goes into the area where the friction coefficient remains a constant, which is at plain sliding stage, because the pressure that the AFD bears doesn't change, the friction force produced on the friction surface remains a constant, and the stiffness is zero. Until the sliding displacement is zero, the loading process is finished.

2.3 Test of AFD

For the proposed adaptive energy dissipation system, some tests are conducted in order to proposed mechanical model. The performance tests on the regularity of the adaptive friction force and fatigue are conducted. Two test samples of the adaptive energy dissipation system are tested, with a maximum friction force of 30 tons and a design displacement of ±100mm. The loading device employed in these tests is the LETS-2000kN electro-hydraulic servo testing machine.

Table 1 Test Case

Case	Loading displacement (mm)	Loading frequency (Hz)	Loading velocity (mm/s)	Number of loops
1	25	0.1	10	3
2	55	0.1	22	3
3	95	0.1	38	3
4	95	0.1	38	30

The above test cases are conducted, the hysteresis curve of the AFD is obtained as shown in Figure 2.

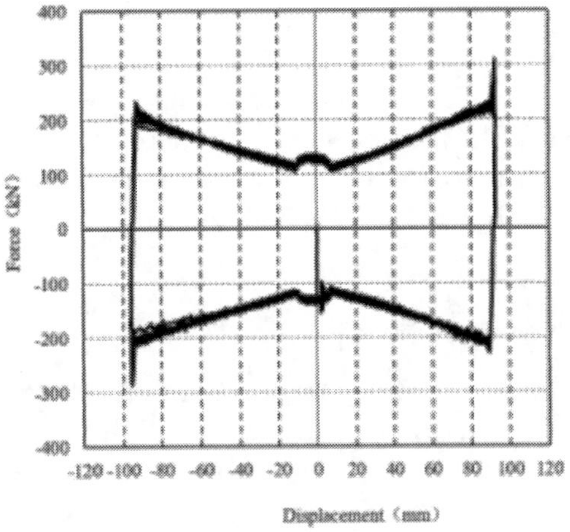

Figure 3 Test performance curve of AFD

According to the test results, the AFD can provide a maximum friction force of 300kN, and the displacement is approximately 100mm. While the displacement is small (±10mm), the friction force is low; after exceeding the limit displacement, the friction force increases linearly, which implies the friction force provided by this system can vary with the displacement. In the case where the maximum displacement is reached, under the 30-loop reciprocating load, the performance of the AFD remains stable, and no significant attenuation of the mechanical performance and damage of members are observed, which implies the good fatigue performance of the AFD.

3. OPTIMAL DESIGN METHOD

3.1 ETH method

In the Endurance Time Method, the intensity of the ground motion increases linearly with the time-history. According to the Endurance Time Method, the target acceleration response spectra is proportional to duration of the time-history at any moment. The target acceleration spectra can be defined as Eq. (3):

$$S_{aT}(T,t) = \frac{t}{t_{Target}} S_{aC}(T) \tag{3}$$

Where: t_{Target} is the target time; $S_{aC}(T)$ is the scaled spectra; T is the natural period of structures; t is the part time of time history; $S_{aT}(T,t)$ is the target response spectra at time t. Displacement spectra is closely related to the acceleration spectra, as shown in Eq.(4):

$$S_{uT}(T,t) = \frac{t}{t_{Target}} S_{aC}(T) \times \frac{T^2}{4\pi^2} \tag{4}$$

Where $S_{uT}(T,t)$ is the target displacement response spectra at time t. The problem to find an acceleration that can satisfy Eq. (3) and Eq. (4) is complicated. For simplicity, the problem is converted into an unconstrained optimization problem as follows:

$$\min F(\ddot{u}_g) = \int_0^{T_{max}} \int_0^{t_{max}} \left\{ \left[S_a(T,t) - S_{aT}(T,t) \right]^2 + \alpha \left[S_u(T,t) - S_{uT}(T,t) \right]^2 \right\} \mathrm{d}t\mathrm{d}T \tag{5}$$

where $\ddot{u}_g$ is the acceleration endurance time history; $S_a(T,t)$ is the acceleration spectra of $\ddot{u}_g$ at time t; $S_u(T,t)$ is the displacement spectra of $\ddot{u}_g$ at time t; α is the weight parameter of displacement spectra.

3.2 GA method

GA is a search heuristic that mimics the process of natural selection. GA is usually used to generate useful solutions to optimization and search problems. In a GA, a population of candidate solutions to an optimization problem is evolved to better solutions. Each candidate solution has a set of properties which can be mutated and altered. GA contains these procedures: initialization, selection, crossover, mutation and termination.

4. DYNAMIC MODEL

4.1 SDOF system

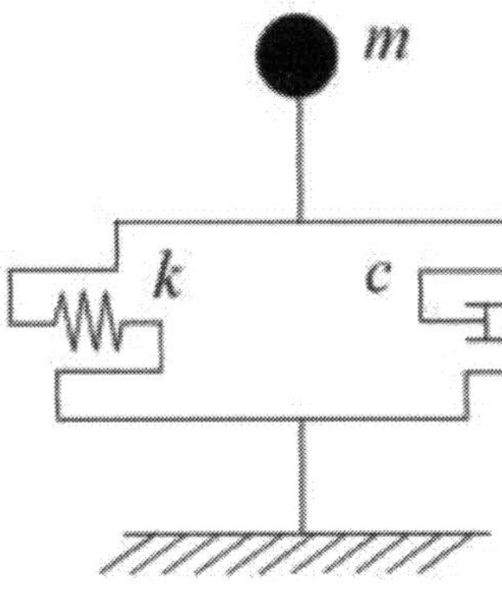

Figure 4 Analytical model of SDOF structure (ST0)

The analytical model of SDOF structure is shown as Figure 3. According to structural dynamic theory, the governing equation of the SDOF structure excited by the ground motion $\ddot{x}_g$,

$$m\ddot{x}+c\dot{x}+f_s(x,\dot{x})=-m\ddot{x}_g \tag{6}$$

where m and c are respectively the mass and damping coefficient of the SDOF structure, and $f_s(x,\dot{x})$ is the restoring force, and x, $\dot{x}$ and $\ddot{x}$ are respectively the relative displacement, velocity and acceleration.When elastic analysis is conducted, the restoring force can be obtained from $f_s(x,\dot{x})=kx$, where k is the structural stiffness, then the Eq. (3) can be expressed as follows,

$$m\ddot{x}+c\dot{x}+kx=-m\ddot{x}_g \tag{7}$$

Dividing the two sides of Eq. (4) with m, the following expression is able to be obtained,

$$\ddot{x}+\frac{c}{m}\dot{x}+\frac{k}{m}x=-\ddot{x}_g \tag{8}$$

Defining $\omega=\sqrt{\frac{k}{m}}$ and $\xi=\frac{c}{2m\omega}$, the Eq. (5) can be written as,

$$\ddot{x}+2\omega\xi\dot{x}+\omega^2x=-\ddot{x}_g \tag{9}$$

where ω and ξ denote the frequency and damping ratio of the structure.

4.2 SDOF isolated system

When base isolation is employed, the isolated structure can be idealized as the analytical model, shown in Figure 4. When the AVD is placed in the isolation layer, the horizental equivalent stiffness and equivalent viscous damping of the isoaltion layer can be seen as the horizental stiffness k and the damping c of the SDOF analytical model, then the analytical modal of the controlled system can be represented as shown in Figure 4, where k_f denotes the stiffness of AVD.

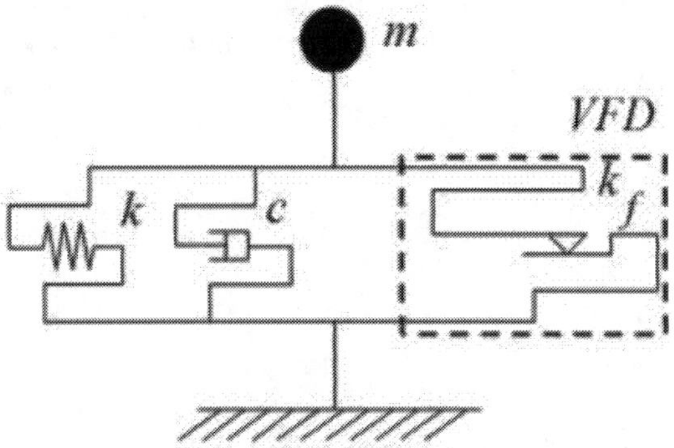

Figure 5 Analytical model of SDOF structure with AFD (ST1)

Correspondingly, the governing equation of the controlled system can be expressed as follows,

$$m\ddot{x}+c\dot{x}+f_s(x,\dot{x})+f(x)=-m\ddot{x}_g \tag{10}$$

Where $f(x)$ is the restoring force provided by AVD, which can be the function of displacement mentioned above, $f(x)=k_f(x)x$, where $k_f(x)$ denotes the stiffness varying with the sliding displacement.

Then the Eq. (9) can be written as follows,

$$m\ddot{x}+c\dot{x}+(k+k_f(x))x=-m\ddot{x}_g \tag{11}$$

4.3 Optimization study

4.3.1 Parameters of optimal model

A SDOF original model (ST0) is employed in this section. The parameters of this SDOF is shown in Table 2. To compare the effect of AVD on the vibration control of structures, the SDOF system, the structural parameters of which is the same as those of ST0, with an AVD (ST1) is also obtained, as shown in Figure 2. The ETH according to Shanghai Code, the peak of ground acceleration of which is adjusted as 200gal, is employed as ground motion subjecting to the structure, and the damping ratio is set to 0.02.

Table 2 Model information of isolated SDOF structure (ST0)

Mass m (kg)	Stiffness k (kN/m)	Damping coefficient C (kN/(m/s))
37331680	419200	5003909

To realize the optimal design of the structure with an AVD, the optimum location, number and parameters of the AVD should be determined to make the vibration response of the structure minimum. The parameters of AFD include the initial sliding force f_p, plain sliding length l_{fp}, initial sliding stiffness k_{f1}, loading stiffness k_{f3}, transition stiffness k_{f4}, unloading stiffness k_{f5}. The plain sliding length l_{fp} can be adjusted according to the actual requirement. The loading stiffness k_{f3} and unloading stiffness k_{f5} are both related to the friction surface between the boards, to analyze simply, we assume the unloading stiffness k_{f5} to be the reserve value of the loading stiffness k_{f3} (k_{f5}= -k_{f3}). The initial stiffness k_{f1} and transition stiffness k_{f4} are both related to the axial deformation of the member, which are much larger than the loading stiffness. In this study, we assume $k_{f1}=10k_{f3}$.

Based on the above analysis, the optimal variables of the AFD include the initial sliding force f_p, plain sliding length l_{fp} and loading stiffness k_{f3}. In this section, we set the plain sliding length l_{fp} to 6mm, and consider the two optimal variables, f_p and k_{f3}, and use genetic algorithms to obtain the optimum solutions. To ensure both the displacement and acceleration of the system effectively controlled, the linear combination of the maximum displacement ratio and maximum acceleration ratio of the ST1 to the ST0 is taken as target function. The restriction conditions is the corresponding specified ranges. In all, the optimal model of AFD is able to be described as,

Optimal variables: $$Variable=[f_p, k_{f3}] \tag{12}$$

Target function: $$f=\frac{1}{1+\gamma}\frac{d_{st1}}{d_{st0}}+\frac{\gamma}{1+\gamma}\frac{a_{st1}}{a_{st0}} \tag{13}$$

Restriction conditions: $$0 \le f_p \le f_{\max}\ ,\ 0 \le k_{f3} \le k_{\max} \tag{14}$$

where $f_{\max}$ is the maximum initial sliding force, $k_{\max}$ is the specified maximum of the loading stiffness; d_{st1} and a_{st1} are respectively the maximum displacement and acceleration of the ST1, d_{st0} and a_{st0} are respectively the maximum displacement and acceleration of the ST0; and γ is the weighting factor, which is set as 0.5 in this study.

4.3.2 ETH analysis and genetic parameters selection

In this paper, the binary encodings with a length of 10 and 16 are respectively used to represent the optimal variables f_p and k_{f3}. According to correspondence between encoding and parameters introduced above, the longer the coding, the higher accuracy of calculation, while the lower efficiency. In order to ensure the accuracy, while improving computational efficiency, we divide the optimal variables f_p and k_{f3} with the corresponding cardinal numbers as follows,

$$N_f = \frac{f_p}{{}_m f_p} \tag{15}$$

$$N_k = \frac{k_{f3}}{{}_m k_{f3}} \tag{16}$$

where N_f and N_k are respectively the direct optimal variables used in genetic algorithms, and ${}_m f_p$ and ${}_m k_{f3}$ are respectively the cardinal numbers of the optimal variables f_p and k_{f3}.

According to the optimal model, the control parameters of genetic algorithms are shown in Table 3.

Table 3 Parameter of GA

Termination generation	Population size	Crossover probability	Mutation probability	Generation gap
100	30	0.7	0.05	0.7

5.2.4 Optimization results analysis

A endurance time history curve is shown in Figure 7. The optimal variables of AFD are shown in Table 4, and the analytical results of the ST0 and ST1 are shown in Table 5. The hysteresis curve of AVD is shown in Figure 7, and the response time histories of displacement and acceleration are respectively shown in Figure 8 and Figure 9.

Table 4 Parameters of optimal AFD

l_{fp} (mm)	f_p (kN)	k_{f1} (kN/m)	k_{f2} (kN/m)	k_{f3} (kN/m)	k_{f4} (kN/m)	k_{f5} (kN/m)
6	342	975000	0	97500	975000	-97500

Table 5 Analytical results

	Max dis. (mm)	(ST0-ST1)/ST0	Max acc. (m/s^2)	(ST0-ST1)/ST0
ST0	278	53%	3.21	39%
ST1	131		1.97	

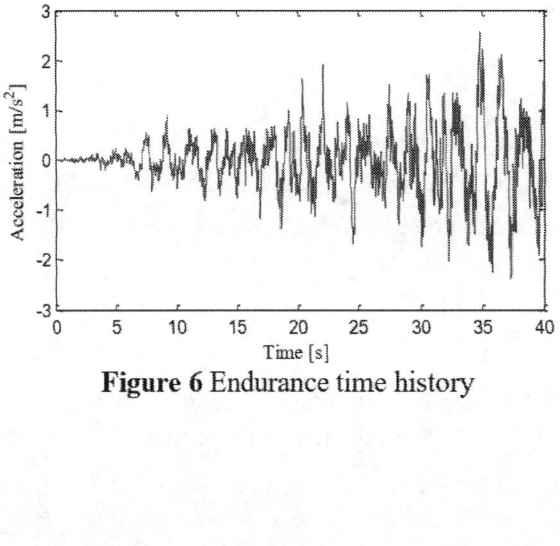

Figure 6 Endurance time history

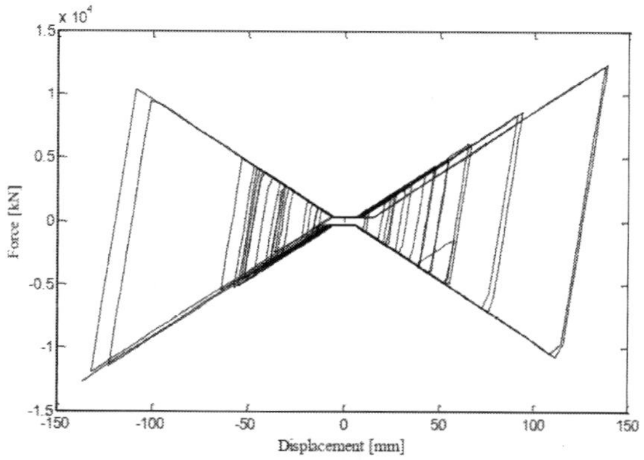

Figure 7 Hysteretic curve of AFD

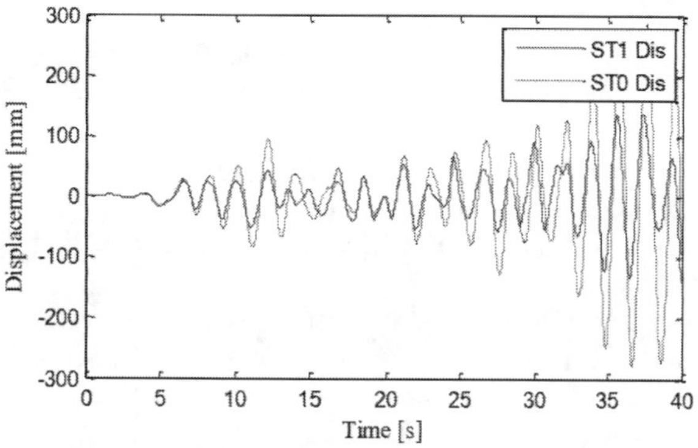

Figure 8 Time history curve of displacement

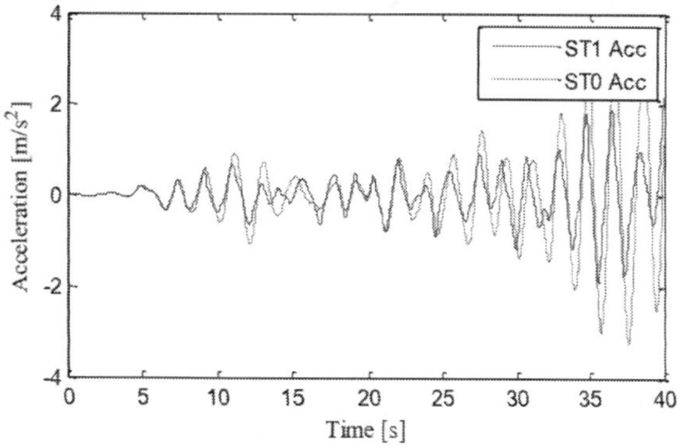

Figure 9 Time history curve of acceralation

5. CONCLUSION

An adaptive energy dissipation system called as AFD proposed in this study. The main conclusions are drawn as follows,

1) An experiment is conducted to verify the shape and properties of the theoretical hysteresis curve of AFD, and the shape of hysteresis cure is almost same as theoretical model;

2) The AFD has the characteristic of variable friction force and variable stiffness which is adapted to different seismic levels;

3) Under the seismic load, by appending structures with an AFD, the maximum displacement of the SDOF is greatly reduced and the maximum acceleration is also reduced.

ACKNOWLEDGES

This work is supported b by the Foundation of Shanghai Engineering Technical Research Centre under Grant No. 10DZ2252000.

REFERENCES

1. Kikuchi, M. and I.D. Aiken, An analytical hysteresis model for elastomeric seismic isolation bearings. *Earthquake Engineering & Structural Dynamics*, 1997. 26(2): p. 215-231.
2. Morgan, T.A. and S.A. Mahin, Achieving reliable seismic performance enhancement using multi-stage friction pendulum isolators. *Earthquake Engineering & Structural Dynamics*, 2010. 39(13SI): p. 1443-1461.
3. Tsai, C.S., T.C. Chiang and B.J. Chen, Finite element formulations and theoretical study for variable curvature friction pendulum system. *Engineering Structures*, 2003. 25(14): p. 1719-1730.
4. Kasai, K. and K. Matsuda, Full-scale dynamic testing of response-controlled buildings and their components: concepts, methods, and findings. Earthquake Engineering and Engineering Vibration, 2014. 13(1 Supplement): p. 167-181.
5. Miyamoto, H.K., et al., Limit states and failure mechanisms of viscous dampers and the implications for large earthquakes. *Earthquake Engineering & Structural Dynamics*, 2010. 39(11): p. 1279-1297.
6. Soong, T.T. and B.F. Spencer, Supplemental energy dissipation: state-of-the-art and state-of-the practice. *Engineering Structures*, 2002. 24(3): p. 243-259.
7. Chien, C.C., L. Lu and Y. Tsai, Analysis of Single-degree-of-freedom Systems Containing Multi-functional Friction Damper. *Procedia Engineering*, 2014. 79: p. 500-505.
8. Bhaskararao, A.V. and R.S. Jangid, Seismic analysis of structures connected with friction dampers. *Engineering Structures*, 2006. 28(5): p. 690-703.
9. Cao, L., et al., Variable friction device for structural control based on duo-servo vehicle brake: Modeling and experimental validation. *Journal of Sound and Vibration*, 2015. 348: p. 41-56.
10. Lu, L.Y., et al., Modeling and experimental verification of a variable-stiffness isolation system using a leverage mechanism. *Journal of Vibration and Control*, 2011. 17(12): p. 1869-1885.
11. Ribakov, Y., Reduction of structural response to near fault earthquakes by seismic isolation columns and variable friction dampers. *Earthquake Engineering and Engineering Vibration*, 2010. 9(1): p. 113-122.
12. Zhou, X. and L. Peng, A new type of damper with friction-variable characteristics. *Earthquake Engineering and Engineering Vibration*, 2009. 8(4): p. 507-520.
13. Yoshioka, H., J.C. Ramallo and B.F. Spencer, "Smart" base isolation strategies employing magnetorheological dampers. *Journal of Engineering Mechanics*-ASCE, 2002. 128(5): p. 540-551.
14. Guo, S.Q., S.P. Yang and C.Z. Pan, Dynamic modeling of magnetorheological damper behaviors. *Journal of Intelligent Material Systems And Structures*, 2006. 17(1): p. 3-14.
15. Lu, L., T. Lee and S. Yeh, Theory and experimental study for sliding isolators with variable curvature. *Earthquake Engineering & Structural Dynamics*, 2011. 40(14): p. 1609-1627.
16. Narasimhan, S. and S. Nagarajaiah, Smart base isolated buildings with variable friction systems: H-infinity controller and SAIVF device. *Earthquake Engineering & Structural Dynamics*, 2006. 35(8): p. 921-942.
17. Estekanchi, H.E., V. Valamanesh and A. Vafai, Application of Endurance Time method in linear seismic analysis. *Engineering Structures*, 2007. 29(10): p. 2551-2562.
18. Estekanchi, H.E., K. Arjomandi and A. Vafai, Estimating structural damage of steel moment frames by Endurance Time method. *Journal of Constructional Steel Research*, 2008. 64(2): p. 145-155.
19. Valamanesh, V., H.E. Estekanchi and A. Vafai, Characteristics of Second Generation Endurance Time Acceleration Functions. *Scientia Iranica Transaction A-Civil Engineering*, 2010. 17(1): p. 53-61.
20. Charmpis, D.C., M.C. Phocas and P. Komodromos, Optimized retrofit of multi-storey buildings using seismic isolation at various elevations: assessment for several earthquake excitations. *Bulletin of Earthquake Engineering*, 2015. 13(9): p. 2745-2768.
21. Kim, H.S. and P.N. Roschke, Design of fuzzy logic controller for smart base isolation system using genetic algorithm. *Engineering Structures*, 2006. 28(1): p. 84-96.
22. Cha, Y. and A.K. Agrawal, Decentralized output feedback polynomial control of seismically excited structures using genetic algorithm. *Structural Control & Health Monitoring*, 2013. 20(3): p. 241-258.

ICAST2015 #072

Implementation of Synchronized Triple Bias-Flip Interface Circuit towards Higher Piezoelectric Energy Harvesting Capability

Yuheng Zhao, Chenbin Zhou, and Junrui Liang*

Mechatronics and Energy Transformation Laboratory (METAL)
School of Information Science and Technology
ShanghaiTech University, Shanghai, China

Abstract

It has been mathematically proved that the harvesting capability of a piezoelectric energy harvesting (PEH) system can be significantly enhanced by taking the synchronized multiple bias-flip (SMBF) strategy. Among all SMBF solutions, the parallel triple bias-flip version (P-S3BF) makes the best compromise between cost and effectiveness. This paper introduces the practical design procedures and implementation details of the P-S3BF interface circuit. The triple bias-flip operation is realized by a current-routing H-bridge, which employs six power MOSFETs for enabling the specified current flows. Experiment result shows that the P-S3BF outperforms the extensively studied synchronized switch harvesting on inductor (SSHI) and standard bridge rectifier (SEH) interface circuits under the same mechanical condition. With the piezoelectric harvester used in this experiment, P-S3BF increases the harvested power by 33%, compared to SSHI; and 300%, compared to SEH.

1. INTRODUCTION

The advancements of ultra-low-power IC technologies and energy harvesting technologies in the last decade are changing the deployment way of wireless sensor networks (WSNs). Future WSN devices can become self-powered by scavenging the energy in their ambience. The power level of existing micro-generators ranges from several microwatts to milliwatts. Three types of micro-generators have been mostly investigated for harvesting the ambient kinetic energy, i.e., electromagnetic, piezoelectric, and electrostatic generators. Among these three electromechanical transducers, the piezoelectric one can be implemented with simple configuration and provides relative high output voltage. Therefore, it is extensively studied for extracting electric power from the ambient vibration sources. There are two important emphases in the research of piezoelectric energy harvesting (PEH) systems: broadening the bandwidth and improving the energy harvesting capability. The former can be effectively achieved by mechanical methods; while the latter requires more capable designs of interface circuits.

The synchronized bias-flip (BF) interface circuits are the most extensively studied circuit solutions for enhancing the piezoelectric power conversion. They were originally designed for the vibration damping purpose [1, 2]. The first BF based energy harvesting interface circuit is called synchronized switch harvesting on inductor (SSHI). It has set a milestone for enhancing the harvesting capability by up to several folds, compared to the standard bridge rectifier (SEH), i.e., the bridge rectifier solution. The general principle of BF solutions is shown in Fig. 1(a). The voltage across the piezoelectric capacitor C_p can be rapidly changed by making use of the electrical resonance of an RLC network. By

*Corresponding author: liangjr@shanghaitech.edu.cn

synchronously flipping the piezoelectric voltage with respect to zero [1] or a dc voltage source [2] at the zero-crossing instants of the vibration velocity, the combination of C_p and the interface circuit behaves like a tunable resistance, as shown in Fig. 1(b). The power consumed by equivalent R_{conv} is regarded as the power extracted (or converted) from the piezoelectric structure. As more power is extracted from the vibrating piezoelectric structure, the electrically induced damping increases as well. The dissipated and harvested powers within the energy extraction has been further distinguished in a more detailed model, as shown in Fig. 1(c) [3]. It implies that the enhancement of energy extracting capability (corresponding to R_h+R_d) does not necessarily lead to the enhancement of energy harvesting capability (corresponding to R_h). The design principle of the PEH interface circuit should be specified as to *maximize the electrically induced damping* (increase R_h+R_d) *at low dissipative cost* (decrease R_d).

After SSHI, various BF based topologies were proposed. Ref. [4, 5] give a study on the evolution of PEH interface circuits. Some researches claimed that the harvested capability can be further enhanced by taking active intervention, i.e., pumping some amount of energy into the system in order to gain more [6, 7]. Yet, the boundary between necessary passive actions and auxiliary active actions were ambiguous until the introduction of the hybrid bias-flip solutions [8, 9, 10].

2. ACTIVE & PASSIVE BF ACTIONS

The bias-flip action can be classified into two types: active BF and passive BF, as shown in Fig. 3. The single supply pre-biasing (SSPB) [9], whose circuit and waveforms are shown in Fig.4, is taken as an example for explaining the concept of these two actions. When the equivalent current from the piezoelectric element crosses zero, two BF actions come into play. The switches S1 and S3 are turned on first. The charge stored in the capacitor C_p then flow through the storage capacitor C_r. During this action, C_r absorbs energy form the piezo-element, or more exactly from C_p. Such BF action is called *passive BF*. After the first action, the voltage across C_p becomes zero, then S1 and S3 are turned off, while S2 and S4 are turned on, which enables the storage capacitor C_r to reversely charge C_p. During this action, C_r injects energy into C_p, such that to pre-bias C_p before the start of the next half vibration cycle. Such BF action is called *active BF*.

In each BF action, the starting voltage across C_p is defined as V_b+U. By shorting C_p to an inductor L_i in series with a dc bias voltage V_b, v_p drops to $V_b+\gamma U$ in tens to hundreds of micro seconds due to the RLC series resonance. The flipping factor γ can be achieved within the range as follows

$$-e^{-\pi/(2Q)} \le \gamma \le 1. \tag{1}$$

Q in (1) is the quality factor of the resonant shortcut. The energy which is absorbed by V_b or injected into the piezoelectric structure from V_b is expressed as follows

$$E_{h/i,BF} = \Delta Q V_b = C_p(1-\gamma)UV_b. \tag{2}$$

ΔQ is the change of charge on C_p in each BF action. When $E_{h/i,BF} > 0$, the bias dc source V_b harvests energy from the piezoelectric system; when $E_{h/i,BF} < 0$, it injects energy into the system. The dissipated energy associated with each bias-flip action is expressed as follows

$$E_{d,BF} = \frac{1}{2}C_p(1-\gamma^2)U^2 \tag{3}$$

3. BEST BIAS-FLIP STRATEGY

Equations (2) and (3) show that, in a BF action, larger voltage change, i.e., $(1-\gamma)U$, linearly increases the extracted energy, while quadratically increases the dissipated energy. The only way to increase the extracted energy at a low dissipation cost is to *make a large voltage change by combining multiple small steps*, which leads to the proposal of SMBF interface circuits [4], as shown in Fig. 5. Mathematical analysis shows that the harvested power (normalized with respect to that in SEH) in terms of voltage variables in SMBF is

$$E_h = 2[\sum_{m=1}^{M}(1-\gamma)U_m V_{b,m} + (2-\Delta U)V_o] \tag{4}$$

Solving the zero derivative point with respect to the intermediate voltages (U_m, m=1,2,...,M) leads to the optimal bias voltages set (normalized with respect to the open-circuit voltage) as follows

$$\begin{cases} U_{m,opt} = \frac{1}{1+\gamma} \\ \Delta U_{opt} = 1 \end{cases}, \quad m = 1,2,3,...,M \tag{5}$$

The normalized bias voltages in optimal P-SMBF can be derived as follows

$$V_{b,m} = \frac{1-\gamma}{1+\gamma}\frac{M-2m+1}{2}, \quad \text{m} = 1, 2, 3, \ldots, M \tag{6}$$

According to the derived best voltage bias-flip strategy, a new method has been discovered for the further enhancement of harvesting capability without pushing $\gamma \to -1$, that is to sophisticatedly exert more bias-flip actions, i.e., increase M, at each synchronized instant.

4. S3BF INTERFACE CIRCUIT

Fig. 6 shows the optimal voltage relays in a synchronized instant (downstairs instant) under different bias-flipping number M. Compared to the pictures of S-SMBF, the bias voltages (designated with cross marks) in the optimal P-SMBF are symmetry with respect to zero volt. In particular, in the triple bias-flips version P-S3BF, there are three bias voltages V_b, 0, and $-V_b$. Since 0 is a free voltage source, V_b and $-V_b$ can be realized with a single source and an H bridge. Therefore, the P-S3BF might be realized by using only one auxiliary voltage source V_b and some power conditioning components. Given a comprehensive consideration on the cost and effectiveness, the P-S3BF solution can make a good compromise between circuit complexity and harvesting capability enhancement. This tells the motivation of our investigation on the P-S3BF interface circuits.

The proposed P-S3BF interface circuit employs six power MOSFETs for executing the triple bias-flip actions at the zero current instants. The circuit topology is shown in Fig. 7. The circuit parameters are listed in Table I. No additional power is required for establishing the bias voltage V_b. It can be self-adaptive across the auxiliary capacitor C_b.

4.1 Operational Principle

The operational principle of P-S3BF are shown in Fig. 9. The triple bias-flip actions take place when the equivalent current i_{eq} crosses zero (or the piezoelectric cantilever is at its displacement extremes). A specific RLC resonant path is turned on at each bias-flip action, providing a shortcut for the charge stored in C_p. Therefore, v_p can be rapidly flipped for three times with respect to the reference voltages V_b, 0, and $-V_b$ in succession. The working phases are described as follows:

- Phase 1: $v_p>0$, *open circuit:* In this phase, the voltage across the piezo element v_p is proportional to the displacement of the piezoelectric cantilever. v_p is not high enough to get through the diode bridge ($v_p<V_o+V_F$).
- Phase 2: $v_p>0$, *constant voltage:* The diode bridge is turned on. v_p is clamped to a certain voltage, which is related to the dc load R_L. The load absorbs energy from C_p (piezoelectric structure).
- Phase 3: *downstairs, 1st bias flip:* M_1 and M_5 are turned on. The current flow is confined in one direction by D_{L2}. The charge in C_p flows through C_b. Since V_b is positive, C_b absorbs energy in this phase. Therefore, it corresponds to a passive bias-flip action.
- Phase 4: *downstairs, 2nd bias flip:* M_4 are turned on. The current flow is confined in one direction by D_{R1}. The charge stored in C_p flows through the resonant path with zero voltage reference. It flips v_p from positive to negative. In this phase, there is no energy change in C_b.
- Phase 5: *downstairs, 3rd bias flip:* M_3 and M_4 are turned on. The current flow is confined in one direction by D_{R3}. In this phase, C_b reversely pumps charge to C_p. Since V_b is negative, C_b injects energy into C_p in this phase. Therefore, it corresponds to an active bias-flip action.

There are other five phases following the aforementioned five phases in the other half vibration cycle. They carry out the upstairs bias-flip actions likewise.

It is worth noting that the energy absorbed by C_b in the 1st bias flip phase and the energy injected from C_b in the 3rd bias flip phase are the same under steady state, which means that there is no energy change in C_b in one vibration cycle. The 2nd bias flip is a similar process as that in P-SSHI. The voltage across C_p changes its sign by flipping with reference to zero volt. There is also no energy change in C_b during this phase.

Enlarging the voltage amplitude across a current source yields more extracted energy; while increasing the bias-flip number and minimizing the voltage change in each step decrease the dissipated energy. These two features can be simultaneously achieved by taking the P-S3BF, therefore it can enhance the energy harvesting capability of a PEH system under the same mechanical vibration.

4.2 Current-Routing H-Bridge

There are totally six bidirectional switches for controlling the conduction or cutoff of each switching path. Each switch is realized by a diode in serial with a MOSFET (M_1 with D_{L1}, M_2 with D_{R2}, M_3 with D_{R3}, M_4 with D_{R1}, M_5 with D_{L2}, and M_6 with D_{L3}). The diode in each switching path not only helps block the voltage, but also confines the current flow in a specified direction in different bias-flip actions. The capacitor C_b in the middle of the bridge serves as the bias voltage source V_b. According to the vibration of the piezoelectric beam, V_b is able to adapt itself towards the optimal value given in (6).

4.3 Driving Circuits

The purpose of the driving circuits is to properly turn on or off the MOSFET switches with a low power consumption. Fig. 7(b) shows the driving circuit in the P-S3BF design. The switch control is delivered by a micro-controller unit (MCU). The MCU is powered by battery (the power consumption of the MCU is not considered in this study). Given the single voltage supply for powering the MCU, it can only generate positive voltage for driving the gate nodes. The low-side N-type MOSFETs (M_3 and M_6) can be directly driven by the positive voltage, yet p-type MOSFETs (M_1, M_2, M_4, and M_5) require negative gate drive voltage. An easy way to solve this problem is to design some voltage level-shifting circuits. For the high-side p-type MOSFET, taking M_1 for example, when the driving pin of the MCU, i.e., PR1 is high, the gate and source voltages approaches V_{con} (a voltage node, rather than a source), given $R_{1b}>>R_{1a}$. When the driving pin is switched to low voltage, the gate voltage falls below V_{con}. Consequently, the negative voltage between the gate and source terminals turns M_1 on.

For the low-side p-type MOSFET, taking M_2 for example, when the driving pin of the MCU, i.e., PR2 is high, the gate and source terminals approach the ground voltage. When the driving pin is switched to low voltage, the gate voltage R2 node goes negative. A negative voltage between the gate and source terminals turns M_2 on.

With these driving circuits, all the six MOSFETs can be properly controlled by the digital I/O pins of an MCU without using the isolating gate-drive chips.

4.4 Control Timing

The control signal is provided by six I/O pins of an ultra-low-power MCU in an energy harvesting evaluation board (eZ430-RF2500, Texas Instrument, Dallas, TX). During most of the time, the MCU is in deep sleeping mode with very low maintaining current (a few μA). It only comes into operation at the bias-flip instants, such that to decrease the power consumption (this external power is not counted into the circuit dissipation in this study). The width of each gate drive pulses is half of an RLC resonance cycle

$$\Delta T = \pi\sqrt{L_i C_p}, \tag{7}$$

The switch timing control, by which the P-S3BF is activated, is shown in Fig. 8(c).

5. EXPERIMENTS

Experiments are carried out with a base excited piezoelectric energy harvester under three cases of interface circuits: SEH, P-SSHI, and P-S3BF. In each case, one of these interface circuits is connected to the piezoelectric cantilever. The actual harvested power is experimentally measured as functions of $\tilde{V}_{rect}$.

5.1 Setup

Fig. 10 shows the experimental setup. The main mechanical structure is a piezoelectric bimorph cantilever. One of its end is fixed on the shaker to form a piezoelectric cantilever. One patch of the bimorph is for energy harvesting while the other is for vibration reference. For the purpose of synchronization in both P-SSHI and P-S3BF, an electromagnetic sensor, which is composed of a coil and a pair of magnets, is installed near the free end for sensing the relative velocity of the cantilever beam. The permanent magnets also act as a proof mass at the same time. It can lower the natural frequencies and increase the displacement of the free end. The output voltage from the coil, which is proportional to the end velocity, is then input to an internal voltage comparator of the MCU. The MCU is coded to generate the switching timing accordingly for driving the six MOSFET switches, such that the synchronized bias-flip actions can be successfully carried out. The mechanical parameters are listed in Table II.

5.2 Results

Experiments are carried out for obtaining the harvested power P_h under a constant displacement magnitude harmonic excitation. Under such excitation condition, the equivalent circuit of a piezoelectric structure can be taken as a current source, thus, the mechanical effect resulting from the backward coupling can be neglected. When different dc load resistors R_L are connected one after another to stabilize the output voltage V_o at different levels, displacement magnitude is kept constant (open circuit voltage V_{oc} keeps unchanged). Once it is changed, as observed from the open-circuit voltage of the referential piezoelectric patch, the excitation amplitude is tuned for keeping the displacement magnitude to constant level.

The experimental harvested power P_h is obtained as follows

$$P_h = \frac{V_o^2}{R_L} \tag{8}$$

The non-dimensional rectified voltage $\tilde{V}_{rect}$ is defined by

$$\tilde{V}_{rect} = \frac{V_{rect}}{V_{oc}} \tag{9}$$

The rectified voltage V_{rect} is the sum of V_o and V_F, where V_F is the forward voltage drop of the bridge rectifier. It is 0.8V in the experimental circuit.

As shown in Fig. 11, the voltage profile of the P-S3BF is quite similar to P-SSHI, except those in the synchronized instants. Three stairs about the same height are observed in the triple bias-flip instants. It makes a good agreement with the theoretical prediction. As we can read from the harvested power illustrated in Fig. 12, under optimal $\tilde{V}_{rect}$, the harvested power of P-S3BF is 33% higher than P-SSHI and 300% higher than SEH in these experiments.

6. CONCLUSIONS

Given the guidance of SMBF strategy, the implementation details of P-S3BF harvesting interface circuit was introduced in this paper. The circuit operation principle were introduced in details, in order to clarify the design considerations. Some essential issues, such as the effectiveness of improvement, circuit complexity, realization of the bias voltage source, current routing, low-power control, etc. were taken into consideration towards the practical design and implementation of the P-S3BF interface circuit. Experiment results show that the harvesting capability of P-S3BF outperforms P-SSHI by 33%, which verified the capability of the SMBF strategy. Future effort should focus on the self-powered solution and the dynamic electromechanical model of the SMBF based PEH systems.

ACKNOWLEDGMENTS

The work described in this poster was supported by the grants from National Natural Science Foundation of China (Project No. 61401277) and Faculty Start-up Grant (Project No. F-0203-13-003) of ShanghaiTech University, Shanghai, China.

REFERENCES

[1] C. Richard, D. Guyomar, D. Audigier, and G. Ching, "Semi-passive damping using continuous switching of a piezoelectric device," in *Smart Structures and Materials 1999: Passive Damping and Isolation*, T. T. Hyde, Ed., vol. 3672, no. 1. SPIE, 1999, pp. 104–111.

[2] H. Ji, J. Qiu, K. Zhu, Y. Chen, and A. Badel, "Multi-modal vibration control using a synchronized switch based on a displacement switching threshold," *Smart Mater. Struct.*, vol. 18, no. 3, p. 035016, 2009.

[3] J. R. Liang and W. H. Liao, "Impedance modeling and analysis for piezoelectric energy harvesting systems," *IEEE/ASME Trans. Mechatron.*, vol. 17, no. 6, pp. 1145–1157, 2012.

[4] J. R. Liang and H. S.-H. Chung, "Best voltage bias-flipping strategy towards maximum piezoelectric power generation," *J. Phys. Conf. Ser.*, vol. 476, no. 1, p. 012025, 2013.

[5] J. Liang, "Synchronized triple bias-flips harvesting circuit: a new solution for piezoelectric energy harvesting enhancement," in *25th International Conference on Adaptive Structures and Technologies*, 2014.

[6] W. Q. Liu, Z. H. Feng, J. He, and R. B. Liu, "Maximum mechanical energy harvesting strategy for a piezoelement," *Smart Mater. Struct.*, vol. 16, no. 6, pp. 2130–2136, 2007.

[7] Y. Liu, G. Tian, Y. Wang, J. Lin, Q. Zhang, and H. F. Hofmann, "Active piezoelectric energy harvesting: general principle and experimental demonstration," *J. Intell. Mater. Syst. Struct.*, vol. 20, no. 5, pp. 575–585, 2009.

[8] M. Lallart and D. Guyomar, "Piezoelectric conversion and energy harvesting enhancement by initial energy injection," *Appl. Phys. Lett.*, vol. 97, no. 1, p. 014104, 2010.

[9] J. Dicken, P. Mitcheson, I. Stoianov, and E. Yeatman, "Power-extraction circuits for piezoelectric energy harvesters in miniature and low-power applications," *IEEE Trans. Power Electron.*, vol. 27, no. 11, pp. 4514–4529, 2012.

[10] D. Kwon and G. A. Rincon-Mora, "Energy-investment schemes for increasing output power in piezoelectric harvesters," in *IEEE 55th International Midwest Symposium on Circuits and Systems*, Aug. 2012, pp. 1084–1087.

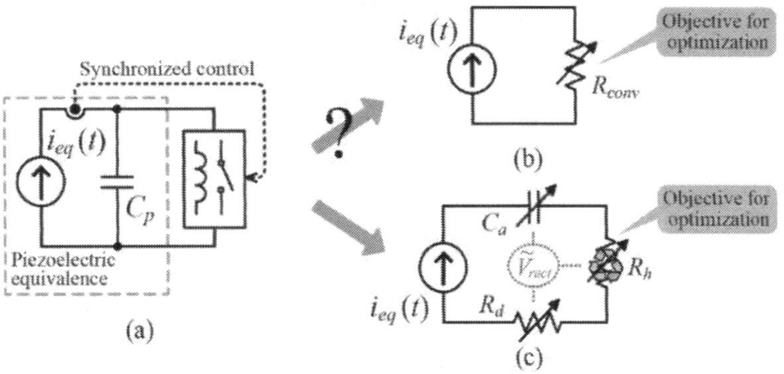

Figure 1. Bias-flip solutions for PEH. (a) Principle. (b) The equivalent targeted at power conversion (extraction). (c) The equivalent targeted at energy harvesting.

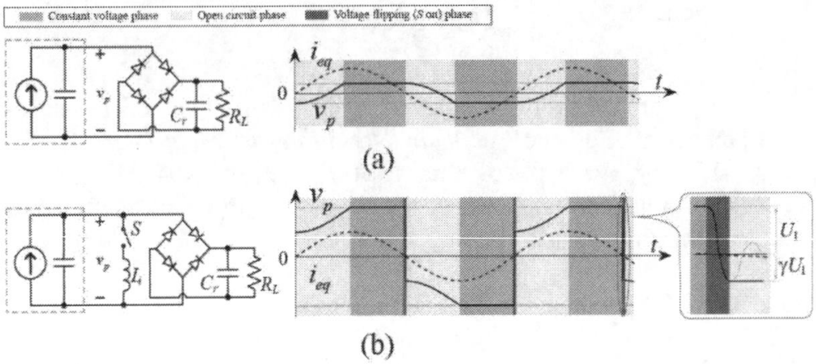

Figure 2. Circuit topologies and waveforms of two PEH interface circuits: (a) SEH. (b) P-SSHI.

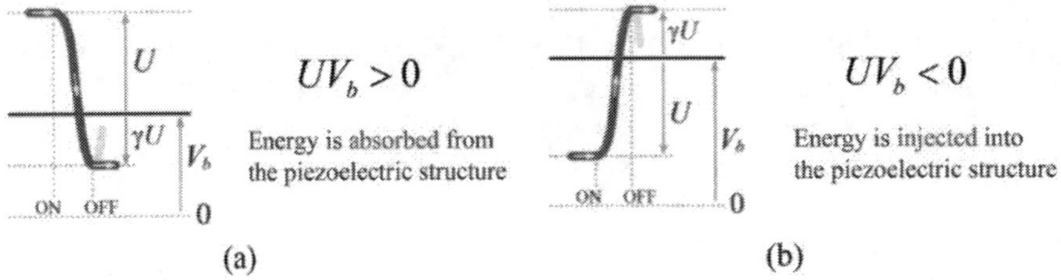

Figure 3. Bias-flip actions. (a) Passive. (b) Active.

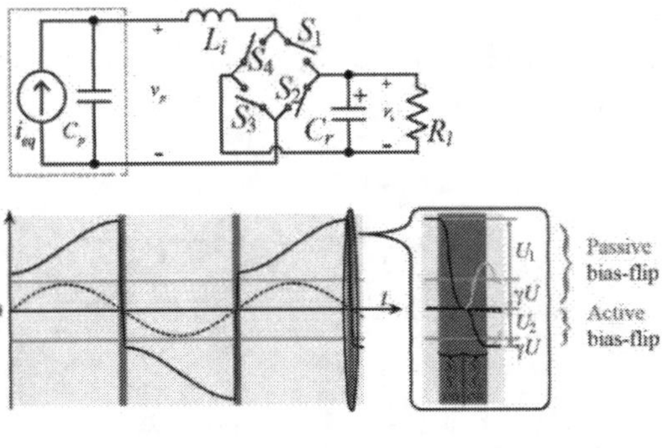

Figure 4. Circuit topology and waveform of SSPB.

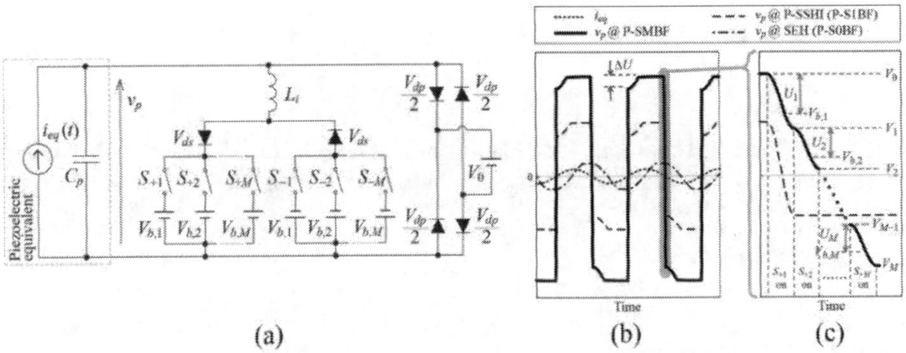

Figure 5. Generalized P-SMBF. (a) Circuit Topology. (b) Operating waveforms. (c) Zoom-in view in the synchronized instant.

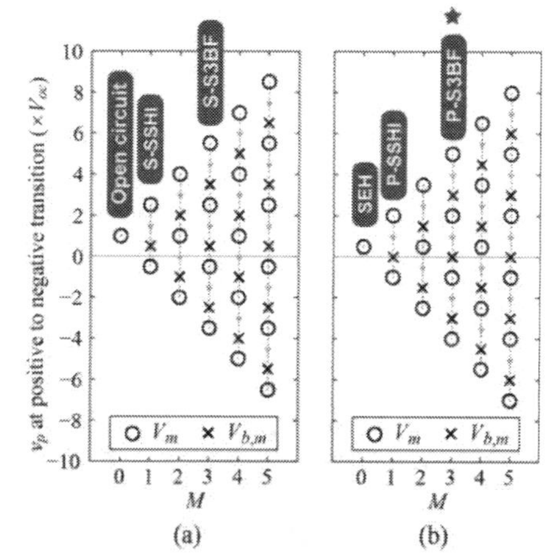

Figure 6. Optimal intermediate and bias voltages in voltage BF relay in (a) S-SMBF; and (b) P-SMBF.

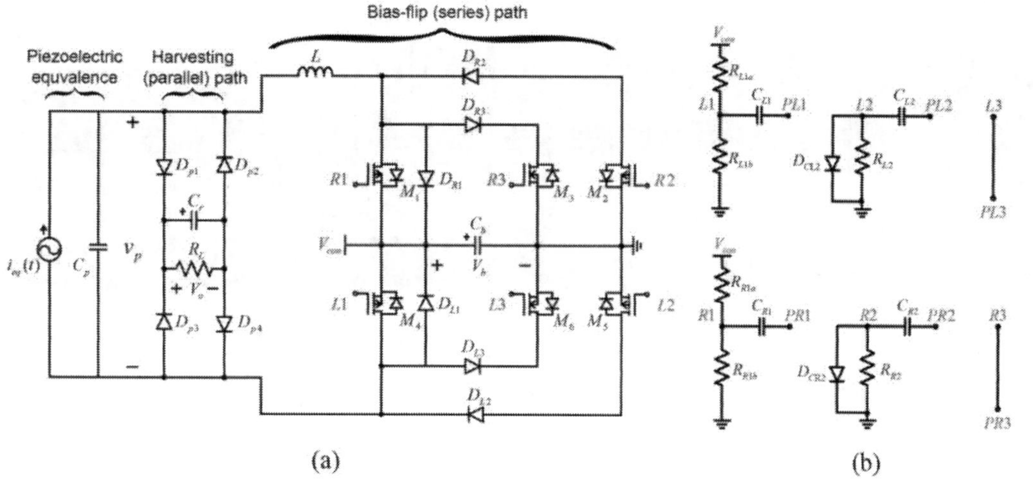

Figure 7. P-S3BF circuit. (a) Circuit topology. (b) Driving circuit.

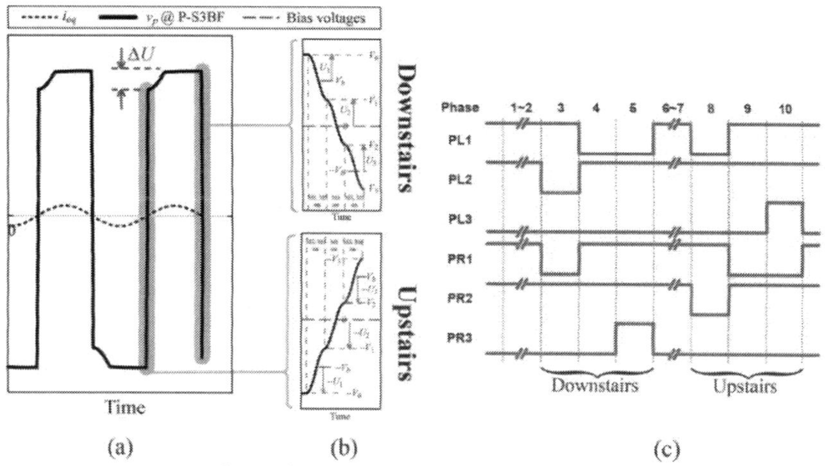

Figure 8. P-S3BF operation principle. (a) Waveforms. (b) Zoom-in view of the synchronized instant. (c) Control timing sequence.

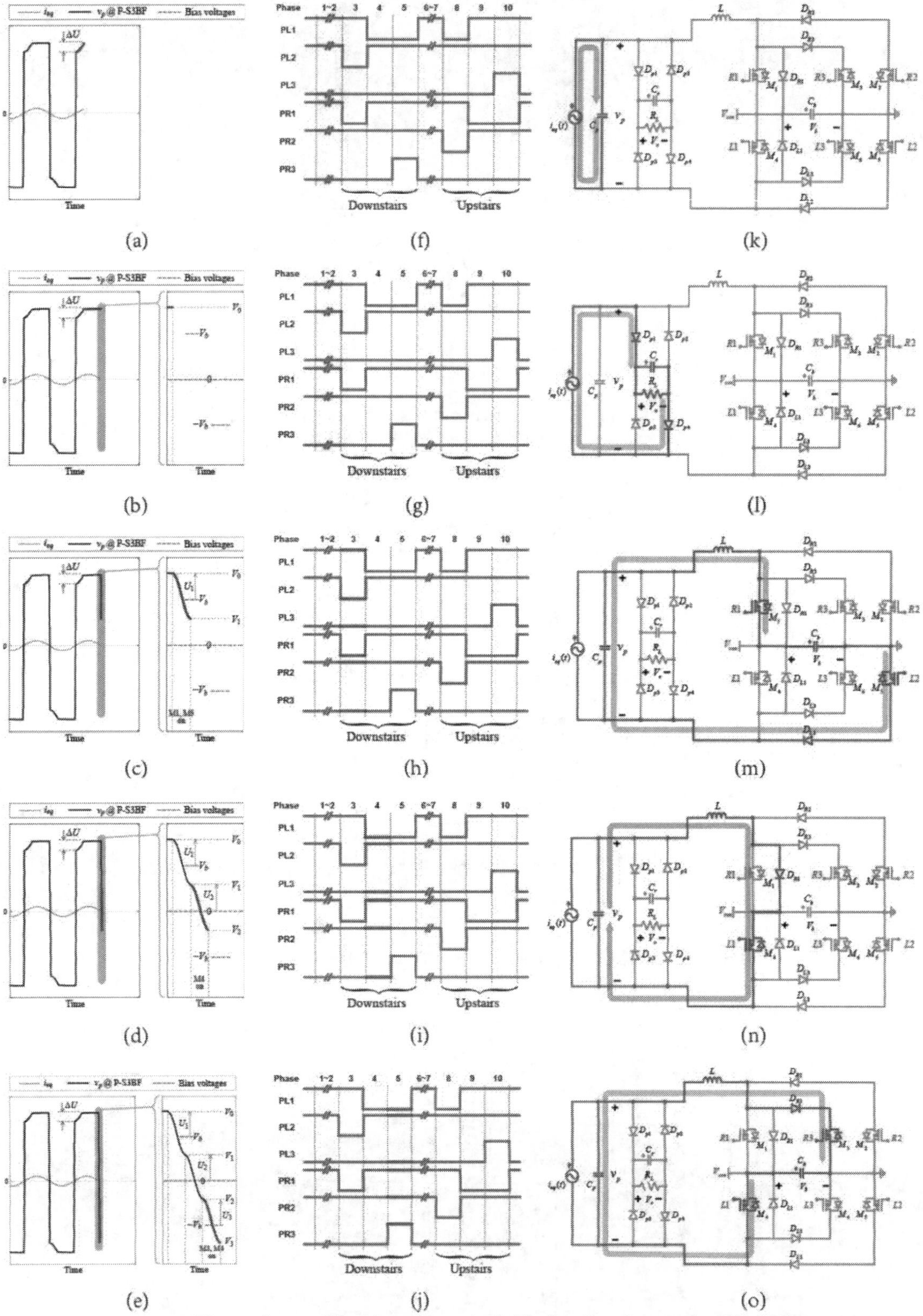

Figure 9. Working phases in half of a vibration cycle of P-S3BF.
(a)-(e) Operation waveforms. (f)-(j) Control timing. (k)-(o) Current flow paths.
(a) (f) (k) Open circuit. (b) (g) (l) Constant voltage. (c) (h) (m) the first bias-flip.
(d) (i) (n) the second bias-flip. (e) (j) (o) the third bias-flip.

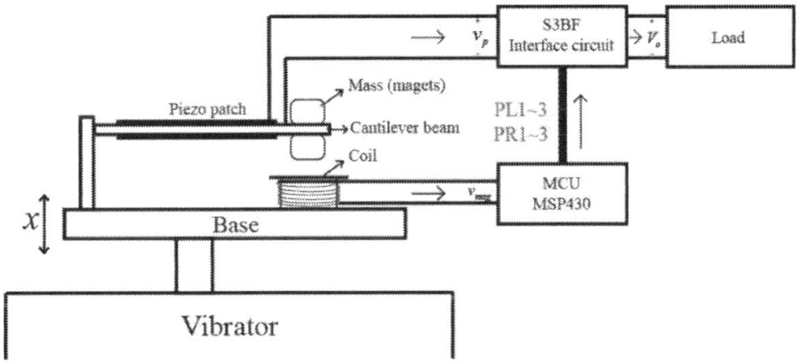

Figure 10. Experimental setup.

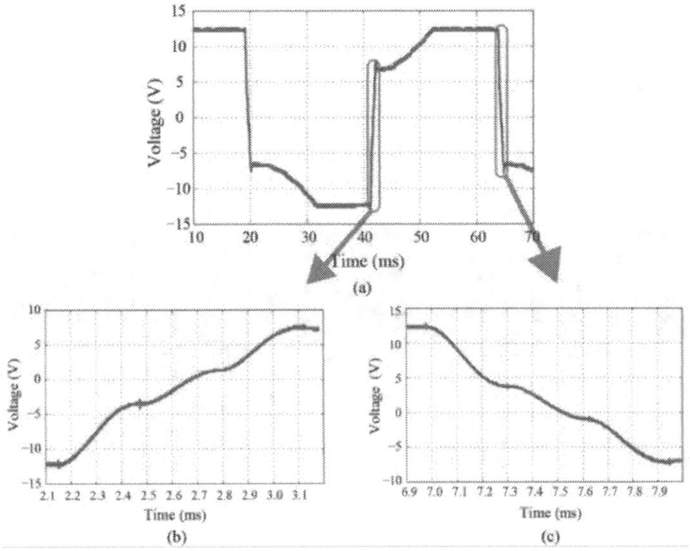

Figure 11. P-S3BF experimental result. (a) Operating waveforms. (b) Upstairs instant. (c) Downstairs instant.

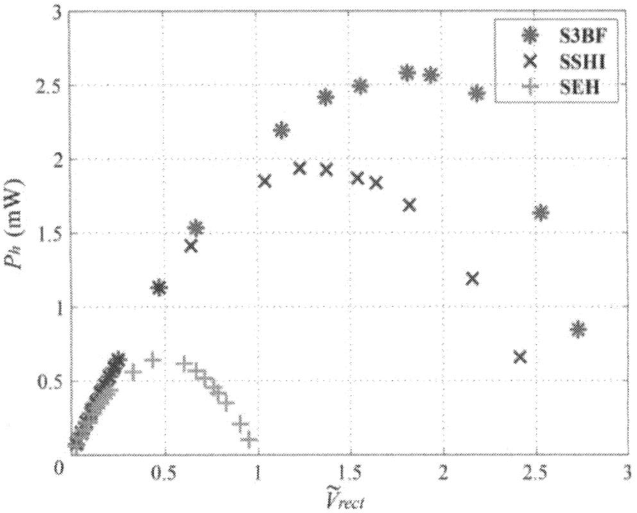

Figure 12. Harvested power by using different interface circuits.

Table 1 Electrical Specifications

Parameters	**Value**
Diodes	1N4004
MOSFETs	VISHAY Si4590DY
L	47 mH
Cr	33 μF
Cb	2.2 μF
R_{R1a},R_{L1a}	6.8 kΩ
R_{R1b}, R_{L1b}, R_{R2}, R_{L2}	680 kΩ
C_{L1}, C_{L2}, C_{R1}, C_{R2}	3.3 μF
γ	-0.64

Table 2 Mechanical & Piezo Specifications

Parameters	**Value**
Piezoelectric patch	50.0×49.7×0.2 (L×W×H) (mm)
Cantilever beam	59.5×53.1×0.2 (L×W×H) (mm)
Material	PZT-5
V_{oc}	10.75 V
C_p	220.75 nF
f_0	24.0 Hz

ICAST2015 #011

Design of an Integrated Semi-active Seat Suspension for both Longitudinal and Vertical Vibration Attenuation

Xian-Xu Bai*, Peng Jiang, and Li-Jun Qian

Department of Vehicle Engineering, Hefei University of Technology, Hefei 230009, People's Republic of China

Abstract

In this paper, an integrated semi-active seat suspension, mainly composed of a switching mechanism and a torque-controllable rotary magnetorheological (MR) damper working in pure shear mode, for both longitudinal and vertical vibration attenuation is proposed and designed. The switching mechanism employs the parallelogram frames as a motion guide which keeps the seat pan moving longitudinally and vertically. Both longitudinal and vertical motions are transformed into a rotary motion which is transmitted to the rotary MR damper after an amplification by a gear mechanism. The torque generated by the MR damper can be tuned through adapting the applied current and thence the seat vibration control could be realized. The mathematical model and the virtual prototype model of the semi-active seat suspension system are established in MATLAB/Simulink and Adams/View, respectively, and vibration control performance of the system is simulated and analyzed. Based on the established experimental test rig, the prototype of the semi-active seat suspension system is tested, and the results of the mathematical model, the dynamics simulation, and the experimental test are compared.
Keywords: Integrated seat suspension, vibration control, MR damper.

1. INTRODUCTION

Vibrations from vehicles, such as motor vehicles, water vehicles, and aircrafts, due to rough roads, waves, and other excitations, are generally smaller than 10 Hz [1], which is the sensitive frequency range of occupant body. Seats are actually a vibration isolation element installed between the occupants and vehicles. When the vibrations transmit to the seats, they would influence riding comfort for occupants and even lead to their health and security issues [2,3]. Two ways or steps could be used to attenuate vibrations transmitted from seats to occupants and improve riding comfort and security performance. On the one hand, vibrations from the vehicles themselves should be controlled so as to control the seat vibrations from the source. On the other hand, the vibration energy could be attenuated on the way of the vibration transmission by the seat suspensions which connect the seats to the vehicles. Suspensions for ground vehicles are a typical application of the former-mentioned approach, because vibrations transmitting from tires to the vehicle body can be to some extent mitigated by the vehicle suspensions [4-6], however, it is difficult to utilize this approach for ships and aircrafts. In recent years, increasing attentions are paid to how to improve the seat suspension performance and control the vibration transmitting to the seats and occupants [7,8].

Seat suspensions could be categorized into passive seat suspensions, active seat suspensions, and semi-active seat suspensions according to control types. Passive seat suspensions composed of springs and hydraulic dampers have advantages of simple configuration and low cost, but the parameters of the

* Corresponding author. Tel: (+86) 189 6379 4953; E-mail: bai@hfut.edu.cn

springs and the hydraulic dampers are determined by experiential or optimal design and cannot be adjusted to adapt to different working conditions. Active seat suspensions provide controllable driving force using hydraulic actuators, pneumatic actuators, or servo motors, and provide excellent vibration control performance. Maciejewski et al. [9] proposed an active seat suspension system based on a controllable pneumatic spring and a hydraulic damper, the active control of driving force is implemented by a controller and a proportional valve which could control expansion and compression of the air-flow in the pneumatic spring. The active seat suspensions show excellent vibration isolation performance, but their practical applications are constrained due to their complex structure, high power consumption, and also high cost [10]. Semi-active seat suspensions mainly consist of stiffness-controllable elastic elements or damping-adjustable elements, they have advantages of simple structure as passive seat suspensions and excellent vibration mitigation performance as active seat suspensions [11,12]. They provide a wide application prospects. Wu et al. [12] developed a semi-active seat suspension using an electrorheological (ER) damper. The damping force of the ER damper could be adjusted in real time. Choi and Wereley [13] showed that semi-active helicopter crew seat suspensions using magnetorheological (MR) dampers could effectively reduce harm to the occupants induced by impact force. Choi et al. [14,15] have done much on the applications of ER dampers, MR dampers, and other semi-active actuators used in the seat suspensions. Sun et al. [16] proposed a seat suspension system based on four MR elastomers installed onto a conventional passive seat suspension that could only attenuate vertical vibration to achieve half semi-active control for the longitudinal vibration and half passive control for vertical vibration at the same time. Up to date, however, most of structural designs of the seat suspension focus on vertical vibration attenuation only, even some seat suspension structures had multi-dimensional vibration attenuation performance, they are all repeated combinations of several one-dimensional hardware and each dimension is decoupled with each other. As we could imagine, if a "functional integration" concept [17-19] could be utilized for seat suspension design to realize an integrated two-dimensional or multi-dimensional controllable seat suspension system, not only the seat suspension structure could be simplified, but also the real-time control of two-dimensional or multi-dimensional seat vibrations could be realized. Moreover, it would also further decrease the system cost and maximize the applications of the controllable seat suspension systems.

In this study, an integrated semi-active seat suspension system for both longitudinal and vertical vibration attenuation is proposed. The system transforms both longitudinal and vertical vibrations of the seat into a rotary motion by a switching mechanism. At the same time, the energy of the rotary motion is transmitted to the transmission shaft of the controllable MR damper and dissipated, so the integrated semi-active control of both longitudinal and vertical vibrations is realized. The semi-active seat suspension system is mathematically analyzed and experimentally tested.

2. STRUCTURE AND PRINCIPLE OF THE SEAT SUSPENSION SYSTEM

2.1 Structure of the seat suspension system

Figures 1 and 2 present the views and the mechanical principle of the seat suspension system, respectively. As shown in figures 1 and 2, the seat suspension system consists of a motion switching mechanism, a damping- or torque-controllable actuator, elastic elements, and a transmission mechanism. The switching mechanism is composed of two parallelogram frames with movable frames, front and rear rockers, and fixed frames. The parallelogram frames will guide the seat to move longitudinally (x-axis) and vertically (z-axis). Meanwhile, the longitudinal and vertical motions of the seat are transformed into a rotatory motion by the parallelogram frames. The damping- or torque-controllable actuator is realized by a rotary MR damper which works in pure shear mode. Four torsional springs with the same property are installed between the hinged point of the four rockers and the fixed frames and work as the elastic elements to support the static weight of the seat and the occupant. The transmission mechanism is realized by a pair of gears, i.e., the driving gear and a driven gear, as shown in figures 1 and 2. The driving gear rotates with the rockers and transmit the seat motion or vibration to the driven gear which is coupled with the transmission shaft of the rotary MR damper. The gear mechanism is also utilized to amplify the rotatory motion of the rockers so as to increase the effective stroke of the rotary MR damper.

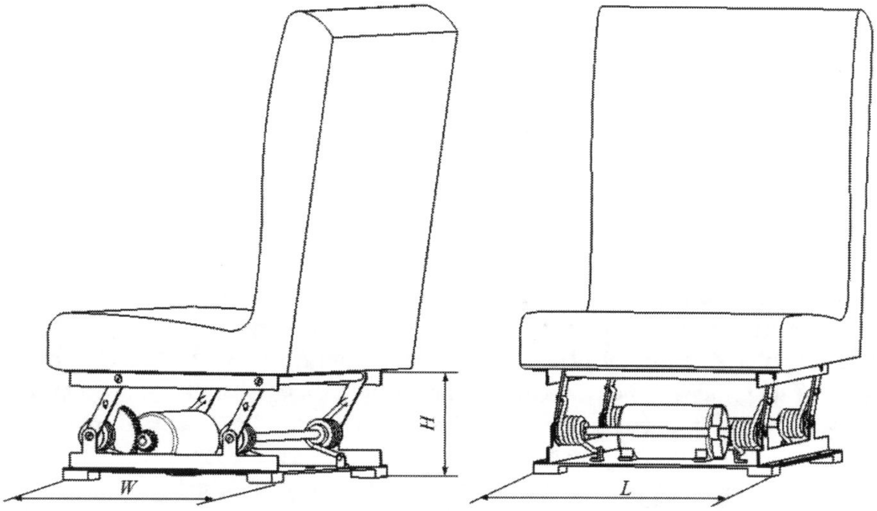

Figure 1. Line drawings of the seat suspension assembly.

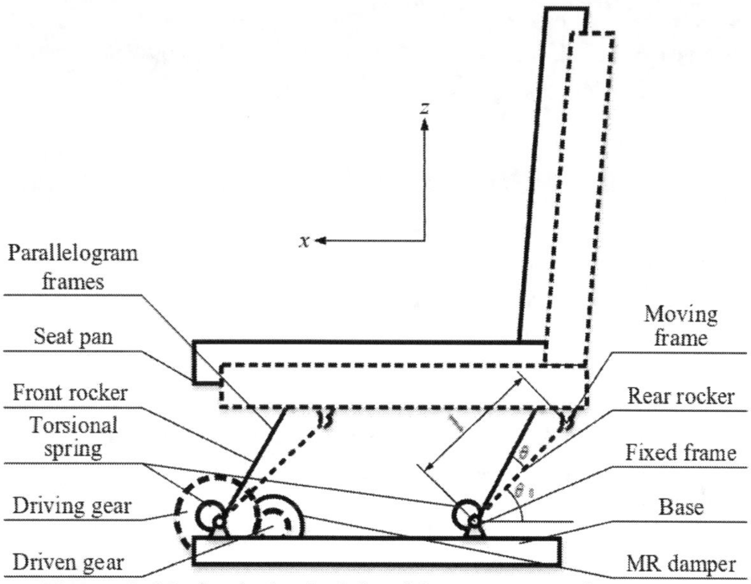

Figure 2. Mechanical principle of the seat suspension system.

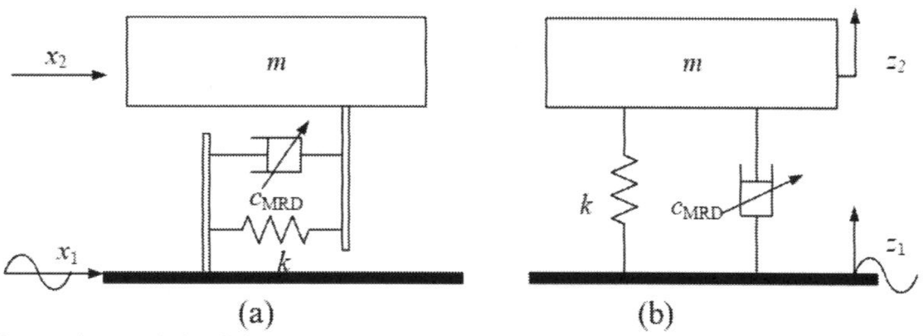

Figure 3. The dynamic model of the seat suspension system: (a) the longitudinal direction and (b) the vertical direction.

2.2 Mathematical model of the seat suspension system

According to figures 1 and 2, the dynamics of the seat suspension system could be decoupled into longitudinal direction (x-axis) and the vertical direction (z-axis), as shown in figures 3(a) and 3(b). The dynamic model of the seat suspension system as shown in figures 3(a) and 3(b) can be expressed as

$$\begin{cases} m\ddot{x}_2 = \frac{k\theta}{l} sin(\theta_0 + \theta) + F_{\mathrm{MR}x} \\ x_2 = x_1 + l\,cos(\theta_0 + \theta) - l\,cos\,\theta_0 \end{cases} \tag{1a}$$

$$\begin{cases} m\ddot{z}_2 = -\frac{k\theta}{l} cos(\theta_0 + \theta) - F_{\mathrm{MR}z} \\ z_2 = z_1 + l\,sin(\theta_0 + \theta) - l\,sin\,\theta_0 \end{cases} \tag{1b}$$

where m is the total mass of the seat and the occupant (Hybrid III 50th male); k is the total stiffness of the torsional springs; l is the length of the rocker; θ_0 is the angle between the rocker and the longitudinal direction when the seat is under the static equilibrium state; θ is the rotation angle of the rocker; x_1 and z_1 are the longitudinal and vertical displacement excitations of the seat suspension system, respectively; x_2 and z_2 are the longitudinal and vertical displacement responses of the seat suspension system, respectively; $\ddot{x}_2$ and $\ddot{z}_2$ are the longitudinal and vertical acceleration responses of the seat suspension system, respectively; $F_{\mathrm{MR}x}$ and $F_{\mathrm{MR}z}$ are the longitudinal and vertical component of the damping force produced by the MR damper, respectively. $F_{\mathrm{MR}x}$ and $F_{\mathrm{MR}z}$ can be expressed as

$$F_{\mathrm{MR}x} = \frac{nc_{\mathrm{MRD}}\dot{\theta}}{l} sin(\theta_0 + \theta) \tag{2a}$$

$$F_{\mathrm{MR}z} = \frac{nc_{\mathrm{MRD}}\dot{\theta}}{l} cos(\theta_0 + \theta) \tag{2b}$$

where n is the transmission ratio of the gear mechanism; c_{MRD} is the equivalent damping of the rotary MR damper; $\dot{\theta}$ is the rotational angular velocity of the rocker.

2.3 Rotary MR damper

Figure 4 shows the structural principle of a rotary MR damper for the proposed seat suspension system. As seen from figure 4, the moving parts, including the transmission shaft, the rotating discs, and the shaft sleeves, rotate relatively to the fixed discs. The bobbins are fixed between two adjacent fixed discs. MR fluid fills the room formed by the fixed discs, rotating discs, bobbins, and shaft sleeves. The fixed discs, the rotating discs, and the cylinder are all made from the material with high permeability, the others are made from non-magnetic materials. The magnetic field generated by the exciting coil with applied current starts from the cylinder, passes through the coil-left fixed disc, the MR fluid between the coil-left fixed disc and the rotating disc, the rotating disc, the MR fluid between the coil-right fixed disc and the rotating disc, and the coil-right fixed disc, and finally closes at the cylinder.

When no magnetic field, MR fluid behaves as Newtonian fluids. Under the effect of the magnetic field generated by the exciting coils with applied current, MR fluid shows a characteristic of Bingham fluids. The torque of the MR damper can be controlled by adjusting the applied current to the exciting coils. The total torque produced by a single pair of rotating-fixed discs can be expressed as

$$T_0 = T_\eta + T_{\mathrm{B}} \tag{3}$$

where T_η is the torque depended on the dynamic viscosity of the MR fluid; T_{B} is the torque due to the magnetic flux density. T_η and T_{B} can be respectively expressed as

$$T_\eta = 2\pi\eta\dot{\gamma} \int_{R_0}^{R_1} r^2 dr \tag{4a}$$

$$T_{\mathrm{B}} = 2\pi\tau_{\mathrm{y}}(B) \int_{R_0}^{R_1} r^2 dr \tag{4b}$$

where η is the dynamic viscosity of the MR fluid; $\dot{\gamma}$ is the shear rate; $\tau_{\mathrm{y}}(B)$ is the shear yield stress related to the magnetic field and is given by

$$\tau_{\mathrm{y}}(B) = k_1 B^\alpha \tag{5}$$

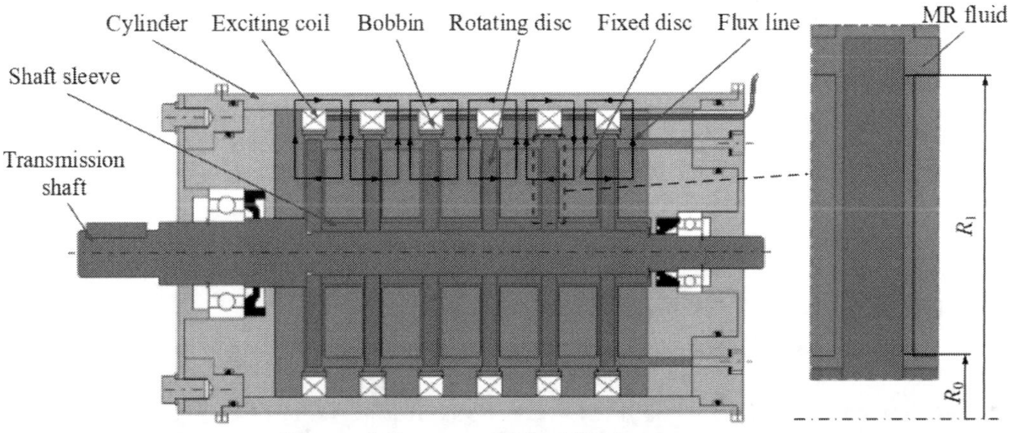

Figure 4. Structure of the MR damper.

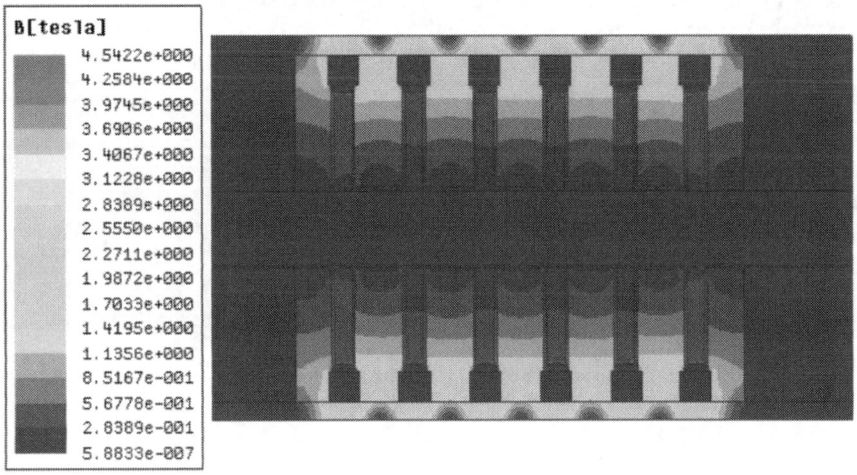

(a)

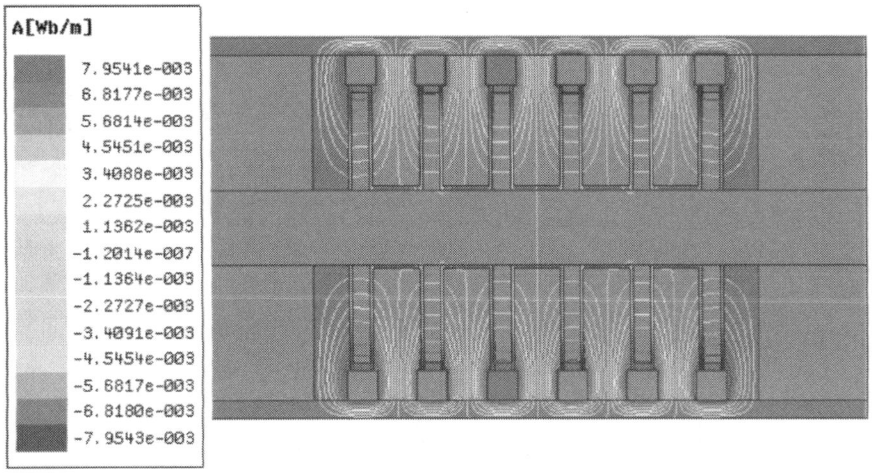

(b)

Figure 5. The magnetic field of the MR damper: (a) the magnetic flux density contours and (b) the magnetic flux lines.

where k_1 and α are the constants; B is the magnetic flux density.

From equations (3), (4), and (5), torque generated by the MR damper will be increased (or decreased) by increasing (or decreasing) the magnetic flux density. Combining equations (1) and (2), dynamic response of the seat could be controlled by tuning the applied current to the exciting coils of the MR damper.

To validate the magnetic principle of the proposed rotary MR damper, the software package ANSYS Maxwell is used to simulate the electromagnetic characteristics of the MR damper. Figures 5(a) and 5(b) present the magnetic flux density contours and the magnetic flux lines when the MR damper is applied a current level of 3 A. As shown in figure 5(a), the magnetic flux density between the fixed discs and the rotating discs is nearly 0.3 Tesla. As seen from figure 5(b), the magnetic flux lines travel through two adjacent fixed discs, the rotating disc between two adjacent fixed discs, and the cylinder, which validates the principle of the MR damper shown in figure 4.

3. NUMERICAL SIMULATION OF THE SEAT SUSPENSION SYSTEM

To theoretically analyze the performance of the proposed seat suspension system, MATLAB/Simulink is utilized to simulate the vibration transmissibility of the seat suspension system when the MR damper provides different equivalent damping (c_{MRD}) values. The excitation to the seat suspension system is sinusoidal swept-frequency displacement signals with an amplitude of 15 mm and frequency range of 0.5-6 Hz.

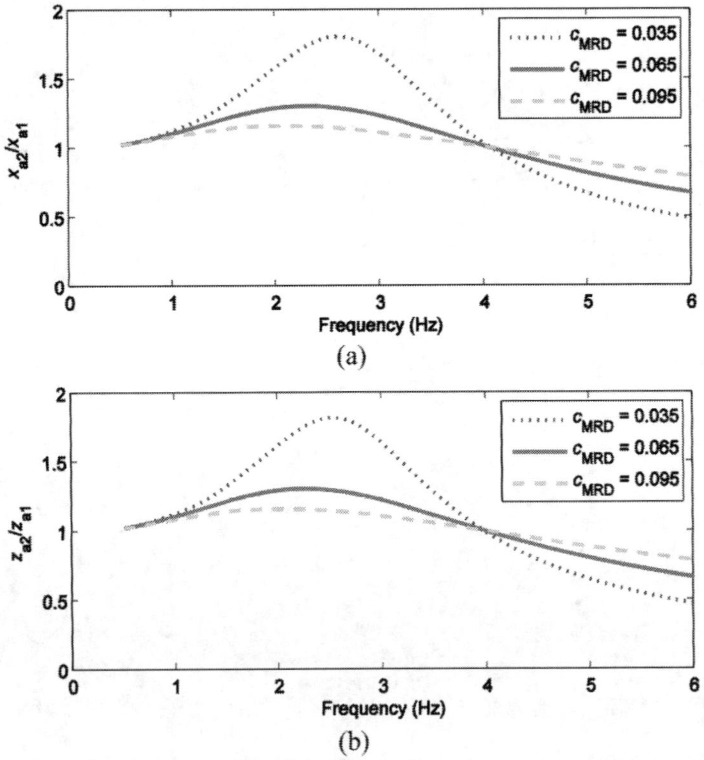

Figure 6. Numerical simulation results of the seat suspension system: (a) transmissibility in the longitudinal direction and (b) transmissibility in the vertical direction.

Figures 6(a) and 6(b) present the transmissibility of longitudinal and vertical vibrations of the seat suspension system, respectively. As shown in figure 6, the resonance frequency (ω_{n1}) of the seat suspension system is nearly 2.5 Hz. When the excitation frequency is lower than the resonance frequency (ω_{n1}), the greater the equivalent damping (c_{MRD}) value is, the smaller the vibration transmissibility will be. When near the resonance frequency, as the equivalent damping (c_{MRD}) value increases, the vibration transmissibility decreases obviously. When the excitation frequency is greater than $\sqrt{2}\omega_{n1}$, the smaller the equivalent damping (c_{MRD}) value is, the smaller the vibration transmissibility will be. It should be noted that, as shown in figure 6, the equivalent damping values (c_{MRD}) of 0.035 Nm s/°, 0.065 Nm s/°, and 0.095 Nm s/° are equivalent to the torque values of 6.58 Nm, 11.04 Nm, and 13.53 Nm provided by the rotary MR damper, respectively. Therefore, by controlling the applied current to the MR damper according to dynamic response of the seated occupants, i.e., controlling the equivalent damping (c_{MRD}) value, the integrated semi-active control of both longitudinal and vertical vibrations of the seat suspension system can be realized.

4. VIRTUAL PROTOTYPE SIMULATION OF THE SEAT SUSPENSION BASED ON ADAMS/VIEW

To further validate the vibration control performance of the seat suspension model, the virtual prototype model of the seat suspension is constructed in the software package Adams/View. The same swept-frequency displacement signals to the numerical simulation are adopted as excitations applied to the base of the seat suspension model in the Adams/View. Figure 7 presents the transmissibility comparison when the equivalent damping (c_{MRD}) values are 0.035, 0.065, and 0.095, respectively. As shown in figure 7, the resonance frequency (ω_{n2}) of the seat suspension system is slightly smaller than 2.5 Hz. When the excitation frequency is lower than the resonance frequency (ω_{n2}), the greater the equivalent damping (c_{MRD}) value is, the smaller the vibration transmissibility will be. Near the resonance frequency of 2.5 Hz, when the equivalent damping (c_{MRD}) value increases, the vibration transmissibility decreases obviously. When the excitation frequency is greater than$\sqrt{2}\omega_{n2}$, the smaller the equivalent damping (c_{MRD}) value is, the smaller the vibration transmissibility will be. Similar results with what are shown in figures 6 and 7, the longitudinal and vertical vibrations can be mitigated by controlling the applied current to the MR damper.

5. EXPERIMENTAL TEST

5.1 Rotary MR damper

Figure 8 presents the established torque-test rig for the MR damper. To obtain the torque controllability of the MR damper, the MR damper under different driving speeds of 15 rpm, 60 rpm, and 180 rpm within a safe current range (0-4 A) is tested.

As shown in figure 9, when no current, the torque generated by the dynamic viscosity of the MR fluid is nearly 1.02 Nm, while with a maximum current of 4 A, the maximum torque generated by the MR damper increases to around 13.53 Nm. The torque approximately increases linearly with the current level. The torque of the MR damper working in pure shear mode is not dependent on the exciting speed, as seen from figure 9.

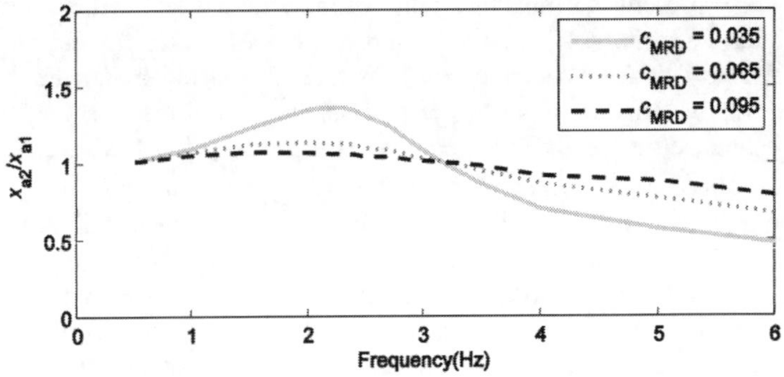

(a)

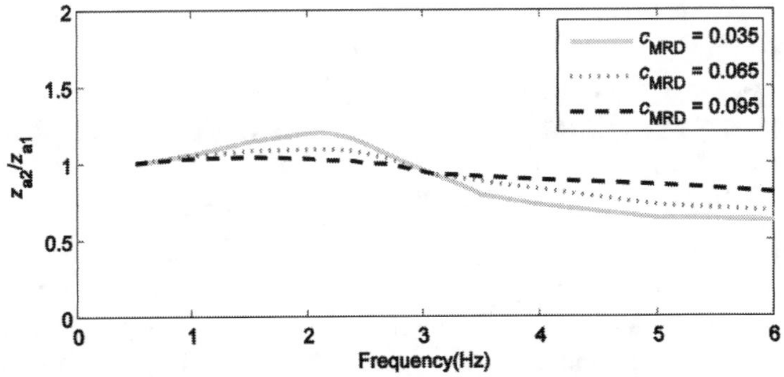

(b)

Figure 7. The simulation results of the virtual prototype of the seat suspension system: (a) transmissibility in the longitudinal direction and (b) transmissibility in the vertical direction.

Figure 8. The torque test rig for the MR damper.

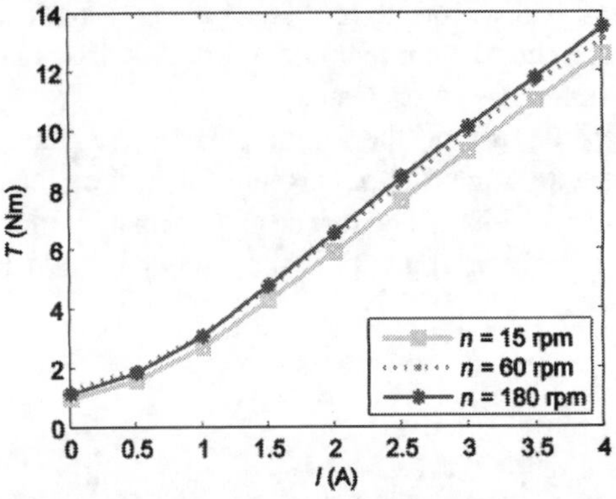

Figure 9. The relationships between the torque and the applied current of the MR damper.

5.2 Vertical vibration test of the seat suspension system

Figure 10 presents the vertical vibration test rig for the seat suspension system. As shown in figure 10, the seat suspension system is installed on the surface of the vibration actuator, and the mass representing the Hybrid III 50th male are fixed on the seat pan. Two accelerometers are respectively installed on the seat pan and the base of the seat suspension to measure vertical acceleration. A data acquisition system (Type: DEWE-501) is utilized to acquire and process the acceleration signals.

Figures 11, 12, and 13 present the time histories of the acceleration responses of the seat suspension when sinusoidal displacement excitations with an amplitude of 15 mm and frequencies of 1.5 Hz, 2.5 Hz, and 5 Hz are applied to the base of the seat suspension and different applied currents are supplied, respectively. Figure 14 presents the time histories of the acceleration response of the seat suspension when a sinusoidal displacement excitation with an amplitude of 15 mm and resonance frequency is applied to the base of the seat suspension and the applied current is 0 A through 4 A.

As shown in figures 11(a) and 11(b), when the excitation frequency (1.5 Hz) is lower than the resonance frequency (2.5 Hz), the vibration acceleration of the seat pan with applied current level of 0 A is nearly the same to that with applied current level of 2 A. As the excitation frequency increases to the resonance frequency (2.5 Hz), the vibration acceleration of the seat pan decreases much with increasing the applied current, as presented in figures 12(a) and 12(b). When the excitation frequency is higher than $2.5\sqrt{2}$ Hz, the seat pan acceleration is smaller than the excitation acceleration for both applied currents of 0 A and 2 A. But the control of the applied current to the MR damper does not work as what shown in figure 13. In order to present in full-scale, figure 14 is shown to present the whole process of dynamic response when increasing the applied current from 0 A to 4 A. As it can be seen from the figure that for the resonance frequency, the bigger applied current is, the better acceleration response of the seat pan will be.

As it can be seen from figures 11-14, by controlling the applied current to the MR damper, integrated semi-active control of vertical vibration of the seat suspension system can be realized. Combining figures 6 and 7, the experimental test results are shown agreement with the results of the numerical simulation and the virtual prototype simulation, and the rotary MR damper can work for the integrated semi-active control of both longitudinal and vertical vibrations of the proposed seat suspension system. It should be noted here, only vertical vibration response of the seat suspension system is tested in this study. The experimental tests for the two-dimensional vibration attenuation performance will be validated.

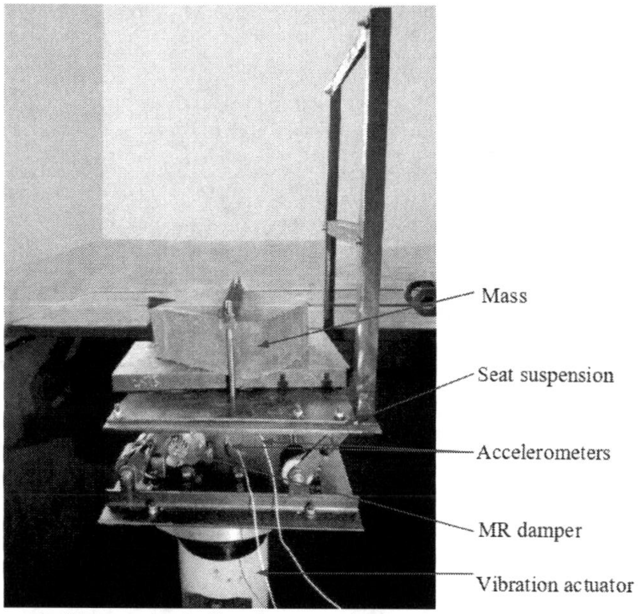

Figure 10. The vertical vibration test rig for the seat suspension system.

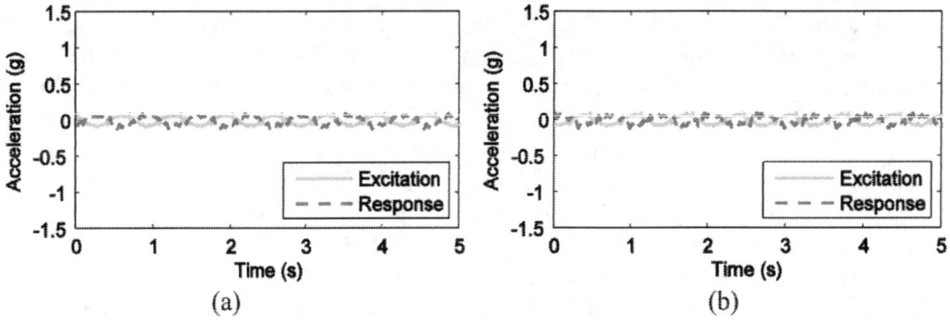

Figure 11. The time histories of the acceleration response of the seat suspension when a sinusoidal displacement excitation with an amplitude of 15 mm and frequency of 1.5 Hz is applied to the base of the seat suspension: (a) $I = 0$ A and (b) $I = 2$ A.

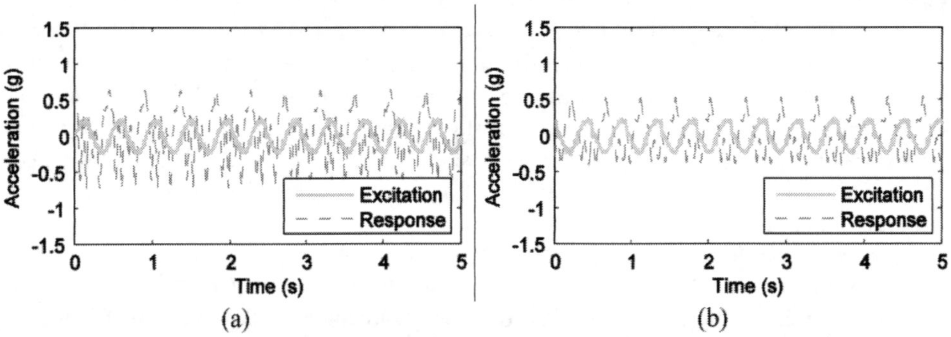

Figure 12. The time histories of the acceleration response of the seat suspension when a sinusoidal displacement excitation with an amplitude of 15 mm and frequency of 2.5 Hz is applied to the base of the seat suspension: (a) $I = 0$A and (b) $I = 2$ A.

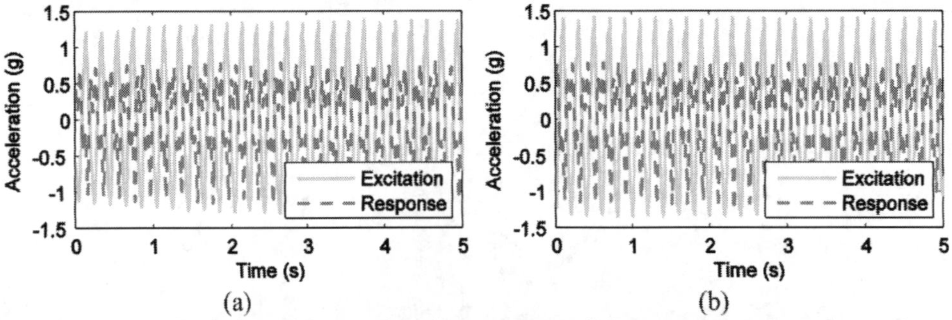

Figure 13. The time histories of the acceleration response of the seat suspension when a sinusoidal displacement excitation with an amplitude of 15 mm and frequency of 5 Hz is applied to the base of the seat suspension: (a) $I = 0$ A and (b) $I = 2$ A.

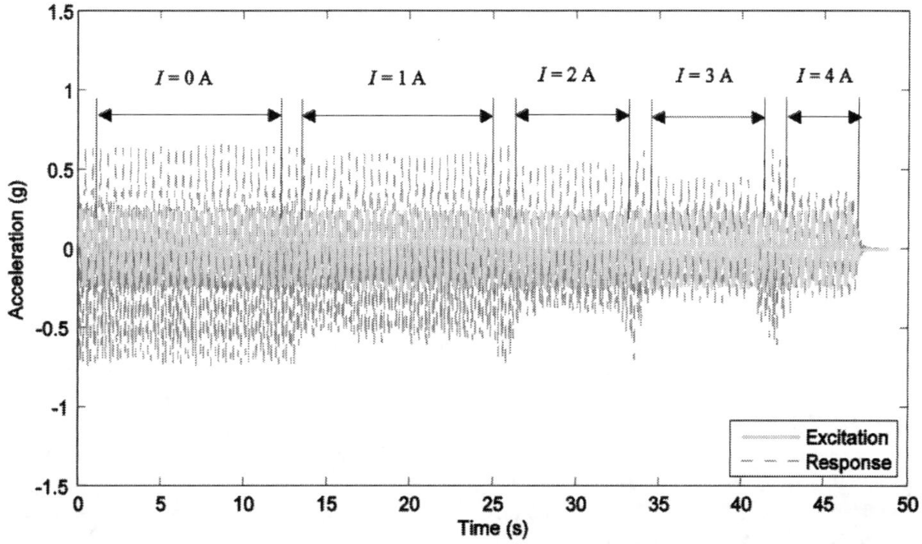

Figure 14. The time histories of the acceleration response of the seat suspension when a sinusoidal displacement excitation with an amplitude of 15 mm and frequency of 2.5 Hz is applied to the base of the seat suspension.

6. CONCLUSIONS

In this paper, an integrated semi-active seat suspension, composed of a switching mechanism transforming both longitudinal and vertical motions into a rotary motion and a real-time damping-controllable rotary magnetorheological (MR) damper working in pure shear mode, for both longitudinal and vertical vibration attenuation was proposed and designed. The mathematical model and the virtual prototype model of the semi-active seat suspension system were established in MATLAB/Simulink and Adams/View, respectively, and the vibration control performance of the system was simulated and analyzed. The vibration control performance of the seat suspension prototype was preliminarily tested. Based on the research results, concluding remarks could be obtained as follows:

(1) When the applied current to the MR damper increases at the resonance frequency of the system, both longitudinal and vertical vibrations could be attenuated significantly. When the excitation frequency is lower than 2.5 Hz or greater than $2.5\sqrt{2}$ Hz, the seat suspension system shows no obvious vibration control effectiveness for the two dimensions.

(2) Within a certain current range, the torque produced by the MR damper increases linearly from 1.02 Nm to 13.53 Nm and is not dependent on the excitation speed.

ACKNOWLEDGMENTS

The authors wish to acknowledge the National Natural Science Foundation of China (Grant No. 51305114), Anhui Provincial Natural Science Foundation (Grant No. 1408085QE91) and Fundamental Research Funds for the Central Universities (No. 2014HGCH0008), for their support of this research. The authors would also like to thank Mr. Hui Pan, a Bachelor student from Hefei University of Technology, for his assistance in design and finite element analysis of the rotary MR damper.

REFERENCES

1. Li, W., Zhang, X. and Du, H., 2012, Development and simulation evaluation of a magnetorheological elastomer isolator for seat vibration control, *Journal of Intelligent Material Systems and Structures*, Vol. 23, No. 9, 1041-1048.
2. Paddan, G. S., Griffin, M. J., 2002, Evaluation of whole-body vibration in vehicle, *Journal of Sound and Vibration*, Vol. 253, No. 1, 195-213.
3. Pope, M. H., Wilder, D. G. and Magnusson, M. L., 1999, A review of studies on seated whole body vibration and low back pain, *Journal of Engineering in Medicine*, Vol. 213, No. 6, 213-435.
4. Gao, H., Lam, J. and Wang, C., 2006, Multi-objective control of vehicle active suspension systems via load-dependent controllers, *Journal of Sound and Vibration*, Vol. 290, No. 3, 654-675.
5. Shafie, A. A., Bello, M. M. and Khan, R. M., 2015, Active vehicle suspension control using electro hydraulic actuator on rough road terrain, *Journal of Advanced Research*, Vol. 9, No. 1, 15-30.
6. Bai, X. X., Hu, W. and Wereley, N. M., 2013, Magnetorheological damper utilizing an inner bypass for ground vehicle suspensions, *IEEE Transactions on Magnetics*, Vol. 49, No. 7, 3422-3425.
7. Ieluzzi, M., Turco, P. and Montiglio, M., 2006, Development of a heavy truck semi-active suspension control, *Control Engineering Practice*, Vol. 14, No. 3, 305-312.
8. Gan, Z., Hillis, A. and Darling, J., 2015, Adaptive control of an active seat for occupant vibration reduction. *Journal of Sound and Vibration*, Vol. 349, 39-55.
9. Maciejewski, I., Meyer, L. and Krzyzynski, T., 2010, The vibration damping effectiveness of an active seat suspension system and its robustness to varying mass loading, *Journal of Sound and Vibration*, Vol. 329, No. 19, 3898-3914.
10. Kela, L., Vähäoja, P., 2009, Recent studies of adaptive tuned vibration absorbers/neutralizers, *Applied Mechanics Reviews*, Vol. 62, No. 6, 060801.
11. Du, H., Sze, K. Y. and Lam, J., 2005, Semi-active $H\infty$ control of vehicle suspension with magneto-rheological dampers, *Journal of Sound and Vibration*, Vol. 283, Nos. 3-5, 981-996.
12. Wu, X. and Griffin, M. J., 1997, A semi-active control policy to reduce the occurrence and severity of end-stop impacts in a suspension seat with an electrorheological fluid damper, *Journal of Sound and Vibration*, Vol. 203, No. 5, 781-793.
13. Choi, Y. T. and Wereley, N. M., 2005, Biodynamic response mitigation to shock loads using magnetorheological helicopter crew seat suspensions, *Journal of Aircraft*, Vol. 42, No. 5, 1288-1295.
14. Choi, S. B., Choi, J. H., Nam, M. H., Cheong, C. C. and Lee, H. G., 1998, A semi-active suspension using ER fluids for a commercial vehicle seat, *Journal of Intelligent Material Systems and Structures*, Vol. 9, No. 8, 601-606.
15. Choi, S. B. and Han, Y. M., 2007, Vibration control of electrorheological seat suspension with human-body model using sliding mode control, *Journal of Sound and Vibration*, Vol. 303, Nos. 1-2, 391-404.
16. Sun, S. S., Yang, J., Deng, H. X., Du, H., Li, W. H., Alici, G. and Nakano, M., 2015, Horizontal vibration reduction of a seat suspension using negative changing stiffness magnetorheological elastomer isolators, *International Journal of Vehicle Design*, Vol. 68, Nos. 1-3, 104-118.
17. Wang, D. H. and Wang, T., 2009, Principle, design and modeling of an integrated relative displacement self-sensing magnetorheological damper based on electromagnetic induction, *Smart Materials and Structures*, Vol. 18, No. 9, Article no. 095025.
18. Wang, D. H., Liao, W. H. and Bai, X. X., 2010, An integrated relative displacement self-sensing magnetorheological damper: prototyping and testing, *Smart Materials and Structures*, Vol. 19, No. 10, Article no. 075001.
19. Wang, D. H. and Bai, X. X., 2013, A magnetorheological damper with an integrated self-powered displacement sensor, *Smart Materials and Structures*, Vol. 22, No. 7, Article no. 075001.

ICAST2015: 26th International Conference on Adaptive Structures and Technologies
October 14-16th, 2015, Kobe, Japan

ICAST2015 #028

PPF-based vibration control of a seat frame of motor vehicle

Keiyu Kadoi[1], Kentaro Takagi[1*], Ichiro Jikuya[2], Satoshi Ishiguro[3], Hirohito Hayashi[4], Hisamitsu Harada[5], and Kazuhiko Oshima[6]

[1] Department of Mechanical Science and Engineering, Nagoya University, Nagoya, Japan.
[2] Department of Aerospace Engineering, Nagoya University, Nagoya, Japan.
[3] 14 Lab, Nagoya, Japan.
[4] Mikawa Sales Office, Kyocera, Anjo, Aichi, Japan.
[5] Harada Vehicle Design, Miyoshi, Aichi, Japan.
[6] Department of Integrated Mechanical Engineering, Daido University, Nagoya, Japan.

Abstract

This paper shows vibration control of small vibration of a seat frame of a motor vehicle by using piezoelectric elements. The idea of Positive Position Feedback (PPF) is applied to propose a method of vibration control, and we call the method by the PPF-based control because the collocation of sensors and actuators are not assumed in our framework. Firstly, we design a PPF-based controller for a single mode. Due to the analogy of PPF to a mechanical tuned-mass damper, we determine a design parameter corresponding to the natural frequency of the controller. We design the controller gain from the stability margin. The remaining parameter, the damping ratio, of the controller is chosen by numerical search in which the maximum of the real parts of the poles of the closed loop transfer function is minimized. Secondly, we propose a design method of a controller which works not only for a single mode, but also for multiple modes simultaneously. The controller for multiple modes is designed by simply adding controllers each of which is designed for a different single-mode. Finally, we demonstrate the effectiveness of the proposed method in the experimental setup. We assemble stack type piezoelectric elements at the bottom of a real seat frame of a motor vehicle. We measure the frequency response from the piezoelectric actuator to the piezoelectric sensor. As a result, the seat frame shows a typical frequency response of a lightly damped flexible structure. We identify a parametric model of the first and the second modes which are approximated as single modes from the frequency response. We design the PPF-based controller for the parametric model and it is shown that the PPF-based controller is effective for the suppression of the first mode, the suppression of the second mode, and the simultaneous suppression of both modes.

1. INTRODUCTION

Motor vehicles are used as a means of transportation all around the world. Vibration takes place in a car because of torsional vibration of the engine, bumps of road and so on. Because the vibration of a seat affects the comfort of occupants, it is important to reduce vibration of seats of motor vehicles. This paper proposes vibration control of a seat frame by using piezoelectric elements to improve the comfort of occupants in cars.

* takagi@nuem.nagoya-u.ac.jp

In general, a piezoelectric element converts the applied deformation to the electrical charge, or the applied voltage to the force in case it is fixed. There are a lot of methods for reducing the vibration of machinery and structures using a piezoelectric element[1-3]. This study focuses on Positive Position Feedback (PPF) as the vibration control method. PPF is a method in which the displacement of the vibration of the structure is measured. PPF is processed through a second order controller and is positively fed back to force. It is well known that the stability of the PPF is guaranteed by a simple condition if collocation of sensor and actuator is satisfied. The idea of Positive Position Feedback (PPF) is applied to propose a method of vibration control. The method is called the PPF-based control because the collocation of sensors and actuators are not assumed in our framework. We consider a simple method for designing a PPF-based controller that can be used for complex structures with non-collocated pair of sensor and actuator. In case the collocation is not satisfied, it is important to guarantee the stability. Firstly, we determine the design parameters of a single-mode PPF-based controller as follows. A PPF controller is a second-order system with three design parameters. We determine one of the parameters, eigen frequency of the controller, to be equal to the natural frequency of the controlled mode due to the analogy of PPF to a mechanical tuned-mass damper. Next, we determine the gain of the controller so that the open-loop gain may have the designated margin from the stability limit. The other parameter, the damping ratio of the controller, is determined by performing the numerical search to minimize the maximum of the real parts of the poles of the closed loop transfer function. Moreover, we design not only the controller which works for a single mode, but also a controller which can suppress multiple modes simultaneously. We propose a design method of the multiple-mode controller by simply adding the controllers designed for each single mode.

In the experiment, firstly, piezoelectric ceramics as the sensor and the actuator are attached at the bottom of the car seat skeleton. We measure the frequency response form the piezoelectric actuator to the piezoelectric sensor to identify a nonparametric model of the controlled object. It is shown that the frequency response of the seat can be regarded as that of a typical lightly damped flexible structure. We obtain parametric models of the first and the second modes from the nonparametric model by using the least squares method. We design PPF-based controllers from the parametric model. The target controlled modes are chosen as three cases, the first mode, the second mode, and both modes. Finally, we show that the small vibration of the seat frame of an automotive is well damped in the experiment.

2. MODELING OF A SEAT FRAME

2.1 Skeletal Seat Frame of a Car

Figure 1 shows the seat frame of a vehicle used in this research. It is the passenger seat frame of an automotive (Aqua, Toyota). The cushion, seat-belt buckles and the seat rail have been removed.

Figure 2 shows the piezoelectric ceramics (AE1414D16DF, NEC Tokin). The seat is placed on four stack-type piezoelectric ceramics which are placed at four corners of the bottom of the seat frame. The stack-type piezoelectric ceramics are placed between the bottom of the seat and the base. The seat and the base are connected by bolts. The size of the piezoelectric element is $14[\mathrm{mm}] \times 14[\mathrm{mm}] \times 20[\mathrm{mm}]$. The bottom of the seat contact with the piezoelectric ceramic which is connected with a metal ball in order to prevent stress concentration of the ceramics. A sheet of paper is inserted between the base and the piezoelectric ceramic to protect the surface of the piezoelectric ceramic.

Figure 1: Photo of the skeletal seat frame of a car

Figure 2: Photo of the stack type piezoelectric element (AE1414D16DF, NEC Tokin) placed between the seat and the base

2.2 Modeling

In this study, we consider the following assumptions for modeling the piezoelectric elements and the seat.

- The seat is modeled as a flexible structure which is expressed by a $N-$ degrees of freedom linear vibration system.
- The piezoelectric actuator generates the force which is proportional to the applied voltage.
- The piezoelectric sensor generates the voltage which is proportional to the deformation.

Figure 3 is a schematic figure of the seat frame and the piezoelectric elements of actuator and sensor. In the figure, $v_a(t)$ is the voltage applied to the actuator piezoelectric element. $v_s(t)$ is the voltage generated by the sensor piezoelectric element.

$G(s)$ represents the transfer function from the piezoelectric actuator voltage, $V_a(s)$, to the piezoelec-

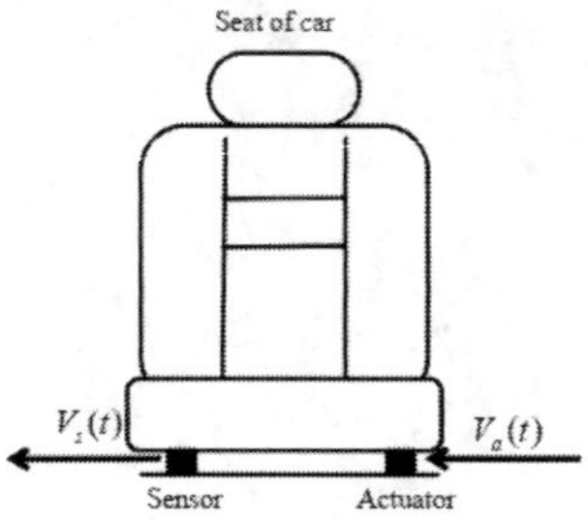

Figure 3: Schematic of a seat frame with piezoelectric elements

tric sensor voltage, $V_s(s)$.

$$V_s(s) = G(s)V_a(s) \tag{1}$$

From the assumption, $G(s)$ can be modeled as follows.

$$G(s) = \sum_{i=1}^{N} \frac{b_i}{s^2 + 2\zeta_i\omega_i s + \omega_i^2} \tag{2}$$

ζ_i is the damping ratio and ω_i is the natural angular frequency. b_i is the constant which expresses the gain from the sensor to the actuator. The subscript, i, of ζ_i, ω_i and b_i denotes the number of mode.

It is difficult to design the controller by considering all modes of the controlled object $G(s)$ which is a $2N$-th order system. For designing the controller, we use the approximated model $\hat{G}(s)$ which has only single target mode ($i = n$) picked out from $G(s)$.

$$\hat{G}(s) = \frac{b_n}{s^2 + 2\zeta_n\omega_n s + \omega_n^2} \tag{3}$$

3. DESIGN METHOD OF THE PPF-BASED CONTROLLER

3.1 PPF-based Controller for a Single Mode

$K(s)$ represents the transfer function of the controller and $V_d(s)$ represents the disturbance voltage. In this study, the disturbance voltage is superposed on the control voltage. The block diagram of the closed loop system is shown in Fig. 4. The relation of $V_a(s)$, $V_d(s)$, $V_s(s)$, $G(s)$ and $K(s)$ is obtained as follows.

$$V_a(s) = V_d(s) - K(s)V_s(s), \tag{4}$$
$$V_s(s) = G(s)V_a(s) \tag{5}$$

where K is the controller which is in the same form of the Positive Position Feedback (PPF) [2]

$$K(s) = -\frac{g_c \omega_c^2}{s^2 + 2\zeta_c \omega_c s + \omega_c^2} \tag{6}$$

where ω_c, ζ_c and g_c are the parameters to be designed. ω_c and ζ_c are chosen as positive.

$$\omega_c > 0 \tag{7}$$

$$\zeta_c > 0 \tag{8}$$

The sign of the g_c is chosen to be equal to b_n. We call $K(s)$ in Eq. (6) as the Positive Position Feedback based (PPF-based) controller because the sensor and the actuator are not assumed to be collocated. From Fig. 4, the closed-loop transfer function is given as follows.

$$\hat{G}_{yd}(s) = \frac{\hat{G}(s)}{1 + \hat{G}(s)K(s)} \tag{9}$$

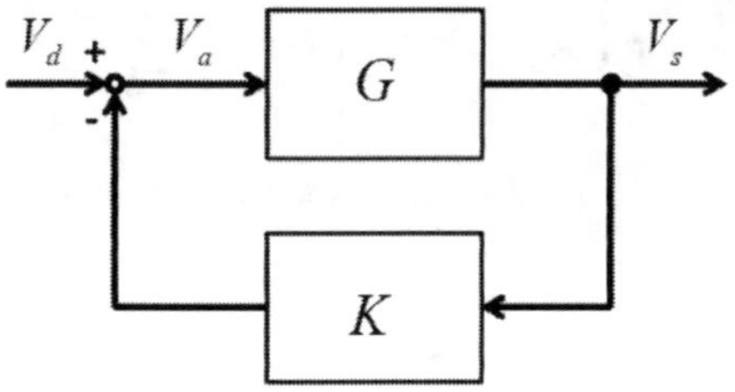

Figure 4: Block diagram of the control system

3.1.1 A Method for Determining the Controller Parameter ω_c

We set ω_c equal to the natural frequency of the controlled mode, ω_n, because the damping principle of PPF is analogous to that of a mechanical tuned mass damper.

$$\omega_c = \omega_n \tag{10}$$

3.1.2 A Method for Determining the Controller Parameter g_c

According to the Routh-Hurwitz criterion, we can show the following condition is the sufficient condition for the stability of the closed-loop system in Eq. (9).

$$0 < \frac{g_c b_n}{\omega_n^2} < 1 \tag{11}$$

Obviously, larger absolute value of the controller gain, g_c, may improve the control performance. However, g_c needs to satisfy the stability condition in Eq. (11). Focusing on the stability limit in Eq. (11), $g_c b_n/\omega_n^2 = 1$, we design g_c so that the system may have stability margin. When the stability margin is expressed as M_{gc}[dB], the condition of the g_c is expressed as below.

$$M_{g_c} = -20\log_{10}\frac{g_c b_n}{\omega_n^2}, \tag{12}$$

or

$$g_c = \frac{\omega_n^2}{b_n}10^{\frac{-M_{g_c}}{20}} = \frac{1}{\hat{G}(0)}10^{\frac{-M_{g_c}}{20}}. \tag{13}$$

It is important to keep a sufficient stability margin because in the design we have used $\hat{G}(s)$ which approximates only a single mode instead of $G(s)$ which considers all modes. After designing the controller, it is also important to check the stability margin of the control system which includes $G(s)$ considering all modes, for example, by using the Nyquist diagram.

3.1.3 A method for Determining the Controller Parameter ζ_c

We denote the pole of $\hat{G}_{yd}(s)$ as $p_i (i = 1, 2, 3, 4)$. The p_i can be located by changing the parameter, ζ_c. In this study, we propose to determine ζ_c by minimizing the maximum of the real part of the poles, as follows.

$$\zeta_c = \arg\min_{\zeta_c}\max_i \mathrm{Re}[p_i] \tag{14}$$

3.2 A PPF-based Controller for Multiple Modes

We denote the PPF-based controller designed for the single n-th mode as $K_n(s)$.

$$K_n(s) = -\frac{g_{cn}\omega_{cn}^2}{s^2 + 2\zeta_{cn}\omega_{cn}s + \omega_{cn}^2} \tag{15}$$

We design multiple-modes PPF-based controller by simply adding single mode controllers when the target vibration modes are n_1, n_2, ... ,n_c.

$$K(s) = \sum_{n=n_1,n_2,\ldots,n_c} K_n(s) \tag{16}$$

4. EXPERIMENTS

4.1 Experimental Setup

Figure 5 shows a schematic diagram of the experimental apparatus. We apply the voltage v_a to the piezoelectric actuator through a piezo amplifier (HJPJ-0.3Px3, MATSUSADA Precision) which has the gain of 30. The piezoelectric element can be applied a only positive voltage and can only be used in the expanding direction. The piezoelectric amplifier can amplify only positive voltage. Therefore, we superpose the offset voltage of 30[V] to the control input. We placed a saturation block such that the output voltage from the DSP is limited to 3.3[V] at the maximum so that the piezoelectric element may not be applied excessive voltage. As the result, the output voltage of the amplifier is limited to 100[V].

The output of the piezoelectric sensor is connected to a data logger to measure the sensor voltage. The piezoelectric element generates charge which is proportional to the displacement. The equivalent circuit of the piezoelectric element is modeled as a parallel connection of a current source and a capacitor, and therefore, voltage which is proportional to the generated charge appears between the terminals. However, the voltage is not proportional to the charge in the very low frequency range because the high-pass filter is formed by the input resistance of the data logger (91.3[kΩ]) and the capacitance of the piezoelectric element (12.13[nF]). Therefore, the cut-off frequency of the CR high-pass characteristics for the sensor signal is 0.903[rad/s], which is sufficiently low. We consider that the input resistance does not affect modeling and control.

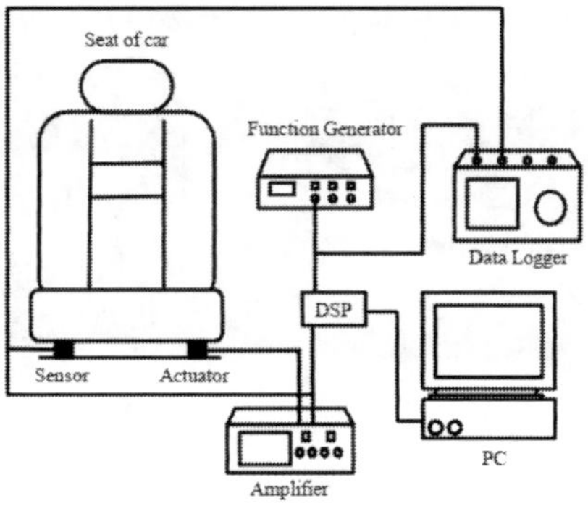

Figure 5: Experimental setup

4.2 Method of the Identification

We measure the frequency response from the piezoelectric actuator to the piezoelectric sensor. In other words, we identify the nonparametric model of the frequency response of $G(s)$. For identifying the frequency response, swept sine wave voltage is applied to the piezoelectric actuator and the voltages of the piezoelectric actuator and sensor are measured. We obtain the parametric models of the first mode, $\hat{G}_1(s)$, and the second mode, $\hat{G}_2(s)$, by parameter estimation from the nonparametric model.

The procedure of the experiment is shown as follows.

1. Add a swept sine wave of amplitude of 0.5[V] to the piezo amplifier during 40[s] from the function generator. That is to say, swept sine wave of 15[V] is applied to the piezoelectric element. The start and stop frequencies are divided into two ranges, from 10 [Hz] to 30[Hz] and from 30[Hz] to 60[Hz], so as not to be swept too fast.

2. Measure the disturbance (actuator) voltage and the piezoelectric sensor voltage. Measured sam-

pling frequency is 200[Hz] and the number of measured points are 8200. The measurement is repeated five times.

3. Calculate the averaged cross and power spectra from the obtained data sets and obtain the frequency response function from the ratio of the averaged spectra.

4. Estimate the single mode transfer function $\hat{G}_i$ of the first and the second modes from the non-parametric model by using the least squares method. `invfreqs` function of MATLAB Signal Processing Toolbox is used for the computation.

4.3 Results of the Identification

The frequency response and the identified transfer function in the experiment are shown in Figs. 6 and 7. Figure 6 shows the gain and Fig. 7 shows the phase. The dots shows the frequency response function. The solid line is the identified transfer function of the first mode. The dashed line is the identified transfer function of the second mode. The identified transfer functions of the first and the second modes match well to the experimental data. Estimated parameters are shown in Table 1.

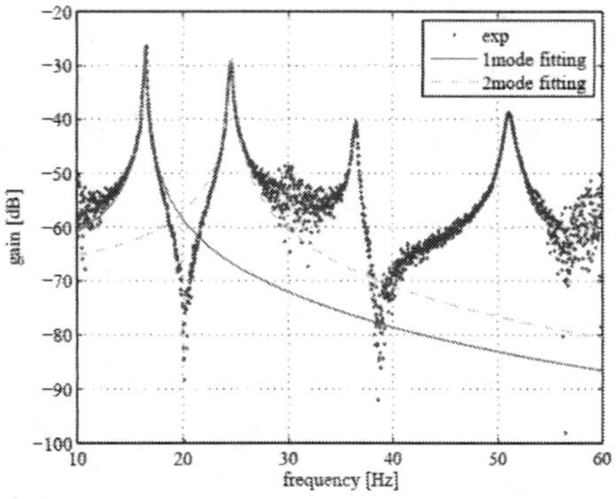

Figure 6: Gain of frequency response and the identified single-mode model

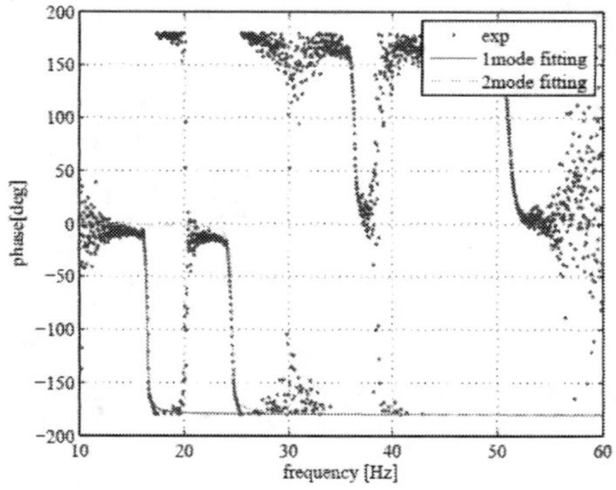

Figure 7: Phase of frequency response and the identified single-mode model

Table 1: Parameters in the experiment

	1st mode ($i = 1$)	2nd mode ($i = 2$)
ω_i	103.24	154.46
ζ_i	0.0060	0.0057
b_i	6.178	11.11

4.4 Design of the PPF-based Controllers

We design the controller from the identified transfer function by the method shown in Section 3.

Firstly, we design the controller K_1 for the first mode. The stability margin M_{gc} is chosen as 30[dB]. Then g_{c1} is calculated by Eq. (13) and the computed result is shown in Table 1. ζ_{c1} is obtained by numerically minimizing the maximum value of the real part of the closed-loop poles. Figure 8 shows the cost function, the max of the real parts of the closed loop poles. The value of ζ_{c1} which achieves the minimum value is found as 0.17. The designed value is shown at the second column in Table 2.

Secondly, we design the controller for the second mode. The stability margin is set as $M_{gc} = 30$[dB], we can obtain g_{c2} by the same procedure as in the case of the first mode controller. ζ_{c2} is designed by minimizing the value of the cost function shown in Fig. 9. The designed controller parameter is shown at the third column in Table 2.

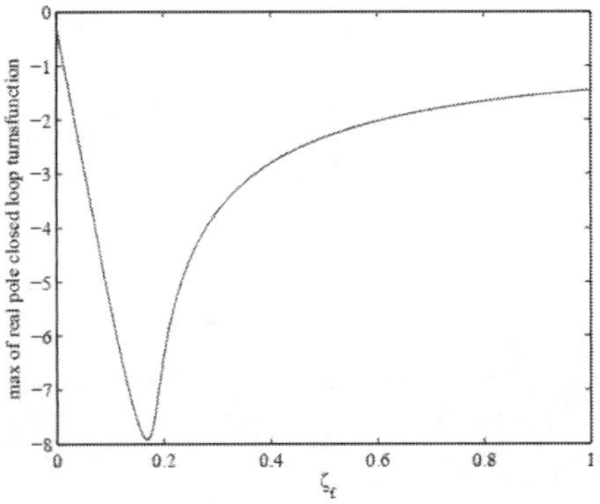

Figure 8: Cost function $\max \mathrm{Re}[p_i]$ for determining ζ_{c1}

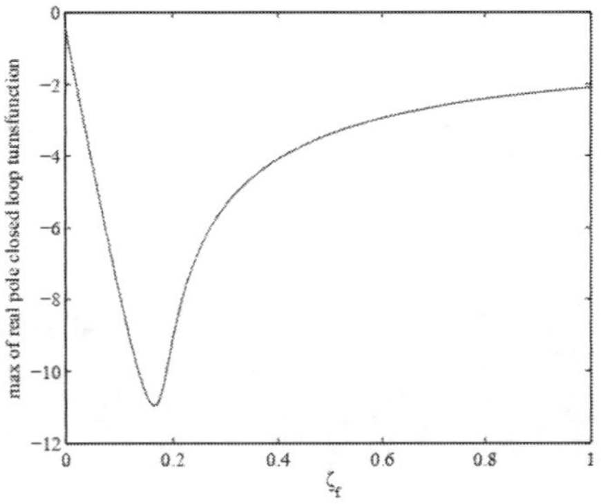

Figure 9: Cost function $\max \mathrm{Re}[p_i]$ for determining ζ_{c2}

Table 2: Parameters of the controllers

	1st mode ($i = 1$)	2nd mode ($i = 2$)
ω_i	103.24	154.46
g_i	56.5645	67.3442
ζ_{ci}	0.1682	0.1662

Finally, we design the multiple-mode controller by simply summing the controllers designed for the first mode and the second mode.

$$K_{12}(s) = K_1(s) + K_2(s) \tag{17}$$

We need to guarantee the stability margin by Nyquist diagram using $G(s)$ because we used the approximated single-mode models, $\hat{G}_i$, instead of G considering the all modes, when we designed the controller.

We confirm the stability margin of the system controlled by K_1 which is designed for damping the first mode. Figure 10 shows the Nyquist diagram of GK_1. We used nonparametric model of G to draw the diagram. Even though the data of the frequencies under 10[Hz] was not measured enough, the system is thought to be stable, because the $|G(0)K_1(0)|$ is guessed to be sufficiently small than 0[dB] and $-1 + 0j$ is out of the loop in the figure. Figure 11 shows the Nyquist diagram for the controller K_2.

The stability can also be confirmed by Fig. 11. Figure 12 shows the Nyquist diagram for the controller $K_1 + K_2$. The system is stable from Fig. 12.

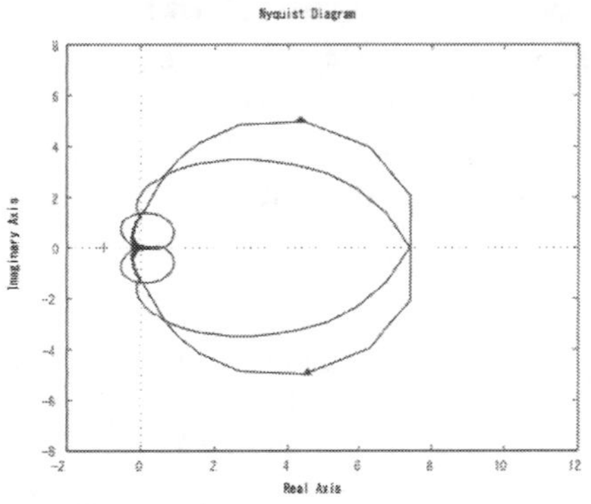

Figure 10: Nyquist diagram of the open-loop transfer function GK_1

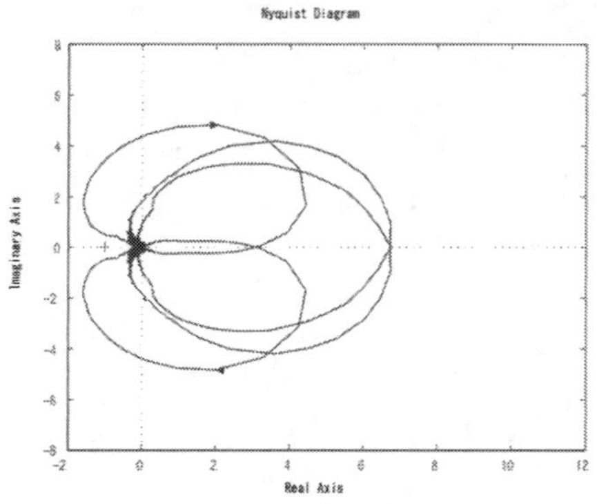

Figure 11: Nyquist diagram of the open-loop transfer function GK_2

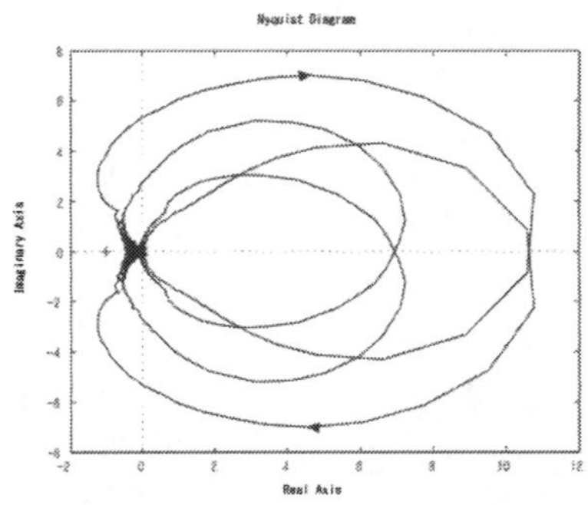

Figure 12: Nyquist diagram of the open-loop transfer function $G(K_1 + K_2)$

4.5 Methods of the Vibration Control

Figure 13 shows the block diagram of the experimental setup. The procedure of the experiment is shown as follows.

1. Program the designed controller to the DSP by using MATLAB/Simulink. The control period is $200[\mu\text{s}]$.

2. Add a swept sine wave of amplitude of 0.5[V] to the piezo amplifier during 40[s] from the function generator to the DSP. That is to say, swept sine wave of 15[V] is applied to the piezoelectric element. The start and stop frequencies are divided into two ranges, from 10[Hz] to 30[Hz] and from 30[Hz] to 60[Hz], so as not to be swept too fast. The disturbance signal which is swept sine wave as stated above is summed to the control signal in the DSP and is outputted to the amplifier.

3. Measure the disturbance voltage and the piezoelectric sensor voltage. Measured sampling frequency is 200[Hz] and the number of the measured points are 8200. The measurement is repeated five times.

4. Calculate the averaged cross and power spectra from the obtained data sets and obtain the frequency response function from the ratio of the averaged spectra.

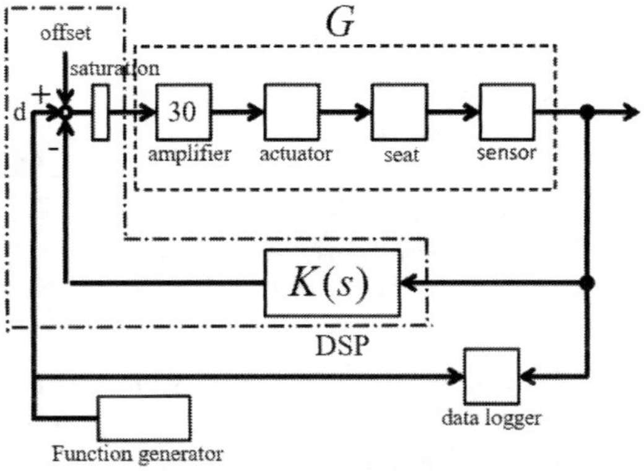

Figure 13: Block diagram of the experiment

4.6 Results and Discussions of Vibration Control

Figure 14 shows the frequency responses derived from the experimental data. In the figure, the blue solid line shows the result without control, the green dashed line shows the result of the first mode controller K_1, the black broken line shows the result of the second mode controller K_2, the red solid line shows the result of the multiple-mode controller K_{12}.

In the case of using K_1 and K_2, the damping effect of the resonance peaks of the first is about 20[dB] and the second modes is about 15[dB]. In the case of using K_2, the controller has an effect not only on the second mode but also of about 10[dB] on the first mode. One possible reason of this result is as follows. The natural frequencies of the first and second modes are close each other and K_2 effects on the first mode. We think a PPF-based controller has damping character in the lower frequency as well as at the target frequency because K_1 does not effect on the second mode.

In the case of using K_{12}, the controller damped both the first and the second modes about 20[dB]. The damping performance of the controller is greater than that of K_1 or K_2 alone.

5. CONCLUSIONS

This study has carried out vibration control of a seat frame of a real automotive using piezoelectric elements. We have proposed a simple method for suppressing a single mode or multiple modes by a controller same to PPF even though the sensor and the actuator were not collocated. Because it was difficult to get a detailed model of the seat, an approximated single-mode model was used to design the

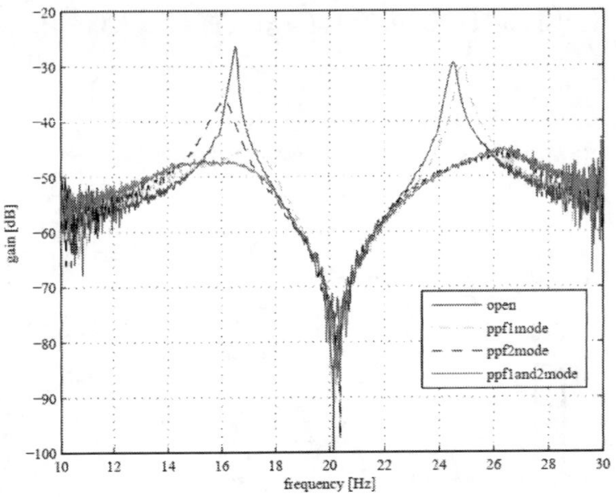

Figure 14: Frequency responses in the vibration control experiment

controller and nonparametric model was used for checking stability. For suppressing multiple modes, we proposed a simple method in which single-mode controllers were just summed to obtain a multiple-modes controller. In the experiment, we showed that the PPF-based controller well suppressed the small vibration of the seat frame.

ACKNOWLEDGMENTS

The authors appreciate Mr. Kenichi Someya (Toyota Motor Company) to his courtesy for supplying the seat.

REFERENCES

1. Balas, M. J., "Direct Velocity Feedback Control of Large Space Structures," *Journal of Guidance, Control and Dynamics*, Vol. 2, No. 3, 1979, pp. 252-253.
2. Fanson, J. L. and Caughey, T. K., "Positive position feedback control for large space structures," *AIAA Journal*, Vol. 28, No. 4, 1990, pp. 717-724.
3. Moheimani, S. O. R. and Fleming, A. J., *Piezoelectric Transducers for Vibration Control and Damping*, Springer, 2006.
4. *Multilayer Piezoelectric Actuators*, Vol.08, Cat.No.9470MPAVOL08E1501H1, Update 2015/07/23, `http://www.nec-tokin.com/english/product/pdf_dl/multilayerpiezoelectricactu.pdf`, NEC Tokin, 2015.

ICAST2015 #068

Input Shaping Methods for Vibration Control of Flexible Structures

Hiroshi Okubo[1*], Manami Yoshimine[1], and Ryosuke Satou[2]

[1] Department of Mechanical Engineering, Kanagawa Institute of Technology, Atsugi, Japan
[2] Former student, Kanagawa Institute of Technology, Atsugi, Japan

Abstract

This study covers the input shaping technique for reducing the vibrations of flexible structures under controlled force inputs applied to the structure. The input shaping profile is investigated for the effective suppression of transient vibrations of modal responses. This is a feedforward control method based on input shaping. It is applied to the input commands to minimize the residual vibrations and suppress the overshoot of the modal responses. The results of numerical simulations, using a simple dynamic model of a flexible spacecraft, show that the input shaping technique is capable of suppressing the residual vibrations caused by attitude maneuvers.

1. INTRODUCTION

One of the problems faced during the attitude control of a flexible spacecraft is unsatisfactory precision control because of residual vibrations in the flexible structures. It is very important to suppress the residual vibrations quickly to meet the mission requirements. The input shaping on the Posicast control proposed by Smith is one of the most effective vibration suppression techniques for a flexible structure [1]-[4]. The input shaping technique has been applied to the concurrent design with feedback control of nonlinear coupled vibration systems such as the trajectory tracking control of a flexible manipulator [5] and the slewing maneuver of a flexible spacecraft [6].

In this study, we apply the input shaping method to control the residual vibrations by convoluting a sequence of impulses in a one-degree-of-freedom vibration system. One of the main advantages of input shaping is that the equation used to reduce the residual vibrations requires only estimates for the natural frequencies and damping ratios. The numerical simulation of a simple model of a flexible spacecraft has demonstrated that the input shaping technique is highly effective in the attitude maneuver control of the spacecraft.

2. INPUT SHAPING TECHNIQUE

Figure 1 shows the block diagram of a servo control system consisting of a feedback loop for pursuing the control objective, such as the attitude maneuver of a flexible satellite, and an input shaper for suppressing the excitation of vibrational responses.

* okubo@me.kanagawa-it.ac.jp

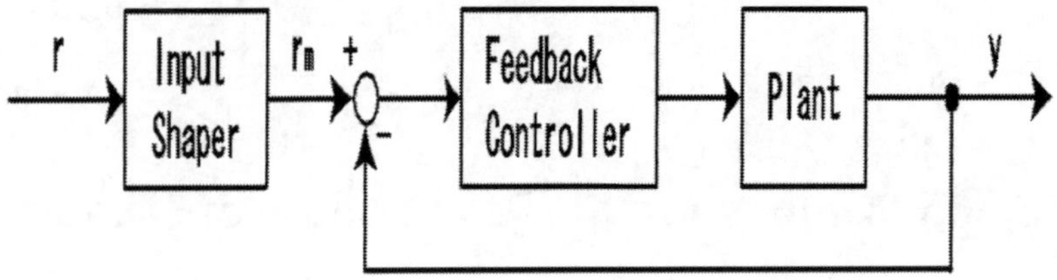

Figure 1. Block diagram of input shaper with a feedback controller

The amplitude of vibration response when N impulses are fed to a one-degree-of-freedom vibration system is given as follows.

$$A_{amp} = \sqrt{\left(\sum_{j=1}^{N} B_j \cos \omega_d t_j\right)^2 + \left(\sum_{j=1}^{N} B_j \sin \omega_d t_j\right)^2} \tag{1}$$

where

$$B_j = A_j \frac{\omega_0}{\sqrt{1-\zeta^2}} exp\left(-\zeta\omega_o\left(t_N - t_j\right)\right) \tag{2}$$

In addition, A_j denotes the amplitude of the jth impulse, t_j is the time at which the impulse occurs, ω_0 is the natural frequency of the system, and ζ, is the damping ratio of the system. To eliminate the vibrational response after the input has been applied, the expression for A_{amp} should be zero at time t_N; when the input ends. In other words, both the squared terms in Eq. (1) should be equal to zero.

$$\sum_{j=1}^{N} A_j \, exp\left(-\zeta\omega_o\left(t_N - t_j\right)\right) \cos \omega_d t_j = 0 \tag{3}$$

$$\sum_{j=1}^{N} A_j \, exp\left(-\zeta\omega_o\left(t_N - t_j\right)\right) \sin \omega_d t_j = 0 \tag{4}$$

For example, consider two impulse inputs at times t_1 and t_2 with amplitudes A_1 and A_2. The responses to the two impulse inputs cancel each other, and the vibrations after time t_2 vanish, only if the natural frequency and damping ratio of the system are known exactly. Figure 2 demonstrates the input shaping technique using two impulses.

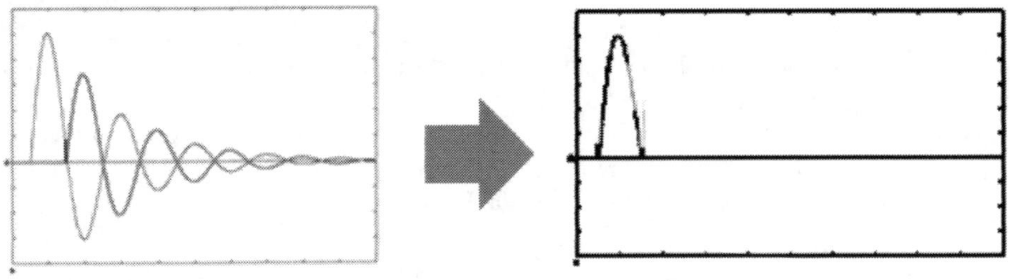

Figure 2. Basic idea of the input shaping technique

For the successful application of the input shaping technique in most real systems, the constraint equations, i.e., Eqs. (3) and (4) must ensure robustness to modeling errors. Singer and Seering [2] have developed a method of robust input shaping by using the condition that the derivative of Eqs. (3) and (4) with respect to the frequency of the residual vibrations equals zero. The resulting shaper is called a zero vibration and derivative (ZVD) shaper. Further details may be obtained by referring to their study.

Figure 3 shows an example of ZVD shaping. The single unit step input is divided into a sequence of three inputs with smaller amplitudes and time shifts satisfying the input shaping conditions.

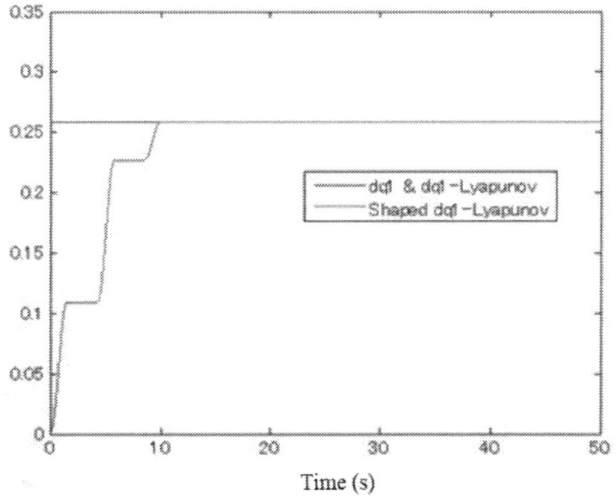

Figure 3. An example of input shaping for a step command [7]

3. MODELING OF FLEXIBLE SPACECRAFT

The dynamic model of a satellite with flexible appendages is shown in Figure 4. The model of the flexible structure is considered as a one-degree-of-freedom-vibration system, as defined in literature [4].

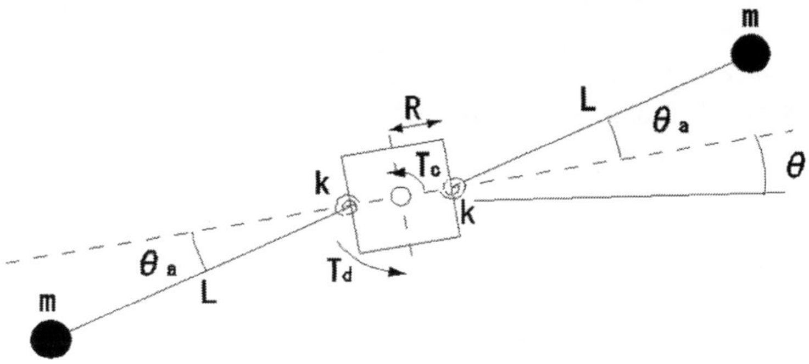

Figure 4. Dynamic model of a flexible spacecraft

where J_b is the moment of inertia of the satellite body, m is the mass of the adjunct, R is the mounting position from the center of the satellite, θ is the attitude angle, θ_a is the angle of the arm, k is the spring constant of torsion, T_c is the torque of the reaction wheel, and T_d is the torque of the thrusters.

The dynamics of the attitude of the satellite with a flexible structure is derived by Lagrange's equations of motion as follows.

$$[J_b + 2m(R+L)^2]\ddot{\theta} + 2m(R+L)L\ddot{\theta}_a = T_d + T_c \quad (5)$$

$$2\mathrm{m}(\mathrm{R}+\mathrm{L})\mathrm{L}\ddot{\theta} + 2\mathrm{m}L^2\ddot{\theta}_a = -2k\theta_a \quad (6)$$

where the angular velocity of the satellite body is ignored, because it is very small when compared to the angular velocity of the reaction wheel.

The torque of the thruster is considered as a pulse in one direction, and it functions with unit amplitude, changing sign at a specified time sequence. The impulses of the thruster torque are shown in Table 1, where A_j denotes the normalized amplitude of a single impulse, t_j is the switching time; $t_1 = 0$, because the first impulse is applied at the initial time.

Table 1 Amplitudes and switching times of thruster torque

A_j	t_j
1	t_1
-1	t_2
1	t_3
-1	t_4
⋮	⋮
-1	t_n

In order to apply the ZVD, Eqs. (3) and (4), and the derivative with respect to the natural frequency are used for the constraint equations. The number of switching times, n, of the impulses to suppress the tangent vibration with frequency robustness is determined by solving the following equations.

$$\sum_{j=1}^{N}(-1)^{j+1} sin\omega_d t_j = 0 \tag{7}$$

$$\sum_{j=1}^{N}(-1)^{j} cos\omega_d t_j = 0 \tag{8}$$

$$\sum_{j=1}^{N}(-1)^{j+1} t_j sin\omega_d t_j = 0 \tag{9}$$

$$\sum_{j=1}^{N}(-1)^{j+1} t_j cos\omega_d t_j = 0 \tag{10}$$

We consider the torque of the reaction wheel. The attitude angle of the satellite is controlled by a simple proportional-integral-derivative (PID) controller.

$$T_c = -I_w\ddot{\theta}_w = -k_p\theta - k_i\int_0^t \theta d\tau - k_d\dot{\theta} \tag{11}$$

where I_w is the moment of inertia of the reaction wheel and $\ddot{\theta}_w$ is the angular acceleration of the reaction wheel.

4. RESULT OF NUMERICAL SIMULATION

The following numerical values are used in simulation.

J_b = 12800 kg$\cdot m^2$, m = 25 kg, R = 2 m, L = 10 m, k = 5685 Nm

The amplitudes and switching times of the thruster are obtained by using the damped natural frequency ω_d. These are listed in Table 2, where the time for the ignition of the thruster is set to $t_1 =$ 0 s and $t_4 = 100$ s.

Table 2 Amplitudes and switching times of thruster torque

A_i	$t_i(s)$
1	0
-1	0.55
1	1.11
-1	100
1	100.55
-1	101.11

The results of numerical simulations are shown in Figures 5-9. Figure 5 shows the time response of the flexible structure when input shaping is not used, and Figure 6 shows the response with input shaping. The vertical axis represents θ_a, which is the amplitude of the angular displacement of the flexible structure, and the horizontal axis represents time. These figures show that the convergence time is shorter, and the maximum amplitude of displacement is smaller, when input shaping is applied for the input torque.

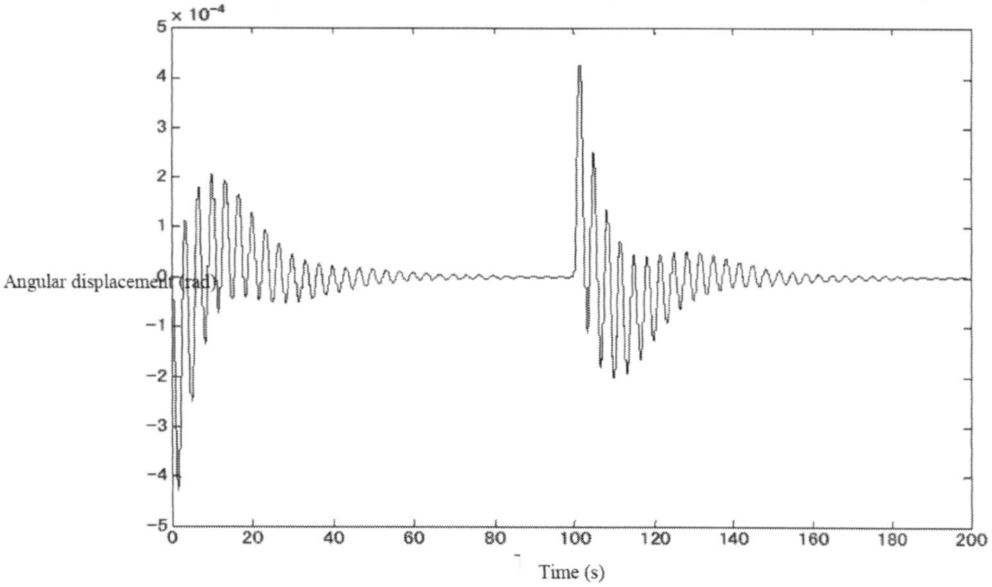

Figure 5. Displacement of flexible structure with input shaping

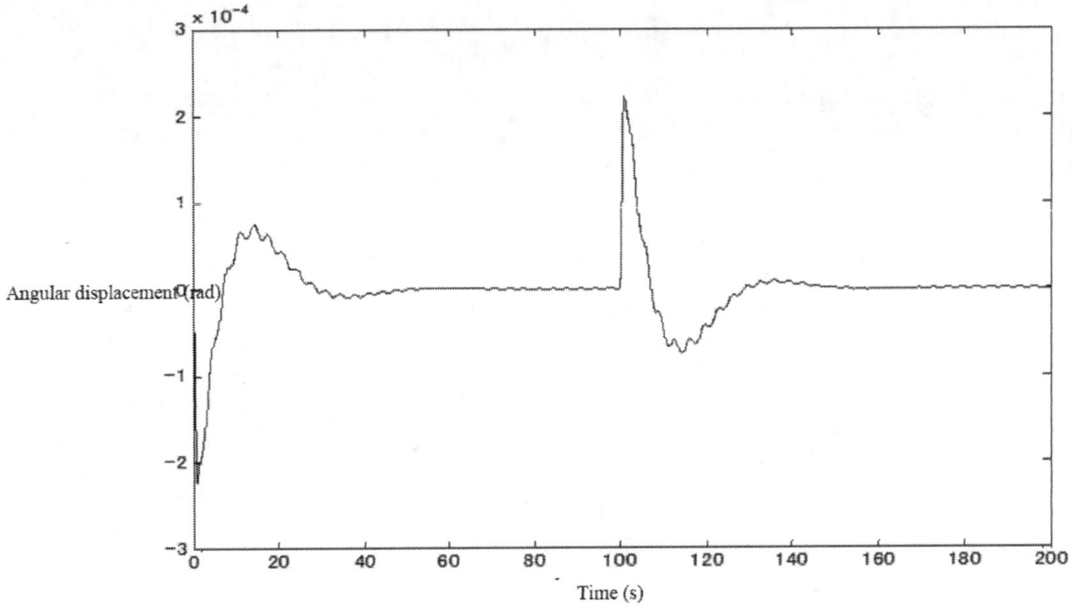

Figure 6. Displacement of flexible structure with input shaping

The input torque T_d of the thruster with input shaping, and the PID feedback control torque T_c of the reaction wheel are shown in Figure 7. Figures 8 and 9 show the enlarged view of Figure 7 around the time of the ignitions, $t_1 = 0$ s and $t_4 = 100$ s. It is observed that the input shaping technique is very effective in suppressing the residual vibrations of the flexible structures excited by the attitude control torque.

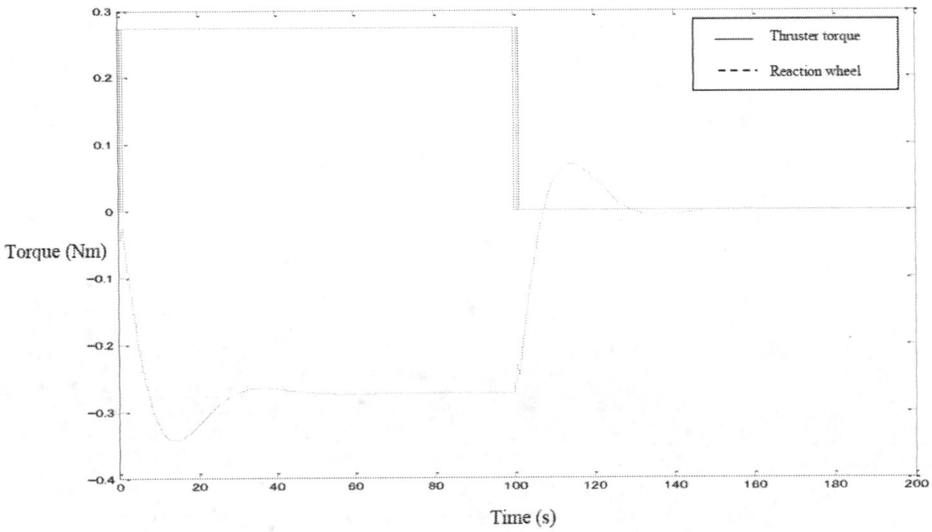

Figure 7. Torque of the reaction wheel and thruster with input shaping

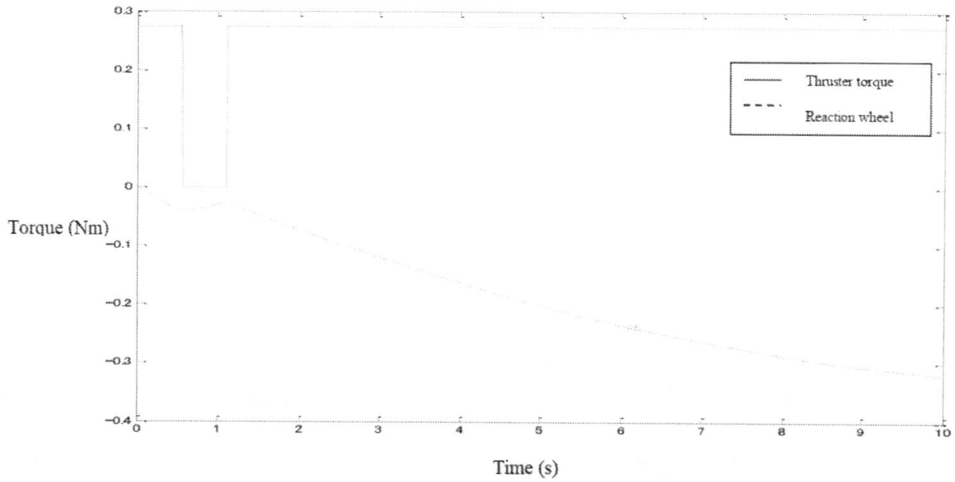

Figure 8. Enlarged view of Figure 7 around time $t_1 = 0$

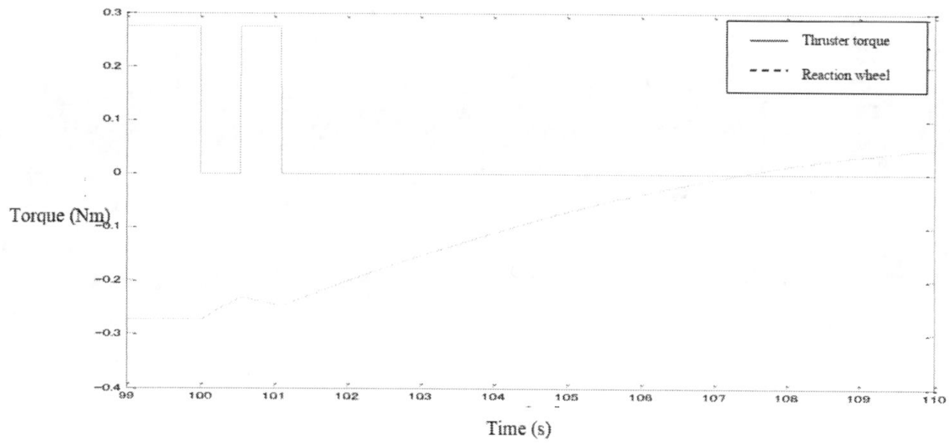

Figure 9. Enlarged view of Figure 7 around time $t_4 = 100$ s

5. CONCLUSIONS

Input shaping is very effective in suppressing the residual vibrations of flexible structures, as well as in completing the agile attitude maneuvers of a flexible spacecraft. Future work involves comparing the input shaping method with other command shaping methods (such as nill-mode excitation (NME) method), and experimental demonstration of the input shaping method with a simple laboratory model.

REFERENCES

1. Smith, O.J.M., *Feedback Control Systems*, McGraw-Hill, New York, 1958, p.338.
2. Singer, N.C., Seering, W.P., "Preshaping command inputs to reduce system vibration," *J. Dynamics Systems, Measurement and Control*, 112, 1, 1990, pp.76-82.
3. Singhose, W.E., "Extra-Insensitive Input Shaper for Controlling Flexible Spacecraft," *Journal of Guidance, Control and Dynamics*, Vol. 19, No. 2, 1996, pp. 385–391.
4. Singhose, W.E., Banerjee, A.K., and Seering, W.P., "Slewing Flexible Spacecraft with Detection-Limiting Input Shaping," *Journal of Guidance, Control and Dynamics*, Vol. 20, No. 2, 1997, pp. 291–298.
5. Hasegawa, Y. and Okubo, H., "Input Shaping for Fast Trajectory racking Control of a Flexible Manipulator," *Proceedings of the MOVIC*, Paper 327, 10 pages, 2008.
6. Kashima, M., "Attitude Control of Flexible Spacecraft Using Control Momentum Gyros", *Master Thesis, Graduate School of Engineering*, Osaka Prefecture University, February 2010.
7. Okubo, H. and Kuwamoto, S., "Integrated Feedback/Feedforward Control of Flexible Spacecraft for Agile Attitude Maneuver and Vibration Suppression," *25th International Conference on Adaptive Structures and Technologies*, Oct. 6-8th, 2014, The Hague, The Netherlands.

ICAST2015: 26th International Conference on Adaptive Structures and Technologies
October 14-16th, 2015, Kobe, Japan

ICAST2015 #071

Dynamic Characteristics of Hanging Truss Having SMA Wires (Vibration Isolation and Absorption Points of View)

Xuan Zhang[1*], Kazuyuki Hanahara[1], and Yukio Tada[1]

[1] Graduate School of System Informatics, Kobe University, Kobe, Japan.

Abstract

There are two viewpoints for reducing the influence of environmental vibration. One is the vibration isolation that aims to cut off the transmission of vibration motion to the target object in consideration. The other is the vibration suppression that aims to reduce or attenuate the vibration itself. The former is well demonstrated by a pendulum; its behavior is not immediately affected by horizontal vibration motion of its support due to the inertia. The latter is a typical effect usually achieved by means of a damping device. In this study, we adopt a truss structural system having shape memory alloy (SMA) wires as a part of its truss members. On the basis of the nonlinear characteristics of SMA that includes the pseudoelasticity and the hysteretic characteristics, such truss structural system can be a promising candidate of a mechanical system that has both capabilities of isolation and absorption of vibration. We deal with the truss structural system in a hanging configuration; that is, the fixed nodes of the truss are assumed to be placed at the support ceiling. Taking into account the effect of gravitational force, the mechanical stability of the overall configuration of the truss is expected to be guaranteed in many situations.

1. INTRODUCTION

Shape memory materials have been studied extensively including various compositions of that kind of material since 1930s[1]. The family of shape memory material is attractive to be applied in modern adaptive structures and intelligent structures due to the relevant properties, the shape memory effect and the pseudoelasticity. The excellence of pseudoelasitic property is that in a relatively high temperature environmental situation, such kind of material is able to experience large strain deformation; however, with removing the external force or energy, they can return to their initial configuration. For a specific application of the material, the relevant parameters such as critical phase transformation temperatures should be given, so as to understand at which temperature range this material shows the pseudoelastic characteristic.

The characteristic of psudoelasticity of shape memory material has been applied to various commercial products. Nowadays, SMA is utilized in high-end eyeglass frames owing to great resistance to accidental damage. The cellular phone antenna at present is extensively manufactured from shape memory alloy; that is also taking into account the resistance to folding damage[2].

* tyou@opt.cs.kobe-u.ac.jp

Pseudoelasticity of SMA material attracts great attention for researchers in the structural vibration control area and in structural vibration isolation area in recent years. For the purpose of dealing with vibration control, the shape memory effect plays an important role, and for the purpose of dealing with vibration isolation, the pseudoelasiticity plays a significant role. Some research results according to experimental and theoretical studies have recently shown that SMA-based structures and devices are practically applicable[3-6].

The first research on the application to seismic area using SMAs traces back to the work of Graesser and Cozzarelli[7]. The main achievement of their work is the influence of frequency and history on the energy dissipation characteristics of SMA wires. This result also tells us that the arrangement of different states of SMA materials in structures attains better performance. Except for the pseudoelasticity and the shape memory effect, the family of SMA material also shows high-fatigue properties and the ability to provide hysteretic damping[8]. A suitable pseudoelastic constitutive mathematical model is developed considering residual martensite strain effect and the behavior of the simulation result agrees with the experiment data. Omar[9] has compared nonlinear seismic performance of braced steel frames with two different bracing systems by means of nonlinear time history analyses, and examined the effectiveness of these two systems.

In this study, we deal with the truss structure in the hanging configuration with an objective apparatus at the end of the truss. Owing to the gravitational force, stable state of the truss structure with target apparatus can be guaranteed in many situations. Therefore, this kind of truss structure system can be expected to demonstrate vibration isolation capability. In addition, because of the pseudoelastic characteristic of SMA material, this kind of truss structure system is expected to absorb the vibration energy when ceiling vibration excitation ceases. We formulate the dynamic equation for the truss structure in hanging configuration and discuss that vibration isolation and vibration absorption ability of the truss structure system through numerical calculation.

2. CONSTITUTIVE MODEL OF SMA MATERIAL

The major loop of pseudoelastic characteristic of SMA is dealt with in the literature[11]. In the current study, since SMA is employed in the form of wire truss members, its compressive stiffness is considered to be zero. We have to deal with this situation. We adopt the reuss scheme[12] in order to deal with the nonlinear relationship between the Young's modulus and the martensite volume fraction of SMA. The minor loop of hysteretic characteristics is dealt with in the manner in [13]. The constitutive model of SMA is briefly summarized as follows.

2.1 Tangent Stiffness

Total strain for SMA material can be expressed as:

$$\epsilon = \epsilon_e + \Lambda\lambda \tag{1}$$

where we denote ϵ_e as elastic strain, λ as martensite volume fraction, and Λ as maximum phase transformation strain.

Taking into account the volume fraction λ, the relationship between stress and strain is expressed as follows:

$$\sigma = E\epsilon_e = E[\epsilon - \Lambda\lambda] \tag{2}$$

On the basis of the linearization, we obtain:

$$d\sigma = E[d\epsilon - \Lambda d\lambda] \tag{3}$$

where Young's modulus is expressed as follows at the stage of phase transformation:

$$E = \frac{E_A E_M}{E_M + (E_A - E_M)\lambda} \tag{4}$$

We denote parameter E_A as Young's modulus at pure austenite phase, and E_M as Young's modulus at pure martensite phase.

2.2 Minor Loop

There are three situations for loading about minor loop classified by the position of start point in the constitutive relation as in Figure 1:

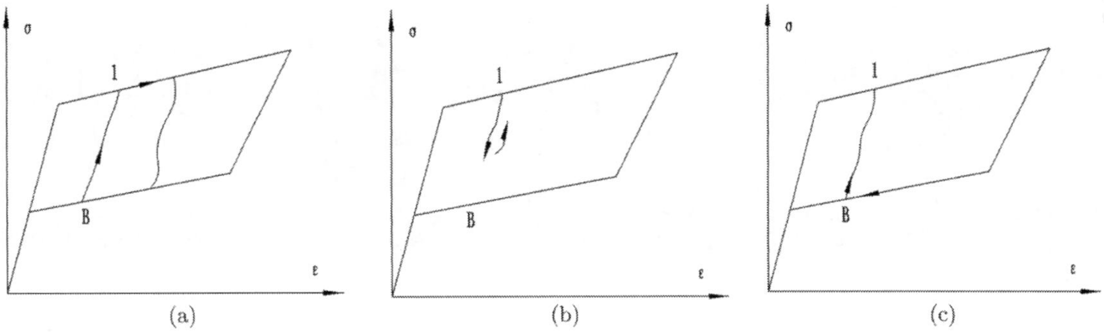

Figure 1: Schematic figure of hysteretic loop

(1). Start point is at the stage of forward transformation as in Figure 1(a).

(2). Start point is at the inner loop where the direction is changing from unloading to loading as in Figure 1(b).

(3). Start point is at the reverse transformation where direction is changing from unloading to loading as in Figure 1(c).

The analytical method and calculation process for unloading is the same as loading and is omitted here.

3. DYNAMIC SIMULATION OF HANGING TRUSS HAVING SMA WIRES

3.1 Equation of Motion

In order to take into account the nonlinear characteristic of SMA wire, we use the tangential stiffness of the SMA material. The tangential stiffness k_i of ith member is expressed as:

$$k_i = \begin{cases} \dfrac{E^R A_i}{l_i} & \text{(rigid member } i\text{)} \\ \dfrac{E_i^{SMA} A_i}{l_i} & \text{(SMA wire member } i\text{)} \end{cases} \tag{5}$$

where A_i and l_i are the current cross-sectional area and length of the ith member, E^R is the Young's modulus of the material of rigid member and E_i^{SMA} is the Young's modulus of the SMA member i corresponding to the current state obtained in the manner as discussed in the previous section. It should be noted that E^R is assumed to be constant in general. On the basis of the diagonal tangential member stiffness matrix $\boldsymbol{K}^R = \text{diag}[k_1, ..., k_M]$, the tangential nodal stiffness matrix $\boldsymbol{K}$ is expressed as [14]

$$\boldsymbol{K} = \boldsymbol{J}^T \boldsymbol{K}^R \boldsymbol{J} \tag{6}$$

where

$$\boldsymbol{J} = \left.\frac{\partial \boldsymbol{L}}{\partial \boldsymbol{X}}\right|_{\boldsymbol{X}=\underline{\boldsymbol{X}}+\boldsymbol{U}} \tag{7}$$

is the Jacobian matrix of the relation $\boldsymbol{L} = \boldsymbol{L}(\boldsymbol{X})$ between the member lengths and nodal positions. The current nodal position $\boldsymbol{X}$ is given as $\underline{\boldsymbol{X}} + \boldsymbol{U}$, where $\underline{\boldsymbol{X}}$ denotes the initial natural nodal positions.

Taking into account the gravitational nodal force $\boldsymbol{G}$ and nodal inertia matrix $\boldsymbol{M}$ which can be obtained in the similar manner as the matrix structural analysis approach, the equation of motion can be formulated in the following general form:

$$\boldsymbol{M}\Delta\ddot{\boldsymbol{U}} + \boldsymbol{K}\Delta\boldsymbol{U} = \boldsymbol{F} + \boldsymbol{G} \tag{8}$$

where $\boldsymbol{F}$ is the external nodal force causing vibration. It should be noted that the above equation is based on the tangential stiffness at the current deformation $\boldsymbol{U}$; we have to update the deformation $\boldsymbol{U}$ as well as the stiffness $\boldsymbol{K}$ in accordance with the realized dynamic motion.

3.2 Numerical Integration

We solve the dynamic equation of motion by means of Newmark-β integration scheme. On the basis of Newmark-β method, the increment of nodal displacement vector, nodal velocity vector and nodal acceleration vector from time step $n-1$ to n are obtained as follows:

$$\Delta\boldsymbol{U}^{(n)} = (\boldsymbol{M}\frac{1}{\beta\Delta t^2} + \boldsymbol{K})^{-1}[\boldsymbol{F}^{(n)} + \boldsymbol{G} + \boldsymbol{M}(\frac{1}{\beta\Delta t}\dot{\boldsymbol{U}}^{(n-1)} + \frac{1}{2\beta}\ddot{\boldsymbol{U}}^{(n-1)})] \tag{9}$$

$$\Delta\dot{\boldsymbol{U}}^{(n)} = \frac{\gamma}{\beta\Delta t}\Delta\boldsymbol{U}^{(n)} - \frac{\gamma}{\beta}\dot{\boldsymbol{U}}^{(n-1)} + (1 - \frac{\gamma}{2\beta})\ddot{\boldsymbol{U}}^{(n-1)}\Delta t \tag{10}$$

$$\Delta\ddot{\boldsymbol{U}}^{(n)} = \frac{1}{\beta\Delta t^2}\Delta\boldsymbol{U}^{(n)} - \frac{1}{\beta\Delta t}\dot{\boldsymbol{U}}^{(n-1)} - \frac{1}{2\beta}\ddot{\boldsymbol{U}}^{(n-1)} \tag{11}$$

The nodal displacement vector, nodal velocity displacement vector and nodal acceleration vector are updated according to the following equations:

$$\boldsymbol{U}^{(n)} = \boldsymbol{U}^{(n-1)} + \Delta\boldsymbol{U}^{(n)} \tag{12}$$

$$\dot{\boldsymbol{U}}^{(n)} = \dot{\boldsymbol{U}}^{(n-1)} + \Delta\dot{\boldsymbol{U}}^{(n)} \tag{13}$$

$$\ddot{\boldsymbol{U}}^{(n)} = \ddot{\boldsymbol{U}}^{(n-1)} + \Delta\ddot{\boldsymbol{U}}^{(n)} \tag{14}$$

where we denote the time step as Δt. The parameters β and γ are for the Newmark-β method.

The numerical calculation process is:

(1). Initialization: nodal position vector $\boldsymbol{U}$, member length vector $\boldsymbol{L}$, member strain vector $\boldsymbol{\epsilon} = [\epsilon_1, ..., \epsilon_M]$.

(2). Exert boundary conditions, in which the longitudinal degree of freedom of the support nodes of the truss structure are fixed.

(3). Calculate Jacobian matrix $\boldsymbol{J}$, inertia matrix $\boldsymbol{M}$ and stiffness matrix $\boldsymbol{K}$.

(4). Calculate the vectors $\Delta\boldsymbol{U}^{(n)}$, $\Delta\dot{\boldsymbol{U}}^{(n)}$ and $\Delta\ddot{\boldsymbol{U}}^{(n)}$ and then obtain the nodal displacement increment.

(5). Update nodal positions on the basis of $\boldsymbol{X} = \underline{\boldsymbol{X}} + \boldsymbol{U}^{(n)}$ and to get member length vector $\boldsymbol{L}$ based on $\boldsymbol{X}$ so as to obtain member strain vector $\boldsymbol{\epsilon}$.

(6). Calculate the constitutive equations for SMA to determine current Young's modulus of the SMA members respectively.

(7). Repeat step (2) through step (6).

There is one point should be noted that the initial displacement is zero as $\boldsymbol{U}^0 = \boldsymbol{0}$. In addition, it is noted that the slack condition

$$l_i < \underline{l}_i \tag{15}$$

has to be taken into consideration for SMA wires, where $\underline{l}_i$ is the initial natural length of the ith member. We should take into consideration that the strain value of the initial length of ith member is assumed to be zero in the current study. For the member i with the condition expressed as in formula (15) a small value k_ϵ is adopted as the stiffness of such member in order to guarantee the stability of numerical integration. The strain values for the slack members are thought of as zero.

4. SIMULATION RESULTS

Example structures with different configurations are shown as in Figure 2. The thick lines stand for rigid member, the thin line stand for taut SMA wires and the broken lines stands for slack SMA wire. The density of the rigid member is assumed as 7.86×10^3kg/m^3, and their Young's modulus is assumed as 200GPa. The length of the member of rigid member is supposed to be 1m, and the cross-sectional area is supposed to be 300mm^2. The mass of the red object is assumed as a target apparatus, the mass of which is 117.9kg. The adopted material parameters of the SMA wire members are shown in Table 1. The excitation of the structure is assumed to be due to the horizontal vibration of the ceiling with amplitude of 0.03m and frequency of 2Hz between 0s and 2.5s.

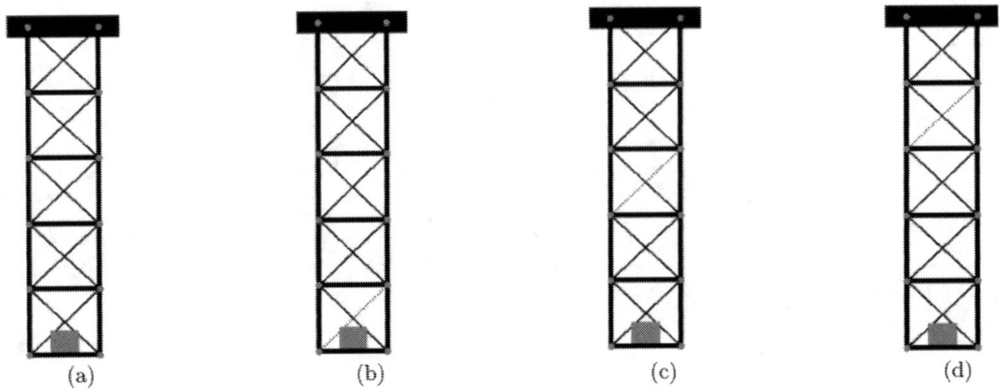

Figure 2: Examples for truss structures with different configurations; (a): having 10 taut SMA wires; (b), (c), (d): having 9 taut SMA wires

The truss structures in Figure 2 is considered to possess the capability of vibration absorption as the hysteretic effect of SMA wires. The corresponding simulation of the dynamic behaviors of the target apparatus on the truss structure are given in Figure 3. The environmental temperature is assumed to be above the austenite phase transformation finish temperature of SMA material; thus, the phase transformation from martensite phase to austenite phase is complete [11]. Therefore, all the SMA wire members show the characteristic of pseudoelasticity.

Table 1: Material parameter of SMA

SMA density	ρ_{SMA}	6500kg/m^3
Cross-sectional area of SMA	A_{SMA}	$1 \times 10^{-6}\text{m}^{-2}$
Young's modulus (austenite)	E_A	70GPa
Young's modulus (martensite)	E_M	30GPa
forward transformation finish stress	σ_f^{MA}	287MPa
forward transformation start stress	σ_s^{MA}	161MPa
reverse transformation start stress	σ_s^{AM}	131MPa
reverse transformation finish stress	σ_f^{AM}	0MPa
forward transformation start stress	ϵ_s^{MA}	0.0023
forward transformation finish stress	ϵ_f^{MA}	0.0596
reverse transformation start stress	ϵ_s^{AM}	0.0544
reverse transformation finish stress	ϵ_f^{AM}	0
maximum phase transformation strain	Λ	0.05

In regard to the simulation, the amplitude of the support ceiling vibration are all 0.03m and the vibration frequency are all 2Hz. Through the comparison of dynamic behaviors of the target apparatus

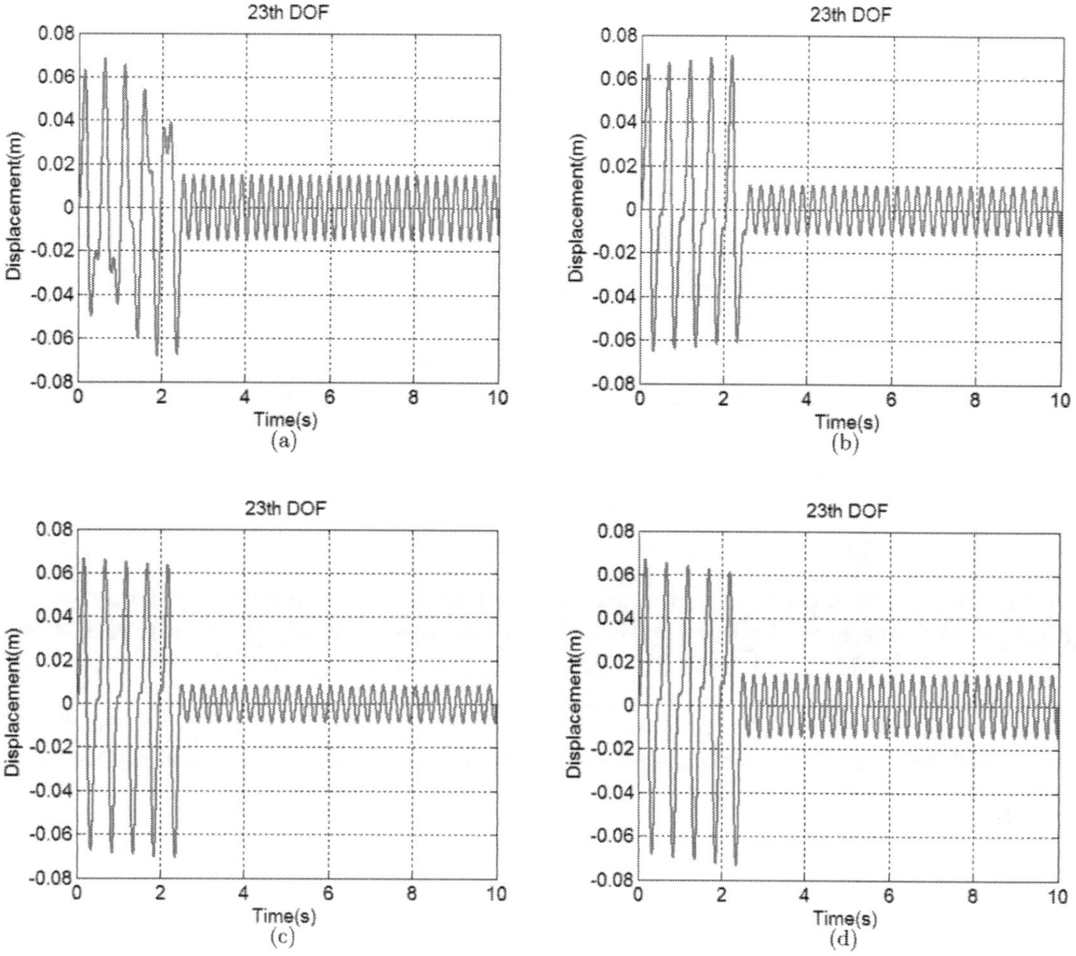

Figure 3: Dynamic behavior of truss structures with vibration frequency of 2Hz and amplitude of 0.03m: (a) is for Figure 2(b); (b) is for Figure 2(c); (c) is for Figure 2(d)

in Figure 3 we can find that the vibration amplitude of the truss structure having the configuration as in Figure 2(b) is smaller than the other configurations with 9 taut SMA wires when support ceiling is vibrating. This shows that the truss structure with configuration as in Figure 2(b) demonstrates better vibration isolation effect than the others structures having 9 SMA wires. As the vibration of the support ceiling ceases, the vibration of the target apparatus is significantly suppressed. In addition, we can find that the vibration amplitude of the truss structure having configuration as in Figure 2(a) is smaller than the vibration amplitude of the truss structure having configurations as in Figure 2(b), 2(c) and 2(d) as the support ceiling vibrates. Therefore, the structure having configuration as in Figure 2(a) demonstrates better vibration isolation effect than the structures having configuration as in Figure 2(b), 2(c) and 2(d). as the support ceiling vibrates.

The following point should be noted in the current simulation study. Since the minor hysteretic

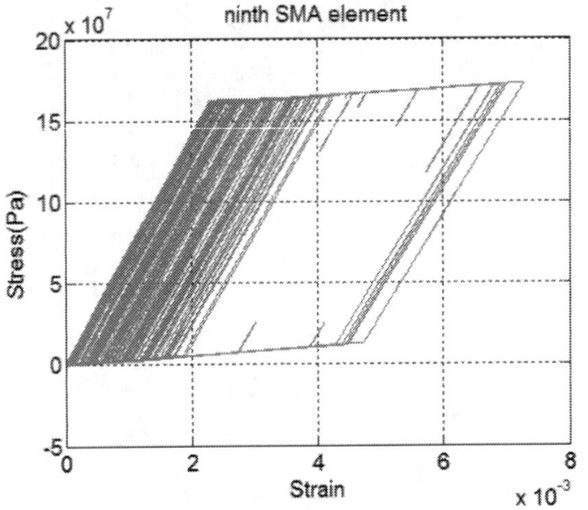

Figure 4: Example time history of behavior of SMA wire: the 9th member of truss shown in Figure 2(a)

loop of the constitutive relation of the SMA material is not taken into consideration, the time history of behavior of the simulated SMA wire in stress-strain space becomes the piece-wise linear path as shown in Figure 4. Owing to this piece-wise linearity without minor hysteretic loop, the vibration energy cannot be completely absorbed in these simulations as can be seen in Figure 3.

5. CONCLUDING REMARKS

For a structural system, there are two types of approaches in order to suppress the influence of vibration of its basis on the supporting object. One is the vibration isolation that cut off the transmission, and the other is the vibration absorption to reduce the amplitude of the vibration itself. In this study, we have dealt with a truss structural system having SMA wire members, which is of hanging configuration from the basis ceiling.

We have developed a constitutive equation of the pseudo-elastic characteristics of SMA with the hysteretic property. The conspicuous characteristic of wire that it cannot take compressive force has been taken into account. The formulated equation of motion is based on the tangential stiffness in order to deal with the nonlinearity.

Dynamic simulations of the hanging truss with SMA wires have been carried out. The obtained results demonstrate the vibration reduction capability of the proposed truss structural system. We can observe the difference between the simulated dynamic response in accordance with the SMA-wire setup of the truss. On the basis of the obtained results, the design of placement of SMA wires and the adjustment of their characteristic parameters are considered to be an important part of the future work; an optimal design approach has to be taken into account.

REFERENCES

1. Lagoudas, D. C. eds., *Shape Memory Alloys (Modeling and Engineering Applications)*, Springer, 2008.
2. Wu, M. H., Schetky, L. M., "Industrial applications for shape memory alloys." *Proceedings of the international conference on shape memory and superelastic technologies, Pacific Grove, California*. Vol. 44 No. 1. 2000. *Computational materials science*, 25.1 (2002): 218-227.
3. Han, Y. L., et al., "Structural vibration control by shape memory alloy damper." *Earthquake engineering & structural dynamics*, 32.3 (2003): 483-494.
4. Khan, M. M., Lagoudas, D. C., "Modeling of shape memory alloy pseudoelastic spring elements using Preisach model for passive vibration isolation." *SPIE's 9th Annual International Symposium on Smart Structures and Materials*, International Society for Optics and Photonics, 2002.
5. Hanahara, K., Tada, Y., "Vibration suppression by means of SMA-Wire Truss." *The national Congress of Theoretical and Applied Mechanics*, 62.0 (2013): 158.
6. Song, G., Ma, N. and Li, H. N., "Applications of shape memory alloys in civil structures." *Engineering Structures*, 28.9(2006): 1266-1274.
7. Mortazavi, S. M. R., Ghassemieh, M. and Motahari, S. A., "Seismic Control of Steel Structures with Shape Memory Alloys." *International Journal of Automation and Control Engineering*, 2013.
8. McCormick, J., et al., "Seismic vibration control using superelastic shape memory alloys." *Journal of engineering materials and technology*, 128.3 (2006): 294-301.
9. Omar, M., "seismic Response of Braced Steel Frames with Shape Memory Alloy and Mega Bracing Systems." *International journal of Civil Architecture, Structural and Construction Engineering*, Vol. 8. No. 2. 2014.
10. Parulekar, Y. M. and Reddy, G. R., "Nonlinear Model of Pseudoelastic Shape Memory Alloy Damper Considering Residual Martensite Strain Effect." *Adv. Acoust. Vib.*, Article ID 231896, 2012, pp11
11. Hu, J. W., "Numerical simulation for the behavior of superelastic shape memory alloys." *Journal of Mechanics Science and Technology*, 27.2 (2013): 381-386.
12. Auricchio, F. and Sacco., "A one-dimensional model for superelastic shape-memory alloys with different elastic properties between austenite and martensite." *International Journal of Non-Linear Mechanics*, 32.6 (1997): 1101-1114.
13. Lagoudas, D. C., et al., "Simplified shape memory alloy (SMA) material model for vibration isolation." *SPIE's 8th Annual International Symposium on SMart Structures and Materials*, International Society for Optics and Photonics, 2001.
14. Hanahara, K. and Tada, Y., "Variable Geometry Truss with SMA Wire Actuators (Based Consideration on Kinematical and Mechanical Characteristics)." *Proceedings of 19th International Conference on Adaptive Structures and Technologies, October 6-9, Ascona, Switzerland*, 2008.
15. Bandeira, E. L., Savi, M. A., et al., "Finite element analysis of shape memory alloy adaptive trusses with geometrical nonlinearities." *Archive of Applied Mechanics*, 76.3-4 (2006): 133-144.
16. Toi, Y. and Tsukamoto, K., "Computational modeling of adaptive trusses with shape memory alloy members." *Engineering Letters*, 19.1 (2011): 31-37.
17. Hanahara, K. and Tada, Y., "Dynamic Characteristics of Variable Geometry Truss with SMA Wire Actuators." *13th Asia Pacific Vibration Conference*, 2009.
18. Narita, R. and Kojima, T., et al., "Evaluation of Performance of Multi-suspended Pendulum Isolation System with Spatial Truss Structure." *12th World Conference Earthquake Engineering*, 2000.

ICAST2015 # 074

Hybrid adaptive tuned vibration absorber and its control plan

Ning Han[1], Weizhi Song[2], Qiang Gao[1,*], and Chendong Duan[3]

[1] School of Automobile, Chang'an University, Xi'an, 710064, China
[2] School of Mechanical Engineering, Luoyang Institute of Science and Technology, Luoyang, 471000, China
[3] School of Electronic and Control Engineering, Chang'an University, Xi'an, 710064, China

Abstract

The tuned vibration absorber (TVA) has been widely used to suppress the vibration of machines, structures, buildings and aerospace equipments in engineering applications since it was invented. The conventional tuned vibration absorber can reduce the primary system vibration significantly when it is tuned to the excitation frequency of the primary system. However, the effective frequency bandwidth of the conventional tuned vibration absorber is very narrow, and can increase the vibration of the primary system when it is mistuned. In recent years, many configurations of the adaptive tuned vibration absorber have been developed to extend the frequency bandwidth of the vibration absorber. In all these configurations, researchers designed a great deal of devices to change absorber stiffness or mass to track the varying excitation frequency. Nevertheless, using variable stiffness or variable mass method can not increase the effective bandwidth of the tuned vibration absorber evidently, and probably can not satisfy the requirements of some practical applications, in which the excitation frequency changes over a wider frequency range. In order to deal with this problem, in this paper, a hybrid adaptive tuned vibration absorber, in which both the stiffness and the mass of the absorber can be varied, is designed. Meanwhile, a new control plan for the designed hybrid adaptive TVA is proposed. A number of numerical simulations were carried out to investigate the performance of the suggested adaptive tuned vibration absorber and the control plan. The results show that the new hybrid absorber and the control plan can widen the effective frequency bandwidth of the absorber greatly and reduce the vibration of the primary system significantly over this bandwidth. The defect of the control plan is analysed as well.

1. INTRODUCTION

The tuned vibration absorber (TVA) [1] is a vibration control device, which includes an additional mass attached to the primary system via a damper and a spring. Because of having the advantage of simple structure, economy and stable performance, the TVA has been widely used in the field of engineering vibration control. Although the TVA can greatly reduce the vibration of the primary system when it is tuned to the excitation frequency, its effective bandwidth is very narrow. In practical applications, the excitation acting on the primary system is usually complicated and the excitation frequency is often changed with time so that the TVA becomes ineffective. This defect limits the application of the TVA. To widen the effective bandwidth, active vibration absorber and adaptive tuned vibration absorber (ATVA) have been put forward in the last two decades. The active vibration absorber often produce an active force acting on the primary system to neutralize the external excitation such that in the ideal case, the two inputs is offset and no net vibration is generated. Actuators are required in the active vibration absorber to produce active force, meanwhile, sensors and controllers are used to measure the motion condition of the primary system and the absorber, and control the absorber. Electromagnetic and piezoelectric actuators have been studied [2, 3]. However, the active vibration absorber often works in a complicated way because controllers, sensors and actuators need to cooperate to achieve vibration reduction, and can cause instabilities of the system.

The ATVA can change the natural frequency by modifying the absorber parameters to match the excitation frequency. So, the ATVA has a wider effective bandwidth, and is suitable to suppress the vibration of the primary system when the excitation frequency varies with time. On this account, many researchers studied into the adaptive tuned vibration absorber in recent years and a number of ATVAs have been designed. Thus far, mechanical structure [4], electromagnetic device [5] and smart material [6] have been applied to develop high performance ATVAs of variable natural frequency, and most ATVAs vary their natural frequency by adjusting the stiffness of the absorber. Another type of ATVA, variable mass adaptive tuned vibration absorber was suggested in Ref. [7], in which the absorber natural frequency is changed by varying ATVA mass. In the continuous studies, a control strategy of the variable mass ATVA was investigated, which can effectively widen the bandwidth of the absorber, meanwhile, reduce the resonance of the primary system due to the introduction of the absorber. The results show that the control plan has a high performance on resonance reduction of the primary system [8].

However, using neither variable stiffness nor variable mass method can increase the effective bandwidth of the tuned vibration absorber evidently, and can not satisfy the requirements of some practical applications, in which the excitation frequency often changes over a wider frequency range. In order to deal with this problem, in this paper, a hybrid adaptive tuned vibration absorber with adjustable natural frequency is suggested, as well a new control strategy for the new absorber is proposed. In the hybrid ATVA, both the stiffness and the mass of the tuned vibration absorber can be adjusted [9]. The effective frequency bandwidth of the absorber can be much wider and the vibration of the primary system can be reduced significantly.

2. MODEL OF HYBRID ADAPTIVE TUNED VIBRATION ABSORBER

Figure 1 shows the model of traditional TVA applied to the primary system, where m_1, m_2 are the mass of the primary system and the absorber, respectively, and k_1, k_2 are their stiffness, c_1, c_2 are their damping, respectively. x_1, x_2 are the displacement of the primary system and the TVA, respectively, F_0, ω are the amplitude and the frequency of the external excitation, respectively. Figure 2 shows the frequency responses of the primary system with and without the TVA attached. From Figure 2, we can see that TVA can effectively reduce the vibration of the primary system while the natural frequency of the TVA is equal to the excitation frequency. However, the parameters of the traditional TVA is fixed after it is manufactured, namely the absorber natural frequency is unchangeable. When the excitation frequency is deviated from the natural frequency, the vibration reduction performance is greatly attenuated. Therefore, the effective bandwidth of the TVA is very narrow. Another main defect of the conventional TVA is that a new resonance peak is introduced to the new system since the absorber-host system becomes a two-degree-of-freedom system when an absorber is attached to a primary system [10]. Consequently, using the TVA can cause the resonance of the primary system.

In order to extend the effective bandwidth of the conventional tuned vibration absorber, many designs of ATVA have been proposed. Some ATVAs were designed based on method of varying stiffness, and others based on varying mass. Figure 3(a) shows the model of the adaptive tuned vibration absorber with variable stiffness element, in which the stiffness of the absorber can be adjusted somehow (for example, by using mechanical, electromagnetic method, or new materials), then the natural frequency of the TVA can be changed to match the time-varying frequency of the excitation. Another type of the adaptive tuned vibration absorber is variable mass absorber, in which a liquid tank working as the variable mass element is introduced to the common TVA, as shown in

Figure 3(b). When the liquid volume in the tank increases or decreases, the mass of the absorber will be changed as well. It is well known that the natural frequency of the absorber can be calculated by using the equation below:

$$\omega_{2n} = \sqrt{\frac{k_2}{m_2}} \tag{1}$$

where ω_{2n} is the natural frequency of the TVA. We can see that the variation of the TVA natural frequency is not proportional to that of the absorber dynamic parameters since there is an extraction of square root in the equation. For example, when $k_{2\max} = 4 * k_{2\min}$, the $\omega_{2\max} = 2 * \omega_{2\min}$. In practical applications, the dynamic parameters of the absorber usually can not vary in a very wide range, consequently, the effective bandwidth of the ATVA probably will not become wide enough to satisfy the requirements of the applications, no matter variable stiffness or variable mass ATVA is applied. In order to deal with this problem, the hybrid adaptive tuned vibration absorber is suggested in this paper. Both stiffness and mass element of the hybrid adaptive tuned vibration absorber can be changed to adjust the natural frequency of the absorber, as shown in Figure 4. It is obvious that the new absorber is the combination of the previous ATVA plans, and can suppress the vibration of the primary system effectively over a wider frequency band.

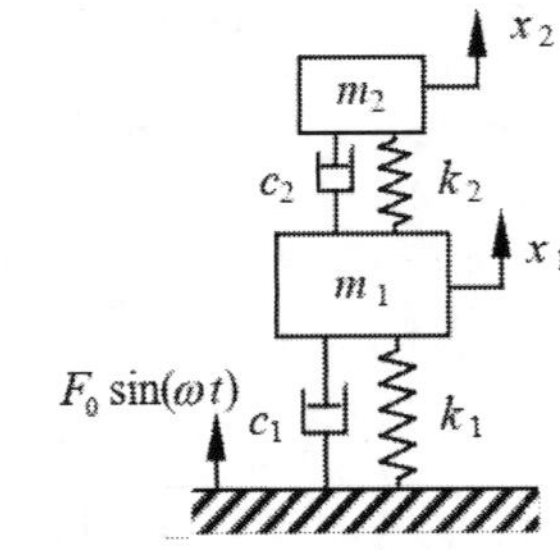

Figure1. Model of TVA to a primary system

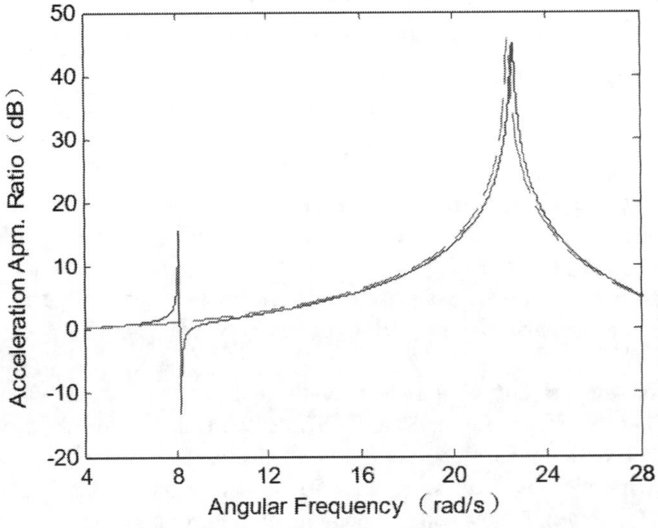

Figure2. Frequency responses of the primary system.
— (blue): with a TVA attached; – – (green): without TVA.

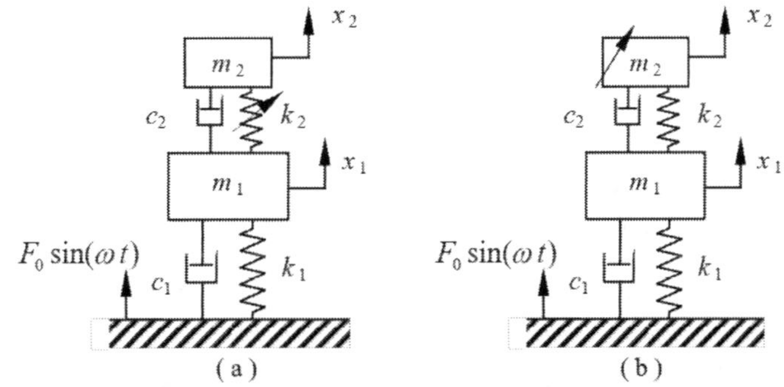

Figure 3. Application of an ATVA to a primary system:
(a) variable stiffness ATVA; (b) variable mass ATVA.

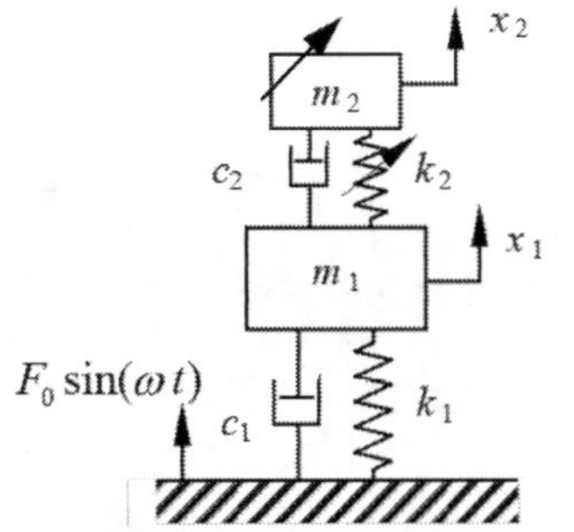

Figure 4. Application of a hybrid adaptive tuned vibration absorber to a primary system

3. CONTROL PLAN OF HYBRID ADAPTIVE TUNED VIBRATION ABSORBER

The previous ATVA designs can be used to implementation the hybrid ATVA. For example, a magnetorheological elastomer, whose elastic modulus (and then stiffness) can be changed by the intensity of magnet field applied on it, works as the stiffness element, and a liquid tank works as the variable mass element. The next problem is how to control the hybrid ATVA. The control plan proposed in this paper is as follows:

The effective frequency band of the hybrid ATVA to reduce the vibration of the primary system is over the range of [$\sqrt{\frac{k_{2\min}}{m_c + m_{v\max}}}$, $\sqrt{\frac{k_{2\max}}{m_c + m_{v\min}}}$], where $k_{2\min}$, $k_{2\max}$ are the minimum and maximum stiffness of the absorber; m_c is the constant mass of the absorber mass, and m_v is the variable mass of the absorber, namely, m_v is the mass of liquid in the tank, $m_{v\min} = 0$, and $m_{v\max}$ is the maximal value of m_v . To design the control plan, the band [$\sqrt{\frac{k_{2\min}}{m_c + m_{v\max}}}$, $\sqrt{\frac{k_{2\max}}{m_c + m_{v\min}}}$] is divided into four parts. The end-point frequencies of these four band are ω_1 , ω_2 , ω_3 , ω_4 , ω_5 . The first frequency band is [ω_1 , ω_2], namely [$\sqrt{\frac{k_{2\min}}{m_c+m_{v1}}}$, $\sqrt{\frac{k_{2p}}{m_c+m_{v1}}}$]; The second frequency band is [ω_2 , ω_3], namely [$\sqrt{\frac{k_{2\min}}{m_c+m_{v2}}}$, $\sqrt{\frac{k_{2\max}}{m_c+m_{v2}}}$]; Similarly, the third frequency band is: [ω_3 , ω_4], namely [$\sqrt{\frac{k_{2\min}}{m_c+m_{v3}}}$, $\sqrt{\frac{k_{2\max}}{m_c+m_{v3}}}$], the fourth frequency band is: [ω_4 , ω_5], namely [$\sqrt{\frac{k_{2\min}}{m_c}}$, $\sqrt{\frac{k_{2\max}}{m_c}}$], where m_{v1} is the liquid mass in the tank over the first frequency band, m_{v2} is that over the second band, and m_{v3} is that over the three band. It is obvious that over the first band, the liquid mass m_{v1} is equal to $m_{v\max}$, and over the fourth band, the liquid mass is zero, that is, all liquid in the tank is released.

In addition, in order to ensure that the whole frequency band is continuous, let

$$\begin{cases} \sqrt{\frac{k_{2p}}{m_c+m_{v1}}} = \sqrt{\frac{k_{2\min}}{m_c+m_{v2}}} \\ \sqrt{\frac{k_{2\max}}{m_c+m_{v2}}} = \sqrt{\frac{k_{2\min}}{m_c+m_{v3}}} \\ \sqrt{\frac{k_{2\max}}{m_c+m_{v3}}} = \sqrt{\frac{k_{2\min}}{m_c}} \end{cases} \tag{2}$$

Then, the liquid mass can be calculated by using the above equation, as following:

$$\begin{cases} k_{2p} = \frac{k_{2\min}^3 \times (m_c + m_{v\max})}{k_{2\max}^2 \times m_c} \\ m_{v2} = \frac{k_{2\max}^2 \times m_c}{k_{2\min}^2} - m_c \\ m_{v3} = \frac{k_{2\max} \times m_c}{k_{2\min}} - m_c \end{cases} \tag{3}$$

And then the end-points can be obtained:

$$\begin{cases} \omega_1 = \sqrt{\frac{k_{2\min}}{m_c+m_{v1}}} = \sqrt{\frac{k_{2\min}}{m_c+m_{v\max}}} \\ \omega_2 = \sqrt{\frac{k_{2p}}{m_c+m_{v1}}} = \sqrt{\frac{k_{2\min}}{m_c+m_{v2}}} \\ \omega_3 = \sqrt{\frac{k_{2\max}}{m_c+m_{v2}}} = \sqrt{\frac{k_{2\min}}{m_c+m_{v3}}} \\ \omega_4 = \sqrt{\frac{k_{2\max}}{m_c+m_{v3}}} = \sqrt{\frac{k_{2\min}}{m}} \\ \omega_5 = \sqrt{\frac{k_{\max}}{m}} \end{cases} \tag{4}$$

According to equations (3) and (4), k_{2p}, m_{v2} and m_{v3} can be computed, and the frequency band can be divided accurately. Then, the control plan for the hybrid ATVA can be described as following equation:

$$F(m_v,k_v)=\begin{cases} m_v=m_{v1},k_v=\omega^2(m_c+m_{v1})\ldots\ldots\ldots\ldots\omega_1\le\omega<\omega_2 \\ m_v=m_{v2},k_v=\omega^2(m_c+m_{v2})\ldots\ldots\ldots\ldots\omega_2\le\omega<\omega_3 \\ m_v=m_{v3},k_v=\omega^2(m_c+m_{v3})\ldots\ldots\ldots\ldots\omega_3\le\omega<\omega_4 \\ m_v=0,k_v=\omega^2 m_c\ldots\ldots\ldots\ldots\ldots\ldots\ldots\omega_4\le\omega\le\omega_5 \end{cases} \tag{5}$$

where k_v is the absorber stiffness when the excitation frequency is varying with time.

So, considering the external excitation frequency acting on the primary system, the liquid supply device injects liquid in or releases liquid from the tank to satisfy the relationship between excitation frequency and variable mass by using equation (5). Meanwhile, the control system as well adjusts the stiffness of the hybrid ATVA according to the excitation frequency. In this way, the hybrid ATVA can track the excitation frequency in real-time.

3. SIMULATIONS OF THE CONTROL PLAN

To analyze the performance of the hybrid ATVA and the proposed control plan, a number of simulations were conducted. Table 1 shows the parameters of the primary system and the absorber. The frequency responses of the primary system are shown in Figure 5. We can see that the vibration of the primary system is significantly suppressed over the frequency band from 5.96 rad/s to 13.42 rad/s by using hybrid ATVA and the control plan. As the counterpoint, the frequency responses of the primary system when the variable mass ATVA, and the variable stiffness ATVA is applied respectively, are presented in Figure 6 and Figure 7. We can see that the variable mass absorber can effectively reduce the primary system vibration over the frequency range from 5.96 rad/s to 10.95 rad/s, and the variable stiffness absorber can do from 10.95 rad/s to 13.42 rad/s. It is evident that the hybrid ATVA has a wider effective bandwidth comparing with both variable stiffness absorber and variable mass absorber.

Table 1. Parameters of The Primary System and ATVA

Parameters of The Primary System	Values	Parameters of Hybrid ATVA	Values
m_1	1	m_c	0.05
k_1	225	$m_{v\min}$	0
c_1	0.01	$m_{v\max}$	0.1185
		$k_{2\min}$	6
		$k_{2\max}$	9
		c_2	0.001

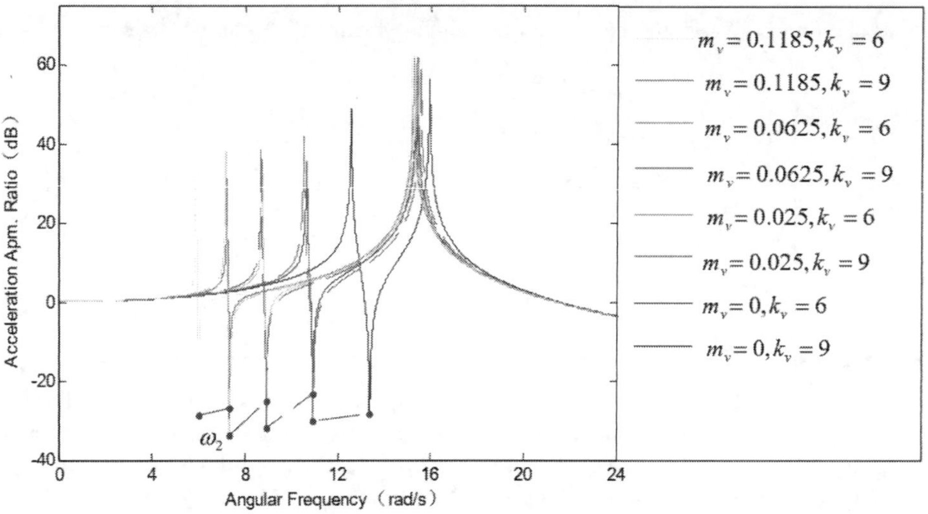

Figure 5. Frequency response of the primary system with a hybrid ATVA attached

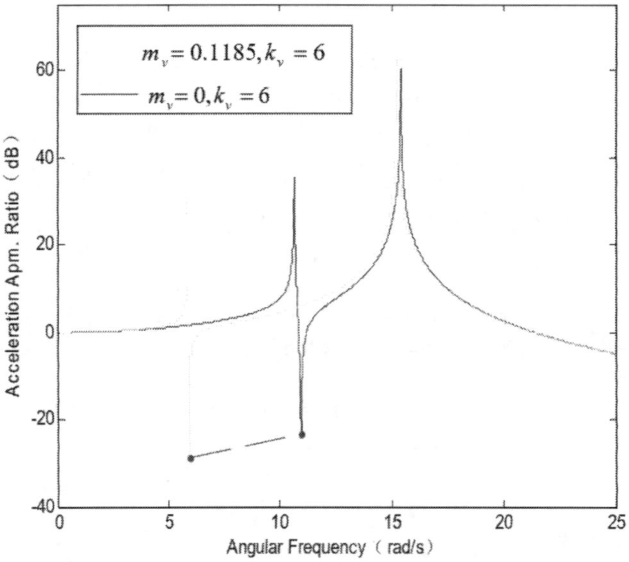

Figure 6. Frequency response of the primary system with a variable mass ATVA attached

However, we can see that there is a number of "jump" at the frequency of the end-points of each band when the hybrid ATVA and the proposed control plan is applied. For example, when the excitation frequency is close to the second end-point, i.e. the point at frequency ω_2, in band of [ω_1, ω_2), the frequency response of the primary system is close to the point (7.3 rad/s, -26.82 dB), which is the lowest point of the red curve, as shown in Figure 5. Meanwhile, when it is close to the point at ω_2 in band of [ω_2, ω_3), the frequency response is close to the point (7.3 rad/s, -33.86 dB), which is on the cyan curve. This implies that there is a abrupt change of 7.04 dB on the primary system vibration when the suggested control strategy is used to control the hybrid ATVA to cross the frequency of ω_2. This situation occurs because we pay more attention on ensuring the frequency band to be successive when we design the control plan, but the frequency response is considered less. It is a defect of this control plan, which probably causes undesirable sudden varying on motion of the primary system. In subsequent study, we will make efforts to solve this shortcoming and improve the performance of the control plan.

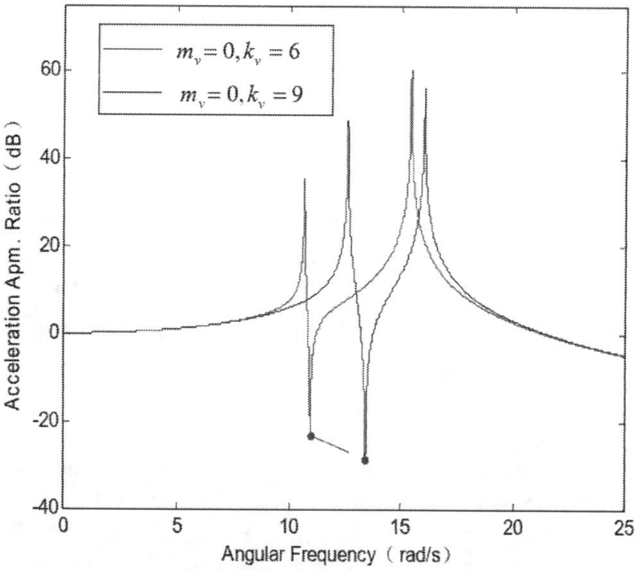

Figure 7. Frequency responses of the primary system with a variable stiffness ATVA attached

4. CONCLUSION

In this paper, a new type of hybrid adaptive tuned vibration absorber and the control plan for it are proposed. Both the absorber stiffness and mass can be changed in this ATVA. Some simulations were carried out to test the performance of the hybrid ATVA and the control plan, and the results indicate that the new hybrid ATVA has a wider effective bandwidth and show valuable in practical applications of vibration control, compared with the previous adaptive absorber, such as variable stiffness absorber, and variable mass absorber. However, there is obvious fluctuation in the frequency response of the primary system at the end-point of the divided band when the control plan is applied. This is the defect of the control strategy. It will be improved in our coming study.

ACKNOWLEDGMENTS

This study is supported by the National Natural Science Foundation of China (Nos. 51175049 and 51275380) and the Special Fund for Basic Scientific Research of Central Colleges, Chang'an University (CHD2012ZD007).

REFERENCES

1. Frahm, H., "Device for damping vibrations of bodies,'',US Patent, 1911, 989, 958
2. Liu, J., He, H. P., Luo, S., " Design and Performance Analysis of New-style and High-powered Electromagnetic Counter-force Actuator,'', Journal of China Mechanical Engineering, Vol. 17,No.18, 2006 , pp.1892-1894
3. Jalili, N., Knowles, D. W., "Structural Vibration Control Using an Active Resonator Absorber: Modeling and Control Implementation,'' Journal of Smart Materials and Structures, Vol. 13, No.5, 2004 ,pp998-1005.
4. Walsh, P. L., Lamancusa, J. S., "A Variable Stiffness Vibration Absorber for Minimization of Transient Vibrations,'' Journal of Sound and Vibration,'' Vol.158, No. 2,1992, pp195-211.
5. Liu, J., Liu, K. F., "A Tunable Electromagnetic Vibration Absorber: Characterization and Application,'' Journal of Sound and Vibration,'' Vol.295, No.3- 5, 2006, pp708-724.
6. Deng, H.X., Gong, X. L., Wang, L. H., "Development of an Adaptive Tuned Vibration Absorber with Magnetorheological Elastomer,'' Journal of Smart Materials and Structures, Vol.15, No.5, 2006, pp111-116.
7. Gao, Q., Duan, C. D., Fang, X. B., Zhao, Y.Q., "Resonance Reduction of Primary System Using Variable Mass Tuned Vibration Absorber," Journal of Noise Control Engineering, Vol.62,No.3, pp 138-144
8. Han, N., Zhao, Y.Q., Gao, Q., Duan, C. D., "Simplified Control Strategy for Resonance Reduction of Primary System,'' Journal of Advances in Mechanical Engineering, Vol. 7, No.9, 2015, pp 1–9.
9. Gao, Q., Song, W. Z., Zhao, Y. Q., "A Hybrid Adaptive Tuned Vibration Absorber and the Control Strategy," China patent
10. Jin, X. X., "Automobile Vibration Analysis," Tongji University Press, 2002. Pp.81-88.

ICAST2015 #042

Improvement of deployment repeatability by the vibration produced by macro fiber composite actuator

Ryotaro Sakamoto[1*], Hiroaki Tanaka[1], and Yoshiro Ogi[2]

[1] Department of Aerospace Engineering, National Defense Academy of Japan, Kanagawa
[2] Institute of Industrial Science, the University of Tokyo, Tokyo

Abstract

In this study, a method of improving the positioning accuracy of deployable space structures was developed and its effectiveness demonstrated. When a space structure expands, friction is generated at the joints between different parts. This friction has adverse effects on the positioning accuracy and produces deviations from the expected position during operation. To achieve high positioning accuracy, it is necessary to not only minimize the friction at the joint but also correct the deviation after it occurs. We apply vibration produced by a macro fiber composite (MFC) actuator to correct the deviation caused by friction. A portion of the structure including the joint is simulated by a simple beam and a ball bearing joint. In the results of our experiments, we observed the effect of applying vibration on the correction of deviations. We also determined the effective vibration based on the natural frequency of the beam and the necessary vibration duration. This method may contribute to improving the accuracy of deployment operations.

1. Introduction

At the joints of large space structures, friction, looseness, and other factors cause adverse effect, such as shape deformation and reduced positioning accuracy [1, 2]. These adverse effects are severe problems that affect the performance of such structures. In particular, the antennas on satellites must have highly precise shapes and positions. This is why several studies have investigated how to reduce friction at the joint and improve deployable repeatability [3]. However, when objects are in contact, it is difficult to reduce the friction to zero, and thus adverse effects or deviations are unavoidable. Therefore, we conducted experiments and analysis on an experimental device that is representative of a portion of the structure to determine how to minimize such adverse effects. This experimental devise can reproduce friction and deviation to examine the effect of vibration. Using this device, we determined the relationship between the friction and torque applied to the joint of the structure. The natural frequency of the device, which is important in this study, is also revealed by a hammer test and the finite element method (FEM). Demonstrating the effectiveness of different vibration frequencies and durations, we established a method of correcting the deviation in the deployment of the space structure by applying vibration with a macro fiber composite (MFC) actuator.

* r.sakamoto6345@gmail.com

2. Experimental methods

2.1 Outline of experiment

In this study, we simulated a deployable structure composed of radial ribs and hoop cables [4, 5] with a simple aluminum beam with dimensions of 40 mm (width) × 2 mm (height) × 700 mm (length) and a ball bearing joint. Figure 1 shows the experimental device. The root of the beam is fixed by a joint. Figure 2 shows the hinge of the device. Figure 3 shows a schematic representation of the device. The beam is hung on a spring with a spring constant of 39 N / m; the spring is located 400 mm from the root. And mass of 200 g is hung at the tip of the beam to keep the beam horizontal. A height gauge was used to raise and lower the beam during the measurement. Three laser displacement sensors were used to measure the relative displacement of the beam from the initial position and determine at each measurement point. The measurement points were 130, 400, and 630 mm from the root. We used an IA-030 laser sensor at the 130 mm position and IL-100 sensors at the others (Keyence). The data was transmitted to a data recorder (GL900 midi LOGGER).

This experiment was conducted to reproduce the deployment behavior by moving the beam. The movement of the beam generates friction at the joint, causing positioning deviation to occur. An MFC actuator was attached 50 mm from the root of the beam. Figure 4 shows the MFC actuator that was attached to the beam. The MFC actuator is a type of piezoelectric actuator [6]. MFCs are suitable for shape control via vibration because of their high performance flexibility and durability [7]. They cause vibration after generating the deviation. We drove the MFC and applied vibration using a signal generator (NF Corporation, WF1974) and an amplifier (TREK, Model 2210).

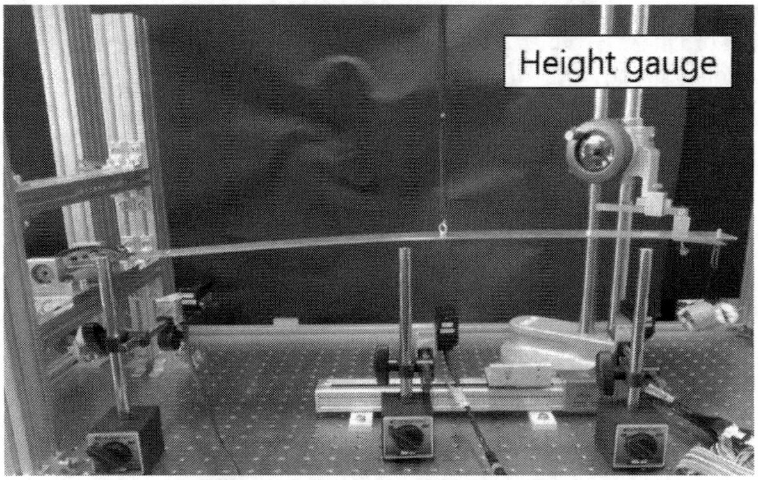

Figure 1. Experimental device

Figure 2. Hinge of the device

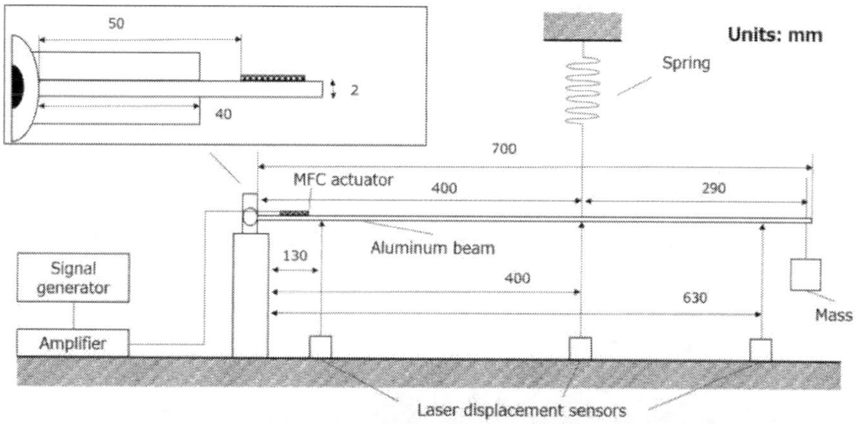

Figure 3. Experimental setup

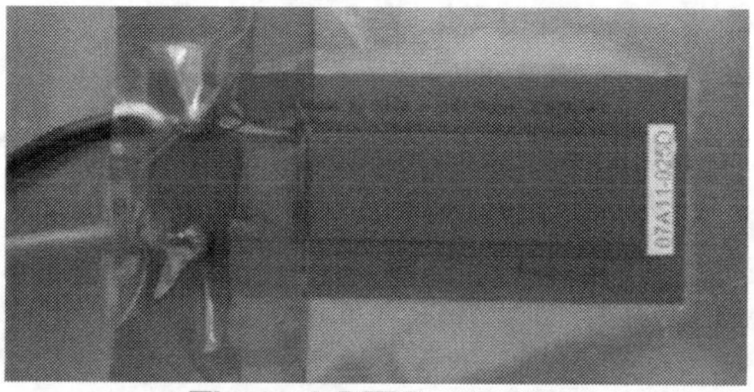
Figure 4. MFC actuator

2.2 Consideration of measurement error caused by temperature

To consider the measurement error caused by the deformation of the jig due to changes in temperature, we measured the tip position 630 mm from the root of the beam and the temperature of the frame using a laser displacement sensor and a thermocouple, respectively. The thermocouple was attached on the frame supporting the beam (Fig. 5). The measurement was conducted over approximately two days. Figure 6 shows the changes in the temperature and the beam position with time. Figure 7 shows the relationship between the temperature and the beam position.

These figures demonstrate the dependence between of the beam position on the temperature. From the results, we obtained approximate expression and reflected in the results.

$$D = 0.028T + 17 \tag{1}$$

where *D* is the amount of deformation and *T* is the temperature

Figure 5. Thermocouple

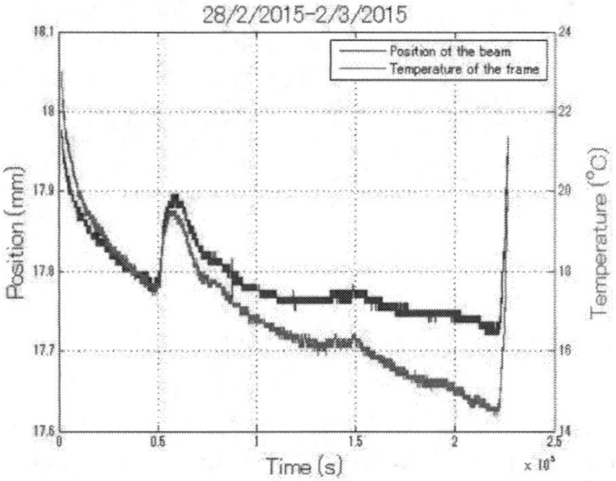

Figure 6. Changes in temperature and beam position with time

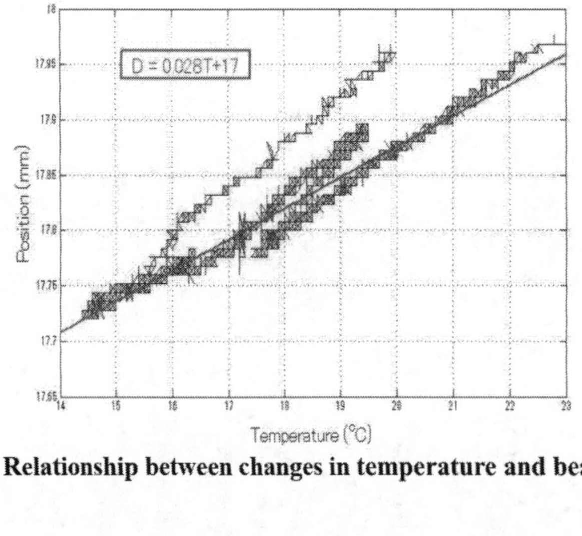

Figure 7. Relationship between changes in temperature and beam position

2.3 Analysis of the natural frequency of the beam

To determine the natural frequency of the experimental device, we conducted a hammer test and finite element analysis. Based on the results, we determined the natural frequency of the experimental device.

In the hammer test, we struck the point midway between the root of the beam and the spring and recorded the position the beam. These data were analyzed by fast Fourier transform. Figure 8 shows the results of the hammer test based on the data of the position between the spring and the tip of the beam. We confirmed peak points at 1.29, 10.4 and 37.9 Hz. These are the natural frequencies from the first, second, and third modes, respectively.

The finite element analysis model is shown in Fig. 9. In the finite element analysis, the device was modeled using beam and truss elements. The shapes of the modes are shown in Fig. 10. Blue limes represent the mode shapes. The obtained natural frequencies are given in Table 1. The natural frequencies of the first five modes obtained using a consistent mass matrix are 1.24, 11.3, 36.9, 80.2, and 140 Hz, respectively.

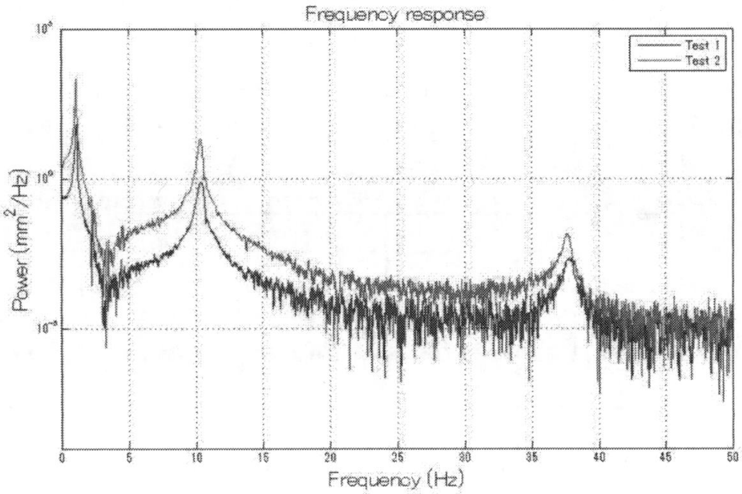

Figure 8. Hammer test results

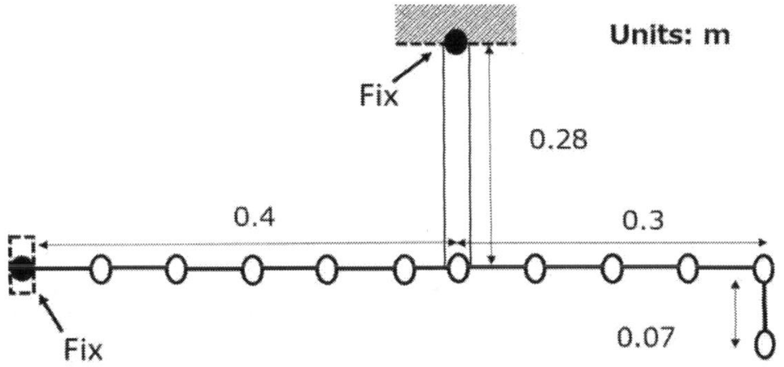

Figure 9. Finite element analysis model

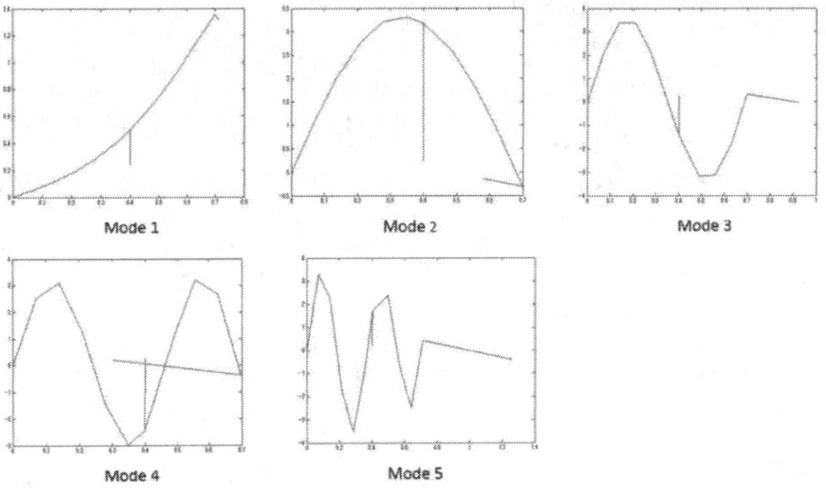

Figure 10. Shapes of the first five modes

Table 1. Natural frequency of experimental device

Mode	Hammer test (Hz)	Finite element method (Hz)	
		Consistent	Concentrated
1	1.29	1.24	1.24
2	10.4	11.3	11.1
3	37.9	36.9	36.6
4		80.2	76.7
5		140	136

3. Experimental results

3.1 Effect of friction

We conducted two experiments to reveal the effect of friction at the joint. The first experiment revealed the difference in the behavior of the beam when torque was and was not applied. We considered two hinge conditions; one is tightening the screw of the hinge by applying a torque of 10 cN·m, and the other is with the screw removed, such that no torque is applied. Figure 11 shows photographs of these two cases. When a torque is applied to the screw, the contact force between the surface of the screw and the plate around it is stronger. The contact surface increases, and friction occurs easily; a schematic representation of this is shown in Fig. 12.

We obtained the relationship between the torque and friction by comparing the displacement from the initial position of the tip of the beam. By moving the tip of the beam up and down by 3, 5, and 10 mm, slight deviation from the initial position was shown to occur when the tip of the beam was returned to its initial location. We measured the position and difference between two values of the before and after moving. Figure 13 shows the results of the experiments with no torque and a torque of 10cN· m applied to hinge. The red dashed and blue lines represent the 10 cN· m and no torque cases, respectively. These lines represent the position of the tip of the beam.

The gap (vertical axis) is defined as the difference between the values of the beam position before and after moving to the initial location. Two conclusions about the influence of friction can be drawn from these results. The first is that when the hinge was not tightened and thus experienced little friction, the movement was highly precise. And the difference of the accuracy between the 10 cN· m and no torque increased when the movement of the tip became large. The second conclusion is that the friction can be reduced, but it cannot be zero. Specifically, even 3 mm movement had a deviation from its original position when the beam was moved.

The second experiment regarding friction at the joint addresses hysteresis. Figure 14 more clearly shows the difference in the friction caused by torque. The red and blue lines represent the 10 cN· m and no torque cases, respectively. The tip of the beam was raised and lowered by displacements ranging from –20 to 20 mm, and the friction moment at the joint was estimated using a force gauge on the spring. When a torque of 10 cN· m was applied, a larger moment was required to start moving the beam than in the free torque case. Furthermore, the repeatability decreased when the movement became large. If no torque was applied to the joint, it experienced a small amount of friction.

From these experiments, to achieve highly accurate deployment, the influence of friction must be minimized.

Figure 11. Torque applied to hinge of experimental device (Left: no torque, Right: 10 cN· m)

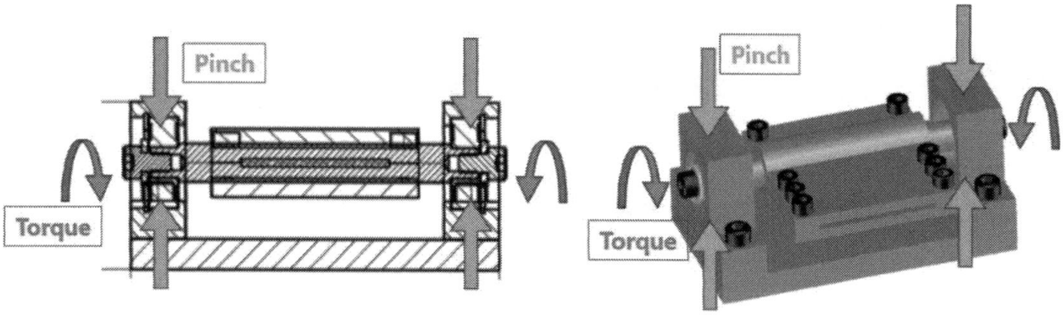

Figure 12. Schematic representation of hinge

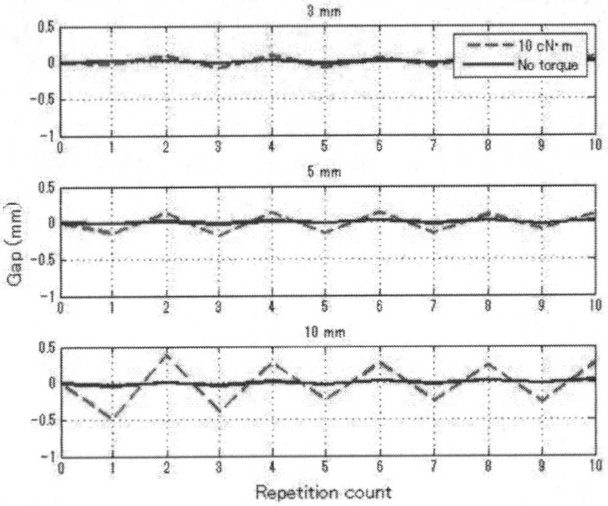

Figure 13. Results for different torques and beam tip displacement

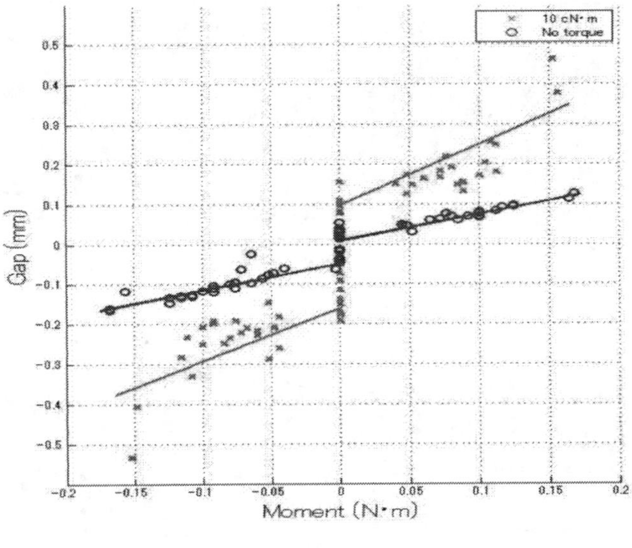

Figure 14. Hysteresis on the joint

3.2 Example of results

Figure 15 shows an example of the behavior of the beam when it is raised or lowered and returned to its initial position. The line represents the position of the tip of the beam. A stable position can be obtained when measurement are taken with repeated up and down movements of the beam. (This average is regarded as the expected position and stable value in this study.) After obtaining the stable value with vertical movement of 3 mm in both direction, the beam was raised by 10 mm and then lowered to its initial position to generate the large deviation. Then, the beam was vibrated at frequencies of 35 and 40 Hz. Different reactions were observed for each vibration. Recovery is defined as the amount of changes before and after vibration.

Comparing the results, we confirmed that a vibration of 40 Hz better corrects the deviation than a vibration of 35 Hz. To numerically evaluate this, we define the correction factor as

$$CF = 100(R - d)/(S - d), \quad (2)$$

where the correction factor *CF* is given as a percentage, *R* is the recovery, *d* is the deviation, and *S* is the sable value.

The calculation results are given in Table 2. The correction factor represents the effect of vibration on the deviation as a percentage. By comparing the correction factors for different frequencies, we confirm the large effect of 40 Hz vibration on the deviation numerically.

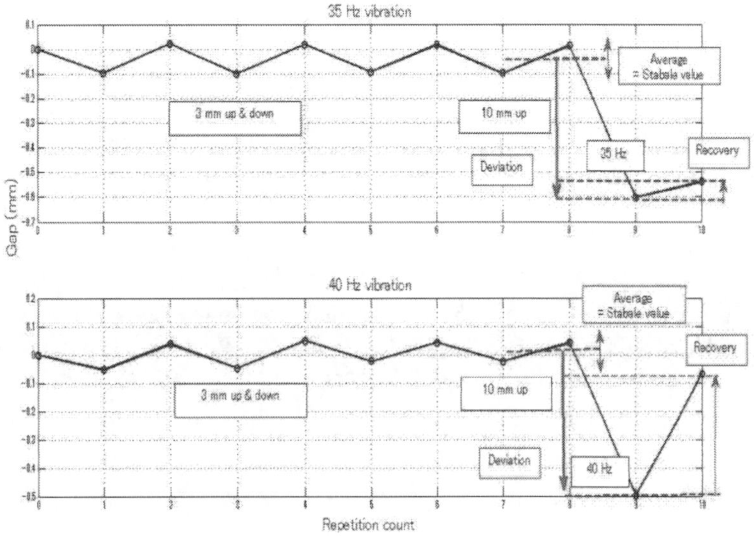

Figure 15. Example of results

Table 2. Calculation results

Frequency [Hz]	Stable value [mm]	Recovery [mm]	Deviation [mm]	Correction factor [%]
35	–0.043	–0.54	–0.60	11.03
40	0.0083	–0.068	–0.50	84.97

3.3 Repeatability under experimental conditions

To consider the dependence of the interference on the time after setting the experimental device, we conducted experiments.

Figure 16 shows the displacement with respect to the elapsed time from setting the device when the beam is moved 3 mm up and down. Regardless of the elapsed time from the setting. The same behavior and stable value obtained.

Furthermore, we hypothesize that repeating up and down movements affects the stable position of the beam. Therefore, we confirmed the effect of vibration makes the beam suitable position from the deviation.

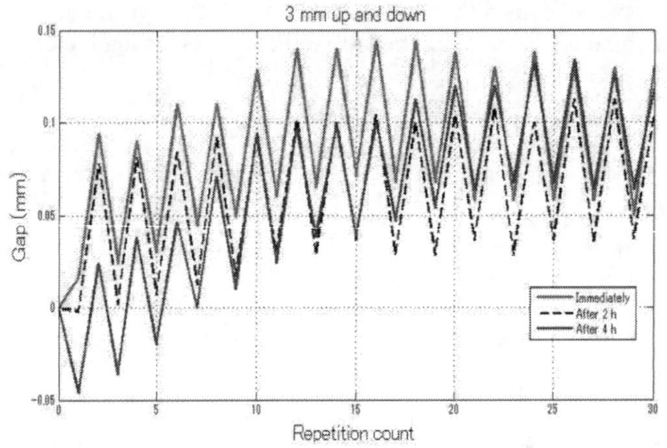

Figure 16. Results of vertical displacement by 3 mm upward and downward

3.4 Vibration duration

We conducted experiments to determine the effective vibration time. The results are shown in Fig. 17. These results demonstrate the effect of the vibration time on the deviation for a vibration of 40 Hz. The times plotted in the figure are 1, 5, 10, 30, 60, and 120 s. When the beam vibration had a sufficiently long duration, a constant value of the correction factor was obtained. Therefore, in this experiment, the vibration duration was set to 1 min as a basis.

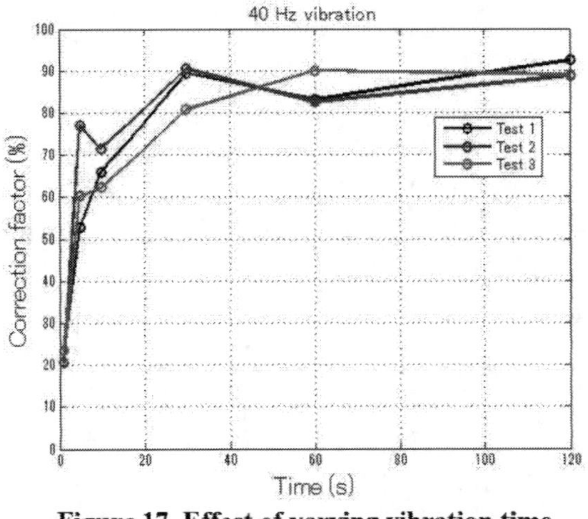

Figure 17. Effect of varying vibration time

3.5 Dependence of natural frequency

The experimental results of the correction when vibrations near natural frequency are applied are shown in Fig. 18. Higher correction factors occur at the natural frequencies rather than at other frequencies. This phenomenon is more pronounced in lower modes. This is because the beam is more affected by the vibration at the natural frequencies.

Furthermore, comparing the higher and lower modes, the effective frequency range in higher modes is larger. For example, in the fifth mode, the correction factor exceeds 50% for frequencies from 125 to 134 Hz. In contrast, in the first mode, correction factor is below 50 % for frequencies differing by more than 0.1 Hz from the frequency at which the peak correction factor is obtained.

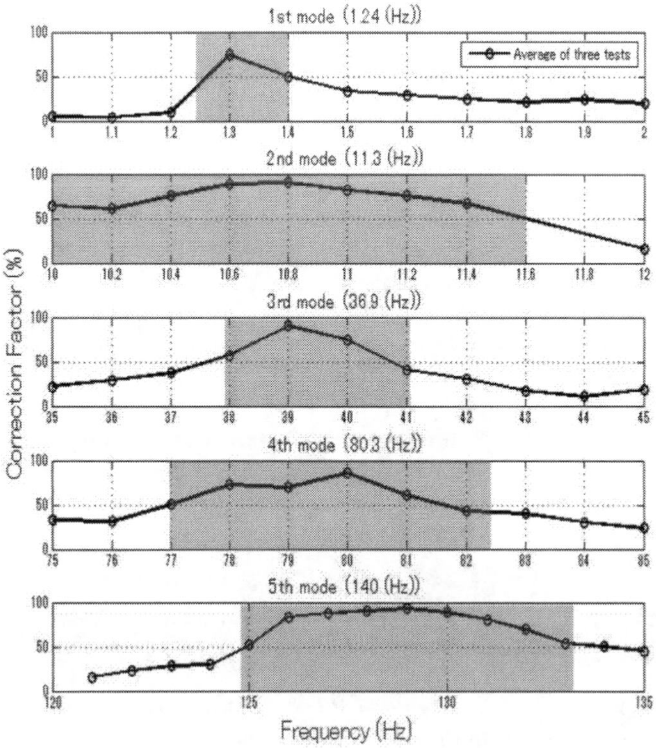

Figure 18. Experimental results of modes 1-5

Figure 18 shows the results when the beam is raised and vibrated. In addition, both the upper side deviation of the beam and the effect of vibration makes the beam suitable position from the deviation of lower side has been confirmed. Figure 19 shows the results of applying a vibration of 40 Hz in both directions.

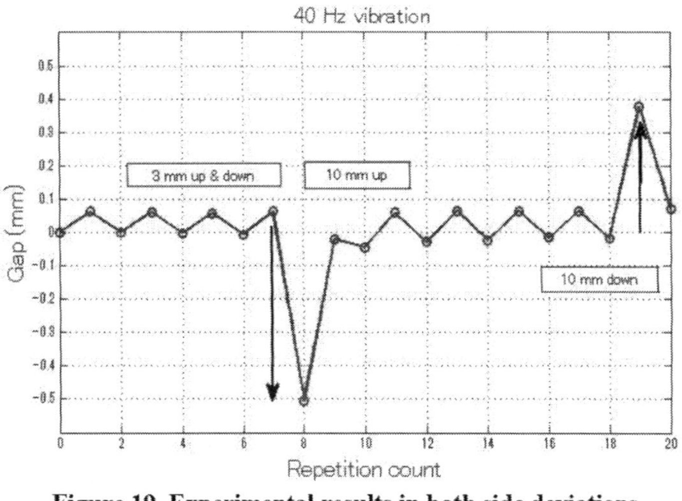

Figure 19. Experimental results in both side deviations

4. Conclusion

In this study, we proposed a method of improving the accuracy of the deployment repeatability of large space structures. When deploying such structures, highly precise deployment is limited by the necessary occurrence of friction at the joint because friction arises when objects are in contact, even if only slightly. Based on the results of our experiments, we confirmed the adverse effect of friction on the movement of such structures. It is necessary to minimize the effect of friction to improve accuracy. Therefore, the proposed MFC vibration method may be able to correct deviation and contribute to improving accuracy.

It was revealed that it is possible to correct the deviation due to friction by applying the vibration at a natural frequency mode. Furthermore the correction factor differed depending on the mode. We also determined the necessary vibration duration for the beam.

In addition, this phenomenon is not only occurs for one-sided deviation but also individual upper and lower deviations of the beam. This study may contribute to achieving high precision in deploying space structures. Additionally, because MFC is very thin and has many applications, this research may also contribute to broadening the applications of MFC.

ACKNOWLEDGMENTS

This work was supported by a grant-in–aid for scientific research 26249131 from Ministry of Education, Culture, Sports, Science and Technology, of Japan

REFERENCES

1. Kojima, Y., Taniwaki, S., and Okami, Y., "Dynamic simulation of stick-slip motion of a flexible solar array," Control Engineering Practice Vol. 16, 2008. Pp. 724-735.
2. Katsumata, N., Higuchi, K., Ookase, Y., and Kaimori, M., "Development on Deployable Space Structures," Memories of the Muroran Institute of Technology Vol. 64. May. 2015. pp69-76.
3. Ogi,Y., Tanaka, H., Tezuka, Y., Miyasaka, A., and Kawaguchi, K., "Effect of Base Hinge on Deformed Shape of Elastic Rib and High Precision Positioning," the 58th Space Sciences and Technology Conference, Nov. 2014.
4. Ishimura, K., Akita, T., Tanaka, H., Goto, K., Yoshiharu, A., Kawano, T., Sakamoto, H., Ogi, Y., Yoshihara, M., and Matsumoto, K., "Stiffness Test of Large Deployable Antenna Composed of Cables, Mesh and Ribs," the 54th JSASS/JSME/JAXA Structures Conference, Aug. 2012.
5. Tanaka, H., Akita, T, Kogiso, N., Ishimura, K., Sakamoto, H., Ogi, Y., and Miyazaki, Y., "A design method for a space antenna structure consisting of radial ribs and hoop cables," the 24th Computational Mechanics Division Conference, Oct. 2011.
6. Robert, G. B., NASA Langley Research Center. "Overview of NASA langley's piezoelectric ceramic packaging technology and applications," the 10th Japan International SAMPE Symposium and Exhibition, Nov. 2007.
7. Henry, S., Daniel, M. P., Elizabeth A. M., Gyuhae, P., and Daniel J. I., "The use of macro-fiber composite in structural vibration applications," the 21st IMAC Conference and Exposition on Structural Dynamics, Feb. 2003.

ICAST2015 #054

High-Precision Positioning of Flexible Deployable Structures Using Kinematic Couplings

Yoshiro Ogi[1*], Yuki Teduka[2], Hiroaki Tanaka[3]
Akihiro Miyasaka[4], and Ken'ichi Kawaguchi[1]

[1] Institute of Industrial Science, the University of Tokyo, Tokyo, Japan
[2] Nissei Plastic Industrial Co., Ltd., Nagano, Japan
[3] Department of Aerospace Engineering, National Defense Academy of Japan, Kanagawa, Japan
[4] Department of Mechanical Systems Engineering, Tokyo City University, Tokyo, Japan

Abstract

This paper presents a high-precision positioning method for flexible deployable structures which have joints like a deployment hinge. Kinematic couplings are applied to joints to restrain necessary degrees of freedom. Results of experiment using a beam regarded as a single rib of a deployable antenna reflector showed that high positioning accuracy can be obtained by properly selecting combination of the kinematic couplings and laying them out.

1. INTRODUCTION

In some space applications, shape accuracy of deployable structures is a key requirement[1)]. One of the factors to disturb the accuracy is deployment repeatability, and that is mainly due to joints where free play and friction affects badly in general. To improve positioning accuracy of flexible deployable structures, this research focuses on deployment hinges and proposes to use kinematic couplings in a proper way.

Hinges are mechanical parts of structure which are designed to restrain translational motion and provide uniaxial rotation for retraction. Typical hinges are plain bearings and rolling bearings, optionally combined with active or passive actuators and latch mechanism. Recent years, a variety of high-performance bearings are available. However, for high-precision large space structures such as a deployable antenna reflector over 10 m, even small friction moment and free play inside hinge's internal parts give non-negligible effect on positioning accuracy.

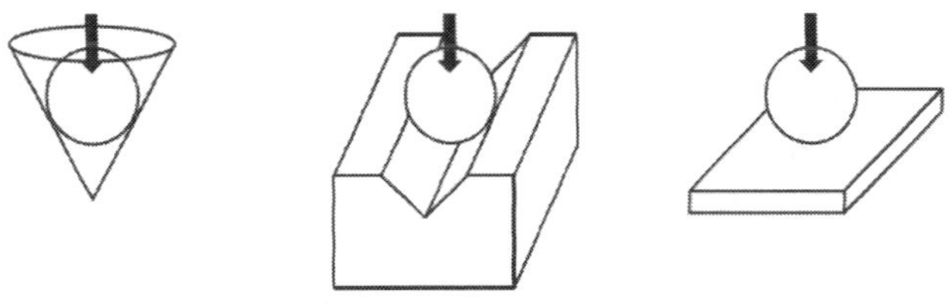

(a) 3DOR (b) 2DOR (c) 1DOR

Figure 1. Example of hoop-rib antenna

* ogi@iis.u-tokyo.ac.jp

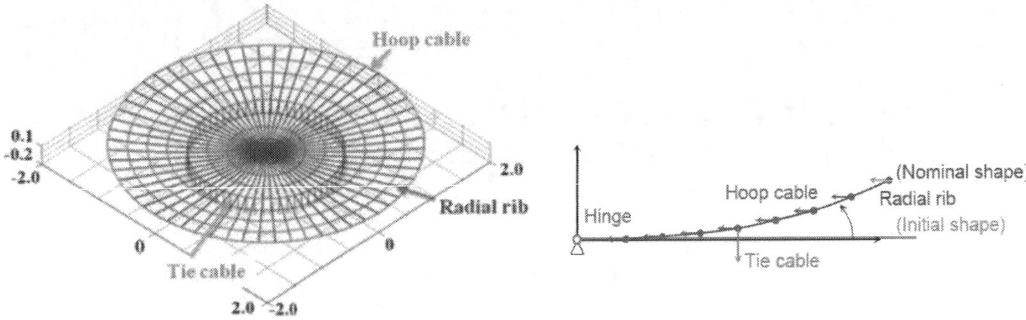

(a) Whole structure (b) Rib

Figure 2. Example of rib-cable type deployable antenna reflector

Kinematic couplings have been widely used for attaining high positioning repeatability in fixtures of machining tools, optical instruments, and so on[2)]. Examples of kinematic coupling are shown in Fig. 1. Just for introduction, in Fig. 1(a), a couple of an outer face of a sphere and an inner face of a cone (or equivalently a set of three flat faces) are fitted under imposing load. If the couple keeps its contact and friction does not exist ideally, the sphere's translational motion is restrained, namely its relative degrees of restraint is three (3DOR). Similarly, degrees of restraint in Fig. 1(b) and Fig. 1(c) are two by a vee-groove (2DOR) and one by a flat face (1DOR) respectively. Popular use of the kinematic couplings is to combine three 2DORs to restrain six degrees of freedom which a free body has.

An application of the kinematic couplings is considered for a space antenna structure[3)]. And some of the authors have been developing hinge-latch mechanism using the kinematic couplings for a solid dish antenna to attain arc-second order deployment repeatability[4)]. The kinematic couplings can be recognized as promising latch mechanism to restrain six degrees of freedom of a free body like a cantilever. However, the principle also should be applied to flexible parts which are supported by other parts and whose necessary degrees of restraint are less than six. The purpose of this study is to explore the possibility of the kinematic couplings for such structures, and to offer one of the methods to realize high-precision large space structures.

The structure considered here is a deployable antenna reflector composed of elastic ribs and tension cables as shown in Fig. 2. In such structure, each rib has a hinge at its one end for retraction. Furthermore, the rib has some extent of stiffness against rotation at the end because the rib is connected to neighboring ribs and backup structure through tension cables. This paper reports experimental results of various joints of a hinge or kinematic couplings, followed by discussion.

2. EXPERIMENTAL SETUP

Figures 3 and 4 show the setup of experiment conducted in this study. An aluminum beam (700 mm×40 mm×2 mm) is used as a single elastic rib. One end of the beam (root) is jointed to a fixture by a hinge or kinematic couplings. Also, the beam is suspended at the middle by a spring whose another end is connected to an aluminum frame fixture through a force gauge. Spring constant of the spring was measured to be 0.039 N/mm. At another end of the beam (tip), two weights of equal mass are hung separately. Mass of the weights and their attachment was 0.2 kg as a whole. The appearance of the experimental setup is like a single rib of a rib-cable antenna structure in cup down configuration, as if the spring is a tie cable connected to backup structure and the weights are hoop cables.

Ideally, if the root is jointed with one degree of freedom for rotation, the structure has a unique equilibrated state under gravity. However, joint's free play or friction is inevitable actually, so the structure should shows multi-states of equilibrium after retraction/deployment operations.

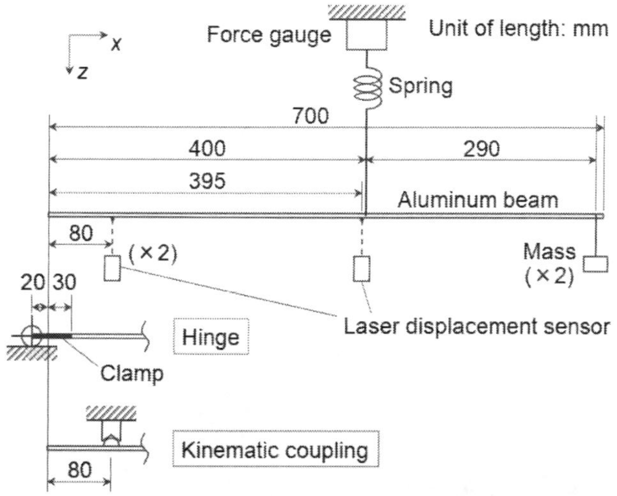

Figure 3. Schematic diagram of experimental setup

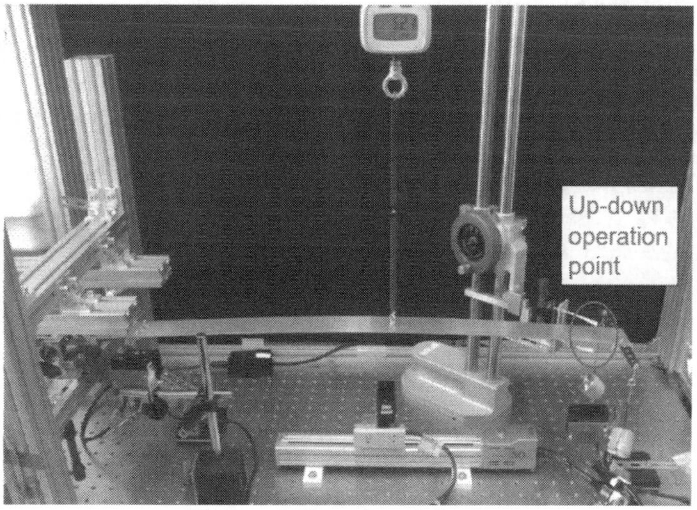

Figure 4. Experimental setup

Root jointing ways examined in this study are listed in Table 1. The hinge is composed of two ball-bearings inside holders attached to a fixture and a shaft holing the beam by a clamp (Fig. 5(a)). Tightness of hinge assembly can be changed by screws at shaft's both ends. Therefore, we can change extent of free play and friction of the hinge, but practically, it is difficult to control and keep those characteristics in high accuracy.

The kinematic couplings used in this experiment are a single or combination of 3DOR, 2DOR, and 1DOR as shown in Figs. 5(b)-(d). Material for all kinematic coupling parts except the 1DOR flat part is SUS440C. Another stainless steel is used for the 1DOR flat part because the 3DOR/2DOR sphere-shape part has a hole at its top (Fig. 5(b)). The radius of the 3DOR/2DOR sphere-shape part is 2 mm. The 3DOR concave part is a cone whose two generatrices in a plane cross at right angles. Similarly, the 2DOR concave part has two flat faces crossing at right angles. No lubricant is used for all the cases.

Kinematic couplings generally require imposing loads to keep contact. In this experimental setup, the root has a tendency to rise up, so that the kinematic couplings are used to push down the root, which means the kinematic couplings naturally have imposing loads.

Table 1 Root jointing ways for experiment

Case	Root jointing way
1	Hinge, tight
2	Hinge, loose
3	3DOR
4	3DOR and 1DOR
5	3DOR and 2DOR

(a) Hinge (Cases 1 and 2)

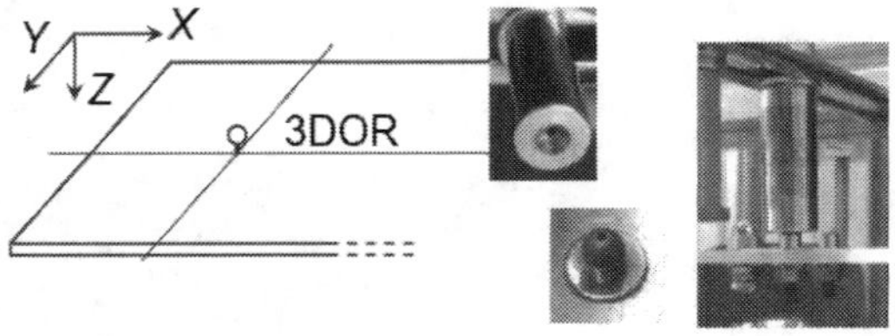

(b) Case 3

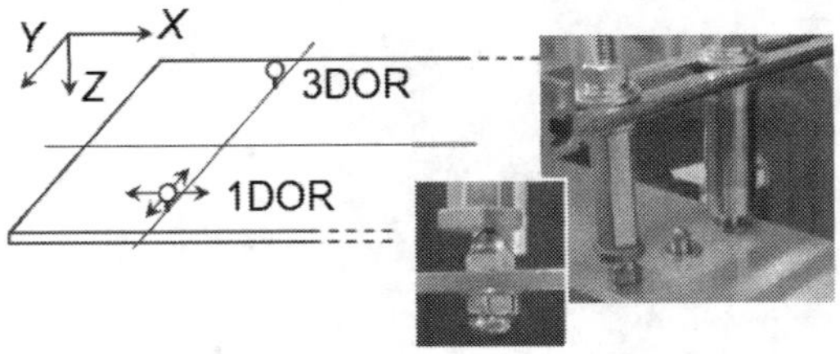

(c) Case 4

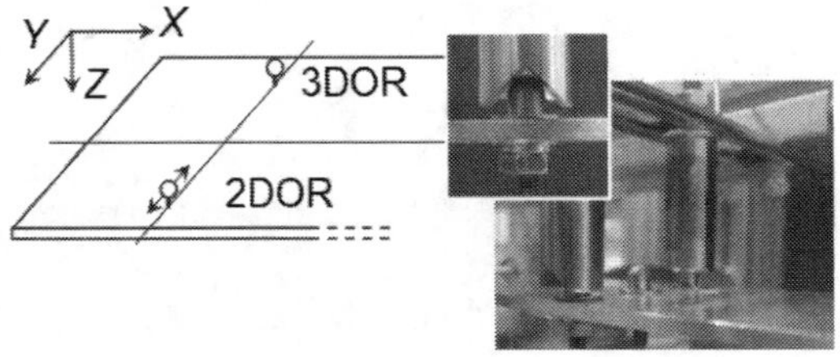

(d) Case 5

Figure 5. Details of root jointing ways

To evaluate positing accuracy of the structure, the tip of the beam is moved up or down by a separated tool as shown in Fig. 4. After that, the tool is detached from the beam and vertical positions at several points are measured by laser displacement sensors whose measurement repeatability are 1.5~4 μm. The magnitude of the movement is 100-mm up for large disturbance such as retraction and deployment, and 5-mm up and 5-mm down for small disturbance. In any experiment cases, even 5-mm movement was sufficient to exceed elastic deformation of the beam and cause its rigid body motion.

Measurement data are recorded by a digital data logger. The operations are conducted by a human paying attention not to induce dynamic behavior. Because equilibrated states are concerned in this paper, mean values of over 400 successive data recorded at 20 ms intervals are used for measurement noise reduction. Temperatures of the experimental setup are monitored using another measurement system using thermocouples.

3. EXPERIMENTAL RESULTS

3.1 Equilibrated position after each operation

For Cases 1-5, equilibrated positions after each operation are shown in Figs. 6-10. The measurement data discussed here is vertical positions in –*Z* direction at the middle of the beam close to the spring suspension point (Fig. 3), and torsion about *X*-axis at the root which is calculated from measurement data of two laser displacement sensors separated along *Y*-axis.

In Figs 6-10, the data after at least three sets of 5-mm up and down movements are used as initial, 100-mm up movement is conducted first, followed by several sets of 5-mm up and down movements. Such a sequence is conducted twice for each experimental case. It should be noted that only variations of the data are concerned in this paper, so that the vertical axis at each graph should be used only for measurement scale. For each case, the temperature variation range was 0.6°C at maximum.

In most cases, graph lines of equilibrated positions show zigzag shape after transient response due to large disturbance of the 100-mm up movement. That is characteristic of free play or friction, and should be minimized for high positioning accuracy.

As for the vertical position of Case 2 (Fig. 7(a)), the results of the 1st and 2nd sequences show different tendency. One of the reasons can be considered that hinge's bearings have several parts inside them, and due to large disturbance, the parts slipped or rotated causing change in hinge's mechanical characteristic.

Difference between the results of the 1st and 2nd sequences can also be observed as for the torsion of Case 5 (Fig. 10(b)), which will be discussed later.

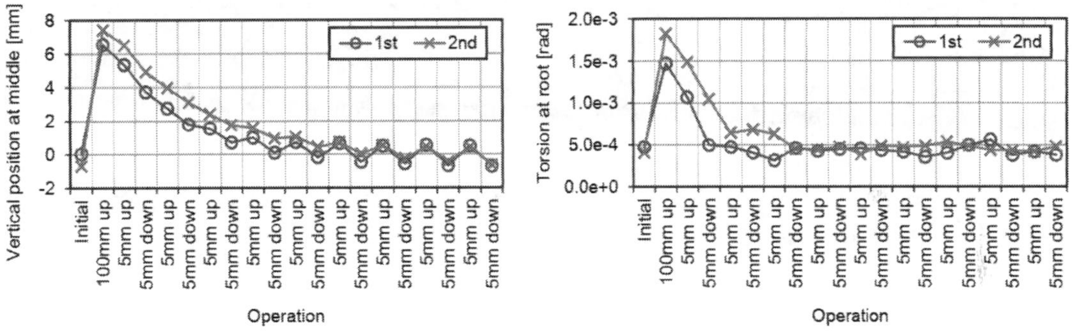

(a) Vertical position in –Z direction at middle (b) Torsion about *X*-axis at root

Figure 6. Equilibrated position after each operation (Case 1)

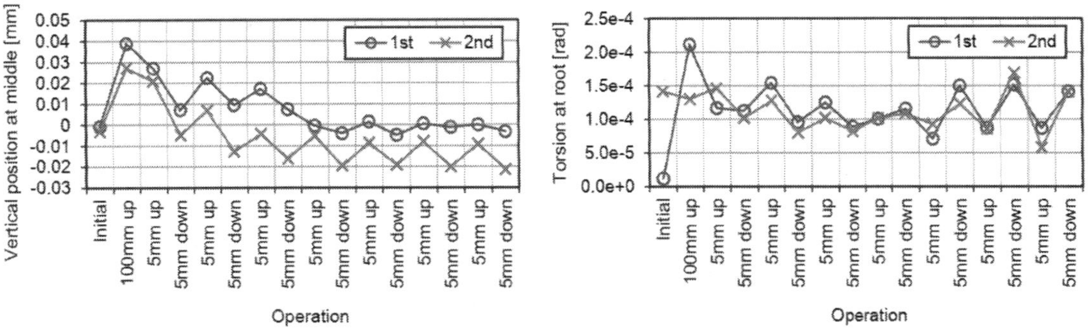

(a) Vertical position in –Z direction at middle (b) Torsion about *X*-axis at root

Figure 7. Equilibrated position after each operation (Case 2)

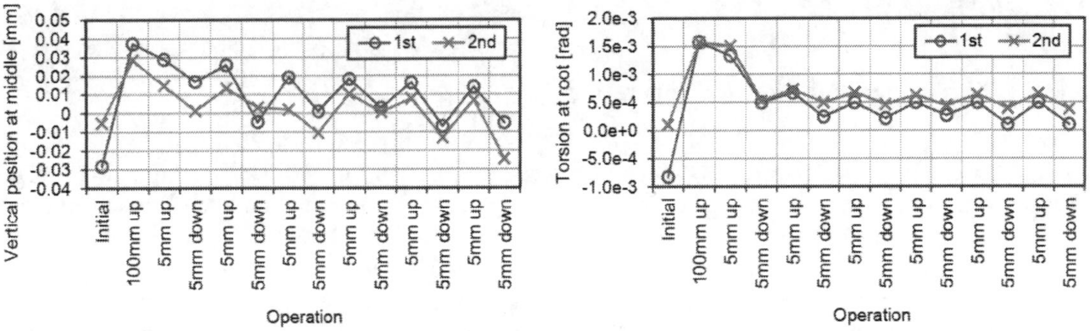

(a) Vertical position in –Z direction at middle (b) Torsion about X-axis at root

Figure 8. Equilibrated position after each operation (Case 3)

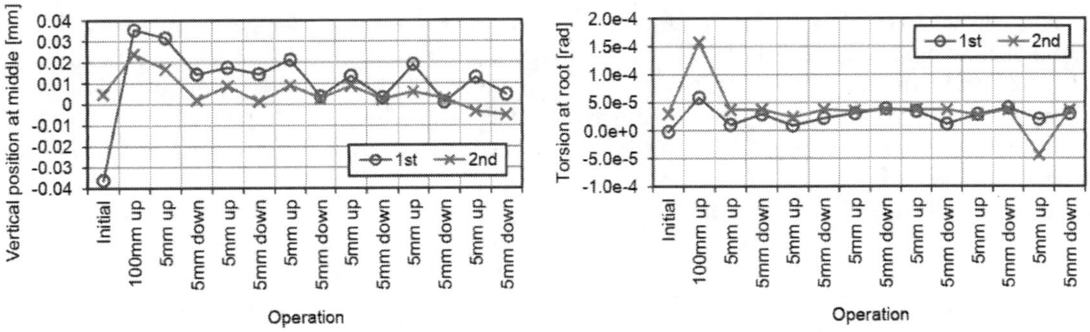

(a) Vertical position in –Z direction at middle (b) Torsion about X-axis at root

Figure 9. Equilibrated position after each operation (Case 4)

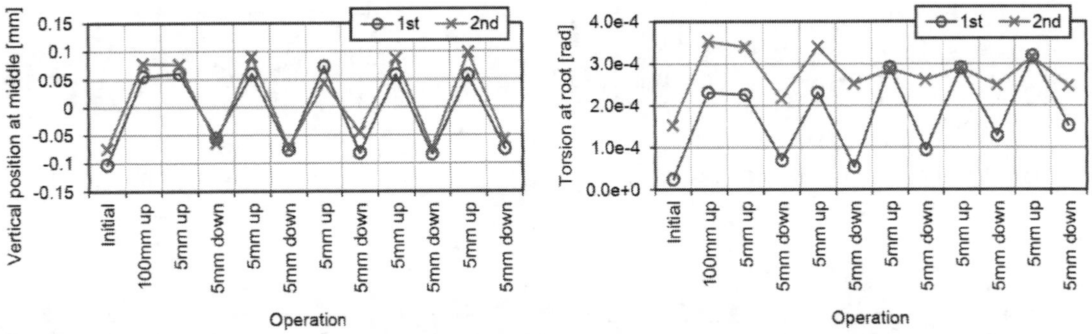

(a) Vertical position in –Z direction at middle (b) Torsion about X-axis at root

Figure 10. Equilibrated position after each operation (Case 5)

3.2 Comparison of root jointing ways

To clarify the difference of the results of all the experimental cases, standard deviations of vertical position at the middle and torsion at the root are shown in Fig. 11. In order to exclude the effect of temperature difference, the standard deviation for each case is calculated for the 1st and 2nd sequence data respectively, and averaged. Although transient phenomena are observed due to the 100-mm up movement, all the data are used for the standard deviation calculation.

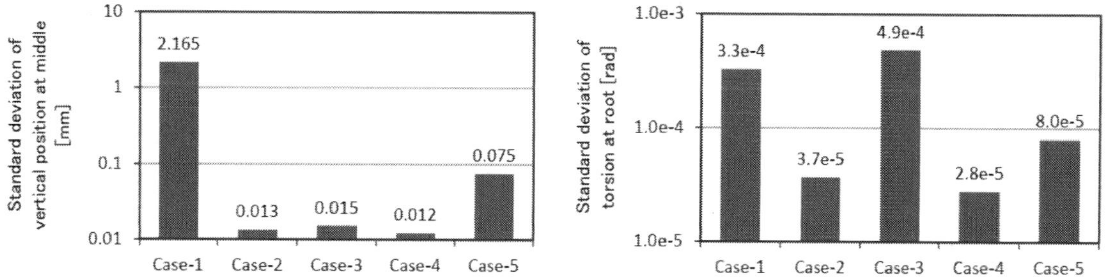

(a) Vertical position at middle (b) Torsion at root

Figure 11. Standard deviation of equilibrated positions

Comparison between the results of Cases 1 and 2 shows that positioning accuracy of the hinge depends on assembly ways of components. That indicates theoretical estimation of mechanical characteristic is difficult in required accuracy and depends on qualification tests. Moreover, as shown in Fig. 7(a), mechanical characteristic of such hinges may change with repetitive tests.

When we focus on the results of kinematic couplings (Cases 3, 4, and 5), additional kinematic coupling of 1DOR or 2DOR to 3DOR is effective for positioning accuracy in torsion because of obvious additional restraint (Fig. 11(b)). As for vertical position, however, the result of Case 5 is the worst of the three. That is because, when the beam rotates about *Y*-axis at the root, 1DOR of Case 4 follows the motion by rolling (Fig. 5(c)) and 2DOR of Case 5 needs slippage at two contacting area being disturbed by their friction (Fig. 5(d)). Therefore, the best combination of kinematic couplings for positioning accuracy in vertical position and torsion is a set of 3DOR and 1DOR (Case 4).

4. CONCLUSIONS

As shown from experimental results, insufficient or excessive restraint using the kinematic couplings worsened positioning accuracy of the beam. In other words, high positioning accuracy can be obtained by properly selecting combination of the kinematic couplings and laying them out. One of the advantages to use the kinematic couplings is that if their contact mechanics can be modeled properly, positioning accuracy can be estimated, and furthermore optimal design can be carried out.

ACKNOWLEDGMENTS

A part of this study was supported by "Study on Development of High-Precision Large Space Structural System" in the "Science and Engineering Experiments of Strategic Development and Research" of Institute of Space and Astronautical Science, Japan Aerospace Exploration Agency.

REFERENCES

1. Ishimura, K., Kii, T., Komatsu, K., Goto, K., Higuchi, K., Matsumoto, K., Iikura, S., Yoshihara, M. and Tsuchiya, M., "Shape Prediction of Large Deployable Antenna Structure on Orbit," *Transactions of the Japan Society for Aeronautical and Space Sciences, Aerospace Technology Japan*, Vol. 10, No. ists28, 2012, pp. Pc_35-Pc_40.
2. Slocum, A.H., "Precision Machine Design," Prentice Hall, 1992.

3. Reynolds, P., Atkinson, C. and Gliman, L., "Design and Development of the Primary and Secondary Mirror Deployment Systems for the Cryogenic JWST," Proc. 37th Aerospace Mechanisms Symposium, USA, May 19-21, 2004, pp. 29-44.
4. Shinohara, K., Ishimura, K., Ogi, Y., Matumoto, K., Tanaka, H., Ito, T. and Kai, K., "Analysis of Latch and Hinge Deployable Structure Using Contact Finite Element Method," 2013-c-37, 29th International Symposium on Space Technology and Science, Nagoya, Aichi, Japan, June 2-9, 2013.

ICAST2015 #56

Deployment Behavior Control using Cables and Bi-SMA-Convex Tape Booms

Nobuhisa Katsumata[1*], Masaru Kume[1], and Ken Higuchi[1]

[1] Department of Mechanical, Aerospace and Materials Engineering, Muroran Institute of Technology, Hokkaido, Japan

Abstract

The purpose of this study is to demonstrate the synchronous and the stable deployment of a newly proposed boom systems consisted of cables, a rotary damper and shape memory alloys (SMA) memorized convex tape shape. Through a shaft, a rotary damper is connected to a reel and cables winded around the reel are connected to the SMA boom tips. The deployment part is consisted of bi-SMA convex tape boom, two SMA convex tape is combined to form a convex lens cross section, and the outside of bi-SMA convex tape is wrapped by a sheet type heater and a polyimide films. The boom deployment is achieved just using the shape recovery force of SMA. Installing cables and a rotary damper, deployment behavior of each boom is controlled and each boom deploy synchronously due to the resistance force of damper to a leading deploy boom. Besides, the structural stiffness control concept of this proposed SMA bi-convex tape boom is discussed because the young's module become almost half on a lower grass transfer temperature conditions.

1. INTRODUCTION

High packing rate, high bending and torsional stiffness, and light weight deployable booms are important and are always required in the field of space development. Besides, the stable deployment and the simple deployment mechanism booms are very useful especially for small satellites because of the limitation of weigh and mounting space. These deployable booms work as actuators for deploying solar array panels and/or membrane, antennas, deorbit membrane, etc. during deployment and work as structural members after deployment. However, each desired property of boom has its contradictive characteristics and it's difficult to satisfy every property at the same high level. Thus, trade-offs or optimum designs have been considered in developing many types of deployable booms.

As for the high packing rate and the light weight aspects, inflatable booms [1, 2] will be the best, compared to any other kind of booms, however a highly pressurized inner condition of the boom or the rigidization of outer inflatable tube material [1-6] is essential to keep and/or get the high bending and torsion stiffness after deployment. STEM booms [7, 8] consisted of Tri-axial Woven Fabrics (TWF) CFRP seems to be one of the solution of high stiffness after deployment, though Single-STEM booms [7, 8] has a low torsion stiffness due to its open cross-section. Bi-STEM booms solve the torsion stiffness program but that are double the weight. As for bending and torsion stiffness, a round-shaped cross section [9, 10] or double Omega-shaped cross section [11-13] booms have high performances. However, the packing rate is not higher than others because these booms are impossible to slide towards the boom's lengthwise direction during packing due to the tight boundary edge between half round-shaped or Omega-shaped

* nobuhisa-k@mmm.muroran-it.ac.jp

sections. The unslidable packing condition leads to a packed configuration with some gaps in the winding roll.

Based on these considerations, the BCON boom [14, 15] design shown in Fig. 1 (a), the combined use of double convex tapes forming elliptical cross section and the slidable boundary condition due to the braid coating, will be optimized. In addition, it is possible to deploy itself by just releasing the elastic energy stored during packing. As shown in Fig. 1 (b), the slidable boundary condition toward lengthwise direction achieve tight winding configuration. The cross section shape of BCON boom on packing configuration is rectangular and the cross section shape recover to the convex lens shape after deployment. This elliptic cross section shape has advantage for bending and torsion stiffness from the view point of after deployment structural member.

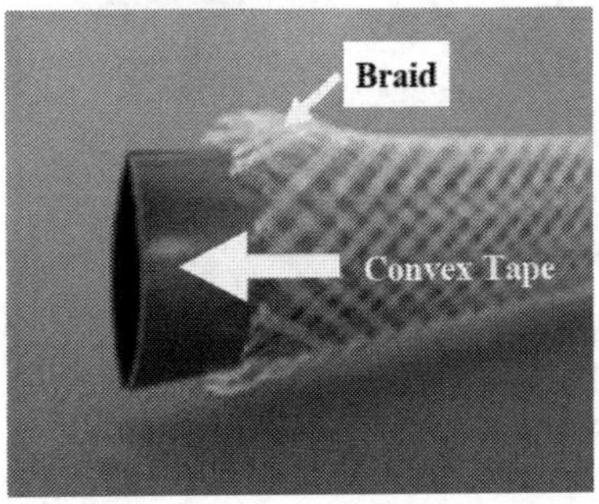

(a) BCON boom configuration

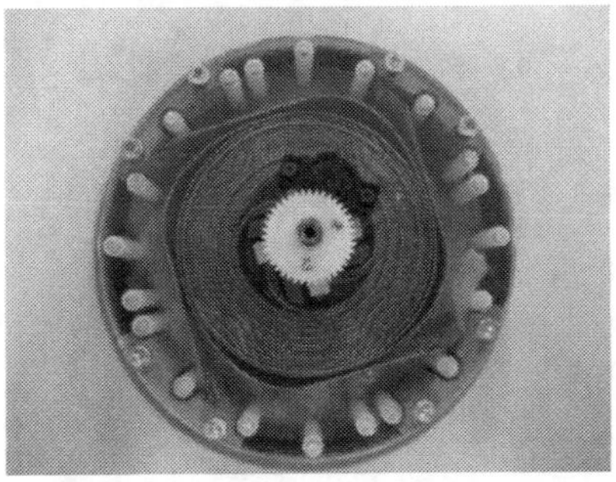

(b) Packed configuration of four BCON booms

Figure 1. Schematic photo of BCON boom and stored configuration

The newly proposed SMA Bi-Convex tape booms (SMA-BiCon) or SMA Braid Coated Bi-Convex tape Boom (SMA-BCON) [16] also have advantage of BCON booms. As for the SMA-BiCon, SMAs memorized convex tape shape is used and the boom deployment is achieved by using shape recovering force of SMA. The boom is stored in the lower temperature of grass transfer, the SMA-BiCon flexibly bend and pack due to the hyper elasticity property of SMA. The most advantage of SMA-BiCon will be a flexibility in the total boom design. For example, if we wind a SMA-BiCon around a polygonal box satellite, a partially different boom design is possible. Bi-SMA convex tape part should be applied on a polygon corner, and a straight and rigid (deformable) boom can be applied to a polygon's side part. To do that, the total boom are consisted of Bi-SMA convex parts and straight rigid boom part. This design flexibility come from the simple deployment mechanism of the SMA-BiCon and the partially different boom design have possibility to the light weight boom design.

Through these structural concept considerations, the deployment experiment of SMA-BiCon are carried out in this paper. Through the past deployment experiment of SMA-BCON [16], synchronous deployment of SMA is important to achieve stable deployment behavior however it's difficult to control the heating condition of SMA by using simple heater. Therefore the simple synchronous deployment mechanism using cable and rotary damper is developed and the demonstration of synchronous deployment is carried out. Besides, the combined use concept of SMA and Shape Memory Polymer (SMP) is discussed to keep a high bending and torsion stiffness after deployment and on martensite phase SMA conditions.

2. SMA BI-CONVEX TAPE BOOM

The conceptual model of SMA Bi-Convex boom (SMA-BiCon) is shown in Fig. 2. Fig. 2 (a) shows the SMA-BiCon boom material, the SMA material is consisted of Ni-Cu alloys. The glass transition point Tg is set as 60 degree centigrade. The memorized shape is convex tape. Fig. 2 (b) shows the conceptual model of SMA-BiCon. Upper figure shows the actual manufactured boom, and lower illustration shows the cross section configurations. As described introduction, the boom consisted of two part, one is the bendable and deployable SMA part, and the other is the straight light weight boom part. The straight light weight boom part is made by punched acrylic plate. The bendable SMA part consisted of three layer toward cross section. SMA convex tape is located in the innermost layer, and the sheet type heater is located in outside of SMA. This sheet type heater is consisted of the polymethylpenten file, daily used food wrap, and the Carbo e-Therm Heater layer. The Carbo e-Therm is a kind of paint containing carbon powder. After spreading Carbo e-Therm on the film, the film works as heater just supplying currents. This heater part is also manufactured ourselves. The outermost layer consisted of polyimide tape. This polyimide tape wrapping work as a kind of insulation material, and increase the heating effect of inner side.

(a) SMA material

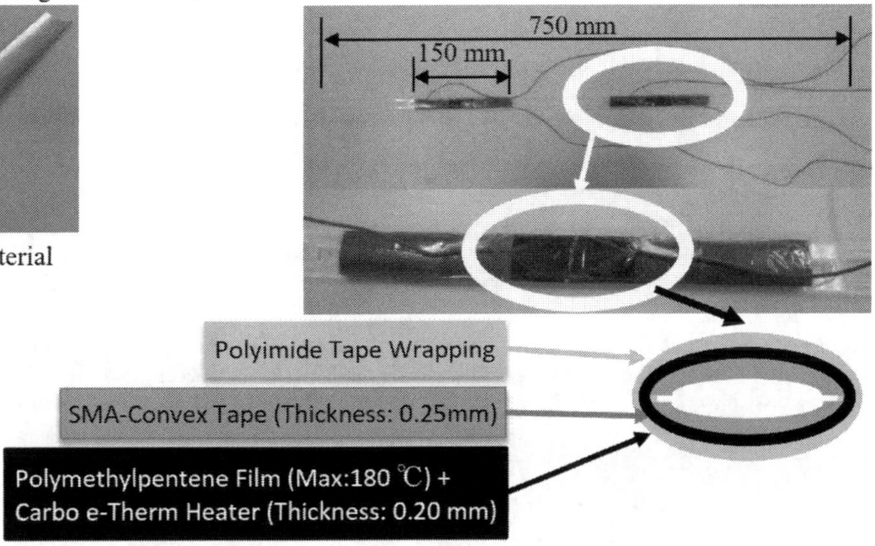

(b) SMA BiCon configuration and schematic illustration of SMA-BiCon cross section

Figure 2. SMA-BiCon material and SMA-BiCon conceptual model configuration

3. SYNCHRONOUS DEPLOYMENT SYSTEM

Supplying one watt, the sheet type heater heat and SMA-BiCon gradually deploy, though the heating condition of SMA is independent. To control the electric power supply, the same watt value is supplied to each heater, however the deployment of SMA is not synchronous due to the contact condition of SMA and heater and/or the individual heat sensitivity difference of SMA. Through try and error examinations, another synchronous deployment method of SMA-BiCon is developed using cable and rotary damper.

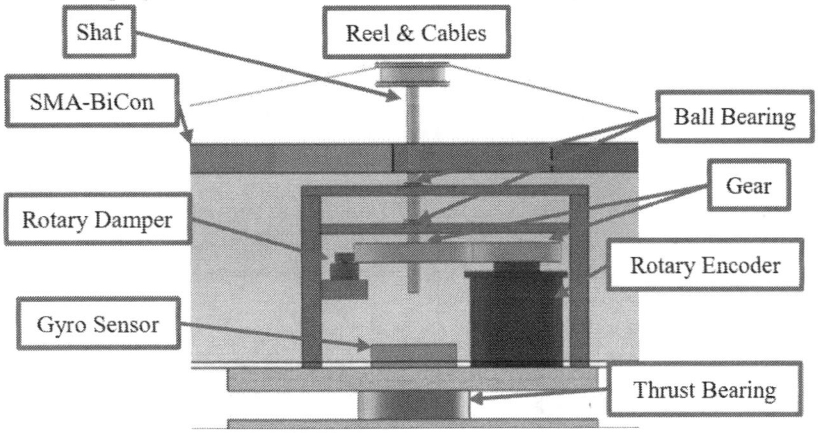

Figure. 3 The outline figure of synchronous deployment system

Fig. 3 shows the outline figure of synchronous deployment system. As shown as Fig. 3, the boom tips and the center shaft are connected through the cables, and the center shaft and the rotary damper are connected through the gears. If one of the boom deploy earlier than others, the deployment is interrupted by the damping force through cables. The system shown in Fig. 3 is still conceptual phase, the cables and the rotary damper system will become more simplify and lightweight in an actual space mission design. To measure the rotational angle and the angular velocity, a rotary encoder and a gyro sensor is also set in the device. A thrust bearing is located in the bottom of box for the free rotational condition.

4. DEPLOYMENT EXPERIMENT

4.1 Experimental Device

The experimental device after first step deployment configuration is shown in Fig. 4. In this study, SMA-BiCon has two deployment parts, and stored around 300mm square box. DC power supply is located under the box, and the heater current is supplied here. In this case, the booms deploy like stepwise manner. It's possible to heat every deployment part and to deploy every boom simultaneously, but stepwise deployment is carried out in this experiment due to demonstrate the synchronous deployment system in any boom length.

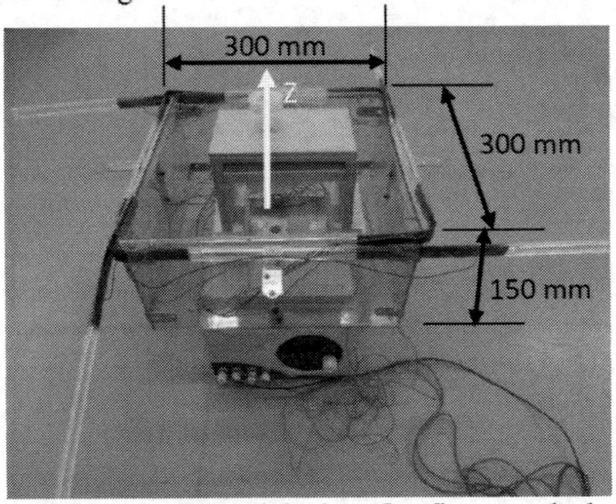

Figure. 4 The experimental device after first step deployment

4.2 Experimental Results

The deployment experiment is examined in two cases, the supplied voltage value is set at 7.5 V and 10.0 V. Deployment behavior is recorded by video camera, and deployment rate is calculated by the captured pictures. The capture pictures are obtained in one second interval. The deployment rate is defined as following: Deployment rate = deployment angle [degree] / 90 [degree]. The deployment angle is measured from the captured pictures using image processing software. The boom number from one to four is set as the manner shown in Fig. 5 (a) and (b). After first step deployment, the stored boom position rotate by 90 degrees clockwise, the number of boom is also set like this manner. Angular velocity value are obtained by the gyro sensor. The z axis angular velocity value, shown as yellow arrow in Fig. 4, is compared in this study.

Fig. 5 and Fig. 7 shows the example of deployment behavior with synchronous deployment system and without the system. Fig. 6 and Fig. 8 shows the deployment rate and angular velocity result in case of supplied voltage is 7.5 V and 10.0 V. The graphs are divided into two part. One is first step deployment results, and the other is second step deployment results.

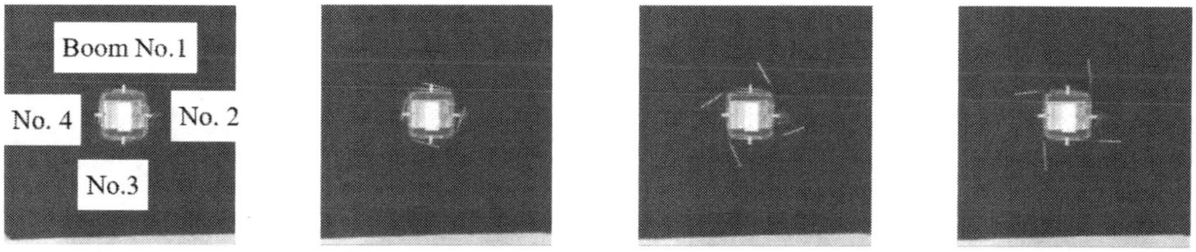

0 s 10 s 20 s 30 s

(a) First step deployment behavior

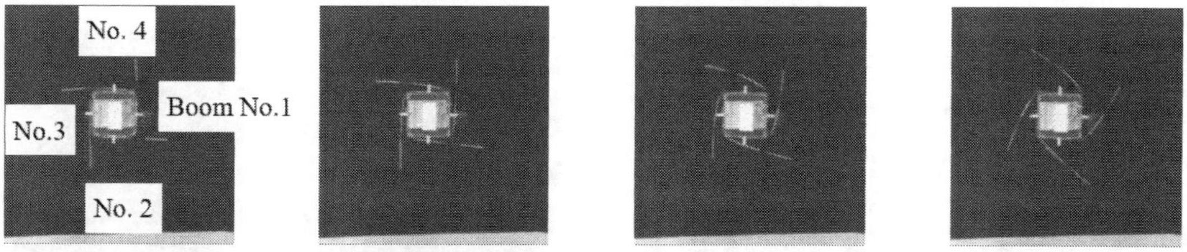

0 s 10 s 20 s 30 s

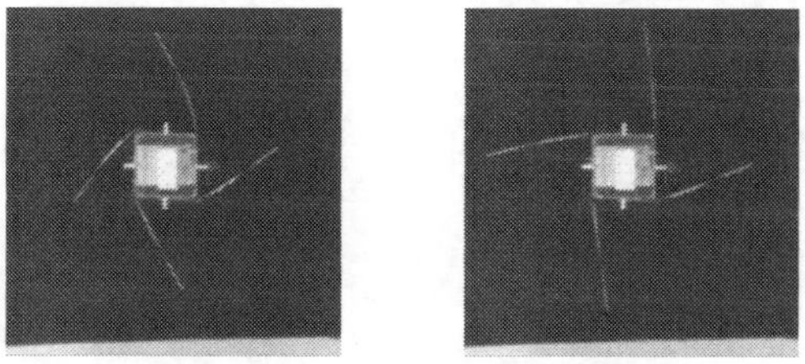

40 s 50 s

(b) Second step deployment behavior

Figure. 5 Example of Deployment behavior of with cables and rotary damper system

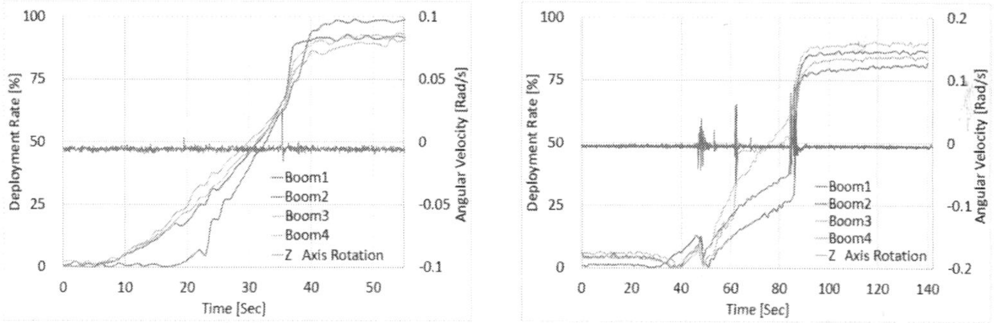

1st step deployment 2nd step deployment

(a) Supplied voltage is 7.5 V

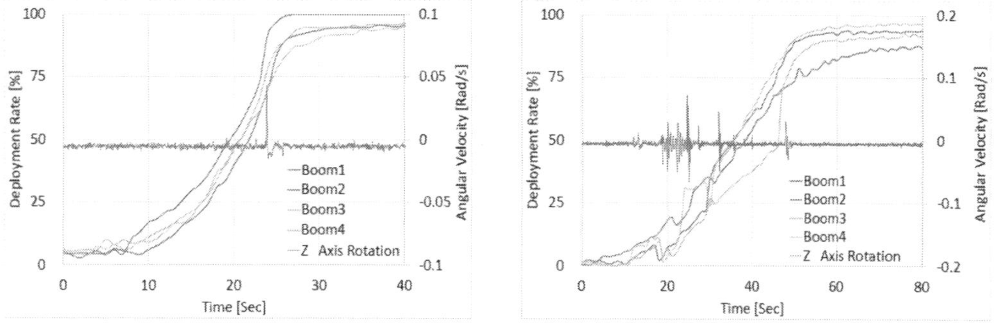

1st step deployment 2nd step deployment

(b) Supplied voltage is 10.0 V

Figure. 6 Deployment rate and angular velocity value with cables and rotary damper system

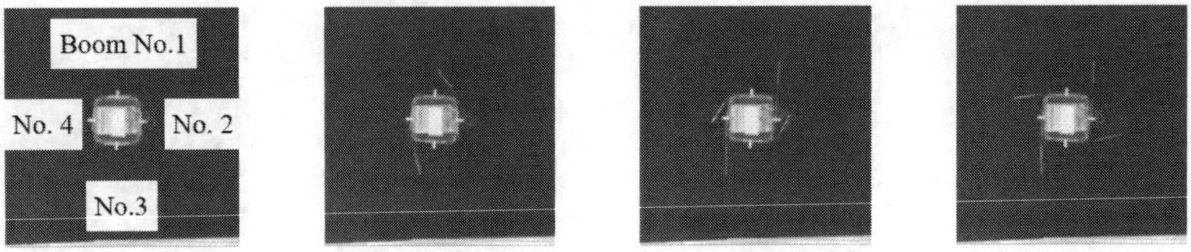

0 s 5 s 10 s 15 s

(a) First step deployment behavior

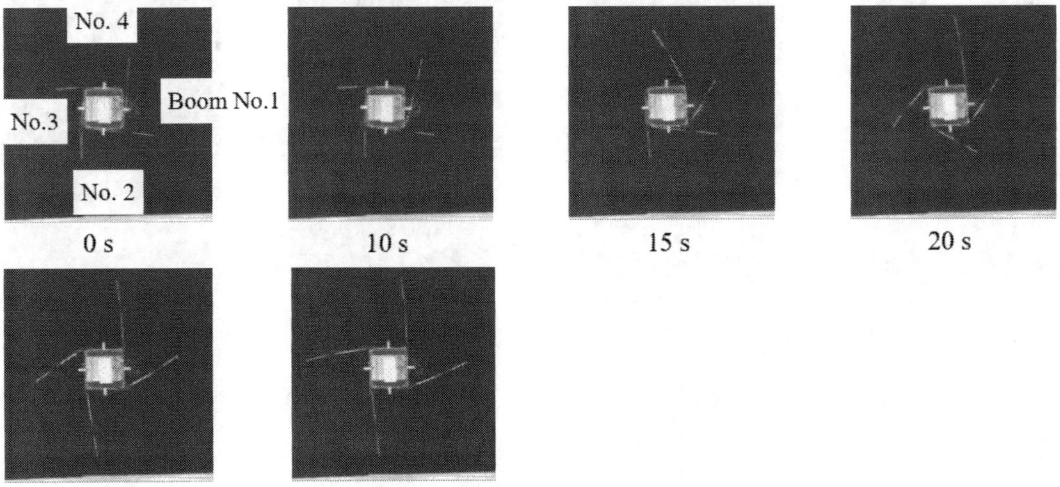

0 s 10 s 15 s 20 s

40 s 60 s

(b) Second step deployment behavior

Figure. 7 Example of Deployment behavior of without cables and rotary damper system

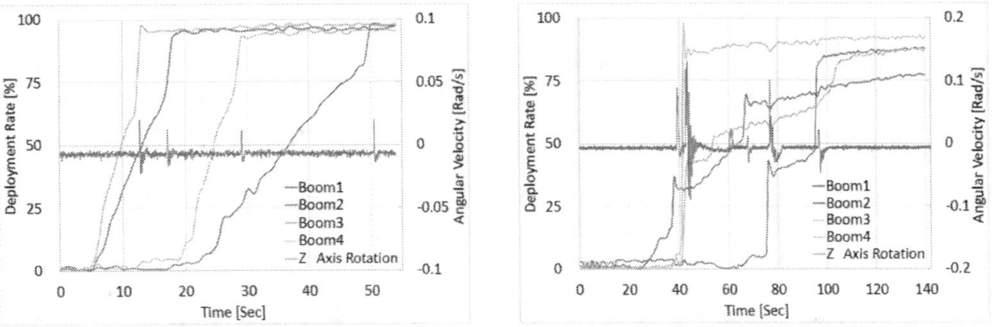

(a) Supplied voltage is 7.5 V

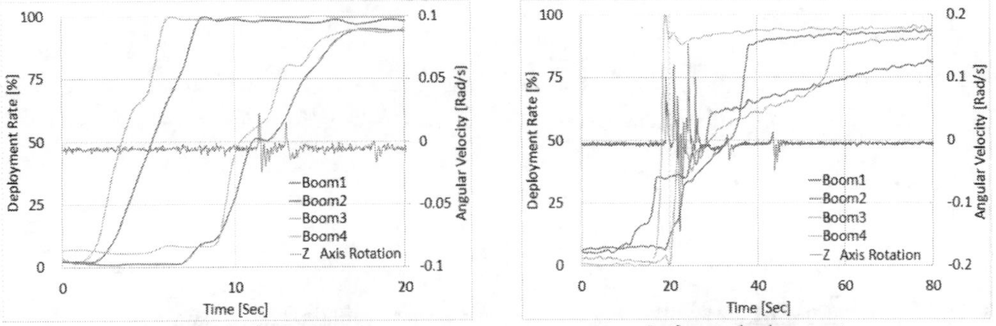

(b) Supplied voltage is 10.0 V

Figure. 8 Deployment rate and angular velocity value without cables and rotary damper system

4.3 Discussion

Comparing to Fig. 5 and Fig. 7, the effect of the synchronous deployment system is clear, and the synchronous deployment behavior is well demonstrated. And in Fig. 6, the deployment rate of each four boom shows a same manner, and synchronous deployment can be confirmed by the graph. Comparing to Fig. 6 and Fig. 8, the angular velocity with cables and rotary damper system shows almost half value than that of without the systems. And the angular velocity have peaks just after completing the deployment or a quick deployment phase. The graph shown in Fig. 8 (a) first step deployment well show the relation between the angular velocity peak and the final phase of deployment, and Fig. 6 (a) second step deployment well show the relation between the angular velocity peak and the deployment speed. Comparing to Fig. 6 (a) and Fig. 8 (a) second step deployment, the deployment of each booms quickly complete around 85 s in Fig. 6, however there has one peak of angular velocity because of the synchronous deployment. Though the boom deployment without the system, shown in Fig. 8 (a) second step, is independently and the boom No. 4, yellow line, and the boom No. 2, blue line, complete the deployment quickly, then the angular velocity shows the many peaks phenomenon, and the deployment of boom becomes unstable. Through these considerations, it is found out that the synchronous and the uniform slow velocity deployment are the key to realize the stable deployment.

5. CONCLUSIN AND FUTURE WORK

The new types of deployable boom consisted of SMA bi-convex tape are proposed, and aiming to the stable and the synchronous deployment, the synchronous deployment system consisted of cables and rotary dampers is also proposed in this study. Through deployment experiments using the conceptual model, the synchronous deployment behavior is achieved by comparing to the model without the system, and the stable deployment is also achieved by comparing the angular velocity value during the deployment.

From the viewpoint of the bending and torsion stiffness after deployment and at the lower grass transfer temperature condition, the martensite phase SMA has almost half young's modulus comparing to the austenitic phase SMA. The important function of a deployable boom is a structural member to keep the shape of a panel, a membrane after deployment. The combined use of SMA and SMP will solve the structural rigidity lowering program. Authors are planning to design the SMA and SMP layer convex tape boom, shown in Fig. 9, and to investigate the structural stiffness, such as the bending, torsion stiffness and natural frequency, in a several temperature conditions.

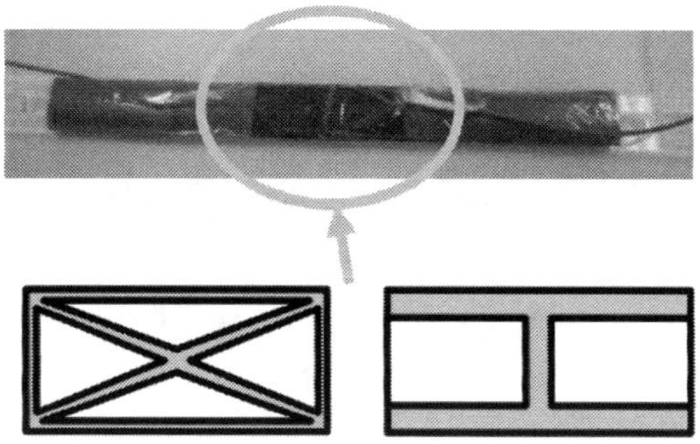

(a) SMP sheet shape idea

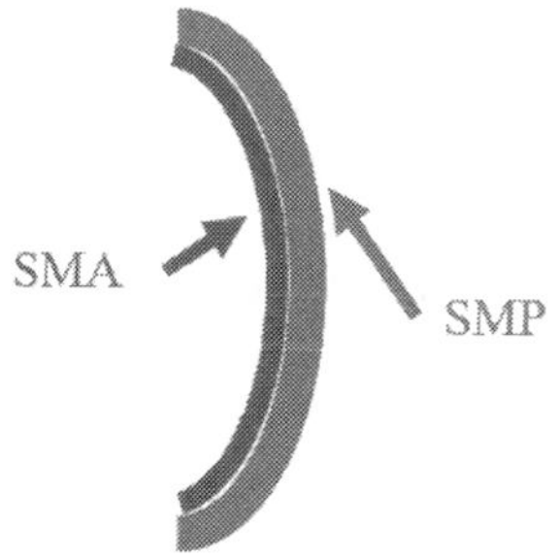

(b) Cross section idea for combining SMA and SMP layer

Figure. 9 Idea for combining SMA and SMP materials for SMA-BiCon boom

ACKNOWLEDGMENTS

This work was supported by JSPS KAKENHI Grant Number 15K18280.

REFERENCES

1. Lichodziejewski D., Derbès B., Slade K. and Mann T., "Vacuum Deployment and Testing of a 4-Quadrant Scalable Inflatable Rigidizable Solar Sail System," 46th AIAA/ASME/ASCE/AHS/ASC Structures, Structural Dynamics & Materials (SDM) Conf., Austin, Texas, 2005, AIAA 2005-2122.
2. Lichodziejewski D., Gordon V. and Derbès, B., "Spiral Wrapped Aluminum Laminate Rigidization Technology," 43rd AIAA / ASME / ASCE / AHS / ASC Structures, Structural Dynamics, and Materials Conf., Denver, Colorado, 2002, AIAA 2002-1701.
3. M. Lou, H. Fang, and L-M. Hsia, "Self-Rigidizable Space Inflatable Tube," J. Spacecraft and Rockets, 2002, Vol.39, No.5, pp.682-690.
4. Watanabe, A. Hori, T. and Tsunoda, H., "Fiber and Fabric Technologies for Space Inflatable Structures," Proc. 47th United Conf. Space Science and Technology (Ukaren), 2G1 (2003), pp. 958-963 (in Japanese).
5. Watanabe, A., Sakai, R., Hori, T. and Ito, H., "Study on Light Curable Space Inflatable Structures Technology by Triaxial Woven Fabric," Proc. 50th United Conf. Space Science and Technology (Ukaren), 3E05 (2006), pp. 1693-1698 (in Japanese).
6. Watanabe, A. Hori, T. and Sakai, R., "Study of Space Inflatable Structures by Chain Curing Polymer," Proc. 53th United Conf. Space Science and Technology (Ukaren), 2H11 (2009), pp. 1509-1512 (in Japanese).
7. Iqbal, K. and Pellegrino, S., "Bi-Stable Composite Shells," 41st AIAA / ASME / ASCE / AHS / ASC Structures, Structural Dynamics, and Materials Conf., Atlanta, GA, 2000, AIAA 2000-1385.
8. Craig, S. H., Keith, R. G., Erik, R. A., Robert, J. D. and Mark, S. L., "Development of a Prototype Elastic Memory Composite STEM for Large Space Structures," 44th AIAA / ASME / ASCE / AHS / ASC Structures, Structural Dynamics, and Materials Conf., Norfolk, VG, 2003, AIAA 2003-1977.
9. Sakamoto, H., Furuya, H., Sato, Y., Natori, M.C. et al., "Origami-Based Membrane Storage and Deployment Technology for De-Orbiting Satellites," 64th International Astronautical Congress, Beijing, China, 2013, IAC-13-B4.6A.4.
10. Ömer, S., "Deployment analysis of a self-deployable composite boom," Composite Structures 89 (2009), pp. 374–381.
11. Herbeck, L., Strcöhlein, T. and Torrez-Torres, J. "Lightwight Deployable Booms: Design, Manufacture, Verification, and Smart Materials Application," 55th International Astronautical Congress, IAF/IAA/IISL, Vancouver, Canada, 2004, IAC-04-I.4.10.
12. Christoph, S., Lars, H. and Elmar, B., "Structural engineering on deployable CFRP booms for a solar propelled sailcraft," Acta Astronautica 58 (2006), pp. 185 – 196.
13. Joachim, B., Marco, S. and Martin, W., "Ultralight Deployable Booms for Solar Sails and Other Large Gossamer Structures in Space," Acta Astronautica 68 (2011), pp. 984–992
14. Watanabe, A. et al., "Study of the extensible structure which coated a braid," Proc. 56th United Conf. Space Science and Technology (Ukaren), 2012, JSASS-2012-4496 (in Japanese).
15. Watanabe, A et al., "Lightweighting examination of the braid coated boom," Proc. 57th United Conf. Space Science and Technology (Ukaren), 2013, JSASS-2013-4679 (in Japanese).
16. N. Katsumata, M. Kaimori, M. Yamasaki, K. Higuchi, M.C. Natori and H. Yamakawa., "Deployment Characteristics of Braid Coated Bi-Convex Tape and Bi-SMA Convex Tape Booms for Deployable Membrane Structures," Journal of Mechanics Engineering and Automation, Vol.4, No.1, January 2014, pp. 52-62.

ICAST2015 #91

An Investigation into the Effect of Freeplay on the Nonlinear Aeroelastic Behavior of a Wing via Higher-Order Spectra

Michael Candon*, Hideaki Ogawa, Robert Carrese and Pier Marzocca
School of Aerospace, Mechanical and Manufacturing Engineering, RMIT University, Melbourne, VIC, Australia

Abstract

Higher-order spectra analysis is used to analyze the nonlinear flutter characteristics of the AGARD 445.6 wing benchmark model, modified to include freeplay structural nonlinearities in pitch and plunge at the wing root. High-fidelity coupled CFD-CSD fluid-structure interaction simulations are conducted and bispectral densities are computed from selected aeroelastic wing dynamic response signals to identify the presence of quadratic nonlinearities within the available time histories. Initially, various freeplay conditions are investigated, before focusing on the intensity of nonlinear behavior at various wing spanwise stations. The considerable effect of freeplay is clearly demonstrated as the wing enters a bounded limit cycle oscillation (LCO) with amplitudes several orders of magnitude larger than that of the case without freeplay. Nonlinear coupling is demonstrated via self-interaction of the LCO frequency and coupling between the LCO frequency and the first torsional mode, this is shown to be strongest at the root where the pitch / plunge mechanism is installed and torsion is most prominent, dissipating spanwise from the pitch / plunge mechanism as the torsional motion becomes less significant. Modal analysis, alongside analysis of the leading-trailing edge phase change and wing deformation shapes supports the findings of the HOS, generating understanding of the underlying physical attributes leading to the observed nonlinear aeroelastic phenomena.

1. INTRODUCTION

Efficient and reliable prediction of nonlinear aeroelastic behavior as well as the expansion of knowledge associated with the underlying physics is currently a crucial and active field of research, drawing on the body of knowledge surrounding linear and nonlinear flutter evaluations [1]. Nonlinear aeroelastic behavior such as limit cycle oscillation (LCO) can compromise the structural integrity of an airframe causing considerable permanent damage and in some cases fatigue or catastrophic failure [2]. LCOs are induced by the presence of structural and aerodynamic nonlinearities within an aeroelastic system. Aerodynamic nonlinearities are generally a result of transonic flow phenomena, *i.e.*, shockwave - boundary layer interactions and shock induced separation or high angle of attack dynamic stall operations. Structural nonlinearities can be either distributed or concentrated. A distributed nonlinearity will affect the entire wing (*i.e.*, nonlinear material properties), whereas concentrated nonlinearities are the result of large concentrated structural loads or freeplay [3]. Generally, freeplay is considered to be inherited from loosened mechanical linkages between the main wing and control surfaces, main wing and store link system or within an all-movable horizontal tail. A review of these various forms of freeplay is presented in [4]. Extending the notion of freeplay occurring in the mechanical linkages of an all-movable tail, variable-incidence and variable-sweep wings pose the risk of suffering the same type of freeplay.

Despite its importance, accurate prediction of nonlinear wing flutter characteristics, such as LCOs and transonic dip, has been a formidable challenge, often hampered by the complex flow structure

* candon.michael@rmit.edu.au

coupling due to aerodynamic and structural nonlinearities. Various experimental techniques have been contrived to achieve thorough understanding and accurate prediction of LCOs [4, 5], but the approaches have significant limitations. For instance experimental methods using wind tunnels or flight tests can be expensive and impractical prior to the design and construction phases. It is therefore urgently required to develop greater understanding and methodologies that can overcome these practical drawbacks.

Higher-order spectra (HOS) analysis is a valuable tool when analyzing nonlinear aeroelastic systems. Their superiority when comparing to traditional linear methods, such as the power spectrum, comes from the ability of the higher-order statistics to predict the presence of nonlinearities. The power spectrum is only able to define second-order statistics, therefore, can only rigorously unveil physics associated with linear phenomena [6, 7]. HOS methods are advantageous in nonlinear aeroelastic analysis as they demonstrate interactions between frequencies which are a result of nonlinearities, hence, HOS can be used to identify the presence of nonlinearities and the transition from linear to nonlinear behavior within an aeroelastic system. It is also worth noting the duality between HOS and Volterra functional series' kernels, where HOS can be represented as the Fourier transform of the Volterra series, *i.e.,* the Volterra series in the frequency-domain [8].

HOS methods have been utilized to analyze various nonlinear aeroelastic studies including the analysis of wind-tunnel and flight test experiments, and fundamental research. Experimental flutter wind tunnel results of a pitch / plunge apparatus were analyzed using HOS in a study conducted by Silva *et al.* [9]. The analysis proved beneficial in identifying regions of linear and nonlinear behavior, and also identifying the transition from linear to nonlinear aeroelastic behavior. Further, HOS has been applied to the results of a wind tunnel flutter experiment for the flexible high speed civil transport (HSCT) semispan model in a study conducted by Hajj and Silva [10]. Nonlinearly coupled frequencies within the aerodynamic forces on the model were identified. HOS is applied to F/A-18 flight flutter test data in a study conducted by Silva and Dunn [11]. It was found that higher-order spectra were valuable in detecting nonlinearities within the experimental data. In another application to flight flutter test data, Hajj and Beran [12] performed HOS analysis on the F-16. The analysis demonstrated areas in which nonlinearities were most prominent and relations between nonlinearities and vibration modes of various components. Finally, Volterra series has been applied in fundamental analytical studies of nonlinear aeroelasticity by Marzocca *et al.* [13-15]. What has not been shown in the literature is a systematic study addressing HOS analysis of high-fidelity nonlinear aeroelastic predictions. The purpose of this paper is two-fold: i) to develop a computational framework for 3D wings capable of providing predictions for 3D nonlinear wings, and ii) apply HOS to the nonlinear aeroelastic response predictions obtained from selected conditions.

In the present study high-fidelity fluid-structure interaction (FSI) simulations are performed on the AGARD 445.6 wing, a classical aeroelastic benchmark case [16]. The AGARD case is selected as experimental flutter data ranges from low-transonic to low-supersonic speeds, hence, validated numerical computations can be performed through subsonic, transonic and supersonic regimes. Typically the AGARD 445.6 displays linear aeroelastic behavior, with a linear structure and with weak nonlinear aerodynamics [17], therefore loosened root connections are installed (in pitch and plunge) to enhance the effect of the nonlinearities on the wing dynamics. Furthermore, higher-order spectra are utilized to analyze the high-fidelity simulations with the objective of identifying the presence and quantifying the effect of nonlinear coupling on the LCO condition. Varying the freeplay parameters, and hence the magnitude of nonlinearities, will allow for relationships to be constructed between the change in HOS and the change in magnitude of nonlinearities. Drawing correlations between the magnitude of nonlinear behavior and the inherent effect on the LCO condition is crucial in gaining greater understanding of nonlinear aeroelastic behavior and in the development of flutter and post-flutter predictive tools. This novel approach to nonlinear aeroelastic analysis (*i.e.,* utilizing HOS to analyze high-fidelity simulation data) enables nonlinear aeroelastic phenomena to be detected, understood and suppression strategies can be assessed in the design phase, reducing timeline delays and minimizing unnecessary financial costs.

2. COMPUTATIONAL FRAMEWORK

2.1 Conditions and Configurations

2. 1.1 AGARD 445.6 Wing

The AGARD 445.6 is a highly swept wing with a NACA 65A004 airfoil section and sweep angle of 45°. The material properties of the wing, presented in Table 1, replicate those of weakened mahogany, consistent with experimental data. The wing model has weakened torsional stiffness, which was implemented experimentally by drilling holes in the wing body and filling with plastic [16]. Figure 1 presents the geometry considered for the present study. The freestream conditions (Table 2) are selected such that in the absence of freeplay the oscillatory motion of the wing is highly stable, not inducing the LCO condition.

Table 1 Wing body material properties [16]

Density (kg/m^3)	Spanwise Elasticity (Pa)	Chordwise Elasticity (Pa)	Poisson ratio
381.98	3.15e+09	4.16e+07	0.31

Table 2 Freestream parameters

Mach No. M_∞	Static Pressure p_∞ (Pa)	StaticTemperature T_∞ (K)	Density ρ_∞ (kg / m^3)	Reynolds Number Re_∞	Angle of Incidence α_0 (°)
0.90	4687.00	257.67	63.39e-02	1.08e+06	0

As aerodynamic nonlinearities, inherent in transonic flow, did not provide sufficient evidence of nonlinear behavior, the model is to be modified by adding freeplay structural nonlinearities in pitch and plunge at the root of the wing, see Figure 1. A similar method for introducing nonlinearities to a simply supported 3-D wing is presented by Kim *et al.* [18] Four cases are investigated with different combinations pitching α_f and plunging h_f freeplay margins, presented in Table 3.

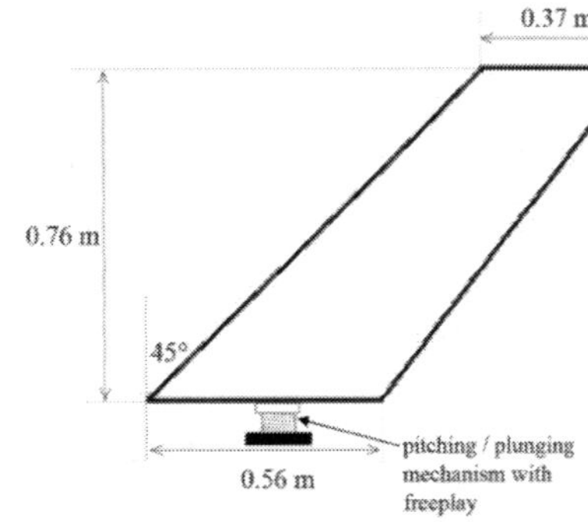

Figure 1 Geometric configuration of the wing with pitching / plunging freeplay

Table 3 Test cases: freeplay selection in torsion and bending at the root.

	Case 1	Case 2	Case 3	Case 4
α_f (°)	0	0	0.50	0.50
h_f (mm)	0	1.00	0	1.00

2.2 Higher-Order Spectra

This study will employ bispectral density analysis to identify nonlinear aspects within the aeroelastic system, *i.e.*, more specifically the nonlinearities introduced through freeplay. The bispectrum determines nonlinear interactions within the aeroelastic system by estimating third-order moments in the frequency-domain [19]. The bispectral density or bispectrum can be defined as

$$B(f_1, f_2) = \frac{1}{M}\sum_{i=1}^{M} X_i(f_1)X_i(f_2)X_i^*(f_1 + f_2) \qquad (1)$$

where M is the number of data segments to be considered. The bispectrum $B(f)$ can be plotted against two frequency variables f_1 and f_2 in a three-dimensional plot. Each point on the plot describes the bispectral energy density of the signal at the bifrequency (f_1, f_2). The bispectrum at any bifrequency (f_1, f_2) measures the level of interaction between the two frequencies f_1 and f_2, which results from quadratic nonlinearities within the system, hence the bispectrum detects the occurrence of quadratic nonlinearities. $X(f)$ is the Fourier transform of the time series $x(t)$ [19].

Whilst the bispectrum detects system quadratic nonlinearities, it does not provide a basis for comparing levels of nonlinearity, hence, as a general practice, the bispectrum is conveniently normalised to give the bicoherence which is bounded between 0 and 1, where spikes in the magnitude of the bicoherence function represent levels nonlinearities.

The bicoherence function b is defined as

$$b^2(f_1, f_2) = \frac{\left|\frac{1}{M}\sum_{i=1}^{M} X_i(f_1)X_i(f_2)X_i^*(f_1 + f_2)\right|^2}{\frac{1}{M}\sum_{i=1}^{M}|X_i(f_1)X_i(f_2)|^2 \cdot \frac{1}{M}\sum_{i=1}^{M}|X_i(f_1 + f_2)|^2} \qquad (2)$$

In estimating the higher-order spectra, the data is split into blocks which are evaluated individually and averaged. It is essential that particular attention be given to the block length M in comparison to the total data length N. A larger the block size will provide a finer resolution, but this comes with greater variance [20]. It is suggested by Dalle Molle and Hinch [21] that when identifying nth-order cumulants, the block length should be $(n-1)th$ root of the sample size. To estimate the bispectrum (bicoherence) numerical codes are implemented based on validation provided in a previous study with pure multiple degrees of freedom mechanical systems [22] which is presented in Reference [23], providing the reader with an example to support the use of HOS and understand its features.

2.3 Numerical Methods

The numerical approach to identify the onset of flutter and LCOs is via two-way fluid-structure interactions (FSI). In this method high-fidelity transient simulations are performed, coupling computational fluid dynamics (CFD) with computational structural dynamics (CSD).

The FSI simulations are computed using the numerical simulation software package ANSYS R15 [23]. The FSI is time-accurate, with implicit coupling of the CFD (ANSYS Fluent) and CSD (ANSYS Transient Structural) at each time-step. The transient time step is set to 0.001 s. Each time step involves a maximum of three coupling iterations, with a maximum of 20 CFD iterations per coupling iteration. The time period for each simulation is 2 s, giving sufficient time for the dynamic oscillations to die-out or enter a bounded limit cycle.

The Navier-Stokes equations for transient flowfields are solved via a coupled pressure-based solver with second order upwind spatial accuracy; higher order term relaxation is utilized to improve convergence stability. Each transient simulation is initialized from a converged steady-state solution. The convergence criteria set to 1e-05 for the locally scaled residuals at each time step. Inviscid flowfields are assumed.

The wing body is composed entirely of mahogany, consistent with the experimental wing [16]. Solid elements are used with 8 nodes per hexahedral element (6 nodes per wedge) which is acceptable as strain gradients are low. Transient structural simulations are conducted via the Newton-Raphson approach. A load (time) step of 0.001 s is set with 5 substeps and a maximum of 15 equilibrium iterations per substep. The wing body contains 4950 elements (96.4% hexahedral and 3.6%wedge). Validation of the structural model is presented in Section 3.2.2.

In order to facilitate the movement of the wing, a dynamic mesh with a smoothing method is utilized. The smoothing method allows the interior nodes of the mesh to deform whilst the number of nodes and their connectivity remain preserved. A diffusion smoothing method is implemented with a diffusion parameter of 1.5 which aims to preserve cells closest to the wall and apply the dynamic motion to the cells in the interior of the far-field. The diffusion smoothing method is chosen as it generally results in a better quality mesh when compared to other smoothing methods such as spring-based smoothing [23].

A C-grid, as presented in Figure 2, is generated via ANSYS ICEM CFD [23]. A mesh convergence study is conducted (Figure 2) considering two grids: nominal comprised of 494,438 hexahedral cells (563,413 nodes) and fine comprised of 3,955,504 hexahedral cells (5,533,360 nodes). The baseline grid is considered of sufficient resolution to accurately represent the flowfield and FSI for this study.

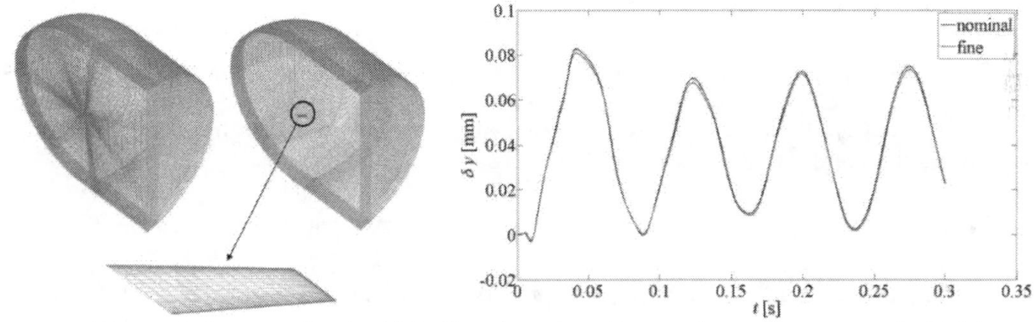

Figure 2 Computational grid (left) and results of the mesh convergence study (right)

The boundaries of the domain (Figure 2) are set such that no interaction will occur with the flow surrounding the wing. The walls of the wing body are set as no slip walls and are specified as dynamic mesh zones to be coupled with the corresponding surfaces on the wing body. A *Pressure Far-Field* is imposed on the freestream inlet (pink) and a *Pressure Outlet* on the outlet (yellow). *Symmetry* is imposed on the wall to which the wing is attached (blue). The wing body is simply supported with a fixed support being imposed on the face of the wing root. The rest of the wing is unconstrained. The wing top, bottom and tip are specified as fluid solid interfaces which are coupled with the corresponding surfaces in the fluid domain [23].

3. RESULTS

In the following section a number of aeroelastic simulations are performed. The numerical tools are validated against experimental data first, then the results of the FSI simulations of the AGARD wing with loosened root connections are presented to analyze the effects of freeplay on nonlinear aeroelastic behavior and the intensity of the nonlinear behavior at different stations on the wing.

3.1 Numerical Tools

3.1.2 Aeroelastic Model

The numerical simulations are validated against experimental results with respect to the flutter frequency ω which is calculated by taking the FFT of the time series plots for the leading edge tip deflection δ_y at freestream Mach numbers ranging from M_∞ =0.499 – 1.141, presented in Figure 3 (a). This demonstrates good correlation between numerical and experimental results. The transonic dip, a term used to describe the drop in the flutter boundary due to compressibility effects in a transonic flight regime, has been accurately predicted at $M_\infty = 0.96$. However there is some variance in the results for $M_\infty = 0.499$ and 1.141.

Figure 3 Comparison between experimental [16] and numerical results for the flutter frequency (a) and first bending (left) and torsional (right) modes of vibration (b)

3.2.2 Structural Model

Modal analysis is conducted for the simply supported wing (no root freeplay) and compared with experimental results in order to validate that the wing body is well modelled in the structural solver. The first bending and torsional modes are obtained as presented in Figure 3 (b). Table 3 demonstrates a very good correlation between the numerical and experimental results, indicating that the wing body is well modelled by the structural solver.

Table 3 Comparison between experimental [16] and numerical data for modal analysis

Mode	Numerical Result (Hz)	Experimental Result (Hz)	Deviation (%)
First Bending	9.62	9.37	3 .00
First Torsional	40.10	39.07	2.60

3.3 Freeplay Parameters

As the aerodynamic nonlinearities present in the traditional simply supported model do not provide sufficient evidence of nonlinear behavior, freeplay in pitch and plunge is installed at the root of the airfoil to induce structural nonlinearities which are detectable via HOS. Case 1 considers the airfoil simply supported (no root freeplay) whilst Cases 2 and 3 consider freeplay in plunge ($h_f = \pm$ 1 mm) and pitch ($\alpha_f = \pm$ 0.5°), respectively. Plots of the transient response for these three cases are presented in Figure 4. Inspection of the common bicoherence plot for these cases indicates that the aerodynamic and induced structural nonlinearities within the system are weak and not detectable. The only noteworthy feature here is that for Case 3 ($\alpha_f = 0.5°$) where the freeplay margin induces an LCO condition, however, the bicoherence plot remains unable to produce evidence of nonlinear behavior.

Figure 5 presents Case 4, where freeplay is installed in both pitch and plunge at the root of the wing. Here, the time history plot demonstrates the considerable effect of nonlinearities on the aeroelastic behavior of the wing, typifying the expected nonlinear phenomena. The oscillations are inevitably guided into a bounded LCO, the amplitude of which is several orders of magnitude larger than the case in the absence of freeplay. The bicoherence and power spectral density (PSD) plots demonstrate a strong peak where the LCO frequency is self-interacting at 12.5 Hz. Further, the LCO frequency couples with a frequency of approximately 37.5 Hz to form a relatively low magnitude peak. The origin of the 37.5 Hz is further investigated in the proceeding section. Finally, an almost negligible peak can be seen at approximately 25 Hz, which is due to superposition, *i.e.,* the frequency of the LCO subtracted from the first torsional mode.

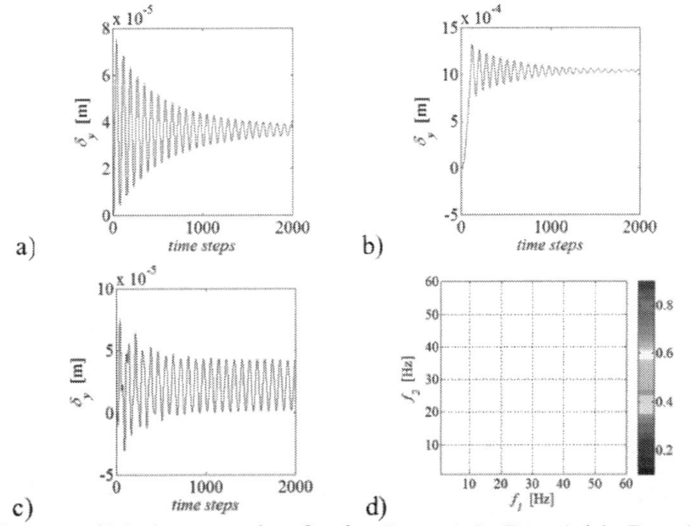

Figure 4 Time history and bicoherence plots for the Cases 1-3. Case 1 (a), Case 2 (b), Case 3 (c) and the common bicoherence plot (d)

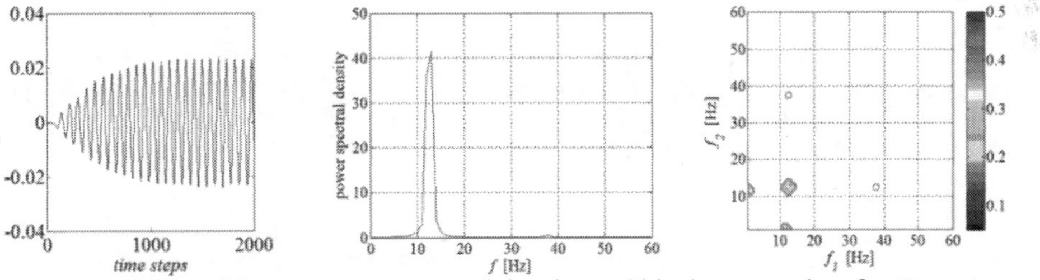

Figure 5 Time history, power spectral density and bicoherence plots for Case 4

3.4 Nonlinear Quantification

To quantify nonlinear behavior on the airfoil, the transient response of the wing is monitored at fifteen stations across the wing planform as presented in Figure 6. These stations are positioned evenly spanwise from root to tip at the leading edge, trailing edge and elastic axis (where the freeplay mechanism is installed). Figure 7 presents the transient response at each station. To analyze the response one time period (in LCO condition) at each of the fifteen stations is presented in Figure 8. Inspection of the response at the root demonstrates the considerable 'step' in the motion which is caused by the pitch / plunge mechanism hitting the freeplay constraints. Further, within each step multiple small (higher frequency) oscillations are observed, these are small rebounding oscillations which are occurring each time the freeplay mechanism hits a constraint. Inspection of the phase relationship at each spanwise position indicates that torsion is most significant at the root where the pitch / plunge mechanism forces the leading and trailing edges to be out of phase. Torsion becomes less prominent towards the tip as the leading-trailing edge oscillations become nearly in phase. It is interesting to note that torsion is least prominent at the three-quarter span region. Figure 11 presents the deformation of the wing for one time period. Here the presence of both bending and torsion can be observed, as to be expected. This suggests that the 37.5 Hz frequency is due to the activation of the first torsional mode (40 Hz) which is presented in Figure 3 (b), the slight dampening of the magnitude being a result of freeplay and aerodynamic dampening.

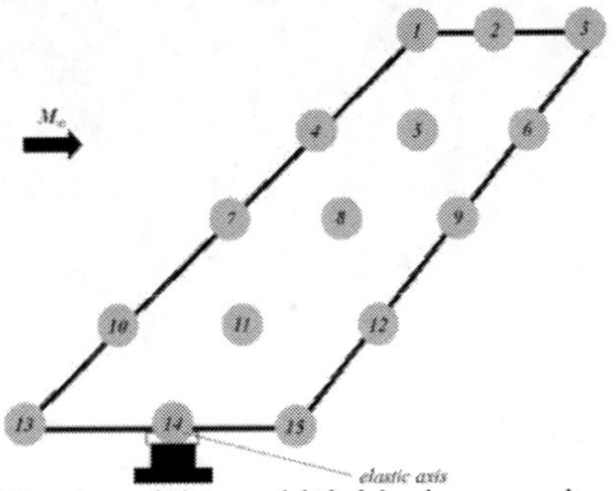

Figure 6 Stations (S) at which bicoherence is calculated

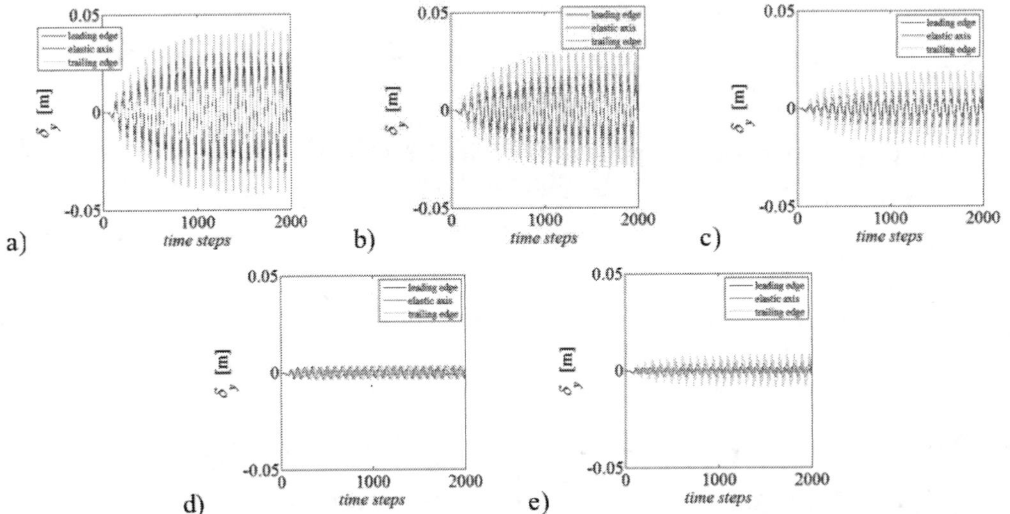

Figure 7 Transient response in bending at each spanwise station for the tip a) S 13-14-15, three quarter span b) S 10-11-12, half span c) S 7-8-9, quarter span d) S 4-5-6 and root e) S 1-2-3

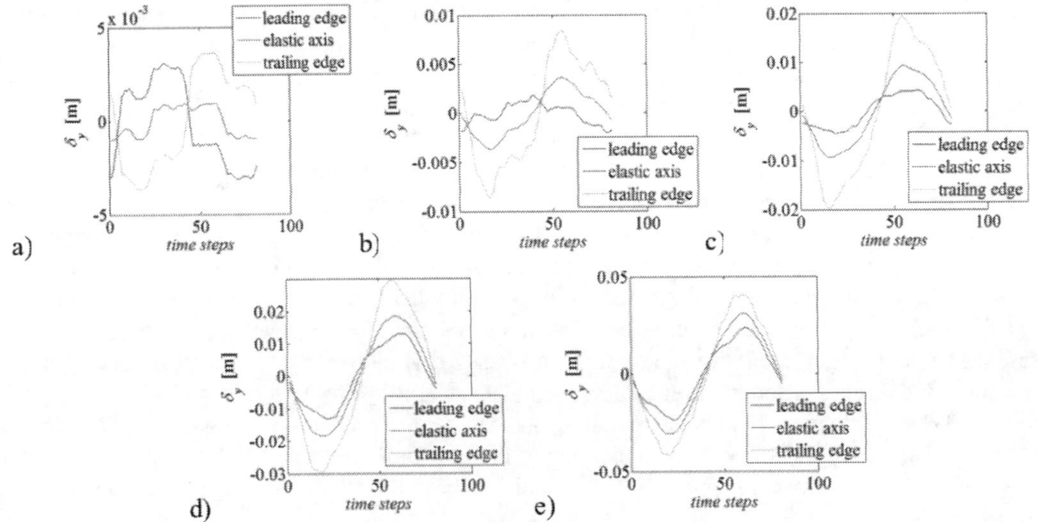

Figure 8 One time period at each spanwise station for the tip a) S 13-14-15, three quarter span b) S 10-11-12, half span c) S 7-8-9, quarter span d) S 4-5-6 and root e) S 1-2-3

Figure 9 presents the PSD plots at each station. In general the PSD plots indicate that the ratio between the LCO frequency and the first torsional mode is most intense at the root of the wing, becoming indiscernible towards the tip, this correlates with Figure 8. Further it can be seen that the presence of the first torsional mode dissipates at a greater rate when moving spanwise from the elastic axis than the leading or trailing edges.

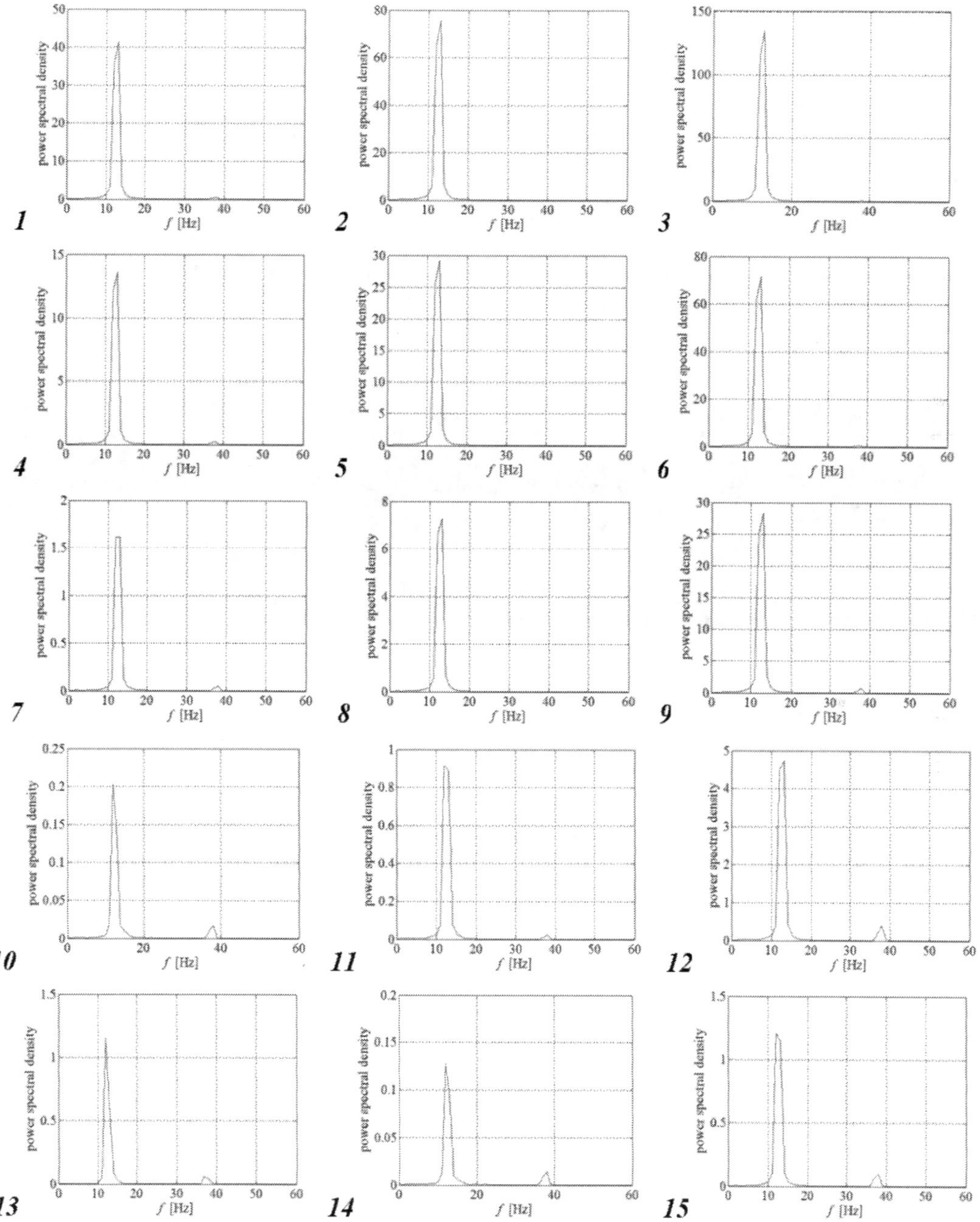

Figure 9 Power spectral density plots for Case 4 at 15 stations across the wing planform

Figure 10 presents the bicoherence plots at each station. In general the bicoherence plots demonstrate behavior similar to that of the PSD plots whilst also providing further validation and understanding of the underlying physical attributes observed. Station 14 (elastic axis) is clearly demonstrating the most intense nonlinear coupling, an expected finding as this is the source of the nonlinearity. Generally, quadratic coupling between the LCO frequency and first torsional mode is far more prominent on the leading edge than the trailing edge. This is explained when observing Figure 11 where it can be seen that deformation in torsion is greater on the leading edge, favoring quadratic coupling between the LCO frequency and first torsional modes. Correlations can be drawn between the leading-trailing edge phase relationships and bicoherence plots, *i.e.,* coupling between the LCO frequency and torsional mode is most prominent in spanwise regions where the leading and trailing edges are most out of phase (S 10-15). Further, all stations in the three-quarter span region (S 4-5-6) where the leading and trailing edges are found to be nearly completely in phase do not indicate the presence of any quadratic coupling between the torsional mode and the LCO frequency. Finally, in regions where the nonlinear interaction between the LCO frequency and first torsional mode are strong (S 10-15), weak peaks are observed at 25 Hz and 50 Hz, which is a result of superposition between the LCO frequency and first torsional mode.

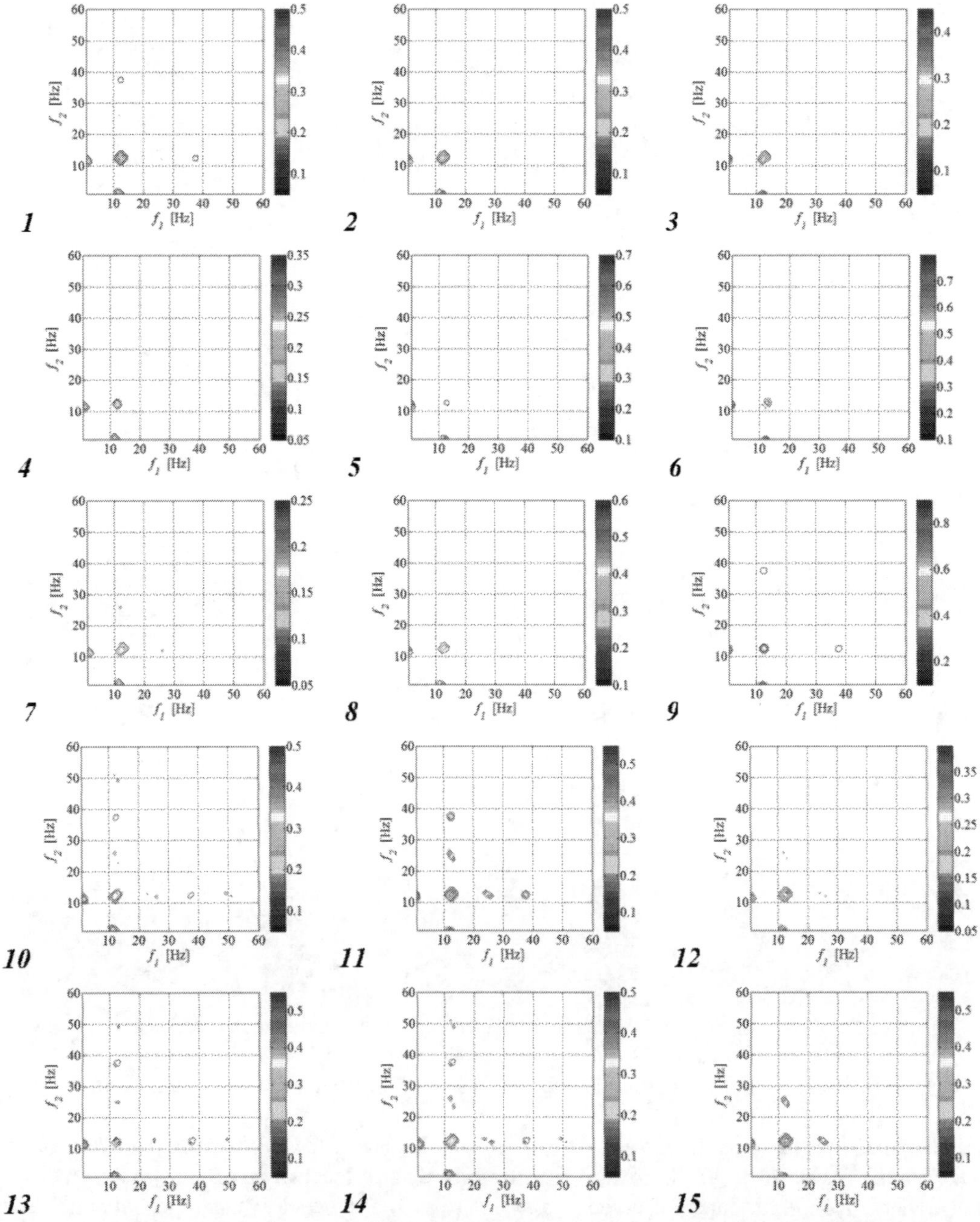

Figure 10 Bicoherence plots for Case 4 at 15 stations across the wing planform

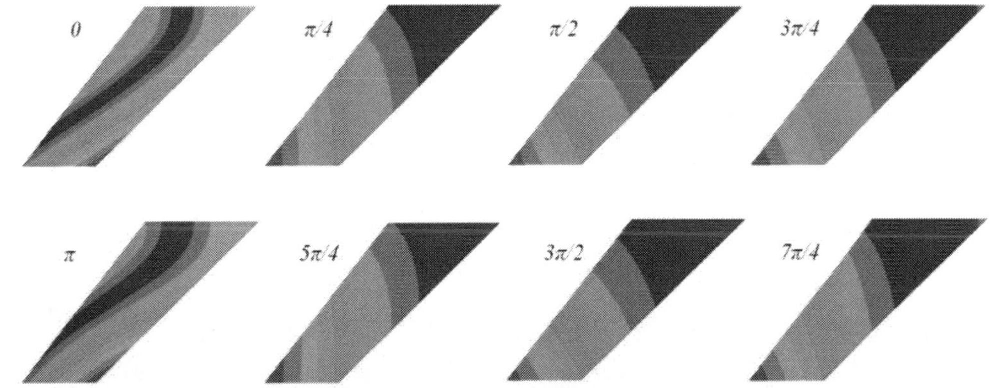

Figure 11 Total deformation shapes for one time period

4. CONCLUDING REMARKS

High-fidelity FSI simulations have been conducted on the benchmark AGARD 445.6 wing, modified to include pitching and plunging freeplay nonlinearities at the root. Analysis of the nonlinear system has been conducted utilizing higher-order spectra to identify the presence of quadratic nonlinearities and quantify nonlinear behavior at various stations across the span of the wing. Initially the FSI model is validated against experimental data. It is shown that the inclusion of both pitching and plunging freeplay margins at the root induces an LCO condition with an amplitude several orders of magnitude larger than that of the cases where nonlinear behavior was not discernable. This is attributed to nonlinear coupling via self-interaction of the LCO frequency plus interaction between the LCO frequency and frequency of the first torsional mode. It is demonstrated that the intensity of nonlinear interactions between the LCO frequency and first torsional mode peaks in the region of the nonlinear source and is dissipating in the spanwise. Further understanding of the underlying physical phenomena is generated by analyzing leading-trailing edge phase relationships and the deformed wing shape over one time period. It is shown that the intensity of coupling between the LCO frequency and first torsional mode is highly dependent on the phase relationship (which in the illustrated case depends on the presence of torsion) and the amplitude of torsional deformation. In regions where coupling is strong between the LCO frequency and first torsional mode, further nonlinear harmonics appear as a result of superposition of the LCO and first torsional mode frequencies. Future work will include further investigations into various freeplay and freestream parameters, drawing correlations as to how these parameters affect the nonlinear behavior of the wing. Further, predictive methods will be developed to alleviate the large computational effort of fully coupled solutions.

ACKNOWLEDGMENTS

The authors would like to thank Dr. Michele Pasquali from University of Rome Sapenzia for supplying his version of the higher-order spectra tool that was used for validation purposes. The authors would also like to thank LEAP Australia for their support with regard to the ANSYS software package.

REFERENCES

1. Dowell, E. H., and Ilgamov, M., Studies in Nonlinear Aeroelasticity, Springer-Verlag New York, New York, USA, 1988.
2. Lee, B. H. K., Tron, A., "Effects of Structural Nonlinearities on Flutter Characteristics of the CF-18 Aircraft," *Journal of Aircraft,* Vol. 26, No. 8, Aug. 1989, pp. 781-786.
3. Dowell, E.H., Edwards, J. W., and Straganac, T., "Nonlinear Aeroelasticity," Journal of Aircraft, Vol. 40, No. 5, Sep. 2003, pp. 857-874.
4. Tang, D., and Dowell, E.H., "Aeroelastic Response Induced by Freeplay, Part 1: Theory," *AIAA Journal,* Vol. 49, No. 11, Nov. 2011 pp. 2532 – 2542.

5. Virgin, L. N., "Introduction to Experimental Nonlinear Dynamics," *Measurement Science and Technology*, Vol. 12, No. 3, 2001, pp.353.
6. Nikias, C.L., and Petropulu, A. P., Higher-Order Spectra Analysis A Nonlinear Signal Processing Framework, PTR Prentice-Hall, Inc., New Jersey, USA, 1993.
7. Bendant, J. S., and Piersol, A. G., Random Data Analysis and Measurement Procedures, John Wiley and Sons, Inc., New York, USA, 2000.
8. Silva, W. A., "Identification of Nonlinear Aeroelastic Systems Based on the Volterra Theory: Progress and Opportunities," *Journal of Nonlinear Dynamics*, Vol. 39, No. 1-2, Jan. 2005.
9. Silva, W. A., Straganac, T., and Muhammmmad, R.J., "Higher-Order Spectral Analysis of a Nonlinear Pitch and Plunge Apparatus," *46th AIAA/ASME/ASCE/AHS/ASC Structures, Structural Dynamics and Materials Conference,* AIAA-2005-2013, Austin, TX, Apr. 2005.
10. Hajj, M. R., and Silva, W. A., "Nonlinear Flutter Aspects of the Flexible High-Speed Civil Transport Semispan Model," *Journal of Aircraft*, Vol. 41, No. 5, 2004, pp. 1202-1208.
11. Silva, W.A., and Dunn, S., "Higher-Order Spectral Analysis of F-18 Flight Flutter Data," *46th AIAA/ASME/ASCE/AHS/ASC Structures, Structural Dynamics and Materials Conference,* AIAA-2005-2014, Austin, TX, Apr. 2005.
12. Hajj, M. R., and Beran, P. S., "Higher-Order Spectral Analysis of Limit Cycle Oscillations of Fighter Aircraft," *Journal of Aircraft,* Vol. 45, No. 6, Nov. 2008, pp. 1917-1923.
13. Marzocca, P., Librescu, L., and Silva, W. A., "Nonlinear Open-/Closed Loop Aeroelastic Analysis of Airfoils via Volterra Series," *AIAA Journal*, Vol. 42, No. 4, 2004, pp. 673-686.
14. Marzocca, P., Librescu, L., and Silva, W. A., "Time-Delay Effects on Linear/Nonlinear Feedback Control of Simple Aeroelastic Systems," *Journal of Guidance, Control, and Dynamics*, Vol.28, No.1, 2005, pp. 53-62.
15. Marzocca, P., Nichols, J. M., Seaver, M., Trickey, S., and Milanese, A., "Second Order Spectra for Quadratic Non-linear Systems by Volterra Functional Series: Analytical Description and Numerical Simulation," *Mechanical Systems and Signal Processing,* Vol. 22, 2008, pp. 1882-1895.
16. Yates, E. C., "AGARD-R-765 AGARD Standard Aeroelastic Configurations for Dynamic Response I-Wing 445.6," *61st Meeting of the Structures and Materials Panel*, Oberammergau, Germany, Sep. 1985. Subjected
17. Silva, W. A., Chwalowski, P., and Boyd, P., "Evaluation of linear, inviscid, viscous, and reduced order modelling aeroelastic solutions of the AGARD 445.6 wing using root locus analysis," *International Journal of Computational Fluid Dynamics,* Vol. 28, No. 3 – 4, Aug. 2014, pp. 122 – 139.
18. Kim, D.H., Lee, I., and Paek, S. K.., "Nonlinear Aeroelastic Instability of a Supersonic Missile Wing with Pitch Axis Freeplay," *International Journal of Aeronautical and Space Science*, Vol. 4, No. 1, May 2003, pp. 53-62.
19. Choudhury, S. M. A. A., Shah, S. L., and Thornhill, N. F., Diagnosis of Process Nonlinearities and Valve Stiction, Springer-Verlag Berlin Heidelberg, Germany, 2008.
20. Subba, T., and Gabr, M. M., "An Introduction to Bispectral Analysis and Bilinear Time Series Models," Lecture Notes in Statistics, Springer-Verlag 24, Berlin 1984.
21. Dalle Molle, J. W., and Hinich, M. J., "The Trispectrum," *Proceedings of the Workshop Higher Order Spectral Analysis*, Vail, 1989, pp.68-72.
22. Pasquali, M., Lacarbonara, W., and Marzocca, P., "Detection of Nonlinearities in Plates Via Higher-Order-Spectra: Numerical and Experimental Studies," *Journal of Vibration and Acoustics*, Vol. 136, No. 4, 2014, pp. 1-13.
23. Hickey, D., Worden, K., Platten, M. F., Wright, J. R., and Cooper, J. W., "Higher-Order Spectra for Identification of Nonlinear Modal Coupling," *Mechanical Systems and Signal Processing*, Vo.23, No. 4, May 2009, pp.1037-1061.
24. ANSYS® Academic Research, Release 15.0, Help System, Workbench Users Guide, ANSYS, Inc.

ICAST2015 #93

Experimental Investigations on High-Lift Wing Stall Control Strategies with Synthetic Jet Actuators

Danilo Andreoli[1], Mario Cassaro[1*], Manuela Battipede[1],
Pier Marzocca[2], and Goodarz Ahmadi[3]

[1] Department of Mechanical and Aerospace Engineering, Politecnico di Torino, Turin, Italy
[2] School of Aerospace, Mechanical and Manufacturing Engineering, RMIT University, Bundoora, Australia
[3] Wallace H. Coulter School of Engineering, Clarkson University, Potsdam, NY, US

Abstract

This paper contributes to the development of two different open/closed-loop strategies for actively control wing aerodynamic stall by means of coordinated arrays of synthetic jet actuators. The control of the flow over aerodynamic shapes in order to achieve performance enhancements, such as improved aircraft maneuverability, has been a lively area of research for the last two decades. Active flow control can produce significant performance improvement when combined in a closed-loop control system. Synthetic jet actuators (SJAs) are devices able to interact actively with the flow around a hosting structure by providing ejection and suction of fluid from an orifice. The research presented in this paper concerns the implementation of Zero-Net-Mass-Flux SJAs airflow control system on a NACA0015 low aspect ratio (LAR) wing section prototype developed by Clarkson University under the Advancement of Intelligent Aerospace Systems (AIAS) AFOSR Grant FA9550-09-1-0051. Wind tunnel experiments, on the uncontrolled wing, are primarily performed for system identification purpose. The open-loop control of the wing is implemented and tested, obtaining a stall delay of about 2.8 degrees of angle of attack. A strategy based on the comparison of several acoustic pressure sensors PSD profiles is adopted to characterize the forthcoming boundary layer detachment. This allows for triggering the closed-loop control in stall proximity only for sake of energy optimization. Pertinent results and discussion will be provided.

1. INTRODUCTION

The main objective of the present research, performed under the Air Force Office of Scientific Research (AFOSR) grant number FA9550-09-1-0051 at Clarkson University, is the boundary layer active control by Synthetic Jet Actuator (SJAs) for stall delay purpose. A closed-loop control strategy is attempted on a Low Aspect Ratio (LAR) wing prototype characterized by low Reynolds number (Re) flight condition. Theoretical and experimental investigation on active control method applied to LAR wings is common practice and returns valuable results (Z.Wang, T.Hu, & I.Gursul, 2014). Interesting findings are related to the aircraft aerodynamic performance improvements, at relatively low costs projects, thanks to actuator energy-saving feature (Tovar, et al., 2012). Scientific literature is populated by several studies conducted on wing assemblies with the intent of enhancing aerodynamic performance (Tang, Salunkhe, Du, & Wu, 2014), (Stolk, et al., 2007), (Hong, 2007) and (Millar, 2005). In particular, SJAs are usually organized in arrays of several elements placed along the

* mario.cassaro@polito.it

wingspan (Hubbard, Flatau, & Clingman, 2014). In the presented study, the wing section prototype is provided with two arrays of SJAs located in proximity of the leading edge the first and close to the flap shoulder the second. This solution is chosen as it returned positive results in previous researches, (Alstrom, 2013) and (Melton & Yao, 2006). In addition, the present work provides interesting experimental outcomes on LAR wings at low Re, which are dominating the current aeronautical scenario as the small and micro unmanned aerial vehicles (UAVs, MAVs) are concerned (Inman, Kochersberger, Kurdila, Mason, & Wilson, 2011). SJAs application is not limited to small and micro aerial vehicles, but it represents an attractive solution to be investigated for large-scale air vehicles fuel saving and pollutants' emission reduction (Jabbal, S.C.Liddle, & W.J.Crowther, 2010). In fact, SJAs implementation and optimization could be critical for the evolution of the next aircraft generation green operation, as they can positively affect take-off/landing distances and global thrust demand. For UAV applications, SJAs have been proven so effective to represent a valid alternative to classical moving surfaces for flight control purpose (Zeiger & Rediniotis, 2004), (Golubev & MacKunis, 2014). An example of SJAs application as hingeless flight control devices is described in (Neidhoefer & Ryan, 2009), where the actuators are acting on both wing sides, in the trailing edge proximity and are integrated with Coanda effect surfaces.

The present study intends to verify and validate the efficacy of a particular SJAs configuration coupled with open/closed-loop controllers through experimental activities. The actuators have been previously designed, optimized by numerical simulation and assembled. The complexity of the work is mainly related to the three-dimensional flow field which characterizes the LAR wing prototyped (Cosyn & Vierendeels, 2006). Two arrays of custom made SJAs, installed in the proximity of the leading edge and of the trailing edge flap hinge, respectively 10% and 65% of the chord length, make up the actuation system. The sensing system consists of different components, each installed for a specific purpose. Eleven acoustic pressure transducers distributed in the wing upper surface, 8 in the wing body and 3 on the flap, serves for the closed-loop trigger in stall proximity. An accelerometer, placed near the wing c.g., serves for system identification purpose. A six-axis force balance serves for the wing open-loop control. A dSPACE hardware system connected to the software environment Matlab/Simulink/ControlDesk complete the experimental set-up.

The paper is organized as follows. SJAs and LAR-wing physics and integration are described respectively in Section 2. The experimental set-up description, in terms of hardware and software architecture, jointly with the SJAs' parameters' analysis are provided in Section 3. Experiment results and discussion are reported in Section 4. Conclusions complete the paper.

2. SJA AND LAR WING

Synthetic jet actuators belong to a wide category of devices employed for boundary layer active control purpose. In particular, they are classified among the fluidic type actuators as their way to control the boundary layer evolution is based on the synthesis of fluid jets that interact with the external flow field, without the need of an external air source; for this reason they are referred to as zero-net-mass-flux actuators. The synthetic jet is created by a periodic suction and injection of the fluid in the boundary layer, through the orifice slot of the actuator assembly. The other elements of the SJA are a housing, which stands for the actuator's cavity and a vibrating diaphragm (the active element), which is integrated in one of the cavity walls to create a variable-volume chamber. In fact, under an external electrical source, which is applied to the membrane, its deflection causes the volume of the housing to increase and decrease periodically, accordingly to the shape of the electrical signal. As a consequence, the air enters and leaves the actuator cavity through its orifice which is the connection element between the SJA and the external environment. The membrane is generally made of piezo-electric material which allows the electrical control of its displacement at a low energy condition. The orifice slot is designed with a proper shape and orientation which define the synthetic jet direction into the flow field and deeply characterize the actuator efficiency on the aerodynamic behavior influence. The solid and confined assembly of the actuator is externally connected through electrical wires which carry the command signal designated to excite the piezo-electric membrane forcing it to vibrate. The controlled vibration has to be optimized by taking into account for several factors such as the size of the chamber (volume and shape) and its resonance frequency (Helmholtz frequency) (Ikhlaq, Ghaffari, & Arik, 2014).

The mechanism by which the actuator jet is synthetized lies in the creation of a shear layer during the phase of air ejection from the actuator cavity, between the jet itself and the external air flow. This shear layer vorticity rolls up driven by its own momentum and then it forms a vortex ring structure carried downstream by the air flow. During the suction phase, a low pressure zone is created in the proximity of the actuator orifice, which calls back the portion of fluid close to the previously generated vortex that, in turns, is at that moment too far from the actuator's cavity to be able to come back inside it. The repeating of this elementary process led to the creation of the synthetic jet which derives, at the end, from a particular alteration of the free-stream flow when it cross a point in the wall characterized by a pulsating local pressure: the actuator's orifice, as it is showed in **Error! Reference source not found.**.

The SJAs employed in the present study have been designed and built during previous research activities at Clarkson University (Alstrom, 2013), (Kabiri, 2012), during which the chamber and orifice shape and dimension, as well as the membrane piezo-electrical characteristics have been optimized. This led to obtain two different SJA elements per array, as in **Error! Reference source not found.**, of which the main features are summarized in **Table 1**. Each array is composed by ten SJA elements. A brass round of 32 mm diameter and 0.10 mm thick, coupled with a piezo-ceramic element (PZT-5h) of 32 mm diameter and 0.22 mm thick, composes the membrane.

The test case analyzed consists of a dual SJAs strip driven by a single control signal, which is optimized in terms of wave shape, mean voltage, frequency and amplitude.

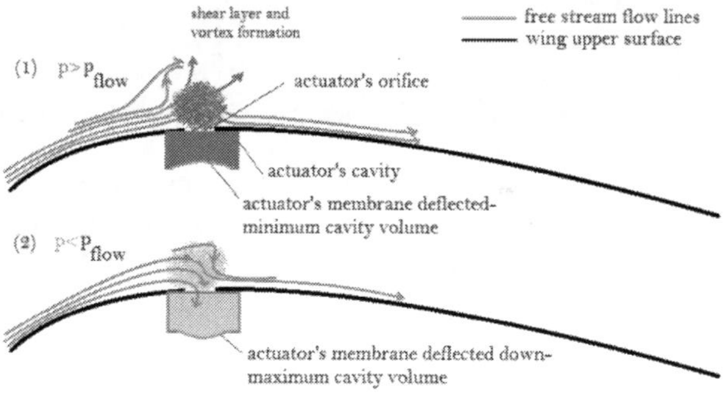

Figure 1. Schematic representation of the streamlines locally within the actuator's orifice proximity, caused by pulsating local pressure between its highest (1) and lowest (2) values.

Table 1 Leading Edge and Flap Shoulder SJAs' Features

	Leading Edge SJA	Flap Shoulder SJA
Cavity Volume [mm³]	4650	10738
Orifice Orientation Angle [deg]	30	

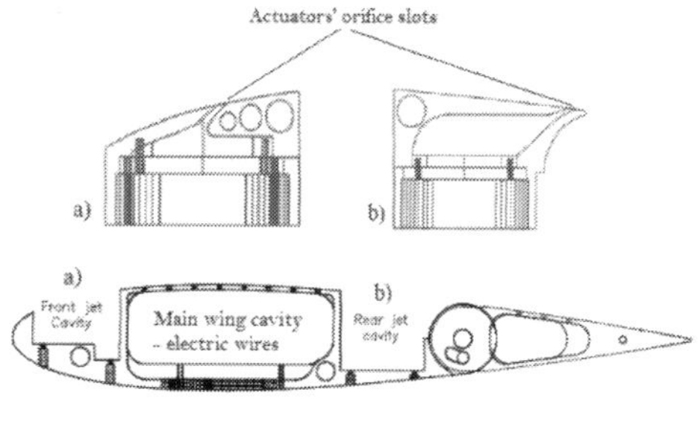

Figure 2. CU SJAs sections and wing locations

a)Leading edge b) Trailing edge (Kabiri, 2012)

In addition, the usage of a LAR wing as SJAs hosting structure allows the investigation of the fluidic actuators effectiveness in presence of complex flow-field. In fact, the three-dimensional flow structure of LAR wings is strongly characterized by wing-tips vortices and separation bubbles at relatively small angles of attack (Noughabi & Tadjfar, 2013). The most relevant effect of having wing-tip vortices of similar scale of the chord length is the streamline separation delay caused by the downwash component, as shown in **Error! Reference source not found.**, (Arivoli, Dodamani, Antony, Suraj, Ramesh, & Ahmed, 2011), (Mueller & Torres, 2001). While the separation bubble phenomenon, induced by the low Re flow regime (Yang & Hu, 2008), causes drag increment and lift distribution disturbance leading to stall anticipation. These phenomena have been identified during the wind tunnel test campaign and helped to analyze the sensors' signal.

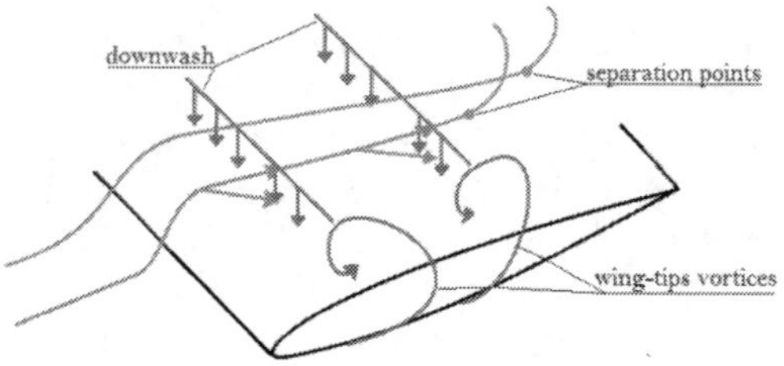

Figure 3. Downwash induced by the wing-tips vortices and downstream shift of the separation point. In green the resulting velocity vector is represented.

3. EXPERIMENT SETUP

The LAR wing hosting the SJAs and sensors' is a characterized by an aspect ratio of 1.33 and a chord length of 0.3 m. It is a modular wing assembly, and the wing airfoil is a NACA0015. The wing is composed by three main parts to guarantee a high level of maintainability and reliability. Sensors and actuators are located inside the wing assembly. The two stripes of ten SJAs are located at 10% and 65% of the chord length, from the leading edge. The sensor system is made up by eleven PCB 103B01 high sensitivity acoustic ICP pressure sensors, grouped as follows: eight in the main wing cavity (between the 19% and the 45% of the chord length) and three inside the flap compartment (between the 75% and the 82% of the chord length). Each pressure transducer has its sensitive element located underneath a pinhole orifice of 1.2 mm of diameter, on the upper surface of the wing. The line of transducers is positioned at 14.4 cm from the right wing tip. Sensors and actuators disposition are shown in **Error! Reference source not found.**.

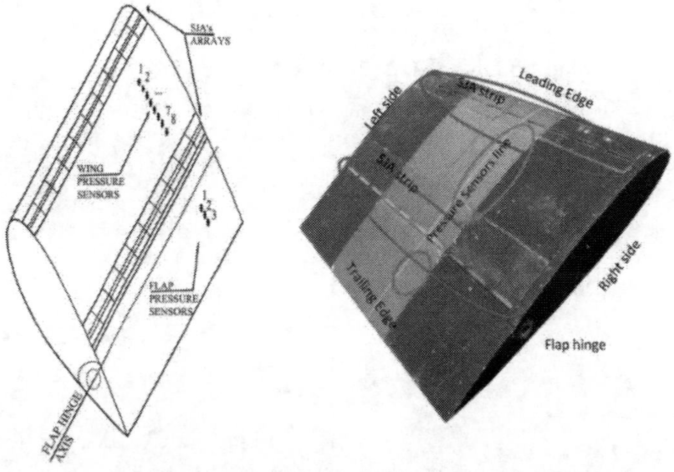

Figure 4. SJAs (actuators) and pressure transducers location

The experimental set-up includes the following hardware and software components required for actuation and data acquisition purpose. Two control computers, a dSPACE DS1103 System, an amplifier Trek350A and three PCB Signal Conditioners for the pressure transducers power supply, all supported by Matlab®/Simulink®/Control Desk® and Labview®. One of the two computers is equipped with the Matlab®/Simulink®/Control Desk® environment and communicates with the dSPACE system. It is used for controlling the SJA arrays, receiving and recording the acoustic pressure transducers signals, and implementing the closed-loop controller. The second computer controls and senses the wind tunnel force balance through a Labview® interface and NI hardware. The dSPACE DS1103 system is a Real-Time-Interface (RTI) and it is made up by two components, a real-time processor (the PPC Controller Board) and a connector panel (CP1103). It is used to guarantee real-time during the control experiment. Both actuators and sensors work at the same frequency as they share the same management system.

The wing model has been tested at the Clarkson University High Speed wind tunnel (CUHS) facility, which is an open-circuit, suction type system. In correspondence of the test section of the wind tunnel, the six-axis AEROLAB TM balance is mounted. The wind speed inside the tunnel is measured using a pitot tube connected to a CPG Digital Pressure Gauge 2500 instrument (Mensor Corporation). The Pitot tube is located in the proximity of the test chamber inlet section. The experiment set-up architecture schematic is represented in Figure 5**Error! Reference source not found.**, in which the input signal paths, referred to the system 'wing / balance', are highlighted.

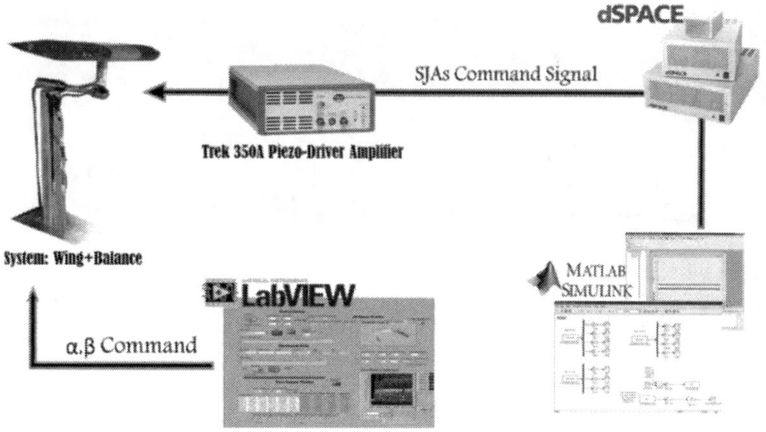

Figure 5. Experimental set-up - System inputs

The SJAs control signal parameters are optimized through open-loop experiment. The same electric signal is used to drive both SJA arrays. This optimization process emphasizes the strong correlation between the actuators power signal and their interaction with the external flow field (Mansor, M.N.Dahalan, Z.K.Lee, & A.Ali, 2014). The carrier wave is sinusoidal and the experimentally tuned parameters are the frequency and the mean voltage (bias), while the amplitude is kept fixed to the value of *5 V* because of the employed piezo membrane physical limitation. In addition, a unitary square wave is added to the previous wave to allow a finest tuning of the actuators' overall performance and its duty cycle and period are experimentally tuned as well. The optimization process starts by exciting the SJAs with a sine wave with a frequency of *460 Hz*, a bias of *25 V* and a duty cycle of *0%*. The optimization sequence consists of separately tuning the parameter values in the following order: sine wave frequency, square wave duty cycle and frequency, sine wave bias. The objective function of the tuning activity is the wing stall angle of attack. During this process the wind speed within the tunnel is kept at U=5 m/s, and the lift coefficients (CL-α curve) is monitored and used as aerodynamic performance index. The sine wave frequency is optimized in a range between *0 Hz* and *1300 Hz*. Figure 6 shows how the stall angle of attack trend with frequency. In particular, the maximum value is experienced in correspondence of the frequency range enclosed between *860 Hz* and *110 Hz*, and the optimal solution is f*=1050 Hz,* as in Figure 7. Consequently, the square wave duty cycle and period are optimized. In detail, the duty cycle has been kept fixed to *50%* and its frequency f_{DC} is tested in a range between *0 Hz* and *200Hz*. The considerable effect of this parameter tuning on the stall angle of attack is of about one degree for f_{DC}*=160 Hz*. Finally, the bias voltage of the carrier wave is raised from *V=25V* to *V=30V*. The bias voltage does not strongly affect the stall angle of attack because the membrane is working close to its operational limit (Figure 8**Error! Reference source not found.**). The lift coefficient drop results delayed of about one degree and this makes the overall $\Delta\alpha_{stall}$ equal to about *3 degrees* from the clean (SJAs not activated) configuration. In ; Table 2 the optimal control signal parameters are summarized.

The effort to correlate the synthetic jets and the wing aerodynamics proper frequency was unsuccessful. The Strouhal number, which is related to the vibration frequency due to the vortex shedding, is computed by using a temporal window of the balance raw voltage measurements and performing the FFT. The Strouhal number referred to the chord length is *St=fc/U=0.3*, where *f=5 Hz* is the vortex shedding frequency. As the synthetic jets and the wing aerodynamics are two different scaled phenomena, it was not possible to directly relate them each other and the experimental analysis has been preferred.

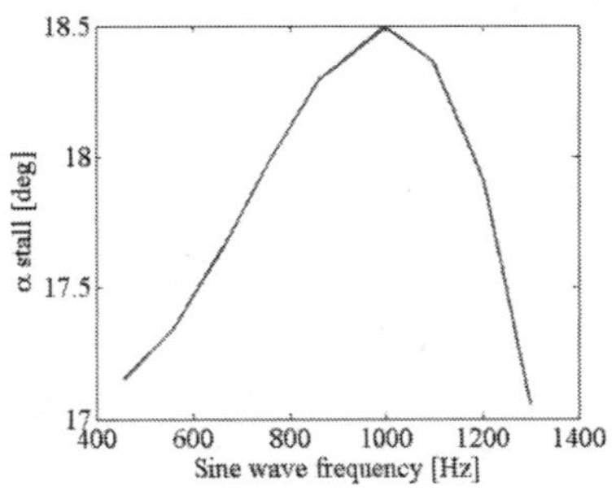

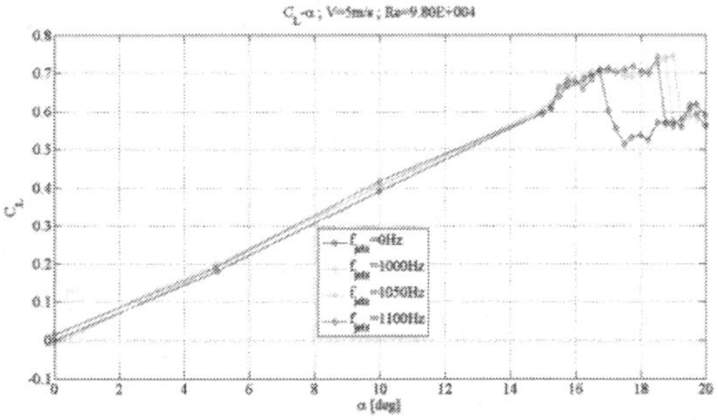

Figure 6. Stall angle of attack dependency with the sine wave frequency **Figure 7.** Sine wave frequency optimization tuning

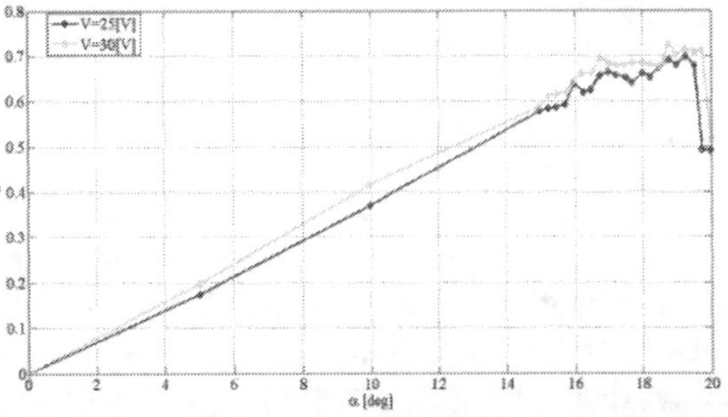

Sine Wave Mean value [V]	30
Sine Wave Amplitude [V]	5
Sine Wave Frequency [Hz]	1050
Square Wave Duty Cycle	50%
Square Wave Frequency [Hz]	160

Figure 8. Lift coefficient vs. AOA @ U=5m/s, for different bias voltage: V=25+5V, V=30+5V; **Table 2** SJAs wave's command signal parameters

5. WIND TUNNEL TEST AND EXPERIMENTAL DATA ANALYSIS

The primary purpose of the wind tunnel test campaign is to derive the wing aerodynamic coefficients in the longitudinal plane for designing the controller. The other important task is the wing flow-field characterization in stall proximity to correlate it to the pressure transducers readings. The objective is to use the transducers' signals for the online real-time flow control. The data acquisition test is performed keeping constant the wind tunnel speed at *U=5 m/s,* which corresponds to a

Reynolds number of $9.53 \cdot 10^4$. The results for the *U=5 m/s* test case are shown in Figure 9 in terms of lift and drag coefficients. The blue and magenta curves are referred to the un-actuated and actuated configuration respectively. The aerodynamic improvements consist in a lift coefficient increment of $\Delta C_{Lmax} \approx 0.06$ and a stall angle of attack postponement of $\Delta \alpha_{stall} \approx 3$ *deg* confirming the optimization process results. The synthetic jets seems to affect the aerodynamic behavior also in terms of C_{L0}, while the lift coefficient curve slope $C_{L\alpha}$ remains constant with the SJAs activation. A SJAs efficiency sensitivity analysis to Re variation has been also conducted. The results for the *U=10 m/s*, shown in Figure 10, suggest that the effects of the higher Reynolds number are a noticeable increment of the clean wing maximum lift coefficient value and a lighter effect of the SJAs on the stall postponement mechanism. In this case, the absence of C_{L0} variation suggests that the effect, previously observed, is to be considered a measurement error due to the very low Reynolds number condition. The aerodynamic drag coefficients plots show that the SJAs act postponing the wing drag abrupt increment due to flow separation.

Focusing on the post-stall region, for the *U=10 m/s* test case (Figure 10 – right) the drag coefficient is higher when the SJAs are activated: the swirling flow field on the wing upper surface seems to be amplified by the synthetic jets action. At the same time, the post-stall lift coefficient drop is less severe than the clean wing case, indicating a benefic effect of the synthetic jets. Anyway, the macroscopic results as those provided by the lift and drag coefficients aerodynamic curves, are not fully satisfactory for explaining the complex interaction mechanism between the synthetic jets and the external flow field. Thus, further studies are required to evaluate the convenience in keeping the SJAs actuated after the wing stall occurrence.

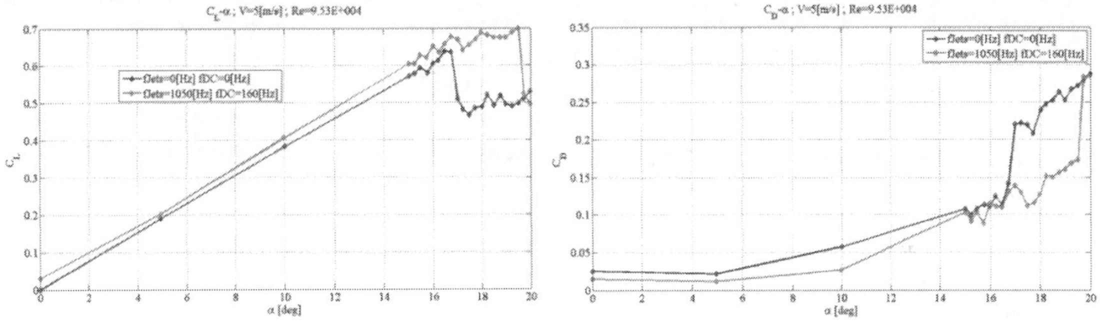

Figure 9. Lift and drag coefficients vs. AOA, actuated and un-actuated configuration @ U=5 m/s

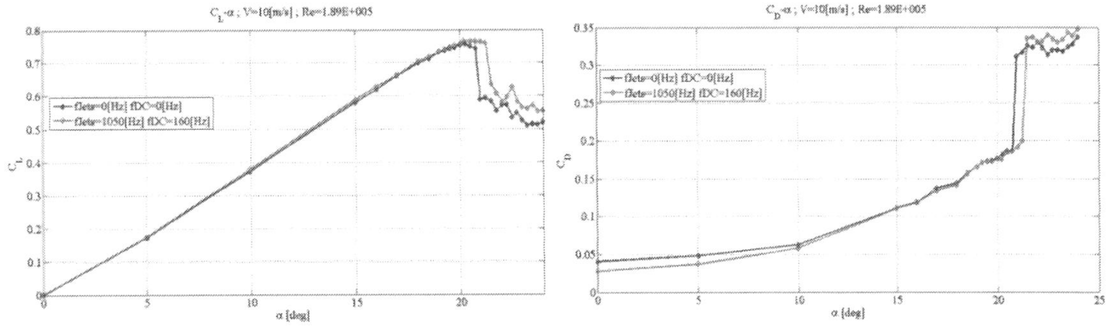

Figure 10. Lift and drag coefficients vs. AOA, actuated and un-actuated configuration @ U=10 m/s

During the entire wind tunnel test campaign, data from the eleven acoustic pressure transducers have been recorded and stored on the monitoring computer in real-time. A data analysis has been conducted to find a signature in the acoustic pressure spatial distribution that indicates the flow separation on the wing upper surface. In fact, this approach could be easily extended and implemented to a real-case application outside the wind tunnel, where the use of a force balance is obviously impracticable. The main objective is to define a trigger event to activate the SJAs controller.

The operative procedure consists of recording *30 seconds* from the eleven pressure sensors, at each angle of attack, from $\alpha=0$ *deg* up to the post-stall region. The transducers sampling is set to $f_S=8$ *kHz*, to allow the analysis of pressure fluctuations in the range of frequency between *0 Hz* and *4000 Hz*. After each recording, the angle of attack is increased by the force balance motion, and a certain time is waited to let the aerodynamic transient in the wind tunnel to extinguish. The primary objective is to determine the pressure fluctuation intensities. To this purpose, the RMS of the pressure signal time histories is computed for each angle of attack, across the chord length. The results show that the acoustic pressure transducers are deeply affected by the external noise of the working environment and the pressure fluctuation intensity sensed by the transducers increases even more when the wind tunnel fan is turned on, as a consequence of the noise and vibration it produces. The identification of the noise contribution in the transducers' recordings is a complex process which in this experiment has been made comparing the FFT computations of the transducers' signals, captured in different configurations: wind tunnel turned off / wind tunnel turned on, both cases with the actuators activated and deactivated. An accelerometer mounted in the proximity of the wing center of gravity has been also used for the identification and filtering off the mechanical vibrations transferred to the wing assembly through the force balance. The difficulties encountered in performing the filtering post-processing of the raw recordings, suggested focusing on the frequency domain analysis. The frequency analysis is performed computing the MSC (Magnitude Squared Coherence) looking for some correlation between the frequency content of different sets of data. Again, the coherence function is applied to the SJAs' on and off configurations. The results reveal that the flow field on the wing upper surface is characterized by frequency content which follows a particular spatial pattern: the sensors' signals are correlated in pairs or triplets at the same relative distances. The frequency coherence study suggests that the flow structure, on the upper surface, is made by a macro vorticity pattern whose scale is about *4 cm*. However, this information is not particularly useful for a stall identification purpose as no evident relation between the vortex scale and the AOA variation has been detected.

Finally, the incoming stall signature has been identified by computing the mean pressure sensed by each transducer along the chord length. Figure 11 (left) shows the pressure distributions along the chord length, normalized using the static ambient pressure value p_{amb}. Focusing on the first eight pressure transducers' readings, which are those located in the main wing body, a pressure distribution trend inversion is observed increasing the angle of attack. This is better observable in Figure 11 (right), in which a linear regression has been applied to the raw pressure distributions. The pressure trend changes its slope sign between the AOA values $\alpha_1=15.75$ *deg* and $\alpha_2=16$ *deg:* according to the force balance results (C_L *vs. AOA* curves) these values of AOA are related to the stall occurrence. Furthermore, for $\alpha<\alpha_{stall}$ the averaged pressure distributions are below one, which means that the averaged pressure p_{avg} sensed by the transducers is lower than the ambient pressure p_{amb}. This correctly characterizes the suction side of the wing upper surface. The separation bubble phenomenon is associated to the pressure discontinuities that affects the raw pressure distributions, and it is typical at low Reynolds numbers (Yang & Hu, 2008).

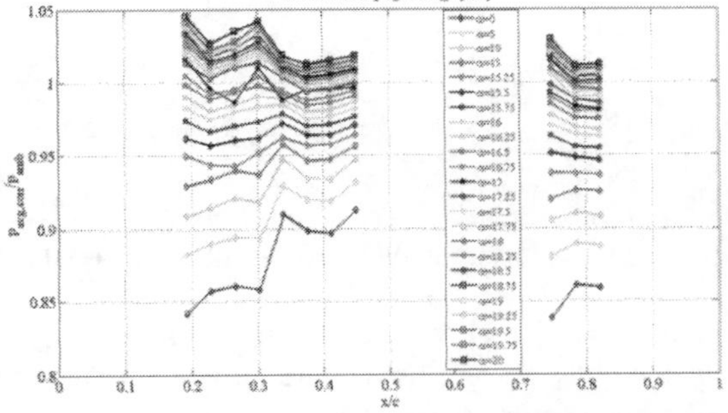

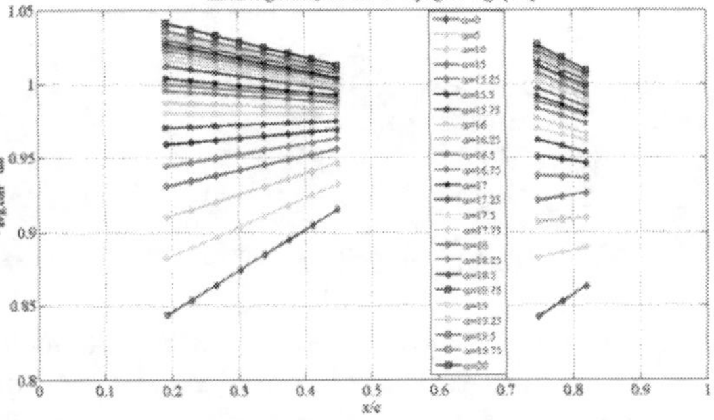

Figure 11. Normalized mean pressure chord-wise distribution sweeping AOA (left); linearization of the pressure distribution (right)

The main reason to develop a closed-loop control strategy is to optimize the SJAs overall system performances in terms of energy consumption. Exactly, the idea is to avoid the continuous SJAs activation but only in stall proximity, which is useful for small aerial vehicle application, where the electric power storing system is limited. For the proposed configuration, made up by only 20 SJA elements, the power required for the activation is *3W*, considering a mean voltage value of *30V* and a current flux of *100 mA*. This value will be much higher for a real wing application.

The proposed closed-loop control scheme is based on an AOA feedback signal provided by the force balance motion encoder. The dSPACE interface uses the AOA input information for triggering the actuation logic in the Matlab/Simulink® control code: this is set in order to activate the SJAs when the AOA value crosses the value of $\alpha=16$ *deg* in the forward direction and vice versa. Then, the angle of attack discrete derivative is necessary to distinguish the forward from the backward triggering event. Figure 12 shows the results of the closed-loop experiment and the comparison with the open-loop curve (red curve in the figure). The stall delay effect is maintained where required and the linear region is not affected by any energy loss.

Further investigations, in the attempt of activating the SJAs using the pressure readings information on the wing upper surface, are necessary as the AOA information as feedback signal is not easily accessible in a real operative scenario.

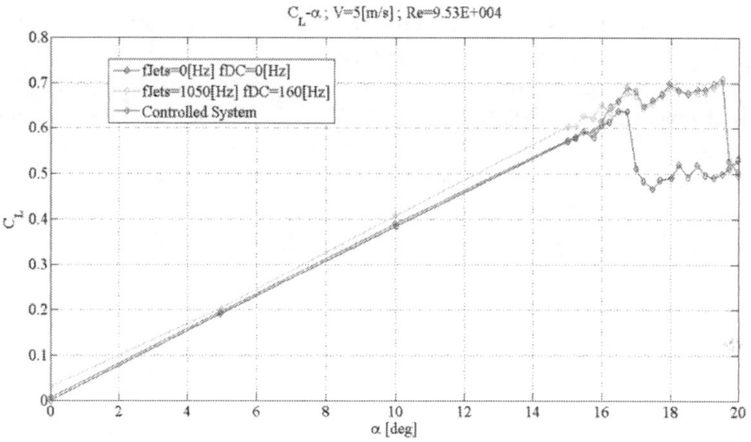

Figure 12. Lift coefficient vs. AOA, Closed-loop / Open-loop comparison

6. CONCLUSIONS

The flow active control on a wing section NACA0015 using two synthetic jet actuator arrays is presented in this article. The symmetric airfoil wing section is characterized by low aspect ratio. Wind tunnel test have been performed at a flow speed of $U=5$ *m/s*, which results in a low Reynolds number condition. The stall delay has been the main objective of the actuators application and to this purpose the SJAs control signal has been experimentally tuned and optimized. By the optimization process, it has been observed that the carrier wave sinusoidal frequency is the most important parameter in defining the actuators effectiveness, compared to the bias voltage, amplitude and the square wave parameters. A second set of wind tunnel experiments have been performed, using the previously obtained SJA input signal, for aerodynamic data acquisition purpose, for both the continuously actuated and non-actuated wing configurations. The effect of the Reynolds number has been also investigated, by increasing the tunnel wind speed up to $U=10$ *m/s*. The results indicate that at higher Reynolds numbers, the SJAs effect on the overall aerodynamic behavior is reduced, and the LAR wing aerodynamic is dominated by the three-dimensional flow field mainly induced by the wing-tips vortices. In the last part of the experimental test, the stall event identification process has been performed by analyzing the recordings of an array of acoustic pressure transducers located on the wing upper surface, along the wing chord; the analysis has been conducted in both time and frequency domains. Three different methodologies are applied and interesting outcomes emerged: the root mean square analysis proved the presence of a low fluctuation zone centered at about 38% of the chord which is probably related to the laminar separation bubble phenomenon; the magnitude square coherence demonstrated that the boundary layer on the wing upper surface is organized in a well-defined structure of pressure fluctuations, in the chord-wise direction. The normalized mean pressure analysis demonstrated the feasibility of identifying the stall approach by using the pressure transducers. A closed-loop control system based on the AOA feedback signal coming from the force balance encoder has been designed and real-time implemented, with the aim of limiting electrical energy consumption associated to the open-loop continuous operation of the SJAs. The result, in terms of lift coefficient – AOA curve, shows that the stall delay is still obtained avoiding energy losses in the linear region of the aerodynamic characteristic. Further experimental studies are necessary on the active control based on the acoustic pressure transducers measurements only.

REFERENCES

1. Alstrom, R. B. "Aerodynamic Flow Control of a High Lift System with Dual Synthetic Jet Aarrays", PhD Thesis, Clarkson University, Potsdam, NY, USA, 2013.
2. Arivoli, D., Dodamani, R., Antony, R., Suraj, C., Ramesh, G., & Ahmed, S. "Experimental Studies on a Propelled Micro Air Vehicle", 29th AIAA Applied Aerodynamics Conference. Honolulu, Hawaii, 2011.
3. Cosyn, P., & Vierendeels, J., "Numerical Investigation of Low Aspect Ratio wings at low Reynolds Numbers", Journal of Aircraft, 43(3), 2006, pp. 713-722.
4. Golubev, V. V., & MacKunis, W., "On UAV Robust Non-Linear Control" ICAS. St. Petersburg, 2014.
5. Hong, G., "Effectiveness of micro synthetic jet actuator enhanced by flow instability", University of Technology Sydney (UTS), Sidney, 2007.
6. Hubbard, J., Flatau, A., & Clingman, D., "Phased array synthetic jets for influencing dynamics of complex flows", University of Maryland College Park, 2014.
7. Ikhlaq, M., Ghaffari, O., & Arik, M., "Effect of actuator deflection on heat transfer for low and high frequency synthetic jets", IEEE Thermal and Thermomechanical Phenomena in Electronic Systems. Orlando, Florida, USA, 2014.
8. Inman, D. J., Kochersberger, K. B., Kurdila, A. J., Mason, W. H., & Wilson, S. B., "Ducted Fan Aerodynamics and Modeling, with Applications of Steady and Synthetic Jet Flow Control", PhD Thesis, Virginia Polytechnique Institute, Blacksburg, VA, US, 2011.
9. Jabbal, M., S.C. Liddle, and W.J. Crowther., "Active Flow Control Systems Architectures for Civil Transport Aircraft", Journal of Aircraft, 47(6), 2010.
10. Kabiri, P., "Active flow control over a NACA0015 airfoil by synthetic jet actuators", PhD Thesis, Clarkson University, Potsdam, NY, USA, 2012.
11. Mansor, S., M.N. Dahalan, Z.K. Lee, and A.Ali, "Flow Visualization and Measurement of a Synthetic Jet Actuator Performance Using PIV System", Journal Technology (Sciences & Engineering), 71(2), 2014, pp. 13-117.
12. Melton, L. P., & Yao, C.-S. S., "Active Control of Separation from the Flap of a Supercritical Airfoil", AIAA Journal, 44(1), 2006, pp. 34-41.
13. Millar, A. C., "Flow Control Via Synthetic Jet Actuation", Master of Science Thesis, Texas A&M University, Texas, 2006.
14. Mueller, T. J., & Torres, G. E., "Aerodynamics of low aspect ratio wings at low Reynolds numbers with applications to micro air vehicles design and optimization", University of Notre Dame. Notre Dame, Indiana, 2001.
15. Noughabi, A. K., and Tadjfar, M., "Cross-Wind Influence on Low Aspect Ratio Wings at Low Reynolds Numbers", ASME 2013 Fluids Engineering Division Summer Meeting, 2013.

16. Stolk, M., Patel, D., O'Donnel, K., Marzocca, P., Breuker, R. D., Abdalla, M., et al., "Active Flow and Aeroelastic Control of Lifting Surfaces Using Synthetic Jet Actuators", Aerospace Technology Conference and Exposition, 2007.
17. Tang, H., Salunkhe, P., Du, J., and Wu, Y., "Active Flow Control Over a Wing Model Using Synthetic-Jet-Actuator Arrays", Fluid-Structure-Sound Interactions and Control, Vol. I, 2014, pp. 115-122.
18. Tovar, E., Pereira, N., Bate, I., Indrusiak, L., Negrao, J., Viana, J. C., et al., "Networked Embedded Systems for Active Flow Control in Aircraft", 11th International Workshop on Real Time Networks RTN, Pisa, 2012.
19. Yang, Z., & Hu, i. H., "An experimental study of the laminar flow separation on a low-Reynolds-number airfoil", Thesis, Iowa State University, Iowa, USA, 2008.
20. Z. Wang, T.Hu, & I.Gursul., "Active control of self-excited roll oscillations of LAR wings", Fluid-Structure-Sound Interactions and Control, Vol. III, 2014, pp 357-362.
21. Zeiger, M. D., & Rediniotis, O. K., "Reconfigurable Synthetic Jet Actuation for Closed-Loop Hingeless Flow Control". Department of Defense, Air Force, 2004.

ICAST2015 #079

Adaptability in Various Structures of Marine Plankton Skeleton

Naoko Kishimoto

Department of Mechanical Engineering, Setsunan University, Osaka, Japan

Abstract

In order to investigate what structures are optimized and adapted under microgravity conditions, we focus on shapes of marine plankton with calcareous or silicate skeletons because they float in water like under microgravity owing to buoyant forces. Various shapes of fossils and modern species also proved that they have varied their shapes to fit to environmental changes in their evolution process for over several hundred million years. The skeleton structures measure several hundred micrometers and usually we should use an electron microscope to observe details of the structures but we cannot observe invisible parts. In this study, we can acquire three dimensional information of the whole structures using a micro X-ray computer tomography. Analysis on acquired three-dimensional information provides geometric and mechanical features of the structures. In this paper, we show some analysis results which indicate their adaptability to environmental changes.

1. INTRODUCTION

Space environment around space structure systems such as rockets, satellites, and the International Space Station is characterized by high vacuum and microgravity. To investigate behavior of a space structure in vacuum environments, we can use a vacuum chamber of which inside is made into a high vacuum state under 10^{-5}Pa equivalent to atmospheric pressure at an altitude over 250km. However, in case of microgravity, only an incomplete test environment can be realized, even if using drop tower or parabolic flight, so that it is difficult to test an optimized structure under microgravity condition.

In this research, we focus on shapes of marine plankton with calcareous or silicate skeletons in order to know what shapes are optimized or adaptive to real microgravity conditions, because they float in water as if under microgravity. Moreover, they have several hundred million years evolutionary history which provides "actual" adaptive process to environmental changes. In usual paleontologists or biologists classify and analyze these micro fossils or skeletons of marine plankton according to two-dimensional electron microscopy images, so that three-dimensional information including thickness and internal details is very limited.

Then we introduce using micro X-ray computer tomography (micro X-ray CT) with high resolution in order to acquire three-dimensional information of the marine plankton skeletons less than 300 micrometers in diameter. Some ingenuity and improvements have enabled to observe three dimensional structures of the skeletons in detail on the same level as using an electron microscopy[1)]. Based on the acquired information, we conducted geometrical and mechanical analysis of the skeleton shapes. And we also produced three-dimensional enlarged models using a three-dimensional laminating printer. These models provide great improving the shape recognition of marine planktons, especially for many people who have been not familiar with the invisible marine plankton. Figure 1 shows outline and flow of our research.

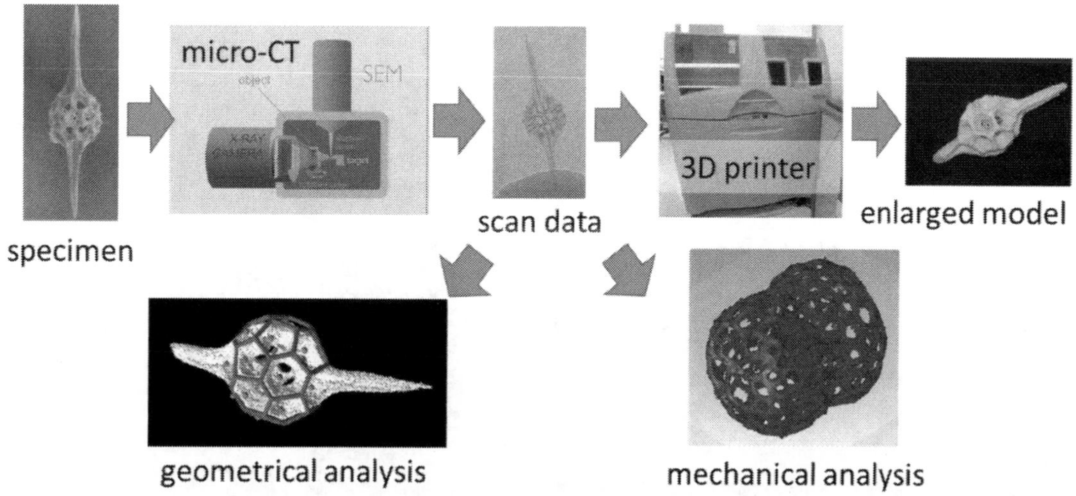

Figure 1. Outline of our research on marine plankton skeletons

2. RESEARCH TARGET: FORAMINIFERA AND RADIOLARIA

Our research targets are marine plankton skeletons[2)] generally smaller than one millimeter, the study of which requires the use of light or electron microscopy. Figure 2 shows representative skeleton shapes of marine plankton, which can be divided into four groups. They are divided two groups from a material aspect: carbonate and silicate, and also divided two groups from a nutrient source: phytoplankton and zooplankton. In general, the dimension of phytoplankton is rather smaller than that of zooplankton. It is difficult for a micro X-ray CT to obtain higher resolution below a submicron level, and we treat mainly zooplankton with 100-300um in size: foraminifera and radiolaria.

Foraminifera are members of a phylum or class of amoeboid protists characterized by streaming granular ectoplasm that among other things is used for catching food, and commonly by an external shell or "test" made of various materials and constructed in diverse forms. Foraminifera typically produce a test, or shell, which can have either one or multiple chambers, some becoming quite elaborate in structure. These shells are commonly made of calcium carbonate ($CaCO^3$) or agglutinated sediment particles. Over 10,000 species are recognized, both living and fossil. They are usually less than 1 mm in size, but some are much larger, the largest species reaching up to 20 cm.

Radiolaria are members of protozoa characterized by intricate mineral skeletons, typically with a central capsule dividing the cell into the inner and outer portions of endoplasm and ectoplasm. They are found throughout the ocean, and their skeletal remains make up a large part of the cover of the ocean floor as siliceous ooze. Due to their rapid turn-over of species, they represent an important diagnostic fossil found from the Cambrian onwards. Radiolarians have many needle-like pseudopodia supported by bundles of microtubules, which aid in the radiolarian's buoyancy. The cell nucleus and most other organelles are in the endoplasm, while the ectoplasm is filled with frothy vacuoles and lipid droplets, keeping them buoyant. The radiolarian can often contain symbiotic algae, especially zooxanthellae, which provide most of the cell's energy. Some of this organization is found among the heliozoa, but those lack central capsules and only produce simple scales and spines. They are usually 0.1-0.2 mm in diameter.

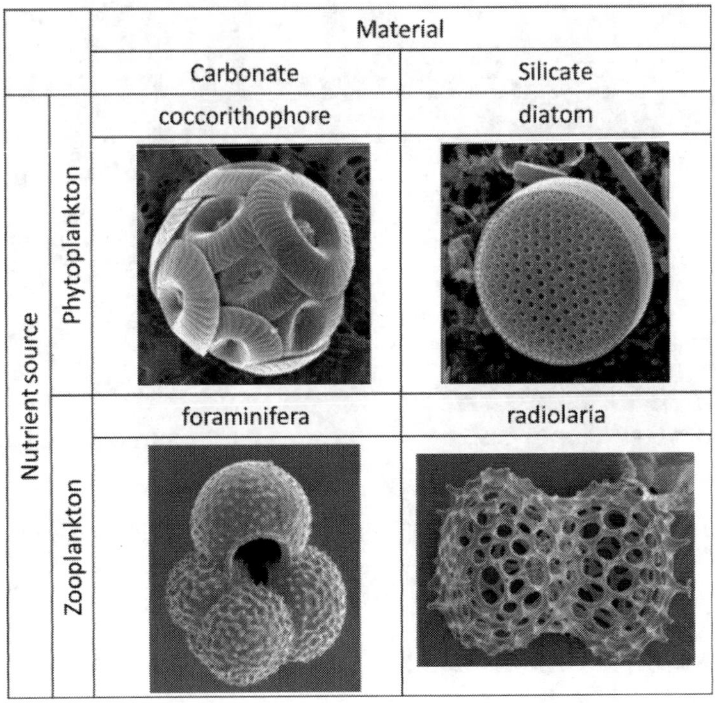

Figure 2. Four types of marine plankton

3. METHOD

Our equipment of micro X-ray computer tomography, micro CT in SEM of Brucker-Skyscan, is attached to an electron microscope, JSM-6510. This system uses the X-rays generated from a metal target irradiated by an electron beam of the microscope. A different metal type of the target generates a different distribution of spectrum of X-rays, and we can adjust the contrast of transmission images changing metal types of the target.

Before observation, using an optical microscope a specimen is glued on a sharpened metallic holder of which diametric size of a tip is 100 or 300 um. It is most important to set the specimen on the exact center of the holder tip to keep its high resolution. To construct a distinct three dimensional image, it is significant to capture transmission images as clearly as possible. Paying the closest attention to adjust some conditions of the electron microscope, we can observe tiny frame structures with submicron diameters shown in Fig.3. *Pterocorys zankuleus* (Haeckel) is a kind of recent radiolarian, which has more thin and delicate frames than fossils have. A thickness of a narrowest frame is 600 or 700 nm. These images show our observation system provide sufficient resolution for radiolarian classification[1)].

Once the three-dimensional information is obtained, we can easily observe an arbitrary section or inside of the structures using the image processing software as shown in Fig.3. And based on the information, we can produce enlarged models by the 3D laminating printer (Fig.4). These models provide a hands-on experience for viewing invisible marine planktons[3)].

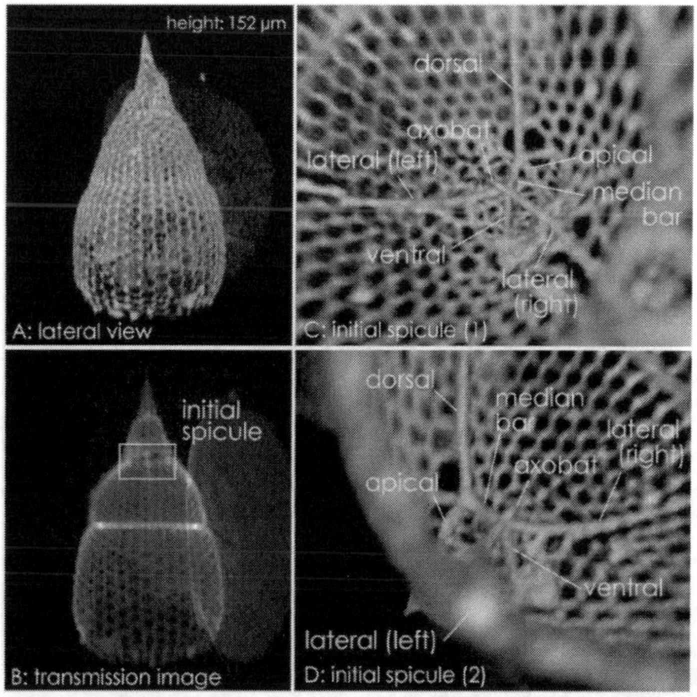

Figure 3. Acquired three dimensional information
Recent radiolaria *Pterocorys zankuleus* (Haeckel)

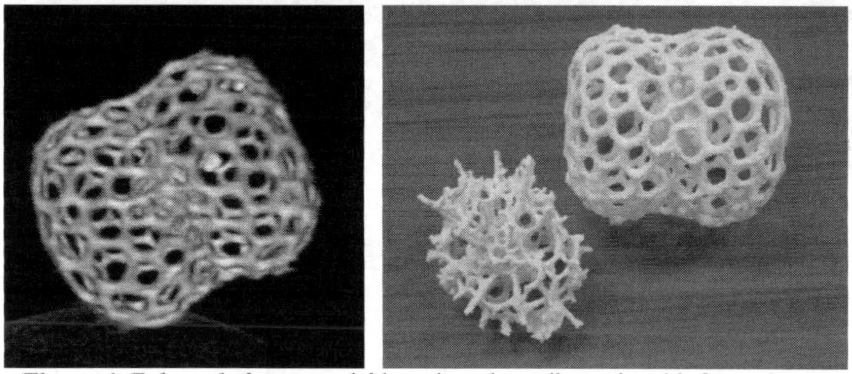

Figure 4. Enlarged plaster model based on three dimensional information
Didymocyrtis tetrathalamus (Haeckel)

4. ANALYSIS USING THREE DIMENSIONAL INFORMATION

Using three dimensional information, we can conduct various efficient evaluations of shapes and structures we have not been able to treat using only two dimensional images relating marine plankton skeletons.

4.1 GEOMETORICAL ANALYSIS

Examples of geometrical analysis include quantification of connection rules among sequential chambers of foraminifera and determination of number and shapes of pores of spherical radiolaria.

4.1.1 Quantification of connection rule among sequential chambers of foraminifera

Foraminifera have calcareous spherical chambers sequentially connected and we modeled their shapes as series of spheres mathematically[4)]. Based on appearance of two dimensional electron microscopy images, we proposed some connection rules between adjacent spherical chambers. Then in order to evaluate the proposed rules, we approximate sequential spheres based on acquired three dimensional information obtained by micro-CT. Figure 5 shows a computer graphic image of acquired 3D information (Fig.5 left) and an approximation of sequential spherical chambers derived from the 3D information (Fig.5 right). Using the approximation models, we can recognize quantitative parameters of foraminifera morphology such as chamber diameters, distances and revolving angles between adjacent chambers, and so on.

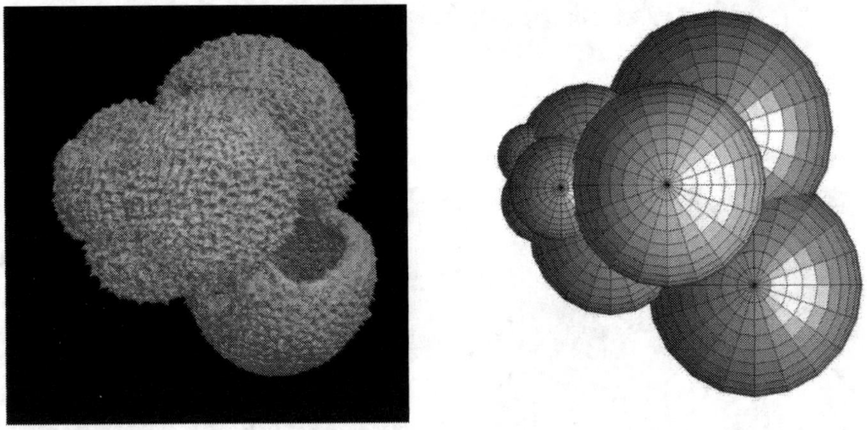

Figure 5. Approximated sequential spherical chambers based on acquired 3D data

4.1.2 Determination of number and shapes of pores of spherical radiolaria[5),6),7)]

Spherical radiolaria have many pores on its sphere. They are classified according to a number and shapes of the pores adding to its whole shape and size. Actually it is difficult to count the exact number of pores only from two dimensional electron microscope images shown in Fig.2. Then we developed the way to automatically count the number and determine the shapes of the pores based on the acquired three dimensional information.

We used a set of grid points on a spherical surface to approximate the skeletal structure, and counted the number of pores by using an algorithm for counting clusters on a grid. Finally, we used Voronoi tessellation to determine the distribution of the pores. Now our collected data concerning Mesozoic radiolaria *Pantanellium* have identified that number of pores takes some discontinuous values owing to geometrical restrictions.

4.2 MECHANICAL ANALYSIS

We can construct a voxel finite element model base on acquired 3D information and calculate mechanical stress distribution of plankton skeleton under some appropriate boundary conditions. Figure 6 shows computer graphic image of *Didymocyrtis tetrathalamus* (Haeckel), its STL format image and constructed voxel finite element model based on the STL format image. This voxel model has 460458 rectangular voxels. However larger number of voxels allows accurate modeling and analysis, it increase computational costs.

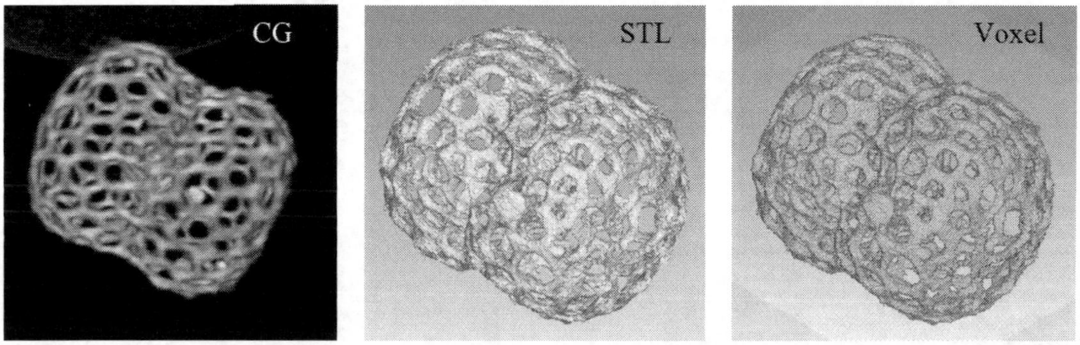

Figure 6. Images and voxel finite element model of *Didymocyrtis tetrathalamus* (Haeckel)

Table 1. Analysis condition on voxel finite element models

Material	Silica glass (Quartz)
Young's modulus	74GPa
Poisson's ratio	0.18
Boundary condition	Free-free
Loading condition	Simulated water pressure (depth of water : ~1m)
Analysis software	VOXELCON 2010

Table 1 shows analysis condition on voxel finite element model as shown in Fig.6. Considering they float in the water, we adopt simulated water pressure around it. Figure 7 shows Mises stress distribution result of *Didymocyrtis tetrathalamus* (Haeckel). Figure 8 also shows Mises stress distribution result of another type of radiolarian *Nasselaria* sp. These figures indicate that no stress concentration is occurred, which means these skeleton structures are some kind of optimized structures in engineering sense.

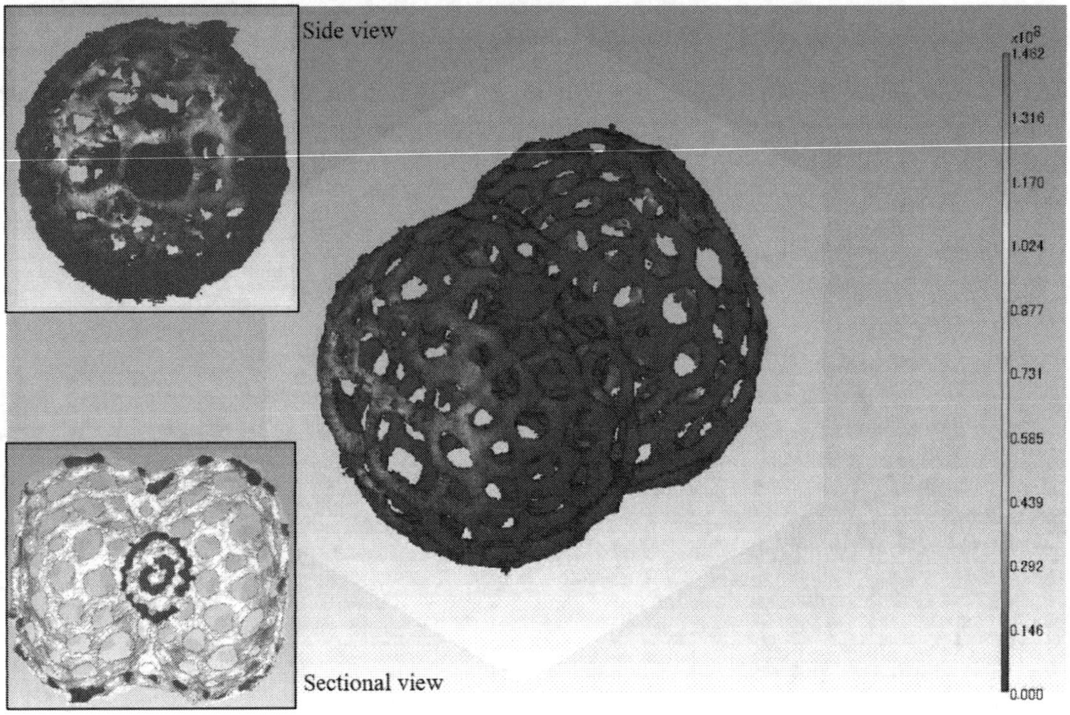

Figure 7. Analysis result (*Didymocyrtis tetrathalamus* (Haeckel)) ; Mises stress distribution

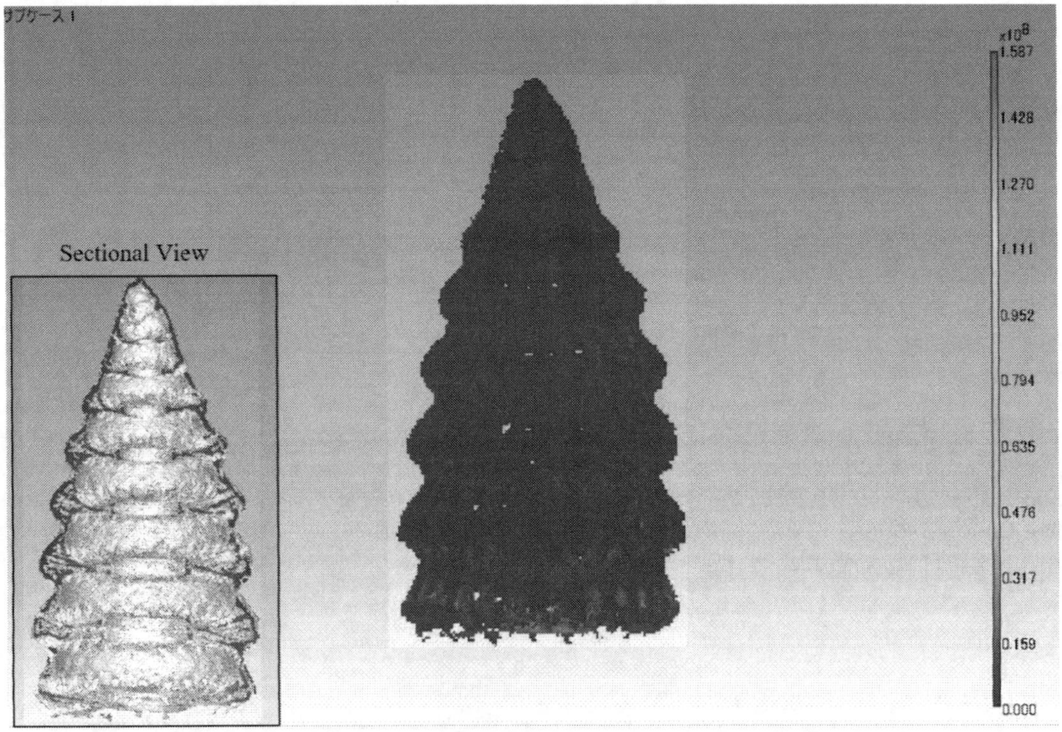

Figure 8. Analysis result (*Nassellaria sp.*) ; Mises stress distribution

4. CONCLUSION

In order to investigate what structures are optimized and adapted under microgravity conditions, we focus on shapes of marine plankton with calcareous or silicate skeletons because they float in water like under microgravity owing to buoyant forces. In this study, we acquired three dimensional information of the whole structures using a micro X-ray computer tomography and conducted geometrical and mechanical analysis on the acquired three-dimensional information. Especially stress distribution under uniform pressure indicated that there is no stress concentration, which means their structures are some kind of adaptive to the boundary condition.

ACKNOWLEDGMENTS

We thank Dr. Naoto Ishida, Meiji Univ., Prof. Takashi Yoshino, Toyo Univ., Atsushi Matsuoka, Niigata Univ., Dr. Katsunori Kimoto, JAMSTEC, and Toshiyuki Kurihara, Niigata Univ. for helpful discussion.

REFERENCES

1. Ishida, N., Kishimoto, N., Matsuoka, A., Kimoto, K., Kurihara, T., Yoshino, T., "Three-dimensional image of the Jurassic radiolarian *Protunuma? Pchiensis* Matsuoka: an experimental study using high resolution X-ray micro-computed tomography," *VOLUMINA JURRASICA*, XIII(1), 2015, pp.77-82.
2. Armstrong, H.A., Brasier, M.D., Microfossils, Second Edition, 2009, Blackwell Publishing.
3. Matsuoka, A., Yoshino,T., Kishimoto, N., Ishid, N., Kurihara, T., Kimoto, K., Matuur, S.," Exact Number of Pore Frames and Their Configuration in the Mesozoc radiolarian *Pantanellium*: An application of X-ray micro-CT and Layered Manufacturing Technology to Micropaleontology," *Marine Micropaleontology*, 88/89, 2012, pp.36-40.
4. Yoshino, T., Kimoto, K., Kishimoto, N., Matsuoka, A., Kurihara, T., Ishida, N., Matuura, S., "A Simple Mathematical Model for Chamber Arrangement of Planktic Foramnifera," *Forma*, 24, 2009, pp.87-92.
5. Yoshino, T., Matsuoka, A., Kurihara, T., Ishida, N., Kishimoto, N., Kimoto, K., Matuura, S., "Application of Voronoi Tessellation of Spherical Surface to Geometrical Models of Skeleton Forms of Spherical Radiolaria," *Forma*, 27, 2012, pp.45-53.
6. Yoshino, T., Kishimoto, N., Matsuoka, A., Ishida, N., Kurihara, T., Kimoto, K., "Pores in Spherical Radiolarian Skeletons Directly Determined from Three-Dimensional Data," *Forma*, 29, 2014, pp.21-27.
7. Yoshino, T., Matsuoka, A., Kurihara, T., Ishida,N., Kishimoto, N., Kimoto, K., "Polyhedron geometry of skeletons of Mesozoic radiolaria *Pantanellium*," *Revue de micropaleontologie*, 2015

ICAST2015: 26th International Conference on Adaptive Structures and Technologies
October 14-16th, 2015, Kobe, Japan

ICAST2015 #052

Airfoil Design for Morphing Wing Using Robust Multiobjective Optimization

Nozomu Kogiso[1*], Masahiro Toyoda[1], Takashi Funahashi[1] and Tomohiro Yokozeki[2]

[1] Department of Aerospace Engineering, Osaka Prefecture University, Sakai, Japan.
[2] Department of Aeronautics and Astronautics, The University of Tokyo, Tokyo, Japan.

Abstract

This study considers the robust optimum airfoil shape of the morphing wing using the corrugated structures arranged in the rear side of the airfoil. The rear side of the wing is deformed like the plain flap by bending the corrugated structure. As the first step, this study considers the multiobjective optimization of the airfoil shape with different objectives corresponding to the flapping angle. In addition, uncertainty of the airfoil shape is considered, thus, the robust multiobjective optimization method is applied. The morphing part is modeled by nonuniform rational basis spline (NURBS) curve and the passing points of the NURBS curve are treated as design variables. In addition, the NURBS curve is smoothly connected by appropriately arranging the passing points to the front side of the airfoil where is not morphed. The aerodynamic analysis is performed by the free computational tool for the wing operating at low Reynolds number range. Then, the surrogate model is constructed using the radial basis function (RBF) network for computationally efficient optimization. Through numerical calculation, the validity of the robust multiobjective optimization method for the airfoil design is demonstrated.

1 INTRODUCTION

A morphing wing can be defined as one which has the ability to either alter its shape in a seamless change along the chord or spar by warping or bending [1, 2]. Yokozeki et al., one of the authors, have developed the variable camber morphing airfoil arranging corrugated structures at the rear side in airfoil as shown in Fig. 1 [3]. Since the corrugated structure has flexibility to bending to the wavy direction, the rear part of the airfoil can morph as a plain flap. The flap has several functions to increase the lift-to-drag ratio at taking off or to increase the lift at approaching. Feasibility of the corrugated morphing airfoil was demonstrated through wind tunnel test using the two-dimensional model [3]. In addition, it is shown that the morphing shape can be designed in terms of the corrugation pitch or the thickness [4]. However, the optimized corrugate arrangement is not studied yet.

Thus, this study applies the robust multiobjective optimum design method proposed by the authors [5] to the airfoil design of the corrugated morphing airfoil. For the formulation of the multiobjective optimization, the satisficing trade-off method (STOM) [6] is applied. STOM is known to be an interactive optimization method and converts a multiobjective optimization problem into the equivalent single-objective optimization problem by introducing an aspiration level that corresponds to the user's preference for each objective function value. In the robust multiobjective optimization, the Pareto solutions are

* kogiso@aero.osakafu-u.ac.jp

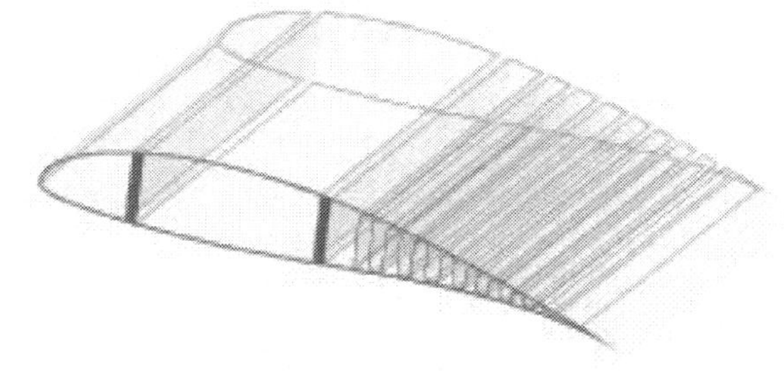

Figure 1. Morphing airfoil using corrugated structure

evaluated to visualize the trade-off between the mean performance and the variation by formulating them as independent objective functions. In previous study, the method was applied to the high high-precision space reflector design to investigate the effect of variations on the applied load on the shape accuracy [7].

As the first step of the optimum design study, this study concentrates on the airfoil shape design considering the aerodynamic performance. The corrugation arrangement design is not considered yet. That will be possible to design the arrangement according to the desired airfoil shape. The morphing wing warped the corrugated section such that the length of the upper curve is unchanged, while that of the lower curve is shrinked [3]. Accordingly, the upper curve is modeled by using the non-uniform rational basis spline (NURBS) curve [8] and the passing points are treated as design variables. The NURBS curve is widely used in the shape optimization such that the airfoil optimization [9] and the membrane design [10]. The curve is modeled under following constraints that the length of the upper curve is unchanged and the morphing part is smoothly connected to the unmorphed front part, where the front part is not modeled as the NURBS curve. In addition, the lower curve of the morphed part is also modeled as the NURBS curve that the passing points are modeled as dependent variables such that the airfoil thickness is unchanged.

The aerodynamic performance is evaluated by using the airfoil analysis tool XFLR5 [11] that is widely adopted to the design of wings operating at low Reynolds numbers. In this study, the numerical analysis tool is not directly used in the optimization. For computational efficiency, the surrogate model [12] is constructed by using the radial basis function (RBF) network [13]. The RBF network constructs the approximation model from several sample points and updates the model by adding the sample properly determined to improve the approximation accuracy. In this study, the RBF network method proposed by Kitayama et al. [14] is adopted for approximation accuracy.

Through numerical examples, it is first demonstrated the efficiency of the morphing airfoil modeling using the NURBS curve proposed in this study. Then, the approximate accuracy of the RBF network for the aerodynamic performance in terms of the design variables, the passing points of the NURBS curve, is discussed. Finally, the usefulness and the efficiency of the robust multiobjective optimization method to the morphing airfoil design are discussed.

2 MORPHING AIRFOIL SHAPE BY NURBS CURVES

The morphing airfoil with corrugation [3] is morphed as the plain flap by bending the corrugation such that the length of the upper curve is unchanged, while that of the lower curve is shrinked as shown in Fig. 2. In addition, the airfoil thickness is unchanged during the morphing.

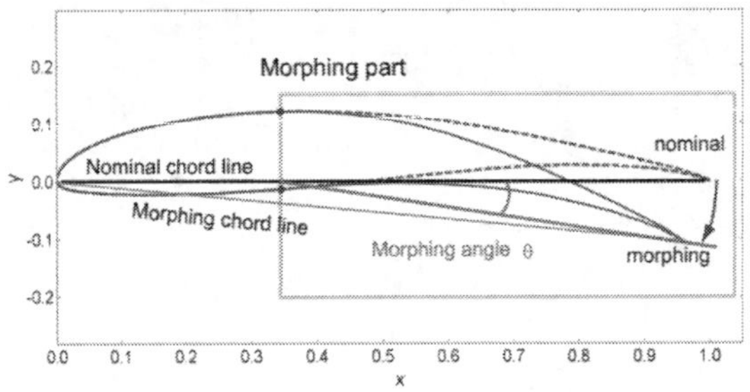

Figure 2. Deformation of morphing wing like plain flap

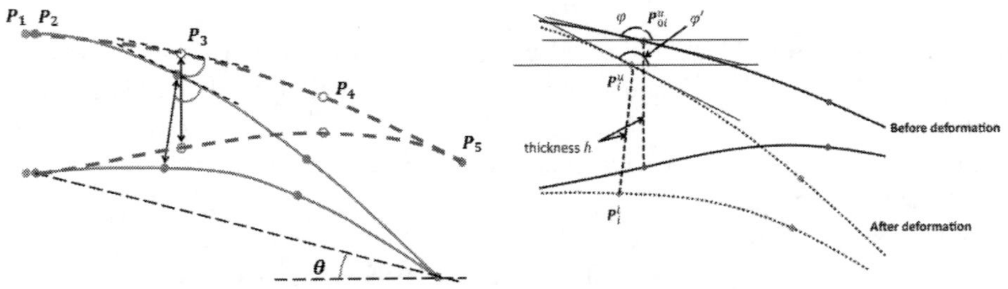

(a) NURBS curve in upper side (b) NURBS curve in lower side

Figure 3. NURBS curve model of morphing part in airfoil

The NURBS curve is adopted to model the morphing section of the rear part of the airfoil. While, the front part without morphing is not modeled as the NURBS curve. Since the total airfoil shape is modeled with a set of nodes passing on the airfoil and the shape is used as an input model to the aerodynamic analysis tool, the nodal points of the rear side are obtained from the NURBS curve and those of the front side are given as the constant values from the airfoil section data. Therefore, the different modeling method for the ordinal airfoil shape optimization using the NURBS curve is required. Especially, the smooth connection is required at the NURBS starting point.

In this study, the NURBS curve passing points are treated as design variables. For efficient optimization, smaller number of the design variables is desirable. The minimum number of the NURBS curve passing points is set to five for sufficient freedom of the airfoil shape and satisfy the smooth connection requirement at the morphing starting point.

At first, consider the upper curve of the airfoil. The NURBS passing point $\boldsymbol{P}_2^u$ is set at the morphing starting point on the upper curve and the another point $\boldsymbol{P}_1^u$ is set at the closed leading edge side on the upper curve as shown in Fig. 3(a), where $\boldsymbol{P}_1^u$ is a starting point of the NURBS curve. The unmoved two points will act such that the change on the curvature of the NURBS curve will be continuous during the morphing for the smooth connection between the front and the rear side. Then, $\boldsymbol{P}_5^u$ is set at the trailing edge point that is the end point of the NURBS curve. The position of the end point $\boldsymbol{P}_5^u$ is determined to satisfy the two conditions as the morphing angle θ and the unchanged length of the upper curve. Where, the morphing angle is defined to the angle between the unmorphed chord line and the line connecting the morphing starting point at the lower curve and the trailing edge as shown in Fig. 2.

The two passing points $\boldsymbol{P}_3^u$ and $\boldsymbol{P}_4^u$ between the starting and the end points on the upper curve are

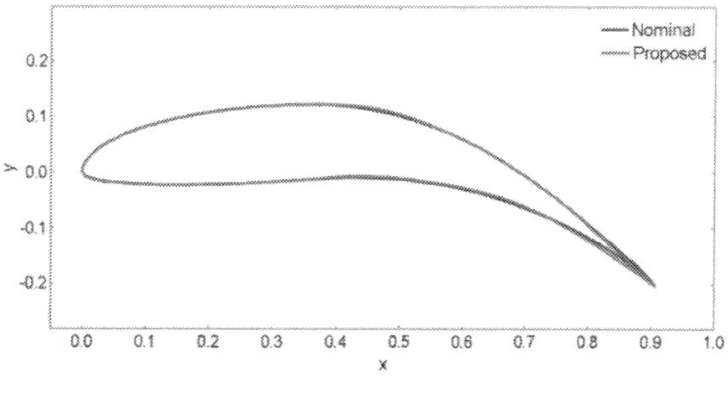

Figure 4. Comparison between NURBS model and original morphing airfoil [3] for 20 degree of morphing angle

freely moved to change the airfoil shape. For sufficient freedom of the airfoil shape and simplicity, the y-coordinates of the points are treated as design variables.

Then, the morphing shape of the lower curve is defined. At first, $\boldsymbol{P}_1^l$ and $\boldsymbol{P}_2^l$ are set at the points on the lower curve shifted perpendicular to the unmorphed chord line that is identical to the x-axis. The end points $\boldsymbol{P}_5^l$ is set as the trailing edge that is identical to $\boldsymbol{P}_5^u$ as the end point of the upper NURBS curve. Then, the points on the lower curve $\boldsymbol{P}_3^l$ and $\boldsymbol{P}_4^l$ are defined as shifting the wing thickness unchanged from the points $\boldsymbol{P}_3^u$ and $\boldsymbol{P}_4^u$ on the upper surface, respectively, as shown in Fig. 3(b). The wing thickness h_j is first defined for the unmorphed wing as follows:

$$h_j = y_{0j}^u - y_{0j}^l \qquad (j = 3, 4) \tag{1}$$

where y_{0j}^u and y_{0j}^l are the y-coordinate of the NURBS passing points for the unmorphed airfoil. As shown in Fig. 3(b), the slope angle of the upper curve at the point is denoted as φ. Then, the passing point coordinate for the morphed airfoil is denoted as $\boldsymbol{P}^u = [x^u, y^u]$. The coordinate of the corresponding points $\boldsymbol{P}^l = [x^l, y^l]$ in the lower curve for the morphed wing is evaluated as follows:

$$x^l = x^u - h\sin(\varphi - \varphi') \tag{2}$$

$$y^l = y^u - h\cos(\varphi - \varphi') \tag{3}$$

where φ' is the slope angle at the upper curve for the morphed wing.

Figure 4 compares the morphing airfoil with 20 degree of the morphing angle between the NURBS modeling by red curves and the original shape in reference [3] by blue curves. It is found that the two curves are almost identical, though small differences appear around the trailing edge.

3 AIRFOIL ANALYSIS AND SURROGATE MODEL

3.1 Airfoil analysis

This study uses the airfoil analysis tool XFLR5 [11] that is widely adopted to the design of wings operating at low Reynolds numbers. The tool provides the change of the lift coefficient, the drag coefficient and the moment coefficient with respect to the angle of attack based on the pressure distributions evaluated by the two-dimensional viscous formulation for the given airfoil shape, the velocity, and the Reynolds number.

3.2 Surrogate Model by RBF Network

A radial basis function (RBF) network is a three-layer feed-forward network using the RBF. The function $f(\boldsymbol{x})$, the output of the network, is approximated as follows:

$$f(\boldsymbol{x}) = \sum_{j=1}^{m} w_j h_j(\boldsymbol{x}) \tag{4}$$

where m indicates the number of sampling points, w_i is the weighting factor, and $h_j(\boldsymbol{x})$ is the j-th basis function defined as follows:

$$h_j(\boldsymbol{x}) = \exp\left[-\frac{(\boldsymbol{x}-\boldsymbol{x}_j)^{\mathrm{T}}(\boldsymbol{x}-\boldsymbol{x}_j)}{r_j^2}\right] \tag{5}$$

where $\boldsymbol{x}_j$ is j-th sampling point,and r_j is the width of the basis function. In this study, r_j is defined by using Kitayama's method [14] as follows:

$$r_j = \frac{d_{j,\max}}{\sqrt{n}\sqrt[n]{m-1}} \tag{6}$$

where n is the number of design variables, m denotes the number of sampling points, and $d_{j,\max}$ is the maximum distance between the j-th sample point and the other sampling points.

The RBF network is usually accomplished by solving the following function to estimate the weighting factor $\boldsymbol{w}$ in Eq. (5).

$$\text{Minimize}: \quad \sum_{i=1}^{p}(\hat{y}_i - f(\boldsymbol{x}_i))^2 + \sum_{j=1}^{m}\lambda_j {w_j}^2 \tag{7}$$

where the first term is sum of squares of the error between the network output $f(\boldsymbol{x}_i)$ and the true value $\hat{y}_i$ and the second term is introduced for the regularization [13]. It is recommended that λ_j is seta as a small value, e.g., 1×10^{-3}.

The necessary condition of Eq. (7) yields the following equation:

$$\boldsymbol{w} = (\boldsymbol{H}^{\mathrm{T}}\boldsymbol{H} + \boldsymbol{\lambda})^{-1}\boldsymbol{H}^{T}\hat{\boldsymbol{y}} \tag{8}$$

where $\boldsymbol{H}$ is given as follows:

$$\boldsymbol{H} = \begin{bmatrix} h_1(\boldsymbol{x}_1) & h_2(\boldsymbol{x}_1) & \cdots & h_m(\boldsymbol{x}_1) \\ h_1(\boldsymbol{x}_2) & h_2(\boldsymbol{x}_2) & \cdots & h_m(\boldsymbol{x}_2) \\ \vdots & \vdots & \ddots & \vdots \\ h_1(\boldsymbol{x}_p) & h_2(\boldsymbol{x}_p) & \cdots & h_m(\boldsymbol{x}_p) \end{bmatrix} \tag{9}$$

and the following matrix $\boldsymbol{A}$ is introduced:

$$\boldsymbol{A} = (\boldsymbol{H}^{\mathrm{T}}\boldsymbol{H} + \boldsymbol{\lambda}) \tag{10}$$

The learning of the RBF network reduces to obtain $\boldsymbol{A}^{-1}$. In this study, the leave one out cross validation(LOOCV) [12] is adopted for efficient learning.

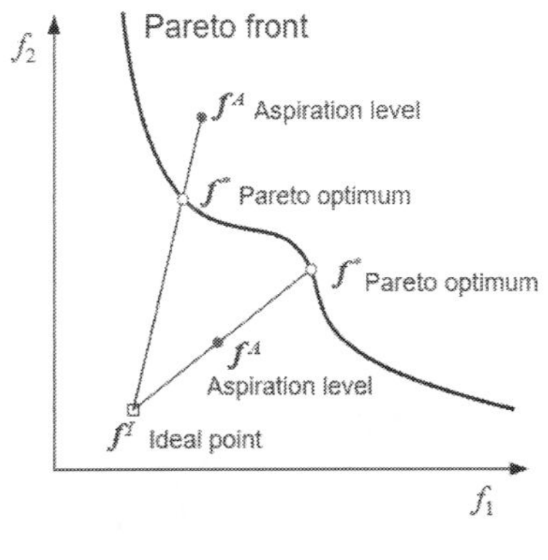

Figure 5. Pareto optimal solution searching by STOM

4 ROBUST MULTIOBJECTIVE OPTIMIZATION

Designing the morphing airfoil requires the robust design that considers the effect of uncertainties of several parameters such as velocity, the angle of attack, the airfoil shape and so on.

The robust optimization can be defined as a multiobjective optimization problem to minimize both the expected value and the variation of the objective function. However, it has been generally formulated as the minimization of the weighted sum of the expected value and the variation as follows:

$$F_{robust}(\boldsymbol{x}, \boldsymbol{z}) = \mathrm{E}[f(\boldsymbol{x}, \boldsymbol{z})] + \alpha\sqrt{\mathrm{Var}[f(\boldsymbol{x}, \boldsymbol{z})]} \tag{11}$$

where $\boldsymbol{x}$ and bmz denotes design variables and random variables, respectively, and $\alpha > 0$ is a weight coefficient. The single-objective optimization using Eq. (11) sometimes fails to obtain the desired solution, especially when the Pareto set has a non-convex shape.

In this study, the robust design optimization is formulated as the following multiobjective optimization problem:

$$\text{Minimize:} \quad f_1(\boldsymbol{x}, \boldsymbol{z}) = \mathrm{E}[f(\boldsymbol{x}, \boldsymbol{z})] \tag{12}$$

$$f_2(\boldsymbol{x}, \boldsymbol{z}) = \sqrt{\mathrm{Var}[f(\boldsymbol{x}, \boldsymbol{z})]} \tag{13}$$

where the expected value and the standard deviation are adopted as individual objective functions.

Part of the authors [5] are proposed to use STOM [6] to the robust multiobjective optimization problems. STOM is known to be an interactive optimization method and converts a multiobjective optimization problem into the equivalent single-objective optimization problem by introducing an aspiration level that corresponds to the user's preference for each objective function value. As shown in Fig. 5, the Pareto optimal solution is usually located on the line connecting the ideal point and the aspiration level in the objective function space, regardless of whether or not the aspiration level lies in the feasible region. STOM is formulated as the following min-max problem:

$$\text{Minimize :} \quad \max_{i=1,2,\cdots,r} w_i(f_i(\boldsymbol{x}) - f_i^I) \tag{14}$$

$$\text{subject to :} \quad g_j(\boldsymbol{x}) \geq 0 \quad (j = 1, 2, \cdots, m)$$

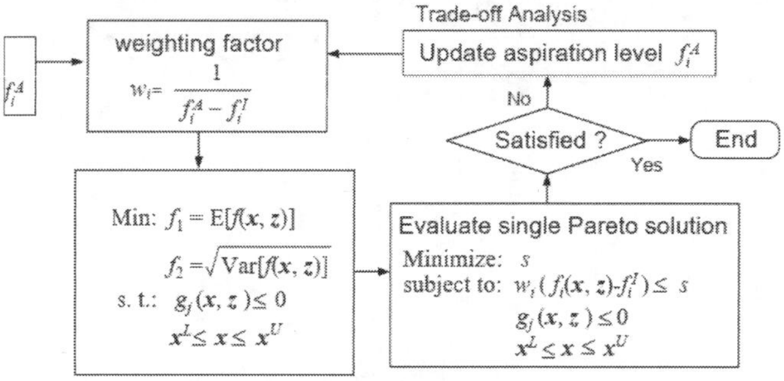

Figure 6. Flowchart of robust multiobjective optimization using STOM

where f_i^I is an ideal point and w_i is a weight coefficient that is evaluated by using an aspiration level f_i^A as follows:

$$w_i = \frac{1}{f_i^A - f_i^I} \qquad (i = 1, \cdots, r)$$

Introducing a slack variable s, the equivalent single-objective optimization problem is defined:

$$\begin{aligned} \text{Minimize:} \quad & s \\ \text{subject to:} \quad & w_i\big(f_i(\boldsymbol{x}, \boldsymbol{z}) - f_i^I\big) \leq s \quad (i = 1, \cdots, r) \\ & g_j(\boldsymbol{x}, \boldsymbol{z}) \leq 0 \quad (j = 1, \cdots, m) \end{aligned} \tag{15}$$

The Pareto solution corresponding to the aspiration level is obtained. When the user is not satisfied with the obtained solution, the user repeats the process by changing the aspiration level. On the other hand, the Pareto surface can be obtained by parametrically changing the aspiration level.

The flow of the robust multiobjective optimization is illustrated in Fig. ??. In this study, the sequential quadratic method (SQP) is adopted to solve the equivalent optimization problem (15).

5 NUMERICAL EXAMPLES

The airfoil adopted in this study is Fx63-137 (maximum wing thickness 13.7%, maximum camber 5.97%) that motor gliders widely adopt, where the Reynolds number is set as $Re = 7.94 \times 10^5$ [3]. Here, the lift maximization design under the case of 10 degree of the morphing angle is illustrated.

The design variables are the y coordinates of the NURBS passing points y_3 for $\boldsymbol{P}_3^u$ and y_4 for $\boldsymbol{P}_4^u$ and the angle of attack for the nominal airfoil α'. Note that α' is different from the angle of attack for the morphed airfoil. In order to avoid the ill-shaped airfoil that has ill effects on the accuracy of the surrogate model, the design variables for 10 degree of the morphing angle are constrained as follows:

$$0.06355 \leq y_3 \leq 0.11855 \tag{16}$$
$$-0.07165 \leq y_4 \leq 0.02835 \tag{17}$$
$$-6.0 \leq \alpha' \leq 5.0 \tag{18}$$

Note that the lower and upper bounds are varied for the morphing angle. In addition, the angle of attack α' is also constrained to avoid numerical errors beyond the stall angle for the aerodynamic analysis.

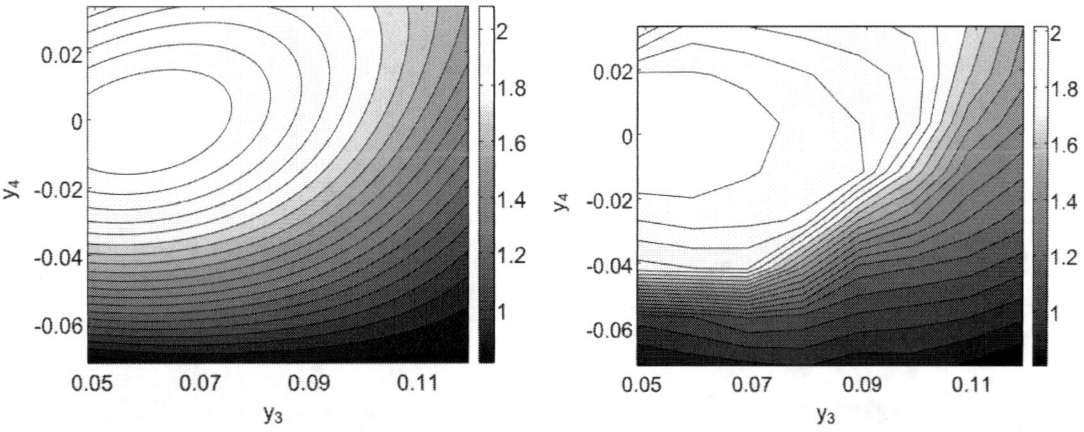

(a) Approximation by RBF network (b) XFLR5

Figure 7. C_L distribution for $\alpha' = 2°$ in design variable space

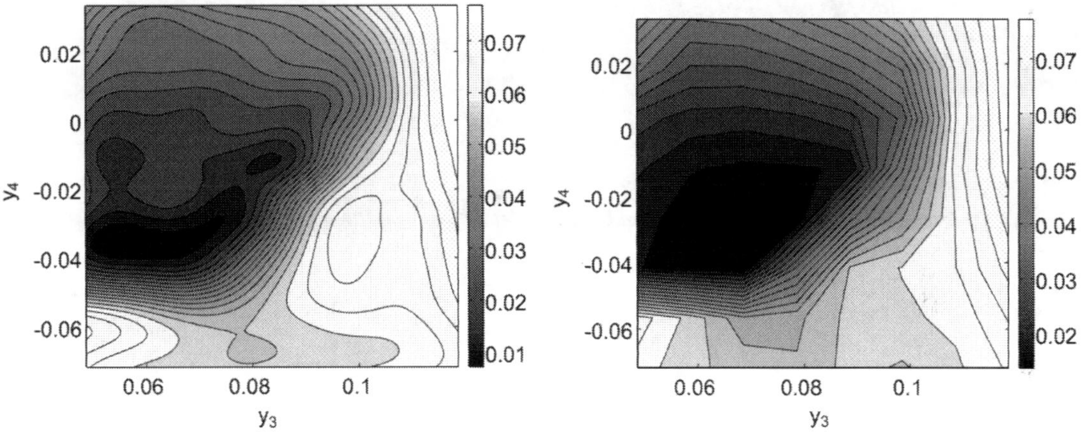

(a) Approximation by RBF network (b) XFLR5

Figure 8. C_D distribution for $\alpha' = 2°$ in design variable space

5.1 Approximation Accuracy of RBF Network

The approximation accuracy of the RBF network is compared for the airfoil under the case of 10 degree of the morphing angle. The lift and the drag coefficient distribution in terms of the design variables y_3 and y_4 for the angle of attack to the nominal airfoil $\alpha' = 2°$ are illustrated in Figs. 7 and 8, respectively. In the figures, (a) indicated the approximated contour curve evaluated by the RBF network, and (b) is a contour plot made from the XFLR5 results at the lattice points. Because of the lattice points is small in number, the contour curves in (b) are shaky. It is found from the figures that the RBF network has sufficient approximation accuracy for the lift coefficient. However, the accuracy for the drag coefficient is insufficient, though the overall trends are very similar from each other. This is because the change of the drag coefficient is much larger for the stall regions in comparison of that of the lift coefficient.

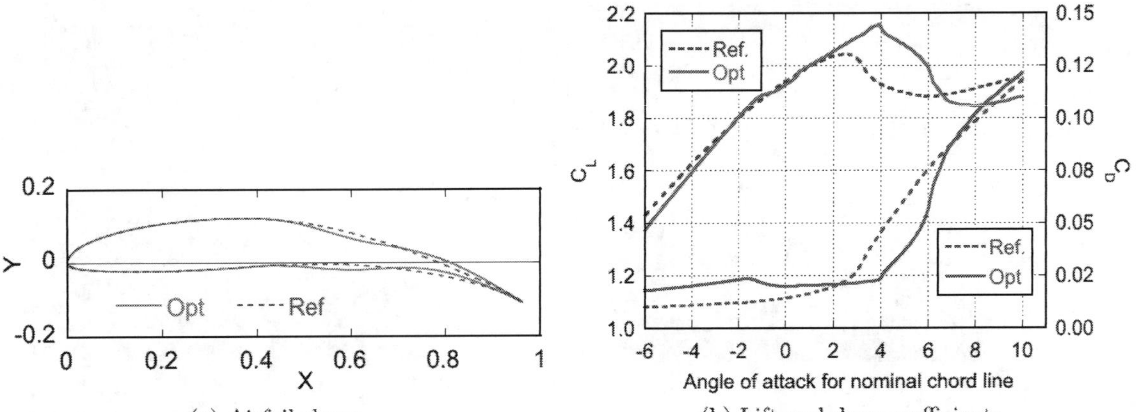

(a) Airfoil shape (b) Lift and drag coefficients

Figure 9. Comparison of C_Lt maximized airfoil shape and the referred airfoil [3] for 10 degree of morphing angle

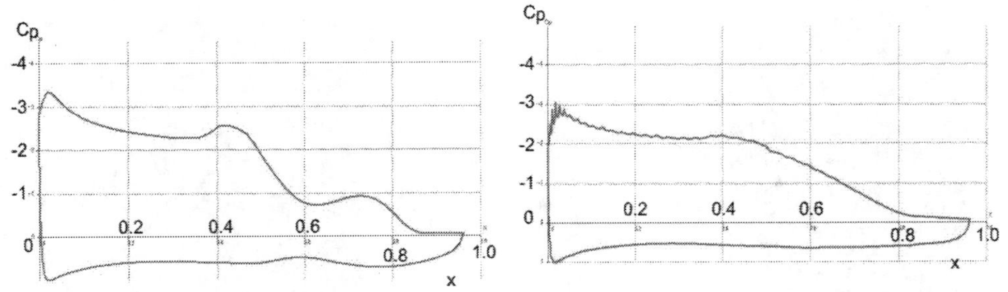

(a) Optimized airfoil for $\alpha' = 3.9°$. (b) Referred airfoil for $\alpha' = 2.5°$

Figure 10. Comparison of pressure distributions at the maximum lift coefficient

5.2 Lift Coefficient Maximization as Deterministic Design

The lift coefficient maximization design for 10 degree of the morphing angle is illustrated in Fig. 9(a). In this figure, "Opt" curve corresponds to the optimized design and "Ref" to the airfoil shape in the reference [3]. Though the optimized airfoil has wavy shape, the maximum lift coefficient is larger than that of the reference airfoil as illustrated in Fig. 9(b), where the horizontal axis indicates the angle of attack for the nominal airfoil (α').

The maximum lift coefficient of the optimum airfoil is 2.157 at $\alpha' = 3.9°$. It is better than that of the referred airfoil, 2.0443 at $\alpha' = 2.5°$. On the other hand, the drag coefficient is about twice as that of the referred wing because of the wavy shape. However, the optimum airfoil has smaller drag coefficient at the stall angle. The pressure distribution is compared in Fig. 10. The horizontal axis indicates the chord line of the nominal airfoil and the lateral axis indicates the pressure coefficient, where the negative value is the upper side. That is, upper curve corresponds to the upper airfoil side. In the optimized airfoil, the negative pressure decreases at the wavy point, but the value is recovered at the backward. That's why the optimum airfoil has larger lift coefficient.

The approximation accuracy is shown in Table 1, where the approximated and the analyzed values

Table 1 Approximation accuracy at the optimum design

Design	Method	$C_{L\max}$	α' (deg)
Optimum	RBFN	2.158	3.09
	XFLR5	2.157	3.90
Reference [3]	XFLR5	2.044	2.50

are compared for the optimized and the referred airfoil. Though the difference of the maximum lift coefficient is small, the stall angle has large difference. It indicated that some improvements are required to establish the approximation accuracy.

5.3 Robust Multiobjective Optimization

Here illustrates the Pareto solutions for the robust multiobjective optimization problem to maximize the lift coefficient, while to minimize the variation of the lift coefficient with respect to the airfoil shape. Considering variations of the airfoil shape, y_3 and y_4 are set as random variables and the standard deviations are set as follows:

$$\sqrt{\mathrm{Var}[y_i]} = 0.01 \quad (i = 3, 4) \tag{19}$$

The design variables are set as the mean values of the coordinates, $E[y_3]$, $E[y_4]$ and the angle of attack for the nominal airfoil, α'. The design problem is formulated as the robust multiobjective optimization problem as follows:

$$\begin{aligned} \text{Maximize:} \quad & f_1 = C_L(E[y_3], E[y_4], \alpha') \\ \text{Minimize:} \quad & f_2 = \sqrt{\mathrm{Var}[C_L(E[y_3], E[y_4], \alpha')]} \\ & y_i^L \leq E[y_i] \leq y_i^U \quad (i = 3, 4) \\ & \alpha'_L \leq \alpha' \leq \alpha' \end{aligned} \tag{20}$$

where the same side constraints as the deterministic design problem are used.

Pareto solutions obtained by parametrically changing the aspiration levels are illustrated in Fig. 11(a). The ideal point is located at the lower right corner, because f_1 (mean value of C_L) will be maximized and f_2 (standard deviation of C_L) will be minimized, where the ideal point is set for both solutions for the single objective optimizations. It is found that the obtained Pareto solutions are located on the line connecting the ideal point and the aspiration levels.

The mean airfoil shapes corresponding to A, B and C denoted in Fig. 11(a) are compared in Fig. 11(b). The solution C that has large mean value of C_L and large variance of C_L is a wavy shape as the deterministic design shown in Fig. 9. On the other hand, the smaller variance designs like B and A have smaller wavy shape. Therefore, the STOM obtains reasonable Pareto solutions.

Finally, the approximation accuracy of the surrogate model is presented. Table 2 compares the approximated C_L by the RBF network with the numerical value by XFLR5 for the obtained airfoil shapes, A, B and C. It is found that the approximation has the sufficient accuracy as the previous examples.

6 CONCLUSIONS

This study evaluates the robust optimum airfoil shape of the morphing wing by the robust multiobjective optimization method by using STOM. Through numerical examples, the following conclusions are remarked.

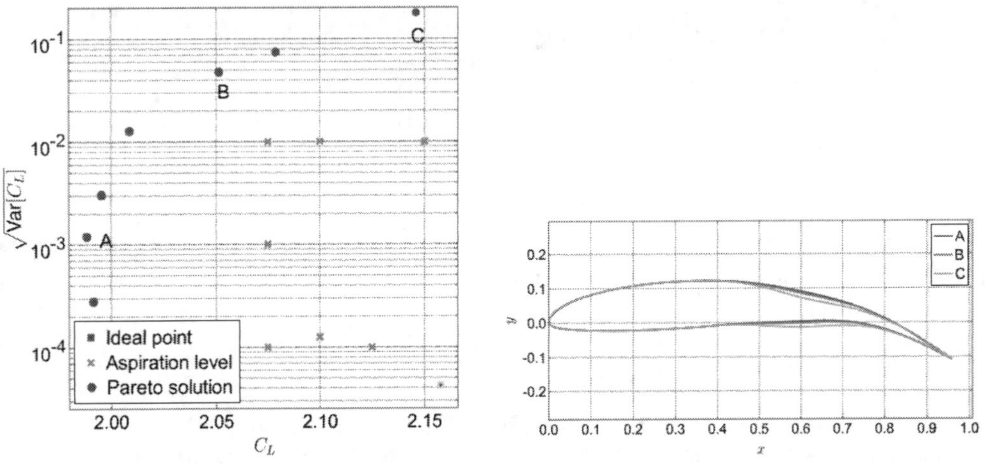

(a) Pareto solutions (b) Airfoil shape for the solution A, B and C

Figure 11. Pareto solutions for robust multiobjective optimization

Table 2 Approximation accuracy by RBF network for the Pareto solutions A, B and C in Fig. 11(a)

Pareto solution	RBF network		XFLR5	
	$E[C_{L_{\max}}]$	α' (deg)	$E[C_{L_{\max}}]$	α' (deg)
A	1.989	3.48	2.007	3.30
B	2.051	3.71	2.056	3.70
C	2.146	3.30	2.132	3.90

- The morphing section is modeled by using the NURBS curve in order to reduce the number of design variables, where the passing points of the NURBS curve are treated as design variables. The smooth connection of the leading edge side and the morphing section at the trailing edge side at the morphing starting point is established by setting the double starting points of the NURBS curve at the morphing beginning points.
- The aerodynamic performance is evaluated by using the analysis tool XFLR5 for low Reynolds number region. To avoid the direct calls of the analysis tool during the optimization, the surrogate model is constructed by using the RBF network. Through numerical examples, the approximation accuracy of the lift coefficient is demonstrated to be sufficient, though some improvements will be required for the drag coefficient approximation.
- The robust multiobjective optimization considering variation of the airfoil shape is demonstrated, where the mean value and the standard deviation of the lift coefficient are formulated as individual objective functions. The Pareto solution obtained by STOM is illustrated to be effective to visualize the trade-off between the mean value and the standard deviation.

The other performance index such as the lift-to-drag ratio should be considered. Then, design methodology for the corrugation arrangement corresponding to the optimized airfoil shape should be established in the future that requires the multidisciplinary coupling analysis with aerodynamics and structure.

REFERENCES

1. Tamayama, M., "Survey on Morphing Activities", *Journal of Japan Society of Fluid Mechanics*, Vol. 28, No. 4, 2009, pp. 277-284, (*in Japanese*)
2. Barbarino, S., Bilgen, O., Ajaj, R. M., Friswell, M. I., and Inman, D. J., "A Review of Morphing Aircraft", *Journal of Intelligent Material Systems and Structures*, Vol. 22, No. 9, 201, pp. 823-877.
3. Yokozeki, T., Sugiura, A., and Hirano, Y., "Development of Variable Camber Morphing Airfoil Using Corrugated Structure", *Journal of Aircraft*, Vol. 51, 2014, pp. 1023-1029.
4. Yokozeki, T., Takahashi, H., and Hirano, Y., "Development and Demonstration of Morphing Wing Using Corrugated Structures", *Proceedings of 46th JSASS Annual Meeting*, 2015, JSASS-2015-1008, pp. 1-5, (*in Japanese*)
5. Toyoda, M., and Kogiso, N., "Robust Multiobjective Optimization Using Satisficing Trade-Off Method", *Journal of Mechanical Science and Technology*, Vol. 29, 2015, pp. 1361-1367.
6. Nakayama, H. and Sawaragi, Y., "Satisficing Trade-Off Method for Multiobjective Programming", *Lecture Notes in Economics and Mathematical Systems*, Vol. 229, 1984, pp. 113-122.
7. Kodama, R., Kogiso, N., Toyoda, M., and Tanaka, H., "Trade-Off analysis for Structural Design of High-Precision Space Reflector Using Multiobjective Optimization Method", *Mechanical Engineering Journal*, Vol. 2, No. 4, 2015, 15-00058, pp. 1-14.
8. Piegl, L., and Tiller, W., *The NURBS Book*, 2nd ed., Springer, 1997.
9. Oyama, A., and Fujii, K., "A Study on Airfoil Design for Future Mars Airplane", *44th AIAA Aerospace Sciences Meeting*, 2006, AIAA-2006-1484.
10. Kogiso, N., Hirajo, D., and Akita, T., "Shape Optimal Design of Membrane Wrinkling Minimization Using NURBS Curves", *Journal of the Japan Society for Aeronautical and Space Sciences*, Vol. 56, 2008, pp. 72-79, (*in Japanese*)
11. "XFLR5, An Analysis Tool for Airfoils, Wings and Planes", http://www.xflr5.com/xflr5.htm, (access on 2015. 10)
12. Forrester, A., Sobester, A., and Keane, A., *Engineering Design Via Surrogate Modeling: A Practical Guide*, Wiley, 2008.
13. Orr, M. J. L., *Introduction to Radial Basis Function Networks*, University of Edinburgh, http://www.anc.ed.ac.uk/rbf/rbf.html, 1996, pp. 1-67, (access on 2015. 10).
14. Kitayama, S., Arakawa, M. and Yamazaki, K., "Sequential Approximate Optimization Using Radial Basis Function Network for Engineering Optimization", *Optimization and Engineering*, Vol. 12, No. 4, 2011, pp. 535-557.

ICAST2015 #057

Morphing beam structure consisting of open- and closed-section members: FEA modeling and demonstration

Atsushi Aso[1*], and Hiroaki Tanaka[1]

[1] Department of Aerospace Engineering, National Defense Academy of Japan, Kanagawa, Japan

Abstract

We propose a new morphing beam concept and develop its design method. The validity of the design method is investigated through numerical simulations and an experiment. In the proposed morphing beam concept, a specially designed beam consisting of two pipes is employed. One pipe is a thick closed-section member, and it carries bending moments. The other pipe is a variable-shape member having open and closed sections and is designed to achieve an objective distribution of twist angles. The ends of these pipes are connected to each other at the tip side, and the other ends are connected to the base body or a motor that provides torque to the member at the root side. In general, the torsional rigidity GJ of the open-section member is smaller than that of the closed-section member. Therefore, the open-section part twists more in comparison with the closed-section parts. The objective distribution of the twist angle can be achieved by designing the appropriate length proportions of the open- and closed-section parts. This morphing beam concept is applicable to morphing wings or deformable space structures.

A numerical model of the beam consisting of two pipes was developed, and the performance was investigated through finite element analysis (FEA). We also manufactured an experimental model using a 3D printer and performed an experiment. The feasibility of the morphing beam concept and the validity of the design method were demonstrated through the numerical simulations and experiments. The results clearly indicate that the morphing beam concept is useful in many fields. For instance, it could bring about a new airplane wing system that includes an aileron by realizing a morphing wing on a seamless rudder that twists like the wings of a bird.

1. INTRODUCTION

Morphing beam concept is applicable to morphing wing or deformable space structures. In both cases, this concept could be applied in order to adapt to their circumstances by means of changing their shape. Therefore many researchers have been studying morphing wings and different ways of morphing, namely span change, chord length change, sweep change, twisting, chord-wide bending, and span-wise bending (Fig. 1) [1]. In any morphing wing, the common purpose is to enhance flight performance without increasing wing weight (if possible). A hinged rudder surface is necessary for flying objects to control the lifting power and to perform take-off, landing, and turning. However, a hinged rudder surface works only over a short period of time and may be heavy due to the mechanical systems. On the other hand, bird wings are not like hinged rudder surfaces but are seamless and can control the lifting power and perform take-off, landing, and turning (Fig. 2) [2]. In general, the seamless surface observed in the

* info@icast2015.com

twist of a bird's wing is lighter thanks to its simple surface structure and can enhance aerodynamic performance.

There are great benefits to wing twist; it was first applied in planes by the Wright Brothers but was soon removed. Because of the demand for enhanced performance and higher airspeed, applying wing twist has not been feasible [3]. To make matters worse, very dangerous phenomena such as aileron reversal and divergence flutter were later found. These crises occurred because of the low stiffness, and generally, a twisting wing is known to have a critical defect in stiffness. Nowadays, thanks to advancements in materials and actuators and the changing demand, many researchers have resumed studying wing twist and various methods of achieving it [3],[4].

In this study, we propose a new morphing beam concept consisting of two pipes. One pipe is a thick closed-section member and carries the bending moments. The other pipe is a variable-shape member with open- and closed-section parts. The combination of the open- and closed-sections is designed to achieve an objective distribution of twist angles. We establish a theory for the twist angle of this specially designed beam and validate it by finite element analysis (FEA) using commercial FEA software (ANSYS Workbench 15). This morphing beam structure can make conventional airplanes lighter and more adaptive to changes in flight conditions. In this study, we focus on the morphing beam for a wing twist type morphing wing.

To accomplish this objective, this paper first derives the theory for the twist angle of the specially designed beam. Next, we analyze the behavior of the beam through numerical simulations using FEA. Finally, we demonstrate the proposed morphing beam for wing twist and validate the theory.

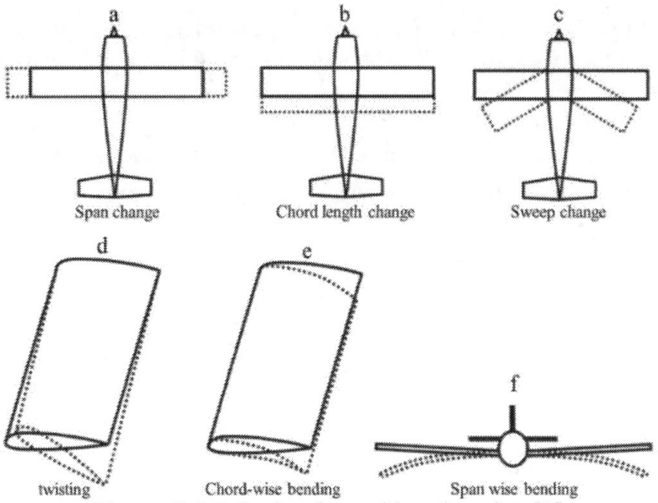

Figure 1. Methods of morphing the wing [1].

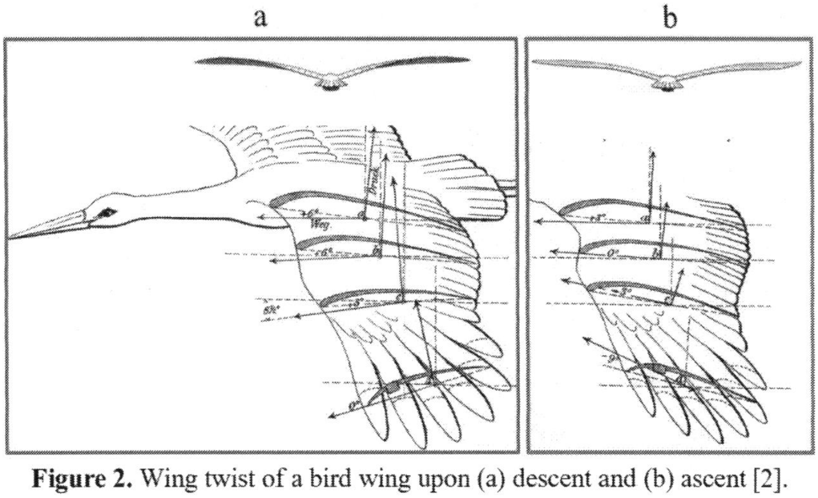

Figure 2. Wing twist of a bird wing upon (a) descent and (b) ascent [2].

2. THEORY OF WING TWIST USING TWO PIPES [5]

In general, when we twist a rod whose cross section is homogeneously hollow and round, the rate of the twist angle θ [rad/m] is a constant value. However, when we twist a rod that consists of an open- and closed-section member, the rate of the twist angle is different in each section. This means that we can design a distribution of twist angles arbitrarily. One example of wing twist with a different attack angle is shown in Fig. 3. The wing tip side has a larger rate of twist angle than the wing root side in the figure on the right (Fig. 3). We can select the rate of the attack angle according to its purpose.

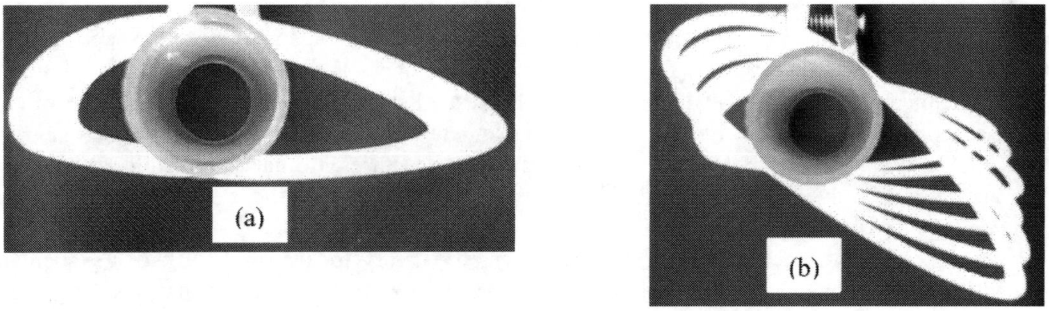

Figure 3. Example of wing twist: (a) untwisted and (b) twisted wing.

In this study, we propose a new morphing beam concept for wing twist using a specially designed beam consisting of the two pipes shown in Fig. 4. One pipe is a thick closed-section member and carries the bending moments. The other pipe is a variable-shape member with open- and closed-section parts. The ribs connect to the beam at the closed sections. To carry out wing twist, the attack angles of the ribs are changed by twisting the beam. The combination of open and closed sections is designed to achieve an objective distribution of twist angles. The ends of these pipes are connected to each other at the wing tip side, and the other ends are connected to the airplane body or motor, which provides a torque to the member at the wing root side.

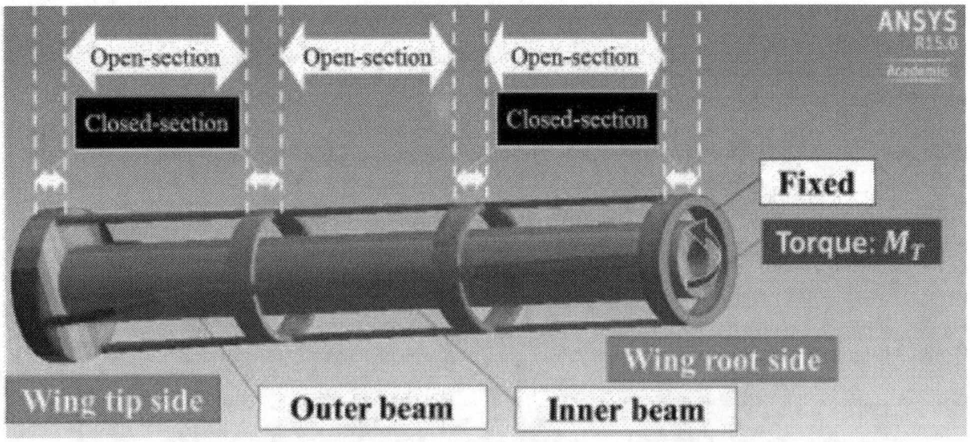

Figure 4. Image of a beam composed of an open- and closed-section member.

2.1 Derivation of twist theory

Now, we consider the torque at the open-section member, which consists of several beams. We can assume that the torque is working as shown in Fig. 5 at the open-section member. In order to consider the deformation of an open section, the torque is divided into horizontal and vertical components, as shown in Fig. 5(b). The forces and moments represent the components (Fig. 5(c)).

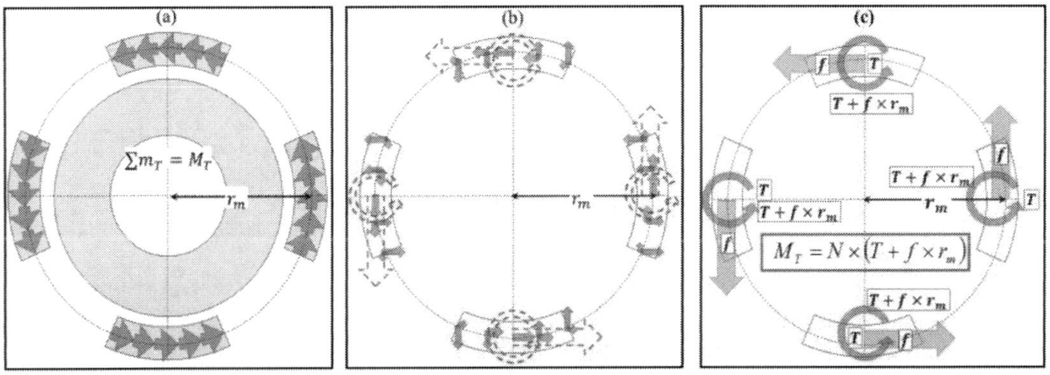

Figure 5. (a) Torque at open-section member, (b) analysis of torque in vertical and horizontal directions, and (c) corresponding forces and moments.

Therefore, the torque M_T is composed of the moment caused by the forces and the twist moment of each beam as follows:

$$M_T = N \times (T + f \times r_m) \tag{1}$$

Here, T denotes the twist moment, f is the total force in each circumferential direction, r_m is the radius of the morphing beam, and N is the number of beams. Next, we consider the boundary conditions shown in Fig. 6.

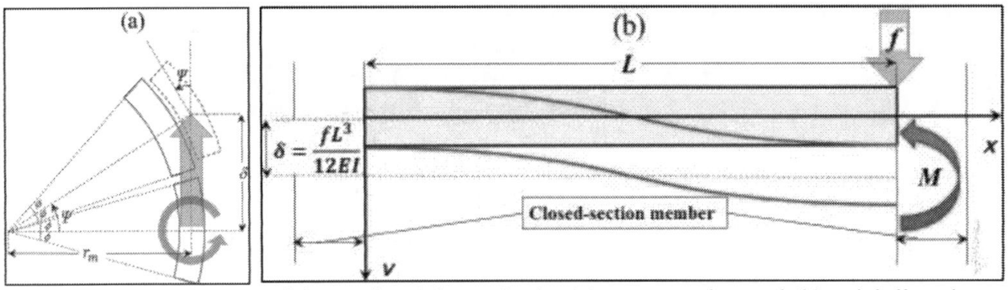

Figure 6. Boundary conditions of each beam in the (a) cross section and (b) axial direction.

Figure 6 indicates that the twist angle ψ of each beam is the same as the twist angle of the open section, which occurs by the rotation shown in Fig. 5. In addition to this constraint, the deflection of the beam tip δ must correspond to the twist deformation. Therefore, the following condition is obtained:

$$\psi = \tan^{-1}\left(\frac{\delta}{r_m}\right) \cong \left(\frac{\delta}{r_m}\right) \tag{2}$$

$$\delta = \frac{fL^3}{3\mathrm{EI}} - \frac{ML^2}{2\mathrm{EI}} = \frac{fL^3}{12\mathrm{EI}} \tag{3}$$

Here, L is the length of the open section, E is the Young's modulus, and I is the moment of inertia. Then, by substituting Eqns. (2) and (3) into the Saint Venant torsion formula, the following is obtained:

$$T = \frac{GJ\psi}{L} = \frac{GJL^2 f}{12\mathrm{EI}r_m} \tag{4}$$

Here, GJ is the torsional rigidity . According to Eqns. (1) through (4), the twist angle of the open-section member can be obtained. Additionally, the twist angle at the closed section can be approximated by the

Saint Venant torsion formula. Therefore, the twist angle of the special beam composed of an open- and closed-section member is calculated as follows:

$$\psi_{tip} = \sum \psi_o + \sum \psi_c = \sum_{i=1}^{k} \frac{L^3}{12EIr_m} \times \frac{M_T}{N\left(r_m + \dfrac{GJL^2}{12EIr_m}\right)} + \sum \psi_c = \frac{M_T L_{oi}^3}{N\left(12EIr_m^2 + GJ_o L_{oi}^2\right)} + \sum_{i=1}^{k+1} \frac{M_T L_{ci}}{GJ_{ci}} \quad (5)$$

Here, the subscript o denotes the open-section member, and c denotes the closed-section member. Next, we examine the rigidities of the members. One beam of the open section is shown in Fig. 7. The polar moments of inertia of the closed-section member J_c and open-section member J_o are as follows:

$$J_c = \frac{\pi}{32}\left(d_2^4 - d_1^4\right), \; J_o = 2\phi r_m t^3 k_2 \quad (6)$$

Here, ϕ is the angle of each open-section member. By approximating the cross section of the beam (part of the ring) as a rectangle, as shown in Fig. 7, k_1 can be calculated by the following equation, and the parameter k_2 is defined as follows [6]:

$$k_2 = \frac{1}{3}\left(1 - \frac{192 \tanh \dfrac{\pi}{2} k_1}{\pi^5 k_1}\right), \; k_1 \equiv \frac{b}{a} = \frac{2r_m \sin\phi}{t} \quad (7)$$

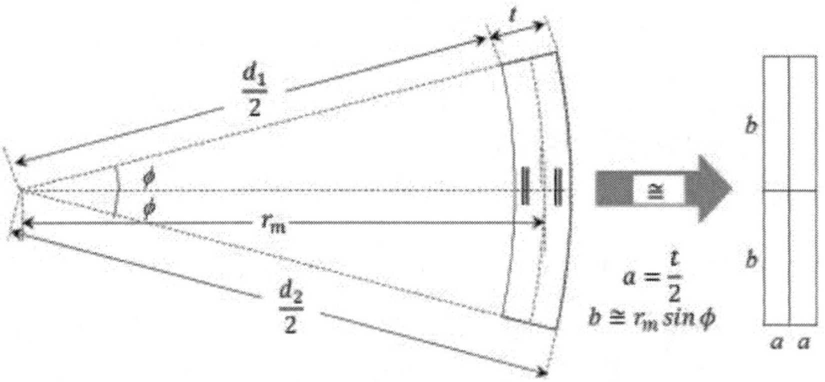

Figure 7. Approximating cross section of the beam (part of the ring) as a rectangle.

In general, the moment of inertia of a ring part is known as follows [6]:

$$I = r_m^3 t(\phi - \sin\phi\cos\phi) \quad (8)$$

2.2 Validation of theory through numerical analysis

In order to validate the derived theory, numerical simulations were performed. The twist angle of the rod was investigated through FEA while changing the open-section member. The results were compared with those calculated by the derived theory. The results are shown in Fig. 8. We set the closed-section member to 5 [mm] in length at both ends of the morphing beam, r_m=9.5 [mm], and t_o =1 [mm]. The size of the mesh used in the FEA was set to 2 x 10^{-4} [m], and the torque applied to the model was 10^{-3} [Nm]. The material properties are obtained from tensile tests as follows: ρ=834.0 [kg/m^3], E=1650 [MPa], ν=0.441, and G=572.8 [MPa].

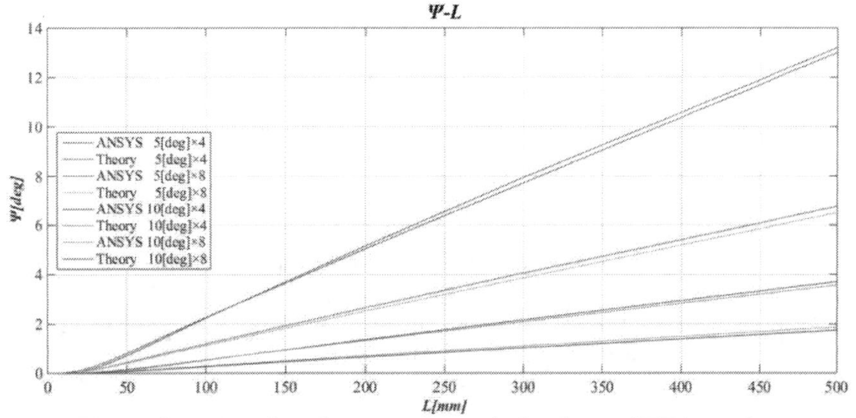

Figure 8. Comparison between theoretical value and FEA results.

Figure 8 shows a comparison between the theoretical value and the FEA results. According to Fig. 8, we can make the following remarks:

(1) The results of the numerical simulations are in good agreement with the results calculated by the derived theory. This indicates that the derived theory is applicable for the design of a morphing beam that achieves wing twist.
(2) The twist angle increased linearly (conforming to Saint Venant twist) for the case where the beam is long.
(3) The twist angle will be a monotonically increasing cubic function (conforming to bending) for the case where the beam is short.
(4) Not only the total angle of the open-section member but also the number of partitions (N) at the open-section member are related to the twist angle.

3. DESIGN METHOD

In the morphing beam system for wing twist, the distribution of the twist angle could be set arbitrarily to meet its purpose. Therefore, we developed a design method for the morphing beam to achieve the target twist angles based on the theory derived in the previous section. According to the previous section, the twist angle of the special beam composed of an open- and closed-section member depends on many parameters. Therefore, there are ways to determine the parameters, and in this study, the authors developed the design method as follows:

(1) The target twist angle (ψ_{tip}) at the wing tip side and the distribution of twist angles are set.
(2) The number of open-section members is determined by considering the size of the wing.
(3) The difference in the twist angles in the open sections is calculated from the distribution of the twist angles.
(4) The applied torque is determined by considering the performance of the motor used in the morphing wing system.
(5) The length of each section L_i, the moments of inertia of each section I_i, the number of beams N, and the angle of each open-section member ϕ_I can be the design variables. Here, we employ the angles of the open-section members as the design variables and set the other parameters by considering the size and weight of the wing and the ease of manufacture.
(6) The angle of each open-section member is calculated in order to achieve the target twist angles with the predetermined design variables.

4. EXPERIMENTAL DEMONSTRATION

To validate the twist angle of the specimen, we demonstrated the twisting of the beam designed using the developed method. Figure 9 shows the experimental apparatus of this demonstration, and an image showing the details of the specimen and a demonstration of the twisting is presented in Fig. 10. The functions of the devices are as follows:

(1) Camera: Monitors and records the behavior of the beam
(2) Beam: The object to be observed
(3) Driver: Interfaces stepping motor with controller and power supply
(4) Vise: Fixes the motor so that it does not slip out of place
(5) Controller: Interfaces the driver with PC2 and transmits commands to the stepping motor
(6) Stepping motor: Applies torque to the beam
(7) Power supply: Supplies electricity to the stepping motor
(8) PC1: Remotely controls the camera to take pictures
(9) PC2: Provides commands to the controller

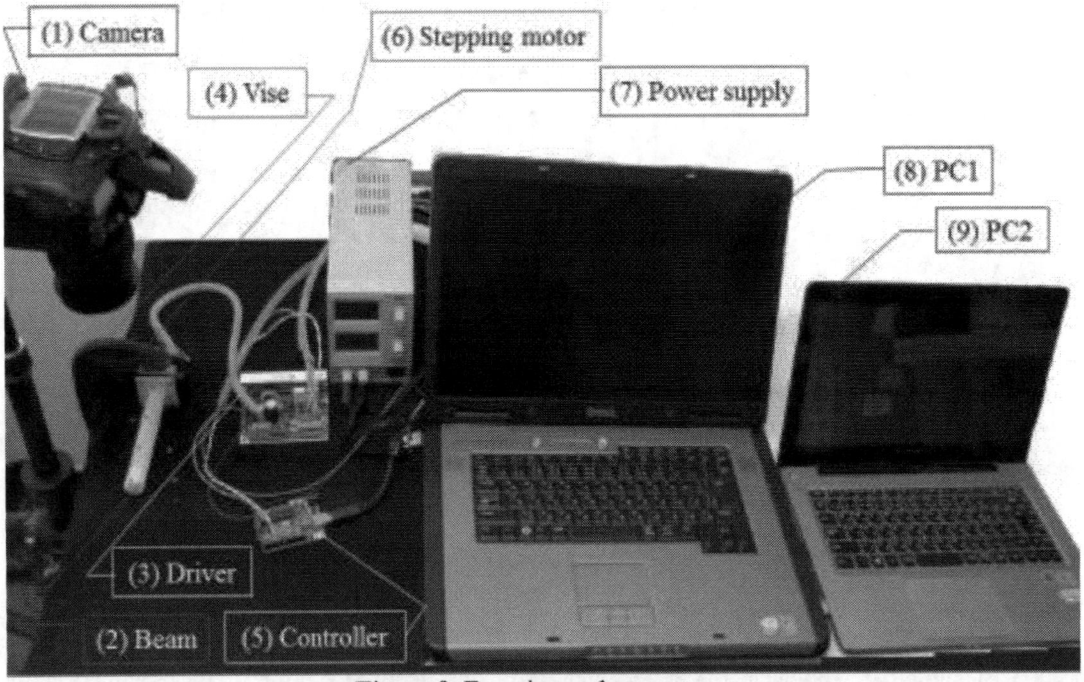

Figure 9. Experimental apparatus.

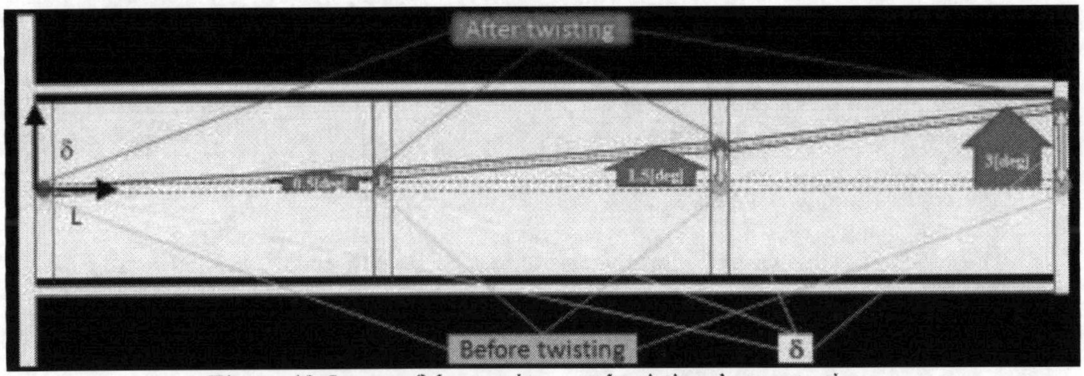

Figure 10. Image of the specimen and twisting demonstration.

4.1 Parameters of specimen

To validate the proposed design method, a morphing beam is designed using the proposed method, and the twist angles are evaluated through numerical simulations and experimental demonstration. The design constraints are as follows:

(c-1) Applied torque: $M_T < 0.1$ [Nm]
(c-2) Model size: less than 120 [mm] x 120 [mm] x 130 [mm]
(c-3) Thickness of the beam: $t \geq 0.8$ [mm]
(c-4) Twist angle at the wing tip: $\psi_{tip} \geq 3$ [deg]
(c-5) Total twist angle at the open-section member is 3 [deg], and those of the open-section parts are 1/2 [deg], 1 [deg], and 1.5 [deg] from the wing root side to the wing tip side.

Based on the above constraints, we designed the morphing beam as follows:

(1) Applied torque: $M_T = 0.02$ [Nm]
(2) The morphing beam has three open-section members and four closed-section members.
(3) There are four beams (N) in the open-section part.

The dimensions of the pipes are determined as follows by considering the manufacturing machine performance:

(4) Outer pipe: $d_1 = 20$ [mm], $d_2 = 18$ [mm], $\Sigma L = 120$ [mm] ($\Sigma L_O = 35$ [mm] x 3, $\Sigma L_C = 5$ [mm] x 3)
(5) Inner pipe: $d_1 = 16$ [mm], $d_2 = 5.45$ [mm], $L = 120$ [mm]
(6) Connect part: $d_1 = 20$ [mm], $d_2 = 5$ [mm], $L = 5$ [mm]

In order to meet these conditions, we designed each open-section member size ϕ to be 16.37 [deg], 12.86 [deg], and 11.15 [deg] from the wing root side to the wing tip side using the developed design method. The specimen we made using this method is shown in Fig. 11. The specimen consists of two parts: the outer pipe and the inner pipe (which is inserted into the outer pipe). They are connected at tip side.

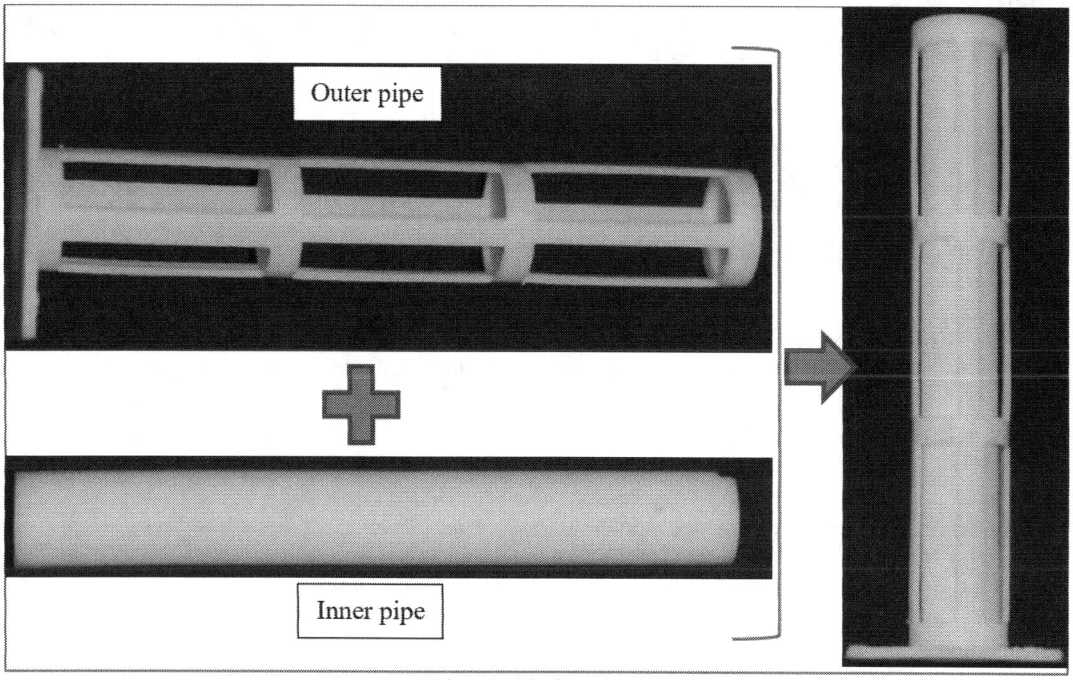

Figure 11. Specimen

4.2 Twist demonstration

To calculate the twist angle at each point, we marked four red points and observed the behavior of the twisting beam. The details of the results are as follows:

(1) Set up stepping motor
In order to twist 3 [deg] at the open-section member, according to Eqn. (5), the stepping motor should in theory rotate 3.045 [deg].

(2) Image processing -
Image processing method is shown in Figs. 12 and 13.

a. Import the original image to PC2.
b. Select one red point and remove the others.
c. Make a grayscale image by extracting the red elements and removing the blue elements from the image and binary image to prepare for calculating the center of gravity of the point.
d. Return to step a. and restart with another red point until the positions of all points are obtained.

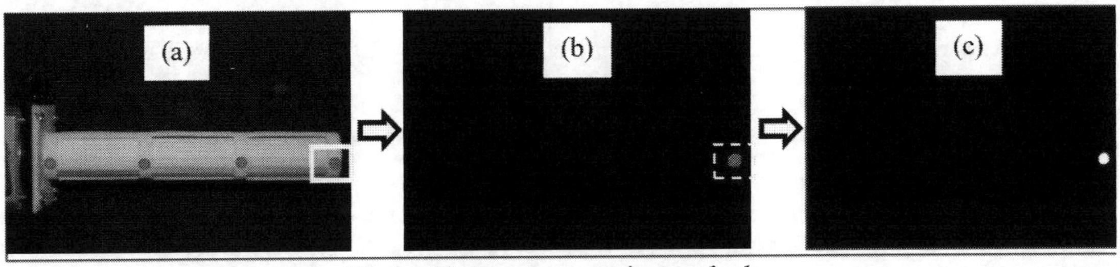

Figure 12. Image processing method:
(a) original image, (b) trimmed image, and (c) binary multilevel image.

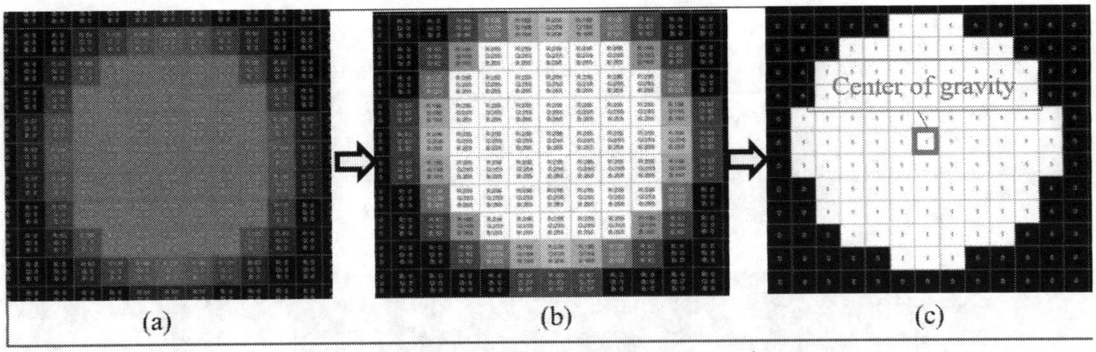

Figure 13. Details of binary image processing:
(a) multilevel image, (b) grayscale image, and (c) binary grayscale image.

(3) Calculate the twisting angle -
After this process, the authors calculated the twist angle as follows:
a. Calculate the center of gravity of each point using the binary image.
b. Compare the deflection of the center of gravity before twisting and after twisting, and then calculate δ (Fig. 10).
c. Calculate the twist angle at each point based on Eqn. (13).

$$\psi = \tan^{-1}\left(\frac{\delta}{d_2/2}\right) \tag{13}$$

4.3 Comparison of the twist angle

A comparison of the twist angle results for the experiment, theory, and numerical analysis is shown in Fig. 14. Furthermore, the authors analyzed four patterns. They are different from the results for the closed-section member.

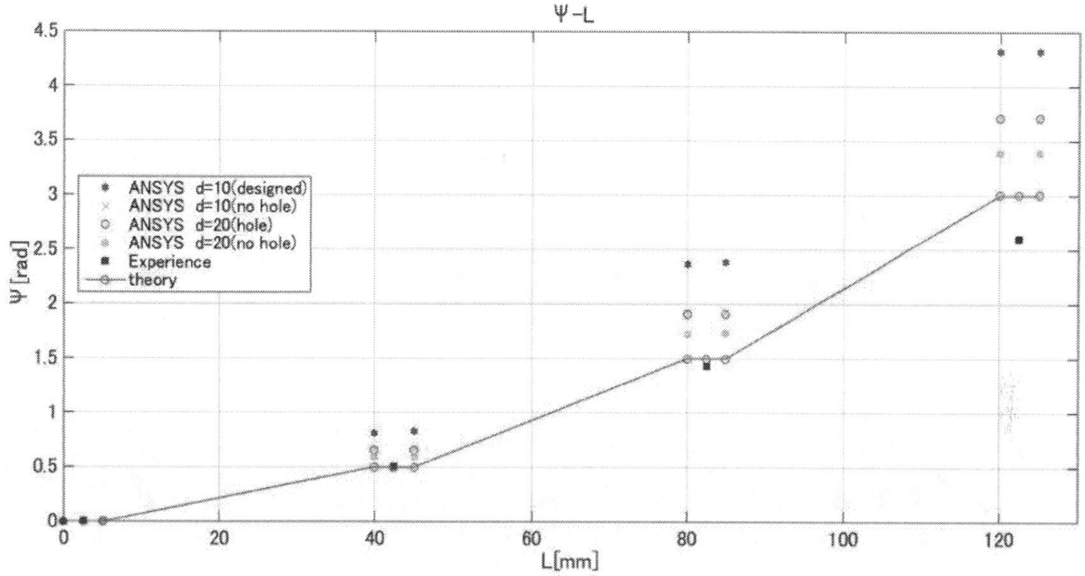

Figure 14. Comparison of twist angle results for the experiment, theory, and analysis.

According to Fig. 14, we can make the following remarks:

(1) The results of the experiment are in good agreement with the results calculated by the derived theory. This indicates that the derived theory is applicable for designing a morphing beam that achieves wing twist.

(2) Although the results of the numerical simulations tend to have larger values than the theory, the tendency of the behavior is similar to that of theory.

(3) The behavior of the numerical simulations depends on the closed-section member, and the higher the rigidity, the closer it behaves to theory.

These results clearly indicate the effectiveness of the developed method for designing a morphing beam for a wing twist system.

4.4 Consideration of the twist angle comparison

We can consider two points from the comparison of the twist angle results for the experiment, theory, and analysis.

(1) Accuracy of the specimen dimension of the specimen.
We designed the specimen according to the design method, but in reality, the dimensions of the specimen are not accurate. A comparison of the parameter 2ϕ is shown in Fig. 15.

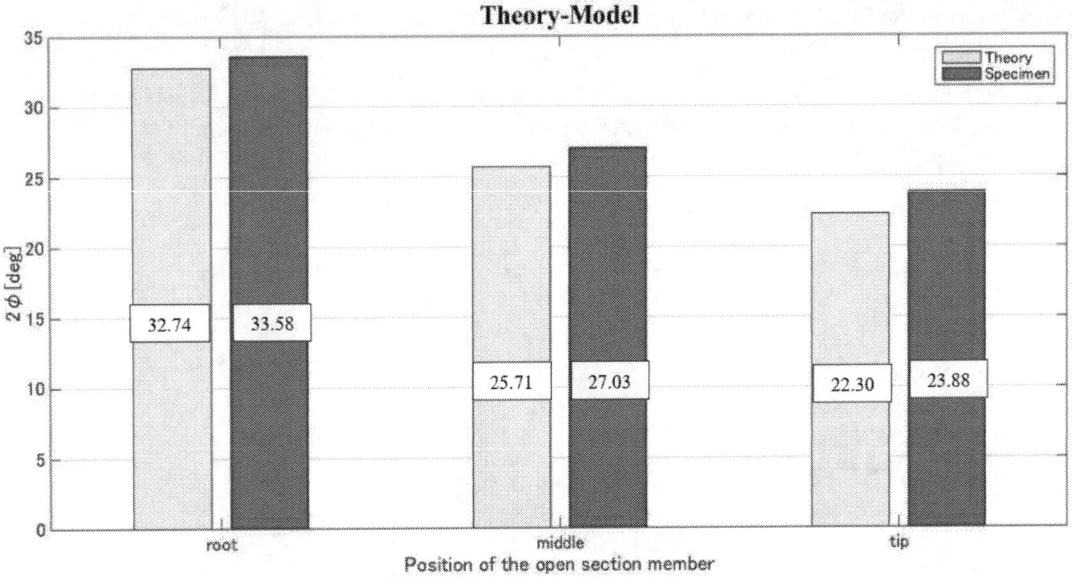

Figure 15. Comparison of the parameter 2ϕ

Figure 15 indicates that the specimen is thicker than the designed value at each open-section member. According to Eqns. (5) to (8), this means that the moment of inertia I and the polar moments of inertia at the open-section member J_o are larger than predicted by theory, and the twist angles ψ_o are smaller. This hypothesis also agrees with the twist experiment results.

(2) Effect of the closed-section member.
We developed our hypothesis based on Fig. 6(b) with the underlying premise that the closed-section member is sufficiently rigid. However, in reality, it differs somewhat from the ideal condition. Therefore, we could consider that this is the reason why numerical analysis shows a larger value than theory.

5. Conclusion

A new morphing beam concept consisting of two pipes was developed, and the effectiveness of the concept was demonstrated through numerical simulations and experiments. The details of the study are as follows:

(1) The theory for calculating the twist angle of an open section was derived.
(2) Some numerical simulations were performed, and the applicability of the theory to the design of a morphing beam that achieves beam twist was demonstrated through numerical simulations.
(3) According to the design method, the model was created and evaluated with a twist experiment. We validated the theory for the twist angle and the design method.

From these three points, a design method for achieving the target twist angles of the morphing beam was developed, and the effectiveness of the method was demonstrated by numerical simulation and experiment. This morphing beam concept is applicable to morphing wing or deformable space structures and this design method can contribute to wing design with the appropriate attack angles leading to enhanced flight performance. The developed morphing beam system can potentially make conventional airplanes lighter and more adaptive to changes in flight conditions.

ACKNOWLEDGMENTS

This work is partially supported by JSPS KAKENHI Grant Number 26249131.

REFERENCES

1. Sofla, A.Y.N., Meguid, S.A., Tan, K.T. and Yeo, W.K., "Shape morphing of aircraft wing: Status and challenges", *Materials and Design*, Vol. 31, 2010, 1284-1292.
2. Otto Lilienthal, Der Vogelflug als Grundlage der Fliegekunst.
3. Barbarino, S., et al., "A Review of Morphing Aircraft," Journal of Intelligent Material Systems and Structurs, vol.22, 2011.
4. Mistry, M., and Gandhi, F., "Design, fabrication, and benchtop testing of a helicopter rotor blade section with warp-induced spanwise camber variation," Journal of Inteligent Material Systems and Structures, vol. 26, 2015.
5. Atsushi, A., and Hiroaki, T, "Study on a morphing wing structure Using open section member", APISAT 2015
6. Terazawa, K. and Matsuura, Y., "Strength of materials (last part)", Kaibundo Ltd, Tokyo, 1967.

ICAST2015#073

Investigation of Relaxation Phenomenon in Piezoelectric Benders

Lanwei Zhou[1, 2*], VikramHrishikeshavan[2], Inderjit Chopra[2]

[1]State Key Laboratory of Mechanics and Control of Mechanical Structures, Nanjing, China
[2]Department of Aerospace Engineering, University of Maryland, College Park, MD, USA

Abstract

Swashplateless control of rotorcraft has motivated the development of piezoelectric actuators for primary and vibration control. For this, the piezoelectric actuator will have to operate under DC and AC excitation fields. It was observed the control authority of the piezoelectric bender varied significantly between DC and AC excitation fields. The mean tip displacement of clamped piezo bimorph with a steady tip load under AC excitation increases when compared with a short circuit or DC excitation condition. This phenomenon of relaxation under AC field is explored in this paper. Three kinds of hypotheses concerning Young's modulus, electrical condition and dynamic stiffness were introduced to explain the phenomenon, respectively. Experiments conducted showed that the change of stiffness is the main reason. The excitation field has a major effect on the stiffness, while excitation frequency not. To verify this hypothesis, theoretical model was established to predict such change. The simulation results showed that the stiffness of clamped bimorph decreased about 55% with an AC excitation of 3.94 KV/cm. The theoretical model could predict the change of stiffness well and had a good agreement with the experimental results. The model also indicates that the larger excitation may lead to negative stiffness and an unstable system.

1. INTRODUCTION

Due to swashplateless control of rotorcraft, piezoelectric actuators for primary and vibration control have made great progress [1-5]. Recent work investigated the feasibility of operating piezoelectric benders to actuate trailing edge flaps on a mach scaled rotor [6-7]. It was found that the control authority of the piezoelectric bender varied significantly between DC and AC excitation fields. An undocumented phenomenon called relaxation was observed in the experiments of piezoelectric benders where benders relax with time in the direction of an applied load. This has implications towards cyclic and collective control of a rotorcraft and hence, it is important to investigate this phenomenon.

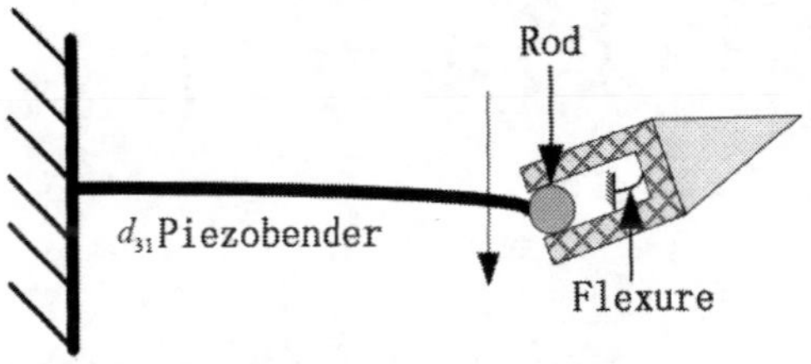

Figure 1. Piezoelectric bender with a steady load

*zhoulanwei@nuaa.edu.cn

When a low frequency signal was used to drive a piezoelectric bender (bimorph) with a steady load applied to the tip as shown in Fig. 1. The amplitude of the motion remained unchanged with load, while the mean tip displacement changed by around 100% more than expected from the measured stiffness with no electric field, as shown in Fig. 2. The change of the mean becomes larger with larger electric field.

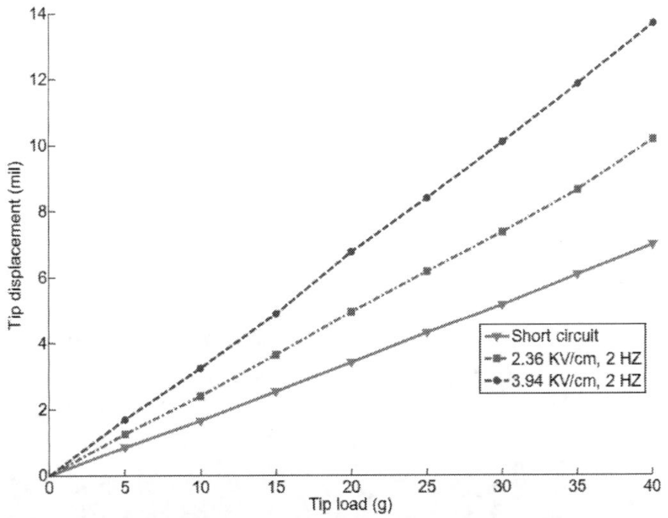

Figure 2. Commercial PZT-5H2 bimorph compliance with live cantilever

There may be three reasons that lead to the above phenomenon. Hypothesis 1 is that Young's Modulus of the PZT material changes with applied electric field. However, tests conducted on single plates and bimorph with DC voltage showed that this is not the case as shown in Fig. 3 [7]. Hypothesis 2 is that the electrical condition of the cantilevered portion of the bender. Inside the cantilever support, each half of the bender can be thought of as a d_{33} stack actuator and has a large blocked force and a small stroke. Ideally these forces cancel out. However, local variations may cause the root slope of the beam change. This change would relieve the large stress caused by not allowing the stack actuator to expand and therefore result in a lower perceived stiffness of the bender. In the dead cantilever hypothesis, electrical isolation of the cantilevered portion of the bender may eliminate relaxation. Hypothesis 3 is that the stiffness of piezoelectric bender with AC excitation changes with Young's modulus being invariant. The stiffness changes due to the vibration of piezoelectric bender under AC electric field.

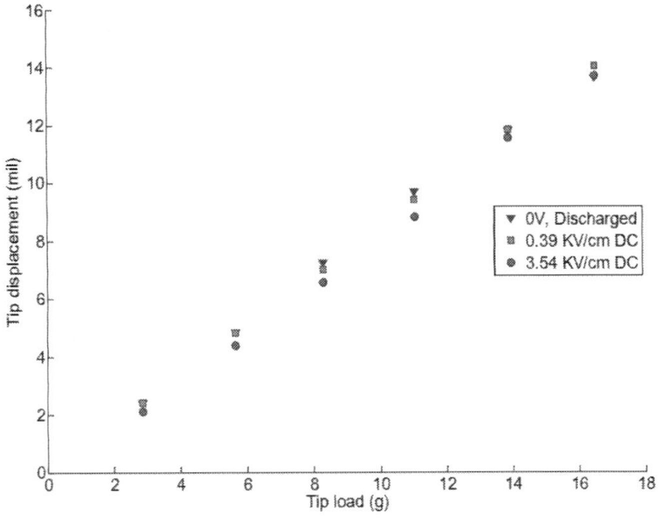

Figure 3. Single Cantilevered PZT-5K4 Plate Stiffness with High DC Electric Field [7]

To verify the latter two kinds of hypotheses, a couple of experiments were carried out. The measurement of stiffness of commercial bimorph under DC and AC excitation fields was conducted. These tests were with different cantilever condition: dead cantilever and live cantilever as shown in Fig. 4. The purpose of the paper is to fully investigate the phenomena shown in Fig. 2 by questioning its repeatability and exploring its causes.

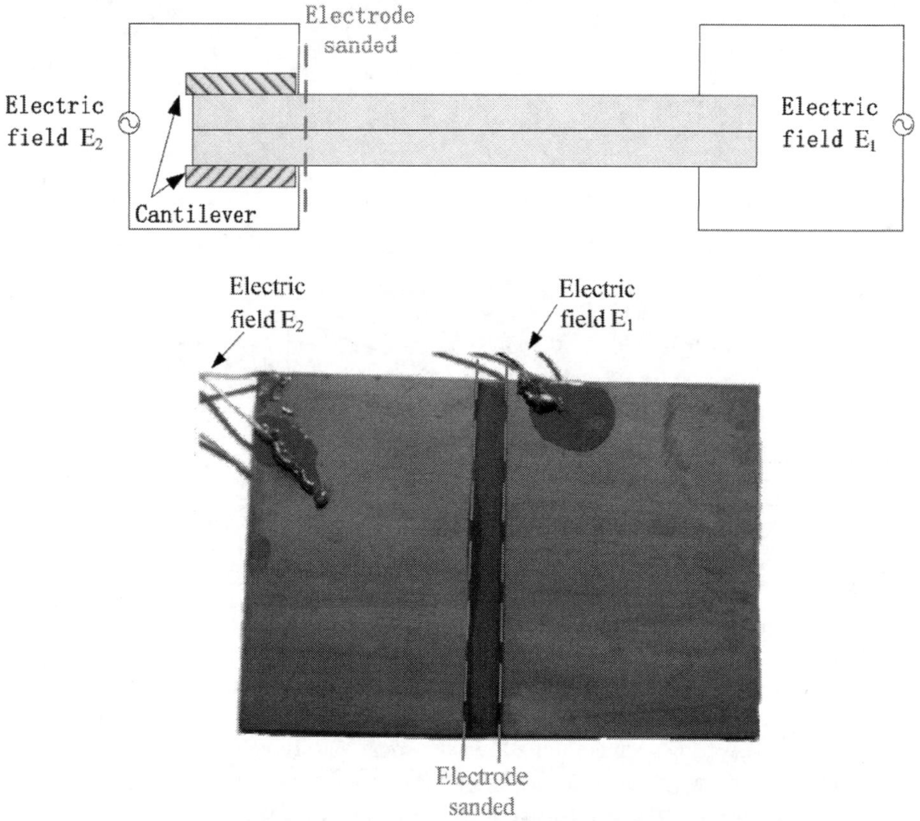

Figure 4. Cantilever condition of bimorph
(E_1=E_2, live cantilever; $E_1 \neq 0$, E_2=0, dead cantilever)

This paper is organized as follows. Section 2 presents the experiments conducted to verify the hypotheses and the preliminary conclusions. A new model for the stiffness of vibrating piezoelectric benders under AC excitation is given in Section 3. Finally, concluding remarks are given in Section 4.

2. EXPERIMENTS UNDER DIFFERENT CONDITION

Experiments on commercial bimorph under DC and AC excitation are conducted as shown in Fig. 5. A 1.5×1×0.0235 inch PZT-5H2 bimorph produced by Morgan Technical Ceramics was used in the experiment. Function generator and power amplifier were used to provide DC or AC excitation to the bimorph. The data obtained from laser displacement sensor was acquired through a NI Instruments data acquisition unit.

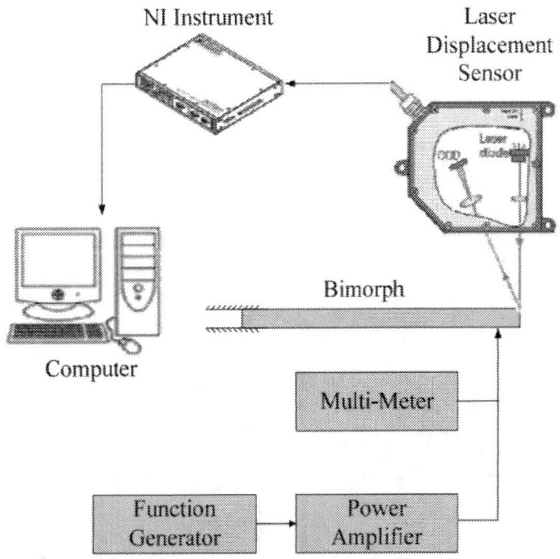

Figure 5. Schematic diagram of experiment

To verify the second hypothesis of electrical condition of the cantilevered portion, we tested the commercial bimorph under live and dead cantilever conditions, respectively. We first obtained a tip displacement of the bimorph in short circuit condition under tip load. Each measurement was taken after the electric field was applied for 5 seconds with the weight suspended from the tip. The tip response of the bimorph under low frequency AC excitation was recorded as the change in mean of the tip motion before and after the tip load was applied. The results showed that the tip displacement with live and dead cantilever conditions increased about 100% and 42% under 3.94 KV/cm, 2 Hz compared with the short circuit condition as shown in Fig.2 and Fig.6. This suggests that the dead cantilever hypothesis may not be responsible for the compliance change. Based on this hypothesis, electrical isolation of the cantilevered portion of the bender should have eliminated relaxation and the compliance of clamped bimorph with dead cantilever should have remained unchanged. The experiment proved that the cantilever condition was not the main reason for the increase of the compliance of the bimorph under AC excitation.

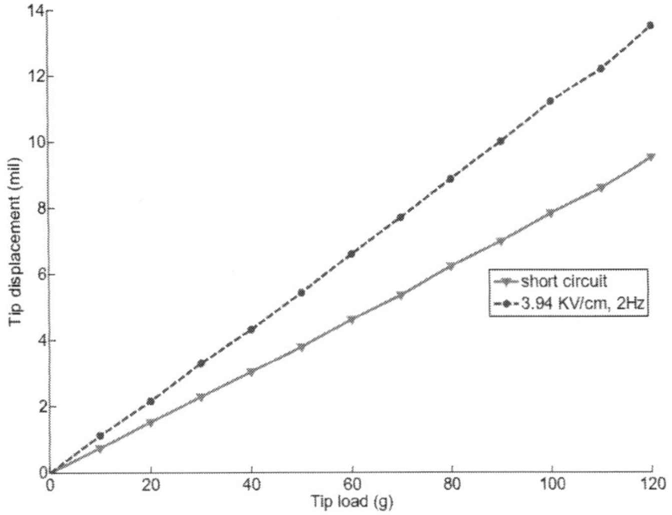

Figure 6. Commercial PZT-5H2 bimorph compliance with dead cantilever

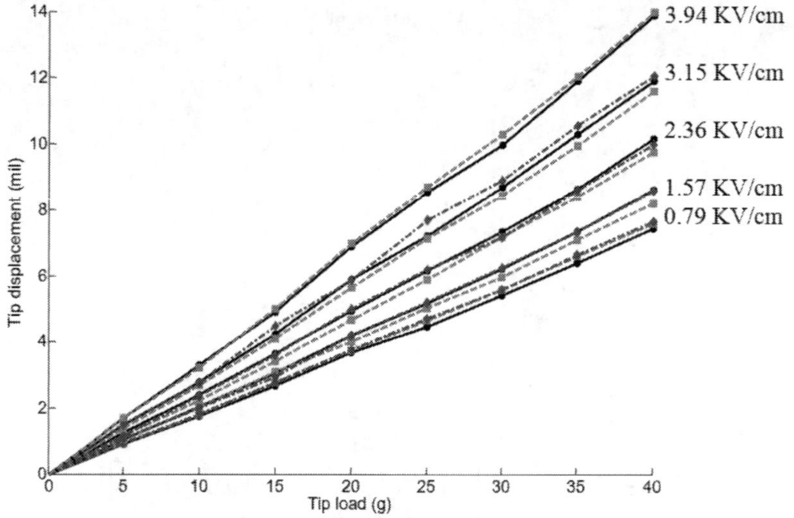

Figure 7. Commercial PZT-5H2 bimorph compliance with live cantilever (——— 2Hz, - - - - - - - -10Hz, - · - · - · -20Hz)

Additionally, as shown in Fig. 7, we tested the bimorph under AC excitation under live cantilever condition with different frequencies (2 Hz, 10 Hz and 20 Hz). Similar to the previous, the compliance grows with increasing the electric field. However, the compliance remains almost the same with different frequency. The results clearly showed that frequency of excitation has little effect on the compliance of clamped bimorph and that the compliance of the bimorph is only dependent on the magnitude of electric filed.

A comparison of stiffness of the bimorph with live cantilever under DC and AC excitation was presented in Fig. 8. The stiffness of bimorph under DC excitation changes little and remains almost the same with increasing DC electric field, while the stiffness of bimorph decreases with increasing AC electric field.

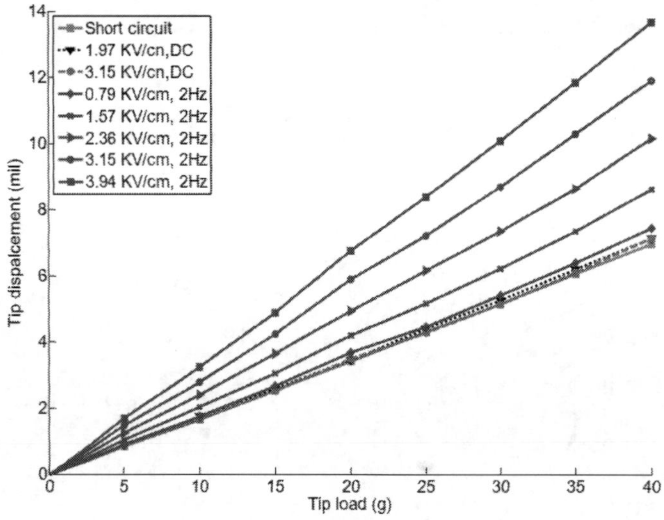

Figure 8. Commercial PZT-5H2 bimorph compliance with live cantilever

In addition, the HPP (half peak to peak) tip displacements of live-cantilevered bimorph under AC excitation with different frequencies were measured as shown in Fig.9. The results showed that the HPP displacement was nonlinear, which means that the stiffness changes under different electric field. Theoretically, the HPP tip displacement should be linear while the stiffness of the bimorph is constant under AC excitation. The HPP tip displacement increases nonlinearly proved that the stiffness decreases with the increase of electric filed. Interestingly, the growth rate of HPP tip displacement matches that of tip displacement under tip load with AC excitation. It means that the third kind of hypothesis should be right. The third kind of Hypothesis could explain the phenomenon shown in Fig.1 well and had no conflicts with the results got from former experiments.

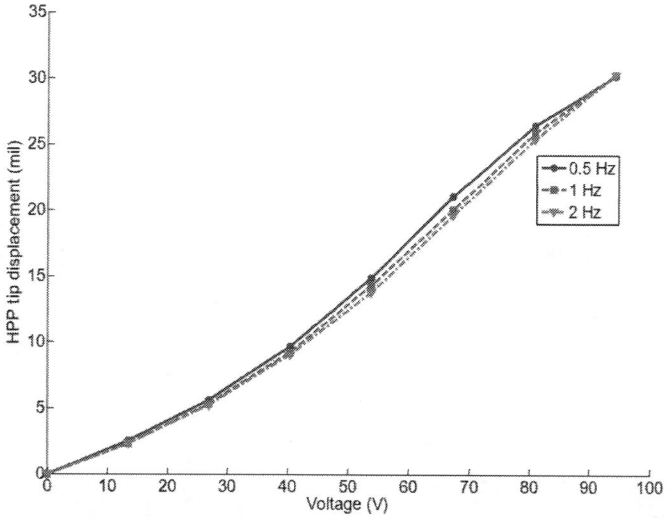

Figure 9. HPP tip displacement of clamped bimorph with live cantilever under AC excitation

To verify the third kind of hypothesis better, a theoretical model of clamped bimorph was established based on Hamilton's Principle in Section 3. Kinetic equation is given and the stiffness obtained from the model will be compared with the experimental results.

3. THEORETICAL MODEL

A clamped bimorph under AC excitation is shown in Fig. 10. A 1.5-inch-long by 1-inch-wide by 0.0235-inch-thick bimorph was bonded at one end with the other end free.

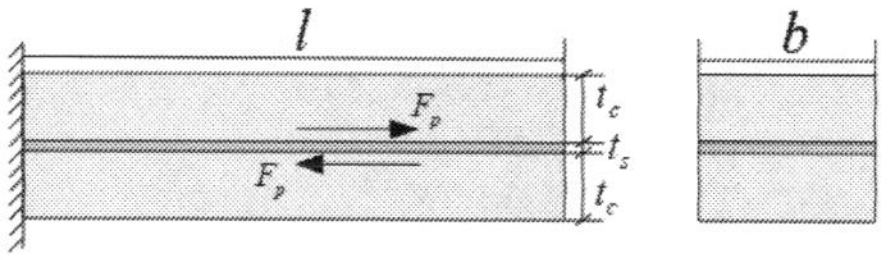

Figure 10. Schematic diagram of clamped bimorph

The axial deflection at the top surface of the metal layer is given by [8]

$$\Delta l_s = \epsilon^s l = -\frac{M}{EI}\frac{t_s}{2} l \tag{1}$$

where M is the induced moment under electric field, EI is the flexural stiffness of the bimorph. They are given as follows

$$M = E_c b_c t_c (t_c + t_s)\Lambda \tag{2}$$

$$EI = \frac{1}{3}\sum_1^4 b_k E_k (h_{k+1}^3 - h_k^3)$$
$$= 2\left\{\frac{1}{3}E_c b_c[(t_c + t_s/2)^3 - (t_s/2)^3] + \frac{1}{3}E_s b_s (t_s/2)^3\right\} \quad (3)$$

where E_c is the Young's modulus of PZT-5H2 bimorph. l, b_c, t_c are the length, width and thickness of single film in the bimorph, respectively. t_s is the thickness of metal layer in the bimorph.

The change in length of each piezoelectric actuator can be found by

$$\Delta l_c = \left(-\Lambda + \frac{F_p}{E_c A_c}\right) l = \left(-d_{31}\frac{V}{t_c} + \frac{F_p}{E_c A_c}\right) l \quad (4)$$

where V is the voltage applied on the bimorph, $E_c A_c$ is the extensional stiffness of a single piezoelectric film, F_p is extensional force on either surface produced by piezos.

For displacement compatibility, $\Delta l_s = \Delta l_c$.

$$-\frac{E_c b_c t_c (t_c + t_s)\Lambda}{EI}\frac{t_s}{2} l = \left(-d_{31}\frac{V}{t_c} + \frac{F_p}{E_c A_c}\right) l \quad (5)$$

This results in a net actuator force

$$F_p = \left(d_{31}\frac{V}{t_c} - \frac{E_c b_c t_c (t_c + t_s)\Lambda}{EI}\frac{t_s}{2}\right) E_c b_c t_c \quad (6)$$

where $\Lambda = d_{31} V/t_c$.

Total potential energy of the bimorph could be expressed as [9]

$$U(t) = \frac{1}{2} EI \int_0^l \left(\frac{\partial^2 w}{\partial x^2}\right)^2 dx + F_p \times \frac{1}{2}\int_0^l \left(\frac{\partial w}{\partial x}\right)^2 dx \quad (7)$$

Total kinetic energy of the bimorph is given by

$$T(t) = \frac{1}{2}(\rho A)_{eq} \int_0^l \left(\frac{\partial w}{\partial t}\right)^2 dx \quad (8)$$

where $(\rho A)_{eq} = b(\rho_s t_s + 2\rho_c t_c)$.

Inserting Eq. (7) and (8) into Lagrange Equation

$$L(t) = T(t) - U(t) \quad (9)$$

Inserting Eq.(7-8) into Eq. (9) and using

$$\frac{d}{dt}\left(\frac{\partial L}{\partial \dot{w}}\right) - \frac{\partial L}{\partial w} = 0 \quad (10)$$

The dynamic equation of vibrating bimorph is

$$(\rho A)_{eq}\frac{\partial^2 w}{\partial t^2} + EI\frac{\partial^4 w}{\partial x^4} - F_p\frac{\partial^2 w}{\partial x^2} = 0 \tag{11}$$

When solving vibration equation of Eq. (11) using assumed mode method, the tip displacement could be expressed as

$$w(x,t) = \sum_{i=1}^{n} \Phi_i(x) q_i(t) \tag{12}$$

$$\sum_{j=1}^{n} M_{ij}\, \ddot{q}_j(t) + \sum_{j=1}^{n} K_{ij}\, q_j(t) = 0 \tag{13}$$

Where $M_{ij} = (\rho A)_{eq}\int_0^l \Phi_j(x)\Phi_i(x)dx$, $K_{ij}=EI\int_0^l \Phi_j(x)\Phi_i^{(4)}(x)dx - F_p\int_0^l \Phi_j(x)\Phi_i^{(2)}(x)dx$.

For cantilever bimorph, $\Phi_i(x)$ could be expressed as $\Phi_i(x) = \sin(i\,\pi\, x/2l)$.Thus the ratio of stiffness with and without AC excitation is

$$\frac{k}{k_s} = 1 - \frac{\sum_{i=1}^{n}\sum_{j=1}^{n} Fp\int_0^l \Phi_j(x)\Phi_i^{(2)}(x)dx}{\sum_{i=1}^{n}\sum_{j=1}^{n} EI\int_0^l \Phi_j(x)\Phi_i^{(4)}(x)dx} \tag{14}$$

From Eq. (14), we can see that the stiffness of the clamped bimorph changes with and without AC excitation and the ratio of stiffness with and without AC excitation is dependent on amplitude of the excitation,

Expanding Eq. (12) for $n = 1$, we could obtain the ratio of stiffness with and without AC excitation

$$\gamma = 1 - \frac{Fp\int_0^l \Phi_1(x)\Phi_1^{(2)}(x)dx}{EI\int_0^l \Phi_1(x)\Phi_1^{(4)}(x)dx} \tag{15}$$

When the load is applied to the tip of the bimorph with AC excitation, the tip displacement is given as

$$w = \frac{P}{K_{11}} = \frac{P}{EI\int_0^l \Phi_1(x)\Phi_1^{(4)}(x)dx - Fp\int_0^l \Phi_1(x)\Phi_1^{(2)}(x)dx} \tag{15}$$

As an illustration, the clamped bimorph with PZT-5H2 piezoelectric actuator attached is considered with the geometric and material constants listed in Table 1.

Table 1 Geometrical and material properties of the bimorph

	Piezoelectric film	**Metal layer**
Effective length(inch)	1.3125	1.3125

Width (inch)	1	1
Thickness (inch)	0.02	0.0035
Young's Modulus (GPa)	66	100

Table 2 Ratio of stiffness with and without AC excitation under live cantilever

	Short circuit	20V	40V	60V	80V	100V
Stiffness	1685.52	1593.25	1380.07	1172.54	1009.16	816.03
Ratio of stiffness by experiment	1.0000	0.9453	0.8188	0.6957	0.5987	0.4841
Ratio of stiffness by theoretical model	1.0000	0.8911	0.7823	0.6734	0.5646	0.4557

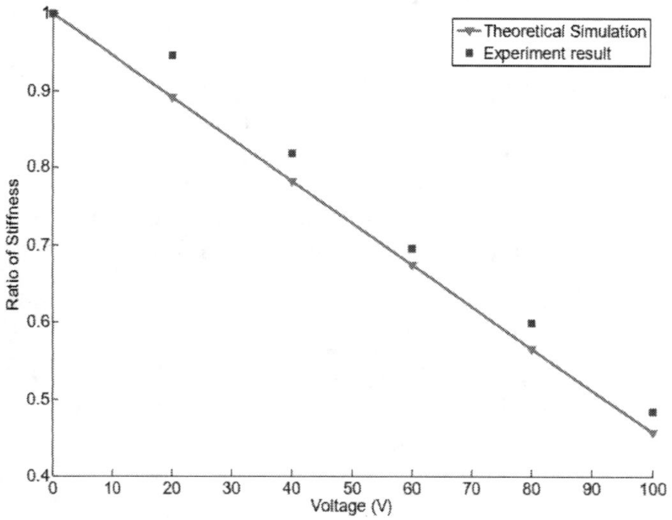

Figure 11. Ratio of stiffness with/without AC excitation under live cantilever

The experiments were conducted with different electric fields. To avoid problems with de-polarization of PZT, [10], the maximum voltage applied in the experiments was 100 volts. The results are included in Table 2 and Fig. 11. The results showed that the stiffness of theoretical model has a good agreement with that obtained from experiments. When the voltage applied to the bimorph increases, the stiffness decreases, which results in a growth of the tip displacement. It also means that when the excitation increases continually, the stiffness would decrease to less than zero. The vibrating piezoelectric bender would become unstable.

4. SUMMARY AND CONCLUSIONS

Piezoelectric actuators for primary and vibration control play a significant role in swashplateless control of rotorcraft. The control authority of the piezoelectric bender varied significantly between DC and AC excitation fields. An increase of compliance of clamped bimorph under low frequency AC excitation was observed during the experiments. Several kinds of hypotheses were introduced to explain the phenomenon. To verify the proposed hypotheses, a series of experiments were designed and conducted. The stiffness was measured under different excitation condition. The results showed that the compliance of clamped bimorph increased because of vibration of piezoelectric bender caused by AC excitation. A theoretical model was established to predict the change of stiffness under AC excitation. Good agreement with test results confirmed the validity of model.

ACKNOWLEDGMENTS

This work was funded by China Scholarship Council and Funding of Jiangsu Innovation Program for Graduate Education (Grant No.CXZZ13_0148) and the Fundamental Research Funds for the Central Universities. The authors would like to appreciate the valuable advice and help of Dr. Anand Saxena in Department of Aerospace Engineering, University of Maryland, College Park.

REFERENCES

1. Spangler, R. and Hall, S., "Piezoelectric Actuators for Helicopter Rotor Control", Proceedings 33rd Structures, Structural Dynamics and Materials Conference, Long Beach, CA, 1990.
2. Fulton, M. and Ormiston, R., "Hover Testing of a Small-Scale Rotor with On-Blade Elevons", *Journal of the American Helicopter Society*, 46(2):96- 106, April 2001.
3. Straub, F, et. al, "Smart Material-Actuated Rotor Technology – SMART". *Journal of Intelligent Material Systems and Structures,* vol. 15, no. 4, 1996. pp. 249-260
4. Straub, F.K., Ngo, H.T., Anand, V. and Domzalski, D.B. "Development of a Piezoelectric Actuator For Trailing Edge Flap Control of Full Scale RotorBlades", *Smart Materials and Structures*, vol. 10, no. 1, 2001, pp. 25-34.
5. Roget, B. and Chopra, I., "Wind-Tunnel Testing of Rotor with Individually Controlled Trailing-Edge Flaps for Vibration Reduc-tion", *Journal of Aircraft*, Vol. 45, No. 3, May-June 2008.
6. Chopra, I., and P. Copp. "Continued development of a mach scale swashplateless rotor with integrated trailing edge flaps.", *Annual Forum Proceedings-American Helicopter Society*. Vol. 64. No. 1. American Helicopter Society, May 2008.
7. Copp, P., "Swashplateless Helicopter Experimental Investigation: Primary Control with Trailing Edge Flaps Actuated with Piezobenders", Ph.D. Dissertation in Aerospace Engineering, University of Maryland, 2012.
8. Chopra, Inderjit and Jayant Sirohi. *Smart Structures Theory*. 1st ed. Cambridge: Cambridge University Press, 2013. *Cambridge Books Online*. Web. 08 September 2015.
9. G. Rezazadeh, M. Fathalilou, R. Shabani. Static and dynamic stabilities of a microbeam actuated by a piezoelectric voltage, *Microsystem Technologies*, 15 (12) (2009), pp. 1785–1791
10. Technical Publication TP- 245 Cantilever Mounted PZT 5A Bimorphs, *http://www.morgante chnicalceramics.com.*

ICAST2015 #01

Fiber Optic Shape Sensor for Morphing Wing Trailing Edge

M. Ciminello[1], S. Ameduri[1], I. Dimino[1], A. Concilio[1]

[1] Smart Structures Laboratory, Centro Italiano Ricerche Aerospaziali, CIRA, Capua (CE), Italy

Abstract

The FBG-based transducers, herein presented, aim at monitoring the actual shape of adaptive trailing edge device (ATED). Compared to conventional systems, the transducers allow a more fail-safe measurements, by taking advantage of a supporting structure, hosting FBG, whose properties may be tailored depending on the architectural requirements and structural constraints, acting as strain modulator. Measuring direct strain may, in fact, be difficult because of the large deformations occurring in morphing elements. A modulation transducer is then necessary to keep the measured strain inside the allowed range. In this application, chord-wise transducer device is a cantilevered beam sliding trough the spars and copying the camber-line of the ribs. A theoretical model describing the system behavior is implemented through a dedicated experimental campaign.

1. INTRODUCTION

Improving aircrafts efficiency is one of the key elements of Aeronautics. Modern aircraft possess many advanced functions, such as good transportation capability, high Mach number, high flight altitude and increasing rate of climb. However no aircraft has possibility to reach all of this optimized performance in a single airframe configuration. The aircraft aerodynamic efficiency varies considerably depending on specific mission and on environmental conditions within which the aircraft must operate. Structures that morph their shape to fit the best environmental condition, may at first seem like the stuff of science fiction, but take a look at nature and lots of examples of plants and animals that adapt to their environment would arise. In order to ensure both the controllable and the static robustness of such complex structural systems, a monitoring network has to verify the effectiveness of the given control commands together with the elastic response. In order to achieve this kind of information, the use of FBG sensors network is, in this work, proposed.

The sensor network is able to measure morphing structures shape which may show large, global displacements due to non-standard architectures and materials adopted. Chord-wise variations may allow setting and chasing the best lay-out as a function of the particular and evolving reference state, always targeting best aerodynamic performance. A good solution for this scope are optical sensors; in fact while keeping a few of drawbacks of the classical systems (like cabling, continuous deployment, and so on), fibre optic sensors may lead to a dramatic reduction of the wires mass and weight thanks to an extreme multiplexing capability. Furthermore, the use of the "light" as "information carrier", permits dealing with nimbler, non-shielded wires and avoids any kind of interference with the on-board instrumentation.

The herein presented FBG system, aims at monitoring the actual shape of a morphing wing trailing edge, conceived and developed in Saristu (Smart Intelligent Aircraft Structures) European project. The original network is constituted of two different solutions, running a morphing wing section both chord-wise and span-wise. While span-wise sensors usually undergo traditional deformation values (10^3-10^4 microstrains), chord-wise sensors can easily attain 10^4–10^5 microstrain. Direct strain may be then difficult to be measured because of the large achieved levels [1]. In the reported application, the modulator device is a cantilevered beam coupled with the camber line of an ATED [2]. In this paper, the design of the system is firstly dealt with. A theoretical model is then introduced, intended to be the base for a shape reconstruction algorithm, able to extrapolate the ATED displacement field from the FBG information.

2. PROBLEM DESCRIPTION

The sensors network herein described is aimed at monitoring the strain field arising inside a 2bay mock-up of an Adaptive Trailing Edge Device (ATED). The demonstrator total dimensions are approximately 800mm x 430mm. The system is made of finger-like, three-hinge ribs, driven by load-bearing actuators; a smooth external surface is guaranteed by the presence of a special deformable skin [3]. A first representation of the target device and is appearance is reported in Figure 1(a).

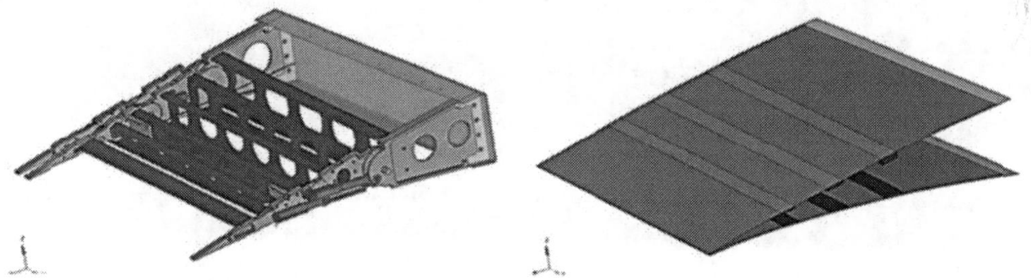

(a) (b)

Figure 1. Saristu dummy adaptive trailing edge (a) without skin, (b) upper and lower skin

Two systems were set up to monitor the ATED shape variations: span-wise and chord-wise networks. While the span-wise bending gives rise to usual deformations (magnitude between thousands and ten thousands of micro-epsilon), typical chord-wise values for morphing applications, range between 1 and 10%. In the studied case, design parameters may span around 5%. These values are concentrated into the morphing skin of the device, that is the element aimed at absorbing the major part of the occurring deformation while stretching and shrinking, Figure 1(b).

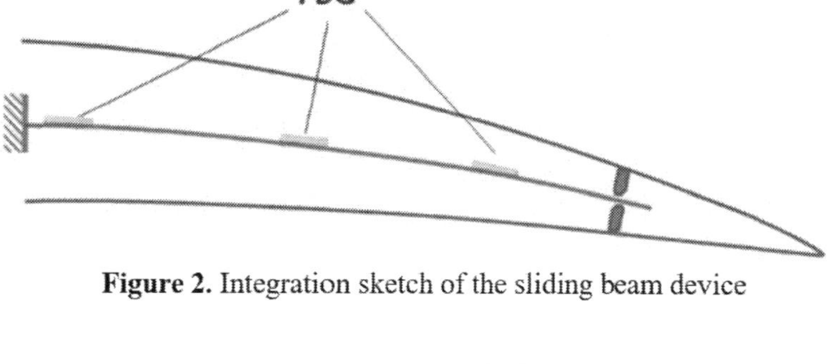

Figure 2. Integration sketch of the sliding beam device

An independent device is herein proposed to measure the camber variation through a non-direct measure of the structural deformation. It is made of a sliding (cantilever) beam hosting an integrated FO array sensors, free to slide trough the wing section spars (Figure 2).

3. FBG-BASED TRANSDUCER

The shape sensor devise is designed: it is made of a cantilevered beam able to slide through spar holes, purposely manufactured. The very ATED tip is hosted into a sized spar cavity, aimed at driving it during morphing. Under the assumption of zero friction between the ATED structure and the sliding beam, ensured in the practice by a teflon cover, only a vertical tip force acts on the latter.

3.1 Test article preliminary characterization

In order to evaluate the accuracy of the transducer system, according to simplified model, the beam is described according to the following equations [4]:

$$\varepsilon(x)=\frac{L_{ext}-L}{L}=\frac{\Delta R}{R} \tag{1}$$

$$\Delta R=\frac{h}{2}; \quad L=R\alpha \tag{2}$$

$$\varepsilon=\frac{h}{2L}\alpha \tag{3}$$

On the basis of Eq.(3), to ensure a deflection resolution of 0.5deg a 22με deformation accuracy is needed, leading to 1.5mm thickness and 300mm length. When the actuator moves downward, the beam bends and the FBG sensors detect the strain due to the stretching of the upper surface of the beam. The baseline configuration (*i.e.* 0 deg) is used for calibration reference signal. On the other hand, when the ATED moves upward, the beam is enforced to copy its shape and thus the sensors detect the compression of the beam upper surface.

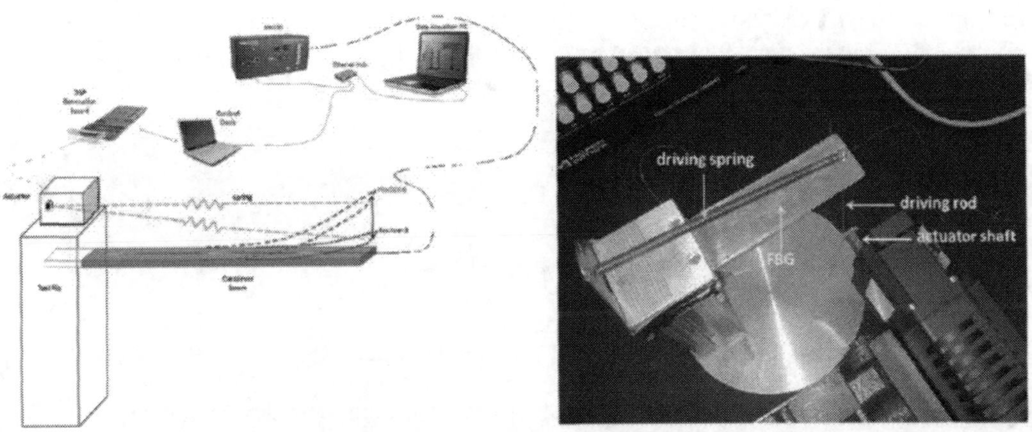

Figure 3. Sliding beam demo setup

A conceptual setup was realized with the aim of demonstrating the functionality of the just mentioned system, integrated with a dedicated control logic (see Figure 3). The elastic reaction given by the morphing surrounding structure is here conceptually represented by a 1D spring. The control action is realized by means of a spring driven by the actuator. The rotary actuator is driven by a target rotation. The beam is enforced to reach a pre-defined deflection and the corresponding strain distribution is measured by an FBG sensor. The FBG calibration curve is reported in Figure 4. The strain time evolution corresponds to a virtual ATED deflection of about +10deg in morphed up status and -10deg in morphed down status respectively.

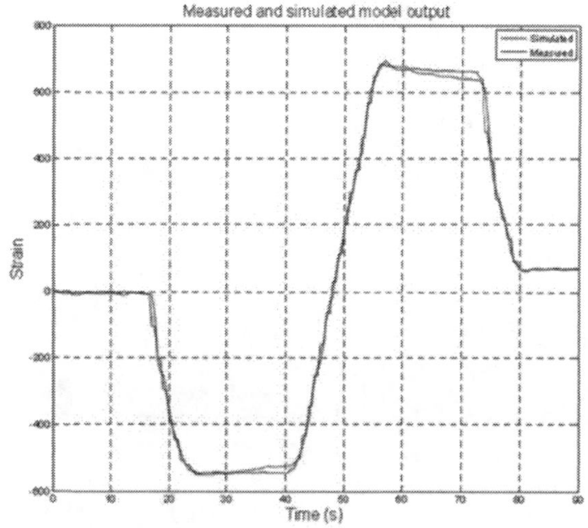

Figure 4. Control system performance

3.2 Two-Bay demonstrator characterization

The equations, describing the beam deformation as a function of the measured strain (ε_{cw}) are following written:

$$\varepsilon_{cw}(x) = ax^2 + bx + c \tag{4}$$

$$\begin{vmatrix} x_1^2 & x_1 & 1 \\ x_2^2 & x_2 & 1 \\ x_3^2 & x_3 & 1 \end{vmatrix} \begin{vmatrix} a \\ b \\ c \end{vmatrix} = \begin{vmatrix} \varepsilon_1 \\ \varepsilon_2 \\ \varepsilon_3 \end{vmatrix} \tag{5}$$

Eq.(4) describes the strain function through a parabola whose unknown coefficients, *a*, *b* and *c*, can be determined through the system (5), once measured the strain ε_1, ε_2 and ε_3. Integrating two times the following relation between the second derivative of the transversal displacement w_{cw} and the deformation:

$$\frac{d^2 w_{cw}}{dx^2} = \frac{2}{t}\varepsilon(x) \tag{6}$$

and imposing the boundary conditions (zero displacement and zero 1st derivative at the clamped edge of the beam):

$$w_{cw}(x)=\frac{2}{t}\left[a\frac{x^4}{12}+b\frac{x^3}{6}+c\frac{x^2}{2}\right] \tag{7}$$

Substituting values (5) in (7) the neutral axis shape can be reconstructed. The simulation of the analytical and numerical reconstruction from 3 acquisition points is reported in Figure 5. The positions of the sensors along the chord of the beam are: 1cm from the clamping, 12cm and 24cm, respectively.

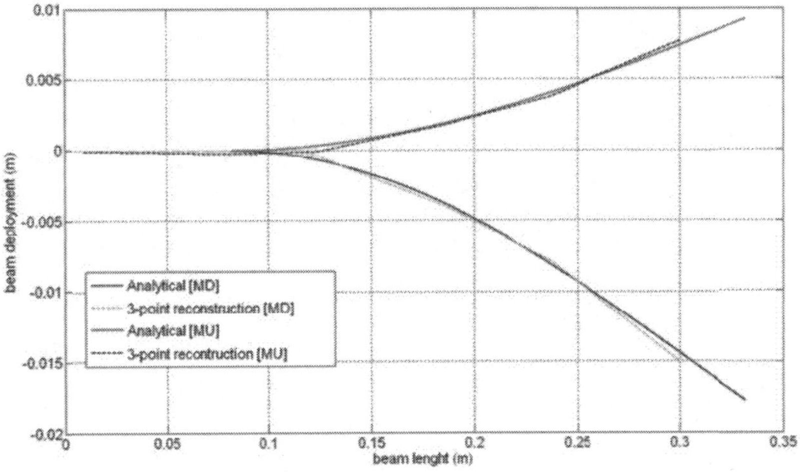

Figure 5. Numerical reconstruction of the morphed up (MU) and morphed down (MD) sliding beam.

4. SHAPE RECONSTRUCTION

A key technology to enable morphing aircraft for enhanced aerodynamic performance is the design of an adaptive control system able to emulate target structural shapes. A first phase of the work involved the determination of optimized airfoils of ATED at a single flow condition (Mach number 0.75) and different angle of attack combinations. For such conditions, theoretical smooth trailing edge shapes were generated and then aerodynamically studied to calculate the optimal performance. The experimental acquisition of the chord-wise sensor system is reported in Figure 6.

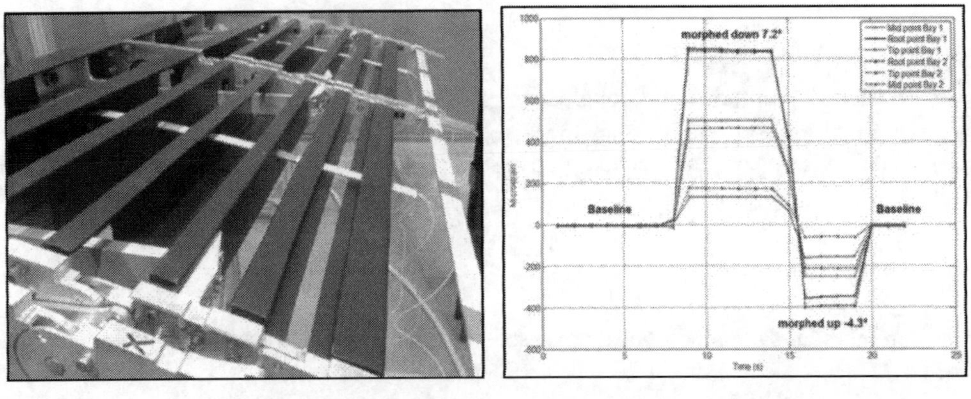

Figure 6. Two-bay demonstrator and sensor system acquisition curves

A continuous wing camber variation corresponding to a rigid TE deflection from -4.3° to 7.2° was investigated. These optimized airfoils were used as a starting point in the actuation and control systems design for determining both the aerodynamic loads fed into the actuation kinematics design and the operative strain distribution over the structure at the target shapes.

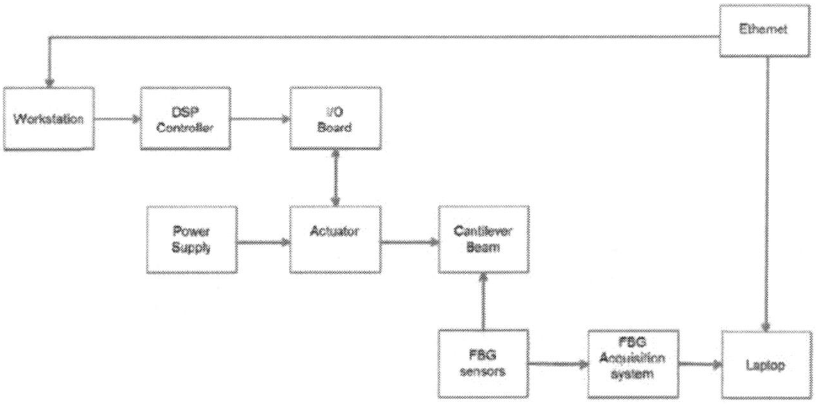

Figure 7. Control system tests: block diagram

The servo actuator model driving the adaptive wing trailing edge was developed in MATLAB/Simulink. The servo motor was designed to be driven from a pulse-width modulation (PWM) driver. The length of the pulse determines where the motor will try to position itself. This feature was simulated through a programmable pulse width modulator whose output is automatically controlled (Figure 7).

The structural camber lines for the 2-bay demonstrator is reconstructed from the measured strain values (Figure 8). The baseline, morphed up, and morphed down configurations are taken into consideration.

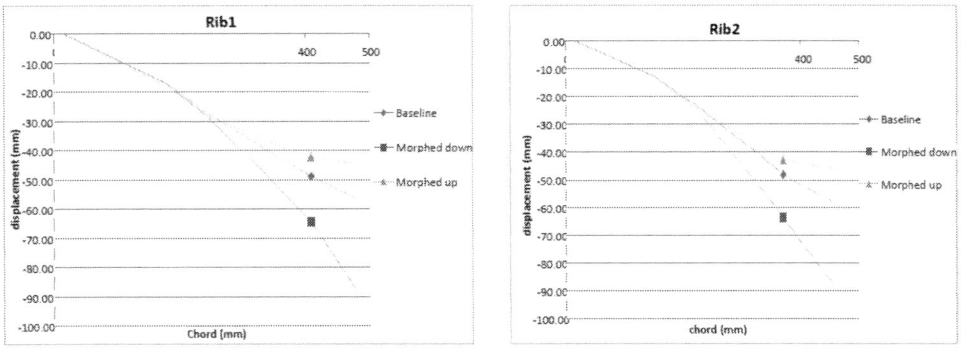

Figure 8. Measurement of chord-wise displacement

A shape reconstruction algorithm is able to reproduce the surfaces generated by the camber lines envelope. In figure 9 the morphed up and morphed down shapes are reported.

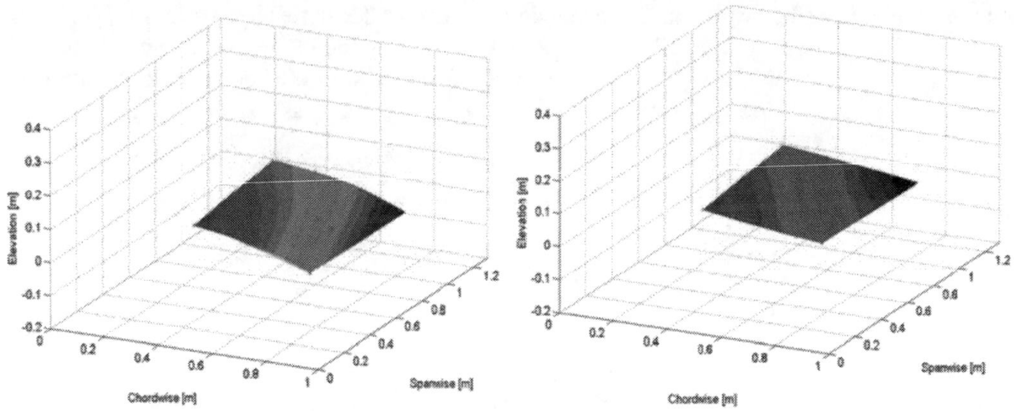

Figure 9. ATED camber line surfaces shape reconstruction

5. CONCLUSION

ATED measurement task herein described is to reconstruct its shape from strain data, retrieved from chord-wise sections, using tailored solutions based on FO technologies. Based on a mechanical device, integrated with an optical fibre sensor, the shape information is translated into strain information and properly modulated. By specific data processing mathematics, these strain values are converted back into an approximation of the produced shape. The shape reconstruction algorithm has been in house developed using commercial Matlab environment. The achieved results proved the effectiveness of this technique to identify and quantify the current deflection level of a morphing device. Further improvement can be obtained, in terms of measurement precision. For instance the adoption of distributed sensors could guarantee a more detailed description of the strain field and thus of the displacement. Also sensors withstanding larger deformation can enable the measure of the strain directly on the skin, this way achieving detailed information, not only on the displacements of the flap mean line, but also on its external shape. This step plays a fundamental role for the reconstruction of the aerodynamic performance attainable through the morphing device.

ACKNOWLEDGMENTS

The research has gratefully received funding from the European Union Seventh Framework Programme (FP7/2007 – 2013) under Grant Agreement n° 284562 SARISTU.

REFERENCES

1. M. Ciminello, S. Ameduri, D. Flauto. Design of FBG based-on sensor device for large displacement deformation, IMOC2013, 4-7 August Rio de Janeiro. Proc. IEEE 113971.
2. Dimino I, Flauto D, Diodati G, Schüller M, Gratias A, Adaptive Shape Control Architecture for Morphing Wings. Proceedings of the 6th ECCOMAS Conference on Smart Structures and Materials. SMART2013, Torino, June 2013.
3. Schorsch O, Lühring A, Nagel C, Pecora R, Dimino I, Polymer Based Morphing Skin For Adaptive Wings, 7th ECCOMAS Thematic Conference on Smart Structures and Materials, SMART 2015
4. Theory of Elasticity (McGraw-Hill International Edition, October 1, 1970 by S. Timoshenko

ICAST2015 #048

Self-powered Wireless Temperature Sensor with Piezoelectric Energy Harvester Fabricated with metal-MEMS Process

Y.C.Hsieh[1], T.K.Lin[1], J.J.Chen[1] and W.J.Wu[1*]

[1] Department of Engineering Science and Ocean Engineering, National Taiwan University, Taiwan

Abstract

In this paper, we present a self-powered wireless temperature sensor powered by a micro piezoelectric energy harvester. The piezoelectric energy harvester was designed to be a cantilever beam structure fabricated with metal-MEMS process. The resonance frequency of the energy harvester was tuned to be around 120Hz, and the bimorph version of our harvester can generate more than 400 μW output power under 1.5g acceleration at its resonance frequency. We used a speaker to simulate the vibration generated by common air conditioning motors, which holds a vibrational frequency around 120Hz and magnitude around 0.5g. Under this level of excitation, our energy harvester is capable to output power around 50μW. An interfacing circuit was used for power management and voltage regulation, which can produce a steady DC output for our sensor. The device was aimed to power a wireless temperature sensor which sends temperature readings to a smart phone via Bluetooth intermittently. When powered by our energy harvest system, our sensor module can be turn on for around 27 seconds to send readings to a smart phone every 19 seconds respectively.

1. INTRODUCTION

As low power microcontroller, sensor and network's technologies improved in past decades, "Internet of Things (IoT)" has become a hot topic nowadays [1]. In order to realize the vision of IoT, it is necessary to obtain different kinds of data from numerous wireless sensors [2]. However, most wireless devices are powered by batteries. Therefore the reliability and performances of the device are limited by the use of batteries and battery lifetime. Moreover, batteries contain toxic materials and can cause serious pollution if it is not recycled properly. The IoT devices and sensors in the future are better to be environmental friendly and maintenance free. Therefore IoT devices with energy harvesting technologies acquiring energy from ambient are definitely the best solution.

Solar, vibration, wind and thermal energies are energy sources which can be scavenged from environment. Our researches focus on using ambient vibration energy as our energy source because of its high power density and easiness of shrinking the size of the energy harvester. With piezoelectric material's great performance on transferring mechanical energy into electric energy, using piezoelectric transducer is one of the best ways to achieve the goal of harvesting vibration energy.

We present a self-powered wireless temperature sensor with piezoelectric energy harvester fabricated with metal-MEMS process in this paper. The design and fabrication of our piezoelectric energy harvester, the interface circuit, and the Bluetooth sensor module will be specified throughout the paper.

2. PIEZOELECTRIC ENERGY HARVESTER

The schematic diagram of the piezoelectric energy harvester and prototype device are shown in Figure 1. The cantilever structure design is working in {3-1} mode; therefore we retrieve electricity form top and bottom electrodes when the structure is vibrating. We chose 301 stainless steel as the substrate because its flexibility and its mechanical properties lead to a better reliability and durability in performance compared to silicon substrate. Moreover, stainless steel device's fabrication process is much simpler than process with silicon substrate [3].

The dimension of our device is 6mm × 8mm. The thickness of the stainless steel substrate is 60μm. The PZT layers, which have a thickness of 15μm, were deposited on both side of the substrate by aerosol deposition method. PZT layers are patterned by lift-off method, and we deposited 200nm thick Pt-Ti electrode on top of PZT layers by e-beam evaporator on both side. The whole device was then annealed at around 650°C in order to improve the piezoelectric layer performance. We can choose connecting top and bottom side PZT layer by either parallel or serial connection to control different poling direction. The poling voltage was DC 15V/μm. Finally, a proof mass was mounted on the end-side of cantilever to tune the resonance frequency. The processing flow is shown in Figure 2. As the result, the serial bimorph version of our piezoelectric energy harvester can generate more than 400μW under 1.5g acceleration's vibration at resonance frequency [4].

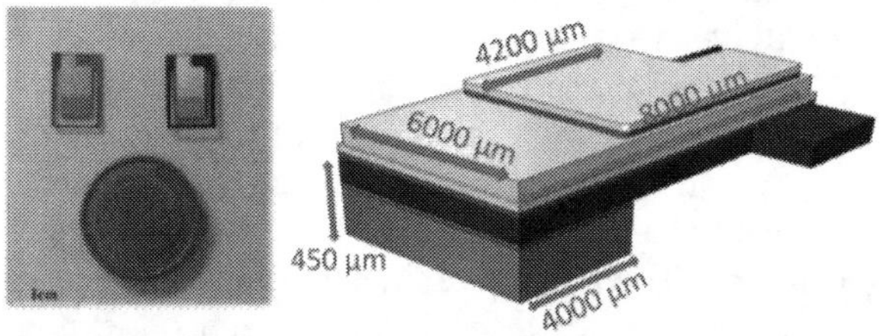

Figure 1. Schematic diagram and prototype device of metal-MEMS piezoelectric energy harvester

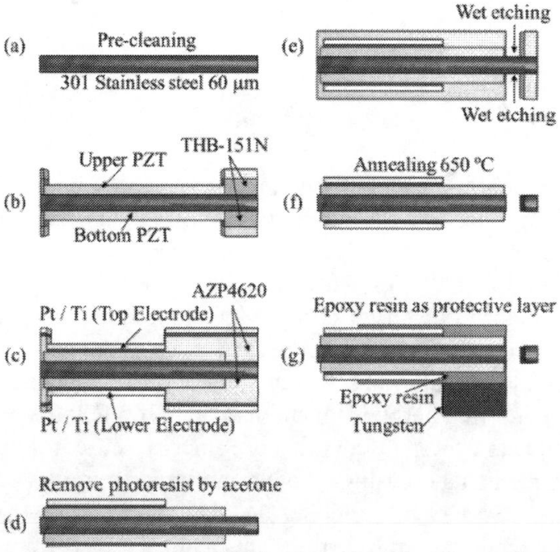

Figure 2. Fabrication flow of metal-MEMS piezoelectric energy harvester: (a) pre-cleaning stainless-steel, (b) use photoresist THB-151N to pattern and deposit PZT layer on both side with aerosol, (c) use photoresist AZP4620 to pattern electrodes, then deposit Pt/Ti electrode on both side, (d) remove photoresist by acetone, (e) use photoresist AZP4620 to pattern the cantilever, then use wet-etching to release the cantilever, (f) annealing at 650°C, (g) paste epoxy as protective layer, then bond a tungsten block as proof mass.

3. THE INTERFACING CIRCUIT WITH ENERGY BUFFER

3.1 Interfacing Circuit

An interfacing circuit was used to accumulate the electrical energy produced by the energy harvester, acting as an energy buffer. The circuit utilizes Linear Technology LTC3588 IC and capacitors for power management and voltage regulator. Figure 4 shows the circuit's schematic diagram. Current source and capacitor inside the red box represents our piezoelectric energy harvester's equivalent circuit. The energy produced by our harvester is first stored into two parallel large (100μ F) capacitors C3 and C4. After the energy in the capacitor reached a certain level, it will then be used by the buck convertor inside LTC3588, which will eventually generate a 1.8V output for our applications. As the energy is used by the load, voltage of capacitors C3 and C4 will ramp down to a lower threshold. After the voltage of the capacitors declined to the lower threshold, the buck convertor inside LTC3588 will shut down and have null output until the next cycle begin.

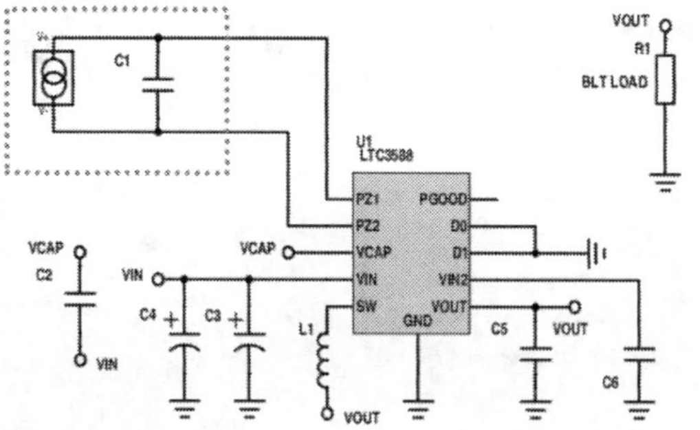

Figure 4. Interfacing circuit schematic diagram

3.2 Bluetooth sensor module

Figure 5 shows the conceptual structure of our Bluetooth sensor module. It is a Bluetooth Low Energy (BLE) module based on Nordic nRF51822. The nRF51822 SoC can be divided into two major parts, the wireless transceiver and the microcontroller. The microcontroller, which is an ARM Cortex-M0 processor, is the smallest ARM processor available. Because of its low power and minimal code footprint, it is very suitable for our energy harvesting system. The processor first collects temperature data from the internal temperature sensor, than it packs the data into wireless packets and sends the collected data to our mobile device through the BLE wireless transceiver. Latest iOS and Android phones or tablets which support BLE protocol are able to receive the broadcasted temperature readings with a simple App.

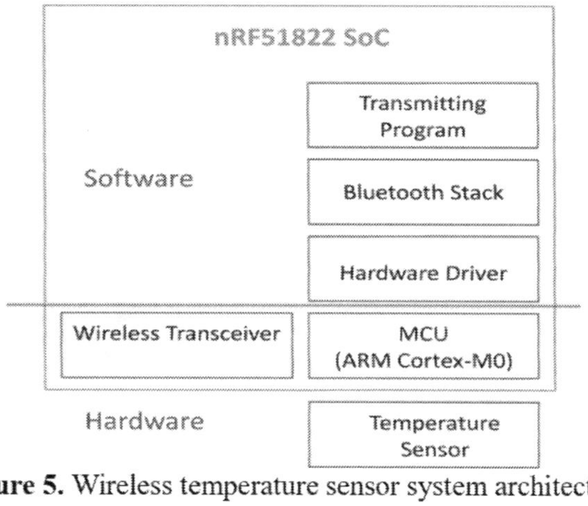

Figure 5. Wireless temperature sensor system architecture

4. EXPERIMENTAL SETUP AND RESULTS

The entire experimental setup is shown in Figure 6. The smart phone on the right side generated a sine wave signal to Bluetooth speaker in the middle of the figure. The frequency of the sine wave signal was tuned to match our energy harvest device's resonant frequency 120Hz, and the magnitude was tuned to be around 0.5g. We soldered our wireless temperature sensor module and interfacing circuit on a 17mm × 29mm double layer printed circuit board (PCB). We then connect the PCB onto our energy harvester, and placed the whole device on top of the speaker to harvest the vibration energy produced by it. The Android tablet on the left was used to receive temperature data from the sensor module via Bluetooth. Whole experiment setup required very little space and equipment, and could be set up easily.

Figure 7(a) shows the output waveform of LTC3588 in our circuit under the condition mentioned above. It can be seen that when the convertor is on, the output voltage is regulated at 1.8V, which can be used by our sensor module. The result also indicates that our sensor module can work for 27 second continuously to send one temperature data per second after a 19 second charging time. This performance is sufficient for many sensing application. Figure 7(b) shows the screenshot of our Bluetooth sensor APP. It shows the ID number of the sensor module on top, and the temperature data reading and the data's received time in the main screen.

Figure 6. Overall experimental setup

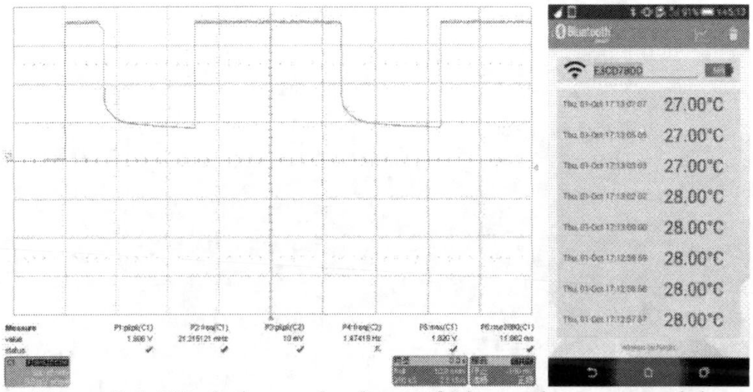

Figure 7. (a)Work time and voltage of Bluetooth sensor module
(b)Temperature data sent from Bluetooth sensor module

5. CONCLUSIONS

This paper demonstrates a fully autonomous Bluetooth wireless sensor. It can be powered by a vibration energy harvester to scavenge power from the environment and need no battery. The entire setup is self-powered and without any external power source. This system can be installed on vibrating devices in household or factory equipment, such as motors, air conditioners, engines, and other rotatory or vibrating devices. This can solve the issues of routinely changing and maintaining batteries, especially for the sensors to be installed in remote area or on rotating machines. In the future we wish to turn the interface circuit into an integrated circuit chip with nonlinear synchronized switching circuit to further boost the output power. By integrating the energy harvester and the circuit into a single package, we can simply put the sensor on any vibration source to directly send data to the internet.

ACKNOWLEDGMENTS

This research is supported by Intel and Ministry of science and Technology.

REFERENCES

1. L. Atzori, A. Iera, and G. Morabito, The Internet of Things: A survey, *Comput. Netw., vol. 54, no. 15*, pp. 27872805, Oct. 2010.
2. Edgecombe,"Discovering Activities to Recognize and Track in a SmartEnvironment" IEEE Trans. Knowledge & Data Eng., vol. 23, no. 4, pp.P. Rashidi, D.J. Cook, L.B. Holder, and M. Schmitter-527-539, Apr. 2011.
3. S.C. Lin and W.J. Wu, "Fabrication of PZT MEMS energy harvester based on silicon and stainless-steel substrates utilizing an aerosol deposition method", *J. Micromech. Microeng.* 23 125028 (2013).
4. S.C. Lin and W.J. Wu, "Piezoelectric micro energy harvesters based onstainless-steel substrates", *Smart Mater. Struct.* 22 405016 (2013).

ICAST2015 #053

Connection Accuracy of Multiple 3D Shape Data Sets due to Feature-Point Matching with Virtual Targets

Taku Harada[1] and Takashi Iwasa[1*]

[1] Department of Mechanical and Aerospace Engineering, Tottori University, Tottori, Japan

Abstract

In this study, we investigated the connection performance of feature-point matching with virtual targets. This is a method to allow precise connection of three-dimensional surface without including any corresponding points or characteristic shapes. To make feature-point matching available for these surface shape data, several fixed targets were virtually created from a preliminary surface shape measurement for dedicative concave models. We conducted a connection experiment using two sets of surface shape data captured by photogrammetry using a grating projection method. The effect of the number of virtual targets on the connection accuracy was studied. The results show that the connection accuracy of the feature-point matching is roughly constant at 70×10^{-6} m root mean square when more than six virtual targets are used. This suggests that at least six virtual targets are needed to provide satisfactory connection performance in feature-point matching with virtual targets.

1. INTRODUCTION

A precise, large, deployable space antenna reflector system based on the smart structure concept has been discussed for future advanced space science missions[1, 2]. The application of the smart structure concept to a space antenna reflector is meant to maintain high mirror accuracy of the reflector surface in orbit by actively controlling its surface shape. However, this requires a shape measurement method capable of capturing the full-field surface shape of a large reflector with high accuracy and spatial resolution, to determine the appropriate control output.

In a previous study[3], we investigated a surface shape measurement method for a large space structure and proposed a method for capturing the full-field surface shape of the structures. In our proposed method, the partial surface shapes of the large space structure are initially measured by photogrammetry using a grating projection method and are connected to represent the full-field surface shape without degrading the measurement accuracy and spatial resolution of the photogrammetry. The connection is performed by a feature-point matching with virtual targets to precisely connect the partial surface shape data without including any corresponding points or characteristic shapes. A verification experiment showed that the proposed method successfully connects two sets of the partial surface shape data to within 100×10^{-6} m root mean square (RMS) accuracy[4].

However, the connection accuracy of the proposed method depends on the number of virtual targets used in the feature-point matching. To apply the proposed method experimentally, the effect of the number of virtual targets on the connection accuracy must be clarified and the number of virtual targets necessary to provide high connection performance must be ascertained. In this study, a

* iwasa@mech.tottori-u.ac.jp

verification experiment on the connection accuracy of the feature-point matching with virtual targets was conducted by connecting two sets of photogrammetry three-dimensional (3D) shape data, using a grating projection method. The effect of the number of virtual targets on the connection performance of the proposed method was also investigated.

2. FEATURE-POINT MATCHING USING VIRTUAL TARGETS

Figure 1 shows an outline of the proposed connection method[4)]. As shown in the figure, the proposed method initially measures partial surface shapes of the target structure by photogrammetry using a grating projection method. The size of the partial surface was set so that the required measurement accuracy and spatial resolution was satisfied in the photogrammetric measurement. The captured partial surface shapes were then connected so that a full field surface shape of the target structure was reconstructed, while preserving both measurement accuracy and spatial resolution. The connection of the partial surface shape data was achieved by transforming the spatial coordinate system of the partial surface shape data onto the reference frame. The coordinate transformation matrix was calculated using feature-point matching with virtual targets. This was used to allow the partial surface shape data to be successfully connected without any corresponding points or characteristic shapes.

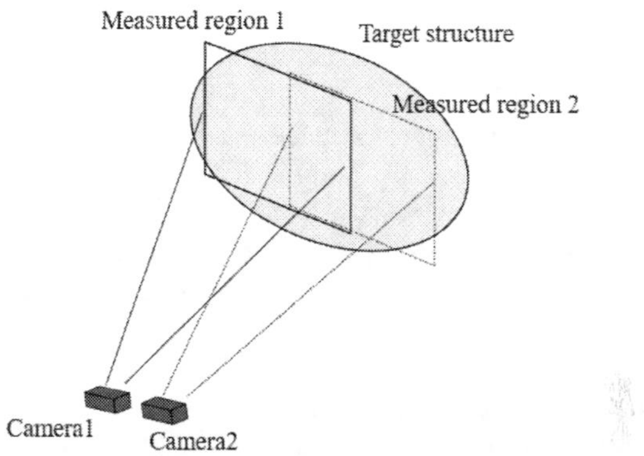

Figure 1. Outline of the proposed connection method

Figure 2 shows the outline of the virtual targets used in the feature-point matching. Several dedicative concave models, whose surfaces were processed to be a part of the ideal spherical surface, were placed in the joint region of the two measurement regions before the surface shape measurement of the target structure was performed. The surface shape of the concave models was measured for both regions. An ideal spherical surface was curve fitted onto the measured surface shape data of the concave models. Since this curve-fitting calculation provides the center points of the ideal sphere, which are fixed in space, the coordinate transformation matrix between the two measured regions was calculated using the traditional feature-point matching for the virtual targets. The calculation of the feature-point matching was conducted using the Singular Value Decomposition (SVD) method.

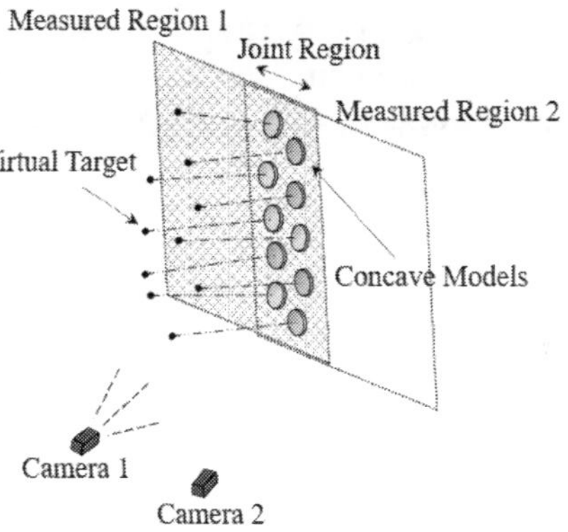

Figure 2. Outline of virtual targets

3. EXPERIMENT

3.1 Measurement system

3.1.1 Photogrammetry using the grating projection method

The grating projection method is one of the active photogrammetric measurement methods. It captures the 3D position coordinates of an object's surface on a pixel-to-pixel basis, using one digital camera and one projector[5-7)]. The grating projection method used in this experiment was able to capture the object surface shape without any lens aberration effects by using a distinctive calibration with two reference planes[8)].

Figure 3 illustrates the measurement system constructed in our laboratory. Two digital cameras were installed in parallel on the measurement jig, so that the object surface shape was measured in two regions. The projector was placed on the underside of the digital cameras at a distance of about 1.05 m. The measurement object was installed in front of the measurement system at a distance of about 1.20 m. To eliminate the effect of any vibration noise transmitted from the floor during the experiment, the measurement object was placed on a vibration isolation table. The specifications of the measurement equipment are given in Table 1.

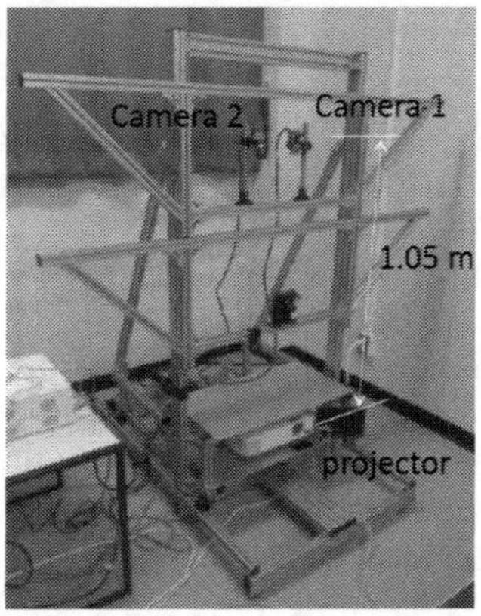

Figure 3. Measurement system

Table 1 Specification of measurement equipment

Equipment	Product	Specification
Digital cameras	IDS UI-3480CP-M-GL	Resolution: 2560×1920
Projector	NEC NP-M300WJL	Resolution: 1280×1800 Luminance: 3000 lm

3.1.2 Measurement objects

Figure 4 shows the two measurement objects: a concave mirror model, used to create the virtual targets, and a spherical mirror model. The concave mirror model had a diameter of 0.05 m, and a thickness of 0.02 m. The object surface was processed at an accuracy 25×10^{-6} m RMS to match the surface shape of an ideal sphere whose radius is 0.1 m. The surface of the model was matted with a white spray to remove any glossy effects on photogrammetric measurement using a grating projection method. Ten concave models were used in this study.

The 0.3 m diameter spherical mirror model was used to investigate the performance of the proposed connection method. The surface shape was processed with 10×10^{-6} m RMS accuracy to match the surface shape of an ideal sphere whose radius is 2.981 m. A detailed description of the model was given in a previous paper[8)].

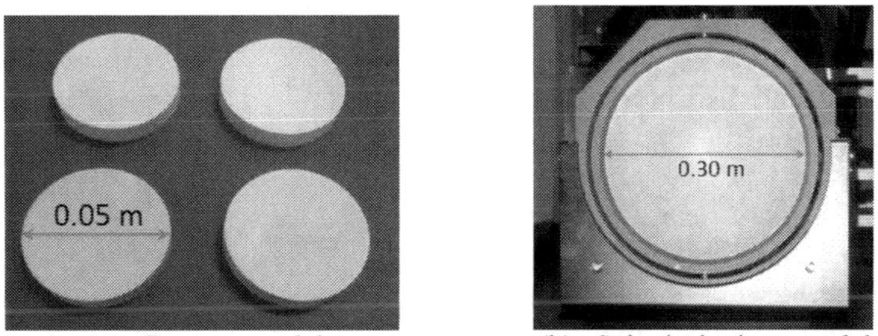

(a) Concave model (b) Spherical mirror model

Figure 4. Measurement objects

3.2 Experiment procedures

Figure 5 shows the experiment procedure. First, we calibrated the photogrammetry using the grating projection method, and the two measurement spaces were arranged so as to be partially overlapping (Figure 2). The size of each measurement space was 0.5 m × 0.7 m × 0.01 m, and the width of the joint region was about 0.20 m.

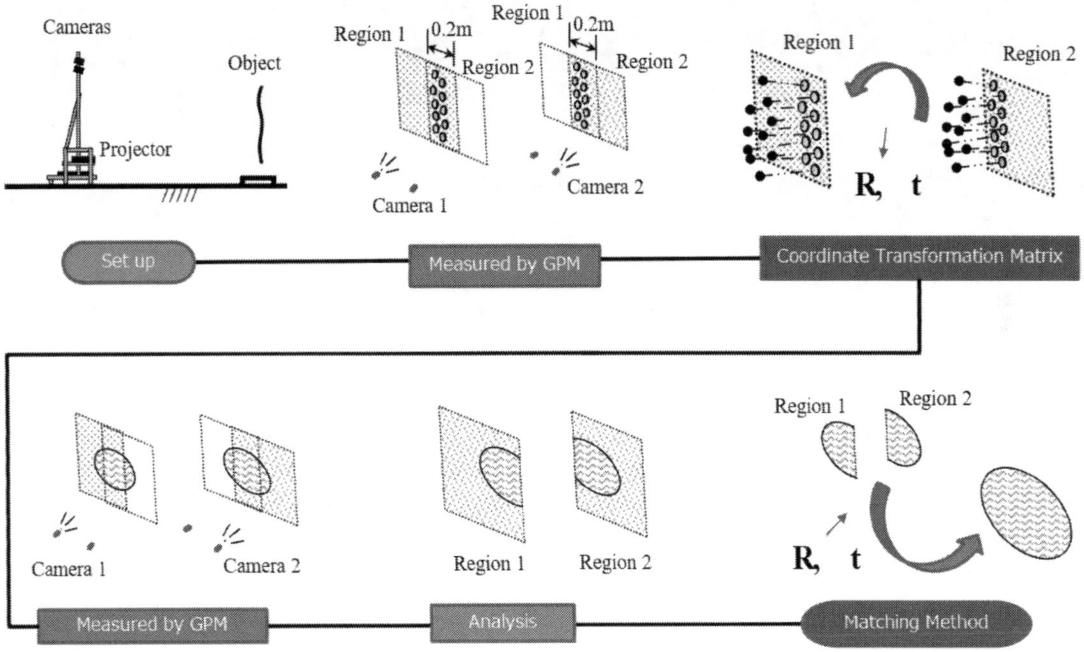

Figure 5. Flow of experiment procedure

The ten concave models were installed in the joint region (Figure 6), and the surface shapes for each region of the concave models were measured by photogrammetry.

By fitting the ideal spherical surface onto the measured surface shape of the concave models, ten virtual targets were created for each measurement region. Figure 7 shows the analysis region applied to the curve-fitting calculation.

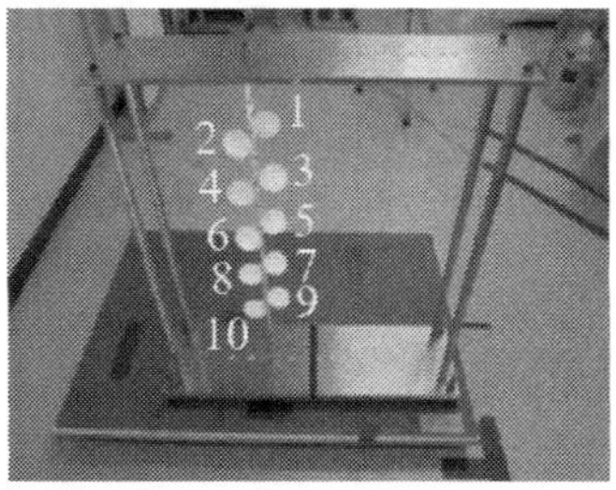

Figure 6. Configuration of ten concave models

Three different patterns of analysis region were used: 20 × 20 pixel^2, 30 × 30 pixel^2, and 40 × 40 pixel^2. This allowed us to analyze the effect of the different sizes of analysis region on the connection performance of the feature-point matching with the virtual targets.

The coordinate transformation matrix between the spatial coordinate systems of the two measured regions was then calculated from the feature-point matching, using the virtual targets created. The SVD algorithm was applied in this calculation. The number of virtual targets was varied from four to ten, in order to investigate the effect of the number of virtual targets on the connection performance. The virtual targets used in the investigation were selected beginning at the top in Figure 6.

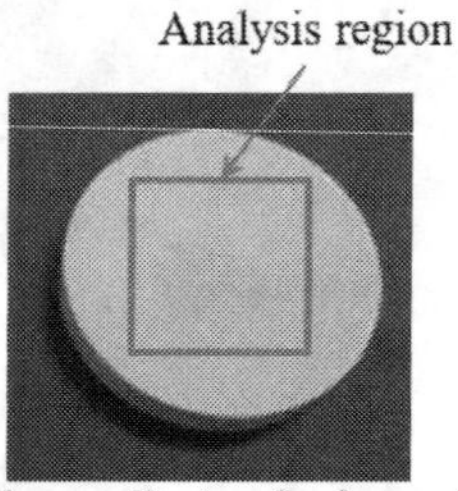

Figure 7. Analysis region

The spherical mirror model was installed in the joint region of the two measurement spaces, and the partial surface shape of the spherical mirror model was captured for the two measurement regions by photogrammetry. Finally, the spatial coordinate system of the partial surface shape data measured in region 1 was transformed onto the spatial coordinate system of the region 2, and the connection performance of the feature-point matching with the virtual targets was analyzed by calculating the differences between the two partial surface shapes.

4. RESULTS AND DISCUSSION

4.1 Effects of number of virtual targets on connection performance

4.1.1 Surface shape measurement of concave models

Figure 8 shows the measurement results of the concave models from the photogrammetry using the grating projection method. The values on the x and y axes indicate the pixel value, and the contour color represents the position coordinate of the object surface in the z axis direction. From the figure, it can be seen that the surface shape of the concave models was successfully captured within the measurement space, and that the concave models were installed at a tilt, relative to each other. Table 2 lists the measurement accuracy for each concave model defined as the RMS value of the difference between the measurement results of the concave model and the ideal spherical surface fitted onto the measurement results. The curve-fitting calculation of the ideal spherical surface was conducted using the square region with 40 pixels on a side in the measured data. The surface shape of the concave models was measured to within an accuracy of 20 × 10^{-6} m RMS.

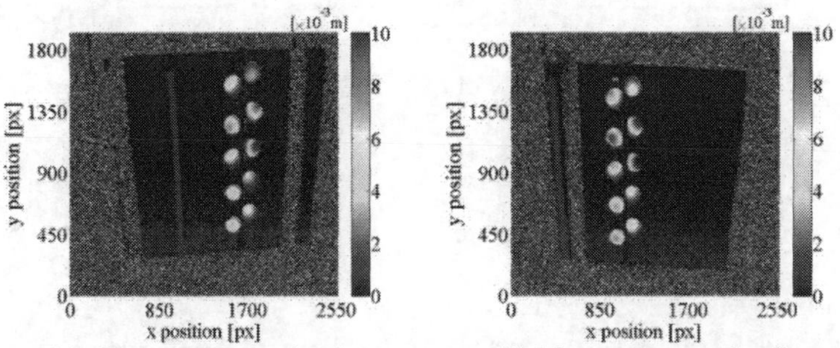

(a) Measurement region 1 (b) Measurement region 2

Figure 8. Measurement results of the concave models

Table 2 Measurement accuracy of concave models [$\times 10^{-6}$ m RMS]

Model number	1	2	3	4	5	6	7	8	9	10
Measured region 1	14	17	14	13	11	15	18	11	12	12
Measured region 2	17	12	16	12	12	13	18	12	10	12

4.1.2 Calculation of virtual targets

The virtual targets were calculated using the measurement results of the concave models shown in Figure 8. Figure 9 shows the analysis region in the surface shape data of the concave model. The size of the analysis region shown by the dotted red line was 40 × 40 pixel^2. Since the measured surface shape data of the concave model fell partly outside the measurement space, the analysis region was manually adjusted to exclude these outlier values.

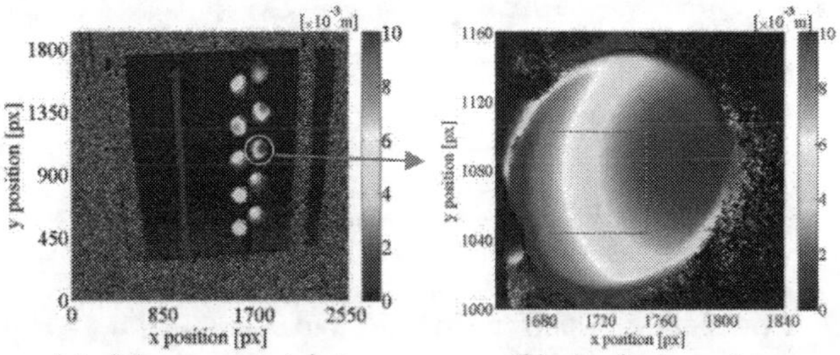

(a) Measurement data (b) Analysis region

Figure 9. Analysis region used in virtual target calculation

Figure 10 shows the 3D position coordinates of the calculated virtual targets. Ten virtual targets were arranged in the vertical direction (y axis) for each region. Since the concave models were installed at a tilt, the virtual targets were not arranged in a perfectly straight line. Using these virtual targets, the coordinate transformation matrix between the measured regions 1 and 2 was calculated with feature-point matching. This calculation was performed while changing the number of virtual targets from four to ten, and the effects of the number of virtual targets on the connection performance of the proposed connection method was investigated.

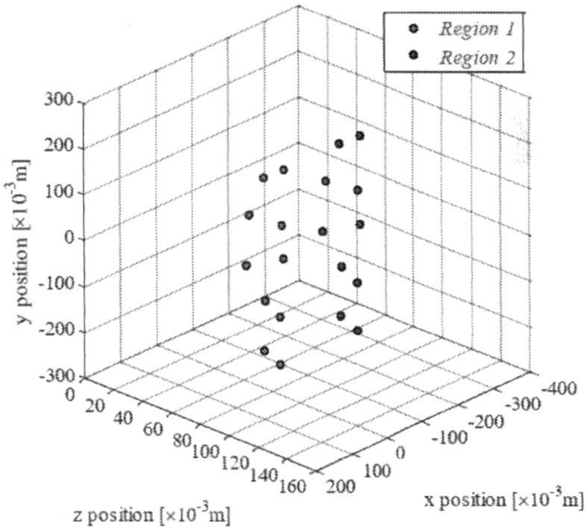

Figure 10. Configuration of calculated virtual targets

4.1.3 Surface shape measurement of the spherical mirror model

Figure 11 shows the measurement results of the spherical mirror model by photogrammetry using the grating projection method. The values on the x and y axes indicate the pixel values, and the contour color represents the position coordinate of the object surface in the z axis direction. The surface shape of the spherical mirror model was successfully captured within the measurement space, with a measurement accuracy of 27.4×10^{-6} m RMS in region 1 and 27.7×10^{-6} m RMS in region 2. These values were derived from the differences between the measurement results of the spherical mirror model and an ideal spherical surface fitted onto the measurement results. The curve-fitting calculation of the ideal spherical surface was conducted for the measured data in the squared region at 450 pixels on a side. From Table 2, the measurement accuracy of the spherical mirror model was lower than that of the concave models, because the photogrammetry with grating projection is sensitive to the optical properties of the surface of the measured object.

(a) Measured region 1

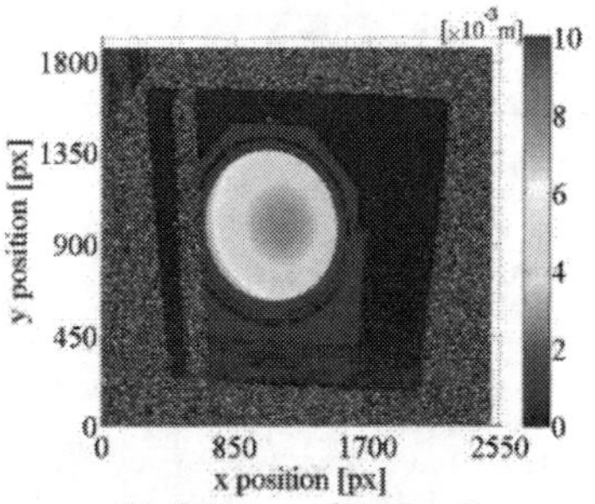

(b) Measured region 2

Figure 11. Measurement results of spherical mirror model

4.1.4 Connection of partial surface shape data of spherical mirror model

Figure 12 shows the partial surface shape data for the spherical mirror model selected for the connection. The size of the selected region from the measured surface shape data was 450×450 pixel^2. The values on the x and y axes indicate the actual position coordinates, and the contour color represents the position coordinate of the object surface in the z axis direction.

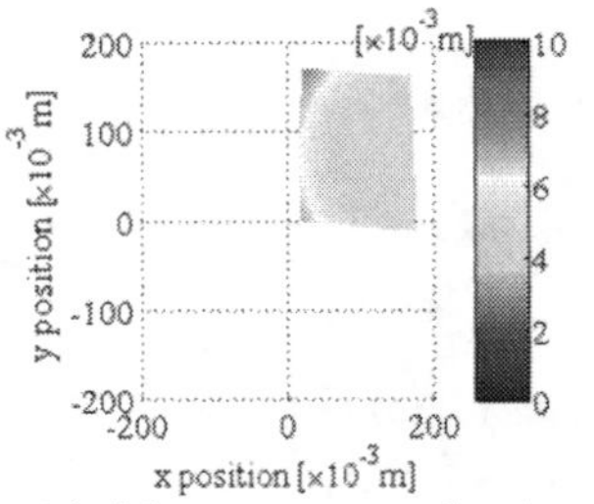

(a) Measurement region 1

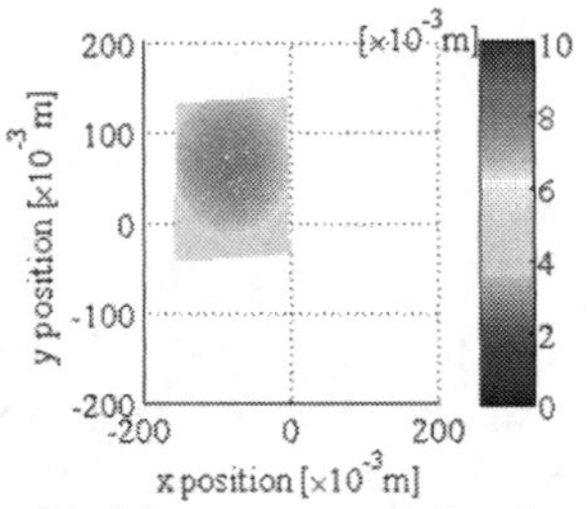

(b) Measurement region 2

Figure 12. Partial surface shape data of spherical mirror model

Figure 13 shows the connection results for the partial surface shape data of the spherical mirror model. These are the results of the feature-point matching, which was conducted with ten virtual targets. Figure 13 (a) and (b) represent the 3D view and the top view of the connection results, respectively. Figure 13 (c) shows the cross-sectional shape of the connection results along the y axis (x = 0 m: vertical line depicted in Figure 13(b)). From the figure, it can be seen that the two sets of partial surface shape data for the spherical mirror model were appropriately connected, with an accuracy of about 72.2×10^{-6} m RMS. This value is given by the difference between the ideal spherical surface fitted onto the measurement results in region 2 and the transformed measurement results in region 1. As this connection accuracy includes the region 1 measurement accuracy, i.e., 27.4×10^{-6} m RMS in the surface shape measurement, the two sets of partial surface shape data for the spherical mirror model can be connected with roughly the same accuracy as that of the photogrammetry using the grating projection method.

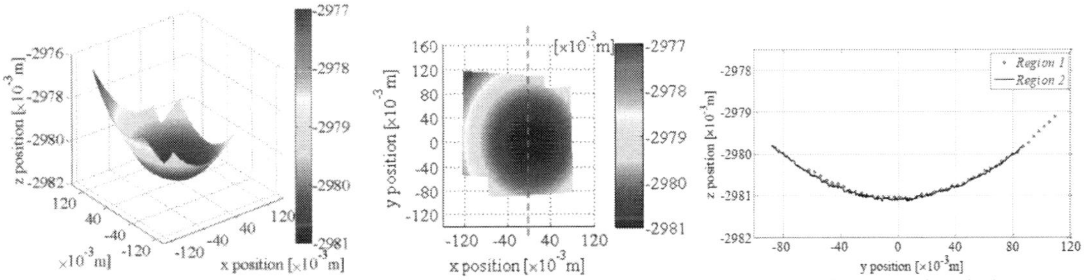

(a) 3D view (b) Top view (c) Cross-sectional view

Figure 13. Connection results of the spherical mirror model

Figure 14 shows the effect of the virtual target number on the connection accuracy. The abscissa indicates the number of virtual targets used in the feature-point matching, and the ordinate indicates the connection accuracy for the spherical mirror model. The partial surface shape data of the spherical mirror model was connected at about 70×10^{-6} m RMS, except in the five virtual target case.

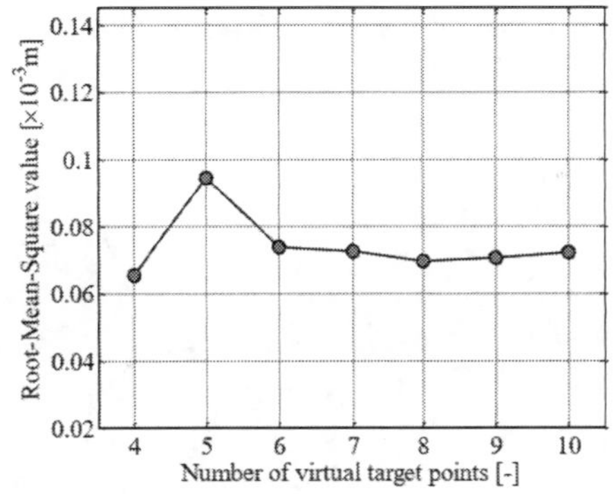

Figure 14. Effects on virtual target number

Figure 15 shows the connection accuracy for the measurement results of each concave model. The abscissa indicates the number of virtual targets, and the ordinate represents the connection accuracy. This figure gives the spatial variation of the connection accuracy of the feature-point matching with the virtual targets, as the ten concave models were separately installed in the joint region, as shown in Figure 8. These results suggest that the connection accuracy varies with location, and that this variation becomes small when the number of the virtual targets is larger than six. The connection performance of the feature-point matching with the virtual targets therefore becomes stable when the number of targets is larger than six. However, when the number of the targets is lower than six, the variation in the connection accuracy is large, and cases are observed where the connection accuracy degraded even when the number of targets increased. The same tendency is also observed in the results for the spherical mirror model, as shown in Figure 14. These results suggest that at least six virtual targets are necessary to allow the connection of 3D surface shape data sets with a stable connection accuracy when using feature-point matching with virtual targets.

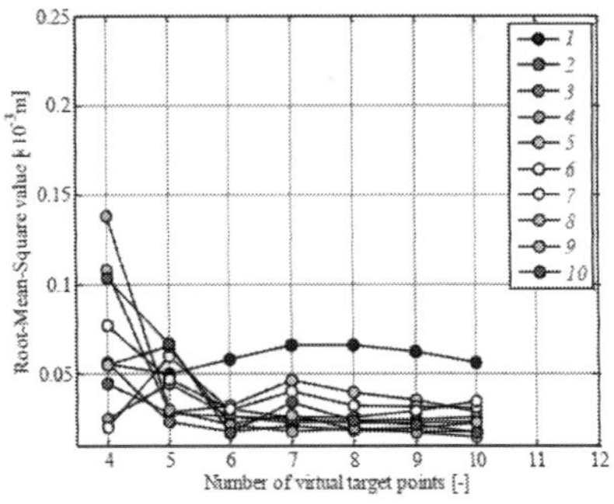

Figure 15. Spatial variation of connection accuracy

4.2 Effects of size of analysis region on connection performance

Figure 16 compares the effect of the number of the virtual targets, against the size of the analysis region selected for the virtual target calculation, on the connection accuracy of the spherical mirror model. The abscissa indicates the number of virtual targets, and the ordinate represents the connection accuracy. No effects are apparent. The differences between these three cases are within 20×10^{-6} m RMS. This is lower than the measurement results for the spherical mirror model.

Figure 17 compares the spatial variation of the connection accuracy with the size of the analysis region selected for the virtual target calculation. Figure 17 (a) indicates the result for the 20 × 20 pixel^2 region, and Figure 17 (b) for the 30 × 30 pixel^2 region. The abscissa indicates the number of virtual targets, and the ordinate represents the connection accuracy for the measured data of the concave models. No clear differences are evident, and the variation of connection accuracy becomes small when the number of the virtual targets is larger than six. This suggests that when the number of the applied virtual targets is larger than six, the 3D shape data sets are appropriately connected by the proposed method at the analysis region sizes.

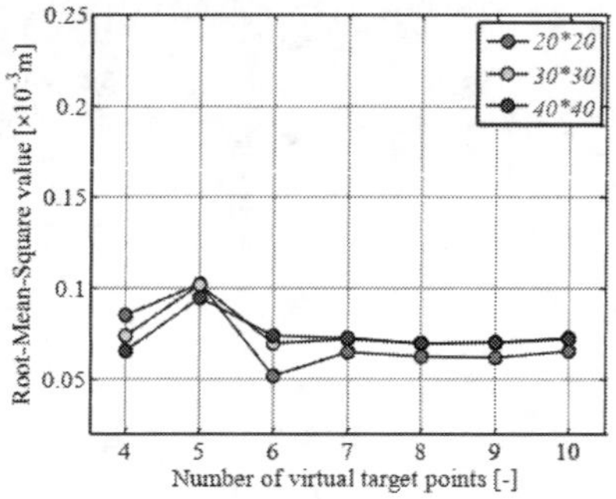

Figure 16. Effects on virtual target number

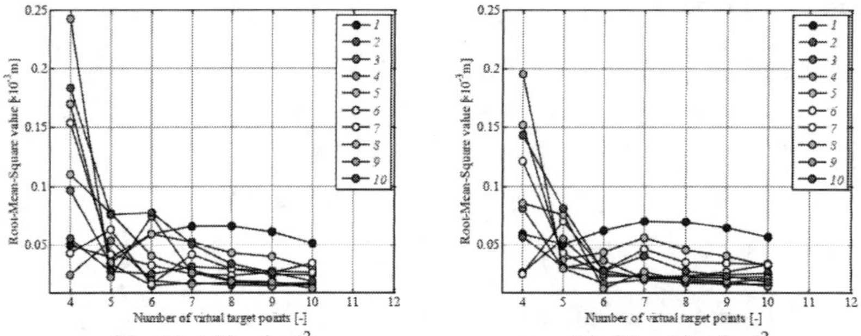

(a) 20 × 20 pixel^2 (b) 30 × 30 pixel^2

Figure 17. Spatial variation of connection accuracy

5. CONCLUSIONS

The connection performance of feature-point matching with virtual targets was investigated. The major findings are as follows.

(1) The connection performance of feature-point matching with virtual targets is roughly constant at 70×10^{-6} m RMS when the number of applied virtual targets is larger than six. Accordingly, at least six virtual targets are necessary to connect 3D surface shape data sets with a constant connection accuracy, when using feature-point matching with virtual targets.

(2) The connection performance is not strongly affected by the size of the analysis region selected for the calculation of the virtual targets. In this experiment, stable connection performance was observed across a range of 20 × 20 pixel^2 to 40 × 40 pixel^2.

ACKNOWLEDGMENTS

This study is partially supported by KAKENHI No. 26249131 and 15K06594.

REFERENCES

1. Santiago-Prowald J and Baier H (2013), Advances in deployable structures and surfaces for large apertures in space, *CEAS Space Journal* **5**(3_4): pp. 89_115.
2. Puig L, Barton A, and Rando N (2010) A review on large deployable structures for astrophysics missions, *Acta Astronautica,* **67**(1_2):, pp. 12_26.
3. Iwasa T, et al. (2015) Proposal of Precise Shape Measurement Method Synthesizing Surface Measurement and Point Measurement for Large Space Structures, *AEROSPACE TECHNOLOGY JAPAN, THE JAPAN SOCIETY FOR AERONAUTICAL AND SPACE SCIENCES,* **14**: pp. 95_103.
4. Harada T and Iwasa T (2015) Proposal of Connection Method for Precise Shape Measurement Data using Virtual Targets, In: *Proceedings of 30th International Symposium on Space Technology and Science.*
5. Tay C J, Thakur M and Quan C G (2005) Grating projection system for surface contour measurement, *Applied Optics* **44**(8): pp. 1393_1400.
6. Xie H et al.(2004) Phase shifting SEM moiré method, *Optics & Laser Technology* **36**(4): pp. 291_297.
7. Fujigaki M, Morimoto Y and Yabe M (2002) Real-time measurement of nanometer displacement distribution by integrated phase-shifting method, *JSME International Journal Series a-Solid Mechanics and Material Engineering,* **45**(3): pp. 448_452.
8. Iwasa T, et al. (2014) An Error Elimination Method for Surface Shape Measurement using the Grating Projection Method, *Transactions of the Japan Society for Aeronautical and Space Sciences, Aerospace Technology Japan* **12**: pp. 81_88.

ICAST2015 #087

Damage depth reconstruction based on genetic algorithm using laser ultrasonic system

Zhan Wang, Chao Zhang, Jinhao Qiu* and Hongli Ji

State Key Laboratory of Mechanics and Control of Mechanical Structures, Nanjing University of Aeronautics and Astronautics, Nanjing, China

Abstract

The safety and reliability of engineering structures cannot be over-emphasized. Along with it is the ever-increasing need for effective and reliable non-destructive evaluation (NDE) techniques. Among various NDE techniques, laser ultrasonic technique provides a new paradigm to detect the damage with non-contact ultrasonic generation, which is more suitable for the damage evaluation in the complex structures with curved surfaces. This paper presents an inverse method for the depth reconstruction of local damage on a 2-D isotropic plate using laser ultrasonic system. In the experiment, a fixed-point acoustic emission (AE) sensor is used to measure the Lamb waves generated by the scanning pulse laser through thermo-elastic effect. The time-space wavefield can be obtained to calculate the phase velocity curve on the unknown structure through two-dimensional Fourier transform (2D-FT). To reconstruct the depth of the damage, two-dimensional short space Fourier transform (2D-SSFT) converts the local time-space wavefield to the local frequency-wavenumber wavefield, giving the phase velocity curve on the local region. Furthermore, an inverse method based on genetic algorithm (GA) is carried out to estimate the thickness of the local structure by comparing the local phase velocity curve with the global phase velocity curve. Finally, the shape of the damage can be detected point-to-point by moving the window function in 2D-STFT. To validate the proposed method, three damages with different depths on an aluminum plate are estimated experimentally. The phase velocities within a broadband are calculated in the frequency-wavenumber wavefield by applying polynomial fitting. The reconstruction results demonstrate that the proposed method is able to estimate the damage depth through the local phase velocity curves, whilst providing a damage detection method without any prior knowledge on material parameters.

1. INTRODUCTION

In recent decades, there have been increasing demands for NDE techniques to detect possible damages and evaluate the safety and reliability of various structures [1-4]. Among them, Lamb wave-based damage detections have been widely used with many advantages, such as high sensitivity to many kinds of damage types, long propagation distance with little attenuation and so on [5-8]. Generally, a large amount of piezoelectric sensors are distributed in the structures to construct a sensing network. Then, one of these sensors generates Lamb waves and the others obtain the damage information about the structure by acquiring the ultrasonic signals [9, 10]. By means of signal processing and data fusion, the damage of structures can be recognized and located. However, in most of the cases, the signals in health situation should be acquired in advance as the baselines to diagnose

* qiu@nuaa.edu.cn

the existence of damage. Furthermore, to improve the quality of inspection, a large number of sensors are required, which hamper their use in practical applications.

To deal with these problems, laser-generation based imaging method has been recently developed [11-15]. By using a movable pulse laser to generate ultrasonic waves and a transducer to detect them, a series of snapshots can be obtained to display the wave propagation from a fixed point to the inspection region. By observing the anomalous waves in the snapshots, the damage positions can be detected directly. To detect the damages precisely, some signal processing methods have been proposed [16-19]. An et al. [18] used three-dimensional Fourier transform to filter the incident waves and reflected waves, and the standing waves which occur in the damaged areas can be obtained after the filter. By calculating the energy of standing wave, the size and location of the notch can be detected. Flynn et al. [19] used frequency–wavenumber analysis to obtain the variation of wavenumbers induced by delamination damage. The new wavenumbers are used to detect and quantify the delamination damage. Finally, the shape of the delamination damage can be detected. Furthermore, several researches have been developed to estimate the depth of damage. Tian et al. [20] used frequency–wavenumber analysis to identify the damage position in the thickness-direction. However, the prior knowledge of the material parameters of the composite plate should be known, which restricts its applications.

This paper presents an inverse method for the depth reconstruction of local damage on a 2-D isotropic plate using laser ultrasonic system without prior knowledge of the material parameters of the material. A fixed-point acoustic emission (AE) sensor is used to measure the time-space wavefield of guide waves, which are generated by a movable pulse laser. Two dimensional Fourier transform is used to convert the wavefield date from the time-space domain to frequency-wavenumber domain, by which the phase velocity dispersion curve can be obtained. To detect the location and size of the damage, a two dimensional short space Fourier transform (2D-SSFT) has been used. The phase velocity in damaged region can be acquired by selecting one segment of the wavefield from the damaged area. Then, the detection of the depth of damage is carried out through a genetic algorithm by minimizing the error function.

2. EXPERIMENTAL SETUP

Figure 1 shows a schematic illustration of the scanning path on an aluminum plate with 2 mm in thickness. A pulsed laser beam is used for Lamb wave generation and a fixed point AE sensor is used to sense the signal of Lamb waves. By scanning the laser point in a line through a laser scanning mirror, the response signals from different laser points and the fixed sensor point can be acquired. According to the reciprocity theorem [21], the wavefield that is generated at the sensor point and propagates to the inspection region can be visualized.

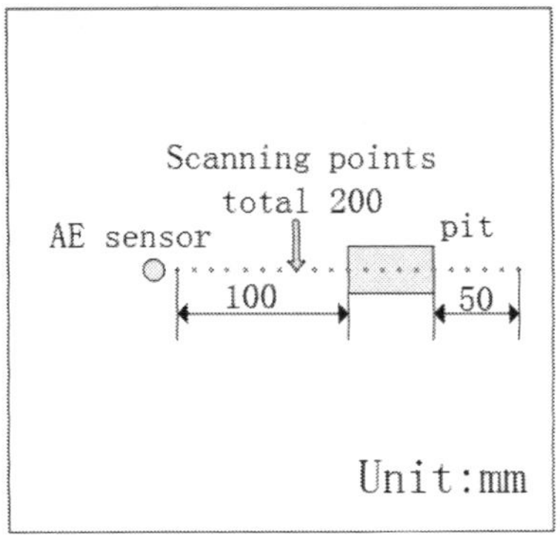

Figure 1. Scanning schematic representation of the test specimen

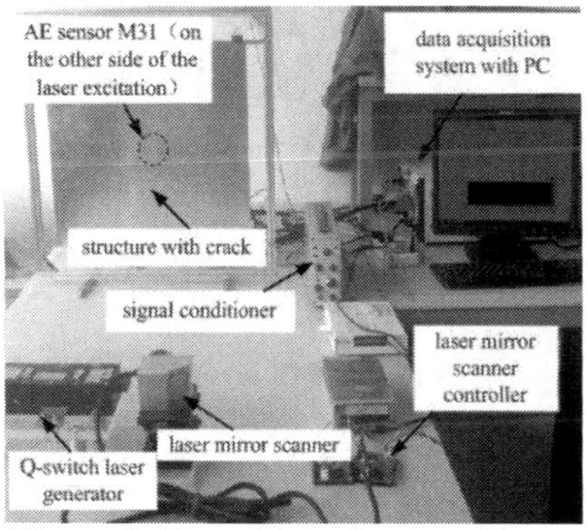

Figure 2. Experimental setup of laser ultrasonic scanning technique

Figure 2 shows the experimental system of the laser ultrasonic scanning technique. The system consists of three units, excitation, sensing and control units. The excitation unit is composed of a lamp-pumped pulse Nd:YAG laser and a two-dimensional laser mirror scanner. The sensing unit is composed of a micro-miniature AE sensor and a signal conditioner. The control unit manages the entire system and synchronizes each unit with a controller (PXIe-8133). The laser mirror is controlled by an analog output module (cRIO-9263) and the signals are sampled by a high-speed digitizer (PXI-5105).

3. DAMAGE EVALUATION WITH FREQUENCY-WAVENUMBER ANALYSIS

As shown in Figure 1, the scanning path starts from 10 mm away from the AE sensor to 210 mm with a spatial resolution of 1 mm, corresponding to 201 measurement points. The pit locates from 110 mm away from the AE sensor to 160 mm with the length of 50 mm. In this study, three pits with different depths on an aluminum plate have been tested. To obtain the signals of the waves propagating though the inspection region, the sampling frequency is 10 MS/s.

3.1 Ultrasonic propagation visualization

The time-space wavefield including a pit with 0.6 mm in depth can be obtained with the signal collected from the scanning line as shown in Figure 3. A faster S_0 mode and a slower A_0 mode can be found in time-space domain. Due to that the AE sensor is response to the out-of-plane velocity of the structural vibration, the amplitude of the A_0 mode Lamb wave is much stronger than S_0 mode in Figure 3. It should be mentioned that the reflection phenomena can be found at both the beginning (x = 100 mm) and the end (x = 150 mm) boundaries of the damage. The interaction of the incident waves and the reflected waves, which is called "trapped wave" phenomenon, exhibits distinctly during the damaged area, showing the position of the damage roughly.

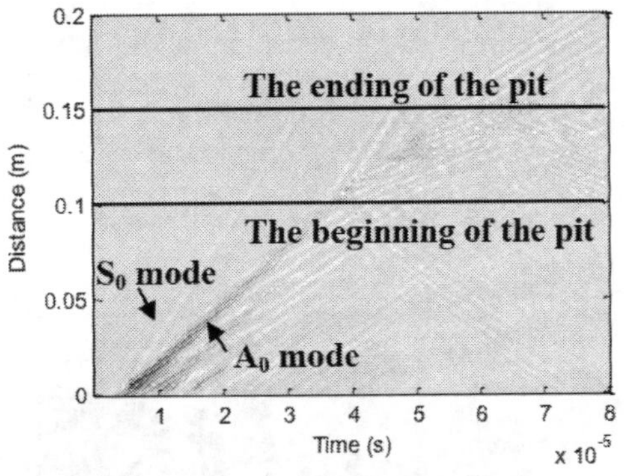

Figure 3. The time-space wavefield of the Lamb wave

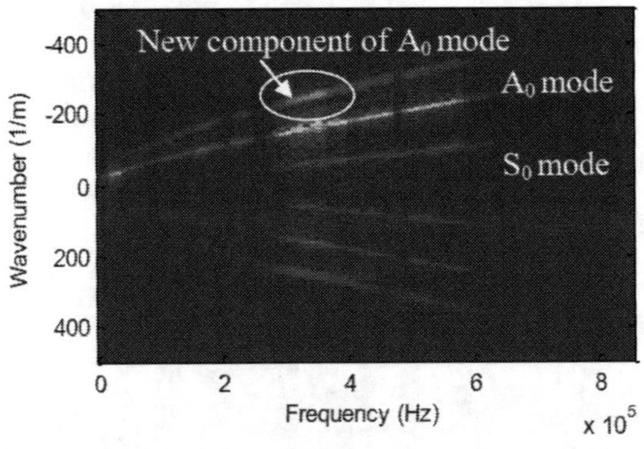

Figure 4. The frequency–wavenumber spectra of the time-space wavefield

3.2 Frequency-wavenumber analysis

In order to detect the depth of the pits, the phase velocity, which is a function of frequency and thickness of the structure, is calculated through the frequency–wavenumber spectra using 2D-FT [22].

The wavefield propagating in an aluminum plate along the x direction can be described by a general analytic expression as

$$w(x,t) = A(\omega)e^{j(\omega t - kx - \theta)} \tag{1}$$

where $A(\omega)$ is the frequency-dependent amplitude, θ denotes the phase, k is wavenumber as the inverse proportional to the wavelength, and ω is the angular frequency. Therefore, the wavefield in frequency-wavenumber domain can be determined by 2D-FT through a series of responses at the positions with an equal interval. The 2D-FT can be given as:

$$W(k,\omega) = \iint w(x,t)e^{-j(kx-\omega t)}dxdt. \tag{2}$$

where $W(k, \omega)$ is the amplitude of the component of the Lamb wave with the wavenumber k and the angular frequency ω. The phase velocity can be acquired from the slope of angular frequency to wavenumber, ω / k.

Figure 4 shows the wavefield in frequency-wavenumber domain, converted from the time-space wavefield in Figure 3. The A_0 mode and S_0 mode of the Lamb wave can be recognized directly, as well as a new component of A_0 mode caused by the damage. As the healthy area is larger than the damaged area, the amplitude of A_0 mode is stronger than the new component. Hence, the dispersion curve of A_0 mode wave in the healthy structure can be obtained by extracting the points with the max amplitude at different frequencies. As shown in Figure 5, by applying polynomial fitting, the phase velocities within a broadband can be calculated.

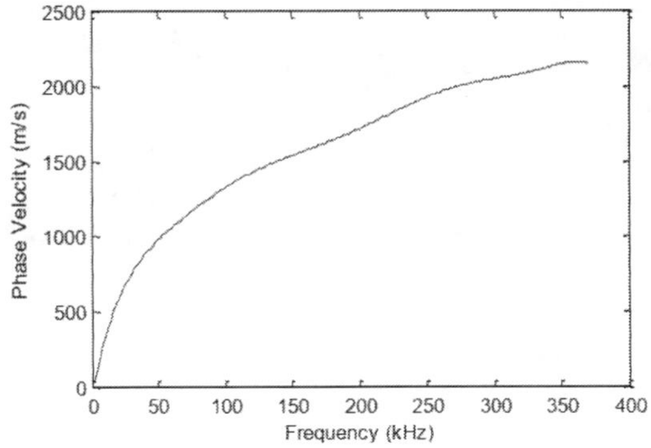

Figure 5. Lamb wave dispersion of A_0 mode

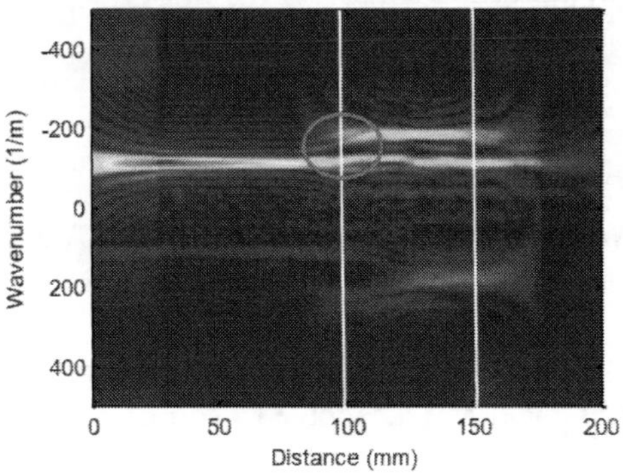

Figure 6. The space-wavenumber figure at frequency of 195 kHz

The space information is ignored after the 2D-FT. In order to retain the space information and get the phase velocity variation along the scanning line, the 2D-SSFT has been developed to obtain the local variation of the phase velocity. A fixed size window function, which is non-zero for only a short period in space, has been used to divide the time-space wavefield into several segments over the space dimension. The 2D-SSFT can be given as

$$SW(\bar{x},k,\omega) = \iint w(x,t)\Omega(x-\bar{x},t)e^{-j(kx-\omega t)}dxdt \tag{3}$$

where $\bar{x}$ is the retained space variable and $\Omega(x,t)$ is a window function. By sliding the window along the space dimension, the abnormal wavenumber in the damaged location can be found. To obtain the location of the damaged areas and the phase velocities in the scanning points, a rectangular window with 50 mm in length has been used in the 2D-SSFT. Figure 6 shows the space–wavenumber slices at frequency of 195 kHz. The space–wavenumber spectrum is significantly changed in the damage region. Therefore, the location of the damage can be detected by the variation of the wavenumbers. Figure 7 and Figure 8 show the local frequency-wavenumber wavefield at 60 mm (in healthy region) and 125 mm (in damage region) respectively.

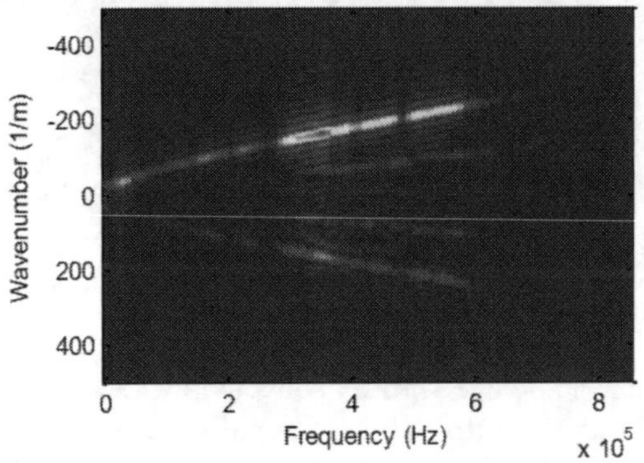

Figure 7. The local frequency-wavenumber wavefield at 60 mm

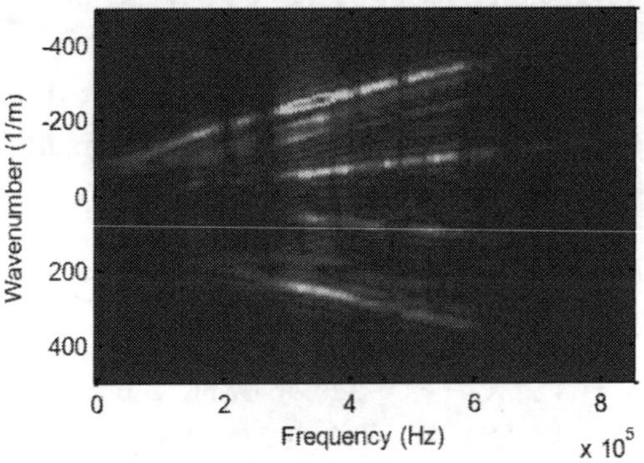

Figure 8. The local frequency-wavenumber wavefield at 125 mm

3.3 Damage depth reconstruction

With the phase velocity on the local region obtained by selecting one segment of the wavefield after 2D-SSFT, a genetic arithmetic is used to reconstruct the depth of damage by minimizing the error function, as:

$$Minimize \quad \mathrm{err} = \sum_{i}^{i=n}\left(v_i^{theory} - v_i^{experiment}\right)^2 \tag{4}$$

where v_i^{theory} is the calculated velocity , $v_i^{experiment}$ is the experimentally measured velocity, and i is the index specifying the sequence of the velocity.

The phase velocity is a function of the product of frequency and thickness, as:

$$v = g(f \cdot d) \tag{5}$$

where v is the phase velocity, g is the function, f is the frequency and d is the thickness of plate. The expression of the function g has been confirmed by applying polynomial fitting the dispersion curve in the healthy structure. With the estimated $g(f\cdot d)$, the phase velocity in any thickness can be calculated. Furthermore, the phase velocity in damaged location can be obtained from the local frequency-wavenumber wavefield at 125 mm in Figure 8. The genetic arithmetic starts with a population of randomly guessed candidate values of the depth, each of which are then evaluated using the error function. A few of them with the lowest error function are preserved, while the others are placed with new trials. After some generations, the arithmetic converges to the value with the lowest error, and the depth can be defined as this value. Finally, three pits with different depths have been detected and the detected depths of the pits are shown in Table 1. It is seen that for any pits, the error between the reconstructed depths and the true depths is below 7%, demonstrating that the proposed inverse method has a better accuracy for the depth reconstruction.

Table 1 Detected Depths and Percentage Error

	Pit 1	Pit 2	Pit 3
true depth (mm)	0.4	0.6	1.4
reconstructed depth (mm)	0.397	0.615	1.486
percentage error	-0.65%	2.43%	6.13%

By repeating the above method in all scanning points, the depths of the plate along the scanning line can be reconstructed as shown in Figure 9. The reconstructed depths coincide well with the true depths in most points along the scanning line. However, due to the influence of the length of the rectangular window, the reconstructed depths near the boundary of the pits do not match well with the true depths. When the rectangular window moves to the boundary of the pits, both the time-space wavefield of the healthy region and damaged area have been taken into account concurrently. The wavenumbers in the pit cannot be extracted by picking the peak value in the wavenumber spectra. This phenomenon can be seen distinctly in Figure 6 marked in red circle. Thus, the length of the window and the method to extract the wavenumbers in damaged area should be considered prudently when using the frequency–wavenumber analysis.

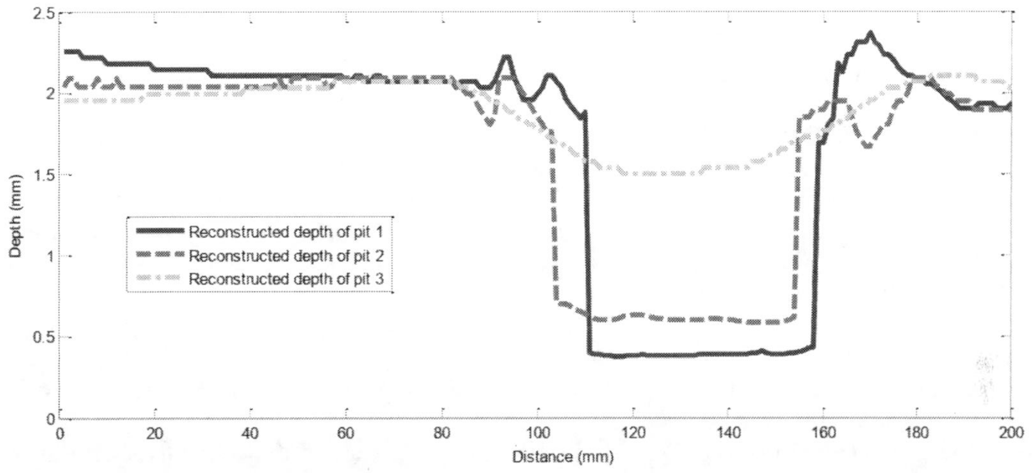

Figure 9. The reconstructed depths of three pits

4. CONCLUSION

This paper presents a method to reconstruct the depth of damage in the aluminum plate based on genetic algorithm using the laser ultrasonic system. In this method, time-space wavefield is obtained from the signals collected from the scanning line. And then, the phase velocity dispersion curve is calculated through the 2D-FT and polynomial fitting. To detect the phase velocities on local region, the 2D-SFFT is used to convert the local time-space wavefield to the local frequency-wavenumber wavefield. Finally, by minimizing the error between the velocities calculated from the formula of velocity dispersion curve and the phase velocities obtained from the local frequency-wavenumber wavefield, the depth of damage can be detected through a genetic algorithm. The error between the reconstructed depths and the true depths is below 7%, which demonstrates that the proposed method is able to achieve good detection accuracies. As a future work, we are willing to extend the present methodology to more complex and practical structure.

ACKNOWLEDGMENTS

This work is supported by A Project Funded by the Priority Academic Program Development of Jiangsu Higher Education Institutions (PAPD), The State Key Laboratory Program under Grant (Nos.MCMS-0514K01), China Postdoctoral Science special Foundation (Nos.2014T70514), and the Fundamental Research Funds for the Central Universities (Nos.NE2015001 &NE2015101).

REFERENCES

1. Drinkwater, B. W., Wilcox, P. D., "Ultrasonic arrays for non-destructive evaluation: A review," *NDT & E International*, Vol. 39, No.7, Oct. 2006, pp. 525-541.
2. Bull, D., Helfen, L., Sinclair, I., Spearing, S., Baumbach, T., "A comparison of multi-scale 3D X-ray tomographic inspection techniques for assessing carbon fibre composite impact damage," *Composites Science and Technology*, Vol. 75, No.7, 2013, pp. 55-61
3. Maier, A., Schmidt, R., Oswald-Tranta, B., "Non-Destructive Thermography Analysis of Impact Damage on Large-Scale CFRP Automotive Parts," *Materials*, Vol. 7, No.1, 2014, pp. 413-429.
4. Jun, C., Ji, H., Qiu, J., Takagi, T., Uchimoto, T., "Novel electromagnetic modeling approach of carbon fiber-reinforced polymer laminate for calculation of eddy currents and eddy current testing signals," *Journal of Composite Materials*, Vol. 49, No.5, 2015, pp. 617-631.
5. Dalton, R. P., Cawley, P., Lowe, M. J. S., "The Potential of Guided Waves for Monitoring Large Areas of Metallic Aircraft Fuselage Structure," *Journal of Nondestructive Evaluation*, Vol. 20, No.1, 2001, pp. 29-46.
6. Su, Z., Ye, L., Lu, Y., "Guided Lamb waves for identification of damage in composite structures: A review. ," *Journal of Sound and Vibration*, Vol. 295, No. 3-5, 2006, pp. 753-780.
7. Raghavan, A., Cesnik, C. E. S., "Review of Guided-wave Structural Health Monitoring ," *The Shock and Vibration Digest*, Vol. 39, No. 2, 2007, pp. 91-104
8. Zhang, C., Qiu, J., Ji, H., "Damage localization using warped frequency transform in active structural health monitoring," *International Journal of Applied Electromagnetics and Mechanics*, Vol. 47, No. 4, Jun. 2015, pp. 897-909.
9. Yeum, C. M., Sohn, H., Ihn, J. B., Lim, H. J.,"Instantaneous delamination detection in acomposite plate using a dual piezoelectric transducer network ," *Composite Structures*, Vol. 94, No. 12, 2012, pp. 3490-3499.
10. Yu, L., Victor, G., "Piezoelectric Wafer Active Sensors in Lamb Wave-Based Structural Health Monitoring ," *JOM*, Vol. 64, No. 7, 2012, pp. 814-822.
11. Sohn, Y., Krishnaswamy, S., "Interaction of a scanning laser-generated ultrasonic line source with a surface-breaking flaw," *Journal of the Acoustical Society of America*, Vol. 115, No. 1, 2004, pp. 172-181.
12. Takatsubo, J., Wang, B., Tsuda, H., Toyama, N., "Generation laser scanning method for the visualization of ultrasounds propagating on a 3-D object with an arbitrary shape," *Journal of Solid Mechanics and Materials Engineering*, Vol. 1, No. 12, 2007, pp. 1405-1411.
13. Yashiro, S., Takatsubo, J., Miyauchi, H., Toyama, N., "A novel technique for visualizing ultrasonic waves in general solid media by pulsed laser scan," *NDT and E International*, Vol. 41, No. 2, 2008, pp. 137-144.
14. Chia, C. C., Jeong, H. M., Lee, J. R., Park, G., "Composite aircraft debonding visualization by laser ultrasonic scanning excitation and integrated piezoelectric sensing," *Structural Control and Health Monitoring*, Vol. 19, No. 7, 2012, pp. 605-620.
15. Sun, G., Zhou, Z., Chen, X., Wang, J., "Laser ultrasonic propagation imaging method in the frequency domain based on wavelet transformation," *Optics and Lasers in Engineering*, Vol. 49, No. 1, 2013, pp. 6481-6486.
16. Ruzzene, M., "Frequency-Wavenumber Domain Filtering for Improved Damage Visualization," *AIP Conference Proceedings*, Vol. 894, No. 1, 2007, pp. 1556-1563.
17. Sohn, H., Dutta, D., Yang, J. Y., DeSimio, M., Olson, S., Swenson, E., "Automated detection of delamination and disbond from wavefield images obtained using a scanning laser vibrometer," *Smart Materials and Structures*, Vol. 20, No. 4, 2011, pp. 45017-45026.
18. An, Y.-K., Park, B., Sohn, H., "Complete noncontact laser ultrasonic imaging for automated crack visualization in a plate," *Smart Materials and Structures*, Vol. 22, No. 2, 2013, pp. 025002.

19. Flynn, E. B., Chong, S. Y., Jarmer, G. J., "Structural imaging through local wavenumber estimation of guided waves," *NDT and E International*, Vol. 59, 2013, pp. 1-10.
20. Tian, Z., Yu, L., Leckey, C., "Delamination detection and quantification on laminated composite structures with Lamb waves and wavenumber analysis" *Journal of Intelligent Material Systems and Structures*, 2014, 1045389X14557506.
21. Fink, M., "Acoustic time-reversal mirrors," *Topics in Applied Physics*, Vol. 84 , 2002, pp. 17-43.
22. Ambrozinski, L. Packko, P., Pieczonka, L., Stepinski, T., Uhl, T., Staszewski, W. J.,"Identification of material properties–efficient modelling approach based on guided wave propagation and spatial multiple signal classification," *Structural Control and Health Monitoring*, Vol. 22, No. 7, 2015, pp. 969-983.

ICAST2015 #045

Performance Analysis of A Multi-Axes Micro Precision Positioner under Static Loads

Jau-Liang Chen*, and Wan-Lin Yu

Department of Mechanical Engineering, National Chung University, Taichung, Taiwan

Abstract

The purpose of this paper is focusing on the performance of a six DOF micro precision positioner under different loadings (248 g and 978 g). The positioner is composed of an X-Y bottom plate with flexure mechanism, three Z-axis flexure mechanisms, and a moving plate on top of the Z-axis mechanisms. Three *Piezomechanik* Pst150/5×5/20 PZT actuators (PZT1, PZT2, and PZT3), are used to provide the displacement along X-axis and Y-axis together with rotation θ about Z-axis. Another three *PhysikInstrumente* P-840.10 PZT actuators (PZT4, PZT5, and PZT6), are used to provide the displacement along Z-axis together with the rotation ϕ about X-axis and φ about Y-axis. An optical measuring system consists of three set of laser diodes, quadrant photodiodes (QPD), and modulation/demodulation signal amplifiers has been applied. From the experimental results, it is found that the maximum displacements under 248 g (978 g) loading are: 12.876 (12.140) μm, 15.771 (15.281) μm, and 13.890 (13.347) μm along the X-, Y-, and Z-axis respectively; while, the maximum rotation angles are: 76.961 (75.885) arcsec, 70.997 (70.172) arcsec, and 36.232 (34.236) arcsec along the ϕ -, φ-, and θ-axis respectively. It is also found that the standard deviations on the linear and rotational axis are less than 0.140 μm and 1.616 arcsec, and the error of the measuring system is less than 0.25 μm and 0.55 arcesec on linear and rotational axis.

1. INTRODUCTION

Along with the development of micro-precision machining technology, the precision of machining already reach nanometers requirement. To achieve this requirement, the precision of the positioning system becomes a very important issue in a machine tool. In order to fulfill the high precision requirement for positioning system, various actuators have been studied for increasing the precision and stability of the stage. For examples, Gao *et. al.* [1] used linear motor to drive a 3DOF positioning stage. Sato *et. al.* [2] in their research used voice coil motor (VCM) to drive the positioning stage. Several researchers [3, 4] have used Piezo-electric (PZT) actuator to drive the stage. It is because the PZT has the following advantages: (1) unlimited resolution, (2) large force generation, (3) no magnetic field, (4) rapid response, (5) low power consumption; (6) operating at cryogenic temperature, (7) vacuum and clean-room compatible, and (8) no wear and tear [5]. According to these advantages, PZT actuators are selected in this study for driving the positioning stage.

To achieve high accuracy, small volume, multi-DOF moving and coplanar mechanism, Jywe *et. al.* [6] in their research used 8 PZT actuators, with cylindrical and arc shape flexure hinges, capacitive sensors, in design of a 5-DOF nanometer-scale stage. Fung *et. al.* [7, 8] have used two 3-DOF

* jlchen@dragon.nchu.edu.tw

precision stages to construct a 6-DOF micro-positioning stage, which is also driven by PZT actuators. Wang [9] in his research has combined a 3-DOF parallel mechanism and a 3-DOF serial mechanism to produce a hybrid 6-DOF positioning stage.

The objective of this research is trying to analyze the performance of the positioner (developed by Wang [10]) under varies static loading, as to be used in a micro-precision milling machine process. The procedures of this research are divided into the following four steps: (1) fundamental formulation derivation, (2) modal analysis, (3) optical measurement system construction, and (4) experiment verification.

2. POSITIONING SYSTEM

The positioning system is constructed by three subsystems (1) positioner, (2) control system, and (3) measuring system (as shown in Fig. 1). The command inputs from control system to the positioner, while the measuring system will feedback the actual displacement to the control system for further command.

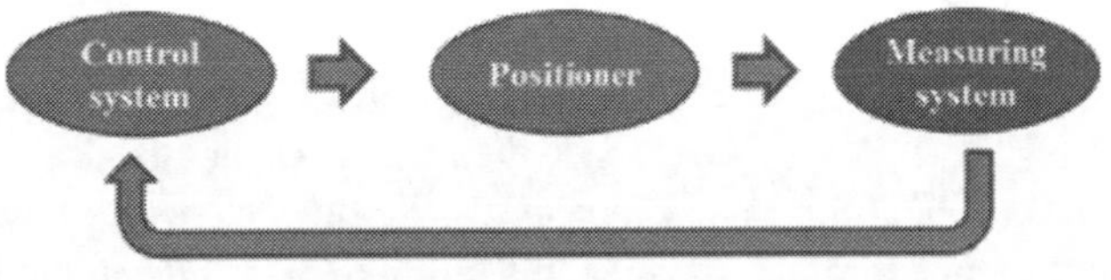

Figure 1. The positioning system

2.1 The Positioner

In this research, the positioner is developed by Wang and is composed of three different parts: (1) XY bottom plate, (2) Z-axis flexure mechanism, and (3) moving stage (as shown in Fig. 2).

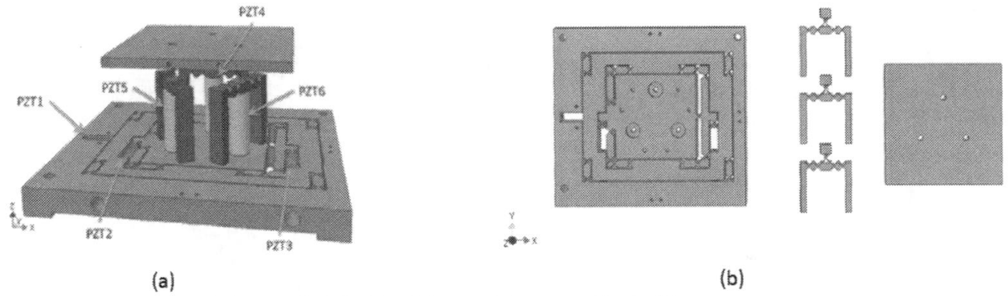

Figure 2. The Positioner (a) assembled, (b) components

The application of flexure structure is the key issue in the design of precision positioning stage. It is using the deformation of a material to achieve a moving motion. It has the advantages of (1) no wearing, need no lubrication, and high stability, (2) with smooth and continuous motion, no gap, and (3) high repeatability. Due to these advantages, it can achieve nanometer positioning accuracy. The flexure structure can be categorized into (1) notch hinge type, and (2) leaf plate type. In notch hinge type structure, it is constructed by two notch, and it acts as a rotational joint. According to Paros and Weisbord's equation [11], the stiffness K_n and maximum rotation angle θ_{max} of notch hinge are given as follows:

$$K_n = \frac{2Ebt^{5/2}}{9\pi r^{1/2}} \tag{1}$$

$$\theta_{\max} = \frac{3\pi r^{1/2}}{4Et^{1/2}}\sigma_Y \tag{2}$$

In this research, two different flexure hinges: (a) notch hinges (as shown in Fig. 3(a)), and dual-axes hinges (as shown in Fig 3. (b)) are being used. The dimensions of this positioner is 164 × 148 × 76 m^3.

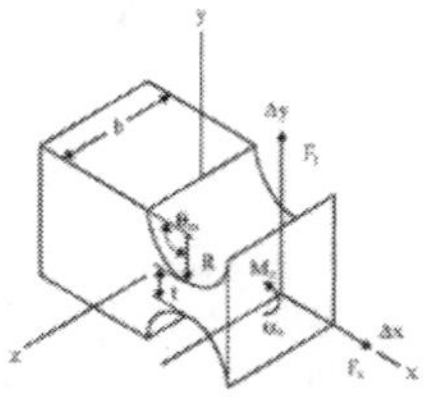

(a)

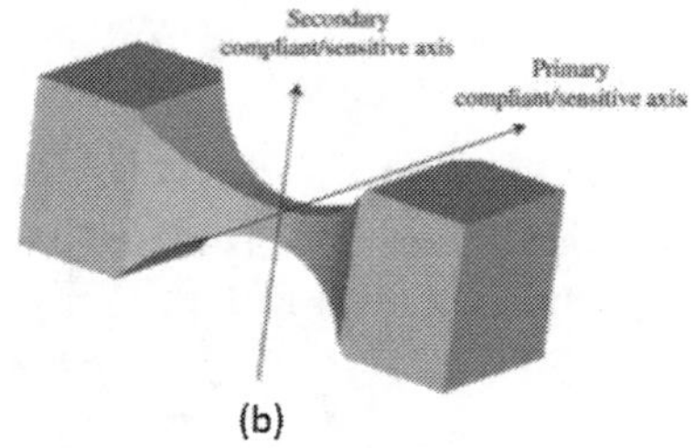

(b)

Figure 3. Flexure hinges (a) notch, (b) dual axes

Three PZT actuators, (PZT1, PZT2, and PZT3) as shown in Fig. 2, are used to provide the displacement along X-axis and Y-axis together with rotation θ about Z-axis. Another three PZT actuators, (PZT4, PZT5, and PZT6) are used to provide the displacement along Z-axis together with the rotation ϕ about X-axis and φ about Y-axis. In this research, *Piezomechanik* PSt 150/5×5/20 piezo-actuators are used for PZT1, PZT2 and PZT3. While, three *PI* P-840.10 piezo-actuators are used for PZT4, PZT5 and PZT6. These actuators are stacked type piezo-actuators.

2.2 The Optic Measuring System

In this study, the optic measuring system is being used to measure the displacement of our positioner. This measuring system is composed of three measuring units, and can measure the six axes displacement simultaneously. Each unit is combined of a quadrant photodiode (QFD) and a laser diode (as shown in Fig. 4). Figure 5 shows the arrangement of the measuring system with the positioner.

(a)

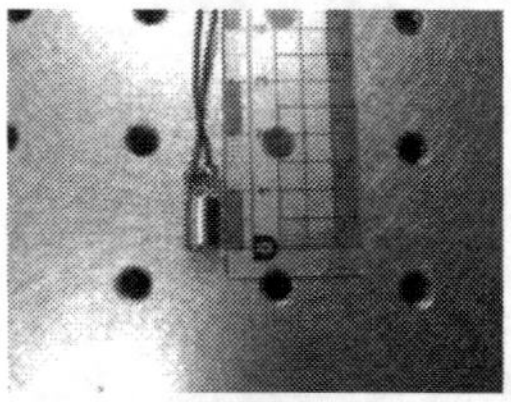

(b)

Figure 4. Measuring devices (a) quadrant photodiode (QPD), (b) laser diode

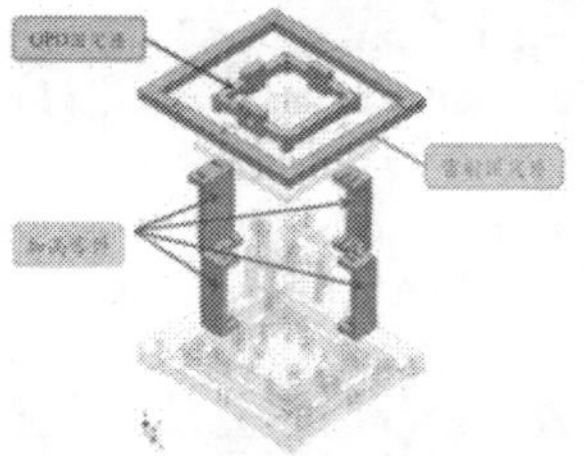

(a)

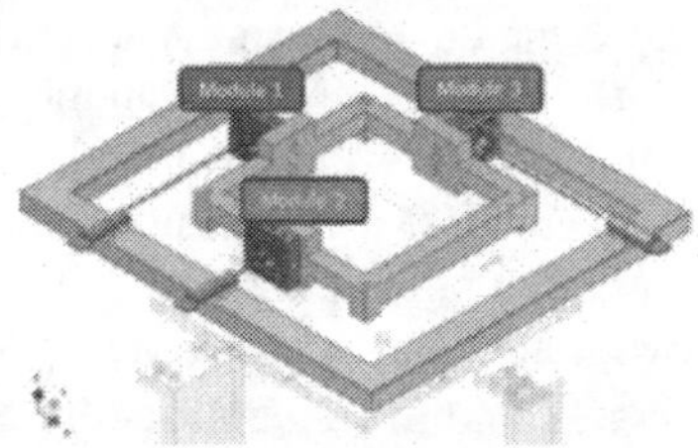

(b)

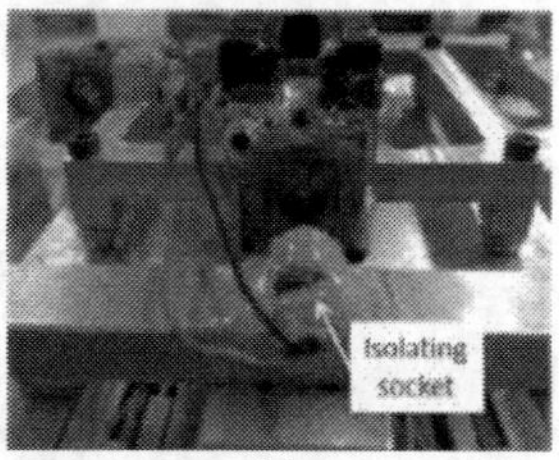

(c)

Figure 5. Arrangement of the measuring system (a) with positioner, (b) arrangement of each unit, and (c) the measuring unit

3. MODAL ANALYSIS OF THE POSITIONER

In this section, the modal analysis is being conducted. Lagrange's equation and virtual work theorems are used to find the dynamic equation and natural frequencies of the positioning stage. Next, ANSYS workbench was used to simulate the mode types and frequencies. To compare with the impact test result, it is assumed to be freely-supported. During the analysis, it is assumed that the monolithic mechanism is a rigid body and only notch hinges will be deformed.

3.1 X-, Y-, and θ-Axis Modal Analysis

Figure 6 shows the sketch diagram of the XY bottom plate, where K_n is the stiffness of the notch hinges on X-axis and Y-axis. When conducting the X-, Y-, and θ-axis modal analysis, beside the bottom plate, the effect of the Z-axis flexure mechanisms and moving plate should also be considered.

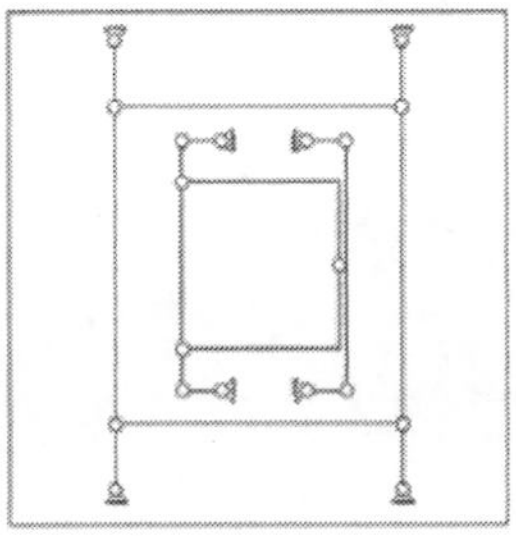

Figure 6. Schematic diagram of the XY bottom plate

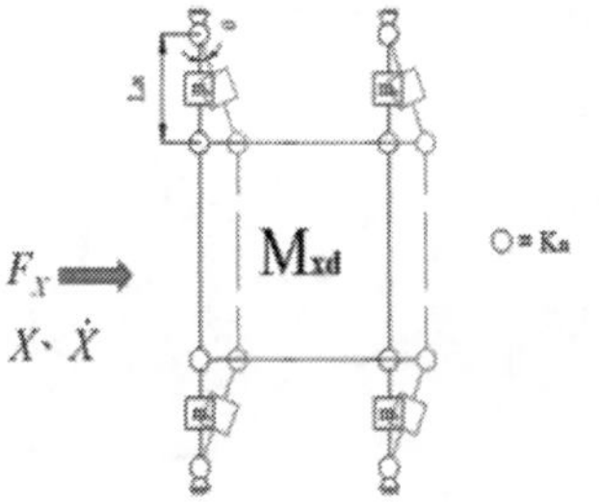

Figure 7. Lumped model in X-axis

3.1.1 X-axis modal analysis

Figure 7 shows the lumped model of the bottom plate on the X-axis, where M_{xd} is the mass of the plate, m_x is the mass of hinge, I_x is the moment of inertia, L_x is the length of hinge. A force F_x is applied by PZT1 along the X-direction.

From virtual work theorem, the kinetic energy T_1 and potential energy U_1 of the bottom plate in X-axis are given by:

$$T_1 = \frac{1}{2}M_{xd}\dot{X}^2 + 4\left[\frac{1}{2}I_x\dot{\theta}^2\right] = \frac{1}{2}M_{xd}\dot{X}^2 + 4\left[\frac{1}{2}I_x(\frac{\dot{X}}{L_x})^2\right] \tag{3}$$

$$U_1 = 8\left[\frac{1}{2}K_n\theta^2\right] = 8\left[\frac{1}{2}K_n(\frac{X}{L_x})^2\right] \tag{4}$$

By using Lagrange's equation $\frac{d}{dt}\left(\frac{\partial L}{\partial \dot{X}}\right) - \frac{\partial L}{\partial X} = 0$, where $L = T_1 - U_1$, we then have

$$\left(M_{xd} + \frac{4I_x}{L_x^{\,2}}\right)\ddot{X} + \left(\frac{8K_n}{L_x^{\,2}}\right)X = 0 \tag{5}$$

Let $M_{x1} = M_{xd} + \frac{4I_x}{L_x^{\,2}}$, and $K_{x1} = \frac{8K_n}{L_x^{\,2}}$.

Next, the same analysis processes are applied for Z-axis flexure mechanism and moving plate, we thus have

$$(M_u)\ddot{X} + (3K_{tx})X = 0 \tag{6}$$

where M_u is the mass of moving plate, and K_{tx} is the stiffness of the Z-axis flexure mechanism. Let $M_{x2} = M_u$, and $K_{x2} = 3K_{tx}$. Now, consider the frame of the base. M_{xx} is the mass of the frame, which is equal to the total mass of the base minus the mass of bottom plate. K_{xx} is the stiffness of the flexure hinge. We have

$$(M_{xx})\ddot{X} + (K_{xx})X = 0 \tag{7}$$

The whole system in X-axis can thus be decomposed as the bottom plate, the frame, and the Z-axis mechanism with moving plate are in serial (as shown in Fig. 8). The natural frequency of this system can thus be solved by the following equation:

$$\det\left([\bar{K}] - \omega_j^2[\bar{M}]\right) = 0 \quad ,(j = 1,2,3) \tag{8}$$

where $[\bar{K}] = \begin{bmatrix} K_{XX} & 0 & 0 \\ 0 & K_{x1} & 0 \\ 0 & 0 & K_{x2} \end{bmatrix}$, and $[\bar{M}] = \begin{bmatrix} M_{XX} & 0 & 0 \\ 0 & M_{x1} & 0 \\ 0 & 0 & M_{x2} \end{bmatrix}$

Thus

$$\begin{vmatrix} K_{x1} - \omega^2 \times M_{xx} & -K_{x1} & 0 \\ -K_{x1} & K_{x1} + K_{x2} - \omega^2 \times M_{x1} & -K_{x2} \\ 0 & -K_{x2} & K_{x2} - \omega^2 \times M_{x2} \end{vmatrix} = 0 \tag{9}$$

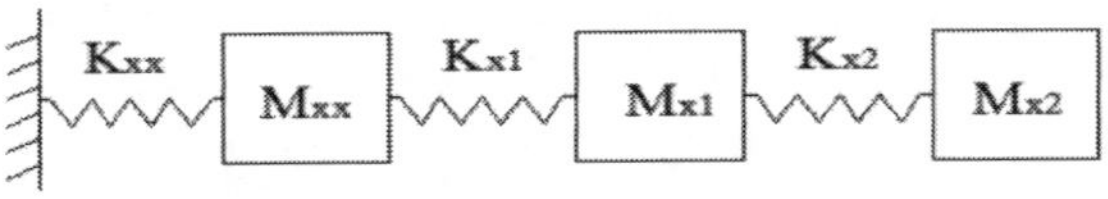

Figure 8. The X-axis system

3.1.2 Y-axis modal analysis

The lumped model in Y-axis of the bottom plate is shown in Figure 9. Force F_y is applied by PZT2 and PZT3 together along the Y-direction. Similar procedure is performed. We then have:

$$(M_{yy})\ddot{Y} + (K_{yy})Y = 0 \tag{10}$$

$$(M_{y1})\ddot{Y} + (K_{y1})Y = 0 \tag{11}$$

$$(M_{y2})\ddot{Y} + (K_{y2})Y = 0 \tag{12}$$

where $M_{y1} = M_{yd} + \frac{4I_y}{L_y{}^2}$, $M_{y2} = M_u$, $K_{y1} = \frac{8K_n}{L_y{}^2}$, and $K_{y2} = 3K_{ty}$. Substitute these values into Eqn. (8), the natural frequency in Y-axis can thus be calculated.

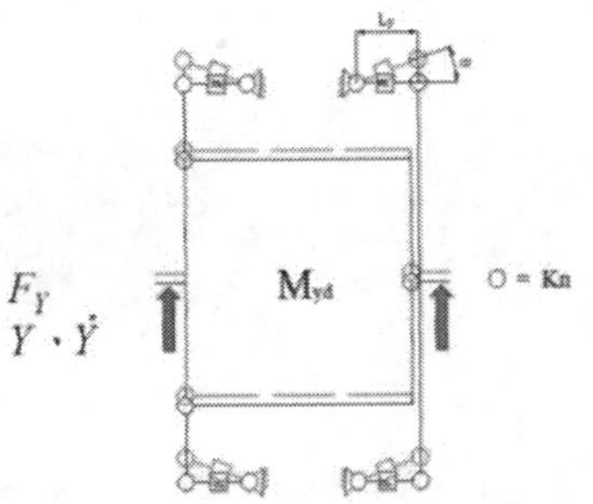

Figure 9. Lumped model in Y-axis

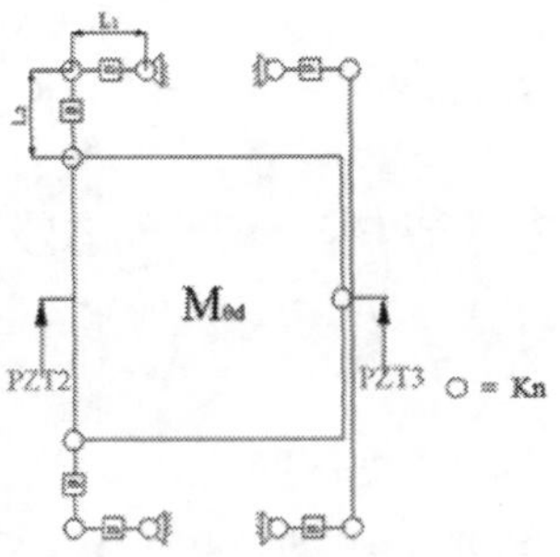

Figure 10. Lumped model in θ-axis

3.1.3 θ-axis modal analysis

Figure 10 shows the lumped model in θ-axis. The rotation is achieved by the elongation difference between PZT2 and PZT3. When the bottom plate rotate an angle θ, the deformation on the notch hinges and parallel mechanism are shown in Fig. 11. L_1 and L_2 are the length of each hinge, while m_1 and m_2 are the masses, a is half the length of bottom plate. When a small rotation occurs, we have:

$$L_1 \sin\theta_1 = a\sin\theta \tag{13}$$

$$L_2 \sin\theta_2 = a\sin\theta \tag{14}$$

$$\theta_1 = \frac{a}{L_1}\theta \tag{15}$$

$$\theta_2 = \frac{a}{L_2}\theta \tag{16}$$

(a) (b)

Figure 11. Deformation model (a) L-type hinge, (b) parallel mechanism.

From virtual work theorem, the kinetic energy T_1 and potential energy U_1 of the bottom plate in X-axis are given by:

$$\begin{aligned} T_1 &= \frac{1}{2} I_{\theta d}\dot{\theta}^2 + 4 \times \frac{1}{2} I_1 \dot{\theta}_1^{\,2} + 2 \times \frac{1}{2} I_2 \dot{\theta}_2^{\,2} \\ &= \frac{1}{2}\left[\frac{1}{6} M_{\theta d}(2a)^2\right]\dot{\theta}^2 + 4 \times \frac{1}{2}\left(\frac{1}{3} m_1 L_1^{\,2}\right)\left(\frac{a}{L_1}\dot{\theta}^2\right)^2 \times \frac{1}{2}\left(\frac{1}{3} m_2 L_2^{\,2}\right)\left(\frac{a}{L_2}\dot{\theta}^2\right)^2 \end{aligned} \tag{17}$$

$$\begin{aligned} U_1 &= 8 \times \frac{1}{2} K_n \theta_1^{\,2} + 4 \times \frac{1}{2} K_n \theta_2^{\,2} + \frac{1}{2} K_n \theta^2 \\ &= 8 \times \frac{1}{2} K_n \left(\frac{a}{L_1}\theta\right)^2 + 4 \times \frac{1}{2} K_n \left(\frac{a}{L_2}\theta\right)^2 + 3 \times \frac{1}{2} K_n \theta^2 \end{aligned} \tag{18}$$

By using Lagrange's equation $\frac{d}{dt}\left(\frac{\partial L}{\partial \dot{X}}\right) - \frac{\partial L}{\partial X} = 0$, where $L = T_1 - U_1$, we then have

$$\left(\frac{2}{3} M_{\theta d} a^2 + \frac{4}{3} m_1 a^2 + \frac{2}{3} m_2 a^2\right)\ddot{\theta} + \left(8K_n\left(\frac{a}{L_1}\right)^2 + 4K_n\left(\frac{a}{L_2}\right)^2 + K_n\right)\theta = 0 \tag{19}$$

Let $I_{\theta 1} = \frac{2}{3} M_{\theta d} a^2 + \frac{4}{3} m_1 a^2 + \frac{2}{3} m_2 a^2$, and $K_{\theta 1} = 8K_n\left(\frac{a}{L_1}\right)^2 + 4K_n\left(\frac{a}{L_2}\right)^2 + 3K_n$.

Next, the same analysis processes are applied for Z-axis flexure mechanism and moving plate, we thus have

$$(I_u)\ddot{\theta} + \left(3K_{t\theta} \times L_\theta^{\,2}\right)\theta = 0 \tag{20}$$

where I_u is the moment of inertia of moving plate, $K_{t\theta}$ is the stiffness of the Z-axis flexure mechanism, and $L\theta$ is the distance from the center of moving plate to the dual axis hinge. Let $I_{\theta 2} = I_u$, and $K_{\theta 2} = 3K_{tx} \times L_\theta^{2}$. While, for the frame, we have

$$(I_{\theta\theta})\ddot{\theta} + (K_{\theta\theta})\theta = 0 \tag{21}$$

Substitute $I_{\theta\theta}$, $I_{\theta 1}$, $I_{\theta 2}$, $K_{\theta\theta}$, $K_{\theta 1}$, and $K_{\theta 2}$ into Eqn. (8), the natural frequency in θ-axis can thus be calculated.

3.2 Z-, ϕ-, and φ-Axis Modal Analysis

When conducting the Z-, ϕ-, and φ-axis modal analysis, the effect of the bottom plate, the Z-axis flexure mechanisms and moving plate should be considered simultaneously.

3.2.1 Z-axis modal analysis

The lumped model of the one single Z-axis flexure mechanism is shown in Fig. 12. Use the same procedure as discussed in section 3.1.1, we have:

$$(M_{zz})\ddot{Z} + (K_{zz})Z = 0 \tag{22}$$

$$(M_{z1})\ddot{Z} + (K_{z1})Z = 0 \tag{23}$$

$$(M_{z2})\ddot{Z} + (K_{z2})Z = 0 \tag{24}$$

$$(M_{z3})\ddot{Z} + (K_{z3})Z = 0 \tag{25}$$

$$(M_{sz})\ddot{Z} + (K_{sz})Z = 0 \tag{26}$$

Substitute these values (M_{zz}, M_{z1}, M_{z2}, M_{z3}, M_{sz}, K_{zz}, K_{z1}, K_{z2}, K_{z3}, and K_{sz}) into Eqn. (8), the natural frequency in Z-axis can thus be calculated.

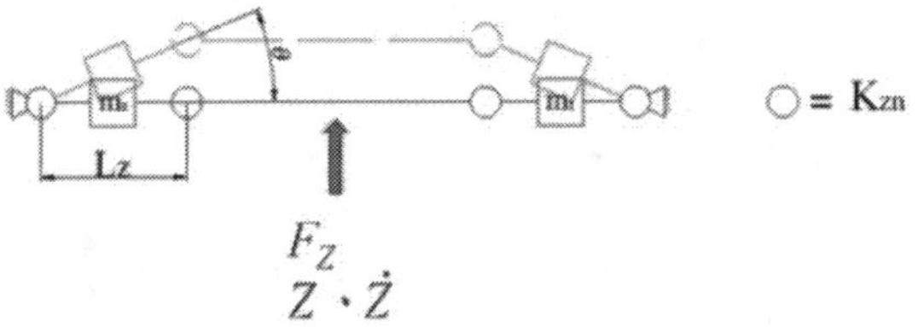

Figure 12. The lumped model of a single Z-axis mechanism

3.2.2 ϕ-axis modal analysis

Figure 13 shows the lumped model in ϕ-axis. The rotation is achieved by the elongation difference among PZT4, PZT5 and PZT6. The arrangement of these three PZTs is shown in Fig. 14. L_1 is the side length of triangle. L_{sz} and b_{sz} are the length and thickness of moving plate respectively. K_{zn} and $K_{t\phi}$ are the stiffness of notch and dual hinge. Using the same procedure in Section 3.1.3, we can then calculate the nature frequency in ϕ-axis.

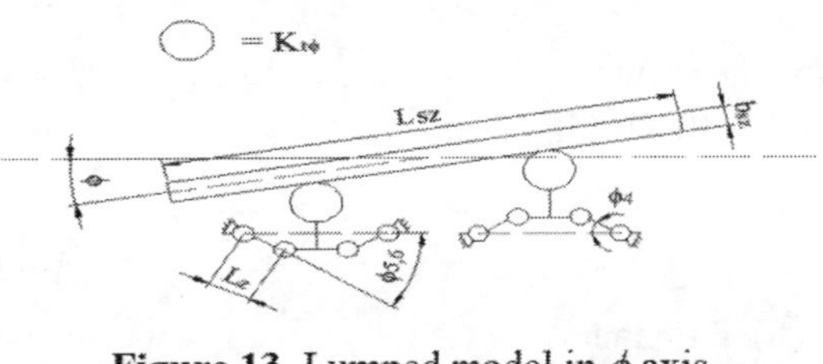

Figure 13. Lumped model in ϕ-axis

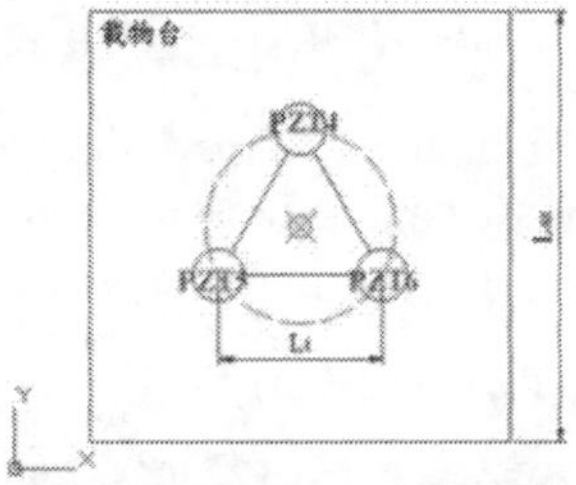

Figure 14. Arrangement of PZT4~PZT6

3.2.3 φ-axis modal analysis

Figure 15 shows the lumped model in φ-axis. The same terminology is applied to calculate the natural frequency.

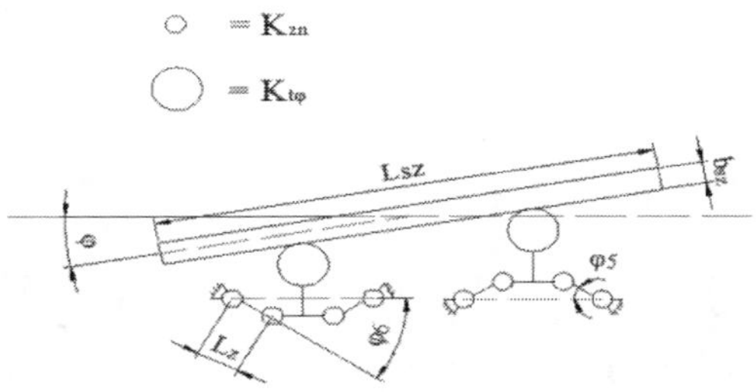

Figure 15. Lumped model in φ-axis

3.3 Finite Element Analysis

In order to verify the accuracy of results from theoretical analysis, FEA by using ANSYS software is conducted. Figure 16 show the simulation results from modal analysis. The comparison between theoretical analysis and ANSYS modal analysis results is shown in Table 1. From this table, it is found that the difference is less than 15%.

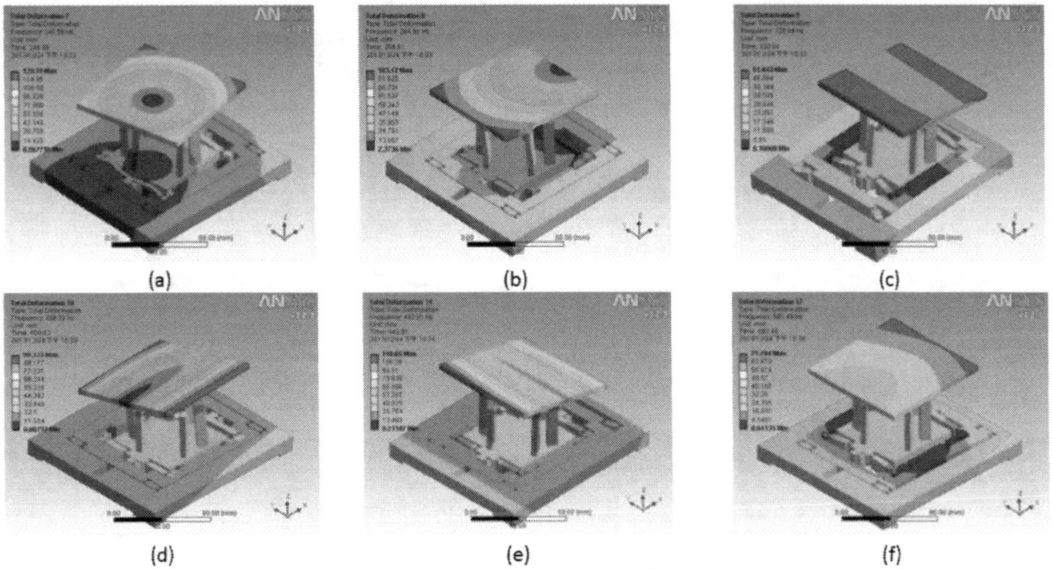

Figure 16. Modal analysis, (a) mode 1, (b) mode 2, (c) mode 3, (d) mode 4, (e) mode 5, (f) mode 6

Table 1 Natural frequency

Mode	1 (θ)	2(Y)	3 (X)	4 (ϕ)	5 (φ)	6 (Z)
Lagrange (Hz)	238.3	291.8	324.0	383.6	385.2	582.2
ANSYS (Hz)	246.6	284.8	320.6	408.0	443.9	583.5
Error (%)	3.4	2.5	1.1	6.0	13.2	0.22

4. EXPERIMENT RESULTS

Three different experiments were conducted in this study to verify our analysis results. The three experiments are (1) impact test, (2) minimum displacement test, and (3) maximum displacement test. The setup of the experiment is shown in Fig. 17.

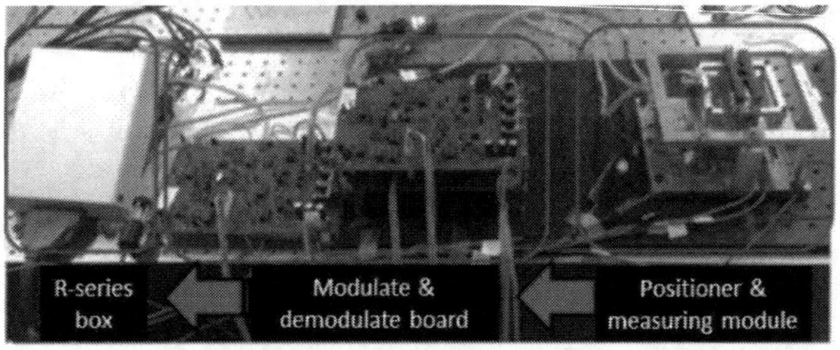

Figure 17. Experiment setup

4.1 Impact Test

The impact test was performed by using a PCB086C04 hammer to strike the stage (as shown in Fig. 18). An 8690C50 3-axes accelerometer was used to collect the vibration signal. And finally, the Oros OR25 vibration analyzer was used to find the natural frequencies of the system. The test results are shown in Table 2. It is found that errors, theoretical values and finite element analysis compare with impact test, are all within 10 %.

(a)

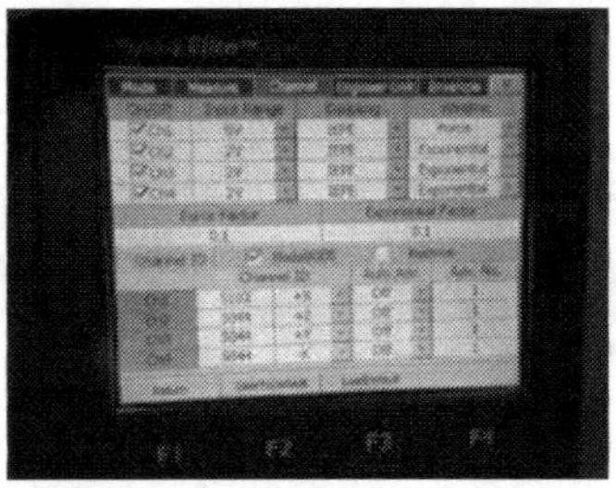

(b)

Figure 18. Impact test (a) experiment setup, (b) analyzer setup

Table 2 Comparison of natural frequency

Mode	1 (θ)	2(Y)	3 (X)	4 (ϕ)	5 (φ)	6 (Z)
Theorems (Hz) (error)	238.3 (5.4%)	291.8 (6.5%)	324.0 (3.8%)	383.6 (3.4%)	385.2 (9.4%)	582.2 (1.3%)
ANSYS (Hz) (error)	246.6 (9.1%)	284.8 (3.9%)	320.6 (2.8%)	408.0 (2.8%)	443.9 (4.4%)	583.5 (1.1%)
Experiment (Hz)	226	274	312	397	425	590

4.2 Minimum Displacement Test

The purpose of this test is trying to evaluate the minimum step that can be achieved in each axis for this positioning stage under different loading. Figure 19 shows the test results with

248g loading. The minimum step on each axis with 248g and 978g loading is shown in Table 3. From this table, it is found that the positioning stage has achieved nano-scale resolution.

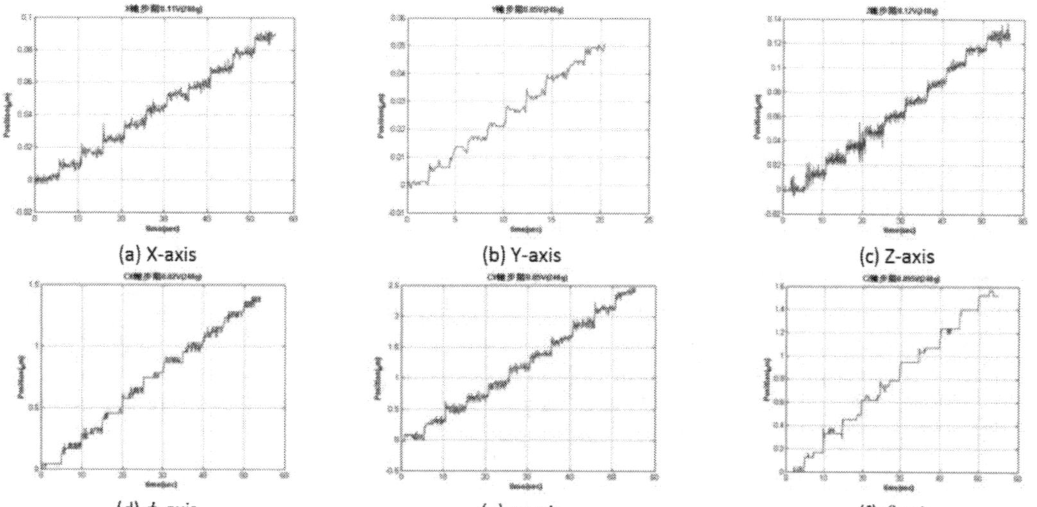

(a) X-axis (b) Y-axis (c) Z-axis

(d) ϕ-axis (e) φ-axis (f) θ-axis

Figure 19. Minimum step in each axis (with 248g loading)

Table 3 Minimum step test results

Load		X	Y	Z	ϕ(arcsec)	φ(arcsec)	θ(arcsec)
248g	Voltage (V)	0.11	0.05	0.12	0.02	0.05	0.055
	Step size	8 nm	6 nm	12 nm	0.14	0.24	0.14
978g	Voltage (V)	0.35	0.25	0.18	0.08	0.15	0.06
	Step size	27 nm	25 nm	17 nm	0.60	0.62	0.21

4.3 Maximum Displacement Test

A maximum displacement experiment was conducted. The procedure of this experiment is simply applied voltage to the PZT actuator from 0 V to 100 V, then reduced the voltage from 100 V to 0 V, and repeated the process for each axis. The results are shown in Table 4. It is found that the maximum displacement is within 12 to 16 μm with standard deviation less than 0.140 μm and residual error less than 0.24 μm in linear translation. While, in rotation, the maximum displacement is around 75 arcsec in ϕ- and φ-axis, 35 arcsec in θ-axis. The standard deviation is less than 1.65 arcsec and residual error is less than 0.55 arcsec.

Table 4 Maximum displacement results

Load		X(μm)	Y(μm)	Z(μm)	ϕ(arcsec)	φ(arcsec)	θ(arcsec)
248g	Displacement	12.876	15.771	13.890	76.961	70.977	36.232
	Standard deviation	0.045	0.140	0.135	0.901	0.623	0.386
	Residual error	0.066	-0.109	0.040	-0.539	0.417	-0.208
978g	Displacement	12.140	15.281	13.347	75.885	70.172	34.236
	Standard deviation	0.045	0.108	0.104	1.616	1.094	0.292
	Residual error	0.049	-0.234	-0.093	-0.365	0.472	-0.064

5. CONCLUSIONS

In this study, the performance of a six DOF micro-precision positioner under different static load has been analyzed. Through theoretical analysis, the mathematical model of this stage was formulated. Finite element analysis with ANSYS has simulated the natural frequency on each axis. Finally, experiments have been conducted to verify the analysis results.

From the experimental results, it is found that the maximum displacements under 248 g (978 g) loading are: 12.876 (12.140) μm, 15.771 (15.281) μm, and 13.890 (13.347) μm along the X-, Y-, and Z-axis respectively; while, the maximum rotation angles are: 76.961 (75.885) arcsec, 70.997 (70.172) arcsec, and 36.232 (34.236) arcsec along the ϕ -, φ-, and θ-axis respectively. It is also found that the standard deviations on the linear and rotational axis are less than 0.140 μm and 1.616 arcsec, and the error of the measuring system is less than 0.25 μm and 0.55 arcesec on linear and rotational axis.

While, in the impact test, it is found that errors, theoretical values and finite element analysis compare with impact test, are all within 10 %.

REFERENCES

1. Gao, W., Dejima, S., Yanai, H., Katakura, K., Kiyono, S., and Tomita, Y., "A surface motor-driven planar motion stage integrated with an XYθZ surface encoder for precision positioning," ***Precision Engineering***, Vol. 28, 2004, pp. 329-337.
2. Sato, K., Nakamoto, K., and Shimokohbe, A., "Practical control of precision positioning mechanism with friction," ***Precision Engineering***, Vol. 28, 2004, pp. 426-434.
3. Chen, D., and Wei, Y., "Study on non-linearity control of a piezoelectric nanoactuator," ***Proceeding of the Fifth World Congress on Intelligent Control and Automation***, China, 2004, pp. 1236-1240.
4. Ku, S.S., Pinsopon, U., Cetinkunt, S., and Nakajima, S., "Design, fabrication, and real-time neural network control of a three-degrees-of-freedom nanopositioner," ***IEEE/ASME Transactions on Mechatronics***, Vol. 5, 2000, pp. 273-280.
5. Physik Instrumente, ***NanoPositioning***, Catalog, US Edition, 2012, pp. 4.5 ~ 4.29.
6. Jywe, W.Y., Jeng, Y.R., Liu, C.H., Teng, Y.F., Wu, C.H., Wang, H.S., and Chen, Y.J., "A novel 5DOF thin coplanar nanometer-scale stage," ***Precision Engineering***, Vol. 32, No. 4, 2008, pp. 239-250.
7. Fung, R.F., and Lin, W.C., "System identification of a novel 6-DOF precision positioning table," ***Sensors and Actuators, A: Physical***, Vol. 150, No. 2, 2009, pp. 286-295.
8. Fung, R.F., Fan C.T., and Lin, W.C., "Design and analysis of a novel six-degrees-of-freedom precision positioning table", ***The Institution of Mechanical Engineers, Part C: Journal of Mechanical Engineering Science***, Vol. 223, No. 5, 2009, pp. 1203-1212.
9. Wang, F.S., ***Design Analysis and Manufacturing of a Hybrid Positioning Stage***, Doctor Dissertation, National Cheng Kung University, Taiwan, 2008.
10. Wang, J.L., ***Design Analysis of A Multi-Axes Micro Precision Positioner***, Master Thesis, National Chung Hsing University, Taiwan, 2011.
11. Paros, J.M., and Weisbord, L., "How to design flexure hinge," ***Machine design***, Vol. 37, 1965, pp.151-157.

ICAST2015 051

Shape Control of the Stewart Platform with Elastic Hinge Utilizing Artificial Thermal Expansion

Ryo Koyama[1*], Kosei Ishimura[2], Akira Iino[1], Yusuke Funakoshi[3], and Hiroshi Yamakawa[1]

[1] Department of Modern Mechanical Engineering, Waseda University, Tokyo, Japan
[2] Institute of Space and Astronautical Science, JAXA, Sagamihara, Japan
[3] Kagoshima Space Center, JAXA, Kagoshima, Japan

Abstract

Recently, structural requirements for observation accuracy of satellites' instruments become increasingly high. This research proposes a new pointing control system which uses Stewart platform with elastic hinge. As a linear actuator, artificial thermal expansion of rod is utilized. This system is highly reliable and enables precise control with 6 DOF. We designed the shape control system whose size is about 300mm on each side. By finite element analysis, it is shown that the shape control system can generate translational motion which is over 100μm in the horizontal direction and 20μm in the vertical direction without rotation. Finally, it is confirmed that relative error between analysis and experimental values is within 15% or less when the displacement is more than 100μm from experiments.

1. INTRODUCTION

Recently, structural requirements for observation accuracy of satellites' instruments become increasingly high. To achieve advanced mission, structural elements such as antennas, telescope aperture, and solar paddles tend to become larger because the larger area they have, the higher performance they can exhibit. In addition to it, introducing redundancy for high reliability and robustness also tend to make satellites larger and complex. Fig. 1 shows that the Japanese astronomical satellites become larger and larger year by year [1].

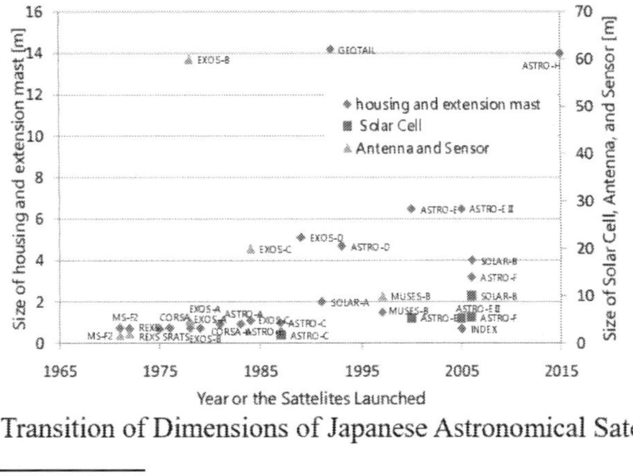

Figure 1. Transition of Dimensions of Japanese Astronomical Satellites [1]

* koyama3553@toki.waseda.jp

Shape accuracy of large space structures are typically influenced by various disturbances. The major errors in shape accuracy are classified into design error, assembly and adjustment error including manufacturing error, and deformation on orbit. For example, cable mesh antennas are usually designed to form approximate parabola surface. Therefore, there is approximation error in such cable mesh antenna as design error. These errors degrade the observation accuracy of satellites' instruments. In order to meet the high scientific requirements, space structures with high shape stability are required.

ASTRO-H, which is X-ray astronomy satellite being developed by JAXA, is given as an example. ASTRO-H is 14 meters long and weighs 2.7 tons — one of the largest Japanese scientific satellites. Required accuracy for optical bench supporting X-ray observation device is very high. In the case of hard X-ray imager system, pointing error must be within 69 arcsec against 12 meters focal length. This means that the displacement between top and bottom of optical bench shall be kept within 4 millimeters [2].

To achieve such high pointing accuracy on orbit, suppression of thermal deformation is one of the most important problems. Conventionally, we have chosen materials with small CTE and have carried out thermal design to keep both spatial temperature distribution and temporally temperature fluctuation to the minimum. However, it is difficult for the conventional methods to realize the accuracy less than millimeters for large space structure more than 10 meters. Therefore, to realize both large and precise space structure, the shape control of structures has been investigated [3-6].

In this study, an active shape control system is proposed and studied. For the shape control of space structures, actuators which satisfy the following requirements are need:

(i) The actuators shall be compatible to space environment.
(ii) The actuators shall achieve position control with high accuracy.
(iii) The actuators shall be simple mechanism for the high reliability.

From the above points of view, we utilize artificial thermal expansion of structural member as an actuator in this paper. Utilizing thermal expansion makes the mechanism of the system remarkably simple and reliable. For example, the longitudinal displacement of the rod member is simply achieved by thermal control of the member. As a result, we can use such a rod member with thermal control as a linear actuator. The linear actuators, where artificial thermal expansion is used to generate longitudinal displacement, are called thermal expansion unit in the following (Fig. 2). Among mechanisms using linear actuator, Stewart platform can control both position and orientation accurately by six linear actuators [7]. Stewart platform is a kind of parallel link mechanism. Generally, parallel link mechanisms can be controlled more precisely than serial link mechanisms, because errors in actuator do not accumulate in parallel link mechanism. In the field of structural mechanics, a lot of researches about Stewart platform has been carried out for pointing control and vibration control [8, 9]. In this research, the Stewart platform is used for precise pointing control.

In general Stewart platform, rotational joints are used at both ends of linear actuators. However, using rotational joints on orbit is not preferable from the tribological point of view because rotational joints may cause cold welding, nonlinearity of the system including backlash, and hysteresis due to friction. Therefore, we propose to replace rotational joints in Stewart platform to elastic hinges (Fig. 2). This change is expected to improve the reliability of control system remarkably.

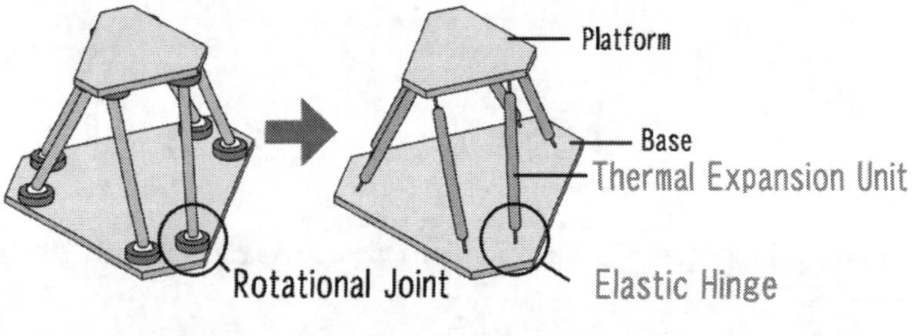

Figure 2. Replacing rotational joints to elastic hinge

2. MODELING OF STEWART PLATFORM WITH ELASTIC HINGE

This section describes how to model the Stewart platform. Generally, the shape of Stewart platform is determined kinematically. However, introducing elastic hinge into Stewart platform requires taking account of elastic deformation. In this paper, we use two models; (a) kinematic model and (b) finite element model. Kinematic model is used to evaluate to rough manipulability area determined by geometrical parameters such as length of linear actuators and to determine the valued of geometrical parameters. Then, finite element model is used to evaluate the influence of elastic hinges and to design the elastic hinges. Using kinematic model saves computational cost to search the desirable geometrical parameter set.

2.1. Kinematic Modeling and Geometrical Parameters of Stewart Platform

In this subsection, kinematic equation between platform position/orientation and displacement of linear actuators are formulated.

The position of both ends of six linear actuators is determined by (a) geometrical parameters $(b_r, \theta_b, s_r, \theta_p)$ which determine the position on base or platform linked with linear actuator (Fig. 3), (b) the position of center of platform against base(${}^b\boldsymbol{p}$), and (c) the orientation of platform against base (Fig. 4). In this research, we set that the initial position of platform $\boldsymbol{\Sigma_p}$ is right above that of base $\boldsymbol{\Sigma_b}$ (${}^b\boldsymbol{p} = (0,0,\ {}^b p_z)$) and there is no rotational displacement between the platform and base. As a result, initial shape of Stewart platform is determined by 5 parameters $(b_r, \theta_b, s_r, \theta_p.\ {}^b p_z)$. As shown in Fig. 4, each vector ${}^b\boldsymbol{L}_i$ corresponding to each linear actuator is given as follows.

$$ {}^b\boldsymbol{L}_i = {}^b\boldsymbol{p}_i + \boldsymbol{R}\,{}^p\boldsymbol{s}_i - {}^b\boldsymbol{b}_i \tag{1} $$

The superscript such as b and p in Eq.(1) means the base and platform coordinate systems, respectively. $\boldsymbol{R}$ is rotational matrix between the base and platform coordinate systems. In this research, displacement of linear actuator is generated by artificial thermal expansion. Therefore, control input is temperature change of thermal expansion unit (Fig.2). The output of the control is translational and rotational displacements. Assuming that the displacement of platform is small enough, we can obtain the relationship between input $\boldsymbol{\Delta T}$ and output $\boldsymbol{\delta x}$ Eq.(3,4) by linearizing Eq.(1) [10].

$$ \boldsymbol{J}_1 \boldsymbol{\delta x}_c = \boldsymbol{J}_2 \boldsymbol{\Delta T} \tag{2} $$

$$ \boldsymbol{\Delta T} = [\Delta T_1 \quad \Delta T_2 \quad \cdots \quad \Delta T_6] \tag{3} $$

is temperature change of each thermal expansion unit.

$$ \boldsymbol{\delta x}_c = \begin{bmatrix} \boldsymbol{\delta p}_c \\ \boldsymbol{\delta \Omega}_c \end{bmatrix} \tag{4} $$

($\boldsymbol{\delta p}_c$: translational displacement, $\boldsymbol{\delta\Omega}_c$: rotational displacement)

is the displacement of the center of the platform.

$$ \boldsymbol{J}_1 = \begin{bmatrix} \boldsymbol{z}_1^T & (\boldsymbol{R}\,{}^p\boldsymbol{s}_1 \times \boldsymbol{z}_1)^T \\ \boldsymbol{z}_2^T & (\boldsymbol{R}\,{}^p\boldsymbol{s}_2 \times \boldsymbol{z}_2)^T \\ & \vdots \\ \boldsymbol{z}_6^T & (\boldsymbol{R}\,{}^p\boldsymbol{s}_6 \times \boldsymbol{z}_6)^T \end{bmatrix} $$

$$ \boldsymbol{J}_2 = \mathbf{diag}\left(\boldsymbol{z}_1^T \frac{\partial \boldsymbol{L}_1}{\partial T_1} \cdots \boldsymbol{z}_6^T \frac{\partial \boldsymbol{L}_6}{\partial T_6}\right) \tag{5} $$

From Eq. (5), $|\boldsymbol{J}_2| \neq 0$ as long as $\frac{\partial \boldsymbol{L}_i}{\partial T_i} \neq \boldsymbol{0}$. When $|\boldsymbol{J}_1| = 0$, the state of Stewart platform is singular. This corresponds to the configurations in which the platform is locally movable even when all the actuated joints are locked [11]. When we determine the initial state of Stewart platform, we must be careful not to set the configuration in singularity.

At first, free thermal expansion of i th thermal expansion unit is assumed. Let CTE as $\alpha[1/\mathrm{K}]$ and length of ith thermal expansion unit as L_i, the following equation is derived.

$$\frac{\partial \boldsymbol{L}_i}{\partial T_i} = \alpha \boldsymbol{L}_i \tag{6}$$

The relationship between input of thermal change $\boldsymbol{\Delta T}$ and output of displacement of platform $\boldsymbol{\delta x}$ is given as follows.

$$\boldsymbol{\delta x}_c = \boldsymbol{J \Delta T} \quad (\boldsymbol{J} = \boldsymbol{J}_1^{-1}\boldsymbol{J}_2) \tag{7}$$

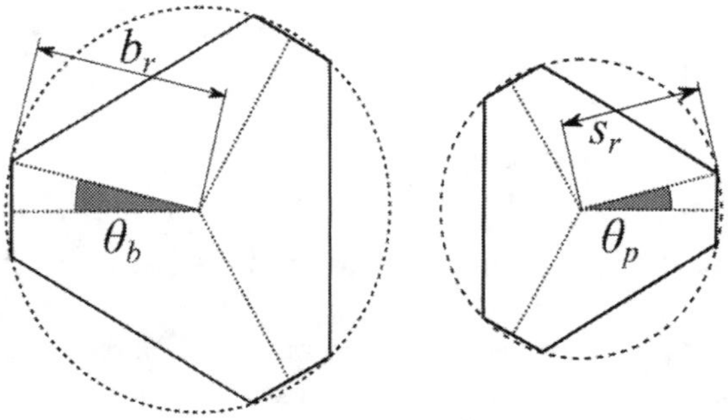

Figure 3. Geometrical parameters of base(left) and platform (Right)

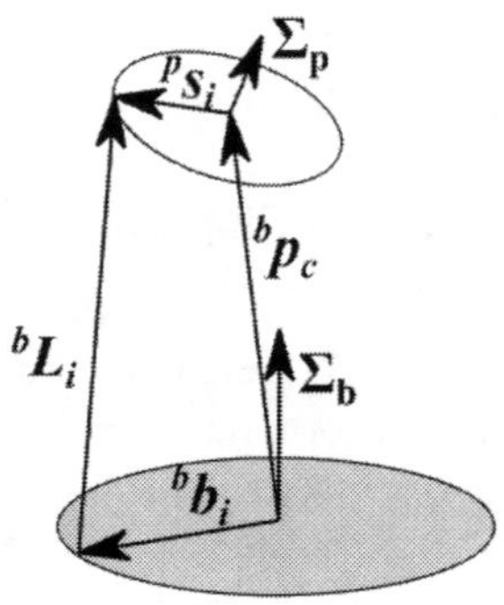

Figure 4. Kinematics of Stewart platform

2.2. Finite Element Model of Stewart Platform with Elastic Hinge

In this subsection, a finite element model of Stewart platform with elastic hinge is introduced. Elastic hinge is a rod member with small diameter. In our model, the platform and base are connected by diagonal liner actuators which consist of thermal expansion unit and hinges as shown in Fig.2. Thermal expansion unit expands in longitudinal direction when it is heated. Each member is modeled as follows.

(i) platform: one rigid plate
(ii) base: one rigid plate
(iii) thermal expansion unit: six elastic rods
(iv) elastic hinge: twelve elastic rods attached at the both ends of thermal expansion units

The parameters introduced in the finite element model are shown in Fig. 5. This system consists of 24 nodes and each node has 6 degrees of freedom. Nodes are connected to each other by beam elements. Beams which share the same node preserve relative angle even if these elements deform. One beam connecting node i and j is modeled as shown in Figs. 6 and 7. We consider that the shear of beams is negligible compared with other deformations such as bending, extension and torsion and that bending, extension and torsion of beams occur independently.

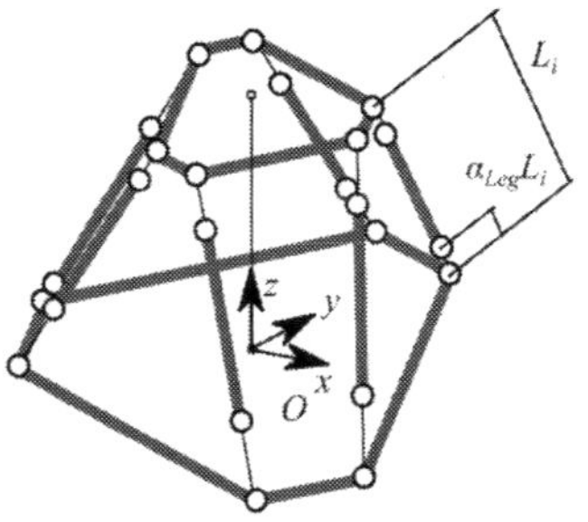

Figure 5. Additional dimensional parameter α_{Leg}

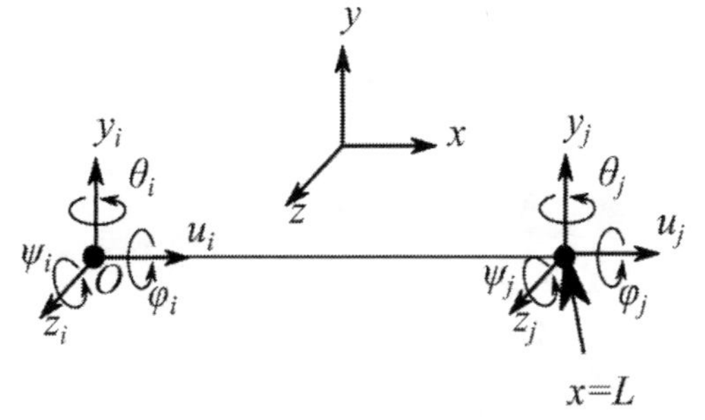

Figure 6. Displacements of nodes

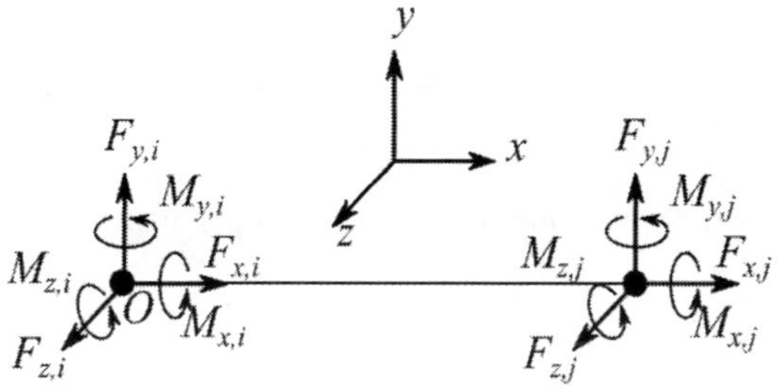

Figure 7. Forces and torques on nodes

2.3. Jacobian Matrix of Finite Element Model

Applying finite element analysis, we can evaluate the deformation of the system when the temperatures changes of each thermal expansion unit are given. This subsection describes how to get Jacobean Matrix. Let $\boldsymbol{\delta x_i} = [\boldsymbol{\delta p_i}\ \boldsymbol{\delta \Omega_i}]^T$ as a displacement of platform when only i th thermal expansion unit is heated by 1K. $\boldsymbol{\delta p_i}$ is translational displacement and $\boldsymbol{\delta \Omega_i}$ is rotational displacement. Because Stewart platform with elastic hinge is modeled as a linear system, we can superpose the displacements of platform for given $\Delta T_i (i = 1,2, \dots ,6)$ which is thermal change in i th thermal expansion unit.

$$\boldsymbol{\delta x_c} = \sum_{i=1}^{6} \Delta T_i \boldsymbol{\delta x_i} = [\boldsymbol{\delta x_1}\ \boldsymbol{\delta x_2} \cdots \boldsymbol{\delta x_6}] \begin{bmatrix} \Delta T_1 \\ \vdots \\ \Delta T_6 \end{bmatrix} = \boldsymbol{J \Delta T} \tag{8}$$

Therefore, we can obtain Jacobean Matrix as Eq. (9).

$$\boldsymbol{J} = [\boldsymbol{\delta x_1}\ \boldsymbol{\delta x_2} \cdots \boldsymbol{\delta x_6}] \tag{9}$$

3. DESIGN OF SHAPE CONTROL MECHANISM: STEWART PLATFORM WITH ELASTIC HINGE

3.1. Evaluation Indexes in Design

In order to evaluate designed shape control mechanism, we use von Mises stress and manipulability ellipsoid. Von Mises stress is used to check whether each member of the system does not deform plastically. Manipulability ellipsoid is used to evaluate the reachable area of the platform center. Manipulability ellipsoid is the ellipsoid where the platform center is reachable under the condition

$|\boldsymbol{\Delta T}| \leq 1$. In this research, we use following four typical manipulability ellipsoid, and lay emphasis on the second one. The symbol in the parentheses distinguishes each ellipsoid.

(p1) Reachable area of translational displacement of platform center
(p2) Reachable area of translational displacement of platform center without changing the orientation of platform
(w1) Reachable area of rotational displacement of platform center
(w2) Reachable area of rotational displacement of platform center without changing the position of platform

Because displacement vector of platform center is given as Eq.(4), manipulability ellipsoid (p1) ~ (w2) are also described as follows.
(p1) manipulability ellipsoid of $\boldsymbol{\delta p_c}$
(p2) manipulability ellipsoid of $\boldsymbol{\delta p_c}$ under the condition that $\boldsymbol{\delta \Omega_c} = \mathbf{0}$
(w1) manipulability ellipsoid of $\boldsymbol{\delta \Omega_c}$
(w2) manipulability ellipsoid of $\boldsymbol{\delta \Omega_c}$ under the condition that $\boldsymbol{\delta p_c} = \mathbf{0}$
The size of manipulability ellipsoid is evaluated by the length of major axis of manipulability ellipsoid r_x, r_y, r_z.

3.2. Optimizing Geometrical Parameters Using Kinematic Model

Under the constraints shown in Table 1, the geometrical parameters of kinematic model is changed, and one set of parameters is selected which satisfying the following guidelines.

(i) Thermal expansion units shall not collide each other.
The distance between every thermal expansion unit shall be more than 20 mm at initial state. (The diameter of thermal expansion unit is set to 15 mm.)
(ii) The initial state of system shall not be in singularity.
The relationship between input $\boldsymbol{\Delta T}$ and output $\boldsymbol{\delta x_c}$ are given by Eq. (2).
To avoid the singular state, remove the design where $|J_1|$ is less than 10^{-40}.
(iii) Length of vertical major axis of manipulability ellipsoid shall be more than 15 μm ($|\boldsymbol{\Delta T}| < 1$[K]).
(iv) Maximize the length of horizontal major axis of manipulability ellipsoid (p2) ($|\boldsymbol{\Delta T}| < 1$[K]).

The geometrical parameters selected by optimization above are also shown in Table 1 and the size of manipulability ellipsoid is shown in Table 2.

Table 1 Geometrical parameters

Parameters	min	max	step	optimized
s_r[m]	0.015	0.135	0.015	0.105
θ_b[deg]	10	50	5	10
b_r [m]	0.015	0.135	0.015	0.105
θ_b[deg]	10	50	5	10
${}^bp_z$ [m]	0.03	0.27	0.03	0.27

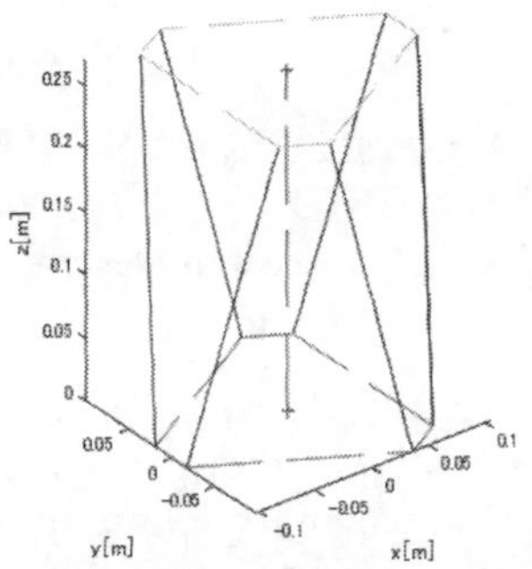

Figure 8. Appearance of optimized design

Table 2 Manipulability ellipsoids with optimized parameters ($|\boldsymbol{\Delta T}| < 1[\mathrm{K}]$)

Condition	Radius of ellipsoid r_x	Radius of ellipsoid r_y	Radius of ellipsoid r_z
p1	16 μm	16 μm	3 μm
p2	15 μm	15 μm	3 μm
w1	8.4 arcsec	8.4 arcsec	22.2 arcsec
w2	7.9 arcsec	7.9 arcsec	22.2 arcsec

3.3. Evaluation of the Effect of Elastic Hinge by Finite Element Analysis

By using optimized geometrical parameters shown in Table 1, a finite element model is constructed to evaluate the effect of elastic hinge. In the FEM, we used parameters shown in Tables 3 and 4 for the elastic hinge and thermal expansion unit. In this condition, the size of manipulability ellipsoid is calculated as shown in Table 5. We achieved more than 10μm manipulability in horizontal plane with input vector $\boldsymbol{\Delta T}$ under $|\boldsymbol{\Delta T}| < 1[\mathrm{K}]$. If the temperature of each thermal expansion unit is controlled in the range of ± 10K, the platform center can be moved about 120μm in horizontal plane and 20μm in vertical direction.

The von Mises stress of hinges is also evaluated. Fig. 9 shows the von Mises stress appears in each member when one thermal expansion unit is heated to 1K. Even if the temperature of thermal expansion unit changes ±50K, maximum von Mises stress is about 30MPa. Therefore, there is no worry of plastic deformation of elastic hinges.

Table 3 Parameters of elastic hinge and thermal expansion unit

diameter of hinge d_1[mm]	3.2
diameter of thermal expansion unit d_2[mm]	15
ratio of hinge length α_{Leg}	0.1
length of thermal expansion unit [mm]	223.5
length of hinge [mm]	27.9

Table 4 Physical properties used in analysis

hinge (SUS)	Young's modulus E_1[GPa]	210
	Poisson's ratio ν_1	0.28
thermal expansion unit A2017	Young's modulus E_2[GPa]	72.4
	Poisson's ratio ν_2	0.33
	CTE α[/K]	24×10^{-6}

Table 5 Manipulability ellipsoid in the case of Finite Element Model($|\boldsymbol{\Delta T}| < 1[\mathrm{K}]$)

Condition	Radius of ellipsoid r_x	Radius of ellipsoid r_y	Radius of ellipsoid r_z
p1	13 μm	13 μm	2 μm
p2	12 μm	12 μm	2 μm
w1	6.7 arcsec	6.7 arcsec	17.8 arcsec
w2	6.3 arcsec	6.3 arcsec	17.8 arcsec

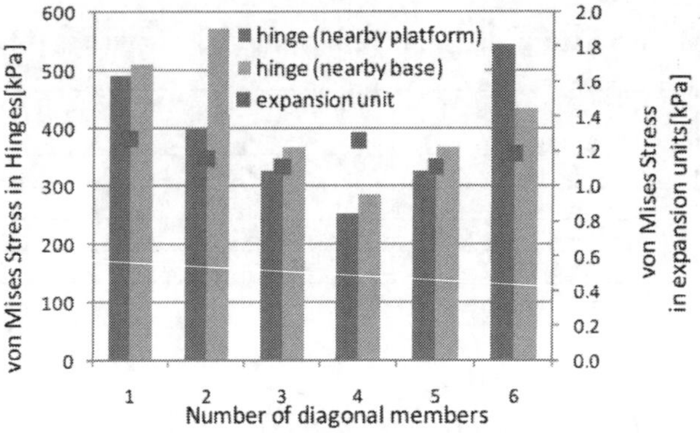

Figure 9. von Mises stress in each member
($|\Delta T|$ of thermal expansion units is 1 K)

4. EVALUATION OF CONTROL PERFORMANCE BY EXPERIMENT

We compare the experiment results with the results of analysis obtained in section 3. Changing the temperature of expansion unit, we measured the translational and rotational displacements of the platform.

4.1. Specification of Stewart Platform used in Experiment

We manufactured the experimental system shown in Fig. 10. The distance between the base and the platform is 270 mm, and the thickness of the plates used in both base and platform is 15 mm. Therefore, the height of Stewart platform is 300 mm. In the experiments, we evaluate the translational displacements (3 degrees of freedom) and rotational displacements except for torsion around the z-axis (2 degrees of freedom) by measuring the displacements at 5 points on the platform (Fig. 11).

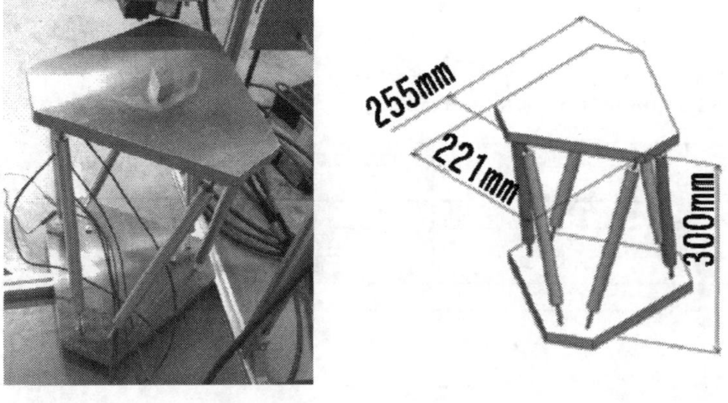

Figure 10. Appearance of experimental system (left) and the size (right)

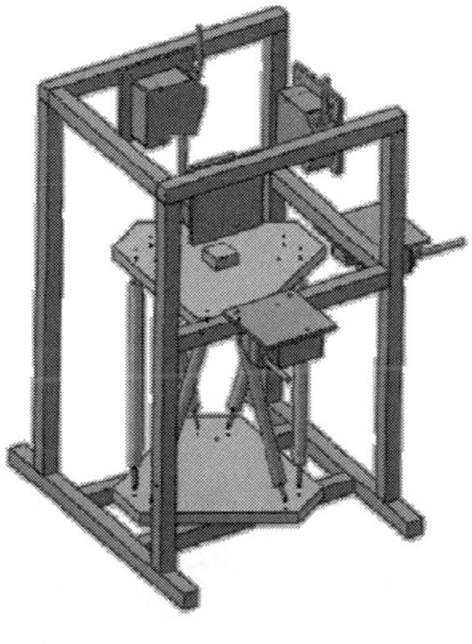

Figure 11. Appearance of measurement system

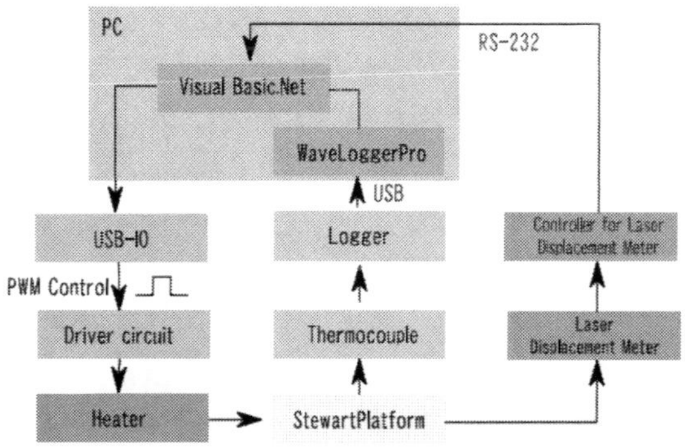

Figure 12. Block diagram of experimental system*

*In experimental system, the voltages of heaters on thermal expansion units are determined by PI controller to control the temperature of thermal expansion unit.

4.2. Experimental Results and Discussion

In this experiment, the temperature of thermal expansion unit was raised by 50K, and dropped by 50K after that. The temperature was changed by 1K after 30 seconds waiting time, so as to realize thermal equilibrium state. The temperature and displacement are sampled at 1sec period. Obtained relationship between position/orientation of the platform and temperature of expansion unit are shown in Figs. 13 and 14. From the result, it is confirmed that the relative errors between analysis and experimental values are within 15% or less when the displacement is more than 100μm.

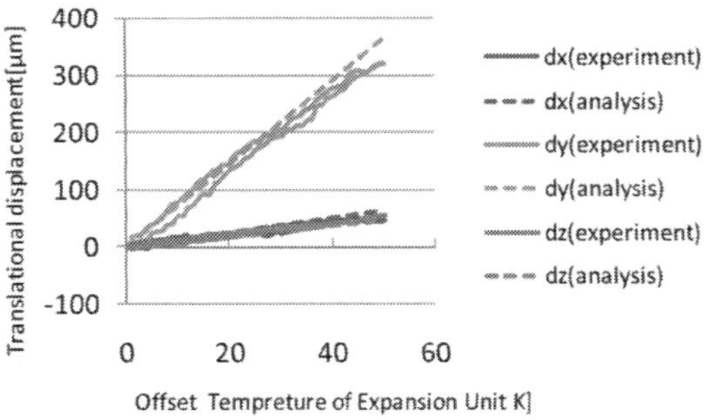

Figure 13. Relationship between temperature of expansion unit and translational displacement

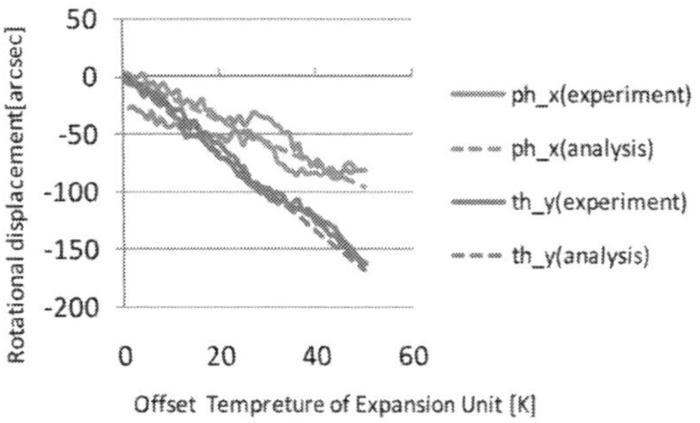

Figure 14. Relationship between temperature of expansion unit and rotational displacement

5. CONCLUSION

We designed and manufactured the Stewart platform with elastic hinge, and evaluate how well we can control the system utilizing artificial thermal expansion by heater. Because the system has no sliding portion, the proposed system is reliable in space. By finite element analysis, it is shown that the designed system can generate translational motion which is over 100μm in the horizontal direction and 20μm in the vertical direction without rotation. From the experiments, we confirmed that relative error between analysis and experimental values is within 15% or less when the displacement is more than 100μm. The result is in good agreement with the finite element analysis. As a result, it is shown that the designed Stewart platform with elastic hinges has useful performance for the shape control of large space structures.

REFERENCES

1. Watanabe, K, "Complete Explanatory Diagram of Space Technology : Space Development in the World," *Kodansya*, 2012. pp. 436-490 (in Japanese).
2. Kawano, T., Ishimura, K., Mnesugi, K., Omagari, K., and Tanaka, K., "A technique to evaluate on-orbit thermal deformation for large precise structures in ASTRO-H," *AIAA SciTech*, 56th AIAA/ASCE/AHS/ASC Structures, Structural Dynamics, and Materials Conference, 2015.
3. Agnes, G., Dooley, J., "Precision deployable structures technology for NASA large aperture missions.," *Space 2004 conference and exhibit* ,Oct. 2004, pp 5899-1–5899-7.
4. Lake, M., Peterson, L., Hachkowski, M., et al., "Research on the problem of high-precision deployment for large-aperture space-based science instruments," *Space technology and applications international forum*, 1998.
5. Miura, K., "Adaptive Structures Research at ISAS," *Journal of Intelligent Material Systems and Structures*, 1992.
6. Korkmaz, S., "A review of active structural control: challenges for engineering informatics," *Computers & Structures*, 2011.
7. Stewart, D., "A PLATFORM WITH SIX DEGREES OF FREEDOM," *Proceedings of the Institution of Mechanical Engineers 1965*, 1965, pp 371-386.
8. McInroy, J., O'Brien, J., Neat, G., "Precise, Fault-Tolerant Pointing Using a Stewart Platform," *IEEE/ASME Transactions on Mechatronics*, VOL. 4, NO. 1, 1999, pp 91-95.
9. Hanieh, A., Preumont, A., Loix, N., "Piezoelectric Stewart Platform for General Purpose Active Damping Interface and Precision Control," *Proceedings of the 9th European Space Mechanisms and Tribology Symposium*, 2001, pp.331-334.
10. Handbook of Robot Engineering, Ed. by Japan Robot Society, 1990, pp.299.
11. Gosselin, C., Angeles, J. "Singularity Analysis of Closed-Loop Kinematic Chains", *IEEE Transactions on Robotics and Automation*, 1990, Vol. VOL. 6, NO. 3, pp 281-290.

ICAST2015 #055

Design and Analysis of Smart Space Reflectors Driven by Piezoelectric Actuators

Kazuaki Hayasaki[1*], Hiraku Sakamoto[1], Hiroaki Tanaka[2], Kosei Ishimura[3], and Masaaki Okuma[1]

[1] Department of Mechanical and Aerospace Engineering, Tokyo Institute of Technology, Tokyo, Japan
[2] Department of Aerospace Engineering, National Defense Academy of Japan, Kanagawa, Japan
[3] Japan Aerospace Exploration Agency, Kanagawa, Japan

Abstract

This paper discusses the configuration of design of smart space reflector system for realization of future high-accuracy antenna in space. This study aims at establishing of design methods for a smart secondary reflector to correspond to deformation mode which caused on the primary reflector surface in space. Authors have investigated effects that perturbation of Kinematic Coupling (KC), which restricted the primary reflector and a back structure, affecting to the primary reflector surface by using finite-element method (FEM). Furthermore, the number of piezoelectric actuators of the back of secondary reflector surface is varied to compare deformation modes of secondary reflector surface with that of primary reflector. Through correlations of deformation of the primary reflector and configuration of the smart secondary reflector, new knowledge of antenna system while considering two reflectors at the same time is obtained.

1. INTRODUCTION

Large and deployable antenna systems (φ : 10m) to observe celestial objects over 100GHz are required for future scientific space antennas. Active shape-control of space reflector have been studied in order to realize future high-accuracy antenna in space[1]; however, a design method is needed for the smart space secondary reflector in consideration of the characteristic of the primary reflector. In order to make a success of space mission, it is important to consider the configuration of design of smart secondary reflector to correct deformation modes of primary reflector effectively.

The precision of the primary reflector surface, which is configured by solid mirror segments, depends on the relative positions of primary reflector segments after developments as a major deformation of primary reflector to correct by smart secondary reflector. Figure 1 illustrates the space-borne radio astronomy observatory system where the secondary reflector corrects the wavefront error due to the shape errors of the large primary reflector. Reflector segments are expected to fix in a deployable back structure by Kinematic Coupling (KC) in figure 2, however, the effect that uncertainties of the fixed position of KC affecting to the mirror surface is not known. In this study, examinations are performed to evaluate deformations of the primary reflector caused by deviation of primary reflector's rigid body position.

* hayasaki.k.aa@m.titech.ac.jp

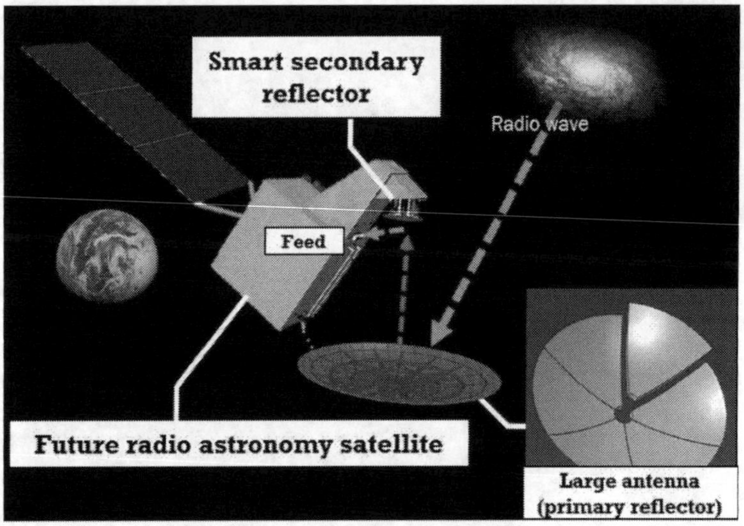

Figure 1. High-accuracy space antenna system with smart secondary reflector.

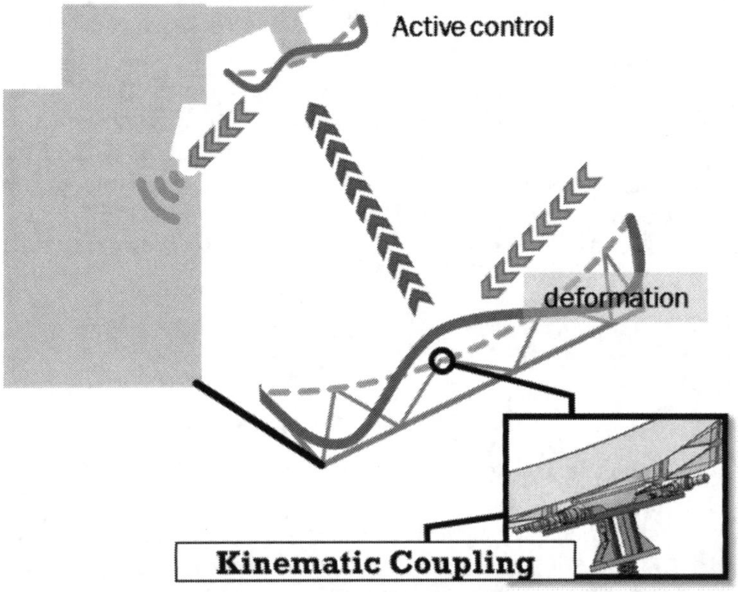

Figure 2. Active shape control by smart secondary reflector concept.

This study aims at clarifying the effect of uncertainties of the KC position to the primary reflector's shape and the configuration of the smart secondary reflector to correct the distortion. In addition, this study aims at obtaining knowledge for simultaneous design optimization for KC positions and the smart secondary reflector.

In order to achieve the research objective, this study exploits the prototype of smart secondary reflector in figure 3. Procedure of analysis is described in section 2. In section 2, primary reflector model and smart secondary model are described as well. In section 3, results of analysis are described and discussed.

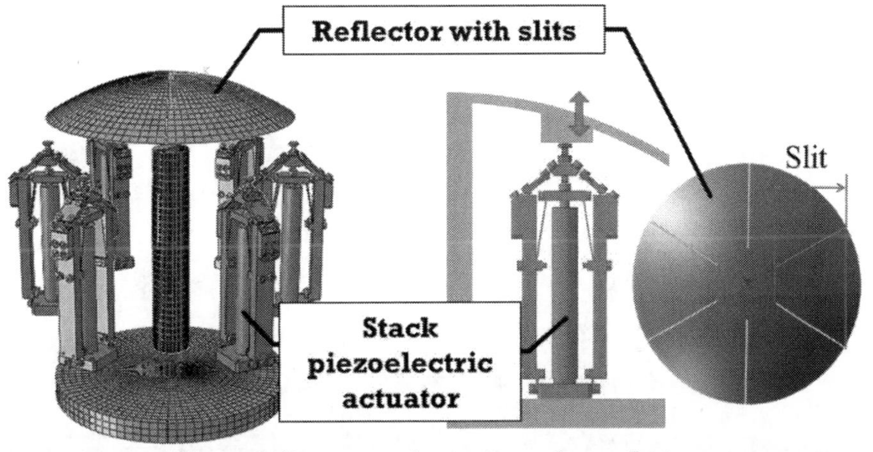

Figure 3. Prototype of smart secondary reflector.

2. PRPCEDURE AND SIMULATION MODEL

This section described procedure of analyses in the following section while reviewing primary reflector model and smart secondary reflector model.

2.1 Perturbation by Kinematic Coupling (KC)

In this research, small displacements are applied each Kinematic Coupling (KC) as perturbation. Figure 4 illustrates about KC. KC consists of three balls in the primary reflector and three radial v-grooves in back structures mate with each other[2]. Each sphere has two contact points for a total of six contact points, enough to constrain all six of the part's degrees of freedom. KC is used to fix the primary reflector with high precision in the back structure. Balls are located to be able to slide to direction of each grooves (x), and to absorb the thermal deformation of the main mirror. Author conducts analysis about only KC_A and B because the primary reflector is symmetric. Small displacements are applied to y and z direction in the coordinate axis of each KC as perturbation. Discuss deformations of the primary reflector of 4 types in figure 5.

2.2 Approach to adjust the shape of smart secondary reflector to the shape of the primary reflector

Approach to adjust the shape of smart secondary reflector to the shape of the primary reflector is described in this paragraph. The antenna is composed of the primary reflector of the paraboloid and the smart secondary reflector of the hyperboloid. Figure 6 illustrates conceptual drawing of searching ideal shape of secondary reflector which correspond to deformations of the primary reflector. Distances between corresponding points where the primary reflector and the smart secondary reflector are constant. Thus, ideal shapes of secondary reflector which correspond to deformations of the primary reflector are derived by moving each point on the surface of secondary reflector in the optical axis direction of the antenna for an amount of corresponding points of the primary reflector.

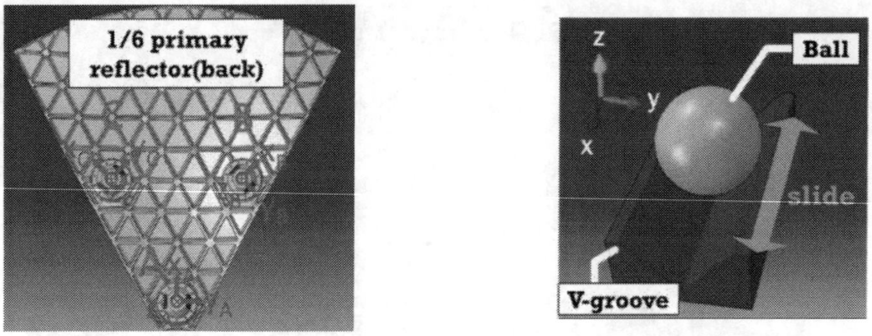

a) b)

Figure 4. a) Places where locate Kinematic Coupling.
b) Conceptual drawing of Kinematic Coupling.

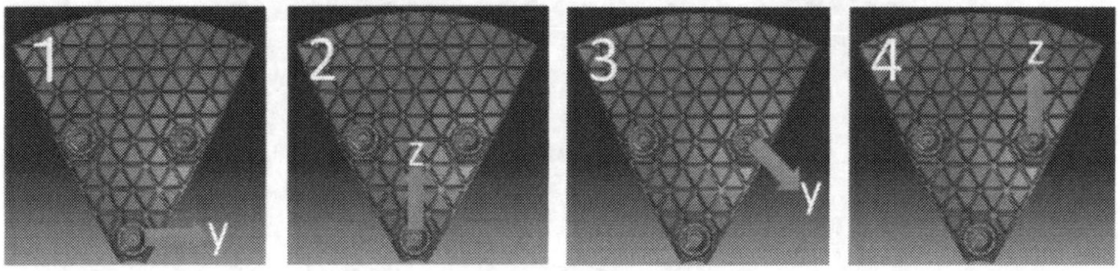

Figure 5. 4 types of applied displacement to investigate primary reflector's deformation mode.

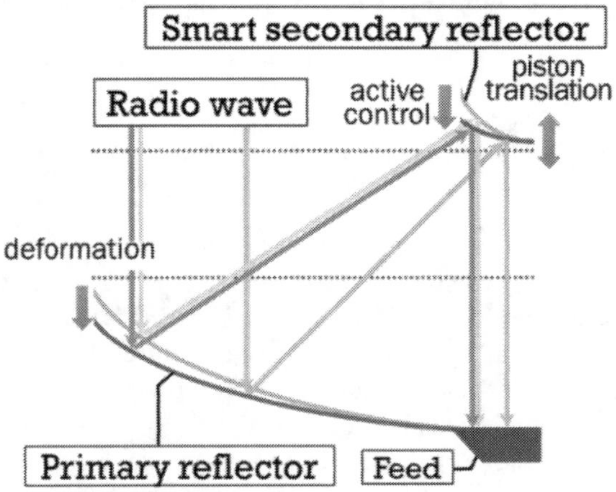

Figure 6. Concept of searching ideal shape of secondary reflector which correspond to deformations of the primary reflector.

2.3 Variation of configuration of smart secondary reflector

This paragraph described variation of configuration of smart secondary reflector to correct deformation of primary reflector. Authors are developing a prototype of smart secondary reflector for observation in the EHF band. This proto model consist of a solid thin surface whose have six slits which extend radially to avoid the occurrence of buckling, supporting members, and surface adjustment mechanism by using piezoelectric actuators. Deformations of smart secondary reflector are caused by adding external forces at back of the surface to the optical axis direction. In this study, assuming that using push/pull type actuator, external forces are enabled to add both directions. Add external forces at positions of red marks in figure 7. Approximate to ideal shapes by actuations of actuators and piston translation of smart secondary reflector.

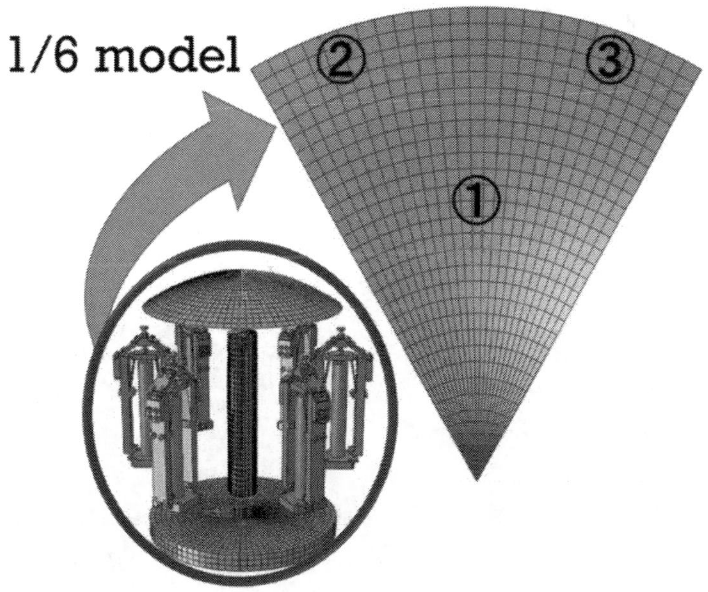

Figure 7. Positions that applies external force.

3. RESULTS AND DISCUSSIONS

In this section, results of analyses are stated. First, 4 types deformations of the primary reflector were investigated. In these analyses, amount of KC's displacements is 1mm, on the grounds that linearity is confirmed between amount of KC's displacements and deformation of primary reflector surface. Obtained surfaces were plotted with 7544 points, and show displacements from original shape with color bar at figure. Second, each deformations of the smart secondary reflector were investigated while changing configuration of it. Variations of configuration are 5 patterns; only piston translation of secondary reflector, 1, 2, and 3 actuators, and no control for purpose of comparisons. Obtained surfaces were plotted with 676 points, and show displacements from ideal shape with color bar at figure. In addition, residual error between obtained surface and ideal shape was showed by Root Mean Square (RMS).

Figure 8 shows results of analyses. Result of experiments clearly following things. Deformation patterns are seen in the primary reflector by perturbations of positions of KC. Piston translation is effective if the smart secondary reflector moved in the optical axis direction of the antenna. Central actuator is effective if the surface distort to the radial direction. Both side actuators are effective if the surface distort to the circumferential direction. In most cases, three actuators model is more effective than two and one model. By contrast, three actuators model is not so effective than two and one model at some deformation modes. These results become the important index in consider the configuration of the smart secondary reflector. Since reduction of the number of actuators in a specific deformation pattern leads to reduction of the cost and improvement of the reliability.

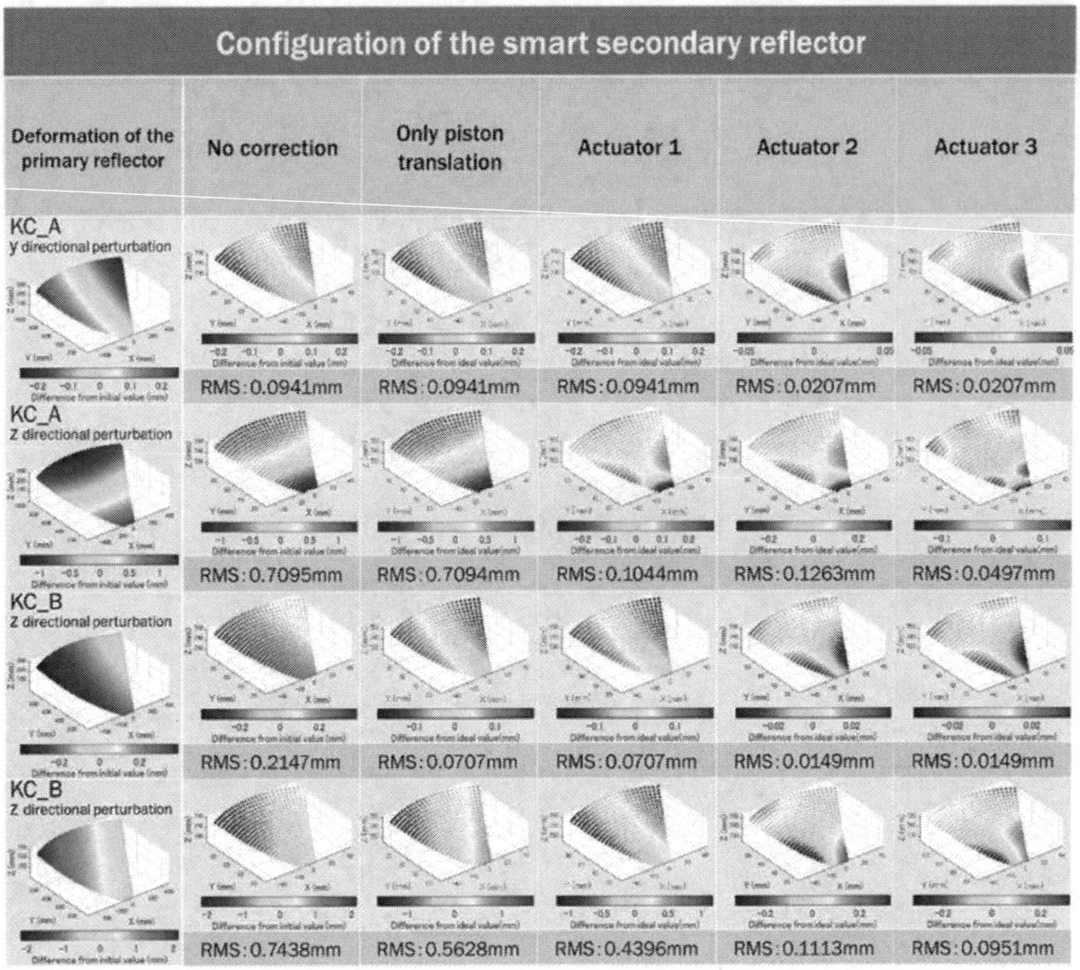

Figure 8. Deformation modes of primary reflector and residual error modes of smart secondary reflector between obtained surface and ideal shape with RMS.

4. CONCLUSION

In this study, a series of analyses of the primary reflector and the smart secondary reflector has been conducted to obtain new knowledges of antenna system while considering two reflectors at the same time. As a result of these analyses, following things were clarified.

i. The effect that perturbation of each KC gives in the primary reflector surface.
ii. Configurations of the smart secondary reflector to correct each deformation mode of the primary reflector.

This study has demonstrated analysis and design method for future reconfigurable structural systems.

ACKNOWLEDGMENTS

This research was partially supported by the ISAS/JAXA strategic research and development grant on "research and development of high accuracy large space structure systems." The authors thank Yoshitaka Sato at JAXA who offered the primary reflector FEM model; Kota Goto and Ken Tsukui at Tokyo Institute of Technology, David Valentin at Imperial College London who helped the experiments and analyses; and the research group members, especially Yoshiro Ogi at University of Tokyo, for their helpful comments on this work.

REFERENCES

1. Inagaki, A., Sakamoto, H., Tanaka, H., Ishimura, K., Okuma, M., "Experimental Study of Reflector Shape Control under Various Thermal Conditions," 2nd AIAA Spacecraft Structures Conference, No. 1399, Jan.2015.
2. Reynolds, P., Atkinson, C., Gliman, L., "Design and Development of the Primary and Secondary Mirror Deployment System for the Cryogenic JWST," Proc. 37th Aerospace Mechanisms Symposium, May 19-21, 2004, pp. 29-44.

ICAST2015: 26th International Conference on Adaptive Structures and Technologies
October 14-16, 2015, Kobe, Japan

ICAST2015 #089

COMPARATIVE STUDY OF PARAMETRIC MODELS FOR MAGNETO-RHEOLOGICAL DAMPERS

Colin Knox, Amin Fereidooni, Yong Chen and Viresh Wickramasinghe

National Research Council Canada, Ottawa, Canada

Abstract

This paper evaluates the fidelity of several parametric models for magneto-rheological dampers using measurement data. The measurement data is obtained through a series of vibration tests on a magneto-rheological damper at various frequencies and displacement amplitudes while the input current to the damper changes within an operational range. The data is then used to estimate the parameters of several models, namely, Bingham-plastic model, Bouc-Wen model, and Kwok and Gu Huo algebraic models. In each case, the parameter estimation is performed by constructing a nonlinear grey box model, and the predicted result is compared with the other models through normalized error metrics. In general, the models show best prediction values for relatively high frequency and displacement inputs. Moreover, the Kwok algebraic model demonstrated better performance, compared to other models evaluated in this study, in predicting the behaviour of the magneto-rheological damper.

1. INTRODUCTION

Magneto-rheological (MR) dampers are semi-active control devices that consist of a hydraulic cylinder with magneto-rheological fluids (MRF). MRFs are field controllable fluids where their rheological properties change when subjected to a magnetic field [2]. This property of MRF's lends itself well to vibration control applications where the damping of the system needs to be controlled. In order to control these dampers effectively, models that adequately characterize the MRF damper behaviour should be incorporated. Two existing categories of these models are parametric and non-parametric [12]. Parametric models have been reported to be more favourable because they provide a better representation of the behaviour of the MR damper [2]. Therefore, in this study, four parametric models have been selected and their performances are compared to select the most suitable model to characterize the MRF damper.

The four parametric models selected for this study are the Bingham, Bouc Wen, Kwok and Gu Huo models. The Bingham model is a simple parametric model first proposed by Stanway *et al.* [8]. Stanway initially developed the Bingham model to characterize an electro-rheological damping mechanism and then extended the model to include MR damper behaviour [8]. It is reported that the Bingham model demonstrates remarkable performance in predicting the damping force in the post yield region; however, it lacks precision in the pre yield region [6]. A parametric model which shows a relatively better prediction in both regions is the Bouc Wen model [6]. The Bouc-Wen model consists of hysteresis and non-hysteresis components, leading to large number of parameters and high computational costs. To alleviate this issue, parametric algebraic function models with relatively less parameters are used in this study. As an example, the Kwok model was developed as an algebraic hyperbolic tangent function-based model for MR dampers [10]. Another simple algebraic model is the Gu Huo model [11]. The distinctive characteristic of the Gu Huo model is the fact that, in addition to velocity, it also considers the acceleration in predicting the damping force.

In order to compare the effectiveness of the above mentioned parametric models, a sequence of experiments is conducted to collect suitable data to input into the models. Using this data, a nonlinear grey box estimation toolbox of MATLAB is employed to estimate the parameters of each model. The discrepancy of the estimated output of each model with the experimental results is then quantified through normalized error equations proposed by Spencer *et al.* [9]. Finally, the parametric model which best predicted the behaviour of the MR damper is presented.

2. MAGNETO-RHEOLOGICAL DAMPER MODELS

2.1 Bingham Model

The Bingham model consists of a Coulomb friction element placed in parallel with a viscous damper [6]. According to the Bingham plastic model, the damping force produced by the MR damper can be expressed as,

$$F(t) = c_0\dot{x} + f_c sgn(\dot{x}) + f_0, \tag{1}$$

where c_0 represents the damping coefficient, f_c is the frictional force related to the field-dependent yield stress, $\dot{x}$ presents the MR damper velocity and f_0 denotes the non-zero mean observed in the measured force. In this study, a piecewise version of the Bingham's model proposed by Wereley *et al.* [15] is used. This model is characterized by the yield force (f_y) and viscous damping coefficient (c_{pot}) as given by the following equation,

$$F(t) = c_{pot}\dot{x} + f_y sgn(\dot{x}), \tag{2}$$

where the yield force (f_y) represents the y-intercept of the force-velocity curve where the MR damper enters the post yield region, and (c_{pot}) is the slope of the hysteresis loop in the post yield region referred to as the viscous damping coefficient. The characteristics of this model can be described based on the following three regions [6]:

1. $F(t) = c_{pot} + f_y$ when $\dot{x} > 0$: the model will exceed the yield force. Acting in a Newtonian, linear fashion with a non-zero asymptote.
2. $-f_y < F(t) < f_y$ when $\dot{x} = 0$: in this pre-yield region, the model assumes that the material is rigid and does not flow. Thus, when $F(t) < f_y$, the shaft velocity ($\dot{x}$) is zero.
3. $F(t) = c_{pot} - f_y$ when $\dot{x} < 0$: the model will exceed the negative yield force. Acting in a Newtonian, linear fashion with a non-zero asymptote.

According to Eq.2, the set of unknown parameters of the Bingham model estimated using the grey box approach, are given as,

$$\theta = [\, c_{pot}, f_y].$$

2.2 Bouc-Wen Model

The Bouc-Wen model was created by Bouc [13] as an analytical description of a hysteretic model and then updated by Wen [14]. The force estimated by this model can be decomposed into two components, as given by the following equation,

$$F(x,\dot{x}) = g(x,\dot{x}) + \alpha z(x), \tag{2}$$

where $g(x,\dot{x})$ is a non-hysteretic component that represents a function of the instantaneous displacement and velocity, and $z(x)$ is a hysteresis component scaled by the Bouc Wen scaling factor α. Spencer *et al.* adapted the Bouc-Wen model and proposed the damping force as follows,

$$F(t) = c_0\dot{x} + k_0(x - x_0) + \alpha z - f_0, \tag{2}$$

where the time derivative of hysteresis component z, is given by

$$\dot{z} = -\gamma|\dot{x}||z|z|^{n-1} - \beta\dot{x}|z|^n + A\dot{x}, \tag{3}$$

where x_0 is the initial displacement, k_0 is the stiffness coefficient, and α, γ, β, n and A are hysteresis loop parameters. Note that, according to Eq. 3, when the velocity of the damper ($\dot{x}$) is zero, the hysteresis term (αz) goes to zero and the Bouc-Wen model behaves as a normal dashpot-spring system. In this case, the variables c_0 and k_0 contribute to the behaviour of a conventional damper without hysteresis. However, when $\dot{x} > 0$, the model mimics the behaviour of a non-linear hysteresis system, where the parameters α, γ, β, n and A contribute to more complicated non-linear hysteresis characteristics.

In this study, in order to characterize the Bouc-Wen model, a set of eight characteristic shape parameters are identified as follows,

$$\theta = [c_0, k_0, x_0, \alpha, \gamma, \beta, n, A].$$

Using the grey box approach implemented in MATLAB, the parameter θ is estimated, and the output of the Bouc-Wen model is compared with experimental data. These results will be presented in Section 5. It is also worthwhile pointing out that, due to the large set of parameters included in the Bouc-Wen model, estimating the parameters of such model becomes relatively more computationally expensive when compared to the Bingham model.

2.3 Kwok Model

Kwok *et al.* developed an algebraic hyperbolic tangent function-based model for predicting the damping force generated by MR dampers [10]. The following equation presents the damper force suggested by the Kwok model,

$$F(t) = c_0\dot{x} + k_0 x + \alpha z + f_0, \tag{4}$$

where z is the hysteresis variable defined as a hyperbolic tangent function and given as,

$$z = \tanh[\beta\dot{x} + \delta sgn(x)], \tag{5}$$

where β and δ are the hysteresis loop parameters.

In case of the Kwok model, the set of parameters that are used to construct the grey box model for parameter estimation is defined as,

$$\theta = [\, c_0, k_0, \alpha, \beta, \delta].$$

Note that the model's less complicated hysteresis function with less number of parameters that allow for an efficient and fast computation.

2.4 Gu Huo Model

Gu and Huo proposed an algebraic model to represent the hysteresis of the MR damper [11]. The damping force predicted by this model is give as,

$$F(t) = f_0 + c_b\dot{x} + \frac{2}{\pi} f_y \arctan\{k[\dot{x} - \dot{x}_0 sgn(\ddot{x})]\}, \tag{6}$$

where c_b is a viscous damping coefficient, f_y is the yield force, $\dot{x}_0$ is the hysteretic velocity and $\ddot{x}$ is the MR damper acceleration. Note that the Gu Huo model is the only proposed model that considers acceleration in its hysteresis component. According to Eq.7, the set of parameters which characterizes the Gu Huo model is as follows,

$$\theta = [\, c_b, k, f_y].$$

Similar to the Kwok model, the Gu Huo model is less complicated than the non-linear models, proving to be computationally efficient.

3. EXPERIMENT

3.1 Magneto-rheological Fluid Damper

MR dampers are semi-active control devices that consist of a hydraulic cylinder with MRF. The damper used for the experiment is obtained from Lord Corporation. The purchased damper, LORD RD-8040-1, is a short stroke magneto-rheological fluid damper commonly used for semi-active seat suspension system on large on and off highway vehicles [1].

Magneto-rheological Fluids are field controllable fluids where their rheological properties change when subjected to a magnetic field. MRF's consist of micron sized ferrous particles suspended in a carrier fluid. When the ferrous particles are subjected to a magnetic field, the dispersed particles instantly form into an organized and structured chain of particles. This reaction increases the apparent viscosity of the overall material [2, 3]. When the magnetic field is removed, the particles return to an un-organized state and the apparent viscosity decreases [4]. The fast and reversible reaction of these fluids to magnetic fields offers them as a suitable candidate to interface mechanical systems with electronic control.

The LORD RD-8040-1 damper operates in valve mode. In valve mode, the MR fluid flows between two electromagnets where the direction of the magnetic flux will be perpendicular to the direction of flow. The damper is divided by the piston between an upper and lower chamber. As the piston moves, the fluid flows through an annular orifice from one chamber to another. The gas chambers on the top sites act as the accumulator of the fluid when the fluid is pushed by the motion of the piston [5]. As the strength of the magnetic field around the annular orifice changes, the viscosity of the MRF passing through the orifice changes within a fraction of a second. This change of viscosity provides controllable resistance against the motion of the piston [6]. Table 1 presents the RD-8040-1 properties [7].

Table 1 MR Damper Properties [7]

Max Stroke (mm)	55
Extended (Total) Length (mm)	208
Cylinder (Body) Diameter (mm)	42.1
Shaft Diameter (mm)	10
Tensile Strength (Max) (N)	8896
Max Current (A)	2.0
Resistance (Ohm)	5 (Ambient Temp.), 7 (Max Temp.)
Operating Temperature (^{o}C)	71

3.2 Test Setup

Tests are conducted to obtain the force-velocity characteristics of the RD-1004 damper at varying frequency, displacement and current. The MR damper is positioned in a MTS Series 311 Load Frame operating with a 22 kip actuator and load cell. The MTS actuator oscillates at a controlled sinusoidal frequency and displacement, and the MTS data acquisition system records data at a rate of 43.22 measurements per second. Applied current is used to control the magnetic field strength in the MR damper. The input current of the MR damper is controlled by a RD-3002-03 Wonder Box and measured using a multimeter. In order to ensure the damper did not exceed the maximum operating temperature, temperature of the damper is recorded with a thermocouple. Figure 1 presents the test setup of the MR damper in the MTS actuator.

Four tests are conducted with the following inputs, all at 0.25 Amp:

- Low frequency (1 Hz), low displacement (0.4 mm)
- Low frequency (1 Hz), high displacement (2 mm)
- High frequency (5.4 Hz), low displacement (0.4 mm)
- High frequency (5.4 Hz), high displacement (2 mm)

Moreover, in order to investigate the MR damper characteristics, the damper is exposed to multiple levels of current at: 0 Amp, 0.25 Amp, 0.5 Amp at 5.4 Hz frequency and 2 mm displacement. The results of these tests will be presented in the subsequent sections.

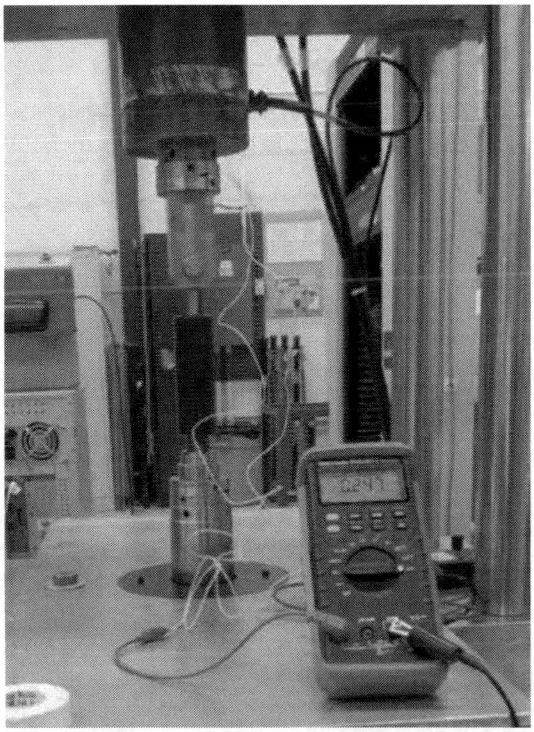

Figure 1 MR Test setup

4. MR DAMPER CHARACTERISTICS

Tests on the MR damper are conducted in order to investigate the damping characteristics of LORD RD-8040-1 device. Figure 2 displays the MR damper's unique hysteresis behaviors at 0, 0.25 and 0.5 Amps when subjected to 5.4 Hz, 2 mm inputs. The following section summarizes the characteristics of the dampers hysteresis and introduces commonly used terms to describe the behavior.

4.1 Hysteretic phenomenon

As the damper strokes between compression and tension, the damper's output force behaves in a unique hysteresis. Figure 2 presents the force-velocity and force-displacement hysteresis respectively, under varying current inputs. The damper's hysteresis progresses along in a clockwise path as time increases. When the damper reaches its maximum stroke, the velocity reaches zero and the force transfers from compression to tension or vice versa. This region of the hysteresis is referred to as the pre-yield region. In contrast to the pre-yield region, the post-yield behavior occurs when the magnitude of the velocity and force increases to its maximum, and the damper approaches its initial displacement. These force-velocity and force-displacement hysteresis behaviors are consistent for each test.

4.2 Current Dependence

The input current affects the viscosity of the MR fluid in the damper. As the current increases, the viscosity also increases. This increase in the viscosity elevates the resistance of the MR fluid against any movement, which eventually results in a greater force required to displace the shaft of the damper. The hysteresis behavior under varying currents is presented in Figure 2. As shown in this figure, increasing input current leads to a greater yield force and viscous damping coefficient.

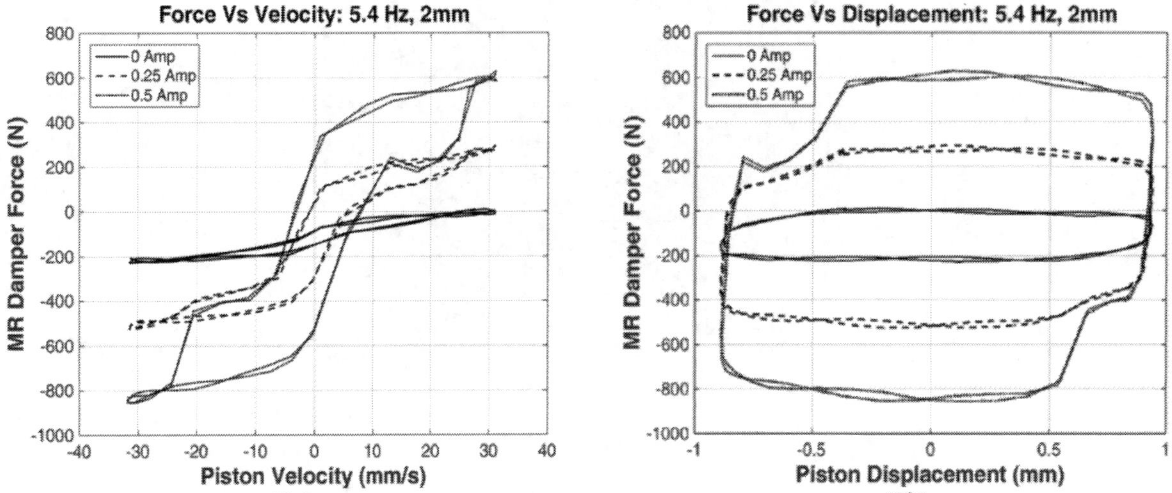

(2a) (2b)

Figure 2 MR damper results under varying currents

5. RESULTS

As mentioned in Section 3, a series of experiments are conducted in this study:

- Low frequency (1 Hz), low displacement (0.4 mm)
- Low frequency (1 Hz), high displacement (2 mm)
- High frequency (5.4 Hz), low displacement (0.4 mm)
- High frequency (5.4 Hz), high displacement (2 mm)

Among these tests, only the results of the last experiment, namely, the high frequency (5.4 Hz) and high displacement test (2 mm), is chosen to be presented in this section. However, in the next section, the entire data sets will be used to compare the efficiency of the models.

Figures 3-6 show the plot of force versus velocity and displacement. In each case, the experimental data is compared with the values obtained from the models. The Bingham model does not accurately replicate the viscous damping coefficient, for this set of data, as shown in Figure 3. In particular, the Bingham model underpredicts the yield force which results in the overprediction of the viscous damping coefficient. Figure 4, which demonstrates the results obtained from the Bouc-Wen model, reveals that this model is able to predict the yield force and the viscous damping coefficient more accurately. The results of the algebraic models, Kwok and Gu Huo models, are depicted in Figures 5 and 6 respectively. In both cases, the predicted results show a better agreement with the experimental data when compared to the previously mentioned models, especially at low velocities, where the models can still estimate accurate force values. However, comparing the two algebraic models, the Kwok model is able to capture the force behavior in the post yield region more accurately than the Gu Huo model.

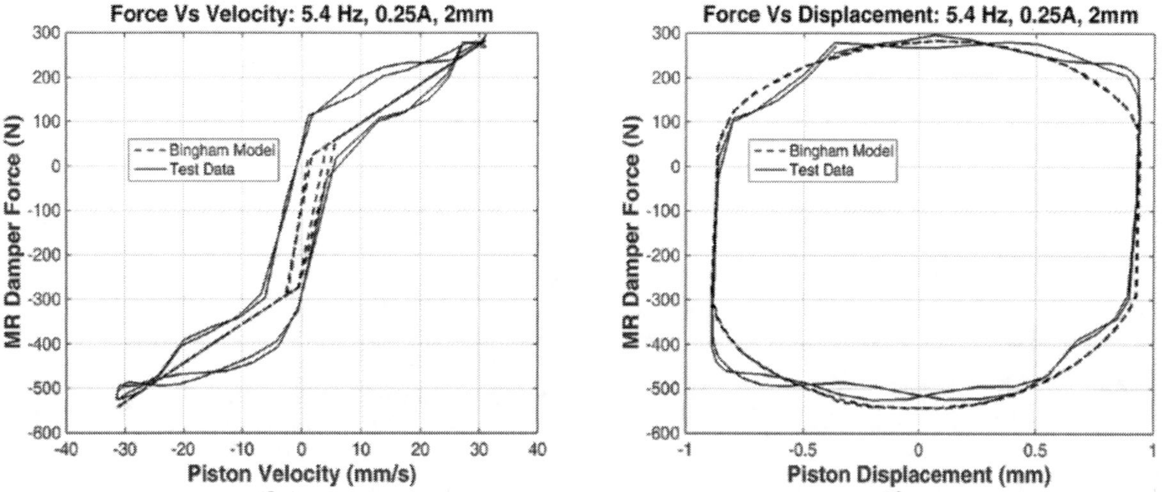

(3a) (3b)

Figure 3 Bingham Model Prediction Vs. Experimental Results

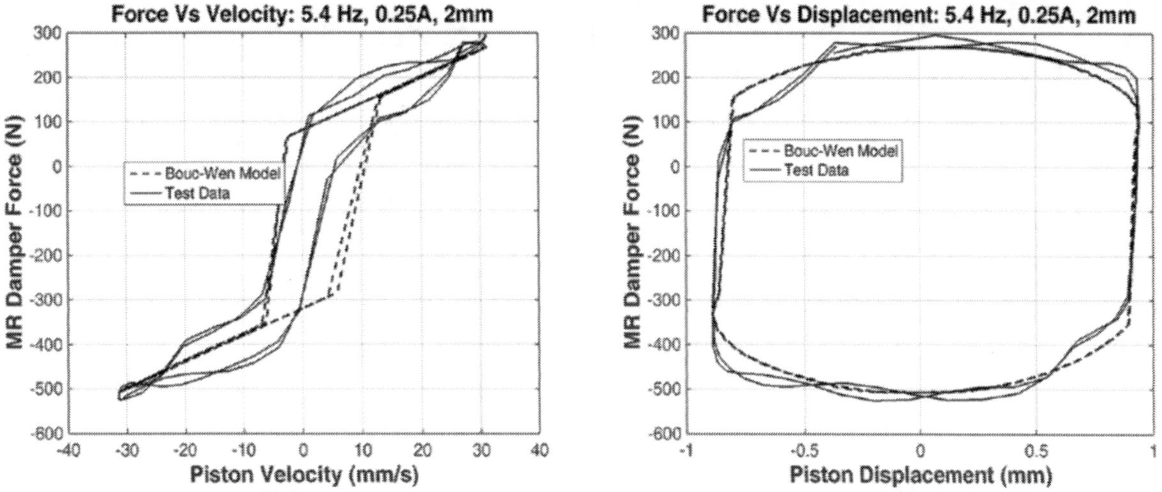

(4a) (4b)

Figure 4 Bouc-Wen Model Prediction Vs. Experimental Results

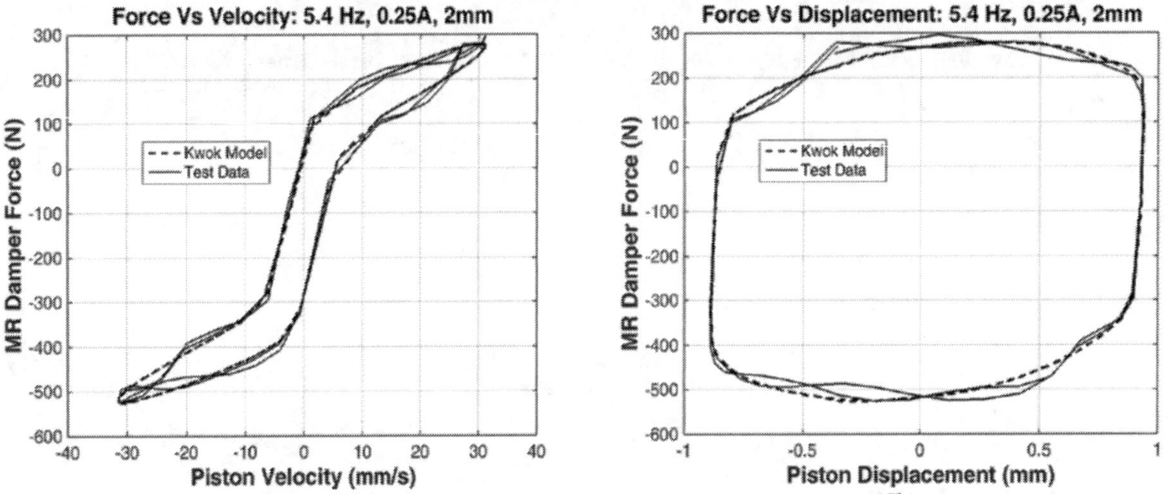

(5a) (5b)

Figure 5 Kwok Model Prediction Vs. Experimental Results

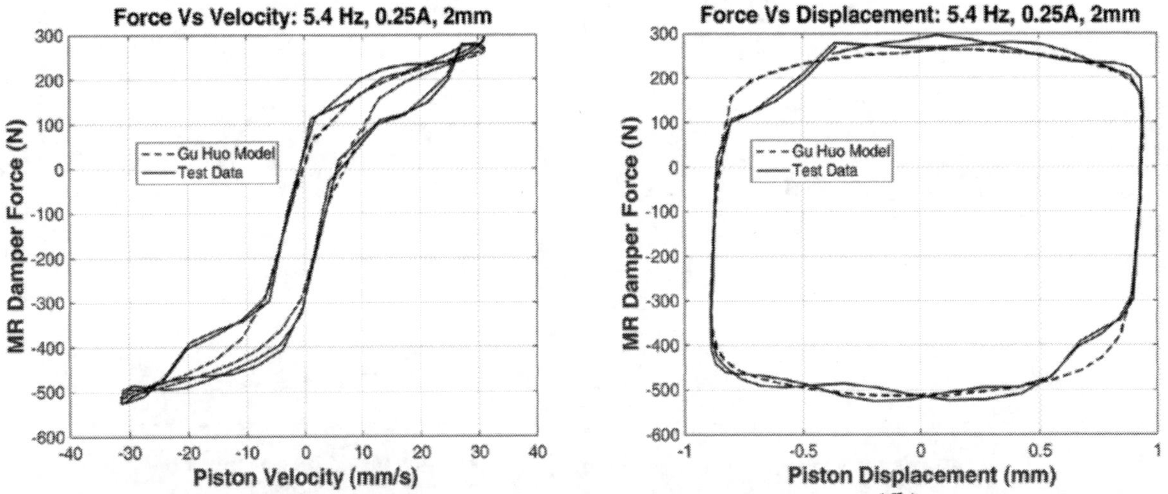

(6a) (6b)

Figure 6 Gu Huo Model Prediction Vs. Experimental Results

6. COMPARISON

6.1 Error Equations

Herein, a set of normalized error equations are used to compare the error between the model prediction and the experimental data. These equations are introduced by Spencer *et al.* as a quantitative comparison method [9]. For the models considered, the error between the predicted force and measured data has been calculated as a function of time, displacement and velocity over two complete cycles, given as,

$$E_t = \frac{\varepsilon_t}{\sigma_F}, \quad (7.1)$$

$$E_x = \frac{\varepsilon_x}{\sigma_F}, \quad (8.2)$$

$$E_{\dot{x}} = \frac{\varepsilon_{\dot{x}}}{\sigma_F}, \quad (8.3)$$

where E_t , E_x and $E_{\dot{x}}$ are time, displacement and velocity based normalized errors respectively, and ε_t, ε_x and $\varepsilon_{\dot{x}}$ are defined by the following equations,

$$\varepsilon_t^2 = \int_0^T (F_{exp} - F_{pre})^2 \, dt, \quad (9.1)$$

$$\varepsilon_x^2 = \int_0^T (F_{exp} - F_{pre})^2 \left|\frac{dx}{dt}\right| dt, \quad (8.2)$$

$$\varepsilon_{\dot{x}}^2 = \int_0^T (F_{exp} - F_{pre})^2 \left|\frac{d\dot{x}}{dt}\right| dt, \quad (9.3)$$

$$\sigma_F^2 = \int_0^T (F_{exp} - \mu_F)^2 dt. \quad (9.4)$$

Note that in the aforementioned equations, F_{exp} is the experimental force, F_{pre} is the model prediction force and μ_F is the average of the experimental force.

6.2 Error Results

Tables 2–5 present the error results obtained from Eq. 8. The four experimental sets mentioned previously are used to present the error: 1. Low frequency (1 Hz), low displacement (0.4 mm), 2. Low frequency (1 Hz), high displacement (2 mm), 3. High frequency (5.4 Hz), low displacement (0.4 mm), 4. High frequency (5.4 Hz), high displacement (2 mm).

Table 2 Low Frequency, Low Displacement Error Comparison

0.25A, 1Hz, 0.4mm	E_t	E_x	$E_{\dot{x}}$
Bingham	0.402	0.486	0.311
Bouc Wen	0.403	0.472	0.339
Kwok	0.261	0.321	0.191
Gu Huo	0.457	0.528	0.389

Table 3 Low Frequency, High Displacement Error Comparison

0.25A, 1Hz, 2mm	E_t	E_x	$E_{\dot{x}}$
Bingham	0.353	0.462	0.255
Bouc Wen	0.402	0.491	0.321
Kwok	0.136	0.151	0.130
Gu Huo	0.347	0.452	0.257

Table 4 High Frequency, Low Displacement Error Comparison

0.25A, 5.4Hz, 0.4mm	E_t	E_x	$E_{\dot{x}}$
Bingham	0.431	0.523	0.286
Bouc Wen	0.340	0.385	0.278
Kwok	0.200	0.233	0.155
Gu Huo	0.320	0.389	0.192

Table 5 High Frequency, High Displacement Error Comparison

0.25A, 5.4Hz, 2mm	E_t	E_x	$E_{\dot{x}}$
Bingham	0.172	0.226	0.103
Bouc Wen	0.189	0.247	0.108
Kwok	0.055	0.056	0.053
Gu Huo	0.092	0.110	0.075

Table 2 presents the error results of low frequency and low displacement inputs. As shown in this table, the Kwok model provides the most accurate predictions, while the Gu Huo model exhibits the greatest error. Similarly, in Table 3, where the error results of low frequency and high displacement inputs are presented, the Kwok model shows the best performance, whereas, the results obtained from Gu Huo and Bingham models show a higher disagreement with the experimental data; however, the performance of the Gu Huo model improves at a higher level of frequency, as presented in Tables 4 and 5. As evident in Table 4, the prediction of both Gu Huo and Bouc Wen models enhance significantly. Finally, as observed in Table 5, each model generated the most accurate prediction results when high frequency and high displacement case is considered.

Overall, the error results presented in this section reveal the fact that the models best predict the measurement data when high frequency and high displacement inputs are applied. Furthermore, it is concluded from the error results that the Kwok model shows a superior performance in all cases when compared to the other models evaluated in this study.

7. Conclusions

Several parametric models suitable for predicting the behaviour of magneto-rheological dampers are compared in this paper, including the Bingham model, Bouc-Wen model, and algebraic models by Kwok and Gu Huo. In order to achieve this, a dynamic vibration test is performed on a magneto-rheological damper to obtain data sets to be use as input into the models. The model prediction results are obtained based on established approaches and they are compared using normalized error equations. Upon comparison using the error equation, it is found that the models predicted the force behaviour of the MR damper more accurately for relatively high frequency and high displacement inputs compared to the low frequency and low displacement case. Out of all models evaluated in this study, the Kwok model best predicted the force behaviour of the MR damper.

8. References

[1] Jolly, Mark R., Jonathan W. Bender, and J. David Carlson. "Properties and applications of commercial magnetorheological fluids." *5th Annual International Symposium on Smart Structures and Materials*. International Society for Optics and Photonics, 1998.

[2] Şahin, İ., T. Engin, and Ş. Çeşmeci. "Comparison of some existing parametric models for magnetorheological fluid dampers." *Smart Materials and Structures*19.3 (2010): 035012.

[3] Weiss, Keith D., and J. David Carlson. "A growing attraction to magnetic fluids." *Machine Design* 66.15 (1994): 61-64.

[4] Ashour, Osama, Craig A. Rogers, and William Kordonsky. "Magnetorheological fluids: materials, characterization, and devices." *Journal of intelligent material systems and structures* 7.2 (1996): 123-130.

[5] Chooi, Weng Wai, and S. Olutunde Oyadiji. "Mathematical modeling, analysis, and design of magnetorheological (MR) dampers." *Journal of Vibration and Acoustics* 131.6 (2009): 061002.

[6] Wang, D. H., and W. Hsin Liao. "Magnetorheological fluid dampers: a review of parametric modelling." *Smart materials and structures* 20.2 (2011): 023001.

[7] *RD-8040-1 and RD-8041-1 Dampers*. Rev. 0. LORD Corporation. June. 2009.

[8] Stanway, R. S. J. L., J. L. Sproston, and N. G. Stevens. "Non-linear modelling of an electro-rheological vibration damper." *Journal of Electrostatics* 20.2 (1987): 167-184.

[9] Spencer Jr, B. F., et al. "Phenomenological model for magnetorheological dampers." *Journal of engineering mechanics* (1997).

[10] Kwok, N. M., et al. "A novel hysteretic model for magnetorheological fluid dampers and parameter identification using particle swarm optimization." *Sensors and Actuators A: Physical* 132.2 (2006): 441-451.

[11] Guo, Dalei, and Haiyan Hu. "Nonlinear stiffness of a magneto-rheological damper." *Nonlinear Dynamics* 40.3 (2005): 241-249.

[12] Boada, M. J. L., et al. "Modeling of a magnetorheological damper by recursive lazy learning." *International Journal of Non-Linear Mechanics* 46.3 (2011): 479-485.

[13] Bouc, R. "A mathematical model for hysteresis." *Acta Acustica united with Acustica* 24.1 (1971): 16-25.

[14] Wen, Yi-Kwei. "Method for random vibration of hysteretic systems." *Journal of the engineering mechanics division* 102.2 (1976): 249-263.

[15] Wereley, Norman M., Li Pang, and Gopalakrishna M. Kamath. "Idealized hysteresis modeling of electrorheological and magnetorheological dampers." *Journal of Intelligent Material Systems and Structures* 9.8 (1998): 642-649.

ICAST2015 #060

Energy-saving shape-retainment using two types of piezoelectric actuators

Tomonori Uchida[1], Tadashige Ikeda[1*], Atsuhiko Senba[2], and Kosei Ishimura[3]

[1] Department of Aerospace Engineering, Nagoya University, Nagoya, Japan.
[2] Department of Vehicle and Mechanical Engineering, Meijo University, Nagoya, Japan.
[3] Institute of Space and Astronautical Science, Sagamihara, Japan.

Abstract

To realize structure systems of future high-precision space observation satellites, smart structure systems which measure and correct their configuration or/and suppress their vibration on obit have been studied. Among sensors and actuators to be used in the smart structure systems, piezoelectric materials are expected to be a good candidate since they have less mechanical parts and high compatibility for space environment. When a shape of a smart space structure is controlled by the piezoelectric actuators attached to its structural element, electric voltage must be continued to be applied to retain a desired shape. To save the amount of electricity usage, a new control method was proposed and its feasibility was examined in the previous papers [Ikeda and Takahashi, Proc. SPIE 8689; Ikeda et al., Trans. JSASS 12 ists29]. In the method the hysteretic behavior of piezoelectric actuators was utilized effectively. The results showed that displacement of a smart beam with a piezoelectric ceramic actuator bonded remained without any voltage applied to the actuators after a pulsed voltage was applied. However, the displacement of the beam overshot a final position. In this paper this overshoot is suppressed. To this end another type of piezoelectric actuator showing almost no hysteretic behavior is bonded on the beam opposing the actuator showing the hysteretic behavior. Result shows that the overshoot can be suppressed by applying a feedback plus feedforward control to the additional actuator while the pulsed voltage is applied to the actuator showing hysteretic behavior.

1. INTRODUCTION

Astronomers require higher mirror accuracy or pointing accuracy for reflectors of space antennae [1-3]. However, their performance is influenced easily by degradation of the materials during the operation and shape errors in folding-deploying process, thermal cycles, disturbance on orbit, manufacturing process, and so on. Accordingly, it is difficult to guarantee their high accuracy during the operation. As a structure systems of future high-precision space observation satellites, smart structure systems which measure and correct their configuration or/and suppress their vibration on obit have been studied [4-8]. In the systems smart materials such as piezoelectric materials and shape memory materials are often used as sensors and actuators, because they have less mechanical parts and hence they are expected to have high compatibility for space environment. Recently, Hill et al. [6] have developed a large lightweight antenna reflector comprised of a flexible Kapton film having polyvinylidene fluoride actuators for surface control. They studied optimal grouping of actuators to reduce number of power supplies. Bradford et al. [7] have developed an active composite reflector

* ikeda@nuae.nagoya-u.ac.jp

comprised of a graphite face sheet and aluminum honey-comb core composite panel having a network of distributed piezoelectric actuators for vibration and shape control. We have also studied a smart antenna reflector system by using piezoelectric materials to realize high mirror accuracy [8].

When shape of an antenna reflector surface is controlled by piezoelectric actuators attached to the structural element, electric voltage must be continued to be applied to keep a desired configuration. The applied electric power itself is low but amount of electricity must be accumulated to become big. However, since available energy is limited especially in space, reduction of amount of electricity usage is necessary to save the energy and to extend lifetime of satellites. Moreover, of course, the shape cannot be retained when the control system is down.

Some researchers have studied shape memory ceramics to keep a position of mechanical relay switches, grippers, etc. without application of electric power. In these studies, the fact that there are two stable states on zero electric filed axis of strain versus electric field curves is applied to alternation of the two positions. These two stable states are caused by hysteresis in phase transition between antiferroelectric phase and ferroelectric phase [9] or hysteresis in domain switching between poled domain state and depoled domain state [10, 11].

Applying the idea of this shape memory ceramics, a new control method was proposed and its feasibility was studied [12]. In the study it was found that (1) about 100 $\mu\varepsilon$ of strain remained after a pulsed voltage, and (2) the residual strain varied depending on the amplitude and period of the pulse. Piezoelectric actuators are often stressed during the operation because they are bonded on a structural element. To examine deformation behavior of piezoelectric actuators under stressed condition, a smart beam model was made, which was comprised of an aluminum beam with a piezoelectric plate actuator bonded on its surface. It was also found from the result of the smart beam that (3) the bending configuration of the beam remained, although the beam vibrated and the tip displacement overshot a final position when the pulse voltage was applied to the actuator. Subsequently, to find an adequate piezoelectric ceramic material for the new shape retainment method, strain response for a pulsed voltage was examined for three types of piezoelectric ceramic actuator plates including soft type and hard type [13]. The coercive field of the hard type piezoelectric actuator is larger than the soft type piezoelectric actuator and the hard type piezoelectric actuator has much less hysteresis in the strain versus electric field relationship than the soft type piezoelectric actuators within the strain range of use. Moreover, it was examined if the vibration of the beam after the pulse could be suppressed by a feedback control or controlling the pulse shape. The results showed that (4) remaining strain of the soft type material was the largest with more than 200$\mu\varepsilon$, (5) the hard type material could not generate enough remaining strain, and (6) the vibration of the beam with the piezoelectric actuator could be suppressed by a feedback control and a "1-cosine" shaped pulse. These results imply feasibility of the shape retainment control with zero or less applied voltage. However, there remain several issues to be solved such as the undesired overshoot, repeatability, reliability, and control method to compensate them. To solve them one by one, in this paper, a suppression method of the overshoot is proposed because the overshoot may damage the structure.

The concept of this shape retainment method and the previous results are reviewed in Sections 2 and 3, respectively. In Section 4 a suppression method of the overshoot is proposed and verified experimentally.

2. CONCEPT OF ENERGY-SAVING SHAPE RETAINMENT CONTROL METHOD

Figure 1 shows a schematic relationship between strain and electric field of a piezoelectric ceramics material. The state where the material has been just fabricated and the dipole moments are randomly oriented is defined as the origin. By applying a strong DC electric field to the material, it polarizes and expands in the electric field direction. Next when the field is removed, the material gradually shrinks. However, residual strain remains even after the field is completely removed as the state (a). Then, if a negative field is applied to the material, it shrinks until the field reaches a coercive field ($-Ec$) shown in the state (b). When the negative field increases beyond the coercive field, the material expands again, and it takes a negative polarized state. Subsequently, when the negative field is removed and then a positive field increases, the material shrinks until the coercive field (*E*c) and then expands. The curve returns to the positive polarized state and shows a so-called butterfly shape.

If the negative electric field is removed from the state (b), the material state goes to the state (c) different from the state (a) because the orientation of the dipole moments in the state (c) is different from that in the state (a). If the negative electric field is removed from a state between the state (a) and the state (b) (e.g. the state (d)), the material state goes to another state between the state (a) and the state (c) (e.g. the state (e)) at zero electric field. Therefore, the strain can take any values between the state (a) and the state (c). By using this concept, if a shape of a smart structure having a piezoelectric actuator becomes a desired shape after removing the electric field, zero electric field is necessary to retain the desired shape. Moreover, even in the case that the shape of the structure deviates slightly from the desired shape due to disturbances or aging, much less electric field is needed to correct the shape than conventional control methods where the electric field must be applied from the state (a).

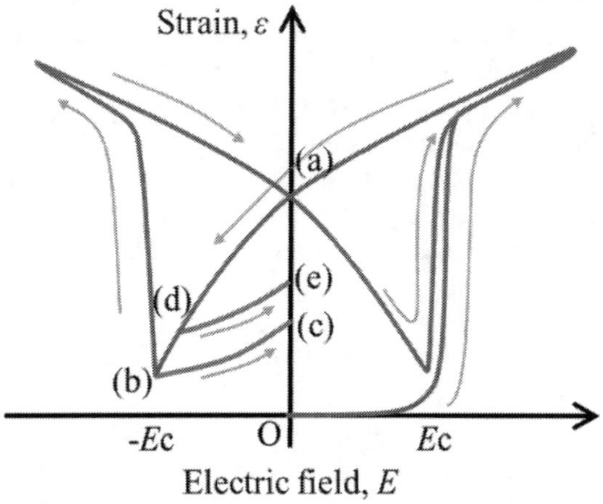

Figure 1. Concept of the proposed shape retainment method on a schematic strain versus electric field curve

3. REVIEW OF THE PREVIOUS EXPERIMENTS [13]

3.1 Strain response of a piezoelectric ceramic plate after pulses of electric field

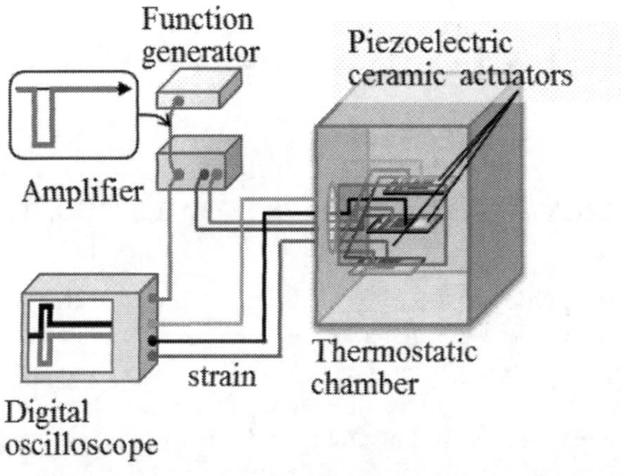

Figure 2. Schematic diagram of an experimental setup for measuring strain responses of piezoelectric ceramic actuators

Schematic diagram of an experimental setup for measuring strain response of a piezoelectric ceramic actuator plate against pulsed voltages is shown in Fig. 2. Input signal of a rectangular pulse was generated by a function generator (NF, WF1974), amplified by an amplifier (Trek, Model 2220), and applied to piezoelectric ceramic actuator plates. Strain on their surface was measured by a strain gauge. The experiment was carried out in a thermostatic chamber because thermal strain was comparable with the strain generated by the electric field.

Figure 3 (a) shows the strain responses of three types of piezoelectric ceramic actuators (Fuji ceramics, type C-91H (soft type, high tan δ), C-6 (soft type), C-202 (hard type)). The specifications of the three types of specimens are listed in Table 1. d_{31}, Y_{11}^{E}, $\varepsilon_{33}^{T}/\varepsilon_0$, tan δ, and T_C denote the piezoelectric constant in the longitudinal direction, Young's modulus in the longitudinal direction, specific permittivity in the thickness direction, loss factor, and Curie point, respectively. Rectangular pulses with amplitudes of –600 kV/m and +600 kV/m and a period of 1 s were applied in series at intervals of 300 s. Figure 3 (b) shows the response of the hard type piezoelectric actuator for the pulses with amplitudes of ±1200 kV/m. It is seen from Fig. 3 (a) that the strains of the soft type piezoelectric actuators after the pulse were different from the ones before the pulse due to the hysteresis in strain versus electric field curve. Approximately 200 με of strain remained for 300 s at least in the soft type piezoelectric actuator C-91H while almost no strain remained in the hard type piezoelectric actuator C-202. This is because that the hard type piezoelectric actuator has much less hysteresis than the soft type piezoelectric actuators within the strain range of use.

Table 1 Specifications of the three types of piezoelectric ceramic actuators examined

Specimen	C-91H	C-6	C-202
Type	Soft, High tan δ	Soft, General Purpose	Hard
d_{31} [pm/V]	–375	2130	–130
Y_{11}^{E} [Gpa]	64	–210	82
$\varepsilon_{33}^{T}/\varepsilon_0$ [-]	5800	62	1600
tan δ [%]	3.7	1.60	0.20
T_C [°C]	170	295	290
Length [mm]	40	50	40
Width [mm]	20	10	20
Thickness [mm]	0.5	0.5	0.5

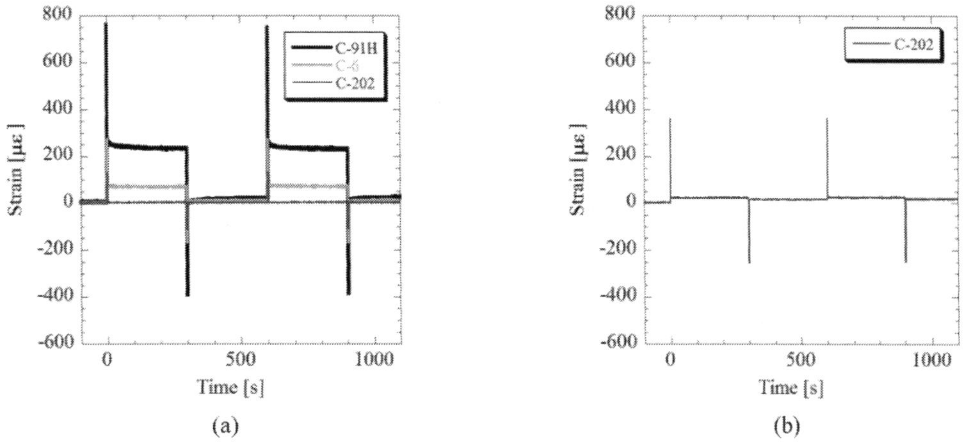

Figure 3. Strain responses of piezoelectric actuators for rectangular pulses with amplitudes of (a) –600 kV/m and +600 kV/m and (b) –1200 kV/m and +1200 kV/m

3.2 Deformation behavior of a smart beam with a piezoelectric actuator

Figure 4 shows an experimental setup for a smart beam. The smart beam model was comprised of an aluminum alloy cantilever beam of 300 mm × 21 mm ×1 mm on a surface of which a piezoelectric ceramic actuator (C-6) of 30 mm × 20 mm × 1 mm was bonded so that the end of the piezoelectric actuator was located at 20 mm from the clamp. Displacement at 20 mm from the tip was measured with a laser displacement meter (Keyence, LB-02/LB-62) and a strain on the piezoelectric actuator was measured by a strain gauge. A rectangular or a "1-cosine" shaped pulse was applied to the piezoelectric actuator.

The applied electric field, the tip displacement of the smart beam, and the strain on the piezoelectric actuator for a pulse of electric field with –600 kV/m amplitude and 1 s period are shown in Fig. 5. Negative pulse input generated positive strain because the strain in the longitudinal direction was induced by d31 effect. It was seen that the strain and the tip displacement remained even at zero electric field after the pulse, although the beam vibrated for the rectangular pulse. This vibration could be suppressed by applying the "1-cosine" shaped pulse. However, the strain and the displacement significantly overshot for both the cases.

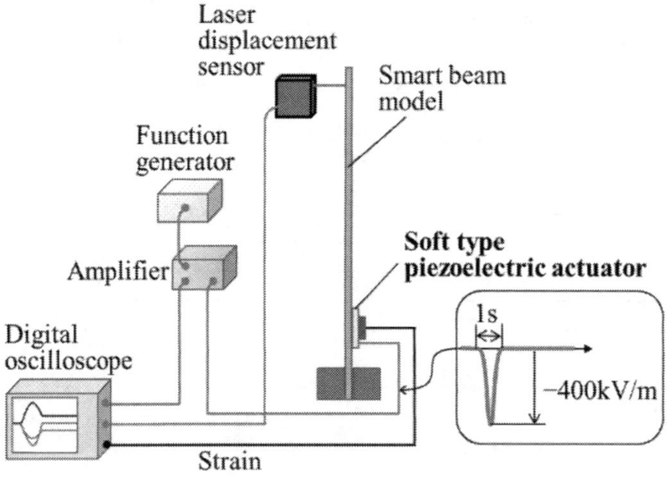

Figure 4. Schematic diagram of an experimental setup for measuring deformation behavior of a smart beam

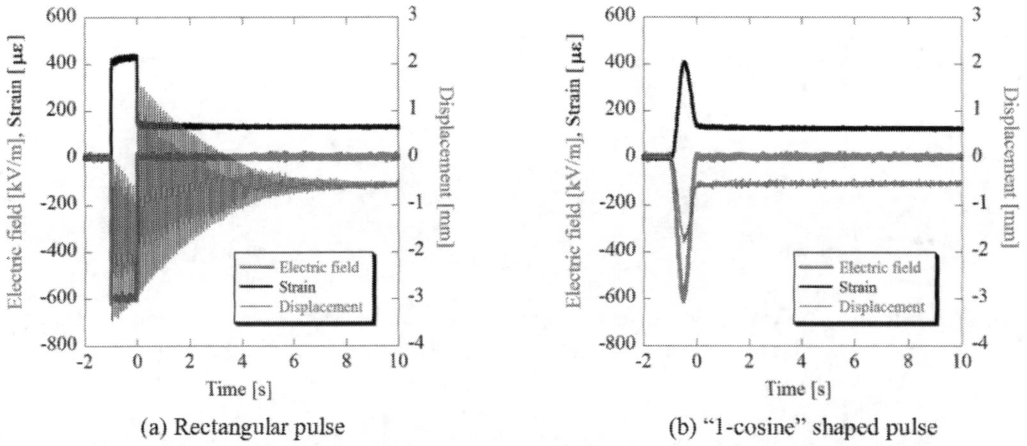

Figure 5. Electric field and responses of the strain of the piezoelectric actuator and the tip displacement of the smart beam

4. SUPPRESSION OF OVERSHOOT USING TWO TYPES OF PIEZOELECRIC ACTUATORS

4.1 Concept of suppression method of the overshoot

It was seen in the previous experiments that (1) the soft type piezoelectric actuators generated much bigger remaining strain than the hard type piezoelectric actuators although both the piezoelectric actuators highly expanded while the pulse was applied, (2) hysteresis of the piezoelectric actuators can be effectively utilized for the shape retainment of the smart beam with no/less energy, (3) displacement of the smart beam significantly overshot the final position although a "1-cosine" shaped pulse was effective for suppressing the vibration of the beam.

Since the overshoot may damage the structure, it must be suppressed. In accordance with the first finding, it is considered that the overshoot can be suppressed using both the soft type piezoelectric actuator and the hard type piezoelectric actuator by applying the electric field to the hard type piezoelectric actuator so that its large strain compensates the large strain of the soft type piezoelectric actuator during the pulse.

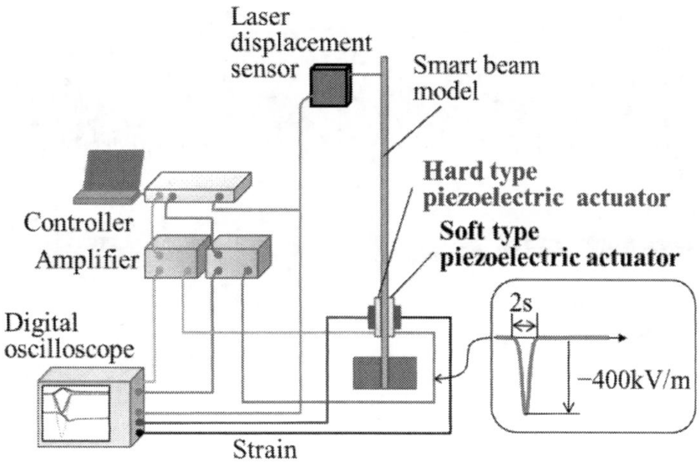

Figure 6. Schematic diagram of an experimental setup for measuring deformation behavior of the improved smart beam

4.2 Method of verification test of the concept

Figure 6 shows an experimental setup to verify the concept described in the previous section. The improved smart beam model was comprised of an aluminum alloy cantilever beam with a size of 300 mm × 22 mm ×1 mm on one surface of which the soft type piezoelectric actuator C-91H with a size of 40 mm × 20 mm × 0.5 mm was bonded in a region between 20 mm and 60 mm from the root and on the other surface of which the hard type piezoelectric actuator C-202 with a size of 40 mm × 20 mm × 0.5 mm was bonded. Displacement at 20 mm from the tip was measured with the laser displacement meter and strains on both the piezoelectric ceramics plates were measured by the strain gauges in a similar way to the previous experiment. A "1-cosine" shaped pulse of the electric field with 400 kV/m amplitude and 2 s period was applied to the soft type piezoelectric actuator. To suppress the overshoot electric field obtained by a feedforward, a feedback, or a feedforward plus feedback control laws was applied to the hard type piezoelectric actuator. In the feedforward control the electric field 2.6 times larger than one applied to the soft type was applied to the hard type piezoelectric actuator. The number of 2.6 was determined through a trial and error process. In the feedback control a PD (proportional-derivative) controller was designed so that the zero displacement was kept while the "1-cosine" shaped pulse was applied to the soft type piezoelectric actuator. To prevent the effect of signal noise or perturbation of the beam, high frequency components of the displacement signal were removed by passage through a low-pass filter. To limit the period while the feedback control was performed, the feedback gain was multiplied by the factor which smoothly increased from 0 % to 100 % for 1 s before applying the "1-cosine" shaped pulse to the soft type piezoelectric actuator and smoothly decreased from 100 % to 0 % for 1 s after the pulse.

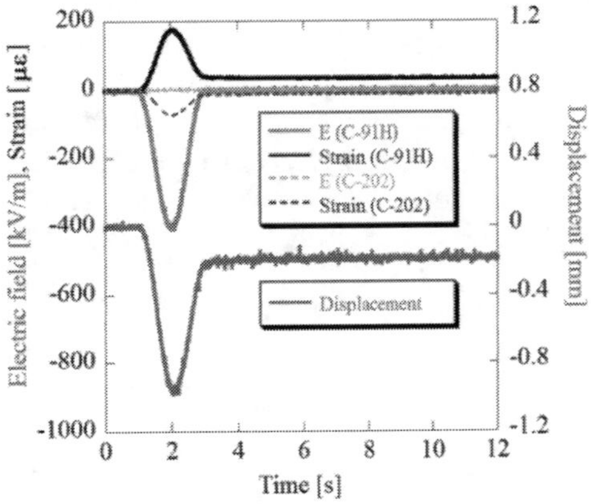

(a) Without control

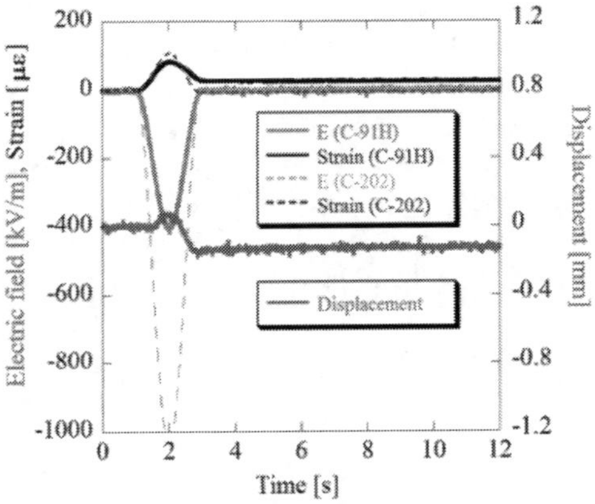

(b) With feedforward control

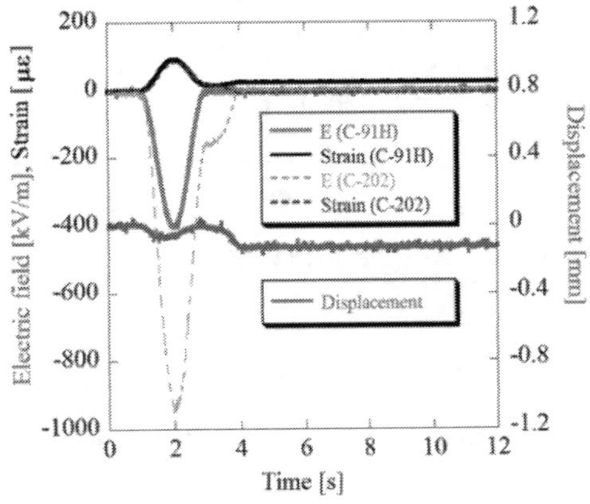

With feedback control

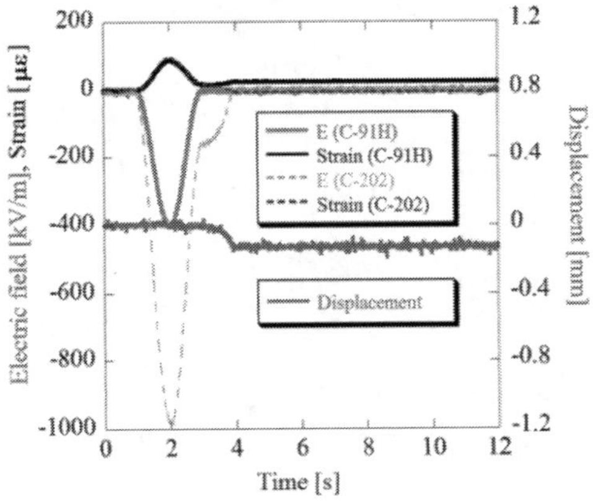

With feedforward + feedback control

Figure 7. Electric fields applied and responses of the strain of the piezoelectric ceramic actuators and the tip displacement of the improved smart beam without and with controls

4.3 Results and discussion

Figures 7 (a) to 7 (d) show the electric fields applied and responses of the strain of the piezoelectric ceramic actuators, and the tip displacement of the improved smart beam without control, with the feedforward control, the feedback control, and the feedforward plus feedback control, respectively. In without control, 0.19 mm of mean tip displacement between 8 s and 10 s remained after the pulse but the displacement overshot by 0.79 mm as shown in Fig. 7 (a). The hard type piezoelectric actuator shrank although no electric field was applied to it. This is because it was compressed passively by bending of the beam due to the expansion of the soft type piezoelectric actuator. In with the feedforward control, it is seen that the large overshoot could be suppressed as shown in Fig. 7 (b). However the tip displacement deflected to the opposite side by 0.08 mm first and still overshot somewhat, by 0.02 mm. This is because the "1-cosine" shaped electric field was also applied to the hard type piezoelectric actuator, which limited performance of the controller. It is noted that the residual deflection also decreased to be 0.13 mm. The reason is thought to be that amount of domain switching of the soft type piezoelectric actuator was reduced due to the stress from the hard type piezoelectric actuator. The feedback control could also suppress the large overshoot, although the zero displacement could not be kept as shown in Fig. 7 (c). This is because the low-pass filter delayed the phase of the displacement signal and that deteriorated the performance. Finally, the feedback control was performed with the feedforward control. Figure 7 (d) shows that zero displacement was kept while the pulse was applied to the soft piezoelectric actuator and after the pulse the displacement smoothly increased. It is seen the effect of the low-pass filter in the feedback control was reduced. It is considered that since the displacement to be suppressed by the feedback control was reduced by adding the feedforward control, the effect was also reduced. The last two controllers could suppress the overshoot almost completely but the residual displacement was also suppressed to be 0.13 mm.

5. CONCLUSION

A suppression method of the overshoot occurring in the shape control of a smart beam utilizing hysteretic behavior of piezoelectric actuators was proposed. In this method two types of piezoelectric actuators were used, one of which was a soft type piezoelectric actuator and the other of which was a hard type piezoelectric actuator. The hard type piezoelectric actuator has much less hysteretic behavior within the strain range of use than the soft type piezoelectric actuator. "1-cosine" shaped pulse of electric field was applied to the soft type piezoelectric actuator so that the strain remained due to the hysteric behavior while the vibration was suppressed. The efficacy of this method was confirmed by an experiment using a smart beam model with a soft type piezoelectric actuator plate bonded on one surface and a hard type piezoelectric actuator bonded on the other surface. The result showed that the controllers could suppress the overshoot almost completely, although the residual displacement decreased a little by the control. This experimental verification indicates that the proposed shape retainment method utilizing the hysteresis edged closer to practical use. There are, however, still several challenges to be solved such as prediction of residual deflection, aging, and so on.

ACKNOWLEDGMENTS

A part of this study was supported by "Study on Large High-Precision Optical Mount" in the "Science and Engineering Experiments of Strategic Development and Research" of Institute of Space and Astronautical Science, Japan Aerospace Exploration Agency.

REFERENCES

1. Higuchi, K., Kishimoto, N., Meguro, A., Tanaka, H., Yoshihara, M., and Iikura, S., "Structure of high precision large deployable reflector for space VLBI (Very Long Baseline Interferometry)," Proceedings of 50th AIAA/ASME/ASCE/AHS/ASC Structures, Structural Dynamics and Materials Conference, AIAA-2009-2609, 2009, pp. 1-9.
2. Agnes, G. S. and Dooley, J., "Precision deployable structures technology for NASA large aperture missions," Proceedings of Space 2004 Conference and Exhibit, AIAA-2004-5899, 2004, pp. 1-7.

3. Takahashi, T., et al., "The NeXT mission," Proceedings of the SPIE, Vol. 7011, 2008, 70110O (14p).
4. Tanaka, H. and Natori, M. C., "Shape control of space antennas consisting of cable networks," Acta Astronautica, Vol. 55, Issues 3-9, 2004, pp. 519-527.
5. Ruggiero, E. J. and Inman, D. J., "Gossamer spacecraft: recent trends in design, analysis, experimentation, and control," Journal of Spacecraft and Rockets, Vol. 43, No. 1, 2006, pp. 10-24.
6. Hill, J. R., Wang, K. W., Fang, W. W., and Quijano, U., "Actuator grouping optimization on flexible space reflector," Proceedings of the SPIE, Vol. 7977, 2011, 797725 (12p)
7. Bradford, S. C., Agnes, G. S., Bach, V. M., and Wilkie, W. K., "An active composite reflector system for correcting thermal deformations," Proceedings of 52nd AIAA/ASME/ASCE/AHS/ASC Structures, Structural Dynamics and Materials Conference, AIAA 2011-1826, 2011, pp. 1-8.
8. Okubo, H., et al., "Demonstration test plan for ultra-high precision smart structural system using high altitude balloon," Proceedings of Balloon Symposium 2009, 2009, pp. 104-107 (in Japanese).
9. Uchino, K., "Shape memory ceramics," in [Shape Memory Materials], Otsuka, K. and Wayman, C. M. (ed), Cambridge University Press, Cambridge, 1998, pp. 184-202.
10. Morita, T., Kadota, Y., and Hosaka, H., "Shape memory piezoelectric actuator," Applied Physical Letter, Vol. 90, 2007, 082909 (3p).
11. Kadota, Y., Hosaka, H., and Morita, T., "Field-induced strain memory with non-180° domain-reorientation control," Journal of the Korean Physical Society, Vol. 57, No. 4, 2010, pp. 902-906.
12. Ikeda, T. and Takahashi, T., "Feasibility study of shape control with zero applied voltage utilizing hysteresis in strain-electric field relationship of piezoelectric ceramics, Proceedings of the SPIE, Vol. 8689, 2013, 86890C (8p).
13. Ikeda, T., Takahashi, T., Uchida, T., and Ishimura, K., "Shape retainment by utilizing hysteresis in piezoelectric ceramics," The ISTS Special Issues of Transactions of JSASS Aerospace Technology Japan, Vol. 12, No. ists29, 2014, pp. Pc_69-Pc_74.

ICAST2015 #006

A PRECISION LIGHT-ACTIVATED ROBOT GRIPPER

Da GUO[1], Huiyu LI[2], Hornsen TZOU[1,2,*]

[1]StrucTronics and Control Lab, School of Aeronautics and Astronautics
[2]The State Key Laboratory of Fluid Power Transmission and Control, School of Mechanical Engineering, Zhejiang University, Hangzhou, Zhejiang 310027, P.R. China
*Corresponding author, Email: hstzou@zju.edu.cn

Abstract

Light-activated shape memory polymer (LaSMP) realizes modulus change function by exposure to ultraviolet (UV) light of different wavelengths. With this noncontact actuation feature, a light-weight and noncontact gripper drived by UV light is presented in this study. Mathematical model of LaSMP's Young's modulus versus exposure time is established first, and then finger models based on the laminated beam structure are presented. With the model of cantilever beam, performance including tip displacement and equivalent force of the finger model are evaluated. In Case studies, laminated beam finger with single LaSMP actuator is presented first. Then the length effect of the LaSMP actuators is studied based on the laminated beam finger with double LaSMP actuators and an optimal length of LaSMP actuators is obtained. According to the result of Case 2, a robotic gripper is designed in Case 3 with optimal length of LaSMP actuators and the gripper is capable of producing large gripping strokes up to 184μm; maximum recommended payload is 0.74mN when 0.5 friction coefficients between finger and target, safety factor of 2 are considered. The performance of the gripper indicates that the grippers have potential use for applications of microelectromechanical system (MEMS).

Keywords: LaSMP, light-weight, noncontact actuation, gripper, optimal length

1. INTRODUCTION

Smart materials have attracted much interest in recent years. Smart materials such as piezoelectric materials, shape memory alloy (SMA), flexoelectric material and magnetostriction material are widely applied in smart structures for vibration control, energy harvesting, actuation, and so on [1-9]. In actuation applications, a lot of mechanisms are actuated by shape memory materials due to their excellent performance. For examples, a beam-shape composite actuator using SMA wires is applied as the active component to provide a large deformation profile and it aims to develop a new locomotion mechanisms based on the deformation profile [10]. Properties of shape memory polymer fibers loaded with carbon nanotubes were studied; the results show that a very large stress was generated in these fibers when they recover their shape after they have been stretched and cooled under tensile load [11].

A deployable hinge was investigated by the application of fiber reinforced shape memory polymer composite and the major advantages of shape memory polymer (SMP) are their extremely high recovery strain, low density and low cost [12].

In this study, a light-weight, noncontact robot gripper is developed based on LaSMP. There are many kinds of robot grippers with diverse functions depending on different actuators. Most of the traditional robot grippers are electric, hydraulic and pneumatic. Control system of contact-type is needed for all of these grippers and their own weight is large versus their payload. For Robotiq adaptive gripper 2-finger model-85, power and communication are established between the gripper and the Robotiq universal controller via a single device cable [13]. Micro grippers based on MEMS have demonstrated their important applications in biological and biomedical researches [14-16]. Hydraulic and pneumatic grippers usually consist of complicated power supply which limits their applications [17-19]. Ionic polymer-metal composites (IPMC), SMA and piezoelectric materials are three most commonly used smart materials in grippers. IPMC grippers have a large bending deflection, thus non-linear large strain response of IPMC to electrical simulation requires more analytical methods to predict their performance based on the material [20,21]. Robot grippers using SMA actuators have been reported to possess large output force and high power density; however actuation frequency is limited by the heating methods and also requires cooling system [22-25]. Piezoelectric gripper offers a great speed, fine positioning and large force, but the displacement generated is small [26-28].

2. LASMP GRIPPER DESIGN

There are two fingers in the LaSMP gripper designed in this section; each finger consists of LaSMP layers and an elastic layer. In gripping mode, the LaSMP finger deforms after exposure to UV light of different wavelengths which results in change of the two LaSMP layers' Young's moduli. LaSMP layers try to recover to the original shape upon exposure but restricted by the elastic layer, thus a load induced by LaSMP layers is applied on the elastic layer. Cooperation of the deformation of these two fingers can pick or place a target. A tiny mass (protective metal surface) is attached at free-end and each finger can be simplified as a cantilever beam without tip mass.

3. MODELING OF LASMP'S YOUNG'S MODULUS

LaSMP's Young's modulus changes when exposure to UV light of different wavelengths. In this section, mathematical model of Young's modulus for LaSMP is established. The mathematical model is divided into two parts, i.e., the forward reaction part and the reverse reaction part.

From Beer-Lambert-Bouguer law, light intensity at the thickness of z in the forward reaction decays exponentially [29]

$$I_r(z) = I_r \mathrm{e}^{-c_r z}, \tag{1}$$

where z is the position of local point along the thickness of LaSMP sample as shown in **Fig.1**; I_r is the incident light intensity on the LaSMP surface in the forward reaction; c_r is the attenuation coefficient in the forward reaction.

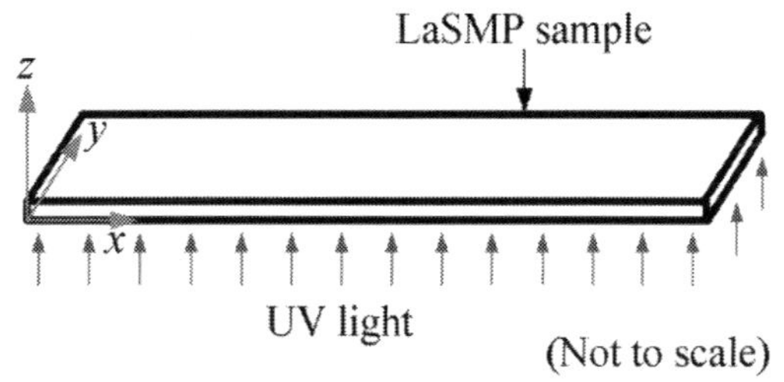

Figure 1. LaSMP sample under UV light exposure

The forward reaction is assumed to be first order and it never reaches static state since light intensity doesn't reach saturation. The concentration of optically activated species P is expressed as

$$\tau_r \frac{\partial P}{\partial t} + P = \frac{1}{I_r} I_r(z) P_{\max}, \tag{2}$$

where t represents the reaction time; τ_r represents the time constant of the forward reaction and it is determined by experiment; $P_{\max}$ is the maximum concentration of optically activated species. Initial condition of Eq.(2) is P=0 when t=0. Solution of the concentration of optically activated species can be obtained

$$P = P_{\max} \mathrm{e}^{-c_r z}\left(1 - \mathrm{e}^{\frac{-t}{\tau_r}}\right). \tag{3}$$

It is assumed that the forward reaction ends at time instant t_1, and the corresponding concentration of optically activated species can be expressed as

$$P_{t_1} = P_{\max} \mathrm{e}^{-c_r z}\left(1 - \mathrm{e}^{\frac{-t_1}{\tau_r}}\right). \tag{4}$$

The local Young's modulus of LaSMP in the sample is assumed to be proportional to the concentration of optically activated species P [30], thus the Young's modulus of LaSMP at the thickness of z is expressed as

$$Y_a(z) = \frac{Y_{\max} - Y_0}{P_{\max}} P + Y_0, \tag{5}$$

where Y_0 is the initial Young's modulus before UV light exposure and $Y_{\max}$ is the maximum Young's modulus when reaction time approaches infinity. Young's modulus of LaSMP actuator can be obtained by integration along the thickness direction and then divided by the thickness

$$Y_a = \frac{\int_0^{h^a} Y(z)dz}{h^a}, \tag{6}$$

where h^a is the thickness of LaSMP actuator. Substituting the concentration of optically activated species into Young's modulus related to the thickness, and substituting the resulting equation into integration gives the Young's modulus of LaSMP actuator in the forward reaction

$$Y_a = Y_0 + \frac{1}{c_r h^a}\left(Y_{\max} - Y_0\right)\left(1 - \mathrm{e}^{-c_r h^a}\right)\left(1 - \mathrm{e}^{\frac{-t}{\tau_r}}\right). \tag{7}$$

Reverse reaction of the light activated species in LaSMP can be obtained when exposure to 248nm wavelength UV light and Young's modulus decreases in this process. The concentration of optically activated species P is expressed as

$$\tau_d \frac{\partial P}{\partial t} + P = 0\,, \tag{8}$$

where τ_d is the time constant of the reverse reaction. For this partial different equation, the initial concentration of optically activated species is equal to the concentration at the end of the forward reaction P_{t1}. Solutions of the concentration of optically activated species in the reverse reaction can be obtained

$$P = P_{\max} \mathrm{e}^{-c_r z} \left(1 - \mathrm{e}^{-\frac{t_1}{\tau_r}}\right) \mathrm{e}^{\frac{-t}{\tau_d}}\,. \tag{9}$$

Substituting the concentration of optically activated species in the reverse reaction into the thickness related Young's modulus and substituting the resulting equation into integration gives the Young's modulus of LaSMP actuator in the reverse reaction

$$Y_a = Y_0 + \frac{1}{c_r h^a}\left(Y_{\max} - Y_0\right)\left(1 - \mathrm{e}^{-c_r h^a}\right)\left(1 - \mathrm{e}^{-\frac{t_1}{\tau_r}}\right) \mathrm{e}^{\frac{-t}{\tau_d}}\,. \tag{10}$$

Young's modulus of LaSMP actuator in the forward and reverse reaction (Eq.(7) and Eq.(10)) depict the Young's modulus of LaSMP actuator under UV lights of different wavelengths and the value of Young's modulus of LaSMP actuator in the forward at time instant t_1 is the initial condition of Young's modulus of LaSMP actuator in the reverse reaction.

4. DESIGN AND ANALYSIS OF AN LASMP LAMINATED BEAM

In this section, analytical solution of a laminated beam is derived based on cantilever beam model. In the design of laminated beam, two pieces of LaSMP layers are glued on the top and bottom surface of an elastic layer. In this study, it is assumed that the LaSMP layers are bonded perfectly with the elastic layer. **Figure 2** illustrates an elastic layer laminated with LaSMP layers.

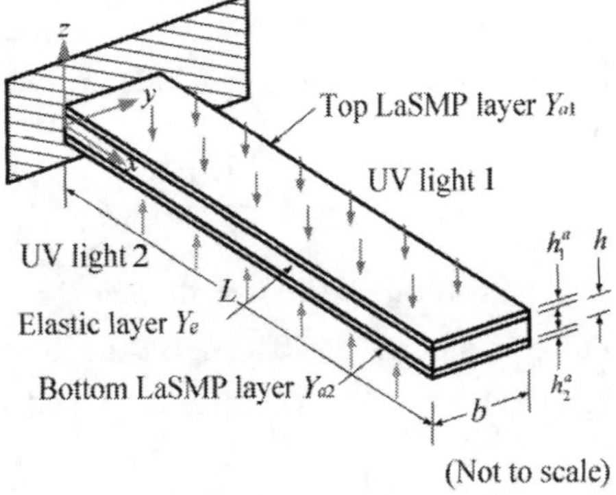

Figure 2. An elastic layer laminated with LaSMP layers

In **Fig.2**, the top layer and the bottom layer are exposured to UV light of two different wavelengths respectively and Young's moduli of these two layers change differently. Note that the Young's modulus of LaSMP were derived based on the local coordinate system and the moments are defined in the global coordinate system. The advantage of this model is that the tip displacement of the finger is bidirectional, i.e., upforward or downforward. Based on this function, the robot gripper could operate in two different ways. The LaSMP layers tend to recover to their original shape when exposured to UV light, but they are constricted by the elastic layer, and they will not be able to recover much of the original shape, thus cause bending moments on the laminated structure. Bending moment induced by the ith LaSMP layer can be obtained

$$M_i^a = \frac{h + h_i^a}{2} S^a Y_{ai} h_i^a \,, \tag{11}$$

where h is the thickness of the elastic layer; h_i^a is the thickness of the ith LaSMP layer; Y_{ai} is the Young's modulus of the the ith LaSMP layer and S is the strain induced by the LaSMP layer. S^a is assumed to be constant value as an approximation in this quasi-static process because it really depends on the elastic layer behavior. For example, S is approximately equal to zero when the elastic layer is very flexible relative to the LaSMP layers since the LaSMP layers will likely recover all of its original shape and very small bending moment will be induced. On the contrary, a bending moment will be induced by the LaSMP layers when the elastic layer is stiff enough and the LaSMP layers are being restrained from recovering the original shape. For laminated beam as shown in **Fig.2**, the LaSMP layers are perfectly glued on the top and bottom surfaces of the elastic layer, resultant bending moment induced by the LaSMP actuators can be expressed as

$$M^a = M_1^a - M_2^a \,. \tag{12}$$

For a beam with a rectangular cross section, only the transverse displacement u_3 dominates the vibration. Thus the system equation can be obtained

$$\rho \tilde{A} \ddot{u}_3 + YI \frac{\partial^4 u_3}{\partial x^4} - \frac{b \partial^2 \left(M^a \right)}{\partial x^2} = bF_3 \,, \tag{13}$$

where ρ is the mass density, $\tilde{A}$ is the cross-section area, Y is the Young's modulus, b is the beam width, F_3 is the mechanical excitation and YI can be obtained as

$$YI = Y_e \frac{bh^3}{12} + Y_{a1} \left[\frac{b\left(h_1^a\right)^3}{12} + \left(\frac{h + h_1^a}{2} \right)^2 bh_1^a \right] + Y_{a2} \left[\frac{b\left(h_2^a\right)^3}{12} + \left(\frac{h + h_2^a}{2} \right)^2 bh_2^a \right], \tag{14}$$

where I is the area moment of inertia and Y_e is the Young's modulus of the elastic layer. For the static structural of a cantilever beam, inertia force and mechanical excitation are neglected, the system equation is reduced to

$$YI \frac{\partial^4 u_3}{\partial x^4} - \frac{b \partial^2 (M^a)}{\partial x^2} = 0 \,. \tag{15}$$

Solution of simplified system equation can be obtained

$$u_3(x) = \frac{1}{6} c_1 x^3 + \frac{1}{2} c_2 x^2 + \frac{1}{2} \frac{bM^a}{YI} x^2 + c_3 x + c_4 \,, \tag{16}$$

where constants can be determined by boundary conditions. At clamped end (x=0)

$$u_3(0)=0; \quad \frac{\partial u_3(0)}{\partial x}=0. \tag{17a,b}$$

At free end ($x=L$)

$$YI\frac{\partial u_3^2(L)}{\partial x^2}=bM^a; \quad \frac{\partial u_3^3(L)}{\partial x^3}=0. \tag{18a,b}$$

Substituting boundary conditions into expression of displacement yields

$$c_1=0; \quad c_2=0; \quad c_3=0; \quad c_4=0. \tag{19a,b,c,d}$$

Thus, the transverse displacement is defined as

$$u_3(x)=\frac{1}{2}\frac{bM^a}{YI}x^2. \tag{20}$$

Thus the detailed expression of the displacement

$$u_3(x)=\frac{1}{4}\frac{(h+h_1^a)bS^aY_{a1}h_1^a-(h+h_2^a)bS^aY_{a2}h_2^a}{Y_e\frac{bh^3}{12}+Y_{a1}\left[\frac{b(h_1^a)^3}{12}+\left(\frac{h+h_1^a}{2}\right)^2bh_1^a\right]+Y_{a2}\left[\frac{b(h_2^a)^3}{12}+\left(\frac{h+h_2^a}{2}\right)^2bh_2^a\right]}x^2. \tag{21}$$

The tip displacement of the laminated beam can be obtained when $x=L$

$$u_3(L)=\frac{1}{4}\frac{(h+h_1^a)bS^aY_{a1}h_1^a-(h+h_2^a)bS^aY_{a2}h_2^a}{Y_e\frac{bh^3}{12}+Y_{a1}\left[\frac{b(h_1^a)^3}{12}+\left(\frac{h+h_1^a}{2}\right)^2bh_1^a\right]+Y_{a2}\left[\frac{b(h_2^a)^3}{12}+\left(\frac{h+h_2^a}{2}\right)^2bh_2^a\right]}L^2. \tag{22}$$

With the tip displacement above, the equivalent force that yields the same tip displacement can be obtained

$$F=\frac{3u_3(x=L)YI}{L^3}. \tag{23}$$

It should be noted that the calculation of the equivalent force is based on the beam without any load. There is not any tip displacement after the beam touching on the target which is assumed to be rigid and the gripping force is obtained by the predicted displacement after touching.

5. CASE STUDIES

Elastic layer integrated with single LaSMP actuator on its top surface is evaluated and length effect of double actuators is studied in this section. Then a gripper is designed based on the optimal length of LaSMP actuators. Two formulations of LaSMP material are available in literature, i.e., EAS-155-93, EAS-155-115 [29]. The main difference between these two materials is the Young's modulus variation range. A wide range of Young's modulus variation from 5 MPa to 123 MPa can be obtained with EAS-155-93. Since wider range of Young's modulus means larger bending moment induced by the LaSMP actuators, formulation of EAS-155-93 is used in Case studies and polypropylen

(PP) is selected as the material of elastic layer. For simplicity, the thickness of each LaSMP layer is identical, i.e., $h^a = h_1^a = h_2^a$. Material properties and geometry of the LaSMP actuator are listed in Table 1.

Table 1. Material properties and geometry of the LaSMP actuator

Property	EAS-155-93
Initial Young's modulus, Y_0 (MPa)	5
Maximum Young's modulus when t approaches infinity, Y_{max} (MPa)	900
Attenuation coefficient in the forward reaction, c_r	82
Time constant of the forward reaction, τ_r	45.45
Time constant of the reverse reaction, τ_d	60
Thickness of the LaSMP layer, h^a (m)	1×10^{-4}

To show the effectiveness of the mathematical model of LaSMP Young's modulus which established in the theoretical section, analytical relusts of Young's modulus of the EAS-155-93 for the forward reaction are compared with the experimental results in **Fig.3** [29]. The error of fitting between analytical results and experimental results is satisfactory as shown in **Fig.3**.

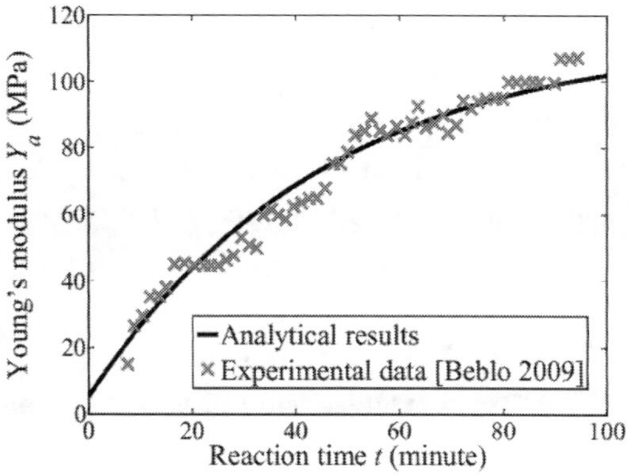

Figure 3. Young's modulus of the LaSMP actuator

Case 1: Laminated Beam Finger of Single LaSMP Actuator

In this Case, single LaSMP layer is glued on the top surface of the elastic PP layer and it is exposured to UV light of two different wavelengths alternately. The advantage of this model is that no cancellation of the stress, less LaSMP material and simpler setup is needed than the situation in which both surfaces of the elastic PP layer are glued with LaSMP layers. This model will be significant when small amount of tip displacement is needed. Material properties and geometry of the laminated beam are listed in **Table 2**.

Table 2. Material properties and geometry of the laminated beam

Property	Laminated beam structure
Young's modulus of the elastic PP layer, Y_e (MPa)	1556
Thickness of the elastic PP layer, h (m)	2×10^{-3}
Width, b (m)	5×10^{-3}
Length, L (m)	0.1
Strain, S^a	0.1%

Eqs.(21-23) can be also applied in this Case by setting M_2^a equal to zero. The Young's modulus of the LaSMP layer is depicted in **Fig.4**. It should be noted that the wavelength of the UV light switches from 325nm to 248nm in the peak in **Fig.4**.Displacement versus horizontal position x at different time instants can be obtained by Eq.(21) and it is presented in **Fig.5**. It should be noted that the deformation of the laminated beam is fast in the beginning and slow down as time goes by, this is because the forward reaction becomes nearly saturated state.

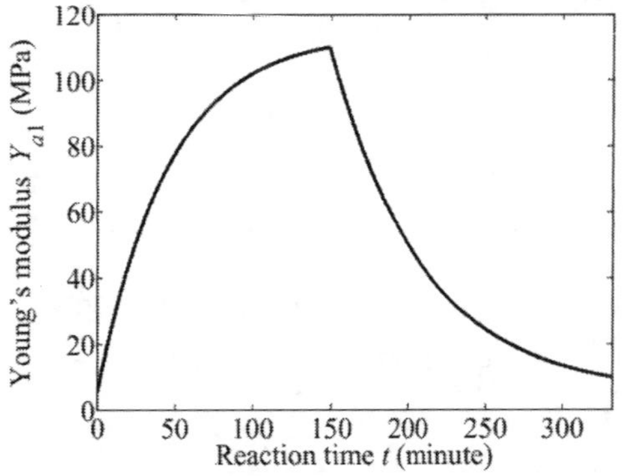

Figure 4. Young's modulus of the LaSMP actuator in Case 1

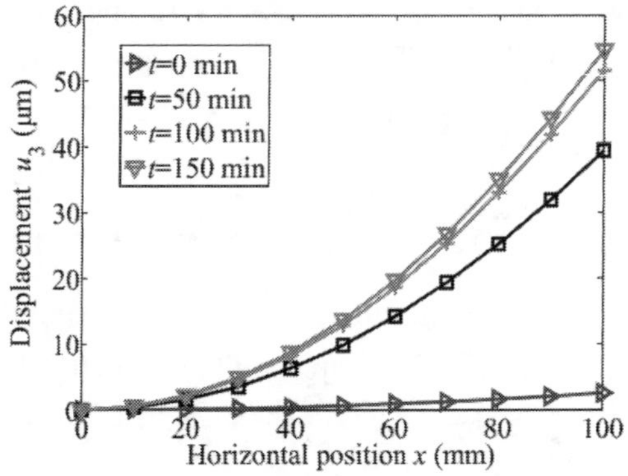

Figure 5. Displacement versus horizontal position x at different time instants

In **Fig.5**, it can be noted that the displacement is not equal to zero when t=0 which caused by the assumption of a constant strain S^a. This result may inconsistent with the fact of gradual change at the beginning, but it makes no matter since the early stage of this process is not the focus of our application.

Case 2: Laminated Beam Finger of Two Variable-Length LaSMP Actuators

In this Case, length effect of the LaSMP layers are evaluated. In addition, double LaSMP layers are glued on the top and bottom suefaces of the PP layer respectively. The other geometry and material parameters are same with those in the first Case.

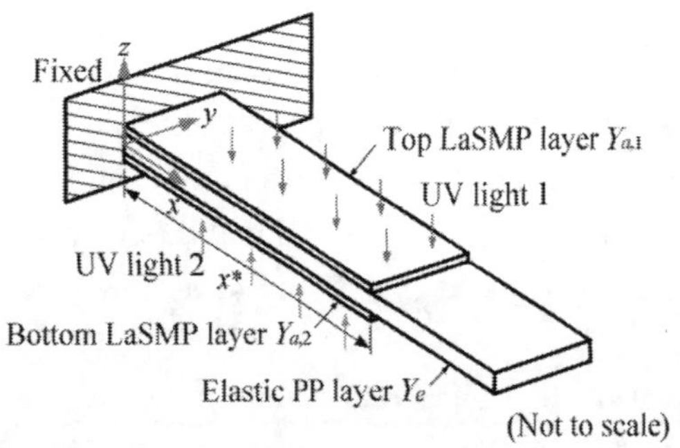

Figure 6. An elastic PP layer laminated with variable-length LaSMP layers

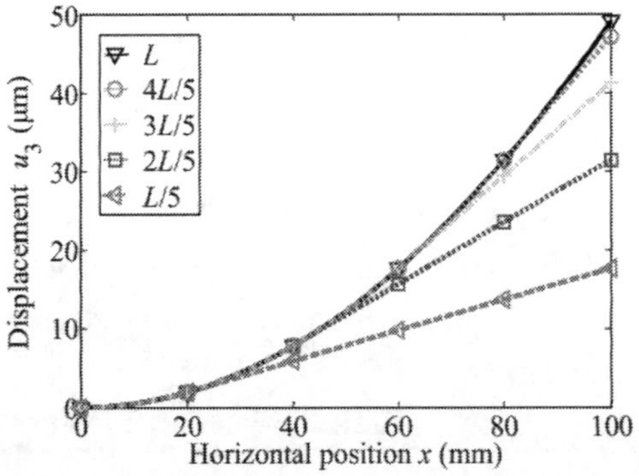

Figure 7. Displacement of laminated beam with variable-length of LaSMP actuators when t=117 minute

The Young's moduli of the two LaSMP layers on the beginning of the cycle is equal to 60MPa; This pretreatment can be obtained by exposure to UV light of 325nm and the purpose of this pretreatment is to increase the Young's modulus of the top LaSMP layer and on the meanwhile decrease the Young's modulus of the bottom LaSMP layer, thus large Young's moduli difference between top and bottom LaSMP layers could be obtained. The Young's modulus of the top and bottom LaSMP layers can be depicted in **Fig.8(a)**.

To evaluate the length effect of the LaSMP layers, the lengths of the LaSMP layers vary in this Case as shown in **Fig.7**. The value of x^* is chosen as L, $4L/5$, $3L/5$, $2L/5$, $L/5$ respectively measured from the fixed end. The part which laminated with LaSMP actuators would deforms after LaSMP actuators exposure to UV light. The remaining part without LaSMP actuators keep its initial shape all the time but its position in the space changes since it is connected with the deformed part. **Fig.7** depicts the shape of the beam with variable-length LaSMP actuators when t=117 minute, i.e., the first peak in **Fig.8(a)**. It is obvious that the tip displacement of the laminated beam reaches its maximum value when the elastic PP layer is fully laminated with LaSMP actuators. This result shows that the relationship between the tip displacement and length of the LaSMP actuators is nonlinear, thus it will be a good reference for designers to keep balance between cost and performance. It can be known from **Fig.7** that the tip displacement only decrease 4% when 80% of the elastic PP layer laminated with LaSMP actuators.

Case 3: Gripper Design Based On the Optimal Length of LaSMP Actuators

Based on the result of Case 2, 80% of the elastic PP layer length is selected as the optimal length for the LaSMP layers. In this Case, a gripper actuated by the optimal length of LaSMP layers is designed and its performance is evaluated.

Young's modulus of the top LaSMP layer Y_{a1} and bottom LaSMP layer Y_{a2} in this Case is same as Case 1. Based on the variation of Young's modulus (**Fig.8(a)**), tip displacement and equivalent force is presented in **Fig.8**. It should be noted that there are four turning points in **Fig.8(a)** and **(b)** respectively. Straight dash line in **Fig.8** shows the relationship between these turning points and variation of Young's modulus, the first and last turning points in **Fig.8(b)** are related to switch of wavelength of UV light irradiating on the top LaSMP layer, the second and third turning points are related to changes of wavelength of UV light irradiate on the bottom LaSMP layer. For example, the UV light irradiating on top LaSMP layer (dash line in **Fig.8(a)**) switches from 325nm to 248nm in the peak and switches from 248nm to 325nm in the trough.Equivalent force that yields the same tip displacement in **Fig.8(b)** can be obtained with Eq.(23) an it is depicted in **Fig.8(c)**. Maximum payload can be calculated when the surface friction coefficient between the target and gripper is considered.

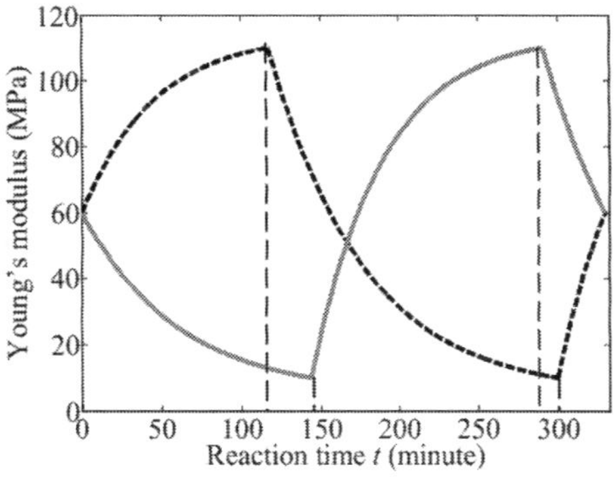

(a)

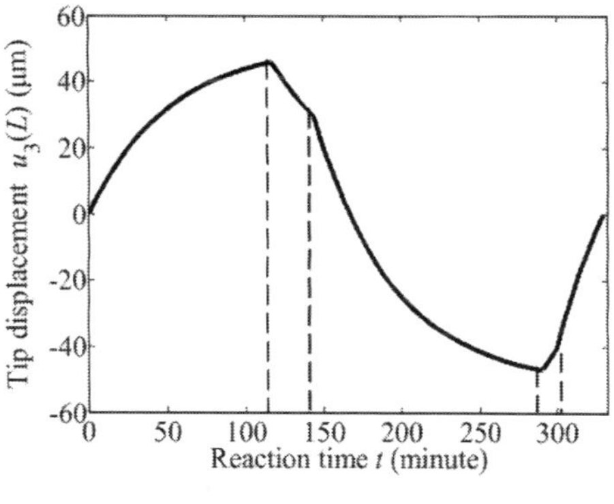

(b)

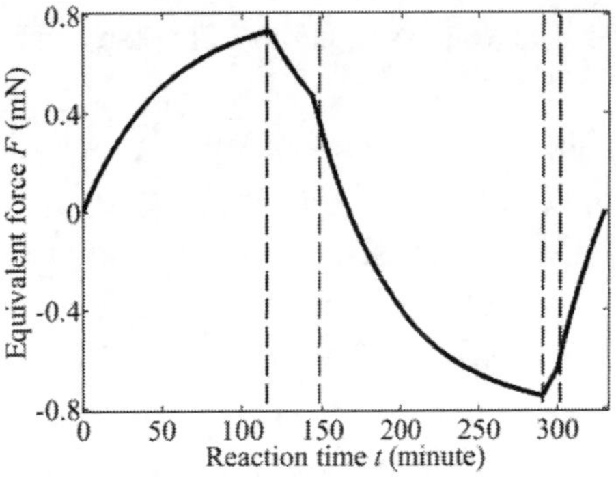

(c)

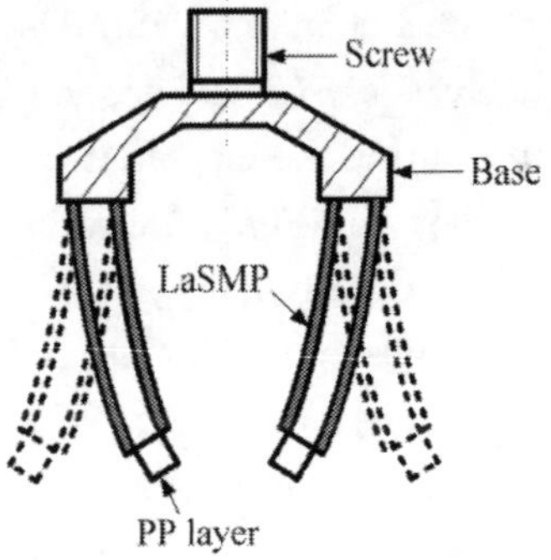

Figure 8. (a) Young's modulus of LaSMP layers (dash: top layer, solid: bottom layer) (b) Tip displacement (c) Equivalent force

Figure 9. Limit positions of the gripper

Figure 9 shows limit positions of the gripper and recalls the tip displacement of one finger, the stroke of the gripper can be calcaulated and it is 184μm here. Based on the result of equivalent force, maximum recommended payload is 0.74mN when 0.5 friction coefficients between finger and target, safety factor of 2 are considered.

6. CONCLUSIONS

A light-weight robor gripper capable of picking and placing target is designed by using elastic PP layer laminated with LaSMP actuators. Besides, based on the LaSMP material, a novel actuation mechanism of noncontact actuation is realized in this robot gripper. The advantages associated with light activated actuation are no electric or electro-magnetic noises and no electric sparking or leakage. Then, mathematical model of LaSMP's Young's modulus versus reaction time is established according to chemical kinetics. Furthermore, laminated beam finger is presented and performances including strokes and force are evaluated.

In the Case studies, robot gripper with variable-length of LaSMP actuators on both its top and bottom surfaces is developed and results indicates that only 4% of the tip displacement is reduced when the length of the actuator is 80% of the length of the elastic PP layer. In the third Case, a gripper is designed based on the optimal length of LaSMP actuators. In addition, the stroke of the gripper is 184μm and maximum recommended payload is 0.74mN (friction coefficients of 0.5 between finger and target, safety factor of 2).

ACKNOWLEDGMENT

This work was supported by the National Natural Science Foundation of China (Nos. 11472241, 11172262, 11321202).

REFERENCES

[1] Dehaven, J. G., and Tzou, H. S., 2008, "Forced Response of Cylindrical Shells Coupled with Nonlinear Shape-Memory-Alloy (SMA) Actuators Regulated by Sinusoidal and Saw-Tooth Temperature Profiles," *Journal of Engineering Mathematics*, 61(2-4), pp.301-313.

[2] Tzou, H. S., 1993, Piezoelectric Shells: Distributed Sensing & Control, Kluwer Academic Publishers, London.

[3] Tzou, H. S., Lee, H. J., and Arnold, S. M., 2004, "Smart Materials, Precision Sensors/Actuators, Smart Structures, and Structronic Systems," *Mechanics of Advanced Materials and Structures*, 11(4-5), pp.367-393.

[4] Zhang, X. F., Hu, S. D., and Tzou, H. S., 2014, "A Generic Double-Curvature Piezoelectric Shell Energy Harvester: Linear/Nonlinear Theory and Applications," *Journal of Sound and Vibration*, 333(26), pp.7286-7298.

[5] Chai, W. K., Tzou, H. S., Arnold, S. M., and Lee, H. J., 2008, "Magnetostrictive Microactuations and Modal Sensitivities of Thin Magnetoelastic Shells," *Journal of Pressure Vessel Technology*, 130(1), pp.011206-1-011206-5.

[6] Li, H., Chen, Z. B., and Tzou, H. S., 2010, "Distributed Actuation Characteristics of Clamped-Free Conical Shells Using Diagonal Piezoelectric Actuators," *Smart Materials and Structures*, 19(11), pp.115015-115026.

[7] Li, H., Hu, S. D., Li, H. Y., and Tzou, H. S., 2014, "Active Vibration Suppression of Payloads with a Conical Isolator," Journal of Intelligent Material Systems and Structures, pp.871-880.

[8] Li, H. Y., Li, H., and Tzou, H. S., 2015, "Frequency Control of Beams and Cylindrical Shells with Light-Activated Shape Memory Polymers," *Journal of Vibration and Acoustics*, 137(1), pp.011006-1-8.

[9] Tzou, H. S., 1989, "Development of a Light-weight Robot End-effector Using Polymeric Piezoelectric Bimorph," IEEE International Conference on Robotics and Automation, pp.1704-1709.

[10] Villanueva, A. A., Joshi, K. B., Blottman, J. B., and Priya, S., 2010, "A Bio-Inspired Shape Memory Alloy Composite (BISMAC) Actuator," *Smart Materials and Structures*, 19(2), pp.025013-025029.

[11] Viry, L., Mercader, C., Miaudet, P., Zakri, C., Derre, A., Kuhn, A., Maugey, M., and Poulin, P., 2010, "Nanotube Fibers for Electromechanical and Shape Memory Actuators," *Journal of Materials Chemistry*, 20(17), pp.3487-3495.

[12] Lan, X., Liu, Y., Lv, H., Wang, X., Leng, J., and Du, S., 2009, "Fiber Reinforced Shape-Memory Polymer Composite and Its Application in a Deployable Hinge," *Smart Materials and Structures*, 18(2), pp.024002-024007.

[13] Rodriguez, A., Mason, M.T., Srinivasa, S.S., 2014, "Manipulation Capabilities with Simple Hands," *Experimental Robotics*, 79, pp.285-299.

[14] Probst, M., Hürzeler, C., Borer, R., and Nelson, B. J., 2009, "A Microassembly System for the Flexible Assembly of Hybrid Robotic Mems Devices," *International Journal of Optomechatronics*, 3(2), pp.69-90.

[15] Colinjivadi, K. S., Lee, J. B., Draper, R., 2008, "Viable Cell Handling with High Aspect Ratio Polymer Chopstick Gripper Mounted on a Nano Precision Manipulator," *Microsystem Technologies*, 14(9-11), pp.1627-1633.

[16] Mackay, R. E., Le, H. R., Clark, S., et al. 2013, "Polymer Micro-grippers with an Integrated Force Sensor for Biological Manipulation," *Journal of Micromechanics and Microengineering*, 23(1), pp.015005-015011.

[17] Michaël, D. V., and Dominiek, R., 2010, "Pneumatic and Hydraulic Microactuators: A Review," *Journal of Micromechanics and Microengineering*, 20(4), pp.043001-043018.

[18] Bos, E. J. C., Bullema, J. E., Delbressine, F. L. M., Schellekens, P. H. J., and Dietzel, A., 2008, "A Lightweight Suction Gripper for Micro Assembly," *Precision Engineering*, 32(2), pp.100-105.

[19] Ahn, K. K., and Nguyen, H. T. C., 2007, "Intelligent Switching Control of a Pneumatic Muscle Robot Arm Using Learning Vector Quantization Neural Network," *Mechatronics*, 17(4), pp.255-262.

[20] Shahinpoor, M., and Kim, K. J., 2005, "Ionic Polymer-Metal Composites: IV. Industrial and Medical Applications," *Smart Materials and Structures*, 14(1), pp.197-214.

[21] Jain, P., Datta, S., Majumder, S., and Dutta, A., 2011, "Two IPMC Fingers Based Micro Gripper for Handling," *International Journal of Advanced Robotic Systems*, 8(1), pp.1-9.

[22] Roch, I., Bidaud, P., Collard, D., 2003, "Fabrication and Characterization of an SU-8 Gripper Actuated by a Shape Memory Alloy Thin Film," *Journal of Micromechanics and Microengineering*, 13(2), pp.330-336.

[23] Teh, Y. H., Featherstone, R., 2008, "An Architecture for Fast and Accurate Control of Shape Memory Alloy Actuators," *The International Journal of Robotics Research*, 27(5), pp.595-611.

[24] Ashrafiuon, H., Eshraghi, M., Elahinia, M. H., 2006, "Position Control of a Three-link Shape Memory Alloy Actuated Robot," *Journal of Intelligent Material Systems and Structures*, 17(5), pp.381-392.

[25] Büttgenbach, S., Bütefisch, S., Leester-Schädel, M., 2001, "Shape memory microactuators," *Microsystem Technologies*, 7(4), pp.165-170.

[26] Brahim, T., Michel, P., Joël, A., Valérie, P., and Nadine, L. F. P., 2012, "Four Dof Piezoelectric Microgripper Equipped with a Smart CMOS Camera," *Journal of Microelectromechanical Systems*, 21(2), pp.256-258.

[27] Götze, H., Pagel, L., 2007, "Development of a Micro-manipulator Based on Piezoelectric-technology," *Microelectronic engineering*, 2007, 84(5), pp.1333-1336.

[28] Tzou, H. S., and Pandita, S., 1987, "A Multi-Purpose Dynamic and Tactile Sensor for Robot Manipulators," *Journal of Robotic Systems*, 4(6), pp.719-741.

[29] Beblo, R. V., and Weiland, L. M., 2011, "Light Activated Shape Memory Polymer Characterization-Part II," *Journal of Applied Mechanics*, 78(6), pp.061016-061024.

[30] Beblo, R. V., 2010, "Characterization and Modeling of Light Activated Shape Memory Polymer," Ph.D., University of Pittsburgh, Pittsbugh PA.

ICAST2015: 26th International Conference on Adaptive Structures and Technologies
October 14-16th, 2015, Kobe, Japan

ICAST2015 #004

Polymer Sensors for Shape Sensing of Adaptive Structures

Mark Melnykowycz[1*], Frank Clemens[1]

[1] Department for High Performance Ceramics, EMPA, Dübendorf, Switzerland

Abstract

Advances in layered manufacturing technologies and materials science are making the manufacture of complex structures with adaptive mechanical characteristics straightforward to produce. Additionally, the ability to print soft and flexible polymers such as Thermoplastic Elastomers (TPE) in consumer-grade Fused Filament Fabrication (FFF) 3D printers is reducing the barriers to producing mechanically adaptive structures. Along with the development of adaptive structures for high-strain applications, Conductive Particle Composite (CPC) piezoresistive sensors based on a Carbon Black (CB) and TPE have been produced in order to measure the deformation of textile bands applied in wearable computing applications. Mechanical testing has shown that the band sensor shows some relaxation during loading to different strain levels, but has very low signal drift over time. Mechanical characterization of the piezoresistive sensor (in 0.3 mm diameter monofilaments) has shown a decoupling between the non-linear force and a linear electrical resistance response during tensile loading. The CB/TPE sensor material can achieve ultimate strains of over 150%, with a working range up to 100%. Pre-straining of the material is required in order to improve the sensor performance. The sensor has been integrated with the FFF printing, to produce an expanding structural design which unfolds along the direction of longitudinal strain to monitor structural deformation.

1. INTRODUCTION

Compliant or adaptive structures can generally be characterized as having the ability to adapt their shape to changing mechanical boundary conditions, their function being related to both their geometry as well as the materials used in their design. By combing adaptive materials with high-strain sensor materials, it would be possible to build structures which sense deformations on irregular surfaces such as the human body, with applications in wearable computing, soft robotics and other new design directions that are not feasible with conventional materials and sensors. In the current investigation, a design for a flexible band sensor based on an elastic substrate and a conductive composite is presented. A design methodology and first results of the integration of 3D printed adaptive structures with polymer sensor material is also shown.

1.1 Adaptive Structures and Soft Sensors

The use of evolutionary design tools such as topology optimization has made it straightforward to investigate design optimizations, but manufacturing such structures was not always feasible with traditional manufacturing methods. The introduction of free-form or layered manufacturing methods (collectively termed 3D Printing), coupled with an ever-expanding library of available printable

* info@icast2015.com

material types is making it easier to produce compliant and adaptive systems. For adaptive structures, the combination of an elastic material with a compliant geometry will yield a design that can experience high strains and conform to complex geometric surfaces not possible with conventional material designs.

Currently many products designed as wearable computing solutions rely on the use of mass-produced sensors including accelerometers, gyroscopes and temperature sensors that can be easily integrated into designs such as fitness trackers and smart watches. While these approaches can measure relative position changes between data points, they do not account for direct strain on a body. Advances in polymer-based sensor technology offer the ability to mount strain sensors on the body for more advanced wearable sensor products. Soft Condensed Matter Sensor (SCMS) systems include flexible sensors for body-mounted applications. The SCMS class of materials defines sensors that are derived from soft materials (as covered in condensed matter physics), with higher mechanical flexibility, low hardness (characteristic of polymers) and potentially non-linear mechanical behavior. These characteristics differs strongly from the behavior of conventional sensors based on hard materials such as bulk silicon and metals that are derived from manufacturing methods of the integrated circuit (IC) industry. Conductive composites can be designed, which incorporate strain sensor behavior with large mechanical deformations (beyond the abilities of conventional strain sensors).

1.2 Piezoresistive Conductive Composite Sensors

To a large extent, polymer strain sensor research has focused on the application of conductive coatings onto polymer monofilament or woven substrates. This has resulted in the development of yarns with conductive fibers [1, 2] or nanotube yarns [3], and carbon or silver-coated polymer fibers [4]. Composite or hybrid materials have been developed via the integration of carbon nanotubes into a polymer matrix [5] or electroactive polymers with carbon fillers [6, 7]. More recently, a pressure sensitive textile has been developed by coating Kevlar fibers with a SBS/AgNP composite, which allows for capacitive based pressure sensing in a woven fabric design [8]. The conventional conductive yarn design encompasses a sensor fiber element (for example a conductive wire) woven or twisted together with load-bearing and electrically insulating fibers. Such an approach appears to have been taken in implementation of Google Project Jacquard [9]. In such a design, the conductive fiber could be metallic or carbon-coated [1, 10], allowing sensor function integration with the structural fibers. The final yarn is a sensor that can support mechanical loads. The current work utilizes polymer composites with carbon conductive particles.

1.3 Conductive Particle Composite (CPC) Sensors

Past investigations focused on the characterization of the mechanical and electrical properties of a piezoresistive polymer monofilament sensor composed of Carbon Black (CB) mixed with a Thermoplastic Elastomer (TPE) matrix. The current work is focused on the integration of this piezoresistive sensor material with high-stretch adaptive structures to measure mechanical strain. The piezoresistive sensor monofilament [11] combines conductive particles together with a flexible polymer matrix, creating a CPC material. This hybrid combination offers advantages in sensor processing, since due to the thermoplastic component, the sensor material can be thermal-formed into various shapes via extrusion. Basic sensor forms include a monofilament or extruded ribbon geometry, which can then be used directly or further processed for specific sensor shape profiles. As the conductive CB content increases the mechanical flexibility decreases while electrical conductivity increases. After various experiments in CB volume fraction, a mixture of 50% CB and 50% TPE was chosen for the ideal balance between electrical and mechanical performance in research work between Empa and ETH Zurich [12], where the ratio of CB to SEBS in the monofilament system was investigated.

2. EXPERIMENTAL

2.1 Band Sensor Design Methodology

The primary goal of body-mounted sensor design in the current work is to create a sensor, which conforms to irregular surface topography, such as of the human wrist, while maintaining the basic geometric shape of the sensor band. As a strain sensor, the sensor would need to undergo large deformations, on the order of 10%-50% strain. The design then required the application of a high-strain sensor material with a substrate that will conform to the body surface. A logical first material choice was the use of an elastic textile band.

2.2 Surface Mounted Integration

In preliminary tests the monofilament fiber sensor was bonded to the elastic textile substrate using a liquid rubber compound. This embodied a simple strain band sensor for tensile testing. The primary finding was that strain shielding occurred in the interface region between the sensor and textile materials. Essentially, under an applied global displacement, the textile band would stretch around the integration area of the sensor would then not see the actual global strain being applied to the band structure. This occurred due to the mismatch in stiffness between the substrate and integration regions (defined largely by the material stiffness of the bonding material).

After further design development, a U-shaped sensor was chosen to surface bond to the elastic substrate, adding stiffening regions at the ends in order to more evenly concentrate the local deformation over the integration region. The U design was chosen in order to minimize the complexity of electrical connection, which would be needed in order to connect the sensor to the data acquisition module when integrated on the body. In order to have better control over the sensor design and substrate, a compliant design approach was investigated based on meta-material designs utilizing an array of tailored geometric cell geometries.

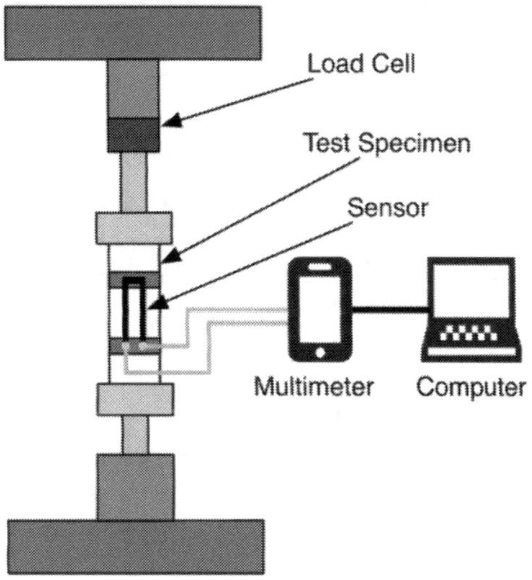

Figure 2.1 Tensile test setup for mechanical cycle testing.

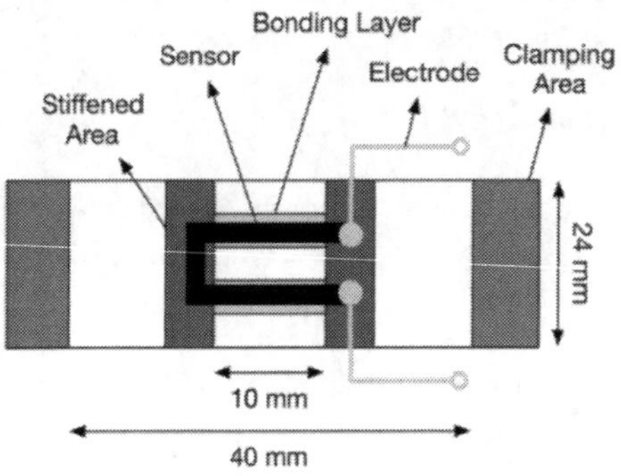

Figure 2.2 Schematic description of the band sensor tensile specimen, electrodes were connected at the ends of the U.

Cyclic tensile loading was used to investigate the electrical-mechanical property coupling of the monofilaments. A tensile testing machine (Zwick / Z005 , Zwick Roell Xforce P 200N load cell, testExpert v11.1 software) was used to apply mechanical load, while a Metex M-3610D multimeter was used to record the resistance of the monofilament in response to applied mechanical strain. ScopeView software was used with the multimeter to log the resistance measurements on a laptop computer. The test setup is depicted in **Figure 2.1**. The specimen dimensions are shown in Figure 2.2, where the tensile gauge length is 40 mm, and the active length of the U-sensor is 10 mm.

A displacement controlled mechanical cycle test consisted of cycling with a 5% range between two different strain levels for 5 cycles with a hold of 1 minute at each strain level. Each sample was subjected from 5% to 50%. The sample would be first brought to the maximum strain of the test (for example 50%), held for 1 minute, and then brought to the strain minimum (45% in this example). After cycling between the maximum and minimum strains the sample was brought back to the test starting point at 0% before the next cycle. The resistance of the sensors was monitored with a Metex M-3610D multimeter (Metex Corporation, Seoul, Korea). A strain step of 50-45% and 30-25% with a 300 second holding time were also used to investigate signal stability over time. Three band sensors were evaluated in the mechanical cycle test. In total, 3 band sensors were produced and tested for the tensile test evaluation.

2.3 Compliant Structure Design

Compliant structural design implies that under a mechanical deformation, a structure will deform in an engineered way such that it conforms to the boundary conditions it is exposed to. Compliant designs offer the ability to tune large geometric deformations by combining the material properties inherent to the geometry along with the material properties of the base material that the geometry is created from. Basic geometric arrays can be based on squares, rectangles, triangles, and hexagonal cells. From a manufacturing perspective, these cell types are easily produced with FFF 3D printers by choosing the corresponding fill pattern algorithm when printing a planar structure such as a rectangle volume.

2.4 Geometric Array Design Approaches

When designing a mechanical cell array, a key design parameter is the relationship between mechanical stiffness and global applied load. An orthogonal or square unit cell will offer the same stiffness in the 0° and 90° directions. In composites design, the +/- 45 layering orientation is often used in order to provide good stiffness multiple directions. If fibers or the structure are all aligned in the direction of loading, then the structure will have a large Poisson's Effect, resulting in a narrowing at the mid-section under tensile loading (as shown in **Figure 2.3**b). This can be improved upon by using a hexagonal geometry, which offers identical geometric stiffness in three principle directions. In order to fulfill the sensor design target of maintaining the geometric form of the array under tensile loading, an auxetic cell was also considered. A material body composed of an auxetic cellular array provides a negative Poisson's Ratio, so that the mid-section of the array will expand in the width direction under tensile loading (**Figure 2.3**c). The auxetic array is initially folded (as shown in **Figure 2.4**a), and unfolds under tensile loading (as shown in **Figure 2.4**b), with an expansion as opposed contraction of the array in the width direction.

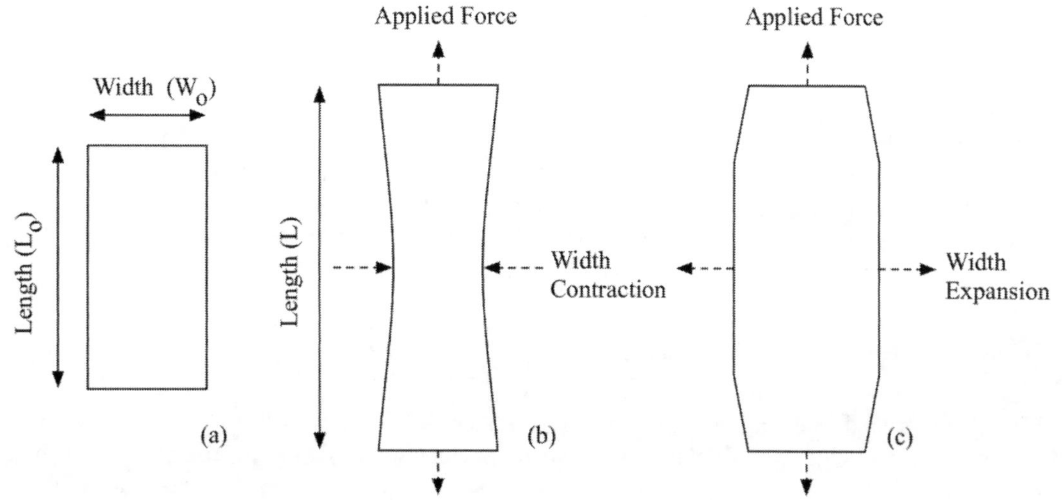

Figure 2.3 Schematic view of body expansion (a) due to a positive (b) or negative (c) Poisson's Effect, resulting in an expansion in the body width as opposed to contraction.

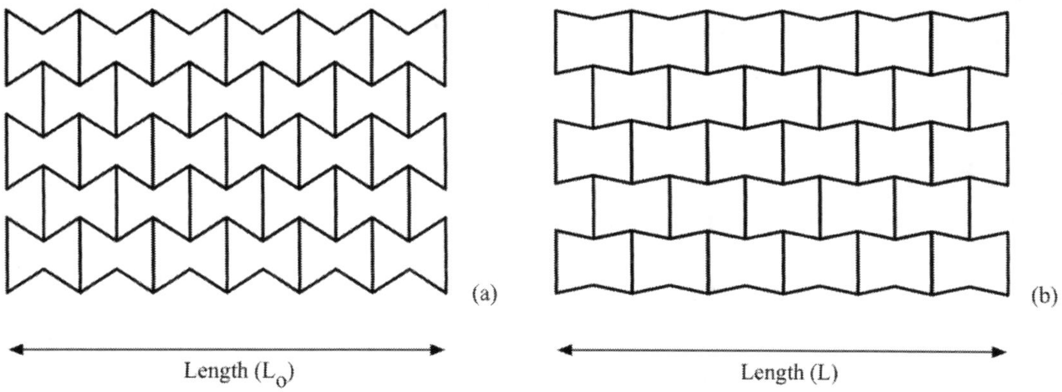

Figure 2.4 Schematic view of an auxetic unit cell structure (a) and near fully expanded (b).

2.5 3D Printing and Process Development

3D printing was done with a Velleman K8200 FFF 3D printer. The open-source design of the K8200 allowed for modification of printing parameters and the addition of a syringe extruder for silicone printing. The commercially available FilaFlex TPE filament was used to produce compliant structures for evaluation as well as casting molds for silicone. The slicing of STL 3D model files and machine control was done with the Simplify3D [13] software.

2.5.1 Printed Silicone

Although not extensively used in 3D printing, methods have been developed for layered manufacturing with silicone. A syringe extruder was installed on the K8200 to print silicone, which was originally developed for chocolate (http://www.open-electronics.org/). A 10 ml syringe was filled with a paste silicone and a plunger mechanism driven by a NEMA stepper motor and gearing system provided force to extrude silicone through the needle of the syringe. A 0.9 mm diameter needle was used with a slow printing speed given the force needed and high viscosity of the silicone.

2.5.2 Silicone Casting

The conventional way to form silicone into shapes is through casting, where a mold is created and filled with silicone paste, which then curies or is dried, depending on the system that is used. The mold then needs to be de-molded easily complex shapes are generally not possible to produce as they may break during demolding. Silicone had been used in previous studies to bond the sensor material to an elastic sport shirt for posture monitoring [14]. As a base material, silicone was an attractive choice as it conforms well to complex surfaces. However, with a Poisson's Ratio near 0.49, a planar silicone structure will neck extensively when stretched over the body surface.

2.5.3 Printed TPE

TPE can be printed using commercially available materials such as FilaFlex (www.recreus.com) or NinjaFlex (www.ninjaflex3d.com), which have been developed specifically for use in FFF machines. TPE was focused on since the CPC sensor is also produced with TPE, which would logically offer similar deformation mechanics as the printed structure. FilaFlex was chosen for this investigation due to its Shore Hardness (84A) and availability. A nozzle diameter of 0.3 mm was used in order to print cell structures with a thin cell wall width in order to reduce the geometric stiffness of the array. The material was extruded at 220° C. A layer height of 0.150 mm was defined to keep the structural stiffness low. Geometric arrays with rectangular, triangular, hexagonal, and auxetic unit cells were printed for evaluation.

2.5.4 Hybrid Compliant Structure Design

After evaluation of printed TPE structures and sensor integration with casted silicone, a hybrid compliant structure was developed by combining a printed TPE auxetic structural array with casted silicone. In this configuration, the auxetic TPE array acted as an exoskeleton for the soft silicone volume in order to stabilize the expansion perpendicular to tensile loading. This allowed the monofilament sensor to be integrated in the silicone volume while the auxetic array provided a negative Poisson's Ratio effect so that the structure remained compliant on the topography of the human body but also didn't excessively deform when stretched along the loading direction.

3. RESULTS / DISCUSSION

3.1 Band Sensor Signal Relaxation Model

The signal relaxation of the sensor material was previously evaluated using the monofilament sensor fibers [11]. Therefore, the current work focused on the band sensor performance characterization. The primary focus was the determination of signal relaxation (mechanical drift). In general, sensor drift is defined as a signal decrease over time (months or years), and is related to physical degradation of the sensor material and therefore a change in sensor performance. For the band sensor, signal relaxation would also be related to changes in the conductive network of CB particles, and also to the mechanical behavior of the elastic band substrate and the bonding materials. The system as a whole can be considered stable so long as the distribution of CB particles does not change. However, given that the band sensor is composed of multiple polymer components, a viscoelastic response to would be expected under mechanical deformation, resulting in a signal relaxation. Since the band was designed with different stiffness zones defined by liquid rubber coatings, there exists the relaxation behavior of the coating areas as a signal relaxation source.

When brought to a new target strain level the sensor signal displays a peak, and then generally sees a relaxation or drift beyond that as shown in **Figure 3.1**a. This initial relaxation is assumed to be related to the intrinsic characteristics of the sensor material, but also to the mechanical relaxation behavior of the elastic band and bonding materials used in the construction of the band sensor design (Figure 2.2). From past research [11] it is known that there generally exists a nonlinear degradation shortly after the target strain is achieved, and afterwards the signal degradation is near-linear. It is expected that as the bonding materials material relax, less strain would be transferred to the sensor region and the signal would accordingly decrease. Alternatively, the sensor material might be slightly under-loaded (**Figure 3.1**b), where the sensor signal rises after attaining the target strain level, this is attributed to loading of the bonding material layer (Figure 2.2). As this material relaxes, more strain would be transferred to the sensor region, resulting in a signal increase.

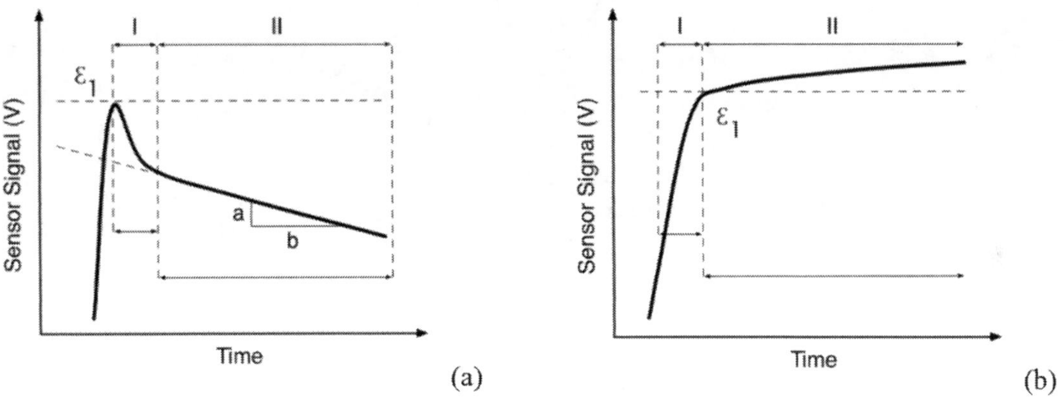

Figure 3.1 General behavior of the conductive polymer when a new strain level is applied (a), where an abrubt peak is seen, followed by a near-linear signal degradation. In a slighlty under-loaded state (b) the signal gradually rises, attributed to the visoelastic nature of the integration materials.

This model for the loading of the band sensor and the signal behavior is represented in the experimental data of Figure 3.2, where a change in the signal character is seen at the higher strain level, while under-loading behavior is seen in the strain step data between 5%-30%, while the higher strains of 40%-50% have a more stable signal at the target strain level. This behavior is not seen in the subsequent applied strain cycles, suggesting that the materials in the band sensor have reached a stable state after the initial strain step. The stable sensor value at the end of each strain step was plotted against the applied strain (**Figure 3.3**), where near-linear behavior is seen throughout the strain range. The linear band sensor behavior mirrors the linear behavior of monofilament sensors previously reported [11]. Given the 50% pre-strain of the current band sensor, these results can be compared with the equivalent monofilament strain range of 50%-100%.

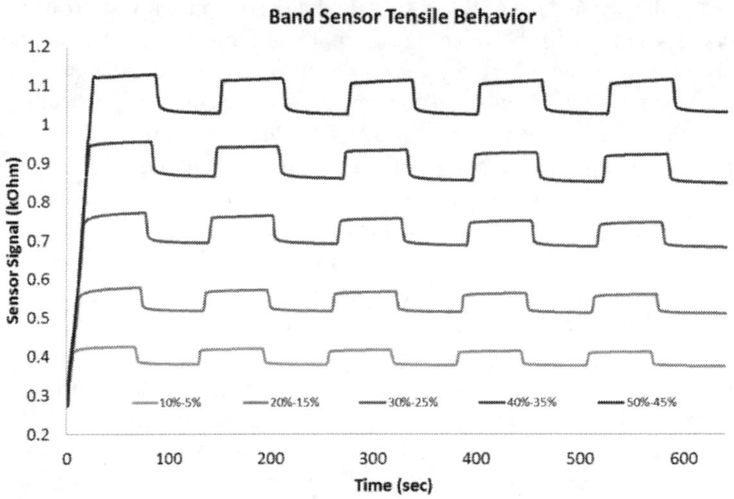

Figure 3.2 Characteristic band sensor response to tensile strain cycle test.

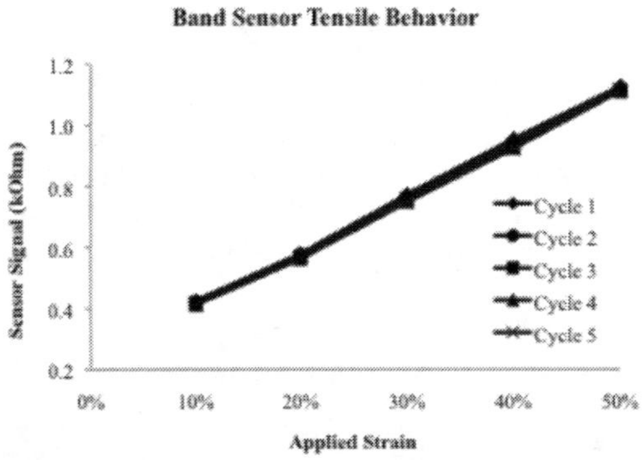

Figure 3.3 Characteristic band sensor tensile strain behavior.

3.2 Band Sensor Drift

The results showed that the absolute drift values were less than .0002 kOhm/sec, over a time of 1 minute. When taken as a percentage of the initial sensor value, the maximum drift loss was 0.1-0.4%/sec for the stretch bands. The Zwick drift investigation showed the stretch bands to have low drift, and when stabilized at a strain level, the drift was near to zero. Although there was room for improvement in the band design, the Zwick tests showed the sensors should perform with low drift when integrated into the STBL-BPM system.

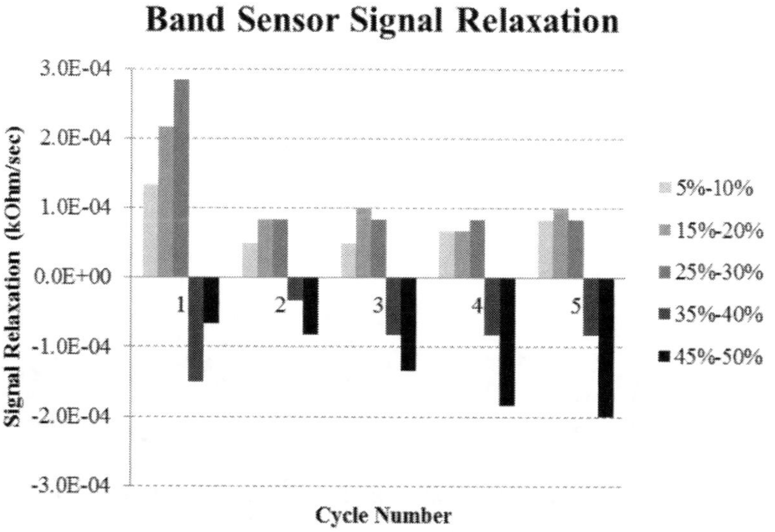

Figure 3.4. Band sensor relaxation as a function of test cycle number and applied strain range.

3.3 Compliant Structure Design and Production

In the exploratory stage, 3D printing tests were conducted different fill patterns including rectangular, triangular, hexagonal, and auxetic patterns. These were printed in FilaFlex TPE at 220° C with a 0.3 mm nozzle and their characteristics evaluated by hand. The printing of silicone was also investigated. **Figure 3.5** shows the results of 3D printing with paste silicone. Although it was possible to extrude silicone, the high viscosity of the paste made it difficult to produce regular continuous material deposition lines. Therefore, TPE was focused on as a printable material.

Figure 3.5 Examples of initial silicone 3D printing tests.

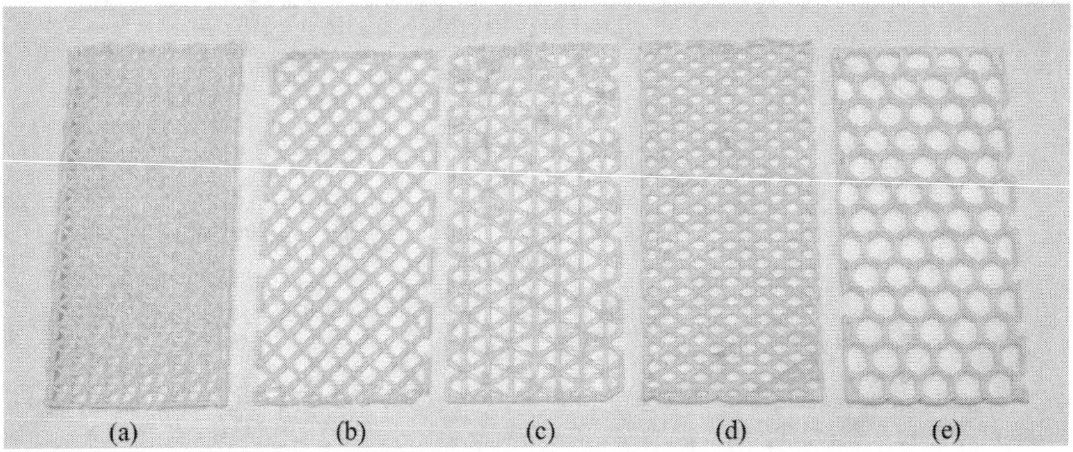

Figure 3.6. Conventional fill patterns printed in TPE including layered +/- 60° triangular (a), +/- 45° rectangual (b), +/- 60° triangular (c), +/- 30° rectangual (d), and hexagonal.

Figure 3.6 shows the results of printing with FilaFlex and different conventional fill patterns. This produced very flexible bodies, which were able to stretch to high strains, but also had large Poisson Ratios (as shown in **Figure 3.8**). Tests were run on printing auxetic arrays, with case cell structures as shown in **Figure 3.7**. Geometric arrays based on auxetic cells have the characteristic of unfolding with an applied mechanical strain along their primary direction. The full cell expansion then defines the strain limit of the full sensor band. In essence, this offers an intuitive way to tune the strain band performance.

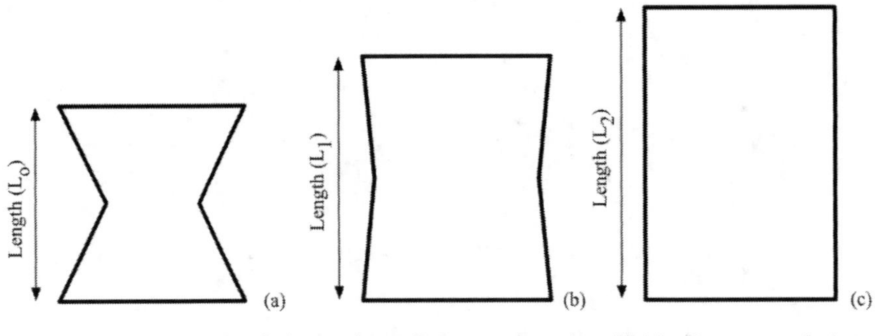

Figure 3.7. Schematic view of the deformation of an auxetic unit cell (a), first expanded to the normal state of geometric expansion (b), and with full extension, with strain occuring in the based material (c).

By combining a compliant geometric array (that has the desired geometric response under loading) with a soft or highly elastic base material, then a band sensor material with tailored high tensile deformation can be attained. Aside from the sensor material design, the integration of sensors into structures presented a difficult design challenge, which required the integration of materials, biomechanics, and mechanical design aspects. The combination of sensor and structural materials requires compliance between the sensor and the structure, in order to provide a high-quality sensor signal as well as design comfort for the person that is wearing the sensor.

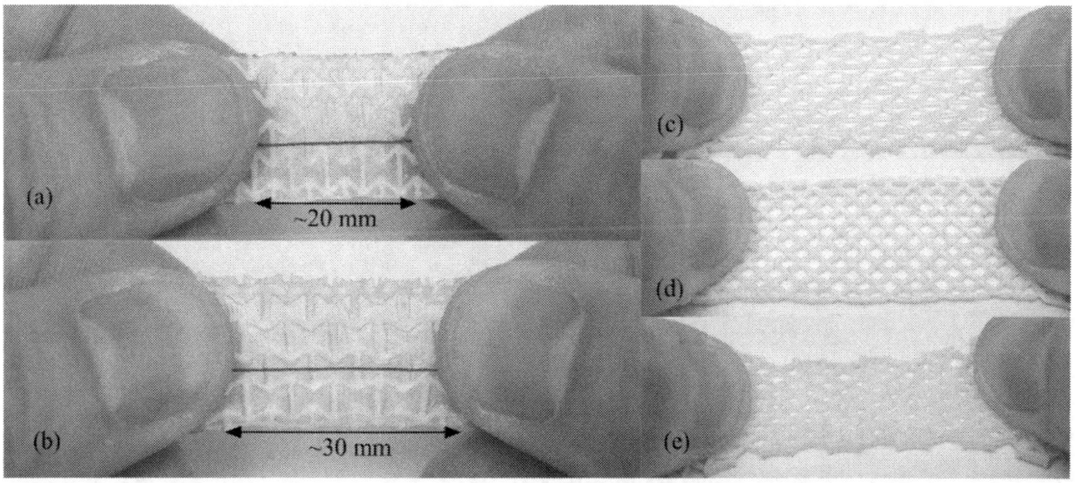

Figure 3.8 Examples of a hybrid band strucutre with a printed TPE auxetic exoskeleton casted with silicone fill and a CPC sensor at rest (a) and under ~50% strain (b), alongside +/- 60° triangular (c), +/- 30° rectangular (d), and +/- 45° rectangualr (e).

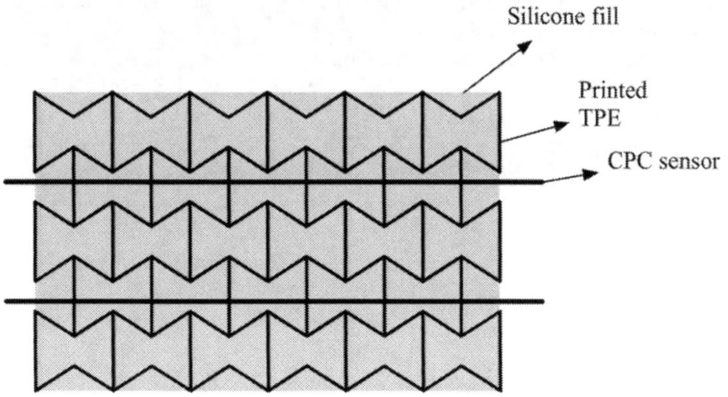

Figure 3.9. Schematic view of the hybrid 3D printed auxetic cell array with casted silicone fill matrix and CPC sensor arrangement.

4. CONCLUSIONS

The development of conductive composites polymers based on CB and TPE offers an attractive material for strain sensing of adaptive structures due to the high strain capability and near-linear response of the resistance of the material. Although a band sensor could be created using a U-shaped sensor configuration bonded to an elastic textile substrate, the combination of freeform fabrication technologies such as FFF combined with silicone casting offers interesting opportunities for creating hybrid sensor structures that utilize the negative Poisson's Ratio effect in order to build sensor structures, which are compliant with complex surface topographies.

5. ACKNOWLEDGMENTS

This work was made possible through a funding grant by the Swiss Commission for Technology and Innovation (CTI Medtech 17614.1 PFLS-LS), and was completed in partnership with STBL Medical Research AG.

6. REFERENCES

1. Huang, C.-T.; Shen, C.-L.; Tang, C.-F.; Chang, S.-H. A wearable yarn-based piezo-resistive sensor. *Sensors and Actuators A: Physical* **2008**, *141*, 396 - 403.
2. Huang, C.-T.; Tang, C.-F.; Lee, M.-C.; Chang, S.-H. Parametric design of yarn-based piezoresistive sensors for smart textiles. *Sensors and Actuators A: Physical* **2008**, *148*, 10 - 15.
3. Zhao, H.; Zhang, Y.; Bradford, P.D.; Zhou, Q.; Jia, Q.; Yuan, F.-G.; Zhu, Y. Carbon nanotube yarn strain sensors. *Nanotechnology* **2010**, *21*, 305502.
4. Gibbs, P.; Asada, H.H. Wearable Conductive Fiber Sensors for Multi-Axis Human Joint Angle Measurements. *Journal of NeuroEngineering and Rehabilitation* **2005**, *2*, 7.
5. Martinez, F.; Obieta, G.; Uribe, I.; Sikora, T.; Ochoteco, E. Polymer-Based Flexible Strain Sensor. *Procedia Chemistry* **2009**, *1*, 915 - 918.
6. Cochrane, C.; Koncar, V.; Lewandowski, M.; Dufour, C. Design and Development of a Flexible Strain Sensor for Textile Structures Based on a Conductive Polymer Composite. *Sensors* **2007**, *7*, 473-492.
7. Shui, X.; Chung, D.D.L. A piezoresistive carbon filament polymer-matrix composite strain sensor. *Smart Materials and Structures* **1996**, *5*, 243.
8. Lee, J.; Kwon, H.; Seo, J.; Shin, S.; Koo, J.H.; Pang, C.; Son, S.; Kim, J.H.; Jang, Y.H.; Kim, D.E.; Lee, T. Sensors: Conductive Fiber-Based Ultrasensitive Textile Pressure Sensor for Wearable Electronics (Adv. Mater. 15/2015). *Advanced Materials* **2015**, *27*, 2409-2409.
9. Google Project Jacquard. **2015**.
10. Flandin, L.; Chang, A.; Nazarenko, S.; Hiltner, A.; Baer, E. Effect of strain on the properties of an ethylene–octene elastomer with conductive carbon fillers. *Journal of Applied Polymer Science* **2000**, *76*, 894–905.
11. Melnykowycz, M.; Koll, B.; Scharf, D.; Clemens, F. Comparison of Piezoresistive Monofilament Polymer Sensors. *Sensors* **2014**, *14*, 1278.
12. Clemens, F.J., B. Koll, T. Graule, T. Watras, M. Binkowski, C. Mattmann, I. Silveira Development of Piezoresistive Fiber Sensors, Based on Carbon Black Filled Thermoplastic Elastomer Compounds, for Textile Application. *Advances in Science and Technology* **2013**, *80*, 7-13.
13. Simplify3D, www.simplify3d.com
14. Mattmann, C.; Clemens, F.; Tröster, G. Sensor for Measuring Strain in Textile. *Sensors* **2008**, *8*, 3719-3732.

ICAST2015 #85

Simulation Method for Performance Evaluation of Structural Deformation Measurements System Using Multiple Cameras

Jae-Hung Han[1*], Jong-Min Yun[1], Ho-Young Kim[1], Hong-Il Kim[2] and Hyuk-Jun Kwon[2]

[1] Department of Aerospace Engineering, KAIST, Daejeon, Korea
[2] Agency for Defense Development, Daejeon, Korea

Abstract

This study presents a performance evaluation method of real-time structural deformation reconstruction technique based on the multiple camera images. Most widely used method to estimate the structural deformation is to convert the measured acceleration and strain data into the deformation shapes. However, the deformation measurement technique using camera images has been studied for this purpose recently, since it is relatively free of sensor installation problem. Among several vision-based structural deformation measurement methods, this paper studies how we can evaluate the performance of SPR (Stereo Pattern Recognition) technique. Structural deformation measurement system utilizes SPR method has many system parameters such as camera specification, camera position and orientation, marker shapes and locations. Also, these parameters should be determined with respect to the motion of target structure. Determination of these related system parameter values in quantitative basis is not feasible. To handle these problems, an integrated simulation S/W called SIMS (SPR Interactive Modeling & Simulation), which can evaluate the shape reconstruction performances considering system parameters and geometry and motion of the target structure, has been developed. Using SIMS, Shape estimation accuracy can be efficiently evaluated under various conditions. Through this simulation, the measurement performance of camera system could be improved for a given structural motion. The decided vision-based structural deformation measurement systems can be used for a real-time wing deflection monitoring, an advanced structural control, and a fast and automated modal testing of structures.

1. INTRODUCTION

Vision-based structural deformation measurement method has its advantage over conventional techniques using installed strain gauge sensors, accelerometers or laser displacement sensors [1-3] since it is not only a non-contact method but is also capable of full-field measurement. Recently, studies regarding vision-based structural deformation measurement method have been performed in wide range of applications. Application of the vision-based structural deformation measurement ranges from laboratory structure model [4, 5] to maneuvering aircraft structures [6, 7] and civil structures. [8-9]

Among several vision-based techniques, Stereo Pattern Recognition (SPR) method is studied in this paper. SPR method uses a set of stereo camera and target keypoints (usually reflective marker) to obtain structural deformation. [10]. SPR method proves to give enough data accuracy compared with other conventional deformation measurement methods with relatively simple configuration [7].

* jaehunghan@kaist.ac.kr

The remaining issue of the SPR method for practical applications is that the measurement performance of it is dependent on various parameters such as camera specifications, camera position and orientation, marker size with its distribution, and the type of motion or dimension of the target object. Simulation is preferred to handle this complex problem because of its difficulty in realizing parameter settings. Keep this in mind, the simulation framework capable of interactively considers structure vibration and the camera system configuration is firstly developed. Then a thin plate with high aspect ratio is considered as an application example. Only the homogeneous set of stereo camera, which is a set with two identical cameras is considered for clarity, in this paper. Finally the simulation result with respect to varying camera altitude is presented to show the usefulness of the developed simulation framework.

2. BACKGROUND

2.1 Obtain 3D position using stereo camera set

By taking an image of an object, it would be projected to the camera image plane. In other words, a 3-dimensional object is mapped into the corresponding 2-dimensional image by camera (Fig. 1).

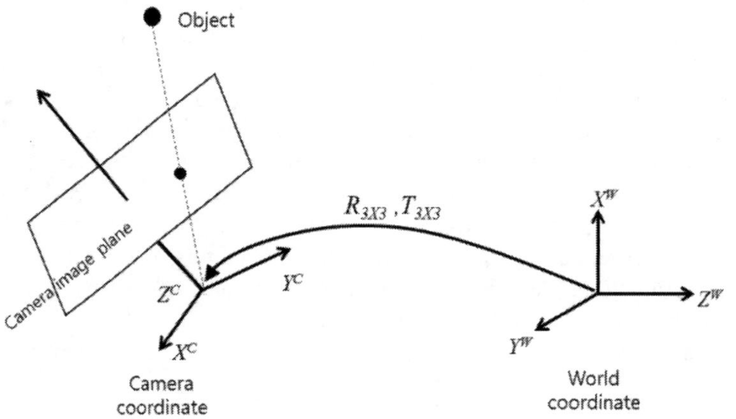

Figure 1. Coordinates geometry

The correspondence between 3D object and 2D image point is obtained as follows; Firstly, the coordinate transform between the world coordinates and the camera coordinates is given by equation (1).

$$\begin{Bmatrix} X^C \\ Y^C \\ Z^C \\ 1 \end{Bmatrix} = \begin{bmatrix} R^T_{3\times3} & 0 \\ 0 & 1 \end{bmatrix} \begin{bmatrix} I_{3\times3} & -T_{3\times1} \\ 0 & 1 \end{bmatrix} \begin{Bmatrix} X^W \\ Y^W \\ Z^W \\ 1 \end{Bmatrix} \tag{1}$$

where $R_{3\times3}$ is the rotation marix and $T_{3\times1}$ is the translation vector, superscript C and W represents camera coordinate and world coordinate, respectively. Rotation matrix is obtained by multiplying rotation matrices in each direction. (as explained in equation (2) and (3))

$$R_{3\times3} = R_z(\psi)R_y(\theta)R_x(\phi), \tag{2}$$

$$R_x(\phi) = \begin{bmatrix} 1 & 0 & 0 \\ 0 & \cos\phi & -\sin\phi \\ 0 & \sin\phi & \cos\phi \end{bmatrix}, R_y(\theta) = \begin{bmatrix} \cos\theta & 0 & \sin\theta \\ 0 & 1 & 0 \\ -\sin\theta & 0 & \cos\theta \end{bmatrix}, R_z(\psi) = \begin{bmatrix} \cos\psi & -\sin\psi & 0 \\ \sin\psi & \cos\psi & 0 \\ 0 & 0 & 1 \end{bmatrix} \tag{3}$$

Second, the correspondence between the object in camera coordinate and projection point on camera image plane is given in equation (4).

$$\lambda \begin{Bmatrix} x^{c,i} \\ y^{c,i} \\ 1 \end{Bmatrix} = \begin{bmatrix} \gamma f & sf & x_0 \\ 0 & f & y_0 \\ 0 & 0 & 1 \end{bmatrix} \begin{bmatrix} 1 & 0 & 0 & 0 \\ 0 & 1 & 0 & 0 \\ 0 & 0 & 1 & 0 \end{bmatrix} \begin{Bmatrix} X^C \\ Y^C \\ Z^C \\ 1 \end{Bmatrix} = K \begin{Bmatrix} X^C \\ Y^C \\ Z^C \\ 1 \end{Bmatrix} \tag{4}$$

where s is the CCD pixel skew , γ is the CCD pixel aspect ratio, (x_0, y_0) is the CCD center point, K is the camera matrix.

In CCD camera model, the K matrix is defined as equation (5).

$$K = \begin{bmatrix} -fm_x & 0 & x_0 m_x \\ 0 & -fm_y & y_0 m_y \\ 0 & 0 & 1 \end{bmatrix} \tag{5}$$

where m_x, m_y are pixel ratios in x and y direction, respectively.

Finally, substituting the equation (1) to the equation (4), the direct relationship between object in the world coordinate and in camera image plane is given as shown in equation (6).

$$\lambda \begin{Bmatrix} x \\ y \\ 1 \end{Bmatrix} = KR_{3\times3}^T[\mathrm{I}_{3\times3} | -T_{3X3}] \begin{Bmatrix} X^W \\ Y^W \\ Z^W \\ 1 \end{Bmatrix} = P \begin{Bmatrix} X^W \\ Y^W \\ Z^W \\ 1 \end{Bmatrix} \tag{6}$$

This P matrix which relates object in world coordinate to camera image plane is called camera projection matrix. The procedure to obtain this P matrix is camera calibration. However, by the projection, all the object lies in the same line from the camera center projects to the single point. In other words, the depth information of object would be missed. Therefore stereo camera is required to obtain 3D position. By stereo camera set, each object projects to both camera image plane. Therefore, to recover 3D position of each object, the correspondence between the two projection points should be established. This process is called stereo matching, and the concept of epipolar geometry is utilized for this process as shown in Fig. 2

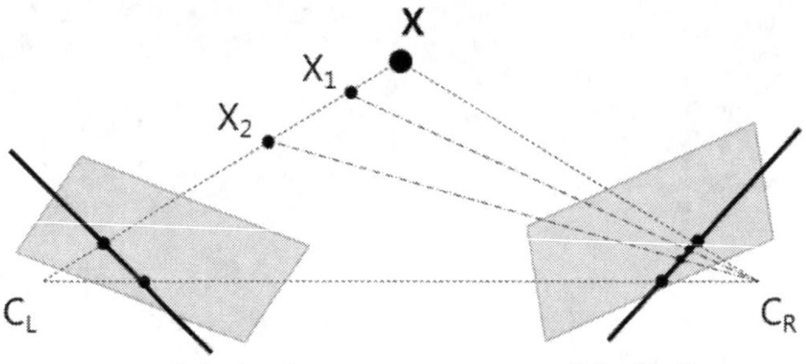

Figure 2. Epipolar geometry concept

2.2 SPR method

The SPR method utilizes a set of stereo camera and keypoints (usually reflective marker) distributed on target structure. After stereo matching is done for every keypoints, 3D positions can be obtained. One of the most widely used method is linear triangulation. As described in section 2.1, the object in world coordinate corresponds to the point on the camera image plane as in equation (7).

$$\lambda x = \lambda \begin{Bmatrix} x \\ y \\ 1 \end{Bmatrix} = PX \tag{7}$$

When the image is taken by n cameras, systems of equation below would be obtained.

$$\begin{bmatrix} x_1 P_{31}^T - P_{11}^T \\ y_1 P_{21}^T - P_{11}^T \\ \vdots \\ x_n P_{3n}^T - P_{1n}^T \\ y_n P_{2n}^T - P_{1n}^T \end{bmatrix} X = 0 \tag{8}$$

where P_{ij} refers i th column of the P matrix of j th camera. The null space of the matrix in LHS of equation (8) gives the recovered 3D position of the object.

3. SIMULATION FRAMEWORK

Simulation framework for quantitative performance estimation of SPR camera system is useful not only because the performance of it dependent on many parameters such as focal length, sampling frequency, resolution, camera position and orientation, marker shape and so on, but also because the direct experimental approach is tedious and not so feasible to perform. Using the developed framework for simulation, it is possible to realize hypothetical SPR camera system with user-specified system parameters. In addition, motion of the target structure can also be generated using the modal data of the structure. Performance of specified SPR camera system is evaluated under specified error criteria. System parameters could be modified to improve the overall system performance. The overall working process of the framework is summarized in Fig. 3.

3.1 Configuration of simulation framework

As it is shown in Fig.3, the simulation framework consists of three parts.
'Structure model part' is to generate vibration data of the target structure. Time-historical vibration data of the target structure is generated on the nodal points which are used for modal analysis of it, via linear time invariant (LTI) simulation based on system of equations of motion. [11] Marker point locations are chosen among these nodal points, shape and dimensions are determined by the user.
'Camera model part' is to produce camera intrinsic and extrinsic parameters. All the intrinsic parameters such as focal length, resolution, sensor dimension, sampling frequency, shutter speed, and the extrinsic parameters such as camera position and orientation are specified. This hypothetical SPR camera system measure the deformation of the target structure in specified vibrating motion. Structural deformation is calculated via recovery of 3D positions in 'SPR algorithm model' part.

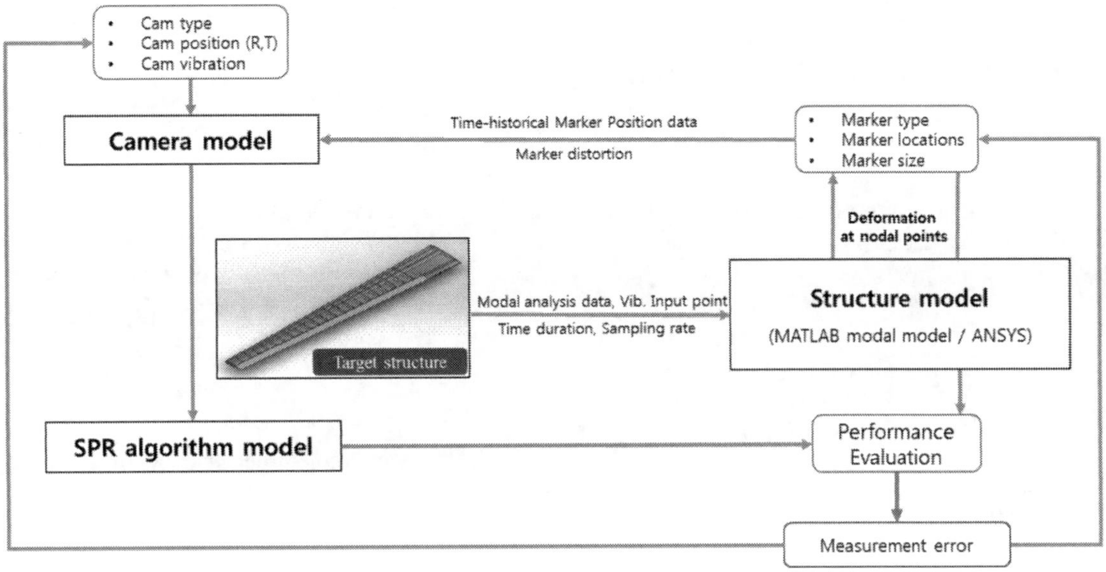

Figure 3. Configuration of the simulation framework

3.2 Error definition for performance evaluation

For quantitative measurement performance evaluation, proper error definition is essential. It starts from the point-wise error on each marker point k, at time step t. (Equation (9))

$$er_{\gamma}(k,t) = H_{\gamma}(k,t) - H_{\gamma}(k,t) \tag{9}$$

where er_{γ} is the point-wise error, H_{γ} is the reference structural shape, H_{γ} is the recovered structural shape.
To get the idea of how this SPR camera system works on the overall target structure, structural root-mean-square (RMS) error in each direction and total direction at time step t is defined as equation (10).

$$E_{RMS,total}(t)=\sqrt{\frac{\sum_{\gamma=1}^{3}\{\mathrm{er}_\gamma(\mathrm{t})\}^T\{\mathrm{er}_\gamma(\mathrm{t})\}}{N}},\quad E_{RMS,\gamma}(t)=\sqrt{\frac{\{\mathrm{er}_\gamma(\mathrm{t})\}^T\{\mathrm{er}_\gamma(\mathrm{t})\}}{N}} \tag{10}$$

where N is the number of the used marker points, and $\{er_\gamma(t)\}$ is an error vector in each direction.

Finally, RMS error would be normalized by the maximum deflection. (Equation (11))

$$E_{NRMS,total}(t)=\frac{E_{RMS,\mathrm{total}}(t)}{\max_t(\sqrt{\sum_{\gamma=1}^{3}\{D_\gamma(\mathrm{t})\}^T\{D_\gamma(\mathrm{t})\}})},\quad E_{NRMS,\gamma}(t)=\frac{E_{RMS,\gamma}(t)}{\max_t(\sqrt{\{D_\gamma(\mathrm{t})\}^T\{D_\gamma(\mathrm{t})\}})} \tag{11}$$

where $\{D_\gamma(t)\}$ is structural deformation at time step t.

This normalized RMS error is called NRE (Normalized RMS Error). NRE would be used as a parameter represents the measurement performance of the SPR camera system in this study.

4. CASE STUDY BY SIMULATION

Using the simulation framework described in section 3, the measurement performance of the SPR camera systems for specified target structure is evaluated.

4.1 Structure model

Fig. 4 shows the simulation model, a thin plate in tapered shape, with aspect ratio of 10. Its dimension and material properties are given in Table 1.

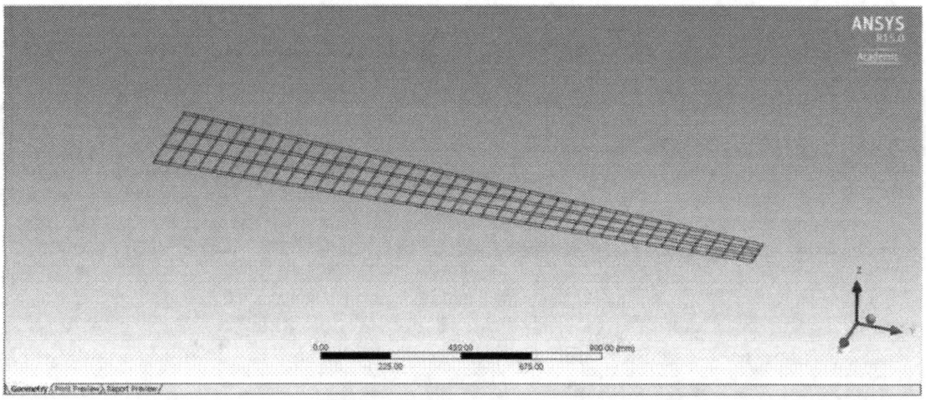

Figure 4. Target structure modelling

Table 1 Structure dimension and material properties

Root chord (a) [mm]	**Tip chord (b)** [mm]	**Span (L)** [mm]
300	100	2000
Thickness (T) [mm]	**Density** [kg/m^3]	**Young's Modulus** [GPa]
10	2770	71.0

The modal analysis for the target structure is performed using ANSYS R15.0 to obtain modal information up to the 8th modes. The result is shown in Fig. 5 and Table 2.

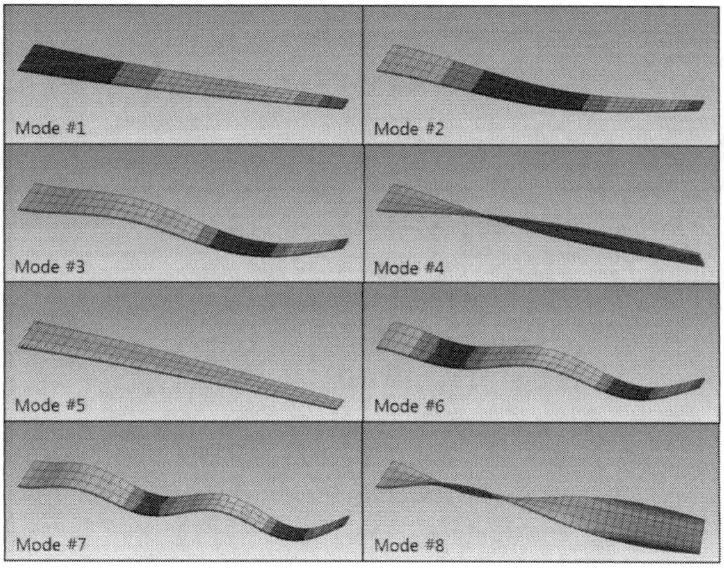

Figure 5. Mode shapes up to 8th mode

Table 2 Natural frequency result by ANSYS analysis

Mode number	**Natural Frequency (Hz)**	**Mode number**	**Natural Frequency (Hz)**
1	2.83	**5**	69.60
2	14.39	**6**	72.80
3	37.81	**7**	119.55
4	56.53	**8**	133.83

4.2 Simulation conditions

The effects of the camera elevation angle are presented using the proposed error criteria. The other parameters are fixed as explained in the subsequent sections.

4.2.1 Camera intrinsic parameter
Table 3 presents the intrinsic parameters of both cameras.

Table 3 Camera intrinsic parameter setting

Sensor type	**Focal length (mm)**	**Resolution (pixels)**
CCD	10	1280×1024
Pixel size (μ m)	**Frame rate (Hz)**	**Exposure time (ms)**
4.8	240	1.0

4.2.2 Marker distribution
Markers are distributed in two rows, each row has four markers. Marker starts from 1100mm away from the root, each column of markers are equally spaced with 200mm, as shown in Fig. 7.

4.2.3 Camera extrinsic parameters
With the spherical coordinate system geometry given in Fig. 6, seven elevation angle cases from 10° to 70° with step size of 10° are considered. Note that camera orientation angles are accordingly changed. The position and orientation of the cameras for seven cases are summarized in Table 4.

Table 4 Camera extrinsic parameter setting

	Position			**Orientation (deg)**		
	r [mm]	θ [deg]	φ [deg]	R_x	R_y	R_z
Camera 1	1000	[10:10:70]	0	[235:2.5:250]	[10:2.5:25]	0
Camera 2	1000	-[10:10:70]	0	[235:2.5:250]	-[10:2.5:25]	0

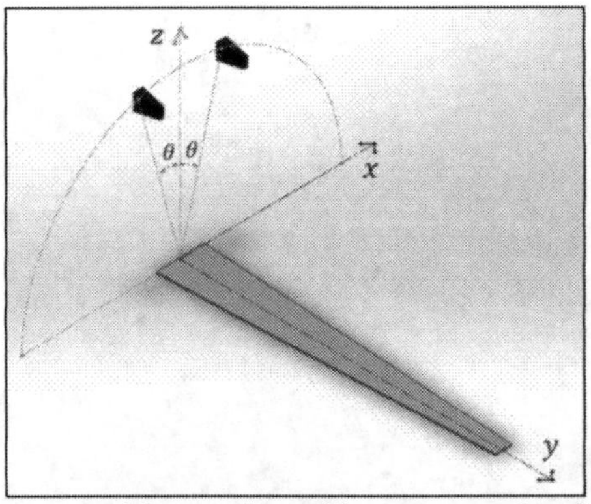

Figure 6. Camera position setting with respect to the elevation angle

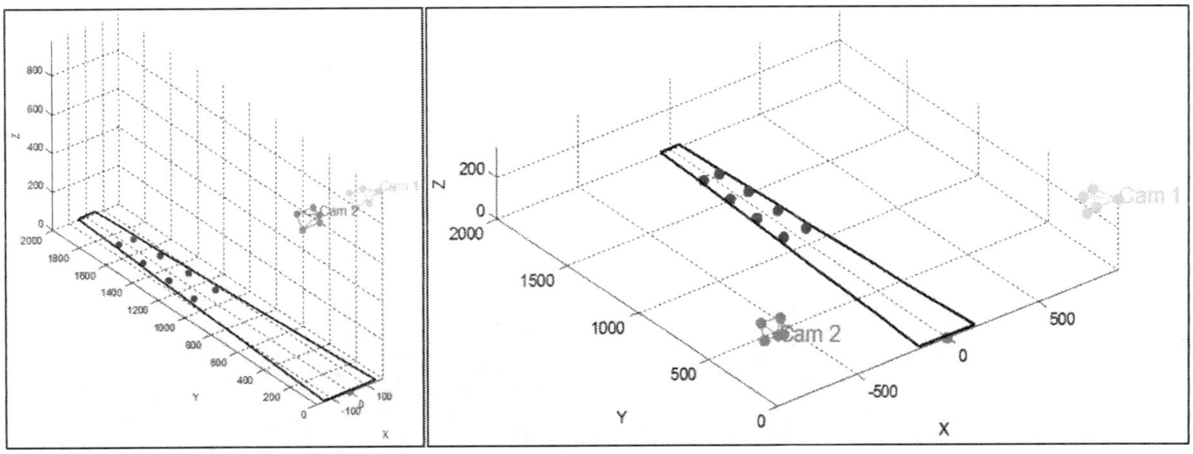

Figure 7. Camera position setting with respect to the elevation angle (10° (left), 70°(right))

4.2.4 Structure vibration case

In this paper, the structure is assumed to vibration with the 1st mode. The vibration amplitude is set to 10 times of the plate thickness. Note that more complex vibration cases are readily evaluated using the developed SIMS.

4.3 Simulation result

To evaluate performance of the SPR camera system, total NRE and z-directional NRE values are calculated for each case. Both total max NRE and z-directional NRE values are smaller than 0.6%, which indicates all of the current camera elevation angles guarantee reasonable accuracy. However, both errors further decrease as the elevation angle increases.

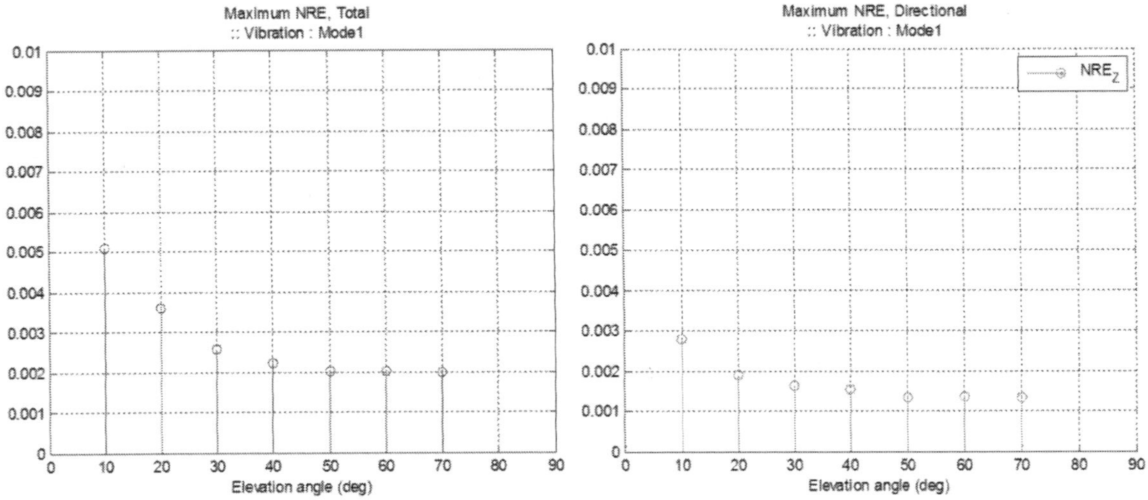

Figure 8. Total maximum NRE (left), Z-directional maximum NRE (right)

5. CASE STUDY BY EXPERIMENT

After the validation of the SPR camera system in structural deformation measurement has done in section 4, the system has applied to the measurement of maneuvering aircraft main wing deformation. The SPR camera system, cameras and markers are installed to the aircraft as shown in Fig. 9.

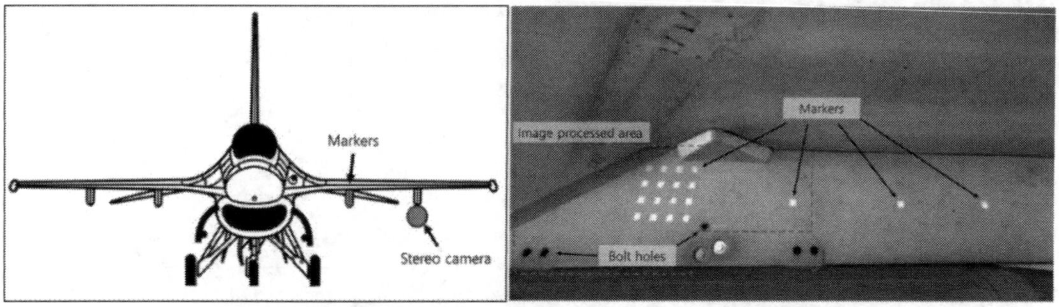

Figure 9. SPR camera system installation to the aircraft

The marker image recorded by stereo camera has processed to give 20 identified keypoints as in Fig. 10.

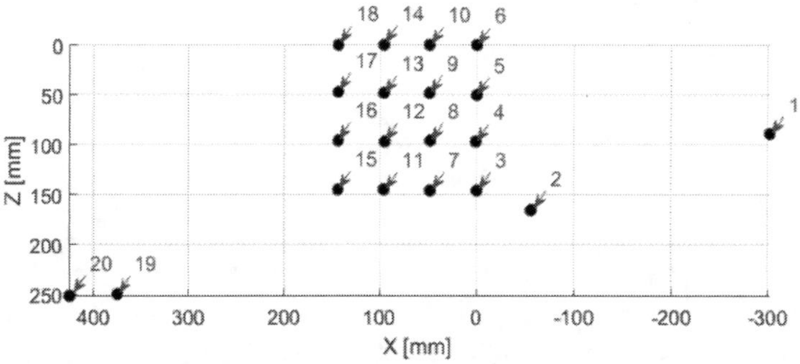

Figure 10. Identification and numbering of 20 keypoints

Here, x axis has defined as wing chord-wise direction, z axis has defined as downward direction of the aircraft. Z-directional displacement of marker 1 has obtained for 400 seconds as shown in Fig. 11.

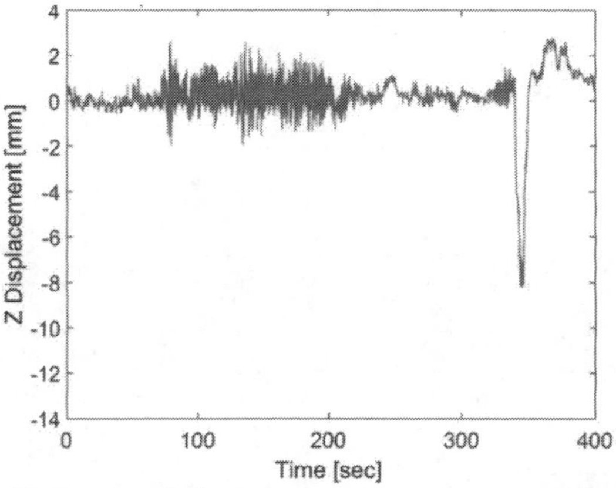

Figure 11. Z-directional displacement measurement result of marker #1

Frequency domain analysis of the above result has done for the time span with continuous vibration, which is [70s, 190s] as shown in Fig. 12.

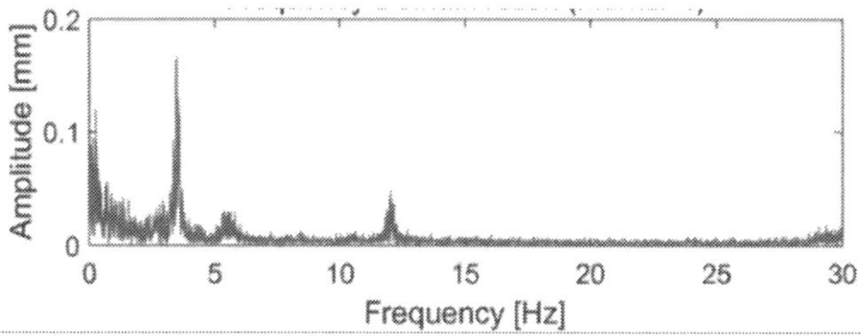

Figure 12. Frequency domain analysis of z-directional displacement of marker #1

It shows that major vibration is limited in low frequency region. Even though the verification of the measurement accuracy couldn't be done because of the limitation in sensor installation, the measurability of the SPR camera system to the maneuvering aircraft has checked through this application example.

6. CONCLUSION

This study presents a performance evaluation method for SPR camera system, and an integrative simulation framework has been presented. The simulation framework is capable of interactive modelling of target structure and SPR camera system. Using this framework, hypothetical camera system without any limitation in specifications or position can be generated. As an application example, vibration of slender tapered shape of thin plate has monitored by user-defined SPR camera system. This paper considers only the first bending mode (1st mode) harmonic vibration as a target structure vibratory motion. To monitor this deformation, SPR camera system with user-defined intrinsic parameters is used with varying altitude. Among seven altitude cases, SPR camera at 70° elevation angle (which is the lowest altitude among test cases) shows the best performance in both total maximum NRE and z-directional NRE sense. In a similar way, the effects of various parameters on SPR performance can be systematically evaluated. With this validation thorough the simulation, the SPR camera system further applied to the measurement of the maneuvering aircraft main wing deformation.

ACKNOWLEDGMENTS

This study is supported by a research program of Agency for Defense Development (ADD) under the contract UD130027JD.

REFERENCES

1. Kim, H.-I., Kang, L.-H., and Han, J.-H., "Shape estimation with distributed fiber Bragg grating sensors for rotating structures", *Smart Materials and Structures*, Vol. 20, No. 3, 2011, 035011.
2. Kim, H.-I., Han, J.-H., and Bang, H.-J., "Real-time deformed shape estimation of a wind turbine blade using distributed fiber Bragg grating sensors", *Wind Energy*, Vol. 17, No. 9, 2014, pp.1455-1467.
3. Korayem, M. H., Shafei, A. M., and Absalan , F., "Estimate a Flexible Link's Shape by the Use of Strain Gauge Sensors", *International Scholarly Research Notices,* Vol. 2013, 2013, pp.1-9.
4. Wu, L.-J., Casciati, F., Casciati, S., "Dynamic testing of a laboratory model via vision-based sensing", *Engineering structures*, Vol. 60, 2014, pp. 113-125.

5. Yun, J.-M., Kim, H.-Y., Kim, H.-I., Han, J.-H., and Kwon, H.-J., "Performance evaluation of homogeneous stereo camera system for full-field structural deformation estimation", *International Journal of Aeronautical and Space Science*, Vol. 16, no. 3, 2015, pp.380-393.
6. Kurita, M., Koike, S., Nakakita, K., and Masui, K., "In-Flight Wing Deformation Measurement", *51st AIAA Aerospace Sciences Meeting including the New Horizons Forum and Aerospace Exposition,* 07-10 January 2013, Grapevine (Dallas/Ft. Worth Region), TX, AIAA 2013-0967
7. Pang, Z. Y., Cesnik, C. E., and Atkis, E. M., "In-Flight Wing Deformation Measurement System for Small Unmanned Aerial Vehicles", *55th AIAA/ASME/ASCE/AHS/SC Structures, Structural Dynamics, and Materials Conference,* 13-17 January 2014, National Harbor, MD, AIAA 2014-0330.
8. Kim, S.-W., and Kim, N.-S., "Multi-point Displacement response measurement of civil infrastructures using digital image Processing", *The Proceedings of the Twelfth East Asia-Pacific Conference on Structural Engineering and Construction,* Vol., 14, 2011, pp.195-203.
9. Lee, J. J., Shinozuka, M., "A vision-based system for remote sensing of bridge displacement", *NDT & E International,* Vol. 39, No. 5, 2006, pp.425-431.
10. Schneider, O., "Analysis of SPR measurements from HART II", *Aerospace Science and Technology,* Vol. 9, No. 5, 2005, pp.409-420.
11. Michael, R. H., *Vibration simulation using MATLAB and ANSYS*, Chapman & Hall, New York, 2001.

ICAST2015 # 065

Adaptive Ensemble Kalman Filter Estimation of Nonlinear Structural Systems with Unknown Noise Covariance

Takeshi Akita[1*], Ryoji Takaki[2], and Nozomu Kogiso[3]

[1] Department of Mechanical Science and Engineering, Chiba Institute of Technology, Narashino, Japan
[2] The Institute of Space and Astronautical Science, JAXA, Sagamihara, Japan.
[3] Department of Aerospace Engineering, Osaka Prefecture University, Sakai, Japan

Abstract

In this paper, an extension of the conventional adaptive Kalman filtering technique to the ensemble Kalman filter is presented. Using the presented method, unknown noise covariances of the filter settings are effectively determined. A simple cantilevered beam vibration problem is provided to verify the effectiveness of the presented adaptive estimation method.

1. INTRODUCTION

Recently, many interests have been attracted for smart structural systems, which are designed to be adjustable via actuators on orbit[1]. One of the main difficulties in constructing effective smart structural systems is to estimate the appropriate current structural states for the actuations because the sensors equipped with spacecraft are usually insufficient to observe whole structural states. To compensate unobserved structural data, numerical simulations based on the finite element (FE) models play an important role.

The FE models usually contains various uncertainties in the systems, such as uncertain structural parameters, and these uncertainties are often identified by using measurement data obtained in validation tests on the ground. The accuracy of the model is limited by how precise the validation test can simulate the orbital spacecraft environment. In future space missions, such as advanced space antennas, the structural design requirements tend to become more and more severe, and the accuracy of the numerical models estimated based on the ground-based validation test could be insufficient.

This research presents an effective model estimation method based on the ensemble Kalman filtering (EnKF) technique[2,3], which can automatically provide optimum estimations of system state variables while assimilating the nonlinear numerical model with the experimental data in sequential manner. Based on the EnKF technique, the structural state estimations can be sequentially obtained on orbit for the actuations in the smart structure system. The accuracies of the EnKF estimations depend on the filter noise settings, that is, the system noise covariance and the measurement noise covariance. These noise settings have to be determined without knowing real state values. In our research, we apply the adaptive estimation technique for the conventional Kalman filter[4,5] to the EnKF, and investigate the effectiveness of the adaptive estimation technique on the EnKF estimations. A simple nonlinear example is provided to verify the effectiveness of the adaptive estimations.

* akita.takeshi@it-chiba.ac.jp

2. NONLINER STRUCTURAL DYNAMIC SYSTEM EQUATIONS

In the finite element analysis, the governing equations for nonlinear structural dynamics problems are given by

$$\mathbf{M}\ddot{\mathbf{u}}_{t+1} + \mathbf{C}\dot{\mathbf{u}}_{t+1} + \mathbf{Q}(\mathbf{u}_{t+1}) = \mathbf{F}(\mathbf{u}_{t+1}) \tag{1}$$

Here, the vectors $\mathbf{u}_{t+1}, \dot{\mathbf{u}}_{t+1}, \ddot{\mathbf{u}}_{t+1}$ indicate the nodal displacement, velocity, and acceleration vectors, respectively, and the matrices $\mathbf{M}$ and $\mathbf{C}$ are the mass and damping matrices, respectively. The internal and external force vectors are denoted by $\mathbf{Q}$ and $\mathbf{F}$, respectively, which are functions of $\mathbf{u}_{t+1}$. The subscript $t+1$ denotes a discretized time step. In this paper, we apply the generalized α method[6] for numerical integration of (1). In the generalized α method, the following governing equations are considered:

$$\mathbf{M}\ddot{\mathbf{u}}_{t+1-\alpha_m} + \mathbf{C}\dot{\mathbf{u}}_{t+1-\alpha_f} + \mathbf{Q}(\mathbf{u}_{t+1-\alpha_f}) = \mathbf{F}(\mathbf{u}_{t+1-\alpha_f}) \tag{2}$$

Here, α_m, α_f are constants that affect numerical damping, and are defined by

$$\alpha_m = \frac{2\rho_\infty - 1}{\rho_\infty + 1}, \alpha_f = \frac{\rho_\infty}{\rho_\infty + 1} \tag{3}$$

where ρ_∞ is the radius of convergence. The discrete time update equations as shown below:

$$\mathbf{\Gamma}\left(\mathbf{u}_t, \mathbf{u}_{t+1}\right) = \mathbf{0} \tag{4}$$

Note that the above equations become nonlinear simultaneous equations and need an iterative method such as the Newton-Raphson method.

3. ENSEMBLE KALMAN FILTER (EnKF)

In this section, we present a basic formulation of the EnKF[2,3], which is a nonlinear extension of the standard Kalman filter. In the EnKF, a priori state estimates are provided by using the Monte Carlo method of nonlinear physical simulations, whereas in the widely used extended Kalman filter (EKF), they are obtained based on linearized systems. Unlike in the EKF, the estimates in the EnKF can inherit many of the nonlinear properties of a priori state estimates.

3.1 Model and observation equations

We define the following model and observation equations:

$$\mathbf{x}_{t+1} = \mathbf{F}_t(\mathbf{x}_t, \mathbf{w}_t) \tag{5}$$

$$\mathbf{y}_{t+1} = \mathbf{H}_{t+1}\mathbf{x}_{t+1} + \mathbf{v}_{t+1} \tag{6}$$

Here, $\mathbf{x}_t$ and $\mathbf{w}_t$ are the state vector and the system noise whose dimensions are n, $\mathbf{F}_t$ is the nonlinear model operator that relates $\mathbf{x}_t$ to $\mathbf{x}_{t+1}$, $\mathbf{H}_{t+1}$ is the transformation matrix from the state vector $\mathbf{x}_{t+1}$ to the measurement $\mathbf{y}_{t+1}$ whose dimension is p, and $\mathbf{v}_t$ is the measurement noise. The covariance matrices of the system noise vector and the measurement noise vector are given by

$$E\left\{\mathbf{w}_t\mathbf{w}_t^T\right\} = \mathbf{Q}_t \tag{7}$$

$$E\left\{\mathbf{v}_t\mathbf{v}_s^T\right\} = \mathbf{R}_t \tag{8}$$

where $E\left\{a\right\}$ denotes the expected value of *a*, the matrices $\mathbf{Q}_t$ and $\mathbf{R}_t$ denote the covariance matrices of the system noise vector and the measurement noise vector, respectively.

In general, the state estimation problem is to find the posterior probability density function (PDF) of the state vector $\mathbf{x}_{t+1}$ given the observation $\mathbf{y}_{t+1}$ with the prior PDF, which is the conditional PDF of $\mathbf{x}_t$ given $\mathbf{y}_{1:t} = [\mathbf{y}_1, \mathbf{y}_2, \cdots, \mathbf{y}_t]$. The prior PDF and the posterior PDF are described as

$$p(\mathbf{x}_{t+1} \mid \mathbf{y}_{1:t}) = p(\mathbf{x}_t \mid \mathbf{y}_1, \mathbf{y}_2, \cdots, \mathbf{y}_t) \tag{9}$$

$$p(\mathbf{x}_{t+1} \mid \mathbf{y}_{1:t+1}) = p(\mathbf{x}_t \mid \mathbf{y}_1, \mathbf{y}_2, \cdots, \mathbf{y}_{t+1}) \tag{10}$$

The state prediction and estimation of $\mathbf{x}_{t+1}$ can be obtained as the expected value of (9) and (10), respectively, as follows:

$$\mathbf{x}_{t+1|t} = E\{\mathbf{x}_{t+1} \mid \mathbf{y}_{1:t}\} = \int_{-\infty}^{\infty} \mathbf{x}_{t+1} p(\mathbf{x}_{t+1} \mid \mathbf{y}_{1:t}) d\mathbf{x} \tag{11}$$

$$\mathbf{x}_{t+1|t+1} = E\{\mathbf{x}_{t+1} \mid \mathbf{y}_{1:t+1}\} = \int_{-\infty}^{\infty} \mathbf{x}_{t+1} p(\mathbf{x}_{t+1} \mid \mathbf{y}_{1:t+1}) d\mathbf{x} \tag{12}$$

3.2 Ensemble approximation

The prior and posterior PDFs usually become non-Gaussian when the model equations are nonlinear, and thus, numerical schemes are needed to evaluate (11) and (12). In the EnKF, the Monte Carlo approach is applied, where the PDFs in (11) and (12) are approximated by generating a number of samples, called particles in the EnKF, as follows:

$$p(\mathbf{x}_{t+1} \mid \mathbf{y}_{1:t}) \simeq \frac{1}{M}\sum_{m=1}^{M} \delta\left(\mathbf{x}_{t+1} - \mathbf{x}_{t+1|t}^{(m)}\right) \tag{13}$$

$$p(\mathbf{x}_{t+1} \mid \mathbf{y}_{1:t+1}) \simeq \frac{1}{M}\sum_{m=1}^{M} \delta\left(\mathbf{x}_{t+1} - \mathbf{x}_{t+1|t+1}^{(m)}\right) \tag{14}$$

Here, M is the total number of particles, $\delta(\bullet)$ is Dirac's delta function, the superscript (m) denotes the index number of the particle set, and $\mathbf{x}_{t+1|t}^{(m)}$ and $\mathbf{x}_{t+1|t+1}^{(m)}$ are the prediction and the estimation, respectively, of the state vector of the mth particle at time step $t+1$. Substituting (13) and (14) into (11) and (12), respectively, we can obtain the state prediction and estimation of $\mathbf{x}_{t+1}$ as the ensemble mean of particles as shown below:

$$\mathbf{x}_{t+1|t} \simeq \hat{\mathbf{x}}_{t+1|t} = \frac{1}{M}\int_{-\infty}^{\infty} \mathbf{x}_{t+1} \sum_{m=1}^{M} \delta\left(\mathbf{x}_{t+1} - \mathbf{x}_{t+1|t}^{(m)}\right) d\mathbf{x} = \frac{1}{M}\sum_{m=1}^{M} \mathbf{x}_{t+1|t}^{(m)} \tag{15}$$

$$\mathbf{x}_{t+1|t+1} \simeq \hat{\mathbf{x}}_{t+1|t+1} = \frac{1}{M}\int_{-\infty}^{\infty} \mathbf{x}_{t+1} \sum_{m=1}^{M} \delta\left(\mathbf{x}_{t+1} - \mathbf{x}_{t+1|t+1}^{(m)}\right) d\mathbf{x} = \frac{1}{M}\sum_{m=1}^{M} \mathbf{x}_{t+1|t+1}^{(m)} \tag{16}$$

Likewise, the covariance matrices of $\mathbf{x}_{t+1|t}$ and $\mathbf{x}_{t+1|t+1}$ are approximated by the particle set as shown below:

$$\mathbf{P}_{t+1|t} \simeq \hat{\mathbf{P}}_{t+1|t} = \frac{1}{M-1}\sum_{m=1}^{M} \left(\mathbf{x}_{t+1|t}^{(m)} - \hat{\mathbf{x}}_{t+1|t}\right)\left(\mathbf{x}_{t+1|t}^{(m)} - \hat{\mathbf{x}}_{t+1|t}\right)^T \tag{17}$$

$$\mathbf{P}_{t+1|t+1} \simeq \hat{\mathbf{P}}_{t+1|t+1} = \frac{1}{M-1}\sum_{m=1}^{M} \left(\mathbf{x}_{t+1|t+1}^{(m)} - \hat{\mathbf{x}}_{t+1|t+1}\right)\left(\mathbf{x}_{t+1|t+1}^{(m)} - \hat{\mathbf{x}}_{t+1|t+1}\right)^T \tag{18}$$

3.3 State estimation

The state estimation is performed through two steps: a prediction step and an estimation step. At the prediction step, all particles are updated through the following equations:

$$\mathbf{x}_{t+1|t}^{(m)} = \mathbf{F}_t(\mathbf{x}_{t|t}^{(m)}, \mathbf{w}_t^{(m)}) \tag{19}$$

where $\mathbf{w}_t^{(m)}$ is the mth realization of system noise vector based on $\mathbf{Q}_t$. At an estimation step, the estimation vector of the mth particle is obtained as follows:

$$\mathbf{x}_{t+1|t+1}^{(m)} = \mathbf{x}_{t+1|t}^{(m)} + \mathbf{K}_{t+1}\left(\mathbf{y}_{t+1} + \mathbf{v}_{t+1}^{(m)} - \mathbf{H}_{t+1}\mathbf{x}_{t+1|t}^{(m)}\right) \tag{20}$$

where $\mathbf{v}_t^{(m)}$ is the realization of measurement noise vector for the mth particle based on $\mathbf{R}_{t+1}$, and $\mathbf{K}_{t+1}$ is the Kalman gain matrix given by

$$\mathbf{K}_{t+1} = \hat{\mathbf{P}}_{t+1|t}\mathbf{H}_{t+1}^T\left(\mathbf{R}_{t+1} + \mathbf{H}_{t+1}\hat{\mathbf{P}}_{t+1|t}\mathbf{H}_{t+1}^T\right)^{-1} \tag{21}$$

Here, $\hat{\mathbf{P}}_{t+1|t}$ is obtained by the ensemble calculations of the particle set in the EnKF.

4. ESTIMATION PROCEDURE FOR NONLINEAR STRUCTURAL SYSTEM

In the previous section, a standard formulation of the EnKF is presented. In this section, we apply it to nonlinear structural model estimation by combining displacement and structural properties into a state vector. The combined state vector for model estimation is defined by

$$\mathbf{x}_t \doteq \begin{bmatrix} \mathbf{s}_t \\ \mathbf{u}_t \end{bmatrix} \tag{22}$$

where $\mathbf{s}_t$ is a k-dimensional parameter vector whose components are various structural parameters such as damping ratios and densities. Likewise, the system noise vector is expressed as the combined vector as follows:

$$\mathbf{w}_t \doteq \begin{bmatrix} \mathbf{w}_{s,t} \\ \mathbf{w}_{u,t} \end{bmatrix} \tag{23}$$

where $\mathbf{w}_{s,t}$ and $\mathbf{w}_{u,t}$ are the system noise in the structural dynamics equations and the system noise in the parameter time update equations, respectively. In the presented approach, the prediction in (19) is performed through two steps. The first step is the prediction of the parameter vector as follows:

$$\mathbf{s}_{t+1} = \mathbf{s}_t + \mathbf{w}_{s,t} \tag{24}$$

By using the updated parameter vector, in the next step, the time state update of the displacement vector is performed as shown below:

$$\boldsymbol{\Gamma}\left(\mathbf{u}_t, \mathbf{u}_{t+1}, \mathbf{s}_{t+1}\right) = \mathbf{0} \tag{25}$$

We consider direct observations of displacement vectors, where the observation equations are given by

$$\mathbf{y}_{t+1} = \begin{bmatrix} {}_p\mathbf{0}_q & \tilde{\mathbf{H}}_{t+1} \end{bmatrix}\begin{bmatrix} \mathbf{s}_{t+1} \\ \mathbf{u}_{t+1} \end{bmatrix} + \mathbf{v}_{t+1} = \mathbf{H}_{t+1}\mathbf{x}_{t+1} + \mathbf{v}_{t+1} \tag{26}$$

where $\tilde{\mathbf{H}}_{t+1}$ is a $p \times n$-dimensional Boolean matrix that relates the observation nodes to the FEA nodes.

The estimation procedure for the nonlinear structural model based on the EnKF is shown below.

(Step 1: Generation of an initial particle set)

The initial particle set is generated by adding initial perturbations to the nominal parameter values in each particle as shown below:

$$\mathbf{s}_{0|0}^{(m)} = \mathbf{s}_0 + \mathbf{w}_0^{(m)}, {}^i w_{s,0} \sim N(0, {}^i\sigma_{s,0}^2) : m = 1, 2, \cdots M \tag{27}$$

(Step 2: Time update of each particle)

In the prediction step, we first produce the realization of $\mathbf{w}_{s,t}^{(m)}, {}^i w_{s,t} \sim N(0, {}^i\sigma_s^2)$ in each particle. Then, we update the parameter vector as follows:

$$\mathbf{s}_{t+1|t}^{(m)} = \mathbf{s}_{t|t}^{(m)} + \mathbf{w}_t^{(m)}, m = 1, 2, \cdots M \tag{28}$$

Next, the displacement vector of each particle is updated by using (25). Consequently, the state prediction vector is given by

$$\mathbf{x}_{t+1|t}^{(m)} = \begin{bmatrix} \mathbf{s}_{t+1|t}^{(m)} \\ \mathbf{u}_{t+1|t}^{(m)} \end{bmatrix}, m = 1, 2, \cdots M \tag{29}$$

In addition, the covariance matrix $\mathbf{P}_{t+1|t}$ of $\mathbf{x}_{t+1|t}$ is calculated by (17).

(Step 3: Calculation of Kalman gain)

The Kalman gain is calculated by using (21) with $\mathbf{P}_{t+1|t}$ obtained in step 2.

(Step 4: Estimation of each particle)

All particles are updated by using (20) with the Kalman gain given in step 3. Note that the parameter estimations are provided by means of the ensemble mean of the particle set as follows:

$$\hat{\mathbf{s}}_{t+1|t+1} \frac{1}{M} \sum_{m=1}^{M} \mathbf{s}_{t+1|t+1}^{(m)} \tag{30}$$

We return to step 2 until the estimation process finishes. Through these steps, we can sequentially obtain the state estimations of the displacement vector and parameters at each observation time step.

5. ADAPTIVE FILTERING FOR THE ENSEMBLE KALMAN FILTER

The settings of system and measurement noise have strong effects on both the efficiency and the accuracy of estimation results. In a real case, we have to choose the noise settings without knowing the real state values. To find appropriate settings, the adaptive Kalman filtering techniques have been proposed for the conventional Kalman filter[4,5]. In the adaptive Kalman filter, a proper cost function related to estimation residuals is minimized with respect to the noise variances. The widely used cost function in the conventional Kalman filter[5] is given by

$$\begin{aligned} J(\mathbf{a}) &= \sum_{t=t_0}^{t_n} \left(\ln |\mathbf{S}(\mathbf{a})| + \mathbf{e}_{t+1}^T \mathbf{S}^{-1}(\mathbf{a}) \mathbf{e}_{t+1} \right) \\ \mathbf{S}(\mathbf{a}) &= \left(\mathbf{H}_{t+1} \mathbf{P}_{t+1|t}(\mathbf{a}) \mathbf{H}_{t+1}^T + \mathbf{R}_{t+1}(\mathbf{a}) \right) \\ \mathbf{e}_{t+1} &= \mathbf{y}_{t+1} - \mathbf{H}_{t+1} \mathbf{x}_{t+1|t} \end{aligned} \tag{31}$$

where $\boldsymbol{a}$ denotes the parameter vector composed of the noise variances as follows:

$$\mathbf{a} = \left[\sigma_{s,1}^2, \cdots, \sigma_{s,n}^2, \sigma_{obs,1}^2, \cdots \sigma_{obs,m}^2 \right] \tag{32}$$

To apply the adaptive Kalman filtering technique to the EnKF, we extend the cost function J to the following equations:

$$\begin{aligned} J(\mathbf{a}) &= \sum_{t=t_0}^{t_n} \left(\ln |\hat{\mathbf{S}}(\mathbf{a})| + \hat{\mathbf{e}}_{t+1}^T \hat{\mathbf{S}}^{-1}(\mathbf{a}) \hat{\mathbf{e}}_{t+1} \right) \\ \hat{\mathbf{S}}(\mathbf{a}) &= \left(\mathbf{H}_{t+1} \hat{\mathbf{P}}_{t+1|t}(\mathbf{a}) \mathbf{H}_{t+1}^T + \mathbf{R}_{t+1}(\mathbf{a}) \right) \\ \hat{\mathbf{e}}_{t+1} &= \mathbf{y}_{t+1} - \mathbf{H}_{t+1} \frac{1}{M} \sum_{m=1}^{M} \mathbf{x}_{t+1|t}^{(m)} \end{aligned} \tag{33}$$

In the adaptive filtering, the tuned noise variances are defined by the solution of the following minimization problem:

$$\min_{\mathbf{a}} J(\mathbf{a}) \tag{34}$$

6. A NUMERICAL EXPERIMENT

A simple cantilevered beam vibration problem is considered to verify the effectiveness of the presented adaptive estimation. The beam is divided into five elements and the y tip displacement is set to the measurement point (Figure 1). The length of the beam is set to 0.4 m, and the base excitation frequency is set to 10 Hz, which is near the 1st eigen frequency of the beam. The base excitation amplitude is set to 5mm, which can induces cubic nonlinearities to the system. In this experiment, the 1st mode damping ratio is set as a model parameter.

The numerical experiment is performed through two processes. In the first process, the structural analysis with the real model parameter value is performed to obtain the displacement history of the cantilevered beam vibration. Then, the artificial measurements are produced by adding the random noise to the calculated observations. In the second process, the EnKF estimation is performed with the artificial measurements. The variance of the measurement noise is set to $\sigma^2_{obs} = 1.0\times10^{-8}m^2$, measurement time step is set as 2ms. The real value of the damping ratio is set to $\zeta = 0.005$, and the initial parameter value in the EnKF is chosen as $\zeta^0 = 0.025$. The total simulation time is 5 s, and the total number of particles is 100.

The solution procedure for the adaptive estimation is summarized in Figure 2. First, we set initial system and measurement noise settings, and perform the EnKF estimation with N_t interval. Then, calculate the cost function J. If the convergence check is OK, the adaptive estimation is finished. Otherwise, we update the noise values, and the EnKF is performed with the new noise values. This process continues until the convergence check is OK. In this paper, we utilize MATLAB® optimization toolbox in the update process. In this experiment, the initial system noise is set $\sigma^{2(0)}_{s,1} = 6.3\times10^{-8}$, and the initial measurement noise is set $\sigma^{2(0)}_{obs} = 1.0\times10^{-8}$. The time interval for adaptive estimation is set $N_t = 5s$.

Figure 3 shows the results of minimization of J. The minimization converges at 10 step, where the system noise is $\sigma^{2(Tuned)}_{s,1} = 4.4\times10^{-6}$ and the measurement noise is $\sigma^{2(Tuned)}_{obs} = 6.6\times10^{-9}$. Figure 4 (a) shows the comparisons of estimation errors of ζ, between the initial system noise setting and the tuned filter setting while the comparisons of y displacement estimation errors in the unobserved nodes (node 4) and the observed nodes (node 6) of the cantilevered beam shown in Figure 1. In these figures, the upper plot shows the estimation errors in the initial setting, while the lower plot shows those in the tuned setting. The 3 standard deviation intervals ($\pm3\sigma$) are also depicted in the figures. The standard deviations are evaluated by calculating the square root of variances of corresponding particle set. As can be seen in Figure 4 (a), the convergence rate of model parameter estimation error with the tuned settings is significantly improved when compared to that with the initial settings. Further, the estimation errors in the initial setting lie outside in the area enclosed by $\pm3\sigma$ around the early time step. In the tuned setting, at almost all the time step, the estimation errors lie in the area. In the practical estimation case, the real sate values are unknown, and thus, the accuracies of error boundaries play an important role to evaluate the estimation results. We can see the same trends in Figures 5 (b) and (c). These results indicate the effectiveness of the presented adaptive estimations.

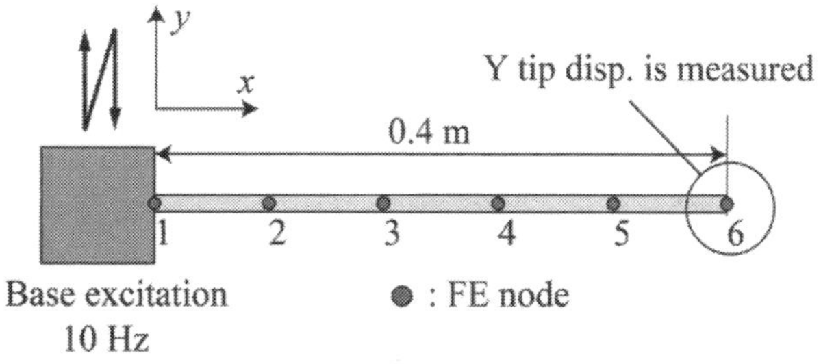

Figure 1. Cantilevered beam model

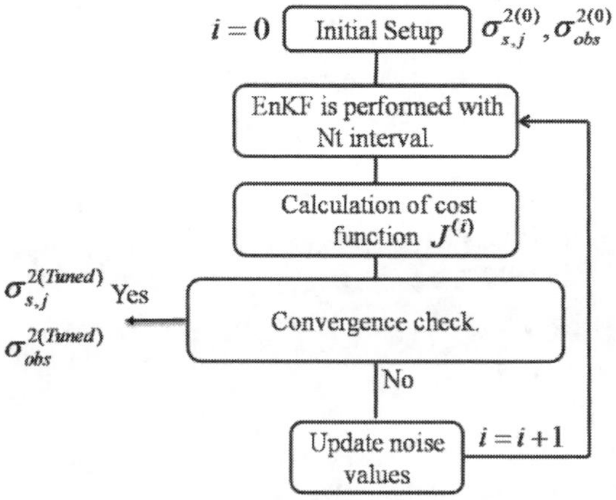

Figure 2. Flow chart in the adaptive estimation

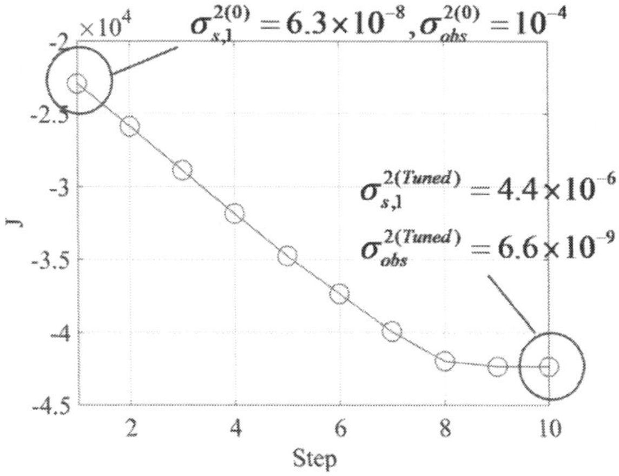

Figure 3. Minimization results

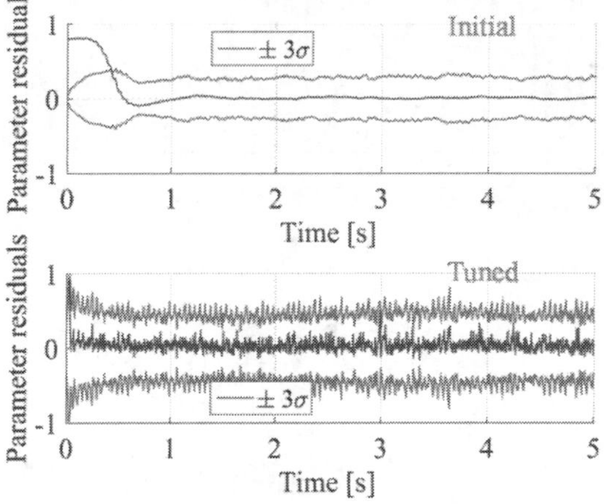

(a) model parameter

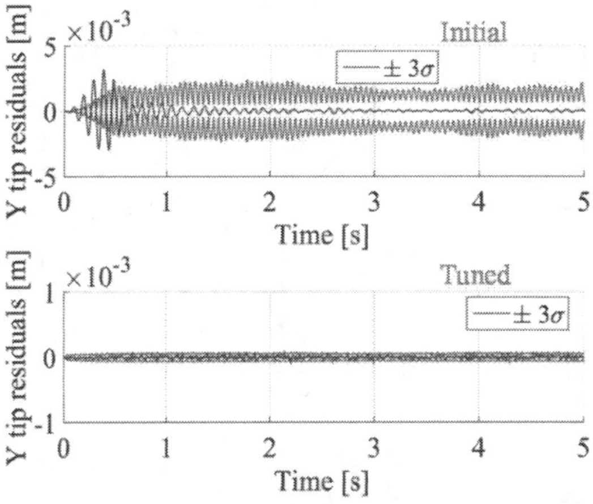

(b) y displacement (node 4, unobserved node)

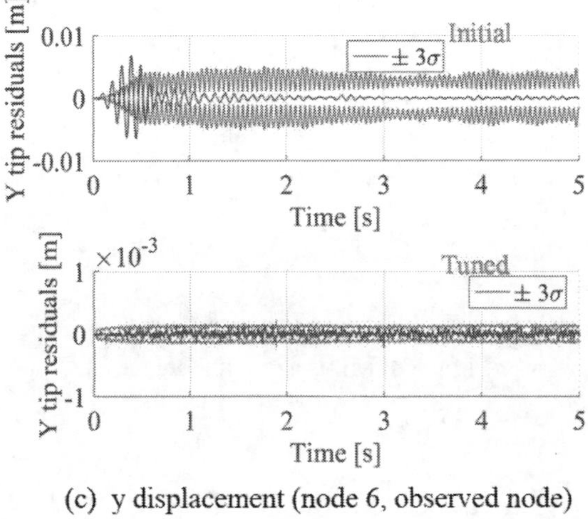

(c) y displacement (node 6, observed node)

Figure 4. Comparisons of estimation errors between the initial noise values ($\sigma^{2(0)}{}_{s,1} = 6.3\times10^{-8}$, $\sigma^{2(0)}{}_{obs} = 1.0\times10^{-8}$) and the tuned noise values ($\sigma^{2(Tuned)}{}_{s,1} = 4.4\times10^{-6}$, $\sigma^{2(Tuned)}{}_{obs} = 6.6\times10^{-9}$)

6. CONCLUSIONS

An adaptive estimation method of nonlinear structural system with the ensemble Kalman filter was presented. The state space equations of the nonlinear structural dynamics model were derived, and the estimation procedure based on the ensemble Kalman filter was presented. A conventional adaptive filtering technique was applied to the ensemble Kalman filtering. A simple cantilevered beam vibration problem was provided to verify the effectiveness of the presented adaptive estimation method.

ACKNOWLEDGMENTS

This work was supported by JSPS KAKENHI Grant Number 26249131.

REFERENCES

1. Tanaka, H., Sakamoto, H., Inagaki, A., Ishimura, K., Doi, A., Kondo, Y., Oyama, K., Watanabe, K., Oikawa, Y., and Kuratomi, T., "Development of a smart reconfigurable reflector prototype for an extremely high-frequency antenna," *Journal of Intelligent Material Systems and Structures*, April 14, 2015.
2. Evensen, G., "Sequential Data Assimilation with a Nonlinear Quasi-Geostrophic Model Using Monte Carlo Methods to Forecast Error Statistics," *Journal of Geophysical Research*, 99, 1994, pp. 10143–10162.
3. Evensen, G., "Data Assimilation," Springer, Berlin, 2009.
4. Maybeck, P.S., "Stochastic Models, Estimation, and Control," vol.2, Academic Press, New York, 1982.
5. Mohamed, A. H., and Schwarz, K. P., "Adaptive Kalman Filtering for INS/GPS," *Journal of Geodesy*, Vol. 43, 1999, pp. 193-203.
6. Chung, J. and Hulbert, G. M., "A time integration algorithm for structural dynamics with improved numerical dissipation the generalized α method," *Journal of Applied Mechanics Transactions of the ASME*, Vol. 60, No. 2, 1993, pp. 371-375.

ICAST2015: 26th International Conference on Adaptive Structures and Technologies
October 14-16th, 2015, Kobe, Japan

ICAST2015 #015

Conceptual Topology Studies on Bi-Stable Composites in Morphing Aerofoils

Izabela K. Kuder[1*], Andres F. Arrieta[2*], and Paolo Ermanni[1]

[1] Laboratory of Composite Materials and Adaptive Structures, ETH Zurich, 8092 Zurich, Switzerland.
[2] School of Mechanical Engineering, Purdue University, West Lafayette, USA.

Abstract

To realise morphing in an efficient manner, multiple contradictory requirements need to be reconciled in a novel technical solution, simultaneously exhibiting significant compliance, adequate load-carrying capacity and low mass. A possible approach is to exploit stiffness variability of bi-stable laminates to impart selective compliance on aerofoil structures. This implies integrating the composites, which feature distinct stiffness properties of each equilibrium configuration, into the distributed topology of a wing profile in a strategic manner, maximising the impact of the bi-stable components on the structural response.

This paper presents a novel, concurrent design methodology aimed at achieving global shape adaptation of such selectively compliant wing profiles. A versatile and computationally efficient representation of the embeddable variable stiffness bi-stable laminates as user-defined finite elements is introduced. These are integrated into the parameterised internal substructure of a NACA 0012 profile. Static aeroelastic simulations coupling Abaqus/Standard® with the aerofoil analysis code XFOIL are performed to explore the global deformation modes achievable with different topologies as regards their aerodynamic characteristics and the contribution to increased compliance provided by the flexible equilibrium configuration. The results confirm the potential of the proposed concurrent design methodology to exploit the variable stiffness of embedded bi-stable laminates to realise selectively compliant morphing structures while preserving an aerodynamically smooth shape.

1. INTRODUCTION

The challenge of realising morphing structures to provide holistic performance benefits over conventional configurations requires reconciling multiple contradictory requirements in a novel technical solution. A compliant, rather than a mechanism-driven, system can allow for efficient shape adaptation [1, 2], yet needs to simultaneously possess adequate load-bearing capacity without suffering mass penalties [3–5]. A promising approach is to base morphing upon selective compliance. This involves adapting the global structural response by selectively activating variable stiffness elements embedded in a tailored internal substructure. In this context, elasticity-based stiffness variability becomes particularly promising as provided by bi-stable laminates [6–10], which exhibit markedly different properties of each equilibrium configuration. Tailoring the spatial stacking sequence distribution of the composites leads

* I. K. Kuder: ikuder@ethz.ch, A. F. Arrieta: aarrieta@purdue.edu

to specific stiffness characteristics while satisfying the integrability objective of clamping two opposite edges within a larger morphing system.
The design space of such embeddable variable stiffness bi-stable laminates, however, features limits in terms of achievable stiffness difference and load-carrying capacity depending on the desired external dimensions [7, 8, 10]. In order to take maximum advantage of the available capabilities, an integrated design and optimization process becomes necessary. This entails finding an optimally distributed internal substructure of a morphing wing section with the embedded variable stiffness components as key elements allowing for selective compliance. A desired deformation mode can then be activated by virtue of the tailored distributed topology responding to the aerodynamic loads in a predefined manner.
This paper presents such a concurrent aero-structural design strategy involving the commercial finite-element solver Abaqus/Standard® weakly coupled with the aerofoil analysis code XFOIL based on an inviscid panel method with an integral boundary layer [11]. First, a versatile and computationally efficient representation of the embeddable variable stiffness bi-stable composites as user-defined finite elements (UEL) is introduced. These are integrated into the parameterised internal substructure of the shape-adaptive profile. Verification against simulations involving full bi-stable laminates demonstrates the capability of the UEL of accurately capturing the structural response, including the morphed shape, while allowing for significant design freedom at reduced computational expense. The global deformation modes achievable with diverse topologies are explored by means of static aeroelastic analyses. Hence insights are gained into the design sensitivity of the proposed morphing structures, with a particular emphasis on positioning the bi-stable components for maximum impact while preserving a smooth morphed shape.
The results confirm the potential of the presented design methodology to fully exploit the variable stiffness of embedded bi-stable laminates to realise selective compliance morphing structures. This provides a basis for further generalisation of the proposed concurrent aero-structural approach to multidisciplinary optimisation.

2. BI-STABLE AEROFOIL CONCEPT

The preliminary concept of the morphing aerofoil based on a 500 mm chord NACA 0012 with integrated variable stiffness bi-stable elements, introduced in [12, 13], is schematically shown in Figure 1(a). The central idea is to achieve global shape adaptation through altering the stiffness characteristics of specific elements embedded within the larger distributed topology of the wing section. For this purpose, bi-stable laminates are employed as a means of providing the desired stiffness variability at reduced actuation cost. The former relies on the tailored, markedly different properties of each equilibrium configuration, which can be accessed through localised actuation between the stable states. The latter exploits the elastic nature of the multi-stability phenomenon combined with the only actuation energy input required to change to the other equilibrium configuration, rather than maintaining these.

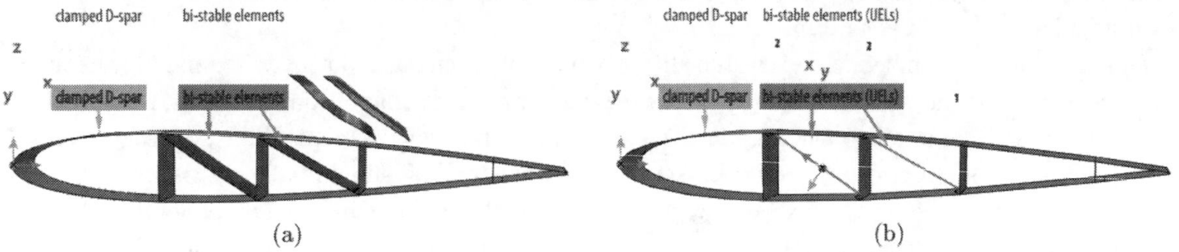

Figure 1. Preliminary morphing aerofoil concept with selectively compliant substructure featuring embedded variable stiffness bi-stable laminates. (a) Full model with bi-stable laminates imported from a separate cool-down analysis. (b) Simplified model with bi-stable components represented by means of UELs.

3. CONCURRENT DESIGN APPROACH

A concurrent methodology following a similar approach to that proposed by Molinari et al [14, 15] and Previtali et al [16, 17] is developed in order to simultaneously incorporate all the relevant dimensions into the design and optimisation process, as schematically shown in Figure 2 and detailed in the following.

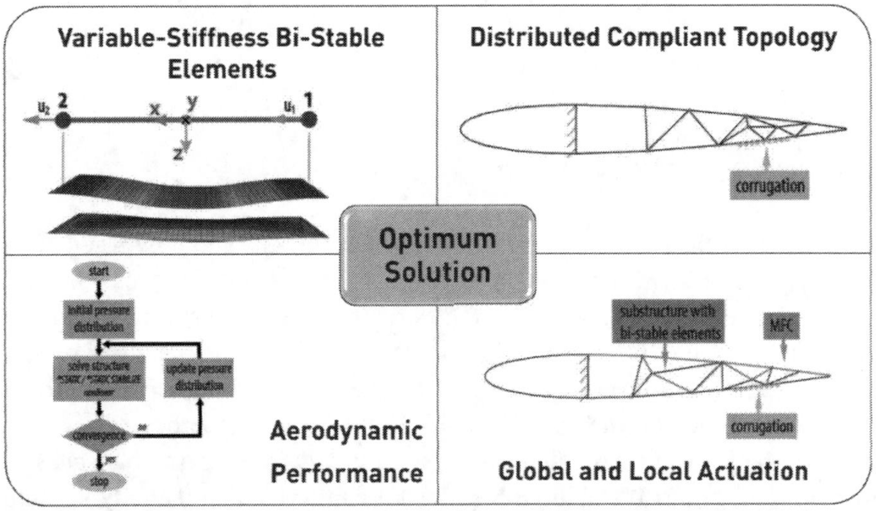

Figure 2. Schematic of the concurrent design and optimisation approach of morphing aerofoils featuring variable stiffness provided by embedded bi-stable laminates.

The interior of the NACA 0012 profile with a clamped D-spar is parametrised to yield a distributed compliant topology, aimed at facilitating camber morphing and guiding aerodynamic loads to achieve a favourable deformed shape. The outer skin possesses a corrugated laminate portion of a simple circular form, ensuring enough extensibility to allow for cambering. The low bending stiffness of the corrugation renders a simultaneous optimisation with the topology necessary to avoid non-smooth deformations. To account for the significant compliance of the proposed wing section, weakly coupled static aeroelastic analyses are performed in an iterative process between Abaqus/Standard® non-linear static step (structural deformations) and the aerofoil analysis code XFOIL (pressure distributions and aerodynamic coefficients). Hence the aerodynamic response of the shape-adaptable profile can be influenced and desired characteristics obtained.

A crucial part of the distributed compliant topology are the monolithically embedded bi-stable laminates. These possess carefully designed two stable configurations: a curved and a flat state of low and high compressive stiffness, respectively, thereby imparting stiffness variability on the wing section. The response of the bi-stable composites is represented by means of a user-defined finite element (`UEL`), detailed in the following section. Thus the efficiency and flexibility of the design process are greatly enhanced. More precisely, a wide range of topologies of the morphing aerofoil can be defined and its individual members selectively replaced with the bi-stable elements to determine the impact on the resulting final morphed shape.

A given global actuation input morphing the profile downwards to increase lift is represented by a thermal analogy of a Macro Fiber Composite (MFC) bi-morph [18] covering the top skin of the aerofoil. Since the bi-stable laminates provide overall stiffness reduction during shape adaptation, it is vital that their flexible equilibrium configuration remain unaffected by the MFC actuation. Hence no tension, which induces the switch to the stiff state, is allowed to act on the composites. This stresses the importance of incorporating the global actuation, topology and bi-stable laminate positioning aspects in a concurrent manner. Finally, the control over the current equilibrium configurations and hence stiffness properties of the integrated bi-stable components is envisaged to be imposed locally, on the laminate level.

4. USER-DEFINED FINITE ELEMENT

4.1 Motivation and Concept Development

The finite element simulation of the morphing aerofoil with embedded bi-stable laminates can be realised as a *full* model, in which accurate displacement and induced stress fields of the bi-stable elements are obtained from separate cool-down analyses and imported into the wing section model [12, 13]. The limitations of the corresponding Abaqus® `IMPORT` option [19], combined with considerable complexity and computational effort, render the method impractical given the intention of proceeding to a fully integrated optimisation process of the selectively compliant profile. A user-defined finite element is therefore developed to represent the axial stiffness behaviour of the bi-stable components when embedded within a larger shape-adaptable system. The Abaqus® user subroutine `UEL` employed to that end allows a general element to be defined by the user and included in the simulations [19, 20]. The approach permits defining an arbitrary number of different `UEL` types, allowing multiple bi-stable laminate designs with distinct stiffness properties to be incorporated into the optimisation process.

Each of the two equilibrium configurations (flexible/curved and stiff/flat) of the bi-stable composites is represented by a distinct user element type, referred to as a *flexible* and *stiff* `UEL`. The force-displacement curves of both stable states of the laminates clamped in-plane are obtained for numerical displacement-controlled stiffness tests in compression and tension, as detailed in [10]. Further, the element is implemented to either accept a particular stiffness curve or a regression equation representing the desired response in a more general manner, as schematically shown in Figure 3(a).

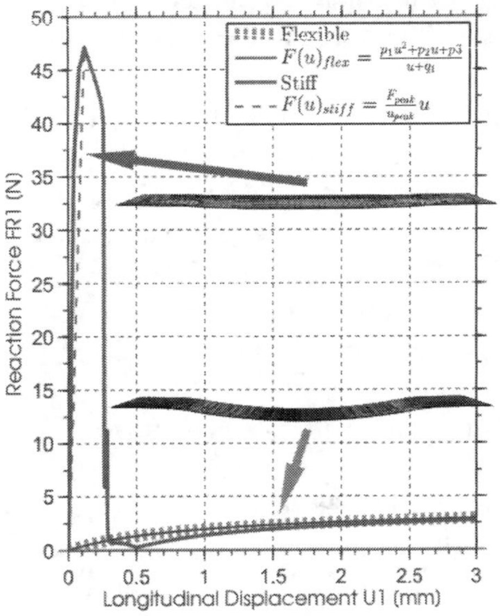

(a)

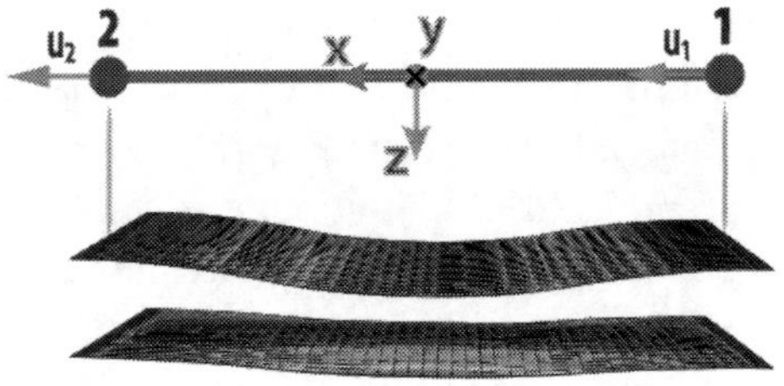

(b)

Figure 3. (a) Example compressive force-displacement curves of the stiff and flexible equilibrium configurations a two-ply bi-stable laminate with regression equation (flexible state) and possible linearisation (stiff state). (b) Schematic of the representation of a variable stiffness bi-stable laminate as a user-defined finite element UEL.

Figure 3(b) indicates the conventions adopted for the derivation of the UEL. The element has a two-dimensional character, assumed to lie in plane (global x-z in the current case, see Figure 1(b)) and therefore possessing only the corresponding translational degrees of freedom. When considered in the global reference frame of the three-dimensional model, as Figure 1(b) illustrates, the orientation of UEL is selected such that both nodes are characterised by the same out-of-plane coordinate (y), which hence does not feature. In the local reference frame of the element, in turn, only axial displacements, u_1 and u_2, are active. This assumption follows from studies into the implications of extending the UEL definition to include more degrees of freedom, which confirm the longitudinal stiffness response as the only component of importance in the case considered.

The tangent stiffness value at the current increment n, k_t^n, is approximated as:

$$k_t^n = \frac{F^n - F^{n-1}}{\Delta u_{1,2}^n} \tag{1}$$

where F^n and F^{n-1} denote the interpolated force values at the current and preceding increments, respectively, and $\Delta u_{1,2}^n$ refers to the relative axial displacement increment during increment n. The subsequent derivation follows standard methods applied to truss elements [21] and is hence omitted for the sake of brevity.

The pronounced anisotropy of the bi-stable laminates, reflected in a distinct structural response in tension and compression, creates the necessity to distinguish between a 180° global rotation of the wing section. By way of illustration, subjecting the shape-adaptable aerofoil presented in Figure 1(b) to an aerodynamic pressure distribution corresponding to a positive angle of attack would preserve the flexible configuration of the bi-stable components, and hence the high compliance of the complete structure, due to the resultant compression acting on the laminates. In contrast, inverting the bi-stable aerofoil under the same loading (curvature of the flexible stable state pointing upwards after a global rotation around the x axis) would force the composites to adopt the stiff equilibrium shape.

4.2 Final Approach

To begin with, basic verification steps are performed to ensure that the element is capable of reproducing exactly the prescribed axial stiffness response described by the force-displacement data. Whereas a significant improvement in computational efficiency accompanies the implementation of the flexible UEL, the numerical difficulties associated with the simulation of the stiff stable state, elaborated in [10], persist due to the highly non-linear characteristics of the problem. Specifically, the snap-through to the curved shape on exceeding a critical load manifests itself as limit points [22, 23] of the corresponding equilibrium path. This poses a considerable challenge for an incremental static analysis, necessitating the employment of stabilisation to achieve convergence. Further, the associated number of iterations and increments of the solution remain high. A potential alternative would be to use the Riks method [19]. This, however, is not pursued herein due to no satisfactory computational set-up being found. By contrast, the static analysis provides a good compromise between accuracy, numerical effort and robustness as regards the solution controls. What is more, the focus of the current approach lies on the selective activation of the deformation modes by virtue of the distinct stiffness characteristics associated with the individual stable states. In this context, the details of the snap-through transition between the equilibrium configurations are not of primary interest and can hence be excluded from the analyses at the present stage. Accordingly, since the post-snap through condition is known (flexible configuration), the simulation of the stiff stable state is only performed until the snap-through threshold corresponding to the force-displacement response peak. The analysis is then interrupted and a message issued indicating the snap-through conditions having been reached. This solution obviates the need for applying a problem-dependent artificial damping to achieve convergence, permitting a standard static analysis, an important aspect in view of automated optimisation implementation.

4.3 Verification of Parametrised Finite-Element Model

In order to assess the performance of the UEL when integrated into the morphing aerofoil, fully parametrised Python routines are employed to generate the finite-element model of the prototype design introduced in [12, 13]. The embedded bi-stable laminates are replaced by the user-defined element, as shown in Figure 1(b). Each end node of a UEL is attached to the corresponding central grid point of the mounting flange by means of of a TIE multi-point constraint, which imposes equal displacements on the common global degrees of freedom of the two nodes [19]. Moreover, the motion of the central grid point of a mounting flange governs the displacements of the remaining flange nodes through a kinematic coupling [19]. This becomes necessary due to the two-dimensional character of the UEL, embedded into a model with a finite third, or spanwise, dimension (Figure 1(b)).

The performance of the UEL is investigated under a distributed aerodynamic load, proving the ability to capture the global stiffness variability of the morphing aerofoil, clear from Figure 4(a). Indeed, the *flexible* mode of the wing section (all the embedded bi-stable elements in the flexible/curved state) features much higher trailing edge displacements than the *stiff* mode (all the embedded bi-stable elements in the stiff/flat state). The slight prediction discrepancy may be primarily explained with the reduction of the bi-stable element response to two dimensions and the corresponding coupling method within a three-dimensional model. Secondly, the full model of the flexible configuration is obtained from the stiff mode by triggering the snap-through of the integrated bi-stable laminates, which induces a slight trailing edge displacement of the entire profile as the initial condition of the flexible full model. By contrast, the UEL is integrated into an undeformed NACA 0012 to directly produce the flexible aerofoil configuration. Nevertheless, the overall difference between the final morphed shapes as obtained on convergence of the static aeroelastic analyses from the full and UEL models is insignificant. This becomes clear from Figure 4(b), which provides a comparison between the final aerodynamic coefficients of the flexible aerofoil mode obtained based on the two models.

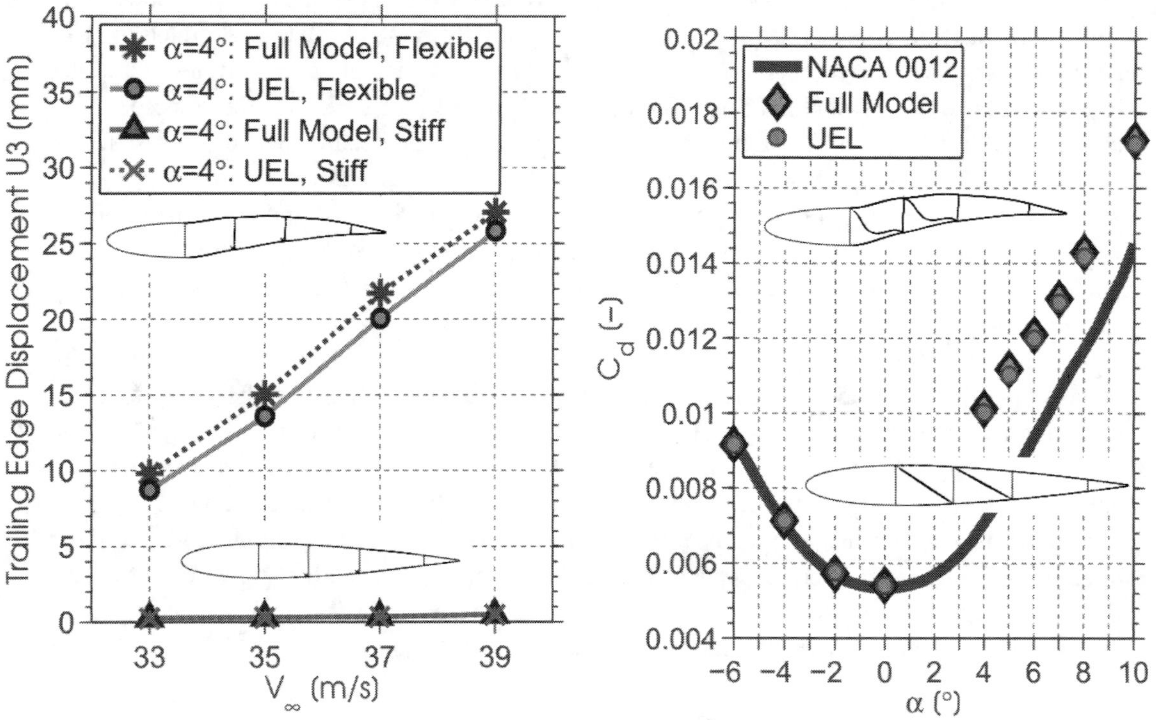

(a) trailing edge vertical displacement

(b) drag coefficient

Figure 4. Verification of the user-defined finite element against a full model in weakly coupled static aeroelastic analyses of the morphing aerofoil. Overlaid final morphed shapes. Please note the lack of the final deformed shapes of the bi-stable elements in (a) compared to the full model (b) due to no visualisation possibility of UELs. (a) Variable stiffness response of the stiff and flexible wing section modes quantified in terms of vertical trailing edge displacement for different freestream velocities and an angle of attack of four degrees. No snap-through of the stiff bi-stable laminates under these conditions. (b) Comparison of the drag coefficients for selected aerodynamic points (freestream velocity of 35 m/s) between rigid NACA 0012 and the flexible aerofoil mode.

5. EXAMPLE TOPOLOGIES

The concurrent approach outlined above is employed to perform conceptual topology studies aimed at determining the most effective positioning of the bi-stable components while preserving a smooth aerodynamic shape. To that end, the trailing edge displacements of the *flexible* and *stiff* modes of the shape-adaptive aerofoil are compared under a specific actuation input (maximum MFC voltage) and aerodynamic loads (20 m/s, 0° incidence). Their ratio allows for quantifying the impact of the stiffness reduction achieved with a particular topology. In the *flexible* aerofoil mode, selected substructural members are replaced with bi-stable elements in the curved state of the laminate configuration used in the experimental demonstrator [12, 13] employed for UEL verification presented above. At this preliminary stage, the size- and layout-dependent stiffness properties of the bi-stable laminates are not considered and no minimum dimensions imposed corresponding to the bi-stability threshold. Accordingly, the *stiff* aerofoil mode is represented by the entire substructure exhibiting the same material properties (polypropylene appropriate for additive manufacturing). The flat equilibrium configuration is therefore not explicitly included due to the associated snap-through corresponding to the neglected geometry-dependent conditions. This simplified approach is adopted to investigate the feasibility of exerting simultaneous control over bi-stable component positioning, internal topology and aerodynamic characteristics of the final morphed shape. The method can subsequently be extended to include the aforementioned constraints, combined with the upscaling possibility of the complete wing section.

It should be noted that only the trailing edge displacements of the actuated stiff aerofoil mode are considered, without the corresponding final aerodynamic coefficients. The reason is the focus on establishing the impact of bi-stable element positioning on the difference in response between the flexible and stiff configurations, without the practical intention of actuating the latter.

As can be seen from Figure 5 showing example topologies, combining a more complex internal substructure covered by a skin with a corrugated portion, and selective stiffness reduction provided by the embedded bi-stable laminates allows an aerodynamically smooth morphed shape to be maintained. This is confirmed by the corresponding aerodynamic coefficients obtained from the static aeroelastic simulations presented in Figure 6, which also provides the characteristics of a rigid NACA 0012 with a discrete flap at an angle of between 5° and 12.5° computed with XFOIL for comparison purposes. It can be seen that promising results are obtained for the lift variation achieved through morphing relative to the associated drag, with all but one topologies predicted to fulfil the goal of performing better than a discontinuous trailing edge device. The aerodynamic coefficients presented in Figure 6 are primarily intended to serve quantification and comparison purposes and therefore the limitations of the two-dimensional XFOIL code should be borne in mind when interpreting the results. First, no influence of the corrugated skin portion of the morphing aerofoil is included in the static aeroelastic simulations, which is expected to lead to a slight drag penalty [24]. On the other hand, the gap drag caused by the trailing edge flap also falls beyond the scope of XFOIL.

The impact of the flexible state of the bi-stable elements ranges from 6.6 to 22.7, as summarised in Table 1. This confirms the considerable contribution of the laminates to the stiffness reduction, and hence a less energy-intensive actuation possibility.

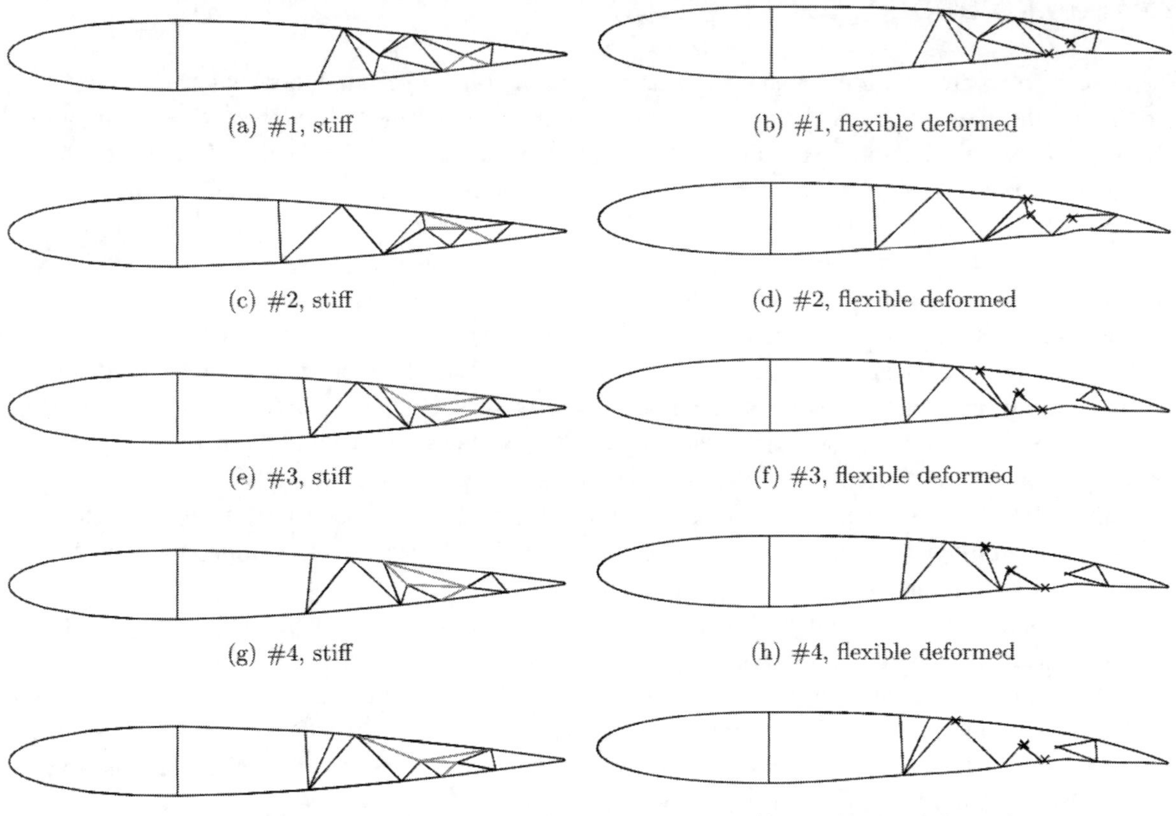

(a) #1, stiff

(b) #1, flexible deformed

(c) #2, stiff

(d) #2, flexible deformed

(e) #3, stiff

(f) #3, flexible deformed

(g) #4, stiff

(h) #4, flexible deformed

(i) #5, stiff

(j) #5, flexible deformed

Figure 5. Example topologies studied with the corresponding final morphed shapes obtained from static aeroelastic analyses for downward MFC actuation at 20 m/s and an angle of attack of 0°. Members replaced with bi-stable elements in the flexible state marked in light blue in the left column. Please note the absence of deformed bi-stable elements in the right column due to lack of post-processing for `UEL`s.

6. CONCLUSIONS AND OUTLOOK

A novel, concurrent design methodology of morphing aerofoils with stiffness variability provided by monolithically embedded bi-stable laminates is presented. The approach involves a complex distributed topology with selected members replaced with bi-stable components, tailored skin stiffness with a corrugated portion, and global actuation inducing the shape adaptation in the low-stiffness state. Further, static aeroelastic response is investigated through a weak coupling between the commercial finite-element solver Abaqus/Standard® and the aerofoil analysis code XFOIL.

Representing the stiffness properties of the bi-stable components by means of user-defined finite elements greatly enhances the design freedom and computational efficiency. The static aeroelastic characteristics of the globally actuated configurations demonstrate that the simultaneous consideration of the aforementioned aspects allows the response of the aerofoil to be tailored to specific requirements.

Specifically, an aerodynamically smooth solution can be obtained while aiming at maximised degree of shape adaptation in the flexible wing section mode.

The results confirm the potential of the presented design methodology to realise selectively compliant morphing structures based on strategically positioned bi-stable laminates providing variable stiffness properties. This allows for further generalisation of the proposed concurrent aero-structural approach to multidisciplinary optimisation.

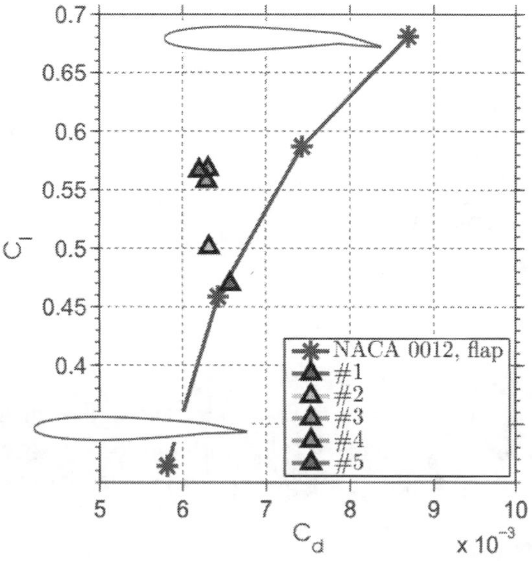

Figure 6. Aerodynamic coefficients of the final morphed shapes of selected topology concepts obtained from static aeroelastic analyses for downward MFC actuation at 20 m/s and an angle of attack of 0°. Characteristics of a rigid NACA 0012 with a downward flap of 5°–12.5°obtained from XFOIL added for comparison.

Table 1 Result summary for selected topologies.

	$u_{3,\mathrm{TE}}^{\mathrm{flex}}$ mm	C_l –	C_d –	$u_{3,\mathrm{TE}}^{\mathrm{stiff}}$ mm	$u_{3,\mathrm{TE}}^{\mathrm{flex}}/u_{3,\mathrm{TE}}^{\mathrm{stiff}}$ –
#1	-9.3	0.4694	0.00657	-1.1	8.6
#2	-11.7	0.5008	0.00632	-0.7	16.3
#3	-13.9	0.5678	0.00631	-2.1	6.6
#4	-13.4	0.5577	0.00629	-0.6	22.7
#5	-13.5	0.5662	0.0062	-0.7	19.7

$u_{3,\mathrm{TE}}^{\mathrm{flex}}$, $u_{3,\mathrm{TE}}^{\mathrm{stiff}}$: vertical trailing edge displacements of actuated flexible and stiff aerofoil modes

ACKNOWLEDGEMENTS

The authors gratefully acknowledge the support of the Swiss National Science Foundation [project number 147059]. Dr. A. F. Arrieta was partially funded through an ETH Zurich Postdoctoral Fellowship.

REFERENCES

[1] S. Kota, J. Hetrick, Z. Li, and L. Saggere. Tailoring unconventional actuators using compliant transmissions: design methods and applications. *IEEE/ASME Transactions on Mechatronics*, 4(4):396–408, December 1999.

[2] L. L. Howell, S. P. Magleby, and B. M. Olsen, editors. *Handbook of Compliant Mechanisms*. John Wiley & Sons Ltd, 2013.

[3] L. F. Campanile. Smart shape control: using compliant and active materials to adapt structural geometry. Challenges and good reasons. In *14th International Conference on Adaptive Structures and Technologies (ICAST), Seoul, Korea, October 2003*, 2003.

[4] L. F. Campanile. Initial Thoughts on Weight Penalty Effects in Shape-adaptable Systems. *Journal of Intelligent Material Systems and Structures*, 16(1):47–56, January 2005.

[5] L. F. Campanile. *Adaptive Structures: Engineering Applications*, chapter 4. Lightweight Shape-Adaptable Airfoils: A New Challenge for an Old Dream, pages 89–135. John Wiley & Sons, Ltd, 2007.

[6] A. F. Arrieta, O. Bilgen, M. I. Friswell, and P. Hagedorn. Passive load alleviation bi-stable morphing concept. *AIP Advances*, 2(3):032118, 2012.

[7] A. F. Arrieta, I. K. Kuder, T. Waeber, and P. Ermanni. Embeddable Variable Stiffness Multi-Stable Composites. In *ICAST2013: 24th International Conference on Adaptive Structures and Technologies, October 7–9, 2013, Aruba*, 2013.

[8] A. F. Arrieta, I. K. Kuder, T. Waeber, and P. Ermanni. Variable Stiffness Characteristics of Embeddable Multi-stable Composites. *Composites Science and Technology*, 97:12–18, 2014.

[9] A. F. Arrieta, I. K. Kuder, M. Rist, T. Waeber, and P. Ermanni. Passive Load Alleviation Aerofoil Concept with Variable Stiffness Multi-stable Composites. *Composite Structures*, 116(September–October):235–242, 2014.

[10] I. K. Kuder, A. F. Arrieta, and P. Ermanni. Design space of embeddable variable stiffness bi-stable elements for morphing applications. *Composite Structures*, 122:445–455, 2015.

[11] M. Drela and H. Youngren. *XFOIL 6.9 User Primer*. Massachusetts Institute of Technology, Cambridge, Massachusetts, USA, 2001.

[12] I. K. Kuder, A. F. Arrieta, M. Rist, and P. Ermanni. Aeroelastic response of a selectively compliant morphing aerofoil featuring integrated variable stiffness bi-stable laminates. *Journal of Intelligent Material Systems and Structures*, under review, 2015.

[13] I. K. Kuder, A. F. Arrieta, M. Rist, and P. Ermanni. Static Design of a Selective Compliance Morphing Wing Profile Based on Integrated Variable Stiffness Bi-Stable Components. In *Proceedings of the ASME 2014 Smart Materials, Adaptive Structures and Intelligent Systems SMASIS2014, September 8–10, 2014, Newport, USA*, 2014. SMASIS2014-7493.

[14] G. Molinari, A. F. Arrieta, and P. Ermanni. Aero-Structural Optimization of Three-Dimensional Adaptive Wings with Embedded Smart Actuators. *AIAA Journal*, 52(9):1940–1951, September 2014.

[15] G. Molinari, M. Quack, A. F. Arrieta, M. Morari, and P. Ermanni. Design, realization and structural testing of a compliant adaptable wing. *Smart Materials and Structures*, 24(10):105027, 2015.

[16] F. Previtali, A. F. Arrieta, and P. Ermanni. Performance of a Three-Dimensional Morphing Wing and Comparison with a Conventional Wing. *AIAA Journal*, 52(10):2101–2113, 2014.

[17] F. Previtali, G. Molinari, A. F. Arrieta, M. Guillaume, and P. Ermanni. Design and experimental characterisation of a morphing wing with enhanced corrugated skin. *Journal of Intelligent Material Systems and Structures*, 2015.

[18] P. Portela, P. Camanho, P. Weaver, and I. Bond. Analysis of morphing, multi stable structures actuated by piezoelectric patches. *Computers & Structures*, 86(3–5):347–356, February 2008.

[19] Dassault Systèmes Simulia Corp. *Abaqus Analysis User's Manual*, 2014. Version 6.14.

[20] Dassault Systèmes Simulia Corp. *Abaqus User Subroutines Reference Guide*, 2014. Version 6.14.

[21] A. Khennane. *Introduction to Finite Element Analysis Using MATLAB and Abaqus*. CRC Press, 2013.

[22] N. I. Obodan, O. G. Lebedeyev, and V. A. Gromov. *Nonlinear Behaviour and Stability of Thin-Walled Shells*. Springer, 2013.

[23] R. Seydel. *Practical Bifurcation and Stability Analysis*. Springer New York, 2010.

[24] G. Molinari, A. F. Arrieta, and P. Ermanni. Numerical and experimental investigation of aero-structural characteristics and performance of distributed compliance morphing wings. In *23rd AIAA/AHS Adaptive Structures Conference, January 5–9, 2015, Kissimmee, FL, USA*, pages 1–14, 2015.

ICAST2015: 26th International Conference on Adaptive Structures and Technologies
October 14-16th, 2015, Kobe, Japan

ICAST2015 041

On an efficient implementation of non-linear structural optimization for the morphing leading edge of an active blown high lift system.

Michael Rose[1*], Anton Rudenko[1], and Hans Peter Monner[1]

[1] Institute of Composite Structures and Adaptive Systems, German Aerospace Center (DLR), Braunschweig, Germany.

Abstract

The permanently increasing demands on performance and noise reduction in today's aircraft high lift systems enforces the study of adaptive technologies to overcome the restrictions of state of the art fully passive technologies. As an alternative to conventional slats and flaps, an active blown Coandă-flap based high lift system is investigated within the German national Collaborative Research Centre 880 to improve on the mentioned limitations. One key feature is provided by an adaptive gapless droop nose with an exceedingly high grade of leading edge morphing. The construction of this component is based on a structural optimization framework, developed at DLR. The framework consists of two hierarchical design steps, an optimization of the skin layout with discrete joints to the inner actuation mechanism and the topology optimization of the latter.

In this paper, the methodology of the second step and its application to the droop nose is discussed. For efficiency reasons, the design of the inner mechanism is completely implemented in the matrix software environment MATLAB. A key component is given by a Finite Element based non-linear solver for the static calculation of the stiffness distribution on a regular grid. As part of the inner loop of this design process, this solver must be implemented as efficient as possible. Due to the lack of sensitivity information in commercial Finite Element programs, the direct implementation of the elasticity equations is further enforced. A main focus of this article concentrates on the underlying assembly approach for the needed system matrices. Finally the application of the tool chain to the droop nose compliant mechanism is presented.

1. INTRODUCTION

The flow control system shown in Figure 1 takes advantage from the Coandă-effect to enhance the flight characteristics of an airplane wing in different flight conditions. It is one of the considered approaches within the framework of the Collaborative Research Center 880 (SFB880). An important component consists of a device capable to adapt its shape according to aerodynamic requirements without any structural gaps or steps. There have been many aerospace research activities in the field of morphing aircraft structures, starting with the flight of the Wright Flayer I. Good reviews are given in [1, 2].

This paper discusses the shape and camber morphing of the airfoil. In constract, the established technique of wing camber changing to enhance the low speed capabilities in commercial aircrafts results in

* info@icast2015.com

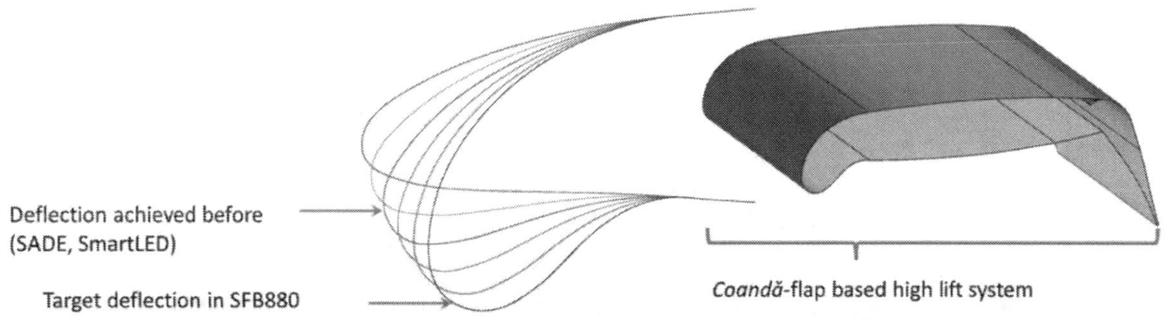

Figure 1. Coandă-flap based high lift system and the deflection of the adaptive droop nose

complex mechanical systems without the ability to provide a stepless aerodynamic shape. The substitution of such systems by less complex adaptive structures without any gaps and steps is very promising to achieve improvements in efficiency. Though such kind of leading edge devices have been successfully presented in [3, 4] as a result of the European project "Smart High Lift Devices for Next Generation Wings" (SADE), the achievable deflection is limited by the considered design. Therefore the considerably higher level of non-linear morphing, which is needed by the Coandă-flap based high lift system with circulation control as described in [5], presents an additional challenge in SFB880. From the structural point of view, this challenge can be separated in some research domains:

- Development of a skin that is flexible enough to sustain the desired shape change and is also stiff enough to carry the aerodynamic loads.
- Design and creation of a morphing mechanism for transferring actuator forces to the skin and aerodynamic loads to the wing structure.
- Definition of an appropriate interface between the skin and the mechanism to provide the desired deformation accordingly to the aerodynamically reasonable target shape.

The skin concept is developed at the Institute of Aircraft Design and Lightweight Structures (Technical University of Brunswick) within the framework of SFB880 and the design relies on input provided by the present work, consisting of the stiffness distribution of the skin. To reach the morphing demand, highly anisotropic materials in combination with enhanced stiffness tailoring is used to overcome the restrictions of conventional composite skins. The morphing mechanism is investigated at the Institute of Composite Structures and Adaptive Systems (German Aerospace Center).

The goal of the topology optimization is to exchange conventional mechanisms like the ones shown in Figure 2 with inner compliant mechanisms to reduce the weight and the need for inner joints simultaneously. The discussion of the methodology and the current status is a central point in this paper.

2. A PROCESS CHAIN FOR STIFFNESS AND TOPOLOGY OPTIMIZATION

Optimization of the skin stiffness distribution has doubtless to be coupled to the optimization of the inner kinematics of the droop nose device, because these are parts of one elastic system and have decisive influence on the shape quality. On the other hand, these two parts have some different requirements

Figure 2. Demonstrator of an adaptive droop nose in normal flight (left) and high lift (right) configuration

towards boundary conditions, applicable optimization methods and parameters: the most important demand on the skin is to provide a reasonable curvature according to the given aerodynamic target shape, but the inward actuating mechanism is more restricted by material loads and weight considerations.

In keeping with this, a two-stage optimization framework shown in Figure 3 (left) was developed to combine these requirements. The first part of the framework handles the stiffness tailoring of the skin and interface elements between the skin and kinematics, described in section 3.

To perform further optimization of the inner kinematics, the skin stiffness data has to be stored and provided to the next optimization step. For this reason, the stiffness matrix of the skin and the force vectors of the actuation nodes have to be stored in so-called "superelements" — substructures, that means in matrices which describe the elastic behavior of the structure in the working point. The substructuring procedure can handle large deflections and rotations of the given problem as long as the local strains remain low (about one percent in the present work). Further description of the substructuring approach for non-linear solutions can be found in [6, 7]. With this structural information, the next optimization step, described in section 4 can be initiated.

3. SKIN STIFFNESS TAILORING

The skin optimization tool is based on a finite element implementation of the droop nose skin with spanwise oriented stringers acting as force application points for the inward kinematic. The skin is modelled with shell elements, allowing the application of composite and hybrid materials. The design parameters of the skin stiffness optimization routine are:

- the stiffness distribution along the airfoil circumference (variable GFRP-layup)
- the locations of the kinematic - skin interfaces
- the direction and amount of the actuation forces at these locations

The objective function is given by the distance between the calculated skin deformation and the target shape provided by a foregone aerodynamic optimization. The optimization constraint is the maximal

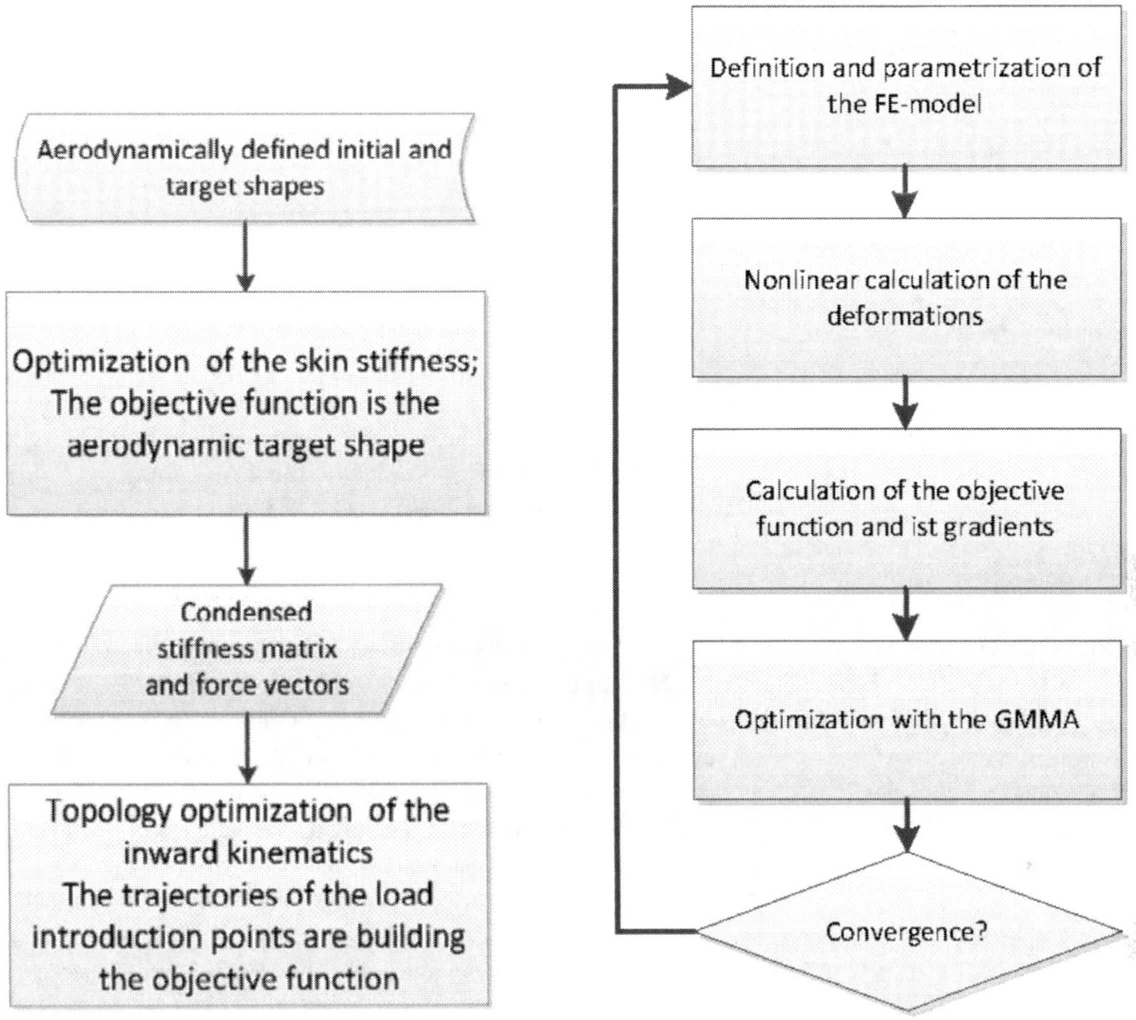

Figure 3. The subdivision of the optimization process chain (left) and the workflow of the topology optimization (right)

elastic strain of the material:

$$\frac{\epsilon_\ell}{\epsilon_{\max}} - 1 \leq 0 \tag{1}$$

Strains and deformations of the skin are calculated with a non-linear ANSYS solver and the optimization is performed with MATLAB-integrated simplex algorithm. An example result of the optimization is presented in Figure 4 (left).

Another important question is the kind and the geometry of the force introduction elements between the skin and the kinematics. In previous projects omega-stringers were chosen mainly because of the robustness and structural strength of this concept. On the other hand, an omega-stringer on a flexible morphing skin results in strong discontinuity of the skin stiffness. A resulting effect on the shape curvature of the morphed airfoil is disadvantageous for the aerodynamic flow and therefore has to be

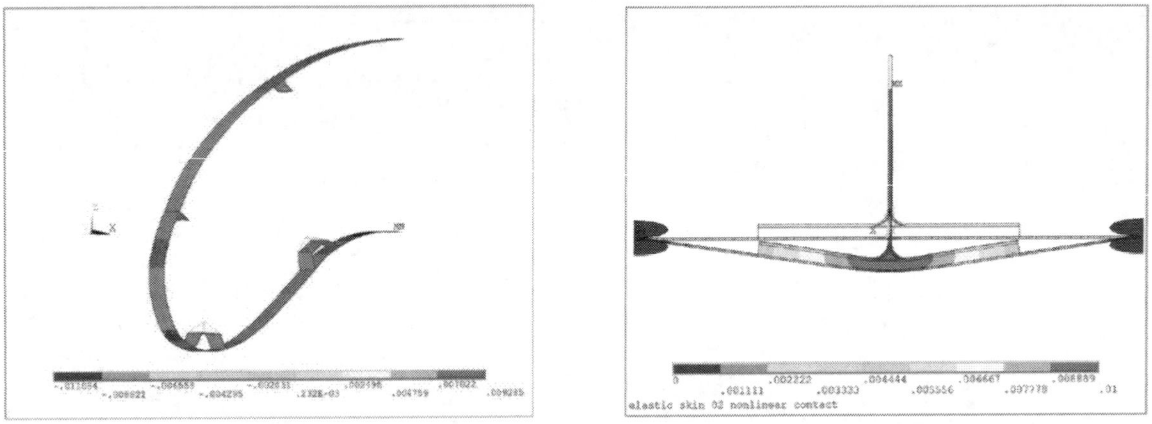

Figure 4. An example of a deformed skin with Omega- and T- stringers, acting as force introduction elements (left) and three-point bending model of a T-stringer on the flexible hybrid elastomer-GFRP skin (right)

minimized.

Based on these considerations a sensitivity analyses was performed to determine an optimal number and type of the force introduction elements to be applied. Additionally to Omega-stringers, T-, L- and I-stringer types were implemented in the optimization loop to investigate the curvature effects.

Some selected results of this analysis are shown in Figure 5, where two optimized droop nose skins with different force introduction concepts are compared regarding the deviation from the target shape and its curvature. It is identifiable that, although the deviations from the target shape are quite small for both concepts, the shapes of the curvature graphs are rather different: The curvature plot of the concept in the Figure 5 (left) shows a generally less continuous behavior with a zero passage (marked with a red circle) by 45 percent of the skin perimeter. Such behavior (especially the zero crossing) of the curvature is well known for disturbance of the aerodynamic flow by disadvantageous pressure gradients. Whereas concept presented in the Figure 5 (right) provides clearly better curvature and can be identified as more appropriate for the morphing skin application.

Having better capabilities as force introduction elements regarding the curvature, T-stringers are less reliable than Omega-stringers regarding the structural strength and stability. To prove the concept, the T-stringer was implemented in a parametric high fidelity finite element model with layered solid elements, shown in Figure 4 (right).

4. TOPOLOGY OPTIMIZATION OF THE INWARD KINEMATICS

For the topology optimization of the complained kinematic mechanism, the Solid Isotropic Material with Penalization (SIMP) was applied. This method is described by Sigmund and Bendsøe in [8, 9] for linear finite element calculations and was extended for non-linear systems. A further step to non-linear analysis like it was applied in this work was presented by Saxena and Ananthasuresh in [10]. The workflow of the optimization routine is described in Figure 3 (right). In order to have the necessary full access to the stiffness matrix and element densities of the compliant kinematic, a proprietary non-linear finite element grid solver was implemented in C/Matlab. The solver allows calculating of grid

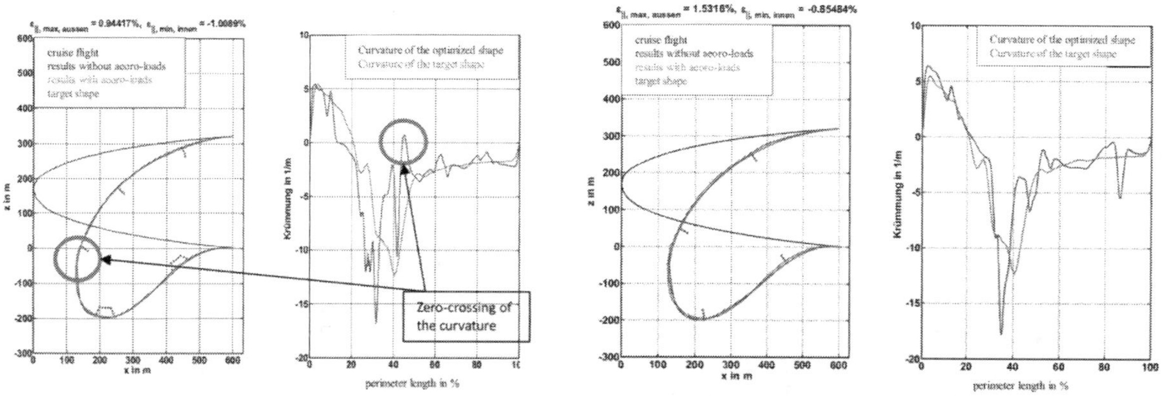

Figure 5. Results of the skin optimization for different force application concepts: 2xOmega- and 3xT- stringer (left), 4xT-stringer (right)

nodes deformations with respect to the actuator forces and has additionally an integrated procedure for determining the sensitivities of topology optimization design parameters presented in equation (5).

The objective function Z for the compliant kinematics is defined as a normed difference between the target position of the force application nodes U_{target} and the nonlinear finite element solution U which is calculated in the first optimization step:

$$Z = ||U_{\text{target}} - U|| = \left(U_{\text{target}} - U\right)^T \boldsymbol{T}\left(U_{\text{target}} - U\right) \tag{2}$$

The matrix $\boldsymbol{T}$ is used here to select the degrees of freedom that are to be used for the formulation of the objective function. These in particular are the degrees of freedom of the elements that contain the force introduction points from the first optimization step. Due to the fact that a SIMP approach prescribes a grid of equal (rectangular) elements, an additional interpolation of the force introduction points on the skin into the elements of the grid has to be performed. Therefore an interpolation matrix with Lagrangian polynomials is implemented according to equations (3) and (4) and also shown in Figure 6.

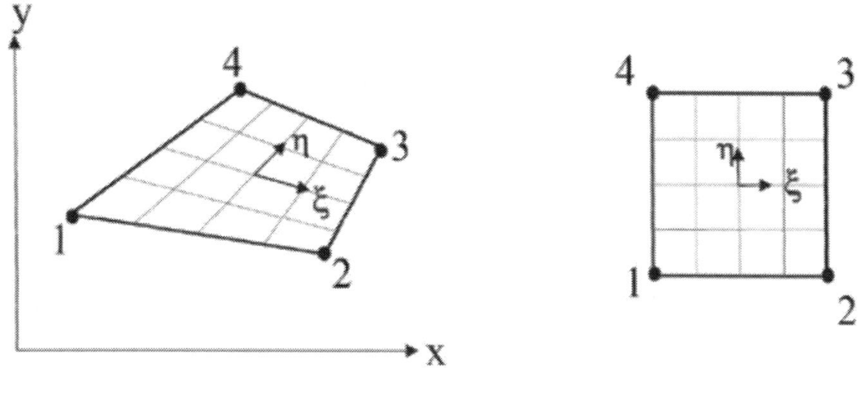

Figure 6. Cartesian and element coordinate systems in a deformed element

$$N_L = L_i^M(\xi)L_j^N(\eta) \tag{3}$$

$$U_i = N_L(\xi,\eta)\hat{x} \tag{4}$$

where U_i describes the position of the force introduction point inside of the element as a function of the preselected element nodes positions $\hat{x} = \boldsymbol{T}U$.

The sensitivities of the objective function Z with respect to the design variables x can be defined as

$$\frac{dZ}{dx} = -2\left[\dot{f} - \sum p\rho^{p-1} f_{\rho,el}(u)\right]^T \cdot \boldsymbol{K}_{TM}^{-1} \cdot \boldsymbol{T}\big(U_{\text{target}} - U\big). \tag{5}$$

Further definitions here are:

- $\boldsymbol{K}_{TM}^{-1}$ as an inverse tangential stiffness matrix of the finite element grid
- $\dot{f}$ as derivative of the actuator forces with respect to the design variables
- ρ as a pseudo density of a finite element
- a penalty factor p
- the non-linear stiffness of an element $f_{\rho,el}(u)$

The following finite element based approach has been developed in order to efficiently derive the matrices for the underlying static non-linear elasticity problem. The elastic constitutive equations consist of the usual linear equations for isotropic material under planar stress conditions

$$\begin{bmatrix}\sigma_x\\ \sigma_y\\ \tau_{xy}\end{bmatrix} = \rho^p \frac{D}{h}\begin{bmatrix}1 & \nu & \\ \nu & 1 & \\ & & (1-\nu)/2\end{bmatrix}\begin{bmatrix}\varepsilon_x\\ \varepsilon_y\\ \gamma_{xy}\end{bmatrix} \quad \text{with} \quad D = \frac{Eh}{1-\nu^2} \tag{6}$$

with the Young modulus E, the Poisson ratio ν, the membrane thickness h to couple the second Piola stress tensor σ with the Cauchy Green strain tensor ε in a geometrically non-linear way. Therefore we have

$$\varepsilon_x = u_{,x} + (u_{,x}u_{,x} + v_{,x}v_{,x})/2, \tag{7}$$

$$\varepsilon_y = v_{,y} + (u_{,y}u_{,y} + v_{,y}v_{,y})/2, \tag{8}$$

$$\gamma_{xy} = u_{,y} + v_{,x} + (u_{,x}u_{,y} + v_{,x}v_{,y}). \tag{9}$$

The virtual stiffness energy is then given by

$$\delta W_{\text{stiff}} = h\int_A \underbrace{\delta\varepsilon_x\sigma_x + \delta\varepsilon_y\sigma_y + \delta\gamma_{xy}\tau_{xy}}_{=\delta w_{\text{stiff}}}\, dA \approx h\sum_{e=1}^{m}\sum_{k=1}^{n_e}\left[\delta w_{\text{stiff}} \underbrace{\frac{\partial(x,y)}{\partial(\xi,\eta)}}_{=J} q_k^e\right]_{(\xi=\xi_k^e,\eta=\eta_k^e)} . \tag{10}$$

For the SIMP approach we use an elementwise constant density $\rho \in [0,1]$ to establish a continuously varying stiffness between no material ($\rho = 0$) and full material ($\rho = 1$). A typical value for the penalty factor would be $p = 3$ in order to tag intermediate stiffness values with a penalty in the optimization process. The approximation in equation (10) uses numerical integration by suitable n_e Gauss points with weights q_k^e in local coordinates (ξ_k^e, η_k^e) for each of the m elements. The usual finite element based approach would be to generate deformation dependent element stiffness matrices, which have to be assembled into the global stiffness matrix $\boldsymbol{K}_{TM}$. In a software environment like *MATLAB* with efficient vector based calculation routines it is hard to obtain an efficient solution for such assembly method. A careful observation of equation (10) after substitution of the constitutive equation (6) and subsequently the strains from (7–9) reveals, that the virtual stiffness energy δW_{stiff} consists of a sum of terms

$$D \sum_{e=1}^{m} \sum_{k=1}^{n_e} q_k^e f_{e,k}^{[1]} \cdot \dots \cdot f_{e,k}^{[r]} \tag{11}$$

where each $f^{[j]}$ is either a function δu, δv, u, v, x, y, ρ^p or a derivative of such a function with respect to ξ or η, evaluated at all Gauss integration points for each element. The whole assembly process can eventually be formulated by the combination of four basic tasks:

- The projection $\mathcal{P}$ of a field u, v, x, y, ρ or its derivative onto the Gauss integration points using the vector of elementwise constant densities $\boldsymbol{\rho}$ and the nodal deformations $\boldsymbol{u}$, $\boldsymbol{v}$.
- The multiplication $\mathcal{M}$ of vectors $\boldsymbol{f}^{[j]}$, which is an efficient pointwise operation.
- The assembly $\mathcal{A}$ of vectors on the Gauss points by a scheme suitable to respect the variations δu, δv or its derivatives. To obtain sensitivity information with respect to densities an assembly of $\delta\rho$ is also needed.
- The sparsity operator $\mathcal{S}$ for a pair of field types as explained in equation (12) to generate sparse system matrices.

$\mathcal{P}$ and $\mathcal{A}$ are adjunct rather than inverse operators. If for example $\boldsymbol{u}$, $\boldsymbol{v}$ are nodal values for the fields u, v and $\boldsymbol{f}$ is a vector of values on the Gauss integration points, we have

$$\left(\mathcal{P}_u[\boldsymbol{u}]\right)^T \boldsymbol{f} = \boldsymbol{u}^T \mathcal{A}_u[\boldsymbol{f}], \quad \left(\mathcal{P}_u[\boldsymbol{u}]\right)^T \operatorname{diag}(\boldsymbol{f}) \mathcal{P}_v[\boldsymbol{v}] = \boldsymbol{u}^T \mathcal{S}_{u,v}[\boldsymbol{f}] \boldsymbol{v}. \tag{12}$$

Because the interpolation of the fields is done on a regular nodal grid for the standard SIMP approach, these projection, assembly and sparsity operators can be implemented very efficiently. We have implemented these rather general operators in the programming language *C* as so called *MEX*-extensions to *MATLAB*. There are several field orders implemented. We currently only use the zero order for elementwise constant density fields and the first order bilinear interpolation for the deformation fields u and v. The implementation is also simultaneously done for one-dimensional, two-dimensional and three-dimensional domains. Again currently only the second case is used. The charme of this approach is its generality. The Ritz/Galerkin matrix equations of a huge class of partial differential equations can be coded based on these basic operations in a simple and efficient way. Even the memory consumption of this approach has the same order as classical finite element codes. And special needs as for example sensitivity information is also easily implementable. One limitation of this method is the use of a regular grid as basic mesh. This is no fundamental restriction as the operators have been successfully generalized in an efficient way for heterogeneous meshes. Another limitation is more severe. We only consider one element type in each single assembly process. Therefore different finite element types have to be assembled seperately. And the vector and matrix operations needed for more complicated elements like beam and shell elements must be expanded into scalar equations, which reduces efficiency and enhances coding complexity simultaneously. Luckily, for topology optimization all these restrictions are usually not present, leading to a very efficient assembly procedure of the needed matrix equations for the elasticity equations.

An evaluation of the objective function and its derivatives is performed by the globally convergent version of the method of moving asymptotes (GCMMA). A detailed explanation of the method and its application in a topology optimization chain can be found in [9]. Depending on the defined convergence criteria (in this case a maximum allowable deviation from the target trajectories), the framework proceeds with updating the design parameter vector x or provides an optimized topology as a pre-design solution for further detailed design of the compliant kinematic. A preliminary result of the topology optimization is shown in Figure 7. In this case a compliant kinematic with one variable actuator force (red arrow) and one target displacement, representative for a force introduction point of the skin was optimized. It is clearly observable that the target position of the control point is reached by the mechanism with quite small deviation.

The implementation of a complete set of boundary conditions representative for a droop nose skin with appropriate constraints (number of force application points, advanced actuation system, installation space) will follow in the next future. For the latter more ambitious goals, the objective function and possibly also the optimization routine must be modified in order to achieve convergence, because the lack of convergence was observed for multi point targets of deflection shapes with highly non-linear deformations.

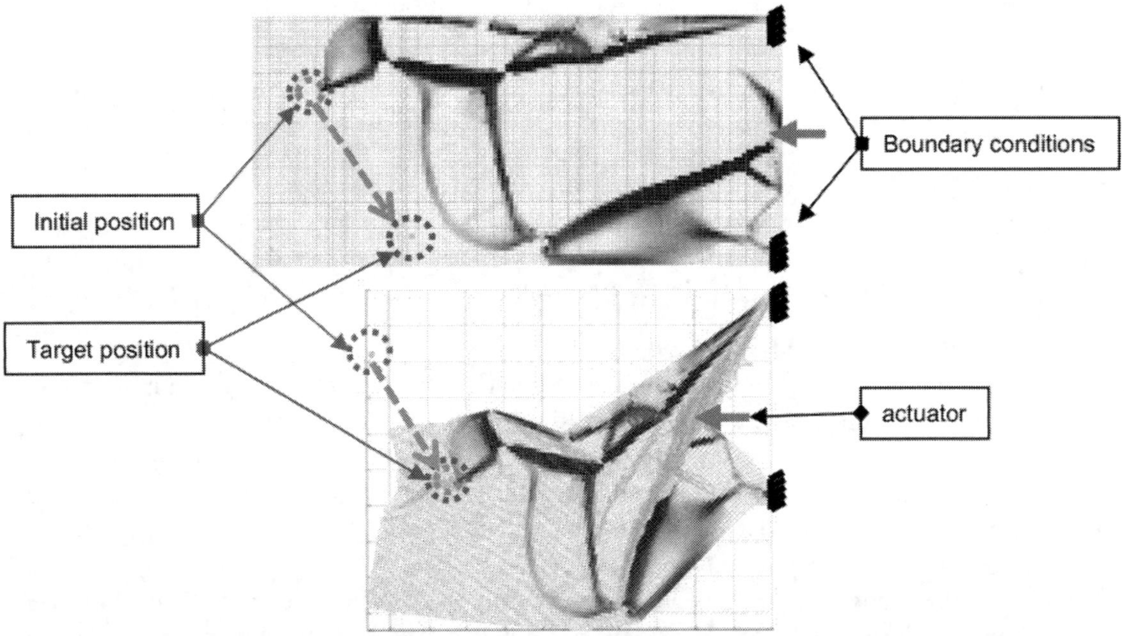

Figure 7. Preliminary results of the topology optimization

5. CONCLUDING REMARKS

A structural optimization framework is set up for an adaptive leading edge device with a high grade of deflection. The maximal deviation from aerodynamic reasonable shapes for different flight (deflection) conditions provides the objective function for the optimization. Two different optimization methods are combined to obtain a morphing solution: a fast and robust simplex optimization of the skin stiffness distribution and a gradient based topology optimization of the non-linear deformed compliant kinematic system. The derivation of the compliant inner mechanism, which currently works for linear elasticity problems, but still has convergence problems for multi point highly non-linear problems must be enhanced regarding the objective function and/or the underlying optimizer, but the efficient underlying solution procedures for the elasticity subproblem have been proven to be mature. The neccessary post processing of the derived mechanisms is another ongoing work. The tool chain itself is already very promising and extensions to three-dimensional problems are also planned.

ACKNOWLEDGEMENTS

Financial support has been provided by the German Research Foundation (Deutsche Forschungsgemeinschaft — DFG) in the framework of the Sonderforschungsbereich 880.

REFERENCES

1. Barbarino, S., Bilgen, O., Ajaj, R. M. and Friswell, M. I., "A Review of Morphing Aircraft", *Journal of Intelligent Material Systems and Structures*, Vol. 22, 2011, pp. 823–899.
2. Sofla, A. Y. N., Meguid, S. A., Tan, K. T. and Yeo, W. K., "Shape morphing of aircraft wing: Status and challenges", *Material & Design - MATER DESIGN*, Vol. 31, 2010, pp. 1284–1292.
3. Kintscher, M., Wiedemann, M., Monner, H. P., Heintze, O. and Kühn, T., "Design of a smart leading edge device for low speed wind tunnel tests in the European project SADE", *International Journal of Structural integrity*, Vol. 2, No. 4, 2011, pp. 383–405.
4. Kintscher, M. and Wiedemann, M., *Design of a smart leading edge device. Adaptive, Tolerant and Efficient Composite Structures*, Springer Verlag, New York 2012, pp.381–390.
5. Burnazzi, M. and Radespiel, R., "Design of a Droopnose Configuration for a Coandă Active Flap Application", *51st AIAA Aerospace Sciences Meeting Grapevine*, Texas, January 2013.
6. Dodds, R. H. and Lopez, L.A., "Substructuring in Linear and Nonlinear Analysis", *International Journal for Numerical Methods in Engineering*, Vol. 15, 2013, pp. 583–597.
7. Hellen, T. K., "Use of substructuring in non-linear material analysis.", *Engineering Computations*, Vol. 1, No. 4, pp. 343–350.
8. Sigmund, O., "On the Design of Compliant Mechanisms Using Topology Optimization", *Mechanics of Structures and Machines: An International Journal*, Vol. 25, No. 4, 1997, pp. 493–524.
9. Bendsøe, M. and Sigmund, O., *Topology Optimization Theory, Methods and Applications*, Springer Verlag, New York, 2nd ed. 2003.
10. Saxena, A. and Ananthasuresh, G. K., "Topology Synthesis of Compliant Mechanisms for Nonlinear Force-Deflection and Curved Path Specifications", *Journal of Mechanical Design*, Vol. 123, 2001, pp. 33–42.

ICAST2015 #014

Application of a New Decoupling Loop Shaping Control Strategy for Load Alleviation of Smart Rotor Blades Equipped with Multiple Trailing Edge Flaps

Terence Macquart[1*], Sachin Navalkar[2] and Roeland De Breuker[3],

[1,3]Faculty of Aerospace Engineering, Delft University of Technology, Delft, The Netherlands.
[2]Delft Center for Systems and Control, Delft University of Technology, Delft, The Netherlands.

Abstract

Wind turbines are fatigue critical machines for which the problem of vibrations is becoming predominant as rotor sizes keep increasing. Whereas the load reduction potential of wind turbine smart blades with trailing-edge control devices is now well-recognised, the design of load alleviation control systems which are efficient, robust and yet simple remains an open problem. The present paper evaluates the efficiency of a new proposed decoupled loop-shaping control (DLSC) strategy for the load alleviation of wind turbine blades equipped with multiple control surfaces. The efficiency of the proposed decoupling control strategy is evaluated against the well-established model predictive controller (MPC). For that purpose, an in-house wind turbine aeroelastic control (WTAC) simulator is employed to simulate the aeroelastic dynamics of the NREL 5 MW wind turbine operating in a fully turbulent environment. WTAC incorporates a wind field generator, an aerodynamic model and a finite element structural blade model. Trailing edge flaps are used as active control surfaces and are dynamically coupled to the aeroelastic wind turbine blade models.

First, a state-of-the-art MPC is designed to address the load alleviation control problem. MPC is specifically chosen on account of its multi-input multi-output (MIMO) nature and constraints handling capabilities. Although high load alleviation efficiency is achieved, the MPC remains a complex time-based state controllers. Next, the proposed frequency-based loop-shaping controller is applied to the same control problem. The DLSC takes advantages of the dynamic response of blades to reformulate the MIMO load alleviation control problem as decoupled single-input single-output (SISO) control loops. Classic loop-shaping control theory is then used to design SISO controllers. The DLSC performance is compared with that of the MPC based on 1P blade root load bending moment reduction. The results of this investigation demonstrate the efficiency of the proposed DLSC strategy with respect to the MPC controller. The advantageous simplicity and insightful features of the DLSC are also highlighted.

* T.B.M.J.Macquart@tudelft.nl

1. INTRODUCTION

Driven by the increasing demand for renewable energy sources, the wind energy market has grown substantially [1]. This growth has triggered the desire for reduced cost of wind energy and improved wind turbine designs, which in turn has revealed the critical need for innovation. Smart rotors are the results of this need for innovation in wind turbine design [2]. With their growing sizes, wind turbine rotors and particularly blades are becoming increasingly prone to fatigue. By actively responding to the flow via a control loop, smart rotors have the capability to reduce fatigue equivalent damage loads, and therefore decrease the cost of wind energy through increased wind turbines lifespan.

The fatigue increase, exacerbated by the growing size of wind turbine rotors, cannot be fully relieved by conventional wind turbine control systems [3]. Numerous load alleviation mechanisms, including individual pitch control and trailing edge flaps, have been proposed [2]. Evaluating and comparing the potential of these concepts has been a major research area over the past decade [4]. Bossanyi, Larsen et al. and, Lackner and van Kuik [4-6] have all significantly contributed to the advance of individual pitch control (IPC). In particular, they have highlighted the effectiveness of IPC for load alleviation as well as potential issues with rapid actuations which may wear excessively on actuators and could adversly affect the reliability and longevity of wind turbines. In comparison, local flow controllers such as control surfaces (CSs) are small and rapid devices distributed along the aeroelastic structure in order to provide control over the local aerodynamic forces. Amongst others, numerical and experimental work by Castaignet et al. [7, 8] have demonstrated the load alleviation potential of trailing edge flaps. Many others have also helped and supported the development and research of active load controllers on wind turbine blades [9-11]. While a necessary step towards the development of smart rotor technology, the design of control strategies for load alleviation remains difficult and more in-depth studies are now required in order to promote the benefits of active load control.

Current research follows the prevailing idea that increasingly complex control architectures such as optimal, multi-input multi-output (MIMO), and predictive control systems will outperform classical controllers for load alleviation [12, 13]. The effectiveness of model predictive control (MPC) in computing control strategies for non-intuitive dynamics such as complex MIMO control problems as well as systems with time delays and constraints is well-recognised [14]. However, issues such as accurate state space estimation, robustness to tuning, model uncertainties and disturbances-dominated dynamics are typical issues which are difficult to address [15]. In this paper, the authors demonstrate that well-designed simple classical controllers can perform as effectively as MPC for wind turbine blade load alleviation employing multiple trailing edge flaps (TEF). In particular, it will be shown that the original MIMO load alleviation control can be decoupled in SISO control problems. As a result, simple, robust and easy-to-implement classical controllers can be employed. The aim of the present paper is to provide supporting evidence to the successful application of the authors' proposed control strategy, namely the decoupling loop shaping control (DLSC) strategy. For that purpose, load alleviation results of the NREL 5 MW wind turbine [16] employing the DLSC strategy are compared against the well-established MPC.

2. NREL 5 MW BLADE LOAD ALLEVIATION - A PROBLEM STATEMENT

The wind turbine model used for this investigation is the variable speed, pitch-controlled 5 MW wind turbine described in the work of Jonkman et al. [16]. Fifteen percent of the blades span is equipped with three trailing edge flaps located in the blades aerodynamic region of efficiency between 75% and 90% of the blade span [17, 18]. The blade displacement is assumed measured through a strain sensor located at 15% of the blade span. Additionally each blade is equipped with two five-hole Pitot tubes used to estimate the local angles of attack and flow velocities in front of the trailing edge flaps as shown in **Figure 1**. The main wind turbine characteristics and control parameters are summarised in Table 1.

Table 1 - Wind turbine model parameters

General Characteristics	Hub height	87.6m	Trailing Edge Flaps	Number of Flaps	3
	Diameter	126 m		Maximum Flap Angle	±5°
	Blade length	61.5m		Covered Span (75% to 90%)	15%
	Blade mass	17 740kg		Maximum Flap Rate	±100°/s
	Number of blades	3			
	Rated speed	12.1rpm			
	Blade structural damping (in % of critical damping)	< 3%			
Sensors	Number of Strain Gages	1	Blade Natural Frequencies	1st Flapwise	0.7056 Hz
	Gages Locations(% of span)	15%		2nd Flapwise	2.0088 Hz
	Number of Pitot Tubes	2		1st Edgewise	1.0943 Hz
	Pitot Locations(% of span)	75% and 90%			

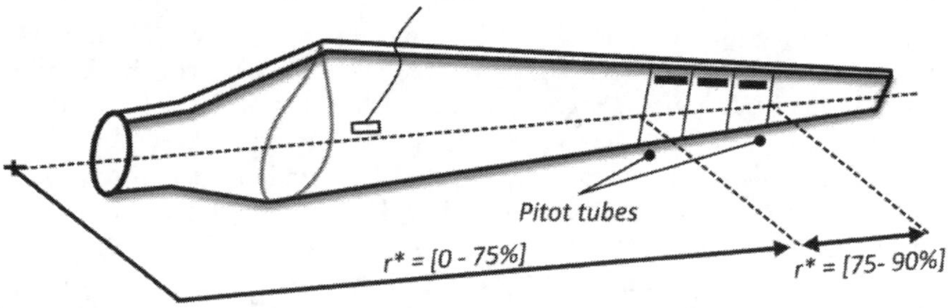

Figure 1 - Illustration of a wind turbine blade equipped with control surfaces, a strain gage sensor and Pitot tubes

Defining wind turbine blade load alleviation as a control problem is the first step towards the design of an active load control system. For that purpose, it is essential to first identify the primary loads responsible for the majority of damage equivalent loading. Wind turbine blades are subjected to various types of loads which mainly include aerodynamic, gravitational and inertia loads. The first (1P) cyclic loads, arising due to the cyclic motion of the blades in a non-axisymmetric wind field, are the prime cause of fatigue [12] as illustrated in **Figure 2**. Accordingly, it is reasonable to define the capability of an active load controller in terms of 1P load reduction. It should, however, be clearly stated that the reduction of 1P loads by the control system should not result in an excessive increase in other frequency bandwidths as shown by the desired spectrum curve in **Figure 2**. Consequently, the load alleviation of wind turbine blades can be defined in the frequency domain as in **Figure 3**. This figure shows the frequency response of an uncontrolled and an ideally-controlled blade structural response where $1N$ and $2N$ denote the first and second natural frequencies and LA_{1P} denote the load alleviation gain for 1P loads. **Figure 3** represents an ideal case which can be used as a loop-shaping target to be achieved while implementating closed-loop control strategies.

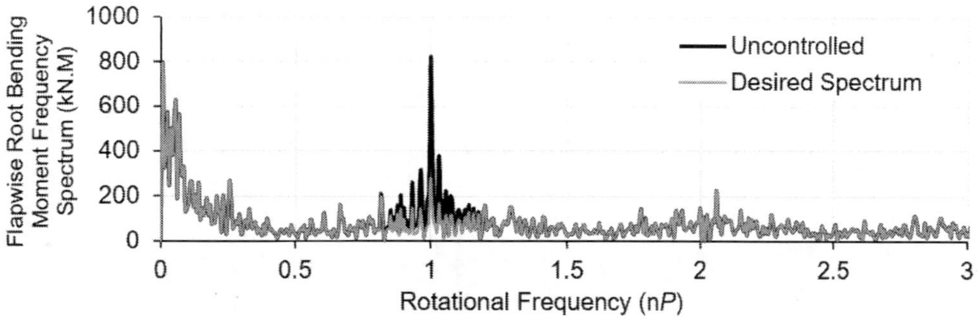

Figure 2 - Typical blade load spectrum (based on the NREL 5 MW wind turbine)

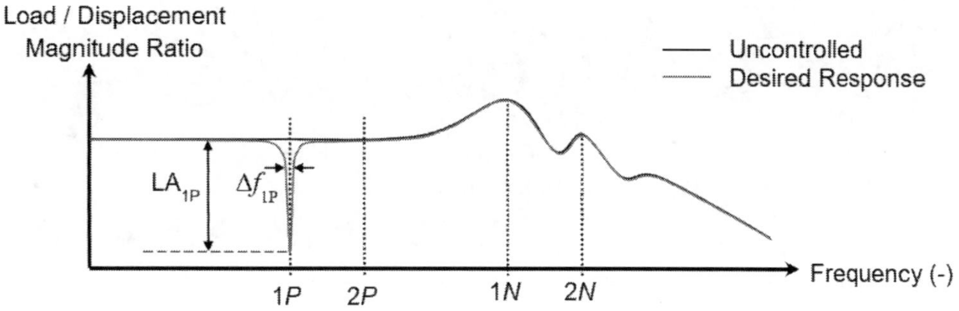

Figure 3 - General load control problem definition

3. SERVO-AEROELASTIC MODEL

The wind turbine aeroelastic performance is obtained using WTAC [19] (Wind Turbine Aeroelastic and Control). WTAC includes a finite element model of wind turbine blades using the blades cross-sectional properties as input. In WTAC the coupled edgewise and flapwise dynamics due to structural twist are described using twisted mode-shapes as shown in **Figure 4**. The out-of-plane (OOP) and in-plane (IP) axis are used as general blade coordinate systems. More details and validations about WTAC can be found in the work of Macquart and Maheri. [19].

Trailing edge flaps are modelled based on thin aerofoil theory according to the original work of Leishman [20]. Under attached flow conditions and for small flap deployment angles δ_F, the Leishman's model predicts a linear relationship between δ_F and the steady state change in lift $\Delta C_{L,ss}(\delta_F)$. In order to account for thicker wind turbine aerofoils, the slope of $\Delta C_{L,ss}$ with respect to δ_F is made a function the blade local angle of attack $\Delta C_{L,ss}(\delta_F, \alpha)$ based on steady state aerodynamic results obtained for each blade aerofoil obtained using XFoil [21]. Denoting the modal coordinate vector $\vec{Q}$, the blade structural system modal form is given as:

$$\left[M_q\right]\ddot{\vec{Q}}+\left[D_q\right]\dot{\vec{Q}}+\left[K_q\right]\vec{Q}=\vec{F}_q \tag{1}$$

where, $\left[M_q\right]$, $\left[D_q\right]$ and $\left[K_q\right]$ are the equivalent modal matrices for mass, damping and stiffness, and $\vec{F}_q$ is the modal force vector.

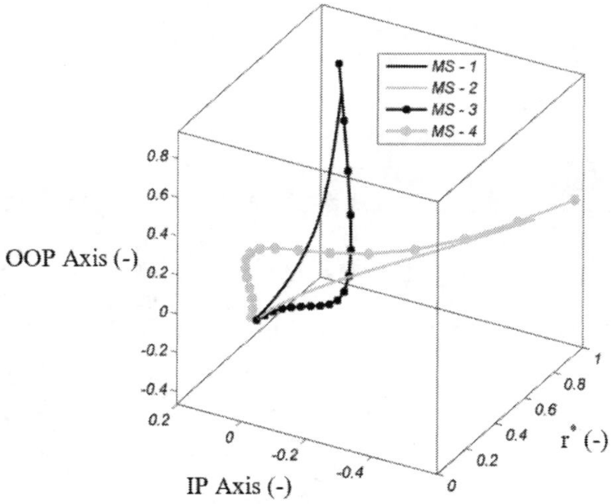

Figure 4 - The NREL 5 MW wind turbine blade coupled in-plane (IP) and out-of-plane (OOP) mode-shapes (MS) as function of the normalised radial coordinate (r*)

Rewriting Equation (1) in a state space form while conserving only the three primary modes and coupling the resulting system with the TEF aerodynamic model (subscript $_{Fc}$), we obtain the aeroelastic system (subscript $_{Ae}$) as follows:

$$\begin{aligned}\dot{\vec{X}}_{Ae} &= [A_{Ae}]\,\vec{X}_{Ae} + [B_{Ae}]\vec{u} + [D_{Ae}]\\ \vec{Y}_{Ae} &= [C_{Ae}]\vec{X}_{Ae}\end{aligned} \tag{2}$$

$$\dot{\vec{X}}_{Ae}=\begin{bmatrix}\dot{Q}_3\\ \dot{Q}_2\\ \dot{Q}_1\\ \ddot{Q}_3\\ \ddot{Q}_2\\ \ddot{Q}_1\\ \dot{\vec{X}}_{Fc}\end{bmatrix}=\begin{bmatrix}0&0&0&1&0&0&0\\0&0&0&0&1&0&0\\0&0&0&0&0&1&0\\r_3&0&0&s_3&0&0&[T_{As3}]\\0&r_2&0&0&s_2&0&[T_{As2}]\\0&0&r_1&0&0&s_1&[T_{As1}]\\0&0&0&0&0&0&[A_{Fc}]\end{bmatrix}\begin{bmatrix}Q_3\\ Q_2\\ Q_1\\ \dot{Q}_3\\ \dot{Q}_2\\ \dot{Q}_1\\ \vec{X}_{Fc}\end{bmatrix}+\begin{bmatrix}0\\0\\0\\0\\0\\0\\ [B_{Fc}]\end{bmatrix}\vec{u}+\begin{bmatrix}0\\0\\0\\F_{q3}\\F_{q2}\\F_{q1}\\0\end{bmatrix} \tag{3}$$

$$\vec{Y}_{Ae}=\left[C_{Ae}\right]\vec{X}_{Ae}=\begin{bmatrix}c_{11}&c_{12}&c_{13}&0&0&0&0\\c_{21}&c_{22}&c_{23}&0&0&0&0\\0&0&0&0&0&0&[C_{Fc}]\end{bmatrix}\begin{bmatrix}Q_3\\ Q_2\\ Q_1\\ \dot{Q}_3\\ \dot{Q}_2\\ \dot{Q}_1\\ \vec{X}_{Fc}\end{bmatrix} \tag{4}$$

The first, second and third modal degrees of freedom are respectively denoted by Q_1, Q_2 and Q_3 for which r_i and s_i describe the unforced dynamics. Each flap dynamic is represented by a third order model represented by the state variables in $\vec{X}_{Fc}$ and a state and a control input matrix denoted by $[A_{Fc}]$ and $[B_{Fc}]$. The interaction of external aerodynamic forces with the structural blade model is described through vectors F_{qi} while the interaction due to flap actuation is described by $[T_{Asi}]$. The unknown external force, included into the vector $[D_{Ae}]$, on each element is assumed to be a uniformly distributed time varying force. With the exception of the last line of $[C_{Ae}]$ that corresponds to the flaps deployment position sensors, the structure displacement at different span locations is a linear combination of the first three modal coordinates.

4. CLOSED LOOP CONTROL SYSTEMS

4.1 Model Predictive Controller (MPC)

A frequency-weighted version of the Model Predictive Control (MPC) methodology will be used to design a reference controller for the load reduction problem. MPC is a time-domain controller design technique that, at each time step, predicts the system loads over a (finite) time horizon as a function of the control input and optimises the control input to minimise blades loading. The main reasons for choosing frequency-weighted MPC control for this application are:

- (i) As it works with state-space realisations, it can directly handle the multivariable nature of the smart rotor control problem.
- (ii) The MPC cost function can be optimised subject to physical constraints, which is desirable for the current application.
- (iii) Frequency-weighting can help focus on the parts of the load spectrum that are primarily responsible for lifetime fatigue damage.

Frequency-weighted MPC has previously been used in [8] for flap control of blade loads and for a more general survey of MPC, the reader is referred to [14]. The control design theory is now briefly summarised. The multi-input multi-output wind turbine system is modelled in the previous section by the linear system (2) - (4) that can be restated in discrete time as:

$$X_{k+1} = [A]X_k + [B]u_k + e_k \tag{5}$$

$$Y_k = [C_1]X_k \tag{6}$$

$$Z_k = [C_2]X_k + [D]u_k \tag{7}$$

Here, X_k is the state vector, u_k is a vector containing the commanded flap deployment angles, and Y_k are the blade loads or deflections measured by the sensors located on the blade. The term e_k refers to the wind-induced loading, whose effect on Y_k is sought to be minimised. Finally, Z_k is a performance measure index and is obtained, in our case, by filtering the output Y_k. The filter used has a bandpass action that emphasises load reductions at the 1P frequency. The matrices $[A]$, $[B]$, $[C_1]$, $[C_2]$ and $[D]$ depend on the system dynamics and the chosen filter and are assumed to be available to the control designer, either from a first-principles model or from system identification experiments. In our case, these matrices are the corresponding discrete equivalent of the continuous matrices (2) augmented with bandpass filter dynamics.

At every time instant k, the controller synthesises a control input sequence that minimises the cost J_k:

$$J_k(u_k, \dots, u_{k+N_c}) = \sum_{j=k}^{k+N} Z_j^T Z_j \tag{8}$$

$$s.t.\ [\phi_k, \dots, \phi_{k+N}] \leq 0 \tag{9}$$

where the following notation has been used:

- N_c is the control horizon, or the period for which the control horizon is active.
- N is the prediction horizon, or the period over which the cost is penalised.
- ϕ_k is the inequality constraint under which the optimisation is to be carried out. Typically, this term is used to enforce a hard limit $u_{\max}$ on the control input, e.g. $\phi_k = u_k - u_{\max}$.

This cost is minimised online with the help of quadratic programming tools at every time instant k to obtain an optimum control input sequence $u_k, \dots, u_{k+N_c}$. Only the first control input command u_k is implemented and the optimisation is then repeated for the next time instant $k+1$, in line with the so-called "receding horizon" principle of MPC.

While MPC is suitable for load alleviation control according to the reasons mentioned above, tuning the parameters N and N_c for satisfactory control performance can be a non-trivial task. The term N_c is usually taken large enough for the predicted impulse response of the closed-loop system to decay to nearly zero. The prediction horizon N is always larger than the control horizon, and is typically large for closed-loop stability. However, increasing N causes a significant increase in computation time. Tuning the performance filter used to synthesise Z_k is also a determining factor in closed-loop stability, and is often required to be done in the frequency domain. The weights on each element of the performance channel have to be defined a priori by the user. Moreover, it should be noted that when the constraints are active in the optimisation, MPC yields a non-linear closed-loop system for which it is typically difficult to characterise closed-loop stability and performance. Finally, current sensors employed for wind turbines do not allow an accurate prediction of the incoming wind field and its corresponding change in aerodynamic forces denoted as e_k. As a result, prediction carried out over large horizon are difficult and likely to incorrectly influence the calculation of Z_k.

4.2 Decoupled Loop Shaping Control (DLSC)

In comparison to MPC, the proposed DLSC aims at achieving similar load alleviation performance with a much simpler control structure. For that purpose, the DLSC strategy is based on two assumptions: (i) First, it is assumed that the blade dynamics are dominated by the first bending mode. This is easily justifiable since large wind turbine blades are slender structures with bending dominated behaviours and modal reduction techniques can be employed in order to accurately capture the dominant blade loads with very few modes (i.e. 2-3). (ii) Second, it is assumed that TEFs installed next to each other have no aeroelastic couplings through the control loop. Two main reasons are given. First, the authors are unaware of any investigation or results about the potential aerodynamic coupling occurring between neighbouring control surfaces deployed on rotating wind turbines. The second reason is related to the limited control effort available to each individual TEF. That is, even if a blade is equipped with TEFs covering 15% of its span, the overall control effort which can be generated by deploying all the flaps simultaneously is generally insufficient to fully alleviate the dominant blade loads [4, 8]. As a result, TEF will not interact with each other through the closed control loop.

Under the above assumptions, the smart wind turbine blade dynamic behaviours can be exploited for control system design. In particular, because of the dominance of the first vibrating mode and because of the limited control effort, multiple TEFs will not interact with each other and the control problem can therefore be decoupled. The domination of the first vibrating mode is illustrated in **Figure 5** in which flapwise tip and root measurements made along the blade span are shown to be strongly correlated. The main idea behind the control scheme proposed by the authors can be expressed as follows: *The dominant vibrating mode and the limited control capabilities and interactions between control surfaces are such that the original MIMO control problem can effectively be decoupled into SISO control problems.* Under the above consideration, we propose to employ the control structure for blade load alleviation as shown in **Figure 6**.

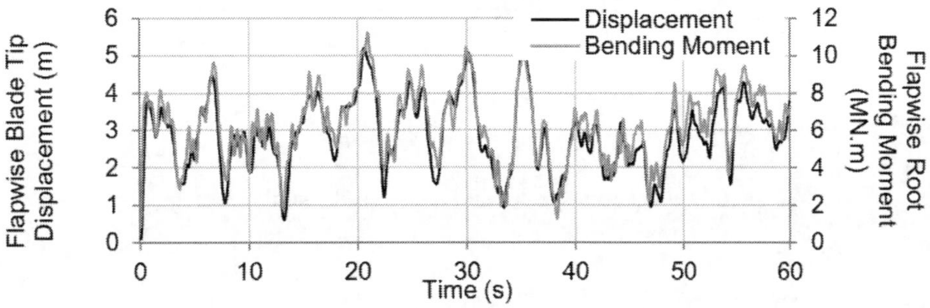

Figure 5 - Superposition of the NREL 5 MW wind turbine blade flapwise tip displacement and root bending moment for a mean wind speed of 15m/s.

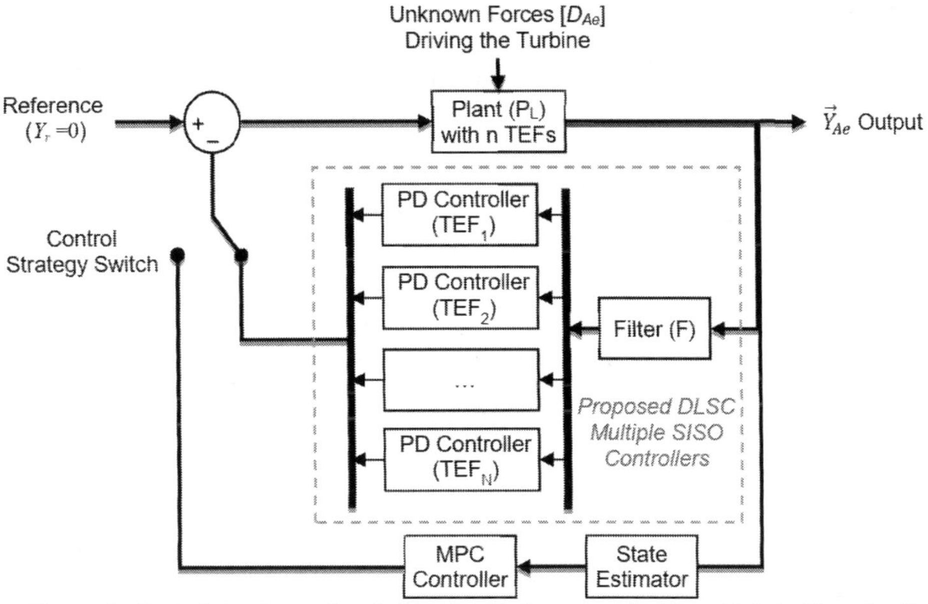

Figure 6 - Control structures of a wind turbine blade equipped with multiple trailing edge flaps

In **Figure 6**, the filter is designed as the combination of a second order band-pass filter and a second order inverted notch filter as illustrated in **Figure 7**. As common practice, the band-pass filter rejects frequency bandwidths below 1P loads which should not be affected by the control as well as high-frequencies. The inverted notch filter amplifies 1P signals measured through the blade strain sensors. As a result, the DLSC control effort will be mostly focused on 1P load rejection. Furthermore, it can be observed (see **Figure 6**) that for similarly tuned individual controllers (TEF_1 to TEF_n) the DLSC control strategy will results in all TEFs deploying in phase with each other.

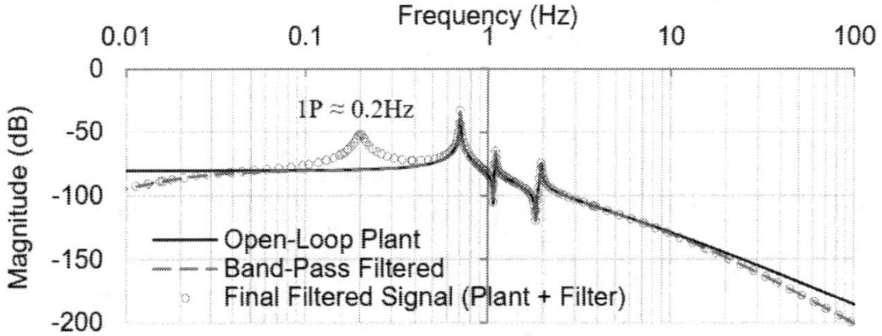

Figure 7 - Frequency spectrum of the original and filtered output signal

5. LOAD ALLEVIATION

We first start by examining the results obtained with the proposed DLSC control strategy compared to the MPC results as shown in **Figures 8** and **9**. These figures compare the load alleviation achieved with both the DLSC and the MPC control strategies. Despite their fundamental differences, it can be observed that both control approaches have similar load alleviation capabilities. As shown by the controlled and uncontrolled load frequency spectrums, the flap activations results in a 1P cyclic loads alleviation of about 600kN.m. Furthermore, it can be seen that the closed loop controls do not results in the increase of higher frequency loads.

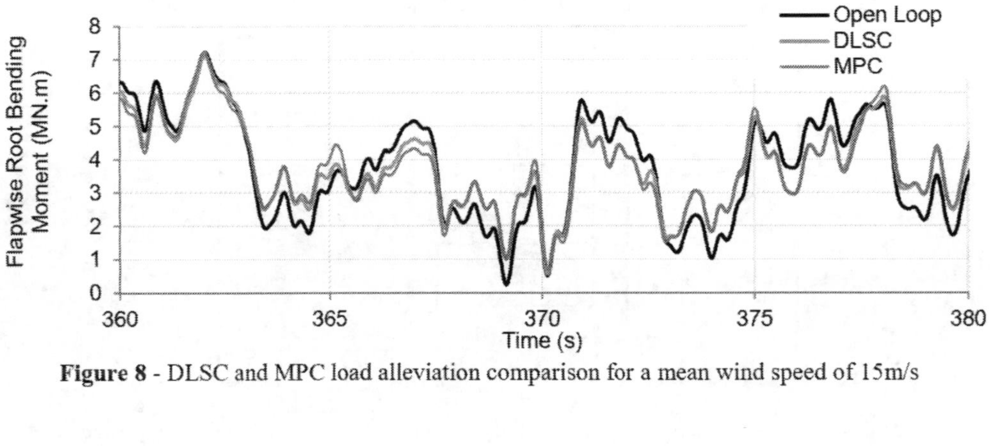

Figure 8 - DLSC and MPC load alleviation comparison for a mean wind speed of 15m/s

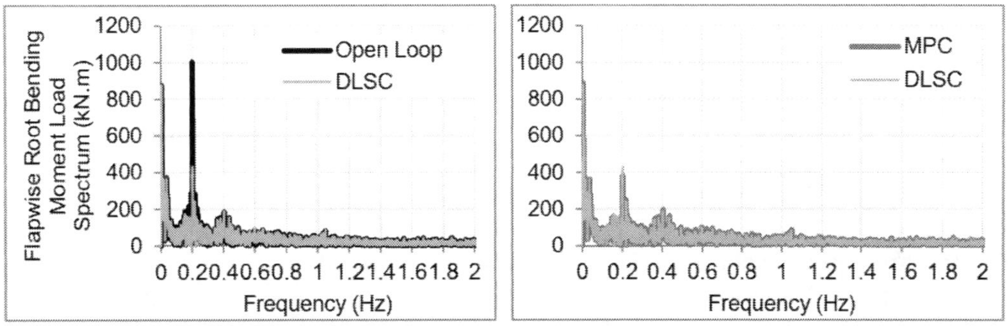

Figure 9 - DLSC and MPC load alleviation frequency spectrum comparison for a mean wind speed of 15m/s

In order to shed light on the load alleviation results presented above, we now compared the controlled flap deployment obtained with both control strategies as shown in **Figure 10**. As previously explained, under DLSC all flaps deploy in phase to each other as all individual controllers share a common input signal and have similar gains (see **Figure 6**). By constrast, the MPC is able to control flaps in any possible configuration in order to minimise the performance criterion Z_k. However, it can be observed that the MPC strategy results in a flap deployment pattern similar to the one achieved by the DLSC. That is, of course, because the 1P loads are dominating the behaviour of the MPC criterion and this is directly reflected onto the flap deployment as shown **Figure 11.** In other words, the assumption made for the DLSC design are correct. As a results, the MPC control strategy converges to an optimal deployment equivalent to the one obtained with the proposed DLSC.

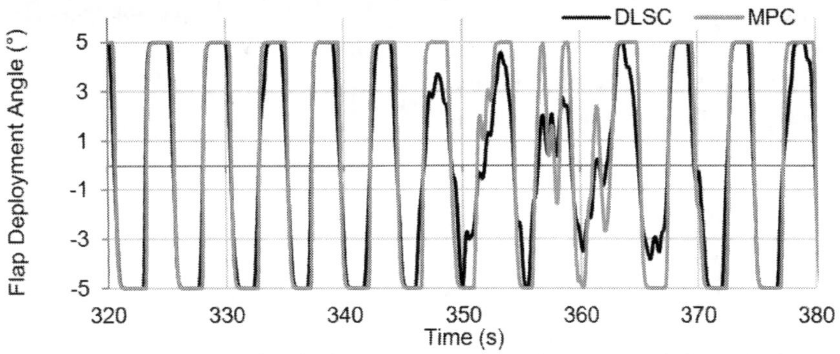

Figure 10 - DLSC and MPC flap deployment comparison for a mean wind speed of 15m/s

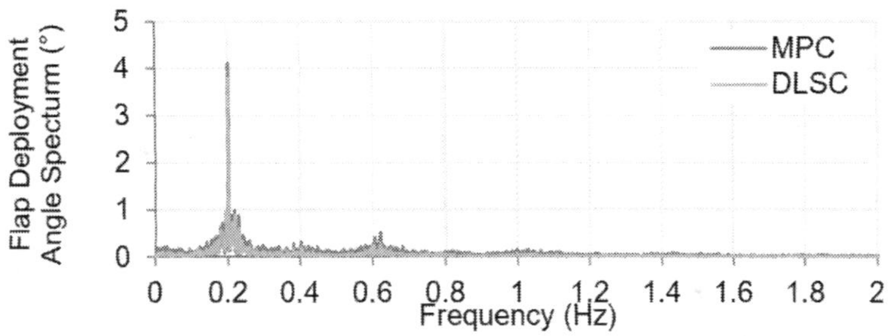

Figure 11 - DLSC and MPC flap deployment spectrum for a mean wind speed of 15m/s

The results presented in this section supports the design of a decoupled loop-shaping control strategy as proposed by the authors. That is, the dominant blade dynamics can be explained in simple behaviours that can be used to design load alleviation controllers based on classical control theory. Efficient load alleviation results have been achieved. The main advantages of using MPC remains the possibility to ensure and exploit hard constraints. While this is not possible using classical control theory, the controller gains can be tuned so as to reach their limits while keeping actuation speed resonable. That is, fast enough for 1P load alleviation but not faster. Furthermore, the implementation of hard boundary constraints for fast deploying devices such as control surfaces should also be handled with great caution in order to avoid excessive actuator wear.

6. CONCLUDING REMARKS

In this paper, the authors successfully employed a classical frequency-based approach to explain the closed-loop system dynamics of wind turbine blades with actively controlled trailing edge flaps. This is in constrast with current trends, attributed in part to predictive control, that promote the increasing complexity required for smart wind turbine control.

The results of this study can be summarised as follows. The deployment of multiple active flow controllers on large scale wind turbine blades was shown to follow a pattern imposed by the dominant first bending vibrating mode. This observation, combined with the limited flap control capabilities, permits the MIMO control problem of smart wind turbine equipped with multiple active flow controllers to be effectively decoupled into simpler SISO control problems. As a result, the proposed frequency-based classical SISO control systems were shown to be highly efficient at reducing 1P loads. Comparisons with the model predictive controller also demonstrated that the proposed approach results in a near optimal flap deployment strategy. The authors believe that the simplicity of their approach, in terms of structure, sensing and tunning, should help the future development and testing of smart controllers for experimental applications.

REFERENCES

1. Greenpeace, and the Global Wind Energy Council. "Global Wind Energy Outlook 2014." Greenpeace and the Global Wind Energy Council, 2014.
2. Barlas, T. K., and van Kuik, G. A. M. "Review of state of the art in smart rotor control research for wind turbines," *Progress in Aerospace Sciences* Vol. 46, No. 1, 2010, pp. 1-27. doi: http://dx.doi.org/10.1016/j.paerosci.2009.08.002
3. Barlas, T. K., and Van Kuik, G. "State of the art and prospectives of smart rotor control for wind turbines," *Journal of Physics: Conference Series*. Vol. 75, IOP Publishing, 2007, p. 012080.
4. Lackner, M. A., and van Kuik, G. "A comparison of smart rotor control approaches using trailing edge flaps and individual pitch control," *Wind energy* Vol. 13, No. 2 - 3, 2010, pp. 117-134.
5. Bossanyi EA. "Individual blade pitch control for load reduction," *Wind Energy* Vol. 6, No. 2, 2003, pp. 119-128.doi: Doi 10.1002/We.76
6. Larsen, T. J., Madsen, H. A., and Thomsen, K. "Active load reduction using individual pitch, based on local blade flow measurements," *Wind Energy* Vol. 8, No. 1, 2005, pp. 67-80. doi: Doi 10.1002/We.141
7. Castaignet D, Poulsen NK, Buhl T, and Wedel-Heinen JJ. "Model Predictive Control of Trailing Edge Flaps on a Wind Turbine blade," *2011 American Control Conference*, 2011, pp. 4398-4403.

8. Castaignet D, Barlas T, Buhl T, Poulsen NK, Wedel Heinen JJ, Olesen NA, Bak C, and Kim T. "Full-scale test of trailing edge flaps on a Vestas V27 wind turbine: active load reduction and system identification," *Wind Energy* Vol. 17, No. 4, 2014, pp. 549-564.
9. Macquart T, Maheri A, and Busawon K. "Microtab dynamic modelling for wind turbine blade load rejection," *Renewable Energy* Vol. 64, 2014, pp. 144-152.
10. Peter Bæk, J.-G. J., Poul Kramer and Mac Gaunaa. "Experimental Comparison of a Miniflap to a Trailing Edge Flap on a Wind Turbine Airfoil."
11. P, B., and Gaunaa M. "Modeling the Temporal Response of a Microtab in an Aeroelastic Model of a Wind Turbine," *AIAA* Vol. 49, 2011.
12. Barlas TK, van der Veen GJ, and van Kuik GAM. "Model predictive control for wind turbines with distributed active flaps: incorporating inflow signals and actuator constraints," *Wind Energy* Vol. 15, No. 5, 2012, pp. 757-771.
13. Castaignet D, Barlas T, Buhl T, Poulsen NK, Wedel Heinen JJ, Olesen NA, Bak C, and Kim T. "Frequency-Weighted Model Predictive Control of Trailing Edge Flaps on a Wind Turbine Blade," *IEEE transaction on control system technology* Vol. 21, No. 4, 2013.
14. Garcia, C. E., Prett, D. M., and Morari, M. "Model predictive control: theory and practice—a survey," *Automatica* Vol. 25, No. 3, 1989, pp. 335-348.
15. Mayne, D. Q., Rawlings, J. B., Rao, C. V., and Scokaert, P. O. "Constrained model predictive control: Stability and optimality," *Automatica* Vol. 36, No. 6, 2000, pp. 789-814.
16. Jonkman, J. M., Butterfield, S., Musial, W., and Scott, G. *Definition of a 5-MW reference wind turbine for offshore system development*: National Renewable Energy Laboratory Golden, CO, 2009.
17. Andersen, P. B., Gaunaa, M., Bak, C., and Buhl, T. " Load alleviation on wind turbine blades using variable airfoil geometry. ," *EWEC 2006*. Athens , Greece, 2006.
18. Macquart, T., Maheri, A., and Busawon, K. "A simple method to determine the optimal location of active flow controllers on wind turbine blades," *Environmental Friendly Energies and Applications (EFEA), 2014 3rd International Symposium on*. IEEE, 2014, pp. 1-5.
19. Macquart T, and Maheri A. "Integrated Aeroelastic and Control Analysis of Wind Turbine Blades Equipped with Microtabs," *Renewable Energy*, 2014.doi: 10.1016/j.renene.2014.09.032
20. Leishman JG. "Unsteady lift of a flapped airfoil by indicial concepts," *Journal of Aircraft* Vol. 31, No. 2, 1994, pp. 288-297.
21. Drela M. "XFOIL: An Analysis and Design System for Low Reynolds Number Airfoilsii," 1989.

ICAST2015 #82

MORPHING ELASTICALLY LOFTED TRANSITION FOR ACTIVE CAMBER CONTROL SURFACES

Benjamin K.S. Woods[1*], Laura Parsons[2], Alexander B. Coles[2], James H.S. Fincham[2], and Michael I. Friswell[2]

[1] Department of Aerospace Engineering, Queens Building, University of Bristol, Bristol UK
[2] College of Engineering, Swansea University, Swansea, UK

ABSTRACT

This paper introduces a compliant morphing flap transition that seeks to address a long-standing source of noise and drag in the design of aircraft wings – the gap present at the spanwise ends of the control surfaces. These gaps create large discontinuities in the flow and allow for pressure leakage from the lower to upper wing surface, generating significant amounts of vorticity, noise, and drag. The concept introduced here seals this gap with a smooth, three-dimensional morphing transition section that elastically lofts between the rigid wing and moving control surface in a passive and continuous manner. Previous transition concepts are first discussed, followed by establishment of an initial desired transition shape. The morphing, elastically lofted transition concept proposed here will then be introduced. In this concept, the complex three-dimensional shape change required is created with a novel structural architecture that combines material and geometric compliance with geometric bend-twist coupling. The concept design and operating principles will be introduced, relevant geometric parameters will be derived, and an initial prototype demonstrator capable of large deflections and smooth transition surfaces will be shown.

1. INTRODUCTION

The gaps present at the ends of aerodynamic control surfaces such as flaps and ailerons are known to be significant sources of noise and drag,[1,2] and researchers have long sought to build structures capable of bridging these gaps. However, success to date has been elusive as the changes in geometry required are quite significant, the aerodynamic load levels are substantial, and the deformations are fully three-dimensional, making mechanical solutions difficult. Figure 1 shows the wing of an Airbus A380 with multiple control surfaces deflected, and the significant size of the resulting gaps can be seen. The relevance of this problem is increasing as multiple segment control surfaces become more common (creating additional gaps), and as control surfaces are increasingly used not just as control effectors for occasional maneuvers but also to actively and continuously manage the lift distribution of a wing. The impact of these gaps on the drag and noise of the aircraft can be significant. In an industry that is facing increasing pressure to improve operating efficiency throughout the full range of flight conditions and to reduce noise generated, particularly on take-off and landing, a device which can help address both problems and which has potential to be retrofit into existing aircraft is particularly attractive.

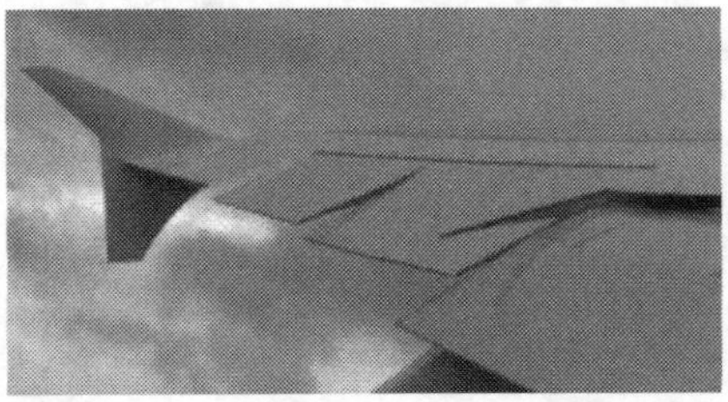

Figure 1. Airbus A380 wing with multiple control surfaces deflected showing the large gaps created.[3]

2. BACKGROUND

Work by various industrial and academic research groups has suggested a number of different approaches to sealing these gaps. In 1984, Kunz proposed a mechanical approach with an auxiliary flap perpendicular to

* ben.k.s.woods@bristol.ac.uk

the main flap and connected to it with a sliding interface, as seen in Figure 2a.[4] It is not known if this device was ever built, but it would seem to require a significant amount of mechanical complexity. Later, in 1998, Diller and Miller of Northrop Grumman proposed a combined mechanical/compliant transition for the spanwise ends of rigid flaps and as a covering for the chordwise gap at the start of the flap.[5] Shown schematically in Figure 2b, their approach covered the gap between flap and wing with compliant silicone elastomer skins which were mechanically reinforced with a large number of embedded rods. These rods were intended to slide into and out of holes in the silicone rubber to allow for the change in length required of the transition, as seen in Figure 2c. The use of so many sliding interfaces passing through the elastomeric material, which would have a high coefficient of friction and which would be susceptible to wear, is problematic and the concept does not seem to have been adopted. Caton *et al.* of The Boeing Company patented a surprisingly similar concept in 2002,[6] which uses the same approach of reinforcing rods sliding inside of an elastomer skin as shown in Figure 2d. The primary difference between their approach and that of Diller and Miller is use of two shorter reinforcing rods instead of a single longer one. This change requires the inclusion of an additional component to cover the joint (which requires even more sliding interfaces), and the design is still likely to suffer from significant tribological issues. In 2013 a second concept from Boeing, shown in Figure 2e, was patented which creates a smooth outer transition surface by covering a series of rigid, mechanical elements in an elastomeric skin.[7] A number of rigid rib segments are mounted to spanwise rods, which slide into and out of the rigid portion of the wing to create the changing length required of the transition section. Of particular note in this design is the use of rotating tips at the trailing edge extents of each rib. This additional mechanical degree of freedom allows the angle of the trailing edge to rotate to follow the smooth shaped desired. This approach requires, however, that the skin is not attached to the rigid rib sections for a significant chordwise length, which raises concerns about out-of-plane deflection under aerodynamic loading. The rotating tips further increase the number of mechanical joints, adding weight and exacerbating concerns about maintainability.

In 2012, a group from NASA Langley research center patented a compliance based approach to flap transitions.[8] Shown in Figure 2f, this concept consisted of a wedge of elastomer bridging the spanwise gap between flap and wing and is significantly simpler than previous approaches. Various configurations were proposed, including solid elastomer, elastomer skins with hollow center, and elastomer skins over foam core. This concept was one of several noise reduction measured tested in a recent wind tunnel campaign by NASA,[9] and was shown to reduce measured noise emissions from a deflected flap by 3 dB or more over a wide frequency range, while also improving the lift to drag ratio of the flap (by an unspecified amount). It appears however, that the device tested was a non-straining elastomeric mockup of the desired shape. This is likely due to the difficulty that was experienced with designing a working version of this transition,[10] wherein the very high strain requirements of ~500% led to unacceptably high force requirements and difficulties designing appropriate fixtures. Attempts to reduce these forces by using the hollowed out configuration led to extensive wrinkling of the transition surface.

Successful transonic wind tunnel testing of a related concept as part of the DARPA smart wing program is overviewed by Kudva.[11] This concept didn't employ transition sections *per se*, at least not in the passive sense being discussed here, but created a similar effect from a series of active camber morphing segments along the span which each had two actuation points. These two degrees of freedom allowed each segment to control the amount of camber and the twist angle of the trailing edge independently. As a result, the camber deflections could be smoothly built up along the span with minimized gaps in between segments. Another compliance based approach was introduced by Pankonien and Inman as a further development of a smart material based active camber concept.[12] This morphing transition concept used an elastomeric skin over 3D printed elastomer honeycomb sub-skin to bridge the gap between different active camber modules along the span of a wing. Low speed wind tunnel testing showed the basic efficacy of the approach, although performance under full-scale aerodynamic loading and energy requirements are unknown. Work by da Rocha-Schmidt and Baier investigated an alternative approach based on shear deformation of a reinforced elastomer skin.[13] The reinforcement in this case consists of a woven metal wire mesh, which is able to deform easily in shear without directly straining the wire filaments. This shear deformable skin is placed over a shear deformable cellular core to create a sandwich structure. Work on this recent concept so far has focused on finite element analysis (FEA) and coupon testing of the skin materials.

One of the more promising concepts shown to date is a morphing transition under development by FlexSys Inc., shown in Figure 2g. This concept has been designed to work in concert with a compliant active camber concept known as FlexFoil™ that has been under development for some time.[14] The transition has recently completed flight testing on a modified Gulfstream III jet aircraft,[15] but little is known about the operating principle or performance as these authors were unable to find any academic publications or even patents. What can be seen though from the picture in Figure 2g is that while the device does indeed span the gap between control surface and rigid wing, it consists of an alternating series of segments which do and don't allow trailing edge rotation, so the net effect is a number of small but discrete steps. The authors suspect that the design is made from a series of thin, spaced rib sections made with the same compliant camber morphing geometry as the control surface that are then covered in an elastomeric skin. In that case, the ribs would be fixed width, and would not rotate, such that the elastomer skin between ribs is forced to bridge the increasing large distance between rib tips, creating the shape seen in Figure 2g. This approach is likely to provide significant noise and drag reduction compared to a control surface with no transition, but perhaps less so than a truly smooth and continuous morphing transition.

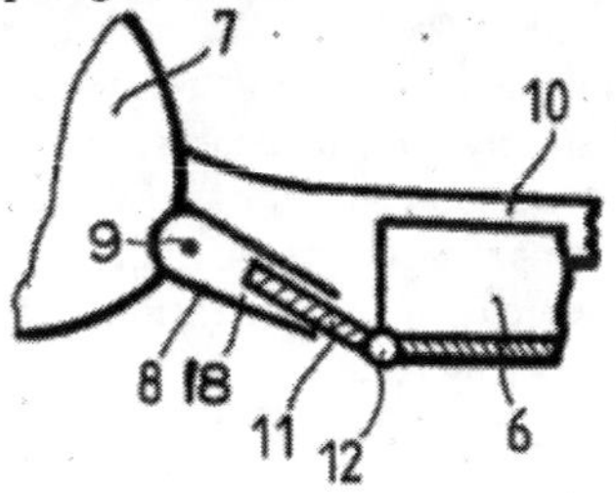

(a) From Kunz [4]

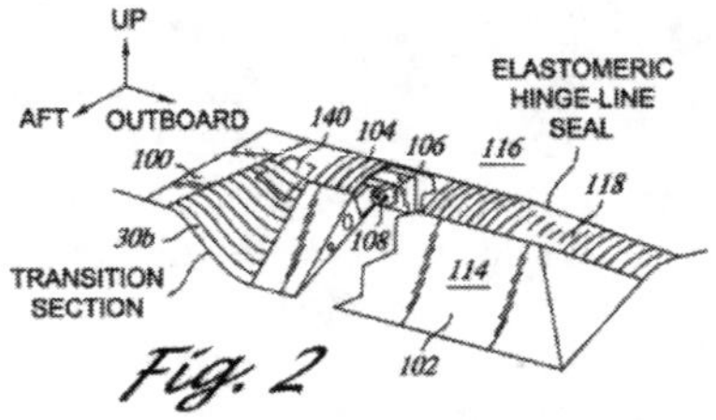

(b) From Diller and Miller [5]

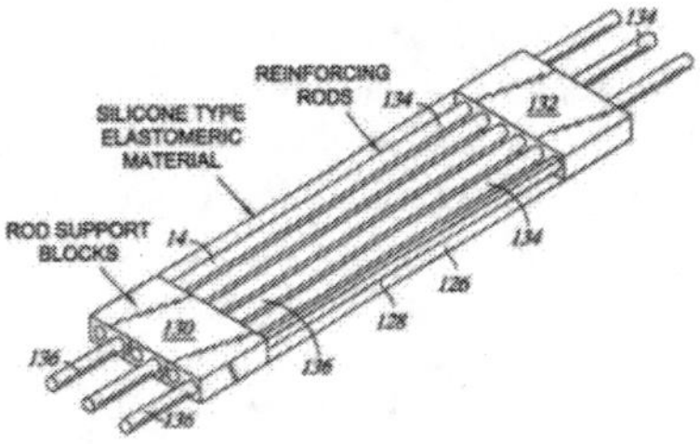

(c) From Diller and Miller [5]

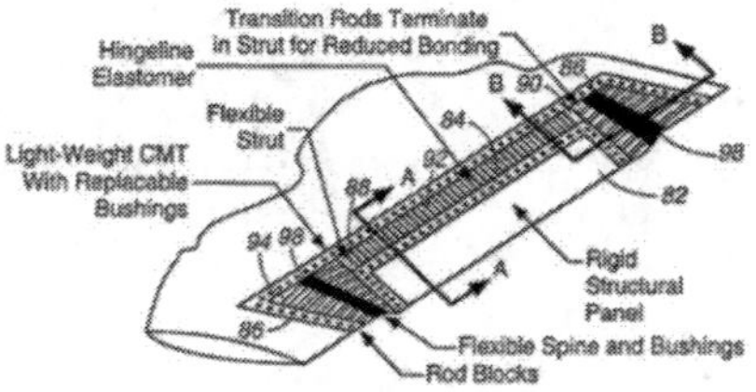

(d) From Caton *et al.* [6]

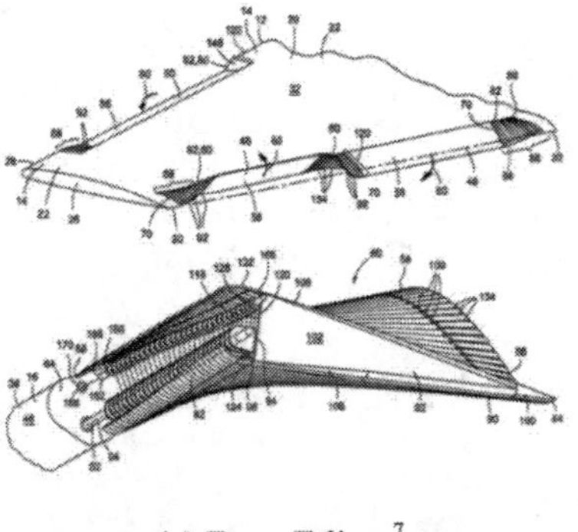

(e) From Etling [7]

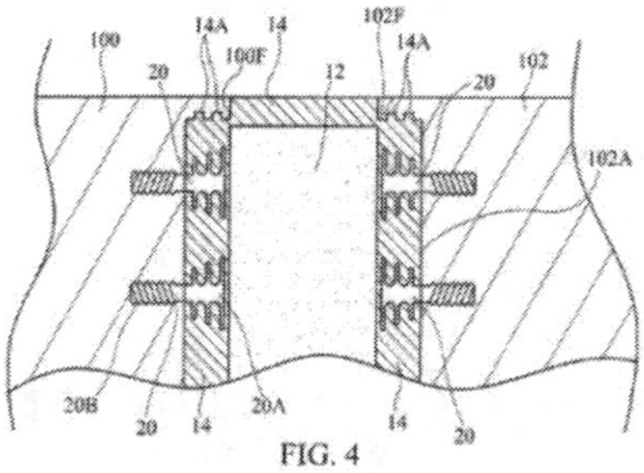

(f) From Khorrami *et al.* [8]

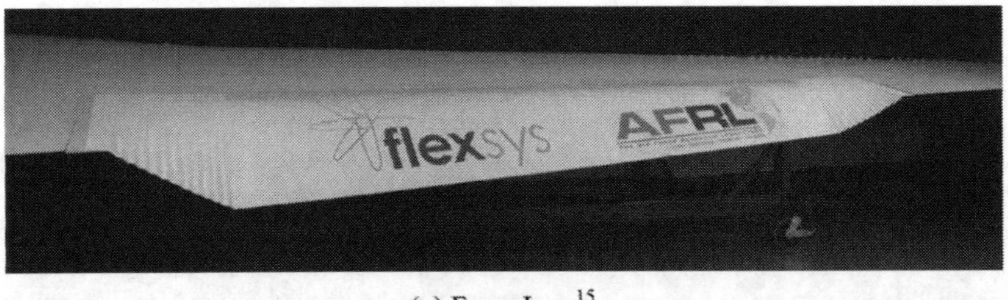

(g) From Lau [15]

Figure 2. A selection of previous flap transition concepts

From this brief review of the existing literature it is clear that engineers have been attempting to solve the problem of flap end gap drag and noise for some time, and that there has been no shortage of ideas. However, much of the work has been carried out by industry and as a result details of the design and performance of these devices is generally unavailable. What's more, with the exception of the newly developed device from FlexSys, none of the proposed solutions appear to have made it to the flight testing stage, and none have been certified for commercial use. It would seem therefore that there is room for the consideration of additional flap end transition concepts.

The previous work shown has been intended for use with a range of control surfaces, including rigid plain flaps, Fowler type flaps, and smooth camber morphing concepts, with the design approach taken in each case and the degrees of freedom achieved determining what types of control surface a given concept is applicable to. The solution proposed here is considered specifically for application to the Fish Bone Active Camber (FishBAC) morphing airfoil, but the basic principles can be readily adapted to other morphing camber concepts. The FishBAC morphing airfoil concept was introduced by Woods and Friswell[16] and has shown promise as a low-drag, high lift authority control surface for application to a wide range of aerodynamic and hydrodynamic surfaces. Wind tunnel testing has shown a 20-25% increase in lift-to-drag ratio at matched lift conditions compared to plain flaps with an achievable change in lift coefficient of $\Delta C_l > 0.7$.[17] The smooth continuous change in camber achieved by the FishBAC (as seen in Figure 3) makes it a good candidate for pairing with an end transition device, and the resulting combined surface would have the potential to operate very efficiently. Furthermore, the lack of any external protuberances from the FishBAC (control horns, flap tracks etc.) and the lack of sharp discontinuities in camber change should lead to a low noise solution as well.

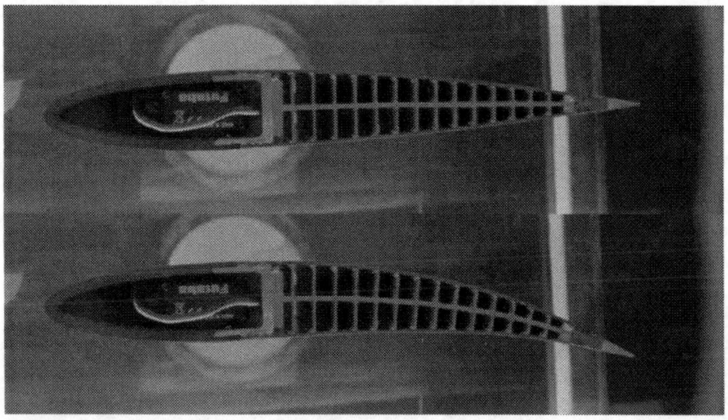

Figure 3. Fish Bone Active Camber morphing airfoil wind tunnel at rest and deflected downwards

Design Requirements

Operating conditions

The conditions under which this control surface transition must operate are demanding aerodynamically, structurally, and environmentally. Ideally, the transition would be usable on a wide range of vehicles, so operating Mach numbers up to the transonic range may be encountered, although the placement of the transition on the relatively lightly loaded trailing edge does reduce the actual aerodynamic pressures the device must support. Furthermore, the aerodynamic loading will vary considerably, increasing significantly with larger control surface deflections. Since all of the shape change in this transition occurs through compliance, the device must undergo significant direct strain; the skin surface in particular will experience large magnitude strains in multiple directions due to chordwise bending and spanwise elongation, on the order of 10%. Creating a structure which can achieve these large strains under the large aerodynamic loadings with minimal actuation requirements and with suitable levels of fatigue life is challenging. What's more, it must work over a wide range of temperatures, on the order of -40°C to +90°C for civil applications and -55°C to +100°C for military applications while being exposed to a host of environmental factors including ultraviolet light, exposure to petroleum products and solvents, ice accumulation, and rain and sand erosion. These environmental factors can be particularly difficult for the types of elastomeric materials typically used for high strain applications. Work with elastomeric erosion barriers[18,19] has led to the development of high performance elastomeric materials which have been flight certified, implying that while a challenging part of the problem, there is reason to believe that material solutions can be found. Having established the demanding operating conditions these devices must operate in, we will now consider the difficulties associated with the design process.

Design Approach

The starting point for the design of a morphing transition such as this is not entirely apparent. The primary reason for having the device is aerodynamic, and so the aerodynamic shapes required to minimize the amount of drag and noise generated at the flap ends would seem to be the obvious starting point. However, it is quite difficult to know what the "best" possible shape is for such a device, particularly at the early stages in the development of a new concept. It is expected that the "optimal" shape for any given application would be sensitive to the particulars of the wing it was being installed on and the operating conditions under which it would be used, with Mach number and Reynolds number being two particularly important parameters that would be expected to impact the fluid flow over and around the device. Given the highly three-dimensional nature of the structure and the expected flow around it, and the importance of viscosity effects and flow separation at the trailing edge of the wing, a full understanding of the flow fields around the transition would require a considerable amount of high-fidelity computational fluid dynamics (CFD) analysis. To then optimize the shape using these tools to analyse a large number of potential geometries under a number of different operating conditions is clearly a significant endeavour. What's more, a reasonable amount of wind tunnel validation of the CFD would be required to establish the validity of the methods, further growing the scope of the problem.

In fact, the high-fidelity optimized shape approach mentioned above, as significant of an effort as it would be, would ultimately not be particularly useful for the case of this device as the compliant nature of the solution proposed implies that the shapes under loading are not the same as the unloaded geometries, and therefore any CFD results which use prescribed, "rigid" shapes would likely deviate significantly from what would exist in flight. Instead, a coupled fluid-structure interaction analysis is needed that can consider the interplay between the compliant structure and the flow field to find stable converged solutions. This of course requires a detailed understanding of the structure and its stiffness properties, which at the start of the design process simply does not exist.

This fundamental design problem exists with all morphing aircraft structures, and indeed increasingly it exists with conventional aerospace structures as well as designers are getting ever more comfortable with larger elastic deflections in wing structures, for example. However, we wish to avoid the full depth of this problem initially to allow us to make some meaningful progress quickly with the new structural architecture proposed. For the sake of this initial development work we will therefore assume some general requirements for the types of shapes obtainable by the structure, and we will design the concept such that there is sufficient control over the design parameters to allow for a number of different desired shapes to be created. If these initial efforts prove successful, there will then be significantly more information on the behavior of the structure available for future fluid-structure interaction analysis.

Nominal Deformed Shapes

With this in mind, we will establish for our purposes here the very simple aerodynamic design guideline that the deformed shape should be smooth and continuous with no discrete changes in shape. This will minimize discontinuities in the surface which would likely create undesirable aerodynamic flow phenomena such as boundary layer tripping, vortex shedding, interference drag, etc. Furthermore, the two fixed sides of the transition, namely the spanwise edge which attaches to the trailing edge of the wing and the chordwise edge which attaches to the rigid portion of the wing trailing edge should have surfaces which remain tangent to the neighboring wing components across the entire deflection range. The same is true for the moving edge of the transition which attaches to the control surface. Enforcing tangency at all attached edges of the transition will ensure that the resulting combined wing surface is also smooth and continuous.

To further simplify the definition of the desired shape for this initial study, we can focus on the profile of the trailing edge of the morphing transition, with the underlying assumption that if the trailing edge is smooth then the continuous compliance of the rest of the structure will help ensure smoothness elsewhere. We will consider the two-dimensional projection of the trailing edge onto a plane aligned with the spanwise and thickness-wise directions, ignoring for the time being the small amount of motion of the trailing edge in the chord direction and considering just its vertical (thickness-wise) displacement. There are a number of shape functions which could be used to describe the desired deflection of the trailing edge along the span of the device, but one particularly simple and useful one is the cosine function:

$$w(y) = h * \cos\left(\frac{\pi y}{l}\right) - h \tag{1}$$

where w is the vertical (thickness direction) displacement of the trailing edge, y is the spanwise distance along the transition, h is the half-amplitude of the control surface deflection, and l is the spanwise length of the morphing portion of the transition. Figure 4 shows an example of trailing edge shapes for the case of a nominal transition that is 200 mm long in span, and has an active chordwise length of 200 mm that undergoes a series of control surface deflections from -60 mm to 60 mm (h = -30 to 30 mm). Note that the sign convention used for the shape function follows that used for trailing edge flaps: positive values of h produce downwards deflections, increasing local camber. It can be seen that this shape function creates a smooth continuous transition between the rigid portion of the trailing edge (left side of the figure) and the morphing control surface (right side) over a wide range of control surface deflections, both upwards and downwards. This bidirectional capability is important for applications such as ailerons where both positive and negative deflections are required. We will consider these desired shapes further and their impact on the structural design in more detail below, but first we will used these nominal deflected shapes in some initial computational fluid dynamics analysis to establish the potential for drag reduction through use of the morphing transition.

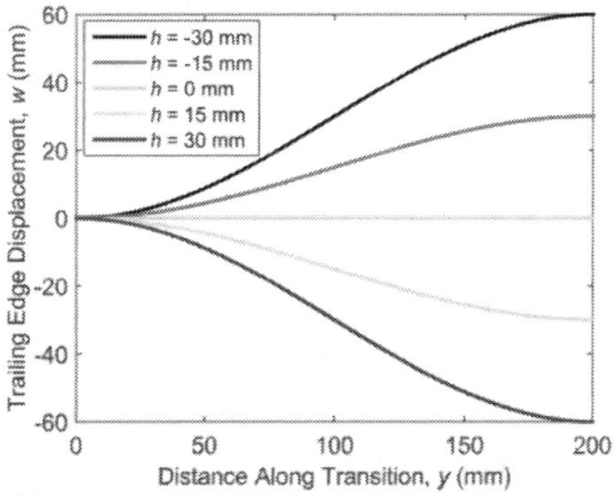

Figure 4. Example desired trailing edge shapes for the morphing transition

3. PROPOSED SOLUTION

To address the demanding design requirements of control surface end transitions, a novel structural architecture is proposed here. This Morphing, Elastically LofteD (MELD) transition combines material and geometric compliance with bend-twist coupling to create a structure capable of large three-dimensional deformations that automatically creates the desired combination of bending and twisting along the span needed to create a smooth and continuous aerodynamic surface.[22] The resulting structure is therefore elastically lofted between the control surface and rigid wing structure in a continuous, passive manner, melding the two structures together. The MELD transition, which is shown schematically Figure 5, is based around a series of bend-twist coupled ribs which are stacked along the span of the module and which are held together by a frame which mounts to the trailing edge of the rigid portion of the wing. The amount of bend-twist coupling in each of the ribs is carefully tailored to achieve the desired distribution of trailing edge angle along the span. At the spanwise extents of the transition there are separate end ribs which are designed to be easily attached to the existing wing structure. One of these ends is rigid, and the other has the camber morphing geometry of the FishBAC (or similar) active camber control surface to which it attaches. The bend-twist coupled ribs, mounting frame, and end ribs are all covered with an elastomeric skin made of silicone or polyurethane rubber sheets, which may or may not be reinforced selectively with fiber, and which are bonded to the upper and lower surfaces. Taken together, these components create a MELD module that can be installed and removed relatively easily into a wing structure, has no mechanical moving parts, and which requires no dedicated actuation; as it is passively driven by the control surface. These features are intended to simplify integration of the transition into new wing designs, and to open up the possibility of retrofitting it into existing designs alongside a retrofit of the FishBAC camber morphing airfoil, which can also be designed as an independent module for ease of integration.

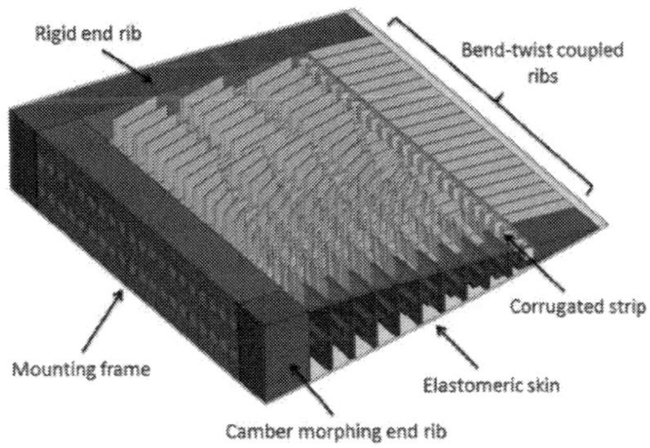

Figure 5. MELD transition concept

Geometric Bend-twist Coupling with Skewed Corrugations

The bend-twist coupled ribs are the key enabling technology for this transition, so their design and operating principle will be discussed in detail here. Driven by the need for high levels of coupling between elastic bending and twisting, the rib sections in this concept use what is believed to be a novel compliant morphing approach that links the two deformations using skewed corrugations. Corrugations are well known in morphing structures and have been proposed in a variety of configurations, albeit most often as an anisotropic extensible skin surface.[23] Corrugated structures have highly direction dependent stiffness, with the direction perpendicular to the corrugations being significantly stiffer in extension and in bending then the direction along the corrugations. Due to their highly anisotropic nature, corrugations are an effective way to create a compliant structure which deforms primarily in one direction. The corrugations in essence act like a series of compliant hinges, allowing for bending deformations to be built up as a series of small deformations in these hinges, which is realized physically as thin plate bending of the segments of the corrugation.

Corrugated members can form the basis of camber morphing ribs, as the corrugated geometry allows for chordwise bending while retaining stiffness in the other loading directions. Figure 6 shows an airfoil trailing edge rib structure built around a corrugated spine.

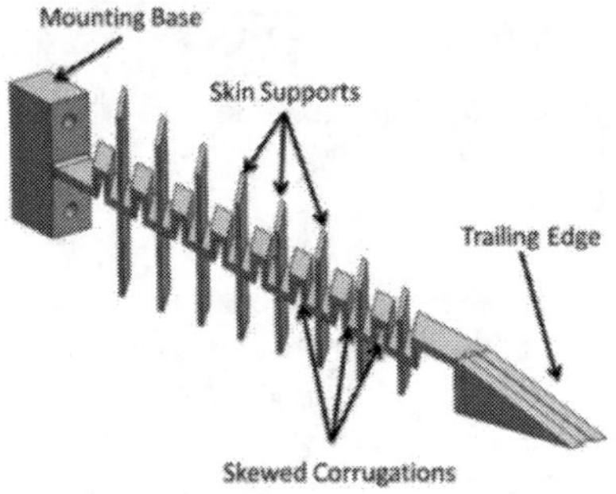

Figure 6. Bend-twist coupled rib with skewed corrugations

If the direction of the corrugation is aligned with the length axis of a rib structure, then the bending of the corrugation will lead to pure bending of the rib, with no induced rotation. If, however, the corrugation is placed at an angle α which is skewed to the long axis of the rib, as in Figure 7, then the bending of the corrugation members will induce twisting in the rib. This is a geometrically induced coupling which unlike other bend-twist coupling concepts does not require asymmetric or unbalanced composite layups or material anisotropy to work.[24,25,26] Composites are indeed preferred materials for the MELD transition as they can provide excellent stiffness and strength properties with low mass, but the concept is not predicated on their use. In addition to allowing for a greater range of materials to be used, the skewed corrugations are also capable of realizing particularly high levels of bend-twist coupling. This is demonstrated in Figure 8 with the finite element analysis results of a skewed corrugation rib under tip load. In this study the mounting base is fixed while the rest of the structure is free to deform under the applied vertical tip load P. The deformed shape shown in Figure 8 is shown at a scale of unity, and the large tip rotation which the tip load induces in clearly visible and is the direct result of the use of skewed corrugations. Having introduced the manner in which bend-twist coupling is created in the MELD transition, we will now consider the required distribution of coupling along the span of the transition.

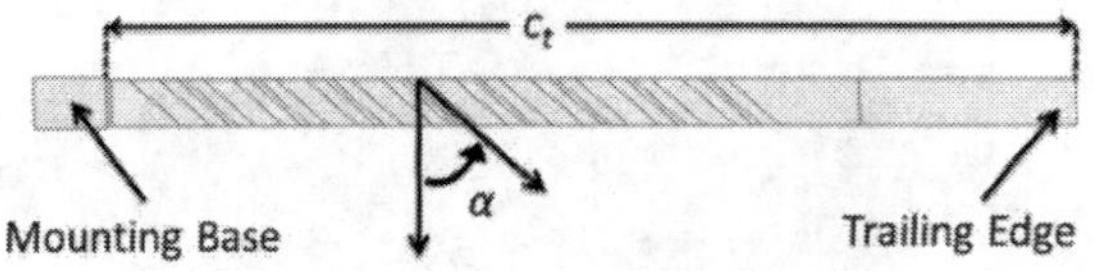

Figure 7. Bend-twist coupled rib, top view showing definition of skew angle α

Figure 8. Finite element analysis result of a skewed corrugation rib ($\alpha = 69°$) under tip load P – note rotation of trailing edge tip relative to the mounting base by angle θ

Bend-Twist Coupling Requirements to Achieve Nominal Shapes

One of the core features of the MELD concept is the ability of the ribs to match both the desired displacement and local surface angle over a range of control surface deflections. The local trailing edge tip angle θ that must be achieved at each point along the span can be found by differentiating the desired deflection:

$$\theta(y) = w'(y) = -\frac{\pi h}{l}\sin\left(\frac{\pi y}{l}\right) \tag{2}$$

For the example trailing edge shapes shown in Figure 4, the corresponding trailing edge angles are shown in Figure 9a.

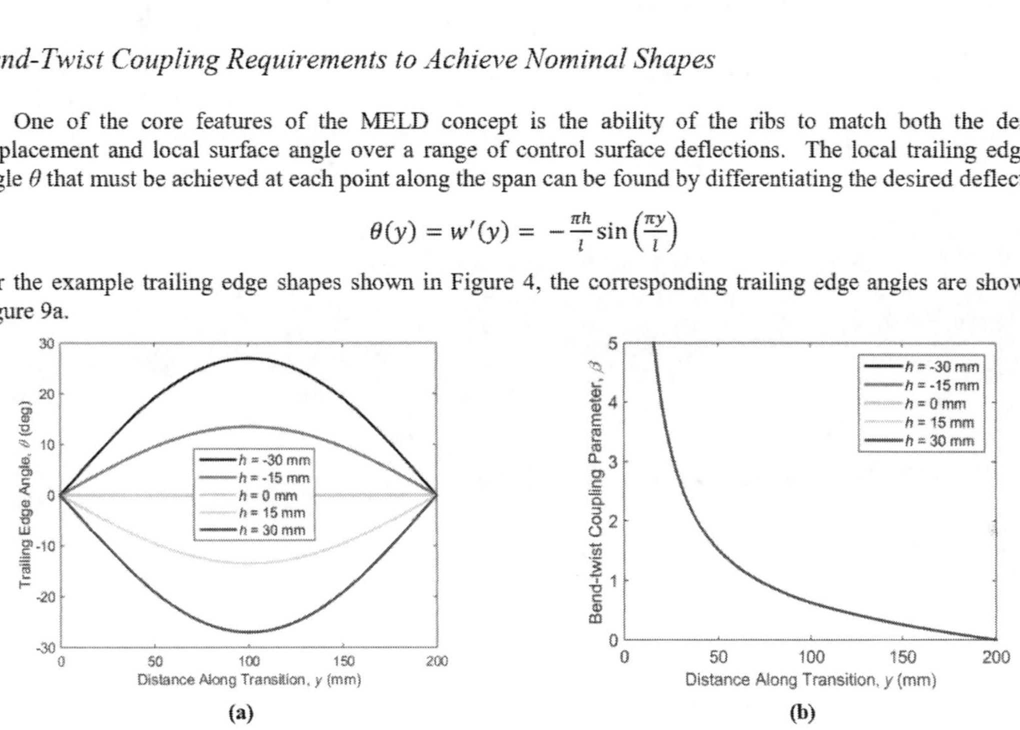

Figure 9. a). Distribution of trailing edge angle for the desired shapes and b). distribution of required bend-twist coupling parameter

We can now define a geometric bend-twist coupling parameter, β, which determines the ratio of rotation to deflection that the tip of each rib must achieve:

$$\beta(y) = \frac{\theta/c_t}{w/c_t^3} = \frac{\theta}{w}c_t^2 \tag{3}$$

The tip angle and deflection are both normalized using the length of the compliant portion of the rib c_t, which is shown in Figure 7, to make β a more generally comparable term that is less beholden to the specifics of the transition geometry. The tip deflection is normalized by c_t to the third power to reflect the distribution of deflection expected under a tip load according to linear elastic beam bending theory (for constant flexural rigidity) and the tip rotation angle is similarly normalized linearly by c_t in line with linear elastic torsional beam theory.[27]

For the example trailing edge shape given in Equation 1, we can evaluate Equation 3 to give the required distribution of bend-twist coupling for the desired shape:

$$\beta(y) = \frac{-\frac{\pi h}{l}\sin\left(\frac{\pi y}{l}\right)}{h*\cos\left(\frac{\pi y}{l}\right)-h}c_t^2 = \frac{\pi c_t^2}{l}\cot\frac{\pi y}{2l} \tag{4}$$

which is plotted in Figure 9b. It is interesting to note that the half amplitude, h drops out of the equation, such that the distribution of coupling required is independent of both sign and amplitude of the desired deflection. This is ideal of course as the control surface to which the MELD transition is attached would be expected to work over a range of upwards and downwards deflections and the value of β of a given rib is an intrinsic geometric property that cannot be changed once manufactured. Note also that the cotangent function approaches infinity at $y = 0$, so the vertical axis of Figure 9b has been truncated to the show the primary region of interest.

The relationship between the values of corrugation skew angle and bend-twist parameter for a given rib is governed by the manner in which the skewed corrugations induce coupling, which in turn is dependent on the detailed geometry and material properties of that rib. For this initial study, a series of ribs with varying skew angle (but equivalent geometry otherwise) were studied using the SolidWorks Simulation finite element analysis program to determine the impact that skew angle α had on the resulting bend-twist parameter β for that specific rib geometry. Both the tip deformation and rotation angle varied linearly with applied load for a wide range of loads, such that a single value of β could be ascribed for a given rib over the full range of deflections of interest for this study. The details of the rib geometry used for the initial demonstrator are discussed in the Concept Demonstrator section. The results of this FEA study are shown in Figure 10. It was found that a simple empirically fit tangent function:

$$\beta = A\tan(\alpha) \tag{5}$$

captured the relationship between β and α quite well when $A = 0.28$, which is plotted along with the FEA data in Figure 10. Note that while α is plotted in degrees, Equation 5 is evaluated in radians.

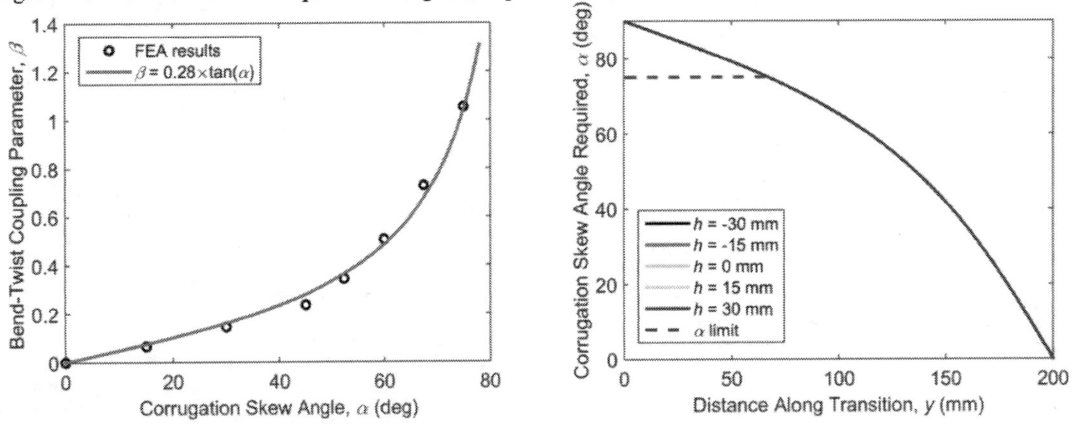

(a) **(b)**

Figure 10. Results from finite element analysis of ribs with an empirically fit tangent function showing the relationship between bend-twist coupling and skew angle, and b.) distribution of required skew angle along span

By combining Equation 5 and Equation 4 we can solve for the distribution of skew angle along the span required to achieve the distribution of bend-twist coupling shown in **Error! Reference source not found.**. While the required bend-twist coupling value approaches infinity as y approaches 0, the corresponding values of skew angle approach a finite value of $\alpha = 90°$, as can be seen in Figure 10b. This is due to the tangent function approaching infinity at 90°. However, it is not feasible to make skewed corrugated ribs with angles that high, as the chordwise length required for even a single segment of corrugation at values of α approaching 90° quickly exceeds the space available. Design work and 3D printing of specimens has shown a maximum practically achievable skew angle of roughly $\alpha = 75°$ for the ribs used here. This restraint is reflected in Figure 10b**Error! Reference source not found.** by the dashed line. The saturation of the skew angle implies that the initial portion of the transition (on the end closest to the rigid portion of the wing) would not have quite as much coupling as desired. However, while the desired coupling is high, the actual deformations and rotations required are quite low, implying that any errors in shape would be small relative to the overall size of the device. The impact of the skew angle saturation is further mitigated by the compliant connections which exist between the ribs which help smooth the shape, as discussed in the next section.

Connections between Rib Members

While the skew corrugated rib members are designed to automatically obtain the desired twist angle when bent, they still require a means of applying the required bending forces. This is accomplished in two ways in the present concept; through the skin and through supplementary corrugated strips. The elastomeric skin which covers the upper and lower surfaces of the MELD transition acts as a compliant connection between adjacent rib members. As it is bonded to both the fixed and moving ends of the control surface, when the control surface is deflected the skin will be directly stretched. Since the skin is also bonded to the ribs, it will to a certain extent bend the ribs along with it, and has some ability to apply force to the ribs. However, in cases where the skin is able to be designed to be quite soft (for lower dynamic pressures for example) or where the ribs are required to be particularly stiff, it is likely that the skin alone will not be able to ensure that sufficient force is applied to the ribs to create the required distribution of bending, leading to discontinuities in the deformed shape. To address this, the MELD concept includes thin spanwise corrugated strips that connect the ribs to the rigid and flexible ends of the transition. A wide range of stiffness properties and actuation energy requirements are achievable through varying the geometry, material properties, number, and chordwise location of the corrugated rib connection members, allowing the MELD transition to be adapted to a wide range of applications and operating conditions.

4. CONCEPT DEMONSTRATOR

To show the structural viability of the MELD transition concept, an initial prototype demonstrator has been designed and built. This demonstrator, the design model of which is shown in Figure 5, was sized for use with a 33% chord FishBAC (i.e. morphing starts at 67% chord) on a 600 mm chord wing section with a NACA 0012 profile. This gives the morphing portion of the demonstrator a chordwise length of 200 mm, with an additional 30 mm of chordwise length included for the non-morphing mounting frame and to provide a bonding surface for the elastomeric skin. The total spanwise length of the demonstrator is 250 mm, with the bend-twist coupled ribs totaling 200 mm and the fixed and morphing end ribs each spanning 25 mm. While these geometry parameters are nominal, they are meant to be representative of a range of wing structures useful for manned and unmanned aircraft, helicopter blades, and wind turbine blade applications. A total of twenty skewed corrugation ribs, each a nominal 10 mm wide, are spaced between the end ribs with corrugation skew angles ranging from 5.06° to 75°. The value of α for each rib was taken from **Error! Reference source not found.**, with the value corresponding to the spanwise center point of each rib being used. Extensions are included off of the corrugated portion of each rib that follow the airfoil contour to provide bonding pads for the silicone skin. The skewed corrugations and skin bonding extensions generally extend to within 60 mm of the trailing edge, although this value varies slightly with skew angle. The bend-twist coupling feature does not extend all the way to the trailing edge due to the limited space available in the sharp trailing edge, and due to the desire to have a rigid region at the end of the rib to create sufficient area to bond the skin to. A single corrugated rib connection member mounted 50 mm in from the trailing edge connects the end ribs to the bend-twist coupled ribs.

This model took advantage of additive manufacturing technology to simplify fabrication of the components, as the combination of skewed corrugations and airfoil profile shaped projections makes for an admittedly complex geometry which would be challenging to make using traditional metal or composite manufacturing methods. While the entire model could have been printed as a single monolithic part, as a means of risk reduction and to allow some ability to replace and change components, the ribs and support pieces were all printed individually. An HP DesignJet 3D Fused Deposition Modeling (FDM) printer was used to print the components in ABS plastic.

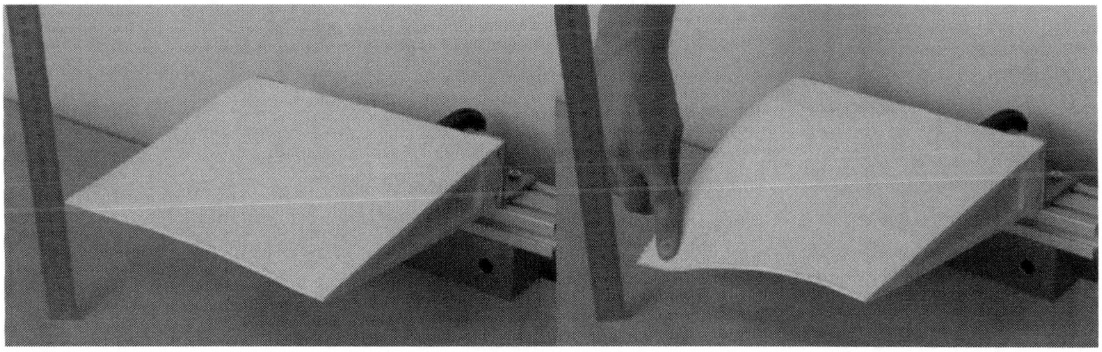

(a) (b)

Figure 11. MELD transition demonstrator as built: a) at rest and b) deforming due to tip load applied to camber morphing end

The silicone skin consists of two separate 1.5 mm thick silicone sheets pre-tensioned and bonded to the structure. Pre-tensioning increases out-of-plane bending stiffness of the skin allowing it to better resist aerodynamic loading, and eliminates buckling of the skin surface in areas which undergo compressive strains upon deformation. This approach is adopted directly from previous analytical and experimental work on the FishBAC, which has shown it to be an effective means of reducing undesired deformations in the compliant skin.

The assembled demonstrator is shown in Figure 11 cantilevered from its mounting frame. This MELD transition behaves as desired, exhibiting smooth and continuous surface shapes when deformed. The importance of the bend-twist coupling to the operation of the device can be seen by the behavior of the portion of the device over which it does not exist. As mentioned above, due to the thinness of the trailing edge, the skewed corrugations stop roughly 60 mm from the trailing edge. The lack of bending compliance and bend-twist coupling in this region leads to the formation of small stepped discontinuities, as can be faintly seen in Figure 11 at the very end of the trailing edge. This effect is small though and can likely be addressed in future designs with geometry changes and/or an additional rib connection member closer to the trailing edge. The lack of any such stepped shape over the rest of the MELD transition demonstrator as it elastically lofts between a prescribed 60 mm tip deflection at the morphing end and no deflection at the fixed end proves the efficacy of the compliant, bend-twist coupled design architecture. Although not shown in the figure, this demonstrator is fully capable of bidirectional deflections as desired.

While the ability of the demonstrator to achieve the desired smoothness in its deformed shape is encouraging, there is clearly more work required to establish the viability of the MELD transition concept. To this end, structural characterization of the demonstrator is ongoing in parallel to finite element analysis of the entire structure. From there, high fidelity modeling of the aerodynamic performance will need to be validated against wind tunnel data, and this coupled fluid-structure interaction problem will need to be solved in a manner which allows for optimization of the design of the MELD transition both aerodynamically and structurally.

5. CONCLUSIONS

In conclusion, this work has introduced a new morphing structure designed to seal the large gaps created at the end of a morphing camber control surface when it is used. These spanwise end gaps are a significant source of drag and noise in current aircraft, and there has been a lot of interest both historically and recently in the design of structures capable of sealing them. Here, a Morphing Elastically LofteD (MELD) transition was proposed which combines material and geometric compliance with bend-twist coupling to create a unique structural architecture capable of deforming in a smooth and continuous manner to connect the moving end of an active camber morphing control surface to the fixed wing structure around it. The inclusion of bend-twist coupling allows the structure to automatically deform in a manner which allows for both the desired local surface displacements and angles to be achieved. Existing work in this area was first introduced, followed by an overview of the design requirements for end transitions. The difficulty of analyzing these structures, particularly for initial design work, due to the strong coupling between aerodynamics and structural deflections and the highly three dimensional nature of the problem was highlighted. This led to the establishment of a nominal target shape as a starting point. The operating principles of the proposed MELD transition were then introduced, with particular attention paid to the design of the skewed corrugation features used in the rib sections as a geometric means of creating the desired bend-twist coupling. The distributions of trailing edge angle and bend-twist coupling required to achieve the target shape were then derived, and a finite element analysis study provided the relationship between corrugation skew angle and bend-twist coupling parameter for the proposed geometry, thereby allowing for the distribution of corrugation skew angle along the span to be determined. An initial technology demonstrator was built with this calculated geometry, and was shown to deform in the desired manner, creating a smooth, elastically lofted shape with minimal surface discontinuities. The basic operating principle of the MELD concept has therefore been demonstrated and the usefulness of the bend-twist coupling shown, although much work remains to be done to develop appropriate analysis tools and to show the efficacy of the concept in realistic operating environments.

6. ACKNOWLEDGEMENTS

The research leading to these results has received funding from the European Research Council under the European Union's Seventh Framework Programme (FP/2007-2013) / ERC Grant Agreement n. [247045].

7. REFERENCES

[1] Macaraeg, M. G., *Fundamental Investigations of Airframe Noise*, NASA Langley Research Center, Hampton, VA, USA: 2004.

[2] Michel, U., Helbig, J., Barsikow, B., Hellmig, M., and Schuettpelz, M., "Flyover noise measurements on landing aircraft with a microphone array," *4th AIAA/CEAS Aeroacoustics Conference*, Toulouse, France: American Institute of Aeronautics and Astronautics, 1998.

[3] Aviation Stack Exchange, "Why do some aircraft have multiple ailerons per wing? - Aviation Stack Exchange" Available: http://aviation.stackexchange.com/questions/921/why-do-some-aircraft-have-multiple-ailerons-per-wing.

[4] Kunz, R., "Apparatus for closing an air gap between a flap and an aircraft. US Patent No. US4471925A. Issued Sep 18, 1984," US4471925 A, September 18, 1984.

[5] Diller, J. B., and Miller, N. F., "Elastomeric transition for aircraft control surface. US Patent No. US6145791A. Issued Nov. 14, 2000.," US6145791 A, November 14, 2000.

[6] Caton, J. H., Hobey, M. J., Groeneveld, J. D., Jacobs, J. H., Wille, R. H., and Jr, L. O. B., "Control surface for an aircraft. US Patent No. US6349903B2. Issued Feb. 26, 2002.," US6349903 B2, February 26, 2002.

[7] Etling, K. A., "Morphing control surface transition. US Patent No. US8342447B2. Issued Jan. 1, 2013," US8342447 B2, January 1, 2013.

[8] Khorrami, M. R., Lockard, D. P., Moore, J. B., Su, J., Turner, T. L., Lin, J. C., Taminger, K. M., Kahng, S. K., and Verden, S. A., "Elastically deformable side-edge link for trailing-edge flap aeroacoustic noise reduction. US Patent No. US8695925B2. Issued Apr. 15, 2014," US8695925 B2, April 15, 2014.

[9] Khorrami, M. R., Humphreys, W. M., Lockard, D. P., and Ravetta, P. A., "Aeroacoustic Evaluation of Flap and Landing Gear Noise Reduction Concepts," Atlanta, GA, United States: 2014.

[10] Sreekantamurthy, T., Turner, T. L., Moore, J. B., and Su, J., "Elastomeric Structural Attachment Concepts for Aircraft Flap Noise Reduction - Challenges and Approaches to Hyperelastic Structural Modeling and Analysis," *55th AIAA/ASMe/ASCE/AHS/SC Structures, Structural Dynamics, and Materials Conference*, National Harbor, Maryland, USA: American Institute of Aeronautics and Astronautics, 2014.

[11] Kudva, J. N., "Overview of the DARPA Smart Wing Project," *Journal of Intelligent Material Systems and Structures*, vol. 15, Apr. 2004, pp. 261–267.

[12] Pankonien, A., and Inman, D. J., "Experimental testing of spanwise morphing trailing edge concept," *Proceedings of SPIE*, 2013, pp. 868815–868815–13.

[13] Da Rocha-Schmidt, L., and Baier, H., "A Shape Variable Gap Cover Concept for Aerodynamic Control Surfaces Based on Shear Deformation," *63. Deutscher Luft- und Raumfahrtkongress*, 2014.

[14] Hetrick, J., Osborn, R., Kota, S., Flick, P., and Paul, D., "Flight Testing of Mission Adaptive Compliant Wing," *48th AIAA/ASME/ASCE/AHS/ASC Structures, Structural Dynamics, and Materials Conference*, Honolulu, Hawaii: American Institute of Aeronautics and Astronautics, 2007.

[15] Lau, S., "Wing morphing experiment takes flight," *Professional Pilot Magazine*, Apr. 2015.

[16] Woods, B. K. S., and Friswell, M. I., "Preliminary Investigation of a Fishbone Active Camber Concept," *Proceedings of the ASME 2012 Conference on Smart Materials, Adaptive Structures and Intelligent Systems*, Stone Mountain, GA: 2012, pp. 555–563.

[17] Woods, B. K. S., Bilgen, O., and Friswell, M. I., "Wind tunnel testing of the Fish Bone Active Camber morphing concept," *Journal of Intelligent Material Systems and Structures*, Feb. 2014, p. 1045389X14521700.

[18] Kreitinger, R. L., and Middleton, D. B., *Aircraft Surface Coatings for Drag Reduction/Erosion Protection*, Warrendale, PA: SAE International, 1981.

[19] Wright, W. ., "Polymers in aerospace applications," *Materials & Design*, vol. 12, Aug. 1991, pp. 222–227.

[20] Woods, B. K. S., Dayyani, I., and Friswell, M. I., "Fluid-Structure Interaction Analysis of the Fish Bone Active Camber Morphing Concept," *Journal of Aircraft*, vol. 52, 2015, pp. 307–319.

[21] Abbott, I. H., and Doenhoff, A. E. V., *Theory of Wing Sections, Including a Summary of Airfoil Data*, New York, New York: Dover Publications Inc., 1959.

[22] Woods, B. K. S., Friswell, M., and Parsons, L., "Morphable Structure," GB1412159.4, July 8, 2014.

[23] Thill, C., Etches, J. A., Bond, I. P., Potter, K. D., and Weaver, P. M., "Composite corrugated structures for morphing wing skin applications," *Smart Materials and Structures*, vol. 19, Dec. 2010, p. 124009.

[24] Jones, R. M., *Mechanics of Composite Materials*, Philadelphia, PA: Taylor and Francis, Inc, 1999.

[25] Chandra, R., and Chopra, I., "Experimental and theoretical analysis of composite I-beams with elastic couplings," *AIAA Journal*, vol. 29, 1991, pp. 2197–2206.

[26] Don Lobitz, and Paul Veers, "Aeroelastic behavior of twist-coupled HAWT blades," *1998 ASME Wind Energy Symposium*, Reno, Nevada: American Institute of Aeronautics and Astronautics, 1998.

[27] Meirovitch, L., *Fundamentals of Vibrations*, New York, NY: McGraw Hill, 2010.

ICAST2015-010

Control and Analysis of a Magnetorheological Elastomer Dynamic Vibration Absorber for Powertrain Mount Systems of Automobiles

Li-Jun Qian, Fu-Long Xin and Xian-Xu Bai*

Department of Vehicle Engineering, Hefei University of Technology, Hefei 230009, China

Abstract

For powertrain mount systems of automobiles using semi-active magnetorheological elastomer (MRE) dynamic vibration absorbers (DVAs), this paper presents a rapid, stable, and effective control algorithm based on state observation. The state-space model for the powertrain mount systems with installation of MRE DVAs is established using the rank criterion method for observable systems. According to the principle of system reconfiguration, a full state observation model of adaptive Kalman filter with Sage-Husa noise estimator is developed. With the state vectors estimated by Kalman filter, the phase difference between the displacement of the dynamic mass of the MRE DVA relative to the powertrain and the absolute displacement of the powertrain is updated continuously based on Simpson's rule. By adjusting the applied current to the MRE DVA with fuzzy logic corresponding to the cosine value of the phase difference, the natural frequency of the MRE DVA could track the excitation frequency of the powertrain well, which results in vibration attenuation of the powertrain mount system.
Keywords: Magnetorheological elastomer (MRE), dynamic vibration absorber (DVA), powertrain mount system, state observation, fuzzy control.

1. INTRODUCTION

Automobile powertrains, composed of engine, transmission, and clutch, is the power source for automobiles and also one of the main vibration sources in automobiles [1]. In order to isolate the vibration transmitting from powertrain to body, engine mount components are usually installed between powertrain and body. The system composed of mount components and powertrain is powertrain mount system. Semi-active mount components are the devices based on smart materials, such as electro- and magneto- rheological materials [2]. Semi-active mount component designed based on principle of dynamic vibration absorbers (DVAs) is one of them. The semi-active DVA could effectively attenuate the vibration of powertrain through adjusting its natural frequency to track the excitation frequency [3]. The elastic element of DVAs are often made of smart materials to have excellent shift-frequency performance, and magnetorheological elastomers (MREs) are one of widely adopted materials [4,5]. MREs have the advantages of rapidly, continuously and reversibly controllable stiffness characteristics, and MREs are suitable to be used as the elastic element of DVAs and have received increasing attention [6-10].

In recent years, several references on structural design of MRE DVAs and vibration control based on MRE DVAs could be found. Xu *et al.* added a voice coil motor which could produce active controllable force to decrease the impact of the damping effect of MRE DVA acting on the vibration

* Corresponding author. Tel: (+86) 189 6379 4953; E-mail: bai@hfut.edu.cn

attenuation performance [6]. In the applications of automobiles vibration attenuation, Hoang *et al.* developed a DVA used for the transmission system, and the test results showed that the transmission's vibration could be significantly reduced by using the MRE DVA [7]. On the control of MRE DVA, Ni *et al.* presented a control algorithm which can realize tuning of the natural frequency of MRE DVA by controlling the compressional force [8]. Kim *et al.* fitted and obtained a mathematical expression between the natural frequency of MRE DVA and the applied magnetic induction intensity [9]. The required applied magnetic field is calculated by measuring the excitation frequency, and the simulation results showed that the proposed algorithm has good performance. Liao *et al.* proposed a phase based stiffness tuning algorithm [10], their research results showed that the control algorithm could efficiently and rapidly attenuate the vibration of the primary system.

However, because of the complex working conditions of powertrains, the excitation of powertrains changes continuously and irregularly, the excitation frequency cannot be identified accurately. In addition, since the shear storage modulus of MRE depends not only on the applied magnetic field, but also the strain and the excitation frequency, it is extremely difficult to build an accurate mathematical model between the shear storage modulus of MRE and the applied magnetic field. Hence, the traditional control algorithms based on frequency identification are difficult to work effectively to attenuate the vibration of powertrain mount systems. Therefore, it is of importance and essential to design a fast and stable algorithm to realize effective control of the MRE DVA for the powertrain mount systems based on the working characteristics of the MRE DVAs.

In this paper, for the MRE DVAs applied in powertrain mount systems of automobiles, a state observation based control algorithm is proposed. The state-space model for the powertrain mount system with installation of the MRE DVA is established. The phase difference between the displacement of the dynamic mass of the MRE DVA relative to the powertrain and the absolute displacement of the powertrain is updated continuously. By adjusting the natural frequency of MRE DVA via fuzzy logic, the vibration of the powertrain mount system could be controlled effectively. In order to validate the effectiveness of the proposed algorithm with fuzzy steps, a MATLAB/Simulink model of the semi-active powertrain mount system based on MRE DVA is established.

2. SEMI-ACTIVE POWERTRAIN MOUNT SYSTEM BASED ON A MRE DVA

2.1 MRE DVA

The conventional DVAs could not support the primary system and extra installation spaces for the DVAs is required, which restrains the applications of DVAs for powertrain mount systems of automobiles. Combining the characteristics of powertrain mount systems and the working conditions of MREs, the principle of the MRE DVA is proposed. The schematic of the MRE DVA [3] is shown in figure 1. As seen from figure 1, the proposed MRE DVA is composed of a semi-active vibration absorbing unit and a passive vibration isolation unit. The vibration absorbing unit is composed of magnetic conductor, shearing sleeve, bobbin core, electromagnetic coil winding, and MRE. The vibration isolation unit is composed of strut, rubber spring, upper housing and lower housing. The magnetic conductor, the bobbin core and the electromagnetic coil winding are supported on the shearing sleeve through the MRE. The MRE is vulcanized between the shearing sleeve and the bobbin core, and the shearing sleeve is fixed to the lower housing. The magnetic conductor moves up and down vertically along the shearing sleeve. The magnetic field produced by the electromagnetic coil winding starts from the bobbin core, and passes through the magnetic conductor and the shearing sleeve, then goes through the MRE and forms a closed loop. The MRE works in pure shear mode. The magnetic conductor, the bobbin core and the electromagnetic coil winding form the dynamic mass of the MRE DVA together.

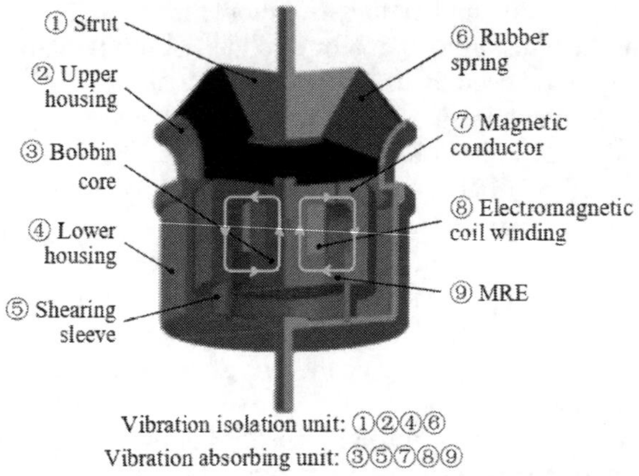

Figure 1. The schematic and configuration of the MRE DVA [3].

When the MRE DVA as presented in figure 1 is installed to the powertrain mount system, the vibration isolation unit is utilized to support the powertrain and used as a passive vibration isolator. The vibration absorbing unit is utilized to absorb the vibration energy. When the excitation frequency changes, by tuning the applied current, the shear storage modulus of the MRE as well as the natural frequency of the MRE DVA can be tuned, which results in vibration attenuation of the powertrain mount systems.

2.2 Vibration characteristics of a MRE DVA based powertrain mount system

During the driving process of automobiles, the main external excitation of the powertrain mount system is the engine's inertia force due to the pistons' reciprocating movement [1]. The main vibration of the powertrain mount system is the vertical one and the powertrain mount system can be simplified to a single degree-of-freedom vibration system. Using the MRE DVA shown in figure 1, the dynamic model of the powertrain mount system could be simplified as a two degree-of-freedom semi-active model as presented in figure 2, and the dynamic model can be expressed as

$$\begin{bmatrix} m_1 & 0 \\ 0 & m_2 \end{bmatrix}\begin{bmatrix} \ddot{x}_1 \\ \ddot{x}_2 \end{bmatrix} + \begin{bmatrix} c_1 & -c_1 \\ -c_1 & c_1 + c_2 \end{bmatrix}\begin{bmatrix} \dot{x}_1 \\ \dot{x}_2 \end{bmatrix} + \begin{bmatrix} k_1 & -k_1 \\ -k_1 & k_1 + k_2 \end{bmatrix}\begin{bmatrix} x_1 \\ x_2 \end{bmatrix} = \begin{bmatrix} 0 \\ f(t) \end{bmatrix} \tag{1a}$$

$$f(t) = F\cos\omega t \tag{1b}$$

where m_1 and m_2 are the dynamic mass of the MRE DVA and the mass of the powertrain, respectively; x_1 and x_2 are the displacement of the dynamic mass of the MRE DVA and the displacement of the powertrain, respectively; k_1 and c_1 are the stiffness and damping of the MRE of the MRE DVA, respectively; k_2 and c_2 are the stiffness and damping of the rubber spring of the MRE DVA, respectively; $f(t)$ is the excitation force of the powertrain; F is the amplitude of the excitation force; ω is the excitation frequency. From equation (1), the displacement of powertrain mount system can be solved

$$\begin{bmatrix} \boldsymbol{X_1} \\ \boldsymbol{X_2} \end{bmatrix} = \frac{F_0}{\det\boldsymbol{Z}}\begin{bmatrix} k_1 + jc_1\omega \\ k_1 - m_1\omega^2 + jc_1\omega \end{bmatrix} \tag{2}$$

where $\boldsymbol{X_1}$ and $\boldsymbol{X_2}$ are the Fourier transforms of x_1 and x_2; det$\boldsymbol{Z}$ is determinant of matrix $\boldsymbol{Z}$, and $\boldsymbol{Z}$ is the displacement impedance matrix of the powertrain mount system, yields

$$\boldsymbol{Z} = \begin{bmatrix} k_1 - m_1\omega^2 + jc_1\omega & k_1 + jc_1\omega \\ k_1 + jc_1\omega & k_1 + k_2 - m_2\omega^2 + j(c_1 + c_2)\omega \end{bmatrix} \tag{3}$$

From equations (1)-(3), the displacement of the powertrain can be expressed as

$$\boldsymbol{X_2} = \frac{F_0\sqrt{(\omega^2 m_1 - k_1)^2 + \omega^2 {c_1}^2}}{\sqrt{C^2 + D^2}} \tag{4a}$$

$$C = -k_1k_2 + \omega^2 m_2 k_1 + \omega^2 c_1 c_2 + \omega^2 m_1 k_1 + \omega^2 m_1 k_2 - \omega^4 m_1 m_2 \tag{4b}$$

$$D = -\omega k_1 c_2 - \omega k_2 c_1 + \omega^3 c_1 m_2 + \omega^3 c_1 m_1 + \omega^3 c_2 m_1 \tag{4c}$$

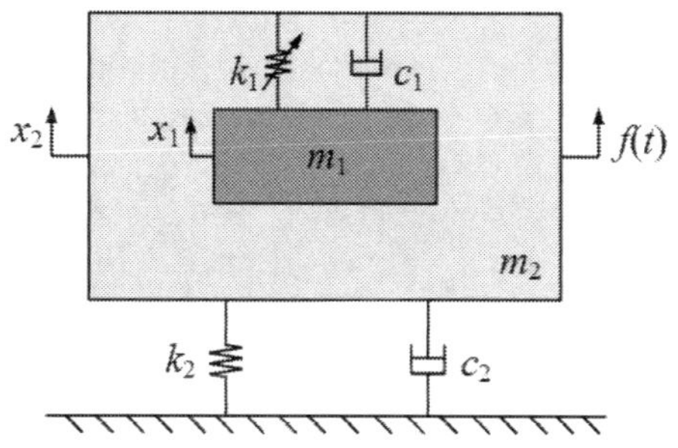

Figure 2. The schematic of a semi-active SDOF powertrain mount system using a MRE DVA.

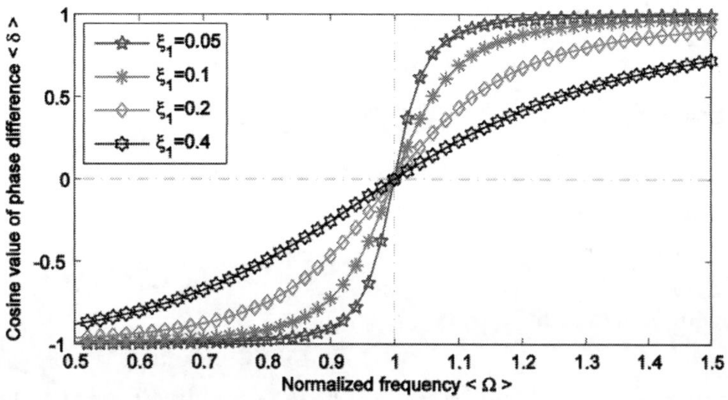

Figure 3. The relationship between the cosine value of phase difference between the displacement of the dynamic mass of the MRE DVA relative to the powertrain and the absolute displacement of the powertrain and the normalized frequency with different damping ratios.

According to equation (4), when the natural frequency of the MRE DVA $\left(\omega_1 = \sqrt{\frac{k_1}{m_1}}\right)$ equals to the excitation frequency (*i.e.*, when $k_1 = \omega^2 m_1$ is satisfied), the displacement of the powertrain achieves the minimum. That is to say, to attenuate the vibration displacement of the powertrain most effectively, the natural frequency of the MRE DVA should track the excitation frequency accurately and real-timely.

Transforming equation (2), yields

$$\frac{\boldsymbol{X}_\mathrm{r}}{\boldsymbol{X}_2} = \frac{1}{(\Omega^2-1)+2j\xi_1\Omega} \tag{5}$$

where $\boldsymbol{X}_\mathrm{r}$ is the Fourier transform of the displacement x_r of the dynamic mass of the MRE DVA relative to the powertrain; Ω is the normalized frequency; ξ_1 is the damping ratio of the MRE of the MRE DVA. $\boldsymbol{X}_\mathrm{r}$, Ω and ξ_1 could be respectively expressed as

$$x_\mathrm{r} = x_1 - x_2 \tag{6a}$$

$$\Omega = \frac{\omega_1}{\omega} \tag{6b}$$

$$\xi_1 = \frac{c_1}{2\sqrt{m_1 k_1}} \tag{6c}$$

From equation (5), the cosine value of the phase difference between the relative displacement $\boldsymbol{X}_\mathrm{r}$ and the absolute displacement $\boldsymbol{X}_2$ can be given by

$$\delta = \begin{cases} -\frac{(\Omega^2-1)}{(\Omega^2-1)^2+(2\xi_1\Omega)^2} & \Omega < 1 \\ 0 & \Omega = 1 \\ \frac{(\Omega^2-1)}{(\Omega^2-1)^2+(2\xi_1\Omega)^2} & \Omega > 1 \end{cases} \tag{7}$$

According to equation (7), the relationship between the cosine value δ and the normalized frequency Ω with different damping ratios ξ_1 is shown in figure 3. As seen from figure 3, no matter what value ξ_1 is, when the normalized frequency $\Omega = 1$ (*i.e.*, $k_1 = \omega^2 m_1$) is satisfied, $\delta = 0$ is always true; when the normalized frequency $\Omega < 1$ (or $\Omega > 1$) is satisfied, $\delta < 0$ (or $\delta > 0$) always holds. Hence, the information of δ could be employed to judge whether the natural frequency of MRE DVA can track the excitation frequency accurately. When $\delta < 0$ (or $\delta > 0$) is satisfied, it means that the natural frequency of MRE DVA is less (or greater) than the excitation frequency, and the applied current should be tuned up (or down) to increase (or decrease) the natural frequency of MRE DVA. When $\delta = 0$ is satisfied, it means that the natural frequency of MRE DVA equals to the excitation frequency, the vibration displacement of the powertrain has already achieved the minimum, there is no need to adjust the applied current.

The displacement $\boldsymbol{X}_{\mathrm{r}}$ of the dynamic mass of the MRE DVA relative to the powertrain and the absolute displacement $\boldsymbol{X}_2$ of the powertrain are harmonic signals with a same frequency. The cosine value δ of the phase difference between $\boldsymbol{X}_{\mathrm{r}}$ and $\boldsymbol{X}_2$ can be calculated by the normalized dot product, according to Simpson's rule [11],

$$\delta = \frac{\boldsymbol{X}_{\mathrm{r}} \cdot \boldsymbol{X}_2}{2\overline{\boldsymbol{X}_{\mathrm{r}}}\,\overline{\boldsymbol{X}_2}} \tag{8}$$

where $\overline{\boldsymbol{X}_{\mathrm{r}}}$ and $\overline{\boldsymbol{X}_2}$ are the modulus of $\boldsymbol{X}_{\mathrm{r}}$ and $\boldsymbol{X}_2$, respectively.

3. CONTROL ALGORITHM

3.1 Adaptive Kalman filtering state observer

Based on the analyses of figure 3, equations (7) and (8) could be applied to judge whether the natural frequency of the MRE DVA can track the excitation frequency. However, it should be noted here, when calculating the value of δ, the real-time signals of $\boldsymbol{X}_{\mathrm{r}}$ and $\boldsymbol{X}_2$ are necessary and at least two sensors are required to guarantee that the control algorithm could work efficiently. It would not only add the cost of the control system, but also increase the complexity of the hardware system. In this study, based on the principle of system reconfiguration, by constructing suitable state observation equations and estimating the state vectors based on the actually measured signals of the selected output vectors, $\boldsymbol{X}_{\mathrm{r}}$ and $\boldsymbol{X}_2$ can be obtained in real time.

According to equation (1), choosing the displacement of powertrain as the output vectors, the dynamic model of the two degree-of-freedom semi-active powertrain mount system, as shown in figure 2, could be converted into state space equation,

$$\dot{\boldsymbol{X}}(t) = \boldsymbol{A}\boldsymbol{X}(t) + \boldsymbol{B}_1 u(t) + \boldsymbol{B}_2 f(t) \tag{9a}$$

$$\boldsymbol{Y}(t) = \boldsymbol{C}\boldsymbol{X}(t) \tag{9b}$$

$$\boldsymbol{X}(t) = [x_1 \quad \dot{x}_1 \quad x_2 \quad \dot{x}_2]^{\mathrm{T}} \tag{9c}$$

$$\boldsymbol{A} = \begin{bmatrix} 0 & 1 & 0 & 0 \\ -\frac{k_{10}}{m_1} & -\frac{c_1}{m_1} & \frac{k_{10}}{m_1} & \frac{c_1}{m_1} \\ 0 & 0 & 0 & 1 \\ \frac{k_{10}}{m_2} & \frac{c_1}{m_2} & \frac{-(k_{10}+k_2)}{m_2} & \frac{-(c_1+c_2)}{m_2} \end{bmatrix} \tag{9d}$$

$$\boldsymbol{B}_1 = \begin{bmatrix} 0 \\ -\frac{1}{m_1} \\ 0 \\ \frac{1}{m_2} \end{bmatrix} \tag{9e}$$

$$\boldsymbol{B}_2 = \begin{bmatrix} 0 \\ 0 \\ 0 \\ -\frac{1}{m_2} \end{bmatrix} \tag{9f}$$

$$\boldsymbol{C} = [0 \quad 0 \quad 1 \quad 0] \tag{9g}$$

where $\boldsymbol{X}$ and $\boldsymbol{Y}$ are the state vector and the output vector, respectively; $\boldsymbol{A}$, $\boldsymbol{B}_1$, $\boldsymbol{B}_2$, $\boldsymbol{C}$ are the system matrixes; $u(t)$ is the active force produced by the MRE DVA; k_{10} is the stiffness of the MRE at field-off state. It should be noted that the MRE DVA at field-off state is assumed as a passive DVA. The natural frequency of the passive MRE DVA equals to the excitation frequency corresponding to the idling conditions of the powertrain mount system. k_{10} and $u(t)$ could be expressed as

$$k_{10} = m_1\omega_0^2 \tag{10a}$$

$$u(t) = (m_1\omega^2 - k_{10})(x_1 - x_2) \tag{10b}$$

where ω_0 is the excitation frequency corresponding to the idling conditions of the powertrain mount system.

Kalman filter could accurately estimate the system state vectors [12]. A Kalman filtering state observer for the powertrain mount system is established and utilized to estimate the system state vectors. As Kalman filter is suitable for the state observation of discrete systems, the continuous system expressed by equation (9) needs to be discretized. According to the rules of discretization, selecting T_s as the sampling period of discretization, the state space equation of discrete system are expressed as

$$\boldsymbol{X}_k = \boldsymbol{\Phi}_{k,k-1}\boldsymbol{X}_{k-1} + \boldsymbol{\Psi}_{k,k-1}u_{k-1} + \boldsymbol{\Gamma}_{k,k-1}f_{k-1} \tag{11a}$$

$$\boldsymbol{Y}_k = \boldsymbol{H}_k\boldsymbol{X}_k + v_k \tag{11b}$$

where $\boldsymbol{X}_k$ and $\boldsymbol{Y}_k$ are the state vector of the discrete system and the output vector of the discrete system, respectively; u_k is the active force of the discrete system; $\boldsymbol{\Phi}_{k,k-1}$, $\boldsymbol{\Psi}_{k,k-1}$, $\boldsymbol{\Gamma}_{k,k-1}$ and $\boldsymbol{H}_k$ are the system matrixes; f_k and v_k are the system noise and measurement noise, respectively, and the covariances of f_k and v_k are Q_k and R_k, respectively. $\boldsymbol{\Phi}_{k,k-1}$, $\boldsymbol{\Psi}_{k,k-1}$, $\boldsymbol{\Gamma}_{k,k-1}$ and $\boldsymbol{H}_k$ can be expressed as

$$\boldsymbol{\Phi}_{k,k-1} = \mathrm{expm}(\boldsymbol{A}T_s) \tag{12a}$$

$$\boldsymbol{\Psi}_{k,k-1} = \boldsymbol{A}^{-1} \cdot (\boldsymbol{\Phi}_{k,k-1} - \boldsymbol{I}) \cdot \boldsymbol{B}_1 \tag{12b}$$

$$\boldsymbol{\Gamma}_{k,k-1} = \boldsymbol{A}^{-1} \cdot (\boldsymbol{\Phi}_{k,k-1} - \boldsymbol{I}) \cdot \boldsymbol{B}_2 \tag{12c}$$

$$\boldsymbol{H}_k = \boldsymbol{C} \tag{12d}$$

where $\boldsymbol{I}$ is the unit matrix.

When automobiles driving in different conditions, as the changes of traffic roads and operation requirements from drivers, the excitation amplitude and excitation frequency of the powertrain always change real-timely. That is to say, the statistical features of the system noise are unknown, which makes it difficult for traditional Kalman filters work effectively. In this study, a full state observation model of adaptive Kalman filter with Sage-Husa noise estimator [13] is built to realize the adaptive observation of the system state vectors. The established adaptive Kalman filtering state observer could estimate the state vectors of the MRE DVA based powertrain mount system accurately, even the statistical features of system noise are unknown.

3.2 Stiffness control with fuzzy steps

Based on the vibration characteristics of the MRE DVA based powertrain mount system as analyzed in section 2.2, by measuring the displacement of powertrain, the displacement of the dynamic mass of the MRE DVA can be estimated by adaptive Kalman filtering state observer. According to Simpson's rule, the cosine value of the phase difference between the displacement of the dynamic mass of the MRE DVA relative to the powertrain and the absolute displacement of the powertrain can be calculated. Based on the information of the cosine value δ, whether the natural frequency of the MRE DVA can track the excitation frequency accurately could be judged and the applied current could be adjusted correspondingly. In the light of the relationship between the cosine value δ and the normalized frequency Ω, as given by equation (7), a control algorithm based on state observation is proposed

$$I_{\mathrm{control}} = -\gamma \int \delta \, dt \tag{13}$$

where I_{control} is the applied current of MRE DVA; γ is the positive integrating amplification factor.

As the range of the excitation frequency of powertrain is wide, and its changing speed varies with the working condition of powertrain, the integrating amplification factor γ must be related to the working condition of powertrain. However, an accurate mathematical model for the relationship between γ and the parameters of powertrain mount system is not easy to develop.

Based on the characteristics of powertrain, the fuzzy control algorithm is utilized to obtain the relationship between γ and the parameters of powertrain mount system [14,15]. The absolute value of the cosine value of phase difference $|\delta|$ and its changing rate ε are selected as the input variables. The integrating amplification factor γ is utilized as the output variable of the fuzzy control algorithm. $|\delta|$ is divided into five subsets: very small (VS), small (S), middle (M), big (B) and very big (VB). ε is divided into five subsets: very small (VS), small (S), middle (M), big (B) and very big (VB). γ is divided into five subsets: very small (VS), small (S), middle (M), big (B) and very big (VB). The degrees of membership of $|\delta|$, ε and γ are shown in figure 4.

As shown in figure 4, as $|\delta|$ changes slowly when closing to 1 and changes fast when closing to 0, the intervals of $|\delta|$ closing to 0 are denser than the intervals closing to 1, correspondingly. The physical meaning of ε is similar with that of $|\delta|$, so the intervals of ε are divided by the same principle as $|\delta|$. The intervals of γ are divided into five even intervals. When $|\delta|$ gets close to 1 and ε gets close to 0, it means that the natural frequency of the MRE DVA is getting far away from the excitation frequency of the powertrain, and a bigger value of γ is needed. When $|\delta|$ gets close to 0 and ε gets close to 1, a smaller value of γ is needed. According to this logical relationship, the fuzzy logic of the integrating amplification factor are established.

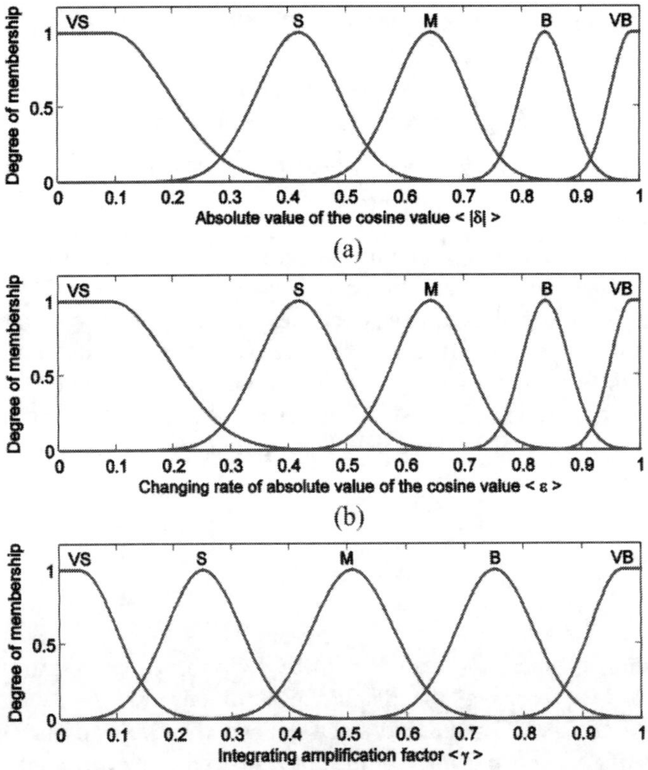

(c)

Figure 4. The degrees of membership of fuzzy subsets: (a) degree of membership of absolute value of the cosine of phase difference; (b) degree of membership of changing rate of absolute value of the cosine value of phase difference; (c) degree of membership of integrating amplification factor.

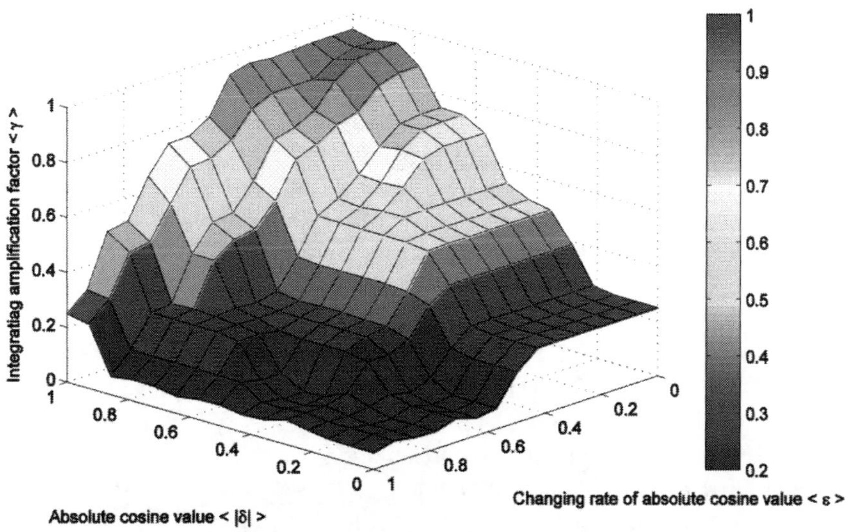

Figure 5. The surface of the fuzzy logic.

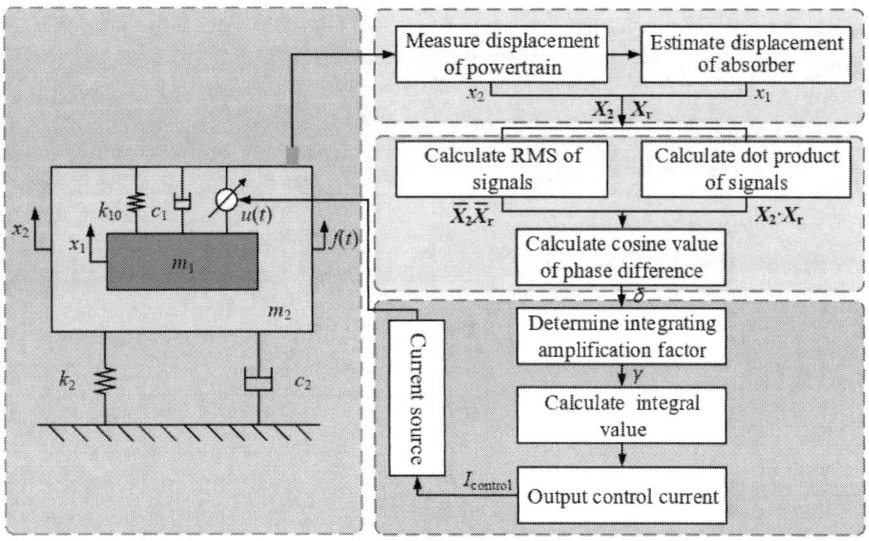

Figure 6. The flowchart of the state observation based control algorithm with fuzzy steps.

Figure 5 presents the surface of the fuzzy logic. As seen from figure 5, γ increases with $|\delta|$ closing to 1 and decreases with ε closing to 1, which means that γ is determined by the working condition of the powertrain. Based on the control algorithm expressed by equation (13), a state observation based control algorithm with fuzzy steps is proposed, and the flowchart of the proposed control algorithm with fuzzy steps is presented in figure 6.

As shown in figure 6, the proposed control algorithm with fuzzy steps could be divided into three processes: state observation process, phase difference calculation process and applied current output process. During the state observation process, by measuring x_2 with the sensor installed on the powertrain, x_1 can be estimated by adaptive Kalman filtering state observer, and x_r can be obtained subsequently. Then the phase difference calculation process starts, by transmitting x_2 and x_r into the Fourier transforms expressed as $\boldsymbol{X}_2$ and $\boldsymbol{X}_r$, δ could be calculated based on Simpson's rule. Subsequently, the algorithm goes into the applied current output process, according to the information of δ, $I_{control}$ as well as γ could be determined by fuzzy logic. At last, the applied current $I_{control}$ is output to the MRE DVA by a controllable current source and the tracking progress of the natural frequency of the MRE DVA to the excitation frequency completes.

4. SIMULATION AND ANALYSIS

In the process of simulation experiments, a typical powertrain with transverse four-cylinder four-stroke gasoline engine is taken as an example. Assuming that the working condition of the powertrain changes timely: the powertrain initially works in a stable process (excitation amplitude and excitation frequency remain unchanged), then changes into an unstable process (excitation amplitude and excitation frequency change continuously), and then experiences a stable process, an unstable process, and a stable process. During these processes, the excitation of powertrain is expressed as [16]

$$f(t) = \begin{cases} F_0 \sin(\Omega_0 t) & 0 \le t \le T_1 \\ [F_0 + \mu(t - T_1)] \sin[\Omega_0 t + \lambda(t - T_1)t] & T_1 < t \le T_2 \\ F_1 \sin(\Omega_1 t) & T_2 < t \le T_3 \\ [F_1 - \mu(t - T_3)] \sin[\Omega_1 t - \lambda(t - T_3)t] & T_3 < t \le T_4 \\ F_0 \sin(\Omega_0 t) & T_4 < t \le T_5 \end{cases} \quad (14)$$

where F_0 and Ω_0 are the excitation amplitude and excitation frequency in the initially stable process, respectively; F_1 and Ω_1 are the excitation amplitude and excitation frequency in the stable process during the running procedure, respectively; μ and λ are the changing rate of excitation amplitude and excitation frequency in the unstable process, respectively.

The vibration attenuation performances of a proposed control algorithm with a fixed step and a traditional FFT control algorithm [8] are presented to illustrate the superiority of the proposed algorithm with fuzzy steps. The simulation results are shown in figure 7. The parameters for the semi-active powertrain mount system are listed in table 1.

Table 1 Parameters for the semi-active powertrain mount system.

Parameter	Value	Parameter	Value
m_1 / kg	5	r_k	0
m_2 / kg	70	R_k	2e-13
ξ_1 / -	0.1	b	0.99
ξ_2 / -	0.05	F_0/ N	20
ω_0 / (rad/s)	2π×10	F_1/ N	100
T_s / s	1e-3	λ/ (rad/s^2)	2π×0.5
Ω_0 / (rad/s)	2π×25	μ/ (N/s)	8
Ω_1 / (rad/s)	2π×30	T_1 / s	10
T_2 / s	20	T_3 / s	30
T_4 / s	40	T_5 / s	50

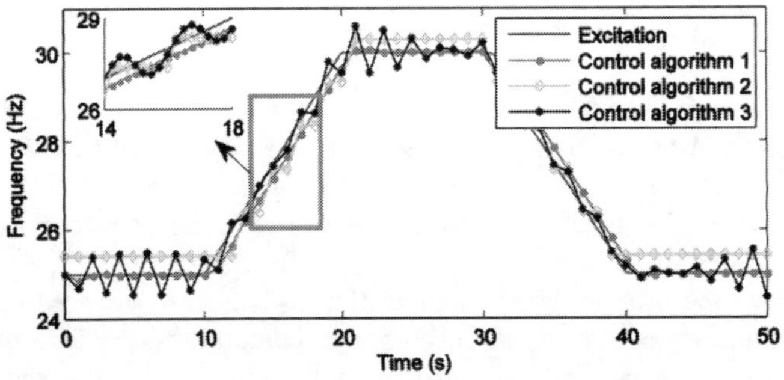

(a)

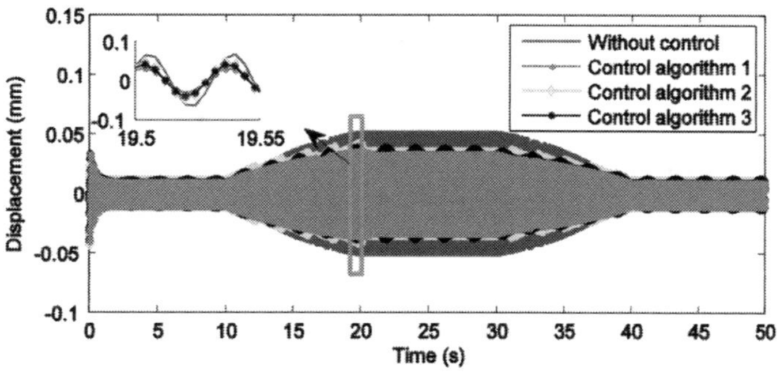

(b)

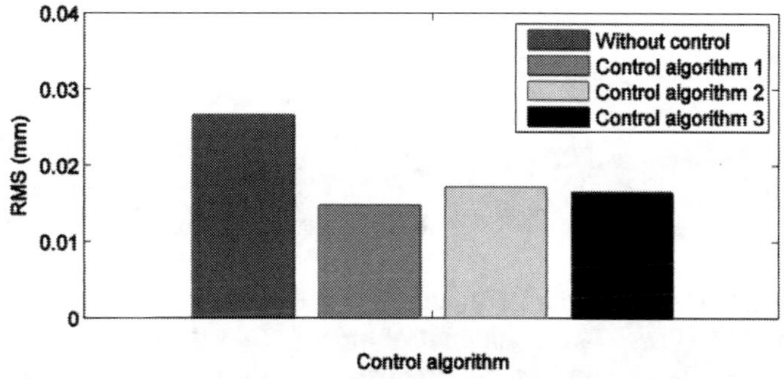

(c)

Figure 7. The vibration attenuation performance of powertrain mount system with the control of the proposed algorithm with fuzzy steps: (a) the tracking process of the natural frequency of the MRE DVA to the excitation frequency with different control algorithms, (b) the displacement of powertrain with different control algorithms, and (c) the RMS of the displacement of powertrain with different control algorithms.

Control algorithm 1: the proposed control algorithm with fuzzy steps; Control algorithm 2: the FFT control algorithm; Control algorithm 3: the proposed control algorithm with a fixed step.

Figures 7(a), 7(b) and 7(c) present the tracking process of the natural frequency of the MRE DVA to the excitation frequency with different control algorithms, the displacement of the powertrain with different control algorithms, and the root mean square (RMS) of the displacement of powertrain with different control algorithms. As shown in figure 7(a), with the control of the proposed algorithm with fuzzy steps, when the engine's excitation frequency changes, the natural frequency of the MRE DVA can follow rapidly and accurately. Though the natural frequency of the MRE DVA controlled by the proposed algorithm with a fixed step and the traditional FFT algorithm could also track the excitation frequency, the track errors are bigger. The natural frequency of the MRE DVA controlled by FFT algorithm tunes in stepped variation. This is because the frequency identification accuracy of FFT method for transient signal is limited, which leads to the jumps among the frequency resolutions corresponding to FFT method. The natural frequency of the MRE DVA controlled by the proposed control algorithm with a fixed step tunes in fluctuant variation due to the relatively large selected integration amplification factor. When the applied current of the MRE DVA is controlled by the algorithm, the natural frequency of the MRE DVA changes too fast and is easy to overshoot the excitation frequency. Then the algorithm repeatedly adjusts the changing orientation of the applied current, which results in the oscillated phenomenon of the natural frequency of the MRE DVA. It can be seen from figure 7(b), with the proposed control algorithm with fuzzy steps, the vibration displacement of the powertrain is attenuated obviously as compared with the other two control algorithms.

To evaluate the effectiveness of the vibration attenuation of different algorithms, the RMS of the displacement of the powertrain is adopted. The RMS of the displacement of the powertrain is calculated by

$$X_{\mathrm{RMS}} = \sqrt{\frac{\sum_{i=1}^{N} x_2^2}{N}} \tag{15}$$

where N is the total sample number.

Observing figure 7(c), the RMS without control reaches 0.0267 mm, the RMS with FFT control algorithm is 0.0172 mm, the RMS with the proposed control algorithm with a fixed step is 0.0165 mm, and the RMS with the proposed control algorithm with fuzzy steps is as low as 0.0148 mm. As compared with the FFT control algorithm and the proposed control algorithm with a fixed step, the RMS with the proposed control algorithm with fuzzy steps decreases by 14.0% and 10.3%, which indicates the proposed control algorithm with fuzzy steps can attenuate the vibration of powertrain more effectively.

4. CONCLUSIONS

Based on the characteristics of powertrain mount systems, a rapid, stable, and effective control algorithm for MRE DVA applied for powertrain mount systems of automobiles was proposed and investigated. A full state observation model of adaptive Kalman filter is built and the state vectors of the powertrain mount system can be estimated accurately. The fuzzy logical relationship between the integrating amplification factor and the parameters of the powertrain mount system is established. Utilizing the established MATLAB/Simulink control model of a semi-active powertrain mount system based on MRE DVA, the effectiveness of the proposed algorithm with fuzzy steps is validated. The results show that the proposed algorithm with fuzzy steps could control the MRE DVA effectively. As compared with the FFT control algorithm and the proposed control algorithm with a fixed step, the RMS of the displacement of the powertrain utilizing the proposed control algorithm with fuzzy steps decreases by 14.0% and 10.3%, respectively.

ACKNOWLEDGMENTS

The authors wish to acknowledge the National Natural Science Foundation of China (Grant No. 51305114), Anhui Provincial Natural Science Foundation (Grant No. 1408085QE91) and Fundamental Research Funds for the Central Universities (No. 2014HGCH0008), for their support of this research.

REFERENCES

1. Yu Y H, Naganathan N G and Rao V D, 2001, A literature review of automotive vehicle engine mounting systems, *Mechanism and Machine Theory*, Vol. 36, No. 1, 123-142
2. Bai X X, Wereley N M and Hu W, 2015, Maximizing semi-active vibration isolation utilizing a magnetorheological damper with an inner bypass configuration, *Journal of Applied Physics*, Vol. 117, No. 17, C711 (4pp)
3. Bai X X, Xin F L, Qian L J and Kan P, 2015, Design and control of a magnetorheological elastomer dynamic absorber for powertrain mount systems of automobiles, *ASME 2015 Conference on Smart Materials, Adaptive Structures and Intelligent Systems (SMASIS2015), Colorado Springs, USA*, Paper No.: SMASIS2015-8893 (7pp), September 21-23
4. Li Y C, Li J C, Li W H and Du H P, 2014, A state-of-the-art review on magnetorheological elastomer devices, *Smart Materials and Structures*, Vol. 23, No. 14, 123001 (24pp)
5. Yang C Y, Fu J, Yu M, Zheng X and Ju B X, 2015, A new magnetorheological elastomer isolator in shear-compression mixed mode, *Journal of Intelligent Material Systems and Structures*, Vol. 26, No. 10, 1290-1300

6. Xu Z B, Gong X L, Liao G J and Chen X, 2010, An active-damping-compensated magnetorheological elastomer adaptive tuned vibration absorber, *Journal of Intelligent Material Systems and Structures*, Vol. 21, No. 10, 1039-1047
7. Hoang N, Zhang N, Li W H and Du H P, 2013, Development of a torsional dynamic absorber using a magnetorheological elastomer for vibration reduction of a powertrain test rig, *Journal of Intelligent Material Systems and Structures*, Vol. 24, No. 16, 2036-2044
8. Ni Z C, Gong X L, Li J F and Chen L, 2009, Study on a dynamic stiffness-tuning absorber with squeeze-strain enhanced magnetorheological elastomer, *Journal of Intelligent Material Systems and Structures*, Vol. 20, No. 10, 1195-1202
9. Kim Y K, Koo J H, Kim K S and Kim S, 2011, Developing a real time controlled adaptive MRE-based tunable vibration absorber system for a linear cryogenic cooler, *IEEE/ASME International Conference on Advanced Intelligent Mechatronics*, Vol. 32, No. 14, 287-290
10. Liao G J, Gong X L and Xuan S H, 2014, Phase based stiffness tuning algorithm for a magnetorheological elastomer dynamic vibration absorber, *Smart Materials and Structures*, Vol. 23, No. 1, 015016 (10pp)
11. Singh P and Singh V, 2012, The duality property of the Discrete Fourier Transform based on Simpson's rule, *Mathematical Methods in the Applied Sciences*, Vol. 35, No. 7, 776-781
12. Liu J and Kalnay E, 2008, Estimating observation impact without adjoint model in an ensemble Kalman filter, *Quarterly Journal of the Royal Meteorological Society*, Vol. 134, No. 634, 1327–1335
13. Kulikov G Y and Kulikova M V, 2015, The accurate continuous-discrete extended Kalman filter for continuous-time stochastic systems, *Russian Journal of Numerical Analysis and Mathematical Modelling*, Vol. 30, No. 4, 239-249
14. Kidner M and Brennan M, 2002, Varying the stiffness of a beam-like neutralizer under fuzzy logic control, *Journal of Vibration and Acoustics*, Vol. 124, No. 1, 90-99
15. Precup R E, Preitl S, Tar J K, Tomescu, M L, Takács M, Korondi P and Baranyi P Z, 2008, Fuzzy control system performance enhancement by iterative learning control, *IEEE Transactions on Industrial Electronics*, Vol. 55, No. 9, 3461-3475
16. Wang M Y, Manoj R and Zhao W, 2001, Gear rattle modelling and analysis for automotive manual transmissions, *Proceedings of the Institution of Mechanical Engineers Part D Journal of Automobile Engineering*, Vol. 215, No. 4, 241-258

ICAST2015 #078

Beetle Inspired Deployable Structures

Kazuya Saito[1*], and Yoji Okabe[1]

[1] Institute of Industrial Science, The University of Tokyo, Tokyo, Japan

Abstract

Foldable hindwings in beetles are one of the ultimate deployable structures. They have sufficient strength and stiffness to tolerate 20–1000 beats per second in the flight position, although they can be folded and unfolded nearly instantly depending on the situation. This study focuses on the hindwing in ladybird beetles and aims to reveal their wing-folding/unfolding mechanisms. These beetles use their wing elasticity for deployment and can therefore quickly take off. First, on the basis of the reports of entomologists, the authors describe the factors related to the actuation of insect wing folding and unfolding. Then, the detailed motions involved in take-off and wing storing in ladybird beetles are observed by a high-speed camera. The analysis of these movies reveals the detailed processes of wing-unfolding and folding. Next, the crease patterns and the transformations of the veins are clarified by using the paper model. Wing-folding/unfolding mechanisms in these insects have various unique capabilities which will provide new possibilities for designing deployable structures.

1. INTRODUCTION

Foldable hindwings in insects are one of the ultimate deployable structures and have large potential for engineering applications, because they have sufficient strength and stiffness to tolerate 20–1000 beats per second in the flight position, although they can be folded and unfolded instantly depending on the situation. In previous works, entomologists reported various types of wing folding. Danforth et al. studied the longitudinally folded forewing of ants and bees (Hymenoptera)[1]. The fan-like folding found in locusts (Orthoptera) and praying mantises (Dictyoptera) are illustrated by Wootton[2]. Among wing folding insects, beetles (Coleoptera) are known to use the most highly diverse wing folding patterns and mechanisms [3-11]. Unlike six legs, the wings of insects are considered to be derived from the tergal exoskeleton. Their folding and unfolding mechanisms involve the elastic behaviors of structures, internal forces driven by muscle or blood pressure and external forces from other body parts. Understanding these mechanisms requires not only analyzing the materials and structures of the wings but also careful observations of how they are actuated in living specimens.

This study focuses on a group of beetles that have elastic-deploying-type wings, and aims to reveal their wing-folding/unfolding mechanisms. These beetles use wing elasticity for deployments and therefore can quickly take off . On the other hand, when it comes to wing storing, they have to fold their wings against elastic forces. Usually, this process requires the supports of other body parts: the abdomen and elytra[5-9]. Note that these folding movements can enable the beetles to achieve two different objectives at the same time: quick wing-folding and efficient storage of elastic energy for subsequent wing deployment. Therefore, detailed investigations of these movements are expected to

*saito-k@iis.u-tokyo.ac.jp

generate useful knowledge in the development of excellent self-deploying systems. This study uses high-speed cameras to reveal the details of these motions. As a typical example of this type wings, this paper focuses on ladybird beetles (Coleoptera: Coccinellidae). First, on the basis of the reports of entomologists and our own observations, the factors relating to the actuation of insect wing folding and unfolding are described. Next the detailed motion involved during take-off and wing storing in ladybird beetles are observed using high-speed cameras. The analysis of these movies reveals the detailed processes of wing-unfolding and folding in ladybird beetles. Finally, the characteristics of wing-folding mechanism are summarized, and the potentials for engineering applications are discussed. The discussion also includes an outline of the future works required on the engineering side to reveal the detailed systems of these excellent deployable structures.

2. ACTUATION FACTORS IN HINDWING UNFOLDING AND FOLDING

Studies by entomologists have included investigations of the mechanisms used in wing-folding/unfolding, as well as the crease patterns of the wings. These mechanisms are influenced by various factors represented by the following list.

Muscle: Muscles are indispensable in wing actuations, but this does not mean that they control all folding/unfolding movements. In foldable hindwings, the veins have many articuli and hinges, but no muscle is observed on these inner bending points. Wing actuating muscles are concentrated on wing bases. They mainly work for flapping in flights, but these basal muscular actions have been shown to be used in wing-folding/unfolding in previous studies[10]. Haas reported on the skeletomuscular apparatus of the metathorax, which is used in wing-unfolding in beetles[11].

Wing elasticity: Many entomologists have pointed out the importance of the intrinsic elasticity of the wings in folding/unfolding. Forbes noted that the wings of certain beetles relax into the folded position at rest[3]. In the same way, the wings of another species of beetles show relaxation to unfolded shapes[5]. These elastic behaviors are observed in wings cut from living or freshly killed specimens. Brakenbury classified the wing-folding mechanism of beetles into two categories depending on the direction of their 'spring back'[6]. In category A, wings are folded automatically and become stable at the resting position as a result of the elastic force. In contrast, wings in category B are unfolded automatically. This study mainly focuses on the latter.

Abdomen: In the folding of unfolding-stable-type (category B) wings, insects must fold against the elastic force acting in the deploying direction. This elastic force is so strong that muscle actuation alone cannot accomplish the folding, and support is needed from other body parts. As discussed in this study, many species are known to use the abdomen in folding. Ladybird beetles fold their wings by "pumping" movements of the abdomen[5]. Rove beetles have freely movable abdomens and fold their wings in a more direct way[7-9]. In many species, the abdomens are equipped the after-mentioned wing-folding spicules[5].

Elytra: The elytra, which are one of the most vital aspects of beetle anatomy, are also used to support wing-folding by holding the mid-folded hindwings in many species[10]. If the elytra are lost, these beetles cannot fold their hindwings. The elytra in some beetles also seem to function as a hold-on release mechanism for elastically-deploying hindwings.

Wing-folding spicules: Many references address the micro spicule patches, which are observed on the abdomens of some species of beetles[5-8]. Wing-atrophied beetles are known to have no such spicules on their abdomens. These spicules are believed to support wing-folding by acting as a barb and helping the abdomen to catch the hindwings in folding movements.

Other potential factors include the blood pressure, which may straighten the hindwings[6,7], but this has not been experimentally confirmed. A similar hydraulic mechanism is known to be used for wing inflation during emergencies[12]. Our observations with the high-speed camera indicate that centrifugal forces caused by wing flapping also play a major role in the wing deployment of some beetles with folding-stable-type wings.

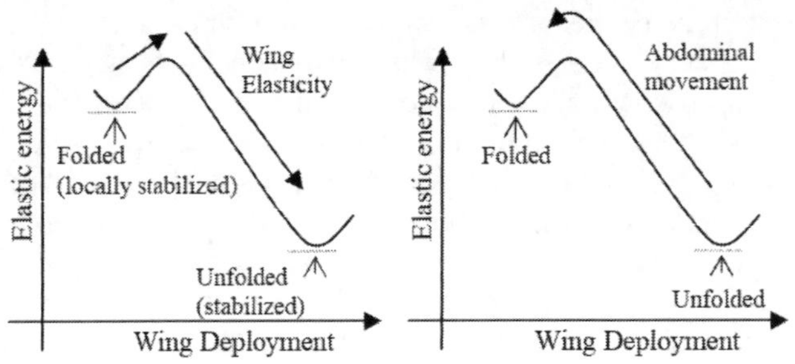

Figure 1. Explanation by Haas's bistable model for unfolding-stable wings.

In wing transformations in unfolding-stable-type wings, driving forces are generated mainly by muscle actuation, wing elasticity, and movements of the abdomen. Although quantitative evaluation of the contributions of each factor requires further engineering studies, Haas proposed convincing explanations of the relationship between intrinsic elasticity and other folding/unfolding mechanisms by using the bistable model[11,13]. He indicated that the wings of beetles may be considered bistable structures with two local stabilization points in both folding and unfolding processes, rather than a simple spring system with a single stable point. This model considers Brackenbury's two categories as states possessing lower elastic energy rather than states possessing a stable point. Figure 1 shows the schematic of the Haas's bistable model in unfolding-stable-type wings. In this wing type, unfolding is easy to achieve because it can be triggered by a small force and automatically progresses by elastic force. Then, muscle actuations are thought to be used mainly to overcome the critical point of elastic energy. On the other hand, the folding process requires a stronger force to overcome the peak of the elastic energy. Therefore, the external force from other body parts is required.

3. WING UNFOLDING MOVEMENTS IN LADYBIRD BEETLES

Ladybird beetles seem to be better at flying than other beetles, and they frequently take off and land, even in captivity. Therefore, they are suitable subjects for research into wing-unfolding/folding mechanisms. Some previous work, including analysis of the crease patterns of hindwings and folding methods[5-7, 13], has been done on this species, but analyses using movies or serially-taken photographs are limited. This study uses high-speed cameras to acquire slow-motion movies of take-off and wing-folding, to clarify the detailed motions of the wings and the abdomen. Specimens were of the species Coccinella septempunctata (Linnaeus), and were captured in the grass fields of Japan. The specimens were put in an acrylic box, and were kept in focus by simultaneously moving the box on the table to keep them in the frame.

Figure 2 shows the take-off process of the ladybird beetle. Ladybird beetles have typical unfolding-stable wings and can deploy their hindwings and quickly take off. Their wing developing elasticity is easily confirmed by the fact that the wings of anesthetized or freshly killed specimens spring back into the unfolded state with mild stimulation. This spring back is activated not only by opening elytra; therefore, it is assumed that the folding state also confers local stability, and that some supporting mechanisms trigger expansion.

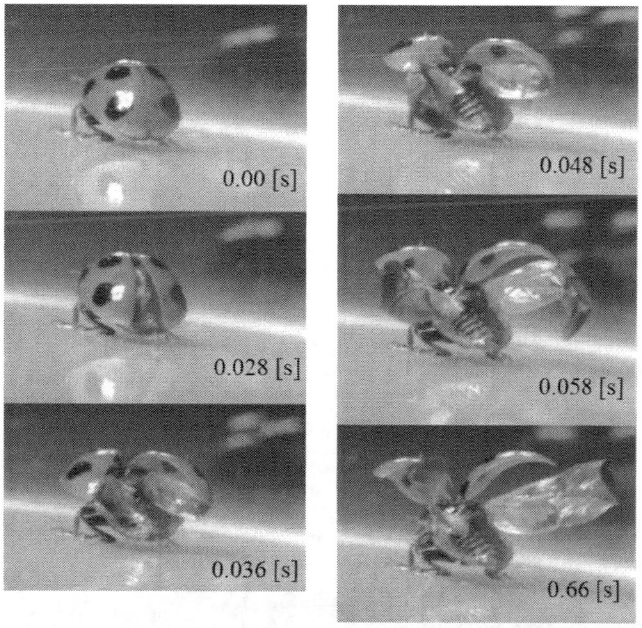

Figure 2. Wing deploying motion of a ladybird beetle. A high-speed camera (HAS-L2, DITECT Co., Ltd., Tokyo, Japan) was used to acquire slow-motion movies at 500 frames per second.

4. CREASE PATTERN AND PAPER MODEL

The crease patterns in most species of beetles have been already clarified in previous works. Representative crease patterns in ladybird beetles can be found in Ref [4] and [10]. However, these crease pattern figures were designed for the purpose of indicating the positional relationship between crease lines and venation, therefore they cannot replicate folded shapes. In order to conform the folded shapes and transformation of the veins, we modified these crease patterns and made a paper model of the hindwings (Fig. 3). This figure is intended to be used to make a paper model that supports understanding of each folding process. This model intentionally ignores the small creases found in adjacent veins and margins of wings and modify the cross angles of crease lines with preference for proper expression of the folded shape and for easy folding.

Unlike in the case of fan-like folding, hindwings in beetles can be folded into both length and width directions. In most large-size beetles, the hindwings have one transverse folding [8] and can be folded once into length direction. Hindwings in ladybird beetles are characterized by having two transverse folding (the principal transverse fold (PTF) and the apical tranverse fold (ATF)) and can be folded into Z-shapes on these folding lines. From an engineering standpoint, this characteristic is beneficial for engineering applications including the development of the space deployable structures and morphing wings because these crease patterns can confer high storing efficiency.

With the support of knowledge obtained from high-speed movies and the paper model, the wing deployment process can be explained as follows. Figure 4 shows the successive pictures of the deformation of the left hindwing during the unfolding process. A high-speed camera (Phantom V1611, Novi Tech Co., Ltd., Tokyo, Japan) was used to acquire slow-motion movies at 3000 frames per second. Focusing on the movements of the veins, it is conformed that two types of movements lead the wing deployment. One is the opening movement in basal two veins (blue arrows in Fig. 4), and the

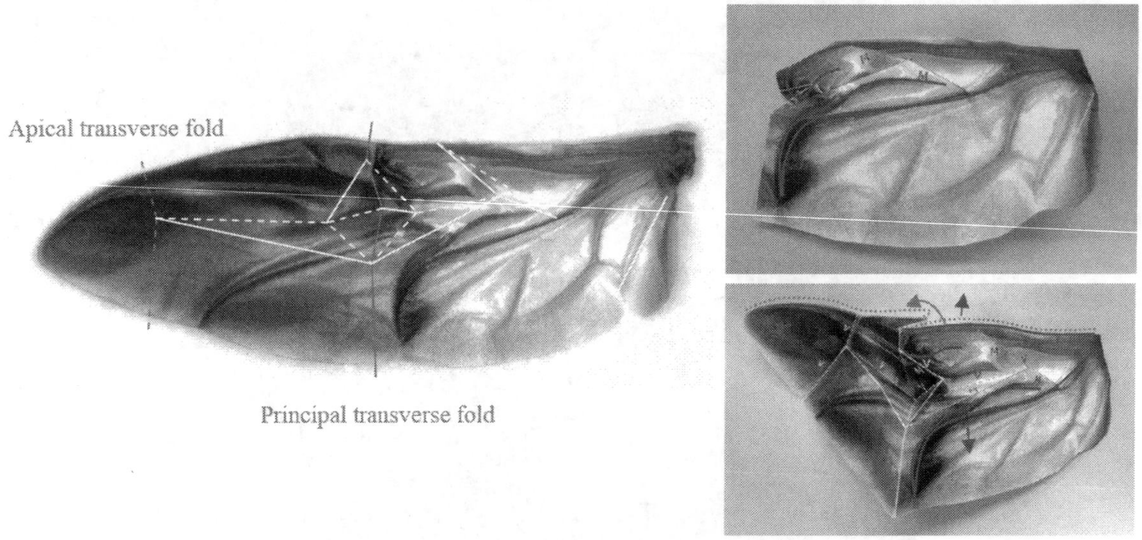

Figure 3. Crease pattern and venation in the hindwing of the ladybird beetle. (Solid line: Mountain fold, dotted line: valley fold).

Figure 4. Successive pictures of left hindwing unfolding in a ladybird beetle.

other is the unbending movement in the anterior margin (red arrows). The former is thought to be actuated by the basal muscles. This type actuation mechanism was already found in the metathorax of other beetles [13]. In the early phase of the deployment, th basal two veins are opened by these muscular actuations, these movements operate simultaneously with the unbending motion on the anterior margin. Based on our microscopic observations, these bending points have no clear joints therefore seem to have some kind of the compliant hinges. The deployment progress at certain points, the wing is deployed automatically by elastic forces in the anterior margin. These movements can be replicated by the proposed paper model (Fig. 3 right).

5. WING FOLDING IN LADYBIRD BEETLES

Figure 5 shows the wing-folding process in the ladybird beetle. After landing, the elytra are folded ahead of the hindwings(Fig. 5A, B). The hindwings are simultaenously aligned backward and strayed out of the closed elytra. In this process, it is seen that the hindwings are slightly curved along the roundness of the abdomen. Next, the ladybird beetle moves its abdomen up and down (Fig. 5B-C and D-E), called "pumping" in previous works. This movement seems to have the effect for rubbing the hindwings on the underside of the elytra. To observe the mid-folded shapes of the hindwings, some specimens had one elytron cut off (Fig. 6). During the folding process, the hindwings are held by the action of the wing-folding spicules and the elytra and folded into Z-shape on the folding lines of the principal transverse fold and the apical transverse fold (Fig. 3). With a few up-down movements, the hindwings are completely folded and stored under the elytra. These wing-folding movements of ladybird beetles are very simple. Nevertheless, they can quickly fold the hindwings (2–3 seconds) into predefined shapes. Note that these folded wings store the elastic energy for the next development, allowing ladybird beetles to easily take off again.

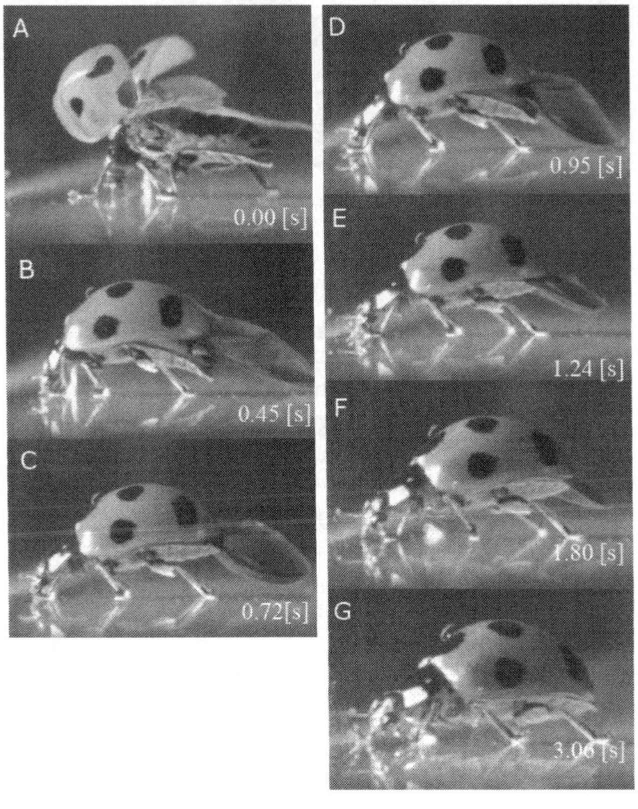

Figure 5. Wing-folding process of a lady bird beetle.

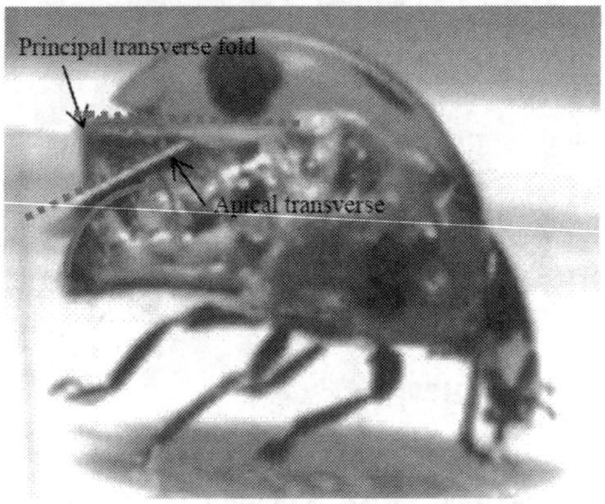

Figure 6. Abdominal movement and mid-folded shape of the hindwing. Right wings are removed in this specimen. The hindwing (dotted line) is folded into Z-shape by the principal transverse fold and apical transverse fold.

5. DISCUSSION & CONCLUSION

This study mainly focused on unfolding-stable-type wings, but the study of folding-stable-type beetles may also give useful knowledge. It is observed that flapping starts before wing unfolding in the take off process of many species of this type beetles. They are believed to use centrifugal forces and aerodynamic forces caused by wing flapping for wing deployments. If the beetle lands on a tree and stops flapping, its hindwings are folded automatically by their intrinsic elasticity. The benefit of this type of wing deployment is that it require no additional muscles or mechanisms for unfolding. Focusing on other group of insects, earwigs (Dermapteran) are known to have folding-stable-type wings. This folding can confer high storage efficiency. The previous studies on earwigs include an interesting hypothesis about the elastically stabilizing or reinforcing mechanisms in wings[13], which is expected to be verified on the engineering research side.

Entomologists have proposed various interesting ideas about the geometry, material, and structure of foldable hindwings of insects. However, despite large potential for engineering applications, few verification studies have been undertaken on the engineering side. In the light of these previous works, future work should include the following three main themes.

Geometry (Crease pattern): Haas proposed simplified geometrical models of crease patterns in beetles based on the analysis of a four-degree vertex[14]. In Ref [8], the pivot folding which consists of two triangle facets and four-degree vertices are described as the representative pattern in beetle wing-folding.

Structure: The structures and mechanisms of wing bases and the articuli of veins have been investigated well, but there are few studies about the mechanical properties of the whole venation. The cross-sectional shape and connecting pattern of veins are thought to contribute to the bistability of the wings. A three-dimensional (3D) scanner and a micro CT can be used to acquire accurate 3D models of wings and veins, and the finite element analysis of these models will enable quantitative analysis on bistability in beetle wings. A 3D modeling system will be also helpful for experimental verification.

Material: Haas et al. investigated the distribution of resilin, a rubber-like protein that contributes to the elasticity of the wings of beetles[15]. This material covers the folding lines and is thought to protect the portion producing large strain during folding/unfolding. Further studies on wing materials will contribute to the development of morphing structures that can tolerate a large number of repeated uses. Investigation of the detailed material properties of the veins and membranes also remains a challenge.

The forms and mechanisms of insect bodies are indivisibly united with their habits. Therefore, detailed observation of living specimens is indispensable. This study is positioned as the starting point for the above-mentioned biomimetic studies. Wing folding/unfolding mechanisms in insects have various unique capabilities, which will provide new possibilities for the design and utilization of deployable structures. Immediate applications include space-deployable structures represented by the solar array paddles and antenna reflectors of satellites. As self-deployment by intrinsic elasticity requires no mechanical actuators, it will contribute to reducing the weight of satellite bodies and improving the reliability of their deployment systems. Moreover, it is expected to broaden the design possibilities for articles of daily use, such as umbrellas and fans

ACKNOWLEDGMENTS

The authors thank Dr. Munetoshi Maruyama and Mr. Shuhei Yamamoto (Kyushu University, Japan) for giving entomological advice. Ms. Akiko Igarashi has supported the capture and upkeep of the ladybird beetles. This work was supported by the special fund of Institute of Industrial Science, the University of Tokyo, and JSPS KAKENHI Grant Number 24860024.

REFERENCES

1. Reichert, G., "Helicopter Vibration Control –Survey," *Vertica*, Vol. 5, No. 1, 1981, pp. 1-20.
2. Lowey, R. G., "Helicopter Vibrations: A Technological Perspective," *Journal of the American Helicopter Society*, Vol. 29, No. 4, Oct. 1984, pp. 4-30.
3. Danforth, B.N., Michener, C.N., "Wing folding in Hymenoptera", *Ann Entomol Soc Am* 81, 1988, pp.342-349.
4. Wootton, R.J., "Geometry and mechanics of insect hindwing fans: A modelling approach", *P Roy Soc B-Biol Sci 262*, 1995, pp.181-187.
5. Forbes, W.T.M., "How a beetle folds its wings", *Psyche 31*, 1924, pp.254-258.
6. Forbes, W.T.M., "The wing folding pattern of the Coleoptera", *J N Y Entomol Soc 24*, pp.42-68, 1926, pp.91-139.
7. Hammond, P.M., "Wing-folding Mechanisms of Beetles, with special reference to investigations of adephagan phylogeny (Coleoptera) ", *CARABID BEETLES, THEIR EVOLUTION, NATURAL HISTORY, AND CLASSIFICATION*, eds Erwin, T.L., Ball, G.E., & Whitehead, D.R., 1979, pp.113-180.
8. Brackenbury, J.H., "Wing folding and free-flight kinematics in Coleoptera (Insecta): a comparative study", *J Zool (Lond) 232*, 1994, pp.253-283.
9. Schneider, V.P., "Flight types and hing wing folding in Coleoptera", *Zool Jahrb Abt Anat Ontog Tiere 99*, 1978, pp.174-210.
10. Blum, V.P., "Phylogeny and ecological significance of elytrareduction and abdomen-mobility of the Staphylinidae (Coleoptera): Comparative and functional morphological studies", *Zool Jahrb Abt Syst Oekol Geogr Tiere 102*, 1979, pp.533-582.
11. Saito, K., Yamamoto, S., Maruyama, M., Okabe, Y., "Asymmetric hindwing folding in rove beetles", *PNAS, 111*(46), 2014, pp.16349-16352.
12. Fedorenko, D.N., *"Evolution of the Beetle Hind Wing, With Special Reference to Folding (Insects, Coleoptera) "*, Pensoft Publishers, Bulgaria. 2009.
13. Haas, F., Beutel, R.G., "Wing folding and the functional morphology of the wing base in Coleoptera", *Zoology 104*, 2001, pp.123-141.
14. Kishimoto, N. et.al., "New Deployable Membrane Structure Models Inspired by Morphological Changes in Nature," 2006, *AIAA*-2006-1898.

15. Haas, F., Gorb, S., Wootton, R.J., "Elastic joints in Dermapteran hind wings: Materials and wing folding". *Arthropod structure & development 29*, 2000, pp.137-146.
16. Haas, F., Wootton, R.J., "Two basic mechanisms in insect wing folding", *Proc R Soc Lond B Biol Sci 263*, 1996, pp.1651-1658.
17. Haas, F., Gorb, S., Blickhan, R., "The function of resilin in beetle wings", *Proc R Soc Lond B Biol Sci 267*, 2000, pp.1375-1381.

ICAST2015 #035

Design of Shape Morphing Cellular Structures and Their Actuation Methods

Robin M. Neville[1*], Fabrizio Scarpa[1]

[1] Advanced Composites Centre for Innovation and Science (ACCIS), University of Bristol, Queen's Building, University Walk, Bristol BS8 1TR, United Kingdom
* rn7318@bristol.ac.uk

Abstract

Kirigami techniques are used to create a new type of "open" honeycomb geometry with shape morphing characteristics. Cables are embedded in the structure to actuate the shape morphing. We present techniques for designing the manufactured shape of the honeycomb, and an analytical model to predict the deformed shapes.

1 INTRODUCTION

In this work we present a family of novel honeycombs produced using Kirigami techniques. Kirigami is the ancient Japanese art of cutting and folding paper. This technique can be used to form a 2D sheet material into a 3D cellular structure. By applying these techniques to sheet materials such as composites and thermoplastics, engineering cellular structures wth complex shapes can be produced. The idea of engineering Kirigami was first explored by Nojima and Saito [1]. Saito, Pellegrino and Nojima developed the mathematical framework linking the 2D cutting pattern to the 3D cellular structure, which enables the creation of specially shaped honeycombs without requiring any machining [2]. They also present a method for applying Kirigami to fibre-reinforced composites [2].

This work builds on that of Nojima and Saito, and introduces the concept of an "open" honeycomb (see Figure 1). We name this architecture "open" honeycomb because it lacks a closed cell topology (we follow the naming conventions of foams, which are also called "open" or "closed" depending on their cell configuration). The folds in the structure give it significantly different mechanical behaviour compared to closed honeycombs. The folds give rise to two new parameters – fold angle α and fold stiffness k. In past work we have characterised the effect of α and k on the mechanical behaviour of open honeycombs, and found that the folds give the structure reduced density, significant anisotropy, and variable stiffness [3], [4].

In this work we exploit two key features which arise from using the Kirigami process.

1. The folds can also function as "hinges" which allow the structure to flex in the Y-direction, while retaining stiffness in the X-direction. This can be expoited to create a morphing or deployable structure.
2. By including circles in the 2D cutting pattern, holes are produced in the walls of the 3D structure. We can then thread cables through these holes to actuate the structure.

The honeycomb studied in this work is an open honeycomb specifically designed for morphing (see Figure 2). We call this a "bridged" honeycomb because of the bridges which link adjacent rows of cells. These bridges are intended to reduce the stress on the folds (by providing two 90° folds instead of one 180° fold), and also to provide a flat surface for potential skin attachment.

In this work we present the following:

1 A method for designing useful shape open honeycombs. This is very similar to Saito's method [2], but is a bit more complicated because there are more variables, and more room for design.
2 A method for modelling the structure as a series of rigid linkages, which can be used to predict the deformed shape of the structure in response to a cable load.

Figure 1. Traditional honeycomb (left) and open honeycomb (right). The folds in the open honeycomb are highlighted yellow.

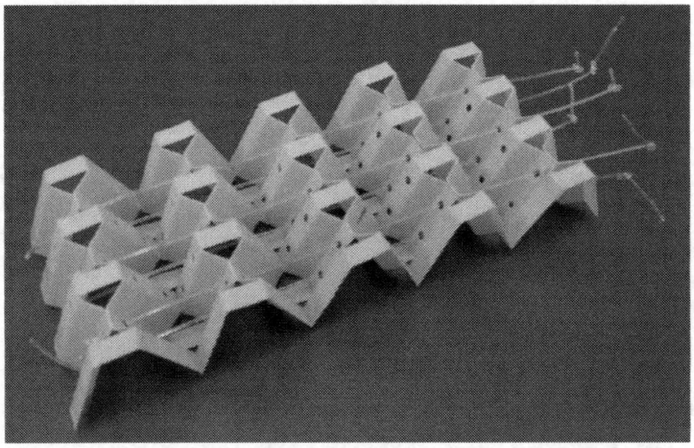

Figure 2. The "bridged" honeycomb used in this work.

2 KIRIGAMI MANUFACTURING

The Kirigami manufacturing process consists of three main steps:

1. **Cutting:** a pattern of slits and holes are cut into the sheet material (see Figure 3a,b,c). The spacing and dimensions of the 2D cutting pattern directly affect the dimensions of the 3D structure.
2. **Corrugating:** the sheet material is thermoformed between hexagonal tooling in a hot press. The slits and holes align with the tool such that the slits describe half a hexagon (see Figure 3d,e). Different shaped tools can be used to produce different shaped cells.

3. **Folding:** the corrugated sheet is folded repeatedly back on itself such that the semi-hexagonal slits open up into hexagonal holes (see Figure 3f,g). The result is an open honeycomb. Normally one would bond together the adjacent cells to form a prismatic hexagonal structure, but in this work we leave the structure un-bonded and allow the hinges to flex.

The bridged honeycomb studied in this work is created using the same process, with a slightly different cutting pattern. Figure 4 shows the location of the bridges on the cutting pattern and in the 3D structure.

PEEK Aptiv film (supplied by Victrex PLC [5]) is used throughout this work because it can be quickly and easily thermoformed. A hot press was used to thermoform the PEEK at 200°C. A Blackman & White Genesis 2100 ply cutter was used to accurately cut the slitting patterns.

It is worth noting that the sheet material itself is of little interest; the Kirigami process can be applied to any sheet material which can be cut and reshaped.

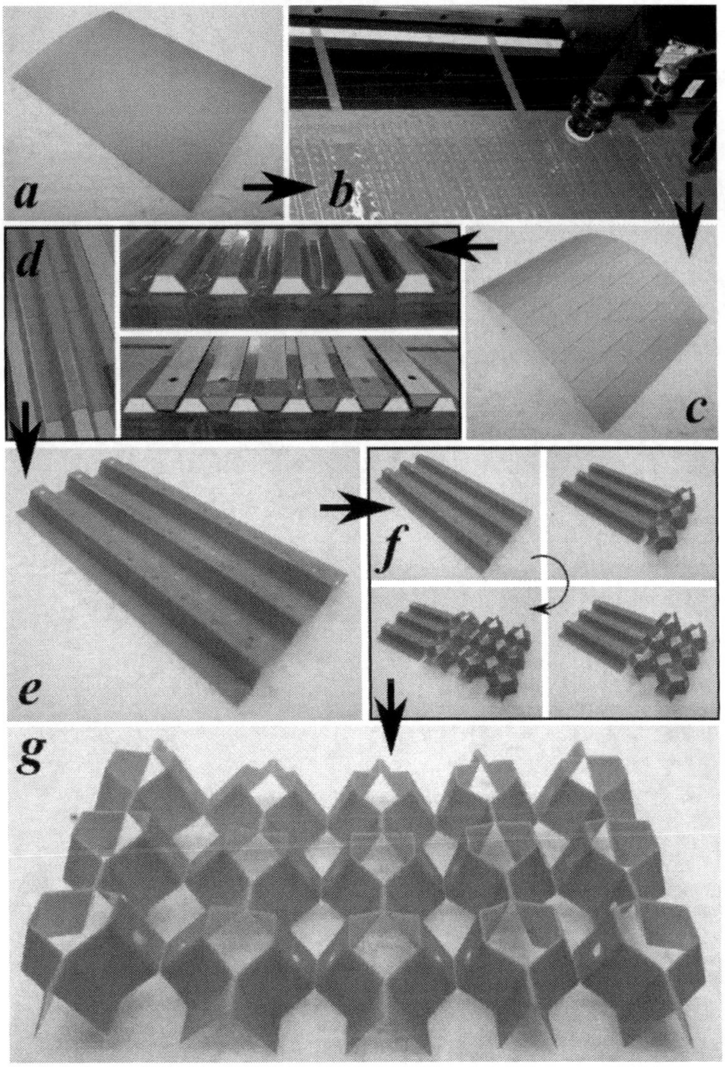

Figure 3. The Kirigami manufacturing process. a) The flat PEEK sheet. b) The cutting patterns are made using a ply cutter. c) The cut sheet with periodically spaced holes. d) The sheet is thermoformed between hexagonal tooling in a hot press to create the hexagonal corrugations. e) The corrugated sheet. f) The corrugated sheet is folded back on itself repeatedly. g) The resulting open honeycomb.

Figure 4. Cutting pattern (left) for bridged honeycomb (right). The bridges are highlighted.

3 GEOMETRY TAILORING

In this section the geometry of the bridged honeycomb is defined, and a method is presented for designing the honeycomb with a specific shape. By changing the cutting pattern we can change the shape of the honeycomb, but the cutting pattern must always line up with the tooling in the X-direction, and the tooling dimensions are fixed. Because the X-dimensions of the honeycomb are fixed, we consider only the shape of the honeycomb in the YZ plane. Figure 5a shows the bridged honeycomb, and Figure 5b shows the bridged honeycomb viewed in the YZ plane, with the unit cell highlighted.

It can be seen from Figure 5b that several of the vertices of the 3D unit cell share the same YZ coordinates. This allows us to use a considerably simplified unit cell in the YZ plane, as shown in Figure 5c (points J and K represent the locations of holes in the cell walls).

We now find the YZ coordinates of the points. We take point A as our starting point and give it coordinates Y_0 and Z_0, as shown in equation (1). From here we find the coordinates of points B to K using the dimensions shown in Figure 5.

$$A = [Y_A, Z_A] = [Y_0, Z_0] \tag{1}$$

$$B = [Y_B, Z_B] = A + \begin{bmatrix} +(c_1 - b_1)\sin\alpha_1 & +(c_1 - b_1)\cos\alpha_1 \\ -\frac{d_1 \sin(\alpha_1+\beta_1)}{\sin\beta_1} \quad , & -\frac{d_1 \cos(\alpha_1+\beta_1)}{\sin\beta_1} \end{bmatrix} \tag{2}$$

$$C = [Y_C, Z_C] = B + [b_1 \sin\alpha_1 \quad , \quad b_1 \cos\alpha_1] \tag{3}$$

$$D = [Y_D, Z_D] = C + \left[\frac{d_1 \sin(\alpha_1+\beta_1)}{\sin\beta_1} \quad , \quad \frac{d_1 \cos(\alpha_1+\beta_1)}{\sin\beta_1}\right] \tag{4}$$

$$E = [Y_E, Z_E] = C + [\chi_1 \sin\phi_1 \quad , \quad \chi_1 \cos\phi_1] \tag{5}$$

$$F = [Y_F, Z_F] = E + [+b_2 \sin\alpha_2 \quad , \quad b_2 \cos\alpha_2] \tag{6}$$

$$G = [Y_G, Z_G] = F + \left[\frac{d_2 \sin(\alpha_2+\beta_2)}{\sin(\beta_2)} \quad , \quad \frac{d_2 \cos(\alpha_2+\beta_2)}{\sin(\beta_2)}\right] \tag{7}$$

$$H = [Y_H, Z_H] = G + [-c_2 \sin\alpha_2 \quad , \quad -c_2 \cos\alpha_2] \tag{8}$$

$$I = [Y_I, Z_I] = G + [\chi_2 \sin\phi_2 \ , \ \chi_2 \cos\phi_2] \quad (9)$$

$$J = [Y_J, Z_J] = B + [e_1 \sin\alpha_1 \ , \ e_1 \cos\alpha_1] \quad (10)$$

$$K = [Y_K, Z_K] = E + [e_2 \sin\alpha_2 \ , \ e_2 \cos\alpha_2] \quad (11)$$

Having defined the dimensions of the unit cell, we can now begin to choose constraints to design a honeycomb. e_1 and e_2 do not affect the cell shape (these specify the location of the holes), and d_1 and d_2 are fixed by the tooling geometry, so there are 12 unknowns in equations (1)-(11).

We can generate 8 additional equations by constraining the vertices to lie on two functions f_1 and f_2 as shown in Figure 6. The functions f_1 and f_2 can be chosen to produce a useful shaped component such as an aerofoil leading edge.

4 more constraints are required to solve the system of equations; the choice of these constraints gives some freedom of design. In this paper we fix α_1, α_2, χ_1, χ_2 to generate honeycombs with regular geometry.

Once the YZ coordinates of the points have been found, the X coordinates are easily calculated using the dimensions of the mould rods. Figure 7 shows an example honeycomb in both 2D and 3D.

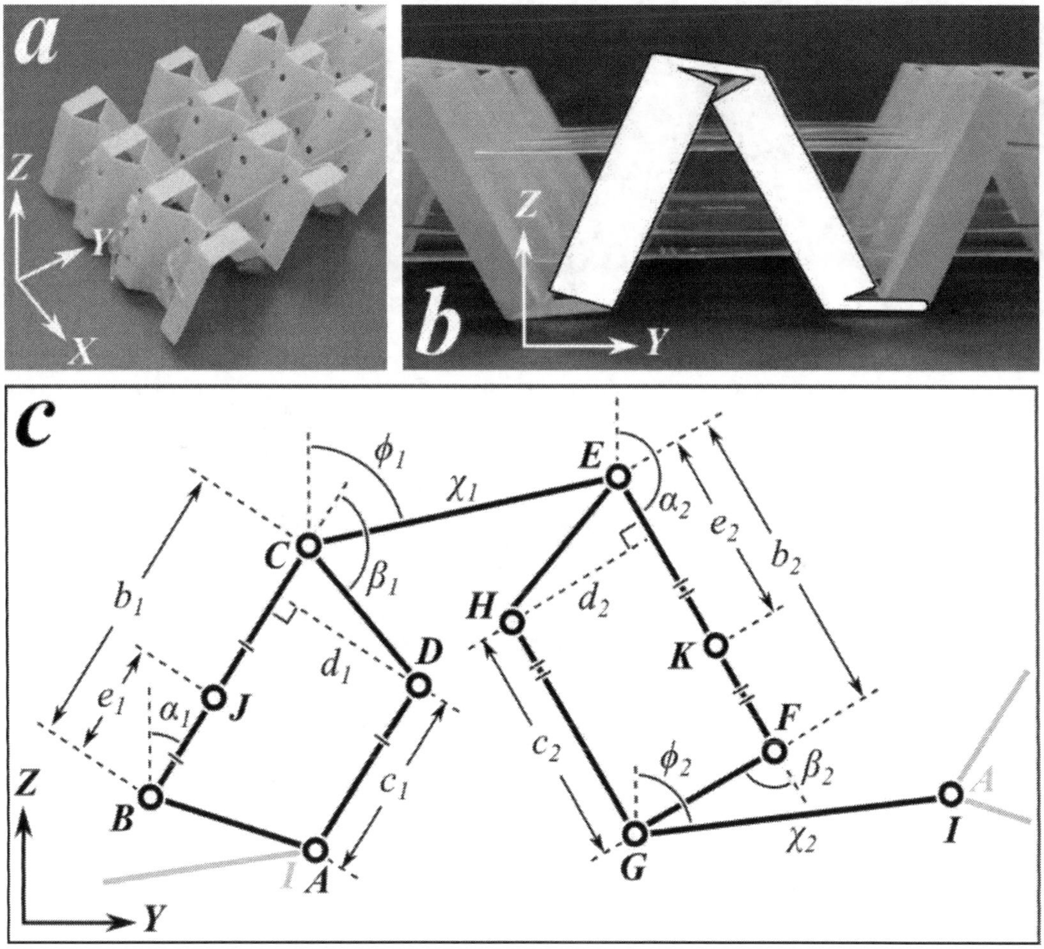

Figure 5. 2D model of the open honeycomb unit cell. a) View of the open honeycomb showing the axes directions. b) Side view of the YZ plane, with unit cell highlighted. c) Unit cell represented by points and lines in the YZ plane, showing dimensions. Parts of the adjacent cells are shown in grey. Points J and K represent locations of holes in the cell walls.

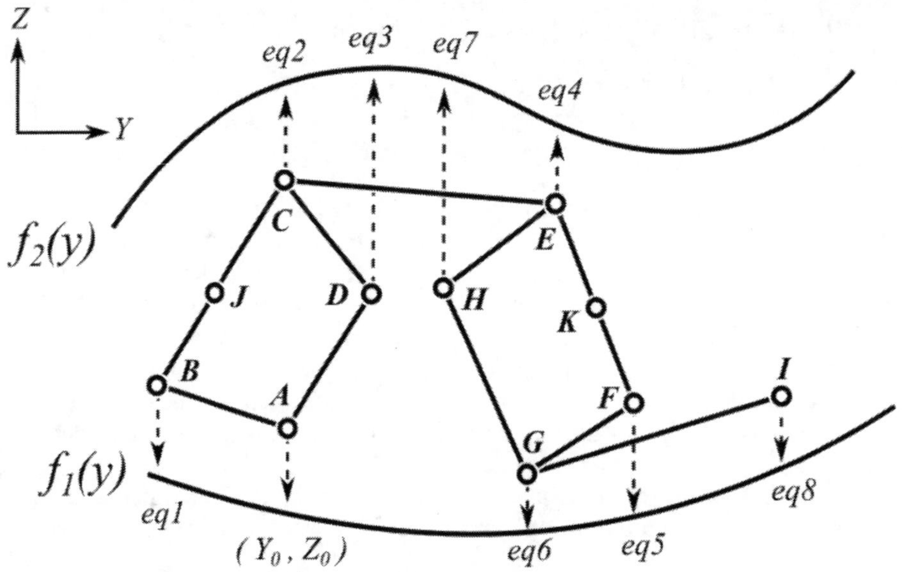

Figure 6. The equations generated by constraining points to lie on functions f_1 and f_2.

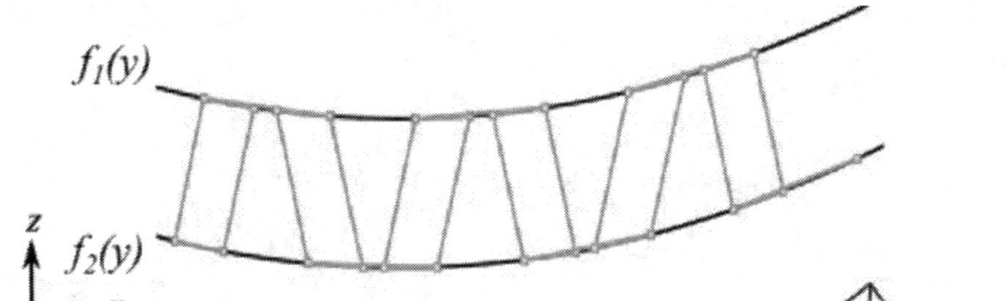

Figure 7. A honeycomb fit between two functions. Functions used were $f_1 = 0.005y^2$, $f_2 = 0.005y^2 + 10$. The four dimensions fixed manually were $\alpha_1 = 15°$, $\alpha_2 = 165°$, $\chi_1 = \chi_2 = 2.5\, l \cos\theta$ (where $l = 5mm$, $\theta = 30°$ as determined by the mould rods).

4 MODELLING THE DEFORMED SHAPE

In this section we demonstrate a method to model the deformation of the structure in response to cable tension. The primary deformation mechanism of the honeycomb is flexing of the folds. Because the folds are all aligned with the X-direction, all the motion of the honeycomb is rotations in the YZ plane. As a result, we can again consider a 2D unit cell.

We assume that the corrugated strips remain rigid, and that the structure deforms purely by rotating about the folds (this is a reasonable assumption because the corrugated strips are much stiffer than the folds). Given this assumption we can idealise the structure as a series of rigid linkages (representing the corrugated strips) joined by rotational springs (representing the folds). Figure 8 shows the idealisation of the unit cell, with forces and moments.

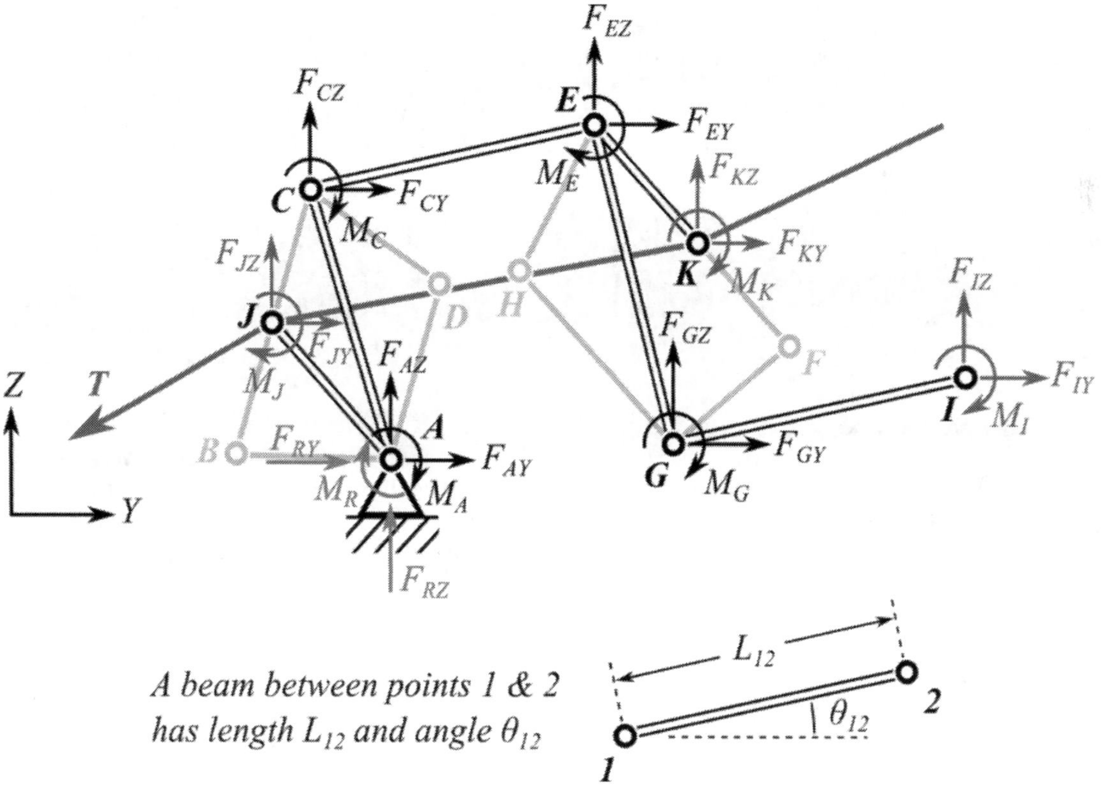

Figure 8. The unit cell idealised as a series of rigid linkages. The large blue line represents the cable with tension T. The cable exerts forces and moments at points J and K – these are shown in blue. Internal fold forces are black. External reaction forces are red. For clarity, lengths and angles of each beam are not shown. They are defined as shown in the bottom insert, with the angle measured positive counterclockwise from the Y axis.

By considering each beam element starting from GI, and working backwards through the structure, we can find the forces and moments at the folds A, C, E, G in terms of the external forces and the lengths and angles of each beam. We represent these in matrix form in equation (12) (i is the cell number, and $j = A, C, E, G$).

$$\{M_j\}_i = [C_M]_i \{F_{ext}\}_i \tag{12}$$

$[C_M]$ is a matrix of coefficients made up of cell dimensions. Treating each fold as a linear rotational spring with relationship $\gamma = M/k$, the fold deflections γ_j are given by:

$$\begin{aligned} \{\gamma_j\}_i &= \{1/k_j\}_i\{M_j\}_i \\ \begin{Bmatrix}\gamma_A\\ \gamma_C\\ \gamma_E\\ \gamma_G\end{Bmatrix}_i &= \begin{Bmatrix}1/k_A\\ 1/k_C\\ 1/k_E\\ 1/k_G\end{Bmatrix}_i \begin{Bmatrix}M_A\\ M_C\\ M_E\\ M_G\end{Bmatrix}_i \end{aligned} \tag{13}$$

We substitute (12) into (13) to obtain:

$$\{\gamma_j\}_i = \{1/k_j\}_i [C_M]_i \{F_{ext}\}_i \tag{14}$$

We now have the deformation of each fold in terms of known parameters (stiffnesses $\{k_j\}$, cell geometry in $[C_M]$, and external forces $\{F_{ext}\}$). We can now find the absolute values of the fold angles measured from the vertical. We denote undeformed dimensions with a superscript "0", and deformed dimensions with a superscript "*". We find the deformed fold angles $\{\theta_j\}_i^* = \{\alpha_1, \phi_1, \alpha_2, \phi_2\}_i^*$ by adding the deformations $\{\gamma_j\}_i$ to the undeformed fold angles $\{\theta_j\}_i^0 = \{\alpha_1, \phi_1, \alpha_2, \phi_2\}_i^0$. The matrix $[C]$ accounts for the fact that point j experiences its deformation plus the deformations of the previous points.

$$\begin{aligned} \{\theta_j\}_i^* &= \{\theta_j\}_i^0 + [C]\{\gamma_j\}_i \\ \begin{Bmatrix}\alpha_1\\ \phi_1\\ \alpha_2\\ \phi_2\end{Bmatrix}_i^* &= \begin{Bmatrix}\alpha_1\\ \phi_1\\ \alpha_2\\ \phi_2\end{Bmatrix}_i^0 + \begin{bmatrix}1&0&0&0\\1&1&0&0\\1&1&1&0\\1&1&1&1\end{bmatrix}\begin{Bmatrix}\gamma_A\\ \gamma_C\\ \gamma_E\\ \gamma_G\end{Bmatrix}_i \end{aligned} \tag{15}$$

This gives us the deformed fold angles in response to cable tension for one cell i. We now expand this to multiple cells. Adjacent cells i and $i+1$ are joined by point I_i meeting point A_{i+1} (see the grey lines in Figure 8). We must account for moments and deformations accumulating throughout the entire structure as well as through the cells. The fold forces and moments at point A_{i+1} act on cell i as the "external forces" at point I_i. For the final cell in the chain, N, external forces at point I_N will be zero. Deformations are summed along the structure in a similar way to how they are summed along a cell. Cell i experiences its deformations plus those of previous cells. This is represented in matrix form for N cells:

$$\begin{Bmatrix}\{\theta_j\}_1\\ \{\theta_j\}_2\\ \vdots\\ \{\theta_j\}_N\end{Bmatrix}^* = \begin{Bmatrix}\{\theta_j\}_1\\ \{\theta_j\}_2\\ \vdots\\ \{\theta_j\}_N\end{Bmatrix}^0 + \begin{Bmatrix}[C]\{\gamma_j\}_1\\ [C]\{\gamma_j\}_2\\ \vdots\\ [C]\{\gamma_j\}_N\end{Bmatrix} + \begin{bmatrix}0&0&\cdots&0\\1&0&\cdots&0\\\vdots&\vdots&\ddots&\vdots\\1&1&\cdots&0\end{bmatrix}\begin{Bmatrix}\Sigma_j\{\gamma_j\}_1\\ \Sigma_j\{\gamma_j\}_2\\ \vdots\\ \Sigma_j\{\gamma_j\}_N\end{Bmatrix} \tag{16}$$

$$\text{Where} \begin{bmatrix}0&0&\cdots&0\\1&0&\cdots&0\\\vdots&\vdots&\ddots&\vdots\\1&1&\cdots&0\end{bmatrix} = D$$

We now have the deformed structure fold angles in terms of fixed geometry dimensions and cable tension T. These can be substituted into equations (1)-(11) to find the deformed coordinates of points A-K. At this point we could input a value for T and find the deformed geometry, but it would be more useful to be able to specify a cable displacement δ_{cable}, since the role of a morphing structure is to assume a specific deflection. We must generate one equation so that we can specify δ_{cable} and solve for T. If we were to pull a section of cable with length δ_{cable} out of the undeformed structure to produce a deformed shape, the length of cable left inside the deformed structure would be given by:

$$L^{*}_{cable} = L^{0}_{cable} - \delta_{cable} \tag{17}$$

Where L_{cable} is the length of cable inside the structure. We can find L_{cable} from the sum of the distances between the various points J and K throughout the cells:

$$L_{cable} = \left(\sum\nolimits_{i}^{2:N} |K_{i-1}J_i| + |J_iK_i|\right) + |J_1K_1| \tag{18}$$

We have J^0 and K^0 because they are fixed by our choice of e_1 and e_2, and we have J^* and K^* in terms of T; thus we can specify a displacement δ_{cable} and solve (17) for T. Figure 9 shows the predicted displacement for a flat honeycomb compared to the displacement of a real honeycomb, and shows that the model produces qualitatively similar results. From Figure 9 we can see differences in the two deformed shapes – this is due to limitations of the model, which are discussed in detail in the next section.

Figure 9. A qualitative comparison between predicted deflection (top) and actual deflection (bottom) when subjected to a similar cable deflection. The transparent plot shows the undeformed structure.

5 DISCUSSIONS & CONCLUSIONS

We have shown methods to design a morphing honeycomb with a functional shape, and predict the deformation of the open honeycomb in response to cable loads. We have shown that the model gives qualitatively sensible results.

5.1 Limitations of the model

The model does not take into account self-weight, and other factors.

These can be easily taken into account in future versions.

The model uses a very simple representation of the fold.

We based this on our observation that the honeycomb always recovers fully from all deformations; this means that the folds are deforming elastically. However, this may not be the case for all materials, and nonlinear stiffness behaviour could be implemented in the model in future.

The folds will repeatedly experience high local stresses which will make them susceptible to breakage. This depends on the material and the thickness, and these will vary depending on the application. We aim here to provide a model for the shape morphing; it is up to the designer to characterise their chosen sheet material and investigate the behaviour of the folds.

The model assumes small displacements.

This means that it does not properly capture the effect of large displacements. The solution to this is to break up a given displacement δ_{cable} into small increments, and run the model over each of these increments. This will then take into account the changing in angles and forces over the course of a large deformation.

The curvature cannot be directly controlled.

Currently we can only control cable tension T or displacement δ_{cable}, and the deformed shape depends purely on the geometry and the fold stiffnesses. The geometry is fixed by the desired component shape, and the fold stiffnesses are fixed by choice of material. It is desirable to influence the shape of the structure in some other way.

5.2 Potential improvements

Options for further curvature control.

Currently we are considering only a simple case with one cable. As shown in Figure 2, many cables can be embedded in the structure. It is also feasible to put multiple knots in a cable, such that it pulls on the structure in several places, and produces compound radii of curvature, as shown in Figure 10a.

It is desirable to limit how much the structure contracts under cable tension. By bonding the structure to an axially stiff plate we could limit contraction and get pure bending (see Figure 10b).

If the cable inside the structure is pulled far enough, the structure will contract until the adjacent strips touch one another; locking it in place. By controlling the length of the bridges between strips it may be possible to tailor this collapsed configuration to a second useful shape (see Figure 10c).

Modelling the effect of a skin

A compliant outer skin could be modelled by spring elements between adjacent points on the upper/lower surface.

6 ACKNOWLEDGMENTS

The Authors acknowledge the Engineering and Physical Sciences Research Council for supporting this work through the Centre for Doctoral Training in Advanced Composites at the University of Bristol, UK (Grant no. EP/G036772/1).

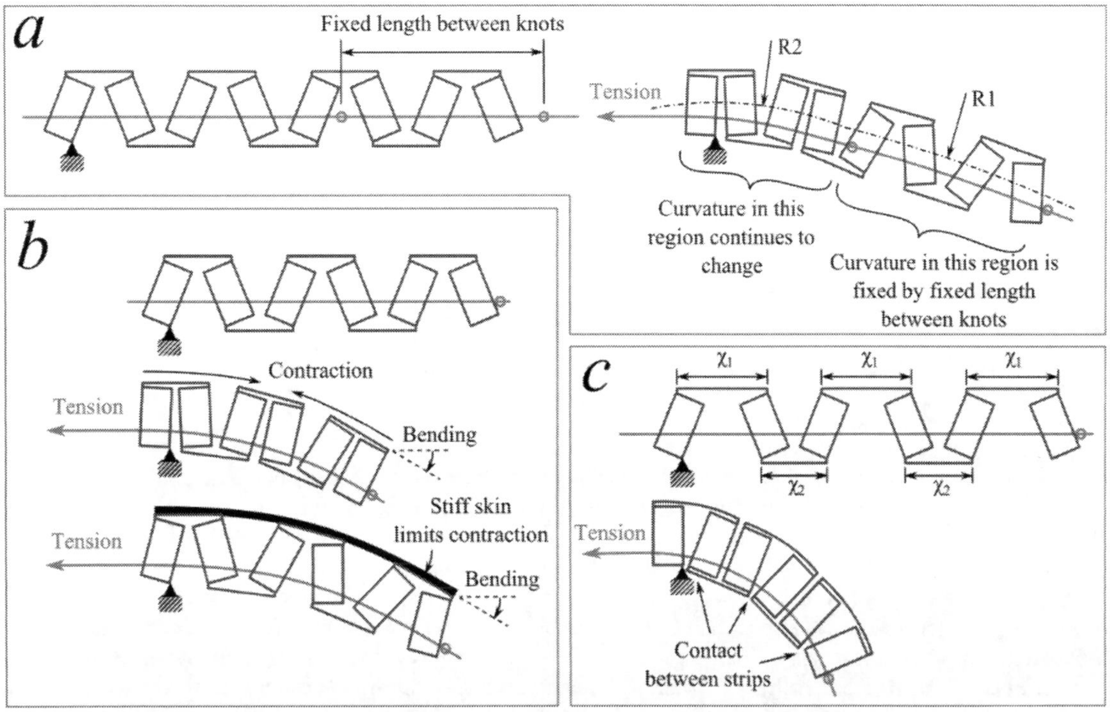

Figure 10. Possibilities for curvature control. a) Multiple radii of curvature produced by a cable with multiple knots. b) An axially stiff skin limits contraction in the Y-direction. c) Bridges with different lengths produce a curved shape when the honeycomb is fully collapsed.

7 REFERENCES

[1] T. Nojima and K. Saito, "Development of Newly Designed Ultra-Light Core Structures," *JSME Int. J. Ser. A*, vol. 49, no. 1, pp. 38–42, 2006.

[2] K. Saito, S. Pellegrino, and T. Nojima, "Manufacture of Arbitrary Cross-Section Composite Honeycomb Cores Based on Origami Techniques," *J. Mech. Des.*, vol. 136, no. 5, p. 051011, Mar. 2014.

[3] R. M. Neville, A. Pirrera, and F. Scarpa, "Open shape morphing honeycombs through kirigami," in *Proceedings of the ASME 2014 Conference on Smart Materials, Adaptive Structures and Intelligent Systems (SMASIS2014)*, 2014.

[4] R. M. Neville, F. Scarpa, and A. Pirrera, "Shape morphing Kirigami mechanical metamaterials (submitted to Scientific Reports)," 2015.

[5] Victrex PLC, "APTIV ® 2000 Series Films."

ICAST2015: 26th International Conference on Adaptive Structures and Technologies
October 14-16th, 2015, Kobe, Japan

ICAST2015 # 017

Wrinkle Generation Mechanisms in Shear-Enforced Membranes

Kei Senda[1*], Mario Petrovic[1], and Kei Nakanishi[1]

[1] Department of Aeronautics and Astronautics, Kyoto University, Kyoto, Japan.

Abstract

The object of this study is to clarify wrinkling behavior of a shear-enforced flat rectangular membrane with free boundaries. For this purpose, an equilibrium path tracking method using a finite element method is developed. This method includes bifurcation path tracking analysis that searches for bifurcation solutions. This method establishes an image of membrane behavior by calculating a series of successive equilibrium states before and after bifurcation buckling. Through detailed analysis of stress, displacement fields, and wrinkle interaction over a load parameter range, the analysis shows how existing wrinkles affect each other and the generation of new wrinkles. As a result, there is wrinkle generation with bifurcation and without bifurcation. The wrinkle generation mechanism without bifurcation is analyzed in detail. Wrinkle generation caused by bifurcations could potentially result in a large number of equilibrium paths. Each equilibrium path represents a specific wrinkle pattern. However, the analyzed results show that significantly fewer equilibrium paths are obtained than expected. These are due to wrinkle generation without bifurcations and to deformation symmetry.

1. INTRODUCTION

Inflatable structures have attracted attention because of light weight and compact storage[1]. This study considers structures using deployable membranes. The structures are initially folded, deployed into their desired shape, unfolded or inflated, and rigidized.

The research goal of this study is to achieve stable deployment along a planned trajectory for a structure folded into a prescribed shape. However, because the structure is a membrane, with many degrees of freedom, the prediction of membrane behavior is difficult. In a deployment experiment using an inflatable tube[2], an origami-like pattern was designed to facilitate predictable deployment. However, small compressive forces cause local buckling and generate wrinkles. Buckling occurs after a bifurcation[3, 4, 5, 6]. When a bifurcation occurs, the material may become unstable and the subsequent state may not be uniquely determined. This makes predicting stable deployment difficult. To solve this problem, an understanding of wrinkle behavior is essential.

In this study, a wrinkle is defined as an out-of-plane membrane deformation that occurs when a compressive load is applied to a membrane. Buckling is defined as a phenomenon generating a finite deformation in a direction different from the load applied to the structure.

An equilibrium state, or an equilibrium point, is defined as the static state of a membrane when a static load or an imposed displacement is applied. By varying a path parameter such as displacement or load, an equilibrium path is obtained, i.e. a sequence of equilibrium points. Bifurcation is a rapid equilibrium point shift and a qualitative change of the equilibrium state when the path parameter varies. A wrinkle, i.e. buckling of a membrane, is generated after a bifurcation. Therefore, in order to deploy a membrane structure along a planned trajectory, understanding of the bifurcation structure and the equilibrium paths is necessary.

The first step in analyzing wrinkling behavior is modeling the membrane structure. Finite element modeling is used to analyze wrinkle behavior. By using shell elements that include bending stiffness, wrinkle amplitude and wavelength are calculated more accurately.

In a straight column or flat plate, the bifurcation after which buckling occurs is a branching bifurcation[3, 7]. For a flat membrane, it is believed that a wrinkle is generated by bifurcation and the number of wrinkles

* senda@kuaero.kyoto-u.ac.jp

increases after bifurcation. A bifurcation is a singularity in the analysis. Therefore, obtaining equilibrium points after bifurcation is difficult and numerous studies have been conducted[8]. A bifurcation path analysis method that is able to search for solutions after bifurcation is needed. In this study, a method of searching for bifurcation solutions is added into the equilibrium path tracking method, of Wagner and Wriggers[9].

In this study, a flat rectangular membrane with free boundary conditions is sheared, see Wong and Pellegrino[4] and Inoue[10]. In Wong and Pellegrino[5, 6] a wide range of numerical results were shown. This study applies relatively small increments to imposed displacement for the bifurcation structure and wrinkle behavior.

Wong and Pellegrino describe wrinkle generation in terms of bifurcations. Bifurcations generate multiple paths for wrinkle generation. Each bifurcation can increase the number of equilibrium paths. There are many possible deformation patterns when the membrane has many wrinkles.

In this study, the results show a membrane has an initial single equilibrium path that branches into 4 paths, then transitions into 2 paths and then again into 4 paths. When the first 4 paths exist, the number of wrinkles increases to 33 or 34 without bifurcations by increasing the prescribed displacement. This study clarifies a new wrinkle generation mechanism, in which wrinkles are generated at regular points on an equilibrium path. Because of this wrinkle mechanism, the branching of the system is reduced and only a small number of paths exist. Two of the first 4 paths undergo bifurcations resulting in only 2 paths.

Additionally, system symmetry is discussed. Symmetry in deformation patterns of a membrane reduces the number of configurations. Finally, a simpler bifurcation structure with fewer equilibrium paths, than expected based on the number of wrinkles is obtained.

The rest of this paper is organized as follows. Section 2 sets up the FEM model of a rectangular membrane using ABAQUS. A custom built subroutine functions as the path analysis method. Section 3 shows the path tracking method for a system with a prescribed displacement. Section 4 introduces the solutions caused by symmetry and the similarity values. Section 5 presents some numerical simulations and discusses the results. Section 6 presents concluding remarks.

2. MODELING ANS PROBLEM

2.1 Analyzed Object and Model

The membrane analyzed in this study is a flat membrane as illustrated in Fig. 1. The material is Kapton and its properties are listed in Table 1. The choice of a Kapton membrane comes from the flat membrane wrinkling data, e.g. Wong and Pellegrino[4] and Inoue[10]. The ratio of the horizontal and the vertical lengths $a : b$ is $3 : 1$. Fixed boundaries are set for the top and bottom edges, and free boundaries for the left and right edges. The fixed top edge is then subjected to a prescribed displacement. Gravity and imperfections are not considered.

The origin of the xyz-coordinate system is fixed at the membrane center of mass in the undeformed state, i.e. when no external loads are applied. The x-axis is parallel to the bottom edge, y-axis is perpendicular to the bottom edge, and z-axis is normal to the neutral plane, as defined in Fig. 1. Coordinates (x, y, z) refer to a position on the membrane prior to deformation, and they do not vary with deformation. Section AA' is defined at $y = 0$[m] and BB' at $y = -0.045$[m].

First, at the top fixed edge, a displacement of $\delta u_y = 30 \times 10^{-6}$[m] is applied in the y-direction and is constant throughout the analysis. This is to maintain consistency with experimental results in Inoue[10]. Then, an imposed displacement δu_x in x direction is applied. In this study, δu_x is increased from 0[m] to 650×10^{-6}[m].

Membrane deformation is described by displacements at all points on the membrane. Point displacements are represented by (u_x, u_y, u_z) in (x, y, z) coordinates. However, the z-displacement of points on the mid-plane is mainly considered, where displacement of a point in z-coordinate $u_z(x, y, 0)$ is simply expressed by $u_z(x, y)$. Stress and strain are given for the mid-plane. The tensile direction is positive for both stress and strain. Considering the plane stress, the $\xi\xi'$-axis corresponds to the first principal stress direction and the $\eta\eta'$-axis corresponds to the second principal stress direction. When wrinkles exist, the $\xi\xi'$-axis in a the wrinkle longitudinal direction is perpendicular to $\eta\eta'$-axis. Both $\xi\xi'$ and $\eta\eta'$ rotate with the wrinkles and remain perpendicular to each other. A wrinkle is a sinusoidal shaped out-of-plane deformation that occurs between two points where curvature in η-direction changes sign.

The problems in this study are defined below. The membrane behavior is analyzed by increasing the imposed displacement δu_x. The behavior includes wrinkle geometry indicated by the displacement $u_z(x, y)$, bifurcation structure of the deformation, and stress. Stress values as well as forces and moments at certain points will be used to explain the behavior in more detail. The wrinkle generation mechanism is described by the following analysis.

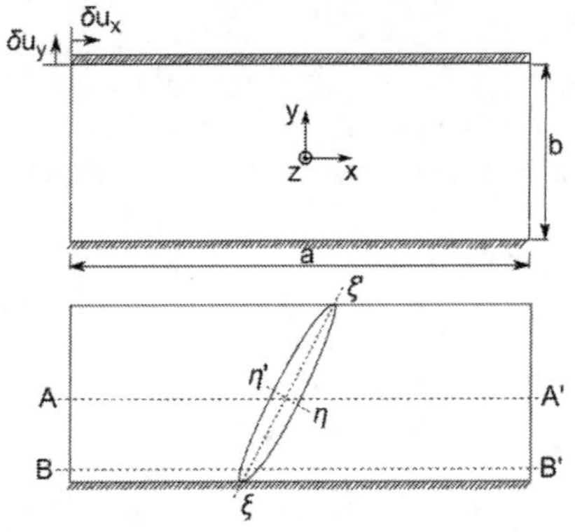

Figure 1. Sheared rectangular membrane. Top and bottom edges have fixed boundaries. Right and left edges have free boundaries.

Table 1 Membrane material properties.

Membrane width	a	0.30 [m]
Membrane height	b	0.10 [m]
Membrane thickness	h	12.5×10^{-6} [m]
Young's Modulus	E	3.0 [GPa]
Poisson's ratio	ν	0.3

2.2 FEM Model

The above geometry, properties and boundary conditions are modeled using ABAQUS. The S4R shell elements in ABAQUS are used. For the mesh, equal divisions in x and y directions are constructed resulting in 360×120 elements. The validity of this mesh density is discussed in Section 5. It should be noted that only geometric nonlinearity is considered. Nonlinearity due to other factors such as material are not considered in the present study.

The static equilibrium of forces, i.e. nonlinear FE equations, $\boldsymbol{F}$, based on the displacement method, is formally written as:

$$\boldsymbol{F}(\boldsymbol{u},\, \boldsymbol{f}) = \boldsymbol{0} \tag{1}$$

where $\boldsymbol{f}$ is the applied force for displacement $\boldsymbol{u}$. The static problem solves the equilibrium equation with δu_x for the equilibrium point $(\boldsymbol{u},\, \boldsymbol{f})$. The imposed displacement δu_x is included in $\boldsymbol{u}$ and the reaction forces for displacements $\boldsymbol{u}$ are included in $\boldsymbol{f}$. Equation (1) is a nonlinear equation for $\boldsymbol{u}$ and $\boldsymbol{f}$, which is difficult to solve. The solution method, e.g. an incremental calculation, is explained in the next section.

3. EQUILIBRIUM PATH TRACKING METHOD

For equilibrium path tracking, a new equilibrium point, near the current equilibrium point, is determined by slightly changing the path parameter. However, if the new point is a bifurcation point, a bifurcation path analysis method is needed to search for post-bifurcation points because a bifurcation point is a singularity in the analysis. A bifurcation diagram and deformation must be obtained in order to understand membrane behavior.

The relationship between the deformed state of the membrane $\boldsymbol{u}$ and the incremental displacement δu_x will be established, where the displacement δu_x is the path parameter. Finally, a method for searching for bifurcation solutions equivalent to that of Wagner and Wriggers[9] is built into the equilibrium path tracking method. This search consists of a stability inspection of the tangent stiffness matrix. Near bifurcation points the relevant eigenvectors are used to obtain bifurcation solutions.

3.1 Successive Solution Calculation

An equilibrium point $(\boldsymbol{u}_0 + \tilde{\boldsymbol{u}},\, \boldsymbol{f}_0 + \tilde{\boldsymbol{f}})$ close to a given equilibrium point $(\boldsymbol{u}_0,\, \boldsymbol{f}_0)$ of Eq. (1) satisfies

$$\boldsymbol{F}(\boldsymbol{u}_0 + \tilde{\boldsymbol{u}},\, \boldsymbol{f}_0 + \tilde{\boldsymbol{f}}) = \boldsymbol{0} \tag{2}$$

Expanding Eq. (2) in a Taylor series and by omitting the higher order terms gives

$$\boldsymbol{K}\tilde{\boldsymbol{u}} = \tilde{\boldsymbol{f}} \tag{3}$$

where $\boldsymbol{K} = \frac{\partial \boldsymbol{F}}{\partial \boldsymbol{u}^T}$ is the tangent stiffness matrix and $-\frac{\partial \boldsymbol{F}}{\partial \boldsymbol{f}^T}$ is the unit matrix.

A solution at the new equilibrium point,$(\boldsymbol{u}_0 + \tilde{\boldsymbol{u}},\, \boldsymbol{f}_0 + \tilde{\boldsymbol{f}})$ of the original nonlinear equation (2) is determined by iterative methods, e.g. Newton-Raphson method. This is the equilibrium path tracking method that seeks successive equilibrium points by gradually increasing $\boldsymbol{f}$ (or δu_x). Also, it is the static solution method that obtains solutions satisfying the static equilibrium equation (1). However, at the structure bifurcation point that is a singular point $(\boldsymbol{u}_c, \boldsymbol{f}_c)$, $\det \boldsymbol{K} = 0$, and the incremental displacement $\tilde{\boldsymbol{u}}$ corresponding to incremental load $\tilde{\boldsymbol{f}}$ cannot be uniquely determined. In this case, a bifurcation path analysis that searches for post-bifurcation paths is needed.

3.2 Bifurcation Path Analysis Method

Bifurcation points may be bifurcation branching points and limit points[11]. When the solution passes through a limit point, snap-through bifurcation occurs and the equilibrium path jumps in a discontinuous manner. If the branching bifurcation points are classified as the symmetric and the asymmetric bifurcation points, they generate symmetric bifurcation buckling and asymmetric bifurcation buckling, respectively. It is possible to predict the type of bifurcation structure using asymptotic theory. In the case of snap-through buckling and symmetric bifurcation buckling, the direction of the incremental displacement from the buckling point is equal to the eigenvector corresponding to a zero eigenvalue. In the case of asymmetric bifurcation buckling, displacement and load should be changed to satisfy a certain relation.

At bifurcation point $(\boldsymbol{u}_c, \boldsymbol{f}_c)$, the tangent stiffness matrix $\boldsymbol{K}$ has a zero eigenvalue $\lambda_c = 0$ (det$\boldsymbol{K} = 0$) and a corresponding eigenvector $\boldsymbol{\phi}_c$. In order to determine the position of the bifurcation point, a pinpointing procedure that is similar to the method to solve the extended system described in Wriggers, et al.[9] is used. The eigenvalue analysis of the tangent stiffness matrix is performed and $(\boldsymbol{u}_c + \epsilon\boldsymbol{\phi}_c, \boldsymbol{f}_c)$ is computed. The perturbed state is in the eigenvector direction from the bifurcation point, where ϵ is an infinitesimal parameter. By assuming the perturbed state is near a post-bifurcation path, a convergence calculation, the Newton-Raphson iteration, is performed. This static analysis searches for a new equilibrium point after the bifurcation. If a post-bifurcation equilibrium point cannot be found, the initial condition is changed by varying ϵ and the search for a new equilibrium point is repeated. This method, which estimates the bifurcation mode as the eigenvector and searches for the bifurcation path, is similar to the method used in Wanger and Wriggers[9].

4. ANALYSIS METHODS OF SOLUTION

4.1 Symmetric Solutions

Based on group theory[12], the equivalence of a system that possesses symmetry is described as:

$$\boldsymbol{M}(\gamma)\boldsymbol{f}(\boldsymbol{u}(\boldsymbol{x}), \mu) = \boldsymbol{f}(\boldsymbol{M}(\gamma)\boldsymbol{u}(\boldsymbol{M}(\gamma)\boldsymbol{x}), \mu) \tag{4}$$

where $\boldsymbol{M}$ is a representation matrix. The observed membrane has symmetry that is invariant with respect to e, s, r and sr, where e is the identity transformation, s is the reflection transformation with respect to xy-plane, r is the rotation transformation by 180° about the z-axis and sr is the combination of s and r. This symmetry group is known as the dihedral group $D_2 = \{e, r, s, sr\}$. The symmetry transformations satisfy the equivalence condition of the governing equations. In addition, the boundary conditions are invariant under the symmetry transformations. As a result, this system has the symmetry that is invariant under all possible transformations.

Deformation $\boldsymbol{u}_0(x,\, y,\, z)$ satisfies the governing equations and boundary conditions. Three deformations $\boldsymbol{u}_s(x,\, y,\, z)$, $\boldsymbol{u}_r(x,\, y,\, z)$, and $\boldsymbol{u}_{sr}(x,\, y,\, z)$ are obtained from $\boldsymbol{u}_0(x,\, y,\, z)$ by means of the reflection transformation, the rotation transformation, and the rotation and reflection transformation, respectively. The three deformations $\boldsymbol{u}_s$, $\boldsymbol{u}_r$, and $\boldsymbol{u}_{sr}$ are the solutions satisfying the governing equations and the boundary conditions.

Symmetry between two solutions is defined through an example. If deformation $\boldsymbol{u}_1(x,\, y,\, z)$ is equivalent to $\boldsymbol{u}_{sr}(x,\, y,\, z)$, there exists a symmetry where $\boldsymbol{u}_0$ agrees with $\boldsymbol{u}_1$ using the rotation and reflection transformation. Also, $\boldsymbol{u}_1$ is equivalent to $\boldsymbol{u}_0$ transformed by the rotation and reflection transformation. In other words, $\boldsymbol{u}_0$ and $\boldsymbol{u}_1$ have the rotation and reflection symmetry. The deformation $\boldsymbol{u}_1$ is a different representation of $\boldsymbol{u}_0$.

Symmetry in a solution can also be defined through an example. If $\boldsymbol{u}_0(x,\, y,\, z)$ is equivalent to $\boldsymbol{u}_r(x,\, y,\, z)$, then $\boldsymbol{u}_0$ has the symmetry that is invariant under the rotation transformation. Also, $\boldsymbol{u}_0$ has the rotation symmetry. In this case, $\boldsymbol{u}_r$ is obtained, which is equivalent to $\boldsymbol{u}_0$ by the rotation transformation. It is said that we obtain the same solution or we do not obtain a new solution by the transformation.

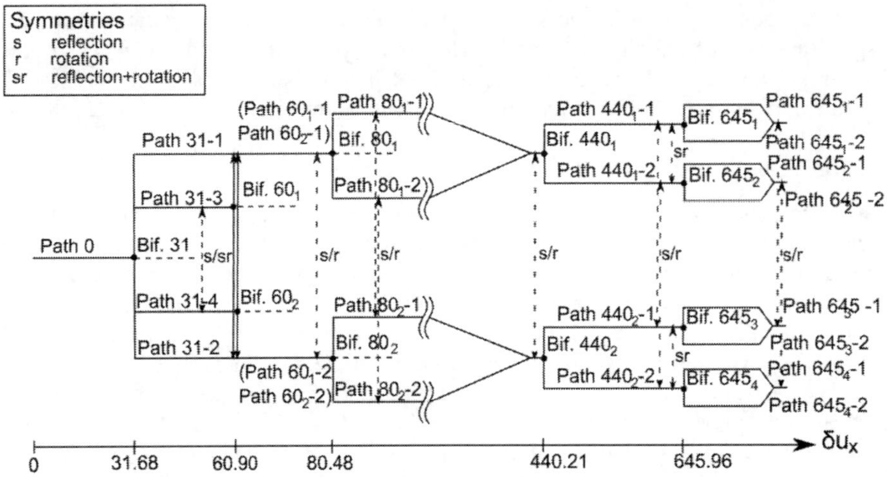

Figure 2: Schematic bifurcation diagram with respect to imposed shear displacement δu_x.

4.2 Similarity Value

In order to identify similarity between the two solutions, the similarity value is constructed as the normalized inner product of solutions. During comparison of different wrinkle patterns, the number of digits to which similarity is obtained is referred (e.g. 0.99999345 achieved similarity within 5 digits). By numerical evaluations, the values after the fifth digit are considered to be numerical error.

5. NUMERICAL RESULTS

5.1 Validity of FEM Modeling

The geometry, properties, and boundary conditions mentioned in section 2 are modeled in ABAQUS. Based on discussions in Wong and Pellegrino[6], the S4R (4 node shell element with reduced Gauss integration) elements are selected. Using this element, numerical results agree best with the experimental results in Wong and Pellegrino[4] and Inoue[10]. The number of elements is $360 \times 120 = 4.23 \times 10^4$, which is the same value used in Wong and Pellegrino[6] for a membrane with an aspect ratio of 3:1. Models with more elements, e.g. 375×125, 390×130, and 432×144 were also used for analysis. The results from these numbers of elements were compared to the result from 360×120 elements to confirm that the results converged and were insensitive above 4.32×10^4 elements.

5.2 Membrane Behavior

As defined in Section 2, the imposed displacement is applied to the membrane and the deformation is calculated. This is done mainly through the observation of the membrane deformation and through the bifurcation structure. Stress states are observed at specific points of interest. For the deformation, the displacement value $u_z(x, y)$ in the z-direction on the mid-plane is shown in the figures. The first and second principal stresses are presented as $\sigma_\xi(x, y)$ and $\sigma_\eta(x, y)$ whose directions are coincident with $\xi\xi'$ and $\eta\eta'$, which are related to the wrinkle shape.

Displacement is imposed on the top edge, first in y-direction as $\delta u_y = 30 \times 10^{-6}$[m]. The incremental displacement in x-direction is imposed δu_x from 0[m] to 650×10^{-6}[m]. The schematic bifurcation diagram is shown in Fig. 2. This bifurcation diagram will be used to illustrate membrane behavior as δu_x is increased.

For notation, bifurcation points are indicated by 'Bif. X_n', where X indicates the micrometer value of the δu_x where the bifurcation point occurs. The subscript n is an identifier used if there are multiple bifurcation points at the same value of δu_x. Paths are indicated by 'Path Y-m', where Y is the value of the δu_x where the path starts, and m is an identifier if there are multiple paths that start at the same point.

5.2.1 Path 0

The equilibrium path is observed for the interval $0 \leq \delta u_x < 31.68 \times 10^{-6}$[m]. The initial imposed displacement is $\delta u_y = 30 \times 10^{-6}$[m] and $\delta u_x = 0$ so the membrane remains flat. This results in a tensile σ_{yy} and, because of the Poisson effect, a tensile σ_{xx}. A state of biaxial tension is maintained throughout most of the membrane. On free boundaries, $\sigma_{xx} = 0$ because no stress is applied from outside of the free boundaries. The σ_{xx} near the free boundaries is smaller than in the central region. As δu_x is increased, the out-of-plane deformation remains 0 and planar shear deformation is observed. This path is followed until the first bifurcation point Bif. 31.

5.2.2 Bif. 31

At Bif. 31 of $\delta u_x = 31.68 \times 10^{-6}$[m], the single equilibrium path branches into four paths, Paths 31-1, 31-2, 31-3 and 31-4. Bif. 31 is a double singular point with two zero eigenvalues. All post-bifurcation paths contain wrinkles that are generated at this point. The out-of-plane deformations of these four paths at $\delta u_x = 31.83 \times 10^{-6}$[m] are shown in Fig. 3. By observing the wrinkles closest to the free boundary, the side with an upward convex wrinkle is identified by + and the side with a downward convex wrinkle by −. The wrinkle patterns of four paths are then indicated as follows: Path 31-1 as (+, −), Path 31-2 as (−, +), Path 31-3 as (−, −), and Path 31-4 as (+, +).

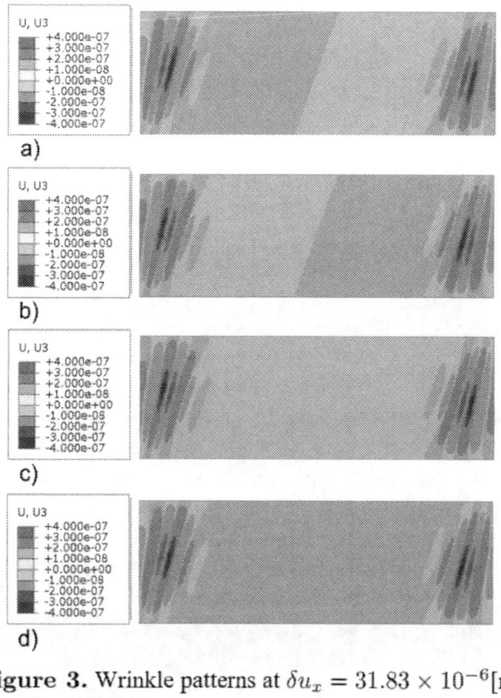

Figure 3. Wrinkle patterns at $\delta u_x = 31.83 \times 10^{-6}$[m] just after Bif. 31.

Relative to the rest of the membrane, the center of the membrane remains flat. A single wrinkle pattern that occupies half of the membrane can be identified. By applying the rotation transformation, the reflection transformation and the rotation and reflection transformation to this one wrinkle pattern occupying half of the membrane and combining the transformations, all full membrane wrinkle patterns are obtained.

Figure 4 a) and b) shows the first principal stress σ_ξ and second principal stress σ_η on the mid-plane at $\delta u_x = 31.83 \times 10^{-6}$ [m]. The initial position of wrinkles near the free boundaries is explained by the compressive stress σ_η near the free boundaries. Therefore, wrinkle formation is expected.

Similarly to Path 0, σ_{xx} near the free boundaries is smaller. As δu_x is increased, the tensile stress to the right and upper direction increases, and σ_ξ becomes as shown in Fig. 4 a) and σ_η becomes as shown in Fig. 4 b). Therefore, the region near the free boundaries takes on the states of uniaxial compression, whereas the other region is in biaxial tension. As δu_x is increased, the compressive regions near the boundaries expand toward the middle, and their compression amplitudes increase.

5.2.3 Path 31

By increasing δu_x until $\delta u_x = 60.90 \times 10^{-6}$[m], all four paths connected from Bif. 31 are observed. The behavior on this path is represented by the generation of new wrinkles one after another on the flat part of the membrane next to existing wrinkles.

This process is shown for Path 31-3 in Fig. 5. Similar behavior occurs along all other paths. A small wrinkle is generated next to existing wrinkles on each side of the flat part of the membrane. By increasing δu_x, this wrinkle grows until it makes contact with the top and bottom boundaries. Once both contacts have been made, the rest of the membrane, up to the middle area, changes its sign in the z-direction, but remains flat. This is followed by the generation of a new wrinkle on each side.

This process is repeated until the flat section of the membrane is filled with wrinkles. Additional shear will increase the number of wrinkles and the amplitude of the wrinkles. The process can be observed for the AA' cross-section in Fig. 6 b). The process for Path 31-1 is shown in Fig. 6 a). The other two paths have reflection symmetry behavior, and are omitted. This behavior will be discussed in a later section.

In Fig. 7 the out-of-plane deformation at $\delta u_x = 60.00 \times 10^{-6}$[m] is shown, just before the upcoming bifurcation. Figures a) to c) correspond to Paths 31-1 to 31-4, respectively. Here, Paths 31-1 and 31-2 have 34 wrinkles and Paths 31-3 and 31-4 have 33 wrinkles. The four paths keep their wrinkle patterns as follows. Paths 31-1, 31-2, 31-3, and 31-4 are identified as $(+,-)$, $(-,+)$, $(-,-)$, and $(+,+)$, respectively.

The key point is that during this process, the eigenvalue analysis of the tangent stiffness matrix showed that all eigenvalues remain positive. Therefore, this wrinkle generation is not a result of bifurcation behavior. This wrinkle generation mechanism is regarded as unique. How this is achieved will be shown later.

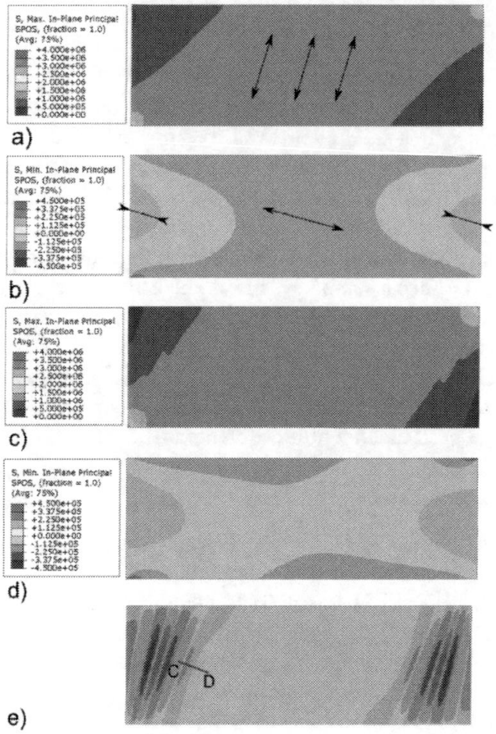

Figure 4. Principal stress values.

5.2.4 Bif. 60

The second bifurcations on this membrane are Bif. 60_1 and Bif. 60_2 at $\delta u_x = 60.90 \times 10^{-6}$[m] on Path 31-3 and Path 31-4, respectively. For Path 31-1 and Path 31-2, all eigenvalues of the tangent stiffness matrix remain positive at larger values of δu_x. At this point, wrinkle generation without bifurcation is complete. No flat sections remain on the membrane.

For Path 31-3, a snap through bifurcation is observed, where one zero eigenvalue is obtained at $\delta u_x = 60.90 \times 10^{-6}$[m], and the path jumps to either Path 31-1 or 31-2. A change in sign occurs near the right free boundary, and it results in a $(-,-)$ configuration becoming a $(+,-)$ configuration. Similarly, when the jump is to Path 31-2, the change is on the right side from $(-,-)$ to $(-,+)$. This transition is the rotational symmetry of the former. Symmetrical transitions occur along Path 31-4. The transition for Path 31-4 with $(+,+)$ to Path 31-2 with $(+,-)$ is the reflection symmetry of the first transition. The transition from Path 31-4 with $(+,+)$ to Path 31-1 with $(-,+)$ is the rotation and reflection symmetry of the first transition.

The conclusion is that after $\delta u_x = 60.90 \times 10^{-6}$[m], there are only two stable equilibrium paths corresponding to Paths 31-1 and 31-2. Their wrinkle configurations are almost the same as shown in Fig. 7 a) and b).

5.2.5 Bif. 440

Path tracking of both Path 80_1-1 and 80_1-2 determines the next bifurcation point at $\delta u_x = 440.21 \times 10^{-6}$[m] as Bif. 440_1. It is a simple singular point and the path branches to either Path 440_1-1 or 440_1-2.

Along Paths 80_1-1 and 80_1-2, collapsed sections increase in size. At $\delta u_x = 440.21 \times 10^{-6}$[m], the collapsed sections of two wrinkles reach critical sizes. One of the wrinkles will split into new wrinkles.

In Fig. 8, the behavior at this point can be observed. The wrinkle pattern at $\delta u_x = 440.21 \times 10^{-6}$[m], before the disturbance vector imposition, is shown in Fig. 8 a). The disturbance vector imposition causes a snap through bifurcation. The above pattern jumps to one of the patterns observed in Fig. 8 b) and c). The collapsed sections near the top and bottom fixed boundaries of one wrinkle expand toward the middle of the wrinkle. By increasing shear to $\delta u_x = 440.50 \times 10^{-6}$[m], the generation of new wrinkles is finalized as in Fig. 8 d) and e). The expanded collapsed sections meet, and the wrinkle amplitude along the center drops until it changes its sign. One wrinkle generates two new wrinkles, one of the same sign and one with the opposite sign.

Comparing the two wrinkle patterns on Paths 440_1-1 and 440_1-2, one is the equivalent of the other by the rotation and reflection transformation. Similarity up to five digits is calculated. Because the two paths are symmetric, only one is tracked until the next bifurcation.

The above behavior and that at Bif. 440_2 on Paths 80_2-1 and 80_2-2 have either reflection or rotation symmetry.

5.3 Wrinkle Generation without Bifurcations

In the previous sections, the membrane behavior were observedfor a range of imposed displacement at $0 \leq \delta u_x \leq 650 \times 10^{-6}$[m]. Now, the focus is on the wrinkles generated without bifurcations. This wrinkle generation occurs on equilibrium paths between the first and second bifurcation point.

5.3.1 Observations

Bifurcation free wrinkle generation is observed in the interval of imposed displacement of $31.68 \times 10^{-6} < \delta u_x \leq 60.90 \times 10^{-6}$[m]. As δu_x is increased, the number of wrinkles increases along all four paths originating from Bif. 31. Along the interval, every path is stable, i.e. the tangent stiffness matrix $\boldsymbol{K}$ on every path is positive definite, and wrinkle generation occurs at stable equilibrium points.

The process of wrinkle generation for a single wrinkle on Path 31-3 is shown in Fig. 5. Figure 5 a) shows the wrinkle pattern at $\delta u_x = 35.00 \times 10^{-6}$[m]. Near the free boundaries, there are three upward convex and three downward convex wrinkles along each boundary. The middle of the membrane remains flat. In Fig. 5 b), the imposed displacement is increased to $\delta u_x = 36.00 \times 10^{-6}$[m]. A small wrinkle is generated on the flat surface of the membrane, next to existing wrinkles. This wrinkle does not stretch from the top to bottom edges of the membrane. In Fig. 5 c) imposed displacement is increased further to $\delta u_x = 37.00 \times 10^{-6}$[m]. The new wrinkle on the right side expands to the top boundary. The wrinkle on left side expands to the bottom boundary. By further increasing the imposed displacement to $\delta u_x = 38.00 \times 10^{-6}$[m], a new wrinkle pattern Fig. 5 d) emerges. Now, the new small wrinkles run from top to bottom. Once the small wrinkles expand from top to bottom, the middle of the membrane changes sign in z-direction. However, the middle remains flat. In this process, where new wrinkles are generated, three new upward convex and four new downward convex wrinkles are observed on each side.

In Figs. 4 c) and d), the two-dimensional stress state on the mid-plane is shown for $\delta u_x = 38.00 \times 10^{-6}$[m]. The first principal stress σ_ξ is tensile throughout the membrane. The second principal stress σ_η is tensile in the middle, on the flat part of the membrane, while it is compressive around the wrinkled areas. However, the compressive stress area does not cover the small wrinkles in Fig. 5 b). Because these small wrinkles are under biaxial tension, to distinguish them from other wrinkles, they are referred to as waves. Wrinkles occur as a result of buckling in a compressive stress area. Waves are generated without buckling in a biaxial tension stress area. The first axis tension and second axis compression area is referred to as the compressive stress areas.

One by one, waves are generated on each side of the flat section of the membrane. As δu_x increases, the compressive stress area expands toward the middle of the membrane. When the compressive stress area covers the wave, by definition the wave becomes a wrinkle. Therefore, a wrinkle is generated without bifurcation buckling.

To show the process in more detail, section CD in the $\eta\eta'$ direction is constructed as shown in Fig. 4 e). The position in section CD is defined by node numbers, projections of finite element nodes onto section CD. The total number of nodes is 30. Figure 9 shows the results for $\delta u_x = 37.58 \times 10^{-6}$[m], $\delta u_x = 37.75 \times 10^{-6}$[m], $\delta u_x = 38.00 \times 10^{-6}$[m], and $\delta u_x = 38.20 \times 10^{-6}$[m]. Depending on the behavior, each figure illustrates an appropriate node range.

Figure 9 d) shows the second principal stress σ_η in section CD. The compressive stress area progresses into the middle. It also shows that there are both the compressive stress area and the biaxial tensile stress areas in section CD. The second principal stress at every node decreases as δu_x increases. Figure 9 e) shows the minimal principal stress in the node interval from 10 to 14 in section CD. The border between the positive and the negative second principal stress areas moves from node 11 to 13 toward point D when the imposed displacement increase from $\delta u_x = 37.58 \times 10^{-6}$[m] to $\delta u_x = 38.20 \times 10^{-6}$[m]. The key point is that the area of node 13 and its right side is the biaxial tensile stress area by $\delta u_x = 38.20 \times 10^{-6}$[m].

Figure 9 a) shows an increase of out-of-plane displacement between nodes 1 and 11 as δu_x is increased. Figure

9 b) shows the out-of-plane displacement more closely between nodes 12 and 19. As the imposed displacement δu_x is increased, the out-of-plane displacement decreases between nodes 14 and 16.

Figure 9 c) shows a change in curvature from positive to negative between nodes 15 and 20 at $\delta u_x = 38.00 \times 10^{-6}$[m]. At $\delta u_x = 38.20 \times 10^{-6}$[m], the wave is convex upward. Recalling Fig. 9 d), the area after node 13 is a biaxial tensile stress area. Therefore, this is not a wrinkle but a wave.

5.3.2 Wrinkle Generation Mechanism

Increasing δu_x from 0, the semicircular regions of uniaxial compression near the free boundaries occur, while the other region remains as a biaxial tension stress area, as indicated by Fig. 4 d). The compressive stress regions gradually expand toward the middle of the membrane with increasing δu_x. This is confirmed by Fig. 9 d).

For the path segment of interest, $37.58 \times 10^{-6} \leq \delta u_x \leq 38.20 \times 10^{-6}$[m], the middle section of the membrane including point D tends to remain flat without out-of-plane displacement due to biaxial tension.

Figure 9 a) shows a downward convex wrinkle near node 3 and an upward convex wrinkle near node 8. Their curvature magnitudes are large. The region from node 1 to node 11 is always a compressive stress area, but its second principal stress σ_η decreases toward node 11. Combining the large curvatures and the decreasing second principal stress results in larger out-of-plane displacements, see the variation between node 1 and 11 in Fig. 9 a).

The out-of-plane displacements of the wrinkles near the free boundary converge to zero toward point D. In Fig. 9 b), while the out-of-plane displacement around node 12 increases, that around node 14 decreases. For this process, the bending moment M_ξ is given in Fig. 9 f) and g) and the section transversal force N_z is given in Fig. 9 h). The M_ξ and N_z are applied to the membrane due to the existence of bending rigidity in the model.

The bending moment changes from positive to negative around node 14. It reaches the minimum value around node 15, and it converges to 0 toward the middle of the membrane. By increasing δu_x, the minimum value also decreases. This negative value of the bending moment results in a downward convex deformation of the membrane.

The transverse force N_z is negative between nodes 8 and 15. This means a downward force is applied in the region. The peak of an upward convex wrinkle occurs near node 8. On this wrinkle, the larger tensile stress σ_ξ and the curvature along ξ causes a negative transverse force. The smaller compressive stress σ_η and the curvature along η cause a smaller positive transverse force. These result in a negative transverse force near node 8. This is confirmed by the fact that a positive transverse force, i.e. an upward force, is applied to a downward convex wrinkle near node 3. As a result, in Fig. 9 b) the deformation is convex upward until node 14, but it becomes convex downward at node 15. In addition, the left side of node 15 is pushed downward. At $\delta u_x = 38.00 \times 10^{-6}$[m], this results in a small region of negative curvature between node 15 to 20. As this region is under biaxial tension, the deformation with varying curvature is a wave.

In summary, initial wrinkles exist in a uniaxial compression region. The biaxial tension region is generally flat. In the biaxial tension region near the existing wrinkles, a wave deformation is generated. This phenomena can be considered as the origin of wrinkles. As the imposed displacement δu_x is increased, the uniaxial compression region spreads to the wave and transforms it into a wrinkle. This process occurs without bifurcations. As the uniaxial compression region spreads, the new wrinkle grows in size and generates the conditions for a new wave. The whole process repeats itself. This is a wrinkle generation mechanism without bifurcations.

6. CONCLUSION

Because wrinkle generation is generally considered to be a result of bifurcations, the expected result was a branching structure at every wrinkle generation. This could potentially result in a large number of equilibrium paths where each represents a specific wrinkle pattern. However, as the results show, significantly fewer equilibrium paths were obtained. This was due to wrinkle generation without bifurcation and to deformation symmetry.

The first bifurcation generates the initial wrinkles along four equilibrium paths. The membrane is divided into the biaxial tension area and the uniaxial compression areas. Wrinkles are only generated in the uniaxial compression area. A wave is generated in the biaxial tension area, near the uniaxial compression area. With increasing shear displacement, the uniaxial compression area expands onto the wave and turns it into a wrinkle. The process of wrinkle generation without bifurcation repeats on all four paths until no section of flat membrane remains. The number of wrinkles increases to 33 and 34. This wrinkle generation without bifurcation results in fewer equilibrium paths than expected for this number of wrinkles.

Increasing shear further causes wrinkle amplitude to increase until the next bifurcation. Different behavior occurs where small sections of existing wrinkles collapsed. With further increases in shear displacement, these collapsed sections govern wrinkle generation by destabilizing existing wrinkles. This is the expected behavior for bifurcations causing wrinkle generation. However, the wrinkle patterns after wrinkle generation show symmetry, resulting in fewer unique wrinkle patterns.

Based on the results of the above analysis, the following conclusions can be made. Based on the number of wrinkles generated, there was a possibility of a complex bifurcation structure with numerous equilibrium paths. Additionally, the existence of symmetry described by a dihedral group indicates three more solutions for each solution obtained, implying a multiplicity of the solution and additional equilibrium points. However, the wrinkle generation mechanism without bifurcations reduces the number of equilibrium paths. Furthermore, the symmetry between deformation patterns reduces the number of possible wrinkle configurations. The number of possible wrinkle configurations is reduced even further by re-coalesce of equilibrium paths occurring after bifurcations. Finally, the membrane bifurcation structure is less complex than it potentially could have been.

The universality of the behavior discussed in this study depends on the universality of the stress field that causes the behavior. Wrinkle generation without bifurcations is caused by the transition between uniaxial compression and biaxial tension. Therefore, any set of conditions that maintain uniaxial compression and biaxial tension would have this type of wrinkle generation.

REFERENCES

1. Chmielewsk,A. B., "Overview of Gossamer Structures," *Gossamer Spacecraft: Membrane and Inflatable Structures Technology for Space Applications*, edited by C. H. M. Jenkins, Vol. 191, Progress in Astronautics and Aeronautics, AIAA, Reston, VA, 2001, pp.1-20.
2. Senda, K., Oda, T., Ota, S., Igarashi, Y., Watanabe, A., Hori, T., Ito, H., Tsunoda, H., and Watanabe, K., "Deploy Experiment of Inflatable Tube using Work Hardening," *Proc. of AIAA Gossamer Spacecraft Forum*, AIAA-2006-1808, Newport, RI, 2006, pp.1-18.
3. Timoshenko, S.P. and Gere, J.M., "Bending of Thin Plates," *Theory of Elastic Stability* 2nd ed., Springer, New York, 2010, pp. 319-346.
4. Wong, Y.W. and Pellegrino, S., "Wrinkled Membranes, Part I," *Journal of Mechanics of Materials and Structures*, Vol. 1, No. 1, Jan. 2006, pp.3-25.
5. Wong, Y.W. and Pellegrino, S., "Wrinkled Membranes, Part II," *Journal of Mechanics of Materials and Structures*, Vol. 1, No. 1, Jan. 2006, pp.27-61.
6. Wong, Y.W. and Pellegrino, S., "Wrinkled Membranes, Part III," *Journal of Mechanics of Materials and Structures*, Vol. 1, No. 1, Jan. 2006, pp.63–95.
7. Bazant, Z.P. and Cedelon, L., "Buckling of Elastic Frames by Equilibrium Analysis," *Stability of Structures*, Oxford University Press, New York, 1991, pp. 53-73.
8. Wang, C.G., Du, X.W., Tan, H.F. and He, X.D., "A New Computational Method for Wrinkling Analysis of Gossamer Space Structures," *International Journal of Solids and Structures*, Vol. 46, No. 6, March 2009, pp. 1516-1526.
9. Wriggers, P., Wagner, W., Miehe, C., "A Quadratically Convergent Procedure for the Calculation of Stability Points in Finite Element Analysis," *Computer Methods in Applied Mechanics and Engineering*, Vol. 70, No. 3, 1988, pp.329-347.
10. Inoue, S., "Prediction Methods of Wrinkling in Thin-Membrane," *Proceedings of International Symposium on Space Technology and Science*, Tsukuba, Japan, 2009, pp.1-7.
11. Endo, A., Hangai, Y., and Kawamata, S., "Post-Buckling Analysis of Elastic Shells of Revolution by the Finite Element Method," *Report of Institute of Industrial Science*, UDC 624.074.4:624.075.4, University of Tokyo, December 1976.
12. Hoyle, R., "A bit of Group Theory," *Pattern Formation* Cambridge University Press, New York, 2007, pp. 52-84.

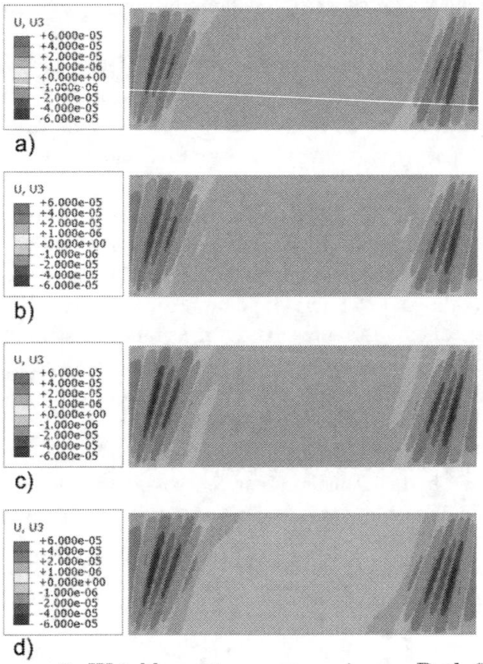

Figure 5. Wrinkle pattern progression on Path 31-3.

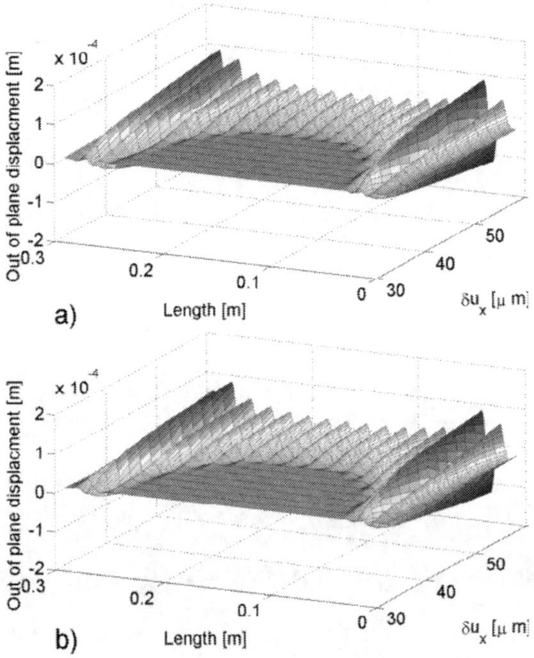

Figure 6. Wrinkle pattern progression in section AA'.

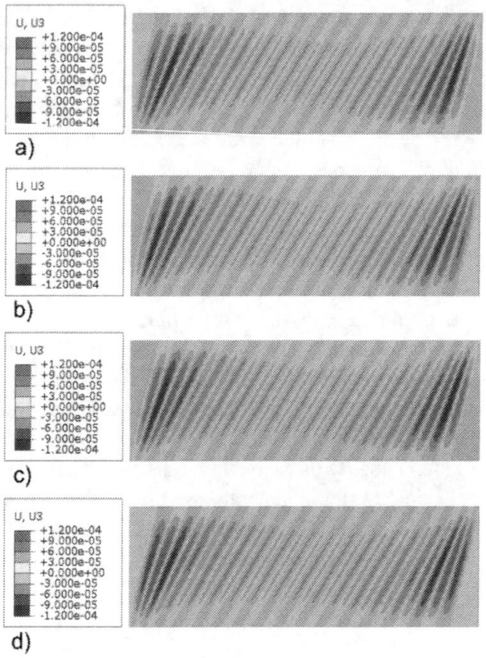

Figure 7. Wrinkle pattern at $\delta u_x = 60.00 \times 10^{-6}$[m].

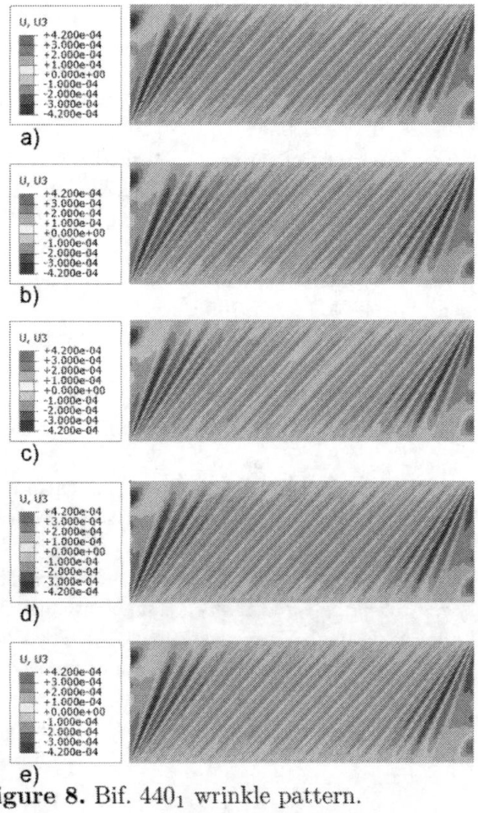

Figure 8. Bif. 440_1 wrinkle pattern.

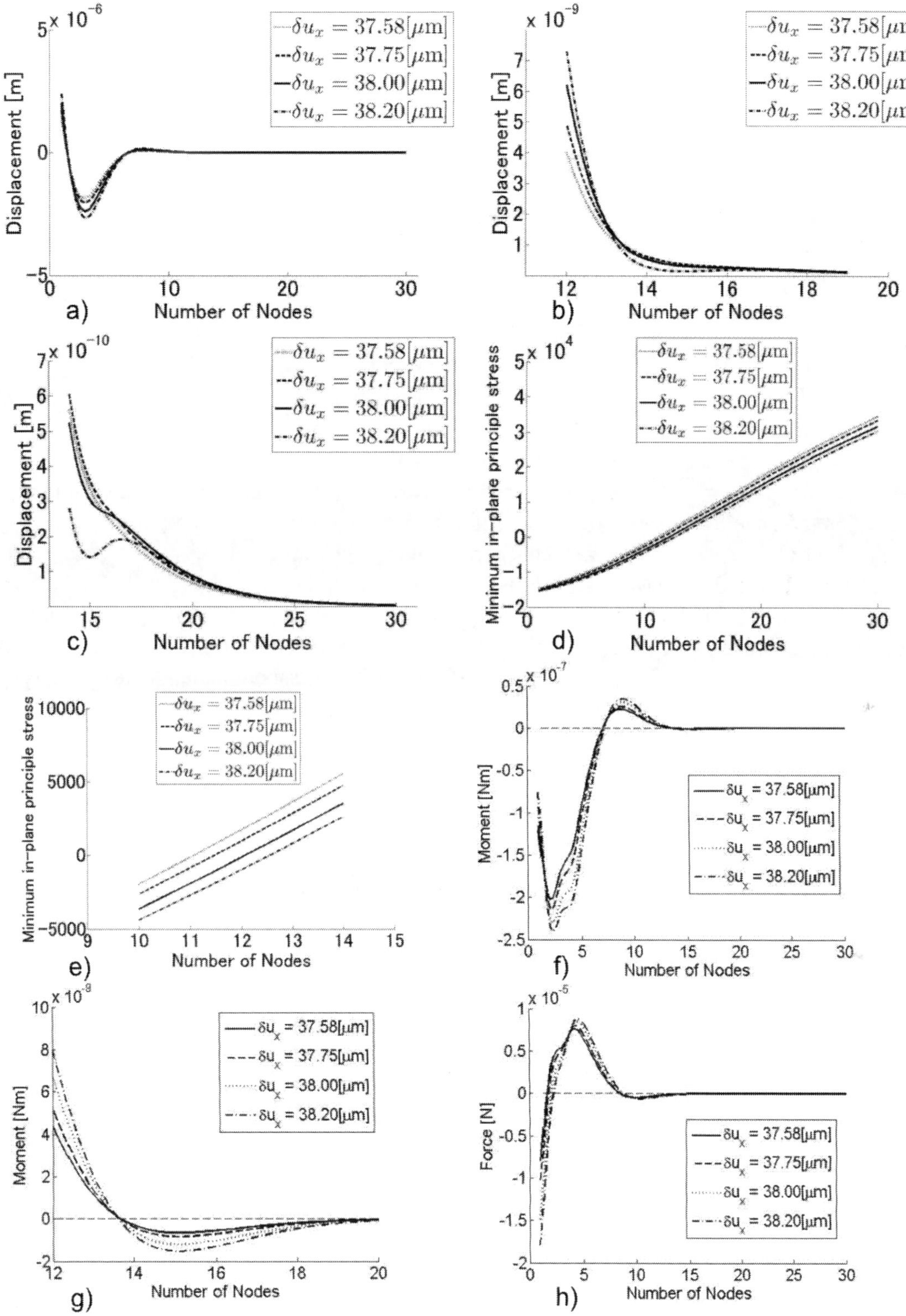

Figure 9. Path 31-3, wave section analysis.

ICAST2015: 26th International Conference on Adaptive Structures and Technologies
October 14-16th, 2015, Kobe, Japan

ICAST2015 #90

Evolutionary Shape Optimization of Transonic Airfoils for Aerodynamic and Aeroelastic Performance

Robert Carrese[1*], Pier Marzocca[1], Michael Candon[1] and Hideaki Ogawa[1]

[1] School of Aerospace Mechanical and Manufacturing Engineering, RMIT University, Bundoora, VIC Australia.

Abstract

The ongoing use and development of optimization frameworks for aircraft design is due to their ability to identify optimal and often non-intuitive shapes pertaining to the multi-disciplinary design criteria. Airfoil design is a continuously revised multi-disciplinary problem in the literature, and is pivotal to illustrate the performance of optimization frameworks involving numerical simulation, flexible shape parameterization, and intelligent evolutionary algorithms. An often overlooked component of this classic problem is to consider the dynamic aeroelastic behavior under trim conditions, which can impose explicit boundaries to the flight envelope. In this paper, a multiobjective particle swarm optimization framework is presented, pertaining to aerodynamic and aeroelastic design criteria at the trim condition. Designer preferences are used to reflect the optimal compromise between the objectives. Results of the optimization process indicate a large spread in design variable influence and interaction, and a subtle yet clear distinction between all objectives is illustrated through the final airfoil shapes obtained.

1. INTRODUCTION

Evolutionary design optimization is a widely used computational technique in modern airfoil design [1]. This state-of-the-art design methodology couples computational fluid dynamics (CFD) solvers with sophisticated optimization algorithms, allowing for the shape of an airfoil section to be optimized for a set of predefined performance criteria known as objective functions. Multiobjective optimization provides a more comprehensive analysis of the design space than single-objective optimization as an array of solutions to $m > 1$ objective functions are provided allowing the designer to investigate the compromise/trade-off between conflicting objectives [2]. The objective functions can be multidisciplinary, where objective functions can stem from a range of engineering disciplines.

Conflicting requirements in transonic airfoil design pose a significant challenge in the conceptual design phase for airfoil robust performance [3]. A significant advantage of adopting evolutionary methods is the level of exploratory capacity of the optimization algorithm, resulting in innovative and often unintuitive geometric profiles for a given set of objectives. While most airfoil design frameworks reported in the literature focus on static objectives, an often overlooked component is to consider the dynamic aeroelastic behavior under trim conditions, which along with providing more realistic constraints on the airfoil profiles, it introduces a significantly strong coupling with objectives governing static performance

* robert.carrese@rmit.edu.au

(e.g. aerodynamic or structural). Airfoil aeroelastic performance is associated with flutter: an aeroelastic event which can cause significant structural damage to the wing and in some cases catastrophic failure [4]. Typically, to avoid the onset of flutter at any trim condition, the velocity at which the system is no longer stable (i.e. damped) should be increased. This increases the flutter boundary for a range of freestream flight conditions giving the airfoil a more robust operational capacity [5].

The conflicting requirements often associated with airfoil design therefore requires optimization techniques which allow the designer to view the tradeoff between solutions to the objective functions and make an intelligent decision on an airfoil shape that is best suited to the design scope [6]. For a solution to be considered Pareto-optimal (i.e. an optimal compromise), it must be identified within the design space for which the value of the objective functions cannot be improved without deteriorating the performance of other objectives. The set of non-dominated solutions which meet the Pareto-optimal criteria are known as the Pareto front. Population-based EMO methods which are able to define the entire Pareto front in one run, thus allowing the designer to view multiple solutions and select one that is most suitable [2]. A significant drawback of this approach however is the large computational cost which can be incurred as a result of the large number of CFD simulations that must be run in order to obtain a converged solution to the objective functions. The computational cost is further augmented by the dynamic aeroelastic objective which entails timely transient CFD simulations. In order to minimize the computational cost, designer-preferences are included to provide guidance in the search with the designers interests (expert opinion) being reflected in the optimal compromise [7]. A reference point is chosen by the designer and projected onto the design space, guiding the swarm towards the preferred solution. The communication of designer preferences via the reference point allows the designer to examine multiple designs within the preferred area without being faced with the unnecessary computational cost that can be inherited when defining the entire Pareto front [8].

2. OPTIMIZATION FRAMEWORK

The feasibility and computational intensity of an airfoil shape optimization process is dependent on the shape parametrization method which is employed, the numerical tools utilized to quantify the objective functions, and the optimization algorithm. This section details the important elements of the optimization framework.

2.1 Airfoil Shape Parameterization

Flexible parameterization methods for airfoil shape optimization are required to provide a complete and representative design space to identify improved and innovative designs [9].

A common method for airfoil shape parameterization is the PARSEC method [10]. It provides high control over important aerodynamic features, and facilitates the implementation of manufacturing constraints. The methodology is characterized by 11 design variables (Figure 1(a)), including leading edge radius (r_{LE}), upper and lower thickness locations (x_{UP}, z_{UP}; x_{LO}, z_{LO}) and curvatures ($z_{xx_{UP}}$; $z_{xx_{LO}}$), trailing edge direction (α_{TE}) and wedge angle (β_{TE}), and trailing edge coordinate z_{TE} and thickness Δz_{TE}. The shape function (describing the thickness) is described by a sixth-order polynomial [10]:

$$z_k = \sum_{i=1}^{6} a_{i,k} \cdot x_k^{i-\frac{1}{2}} \tag{1}$$

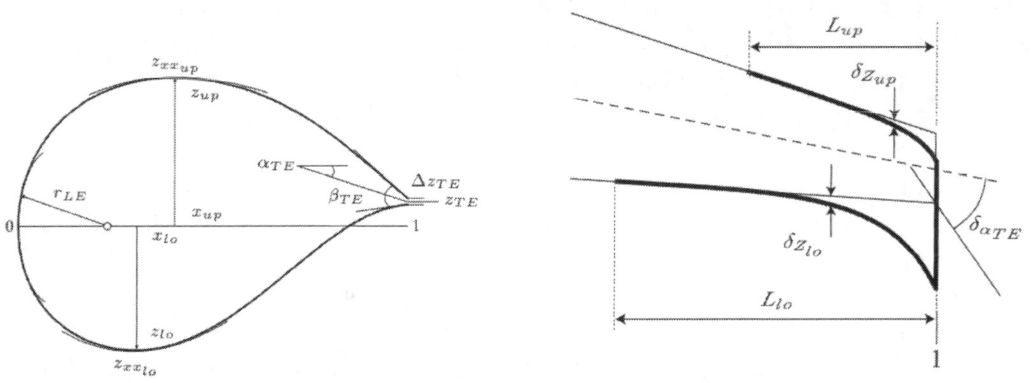

(a) PARSEC shape parameters [10]

(b) Trailing-edge modification [11]

Figure 1. PARSEC airfoil parametrization to represent transonic profiles

where (x, z) are the shape coordinates and k denotes either the upper or lower airfoil surface. The coefficients $a_{i,k}$ are determined from the geometric parameters. For transonic airfoils, a modification is introduced to reduce the probability of downstream boundary layer separation [11]. This is controlled by the parameter $\delta_{\alpha_{TE}}$, which directly influences the additional curvature of the trailing edge. The modification decouples the trailing edge parameterization by first defining a smoother upper surface contour and then constraining the lower surface to intersect the trailing edge coordinate. Figure 1(b) illustrates the modification to the trailing edge curvature. The modification is applied to the upper and lower surfaces as follows:

$$\delta_z = \frac{L \cdot \tan \delta_{\alpha_{TE}}}{2\mu\tau} \left[1 + \eta \cdot x^\tau - (1 - x^\tau)^\mu\right] \quad (2)$$

where the constants η, μ, and τ are set to 0.8, 2, and 6, respectively [11]. Table 1 stipulates the design space boundaries, which result in a wide range of feasible airfoil shapes.

Table 1 Design variable ranges for optimization study

Parameter	**Description**	**Lower bound**	**Upper bound**
r_{LE}	Leading edge radius	0.0063	0.0151
α_{TE}	Trailing edge direction	-0.2405	-0.0026
β_{TE}	Trailing edge wedge angle	0.0655	0.2618
x_{UP}	Upper-crest location	0.3170	0.5250
z_{UP}	Upper-crest ordinate	0.0497	0.0683
$z_{xx_{UP}}$	Upper-crest curvature	-0.5135	-0.2393
x_{LO}	Lower-crest location	0.2835	0.3418
z_{LO}	Lower-crest ordinate	-0.0603	-0.0478
$z_{xx_{LO}}$	Lower-crest curvature	0.2535	0.8405
z_{TE}	Trailing edge ordinate	0.0	0.0
Δz_{TE}	Trailing edge thickness	0.0	0.0
$\delta_{\alpha_{TE}}$	Trailing edge curvature	-0.0080	0.3696

2.2 Flow Solver

Solutions to the objective functions are generated via a flow solver. The fidelity to which the flowfield should be resolved is governed by the aerodynamic complexity of the system and the desired computational accuracy of the designer [1]. For transonic wings, the dynamic motion of the airfoil along with the transonic nature of the flow can have the tendency to generate a highly nonlinear aerodynamic system. Aerodynamic nonlinearities are the product of shock wave boundary layer interaction, inevitably resulting in shock-induced separation. The viscous contributions to the pressure distribution require solution of the Reynolds-Averaged Navier-Stokes (RANS) equations via CFD.

The Navier-Stokes equations for the turbulent flowfields are solved using the commercial finite volume solver ANSYS Fluent R16 [12]. A coupled pressure-based scheme with second-order spatial accuracy is adopted with the turbulence model selected as the one-equation Spalart-Allmaras model [13], assuming turbulent flow over the complete airfoil surface. The steady-state flowfield is initialized via a hybrid initialization scheme which provides a Potential flow solution to determine the momentum field. The C-type grid (as represented in Figure 2) stretches 30 chord lengths aft and normal of the airfoil section. The first grid point normal to the airfoil surface corresponds to an average $y^+ < 120$.

An elastically-mounted rigid airfoil with two-degrees-of-freedom respectively in pitch, α [rad] and plunge, h [m] is considered for a coupled aeroelastic model. The generalized equations of motion for small vibration amplitudes governing this aeroelastic system are [14]:

$$\frac{1}{\omega_\alpha^2}\begin{bmatrix} 1 & x_\alpha \\ x_\alpha & r_\alpha^2 \end{bmatrix}\begin{Bmatrix} \frac{\ddot{h}}{b} \\ \ddot{\alpha} \end{Bmatrix} + \begin{bmatrix} \left(\frac{\omega_h}{\omega_\alpha}\right)^2 & 0 \\ 0 & r_\alpha^2 \end{bmatrix}\begin{Bmatrix} \frac{h}{b} \\ \alpha \end{Bmatrix} = \frac{V^2}{\pi\mu\omega_\alpha^2 b^2}\begin{Bmatrix} c_l \\ 2c_m \end{Bmatrix} \tag{3}$$

where ω is the structural natural frequency in pitch and plunge, x_α and r_α^2 are the static unbalance and radius of gyration about the elastic axis, b is the semi-chord of the section, μ is the mass ratio, and V is the freestream velocity. The array $\{c_l\ 2c_m\}^T$ is the generalized aerodynamic force coefficient vector for the two-dimensional aeroelastic system, comprised of the lift coefficient c_l and moment coefficient about the elastic axis c_m. From a converged static trim result, a dynamic mesh model to capture the induced momentum due to the motion of the wing is introduced based on diffusive smoothing, this

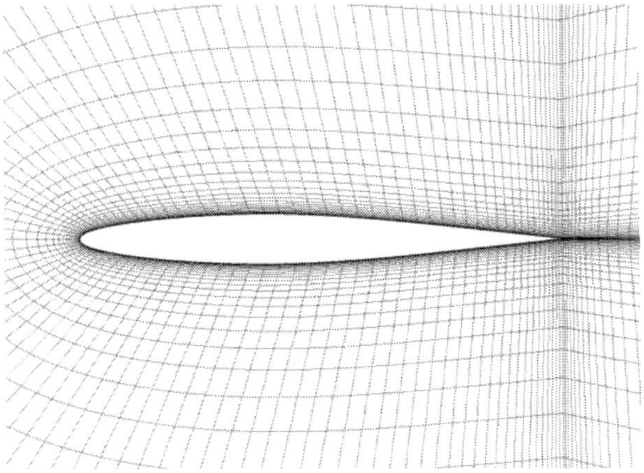

Figure 2. Computational C-grid used for transonic flow solver

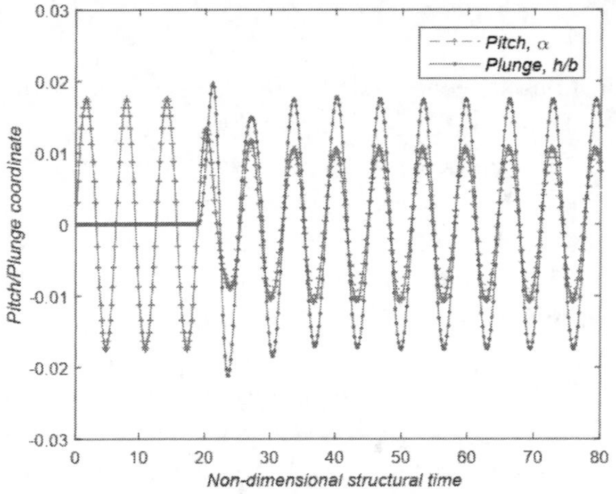

Figure 3. Aeroelastic system exhibiting neutral response $V_f = V_f*$

preserves mesh quality close to the boundary whilst absorbing motion in the far-field. A dual time-stepping scheme is employed with second-order implicit temporal discretization. The time-step selected was based on temporal discretization of ω_α by 36 steps, with approximately 5-10 inner iterations to ensure residual convergence of $\mathcal{O}(10^{-5})$. Within each real time step, the set of ordinary differential is solved using a four-stage Runge-Kutta scheme. Initially a sinusoidal forcing function is applied in pitch about the elastic axis for three complete cycles at ω_α with the airfoil pinned and unable to displace in plunge. At the end of these forced cycles, the imaginary pin is removed, and the system is allowed to evolve forced by its own self-induced loads [14]. The response (see Figure 3, over non-dimensional time $\tau = \omega_\alpha t$) is computed at the trim condition for increasing values of the speed index V_f, to determine the location of the flutter boundary V_f*, which coincides with a neutral response.

2.3 Optimization Algorithm

The limitation of evolutionary multiobjective optimization algorithms is the often significant computational cost associated with identifying all potential compromising solutions on the Pareto front. The incorporation of designer preferences relieves some of the computational cost which would otherwise be associated with exploring the entire Pareto front. Here, a preference-based swarm algorithm is used, where a designer driven distance metric is considered to scalar quantify the success of the solution. This guides the swarm according to the reference point, limiting the search to the preferred region of the Pareto front. Using this methodology, the optimal solution is essentially an optimal compromise of solutions to the objective functions, reflecting the compromise of solutions indicated by the reference point.

The multiobjective evolutionary optimization algorithm (UPMOPSO) is comprehensively described in previous publications, including a discussion on the swarm dynamics and topology, as well as metrics on the performance of the algorithm [6]. For the present article, the discussion is limited to a brief description of the algorithm.

2.3.1 Particle Swarm Optimization

Particle Swarm Optimization (PSO) is derived from the behavioral tendency of individuals to learn from previous experience and emulate the success of others [15]. The model utilizes a swarm of N particles which are flown through $S \in \mathbb{R}^n$. The particles are represented by n-dimensional vectors $\mathbf{x_i}$ and $\mathbf{v_i}$, which are the particle position and velocity respectively. Particles share experiences and follow the most successful particles to explore and exploit promising areas of the design space. The ith particle of the swarm is accelerated towards its personal best position, $\mathbf{p_i}$ and the global best position, $\mathbf{p_g}$ of the swarm. The updated position and velocity vectors at pseudo-time $t+1$ are a resultant of these two components:

$$\mathbf{v}_{i,(t+1)} = \chi\left[\mathbf{v}_{i,(t)} + \mathbf{R}_1[0,\varphi_1] \otimes \left(\mathbf{p}_{i,(t)} - \mathbf{x}_{i,(t)}\right) + \mathbf{R}_2[0,\varphi_2] \otimes \left(\mathbf{p}_{g,(t)} - \mathbf{x}_{i,(t)}\right)\right] \quad (4)$$

$$\mathbf{x}_{i,(t+1)} = \mathbf{x}_{i,(t)} + \mathbf{v}_{i,(t+1)}, \quad (5)$$

where $\mathbf{R}_1[0,\varphi_1]$ and $\mathbf{R}_2[0,\varphi_2]$ are two functions returning a vector of uniform random numbers in the range $[0,\varphi_1]$ and $[0,\varphi_2]$, respectively. The constants φ_1 and φ_2 are set to $\varphi/2$ where $\varphi = 4.1$. The constriction factor χ applies a dampening effect as to how far the particle explores within the search space, given as [16]:

$$\chi = 2/|2 - \varphi - \sqrt{\varphi^2 - 4\varphi}|. \quad (6)$$

Solutions to multi-objective problems are non-dominated or Pareto-optimal, each solution offering a compromising result between all objectives. Since each Pareto-optimal solution is considered equally optimal, there are several global or swarm leaders. To focus the computing effort on only preferred regions of the Pareto front, a designer-preference module is implemented to provide additional guidance to the swarm in parallel with the dominance criterion. This module assists the swarm to converge to preferred regions of the Pareto front to find only solutions of interest to the designer. The guidance mechanism takes the form of a reference point $\bar{z}$ which is a means of drawing on the designer's experience to express a feasible compromise, rather than specific target values. The reference point is used to construct a distance metric to be minimized for $\mathbf{x} \in S$ [7],

$$\text{minimize } d_z = \max_{i=1:m} \{(f_i(\mathbf{x}) - \bar{z}_i)\} \quad (7)$$

where $\bar{z}_i$ is the ith component of the reference point or the aspiration value to the ith objective. As described in Carrese and Li [6], as the algorithm advances in time, a solution is said to be preferred if it satisfies the condition $d_z(\mathbf{z}') = \min(\forall d_z(\mathbf{x})$. The designer has no prior knowledge of the location of the Pareto front, and reference points may be ideally placed in any feasible or infeasible region. The algorithm applies the dominance criteria concurrently with Eq. (7) to find a feasible set of Pareto-optimal solutions near the preferred regions. The spread of solutions along the Pareto front is controlled by δ which is the maximum variation as $d_z(\mathbf{x}) \leq d_z(\mathbf{z}') + \delta$, as shown in Figure 4 [7].

3. COMPUTATIONAL RESULTS

The incorporation of designer preferences to the multi-objective optimization problem relieves some of the computational cost which would otherwise be associated with exploring the entire Pareto front.

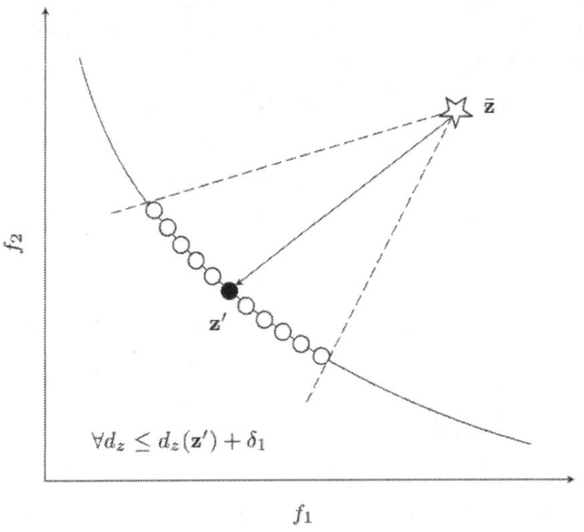

Figure 4. Swarm guidance via the preference module

The algorithm described in the previous section is combined with the flow solver and PARSEC parametrization method for an efficient design framework. The objectives are defined as:

$$\mathbf{f} = \min \left\{ \begin{array}{ll} f_1 & : c_d \\ f_2 & : (1/V_f*) \end{array} \right\} \quad \text{at } M = 0.8,\ c_l = 0.4. \tag{8}$$

While typically the aeroelastic boundary is investigated at zero mean angle of attack, the current study aims to investigate the influence of a mean aerodynamic loading, attributed to the trim condition. For any candidate profile, the trim condition is explicitly determined by allowing the angle of attack to vary until the desired lift coefficient is obtained.

3.1 Designer Preferences

The Isogai Case A pitch / plunge aeroelastic benchmark at $M = 0.8$, $c_l = 0.4$ (see Figure 5) is selected as a suitable reference point to explore the feasibility of this optimization framework [17]. The dimensionless structural parameters of this configuration are chosen to represent the dynamics of the outer section of a swept-back wing, with centre-of-gravity, $x_{cg} = 0.4$, centre-of-rotation, $x_\alpha = 1.8$, and radius-of-gyration about the centre-of-rotation, $r_\alpha = 1.865$. This places the pivot point forward of the leading-edge. The ratio of natural (uncoupled) frequencies $\omega_h/\omega_\alpha = 1.0$, and a mass ratio $\mu = 60$. No structural damping is considered.

The trim condition for the Isogai benchmark, as shown in Figure 5 results in an angle-of-attack $\alpha = 2.077°$ to obtain a lift coefficient of $c_l = 0.4$. A characteristic of this profile is the lightly loaded trailing edge, which noticeably shifts the centre of pressure forward. The objectives for the reference point are calculated as $\bar{z} = [0.0301,\ 0.7928]$.

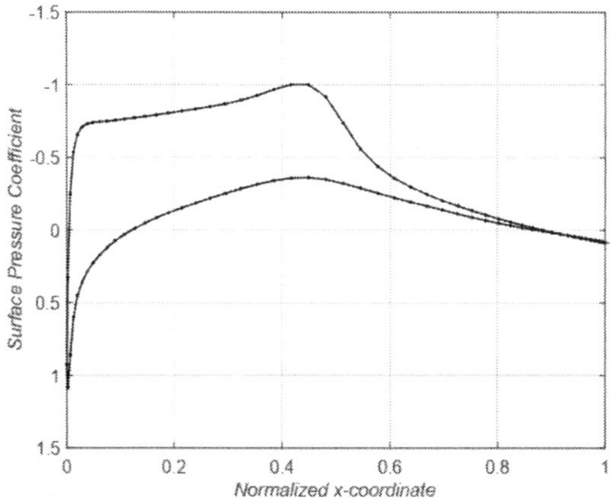

(a) Surface pressure distribution

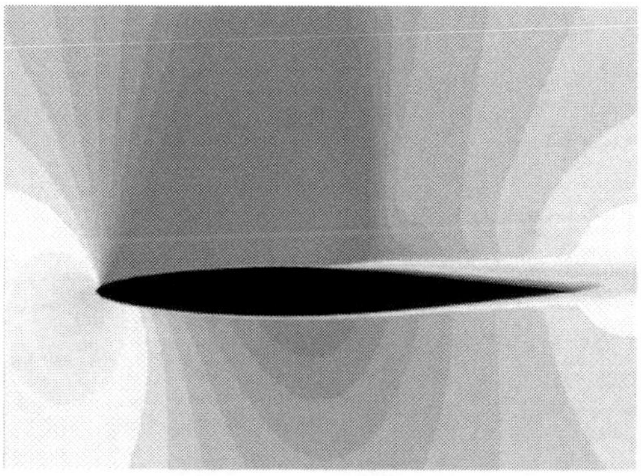

(b) Mach number contours ($0.0 \leq M \leq 1.3$)

Figure 5. Isogai Case A reference configuration at $M = 0.8$, $c_l = 0.4$

3.2 Optimization Results

A swarm population of $N = 50$ particles is progressed for a total of $t = 50$ pseudo timesteps, resulting in 2500 evaluations of the flow solver. The solution spread is limited by $\delta = 1e{-}2$. Figure 6(a) illustrates how the progress of the optimization routine can be intuitively monitored via a single scalar, namely, the progress of the most preferred design. The algorithm is initially exhibiting characteristics of exploration, since there are relatively high performance gains ($\approx 95\%$ of final preferred design) for $t < 30$. After that, the swarm begins to exploit this region to provide a more suitable solution set to the designer. Figure 6(b) shows the final set of (non-dominated) solutions as obtained at $t = 50$. No comment can be made on the optimality of this non-dominated front, since convergence of the optimization framework is dictated solely by the imposed computational budget. Nevertheless, the non-dominated front is well defined, exhibiting characteristics of a convex and discontinuous Pareto landscape. It is also shown, that the Pareto front is highly skewed in favour of the second objective f_2, indicating that the aeroelastic performance is ultimately driving the design.

3.3 Post-processing

To illustrate where the final solutions are located in the design space, a boxplot (see Figure 7(a)) is generated of the non-dominated solutions. The design space is normalized to $[0, 1]^n$ for best visualization. It is observed from Figure 7(a) that the preferred region of the non-dominated front as dictated by the reference point is fairly localized within the design space, with high spatial variation for the trailing-edge wedge angle β_{TE} and the lower thickness co-ordinates (x_{LO}, z_{LO}). This is a very intriguing result and further discussion can be made through visualization of some of the airfoil shapes of interest from the non-dominated solution set. The three airfoils of interest pertain to $\min f_1$, $\min f_2$ and $\min d_z$. The airfoils employ a similar topology, with a relatively high leading-edge radius and location of upper thickness pushed further aft to maintain a relatively low pressure gradient over the upper surface. The subtle variation is evident through the location of the lower thickness co-ordinates, which are pushed further aft for maximum aerodynamic performance.

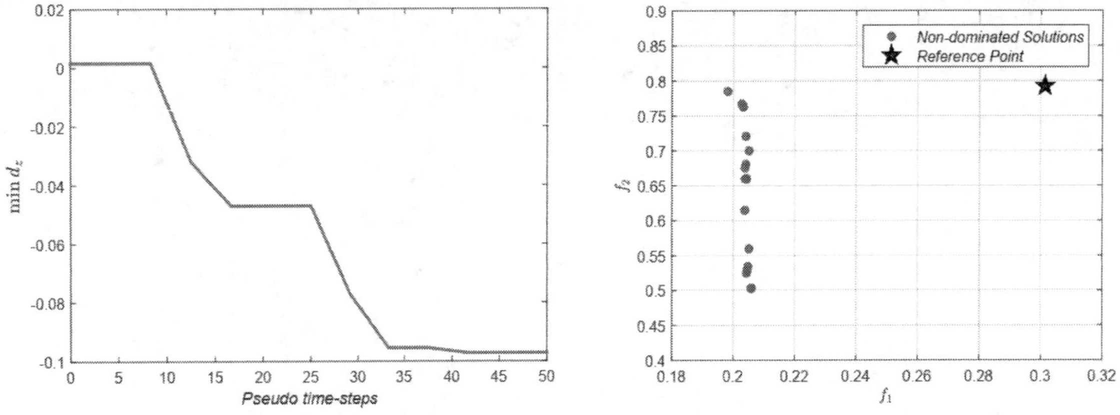

(a) History of most preferred solution

(b) Final non-dominated solution set

Figure 6. Algorithm performance for optimization case-study

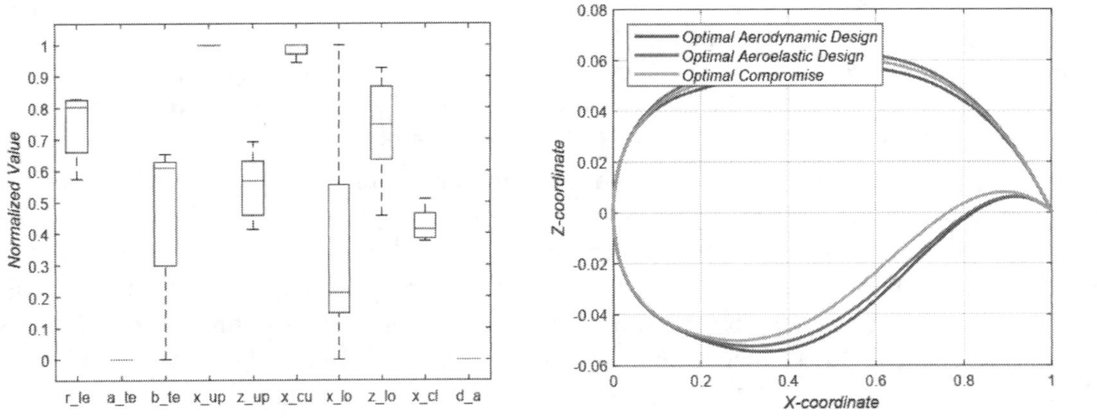

(a) Spread of PARSEC parameters

(b) Optimal airfoil shapes of interest

Figure 7. Shape variation for final catalog of airfoil shapes

Results of the final airfoil shapes are shown in Table 2. Each of the shapes of interest dominate the reference point, and provide performance gains over each objective. Further correlations can be made by visualizing the surface pressure distributions of the candidate airfoils. The optimal aerodynamic design is characterized by a front loading pressure distribution typical of supercritical airfoils, with a very weak shock and a delayed adverse pressure gradient. By contrast, the optimal aeroelastic design has a shock of higher strength, but noticeably employs a lower loading on the forward section of the airfoil. The optimal compromise is shown to mimic the optimal aeroelastic design closely, with the exception that the shock is delayed further downstream, to improve aerodynamic performance. Superior aeroelastic performance is thus dictated by maintaining as much as possible a zero-pressure gradient for the suction and pressure surfaces, resulting in a flat panel appearance.

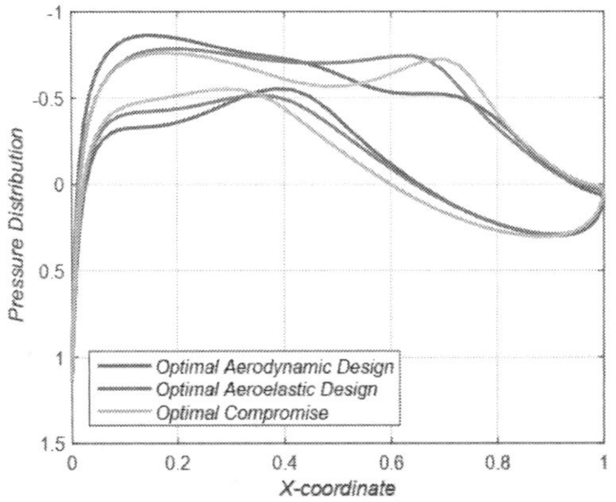

(a) Surface pressure distributions

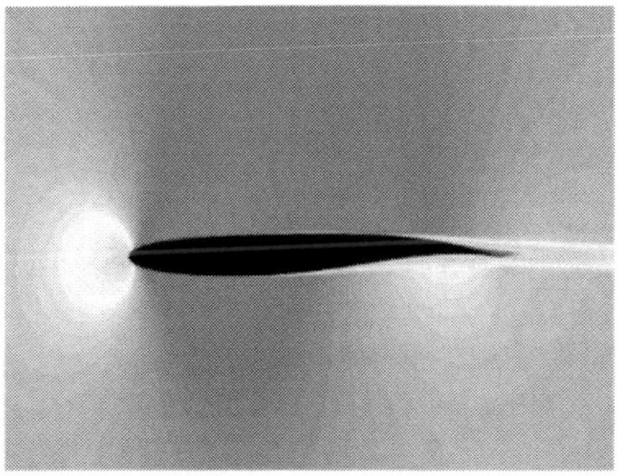

(b) Optimal compromise: Mach number contours ($0.0 \leq M \leq 1.3$)

Figure 8. Trim performance of catalog of airfoil shapes

Table 2 Performance metrics for optimal airfoils of interest

Airfoil	Angle, α	Drag, c_d	Flutter Speed, V_{f*}
Reference Point	2.077°	0.0301	1.261
Optimal Aerodynamic	0.6256°	0.0198	1.273
Optimal Aeroelastic	0.4688°	0.2060	1.987
Optimal Compromise	0.2038°	0.0204	1.514

4. CONCLUDING REMARKS

An initial framework for multi-disciplinary transonic airfoil shape optimization has been presented considering static aerodynamic performance and dynamic aeroelastic performance using computational fluid dynamics. A coupled two-degree-of-freedom pitch/plunge structural model is developed to determine the time-accurate response of the structure at the relevant trim condition as prescribed by the designer preferences. A user-preference multi-objective particle swarm optimization, which employs the concept of a reference point to provide further guidance to the swarm population is used to isolate only the preferred region of the Pareto front to maximise the computational efficiency of the evolutionary algorithm.

Analysis of the final preferred candidates obtained suggested that the dynamic aeroelasticity objective function drives the design and significant improvements for the flutter speed are attainable, at the expense of only slight performance degradation in aerodynamic drag at the trim condition. It is evident from the pressure distributions, that the optimal candidates follow the supercritical airfoil topology, with only slight deviations to redistribute pressure and stabilise the pressure gradient forward of the shock in an attempt to delay the onset of dynamic instability.

The design methodology presented in this study paves the direction for further development in the multi-disciplinary transonic airfoil shape optimization framework, including constraints on the structural integrity of the wing, and to consider a robust operational envelope, for a range of Mach numbers.

REFERENCES

1. Keane, A. J. and Nair, P., *Computational Approaches for Aerospace Design: The Pursuit of Excellence*, John Wiley & Sons, 2005.
2. Deb, K., *Multi-objective optimization using evolutionary algorithms*, John Wiley & Sons, 2001.
3. Li, W., Huyse, L. and Padula, S., "Robust airfoil optimization to achieve drag reduction over a range of Mach numbers", *Structural and Multidisciplinary Optimization*, Vol.24, No.1, 2002, pp. 38-50.
4. Dowell, E. H., *A Modern Course in Aeroelasticity*, Springer, 2015.
5. Thomas, J. P. and Dowell, E. H., "Discrete adjoint approach for aeroelastic design optimization", *15th AIAA/ISSMO Multidisciplinary Analysis and Optimization Conference*, Atlanta, GA, AIAA 2014-2298, 2014.
6. Carrese, R. and Li, A., "Preference-based multiobjective particle swarm optimization for airfoil design", *Springer Handbook of Computational Intelligence*, Springer Berlin Heidelberg, 2015, pp 1311-1331
7. Wickramasinghe, U. K. and Li, X., "Integrating user preferences with particle swarms for multi-objective optimization", *10th Annual Conference on Genetic and Evolutionary Computation ACM Press*, New York, 2008, pp. 745-752.
8. Carrese, R., Sobester, A., Winarto H. and Li, X., "Swarm heuristic for identifying prefered solutions in surrogate-based multiobjective engineering design", *AIAA Journal*, Vol.49, No.7, 2011, pp. 1437-1449.
9. Sobester, A. and Forrester, A., *Aircraft aerodynamic design: Geometry and optimization*, John Wiley & Sons, 2014.
10. Sobieczky, H., "Parametric airfoils and wings", *Notes on Numerical Fluid Mechanics*, Vol.68, 1998, pp 71-88.
11. Shahrokhi, A. and Jahangirian, A., "Airfoil shape parameterization for optimum NavierStokes design with genetic algorithm", *Aerospace Science and Technology*, Vol.11, No.6, 2007, pp 443-450.
12. ANSYS Academic Research, Release 16.2, Help System, Fluent Documentation, ANSYS, Inc.
13. Spalart, P. R. and Allmaras, S. R., "A one equation turbulence model for aerodynamic flows", *La Recherche Aerospatiale*, Vol.1, 1994, pp 5-21.
14. Alonso, J. J. and Jameson, A., "Fully-implicit time-marching aeroelastic solutions", AIAA-94-0056, 1994.
15. Kennedy, J. and Eberhart, R., "Particle swarm optimization", *IEEE International Conference on Neural Networks IV*, 1995, pp 1942-1948.
16. Clerc, M. and Kennedy, J., "The particle swarm - explosion, stability, and convergence in a multidimensional complex space", *IEEE Transactions on Evolutionary Computation*, Vol.6, No.1, 2002, pp 58-73.
17. Isogai, K., "On the transonic-dip mechanism of flutter of a sweptback wing", *AIAA Journal*, Vol.17, No.7, 1979, pp 793-795.

ICAST2015: 26th International Conference on Adaptive Structures and Technologies
October 14-16th, 2015, Kobe, Japan

ICAST2015 #046

Optimal Actuator Layout Design on Circular Plate for High-Precision Smart Space Structure

Tomohiko Naka[1], Nozomu Kogiso[1*], Tadashige Ikeda[2], and Hiroaki Tanaka[3]

[1] Department of Aerospace Engineering, Osaka Prefecture University, Sakai, Japan.
[2] Department of Aerospace Engineering, Nagoya University, Nagoya, Japan.
[3] Department of Aerospace Engineering, National Defense Academy of Japan, Yokosuka, Japan.

Abstract

Space antennas for space exploration misssions using high frequency wave require the precious structural surface shape that will not allow even thermal deformation on orbit. One design candidate to meet such severe design requirements is a smart structural system which the structural shape is actively controlled by actuators. For accurate and flexible shape control, the larger number of actuators is desired. However, as the number of actuators increases, the weight and the required power increase. This study investigates effects of the number of the actuators on the deformation error for a simple circular plate fixed at the center and deformed into the desired shape by the arranged point actuators on the plate. The actuator layout problem is formulated as the optimization problem in terms of the actuator positions and the actuator force that gives the concentrated forcein out-of-plane direction. The deformation is evaluated based on Kirchhoff-Love plate theory. Through the optimization results for some optimum actuator layout designs, efficient actuator layout design strategy is discussed. Then, the optimum design obtained through the FEM model using shell elements is investigated for the modeling verification.

1 INTRODUCTION

Space antennas for space exploration missions require a large aperture areas and high surface shape accuracy as well as light weightness. Generally, the required RMS (root mean square) error of the surface shape is said to about $1/10 \sim 1/20$ of the wavelength of the observation electromagnetic wave. The observation frequency about the extremely high-frequency (EHF) band (30-300GHz) will not allow even thermal deformation on orbit.

One of design candidates to meet the severe design requirement is a smart structure system which can adjust the surface shape by the implemented actuators on orbit. Recently, several studies have developed the smart space reflectors. For example, Fang et al. developed an engineering model of a membrane antenna of 2.4m diameter with adaptive surface control system [1]. The engineering model equips 168 polyvinylidene fluoride (PVDF) actuators to control the surface shape. Since an independent actuator control is impractical for the control system with such many actuators, the actuators are controlled with several groups that are arranged by a group optimization method [2]. Bradford et al. developed the active composite reflector panel with macro fiber composite (MFC) actuators to control the wavefront errors in

* kogiso@aero.osakafu-u.ac.jp

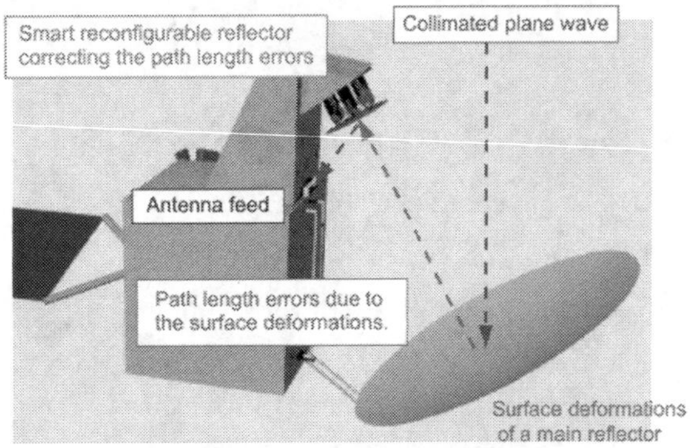

(a) High-accuracy antenna system

(b) Prototype of smart reconfigurable reflector

Figure 1. High-accuracy antenna system with a smart reconfigurable reflector.

the reflector [3]. The system uses 37 actuators in each hexagonal panel with an isogrid array. Datashvili et al. developed laboratory models of reconfigurable reflectors that employed a carbon fiber-reinforced silicon (CFRS) surface with 91 actuators to control the surface shapes [4].

Since such systems with many actuators require more electric power and more complicated control strategy, and hence, reduce the reliability. Tanaka et al., one of the authors, has proposed the different space antenna system that the secondary mirror has the actuators as illustrated in Fig. 1(a) [5]. The smart secondary mirror will adjust the signal error caused by the surface error of a main reflector. As the secondary mirror is sufficient small, the required power of the actuators can be significantly reduced as well as weight.

In order to validate the strategy, the prototype of 200mm diameter secondary mirror with six Piezo actuators as shown in Fig. 1(b) has developed to demonstrate the efficiency through several measurement experiments [5]. However, the number of the actuator and the actuator layout were not optimally determined yet. In addition, the six thin slits introduced to reduce the deformation errors should have been avoided to yields some ill effects such as stress concentration.

Final objective of this study is to reduce the required number of actuators that satisfy the deformation error criteria and to obtain the optimum actuator layout with the smart reflector. As the first step, effects of the number of the actuators that give the out-of-plane concentrated force to the structure on the shape accuracy are investigated for the basic structural element through the optimization methods. In previous study, we investigated the required number of the actuators and the optimum layout for the simple cantilever beam deformed to the ideal deformation shape [6, 7]. In the study, it was found that the two actuators are sufficient to deform into the parabola by arranging the both actuators at the free edge and applying the same magnitude force to the both actuators in the opposite bending direction. That is reasonable for the cantilever problem, because the resultant actuator force is equivalent to the bending moment at the edge.

This study investigate the effects to the circular plate fixed at the center. The optimization problem is defined as minimization of the RMS error between the ideal shape and deformed shape in terms of the actuator position and the actuator force. At first, the axial symmetry condition is considered, that is equivalent to the infinite number of the actuators arranged in the circumferential direction. Then, consider the finite number of the actuator arranged at regular intervals in the circumferential direction and the effect of the number of the actuators is investigated on the RMS error.

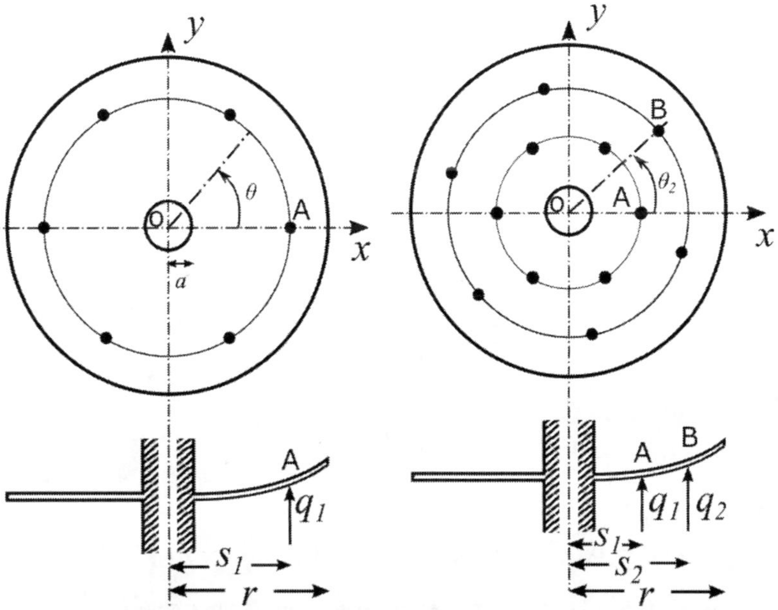

(a) Single type (b) Double type

Figure 2. Circular plate fixed at the center and actuators arranged on concentric circle

As a next step, the strategy is adopted to the FEM model using linear shell element to validate the optimization results between the theoretical model and the numerical model. In the FEM model, the actuators are located at the node that are treated as discrete design variables. On the other hand, the actuator force is modeled as a continuous design variable. For the mixed design variable problem, the evolutionary optimization method (Covariance Matrix Adaptation Evolution Strategy: CMA-ES) [8] is adopted.

Through comparison of the results between the two approaches, efficient actuator layout design is discussed. In addition, numerical result of the nonlinear FEM analysis is compared with that of the linear FEM analysis.

2 PROBLEM FORMULATION FOR BASIC PLATE THEORY

A circular plate with radius r is assumed to be fixed at the small center circle of radius a and free at the edge. Then, n actuators are allocated at n equally divided points on the concentric circle of the radius s_1 and applied the out-of-plane load q_1 as the concentrated load as shown in Fig. 2 (a). The bending deformation is evaluated based on Kirchhoff-Love plate theory [9]. For the optimum actuator layout design, the actuator position s_1 and the applied load q_1 are treated as design variables.

We will discuss the design problem with double circles illustrated in Fig. 2 (b) later. The bending deformation for such actuator allocations can be obtained by superposition of the two single type deformations under the linear deformation assumption. Though the following explanation is limited based on the single type case, it is easily extended to the double type problem.

The desired deformation shape w_I by actuators is set as a paraboloid shape as follows:

$$w_I(s,\theta) = a_I s^2 \qquad (0 \le s \le r, \quad 0 \le \theta < 2\pi) \tag{1}$$

where (s, θ) indicate the polar coordinates. a_I is a coefficient, where the maximum deformation is determined from the design requirement. In this study, it is set as 0.01 times of the plate radius as the small deformation.

2.1 Kirchhoff-Love Plate Theory

When the deformation is small, the plate deformation w by the actuator force is obtained based on Kirchhoff-Love plate theory as follows [9]:

$$w(s_1, q_1; s, \theta) = R_0(s) + \sum_{m=1}^{\infty} R_m(s)\cos(m\theta), \qquad (a \le s \le r, \quad 0 \le \theta < 2\pi) \tag{2}$$

The series terms $R_i(s), (i = 0, \cdots, m)$ are defined as follows.

$$R_0(s) = A_0 + B_0 s^2 + C_0 \log s + D_0 s^2 \log s \tag{3}$$

$$R_1(s) = A_1 s + B_1 s^3 + C_1 s^{-1} + D_1 s \log s \tag{4}$$

$$R_m(s) = A_m s^m + B_m s^{-m} + C_m s^{m+2} + D_m s^{-m+2}, \qquad (m = 2, \cdots, \infty) \tag{5}$$

where A_i, B_i, C_i, and D_i $(i = 0, \cdots, \infty)$ are determined from the boundary conditions [9].

Obviously, as the number of actuators increases, the deformation can approach the desired deformation shape. When the number of the actuators increase to infinity, the actuator load is modeled as the distributed load along the concentric circle of radius s_1. In such a case, the deformation w can be obtained through an axial symmetry model as follows [9]:

$$w_{in}(s_1, q_1; s, \theta) = \frac{q_1 s^2}{8\pi D}(\log\frac{s}{s_1} - 1) - \frac{c_1 s^2}{4} - c_2 \log\frac{s}{s_1} + c_3, \quad (a \le s \le s_1 \le r, 0 \le \theta < 2\pi)$$

$$w_{out}(s_1, q_1; s, \theta) = -\frac{c_4 s^2}{4} - c_5 \log\frac{s}{r} + c_6, \qquad (a \le s_1 \le s \le r, 0 \le \theta < 2\pi) \tag{6}$$

where w_{in} is the deformation inside of the actuator position s_1, w_{out} is that outside of the actuator position and q_1 is the total applied load. The coefficient c_1 to c_6 are determined from the boundary conditions.

2.2 Finite Element Method Model

The finite element model using linear shell elements as shown in Fig. 3 is introduced to validate the actuator layout optimization. Because the numerical model will be required for development of the actual smart reflector. The FEM model consists of 3072 elements, where the plate is equally divided into 96 pieces in the angular direction and unequally into 32 pieces in the radius direction to set the element aspect ratio properly. In this study, a semi-commercial code FEAP (Finite Element Analysis Program) [10] is adopted in this study. As the actuator can be allocated on the node, the optimization problem will be formulated as mixed design variable problem, where the actuator force is set as a continuous design variable.

2.3 Optimum Design Problem

In this study, the optimum actuator layout design problem is formulated to minimize the RMS error between the desired shape and the deformed shape in terms of the actuator position s_1 and the actuator

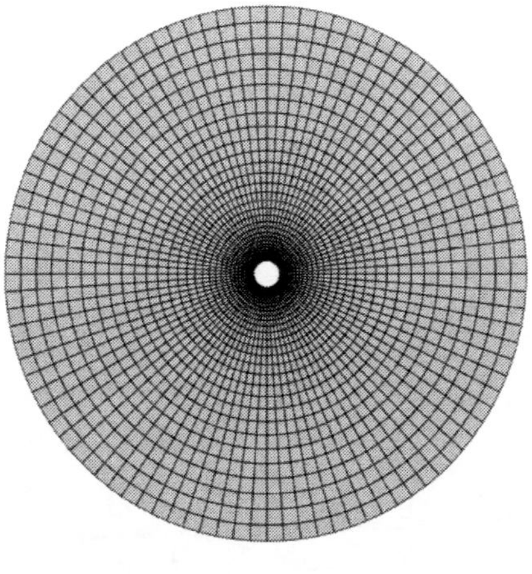

Figure 3. Finite element model of the circular plate

force q_1. The objective function is defined as the normalized RMS error as follows:

$$\text{minimize}: \quad f_D(s_1, q_1) = \frac{f_{rms}(s_1, q_1)}{f_I} \tag{7}$$

$$\text{where}: \quad f_{rms}(s_1, q_1) = \left(\frac{1}{\pi(r^2 - a^2)} \int_0^{2\pi} \int_a^r (w(s_1, q_1; s, \theta) - w_I(s, \theta))^2 s \, ds \, d\theta \right)^{1/2} \tag{8}$$

$$f_I = \left(\frac{1}{\pi(r^2 - a^2)} \int_0^{2\pi} \int_a^r w_I(s, \theta)^2 s \, ds \, d\theta \right)^{1/2} \tag{9}$$

where f_{rms} is the RMS error between the desired and the deformed shapes, that we would like to minimize. f_I in Eq. (9) is the normalized factor that the RMS error between the desired and the undeformed shapes. Therefore, the objective function value approaches zero as the deformation approaches the desired shape. On the other hand, the value is one for the undeformed plate.

In addition, following constraint is imposed such that the actuator should be located on the plate:

$$a \leq s_1 \leq r \tag{10}$$

For the FEM model, the RMS error is evaluated by transforming the integral to the summation with respect to the node as follows:

$$f_{rms}(s_1, q_1) = \left(\frac{1}{\sum_{k=1}^{N_n} u_i} \sum_{k=1}^{N_n} u_i (w_k - w_I)^2 \right)^{1/2} \tag{11}$$

where N_n indicates the number of the nodes and w_k corresponds to the out-of-plane deformation of the kth node. The area around the node u_i depends on the radial position as shown in Fig. 4. That is defined

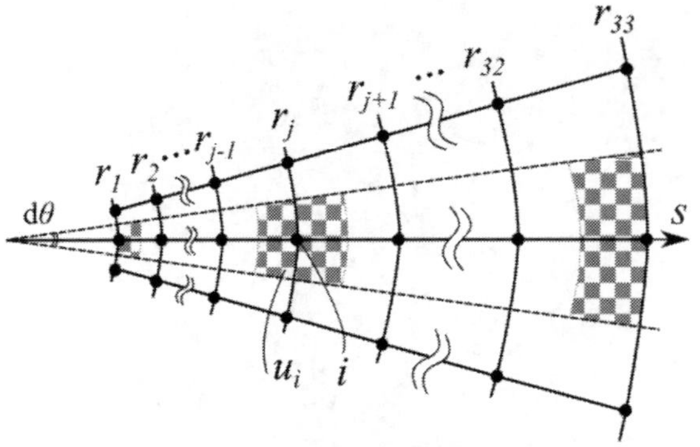

Figure 4. Area u_i around the node.

Table 1 Plate and load parameters

Plate size		Desired deformation	Plate stiffness			Actuator load	
a	r	a_I	D	ν	κ	q	p^*
0.05	1	0.01	1	0.3	$\frac{5}{6}$	$\frac{p}{p^*}$	0.03165

as follows:

$$u_1 = \left(\frac{r_2 - r_1}{2}\right)\left(\frac{3r_1 + r_2}{4}\right) \mathrm{d}\theta \tag{12}$$

$$u_i = \left(\frac{r_{i+1} - r_{i-1}}{2}\right)\left(\frac{r_{i-1} + 2r_i + r_{i+1}}{4}\right) \mathrm{d}\theta \qquad (i = 2, \ldots, 32) \tag{13}$$

$$u_{33} = \left(\frac{r_{33} - r_{32}}{2}\right)\left(\frac{3r_{32} + r_{33}}{4}\right) \mathrm{d}\theta \tag{14}$$

where r_i is the radial position of the node as shown in Fig. 4.

3 NUMERICAL INVESTIGATION AND DISCUSSION

The circular plate and the applied load are modeled as dimensionless form. The radius of the plate is set as $r = 1$ and the hole is set as $a = 0.05$. As the maximum direction of the desired shape is set as 0.01 times of the plate radius, the coefficient in Eq. (1) is set as $a_I = 0.01$. The plate bending stiffness is set as $D = 1$, the Poisson's ratio is set as 0.3 and the shear correction factor is set as $\kappa = 5/6$. The applied load is also normalized by the load applied at the edge point p^* that gives the maximum deformation 0.01 as a_I. These parameter values are summarized in Table 1.

3.1 Actuator Layout for Basic Plate Model

3.1.1 Actuator Layout on Single Concentric Circle

Consider the case that the actuators are allocated equally divided position on the single concentric circle as shown in Fig. 2(a).

Table 2 Optimal layout for actuator allocation on single concentric circle

Number of Actuators	2	3	4	5	6	8	10	12	∞
Arranged radius s_1	0.26	0.81	1.0	1.0	1.0	1.0	1.0	1.0	1.0
Normalized load q_1	2.0	2.3	1.5	1.2	0.98	0.74	0.59	0.49	0.93
Normalized RMS error	0.499	0.184	0.114	0.101	0.0985	0.0978	0.0976	0.0976	0.0971

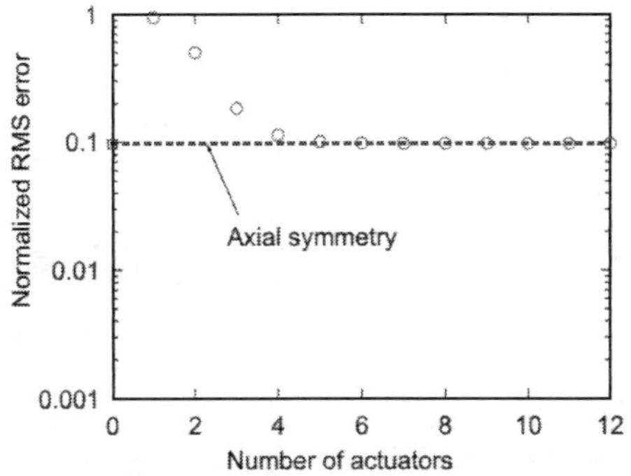

Figure 5. Optimum RMS error for actuator layout in single concentric circle

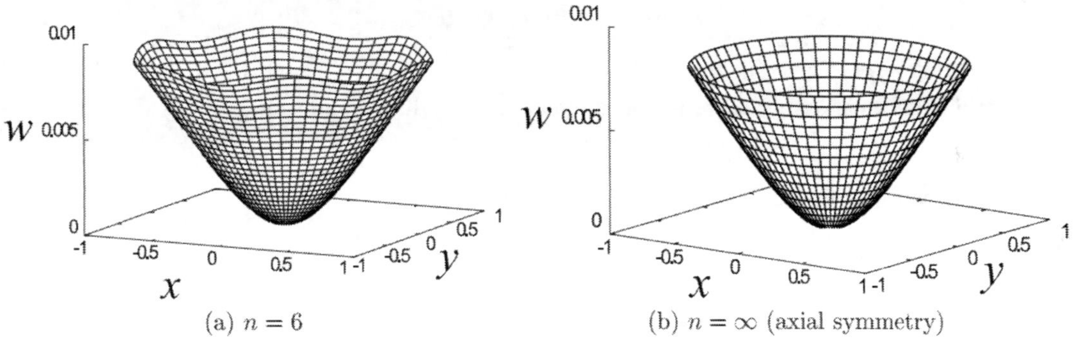

Figure 6. Deformation shape in optimal design in case of single concentric circle

For the deformation evaluation, the series terms in Eq. (2) is set as $m = 15$, because higher terms have found to be negligible. For the integration for evaluating the RMS error in Eqs. (8) and (9), the Gauss-Legendre integration is adopted, where the number of evaluation points is set to 20 in the radius direction and 15 in the angular direction after several preliminary calculations.

The optimization is performed by sequential quadratic method (SQP). The optimal actuator layout designs for several numbers of actuators are compared in Table 2 and Fig. 5, where $n = \infty$ corresponds to the case for the equally distributed load on the circle that is modeled as axial symmetry. As the number of the actuator increases, the actuator position converges to the edge and the RMS error decreases to converge to 0.097. For $n = 6$, the value is almost identical to that for $n = \infty$.

The deformation shapes for $n = 6$ and $n = \infty$ are illustrated in Fig. 6. The deformation for $n = 6$ is found some bumps corresponding to the actuator locations, that is different from the deformation for $n = \infty$. However, the RMS errors of both cases are almost identical with relatively large values. This

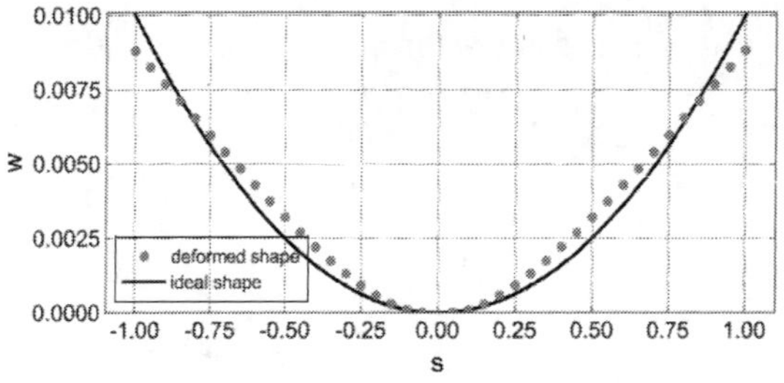

Figure 7. Cross-sectional view of deformation shape for $n = \infty$ optimum design in case of single concentric circles

Table 3 Optimal layout for actuator allocation on double concentric circles

Number of actuators	4	6	8	12	∞
Arranged radius s_1	0.590	0.788	0.860	0.924	0.874
Arranged radius s_2	0.733	0.877	0.926	0.969	0.996
Normalized load q_1	-9.09	-8.98	-8.97	-9.00	-8.97
Normalized load q_2	8.89	9.02	9.03	9.04	9.06
Shift angle θ_2	0.0	0.0	0.0	0.0	0.0
Normalized RMS error	0.883	0.0292	0.0145	0.00681	0.00398

is because the deformation curvature is opposite to the desired shape as shown in Fig. 7. That is, the actuator allocation on the single concentric circle is not sufficient to deform into the desired shape.

3.1.2 Actuator Layout on Double Concentric Circles

As the actuator allocation on the single concentric circle is not sufficient to deform into the desired shape, consider the next case that the actuators are allocated on the double concentric circles as shown in Fig. 2(b).

In this problem, the same numbers of actuators are allocated in both concentric circles with equally divided position. The actuator radial position s_i and the applied load q_i, $(i = 1, 2)$ are treated as design variables. Additionally, as the actuators on the outer circle can be shifted in the angular direction to those on the inner circle, the shift angle θ_2 are also treated as design variable. That is, the total number of design variables is five. The deformation shape is evaluated as superposition of the deformations due to the inner and the outer actuators. Each deformation is evaluated by Eq. (6).

In the optimization, the two constraints are introduced to avoid switching to the inner and the outer circles and to limit the shifting angle as follows:

$$a \leq s_1 \leq s_2 \leq r \tag{15}$$

$$0 \leq \theta_2 \leq \frac{2\pi}{n} \tag{16}$$

As in the previous studies [6, 7], this problem have many local optima in terms of the applied load. First of all, the optimal actuator layout designs for several numbers of actuators are compared in Table 3 and Fig. 8, where $n = \infty$ corresponds to the case for the equally distributed load on the circle. The

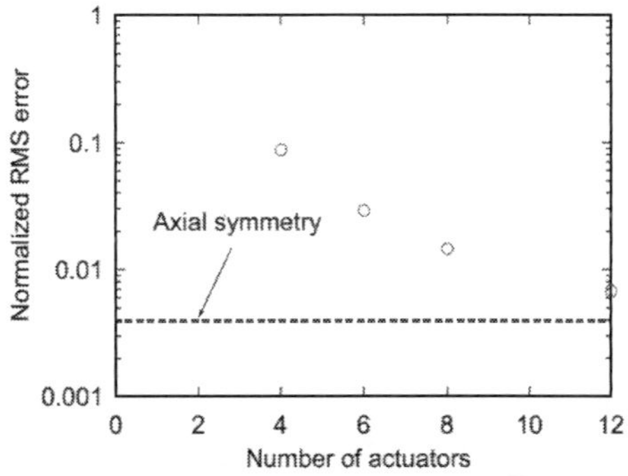

Figure 8. Optimum RMS error for actuator layout in double concentric circle

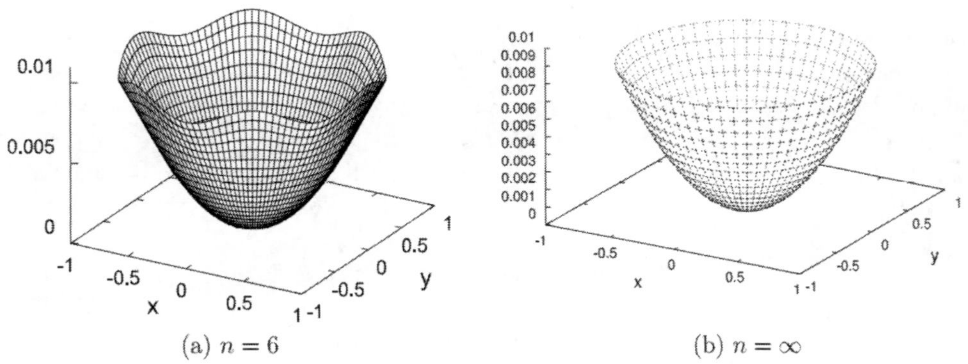

Figure 9. Deformation shape in optimal design in case of double concentric circles

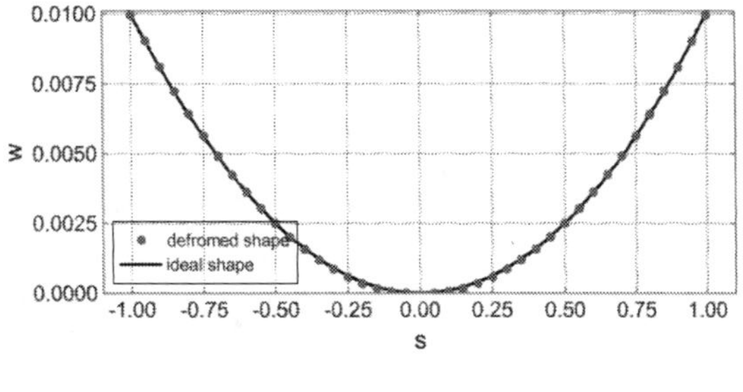

Figure 10. Cross-sectional view of deformation shape for $n = \infty$ optimum design in case of double concentric circles

two actuator positions converge to around $s_1 = 0.9$ and $s_2 = 1.0$, the applied load of the inner circle converges to -9.0 and that of the outer circle converges to 9.0. That is, for the optimum design, the actuators will apply the couple of force to the plate. As the single circle design shown in Fig. 7 has the opposite bending curvature, the design with double circles changes the bending curvature by pulling down the inner actuator.

Table 4 Local optima in Optimization results with concentrated loads

(a) $n = 6$

Position		Applied Load		Shift	Normalized
s_1	s_2	q_1	q_2	θ_2	RMS Error
0.706	0.952	-3.33	3.34	0.0	0.0297
0.749	0.913	-4.94	4.97	0.0	0.0293
0.775	0.890	-6.98	7.02	0.0	0.0292
0.788	0.877	-8.98	9.02	0.0	0.0292

(b) $n = \infty$

Position		Applied Load		Normalized
s_1	s_2	q_1	q_2	RMS Error
0.763	1.00	-4.32	4.40	0.00506
0.824	1.00	-6.02	6.11	0.00415
0.845	1.00	-6.93	7.02	0.00403
0.874	0.996	-8.97	9.06	0.00398

Table 5 Optimization results of $n = 6$ with linear FEM model

Position		Applied Load		Shift	Normalized
s_1	s_2	q_1	q_2	θ_2	RMS Error
0.709	0.940	-3.56	3.57	0.0	0.0325

The RMS error is converged to 0.00398. It means that difference from the desired shape is about 0.4% of the difference between the desired shape and the initial shape. As shown in Fig. 10, the deformation shape is almost identical to the desired shape. However, the convergence is not so good in terms of the number of actuators.

For example, the RMS error for $n = 6$ is about 10 times worse than that for $n = \infty$, though the RMS error value itself is small enough. This is because the effect of the deformation bump along the angular direction is no longer negligible.

As described above, this problem has many local optima as listed in Table 4. All local design has similar tendency that the two loads have opposite directions and the almost same magnitude. For the actual design problem, the load magnitude constraint should be considered.

In addition, the shifting angle θ_2 converged to zero for all cases. It means that the inner and the outer actuators should arrange on the same radius vector.

3.2 Finite Element Method for Double Concentrated Circles

The optimization result using FEM model shown in Fig. 3 is compared with the theoretical model with double concentrated circles. The parameters are set as the same values as listed in Table 1. As the bending stiffness D cannot set directly, Young's modulus E and the thickness h are set as 10.92×10^6 and 0.01, respectively, to $D = 1$.

As the actuators are put on node as discrete design variables, the CMA-ES (Covariance Matrix Adaptation Evolution Strategy) [8] is adopted as optimizer. The optimization result for $n = 6$ is listed in Table 5. This result is almost identical to that of the plated model listed in Table 4. The RMS error is about 10% worse than that of the plate model. This is because the actuator position is limited as nodal point.

The deformation of the optimum design cannot be avoided the bumps as illustrated in Fig. 11. As shown in the previous results, such bumps around the edges cannot be avoided by the finite number of actuators. As the bump deformation is limited to the edge, it will be a good strategy that the actuator is arranged outside of the reflector to resolve the problem [11]. In addition, the optimization of the reflector itself such as thickness distribution is another strategy.

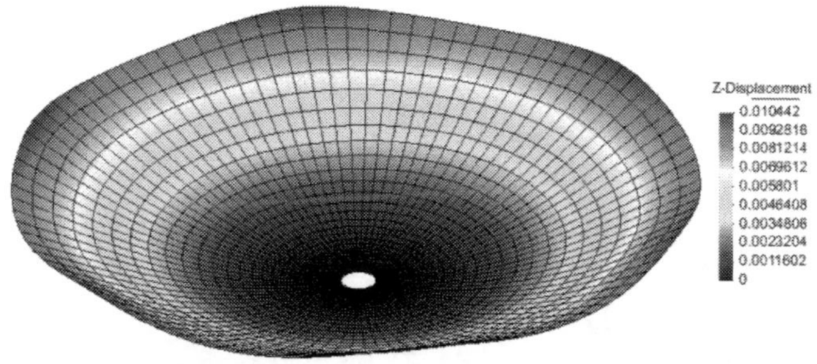

Figure 11. Deformation of the optimum actuator layout for $n = 6$

4 CONCLUSION

In this study, we consider the optimal actuator layout design that gives the out-of-plane deformation to the circular plate as the basic study of the high-precision smart structure. The optimization problem is defined as minimization of the RMS error between the desired shape and the deformed shape in terms of the actuator positions and the actuator forces in out-of-plane direction. The deformation is evaluated based on Kirchhoff-Love plate theory. Then, the optimum design for the FEM model using shell elements required for the actual design problem is investigated for verification.

Through numerical examples, the following conclusions are remarked.

- Considering the two types of the arrangements, it is found that the actuator arrangement on the double concentric circle is useful solution. However, for the limited number of actuators, the deformation with bumps around the edges cannot be avoided. Therefore, other strategy including the structural optimization or the actuator arrangement outside of the reflector should be considered in the future.
- The optimum design using FEM is found to be almost identical to that using the plate bending theory. This kind of structural modeling using FEM can be adopted for the future work to develop the smart reconfigurable mirror system.
- In FEM, the actuator position is limited on the nodal position and is treated as a discrete design variable. Therefore, the CMA-ES method is adopted as optimizer. As the evolutionary method is time consuming in comparison with mathematical programming method, more efficient optimization method or the surrogate modeling should be considered in the future.

ACKNOWLEDGEMENTS

This work is partially supported by JSPS KAKENHI Grant Number 26249131. The authors would like to express our gratitude to Prof. Iwasa, Tottori University for his valuable advice for appropriate FEAP modeling and to Mr. Taku Hagegawa, Osaka Prefecuture University for his valuable advice for the optimization using the CMA-ES algorithm.

REFERENCES

1. Fang, H., Quijano, U., Bach, V., Hill, J., and Wang, K. W. "Experimental Study of a Membrane Antenna Surface Adaptive Control System", *52nd AIAA/ASME/ASCE/AHS/ASC Structures, Structural Dynamics and Materials Conference*, 2011, AIAA 2011-1828, pp. 1-10.

2. Hill, J. R., Wang, K. W., Fang, H. and Quijano, U., "Actuator Grouping Optimization on Flexible Space Reflectors", *SPIE Proceeding*, Vol. 7977, Active and Passive Smart Structures and Integrated Systems, 2011, pp. 1-12, doi: 10.1117/12.880086.

3. Bradford, S. C., Agnes, G. S., Bach, V. M. and Wilkie, W. K., "An Active Composite Reflector System for Correcting Thermal Deformations", *52nd AIAA/ASME/ASCE/AHS/ASC Structures, Structural Dynamics and Materials Conference*, 2011, AIAA 2011-1826, pp. 1-8.

4. Datashvili, L., Baier, H., Wei, B., Hoffman, J., Wehrle, E., Schreider, L., Mangenot, C., Santiago-Prowald, J., Scolamiero, L. and Angevain J.-C., "Mechanical investigations of in-space-reconfigurable reflecting surfaces". *32nd ESA Antenna Workshop on Antennas for Space Applications*, 2010, pp. 1-8.

5. Tanaka, H., Sakamoto, H., Inagaki, A., Ishimura, K., Doi, A., Kono, Y., Oyama, T., Watanabe, K., Oikawa, Y. and Kuratomi, T., "Development of a Smart Reconfigurable Reflector Prototype for an Extremely High-Frequency Antenna", *Journal of Intelligent Material Systems and Structures*, 2015, pp. 1-10, doi:10.1177/1045389X15580660.

6. Kogiso, N., Naka, T., Ikeda, T. and Miyazaki, Y., "Basic Study Towards the Optimal Actuator Layout Design of High-Precision Smart Structure", *56th JSASS/JSME Structures Conference*, 2014, 1A12 (*in Japanese*)

7. Naka, T., Toyoda, M., Kogiso, N. and Ikeda, T., "Basic Study on Actuator Layout Design of High-Precision Smart Structure Using Robust Multiobjective Optimization", *58th Space Science and Technology Conference*, 2014, 3B04 (*in Japanese*)

8. Hansen, N, "The CMA Evolution Strategy: A Comparing Review", *Towards a New Evolutionary Computation*, Springer, 2006, pp. 75-102,

9. Timoshenko, S. P., *Theory of Plates and Shells*, McGraw-Hill, 1959, pp. 51-60, pp. 282-293.

10. Taylor, R. L., *FEAP – A Finite Element Analysis Program User Manual*, Ver. 8.3, http://www.ce.berkeley.edu/projects/feap/ (access on Oct., 2015)

11. Laslandes, M., Hugot, E., Ferrari, M., Hourtoule, C., Singer, C., Devilliers, C., Lopez, C., and Chazallet, F., "Mirror Actively Deformed and Regulated for Applications in Space: Design and Performance", *Optical Engineering*, Vol. 52, No. 9, 091803, 2013, pp. 1-12.

ICAST2015 #024

On-orbit shape control of a reflector using PZT actuators

Lan Lan[1], Yang Zhou[1], Shuidong Jiang[1], Houfei Fang[1*], Jianming Du[2], Zhigang Wu[2]

[1] Shanghai YS Information Technology Co., Ltd., Shanghai, China
[2] State Key Laboratory of Structural Analysis for Industrial Equipment, Dalian University of Technology, Dalian, China

Abstract

With the advancing of space technologies, the radio frequencies used for space observation and communication are continuously increasing. Therefore, the geometry precision of the reflector onboard a spacecraft needs to be increased accordingly. Although many efforts have been made to reduce on-orbit thermal distortion, geometry adaptive control is becoming more and more attractive for high precision reflectors. Using multidisciplinary analytical method, the relationship between the actuation displacement of a PZT actuator and the shape changing of the reflector surface can be studied by simulation. An adaptive reflector is subjected to significant thermal loading during orbiting. For an adaptive reflector on the geosynchronous orbit, the transient temperature fields were simulated by using thermal model, which was established in TMG. Then the deformations of the adaptive reflector at each time step was analyzed using the multidisciplinary analytical method. In this study, the wave-front error control of the reflector during a whole orbiting period was carried out using Least Squares Method and influence matrix approach. The on-orbit wave-front error controllability of the adaptive reflector was determined by this study. The results show that PZT actuators can compensate thermal distortion at all time and this wave-front control technology is very efficient.

1. INTRODUCTION

To meet various applications such as earth science, space science and reconnaissance, large, lightweight and high precision reflectors have been required. Graphite Fiber Reinforced Composite (GFRC) materials are usually used due to their high specific strength and stiffness [1-5]. There exist many environmental factors in space, such as microgravity, thermal cycling, and ultraviolet radiation. In particular, the temperature gradient and bulk changing is the most significant factor to affect the reflector surface [6-10]. Thus, adaptive reflector concepts, coupling with PVDF, MFC and PZT actuators, are proposed for large aperture reflectors used for micrometer to nanometer wave lengths [11-18].

Fig. 1 illustrates the architecture of an adaptive GFRC reflector, it is composed of a GFRC reflector with Piezoelectric Ceramic Transducer (PZT) actuators assembled on the ribs. A multidisciplinary analytical model has been developed for adaptive GFRC reflector design and multi-physics simulations, which includes composite structure, thermal and electrical actuation behaviors [18].

* houfei_fang@yahoo.com

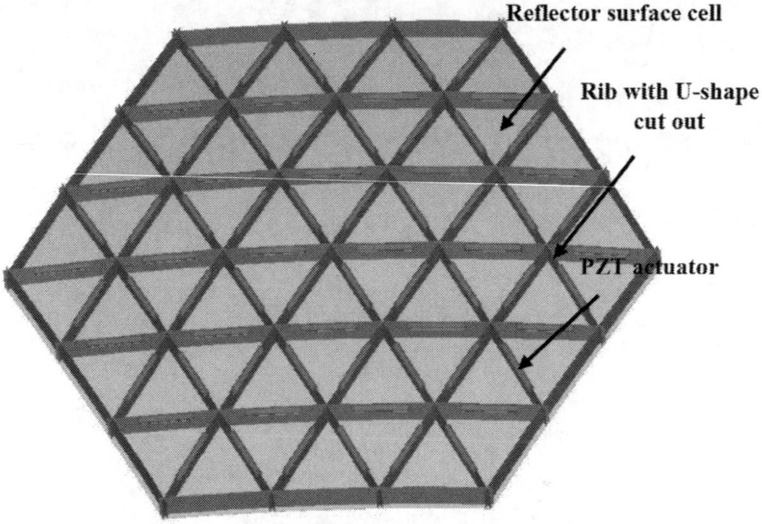

Figure 1. Adaptive hexagonal GFRC reflector model

Using multidisciplinary analytical method, the deformation of the adaptive GFRC reflector can be obtained by simulation. The multidisciplinary analytical method has been validated by experiments. Therefore the thermal-structural behaviors of an adaptive GFRC reflector can be investigated by numerical simulation. Assuming it is on the geosynchronous orbit, the thermal model was set up using TMG and the transient temperature fields of the adaptive reflector were obtained. The deformations of the adaptive reflector at each time step were analyzed by using multidisciplinary analytical method. The wave-front error control of the reflector during the whole orbiting period was discussed by using influence functional matrix approach and Least Squares Method.

2. MULTIDISCIPLINARY ANALYTICAL METHOD

Fig. 1 shows the back of the adaptive reflector. Ribs form triangular lattices that support the face sheet of the reflector. PZT actuators are placed along the edges of the ribs and between rib intersections. When a voltage is applied to an actuator, the actuator expands to bend the rib and to deform the face sheet of the reflector. This can be used to correct the thermal induced deformations of the reflector.

In ref. 18, a multidisciplinary analytical model has been developed. To mimic PZT actuator's behavior, a temperature changing (ΔT) is applied and the actuator element expands or contracts longitudinally. The temperature changing in the embedded actuator causes the actuator to strain until the stress in the actuator balances the stress in the structure.

A test was first conducted on an optical platform under steady temperature and humidity with free-free boundary conditions. Then three PZT actuators were assembled in the triangular GFRC specimen, which is shown in Fig. 2. The test results and the FEA results are displayed in Fig. 3. The FEA analysis uses the multidisciplinary analytical model developed in ref. 18. The length of the top plate is 170 mm. The rib length and the rib height are 140 mm and 60 mm, respectively. The cut out length and height are 84 mm and 50 mm. There is a 5 mm radius circular fillet at each corner of the cut out. The distance between the PZT actuator and the top of the rib is 35 mm. The elastic modulus and the equivalent CTE of the actuator are 27.85 GPa and $4.76E^{-6}$ /°C, respectively. Fig. 3 shows there is a good agreement between the experimental results and numerical results.

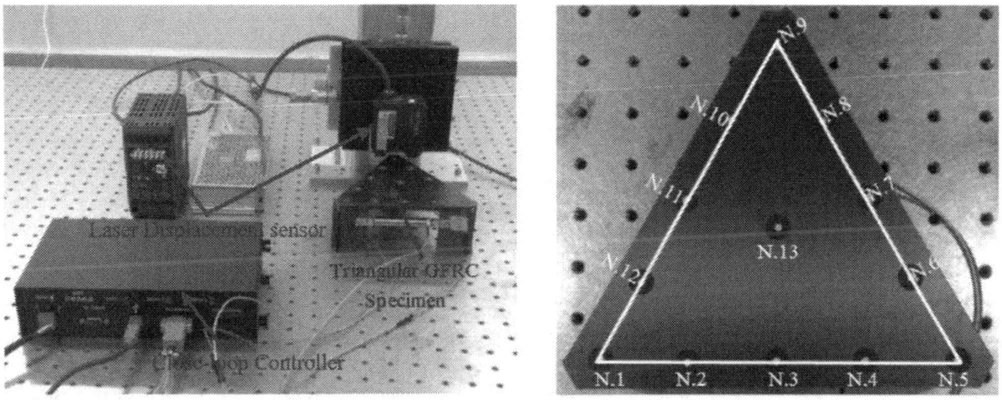

Figure 2. Test set up (left) and measured points (right)

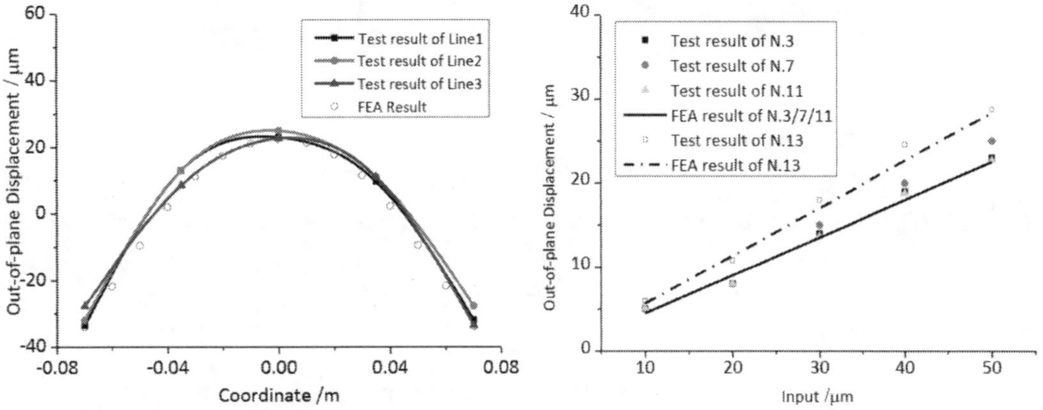

Figure 3. Comparisons between test results and FEA results: three actuators elongate 50 µm (left); displacements of several points under different actuators' elongations (right).

3. ON-ORBIT ANALYSIS OF ADAPTIVE REFLECTOR

There are many factors which can influence the precision of the reflector, such as manufacturing process and on orbit thermal distortion. In space thermal distortions, including temperature gradient and bulk temperature changing, are the most significant sources that influence the surface accuracy. Thus, it is necessary to analyze the thermo-structural behavior of the reflector for space application and wave-front error control.

3.1 Thermal model of adaptive reflector

Thermal load of the adaptive hexagonal GFRC reflector is mainly from solar radiation, additionally due to solar flux reflected from the Earth. It is a complex problem to predict the thermal distortion of a high precision reflector while it is orbiting. The thermal model of the adaptive hexagonal GFRC reflector was created in TMG to calculate temperature distributions at every time step. The inscribed circle diameter of the adaptive hexagonal reflector is 1.0 m and the focal length of the reflector is 2.1 m. On-orbit thermal analysis model of the adaptive reflector didn't include PZT actuators.

Four typical days are chosen by this study, which are the Spring Equinox, the Summer Solstice, the Autumnal Equinox, and the Winter Solstice. The positions and the orientations of the adaptive reflector on a GEO stationary orbit in these four days at noon time are shown in Fig. 4. GEO stationary orbit is a circle and its radius is $4.2E^{+10}$ m. The thermal environment parameters for on-orbit analysis are displayed in Table 1.

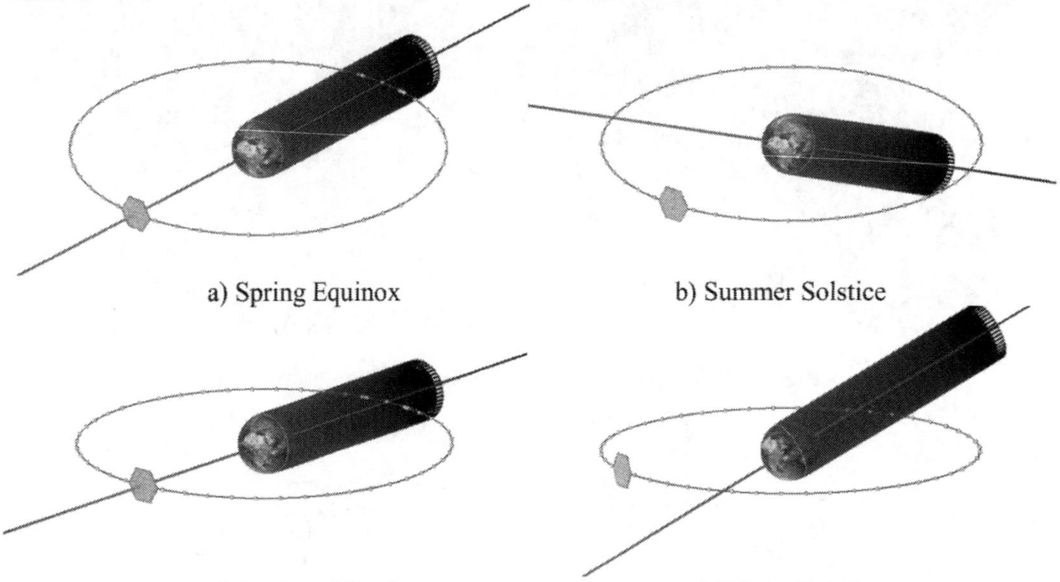

a) Spring Equinox b) Summer Solstice

c) Autumnal Equinox d) Winter Solstice

Figure 4. The positions and the orientations of the adaptive reflector in the GEO stationary orbit

Table 1 Thermal environment parameters

Property	Value			
	Spring Equinox	**Summer Solstice**	**Autumnal Equinox**	**Winter Solstice**
Solar Flux (W/m^2)	1377.8	1323.64	1357.4	1411.56
Earth IR Temperature (K)	237.04			
Albedo Factor	0.306			

For the temperature analysis, the heat flux from the sun, the internal and external radiations of the adaptive GFRC reflector surfaces, and the conductivities of the materials that fabricated the GFRC reflector were taken into consideration. The GFRC reflector was made of laminate plate using eight layers of unidirectional carbon fibers with the thickness of each layer (nominally) to be 0.075mm. The orientations of the layup are $(0°/45°/-45°/90°)_s$. Table 2 shows the properties of the GFRC reflector, which were used for on-orbit thermal analysis and on-orbit structure analysis.

The finite element model of the adaptive GFRC reflector is shown in Fig. 5. The rib length and rib height are 188.6 mm and 40 mm, respectively. The width of the out-side edge is 10 mm. The cut out length and height are 100 mm and 20 mm, respectively. Using the equivalent method proposed in ref. 18, the elastic module and equivalent CTE are 26.53 GPa and $4E^{-6}$ /°C. Boundary conditions for structural analysis are displayed in Fig. 5, three translational degrees of freedom are fixed at three nodes. A finite element model for the thermal analysis is established and the temperature distributions on the adaptive GFRC reflector are calculated every ten minutes during orbiting.

Table 2 Properties of the adaptive GFRC reflector for on-orbit analyses

Property	Structure analysis parameters		Thermal analysis parameters
	Single fiber layer		Quasi-isotropic panel
Elastic Modulus (GPa)	E_1	156	99.46
	E_2	11	
Poison's Ratio	v_{12}	0.25	0.32
Shear Modulus (GPa)	G_{12}	5.28	/
	G_{23}	4.23	
CTE (/°C)	α_1	$-1.01E^{-6}$	/
	α_2	$3.81E^{-5}$	
Thermal Conductivity (W/m.K)	/		0.85
Specific Heat (J/Kg.K)	/		800
Density (Kg/m³)	1800		1800

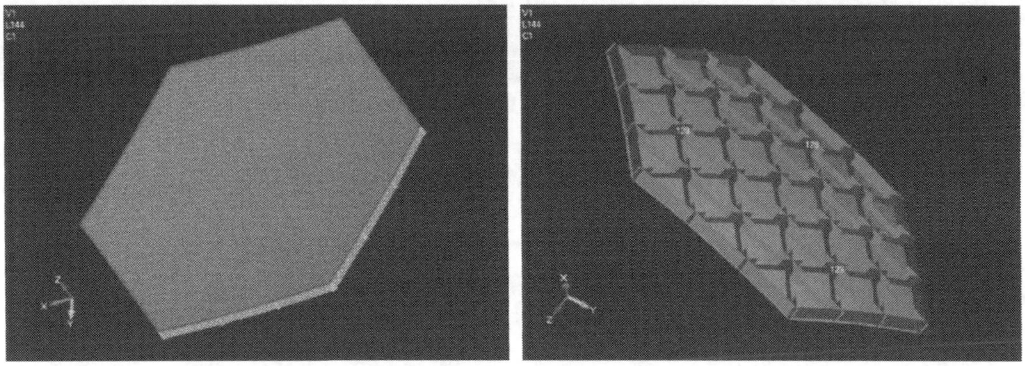

Figure 5. The finite element model of the adaptive reflector for thermal analysis

The maximum, average and minimum temperatures of the adaptive GFRC reflector during a full diurnal orbiting cycle in the four typical days are shown in Fig. 6. Fig. 6 indicates that the maximum, average and minimum temperatures of the adaptive GFRC reflector are very close. In addition, the temperature curves are basically the same between Spring Equinox and Autumnal Equinox as well as between Summer Solstice and Winter Solstice. The temperature levels of the adaptive GFRC reflector alternate greatly in orbital periods, which is about in the range from 30°C to -150°C in Spring Equinox and Autumnal Equinox, as well as from 20°C to -110°C in Summer Solstice and Winter Solstice. Thus it is shown that the Spring Equinox or the Autumnal Equinox is the worst scenario.

Fig. 7 shows the transient temperature fields of the adaptive GFRC reflector at the highest temperature point and the lowest temperature point in Spring Equinox. The highest temperature point and the lowest temperature point correspond to t=600 s and t=45000 s, respectively. The adaptive GFRC reflector fully faces the sun at time t=600 s, corresponding to twelve at noon. At this time the heat flux from the sun is the main part of thermal load acting on the adaptive GFRC reflector. Thus, the temperature of the adaptive GFRC reflector is the highest at this time. The adaptive GFRC reflector is in the earth shadow at time t=45000 s, it doesn't receive any heat flux from the sun, therefore the temperature of the adaptive GFRC reflector is at the lowest at this time.

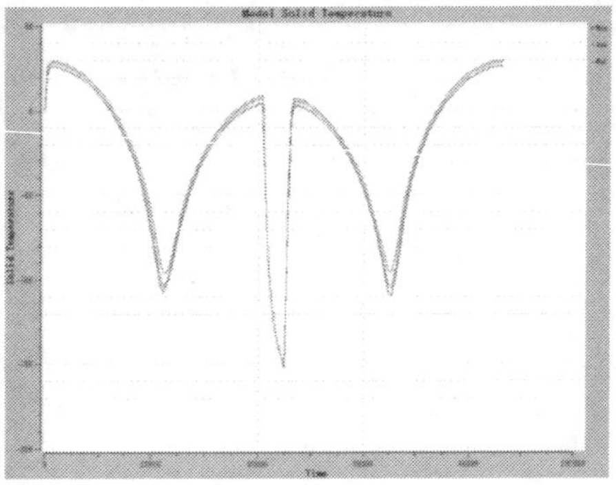

a) Spring Equinox

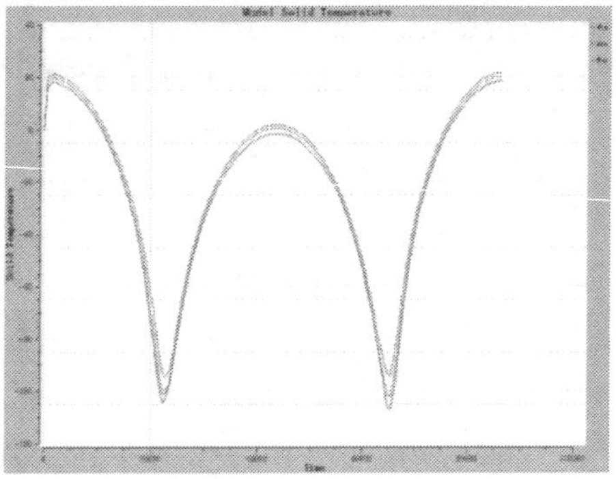

b) Summer Solstice

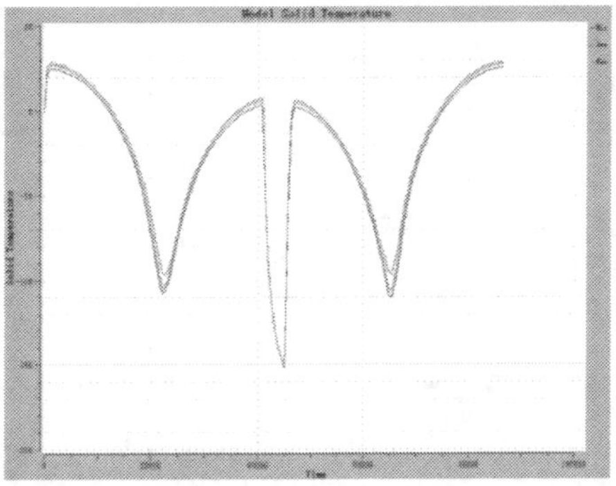

c) Autumnal Equinox

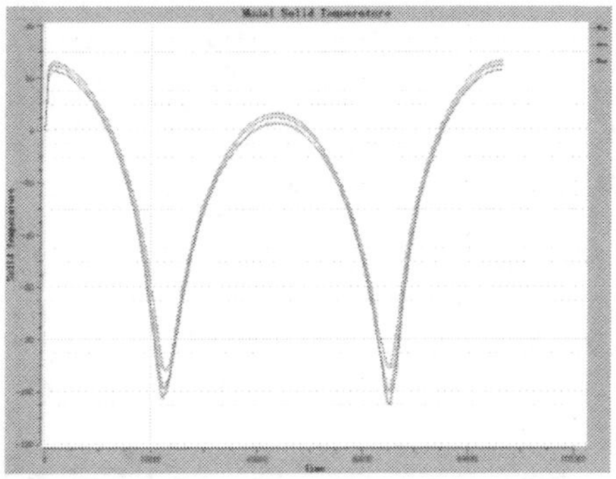

d) Winter Solstice

Figure 6. Maximum, average and minimum temperatures of the adaptive GFRC reflector

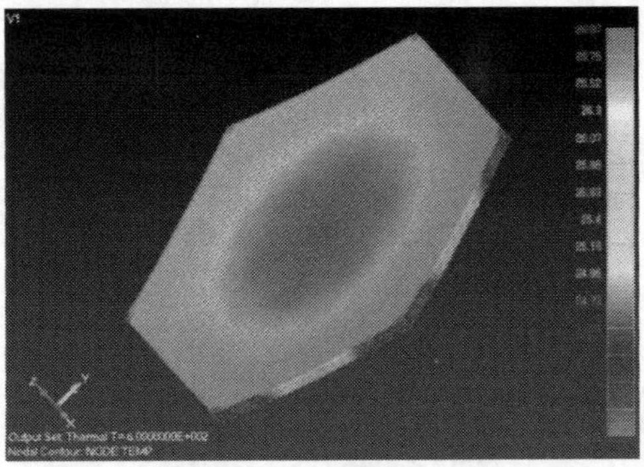

a) t=600 s in Spring Equinox

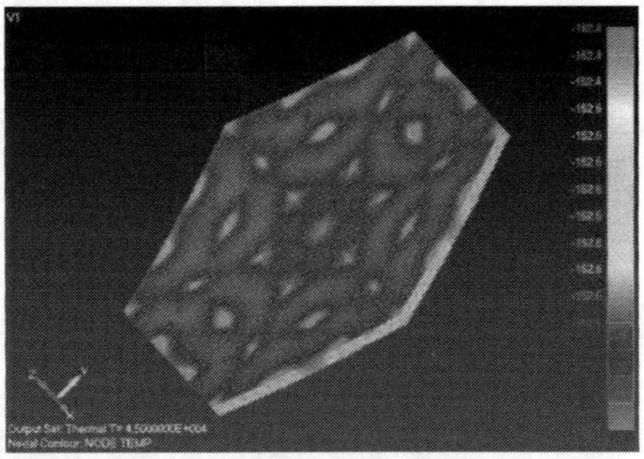

b) t=45000 s in Spring Equinox

Figure 7. The temperature distribution of the adaptive GFRC reflector

3.2 Structural Model of Adaptive Reflector

After the temperature distributions had been obtained, another finite element model that used the temperature distributions as the thermal loads to calculate the reflector deformations was established. Structural deformations of the adaptive GFRC reflector at every time step were then calculated. The wave-front Root Mean Square (RMS) errors [5] of the adaptive GFRC reflector at every time step were then calculated by using the deformation analysis results during the whole orbiting. Fig. 8 shows RMS errors of the adaptive GFRC reflector during that four typical days. The total translations corresponding to the temperature loads in Fig. 7 are shown in Fig. 9.

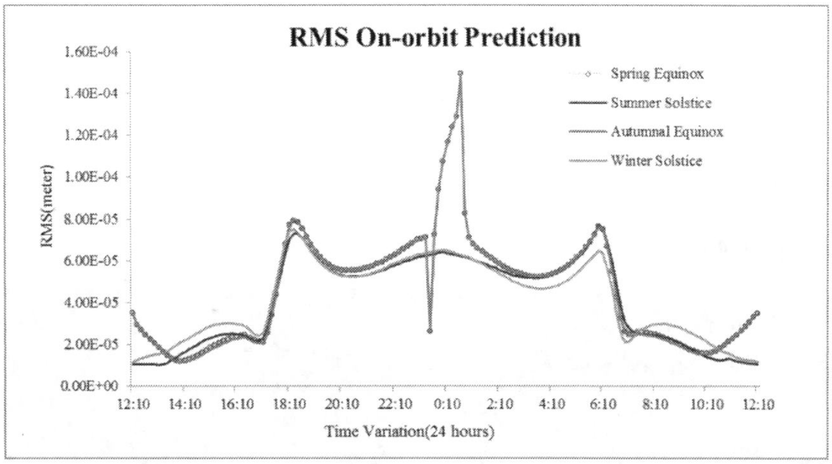

Figure 8. RMS errors as the function of orbiting time

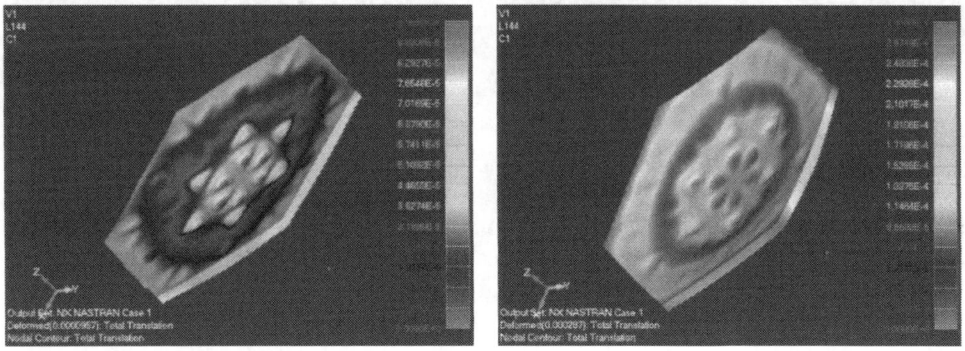

a) t=600 s in Spring Equinox b) t=45000 s in Spring Equinox

Figure 9. The total translations of the adaptive GFRC reflector

4. ON-ORBIT SHAPE CONTROL OF ADAPTIVE REFLECTOR

4.1 Shape Control Method

Influence functional matrix approach is used to conduct on-orbit shape control for the reflector. The equation to obtain the actuation voltage for a given wave-front error is written as:

$$[B_{IF}]\{U\}+Z_e=0\,. \tag{1}$$

Where Z_e represents a known wave-front error in z-direction, $[B_{IF}]$ is the matrix of the actuation influence functions, and $\{U\}$ is the set of actuation voltage. The influence functions represent the influences of actuators to each node on the reflector surface. The number of $[B_{IF}]$ rows is equal to the number of nodes on the reflector surface. On the other hand, the number of $[B_{IF}]$ columns is equal to the number of actuators in the whole reflector shape control system. Generally, the number of equations in Eq. (1) is much more than the number of the unknown variables. So several numerical methods can be used to solve Eq. (1). Least Squares Method and Pseudo Inverse are two examples of these numerical methods. For this study, the Least Squares Method is selected to solve Eq. (1). The matrix of influence functions, $\left[B_{IF}\right]$, is obtained by simulation using the multidisciplinary analytical method.

4.2 Results and Discussions

The out-of-plane translations corresponding to the temperature loads shown in Fig. 7 are given in Fig. 10. The on-orbit deformation shown in Fig. 10 is one of the wave-front errors that needs to be corrected. In order to correct the on-orbit deformations and obtain the actuation voltages, the influence function matrix was numerically calculated.

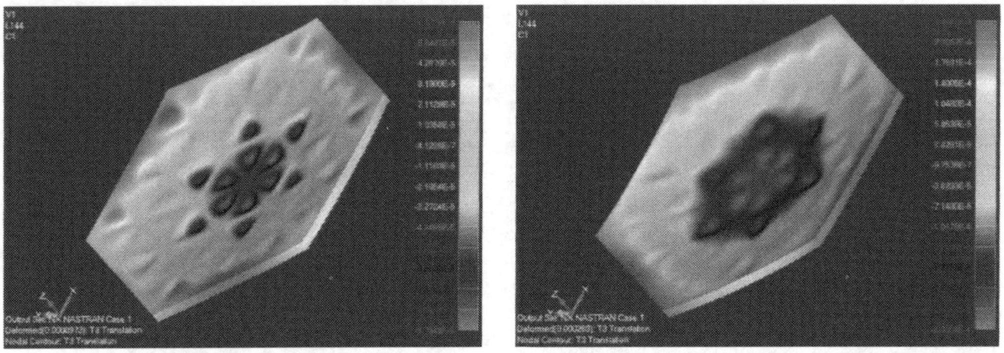

a) t=600 s in Spring Equinox b) t=45000 s in Spring Equinox

Figure 10. The out-of-plane translation contours of the adaptive GFRC reflector

The PZT actuators have an operating range from 0 V to 150 V. After the on-orbit wave-front errors shown in Fig. 10 are controlled, the residual errors and the actuation voltages are shown in Fig. 11 and Fig. 12. By iteration, the RMS errors before and after control for the Spring Equinox are displayed in Fig. 13. From the adaptive GFRC reflector on-orbit analysis, the maximum wave-front RMS error is about 149.34 μm. Using PZT actuators, the reflector shape control system is able to correct approximately 75.34% of the error and the residual maximum RMS error is about 36.83 μm. Fig. 13 indicates that the average value of initial wave-front error in Spring Equinox is about 45.92 μm and the average residual wave-front error is about 15.41 μm. Thus, the controllability of the adaptive GFRC reflector system is around 66.44% for the worst day on-orbit.

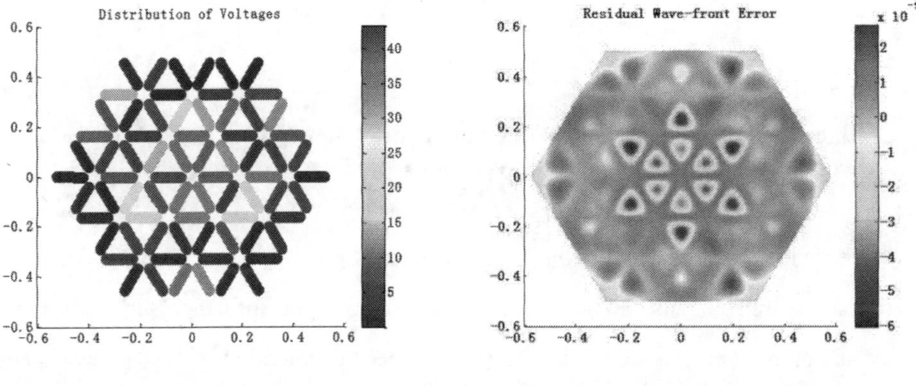

Figure 11. The shape control results of the adaptive GFRC reflector (t=600 s in Spring Equinox)

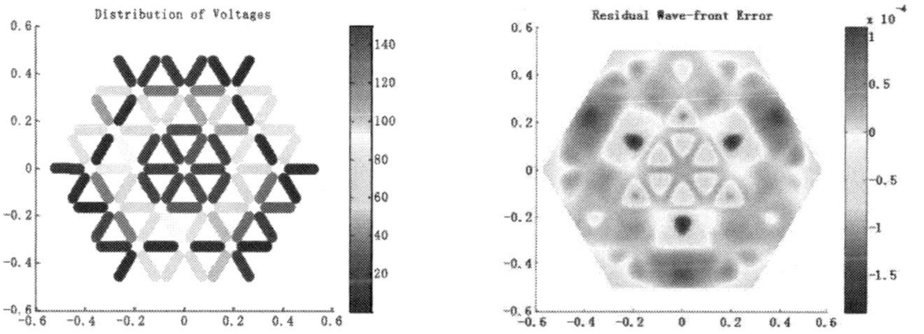

Figure 12. The shape control results of the adaptive GFRC reflector (*t*=45000 *s* in Spring Equinox)

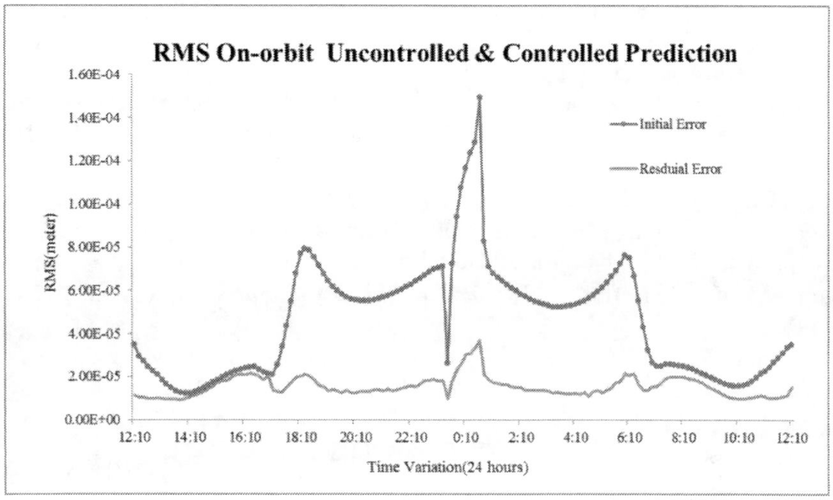

Figure 13. Initial RMS errors and residual RMS errors as the function of orbiting time

5. CONLUSIONS

The GFRC reflector assembled with PZT actuators can be used to effectively correct the on-orbit wave-front error. The PZT actuators equivalent analysis model and the multidisciplinary analytical method for adaptive GFRC reflector has been validated by experiment. Using the multidisciplinary analytical method coupled with thermal-structural analysis method, the temperature distribution and the thermal deformation can be obtained. The influence of each PZT actuator is fairly linear, therefore the influence function matrix and the least squares control method are applicable. These methods are feasible and effective to conduct the on-orbit shape control for a GFRC adaptive reflector.

REFERENCES

1. Abusafieh, A., Federico, D., Connell, S., "Dimensional Stability of CFRP Composites for Space Based Reflectors," *Proc. SPIE*, Vol. 4444, 2001, pp. 9-16.
2. Ozaki, T., Hahn, S., "Composite Materials for Extremely Large Mirrors and Optical Structures," *Proc. SPIE*, Vol. 5382, 2004, pp. 305-312.

3. Kendrew, S., Doel, P., Brooks, D., Dorn, C., Yates, C., Dwan, R., M., Richardson, I., Evans, G., "Development of a Carbon Fiber Composite Active Mirror: Design and Testing," *Optical Engineering*, Vol. 45, No. 3, 2006, pp. 03340-1-7.
4. Lang, M., Baier, H., Ernst, T., "Surface Accuracy Prediction and Optimization of a High Precision Q/V-Band Reflector," *Proc. AIAA*, 2006, pp. 2006-2216
5. Fang, H., Huang, P., Zhou, Y., Liu, L., Liang, Y., Jiang, W., Zhang, S., "Analytical Investigation of a High Precision Reflector," *Proc. AIAA*, 2014, pp. 2014-1508.
6. Dano, M., L., Julliere, B., "Active Control of Thermally Induced Distortion in Composite Structures using Macro Fiber Composite Actuators," *Smart Material Structure*, Vol. 16, 2007, pp. 2315-2322.
7. Sun, H., W., Du, X., W., Tan, H., F., "Investigation of Thermal Distortion and Control of Spacecraft Based on Shape Memory Materials," *Proc. SPIE*, Vol. 7493, 2009, pp. 74934A-1-8.
8. Bradford, S., C., Agnes, G., S., Bach, V., M., "An Active Composite Reflector System for Correcting Thermal Deformations," *Proc. AIAA*, 2011-1826, 2011, pp. 1-8.
9. Sun, H., B., Zhou, J., Zha, Z., L., "Simulation and Experiment for Large Scale Space Structure," *Proc. SPIE*, Vol. 8692, 2013, pp. 86923T-1-7.
10. Ponomarev, V., S., Gerasimov, A., V., Ponomarev, S., V., Shendalev, D., O., "Spacecraft Reflectors Thermomechanical Analysis," *EJP Web of Conferences*, Vol. 82, 2015, pp. 01005-p.1-p.6.
11. Fang, H., Quijiano, U., Bach, V., Hill, J., Wang, K., W., "Experimental Study of a Membrane Antenna Surface Adaptive Control System," *Proc. AIAA*, 2011, pp. 2011-1828.
12. Hill, J., Wang, K., W., Fang, H., "Optimized Gore/seam Cable-actuated Shape Control of Gossamer Membrane Reflectors," *Journal of Spacecraft and Rockets*, Vol. 44, No. 5, 2007, pp. 1122-1130.
13. Hill, J., Wang, K., W., Fang, H., "Advances of Surface Control Methodologies for Flexible Space Reflectors," *Journal of Spacecraft and Rockets*, Vol. 50, No. 4, 2013, pp. 816-828.
14. Fang, H., Pattom, M., J., Wang, K., W., Im, E., "Shape Control of Large Membrane Reflector with PVDF Actuation," *Proc. AIAA*, 2007, pp. 2007-1842.
15. Chen, Q., Natale, D., Neese, D., Ren, K., Lin, M., Zhang, Q., M., Pattom, M., Wang, K., W., Fang, H., Im, E., "Piezoelectric Polymers Actuators for Precise Shape Control of Large Scale Space Antennas," *Proc. SPIE*, Vol. 6524, 2007, pp. 65241P-1-11.
16. Bradford, S., C., Agnes, G., S., Ohara, C., M., Shi, F., Peterson, L., D., Hoffman, M., Wilkie, W., K., "Piezocomposite Actuator Arrays for Correcting and Controlling Wavefront Error in Reflectors," *Proc. AIAA*, 2012, pp. 2012-1743.
17. Bradford, S., C., Agnes, G., S., Ohara, C., M., Green, J., J., Shi, F., Zhou, H. Peterson, L., D., Wilkie, W., K., "Controlling Wavefront in Lightweight Active Reflector Systems Using Piezocomposite Actuator Arrays," *Proc. AIAA*, 2013, pp. 2013-1525.
18. Lan, L., Jiang, S., D., Zhou, Y., Fang, H., F., Tan, S., J., Wu, Z., G., "Geometry Adaptive Control of a Composite Reflector Using PZT Actuator," *Proc. SPIE*, Vol. 9433, 2015, pp. 943305-1-14.

ICAST2015: 26th International Conference on Adaptive Structures and Technologies
October 14-16th, 2015, Kobe, Japan

ICAST2015 #005

Axial Stiffness Variation of Thin Walled Laminated Composite Beams Using Piezoelectric Patches- a New Experimental Insight

H. Abramovich [1*]

[1] Faculty of Aerospace Engineering, Technion, I.I.T., Haifa, Israel

Abstract

Altering the natural frequencies of slender thin walled laminated composite beams, by changing its stiffness has been reported in the literature. This had been performed either using optimal mechanical and thermal induced stresses or by using collocated piezoelectric patches bonded on the structure. It was shown that for the cases of a beam equipped with piezoelectric patches, restraining the beam's axial displacements at its two ends can lead to either softening of stiffening of the beam stiffness, as a function of the voltage supplied to the patches.

However, when those structures behave like a cantilevered beam, or a free-free beam, classical beam theories show that the lateral vibrations of those beams cannot be altered by applying a constant electric voltage to the collocated surface bonded piezoelectric layers or patches. This occurs because the resultant force on each cross section of the beam is always identically zero (due to the fact that axial external forces are not applied on the beam).

It is the aim of the present manuscript, to show that under certain circumstances one can alter also the natural frequencies of a cantilevered piezolaminated beam. The present paper will investigate this phenomenon, and provide a new insight into this topic.

1. INTRODUCTION

The issue of utilizing piezoelectric patches to control and/or improve the natural frequencies and/or buckling behavior of thin walled structures, like beams and plates had been investigated and discussed in the literature.

Waissman & Abramovich [1, 2], published results of stiffening effects of beams, by calculating the vibrations of beams with piezoelectric layers or patches bonded on their surfaces. Their theoretical model was based on Timoshenko's beam theory (first order shear deformation theory), which assumes that plane cross sections of the beam before deformation, remain plane after deformation and in-plane deformations of the cross sections, namely geometric variations of the cross sections, can be neglected.

The investigation of the stiffening phenomenon had been performed in two steps:

a) Initially a constant voltage was applied to the electrodes of the piezoelectric layers or patches. Since the axial displacement at the two ends of the beam was constrained, axial resultant cross sectional forces were developed along it. In the case of asymmetric cross sections bending may also occur, but the present investigation and discussions, for the sake of simplicity, are confined to symmetric cross

* haim@technion.ac.il

sections only. One should note that only the in-plane (axial) boundary conditions are effective, while the lateral boundary conditions have no influence on the phenomenon at this first step.

b) At the second step, small lateral vibrations were superimposed on the initially compressed or tensed beam (depending if the voltage was positive or negative). At this moment the effect of the previous step was introduced solely through the cross sectional resultant axial force that acted along the beam. During the second stage of the investigation, the lateral boundary conditions defined the frequencies and mode shapes of the vibrations, which were not influenced by the axial constraints. Similar results showing the change in the stiffness of a structure with piezoelectric patches had been shown by other studies (see for example Refs. 3-13).

The above presented research [1-2], that supported previous similar results that had been obtained by other researchers [3-8], aroused some interest and lead to various discussions on the nature of the phenomenon. According to the above model the entire phenomenon depends on the resultant axial force that acts along the beam, which is a result of the axial constraints at the beam's boundaries. In the case of a cantilevered beam (clamped /fee boundary conditions) the cross sectional resultant axial force is always equal to zero. Thus it was concluded by various readers of [1-2] that it is definitely impossible to affect the lateral vibrations of a cantilevered beam by applying a constant voltage to piezoelectric layers or patches that are bonded on its upper and lower surfaces.

The purpose of the present study is to show a simple example of a cantilevered beam where the lateral vibrations are affected significantly by applying a constant voltage to piezoelectric patches that are bonded on its surfaces and to further explain the study which has been already communicated in a short note [16]. The effect in this case is due to small variations of the geometry of the beam's cross sections, due to the small lateral vibrations. These small variations of the cross sectional geometry result in bending moments, about the cross sectional center, due to the internal stresses that are created during the above described step one. This bending moment affects the lateral vibrations even though the resultant cross sectional axial force is equal to zero.

2. THE MODEL

2.1 The theoretical model

The employed model is presented in Fig. 1. It consists of a laminated composite beam under lateral and axial loads. The beam includes continuous piezoelectric layers, beside composite structural ones. The assumed displacements are: the lateral displacement, W(x,t), the axial displacement U(x,t) and the rotation of the cross section of the beam (around y axis), Φ (x, t). The equations of motion for this model can be written as (see also Refs. [14, 15]):

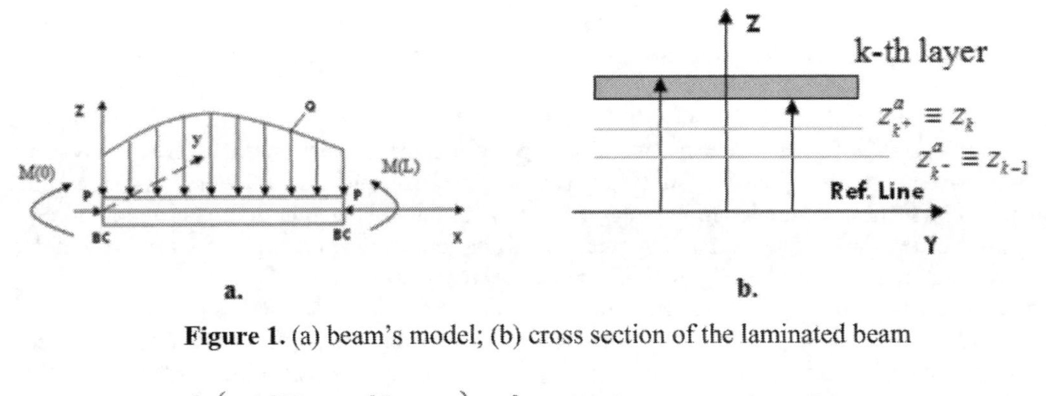

a. b.

Figure 1. (a) beam's model; (b) cross section of the laminated beam

$$\frac{\partial}{\partial x}\left(A_{11}\frac{\partial U}{\partial x}+B_{11}\frac{\partial \Phi}{\partial x}-E_{11}\right)=\frac{\partial}{\partial t}\left[I_1\dot{U}+I_2\dot{\Phi}\right] \quad (1)$$

$$\frac{\partial}{\partial x}\left[A_{55}\left(\Phi+\frac{\partial W}{\partial x}\right)-P\frac{\partial W}{\partial x}\right]=\frac{\partial}{\partial t}\left[I_1\dot{W}\right]+Q \qquad (2)$$

$$\frac{\partial}{\partial x}\left(B_{11}\frac{\partial U}{\partial x}+D_{11}\frac{\partial \Phi}{\partial x}-F_{11}\right)-A_{55}\left(\Phi+\frac{\partial W}{\partial x}\right)=\frac{\partial}{\partial t}\left[I_3\dot{\Phi}+I_2\dot{U}\ \right] \qquad (3)$$

with their associated possible boundary conditions:

$$A_{11}\frac{\partial U}{\partial x}+B_{11}\frac{\partial \Phi}{\partial x}=-p_m \quad \underline{or} \quad U=0 \qquad (4a)$$

$$A_{55}(\Phi+\frac{\partial W}{\partial x})-P\frac{\partial W}{\partial x}=0 \quad \underline{or} \quad W=0 \qquad (4b)$$

$$B_{11}\frac{\partial U}{\partial x}+D_{11}\frac{\partial \Phi}{\partial x}=m_m \quad \underline{or} \quad \Phi=0 \qquad (4c)$$

where p_m and m_m , the generalized axial load and moment, including the influence of the piezoelectric layers, are†:

$$p_m=P-E_{11} \qquad (5a)$$

$$m_m=M+F_{11} \qquad (5b)$$

For uniform properties along the beam, the equations of motion become:

$$A_{11}\frac{\partial^2 U}{\partial x^2}+B_{11}\frac{\partial^2 \Phi}{\partial x^2}=I_1\ddot{U}+I_2\ddot{\Phi} \qquad (6)$$

$$A_{55}(\frac{\partial \Phi}{\partial x}+\frac{\partial^2 W}{\partial x^2})-P\frac{\partial^2 W}{\partial x^2}=I_1\ddot{W}+Q \qquad (7)$$

$$B_{11}\frac{\partial^2 U}{\partial x^2}+D_{11}\frac{\partial^2 \Phi}{\partial x^2}-A_{55}\left(\Phi+\frac{\partial W}{\partial x}\right)=I_3\ddot{\Phi}+I_2\ddot{U} \qquad (8)$$

where $A_{11},B_{11},D_{11},A_{55}$ are the usual stiffness coefficients of a composite beam (including the piezoelectric layers), I_1,I_2,I_3 are the mass moments of inertia and E_{11} and F_{11} are defined as :

$$\left(I_1,I_2,I_3\right)=\int\rho\cdot\left(1,z,z^2\right)\cdot dz \qquad (9)$$

$$E_{11}=b\sum_{k=1}^{N}(\overline{Q_{11}})_k^a V_k(x,t)(d_{31})_k \qquad (10)$$

$$F_{11}=\frac{b}{2}\sum_{k=1}^{N}(\overline{Q_{11}})_k^a V_k(x,t)(d_{31})_k(z_{k^+}^a+z_{k^-}^a) \qquad (11)$$

ρ is the density of each layer of the beam , $V_k(x,t)$ is the voltage applied to the k-th piezoelectric patch, $(d_{31})_k$ is the piezoelectric coefficient of the k-th patch and $\left(\overline{Q_{11}}\right)_k^a$ is the stiffness of the k-th actuator .

Assuming

† Note that in Eqs. (2, 4b) one should use in the non-linear expression the term P rather than p_m.

$$W(x,t)=w(x)e^{i\omega t}\ ;\Phi(x,t)=\phi(x)e^{i\omega t}\ ;\ U(x,t)=u(x)e^{i\omega t} \qquad (12)$$

where ω is the natural frequency of the beam's vibration and substituting in Eqs. (6-8) one obtains:

$$A_{11}u''+B_{11}\varphi''=-\omega^2 I_1 u-\omega^2 I_2\varphi \qquad (13)$$

$$A_{55}(\phi'+w'')-Pw''=-\omega^2 I_1 w \qquad (14)$$

$$B_{11}u''+D_{11}\varphi''-A_{55}\left(\varphi+w'\right)=-\omega^2 I_3\varphi-\omega^2 I_2 u \qquad (15)$$

Decoupling the equations (see [14, 15]) for the general case yields the following expressions for the assumed displacements and angle:

$$u(x)=\alpha_1\sinh(\lambda_1 x)+\alpha_2\cosh(\lambda_1 x)+\alpha_3\sin(\lambda_2 x)+\alpha_4\cos(\lambda_2 x) \qquad (16a)$$

$$w(x)=\beta_1\sinh(\lambda_1 x)+\beta_2\cosh(\lambda_1 x)+\beta_3\sin(\lambda_2 x)+\beta_4\cos(\lambda_2 x) \qquad (16b)$$

$$\phi(x)=\delta_1\sinh(\lambda_1 x)+\delta_2\cosh(\lambda_1 x)+\delta_3\sin(\lambda_2 x)+\delta_4\cos(\lambda_2 x) \qquad (16c)$$

Application of the appropriate boundary conditions, would enable to calculate the natural frequencies of the beam in the presence of induced axial compression.

2.2 The experimental model

The experimental model was mainly made of glass-epoxy, having a uniform rectangular hollow cross section. Two glass-epoxy plates were glued at the root and tip cross sections (see Figs. 2 and 3). The centers of these plates, namely the centers of the root and tip cross sections, were connected by a 0.6 mm steel wire. Three and later on during the tests campaign, six pairs of piezoelectric patches were bonded on the upper and lower surfaces of the beam. To be able to apply relatively high voltages (from -500 up to 1500 V), the piezoelectric patches were MFC (Macro Fiber Composite type M-2814-P1) ones, manufactured by Smart Material Corporation [17]. The MFC was originally developed at NASA's Langley Research Center for aerospace applications, and is an innovative actuator that offers high performance and increased flexibility. It consists of rectangular piezo-ceramic rods, sandwiched between layers of adhesive and electrode polyimide film. The piezoelectric patches were electrically connected in parallel, thus identical voltage was applied to all of them.

To monitor the behavior of the beam, pairs of strain gages were bonded along the upper and lower surfaces of the beam and at the beam's root (see Figs. 2 and 3).

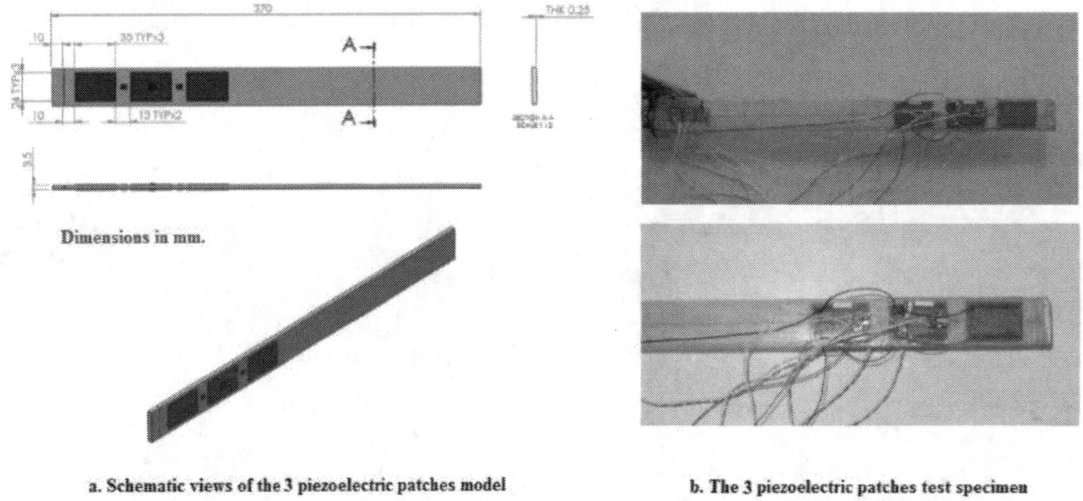

a. Schematic views of the 3 piezoelectric patches model

b. The 3 piezoelectric patches test specimen

Figure 2. The experimental model (3 patches)- a. Schematic views, b. the real specimen

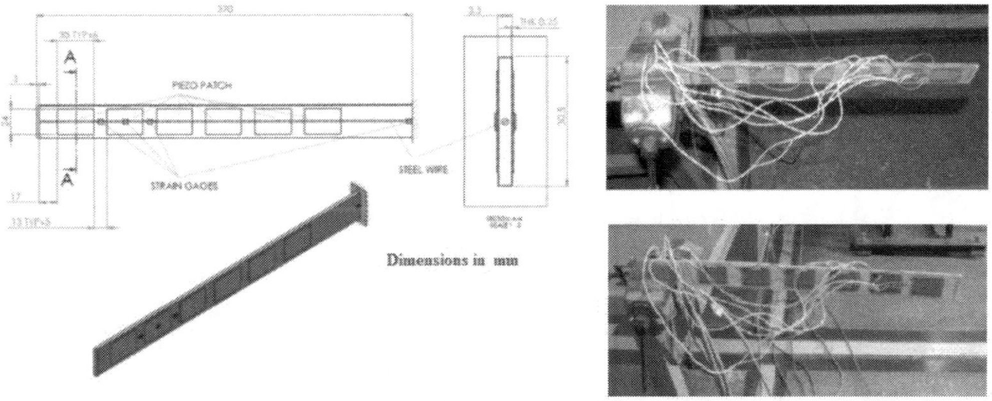

a. Schematic views of the 6 piezoelectric patches model

b. The 6 piezoelectric patches test specimen

Figure 3. The experimental model (6 patches)- a. Schematic views, b. the real specimen

2.3 The finite element model

The ANSYS finite element code [18] was employed to calculate both the static and dynamic response of the beam with both 6 (3 pairs) and 12 (6 pairs) piezoelectric patches bonded on its surfaces. Within this code, a dedicated element enables the user to add to the normal degrees of freedom, the geometric displacements, and another degree of freedom, the voltage (applied or generated) at the nodes of the element. The beam was modeled using 4480 shell type elements; the steel wire was modeled using 112 beam elements, while the piezoelectric patches were modeled using 1160 elements which were described earlier. To increase the accuracy of the model, it was found that the masses of the patches should be included in the calculations.

3. RESULTS

The applied voltage was increased in steps of 50 V. At each voltage level the strains along the beam were recorded. Then the beam was excited using a shaker and the response was measured using Lissajous figures, yielding the first natural frequency of the beam.

First the 3 pairs piezoelectric experimental model was tested (Fig.2). The strain gages reading are given in Fig. 4 for the static case.

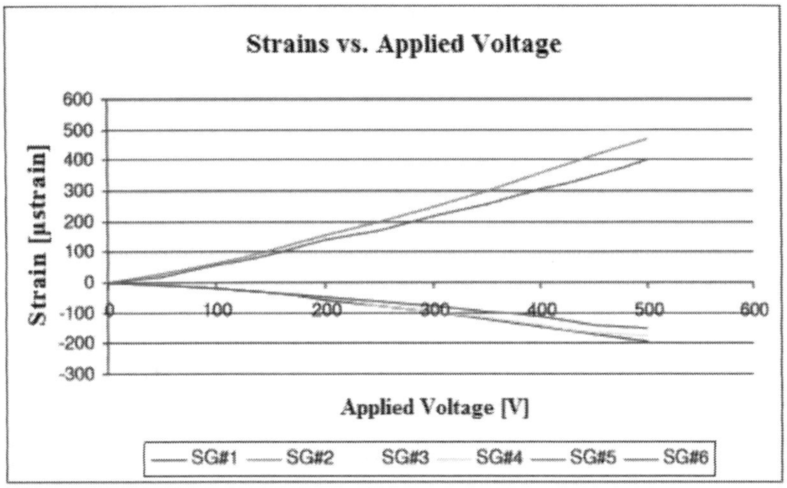

Figure 4. 3 Strains vs. applied voltage for the 3 patches model

One should note that strain gages # 2 , 5 are located on the piezoelectric patches and are in tension, while strain gages # 1,4,3,6 are on the beam and therefore in compression. Figure 5 presents the results of the investigation of whether to include the masses of the patches in the calculation of the natural frequencies or not. It is clear that masses of the piezoelectric patches play an important role in the dynamic behavior of the beam and therefore should be included in the FE calculations. This was done throughout the testing campaign.

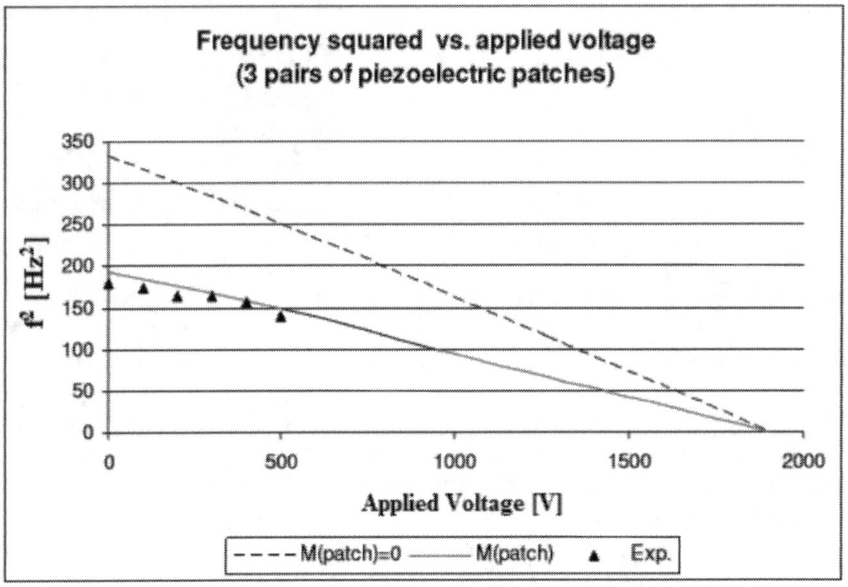

Figure 4. Frequency squared vs. the applied voltage for the case of a beam with 3 pairs of piezoelectric patches.

Another FE investigation was performed to find what is the minimal number of piezoelectric patches capable of inducing buckling in the tested beam in the vicinity of 1000V, the maximal voltage that can be applied to the piezoelectric patches using the laboratory amplifier. Figure 5a presents the results for 6 and 7 pairs of piezoelectric patches bonded on the tested beam. Due to shortage in patches it was decided that the 6 pairs model would be the final testing model.

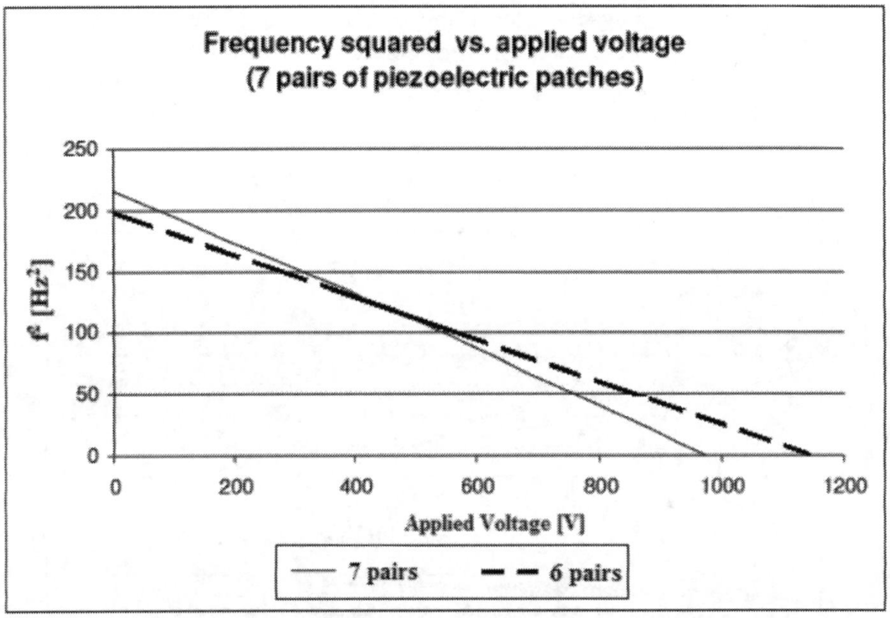

a. A beam with either 6 or 7 pairs of piezoelectric patches

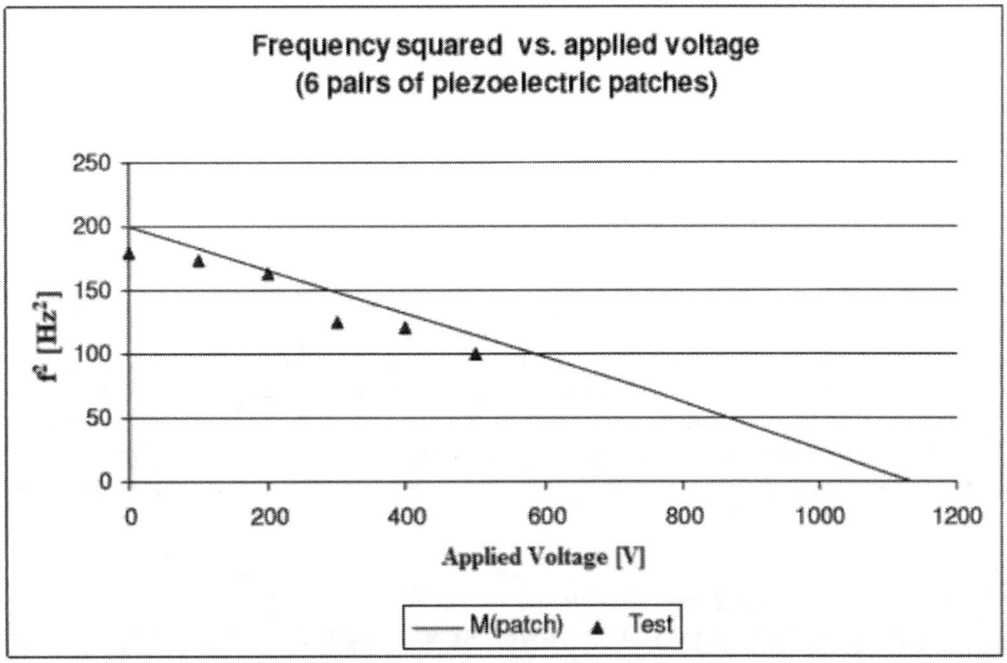

b. A beam with 6 pairs of piezoelectric patches - ANSYS FE results vs test results

Figure 5. Various models used in the present study: a. f^2 vs. applied voltage for 6 and 7 pairs of piezoelectric patches (FE results), b. FE and experimental results for a beam with 6 pairs of piezoelectric patches

Figure 5b, presents a comparison between tests and FE calculations up to an applied voltage of 500 [V]. The accompanying static readings of the strains located at the beam's root are shown in Fig.6.

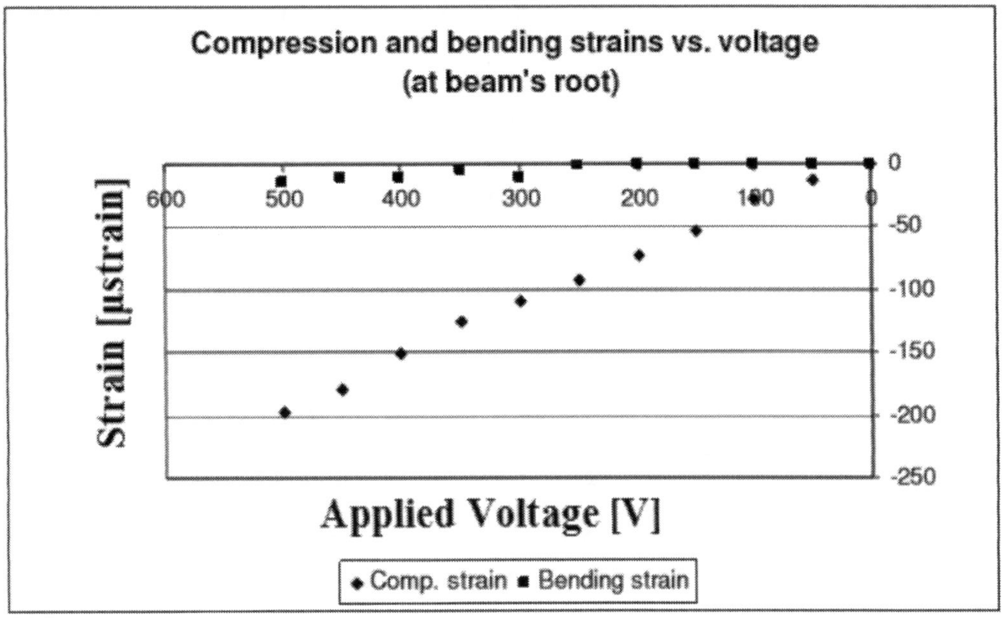

Figure 6. Compression and bending strains as a function of the applied voltage at the root of the beam for low applied voltage

Figure 7 presents a comparison between the frequencies that were measured during the test and calculated values. In general there is a good agreement between both results. Initially the experimental results are lower than the calculated ones. The reasons may be incomplete clamping on one hand and added mass associated with the cables connecting the patches to the voltage source and the strain gages to the strain recorder. At higher voltages the experimental results exhibit a better agreement with the calculations. The square of the frequency is reduced almost linearly with the increase in voltage. The point where the natural frequency becomes zero indicates buckling of the beam. It is well known that because of the initial curvature that always exist in test models (initial imperfections) large bending deformations occur before buckling is reached. The measurement of the natural frequencies stopped at 850 V, as no natural frequencies could be detected above it.

A common method of determining the experimental buckling load (or buckling voltage in the present case) is to use a Southwell plot, as shown in Figure 8. In this figure the bending strain at the root of the cantilevered beam is plotted as a function of the same bending strain divided by the applied voltage. At high enough voltages a straight line is obtained. The slope of this line indicated the buckling voltage. The present value is 1168 Volts, which agrees very well with the calculated value of 1135 Volts (a difference of 2.9 %).

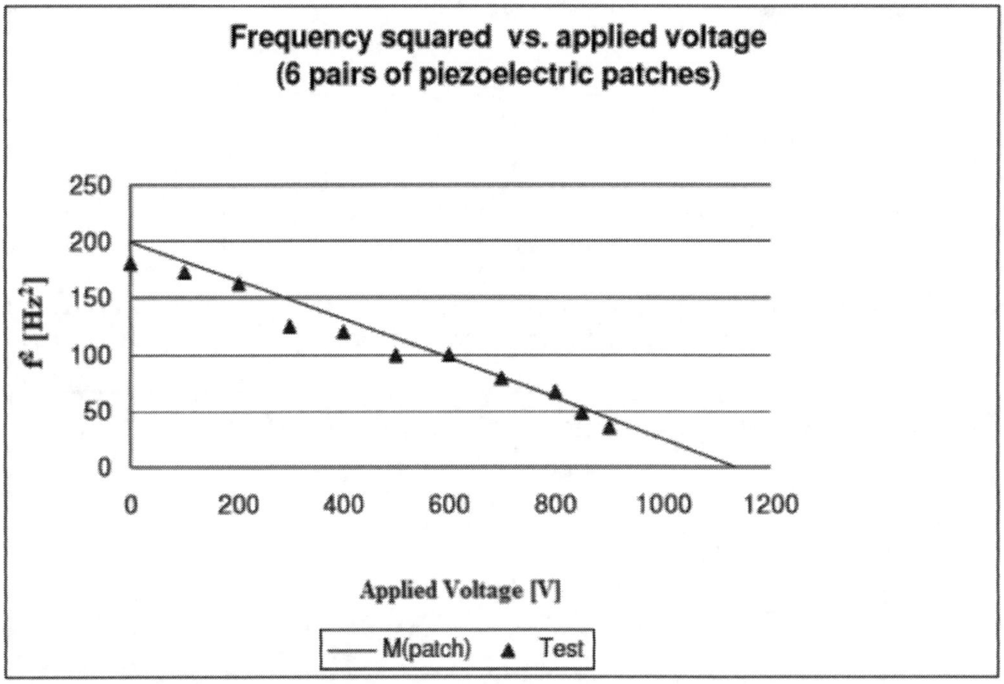

Figure 7. Frequency squared vs. the applied voltage for the case of a beam with 6 pairs of piezoelectric patches till "buckling"

The distribution of the bending strains at the beam's root, (S.G. # 17 and 18) are described in Fig. 9 as a function of the applied voltage, for the static case. Figures 9-10 describe the behavior of the strain gages at the beam's root. It seems from the three graphs that the buckling voltage is about 1150 [V]. A well-defined buckling load can be determined from Fig. 11, yielding similar results to the ones presented in Figs. 7 and 8 for the voltage which will induce buckling in the beam.

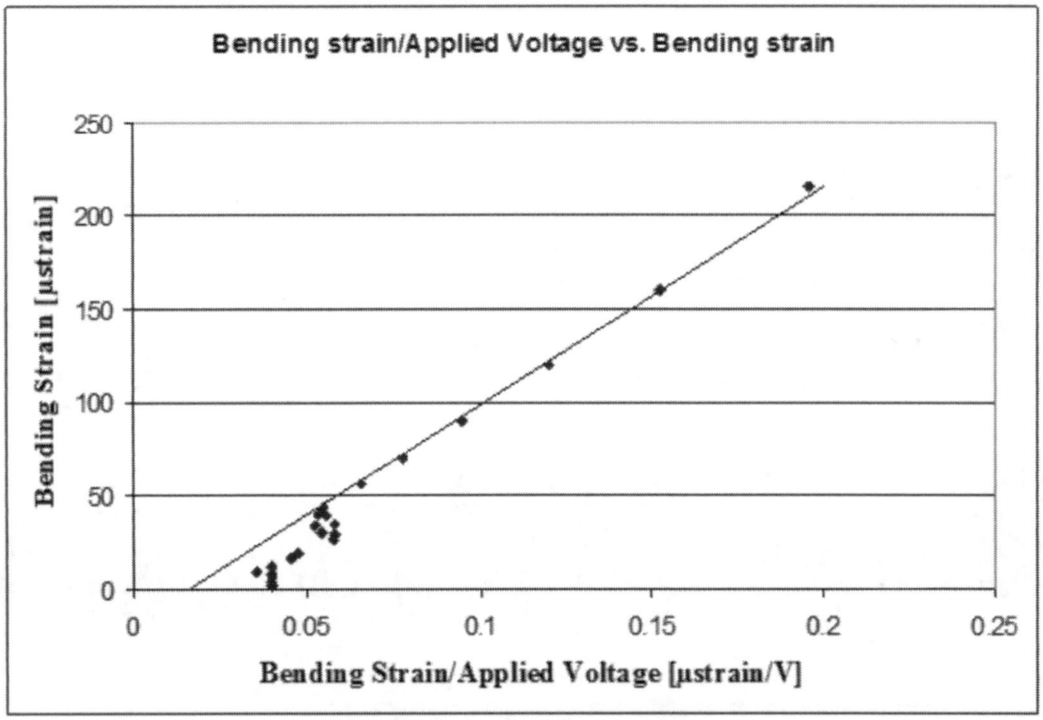

Figure 8. A modified Southwell plot

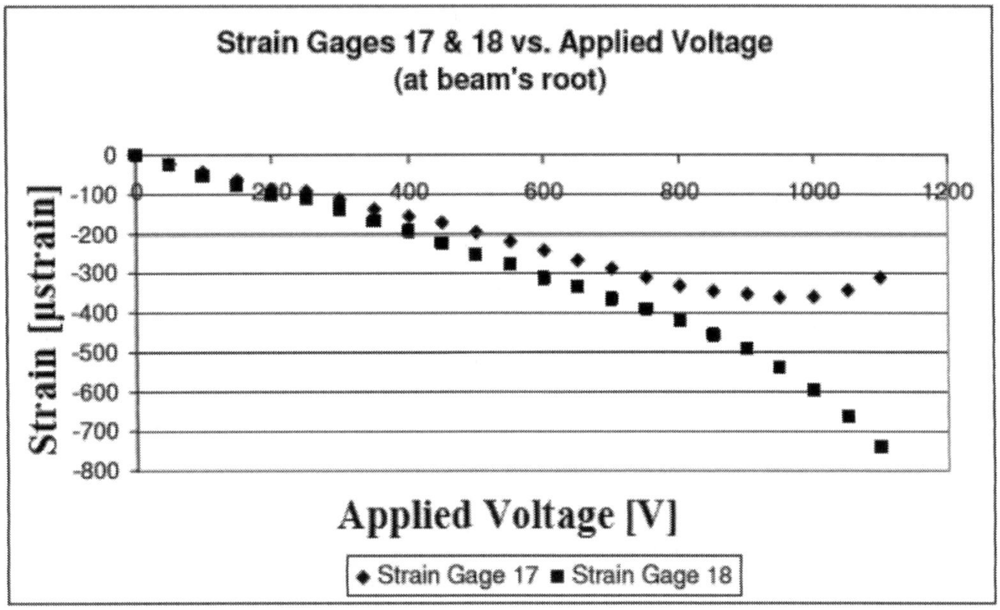

a.

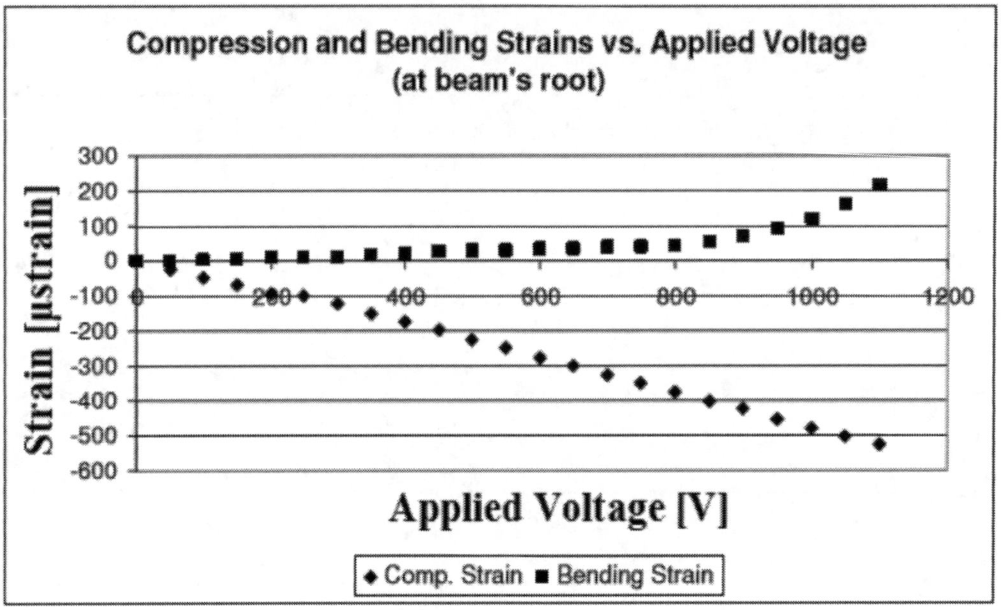

b.

Figure 9. Strains measured at the beam's root vs. applied voltage: a. S.G. #17&18, b. compression and bending strains

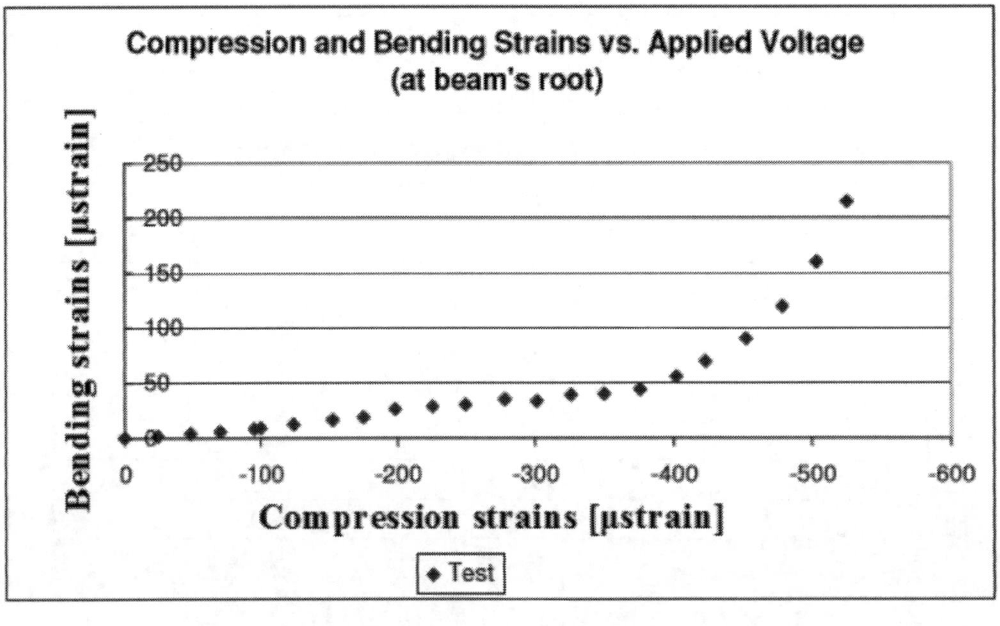

Figure 10. Compression strains vs. bending strains (S.G. 17 & 18)

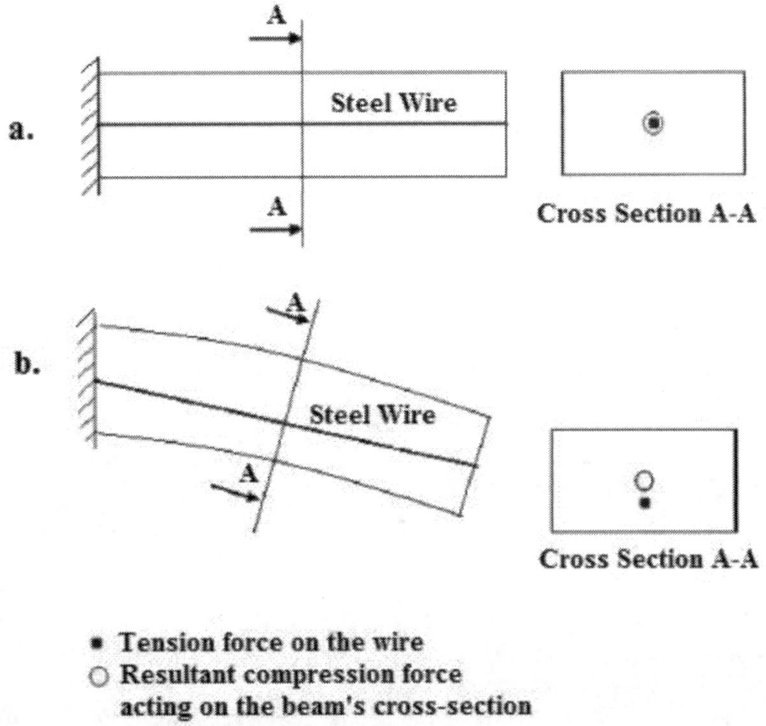

Figure 11. (a) The beam at zero lateral deflection and (b) the beam under lateral deflection

4. DISCUSSION AND CONCLUSIONS

A cantilever composite beam, with bonded piezoelectric patches, was buckled due to application of voltage on the patches. It was proved experimentally, that with the correct restrained ends of a beam, even for a cantilever beam, its natural frequencies can be altered and buckling can be reached, by only applying electrical voltage. Figures 11a-b try to explain schematically the phenomenon involved in the study. The resultant axial force at each cross section of the beam is zero for the cantilever beam. Two opposite components form this resultant: a resultant compressive force acting through the cross-sections centers of the beam and an equal tension force along the wire. When no lateral deflections are experienced by the beam the two components are collinear (Fig. 11a), while in the presence of lateral deflections the two forces would yield a couple which is proportional to the magnitude of the two opposite components of the resultant axial force (Fig. 11b). They are constant during the vibrations. When the magnitude of the lateral deflection (which varies harmonically) is large enough buckling can occur. The buckling of the beam was enabled due to the presence of a steel wire connecting its free end to the clamped side of the beam. The application of voltage to the six pairs of piezoelectric patches bean bonded on the surface of the beam's profile induced tension stresses in the wire, thus causing the buckling of the beam due to the compressive stresses the beam was encountering.

ACKNOWLEDGMENTS

The author would like to thank Mr. A. Kotler for manufacturing the test specimen and for well performing the experimental campaign.

REFERENCES

1. Waisman, H. and Abramovich, H., "Variation of natural frequencies of beams using the active stiffening effect", *Composites Part B: Engineering*, Vol. 33, Issue 6, pp. 415-424, 2002.
2. Waisman, H. and Abramovich, H., "Active stiffening of laminated composite beams using piezoelectric actuators", *Composite Structures*, Vol. 58, Issue 1, pp. 109-120, 2002.
3. Ogauamanam, D.C.D, Almeida, S.F.M. and Hansen, J.S., "Stress stiffening effects in laminated beams with piezoelectric actuators", *J. Intell. Mater. Sys. Struct.*, Vol. 9, pp. 137-145, 1998.
4. Hernande, J.A., Almeida, S.F.M. and Nabarrete, A., "Stiffening effects on the free vibration behavior of composite plates with PZT actuators", *Composite Structures*, Vol. 49, Issue, pp.55-63, 2000.
5. de Faria, A. R. and Donadon, M.V., "The use of piezoelectric stress stiffening to enhance buckling of laminated plates", *Latin American Journal of Solids and Structures*, Vol. 7, No. 2, pp. 167-183, Rio de Janeiro, 2010.
6. Kuo, S.-Y., "Stiffening effects on the natural frequencies of laminated beams with piezoelectric actuators", *Journal of Aeronautics, Astronautics and Aviation, Series A*, Vol.42, No.1, pp. 67-72, 2010.
7. Donadon, M. V., Almeida, S. F. M., and de Faria, A. R., "Stiffening effects on the natural frequencies of laminated plates with piezoelectric actuators," *Composites Part B: Engineering*, Vol. 33, No. 5, pp. 335-342, 2002.
8. Wang, Q. and Quek, S. T., "Enhancing flutter and buckling capacity of column by piezoelectric layers," *International Journal of Solids an Structures*, Vol. 39, No. 16, pp. 4167-4180, 2002.
9. Thompson, S. P. and Loughlan, J., "The active buckling control of some composite column strips using piezoelectric actuators", *Composite Structures*, Vol. 32, Issue 1-4, pp. 59–67, 1995.
10. M. Shariat, M., "Dynamic buckling of imperfect laminated plates with piezoelectric sensors and actuators subjected to thermal-electromechanical loadings, considering the temperature-dependency of the material properties", *Composite Structures*, Vol. 88, Issue 2, pp. 228–239, 2009.
11. Fridman, Y. and Abramovich, H., "Enhanced structural behavior of flexible laminated composite beams," *Composite Str*uctures, Vol. 82, Issue 1, pp. 140-154, 2008.
12. Meressi, T. and Paden, B., "Buckling control of a flexible beam using piezoelectric actuators", *Journal of Guidance, Control and Dynamics*, Vol. 16, No. 5, pp. 977-980, 1993.
13. Mukherjee, A., Joshi, S.P. and Ganguli, A., "Active vibration control of piezolaminated stiffened plates", *Composite Structures*, Vol. 55, Issue 4, pp. 435–443, 2002.
14. Abramovich, H. and Livshits, A., "Dynamic behavior of cross-ply laminated beams with piezoelectric layers", *Composite Structures*, Vol. 25, No.1-4, pp. 371-379, 1993.
15. Abramovich, H. and Livshits, A., "Free vibrations of non-symmetric cross-ply laminated composite beams", *J. of Sound and Vibration*, 176(5), pp. 597-612, 1994.
16. Abramovich, H.," A new Insight on vibrations and buckling of a cantilevered beam under a constant piezoelectric actuation", *Composite Structures*, 93, Issue 2, pp. 1054–1057, 2011.
17. *Smart Material Corporation*-http://www.smart-material.com
18. *ANSYS*- http://www.ansys.com/products/

ICAST2015: 26th International Conference on Adaptive Structures and Technologies
October 14-16th, 2015, Kobe, Japan

ICAST2015 # paper-number 027

Design optimization of generalized coupling coefficient for vibration attenuation in metamaterials with local piezoelectric shunts

Edgar A. Flores Parra[1*], Andrea Bergamini[2], Edoardo Belloni[3], Tommaso Delpero[2] , Paolo Ermanni[2]

[1] Composite Materials and Adaptive Structures Laboratory, ETH Zürich, Switzerland
[2] Materials Science and Technology, Laboratory for Mechanical Integrity of Energy Systems, EMPA, Switzerland
[3] Department of Mechanical Engineering, Politecnico di Milano, Italy.

Abstract

Locally resonant structures have been shown to offer sub-wavelength modifications to the propagation of mechanical waves. Following the pioneering work of Liu, where this principle was demonstrated in a purely mechanical system, the use of locally shunted piezoelectric resonators has been reported for electromechanical systems. The prediction of the properties of a metamaterial comprised of local resonators is based on the calculation of its dispersion curves, assuming periodicity of a medium with infinite spatial extension. However, any practical application of these metamaterials implies a finite size structure comprised of a limited number of unit cells where the authority of the piezoelectric elements on the mechanical medium has a significant impact on the response of a finite sample subject to dynamic loads. In analogy to the generalized electromechanical coupling coefficient, as defined for a piezoelectric damped structure, we investigate the effect of switching the electrical boundary conditions applied to the piezoelectric element of the unit cell, from open to closed circuit, on the dispersion properties of the system, as an expression of its stiffness. Combining the generalized coupling coefficient, calculated for the unit cell, with the optimum resistance and inductance allows for minimization of the frequency response amplitude at a specific harmonic frequency of an infinite medium. Furthermore, the relation between the spatial decay seen in the dispersion curves and the wave attenuation in a finite-size sample of a piezoelectric metamaterial is investigated numerically, using models that have been previously tested against experimental results. Conclusions about the optimal design of these electromechanically coupled metamaterials are drawn and discussed in relation to possible practical applications. In particular, considerations are made relative to the amount of piezoelectric material needed to obtain a desirable level of attenuation in finite structures.

1. INTRODUCTION

In recent years, piezoelectric materials have raised interest as energy converting elements in passive vibration attenuation devices. The piezoelectric material of these devices strains when bonded to a vibrating host structure, resulting in an electric charge produced through the direct piezoelectric effect. The associated electrical power can then be dissipated through multiple different shunt circuits connected to the electrodes of the piezoelectric element.

* edgarf@ethz.ch

The amount of electric charge produced from the mechanical strains is determined by the generalized coupling coefficient. This coefficient is the ratio of energy transformed from mechanical to electrical, which is itself dependent on the intrinsic electromechanical coupling coefficient of the piezoelectric material and the mechanical coupling between the piezoelectric and the host structure. From the latter, it might be expected that the larger the generalized piezoelectric coefficient the greater the attenuation [4]. However, the level of attenuation is also strongly dependent on the type, and the parameters of the shunt circuit. In the case of the RL circuit used in this work, the optimum values of the shunting resistance and inductance, which are a function of the generalized coupling coefficient, were determined by Hagood using the transfer function optimization method [2].

Previous work has focused on methods for evaluating the generalized coupling coefficient [3] as well as optimizing it for applications such as energy harvesting, passive shunt damping, and damage detecion[6],[7]. However, there has been little work in the direction of establishing a correlation between the generalized coupling coefficient of a piezoelectrically shunted unit cell and the wave attenuation seen in a finite sample of the same medium. Spadoni [5] showed that from the dispersion analysis, the frequency range over which wave attenuation in a finite plate with periodic piezoelectric shunts, could be determined. However, this work did not consider the effect the generalized coupling coefficient or the shunt circuit on the wave attenuation.

The objective of this work is to show that a relation exists between the generalized coupling coefficient calculated for a unit cell of an infinite medium, with an optimum RL series shunt, and the wave attenuation achieved in a finite 1D structure with sufficient periodic piezoelectric shunts to approximate the response of an infinite medium. Based on the relation between the unit cell geometry and the wave attenuation of a finite structure the piezoelectric element size can be optimized. Moreover, RL resonators were chosen for their simplicity in modeling and at a later stage for implementing in experimental testing.

2. Generalized coupling coefficient K

Several approaches exist for evaluating the generalized coupling coefficient such as using the modal strain energy, modal stiffness or eigenfrequecies of the system [3].

K quantifies the energy exchange between the mechanical and the electrical domain and can be defined through Eq. (1).

$$K^2 = \frac{\omega_{OC}^2 - \omega_{SC}^2}{\omega_{SC}^2} \tag{1}$$

Where ω_{SC} is the short-circuit resonance frequency and ω_{OC} is the open-circuit configuration. The open circuit state corresponds to the particular case $Z(s) \to \infty$, where $Z(s)$ is the generalized impedance of the shunt, so that the shunting term becomes an additional contribution to the structure stiffness. The latter shows that the open-circuit eigenfrequency, and thus stiffness, is larger than the short-circuit eigenfrequency. The change of stiffness between the open and short circuit configurations is an indicator of the authority of the piezoelectric element on the structure.

By Kirchoffs law, current flowing in the shunt circuit is constant through elements in series. Therefore, a resistor and inductor in series with the PZT can achieve maximum current through the resistor. The optimum resistance and inductance values for the series circuit were calculated for the system at hand, shown in Fig. 1 , following the procedures of Hagood and Von Flotow [2] for minimization of the maximum frequency response amplitude of the structure. These optimum values (see Table 1) which

$t_p[mm]$	K (t_s=1mm)	$C_p[nF]$	$L_{opt}[H]$	$R_{opt}[k\Omega]$
0.2	0.132	6.96	0.14	0.84
0.5	0.107	3.34	0.30	1.43
0.75	0.084	2.18	0.46	1.72
1.0	0.067	1.56	0.65	1.91
2.0	0.037	0.90	1.12	1.87

Table 1: Generalized coupling coefficient, and electrical parameters for coincidence frequency ω_n=5kHz and wavenumber $k_x = \pi/2L_x$.

are a function of the generalized coupling coefficient are given by Eq. (2,3), where ω_n is the electrical resonance frequency and C_p^S the piezoelectric capacitance which, as expected, increases with decreasing thickness of the piezoelectric element. L_x is the lattice constant and k_x is the wavenumber in the X direction.

$$L_{opt} = \frac{1}{\omega_n^2 C_p^s (1 + K^2)} \tag{2}$$

$$R_{opt} = \sqrt{2}\frac{K^2}{C_p^s \omega_n (1 + K^2)} \tag{3}$$

3. Methods

For all transmittance and dispersion case studies, presented in the following section, the host structure was made of *aluminium* with Youngs Modulus E_s =69GPa, density ρ_s = 2700kg/m^3, Poisson's ratio $\nu = 0.33$ and thickness t_s = 1mm. The material of the piezoelectric element was PZT-151 with the open circuit Youngs Modulus E_p^D =77.4GPa, density ρ_p =7760 kg/m^3, coupling coefficient $k_{31} = 0.37$. The superscript, s, refers to the substrate, host structure, while the supercript ,p, refers to the piezoelectric element.

The optimum inductance and resistance values for the five different piezoelectric element sizes utilized in the transmittance and inverse dispersion studies, presented in the following sections, can be found in the Table 1. The values of R_{opt} and L_{opt} were calculated for a 1mm beam host structure and a coincidence frequency of 5kHz and wavenumber $k_x = \pi/(2L_x)$. The value $k_x = \pi/(2L)$ was chosen as it is approximately the wavenumber corresponding to the 5kHz coincidence frequency for the five unit cell configurations considered in this work. The generalized coupling coefficient was obtained from the results shown in Fig. 3b for the 1mm thick substrate. The electrical parameters L_{opt} and R_{opt} were derived for a finite structure with a local resonator shunt. However, we have implemented them in the unit cell (i.e: infinite medium) as the electrical impedance of the RL circuit is independent from the Floquet periodic boundary conditions but can account for the unit cell generalized coupling coefficient.

3.1 Numerical Calculation of Unit Cell Generalized Coupling Coefficient

The generalized coupling coefficient as a function of wavenumber was calculated numerically using

ANSYS. A unit cell model was created, where the host structure featured side lengths $L_x = 15$mm, $L_y = 10$mm, with a thickness that was varied $t_s = [0.5; 1; 1.25; 2; 2.5; 3; 4]$mm. The piezoelectric element of the unit cell had a length, l_p=10mm, and its thickness was varied t_p =[0.2; 0.5; 0.75; 1; 2]mm. Floquet boundary conditions were applied in the X direction and an eigenfrequency analysis was then conducted at each wavenumber over the range $[0...\pi/L_x]$ for both open and closed circuit. The open-circuit configuration was modeled in ANSYS by imposing a zero charge on the terminals of the piezoelectric element, while the short-circuit configuration was modeled by imposing a condition of zero voltage across the terminales of the piezoelectric element. The generalized coupling coefficient of the infinite medium was calculated by applying Eq.(1), using the first eigenfrequency, at each wavenumber, of the dispersion relations for the open and close circuit. The generalized coupling coefficient was then calculated for each wavenumber along the edge Γ-X of the Brillouin zone for each of the 35 different geometric combinations (5 piezo thicknesses x 7 substrate thicknesses) as seen in Fig. 3a, making K wavenumber dependent, Eq.(4).

$$K(k_x)^2 = \frac{\omega_{OC}^2(k_x) - \omega_{SC}^2(k_x)}{\omega_{SC}^2(k_x)} \tag{4}$$

Due to the large number of combinations, (35 geometries x N wavenumbers in the range) ANSYS was chosen as the prefered solver as it allowed much faster computation of the desired eigenvalues compared to COMSOL V4.4.

3.2 Numerical Calculation of the Transmittance of the metamaterial.

For the numerical calculation of the transmittance of the finite metamaterial sample, we used COMSOL Multiphysics V4.4. As seen in Fig. 1 we considered a two-dimensional model of an aluminum beam with a thickness of 1mm, width of 10mm and a total length of 480mm. In the center portion of the structure, a metamaterial with unit cell size, L_x=15mm, was added, consisting of 20 unit cells positioned between 90mm and 395mm from the origin, each a length l_p=10mm piezoelectric elements and thicknesses ranging over $t_p = [0.2; 0.5; 0.75; 1; 2]$mm. Moreover, we defined a 5mm perfectly matched layer at the boundary condition at either end of the structure to suppress the reflection of mechanical waves. We modeled the inductive and resistive components using the electric circuit physics of COMSOL. The selected values for the inductors and resistors, L_{opt} and R_{opt} respectively, as well as for the geometry of the unit cell and the material properties were the same as for the calculation of the dispersion curves in order to establish a direct correlation between the dispersion curves and the transmittance of the finite metamaterial. The mechanical transmittance of the finite metamaterial sample was calculated by taking the ratio of the spatial average of the transverse velocity amplitudes, over a region with 80mm length, before and after the periodic arrangement.

$$Transmittance(\omega) = \frac{v_{a,out}}{v_{a,in}} = \frac{mean(v_a(\omega, x \in [395\text{ mm}, 475\text{ mm}], y \in [0, t_s]}{mean(v_a(\omega, x \in [10\text{ mm}, 90\text{ mm}], y \in [0, t_s]} \tag{5}$$

The system was excited mechanically by a discrete force F(t) applied 7.5mm from the origin, Fig. 1. The steady state frequency response of the system was calculated over the frequency range of interest.

3.3 Inverse Dispersion Calculation

The common approach to evaluate the dispersion relation consists in specifying the wavenumber

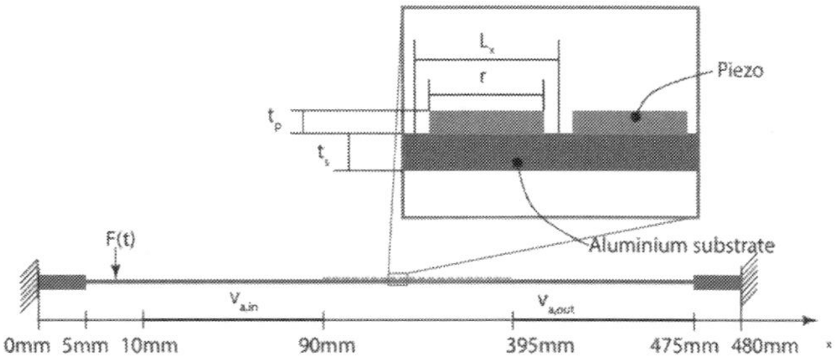

Figure 1: Overview and dimensions of the investigated finite metamaterial sample with detailed view of two unit cells.

and solving for the frequency. However, this approach does not yield the imaginary component of the wavenumber which corresponds to the spatial attenuation. The complex wavenumber can be obtained using transfer matrix technique [10] which allows for the calculation of the inverse solution of the dispersion of the unit cell. The inverse solution consists in specifying the frequency and solving for the wavenumber which in this case is restricted to 1D. The wavenumber is solved for from the eigenvalues of the transfer matrix $\mathbf{T}(\omega)$, calculated from the eigenvalue problem given by Eq.(6), where $\lambda = e^{-ik_x L_x}$.

$$\mathbf{T}(\omega) - \lambda \mathbf{I} = 0 \tag{6}$$

From this expression both the real and imaginary components of the propagation constant can be obtained. The derivation of matrix $\mathbf{T}(\omega)$ and the setup of the eigenvalue problem is described in the following section, where $D(x,\omega), B(x,\omega), K(x,\omega), m(x,\omega)$ are the bending, coupling, axial stiffness and the mass per unit length which are dependent on the position x and the frequency ω. The latter matrices are piecewise constant, describing the pure aluminium region and the region of aluminium plus the piezoelectric shunt as seen in Fig. 2. E_p^{Sh} is the shunted piezoelectric modulus, α is the percentange of the unit cell with piezoelectric element, L and R are the indutance and resistance, respectively.

$$E_p^{Sh}(\omega) = E_p^D\left(1 - \frac{k_{31}^2}{1 + i\omega C_p^S(R + i\omega L)}\right) \tag{7}$$

$$D(x,\omega) = \begin{cases} E_s\frac{b_s t_s^3}{12} & -\alpha L_x<x<0 \\ E_s b_s\left(\frac{t_s^3}{12}+\frac{t_s h_p^2}{4}\right)+E_p^{Sh}(\omega)b_p\left(\frac{t_s^3}{12}+\frac{t_s t_p^2}{4}\right) & 0<x<(1-\alpha)L_x \end{cases} \tag{8}$$

$$B(x,\omega) = \begin{cases} 0 & -\alpha L_x<x<0 \\ \left(\frac{-b_s E_s t_s t_p^3}{2}\right)+E_p^{Sh}(\omega)b_p\left(\frac{t_s^3}{12}+\frac{t_s t_p^2}{4}\right) & 0<x<(1-\alpha)L_x \end{cases} \tag{9}$$

$$K(x,\omega) = \begin{cases} E_s t_s b_s & -\alpha L_x<x<0 \\ E_s t_s b_s+E_p^{Sh}(\omega)b_p t_p & 0<x<(1-\alpha)L_x \end{cases} \tag{10}$$

$$m(x) = \begin{cases} \rho_b t_b b_b & -\alpha L_x<x<0 \\ \rho_b t_b b_b+\rho_p t_p b_p & 0<x<(1-\alpha)L_x \end{cases} \tag{11}$$

$$\mathbf{A}(x,\omega) = \begin{bmatrix} K(x,\omega) & 0 & -B(x,\omega) & 0\,0\,0 \\ 0 & 1 & 0 & 0\,0\,0 \\ B(x,\omega) & 0 & -D(x,\omega) & 0\,0\,0 \\ 0 & 0 & 0 & 1\,0\,0 \\ 0 & 0 & 0 & 0\,1\,0 \\ 0 & 0 & 0 & 0\,0\,1 \end{bmatrix} \quad \mathbf{L}(x,\omega) = \begin{bmatrix} 0 & 0 & 0 & 1\,0\,0 \\ 0 & 0 & 1 & 0\,0\,0 \\ 0 & 0 & 0 & 0\,1\,0 \\ -m(x)\omega^2 & 0 & 0 & 0\,0\,0 \\ 0 & 0 & 0 & 0\,0\,1 \\ 0 & 0 & -m(x)\omega^2 & 0\,0\,1 \end{bmatrix}$$

The coupled system of equations can be written in matrix form, Eq. 12 ,where $\mathbf{A}(x,\omega)$ and $\mathbf{L}(x,\omega)$ are two coefficient matrices.

$$\mathbf{A}(x,\omega)\frac{d}{dx}\mathbf{z}(x) = \mathbf{L}(x,\omega)\mathbf{z}(x) \tag{12}$$

Where the state vector $\mathbf{z}(x)$ contains the unknown displacements, resulting forces and moments:

$$\mathbf{z}(x) = [u(x) \quad \omega(x) \quad \omega^{'}(x) \quad N_x(x) \quad Q_y(x) \quad M_b(x)] \tag{13}$$

Inverting $\mathbf{A}(x,\omega)$ the final form of the system of equations is obtained.

$$\frac{d}{dx}\mathbf{z}(x) = \boldsymbol{\theta}(x,\omega)\mathbf{z}(x) \tag{14}$$

$$\boldsymbol{\theta}(x,\omega) = \mathbf{A}(x,\omega)^{-1}\mathbf{L}(x,\omega) \tag{15}$$

Looking at the difference between a solution at $x = -\alpha L_x$ and $x = 0$ and then at $x = (1-\alpha L_x)$ and $x = 0$, we can write Eq. 16.

$$\boldsymbol{T_1} = e^{\theta_1(\omega)\alpha L_x} \qquad \boldsymbol{T_2} = e^{\theta_2(\omega)(1-)\alpha)L_x} \tag{16}$$

The complete transfer matrix for the unit cell is now defined:

$$\boldsymbol{T} = \boldsymbol{T_1}\boldsymbol{T_2} \tag{17}$$

The next step to obtain Eq. 6 is to combine Floquet's theorem with the Transfer matrix method, see Fig. 2. Applying Floquet's theorem we can obtain Eq. 18.

$$\mathbf{z}((1-\alpha)L_x) = e^{ik_x L_x}\mathbf{z}(-\alpha L_x) \tag{18}$$

Analogously, we can then relate both ends to the unit cell using the Transfer matrix method, Eq. (18).

$$\mathbf{z}((1-\alpha)L_x) = \mathbf{T}\mathbf{z}(-\alpha L_x) \tag{19}$$

Combining Eq.(18,19), we obtain the standard eigenvalue problem, Eq. (20)

$$[\mathbf{T} - e^{ik_x L_x}\mathbf{I}]z(-\alpha L_x) = 0 <=> |\mathbf{T} - e^{ik_x L_x}\mathbf{I}| = 0 \tag{20}$$

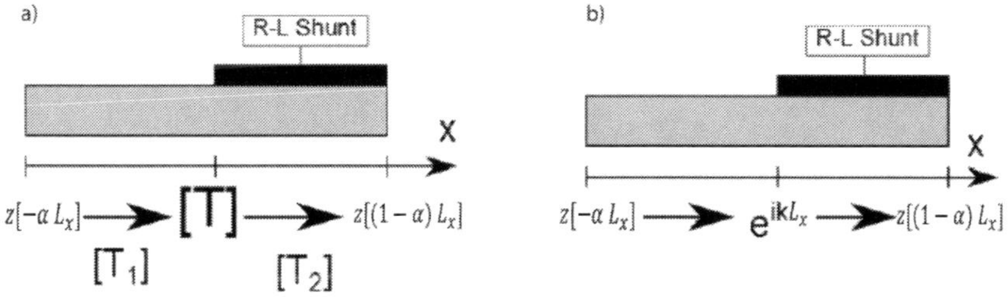

Figure 2: a) Illustrates the two intervals of the unit cell into which the transfer matrix can be divided. b) Floquet's theorem relating degrees of freedom at both ends of the unit cell.

4. Results and Discussion

4.1 Unit cell generalized coupling coefficient

Fig. 3 shows that the generalized coupling coefficient remains almost invariant over the wavenumber range of the $\Gamma - X$ edge, except when $k_x \to 0$ and $k_x \to \pi/L_x$. While not shown in Fig. 3a, since the eigenvalue solver does not clearly resolve the eigenfrequencies near the long wavelength limit, the generalized coupling coefficient tends to a finite value as the wavenumber tends to zero. As the wavenumber tends to the short wavelength limit (wave length equal to twice the lattice constant, L_x), the generalized coupling coefficient tends to zero. The latter implies that towrads this wavenumber limit, the the difference in the modal stiffness of the open and short circuit decreases.

In this work, we are not interested in the limits of $\Gamma - X$ wavenumber range, thus taking a cut-out from the 3D plot at the wavenumber corresponding to the coincidence frequency, $k_x = \pi/(2L_x)$, shows more clearly the depedance of K on the thickness variation of the piezoelectric elements, as seen in Fig. 3b. As the thickness of the piezoelectric element increases the coincident wavenumber increases. However, the variation of the generalized coupling coefficient over this range is minimum compared to the change at the short wavelength limit. Hence, looking at a fixed wavenumber to obtain the generalized coupling coefficient for the various thickness samples considered is sufficient. Moreover, from Fig. 3b we can see that in general as the thickness of the piezoelectric element increases the generalized coupling coefficient decreases. The latter is due to the fact that as the piezoelectric element thickness increases, it becomes harder to strain it, and more of strain energy concentrates on the unit cell substrate interval not attached to the piezoelectric element. Note, as expected that the magnitude of the generalized coupling coefficient is less than the material coupling coefficient k_{31} as not all strain energy is applied on the piezoelectric element. Based on the results seen in Fig. 3b, we expect that for t_s=1mm, the thinnest piezoelectric element t_p=0.2mm to have the highest mechanical to electrical energy conversion ratio and the thickest piezo t_p=2mm will have the least.

4.2 Dispersion of unit cell with optimum RL shunt

Having determined the generalized coupling coefficient for a combination of various substrate and piezoelectric thicknesses over a range of wavenumbers, we studied the dispersion curves of the five

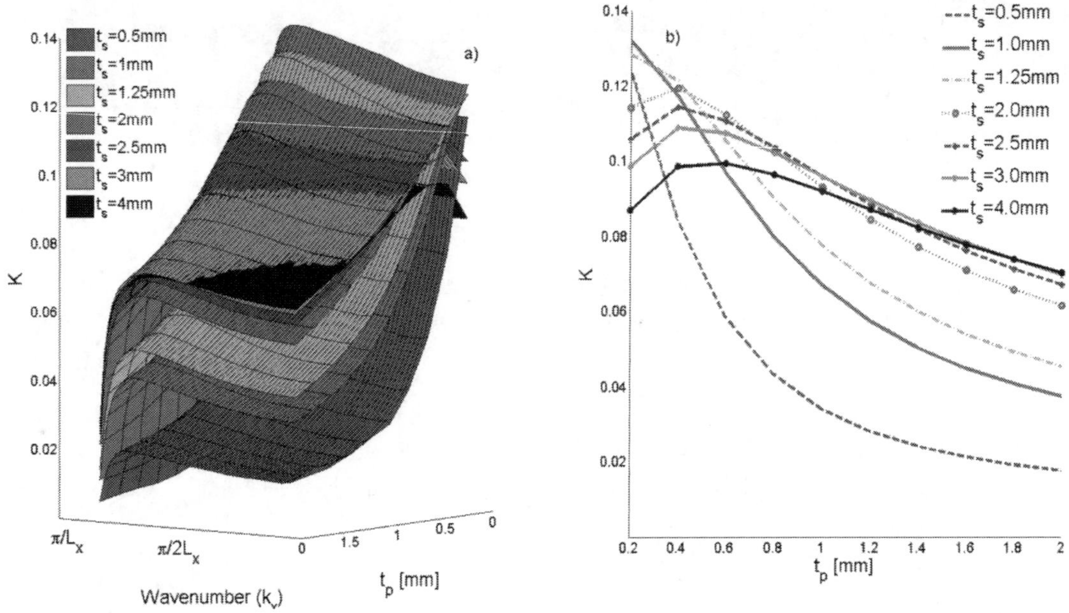

Figure 3: a) Generalized coupling coefficient as function of piezoelectric element thickness, substrate thickness, and wavenumber. b) Generalized coupling coefficient as function of piezo thickness for fixed wavenumber $k_x = \pi/(2L_x)$.

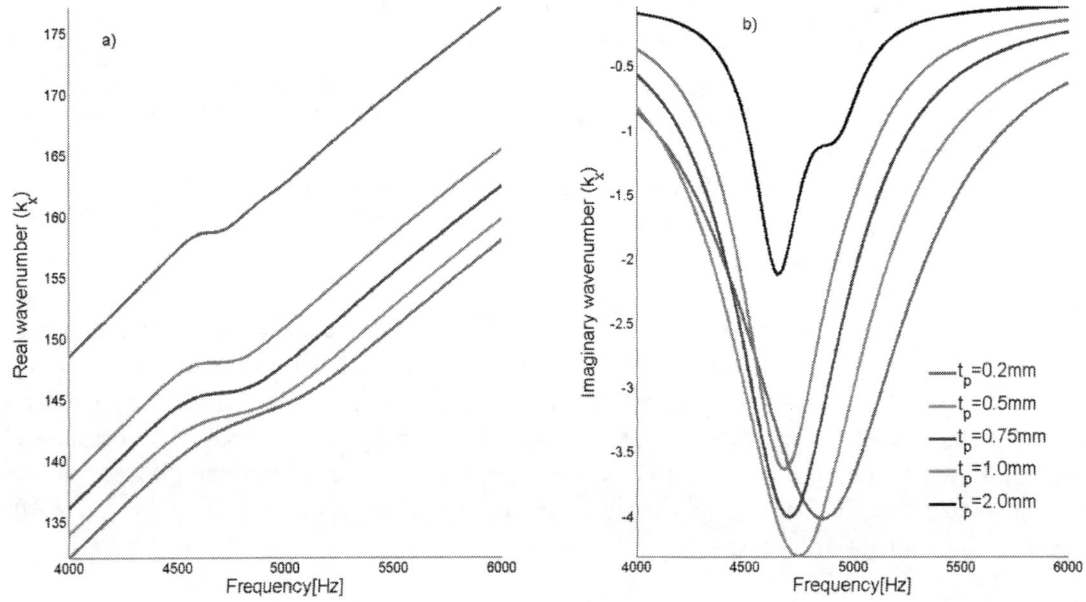

Figure 4: a) Real wavenumber dispersion curves b) Imaginary wavenumber dispersion curves for the five unit cells with t_s =1mm and $t_p = [0.2; 0.5; 0.75; 1; 2]$mm.

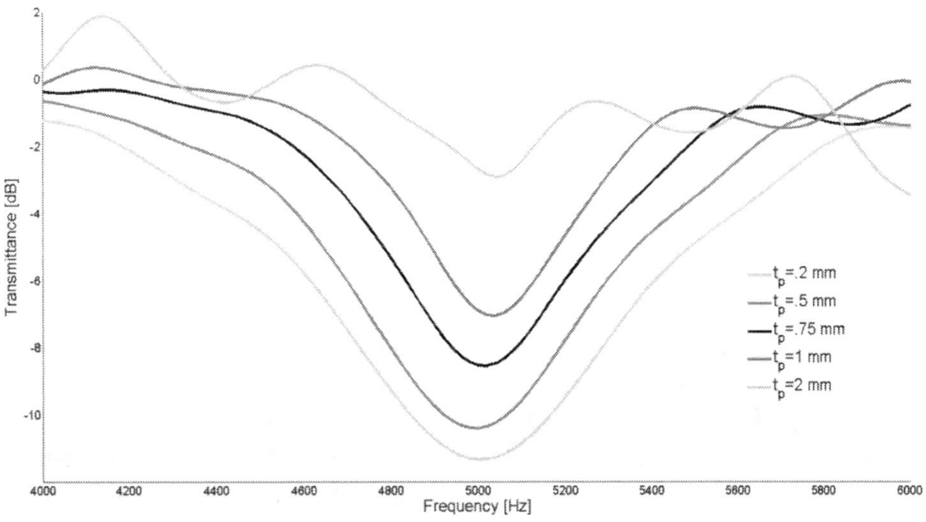

Figure 5: Numerical transmittance of the five finite samples with t_s = 1mm and t_p = $[0.2; 0.5; 0.75; 1; 2]$mm centered around the 5kHz coincidence frequency.

piezoelectric elements described in Table 1, and attached on a 1mm substrate, combined with their corresponding optimum RL circuit. The inverse solution technique described in Section 3 was utilized to investigate the dispersion curves of the five different unit cell configurations. Fig. 4 shows the real and imaginary parts of the wavenumber as a function of frequency. As expected from piezoeletric local resonator shunts, the dispersion curves display an $'S'$ shape corresponding to the coincidence frequency between the mechanical bending and the zero group velocity electrical mode [8]. Fig. 4b shows the imaginary part of the wavenumber, which is a measure of the spatial wave decay and can be paired with the group velocity to compute an equivalent of the temporal decay [9]. From Fig. 4b, we see that the thinner the piezoelectric element the greater the absolute value of the imaginary component of the dispersion, with the exception of the thinnest, t_p=0.2mm, sample. Hence, we expect that the wave attenuation around the coincidence frequency, in the finite samples composed of these unit cell gemoetries, will increase with decreasing thickness of the piezoelectric element.

4.3 Transmittance of the finite metamaterial sample

The transmittance curves of the five finite metamaterial samples, each with 20 unit cells of the five geometries studied in the previous section, were calculated numerically around the coincidence frequency as seen in Fig. 5. Moreover, Fig. 3b indicates that the highest generalized coupling coefficient corresponds to the thinnest piezoelectric element for the chosen substrate, t_s=1mm. Hence, it was expected that the finite metamaterial sample with the highest attenuation would be the one shunted with the thinest piezoelectric elements. This expectation was corroborated in the Fig. 5, where the highest wave attenuation was achived for t_p=0.2mm. However, for this case the prediction from the imaginary dispersion did not match the transmittance results, as we see a lower absolute amplitude of the imaginary wavenumber for t_p=0.2mm than for t_p=0.5mm. The coincidence frequency of t_p=0.2mm is also offset from that of the other samples, thus, we believe that the transfer matrix model does not accurately predict the dispersion for piezoelectric elements that are too thin. However, the model does accurately predict expected trend in the attenuation based on the trend in the magnitude of the imaginary wavenumber for all the other samples ranging from t_p=0.5mm up to t_p=2mm.

5. CONCLUDING REMARKS

In this work we have presented an approach, based on the unit cell generalized coupling coefficient, to optimize the size of the piezoelectric elements for a 1D metamaterial composed of periodic RL shunts, with optimum resistance and inductance values, for minimizing the maximum frequency response amplitude at a specific harmonic frequency. We observed that for an infinite periodic medium, the generalized coupling coefficient tends to a finite value at the long wavelength limit. As the wavenumber tends to the short wavelength limit the generalized coupling coefficient tends to zero indicating a decrease in the difference between the modal stiffness of the open and short circuit configurations. In between the long and short wavelength limits, the generalized coupling coefficient remains mostly constant for the different unit cell geometries. The latter is a useful observation as the response of several geometries that lie within this wavenumber range could be analysed by studying them at a fixed wavenumber, as was done in this work. Moreover, the attenuation achieved in the frequency response amplitude of the finite metamaterial sample, for the considered piezoelectric elements sizes, was compared to the spatial wave decay given by the imaginary wavenumber obtained from the inverse dispersion solution calculated using the transfer matrix method. The results for the finite sample indicate that the highest attenuation was achieved for the piezoelectric element with the highest generalized coupling coefficient and that the maximum attenuation decreased with decreasing values of the generalized coupling coefficient. Moreover, with the exception of the thinnest piezoelectric element, the trend in the attenuation observed for the infinite medium and the finite samples show a good correlation. We conclude that calculating the generalized coupling coefficient at the unit cell level is a powerful and efficient approach for optimizing the piezoelectric element geometry of a metamaterial, and the relative trend in the wave attenuation of different geometries of finite metamaterial samples can be evaluated at the unit cell level.

ACKNOWLEDGEMENTS

The authors thank Sebastian Krödel for the development of the transfer matrix model for asymmetric beams with RL shunts.

REFERENCES

1. Trindade, Marcelo A and Benjeddou, Ayech, "Effective electromechanical coupling coefficients of piezoelectric adaptive structures: critical evaluation and optimization", *Mechanics of Advanced Materials and Structures*, Vol.16, No.3, 2009.
2. Hagood, Nesbitt W and von Flotow, Andreas, "Damping of structural vibrations with piezoelectric materials and passive electrical networks", *Journal of Sound and Vibration*, Vol. 146, No. 2, 1991, pp. 243–268.
3. Trindade, Marcelo A and Benjeddou, Ayech, "Effective electromechanical coupling coefficients of piezoelectric adaptive structures: critical evaluation and optimization", *Mechanics of Advanced Materials and Structures*, Vol.16, No.3, 2009, pp. 210-223.
4. Donald, J Leo and Leo, DJ, *Engineering analysis of smart material systems*, John Wiley & Sons, Inc, 2007.
5. Spadoni, Alessandro and Ruzzene, Massimo and Cunefare, K, "Vibration and wave propagation control of plates with periodic arrays of shunted piezoelectric patches", *Journal of Intelligent Material Systems and Structures*, Vol.20, No.8, 2009, pp. 979-990.
6. Howells, Christopher A, "Piezoelectric energy harvesting", *Energy Conversion and Management*, Vol.50, No.7, 2009, pp. 1847–1850.
7. Hollkamp, Joseph J, "Multimodal passive vibration suppression with piezoelectric materials and resonant shunts", *Journal of intelligent material systems and structures*, Vol.5, No.1, 1994, pp. 49-57.
8. Casadei, Filippo and Delpero, Tommaso and Bergamini, Andrea and Ermanni, Paolo and Ruzzene, Massimo, "Piezoelectric resonator arrays for tunable acoustic waveguides and metamaterials", *Journal of Applied Physics*, Vol.112, No.6, 2012, pp. 064902.
9. Andreassen, Erik and Jensen, Jakob S, "Analysis of phononic bandgap structures with dissipation", *Journal of Vibration and Acoustics*, Vol.135, No.4, 2013.
10. Airoldi, L and Ruzzene, M, "Design of tunable acoustic metamaterials through periodic arrays of resonant shunted piezos", *New Journal of Physics*, Vol.13, No.11, 2011.

ICAST2015: 26th International Conference on Adaptive Structures and Technologies
October 14-16th, 2015, Kobe, Japan

ICAST2015 #069

Damping of Sandwich Panels via Acoustic Metamaterials

George A. Lesieutre*, and Tianliang Yu

Department of Aerospace Engineering, The Pennsylvania State University, University Park, Pennsylvania, USA

Abstract

Sandwich panels are often used in aerospace structures where high stiffness-to-weight is required, such as aircraft fuselage shells. Interior noise reduction in aircraft using such panels is a challenge because acoustic attenuation is reduced for light, stiff composite structures—especially those manufactured to have fewer mechanical joints. Acoustic metamaterials offer an approach to reducing the dynamic response of, and noise transmission through, sandwich panels. The key concept underlying this approach is to consider a metamaterial as a highly distributed system of tuned vibration absorbers that introduces one or more stop bands in which range the response of the global structure is reduced. The resonance frequency (or frequencies) of the absorber system may be tuned to match an excitation frequency and/or a global resonance frequency. Using the assumed-modes method, an initial metamaterial system was designed for integration into the honeycomb core of a representative sandwich panel; this design was refined using finite element analysis. To determine the dynamic response of the global sandwich panel, the metamaterial system was modeled as an effective distributed complex mass density. The cores for two sandwich panels were fabricated using 3-D printing technology, using a stiff polymer for the baseline honeycomb core, and a combination of a stiff and soft/lossy polymers for the metamaterial-augmented core. The cores were characterized statically to determine effective elastic properties, and dynamically to determine the natural frequencies and loss factors of the metamaterial system. Unidirectional carbon-fiber face sheets were bonded to both cores to construct sandwich panels. The sandwich panels were tested dynamically for two different boundary conditions, cantilevered and free-free. Experimental results confirmed that the metamaterial core reduced the peak dynamic response at the natural frequencies of the sandwich panel—typically by about 10 dB—with reasonably good agreement with model predictions.

1. INTRODUCTION

Sandwich panels, comprising stiff face sheets and a lightweight core, find extensive use in aerospace, civil and mechanical structures. Interior noise reduction in structures using such panels is challenging because acoustic attenuation is reduced for light, stiff composite structures, especially those manufactured to have fewer mechanical joints.[1] Damping can help by reducing structural dynamic response at resonance frequencies.

The constrained layer damping (CLD) treatment is a popular passive strategy used to reduce vibration in aerospace, civil and automotive structures.[2] CLD is a strain-based approach in that it requires dynamic strain in the base structure to function effectively. Sato, Shikanai & Tainaka examined the effects of tensile and compressive loads on the transverse vibration of beams with CLD,

* g-lesieutre@psu.edu

and demonstrated, analytically and experimentally, that modal damping could be significantly reduced by in-plane tensile loads.[3] The transverse vibration of pressurized shells and spinning helicopter rotor blades are two aerospace examples of structures that experience substantial tensile loads and reduced modal damping in operation.[4] Non-strain-based damping treatments are needed to reduce the structural dynamic response of pressurized fuselages.

Unlike strain-based treatments, inertial dampers respond to motion of the structure, whether or not that motion produces base strain. The classical inertial damper is a moderately-damped version of a tuned mass vibration absorber.[5] These are discrete devices that are typically effective over a narrow frequency band, which can be broadened in a relative sense by introducing dissipation into the spring. The effective range can be further widened by using multiple absorbers, each tuned to a slightly different frequency.

Several authors have examined inertial dampers distributed either in space or in frequency. Nashif, Jones & Henderson examined the effect of distributed dampers on beams,[5] whereas Smith, Rao & Dyer analyzed the transverse vibration of flat plates with distributed tuned absorbers.[6] Igusa & Xu showed, on a single-degree-of-freedom base structure, how multiple tuned mass absorbers tuned to different frequencies could provide broadband energy dissipation.[7]

The analysis of tuned mass systems, distributed *both* spatially and in frequency is a more recent area of research. In the context of fuzzy structures,[8] Pierce, Sparrow & Russell used a distribution of spring-mass systems to *model* a plate with internal attachments.[9] They showed how, as the number of absorbers becomes large, the effective mass of the absorbers can be approximated by a smooth frequency-dependent function.

Zapfe & Lesieutre considered a distributed (spatially and in frequency) tuned mass system as an inertial, broadband damping treatment.[10] The effectiveness of this non-strain-based approach was shown using a simply supported beam under tension. Kang, Smith & Lesieutre later demonstrated this approach experimentally for helicopter blade lag damping.[11]

Metamaterials are micro-architected materials that can exhibit unusual macroscopic properties—properties not generally observed in nature or in the constituent materials—materials that allow realization of unprecedented behavior. The use of the word "metamaterial" in the technical literature took off around 2000 when Smith *et al.* described a structured material with simultaneous negative permeability and permittivity—and hence negative refractive index—at microwave frequencies.[12] Because the wavelengths of such electromagnetic (EM) waves are very short, manufacturing subunits of appropriate size is challenging.

More recently, the analogy between acoustic and electromagnetic waves motivated research in metamaterials that addresses mechanical waves, which are generally much longer, hence ameliorating manufacturing difficulties.[13, 14] By properly tailoring fine-scale geometry, acoustic metamaterials can exhibit novel properties, including negative effective density and bulk modulus, albeit over limited frequency ranges.[15-18] Previous work on acoustic metamaterial absorbers is mostly theoretical, and is based on so-called "stop band" behavior.[19-21] Other researchers have addressed applications to acoustic enclosures.[22]

With potential for such properties, acoustic metamaterials offer a fresh approach to reducing the dynamic response of, and noise transmission through, sandwich panels. A metamaterial core with vibration absorbers could also be effective in the presence of tensile pressure loads where strain-based approaches would be less effective. As illustrated in Fig. 1, a key concept underlying this approach is to consider the metamaterial as a highly-distributed system of tuned vibration absorbers that introduces frequency bands in which the response of the global structure is reduced. In principle, the resonance frequency (or frequencies) of the absorber system may be tuned to match an excitation frequency (or a range of, or several, excitation frequencies) and / or to match one or more global resonance frequencies.

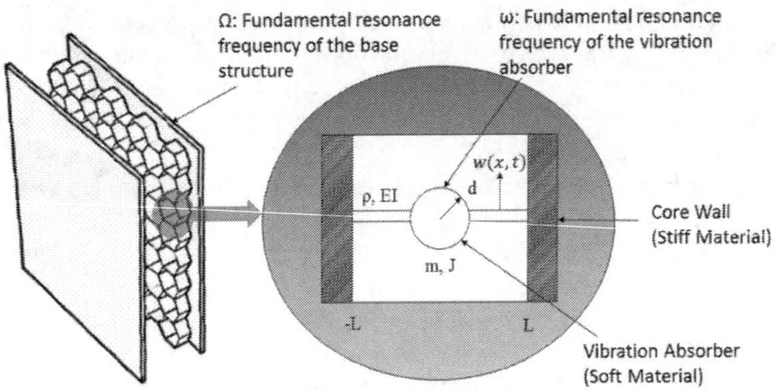

Figure 1. A concept for a honeycomb core with absorber system

The overarching goal of the research described herein was to conceive, design and implement acoustic metamaterials and demonstrate their utility in reducing the dynamic response of sandwich-panel structures.

2. MODELING AND DESIGN

2.1 Sandwich panels with integrated metamaterial cores

A first-order approach to integrating a SDOF metamaterial model into a sandwich panel model is to treat it as a frequency-dependent distributed mass term that affects the lateral translation of points on the panel midplane. This is most reasonable when the dimensions of a vibration-absorber unit are a small fraction of an effective wavelength of concern. For illustration of the approach, consider a rectangular Kirchoff plate with specially-orthotropic material properties, neglecting consideration of transverse shear for convenience, as shown in Fig. 2.

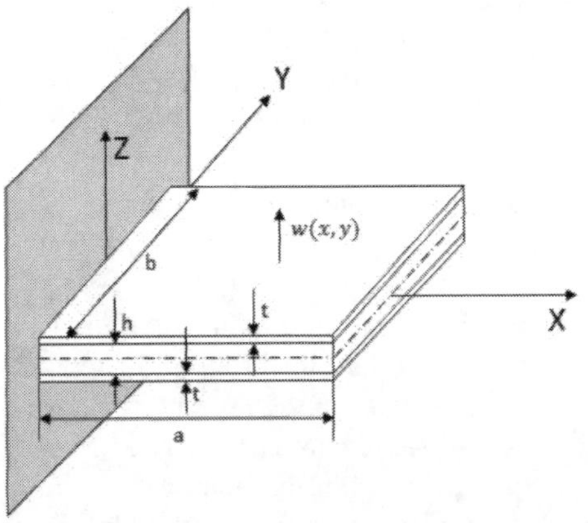

Figure 2. Transverse motion of a cantilevered sandwich panel

The governing PDE is:[23]

$$\rho h \frac{\partial^2 w}{\partial t^2} + \left(D_{11} \frac{\partial^4 w}{\partial x^4} + \left(2D_{12} + 4D_{66}\right) \frac{\partial^4 w}{\partial x^2 \partial y^2} + D_{22} \frac{\partial^4 w}{\partial y^4} \right) = -p\left(x, y, t\right)$$

Considering a system of N identical highly-distributed vibration absorbers having individual mass m, natural frequency ω, and damping loss factor η, as shown in Fig. 3, the equivalent effective mass per unit volume (of core), or effective density, is:[10]

$$\rho_{eff}(\Omega) = \frac{M_{eff}(\Omega)}{abh} = \frac{N}{abh}\frac{m\omega^2(1+i\eta)}{\omega^2(1+i\eta)-\Omega^2}$$

where h is the core thickness, and a and b are its length and width. Note that this effective density is presented as a function of the forcing frequency, Ω. It can be added directly to the density associated with the base sandwich panel when the equations of motion are expressed in the frequency domain.

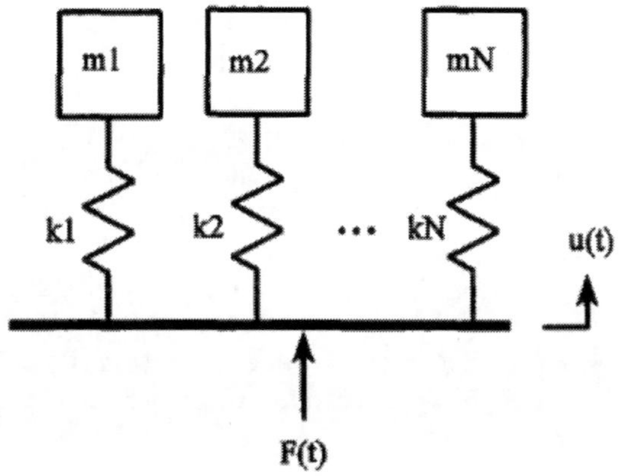

Figure 3. A system of N identical highly-distributed vibration absorbers

Discretized equations of motion for the sandwich panel can be obtained using an energy method such as the assumed modes or the finite element method. The assumed modes method was used in this work, considering a rectangular orthotropic plate of dimensions a x b, as described in (Yu, 2014).[24] Additional displacement degrees of freedom were included to allow for the possibility of an imperfect cantilever and base driving. The resulting equations of motion have the following canonical form:

$$[M]\{\ddot{q}\}+[K]\{q\}=\{F\}$$

If the equations of motion for the sandwich panel are expressed in the frequency domain, the effective mass of the absorber system can be included simply by adding it to that of the base plate. The discretized linear equations of motion are readily expressed in terms of frequency as:

$$\left[[K]-\Omega^2[M(\Omega)]\right]\{\tilde{q}(\Omega)\}=\{\tilde{F}(\Omega)\}$$

2.2 Metamaterial vibration absorber system

A first-order approach considers a vibration absorber as a single degree-of-freedom (SDOF) system. However, a real system will employ elemental units having distributed stiffness and mass, and thus exhibit multiple modes of vibration. In addition to achieving the desired dynamic behavior, key requirements for this study include ready integration with a sandwich panel core as well as co-manufacturability using 3D printing. Fig. 2 shows the focus concept for this investigation: a central mass suspended between opposite walls of a honeycomb cell via "bridging" beam-like springs.[24]

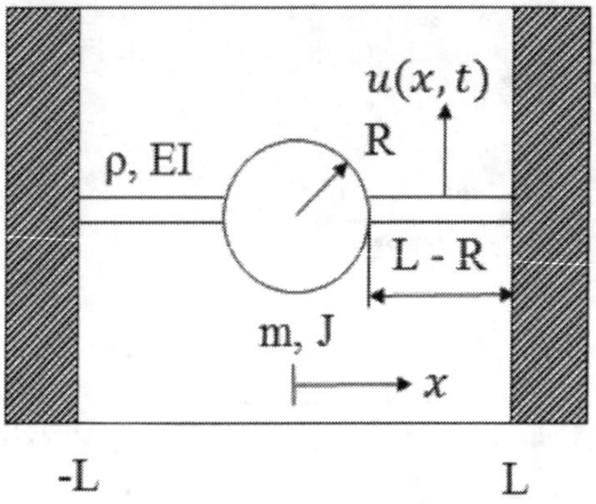

Figure 2: Vibration absorber "bridging beams with central mass"

Assuming material properties for DM 9860, a "digital" 3D-printing polymer developed for Objet PolyJet printers, the modal frequencies of the vibration absorbers were calculated using ANSYS. Fig. 3 shows the first four corresponding mode shapes and frequencies. The first mode involves symmetric bending and has a natural frequency of about 85 Hz. The second mode involves asymmetric bending and significant rotation of the central mass, at a frequency just under 200 Hz. The third and fourth modes are higher-order versions of the first two, at frequencies just below and just above 2000 Hz. The first mode was also estimated using the assumed modes method but, because it neglected transverse shear in the relatively soft "bridging beams," it significantly over-predicted the fundamental frequency.

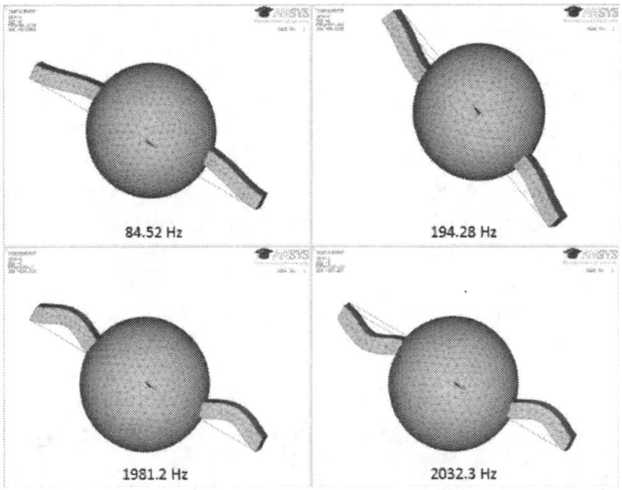

Figure 3. Modal frequencies and mode shapes of a bridging beam vibration absorber

In principle, each of these absorber modes can interact with the dynamics of the base sandwich panel, and all four were included in the system model using a separate effective density for each mode. An *ad hoc* method was used to distribute the absorber mass among the modes, as described in (Yu, 2014).[24] This method should be improved in continuing research.

2.3 Design of metamaterials systems and damped sandwich panels

The models described in the two preceding subsections were used to design representative sandwich panels, cores and metamaterials for potential applications of interest. A key challenge was designing the fine-scale features of the metamaterial—that would typically exhibit high resonance frequencies itself—to match the lowest frequencies of the global structure. At the same time, these designs should exploit known capabilities of 3D manufacturing materials and methods.

Using results from initial exploratory research, the concept shown in Fig. 6 for a metamaterial honeycomb core was developed for manufacturing purposes. It was designed to match an absorber natural frequency to that of one of the lower global panel modes (accepting that predictions of both are imperfect). This was accomplished by using carbon-fiber composite face sheets, a stiff polymer for the honeycomb and, as described in the previous subsection, a relatively soft polymer for the vibration absorbers.

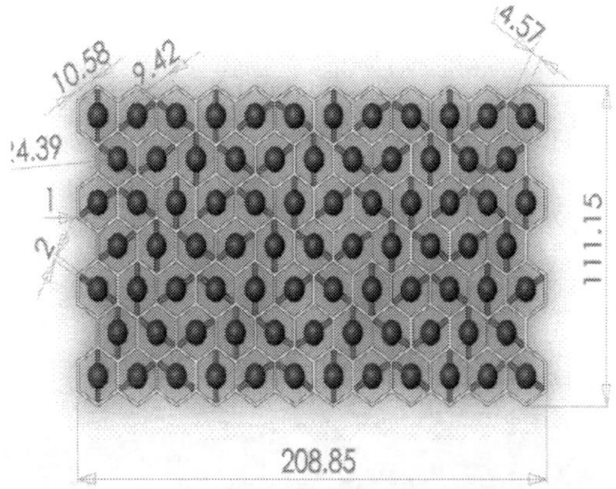

Figure 6. Design drawing of a metamaterial honeycomb core

Face sheets were also designed to complete the definition of a specific sandwich panel, as illustrated in Fig. 7. The face sheets have a very significant effect on the lowest natural frequencies of the global panel, which were specified to be representative of those in aerospace fuselage applications, typically several hundred Hz.

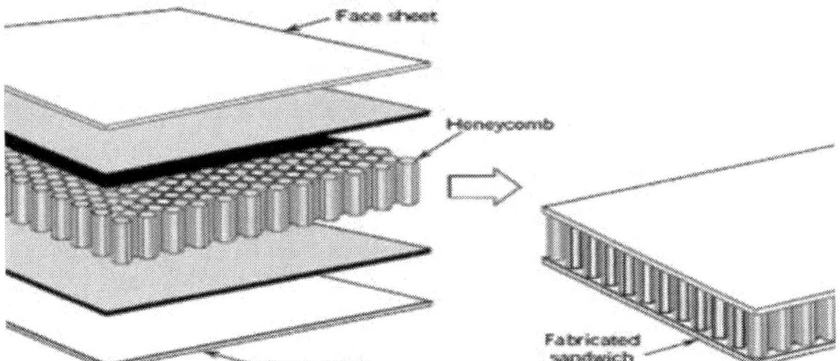

Figure 7. Design drawing of an integrated sandwich panel

Based on representative applications, sandwich panel size and boundary conditions were defined. These are essential elements of the definition of global structural modes and frequencies, which provide targets that guide the design of the metamaterial vibration absorber system and core. The global natural frequencies are especially sensitive to the boundary conditions.

3. FABRICATION AND TESTING

3.1 Metamaterial cores and integrated sandwich panels

The cores for two sandwich panels were 3-D printed using a stiff polymer for the baseline honeycomb core, and a combination of stiff and soft/lossy polymers for the metamaterial-augmented core. The metamaterial core was manufactured using a commercial Objet 260 ConnexTM, which is able to print two different materials simultaneously. Fig. 8 shows the finished specimens; (b) is the baseline core and (a) is the core with the integral metamaterial absorber system. Table 1 lists the material properties of the polymers that comprise the core. Both DM 8510 and DM 9860 are commercial digital materials developed based on Objet PolyJetTM technology.

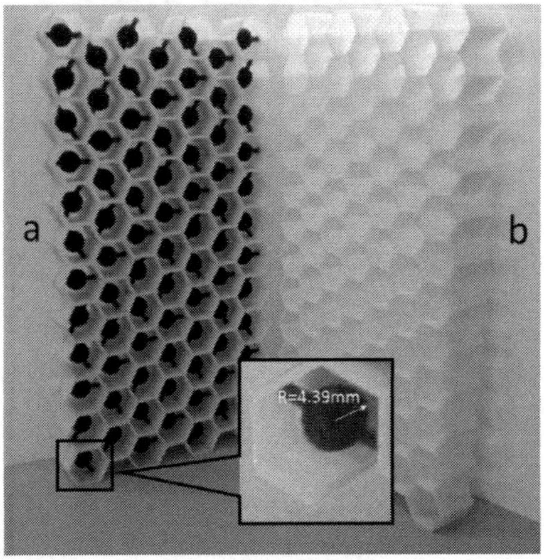

Figure 8. 3-D printed honeycomb cores with and without absorber system

Table 1. Material properties of the honeycomb core and vibration absorbers

	Honeycomb Core	Absorbers
Material	DM 8510	DM 9860
Hardness (Shore A)	N/A	60
Young's modulus (MPa)	2350	3.605
Density (g/cm^3)	1.16	1.13

Unidirectional carbon-fiber composite face sheets were then bonded to both cores to construct sandwich panels, as shown in Fig. 9. Table 2 lists the properties of the unidirectional material used.

Table 2. Material properties of the carbon-composite face sheets (t = 0.030 inch)

Longitudinal modulus, E_1	135 GPa
Transverse modulus, E_2	10 GPa
In-plane shear modulus, G_{12}	5 GPa
Density, ρ	1.6 g/cm3
Poisson's ratio, ν_{12}	0.3

Figure 9. Finished sandwich panels with unidirectional carbon fiber face sheets

3.2 Experimental characterization of metamaterial cores and sandwich panels

The two cores were characterized statically to determine their effective elastic properties, and then dynamically to determine the natural frequencies and loss factors of the integrated core/metamaterial system. The left hand side of Figure 10 shows the setup for the static test, in which the baseline core was treated as a continuous plate to determine the effective Young's modulus (E_{eff} = 18.8 MPa) and Poisson's ratio (υ = 0.32) of the honeycomb core layer. The right hand side of Figure 10 shows the setup for the dynamic test, from which the natural frequencies of the honeycomb cores and the loss factors of the vibration absorber system were determined. Although not shown here due to space limitations, the peak response is generally lower for the metamaterial core.

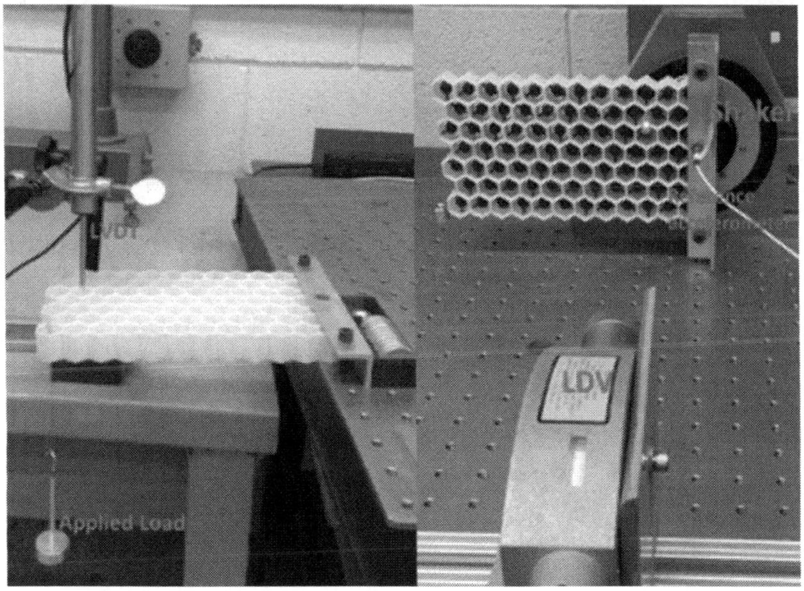

Figure 10. Static and dynamic set-ups for the honeycomb core tests

As shown in Figure 11, one side of a complete sandwich panel was clamped to simulate—to some extent—a cantilevered boundary condition. The center of the clamp was bolted to a shaker, which excited the root of the panel with a white noise signal. Two accelerometers were attached at the tip and the root of the panel respectively. The (transformed) output acceleration (tip) divided by the (transformed) input acceleration (root) yielded an experimental frequency response function (FRF) that could be readily compared to model predictions.

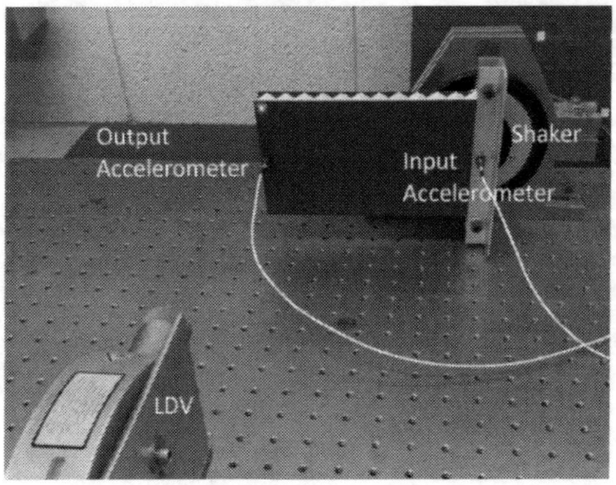

Figure 11. Set-up for sandwich panel dynamic test with cantilevered boundary condition

Since it is difficult to implement a true cantilevered boundary condition in the laboratory, especially for a stiff plate, the sandwich panel model was modified to include a stiff (but not infinite) rotational spring at the boundary, as well as a lateral degree of freedom to represent the excitation condition. The boundary spring value was selected to match the frequency of the fundamental mode of vibration.

4. RESULTS

Figure 12 compares the experimental and model FRFs. The four vertical dashed lines represent the modal frequencies of the absorber system shown in Figure 3.

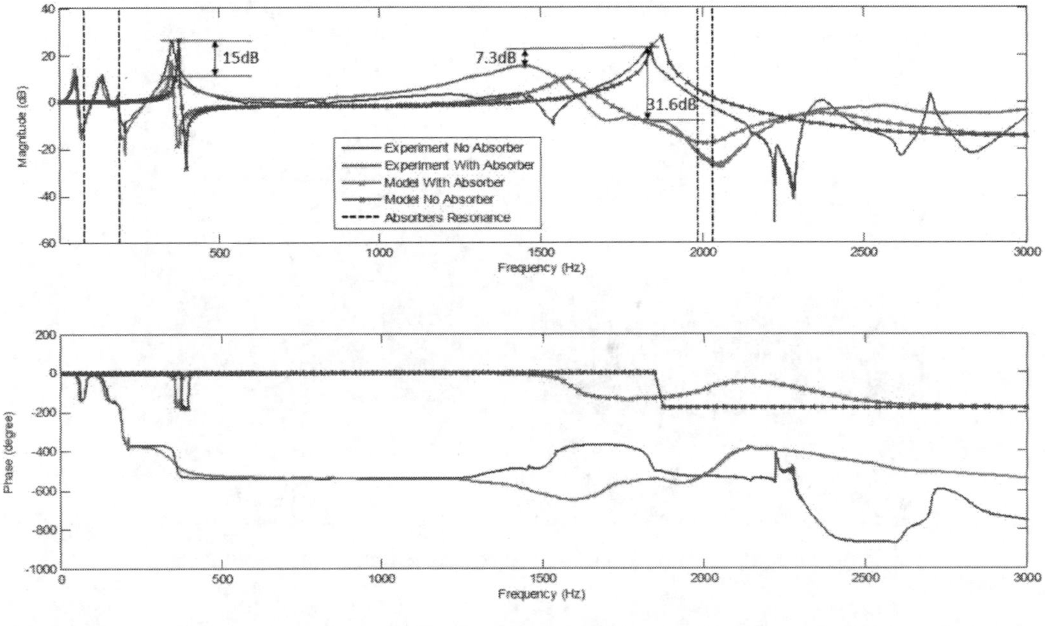

Figure 14. Comparison of the frequency response functions obtained from experiment and model

For the sandwich panel without absorbers, the model matches the experiment reasonably well in the range of 200Hz-2000Hz. The two peaks lower than 200 Hz in the experimental results represent resonance frequencies associated with the test bench; they are not affected by the panel damping treatment.

For the sandwich panel with absorbers, the peak responses near the natural frequencies of the metamaterial system are reduced; the peak response at 355Hz (without absorbers) is reduced by 15 dB, and the peak response at 1835Hz (without absorbers) is separated into two peaks which are 7.3dB and 31.6dB lower than the original one, respectively. For frequencies higher than 2000Hz, no dramatic reduction was observed, but the response curve was much smoother, with virtually no peaks in the response. This positive experimental result suggests the general efficacy of the approach.

The baseline analysis model includes mainly the effects of the fundamental absorber mode of vibration (~85Hz). As previously noted, in recognition that the higher absorber modes affect the measured response, this model was modified in a somewhat *ad hoc* way. In effect, the modeled mass of the absorber system was distributed among the four absorber modes in the frequency range of interest in order to improve the agreement.

5. CONCLUSIONS

Experimental and modeling results demonstrate that a metamaterial-augmented core can be effective in reducing the dynamic response of a composite sandwich panel. 3-D printing enabled the physical realization of a specific "bridging beam" concept for a vibration-absorber metamaterial system. For the structures considered in this study, the peak dynamic response was reduced by about 15 dB at the first mode, and about 5 dB at the second mode; this is consistent with the mode shapes and frequencies of the individual vibration-absorber units. Model predictions agreed reasonably well with experimental results, and there is potential to improve the agreement by incorporating better models of the higher-order modes of the absorber system. As a non-strain-based damping treatment, this approach has special promise for application to flexural structures that carry significant tensile membrane loads, such as pressurized airplane fuselages.

Future research will incorporate improved models of higher-order absorber modes in the sandwich panel model, consider distributions of absorber frequencies, absorber density, and absorber orientation to optimally address vibration reduction, consider the use of active materials for adaptive tuning, and demonstrate the approach in realistic applications, including manufacturing.

ACKNOWLEDGMENTS

This research was pursued without external funding support; the support of The Pennsylvania State University is appreciated.

REFERENCES

1. Liguore, S., Montgomery, J., and Foss, G., "Prediction and Measurement of Structural Loss Factors in Damped Composite Panels under Pressure Loads," AIAA 2008-2236, proc. 49th AIAA Structures, Structural Dynamics and Materials (SDM) Conference, Schaumburg, IL 7-10 April, 2008.
2. Ross, D., Ungar, E.E., Kerwin, E.M., Jr., "Damping of Plate Flexural Vibrations by Means of Viscoelastic Laminae," ASME Monograph on Structural Damping, 48-87, 1959.

3. Sato, K., Shikanai, G., and Tainaka, T., "Damping of Flexural Vibration of Viscoelastic Sandwich Beam Subjected to Axial Force," *Bulletin of Japan Society of Mechanical Engineers*, Vol. 29, No. 253, 1986, pp. 2204-2210.
4. Lesieutre, G.A., "How Membrane Loads Influence the Modal Damping of Flexural Structures," *AIAA Journal*, Vol. 47, No. 7, July 2009, pp. 1642-1646.
5. Nashif, A. D., Jones, D. I. G., and Henderson, J. P., Vibration Damping, Wiley, New York, 1985, pp. 189-257.
6. Smith, T. L., Rao, K., and Dyer, I., "Attenuation of Plate Flexural Waves by a Layer of Dynamic Absorbers," *Noise Control Engineering Journal*, Vol. 26, No. 2, 1986, pp. 56-60.
7. Igusa, T., and Xu, K., "Vibration Control Using Multiple Tuned Mass Dampers," *Journal of Sound and Vibration*, Vol. 175, No. 4, 1994, pp. 491-503.
8. Soize, C., 1986, "Probabilistic Structural Modeling in Linear Dynamic Analysis of Complex Mechanical Systems: I. Theoretical Elements," *La Recherche Aérospatiale*, Vol. 1986-5, pp. 23–48.
9. Pierce, A.D., Sparrow, V.W. and Russell, D.A., "Fundamental Structural-Acoustic Idealizations for Structures with Fuzzy Internals," *Journal of Vibration and Acoustics*, 117, 339-348 (1995).
10. Zapfe, J.A. and G.A. Lesieutre, "Broadband Vibration Damping Using Highly Distributed Tuned Mass Absorbers," *AIAA Journal*, Vol. 35, No. 4, April 1997, pp. 753-756.
11. Kang, H., Smith, E.C., and Lesieutre, G.A., "Experimental and Analytical Study of Blade Lag Damping Augmentation using Chordwise Absorbers," *Journal of Aircraft*, v 43, n 1, January 2006, pp. 194-200.
12. Smith, D. R., Padilla, W. J., Vier, D. C., Nemat-Nasser, S. C., & Schultz, S. (2000). Composite Medium with Simultaneously Negative Permeability and Permittivity. *Physical Review Letters*, 84 (18).
13. Wegener, M. (2013). Metamaterials Beyond Optics. *Science*, 342, 939-940.
14. Kadic, M., Bückmann, T., Schittny, R., & Wegener, M. (2013). Metamaterials beyond electromagnetism. *Reports on Progress in Physics*, 76 (12).
15. Li, J., & Chan, C. T. (2004). Double-negative acoustic metamaterial. *Physical Review E*, 70:055602(R).
16. Yao, S., Zhou, X., & Hu, G. (2008). Experimental study on negative effective mass in a 1-D mass-spring system. *New Journal of Physics*, 10 (4).
17. Huang, H. H., & Sun, C. T. (2009). Wave attenuation mechanism in an acoustic metamaterial with negative effective mass. *New Journal of Physics*, 11 (1).
18. Brunet, T., Leng, J., & Mondain-Monval, O. (2013). Soft Acoustic Metamaterials. *Science*, 342, 323-324.
19. Sun, H., Du, X., & Pai, P.F. (2010). Theory of Metamaterial Beams for Broadband Vibration Absorption. Journal of Intelligent Material Systems and Structures , 21, 1085-1101.
20. Claeys, C. C., Sas, P., & Desmet, W. (2014). On the acoustic radiation efficiency of local resonance based stop band materials. *Journal of Sound and Vibration*, 333, 3203–3213.
21. Pai, P. F., Peng, H., & Jiang, S. (2014). Acoustic metamaterial beams based on multi-frequency vibration absorbers. *International Journal of Mechanical Sciences*, 79, 195-205.
22. Claeys, C., Pluymers, B., Sas, P., Desmet, W. (2014). Design of a resonant metamaterial based acoustic enclosure. Proceedings International Conference on Noise and Vibration Engineering ISMA2014. Leuven, Belgium, 15-17 September 2014 (art.nr. 561) (pp. 3351-3358).
23. Timoshenko, S., & Woinowsky-Krieger, S. (1959). Theory of Plates and Shells. New York: McGraw-Hill.
24. Yu, Tianling (2015). Vibration Reduction of Sandwich Composites via 3-D Manufactured Acoustic Metamaterial Cores, M.S. Thesis, The Pennsylvania State University, August, 2015.

ICAST2015: 26th International Conference on Adaptive Structures and Technologies
October 14-16th, 2015, Kobe, Japan

ICAST2015 #92

An Unconventional Adaptive Flutter Suppression Actuation System: from Modeling to Experimentation

Mario Cassaro[1*], Manuela Battipede[1], Piero Gili[1], and Pier Marzocca[2]

[1] Department of Mechanical and Aerospace Engineering, Politecnico di Torino, Turin, Italy
[2] School of Aerospace, Mechanical and Manufacturing Engineering, RMIT University, Bundoora VIC 3083, Australia

Abstract

This paper contributes to the definition of an unconventional actuation system, intended specifically for slender/highly flexible wings' flutter suppression, and coupled with an adaptive control algorithm to handle post-flutter flight dynamics and uncertainties deriving from unpredictable degradation of the structural properties. The design and validation process of the novel actuation architecture is presented. It is based on a row of multiple small spoiler, located at fifteen percent of the mean aerodynamic chord and coordinated by a modified Model Reference Adaptive Control (MRAC) algorithm. The spoilers' concept design is optimized by Computational Fluid Dynamics (CFD) numerical simulation, afterwards realized and wind tunnel tested to derive the aerodynamic database by means of a six-axes force balance. The mathematical model has served to implement and validate the adaptive control algorithm for a wide range of condition. The modeled system is tested from on-design flutter speed and nominal structural stiffness to post-flutter speed and reduced structural stiffness through the analysis of the wing proper frequencies and phases, which has demonstrated to be very effective in testing the adaptivity of the control architecture. This approach proves the robustness of the proposed architecture before experimentation, which is performed through a custom-made wind tunnel apparatus. The two degree of freedom oscillations during flutter are successfully controlled in all conditions. This paper aims at defining a robust procedure for aeroelastic phenomena control system design, which employs a synergy of modeling, simulation and experimental approaches. Discussions and pertinent conclusions are outlined in the final section of the paper.

1. INTRODUCTION

The aeroelastic instabilities and their control are topics that attracted considerable attention in the aerospace scientific literature in the last decade. In particular, aeroelastic phenomena such as divergence, flutter, and LCOs have been carefully investigated [1-3] and, for many cases, the instability boundaries were determined with reasonable accuracy, using theoretical and experimental techniques for various operation conditions. However, unpredictable structural damages can lead to unexpected aeroelastic behavior within the nominal flight envelope, which explains the necessity to actively control the response of the lifting surfaces to prevent any potential aeroelastic-derived failure. In these circumstances, changes in the structural properties modify the nonlinear system dynamic response, which requires robust adaptive control law to be handled properly. Different research teams have proposed different control laws and architectures, designed and tested both numerically, in

* mario.cassaro@polito.it

simulation environment, and experimentally, with wind tunnel experimentation or flight-testing. In [4-6] the authors developed and tested their control algorithms for the Benchmark Active Control Technology (BACT) Wind-Tunnel Model, whereas in [7-9] the authors designed and tested their control laws for the Nonlinear Aeroelastic Testbed Apparatus (NATA) [10] at Texas A&M University. In [11] the authors tested and validated several adaptive control architectures on similar 2D aeroelastic model. The control systems developed, including the most recent neural network [12], adaptive back-stepping [13] and $\mathcal{L}_1$ adaptive controller [14], have been applied to wing models, in simulation or experimental setup, with either a single trailing edge control surface or a combination of leading and trailing edge control surfaces. Some issues related to these solutions can be highlighted: they range from trailing edge flap saturation and command inversion to the problem of the manufacturing complexity of the leading edge slat actuation system. In detail, command saturation is critical, as the pilot might be unable to control and maneuver the aircraft, if the control surface is operating to suppress flutter. Moreover, command inversion is critical since a control law, designed at nominal condition, and associated to the flap, would not be able to act properly on the system and damp out the oscillation.

The present work reports the investigation on the effectiveness of an original configuration, which would represent a possible alternative and a solution to the previous mentioned issues. Specifically, a novel coordinated multiple spoiler actuation strategy is evaluated for flutter suppression, which, to the best of the authors' knowledge, was not attempted before. The innovative test article consists of a wing section, equipped with multiple-spoilers installed on a strip, located at the 15% of the chord length, with a predefined and coordinated actuation strategy. The proposed architecture, optimized through computational fluid dynamics (CFD) analysis, is manufactured and tested in the wind tunnel to validate the aerodynamic properties of the wing section. The experimentally obtained nonlinear aerodynamic database is used in conjunction with a classical nonlinear plunging/pitching structural model, implemented in a simulation environment to investigate primarily the dynamic response of the proposed wing configuration aeroelastic model. The resulting aeroelastic model exhibits a supercritical Hopf-bifurcation behavior with a stable Limit Cycle Oscillations (LCOs) occurring past the flutter speed. The open and closed loop responses of the system are investigated as to consistently proceed with the flutter suppression wind tunnel test campaign, to validate the novel architecture in a real world application. The experimental setup, consisting of a wing section with multiple spoilers, installed to a plunging and pitching apparatus, similar to the one reported in [10], was designed to exhibit low speed flutter. This provides a solid benchmark for a comparative evaluation of the obtained results. Preliminary control investigations consider a simple PID control strategy exploited to assess the overall capabilities of the wing-spoiler system to suppress flutter and control post-flutter. Consequently, the regulation problem is also formulated, in simulation environment, for a normalized Model-Reference-Adaptive-Controller (MRAC) scheme, modified for performance improvement. The obtained results demonstrate the efficacy and solidity of the overall configuration investigated.

2. SPOILER ACTUATION: FROM CONCEPT TO PROTOTYPE

The novel control surface architecture is conceived primarily to reduce lift and to generate an opposite sign pitching moment with respect to the standard spoiler configuration. From the initial concept, the final actuation configuration is achieved through several optimization steps. CFD analysis has been carried out to optimize the spoiler parameters, evaluating the effect in terms of aerodynamic coefficient variation. Once optimized, the available technology and budget have contributed to the decision of the material and actuator system employed on the final wing prototype. An extensive set of CFD simulations are needed in order to optimize the spoiler effect, evaluated in terms of aerodynamic coefficient. The aerodynamic numerical solver is the CFX module of ANSYS Workbench ver. 14.0. The wing section airfoil is a NACA 0024 with a chord length of 0.2 meters. A thick airfoil is required to accommodate the actuation mechanisms. The three-dimensional wing model has a limited depth, with a spoiler spanning from one side to the other side, compatible with the CFD investigation. The flow field is fully developed in a domain that is 10 chords long and 5 chords height (Figure 1a). The boundaries of the computational domain are far enough from the wing surface to cause negligible effect on the aerodynamic characteristics of the wing. The flow field has been discretized with a numerical mesh adequately dimensioned to obtain reliable results: if for the clean wing section configuration about 850,000 elements must be considered, more than 1 million elements are needed in meshing the domain which includes the wing-spoiler. Even if an unstructured mesh usually results in a lower fidelity than a structured mesh, it provides a computationally efficient solution to the problem at hand. In this investigation a triangular mesh was adopted: it simplifies the mesh generation around the spoiler edges, making the simulation more reliable. Figure 1a and Figure 1b show the generated numerical mesh.

The macroscopic effect of the spoiler location is verified to assess the validity of the assumption on which the forward spoiler idea has grown. A constant setting of wind velocity vector equal to $[\mathrm{u\,v\,w}] = [15\ 0\ 2](\mathrm{m/s})$, which makes the angle of attack (AoA) constant at $7.59\ (\mathrm{deg})$, is used to give a rough estimation of the coefficient trend for three different spoiler positions. Precisely, the spoiler that have the same dimension of $15\ (\mathrm{mm})$ and opening angle of $54\ (\mathrm{deg})$, is analyzed in the rear (80%), middle (50%) and front (10%) positions, measured with respect to the chord length. Results, reported in Figure 2a, demonstrate that the pitching moment coefficient experience sign inversion when the spoiler location is moved forward crossing the zero value when they are located in the middle. The well-known lift reduction and drag increment effect is reproduced. Once the desired effect and the corresponding spoiler position has been obtained, a fine tuning is performed to optimize the spoiler shape.

The spoiler is parametrized as a function of size, opening angle, and position along the chord as shown in Figure 2b. The mutual combination of these parameters is analyzed in the selected range of variation:

- Spoiler position: ranging from 10% to 30% of the chord length;
- Spoiler dimension: ranging from 10 to 25 (mm);
- Spoiler opening angle: ranging from 30 up to 90 degrees.

The boundary conditions are set as in Figure 1d, the bottom and front surface of the domain are considered inlet, the top and rear surface outlet. The velocity vector is maintained constant so as to reproduce an averaged values of the wing AoA during LCOs. The turbulence intensity is set at a medium level (5%) and ambient sea level conditions are used for the flow-field. The outlets static relative pressure was set to be 0 (Pa) and the flow is subsonic in the whole computational domain. The simulation is representative of an infinite wingspan with a two-dimensional flow by imposing symmetry on the two side walls of the computational domain. A standard roughness is set for the airfoil surface. The reference pressure is 101,325 (Pa), and the turbulence model used is the *Shear Stress Transport* model, a modified $k - \omega$ turbulence model. Transitional turbulence option is enabled for more accurate low-speed computation. The convergence is smoothly reached after 350 iterations tipically. The quality of the results is verified by checking the Y+ on the surface of the wing. Although $Y+ < 1$ is preferable, to reach a compromise between accuracy and computational efficiency, a Y+ under 25 was considered acceptable. This is an adequate choice considering the preliminary nature of this parametric study in the spoiler design process. Some examples of the results achieved in terms of pressure distribution are reported in Figure 3. The spoiler optimization process is performed iteratively by evaluating the effect of a single parameter modification per time. The other two parameters are maintained constant meanwhile. The effect on the aerodynamic coefficients of the spoiler length modification is evaluated starting from a standard spoiler configuration, defined as: position at 70% of the chord length, opening angle of 30 (deg) and spoiler length of 0.8 (mm). From the same starting condition, the same process is repeated for the opening angle parameter. Finally, the spoiler position is optimized maintaining constant the other two parameters at their already selected values. Figure 4 reports the main results of the investigation and the chosen final configuration, highlighted by the red square boxes. Despite a negligible effect on ΔC_m provoked by the opening angle variation, the 90 (deg) solution (which will become 85 (deg) when prototyped) is selected. This choice relies on the consideration that the concurrent effect of lift reduction is beneficial during flutter. The same argumentation is valid for the choice of the spoiler position parameter. Once the desired effect on the lift and pitching moment coefficient is obtained, the smallest length considered in the analysis is chosen. This is justified by the necessity of avoiding undesired coupling between flutter and lag motion during tests, caused by excessive drag acting on the opened spoiler.

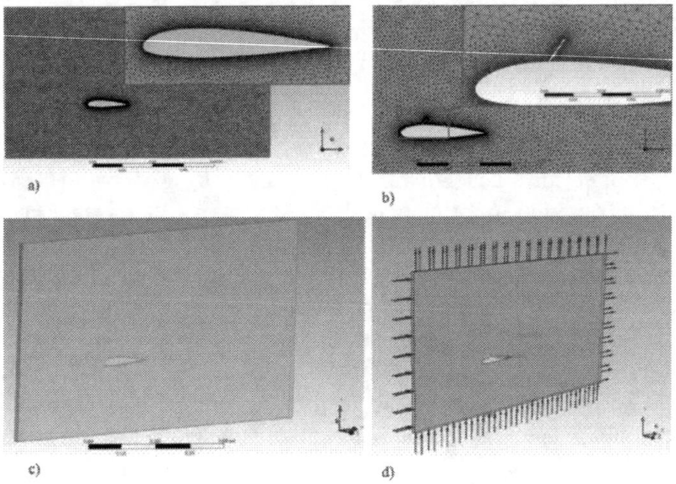

Figure 1. CFD parameter setup: a) Numerical mesh of clean configuration for CFD simulation; b) Numerical mesh of active spoiler configuration for CFD simulation; c) CFD simulation domain; d) CFD boundary conditions.

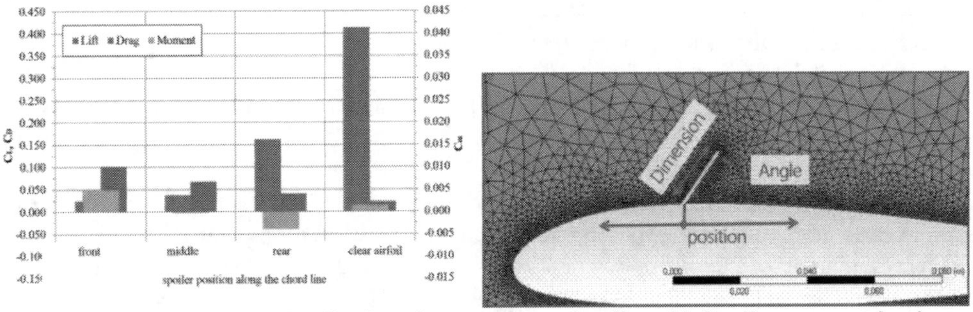

Figure 2. a) Spoiler location macroscopic effect; **b)** Spoiler parametrization

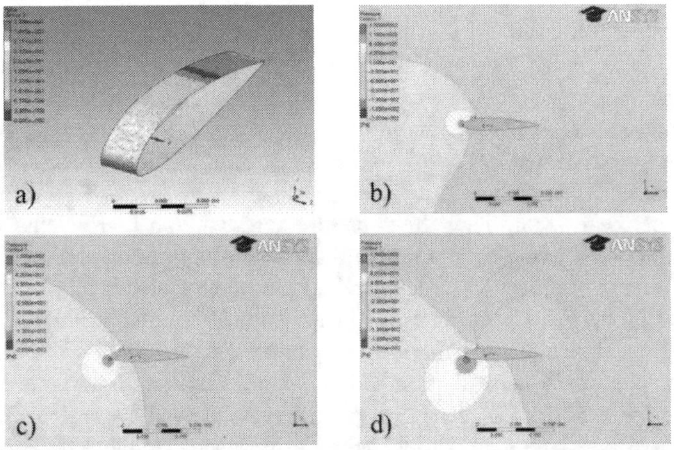

Figure 3. a) Distribution of Y plus on wing surface; **b)** Pressure distribution with active spoiler, AoA 8; **c)** Pressure distribution with active spoiler, AoA 14; **d)** Pressure distribution with active spoiler, AoA 18.

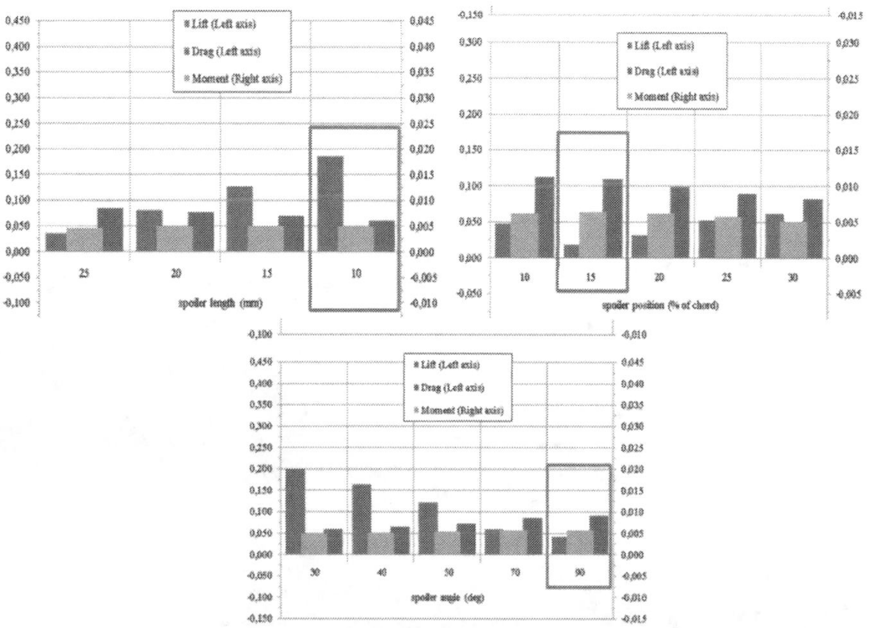

Figure 4. Spoiler optimization results.

Based on the results obtained by CFD simulation, the wing section prototype has been designed and manufactured. The schematic, reported in Figure 5a, shows the overall architecture designed to be prototyped. The geometrical characteristics are:

- Wing section chord length: 0.2 (m);
- Wing section spanwise: 0.3 (m) ;
- Wing section airfoil: NACA 0024.

The materials' choice is operated to meet the structural solidity requirements, during LCOs, and to accommodate appropriately the solenoidal actuators. In particular, the inner structure of the wing prototype is made up by twenty wooden carved wing sections. They are 1 (cm) thick, properly milled, lightened and cut to host the actuation system. Wood has been chosen because of its structural properties, cost and ease of processing. The small wing sections, once polished, are glued and held together by two aluminum bars, located at 20% and 80% of the chord length. The bars are threaded at the tip to obtain compression by tightening the flat bolts acting on the external wing sections. The actuation system is embedded in the wing model. It is composed by push type solenoids, where the strokes acts on an aluminum plate, which is the actual spoiler. When the solenoid is powered, it pushes out the stroke deflecting the spoiler at 85 (deg). The spoiler is realized by two hinged aluminum plates, one of which is glued to the wing surface and the other is free to rotate. A small return spring guarantees closure motion when the solenoid is not powered. The spoiler hinge is located to comply with the CFD optimization process results. It is embedded in the wing surface by an appropriate groove. The wooden wing surface is treated with a thin layer of plaster and then covered by a Monokote® film. This components are employed to guarantee the correct airfoil shape and a low level of surface roughness. When fully deployed the five spoilers operate across the entire wing section span. However, as the control action required is not always maximal, a pre-defined coordinated opening strategy has been implemented. The strategy is intended to preserve pressure distribution symmetry on the wing upper surface, so as to avoid undesirable rolling moment. Numbering the spoilers progressively, 1 to 5, from left to right along the wing section span, as in Figure 5a, the opening strategy sequence is: spoiler #3, #2-4, #1-3-5,#1-2-4-5 and finally #1-2-3-4-5. The final prototype is shown in Figure 5b during wind tunnel static tests.

The predefined opening strategy is realized by implementing a control algorithm on an Arduino board, which is used both for sensing and controlling, in the aeroelastic dynamic experiment. The actuators are wired to a custom made electric board with optical diode, which serves as protection bridge to avoid the higher supplying current to flow back to the Arduino board. The electric wires run inside the wing and are placed to minimize the interference with the wing section aerodynamics during wind tunnel test.

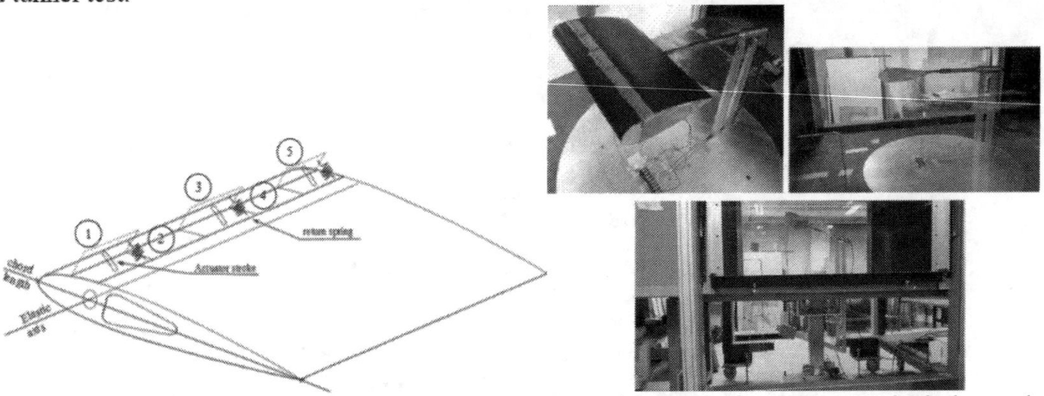

Figure 5. a) Wing-Spoiler configuration, prototype schematic.; **b)** Final prototype and wind tunnel static experiment setup

3. WING AEROELASTIC MODEL

The proposed solution allows the individual actuation of spoilers to provide a range of lift and moment increments, ΔC_L and ΔC_m in terms of coefficients, differently from the operation of a gradually activated single spoiler. The row of small spoiler has also a predefined opening strategy designed to preserve symmetry in the airflow on the wing upper surface [15]. A structural model is associated to the wing for aeroelastic control law design purpose. The schematic view of the aeroelastic configuration described is reported in Figure 6, where the i_{sp} parameter is the number of active spoiler. The plunging and pitching displacements are restrained by a pair of springs attached to the elastic axis (EA) of the airfoil with spring constants, k_h $(N \cdot m^{-1})$ and $k_\theta(\theta)$ $(N \cdot rad^{-1})$, respectively. Specifically, the plunging h (m), is positive downward whereas the pitching θ (rad) is positive nose up and measured from the horizontal, at the elastic axis of the airfoil. Here, $k_\theta(\theta)$ denotes a continuous, linear parameterizable nonlinearity, that is, the aeroelastic system has a continuous nonlinear (3rd-order) restoring moment in the pitch degree of freedom (DOF). As explained afterward, the deduction and incremental values of lift and pitching moment are functionally related to the i_{sp} parameter and the spoiler effectiveness factor $\left(k_{\delta_{sp}}\right)_L$, which takes into account of the selected spoiler opening sequence. Additionally, it should be noted that the actuator working frequencies are one order of magnitude faster than the aeroelastic properties of the model analyzed and for this reason they will be neglected in the model dynamic equations. The aerodynamic model, coupled with the structural one, is referred to the actual wing section prototype. In fact, the additional value of this approach consists of not describing the aerodynamic force and moment by any preexisting analytical unsteady or quasi-steady form [16-20] but rather, using an alternative approach, which is common in flight simulation [21], based on the use of the aerodynamic database experimentally obtained by wind tunnel testing.

This approach allows the inclusion of the stall driven nonlinearities and the modeling of the effects of the spoiler activation strategy on the aerodynamic loads. The equations describing the aeroelastic model in Fig. 1 are:

$$\begin{bmatrix} m_t & m_w x_\theta b \\ m_w x_\theta b & I_\theta \end{bmatrix} \begin{bmatrix} \ddot{h} \\ \ddot{\theta} \end{bmatrix} + \begin{bmatrix} c_h & 0 \\ 0 & c_\theta \end{bmatrix} \begin{bmatrix} \dot{h} \\ \dot{\theta} \end{bmatrix} + \begin{bmatrix} k_h(h) & 0 \\ 0 & k_\theta(\theta) \end{bmatrix} \begin{bmatrix} h \\ \theta \end{bmatrix} = \begin{bmatrix} -L \\ M \end{bmatrix} \tag{1}$$

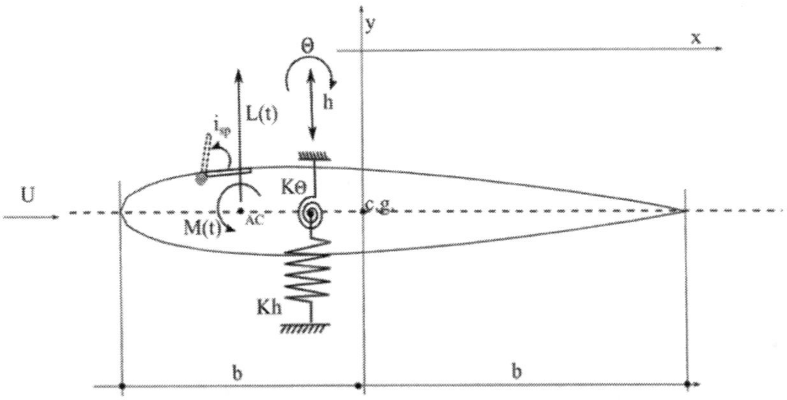

Figure 6. 2D aeroelastic wing-spoiler configuration schematic

where $m_t\ (kg)$ is the total mass of the system, $m_w(kg)$ the mass of the airfoil, x_θ the dimensionless distance measured from the elastic axis to the center of mass, $I_\theta\ (Kg \cdot m^2)$ the mass moment of inertia of airfoil about elastic axis, $c_h\ (N \cdot m^{-1} \cdot s^{-1}), c_\theta(N \cdot s^{-1})$ the structural damping coefficients in plunging and pitching. The external aerodynamic force $L\ (N)$, and moment, $M\ (N \cdot m)$, are expressed as a function of the lift and pitching aerodynamic moment coefficient and are written by the classical formulation per unit of length:

$$L = \rho U^2 b C_L\left(\alpha_e, Re, i_{sp}, w_G(\tau)\right),$$
$$M = \rho U^2 b^2 C_m\left(\alpha_e, Re, i_{sp}, w_G(\tau)\right) \tag{2}$$

where $\alpha_e\ (rad)$ is the effective aerodynamic AoA, Re the Reynolds number and $w_G(\tau)$ the gust velocity profile.

The experimental aerodynamic coefficient dataset, obtained by means of a six axes wind tunnel force balance, are reported in **Error! Reference source not found.**. The measured aerodynamic AoA domain is $\pm 18\ deg$. In the figure, the lines represent configurations going from 0 (clean) up to all active 5 spoilers. As it can be noticed, with this approach, nonlinear stall region of the wing and the nonlinear effect of the spoiler activation are implicitly included in the formulation. The wing section in post-flutter conditions exhibits LCOs, hence it is assumed that the aerodynamic AoA α, used to interpolate the data can be written as:

$$\alpha = \left[\theta + \frac{\dot{h}}{U} + \left(\frac{1}{2} - a\right) b \left(\frac{\dot{\theta}}{U}\right)\right] \tag{3}$$

where θ is the geometrical pitch angle, a and $b\ (m)$ are respectively the dimensionless distance between the mid-chord and the elastic axis and the semi-chord of airfoil, and the contributions of both the plunge and pitching velocities are taken into account. The analytical expressions of the lift and aerodynamic moment, factoring the experimentally evaluated coefficients, are:

$$L = \rho U^2 b C_{L_\alpha}\left[\theta + \frac{\dot{h}}{U} + \left(\frac{1}{2} - a\right) b \left(\frac{\dot{\theta}}{U}\right)\right] - \rho U^2 b \Delta C_{L_{sp}} \tag{4}$$

$$M = \rho U^2 b^2 C_{m_\alpha}\left[\theta + \frac{\dot{h}}{U} + \left(\frac{1}{2} - a\right) b \left(\frac{\dot{\theta}}{U}\right)\right] + \rho U^2 b^2 \Delta C_{m_{sp}} \tag{5}$$

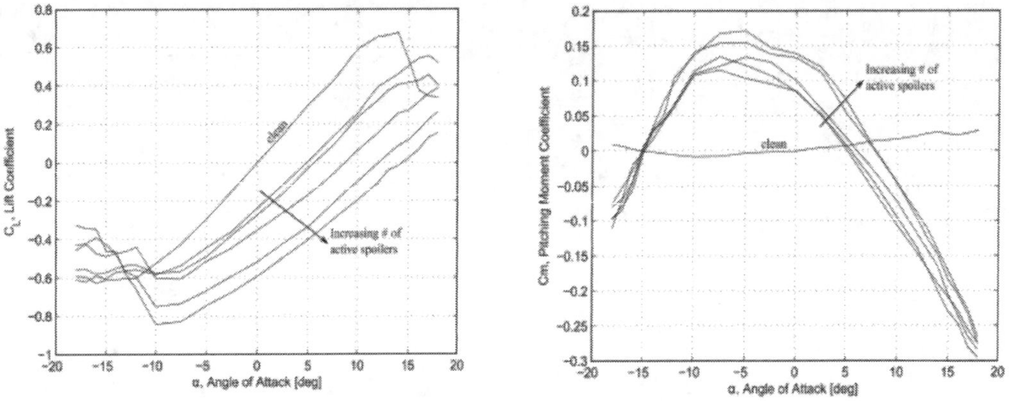

Figure 7 - Aerodynamic database: lift and pitching moment coefficient experimental values. Transition from clean to five open spoiler configuration is specified by the arrows

where U (m/s) is the free stream velocity, c_{l_α} c_{m_α} are lift and moment coefficients of the airfoil and ρ (Kg/m^3) is the air density. Herein, the contribution to lift and pitching aerodynamic moment of the spoiler is written as a function of the wing angle of attack, flight speed and number of opened spoilers. The general formulation describing the spoiler effect on lift coefficient is:

$$\Delta C_{L_{sp}} = \sum_{i=i_{sp}} \left(k_{\delta_{sp}}\right)_i \left(\Delta C_{L_{sp}}(\alpha)\right)_{\delta_{max}} \cdot \frac{(C_L)_M}{(C_L)_{M=0}} \left(\frac{L_E}{L_R}\right)_{sp} \tag{6}$$

where $\left(k_{\delta_{sp}}\right)_L$ is the single spoiler lift effectiveness factor, $\left(\Delta C_{L_{sp}}(\alpha)\right)_{\delta_{max}}$ is the single spoiler lift increment at maximum deflection for given AoA. The term $\frac{(C_L)_M}{(C_L)_{M=0}}$ takes into account the compressibility, whereas $\left(\frac{L_E}{L_R}\right)_{sp}$ considers the flexibility of the wing section. The summation is extended to the number of operating spoiler panels, i_{sp}. As for the prototyped wing configuration, the spoiler works in a binary on/off mode, with maximum and unique allowed deflection at $85\,deg$. The airflow is uncompressible and the wing properties are lumped in a single 2D section, which is rigid by definition. Therefore, equation (6) reduces to

$$\Delta C_{L_{sp}} = \Sigma_{i=i_{sp}} \left(k_{\delta_{sp}}\right)_i \Delta C_{L_{sp}}(\alpha) \tag{7}$$

Analogous considerations are applied to the pitching moment coefficient, for which the final equation has exactly the same form:

$$\Delta C_{m_{sp}} = \sum_{i=i_{sp}} \left(k_{\delta_{sp}}\right)_i \Delta C_{m_{sp}}(\alpha) \tag{8}$$

The nonlinear trend of the spoiler aerodynamic increments is attributed to the nonlinearity of the effectiveness factor, which varies with the number of the open spoilers, their location and consequently to the opening strategy [15]. The proposed aeroelastic model, with the parameters set as in Table 1, exhibits periodic LCOs beyond a critical value of the free stream velocity U in the absence of wing gust, that is $U_{flutter} = 7.5\,m/s$. The uncontrolled plant response in both DOF is reported in Figure 8 in the form of phase diagrams. Obviously, in this case, C_L and C_m are allowed to assume values belonging only to the clean configuration. In line with the main purpose of this work, that is the adaptive control of flutter phenomena, attention is focused on the off-design condition of the plant. Hence, the dynamic response of the plant is analyzed also under failure conditions and post-flutter speed. Figure 8 demonstrates that a stiffness reduction of 50%, $K_{reduced} = K_{nominal}/2$, simulating an unpredictable wing structure failure, affects the dynamic response of the system differently from the post flutter condition. This explain the necessity of validating the control architecture in both configurations.

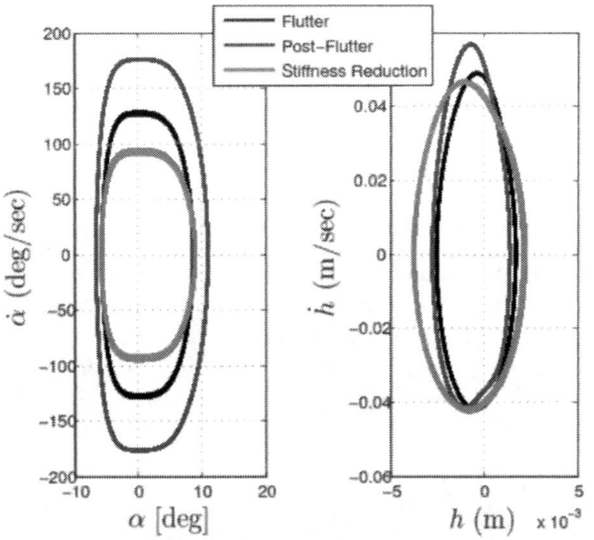

Figure 8. Uncontrolled LCOs response of the aeroelastic plant, pitch and plunge phase diagrams

A Model Reference Adaptive Control (MRAC), modified for performance improvement, is formulated for the regulation problem and implemented. The aeroelastic equation of motion (1) belongs to a class of plant that can be rewritten in an amenable form to model reference control schemes as:

$$\dot{x}(t) = A_m x(t) + B\theta^{*T} x(t) + Bu(t) \tag{9}$$

where $x \in \mathbb{R}^n$ is the state vector, assumed to be measurable, $u \in \mathbb{R}$ is the control input, $\theta^* \in \mathbb{R}^n$ is an unknown parameters vector belonging to a known compact convex set $\Omega \subset \mathbb{R}^n$, $A_m \in \mathbb{R}^{n\times n}$ is Hurwitz and the pair A_m; B, with $B \in \mathbb{R}^n$, is known and controllable. The controller derivation is based on a parametric expression of (9) and the adaptive and control laws work on normalized signals, which increase the robustness of this scheme. Equation (9) can be rewritten as

$$x(t) = W_b(s)[\theta^{*T} x(t)] + W_b(s)[u(t)] \tag{10}$$

where $W_b(s) = (sI - A_m)^{-1}B$ Based on (A_m, B) controllability, a vector $c_0 \in \mathbb{R}^n$ is defined so that $W_m(s) \triangleq c_0^T W_b(s)$ is a strictly proper minimum-phase transfer function. The parametric expression of the model is consequently written as

$$z(t) = \theta^{*T}\phi(t) \tag{11}$$

where $z(t) = C_0^T x(t) - W_m(s)[u(t)]$ and $\phi(t) = W_m(s)[x(t)]$ are available for measurements. The structure and parameters of the unmodeled dynamics are assumed unknown. Performance improvement is obtained by enriching the control signal with an auxiliary input signal u_a, which involves filtering a feedback signal, as follows:

$$u(t) = -\hat{\theta}^T(t)x(t) + k_0 r(t) + u_a(t) \tag{12}$$

where

$$u_a(t) = -Q(s)\left[\epsilon m_s^2 + W_{c0}(s)[\, W_b(s)[x^T(t)]\dot{\hat{\theta}}(t)]\right] \tag{13}$$

Herein, $W_{c_0}(s) = -c_0^T(sI - A_m)^{-1}$, $Q(s) = W_m(s)^{-1}/(\tau s + 1)^{n^*}$, and n^* is the relative degree of $W_m(s)$, while $\tau > 0$ is a design parameter. The associated adaptive algorithm is expressed as:

$$\dot{\hat{\theta}}(t) = \text{proj}\big(P(t)\epsilon(t)\phi(t)\big), \qquad \hat{\theta}(0) = \hat{\theta}_0 \in \Omega \tag{14}$$

$$\dot{P}(t) = -P(t)\frac{\phi(t)\phi^T(t)}{m_s^2}P(t), \qquad P(0) = P_0 \tag{15}$$

where ϵ is the estimation error, defined as $\epsilon(t) = \left(z(t) - \hat{\theta}^T\phi(t)\right)/m_s^2$. The signal $m_s^2(t) = 1 + \phi^T(t)\phi(t)$ is a normalizing term: it is designed to guarantee boundedness of $|\frac{\phi(t)}{m_s(t)}|$, independently whether $\phi(t)$ is bounded or not. Stability and convergence proof of this robust adaptive law are reported in [22]. The spoiler actuation system accept a discrete value ranging from 0 to 5 as input. For this reason a conversion look-up table, acting on the control signal, was introduced to convert the continuous control value in the integer number of open spoilers that produces the corresponding aerodynamic effect.

4. SIMULATION RESULTS AND DISCUSSION

Hereafter only an example will be given of the extensive set of simulation that has been performed to verify and validate the performance of the proposed adaptive flutter suppression actuation system. Results are presented recording and analyzing for each test case the state variable $[\alpha\ h]$, their derivatives $[\dot{\alpha}\ \dot{h}]$ and the control signal time histories. Figure 9 highlights that the problem is well posed and the aeroelastic system belongs to the class of plant controllable by the applied modified MRAC scheme. This is verified by performing a wide set of simulations where convergence is always reached. The regulation takes place with a smooth and constantly decreasing motion. This is a consequence of the single-side action of the spoilers, as the implemented aeroelastic model refers to the wing section designed and prototyped, where the row of multiple spoilers is installed only on the wing upper surface. The regulation time of about 2 seconds remains acceptable, considering the low frequency flutter phenomena for which this actuation solution has been thought. The subplots denoted by the letter *(e)* of Figure 9, show the time histories of the control signal, which is made up of discrete values, between 0 and 5, representing the actual number of open spoilers. This is obtained by bounding the signal with a saturation operator and explains the discontinuities encountered in the time histories. One additional effect of the rounding operation is that for $u < 0.5$ the control law output is floored to zero, which means none of the spoiler is active. This happens when the AoA α is at very small values and close to be fully regulated. For this reason, the actuation dead-zone does not have any impact on convergence, which is only postponed by a little delay. However, the dead-zone actuation issue can be easily overtaken by allowing gradual, and not binary, operation to at least the first open spoiler (that is #3), as explained in Section 3. The results' analysis demonstrates the robustness of the combined spoiler-Modified MRAC control architecture. In fact, the regulation time remains almost constant in all the analyzed test cases. Whereas an unexpected beneficial effect on the control performance due to the stiffness reduction and post-flutter wind speed is detected. This phenomenon derives from the higher effectiveness of each spoiler when higher amplitude oscillations and faster wind condition occurs.

5. CONCLUSION

In this paper, a multiple spoiler configuration, with predefined and coordinated actuation strategy, designed to control flutter and post-flutter limit cycle oscillations is presented. Based on computational fluid dynamics (CFD) studies, the proposed wing-spoiler configuration is fabricated and tested in the wind tunnel to obtain the nonlinear aerodynamic database implemented in the simulation model. A well-known, preexisting plunging/pitching nonlinear structural model is adopted for the aeroelastic system. The same aeroelastic plant is modified to account for trailing-edge control surface effect, which is used for validation and results comparison purpose. A normalized Model Reference Adaptive Control (MRAC) scheme, modified for performance improvement, is applied to both plants. Regulation of the states dynamics to zero is obtained with the same control algorithm, which provides additional validation proofs to the scheme's robustness and adaptation capabilities. Performance analysis demonstrates that the spoiler control architecture represents a viable alternative and can even surpass the standard trailing-edge flap controller by providing energy saving and additional safety. Further analyses, as a multi-input multi-output control of the same aeroelastic system with combined action of spoilers and flap, are planned as future work.

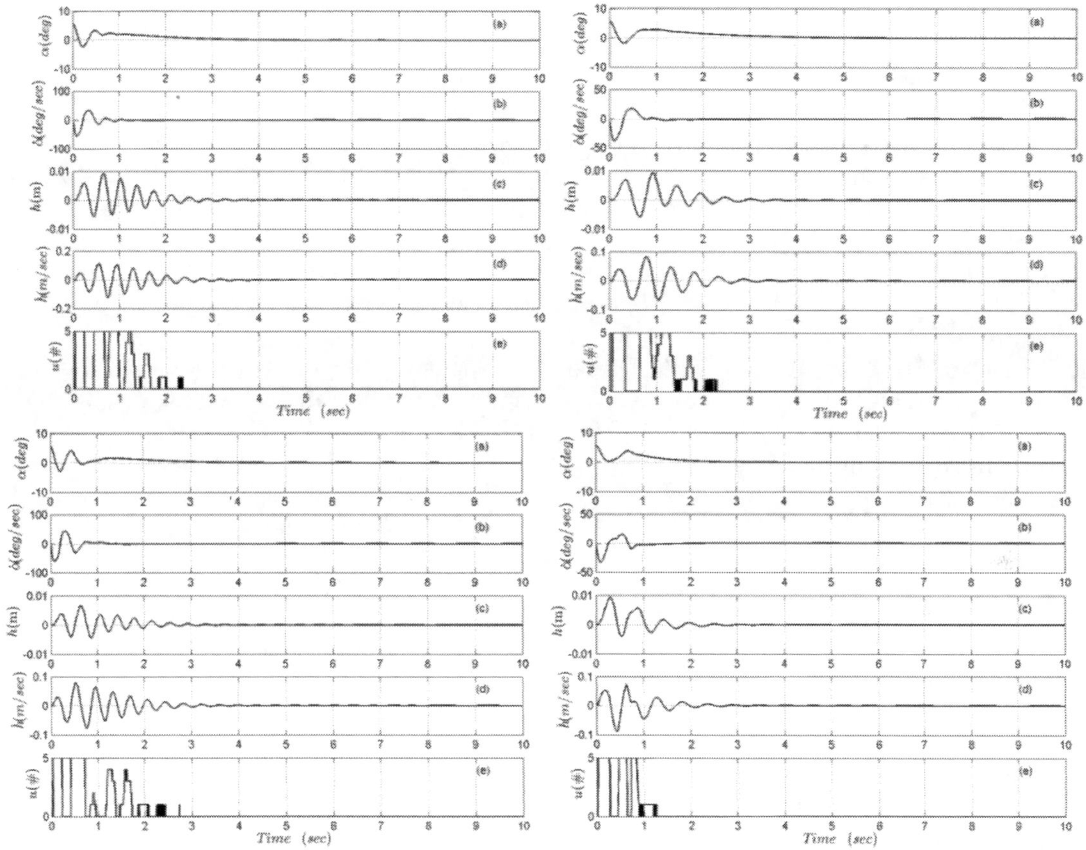

Figure 9. Spoiler/MRAC configuration time response: **a)** $U = U_{flutter}, K = K_{nom}$; **b)** $U = U_{flutter}, K = K_{reduced}$; **c)** $U = U_{post-flutter}, K = K_{nom}$; **d)** $U = U_{post-flutter}$, $K = K_{reduced}$.

REFERENCES

1. Dowell, E.H., A Modern Course in Aeroelasticity, Sijthoff and Noordhoff, 1978.
2. Fung, Y.C., An Introduction to the Theory of Aeroelasticity, Dover, New York, 1955.
3. Bisplinghoff, R.L., Ashley, H., Halfman, R.L., Aeroelasticity, Dover, New York, 1996.
4. Waszak, M.R., "Robust Multivariable Flutter Suppression for Benchmark Active Control Technology Wind-Tunnel Model", Journal of Guidance, Control, and Dynamics, Vol. 24, No. 1, 2001, pp. 147-153. doi: 10.2514/2.4694

5. Mukhopadhyay, V., "Transonic Flutter Suppression Control Law Design and Wind-Tunnel Test Results", Journal of Guidance, Control and Dynamics, Vol. 23, No. 5, 2000, pp. 930-937. doi: 10.2514/2.4635
6. Kelkar, A.G., and Joshi, S.M., "Passivity-Based Robust Control with Application to Benchmark Controls Technology Wing", Journal of Guidance, Control and Dynamics, Vol. 23, No. 5, 2000, pp. 938-947. doi: 10.2514/2.4636
7. Block, J.J., and Strganac, T.W., "Applied Active Control for a Nonlinear Aeroelastic Structure", Journal of Guidance, Control, and Dynamics, Vol. 21, No. 6, 1998, pp. 838-845. doi: 10.2514/2.4346
8. Platanitis, G., and Strganac, T.W., "Control of a Nonlinear Wing Section Using Leading- and Trailing-Edge Surfaces", Journal of Guidance, Control, and Dynamics, Vol. 27, No. 1, 2004, pp. 52-58. doi: 10.2514/1.9284
9. Platanitis, G., and Strganac, T.W., "Suppression of Control Reversal Using Leading- and Trailing-Edge Control Surfaces", Journal of Guidance, Control, and Dynamics, Vol. 28, No. 3, 2005, pp. 452-460. doi: 10.2514/1.6692
10. O'Neil, T., and Strganac, T.W., "Aeroelastic Response of a Rigid Wing Supported by Nonlinear Springs", Journal of Aircraft, Vol. 35, No. 4, July–August 1998, pp. 616-622. doi: 10.2514/2.2345
11. Cassaro, M., Battipede, M., Marzocca, P., Behal, A., "A Comparison of Adaptive Control Architectures for Flutter Suppression", Journal of Guidance, Control, and Dynamics, accepted.
12. Gujjula, S., Singh, S.N., Yim, W., "Adaptive and Neural Control of a Wing Section Using Leading- and Trailing-Edge Surfaces", Aerospace Science and Technology, Vol. 9, No. 2, 2005, pp. 161-171.
13. Behal, A., Marzocca, P., Rao, V.M., Gnann, A., "Nonlinear Adaptive Control of an Aeroelastic Two-Dimensional Lifting Surface", Journal of Guidance, Control, and Dynamics, Vol. 29, No. 2, March-April 2006, pp. 382-390. doi: 10.2514/1.14011
14. Cassaro, M., Battipede, M., Cestino, E., Marzocca, P., Behal, A., "L_1 adaptive flutter suppression control strategy for highly flexible structure", SAE Int. J. Aerosp., Vol. 6, No. 2, December 2013. doi:10.4271/2013-01-226
15. Cassaro, M., Nágy, A., Marzocca, P., Battipede, M., Ahmadi, G., "Novel Active Control Strategy for LCO and Flutter Suppression by a Coordinated Use of Multiple Distributed Surface Actuators", Proceedings of the ASME 2014 International Mechanical Engineering Congress & Exposition, IMEC2014, November 14-20, 2014, Montreal, Quebec, Canada.
16. Wagner, H., "Uber die Entstehung des dynamischen Auftriebes von Tragflugeln", Zeitschrift fur Angewandte Mathematic und Mechanik, 5(1):17–35, 1925.
17. Theodorsen, T., "General Theory of Aerodynamic Instability and the Mechanism of Flutter," NACA Rept. 496, 1935.
18. Sheta, E.F., Harrand, V.J., Thompson, D.E., Strganac T.W., "Computational and Experimental Investigation of Limit Cycle Oscillations of Nonlinear Aeroelastic Systems", Journal of Aircraft, Vol. 39, No. 1, January–February 2002. doi: 10.2514/2.2907
19. Behal, A., Marzocca, P., Rao, V.M., Gnann, A.," Nonlinear Adaptive Control of an Aeroelastic Two-Dimensional Lifting Surface", Journal of Guidance, Control, and Dynamics, Vol. 29, No. 2, March-April 2006, pp. 382-390. doi: 10.2514/1.14011
20. Ko, J., Strganac, T.W., Kurdila, A.J., "Stability and Control of a Structurally Nonlinear Aeroelastic System", Journal of Guidance, Control, and Dynamics, Vol. 21, No. 5, 1998, pp. 718–725. doi: 10.2514/2.4317
21. Hanke, C.R., Nordwall, D.R., The Simulation of a Jumbo Jet Transport Aircraft, D6-30643, Volume II: Modeling Data, The Boeing Company, Wichita Division, September 1970, Wichita, Kansas.
22. Ioannou, P.A. and Sun, J., Robust Adaptive Control, upper Saddle River, NY: Prentice Hall, 1989, Ch. 9.

ICAST2015: 26th International Conference on Adaptive Structures and Technologies
October 14-16th, 2015, Kobe, Japan

ICAST2015 #026

Frequency-Domain Mathematical Model of the Smart Spring Device for Vibration Reduction

Fred Nitzsche[1], Melissa Arras[2*], and Giuliano Coppotelli[2]

[1] Department of Mechanical and Aerospace Engineering, Carleton University, Ottawa, Canada.
[2] Department of Mechanical and Aerospace Engineering, University of Rome La Sapienza, Rome, Italy.

Abstract

In this paper, the capability of a semi-active device for vibration control, named Smart Spring, is analytically investigated. Such a device allows to actively control vibration by modulating the structural properties, such as the mass, stiffness, or damping, of a system, through a piezoceramic actuator. The analytical study of the Smart Spring is performed by developing its mathematical model in the frequency domain, in order to evaluate the response of the system in terms of its harmonic components. Furthermore, an open-loop control law for the modulation of the system stiffness is introduced. It is shown that the displacement solution depends on three dimensionless parameters, representing both structural and control properties of the system, that can be opportunely chosen to reduce the response, then vibration.

1. INTRODUCTION

Reduction of vibration is one of the most challenging problem in many engineering applications, especially for a complex system as the helicopter machine. In steady-state forward flight, the cyclic variation of the aerodynamic loads acting on the blade causes a periodic motion of the blade, thus generating periodic forces and moments which are transmitted to the rotor hub. However, most of these loads are cancelled at the hub, except for their nN_b/rev harmonics, where n is an arbitrary integer and N_b is the number of blades, Ref. [1]. Due to the coupling between the rotor system and the airframe, the vibratory hub loads are transferred throughout the structure thus reducing the comfort of passengers, the crew efficiency, and the fatigue lives of airframe structural components, Ref. [2]. Several approaches for vibration control, either *passive* or *active*, have been proposed over the past years, Refs. [3, 4]. Specifically, passive control systems alter the structure by adding vibration absorbers or isolators, with the advantage to not require an external power source to operate. However, they are effective only at a specific frequency. On the other hand, active control systems act over a broadband of frequencies by using one or more actuators, powered by an external source, that apply forces to the structure adaptively, based on the response of the system measured by physical sensors. The advance of active materials, especially the piezoelectric ones, allowed a wide extent of the active control techniques as these materials directly convert electrical energy into a mechanical deformation that can be used for structural actuation, Ref. [5]. The active control systems are usually classified into direct-active and indirect-active, or semi-active. In direct-active schemes, vibration is controlled by acting directly against the excitation force,

* melissa.arras@uniroma1.it

thus requiring high actuation forces to counteract the vibratory forces. Such an issue is overcome by the semi-active concepts that reduce vibrations by modulating the structural properties, such as the mass, stiffness, or damping of a system. This results in lower actuation forces and displacements with respect to the direct-active schemes. In this paper, a semi-active device named Smart Spring is presented. It allows to control vibration by adaptively changing the structural impedance of a system, e.g. a helicopter blade, through a piezoceramic actuator. Its main advantage, compared to other approaches based on piezoelectric actuators, is that it requires the actuators to produce micro displacements to generate high actuation forces, Refs. [6, 7]. The aim of this paper is to develop the mathematical model of the device in the frequency domain by considering the excitation force and the displacement solution as a combination of harmonics of the multiples of the exciting frequency, being it extremely useful when dealing with rotorcraft applications. Although the Smart Spring could control vibration by simultaneously altering the mass, stiffness, and damping of a system, it is assumed in this study that only the stiffness of the system is modulated, through a square-wave control law. Furthermore, numerical results that show how the amplitude of the response can be varied by acting on the structural and control properties of the system are presented.

2. THE MATHEMATICAL MODEL OF THE SMART SPRING

The conceptual drawing of the Smart Spring is shown in Fig. 1, where it can be noted that it is modeled as two mass-spring systems denoted with subscripts 1 and 2. Specifically, k_1 and m_1 represent the main system stiffness and mass, respectively, whereas the secondary system is characterized by k_2 and m_2. The main spring is attached to a moving structure, that has a built-in sleeve, where the external vibratory force $F(t)$ is considered applied to. The piezoceramic actuator is attached to the secondary system which is inserted into the sleeve. Activating the actuator, the resulting stiffness of the system varies in the range $[k_1, k_1 + k_2]$. When the actuator is off, the sleeve can move freely and only the main system determines the dynamic response of the system. Instead, when the actuator is powered on, a normal force $N(t)$, that engages the two systems, is generated. If the resultant friction force $\mu(t)N(t)$, being $\mu(t)$ the friction coefficient that is, in turn, time dependent, is sufficiently large, the two systems are fully coupled and the system is characterized by an equivalent stiffness, k, given by the "parallel" of the stiffness of the two systems, that is $k = k_1 + k_2$, and the mass of the system increases from m_1 to $m = m_1 + m_2$. The friction force applied to the sleeve can be controlled by an external voltage stimulus to the actuator. Therefore, the voltage applied to the piezoceramic actuator could continuously provide the control of the resulting stiffness, damping, and mass of the Smart Spring system.

The frequency-domain mathematical model of the device is derived by writing the equations of motion of the system. Referring to Fig. 1, they result in the following:

$$\begin{cases} m_1\ddot{x}_1(t) + k_1x_1(t) = F(t) \mp \mu(t)N(t) \\ m_2\ddot{x}_2(t) + k_2x_2(t) = \pm\mu(t)N(t) \end{cases} \tag{1}$$

which can be rearranged by summing the two equations as:

$$m_1\ddot{x}_1(t)\left(1 + \frac{m_2}{m_1}\frac{\ddot{x}_2(t)}{\ddot{x}_1(t)}\right) + k_1x_1(t)\left(1 + \frac{k_2}{k_1}\frac{x_2(t)}{x_1(t)}\right) = F(t) \tag{2}$$

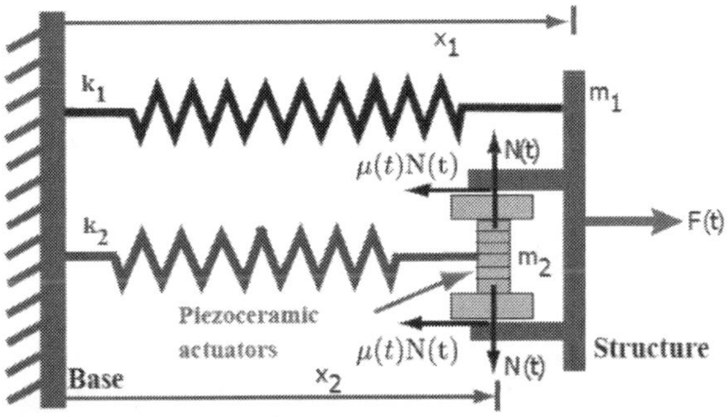

Figure 1. Conceptual drawing of the Smart Spring.

By introducing:

$$m(t) = m_1 \left(1 + \frac{m_2}{m_1}\frac{\ddot{x}_2(t)}{\ddot{x}_1(t)}\right) \tag{3}$$

$$k(t) = k_1 \left(1 + \frac{k_2}{k_1}\frac{x_2(t)}{x_1(t)}\right) \tag{4}$$

Eq. (2) can be rewritten as:

$$m(t)\ddot{x}_1(t) + k(t)x_1(t) = F(t) \tag{5}$$

It is worthwhile specifying that Eq. (5) includes both the cases in which the actuator is off and powered on, achieved by opportunely modulating the mass and stiffness of the system through Eq. (3) and Eq. (4), respectively. In the assumption to modulate only the stiffness of the system, i.e. $|m_2\ddot{x}_2(t)| \ll |m_1\ddot{x}_1(t)|$ (note that usually $m_2 \ll m_1$), Eq. (5) becomes:

$$m_1\ddot{x}_1(t) + k(t)x_1(t) = F(t) \tag{6}$$

The stiffness modulation $k(t)$, which depends on time through the ratio $\frac{x_2(t)}{x_1(t)}$ (see Eq. (4)), between the two theoretical limits k_1 and $k = k_1 + k_2$ is represented by the square-wave law depicted in Fig. 2, where the new variable α is introduced, that is:

$$\alpha = r\psi + \phi \tag{7}$$

in which r and ϕ are the control harmonic and control phase of the Smart Spring, respectively, whereas $\psi = \Omega t$ being Ω the rotor fundamental frequency. The stiffness of the system is then modulated according to:

$$k(t) \rightarrow k(\alpha) = k_1 + k_2 f(\alpha) \tag{8}$$

where $f(\alpha) = \frac{x_2}{x_1}(\alpha)$. The harmonic balance method allows to write the equation of motion, that is Eq. (6), in the frequency-domain by assuming that the excitation force $F(t)$ and the displacement solution $x_1(t)$ are harmonic of the multiples of the exciting frequency, that is:

$$F(t) = \sum_{n=-\infty}^{+\infty} F_n e^{in\psi} \tag{9}$$

$$x_1(t) = \sum_{n=-\infty}^{+\infty} x_{1_n} e^{in\psi} \tag{10}$$

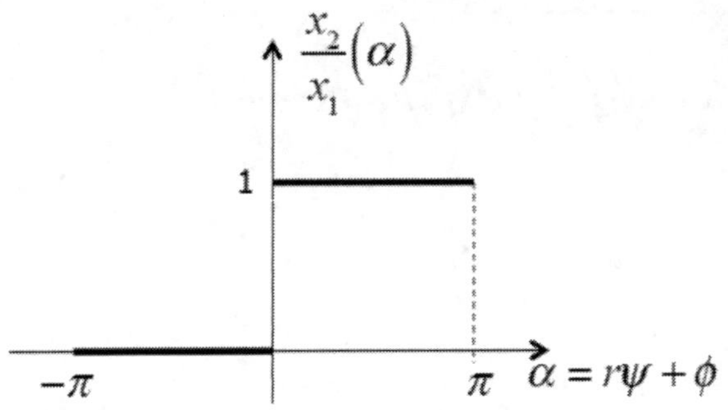

Figure 2. Smart Spring stiffness modulation.

where the coefficients associated with the negative harmonics are the complex conjugates of their positive counterparts. Furthermore, the square-wave control law for the stiffness modulation can be expanded into a complex Fourier series, Ref. [8], to have:

$$k(\alpha) = k_0 + \frac{k_2}{i\pi} \sum_{p=\pm 1, \pm 3, \dots}^{+\infty} \frac{1}{p} e^{ip\alpha} \tag{11}$$

where

$$k_0 = \frac{2k_1 + k_2}{2} \tag{12}$$

By substituting Eqs. (9), (10), and (11) into Eq. (6), the following equation of motion in the frequency domain is obtained:

$$\begin{aligned} \sum_{n=-\infty}^{+\infty} \sum_{p=\pm 1, \pm 3, \dots}^{+\infty} & \left[\left(-m_1 \Omega^2 n^2 + k_0\right) x_{1_n} e^{in\psi} + \right. \\ & \left. + \frac{k_2}{\pi} \frac{1}{p} x_{1_n} e^{i(pr+n)} e^{i(p\phi - \frac{\pi}{2})} \right] = \\ & = \sum_{n=-\infty}^{+\infty} F_n e^{in\psi} \end{aligned} \tag{13}$$

When transformation of the dummy indices p and n in the above equation is performed, that is:

$$\sum_{n=-\infty}^{+\infty} \sum_{p=\pm 1, \pm 3, \dots}^{+\infty} x_{1_n} e^{i(pr+n)\psi} = \sum_{n=-\infty}^{+\infty} \sum_{p=\pm 1, \pm 3, \dots}^{+\infty} x_{1_{(pr+n)}} e^{in\psi} \tag{14}$$

Eq. (13) can be rearranged as follows:

$$\begin{aligned} & \left(-m_1 \Omega^2 n^2 + k_0\right) x_{1_n} + \frac{k_2}{\pi} \sum_{p=\pm 1, \pm 3, \dots}^{+\infty} \frac{1}{p} e^{i(p\phi - \frac{\pi}{2})} x_{1_{(pr+n)}} = F_n \\ & n = 0, \pm 1, \pm 2, \dots \end{aligned} \tag{15}$$

Dividing Eq. (15) by k_1, recalling Eq. (12), and introducing the dimensionless parameter $\kappa = \frac{k_2}{k_1}$, Eq. (15) results in:

$$\left[-\left(\frac{\Omega}{\omega_1}\right)^2 n^2 + 1 + \frac{\kappa}{2}\right] x_{1_n} + \frac{\kappa}{\pi} \sum_{p=\pm1,\pm3,\ldots}^{+\infty} \frac{1}{p} e^{i(p\phi - \frac{\pi}{2})} x_{1_{(pr+n)}} = \frac{F_n}{k_1} \qquad (16)$$

$$n = 0, \pm1, \pm2, \ldots$$

The above equation can be rearranged in the algebraic system $[A]\{z\} = \{b\}$ to be solved for $\{z\} \in \mathbb{C}^{N\times1}$, being $N > n + rp_{max}$, which groups the harmonic components x_{1_n} of the displacement solution. For the sake of brevity, all the mathematical operations that allow to write that system of algebraic equations are not reported in this paper. However, it is important to note that Eq. (16) clearly shows the coupling between the uncontrolled, i.e. $x_{1_n} = \frac{F_n}{k_1}$, and the controlled, i.e. $x_{1_{(pr+n)}}$, harmonic components, leading to a redistribution of the dynamic response spectrum of the system. In addition, it can be observed from the same equation that the solution depends on three dimensionless parameters, namely κ, $\frac{\Omega}{\omega_1}$, and the control phase of the Smart Spring ϕ. In a simple open-loop control approach, the control objective is to adjust those parameters in order to reduce the amplitude of the response, so the vibration of the system.

3. RESULTS

The aim of the numerical analyses presented in this section is to show how the amplitude of the response can be reduced by properly choosing the parameters which it depends on, that is κ, $\frac{\Omega}{\omega_1}$, and ϕ. A constant excitation force has been considered and the control harmonic r, see Eq. (7), has been set equal to 4 simulating to control the critical harmonic of a four blade rotor. The most significative results up to the 8-th harmonic component of the displacement solution are shown in Figs. 3-10, for two operative working conditions identified by the ratio $\frac{\Omega}{\omega_1}$.

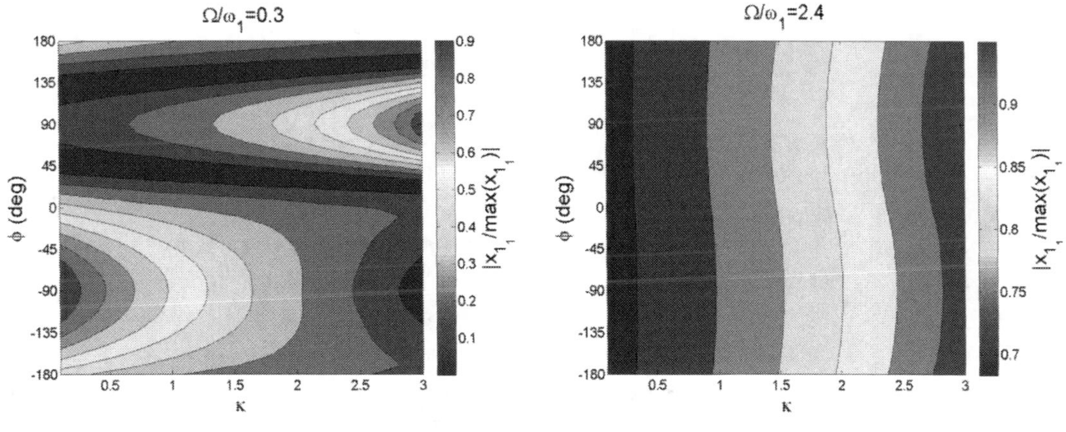

(a) (b)

Figure 3. Parametric study for the 1/rev harmonic component of the displacement solution.

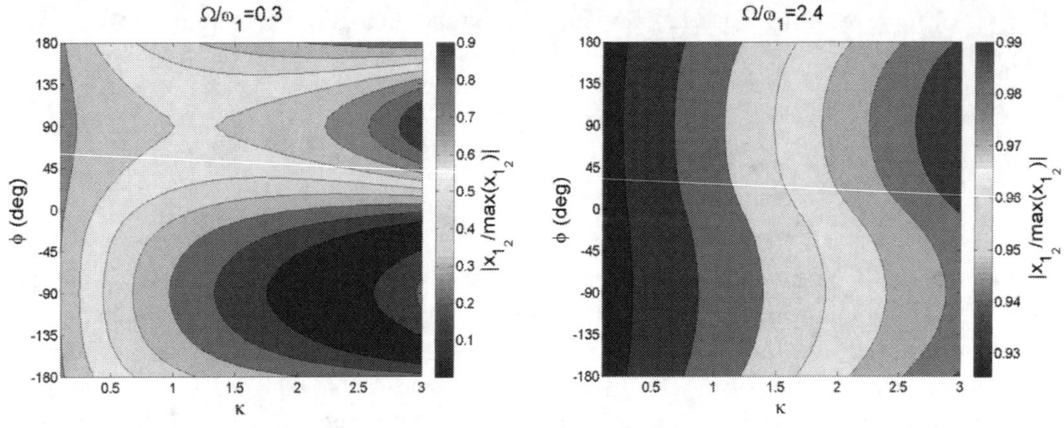

(a) (b)

Figure 4. Parametric study for the 2/rev harmonic component of the displacement solution.

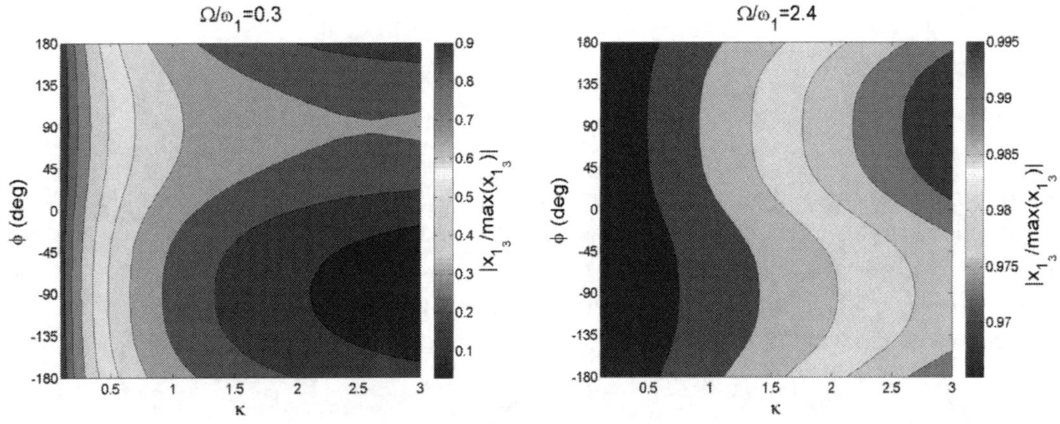

(a) (b)

Figure 5. Parametric study for the 3/rev harmonic component of the displacement solution.

Fig. 3 shows the amplitude of the 1/rev harmonic component of the displacement solution. It can be observed from Fig. 3(a) that for small values of $\frac{\Omega}{\omega_1}$ there is a strong dependence on ϕ, that is given a value of κ it is possible to choose the optimal value of the control phase of the Smart Spring that minimizes the response amplitude by acting in the control law defined by Eq. (11). Such dependence gradually vanishes as $\frac{\Omega}{\omega_1}$ increases, when the amplitude of the displacement becomes almost constant with ϕ, Fig. 3(b). It means that given an operative working condition and identified a system, in terms of $\frac{\Omega}{\omega_1}$ and κ, the amplitude of the response cannot be reduced by modulating the control phase ϕ. However, such a result could suggest to modify the stiffness of the system according to the most suitable value of κ that minimizes the response. A similar behavior is obtained for the amplitude of the 2/rev and 3/rev harmonic components as suggested by Fig. 4 and Fig. 5, respectively. As regards the 4-th, Fig. 6, and 5-th, Fig. 7, harmonic components they show an almost constant amplitude for a wide range of κ for small values of

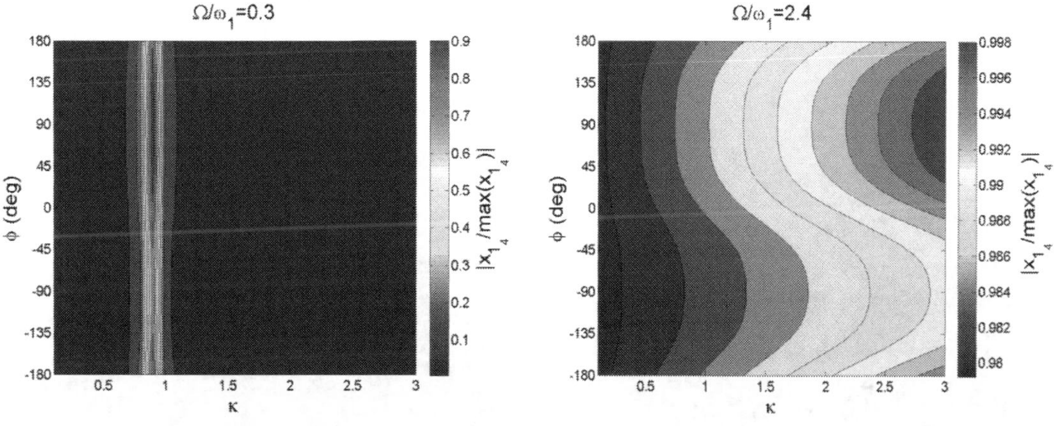

(a) (b)

Figure 6. Parametric study for the 4/rev harmonic component of the displacement solution.

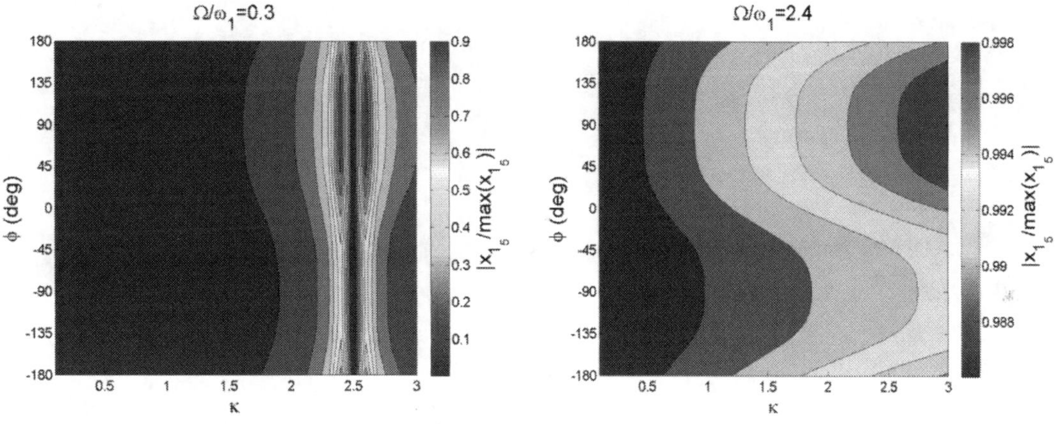

(a) (b)

Figure 7. Parametric study for the 5/rev harmonic component of the displacement solution.

$\frac{\Omega}{\omega_1}$, then assuming the same behavior observed for the lower components. The amplitude of the higher harmonics from the 6-th to the 8-th, reported in Figs. 8-10, tends to increase with κ also for small values of $\frac{\Omega}{\omega_1}$. It is important to note that for a specific harmonic component, the contour plots follow the same behavior but the reduction of the amplitude (obtained by varying κ and/or ϕ) becomes lower as soon as the operative working condition, i.e. $\frac{\Omega}{\omega_1}$, increases, especially for the higher harmonic components. Considering the 7/rev component depicted in Fig. 9 as an example, its amplitude can be reduced by 50% by varying κ when $\frac{\Omega}{\omega_1} = 0.3$, Fig. 9(a), but such a percentage decreases to about 0.7% when $\frac{\Omega}{\omega_1} = 2.4$, Fig. 9(b). This behavior can be justified by Eq. (16), whose terms in the square brackets suggest that as soon as $\frac{\Omega}{\omega_1}$ increases, the amplitude of the n-th harmonic component is less affected by κ.

As a general behavior, it can be noted that the maximum reduction of the amplitude for all the harmonic components is obtained for low values of κ when $\frac{\Omega}{\omega_1} < 1$.

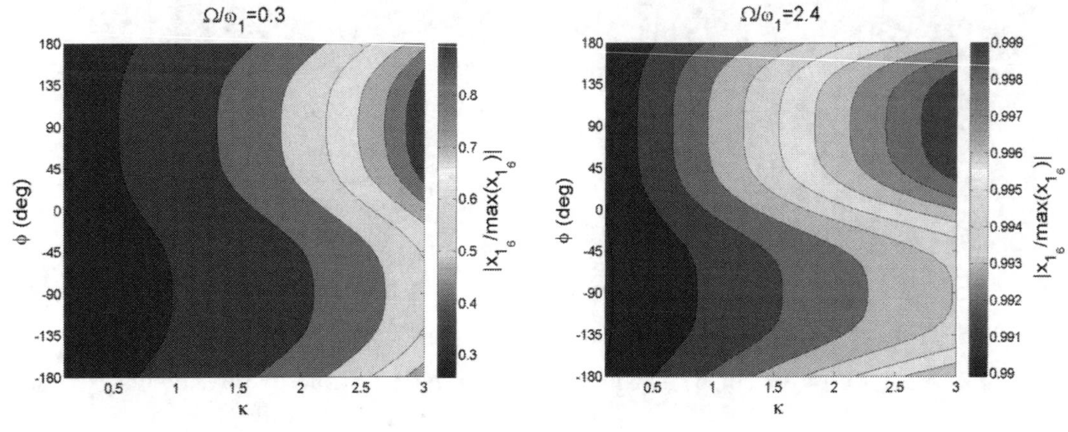

(a) (b)

Figure 8. Parametric study for the 6/rev harmonic component of the displacement solution.

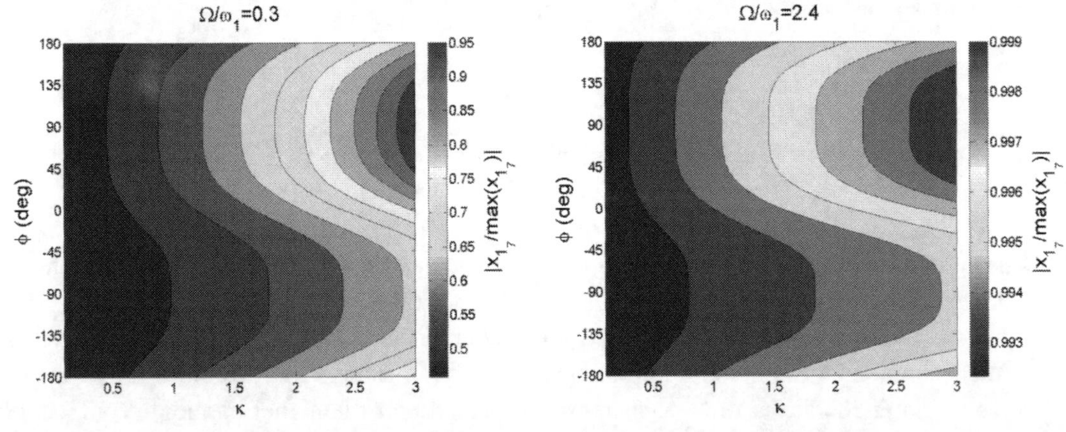

(a) (b)

Figure 9. Parametric study for the 7/rev harmonic component of the displacement solution.

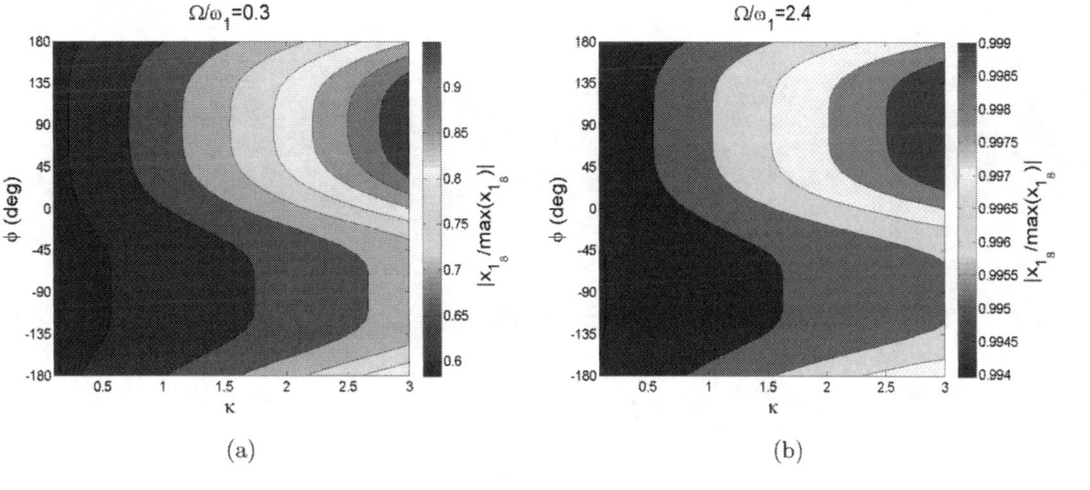

Figure 10. Parametric study for the 8/rev harmonic component of the displacement solution.

4. CONCLUDING REMARKS

In this paper, the effectiveness of the Smart Spring device for vibration control has been analytically validated. Such a device belongs to the category of semi-active approaches, allowing to reduce vibration by modulating the structural properties, such as the mass, stiffness, or damping, of a system through a piezoceramic actuator. The device has been modeled as two mass-spring systems and its mathematical model has been developed in the frequency domain by introducing the harmonic balance method, in order to study the response of the system in terms of its harmonic components. A square-wave control law has been introduced for the stiffness modulation of the system, highlighting the dependence of the system response on both the uncontrolled and controlled harmonics. The capability of the Smart Spring has been finally verified through a numerical study by varying the parameters which the response of the system depends on. Such a study has shown that the amplitude of the system response, then the vibration, can be opportunely reduced for a given operative working condition by modulating the stiffness of the system, thus validating the Smart Spring as a semi-active approach for vibration reduction.

REFERENCES

1. Johnson W., *Helicopter theory*, Dover Publications, 1994.
2. Bramwell A. R. S., Done G. and Balmford D., *Bramwell's Helicopter Dynamics*, Butterworth-Heinemann, 2001.
3. Inman D. J., *Vibration with control*, John Wiley & Sons, 2006.
4. Bielawa R. L., *Rotary Wing Structural Dynamics and Aeroelasticity*, AIAA Education Series, 1992.
5. Jaffe B., Cook W. R. and Jaffe H. L., *Piezoelectric ceramics*, Academic Press, 1971.
6. Wickramasinghe, V. K., Yong, C., Zimcik D. G., Harold T. and Nitzsche, F., "Smart Spring: A Novel Impedance Control Approach for Active Vibration Suppression Applications", *Smart Structures and Materials 2004: Smart Structures and Integrated Systems, Proc. of SPIE*, Vol. 5390, 2004.
7. Yong, C., Zimcik D. G., Wickramasinghe, V. K. and Nitzsche, F., "Development of the Smart Spring for Active Vibration Control of Helicopter Blades", *Journal of Intelligent Material Systems and Structures*, Vol. 15, 2004, pp. 37-47.
8. Shin K. and Hammond P., *Fundamentals of Signal Processing for Sound and Vibration Engineers*, John Wiley & Sons, 2008.

ICAST2015 **#047**

Smart Composite Repair Patches for Aerospace Applications

Lukasz Pieczonka[1*], Tomasz Pruciak[1], Maciej Konieczko[1], Wieslaw J. Staszewski[1], Sofia Pavlopoulou[2] and Constantinos Soutis[3]

[1] Dept. of Robotics and Mechatronics, AGH University of Science and Technology, Krakow, Poland
[2] Jaguar Land Rover, Coventry, UK
[3] Aerospace Research Institute, University of Manchester, Manchester, UK

Abstract

The paper investigates the problem of Structural Health Monitoring of composite patch repairs used in aerospace applications. The vibro-acoustic modulation technique (VAM) is applied to evaluate structural health of test samples with simulated damage. The VAM is a nonlinear ultrasonic technique based on the nonlinear modulations of two wave fields caused by the presence of damage. Test samples are made of Carbon Fiber Reinforced Polymer (CFRP) bonded patch repairs applied on aluminum and CFRP substrates. Damage is simulated by placing Teflon inserts of different sizes and locations during the manufacturing of repair patches. After manufacturing, ultrasonic C-scan and vibrothermography are used to confirm the presence and the locations of debondings caused by Teflon inserts. Subsequently the VAM technique is applied using surface bonded piezo ceramic transducers for excitation and laser Doppler vibrometer for vibration sensing. The paper shows that it is feasible to apply the vibro-acoustic modulation technique to detect the presence of damage in the analyzed samples. The paper discusses technical details of the measurement technique and provides detailed description of the analyzed test samples.

1. INTRODUCTION

Composite materials, especially carbon/epoxy systems, are being increasingly used in the new age commercial and military aircraft structures [1,2]. Until recently the composites were used in aircraft industry mainly for the less critical secondary structure like winglets, radome or necelles as well as for the interior of aircrafts like the ceiling panels, partition walls etc. Currently the composites are being incorporated also into the primary structure of aircrafts like the fuselage frame and skin or engine fan blades. The primary structure is the responsible load-bearing structure of an aircraft and is therefore critical for its performance and safety. Structural composites, which include laminates and sandwich panels, offer many advantages including the high strength-to-weight ratio and flexibility in design. One of the main drawbacks of composite materials is, however, their susceptibility to impact damage. The most common type of damage results from a blunt force impact. In aerospace applications this type of impact damage usually result as a consequence of collision with ground service equipment, bird strike, hail impact or runaway debris during takeoff or landing. Replacement of the damaged areas is not a preferable solution due to the high level of integration and the big size of the components. Therefore there is an increasing need for repair technologies performed on primary structural components of the aircraft. Repair patches are one of the possible solutions. They may be attached to the parent structure by adhesive bonding or mechanical fastening. The former solution is preferable as it offers significantly higher performance, creates minimal damage to the parent structure and facilitates non-destructive inspection [3-6]. The use of composite repair techniques drives the demand for reliable and efficient nondestructive testing (NDT) and Structural Health Monitoring (SHM) techniques [7,8].

This paper demonstrates an application of a nonlinear ultrasonic damage detection technique based on vibro-acoustic wave modulations applied for testing of bonded patch repairs. Aluminum and Carbon Fiber Reinforced Polymer (CFRP) samples with bonded patch repairs containing artificial damage were tested with the nonlinear ultrasonic technique. Reference damage detection data were provided by the classical nondestructive testing techniques – ultrasonic c-scan and vibrothermography. This paper gives details of the test samples and measurement techniques applied to reveal the damage.

2. TEST SAMPLES

The test samples were two square substrate plates with the dimensions of 250x250 mm with applied bonded patch repairs. One plate was made of aluminum and the other was made of carbon fiber reinforced polymer (CFRP). The metallic substrate was made of aluminum grade 2024 T3 with 1.6 mm thickness. The composite substrate consisted of 5 plies with total theoretical thickness of 1.6 mm and ply orientation of [0/90°/0/90°/0]. Both test specimens are illustrated on the photograph in Figure 1. Repair patches were applied on both the aluminum and CFRP substrates in a manual way after surface preparation with sand paper and cleaning with degreasing agent. The repair patches were composed of five plies of *HEXCEL 43280* reinforcement impregnated with *Epocast 52 A/B* epoxy resin following a 20 mm step wise configuration. The overall size of the patch was 160 x 160 mm, as shown in Figure 2a. Impregnation was performed in on a heated table surface (60°C) in order to obtain low viscosity of the resin. The repair patch plies were placed on the substrate with ply orientation of [0/90°/0/90°/0°]. After the 5 plies layup completed, curing was performed for 4 hours in 95°C in oven under vacuum conditions (-700 mbar).

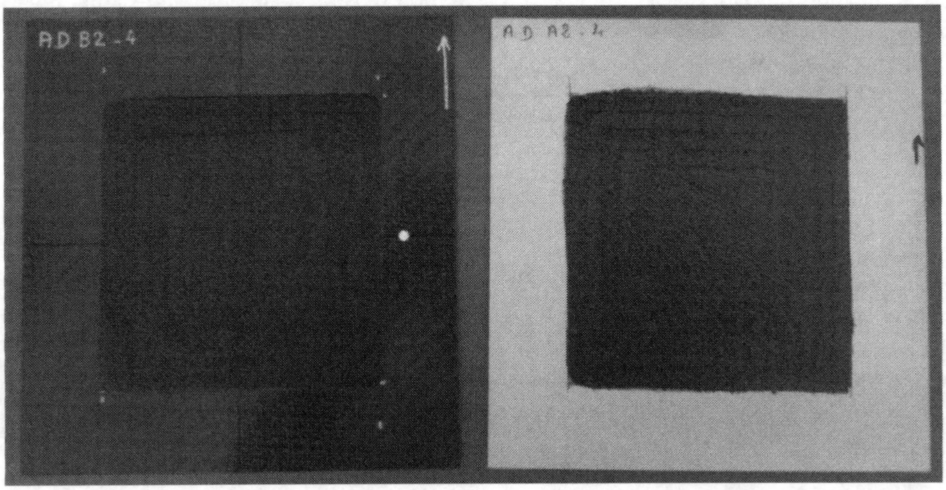

Figure 1. Analyzed test samples. CFRP substrate with CFRP bonded patch repair (left) and aluminum substrate with CFRP bonded patch repair (right).

Four areas of no adhesion were introduced in the applied bonded patch repairs. Four circular Teflon patches with the diameters of 5, 10, 15 and 20 mm were inserted between the substrate and first impregnated ply. Locations of the inserts are shown in Figure 2b.

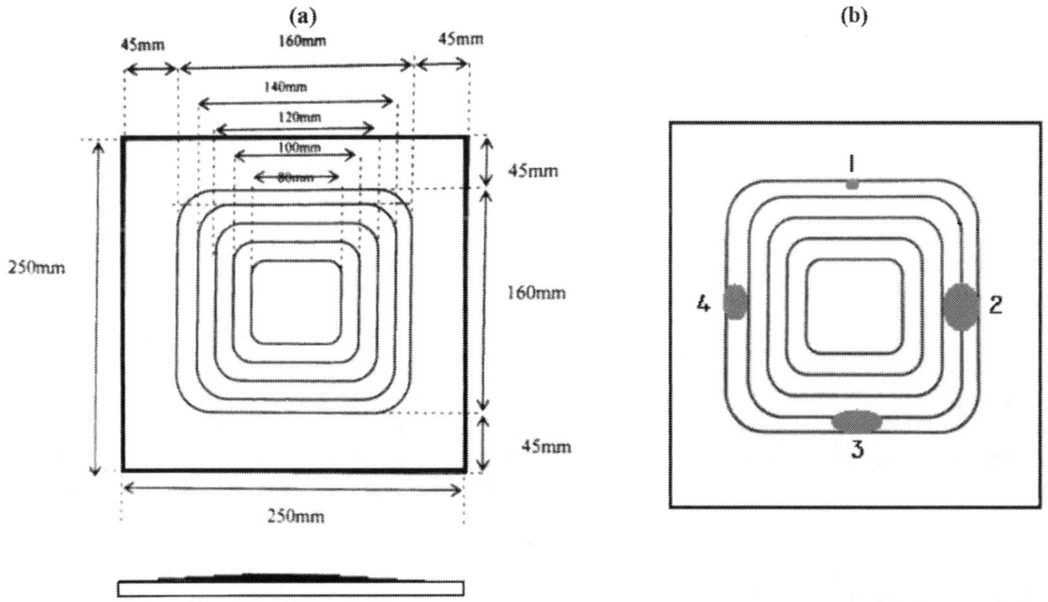

Figure 2. Dimensions of the applied patch repairs (a) and locations of the artificial debondings (b)

3. CLASSICAL NONDESTRUCTIVE TESTING

After the manufacturing of the patch repairs the classical NDT inspection was performed with use of two techniques, namely the ultrasonic C-scan and vibrothermography. Both samples were analyzed using the same test protocol. The goal of NDT inspections was to confirm the presence and location of the debondings in both samples. An ultrasonic transducer working at 5 MHz central frequency was used. The test was performed in the pulse echo mode. The samples were scanned with the spatial resolution of 0.5 mm. Results of the inspection for both the aluminum and composite substrates are shown in Figure 3a and 3b respectively. As can be seen the C-Scan images of both samples show the areas with no adhesion between the repair patch and the substrate as bright spots on image at the locations of Teflon inserts.

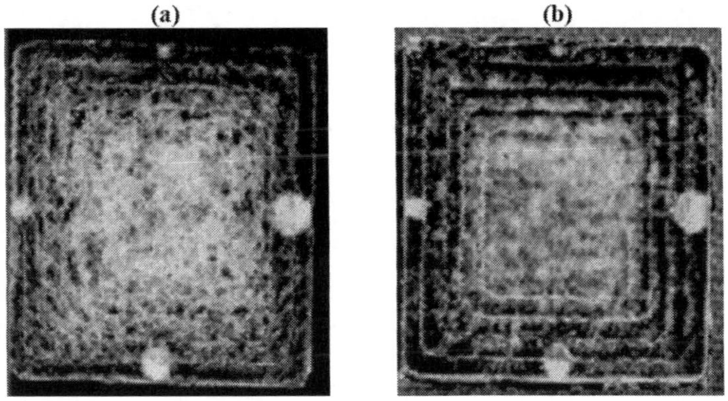

Figure 3. Ultrasonic C-Scan inspection results for the analyzed test samples – (a) aluminum substrate and (b) composite substrate.

The second NDT technique applied to reveal debondings in the test samples was vibrothermography. Vibrothermography, also referred to as ultrasonic thermography or thermosonics, is an active Thermographic Nondestructive Testing (TNDT) method based on the local heat generation [9,10]. The heat is generated selectively at the locations of material discontinuities as a consequence of the externally generated elastic waves (with frequencies typically from 15 to 70 kHz). The mechanical energy is dissipated into heat mainly due to friction. Generated heat propagates to the surface, where it may be detected by an infrared camera. In this study the *Monit SHM* vibrothermographic test system using the 35 kHz ultrasonic excitation column was used to deliver ultrasonic vibration to the test sample. *FLIR Silver 420M* photon detector camera was used to acquire thermal image sequences. The transducer was exciting the plate for 500 milliseconds and thermographic camera was acquiring the signal at 100 Hz frame rate for 3 seconds. Results of the inspection for both the aluminum substrate and the composite substrate are shown in Figure 4a and 4b respectively. In case of the aluminum substrate the locations and sizes of three debondings are clearly visible, while only a slight indication of the smallest debonding at the top edge of the repair patch can be seen. In case of the composite substrate the locations of all four debondings can be easily seen. The size of the simulated debondings is less clearly seen due to the weak bonding of the edge of the first ply of the repair patch and the substrate.

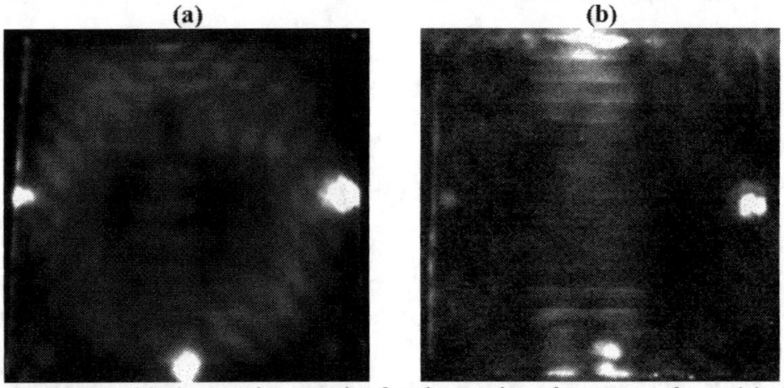

Figure 4. Vibrothermographic inspection results for the analyzed test samples – (a) aluminum substrate and (b) composite substrate.

4. Nonlinear vibro-acoustic modulation (VAM) testing

The nonlinear acoustic damage detection method considered in this study is based on a combined vibro-acoustic modulation (VAM) of an intensive low-frequency pumping wave and weaker high-frequency probing wave [11-16]. These two excitations are introduced to the structure simultaneously, as shown in Figure 5. Typically, one uses a weak High-Frequency (HF) ultrasonic wave as the probe and an intensive Low-Frequency (LF) modal/vibration excitation as the pump wave. In an idealized case when the sample is perfectly linear (undamaged) the response signal spectrum exhibits only the major frequency components i.e. the HF probe and LF pump. However, when the sample is nonlinear (e.g. due to the presence of damage) the spectrum of the response signal reveals additional frequency components such as higher harmonics and modulation sidebands around the high frequency probe component. Other experimental configurations of VAM include the LF impact excitation of natural modes of vibration of monitored structure or the frequency sweep HF excitation [12]. The method is also referred to as the nonlinear wave modulation spectroscopy (NWMS) [16].

Modulation intensity coefficient is one of the most commonly used damage indicators. This coefficient is calculated as the ratio between the sum of amplitude A^i_{LSB} and A^i_{RSB} of the i-th pair of modulation sidebands and the A_{HF} amplitude of the high-frequency component, i.e. :

$$R = \frac{\sum_{i=1}^{n}\left(A^i_{LSB} + A^i_{RSB}\right)}{A_{HF}} \qquad (1)$$

Frequently, only the first pair of sidebands are considered in computation of the R parameter, which has been done in this study. The low frequency excitation LF is frequently selected as one of the resonant frequencies of the test sample in order to increase the level of strains generated in the sample for a given input power. The high frequency excitation HF is selected arbitrarily in the range from few kHz to few hundred kHz.

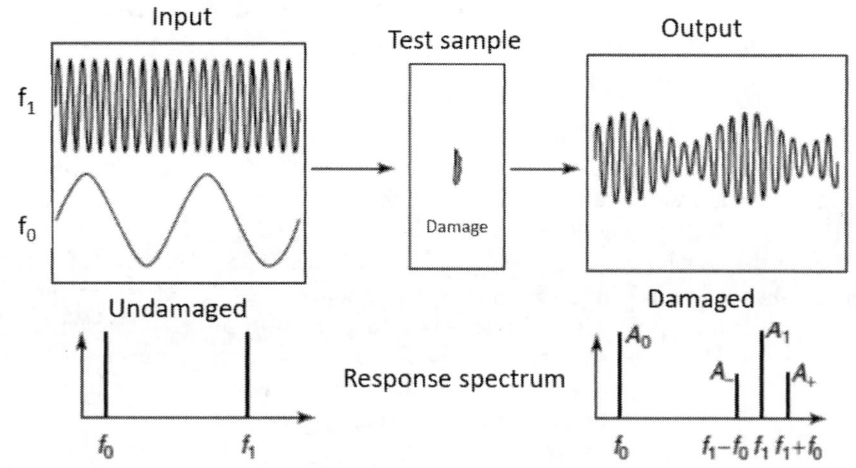

Figure 5. Schematic of the vibro-acoustic modulation (VAM) technique.

In order to apply VAM testing for damage detection one needs to know the baseline level of nonlinearity for the undamaged samples. For this reason two additional samples were added to the test – both were identical to the test samples described above (i.e. aluminum substrate with CFRP repair patch and CFRP substrate with CFRP repair patch) with the only difference being the lack of Teflon inserts that caused debondings in the repair patches. The first step of VAM testing on all considered test specimens was modal analysis. This was necessary in order to find the resonant frequencies of the test samples, that were going to be used as low-frequency pump excitations. The same measurement setup was used for all test samples. The plates were freely suspended using elastic cords to obtain the free-free boundary conditions. The surface-bonded *Noliac CMAP4* stack actuator was used to excite the plate. The white noise excitation signal was generated by a built-in Polytec signal generator and amplified by an *EC Test Systems PAQG* amplifier. A *Polytec PSV-400* scanning laser Doppler vibrometer (SLDV) was used to acquire vibration velocity signals in a non-contact manner. The Frequency Response Functions (FRFs) were calculated from the experimental input and output data using the Polytec PSV software. FRF amplitudes were gathered from the 27×27 measurement grids to obtain the relevant mode shapes corresponding to the selected natural frequencies. The experimental setup is shown schematically in Figure 6.

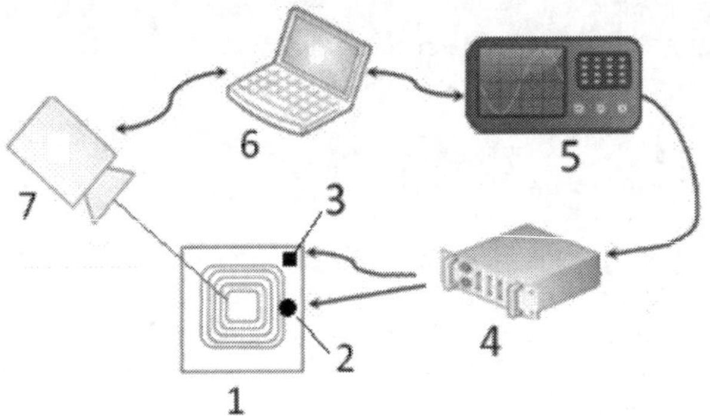

Figure 6. Experimental test setup for modal testing and vibro-acoustic modulation (VAM) measurements: (1) test sample, (2) high-frequency excitation transducer, (3) low-frequency excitation transducer, (4) signal amplifier, (5) arbitrary waveform generator, (6) data acquisition and control unit, (7) scanning laser Doppler vibrometer (SLDV).

Strong vibration peaks have been identified between 700 Hz and 800 Hz in case of the aluminum substrate and composite substrate. The corresponding mode shapes for 763 Hz in case of aluminum substrate and 719 Hz in case of composite substrate are shown in Figures 7a and 8a respectively. These particular mode shapes have been selected based on the measured vibration levels. The arbitrary choice of the mode shapes was possible as it was not essential for the VAM technique to use any particular mode shape for pump excitation. There was a slight shift in resonant frequencies between the damaged and undamaged specimens in both configurations. This could be attributed, however, not only to the presence of debondings, that reduce material stiffness, but also to the natural variability of modal parameters due to the inherent variability of material properties, geometry and boundary conditions.

The experimental setup for VAM testing was the same as the one used for modal testing. It has been only supplemented with an additional, smaller, piezoceramic stack actuator for high frequency ultrasonic excitation, as shown in Figure 6. The frequency of excitation was in this case 45 kHz, and the driving voltage was 80 Vpp. The driving voltage for the low frequency pump excitation was incrementally increased from 10 Vpp to 70 Vpp in 10 Vpp steps. The modulation spectra for each LF voltage were measured at six different points on the test sample. Subsequently the modulation intensity coefficients were computed for each point according to Equation 1.

(a) (b)

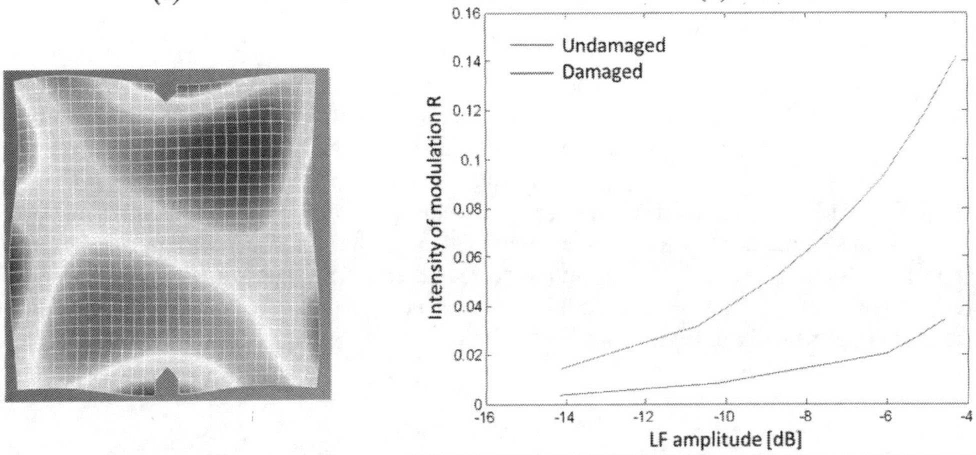

Figure 7. VAM testing results for the aluminum substrate. (a) mode shape corresponding to the LF excitation and (b) intensity of modulation R for damaged and undamaged specimens as a function of LF amplitude.

Results obtained for the aluminum and composite substrates are shown in Figures 7b and 8b respectively. The figures present the dependence of the intensity of modulation coefficient R on the applied low frequency pump wave amplitude for the undamaged samples (blue curves) and damaged samples (red curves). The curves represent the average response for all six measurement points. As can be seen the intensity of modulation increases with the increasing LF signal amplitude. What is more important, however, is that the intensity of modulation is always higher in case of test samples with debondings (red curves). The difference between the observed modulation intensity for the undamaged and damaged specimens increases with the increasing LF amplitude. This indicates that higher amplitudes of LF pump excitation are preferable for the discrimination of damage.

(a)

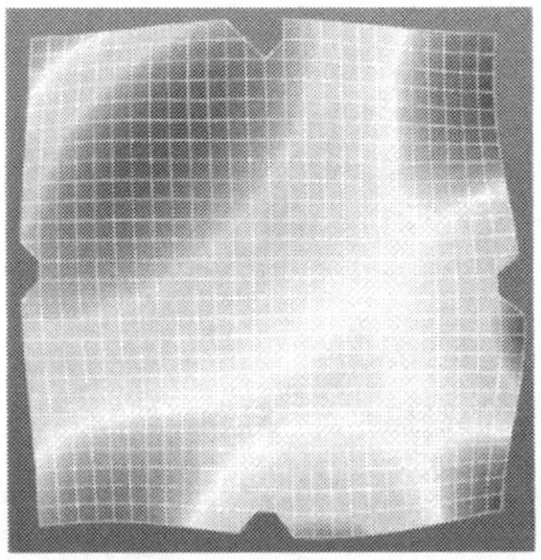

(b)

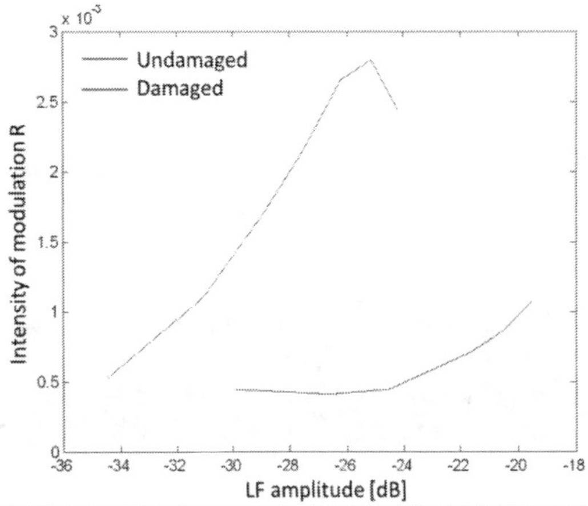

Figure 8. VAM testing results for the composite substrate. (a) mode shape corresponding to the LF excitation and (b) intensity of modulation R for damaged and undamaged specimens as a function of LF amplitude.

5. Conclusions

The paper presented an application of the classical and non-classical damage detection techniques to reveal the presence of artificial disbonds in bonded patch repairs. The classical NDT techniques – ultrasonic C-scan and vibrothermography – have been successfully applied to reveal the presence and the location of disbonds in aluminum and composite substrates repaired with CFRP bonded repairs. In parallel the non-classical nonlinear vibro-acoustic modulation technique (VAM) has been applied. The method has utilized piezo-based low-frequency vibration and high-frequency ultrasonic excitations. The work presented has been focused on the analysis of the intensity of modulation coefficient R. It has been shown that it is possible to discriminate the undamaged and damaged scenarios based on the observed level of modulation intensity. The VAM technique offers the level I of SHM i.e. the capability to detect the presence of damage. Some recent papers suggest, however, that it is possible to extend that capability also to damage localization. The VAM technique has a very simple experimental setup and in principle it is sufficient to use only three piezoceramic transducers: two for excitation of low frequency and high frequency waves respectively and one for the acquisition of vibration responses. The application of permanent monitoring with use of VAM could reduce the number of necessary NDT inspections and to increase the safety of aircrafts.

Further work in this area is underway to test the technique in the industrial environment and to understand the physical mechanisms behind the observed signal modulations that could lead to identification of the type of damage.

ACKNOWLEDGMENTS

The research on nonlinear test techniques was financed by the Foundation for Polish Science (FNP) within the scope of the WELCOME project no. 2010-3/2. The authors would like to acknowledge the EU FP7 project no. ACP8-GA-2009-234333 for providing the test samples.

REFERENCES

1. Baker, A., Dutton, S., Kelly, D., "Composite Materials for Aircraft Structures," *AIAA*, 2004.
2. Hexcel Composites, Aerospace Selector Guide, 2014.
3. Katnam, K. B., Da Silva, L. F. M., Young T. M., "Bonded repair of composite aircraft structures: A review of scientific challenges and opportunities," *Prog. Aerosp. Sci.*, Vol. 61, 2013, pp. 26–42.
4. Pavlopoulou, S., Soutis, C., Staszewski W. J., "Structural Health Monitoring of Composite Scarf Repairs with Guided Waves", *Key Eng. Mater.*, Vol. 518, 2012, pp. 328–337.
5. Fischer, F., Romoli, L., Kling, R., "Laser-based repair of carbon fiber reinforced plastics," *CIRP Annals - Manufacturing Technology*, Vol. 59, No. 1, 2010, pp. 203–206.
6. Hexcel Composites., "Composite repair", Duxford, 1999.
7. Staszewski, W. J., Boller, C., Tomlinson, G. R. (Eds.), "Health Monitoring of Aerospace Structures," *Wiley, Chichester,* 2004.
8. Stepinski, T., Uhl, T., Staszewski, W. J. (Eds.), "Advanced Structural Damage Detection: From Theory to Engineering Applications," *Wiley, Chichester,* 2013.
9. Reifsnider, K. L., Henneke, E. G., Stinchcomb, W. W. , "The mechanics of vibrothermography," *in Mechanics of Nondestructive Testing*, edited by W. W. Stinchcomb (Plenum, New York, 1980), pp. 249–276.
10. Pieczonka, L., Szwedo, M., "Vibrothermography," *in Advanced Structural Damage Detection: From Theory to Engineering Applications*, edited by Stepinski, T., Uhl, T., Staszewski, W. J. (Wiley, 2013), pp. 233–261.
11. Donskoy, D. M., Sutin, A. M., "Vibro-acoustic modulation nondestructive evaluation technique," *J. Intell. Mater. Syst. Struct.*, Vol. 9, 1999, pp. 765–775.
12. Pieczonka, L., Klepka, A., Martowicz, A., Staszewski, W. J., "Nonlinear vibroacoustic wave modulations for structural damage detection: an overview", *Opt. Eng.,* Vol. 55, No. 1, 2015, pp. 011005.
13. Parsons, Z., Staszewski, W. J., "Nonlinear acoustics with low-profile piezoceramic excitation for crack detection in metallic structures," *Smart Mater. and Struct.*, Vol. 15, 2006, pp. 1110–1118.
14. Aymerich, F., Staszewski, W. J., "Impact damage detection in composite laminates using nonlinear acoustics," *Comp. Part A: Appl. Sci. and Manuf.*, Vol. 41, No. 9, 2010, pp. 1084–1092.
15. Klepka, A., Pieczonka, L., Staszewski, W. J., Aymerich, F., "Impact damage detection in laminated composites by non-linear vibro-acoustic wave modulations," *Comp. Part B Eng.*, 2013
16. Van Den Abeele, K., Johnson, P. A., Sutin, A., "Nonlinear Elastic Wave Spectroscopy (NEWS) Techniques to Discern Material Damage, Part I : Nonlinear Wave Modulation Spectroscopy (NWMS)", *Res. Nondestr. Eval.*, Vol. 12, 2000, pp. 17–30.

ICAST2015 #007

Investigation of Aircrew Noise Exposure Levels and Hearing Protection Solutions in Helicopter Cabin

Yong Chen*, Sebastian Ghinet, Andrew Price, Viresh Wickramasinghe and Anant Grewal

Flight Research Laboratory, Aerospace, National Research Council Canada, Ottawa, Canada

Abstract

High noise levels in the helicopter cabin adversely affect aircrew communication and reduce comfort in the short-term, and may lead to hearing loss in the long-term if flight helmets cannot provide sufficient protection to the aircrew. A cabin noise exposure survey has been performed on a Royal Canadian Air Force (RCAF) CH-147F Chinook heavy lift helicopter to evaluate the noise environment and noise protection performance of the flight helmet. Investigation results showed that the low frequency noise attenuation provided by the RCAF flight helmet was marginal in high speed flight conditions that generate loud cabin noise. Therefore, in-canal earphone integrated with active noise cancellation (ANC) capability was investigated to provide enhanced noise protection and improve clarity in voice communication. Simulation and proof-of-concept test results verified that ANC in-canal earphones can serve as a feasible technical solution to provide enhanced noise attenuation to mitigate the low frequency N/rev tonal noise generated by the aerodynamic pressure from the helicopter rotor blades.

1. INTRODUCTION

The helicopter is a versatile air-borne platform that provides the unique capability of vertical taking-off, landing and hovering. However, the ride quality in a helicopter cabin is generally unpleasant due to high vibration and noise levels. The helicopter cabin acoustic environment typically consists of multiple N/rev low frequency tonal noise peaks related to the rotor aerodynamics and numerous meshing tonal noise related to the transmission/hydraulic system at the mid and high frequencies in addition to the broadband aerodynamic noise. In order to reduce noise exposure levels and enhance communication, helicopter aircrew is required to wear flight helmet throughout flight missions. If the flight helmet cannot provide sufficient noise protection or the helmet is improperly worn by the aircrew members, the high noise levels in the helicopter cabin can adversely affect aircrew communication and reduce comfort in the short-term, and may lead to aircrew hearing loss or even loss of flight status in the long-term.

Evaluation of helicopter cabin noise environment is essential in the selection of hearing protection devices to ensure aircrew safety and mitigation of long-term health issues due to exposure to high noise levels. This paper presents a flight test performed on a Royal Canadian Air Force (RCAF) CH-147F Chinook tandem rotor heavy lift helicopter. The Sound Pressure Levels (SPL) were measured at

* Eric.Chen@nrc-cnrc.gc.ca

nine aircrew stations for selected normal and combat level flights and maneuvers to characterize the cockpit and cabin noise environment. The measurement results were analyzed to determine the noise exposure levels for the CH-147F aircrew members in the tested flight conditions based on the acoustic performance of the hearing protection equipment used by the RCAF helicopter aircrew. Due to the presence of high level tonal noise related to the N/rev harmonics of the tandem rotor system, in-canal earphones integrated with active noise cancellation (ANC) capability have been investigated for enhanced hearing protection and improved clarity in the voice communication through numerical simulation and proof-of-concept tests in a reverberant acoustic chamber facility.

2. NOISE MEASUREMENT OF THE CH-147F CHINOOK HELICOPTER CABIN

A RCAF CH-147F Chinook helicopter was instrumented for noise measurement in this investigation. The CH-147F is Canadian military version of the Boeing CH-47F Chinook heavy lift helicopter that is equipped with a twin-engine tandem rotor system. Each rotor is equipped with three blades, and the nominal rotating speed is 215 RPM, or 3.75Hz. The primary roles of this vehicle include troop movement, artillery placement and battlefield resupply.

The test instrumentation and test procedures for this in-flight noise measurement effort were developed based on ISO 5129, ISO 9612, and CSA Z107.56-06 standards [1, 2 and 3]. In accordance with ISO 5129, the interior of the aircraft was unaltered with reference to the normal flight mission configuration. Seat backrests were set to their most upright position, while the number of occupants in the flight was kept to the minimum (6 persons) required for conducting this test. To eliminate interference of sound propagation, no obstruction was placed between the microphone locations and aircrew positions. The selected aircrew stations for noise measurement included the right seat pilot and left seat pilot (copilot) in the cockpit; five empty seat positions and two standing positions in the passenger cabin, as schematically shown in Figure 1.

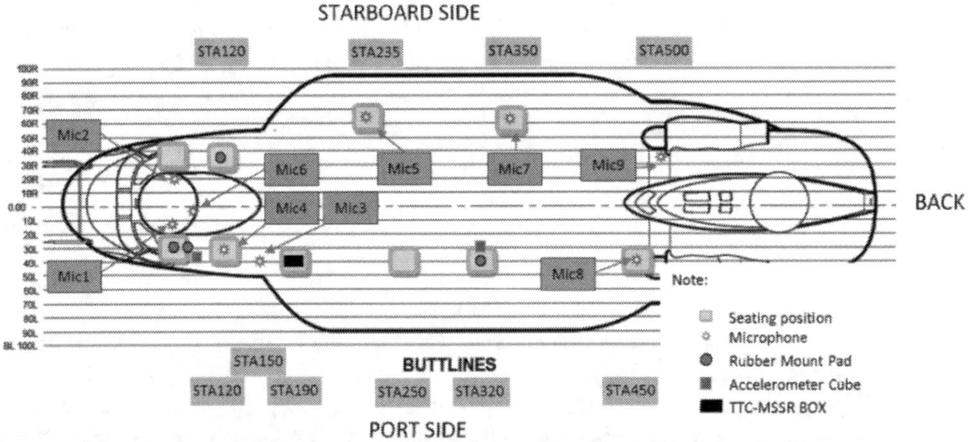

Figure 1. Sensor and aircrew locations in the CH-147F helicopter cabin

Military helicopters are required to operate in various flight conditions and configurations based on their missions. For example, missions such as rappelling, search and rescue, as well as the delivery of payloads and weapons require the helicopter to operate with opened doors. In this configuration the helicopter cabin noise levels increases significantly and may exceed the limit of aircrew hearing protection equipment. Therefore, the flight test sequence was designed to cover three groups of operations, as listed in Table 1. The ground interface maneuvers included ground running, taking-off, and landing on a paved tarmac. Stationary flight involved hovering at two altitudes of 10 feet and 40 feet, respectively, as well as with a sling load of 4000 lbs(f) at two altitudes of 40 feet and 80 feet. Steady airspeed maneuvers included climbing, level, and descending flight. These flight conditions were performed in both doors opened and closed configurations to represent the typical operations of the Chinook helicopter fleet.

Nine ICP type PCB378B02 microphones fitted with preamplifier type 426E01 and windshield were used for acoustic measurement. Each microphone was calibrated according to standard test procedures using a GRAS Type 42AC sound calibrator prior to the flight test. For each seat position, in accordance with MIL-STD-1294A [4], the microphone was oriented vertically upwards at a distance of 0.15 meters from the headrest, and 0.8 meters above the unoccupied seat cushion. In the cockpit, two microphones were located at the seated head height of the two pilots. Each microphone was kept 0.1 meters from the pilot helmet position with the pilots seated for flight operations. For the two standing crew microphone positions, each microphone was positioned 1.65 meters above the floor on secured tripods. Typical arrangements are shown in Figure 2.

Table 1 Flight sequence in measurement runs

Doors Open				Doors Closed			
Run	Flight	Description	Time	Run	Flight	Description	Time
1		EPUSHA ON	60s	15	Hover	Hover 10 ft., door closed	60s
2		Ground Power, Avionics ON	60s	16		Hover 40 ft., door closed	60s
4	Ground	APU ON, Avionics Running	60s	17	Sling	Sling 40 ft., door closed, hatch	60s
6		APU ON, Engine Levers Idle	60s	18		Sling 80 ft., doors closed,	60s
8b		APU ON, Engine Levers Flight	60s	19		30 kts	60s
10		APU OFF, Engine Levers Flight	60s	20	Level	80 kts	60s
11	Hover	Hover 10 ft., doors open,	60s	21		120 kts	60s
12		Hover 40 ft., doors open,	60s	22		140 kts	60s
13	Sling	Sling 40 ft., doors open, hatch	60s	23		150 kts (optional)	60s
14		Sling 80 ft., doors open, hatch	60s	24		80 kts, 30 Deg bank	60s
35		30 kts	60s	25		80 kts, 45 Deg bank	60s
36	Level	80 kts	60s	26		100 kts, 30 Deg bank	60s
37		120 kts	60s	27	Manoeuvre	100 kts, 45 Deg bank	60s
38		140 kts	60s	28		120 kts, Rate 1 turn	60s
39		150 kts	60s	29		140 kts, Rate 1 turn,	60s
40		80 kts, 30 Deg bank	60s	30		70 kts, Descend turn, 60 Deg	60s
41		80 kts, 45 Deg bank	60s	31		70 kts, Descend turn, 60 Deg,	60s
42		100 kts, 30 Deg bank	60s	32	Approach	Normal Approach to Hover	60s
43	Manoeuvre	100 kts, 45 Deg bank	60s	33		Normal Approach to No Hover	60s
44		120 kts, Rate 1 turn	60s	34		Roll on Landing	60s
45		140 kts, Rate 1 turn,	60s				
46		70 kts, Descend turn, 60 Deg	60s				
47		70 kts, Descend turn, 60 Deg,	60s				

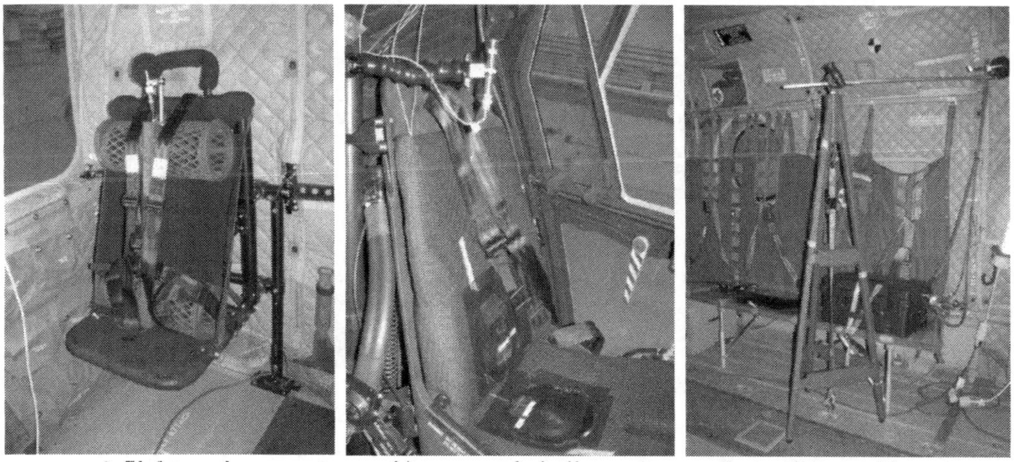

a) flight engineer seat b) unoccupied pilot seat c) standing position

Figure 2. Microphone configuration in the CH-147F helicopter cabin

The HGU-56P-CF flight helmet used by the RCAF helicopter fleet was worn by all aircrew members for voice communication and hearing protection during the flight test. A TTC MSSR-100C series miniature data acquisition system was secured on the cabin bench as a non-essential item for normal flight operations. It is important to note that the noise measurement system was not part of the RCAF CH-147F Chinook helicopter critical flight instrumentation and the system did not interfere with aircraft operations.

3. AIRCREW NOISE EXPOSURE LEVELS

3.1 Flight Helmet Insertion Loss Performance

Aircrew noise exposure level depends on the acoustic Insertion Loss performance of the flight helmet used by the aircrew as well as the noise level of the surrounding environment. Therefore, the helmet Insertion Loss performance of the HGU-56P-CF flight helmet was measured in accordance with procedures specified in ANSI/ASA S12.42-2010 [5] using a GRAS 45CB Acoustic Test Fixture. The HGU-56P-CF flight helmet installed on the acoustic test fixture and average Insertion Loss performance are shown in Figure 3.

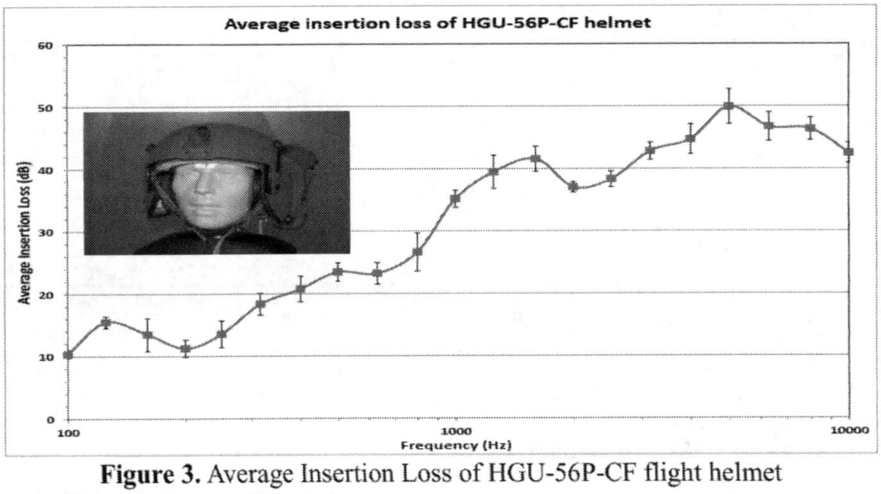

Figure 3. Average Insertion Loss of HGU-56P-CF flight helmet

The HGU-56P-CF helmet Insertion Loss performance showed that the helmet was able to provide more than 30dB reduction to noise above the 1/3 Octave frequency band of 1000 Hz. However, the Insertion Loss performance reduced significantly in the low frequency band. For the 1/3 Octave frequency band of 100 Hz, the insertion loss provided by the HGU-56P-CF flight helmet was approximately 10dB. For the 1/3 Octave frequency band of 400 Hz, the insertion loss was close to 20dB. Considering the fact that helicopter cabin noise is typically dominated by the N/rev harmonic tones of the rotor aerodynamic pressure in the low frequency range, consideration should be taken to ensure the aircrew is provided with appropriate hearing protection to avoid excessive noise exposure in the frequency range of 100Hz and 400Hz.

3.2 Characterization of Helicopter Cabin Noise Environment

The SPLs at the nine selected cockpit and cabin aircrew stations were measured on the CH-147F helicopter operated by the RCAF. The Power Spectral Density plots at the nine aircrew stations for the Closed Doors configuration at 150 knots level flight condition are shown in Figure 4. It can be observed that the sound pressure spectra below 400Hz were dominated by the N/rev harmonics of the helicopter tandem rotor aerodynamic pressure while the sound pressure spectra in the mid and high frequency ranges between 400Hz and 5500Hz were influenced by numerous tonal noise peaks resulting from the helicopter rotor transmission gearboxes and hydraulic pumps.

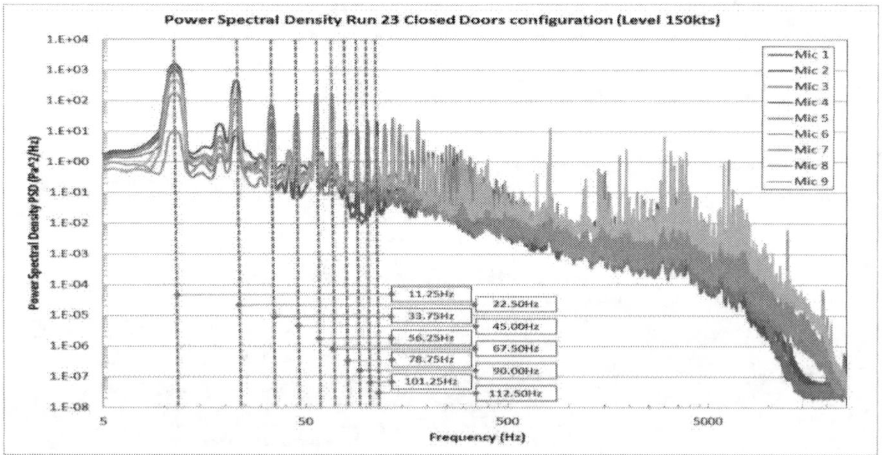

Figure 4. Power spectral density for Closed Doors configuration at 150 knots level flight

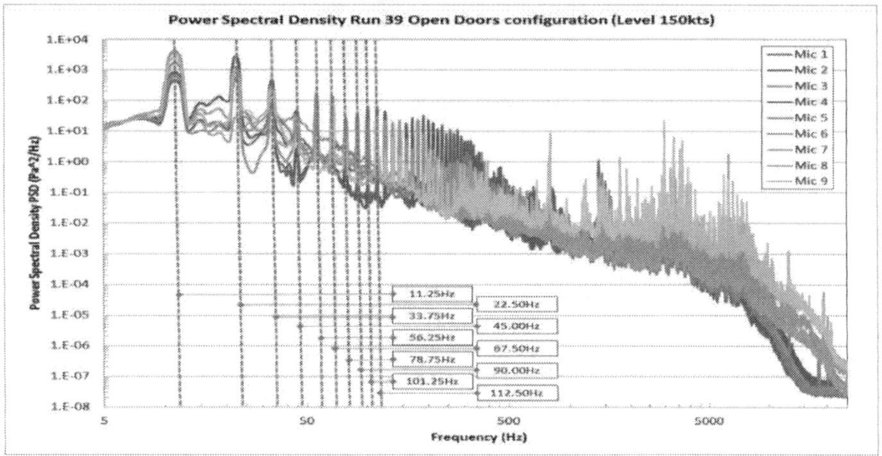

Figure 5. Power spectral density for Open Doors configuration at 150 knots level flight

The power spectral density plots at the nine aircrew stations for the Open Doors configuration at 150 knots level flight condition are shown in Figure 5. Compared to the Closed Door configuration at the same flight speed, the overall SPL generally increased for all locations. For example, the overall SPL of the starboard side flight engineer seat measured by Mic 4 increased from 127.1 dB to 129.9 dB, mainly due to the increase in sound levels between the 1/3 octave bands of 16 Hz and 630 Hz. Consequently, the overall A-weighted sound pressure level at this location also increased from 109.0 dB(A) to 113.1dB(A).

In the Open Doors configuration, the transmission loss provided by the doors and windows was not available. This facilitated the low frequency airborne noise components such as the N/rev aerodynamic pressure to increase significantly in the helicopter cabin and cockpit. Comparatively, no significant difference was identified in the mid and high frequency ranges because the noise within this range was dominantly generated by the meshing tones related to the helicopter rotor gearboxes and hydraulic pumps. The tonal noise was transmitted through helicopter fuselage in the form of structure borne vibration and radiated noise inside the cabin. Among the measured stations, the most severely affected location was the starboard flight engineer seat at the STA120 location which was located between an open door and a window.

To account for the relative loudness perceived by human ears, the A-weighted noise spectra for the nine aircrew stations at 150 knots level flight condition with doors closed are shown in Figure 6. The maximum overall A-weighted SPL was 119 dB(A) which was measured by Mic 9 at the aircrew station located at the STA500 starboard side. Based on the microphone configuration shown in Figure 1, this passenger location was at the back of the cabin and close to the rear ramp door where many hydraulic systems and electronic instrumentation boxes were installed. This equipment emitted high levels of tonal noise in the frequency range between 1 kHz and 5.5 kHz, as shown in Figure 5 and Figure 6 as a light green line (Mic 9). As a result, the A-weighted SPL between the 1/3 Octave band of 1 kHz and 6.3 kHz showed higher SPL than other aircrew stations.

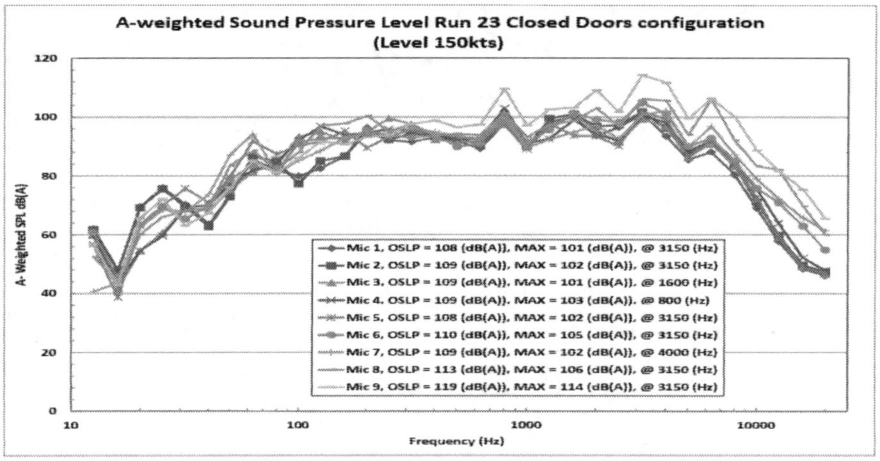

Figure 6. Sound Pressure Level for Closed Doors configuration at150 knots level flight segment

In order to evaluate the impact to aircrew hearing due to the helicopter cabin noise environment, the Insertion Loss performance of the RCAF HGU-56P-CF helmet was used to determine the actual SPL to which the aircrew member was exposed. The estimated A-weighted SPL for the 150 knots level flight condition with helicopter doors open is shown in Figure 7.

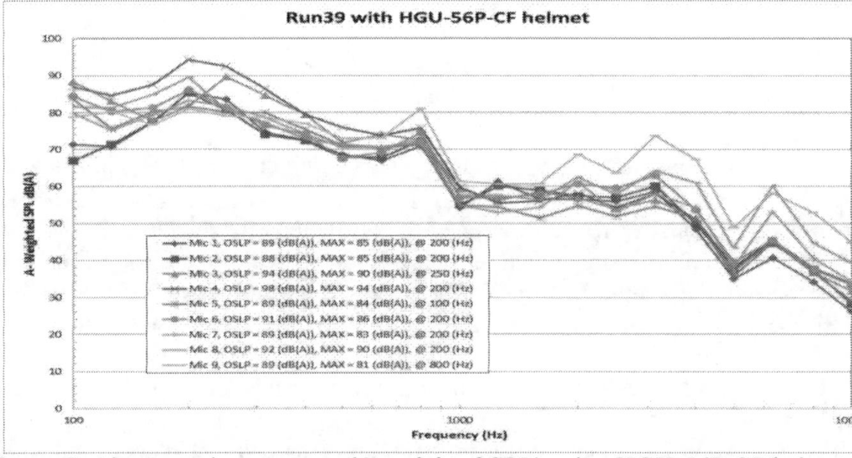

Figure 7. Aircrew noise exposure (A-weighted SPL) using HGU-56P-CF helmet

It should be noted that this flight condition recorded the highest noise level in this flight test. Equipped with a properly worn HGU-56P-CF flight helmet, high sound pressure levels existed in the 1/3 Octave bands between 100 Hz and 400Hz. The highest sound pressure level of 96 dB(A) was identified at the Mic 4 location which was related to the starboard side flight engineer seat at STA120. Other microphone locations such as Mic 3, 6, 8 and 9 also exceeded 85 dB(A) within the 1/3 Octave frequency bands between 100 Hz and 400Hz. The A-weighted SPL exceeded 80 dB(A) in the same frequency range for all measured aircrew stations.

Referencing to the Canada Occupational Health and Safety Regulations on sound levels [6], the protection provided by the HGU-56P-CF flight helmet may be insufficient for a few stations such as the Mic 9 location in long flight missions [7]. Therefore, enhanced sound attenuation between 100 Hz and 400Hz is required to ensure adequate hearing protection for the aircrew members in the RCAF CH-147F helicopter fleet. This is important because the maximum mission duration of the CH-147F Chinook could exceed 9 hours with external fuel tanks.

4. ACTIVE NOISE CANCELLATION FOR ENHANCED HEARING PROTECTION

In the current HGU-56P-CF flight helmet design, two in-canal earphone plugs are used for passive noise attenuation and voice communication purposes. To enhance the noise reduction performance specifically to suppress the high sound pressure level between the frequency range of 100 Hz and 400Hz, a feasible technical approach is to integrate active noise cancellation (ANC) technology into the existing Communication Ear Plugs (CEP). This technical solution is beneficial because it requires minimal design changes to the flight helmets. Compared to the passive in-canal earphone plugs used in the RCAF HGU-56P-CF, ANC earphone plugs can provide enhanced protection in the low frequency range to enhance the protection provided by the current helmets, and also to improve the clarity of the voice communication by reducing background noise in the aircrew ear canals.

4.1 Feasibility Study

A numerical simulation study was performed to investigate the feasibility for ANC technology to reduce the N/rev harmonics tonal noise in the aircrew ear canals. In this study, an energy normalized least mean-square (LMS) algorithm was used as the active noise cancellation algorithm in a Filtered-X feedforward adaptive controller [8, 9]. The earphone response dynamics was assumed to be flat between 20 Hz and 2000Hz.

The sound pressure level measured near the right side pilot seat of the CH-147F helicopter was used as the target signal in this noise suppression study. The sound attenuation provided by the flight helmet was not considered in this numerical feasibility study. The focus of the simulation was to suppress the N/rev harmonic tonal noise between the frequency range of 100Hz and 500Hz. To minimize the control energy required for the noise cancellation, the sound pressure level was A-weighted prior to the adaptation by the ANC algorithm that generated the control output signal for in-canal noise cancellation. The frequency range of the simulation was restricted to the maximum of 1000Hz using a MATLAB SIMULINK model. In this simulation, the noise signal measured in the cockpit was used as the control reference signal because it contained the N/rev harmonic tones that showed high level of correlation with the target noise signal to be suppressed in this ANC application.

It was important to note that the highest non-weighted SPLs came from the lower N/rev harmonics tones such as the 3/rev at 11.25Hz as measured in the CH-147F cockpit and cabin. However, the perceived loudness and impact on human hearing was dominated by frequencies above 100 Hz. Based on flight test results, the target frequency range was required to reduce the N/rev tonal noise between 100Hz and 400Hz while the protection in higher frequency bands would be adequately provided by the HGU-56P-CF flight helmet if properly worn by the aircrew members.

Typical numerical simulation results are shown in Figure 8. In this case, the sound pressure signal from the 150 knots forward flight condition was used for the simulation. Simulation results showed that the amplitudes of major N/rev harmonic tonal noise peaks between 100 Hz to 500 Hz were effectively suppressed by the ANC algorithm, and control spillover effect was not observed. Therefore, the concept of ANC in-canal earphones can serve as a feasible technical solution to provide enhanced hearing protection to the aircrew by significantly reducing the N/rev harmonic tonal noise generated by the aerodynamic pressure of helicopter rotors.

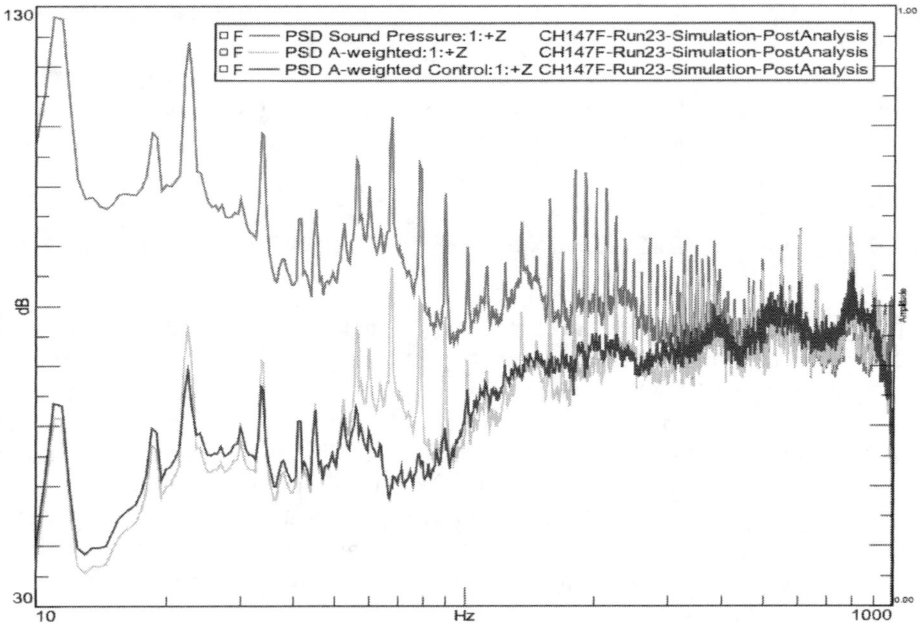

Figure 8. Numerical simulation results

4.2 Proof-of-Concept Experiment

Following the numerical simulation study, proof-of-concept experiments were conducted using the NRC acoustic reverberant chamber facility to further validate the in-canal ANC earphone concept to enhance low frequency noise attenuation for helicopter aircrew.

To provide a representative acoustic environment similar to the CH-147F cabin in the reverberant chamber, the sound pressure signal measured in the CH-147F cabin during the 140 knots forward flight condition was recorded and played back as an analogue audio signal. The audio signal was connected to two independent sets of audio shapers and amplifiers to drive a group of low frequency speakers and a group of high frequency speakers simultaneously to replicate the sound pressure spectrum of CH-147F helicopter cabin in the acoustic reverberant chamber. Due to insufficient actuation power of the low frequency speakers, the replicated SPL was not able to reach the target level in the frequency range below 100 Hz. However, the representative N/rev harmonic tonal noise spectra measured in the helicopter cabin between 100 Hz and 500 Hz was reproduced in the spectral domain and reached sufficient SPL for proof-of-concept experiments. In this experiment, an overall sound pressure level of 120.0dB was generated in the reverberant acoustic chamber.

The experimental configuration in the reverberant acoustic chamber is shown in Figure 9. To evaluate the effectiveness of the in-canal ANC earphones to suppress the N/rev harmonic tonal noise, three separate tests were performed with the input noise level in the acoustic reverberant chamber kept at 120.0dB. A GRAS 45CB acoustic test fixture was used in the test and placed in the middle of the reverberant acoustic chamber for evaluation of the SPL in the ear canals. This acoustic test fixture is equipped with two in-canal microphones to enable high fidelity measurement of SPL as perceived by human ears.

The tests were performed in three configurations. The first configuration was to measure the in-canal sound pressure with the HGU-56P-CF flight helmet properly worn on the GRAS acoustic fixture, as shown in Figure 9b. The second configuration used the flight helmet on the GRAS acoustic test fixture and two BOSE 5FAC in-canal earphones plugged in both ear canals of the GRAS acoustic test fixture without activation of the ANC algorithm. In this case, the in-canal earphones worn by the GRAS 45CB fixture is shown in Figure 9a, while the earphones cannot be seen from outside when the flight helmet was used. With comparison of results obtained in the two configurations, this would enable evaluation of the passive sound attenuation effect provided by the in-canal earphone plugs. The third configuration used the same hardware configuration as the second except the ANC algorithm was activated for in-canal noise cancellation. By comparing the SPLs in the ear canals of the GRAS 45CB acoustic fixture in each configuration, the effectiveness of the ANC in-canal earphones was obtained in terms of both spectral plots and 1/3 Octave SPL values.

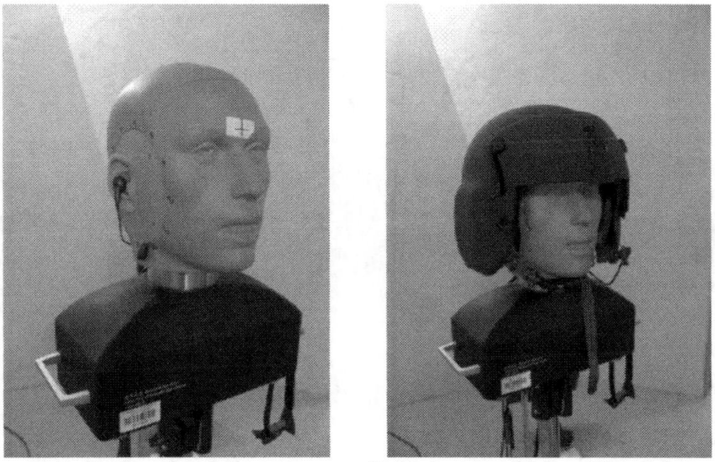

a) In-canal earphones b) Helmet and in-canal earphone

Figure 9. Acoustic test configuration in the reverberant acoustic chamber

Typical test results are shown in Figure 10. The red line indicates the sound pressure spectrum reproduced in the reverberant chamber. The green line indicates the in-canal sound pressure spectrum of the GRAS acoustic fixture with the use of a HGU-56P-CF flight helmet. The blue line indicates the in-canal sound pressure spectrum of the acoustic fixture equipped with two passive earphones and the HGU-56P-CF flight helmet. The pink line indicates the in-canal sound pressure spectrum when the ANC algorithm is activated.

In the first configuration as shown by the green line, the flight helmet demonstrated satisfactory sound attenuation performance of more than 25dB in the frequency range above 500Hz and 40dB above 1000Hz. However, the sound attenuation for frequency ranges below 100Hz was less effective. This was generally consistent with the Insertion Loss data as described in Section 3.1. Consequently, the overall non-weighted SPL was measured as 111.4dB and the A-weighted SPL was 90.3dB(A) in the ear canal locations of the GRAS acoustic test fixture.

With the addition of two passive earphone plugs in the left and right ear canals as tested in the second configuration, as shown by the blue line, the SPL in the frequency range above 1000Hz was observed to reduce significantly. This showed that the passive earphone plugs were able to provide enhanced noise attenuation to further reduce the tonal noise in the mid frequency range above 1000Hz that was generated by the rotor gearboxes and hydraulic pumps. However, the amplitude of low frequency noise below 100Hz was not reduced but marginally increased. This may be related to the break in the earmuff sealing by the wires of the BOSE 5FAC earphone plugs. In this case, the overall in-canal SPL was 110.9dB and the A-weighted SPL was 88.3 dB(A). This demonstrated that passive earphone plugs were not very effective to reduce the aircrew noise exposure.

In the third configuration, as shown by the pink line when the ANC algorithm was activated to control the earphones adaptively for noise cancellation, the SPL in the ear canal was reduced effectively. Similar to that shown in the numerical simulation, the amplitudes of major N/rev harmonic tonal peaks below 500Hz have been suppressed simultaneously. As a result, the non-weighted SPL in the canal location of the acoustic fixture was reduced to 107.8dB and the A-weighted SPL was reduced to 77.3 dB(A). Compared to the previous test case, the A-weighted OSPL was reduced effectively by 11.0 dB(A) through adaptive suppression of the N/rev tonal noise inside the ear canals of the GRAS acoustic test fixture. In this case, the highest non-weighted SPL of the acoustic fixture was measured as 81.7dB at 25Hz. For the frequency range of interest between 100Hz and 400Hz, the highest SPL of 71.0 dB was measured at the 1/3 Octave band of 200 Hz.

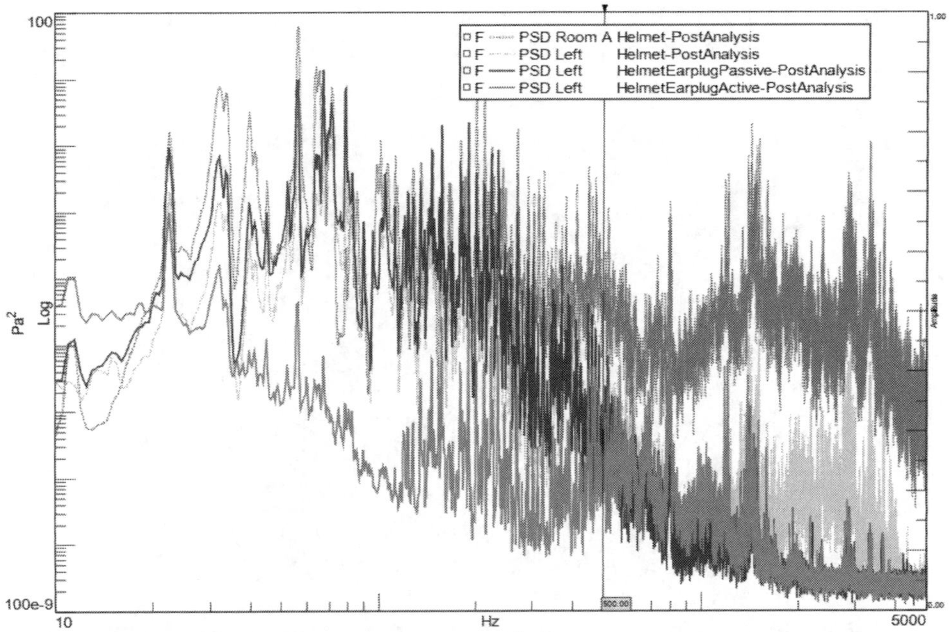

Figure 10. In-canal sound pressure level in the reverberant acoustic chamber

The proof-of-concept experiments verified that the RCAF HGU-56P-CF flight helmet was able to provide effective hearing protection to the aircrew members in the frequency range above 500Hz. The passive earphone plugs were able to enhance noise attenuation in the frequency range above 1000Hz to reduce the noise generated by the gearboxes and pumps. However, it was observed that the passive ear plugs were not effective in reducing low frequency noise such as N/rev harmonic noise. The integration of ANC capability to the in-canal earphones was able to provide significant attenuation to reduce the N/rev harmonic tonal noise in the critical frequency range between 100Hz and 400Hz. Therefore, ANC in-canal earphones can serve as a feasible technical solution to enhance hearing protection performance of helicopter flight helmets.

5. CONCLUSION

A flight test was conducted on a Royal Canadian Air Force CH-147F Chinook heavy lift helicopter to measure the cabin and cockpit noise environment. The Insertion Loss performance of the in-service HGU-56P-CF helmet was able to provide sufficient reduction in noise levels above 1000Hz. However the test results revealed that the sound attenuation performance provided by the flight helmet may be insufficient for certain aircrew stations in the frequency range between 100Hz and 400Hz when performing long duration and high speed flight missions.

To improve the sound attenuation performance of flight helmets within the frequency range of concern, active noise cancellation (ANC) earphones were integrated into the flight helmet to provide enhanced noise protection to the aircrew. Numerical simulation and proof-of-concept tests have been performed using representative sound pressure levels and test configurations inside a reverberant acoustic chamber. The test results verified that the ANC in-canal earphones achieved 11.0 dB(A) reduction in the overall SPL measured at the ear canal location of a GRAS acoustic test fixture. This demonstrated that ANC system was able to reduce the noise through effective suppression of major N/rev harmonic noise generated by the aerodynamic pressure of helicopter rotor blades.

This investigation confirmed that integration of ANC earphones is a feasible technical solution to improve the sound attenuation performance of helicopter flight helmets to mitigate the risks of excessive noise exposure on helicopter aircrew.

ACKNOWLEDGEMENTS

The authors would like to acknowledge the support from the Directorate of Technical Airworthiness and Engineering Support (DTAES) - Human Factors Engineering of the Department of National Defence of Canada and RCAF 450 THS squadron staff. The assistance from Heather Wright, Brent Lawrie and Camile Lebrun, and the Airworthiness Group of the NRC Flight Research Laboratory is also appreciated.

REFERENCES

1. International Standard Organization, “Acoustics - Measurement of Sound Pressure Levels in the Interior of Aircraft during Flight”, ISO 5129:2001.
2. International Standard Organization, “Acoustics - Determination of Occupational Noise Exposure - Engineering Method”, ISO 9612:2009.
3. Canadian Standard Association, “Procedures for the Measurement of Occupational Noise Exposure”, CSA Z107.56-06, 2008.
4. United States Military Standard, “Acoustical Noise Limits in Helicopters”, MIL-STD-1294A.
5. American National Standards Institute, “Methods for the Measurement of Insertion Loss of Hearing Protection Devices in Continuous or Impulsive Noise Using Microphone-in-Real-Ear or Acoustic Test Fixture Procedures”, ANSI/ASA S12.42-2010.
6. Levels of Sound - Canada Occupational Health and Safety Regulations, Part VII - IPG-074.
7. Ghinet, S., Price. A. and Wickramsinghe, V. et al. “Flight test investigation of cabin noise exposure in the Royal Canadian Air Force CH-147F helicopter”, 44th Inter-Noise Congress and Exposition on Noise Control Engineering, San Francisco, USA, August 9-12, 2015.
8. Millot T.A. et al., “Flight test of active gear-mesh noise control on the S-76 aircraft”, 54th American Helicopter Society Annual Forum, Washington DC, USA, 1998.
9. Glover J.R., “Adaptive noise canceling applied to sinusoidal interferences”, *IEEE Transactions on Acoustics, Speech, and Signal Processing*, Assp-25 (6), 1977, pp.484-491.

ICAST2015 #088

Semi-active noise control using non-symmetrical SSD method

Hongli Ji[1,2], Li Cheng[2], Hong Nie[1], and Jinhao Qiu[1*]

[1] State Key Laboratory of Mechanics and Control of Mechanical Structures, Nanjing University of Aeronautics and Astronautics, Nanjing 210016, China
[2] Department of Mechanical Engineering, Consortium for Sound and Vibration Research, The Hong Kong Polytechnic University, Hung Hom, Kowloon, Hong Kong, China.
qiu@nuaa.edu.cn

Abstract

Suppression of noise transmission has received much attention in recent years because of its practical application in many engineering structures such as civil aircraft, passenger trains and cars etc.. Conventional methods start to hit their limit in many applications. Semi-active synchronized switch damping (SSD) approaches show great appealing features in vibration control because no proportional power amplifiers and high-performance digital signal processing units are required, and these systems are much easy to implement for practical applications. Vibration control consequently leads to suppression in the noise transmission. In semi-active synchronized switch damping (SSD) approaches for structural vibration control, the damping effect is achieved by properly switching the voltage on the piezoelectric actuators. The control performance mainly depends on the value of the voltage applied to the piezoelectric elements. Several variants of the SSD control techniques have been proposed to increase the control performance. Due to the inherent properties of piezoelectric materials, the applicable bipolar voltage of a piezoelectric actuator is unsymmetrical. In this paper, unsymmetrical SSD switch circuit is designed to increase the effective voltage range on the PZT actuator for improvement of vibration and noise control performance. First the model of an unsymmetrical SSD system is presented and the working principle is introduced. Next, the switch control logic using modal observer is optimized based on acoustic radiation mode and vibrational mode. Finally, a setup of a plate with unsymmetrical switch circuit is used to demonstrate the control performance. Experimental results show that the proposed control approach exhibits good performance in suppressing noise transmission.

1. INTRODUCTION

Suppression of noise transmission has received much attention in recent years because of its practical implication in many engineering structures such as civil aircraft, passenger trains and cars etc. [1-2]. Conventional methods can be roughly categorized into passive, active and semi-active. In passive methods, absorbing materials are used to improve the effectiveness of noise transmission suppression of double-walled structure [3]. Damping layers can also be used to reduce the vibration level of structures, consequently leading to suppression in the noise transmission [4]. However, these methods start to hit their limit in many applications. Further improvement through purely passive means is usually at the expense of increasing cost, structural weight and size.

* qiu@nuaa.edu.cn

Active methods for noise transmission suppression using structural vibration control have been proposed owing to the progress in electronic and computer technologies to overcome the disadvantage of passive methods. Studies in the active suppression of noise transmission have been reported [5-9]. Jones and Fuller [5] studied the reduction of interior sound fields in flexible cylinders induced by an exterior acoustic monopole using active vibration control. Results indicated spatially averaged noise reductions of above 20 dB over the source plane for acoustic resonant conditions within the cavity. Ho et al. [7] used a piezoelectric actuator and a velocity feedback controller for the reduction of noise transmission through a plate subject to a white noise excitation using vibration as the feedback signal. A new noise isolation system of a composite board without using external noise sensors such as microphones was constructed by Ji et al. [8]. With only one embedded piezoelectric patch in the smart board which functions as both a sensor and an actuator, the system consists of a self-sensing actuator, a neural network identifier and an adaptive feedback controller using Filtered-X LMS algorithm.

Although active methods have the advantages of high control performance and robustness, they also share the drawbacks of requiring high-performance digital signal processors and bulky power amplifiers, which is undesirable in many applications. In order to tackle the problem, several semi-active approaches have been proposed [9-17], among which the pulse switching technique has been receiving much attention in recent years [12-17]. The pulse switching technique consists in a fast inversion of voltage on the piezoelement using a few basic electronics, which is synchronized with the mechanical vibration. In the SSD technique, the control performance mainly depends on the value of the voltage applied to the piezoelectric elements. Several variations of the SSD control techniques have been proposed in the literature to increase the control performance. To increase the voltage on the piezoelectric actuator, an inductance can be connected to the shunt circuit to invert the voltage. This method is called SSDI. To further improve the control performance, a method called SSDV, standing for synchronized switch damping on voltage, has been proposed.

Application of the SSD methods in multimodal vibration control has been of great interest and also of great challenge because defining optimal switching moment in a MDOF system is a very difficult issue. Several research works have been reported on application of the SSD techniques in multimodal vibration control [15-17]. The techniques of SSD method for multimodal vibration control introduced above can be divided into two categories. The first one is that using filter to select modes and the voltage is switched at the displacement extrema of the selected mode. The second one is using some algorithms to find the optimal switching moment in the resultant displacement of multimode vibration to maximize the converted energy. Although the techniques in the second category are effective for simple system with two or three modes, the algorithms to find the optimal switching moment may become very complicated when the number of modes increases. Hence for a complex system with many modes, multi-channel implementation of the techniques in the first category may be simpler because of the low cost of the SSD system.

Although a multitude of outstanding works have been made in SSD control, the generated voltage on the PZT is symmetrical in the original SSD method. But due to the inherent properties of piezoelectric materials, the applicable bipolar voltage of a piezoelectric actuator is unsymmetrical. That is, the voltage applicable in the positive direction (poled direction) is much higher than that in the negative direction because the applicable positive voltage depends on the electrical breakdown voltage of the piezoelectric material, but the applicable negative voltage depends on its depoling voltage. If symmetrical switching circuit is used, the actuator capability will be wasted.

To solve this problem, an unsymmetrical switch circuit was designed for high-voltage semi-active control method based on SSDV by Ji et al [18]. A bypass capacitor and an additional switch are used to realize unsymmetrical bipolar voltage. Experiments were carried out to verify the designed circuit. Experimental results show that the designed circuit can generate a high unsymmetrical voltage on the piezoelectric actuators.

In this study, unsymmetrical SSD method combined with modal observer is applied to suppression of noise transmission through a panel. In this study, the microphone is not used as feedback sensor, but to evaluate the control performance. The layout of the piezoelectric elements on the panel has been optimized base on the mode shapes, responsible for noise radiation. A setup of a plate with unsymmetrical switch circuit is used to demonstrate the control performance.

2. SSD CONTROL METHOD

2.1 Principle of SSD method

SSD control was proposed to provide an alternative to the existing active and passive control methods, and hopefully could alleviate some of the existing drawbacks of these methods. In SSD control, the voltage on the piezoelectric actuator is switched synchronously with the vibration of the host structure to achieve enhanced damping effect. The power consumption of the control system is usually low because the main consumption is in the switch circuit. It is also robust because switch actions are synchronized with vibration so that it can automatically trace the frequency of structural vibration.

With a piezoelectric actuator bonded on a host vibrating structure, a voltage is generated on it in the open-circuit state by the mechanical strain, which is proportional to the deformation of piezo-element. Though mechanical energy is converted to electrical energy as the strain increases, electrical energy is converted back to mechanical energy as the strain decreases. Hence there is no net energy conversion in a full cycle of mechanical vibration. If the voltage on the piezoelectric actuator is processed by a switch circuit by shifting its phase, its amplitude can be magnified. Hence net energy conversion from mechanical to electrical is produced, resulting in an equivalent added damping effect to the host structure. A typical switch circuit for SSDV method is shown in Fig. 1(a) and the corresponding actuator voltage is shown in Fig. 1(b).

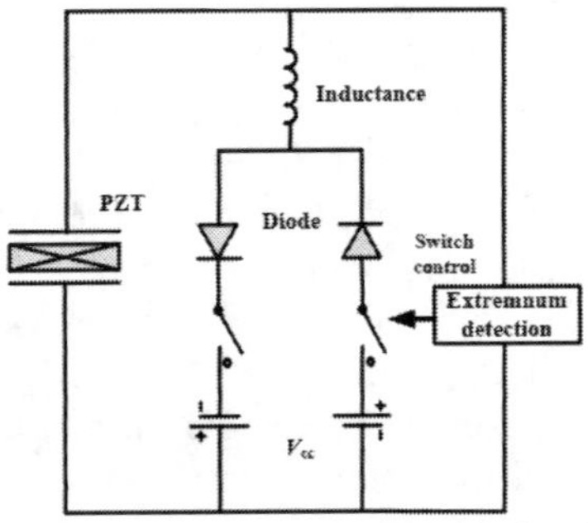

(a) The schematic of a SSDV circuit

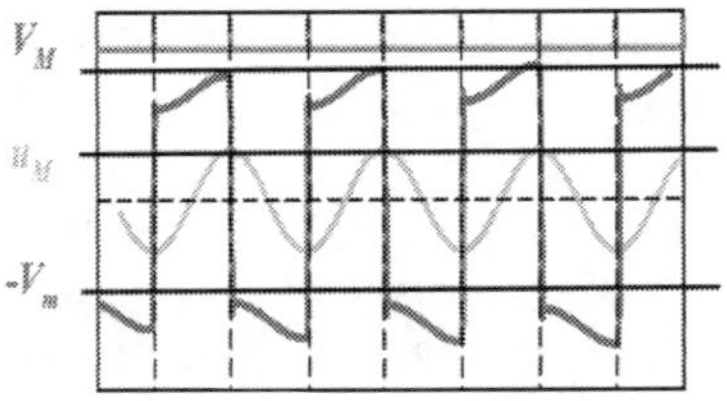

(b) The actuator voltage

Figure 1. The principle of SSDV

2.2 Principle of unsymmetrical SSD method

An unsymmetrical SSD switch circuit to increase the effective actuator voltage range on the PZT has been designed by Ji et al [18]. The schematic diagram of the unsymmetrical SSDV control system is shown in Fig. 2(a). The difference from the classical SSDV is that a unidirectional bypass capacitor C_b and a switch SW3 are connected in parallel to the piezoelectric actuator, which is used to realize the unsymmetrical voltage inversion on the piezoelectric element. In order to obtain unsymmetrical actuator voltage, the switching logic is used to control the three switchers: SW1 is closed at every maximum of displacement and is kept closed for half a period of electrical oscillator; SW2 is closed at every minimum of displacement and is kept closed for half a period of electrical oscillator; SW3 is closed at the same time with SW2, but it is kept closed until the voltage on the piezoelectric actuator equals to 0. The unsymmetrical switched voltage is schematically shown in Fig. 2(b).

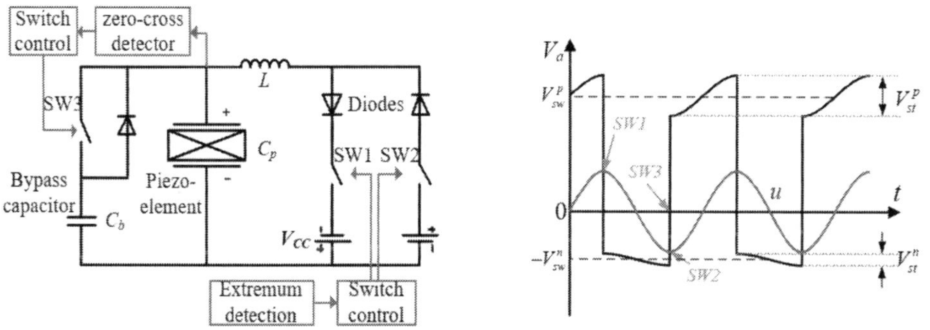

(a) Unsymmetrical SSDV circuit (b) The waveforms of generated voltage and displacement

Figure 2. Principle of unsymmetrical SSDV

3. ELECTROMECHANICAL MODEL OF A VIBRATION SYSTEM

3.1 Electromechanical model

As introduced above, modal displacements are necessary for voltage switching of piezoelectric actuators. Although the sensor signal contains the information of all the modal displacements to be controlled, they are mixed together. Hence, it is necessary to separate them using an observer. In order to identify the modal displacements, the state-space model of structure is necessary.

The vibration displacement $w(x, t)$ at any point of the structure, $\mathbf{x}$, can be written as the summation of the product of the modal coordinates and the eigenmode functions as follows:

$$w(x,t)=\sum_{i=1}^{n}\phi_i(\mathrm{x})u_i(t), \tag{1}$$

where $\phi_i(\mathrm{x})$ is the modal shape of the ith natural mode and $u_i(t)$ is its modal coordinate.

If it is assumed that the modal coordinates have been normalized with respect to modal mass, then the equation of motion in modal space after neglecting the high-order modes can be expressed in the following forms:

$$\ddot{\mathbf{u}}+2[\zeta_i\omega_i]\dot{\mathbf{u}}+[\omega_i^2]\mathbf{u}=-\mathbf{A}_1\mathbf{V}_1-\mathbf{A}_2\mathbf{V}_2+\mathbf{F}_e \tag{2}$$

where $\mathbf{u}$ is the vector of modal coordinates with length n, ω_i is the natural angular frequency of the ith natural mode when all the piezoelectric patches are in short-circuit state, ζ_i is the modal damping, $\mathbf{F}_e$ is the vector of modal force of excitation, $\mathbf{A}_1$ is the voltage coefficient matrix of piezoelectric patches used as actuators, and $\mathbf{A}_2$ is the voltage coefficient matrix of piezoelectric patches used as sensors. $\mathbf{V}_1$ and $\mathbf{V}_2$, having the dimension of k_1 and k_2 (k_1+k_2=m, the total number of piezoelectric patches), are the vectors of actuator voltages and sensor voltages, respectively. They are defined as follows:

$$\mathbf{A}=\left[\mathbf{A}_1\,|\,\mathbf{A}_2\right]=\begin{bmatrix}\alpha_{11} & \alpha_{12} & \cdots & \alpha_{1m}\\ \vdots & \vdots & & \vdots\\ \alpha_{n1} & \alpha_{n2} & \cdots & \alpha_{nm}\end{bmatrix},\quad \mathbf{u}=\begin{Bmatrix}u_1\\ \vdots\\ u_n\end{Bmatrix},\quad \mathbf{F}_e=\begin{Bmatrix}F_{e1}\\ \vdots\\ F_{en}\end{Bmatrix},$$

$$[2\zeta_i\omega_i]=\begin{bmatrix}2\zeta_1\omega_1 & \cdots & 0\\ \vdots & \ddots & \vdots\\ 0 & \cdots & 2\zeta_N\omega_N\end{bmatrix},\ [\omega_i^2]=\begin{bmatrix}\omega_1^2 & \cdots & 0\\ \vdots & \ddots & \vdots\\ 0 & \cdots & \omega_N^2\end{bmatrix}, \tag{3}$$

The electrical equations of the piezoelectric patches can be expressed as

$$\begin{bmatrix} \mathbf{A}_1^T \\ \hline \mathbf{A}_2^T \end{bmatrix} \mathbf{u} - \left[\begin{array}{c|c} \mathbf{C}_{\mathrm{p1}} & 0 \\ \hline 0 & \mathbf{C}_{\mathrm{p2}} \end{array} \right] \begin{bmatrix} \mathbf{V}_1 \\ \hline \mathbf{V}_2 \end{bmatrix} = \begin{bmatrix} \mathbf{Q}_1 \\ \hline \mathbf{Q}_2 \end{bmatrix}. \tag{4}$$

$\mathbf{C}_{\mathrm{p1}}$ and $\mathbf{C}_{\mathrm{p2}}$ are the diagonal matrix of inherent capacitance of the piezoelectric patches used as actuators and sensors, respectively, and $\mathbf{Q}_1$ and $\mathbf{Q}_2$ are the vector of electric charges on the piezoelectric actuators and sensors, respectively. They are defined as follows:

$$\mathbf{C}_{\mathrm{p}} = \left[\begin{array}{c|c} \mathbf{C}_{\mathrm{p1}} & \mathbf{0} \\ \hline \mathbf{0} & \mathbf{C}_{\mathrm{p2}} \end{array} \right] = \left[\begin{array}{ccc|ccc} C_{\mathrm{p1}} & & & & & \\ & \ddots & & & 0 & \\ & & C_{\mathrm{p}k_1} & & & \\ \hline & & & C_{\mathrm{p}k_1+1} & & \\ & 0 & & & \ddots & \\ & & & & & C_{pm} \end{array} \right], \quad \mathbf{Q} = \begin{Bmatrix} Q_1 \\ \vdots \\ Q_{k_1} \end{Bmatrix}, \tag{5}$$

If the input impedance of the sensor interface circuits is larger enough, the sensors can be assumed to be the open-circuit state, that is, the charges on the piezoelectric sensors are zero. This leads to

$$\mathbf{A}_2^T \mathbf{u} - \mathbf{C}_{\mathrm{p2}} \mathbf{V}_2 = 0. \tag{6}$$

Substition of Eq. (6) into Eq. (2) yields

$$\ddot{\mathbf{u}} + 2[\zeta_i \omega_i]\dot{\mathbf{u}} + \left([\omega_i^2] + \mathbf{A}_2 \mathbf{C}_{\mathrm{p2}}^{-1} \mathbf{A}_2^T\right) u = -\mathbf{A}_1 \mathbf{V}_1 + \mathbf{F}_{\mathrm{e}} \tag{7}$$

If we define the state vector,

$$\mathbf{X}(t) = \begin{Bmatrix} \mathbf{u} \\ \hline \dot{\mathbf{u}} \end{Bmatrix}, \tag{8}$$

Then the system can be written in the following form:

$$\begin{aligned} \dot{\mathbf{X}}(t) &= \mathbf{A}_{\mathrm{c}} \mathbf{X}(t) + \mathbf{B}_{\mathrm{c}} \mathbf{u}_{\mathrm{c}}(t) + \tilde{\mathbf{F}}_{\mathrm{e}}(t) \\ \mathbf{y}_{\mathrm{c}} &= \mathbf{C}_{\mathrm{c}} \mathbf{X} \end{aligned} \tag{9}$$

where

$$\mathbf{A}_{\mathrm{c}} = \left[\begin{array}{c|c} 0 & I \\ \hline -\left([\omega_i^2] + A_2 C_{\mathrm{p2}}^{-1} A_2^T\right) & -[2\zeta_i \omega_i] \end{array} \right], \quad \mathbf{B}_{\mathrm{c}} = \begin{bmatrix} 0 \\ \hline \mathbf{A}_1 \end{bmatrix}, \quad \tilde{\mathbf{F}}_{\mathrm{e}} = \begin{bmatrix} 0 \\ \hline \mathbf{F}_{\mathrm{e}} \end{bmatrix} \tag{10}$$

$$\mathbf{C}_{\mathrm{c}} = \mathbf{C}_{\mathrm{p2}}^{-1} \mathbf{A}_2^T, \quad \mathbf{u}_{\mathrm{c}} = \mathbf{V}_1, \quad \mathbf{y}_{\mathrm{c}} = \mathbf{V}_2$$

The term $\mathbf{A}_2 \mathbf{C}_{\mathrm{p2}}^{-1} \mathbf{A}_2^T$ is modification to the angular natural frequencies due to the electromechanical coupling effect of the piezoelectric patches in open-circuit state. For a structure considered in this study, the difference between the natural frequency in short-circuit state and that in the open-circuit state is very small, usually smaller than 0.1%. Hence, this term is neglected in the following discussion.

3.2 Modal observer

In this study, a state-space observer was used to estimate the state vector, which contains the modal displacements as a sub-vector. The modal displacements are necessary in voltage switching of the piezoelectric actuators. The state-space model of the observer can be expressed in the following form:

$$\dot{\hat{\mathbf{X}}} = \mathbf{A}_{\mathrm{c}} \hat{\mathbf{X}} + \mathbf{B}_{\mathrm{c}} \mathbf{V}_{\mathrm{a}} - K(\hat{\mathbf{V}}_{\mathrm{s}} - \mathbf{V}_{\mathrm{s}}) \tag{11}$$

where $\hat{\mathbf{X}}$ is the state variable of the observer, which is used to approximate the state variable the structure, $\mathbf{V}_{\mathrm{a}}$ is the switched voltage on the piezoelectric actuator, $\mathbf{V}_{\mathrm{s}}$ is the output of the piezoelectric sensor, and $\hat{\mathbf{V}}_{\mathrm{s}}$ is the estimated sensor output from the observer. The matrix of gain K was designed using the Kalman filter approach. Since the state variable of the state-space model of the structure is not available, the estimate state variable $\hat{\mathbf{X}}$, which include the modal displacement, was used for voltage switch.

4. EXPERIMENT SETUP AND CONDITIONS

4.1 Experimental setup

The panel used in this study was an aluminum plate of 2 mm thick and 420 mm × 620 mm large. The four edges of the panel were clamped on a box using a rigid aluminum frame as shown in Figure 3, producing an effective vibrating area of 400 mm × 600 mm. The walls of the box were made of PMMA (polymethy methacrylate) with 70mm thick, which could provide more than 70dB sound isolation in the frequency band of interest below 1 kHz.

The material properties of the panel are tabulated in Table 1. Five piezoelectric patches, used as actuators and a sensor, were bonded on the surface of the panel, with locations shown in Figures 3 and 4. Their geometrical and material parameters are shown in Tables 2. The five piezoelectric patches are divided into three sets: the first set, that the one at the center numbered 1 is used as a sensor, the second set, that the upper and down ones numbered 2 and 3 are used as actuators for controlling the first mode, which are switched by the same switch in a synchronized way. The third set, that the other ones numbered 4 and 5 are used as actuators for controlling the second mode, which are switched by the same switch in a synchronized way.

A 6.5 inch loudspeaker inside the box was used to generate noise and an accelerometer shown by a circular mark in Figure 5 was used to measure the vibration of the panel. A microphone installed at 0.4 m way from the center of the plate was used to monitor the sound pressure outside the box, inside a full anechoic chamber, as shown in Figure 3. The schematic of the control system is shown in Figure 5. A source signal from the signal generator was amplified and applied to the loudspeaker to generate noise inside the cavity. The noise transmitted through the panel was measured by the microphone. The signal from the PZT numbered 1 was used for switch control. In two-mode control, an observer was used to separate the modal information of the two dominant modes.

Figure 3. Experimental setup

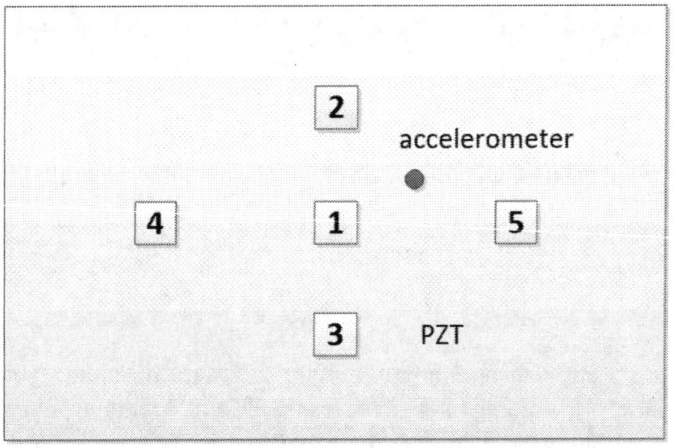

Figure 4. Distribution of piezoelectric patches and accelerometer

Table 1. Parameters of the panel

Components	Parameters
Actual volume of the cavity	420 mm×420 mm×365 mm
Thickness of the aluminum	1 mm
Dimensions of the outer frame	480 mm×480 mm
Dimensions of the inner frame	420 mm×420 mm
Thickness of the PMMA	50 mm
Poisson ratio of the aluminum	0.33
Young's module of the aluminum	70 GPa
Density of the aluminum	2700 kg/m^3

Table 2. Parameters of the PZT patches

Properties	PZT-PMN-51
Poisson ratio	0.39
Density	7600 [kg/m3]
Thickness	1×10^{-3} [m]
Piezoelectric constant d_{31}	-270×10^{-12} [c/N]

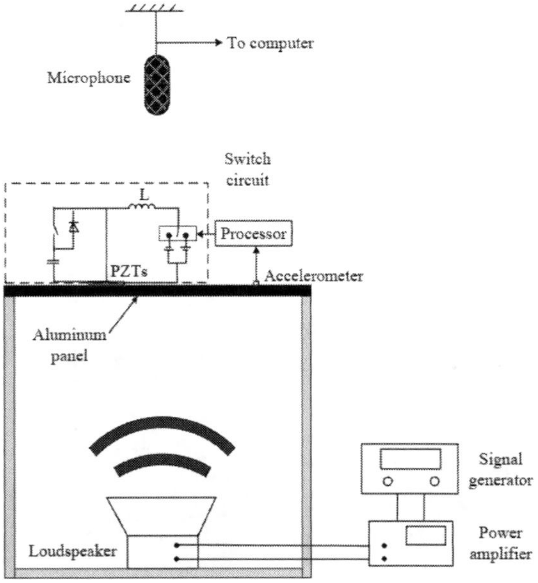

Figure 5. A schematic of the control system

4.2 Characteristics of Vibration and Noise Transmission

The dynamics of the panel were shown by the measured frequency response function using the accelerometer under white noise excitation by the speaker inside the box, shown in the dotted line in Figure 6. The solid line in Figure 6 shows the spectrum of the sound pressure measured by the microphone.

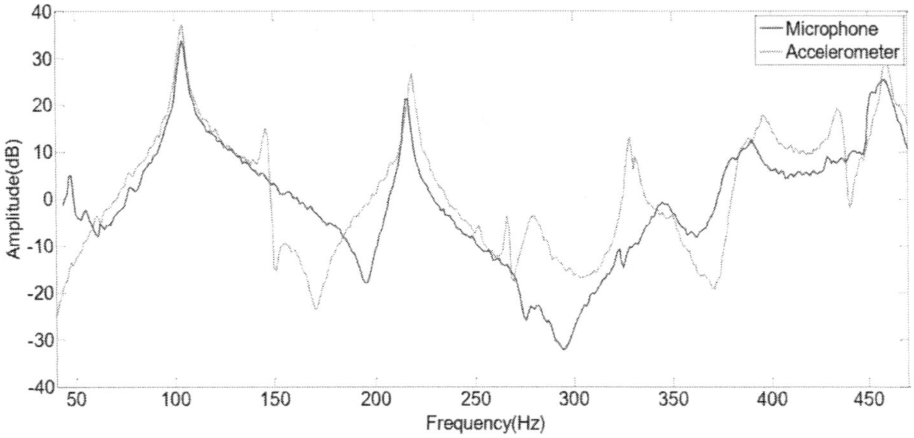

Figure 6. Frequency response of the panel and spectrum of the sound pressure level under white noise excitation

It can be seen that the panel exhibits rather complex dynamics, evidenced by a number of resonance peaks. However, only a few modes contribute to the sound radiation. More specifically, the two peak frequencies of the noise spectrum at 82 Hz and 197 Hz correspond to two structural resonances (1st and 4th modes of the panel). Simulated mode shapes of these two modes are shown in Figure 7.

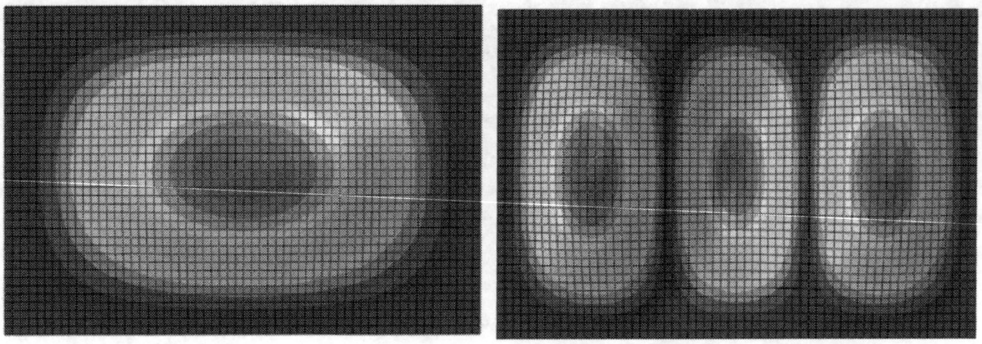

Figure 7. Shape of the 1st and 4th modes of the panel (simulation result)

5. CONTROL RESULTS AND ANALYSES

5.1 Two-Mode Control

The two-mode control scheme attempts to control both structural modes simultaneously. In the experiments, the panel was excited simultaneously by two harmonic sounds at 82 Hz and 197 Hz. The spectrum of the sound pressure is shown by the solid line in Figure 8. The peaks at the two frequencies of excitation are dominant, but there are also small peaks at other frequencies. In order to control the two components of noise dominated by different structural modes, the piezoelectric patches were divided into two sets and switched by two independent circuits as discussed above. The first set of actuator (the one at the center) was used to target the 5th mode and the second set including a cluster of four PZT patches was used to counter the 1st mode. In order to tackle these two modes, the vibration of the panel is needed. The acceleration of the panel was measured by the accelerometer and modal displacements of the two modes were identified by a state observer [19]. The identified displacements were used for action control of the two switches.

The spectrum of the controlled sound pressure is shown in Figure 8, which shows that the peak at 82 Hz is reduced by 21 dB, whist the one at 197 Hz by 5.3 dB. The amplitude of the sound pressure is reduced significantly. To evaluate the overall control performance quantitatively, the overall sound pressure level Lp was calculated by integrating the spectrum from 20 Hz to 500 Hz. An attenuation of 18.4 dB was achieved in the overall sound pressure level.

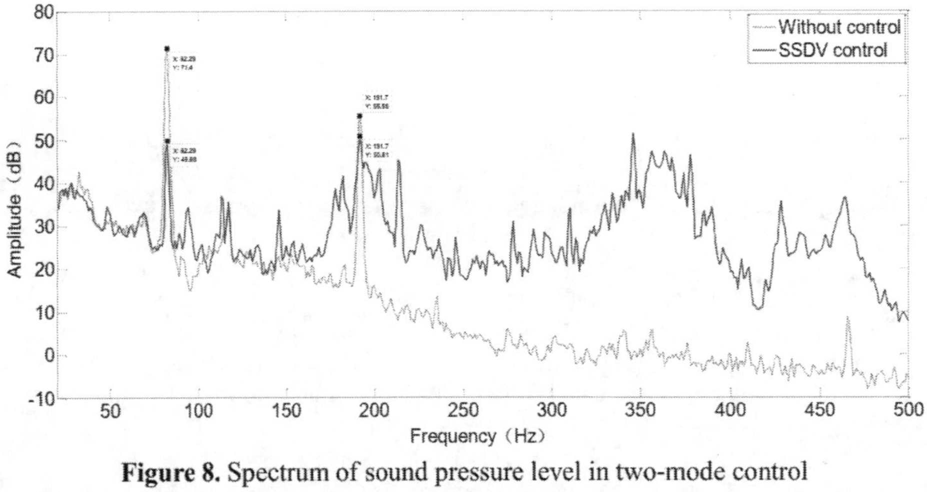

Figure 8. Spectrum of sound pressure level in two-mode control

Having demonstrated the performance of the proposed unsymmetrical semi-active SSDV method, a few remarks are noteworthy. Since noise transmission occurs mainly at the resonant frequencies of the panel, it is assumed that suppression of sound transmission can be attributed to the added damping effect through SSDV control. To verify this assumption, the panel vibration was also measured using the piezoelectric element attached to the panel. Figure 9 shows the spectrum of the vibration signal from PZT 1 without and with control when the two main modes were excited. It can be seen that the vibration at 82 Hz was reduced by about 21.2 dB and that at 197 Hz was reduced by 6.5 dB. While the reduction levels in both vibration and the radiated sound at the first resonance (82 Hz) are very close (around 21 dB in both cases), the ones at 166 Hz are litter different.

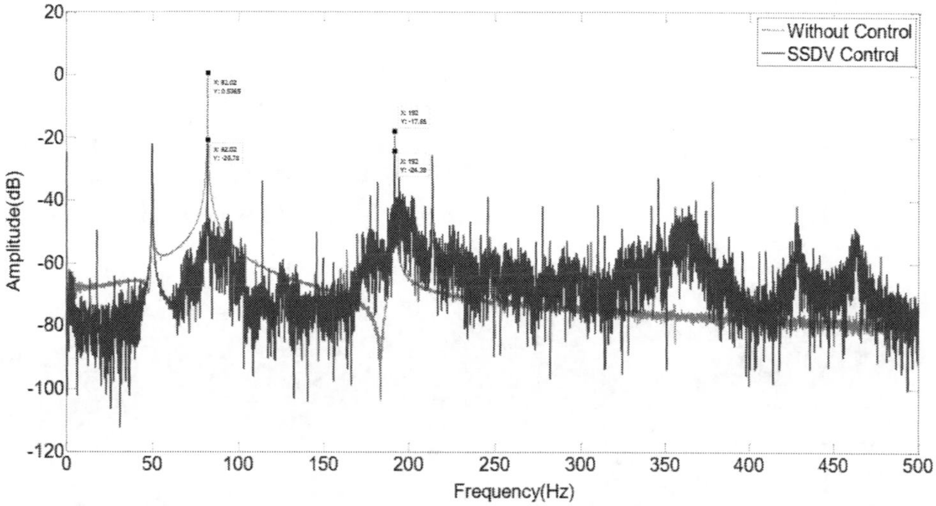

Figure 9. Spectrum of vibration signal from sensor (PZT 1) in two-mode control

5.2 Wide-band Control

In the wide-band control experiment, the panel was excited by a wideband signal with the frequency from 0-500Hz. The solid line in Figure 8 shows the spectrum of sound pressure level. The peaks at the frequencies of 82Hz and 197Hz are dominant. When unsymmetrical semi-active control is deployed, the dominant peak at 82 Hz is reduced from 70.46 dB to 62.89 dB, while the dominant peak at 197 Hz is reduced from 57.2 dB to 48.41 dB, as shown in Fig. 10. Figure 11 shows the spectrum of the vibration signal from PZT 1 without and with control under the white noise excitation. However, the spectrum in Fig.10 and Fig. 11 both show that a few spikes were also produced at the high-order harmonic frequencies. The overall sound pressure level Lp was calculated by integrating the spectrum from 20 Hz to 500 Hz. An attenuation of 3 dB was achieved in the overall sound pressure level.

The advantage of semi-active SSDV control for suppression of noise transmission is that its system can be realized without using acoustic sensors for feedback because the voltage inversion is synchronized with vibration so that it can be implemented with strain sensors such as piezoelectric patches. Since sound transmission is mainly induced by resonant vibration in the experimental system of this study, the switches can be directly controlled by the modal displacements, which are identified from the sensor signal using a modal observer. In some systems, strong noise transmission occurs at non-resonant frequencies. In these cases, the acoustic radiation modes should be identified instead of the vibration modes for switch control. Design of appropriate acoustic radiation modal observer from the mathematical relationship between the mechanical modes and acoustic radiation modes will be topic of future work.

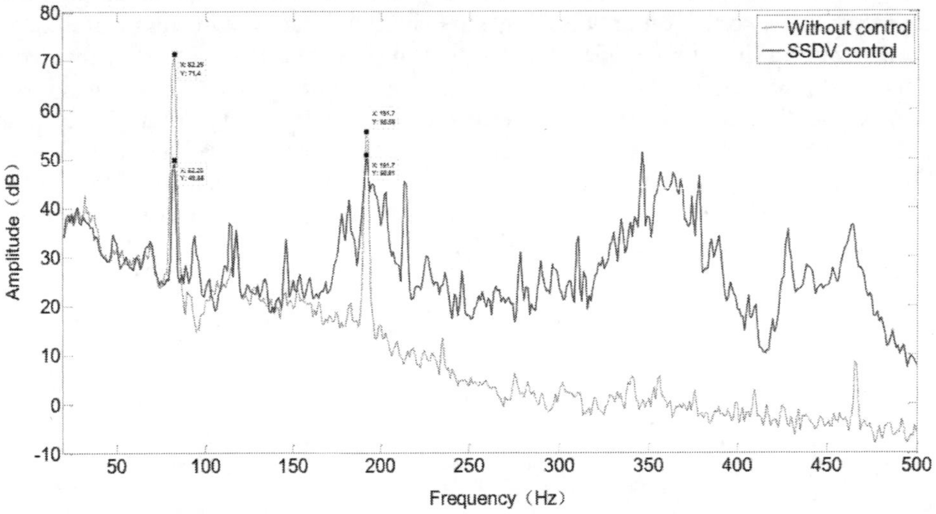

Figure 10. Spectrum of sound pressure level in wide-band control

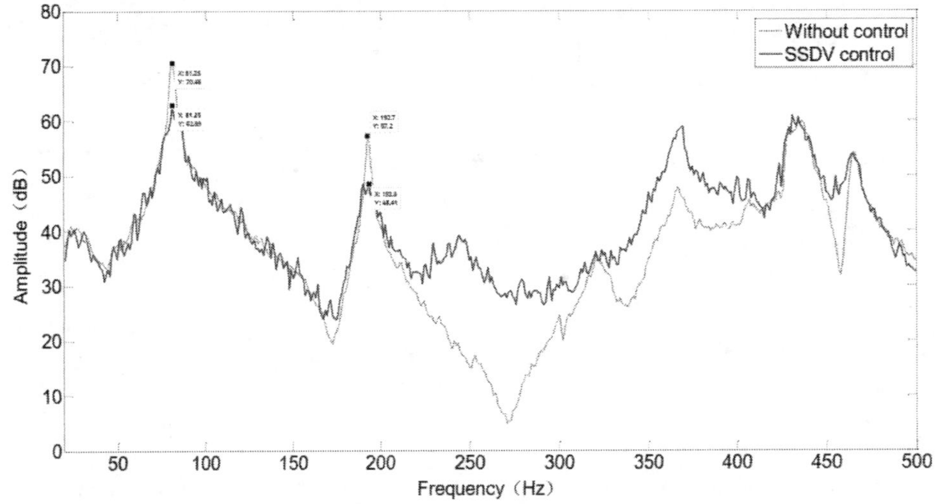

Figure 11. Spectrum of vibration signal from sensor (PZT 1) in two-mode control

5. CONCLUSIONS

The unsymmetrical SSDV method combined with modal observer has been successfully applied to the suppression of noise transmission through an aluminum panel. The four piezoelectric patches bonded on the panel were divided into two sets and switched by two independent circuits to control the two dominant modes of the panel. In the two mode control scheme, modal displacements were identified by an observer and used for switch control. The transmitted noise was measured by a microphone sensor for the evaluation of control performance. The experimental results show that by switching the voltages on the sets of piezoelectric actuators independently, the resonant peaks of the transmitted sound can be successfully attenuated. Significant reduction of overall sound pressure level was also achieved in both single-mode and two-mode controls.

ACKNOWLEDGMENTS

This research is supported by the NSFC under Grant 51375228, NSF of Jiangsu Province under Grant BK20130791, The State Key Laboratory Program under Grant 0515Y02, the Fundamental Research Funds for the Central Universities under Grants NE2015101 and NE2015001, and PAPD.

REFERENCES

1. Wang, X. N., "Hybrid Noise Control in a Duct Using Light Micro-perforated Plate," *J. Acoust. Soc. Am.*, 2012; 132(6): 3778-3787
2. Yuan, M., "Active control of sound transmission using a hybrid/blind decentralized control approach," *Journal of Vibration and Control*, doi: 10.1177/1077546313514758
3. Delany, M. E., "Acoustical properties of fibrous absorbent materials," *Applied Acoustics*, 1970; 3(2): 105-116
4. Ekici, B., "Improving sound absorption property of polyurethane foams by adding tea-leaf fibers," *Archives of Acoustics* 2012; 37(4): 515-520
5. Jones, J. D., "Reduction of interior sound fields in flexible cylinders by active vibration control," *Journal of Modal Analysis*, 1989; (5): 45-50
6. Petitjean, B., "Active control experiments for acoustic radiation reduction of a sandwich panel: feedback and feedforward investigations," *Journal of Sound and Vibration* 2002; 252 (1): 19-36
7. Ho, S., "Active Control of Plate-through Noise Transmission," *Dynamics and Design Conference 2000, JSME*, CD-ROM Proceedings, Paper No. 746
8. Ji, H. L., "An improved system of active noise isolation using a self-sensing actuator and neural network," *Journal of Vibration and Control* 2009; 15(12): 1853-1873
9. Wang, K. W., "Energy-based parametric control approach for structural vibration suppression via semi-active piezoelectric networks," *J VIB ACOUST.* 1996; 115: 505-509
10. Clark, W. W., "Semi-active vibration control with piezoelectric materials as variable stiffness actuators," In: *1000 AIAA/ASME/ASCE/AHS/ASC Structures, Structural Dynamics, and Materials Conference and Exhibit*, 1999, 2623-2629
11. Richard, C., "Semi-passive damping using continuous switching of a piezoelectric device," In: *SPIE Smart Structures and Materials Conference: Passive Damping and Isolation*, 1998, 104-111
12. Onoda, J., "Energy-recycling semi-active method for vibration suppression with piezoelectric transducers," *AIAA Journal*, 2003; 41(4): 711-719
13. Badel, A., "Piezoelectric vibration control by synchronized switching on adaptive voltage sources: Towards wideband semi-active damping," *J ACOUST SOC AM*, 2006; 119(5): 2815-2825
14. Ji, H. L., "Semi-active vibration control of a composite beam using an adaptive SSDV approach," *Journal of Intelligent Material Systems and Structures*, 2009; 20(3): 401-412
15. Yan, L. J., "Multimodal nonlinear damping technique using spatial filtering," *Journal of Intelligent Material Systems and Structures*, 2014; 25(3): 308-320
16. Makihara, K., "Fuselage panel noise attenuation by piezoelectric switching control," *Smart Materials and Structures*, 2010; 19(8): 085022 (10pp)
17. Ji, H. L., "Two-mode vibration control using nonlinear synchronized switching damping based on the maximization of converted energy," *J. Sound Vib.* 2010; 329: 2751-2767
18. Ji, H. L., "Semi-active vibration control based on unsymmetrical synchronized switching damping: Circuit design," *Journal of Intelligent Material Systems and Structures*, 2015, DOI: 10.1177/1045389X15585898
19. Ji, H. L., "Semi-active vibration control of an aircraft panel using synchronized switch damping method," *INT J APPL ELECTROM*, 2014, Vol. 46, pp. 879-893, 2014

ICAST2015 #086

Mixed operation mode of thermoelectric modules for thermal gradient reduction in gas foil bearings

Adam Martowicz[1*], Michał Lubieniecki[1], Jakub Roemer[1], and Tadeusz Uhl[1]

[1] AGH University of Science and Technology, Department of Robotics and Mechatronics,
Al. A. Mickiewicza 30, 30-059 Krakow, Poland

Abstract

The gas foil bearings (GFB) have gained significant increase of the application range due to their advantages, including oil-free operation, reduced weight and maintenance costs. The GFBs can operate nearly frictionless due to hydrodynamic action, which allows to elevate a rotating shaft over the set of supporting foils. The fragility of developing thin fluid gap between the shaft and a top foil, limits the possible applications to those of high rotation speed and to lightly loaded machinery. Normal operation of the bearing strongly depends on the temperature distribution in the bearing foils and, as reported in the literature, favorable operation conditions may be relatively easily lost when the thermal gradients in the top foil unexpectedly rise. This behavior is found to be in contrast to a case when an increase of absolute or average temperature of the whole bearing is identified, which is not considered so critical. The authors propose to apply a local temperature control method to equalize its values within the top foil of the bearing. It should be noted that all recently developed approaches mainly follow a global cooling method. This method is efficient in reduction of an average temperature, however, does not allow for significant thermal gradient decrease. Moreover, the axial cooling method may even lead to a rapid increase of thermal gradients. Only one approach, i.e. radial cooling, offers means for local thermal control, however, due to the bulkiness of the air spreading system its practical application may be questionable.

The work presents an approach for thermal gradient reduction in gas foil bearings carried out with smart materials. Both experimental and numerical analyses are described showing efficiency of the proposed technique. First, the authors show a proof-of-concept experimental test to confirm the capability of thermal gradient reduction in elaborated test stand. Then, a numerical model of the bearing is presented. The enhanced bearing construction consists of 36 thermoelectric modules (TEM) uniformly circumferentially and axially distributed in the bushing of the bearing. Steady state simulations are performed to analyze temperature distribution in the top foil for different currents' configurations. In the study, both cooling and heating are allowed to locally control the temperature. Moreover the Response Surface Method is applied to find the relationships between the control currents and the thermal scatter observed for the top foil. Based on the determined metamodel, a more precise search is performed to find cases exhibiting effective reduction of the thermal gradients. The found configurations are discussed in terms of their efficiency, i.e. power consumption in thermoelectric modules necessary to control the temperature sufficiently.

* adam.martowicz@agh.edu.pl

1. INTRODUCTION

The gas foil bearings (GFB) have gained significant increase of the application range due to their advantages, including oil-free operation, reduced weight and maintenance costs. They are mostly used in high speed and preferably lightly loaded machinery - up to the power 200kW, e.g. turbochargers, compressors and micro-turbines [1]. They operate based on hydrodynamic action, which allows to elevate a rotating shaft. The maintained thin fluid gap between the shaft and a top foil - usually made by a thin air layer - allows for low friction [2]. GFB can be used in high temperatures. However, their normal operation strongly depends on the temperature distribution in the bearing foils. Nominal operation conditions may be relatively easily lost when the thermal gradients in the top foil unexpectedly rise. This may happen irrespectively from the increase of absolute or average temperature of the whole bearing, which is not considered so critical.

As reported in the literature, three general reasons of failures in GFBs are known [3]. These are overheating, thermal runaway and abnormal growth of thermal gradients. Overall overheating of the bearing structure leads to loss of nominal load capacity. It happens due to drop of stiffness of the air film as temperature rises. The bearing, exhibiting a constant operational load, may eventually break the air film completely. A non-uniform distribution of the heat flux is responsible for thermal runaway, in turn [4]. The shaft in GFB expands faster than the foils and whole housing due to uneven heat absorption of the mating bearing parts. It results in increased preload acting on the GFB. A self-sustaining process is activated due to relationships between heat generation, thermal expansion and uncontrolled preload, which inevitably leads to thermal seizure. The last reason for failures in GFB is increasing thermal gradients and is considered the most dangerous source of damages. It may occur unexpectedly, even though the overall temperature is not high and the bearing operates stably [3].

Several methods are used to keep stable condition of GFB at their operation. Different types of cooling methods are developed to effectively address thermal stability issues. All methods listed below involve forced air stream. The air can be directed to either the surface of the journal or inside the hollow shaft, the axial air flow can be forced through the bumps of the foil. Alternatively, nozzles may be spread on the bearing's bushing to allow for radial delivery of the air stream directly on the shaft. These methods are known as direct, indirect, axial and radial cooling respectively and enable global cooling of GFB with significant decrease of the overall temperature [5-8]. However, substantial cooling flows must be ensured for their successful operation. Only the last mentioned method can lead to effective and controllable gradient reduction. On the other hand, while an application of the axial cooling method results in lowering the overall bearing temperature, the temperature distribution may get severely detrimental to the bearing at the same time. Hence, alternative methods to assure thermal stability are demanded, including those based on passive cooling.

The authors of this paper propose the method of an active control of local bearing temperature based on the application of smart materials. Both experimental and numerical analyses (aided with metamodeling technique) are described showing efficiency of the proposed technique. As shown, the adopted approach allowed the efficient equalization of the temperature within the top foil of GFB - in contrast to the recently developed approaches following a global cooling and consuming much energy to provide sufficient air flow. As already mentioned, these methods are efficient in reduction of an average temperature, however, does not enable for significant thermal gradient decrease.

The work provides a continuation of the authors' previous efforts. Simulations for thermal gradient reduction in GFB are described in [5]. However, only cooling is considered in the cited work. An efficient currents' configuration has been also found based on a metamodeling technique. An innovative measurement technique applied to GFB is presented in [9]. Novel thermocouples, which are made of Pt-Inconel junctions, were used for mapping temperatures in the bearing's foils.

The remaining part of the paper addresses the following contributions. Section 2 provides the description of the test stand, Section 3 presents the results of a proof-of-concept experimental test to show the capabilities of the proposed approach and effectiveness of thermal gradient/scatter reduction. Section 4 introduces a model of the bearing, followed by the results of numerical simulations. Final Section 5 summarizes the work and draws the concluding remarks.

2. TEST STAND

The enhanced construction of the bearing investigated during the experiment is shown in Figure 1. Compared to the standard construction the one-piece bushing has been replaced, for technological reasons, by the assembly of the inner bushing (A) and outer bushing (B) in the form of three separate segments. The control currents are delivered for each individual module by separate set of wires (C). The temperature of foils is measured with standard thermocouples (D) by the inspection holes present in the outer sleeve. The setup allows to measure the temperature on both the top and bump foil, for every 120^0 along the circumference, however, the readings on each foil are shifted by 40^0 one with respect to another. The bearing is mounted on the spindle (E) running at 15 000 rpm.

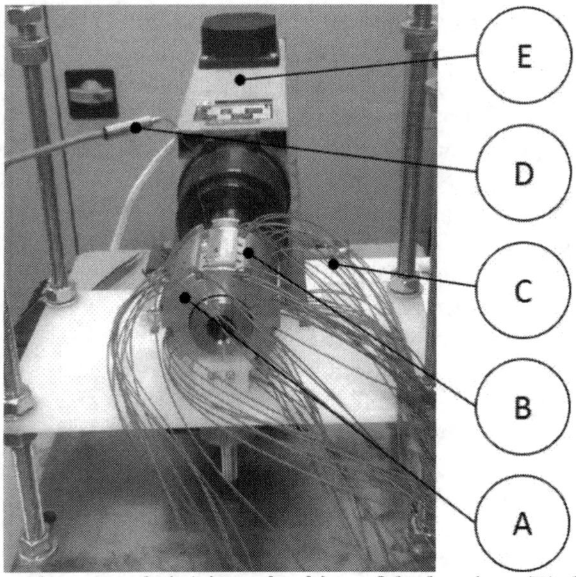

Figure 1. View of the used test stand; (A) inner bushing of the bearing; (B) three-segment outer bearing bushing; (C) set of wires driving currents for each TEM; (D) single thermocouple (other are removed for clarity); (E) electric spindle.

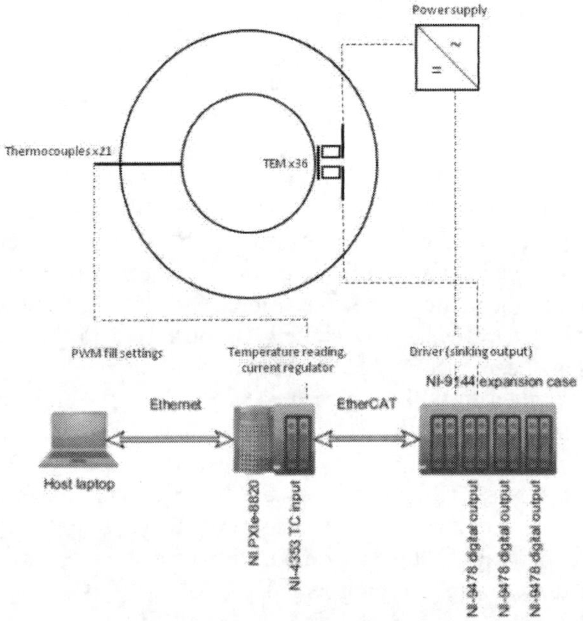

Figure 2. The general scheme of the control system in use.

The control system used for measurements is based on the National Instrument general purpose PXI platform with C-module expansion case. The thermoelectric modules are controlled by sinking digital output modules with maximum current up to 1.5A per module with PWM signal set individually. The general scheme of the control system is shown in Figure 2.

3. PROOF OF CONCEPT – EXPERIMENT

The experiment was conducted under the steady state conditions reached after the spindle start-up. During the proof of concept experiment all modules were powered according to the arbitrarily chosen set of currents (in a range 0.75-0.9A per module). The total current driving the TEM was at the level of 30.6A. The temperatures recorded within a 10-minute interval on the top foil are presented in Figure 3. The temperature scatter under nominal operational condition was measured as 3.2°C. The moment the control currents were applied to the modules, both the recorded temperatures and resulting scatter were reduced. The temperatures in almost all recorded points were immediately reduced but two of them with the smallest currents' values assigned. This is the result of a thermal bridge between the modules' hot sides placed under same segment of the outer sleeve. Low amount of heat dissipated by natural convection on the surface of the bearing bushing strengthens the effect. The low rate of heat dissipation is also a cause of slow but steady process of heating up the whole structure of the bearing. The same behavior is noted for the temperatures measured on the bump foil (Figure 4). It was noted that with respect to the steady state condition with no currents applied to the modules the scatter was reduced by 25% (from 3.2°C to 2.4°C) and 29% (from 2.91°C to 2.1°C) for the top and bump foils respectively.

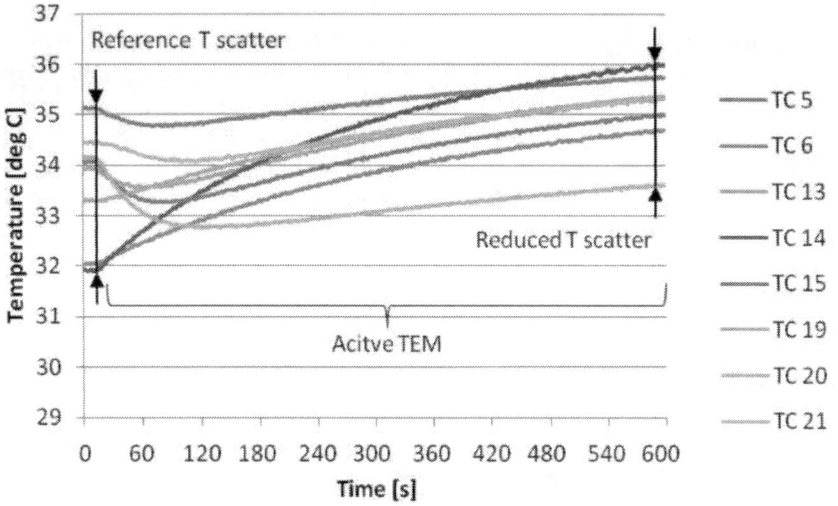

Figure 3. Temperatures recorded for the top foil. It is marked how reference and reduced scatters are determined. The reduced scatter remained stable over the whole observation period irrespectively from the global temperature increase.

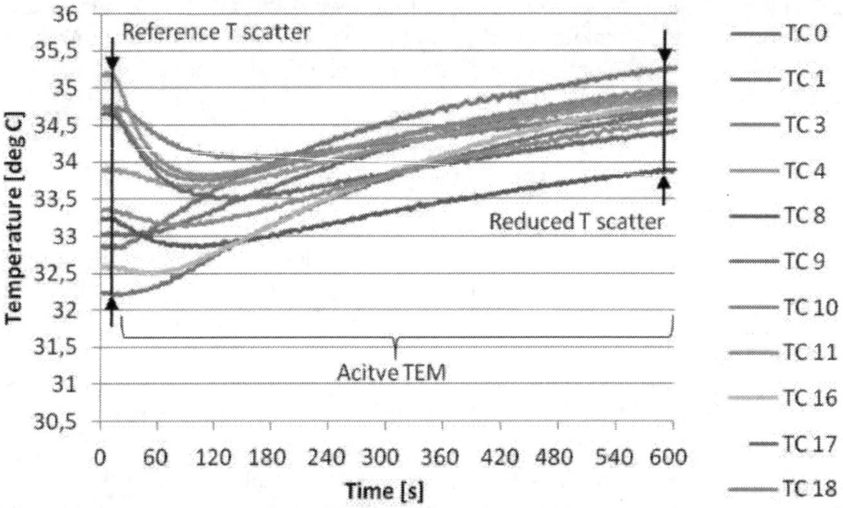

Figure 4. Temperatures recorded for the bump foil. It is marked how reference and reduced scatters are determined. The reduced scatter remained stable over the whole observation period irrespectively from the global temperature increase.

4. NUMERICAL SIMULATIONS FOR MIXED OPERATION MODE FOR THERMOELECTRIC MODULES

The numerical model created in the ANSYS environment consists of 36 thermoelectric modules uniformly circumferentially and axially distributed in the bushing of the bearing (Figure 5). Steady state thermal numerical simulations are performed to analyze temperature distribution in the top foil for different currents' configurations. Thermoelectric modules were implemented analytically [10]. In the study, both cooling and heating are allowed to control the temperature. Figure 6a presents a reference polar plot for the temperature distribution when all thermoelectric modules are not active. All modules are grouped according to their circumferential localizations to simplify the procedure of thermal gradient reduction as shown in Figure 6b.

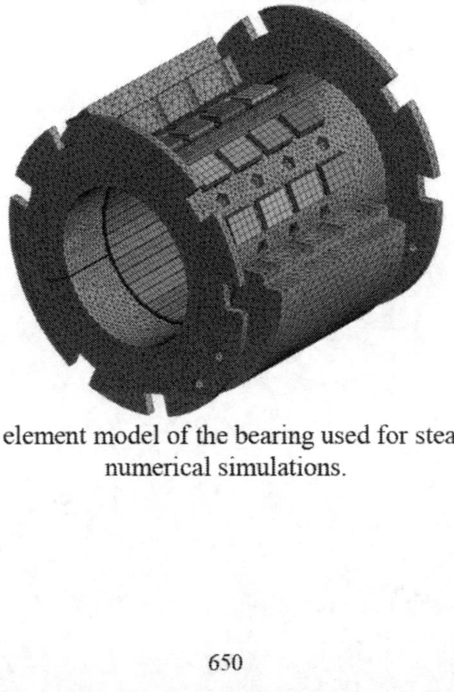

Figure 5. Finite element model of the bearing used for steady state thermal numerical simulations.

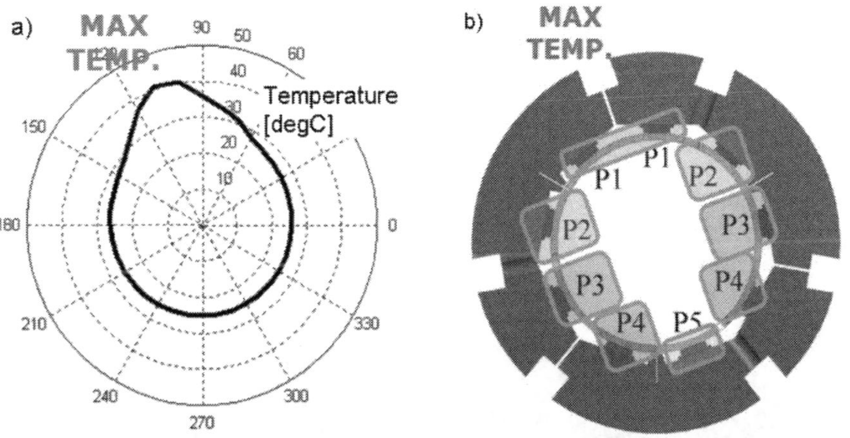

Figure 6. Circumferential temperature profile for the nominal configuration of the modules' currents, when all are set to zero (a) and 5 groups of thermoelectric modules P1, ..., P5 arbitrarily selected according to the temperature profile (b).

The Response Surface Method (RSM) is additionally applied to approximate the relationships between the control currents and the thermal scatter observed for the top foil [11]. There are many effective applications of RSM for variety of structures of different geometric and time scales [12-14], therefore, the authors decided to use the metamodeling technique for more precise search for acceptable configuration of the modules' currents. Table 1 briefly summarizes found configurations in terms of their efficiency, i.e. power consumption in thermoelectric modules necessary to control the temperature sufficiently.

Table 1 Characteristics of selected operation modes for the model of GFB

Case	**Average temperature [^{0}C]**	**Temperature scatter [^{0}C]**	**Reduction of temperature scatter [%]**	**Configuration for currents: I_{P1}, I_{P2}, I_{P3}, I_{P4}, I_{P5} [A]**	**Total current consumption [A]**
Nominal	27.6	15.7	-	0, 0, 0, 0, 0	0
Cooling	38.9	10.6	32.4	1, 0, 0, 0.5, 0	12
Mixed operation mode	45.1	9.07	42.1	1, -0.5, 0, 0.5, -0.5	18
Mixed operation mode aided with metamodel	45.2	9.00	42.6	1, -0.5, -0.2, 0.4, -0.5	18.8

The electric currents for all groups of thermoelectric modules are independently set and take one of the following values: -0.5A (negative values are considered for heating), 0A, 0.5A and 1A (in case of cooling). Hence, 1024 configurations were checked to cover the whole domain of allowed values of the currents. The performance of the proposed method may be assessed comparing the temperature distributions for cooling and mixed operation, which is shown in Figure 7, 8 and 9.

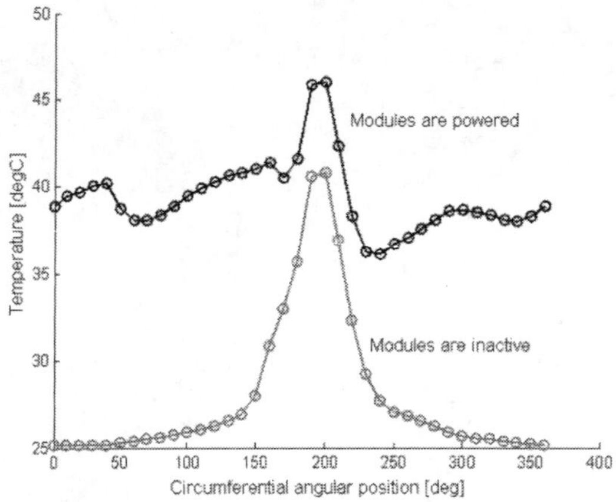

Figure 7. Comparison between temperature distributions achieved for cooling.

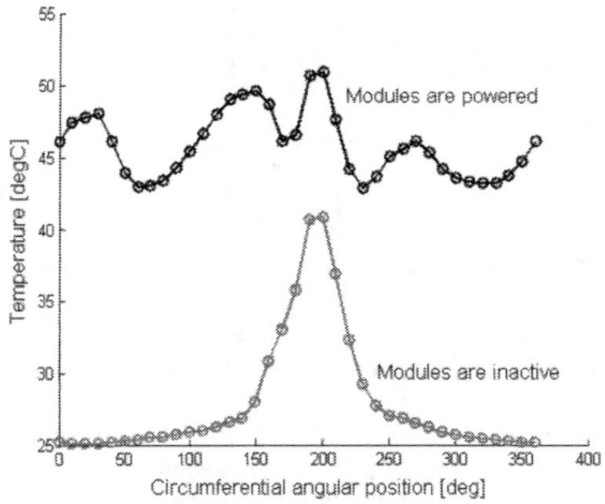

Figure 8. Comparison between temperature distributions for mixed operation mode.

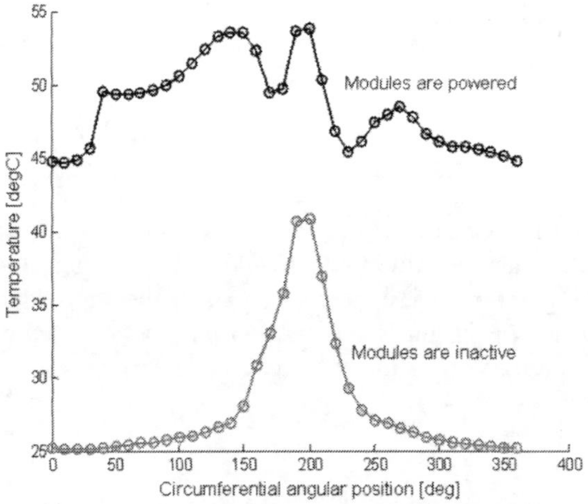

Figure 9. Comparison between temperature distributions for mixed operation mode with the currents' configuration determined with metamodel

The yielded results for cooling mode coincide with the observations on the thermal gradient reduction in experiments. Temperature gradient (represented with temperature scatter) is effectively reduced with the efficiency rate up to 42.6%, which was identified for the mixed operation mode when local heating is also allowed. Due to resultant growth of the power delivered to the bearing with heaters, the overall temperature of the construction rises. This is however acceptable cost of energy redistribution in GFB

5. SUMMARY AND CONCLUDING REMARKS

The paper introduces the first known trials and obtained results regarding mixed operation mode for reduction of critical thermal gradients in GFB, which is carried out with thermoelectric modules. The scope of the work covers both experiments and numerical simulations. First, a proof-of-concept experiment - using laboratory test stand - is performed to demonstrate the efficiency of the proposed approach. Finally, the results of numerical analyses are presented for the model of GFB enhanced by locally controlled cooling and heating.

As found in the study, the gradient reduction is evident when the thermoelectric modules are activated. It is confirmed that they may reduce the temperature scatter efficiently, to at least the same extent as other methods, however, at lower power consumption. For selected configurations only a few out of 36 modules were used to control the temperature field properly. Mixed mode of operation with the currents selection based on metamodeling technique may increase the performance of the control system of additional 10%.

The main implementation difficulty is related to the size of the whole bearing assembly (35mm of bore diameter) and the number of necessary electric connections used for thermoelectric modules and thermocouples. However, the electric wiring can be significantly reduced if best localization for the TEM is found and any redundant module and measurement point is removed. This aspect of enhanced construction of the GFB as well as both axial and circumferential gradient removal and elaboration of the control law based on the metamodel are within the range of the authors' ongoing and future works.

ACKNOWLEDGMENTS

The authors acknowledge the project "The use of thermoelectric materials to improve thermal stability of high-speed rotor bearings systems", no. PBS1/A6/6/2012 financed by the National Centre for Researched and Development.

REFERENCES

1. Kim, D., Lee, A., Choi, B., "Evaluation of Foil Bearing Performance and Nonlinear Rotordynamics of 120 kW Oil-Free Gas Turbine Generator," *Journal of Engineering for Gas Turbines and Power*, Vol. 136, No. 3, 2013.
2. Nalepa, K., Pietkiewicz, P., Żywica, G., "Development of the Foil Bearing Technology," *Technical Sciences*, No. 12, 2009.
3. Radi, K., Batcho, C., "A Novel Thermal Management Approach for Radial Foil Air Bearings," Storming media, 2010.

4. Ryu, K., San Andrés, L., "On the Failure of a Gas Foil Bearing: High Temperature Operation Without Cooling Flow," *Journal of Engineering for Gas Turbines and Power*, Vol. 135, No. 11, 2013.
5. Roemer, J., Lubieniecki, M., Martowicz, A., Uhl, T., "Multi-point control method for reduction of thermal gradients in foil bearings based on the application of smart materials," Proceedings of the SMART 2015 ECCOMAS thematic conference on Smart structures and materials, Ponta Delgada, Azores, 3–6 June, 2015.
6. Shrestha, S., "Experimental feasibility study of radial injection cooling of three-pad radial air foil bearings," PhD thesis, The University of Texas at Arlington, 2012.
7. Heshmat, H., Walton II, J., Tomaszewski M., "Demonstration of a Turbojet Engine Using an air Foil Bearing," Proceedings of 2005 ASME Turbo Expo, 2005.
8. Radil, K., Dellacorte, C., Zeszotek, M., "Thermal Management Techniques for Oil-Free Turbomachinery Systems," *Tribology Transactions,* Vol. 50, No. 3, 2008.
9. Lubieniecki, M., Roemer, J., Martowicz, A., Wojciechowski, K.T., Uhl, T. "A multi-point measurement method for thermal characterization of foil bearings using customized thermocouples," Proceedings of ICT & ECT 2015 Annual International Conference on Thermoelectrics (ICT 2015) and 13th European Conference on Thermoelectrics (ECT 2015), Dresden, Germany, June 28 – July 2, 2015 .
10. Roemer, J., Lubieniecki, M., Polaczek, D., Miodunka, M., Uhl, T., "Eliminating the Risk of Thermal Instability in Foil Bearings by Thermal Gradient Reduction with the Use of Thermoelectric Coolers," XIV Warsztaty Projektowania Mechatronicznego, Kraków, 2014.
11. Myers, R.H., Montgomery, D.C., "Response Surface Methodology process and product optimization using designed experiments," John Wiley & Sons, Inc., New York, 1995.
12. Gallina, A., Martowicz, A., Uhl, T. „Robustness analysis of a car windscreen using Response Surface techniques," *Finite Elements in Analysis and Design*, Vol. 47, No. 1, 2011, pp.46-54.
13. Martowicz, A., Vige', D., Gallina, A., Uhl, T., "Coupling Monte Carlo Simulation with Response Surface Methodology for evaluating dynamic properties of FE model of real-life vehicle's structure," Symposium: NATO/AVT-147, Computational Uncertainty in Military Vehicle Design, Athens, Greece, December 3-6, 2007.
14. Martowicz, A., Kurowski, P., Uhl, T., Lasko, G., "Robust design optimisation of the FE model of rail vehicle", The WCSMO-8: 8th World Congress on Structural and Multidisciplinary Optimization, Lisbon, Portugal, June 1-5, 2009.

ICAST2015 #62

Characteristics of Square Shaped and Hexagonal Shaped Membrane Space Structures with Steric Support of Booms and Cables

Ayako Torisaka[1], Yoshitaka Sato[2], Takeshi Akita [3], M.C. Natori [4],

Hiroshi Yamakawa[5], Tomoyuki Miyashita[5]

[1] Dept. of Aerospace Engineering, Tokyo Metropolitan University, Hino, Japan
[2] Dept. of Modern Mechanical Engineering, Waseda University, Shinjyuku, Japan
[3] Dept. of Mechanical Science of Engineering, Chiba Institute of Technology, Narashino, Japan
[4] Institute of Space and Astronautical Science, Sagamihara, Japan
[5] Dept. of Modern Mechanical Engineering, Waseda University, Shinjyuku

Abstract

This paper discusses and suggests a basic structure of a mixed constitute of lightweight membranes and booms by looking slightly future of space structure which will become larger and larger to be tens of meter class such like future deep space exploration by solar sails or self-power generation by thin solar films etc…Or in short-sighted developing vision, this type of structure is going to be applied to the deorbit hardware for low-orbit satellites to earn much drag for atmosphere entry within smaller years. All of such application need that the structure should have a certain stiffness, so our interest is how to design to get efficient stiffness improvement and also we should be care not to ruin other structural merit such like small wrinkle amount. First, we focused on adding extension force in axial direction of each boom and studied where to put. Then the efficiency of cable connection on the oblique line of a membrane is also verified. After integrated these design factors, we focused on the actual construction phase that may be difficult to realize completely flat surface especially in tens of meters class structure, so we assumed that the manufactured structure will be steric, and we focused on finding the steric stabilized structure by using boom and cables as supportive members. Relation between 1st eigen frequency which is the bending mode of a boom is mainly studied and it becomes obvious that the steric structure improves it more efficiently than other design factor. These examinations are not only carried out on square model, but also on hexagonal model which is also versatile geometry for expanding module structure in the future, and this shows that the optimum boom angle exists to make wrinkle amount becomes minimum. This can be a designing guidepost for future use of boom-membrane structure. Then we consider all the design factor at once and suggest the three dimensional basic structural design for large light weight space structure.

1. INTRODUCTION

Recent success of membrane structure represented by solar sail, IKAROS (Sawada H, Mori O and Okuizumi N, 2012) broaden the possibilities of space exploration and its potential has been developed actively since then. Since it is a tension element, for future large scale space structure systems centrifugal force due to spinning is useful in order to stabilize it, and there already appeared a spinning solar sail spacecraft to demonstrate such basic structure concept.

But for those in rather near future, a corresponding compression element such as a boom should be used from the viewpoint of easy maneuverability on its whole system, and the mechanical characteristics of compound structure systems consisted of membrane elements and booms must be clarified. So one of the prevailing structure aiming for deep space exploration or constructing modular structure is four deployable booms with a square membrane type (Stohlman O.R., Fernandez J. M., Lappas V. J, et al. 2013, H.Sakamoto, H.Furuya, Y.Satou,2013). Detail design study of a square solar sail architecture was carried out by G. Greschik, M. M. Mikulas, (2002), and one of his summarize is that optimal design efficiency is not degraded dramatically by slight to moderate boom cross section variations. Then he also proposing the striped square architecture with diagonally boom and continuously suspension between membranes, but the study about the optimum configuration of suspension layout (connection layout) has not been carried out. So we focused on this point and thought additional detail design method should be introduced by using additional other structural members.

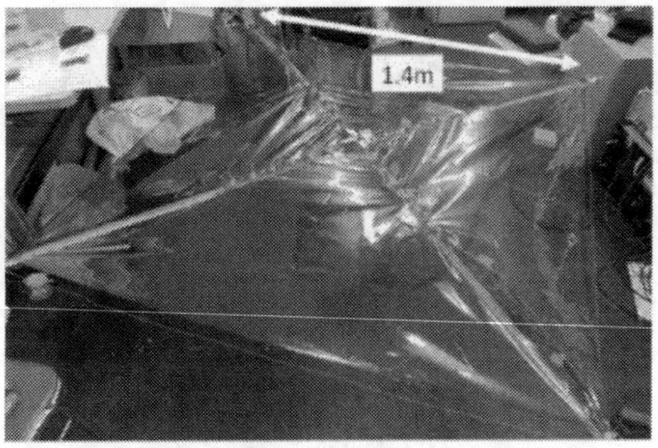

Figure 1 Membrane model suspended by four diagonal Convex booms with cables

Focusing on the use of cable, Sakamoto H. and Park K.C. (2005) proposed surrounding membrane structure with weblike cables along the membrane perimeter to realize alleviating wrinkle in the membrane and reduction in corner cable tensions. When viewed in light of membrane which is suspended by four diagonal boom structure, the tension members are booms in this case, but its elastic deformation may be a load pass which would be disturbance for membrane and would cause wrinkle. So cable should be a supportive role for reducing this and efficient way of using should be found out.

From above, the problems on this research are summarized as below.

- ✓ Are two connections on a boom edge (one edge and other side) are really enough?
- ✓ How and where do you apply cables? And what should the cable aim for?

Then we had constructed the research map for boom-membrane structure as **Figure 2**. Boom-Membrane structure is examined not only with square model but also hexagonal model.

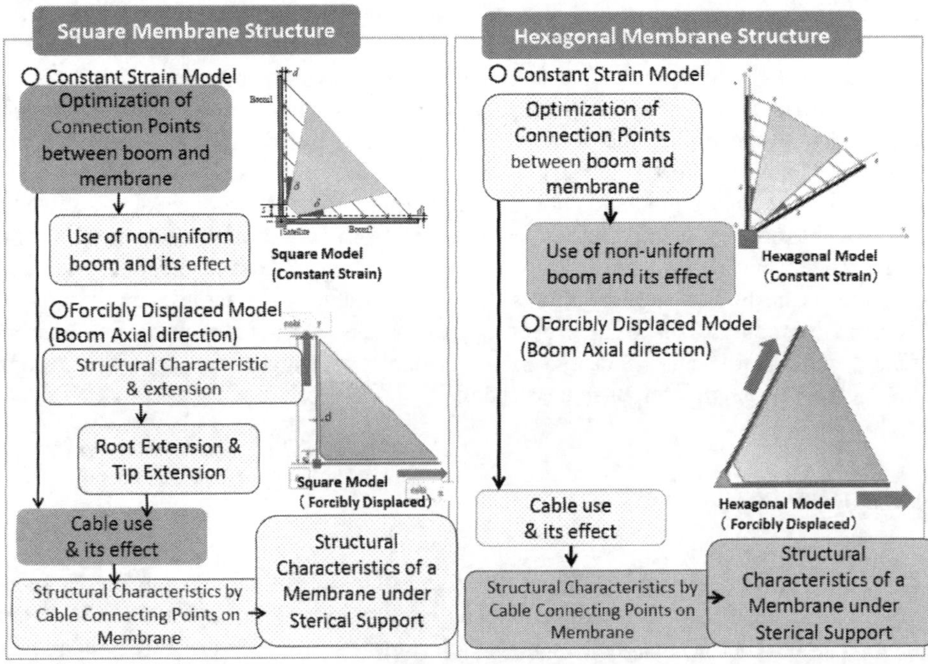

Figure 2 Research Map

Red colored items were already examined and reported before (Torisaka A., Satoh Y., Natori M.C. et al. , 2013). In that paper, we focused on the conventional model as having two connections on a boom edge is not stiff for stress on membrane, so we carried out the optimization of connection points between boom and membrane by using satisficing optimization method, and suggested it as a basic design of a boom-membrane structure. As far as cable use and its effect, we obtained that the cable improves the 2nd mode or higher mode.

Next as a series of research, we examined about yellow and green colored items in **Figure 2**. Here each flow of square model and hexagonal model is almost similar and obtained results are similar, too. So here we mainly focus on and explain about the hexagonal model, while as far as "root extension or tip extension" the obtained knowledge is mentioned by using square model.

First, the effect of non-uniform boom is explained. Then next topic is cable connection on a membrane and how it affect on wrinkle. Last one is new suggestion as a 3D boom-Membrane structure by attaching a boom with some angle toward the out-plane direction.

2. BOOM-MEMBRANE STRUCTURE

This section describes the analytical model of a boom-membrane solar sail as shown in Figure 1 which assume the existing study by Sakamoto, Furuya, and Satou et al., 2013. Four booms are connected to the main satellite and they support the petals at the deployed configuration, and main concept of their hardware can wrap the booms and a membrane together in the same hub by using the rotational skew folding pattern (Furuya, Masuoka, 2004, Furuya, Inoue, Masuoka, 2005). This concept can be developed easily into more ideal one with spiral folding pattern which can absorb boom thickness, and membrane can be stowed in a desired height (Natori, Watanabe, Kishimoto et al. 2007). So this research is significant as a basic study for future actual designing.

Material properties of the structure that is being studied are decided as Table 1 according to our previous study (Tamaru, Akita, Natori, et al. 2012). Reason of taking these values is that to correspond the future comparative study by the laboratory model experiments. Here, the material of the membrane is assumed to be poly-imide film (25μm), the boom material is SMA (Shape Memory Alloy), and the length of the boom is 1.0 [m], 0.0001 [m] in thickness, 0.015 [m] width.

Table 1 Material Property

	Young's modulus [GPa]	Poisson's ratio	Shear modulus [GPa]	Density [kg/m^3]
Membrane(Poly-imide)	3.19	0.34	-	1420
Boom (SMA)	60.0	-	20.6	6500

The initial shape of membrane is an acute-angled triangle truncated near the origin as shown in **Figure 3**. After the membrane is stretched with the angle θ, the membrane is connected to booms through penalty springs, and its tension is dominated by four parameters, θ, d, s and δ. d is the basic distance between membrane and boom elements, δ is the adjustment angle of the additional distance between the elements. Near the origin a small isosceles triangle with the base length s is cut from the original membrane. Here, δ is determined to make strain of membrane uniform, which means δ becomes to be a function of s and d.

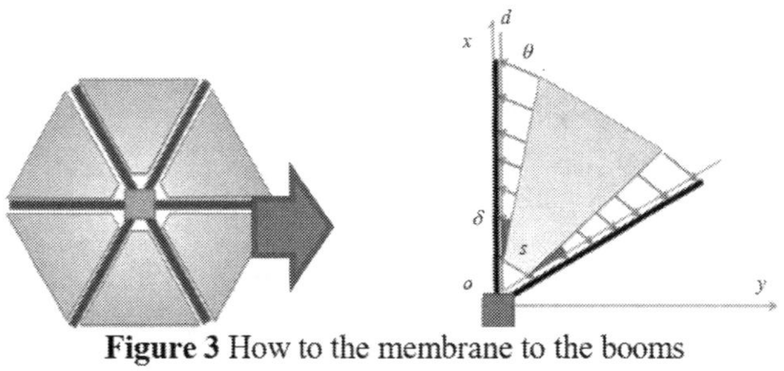

Figure 3 How to the membrane to the booms

Regarding with connecting points, both tips of a boom are always connected as default, and connectivity of the points between these two tips are set as design variables. Expression rule for connection or no-connection is introduced as 1 or 0, so the **Figure 4** is expressed like [0100] for example. Further, the total spring stiffness is treated as constant in this model even the total connecting points are changed.

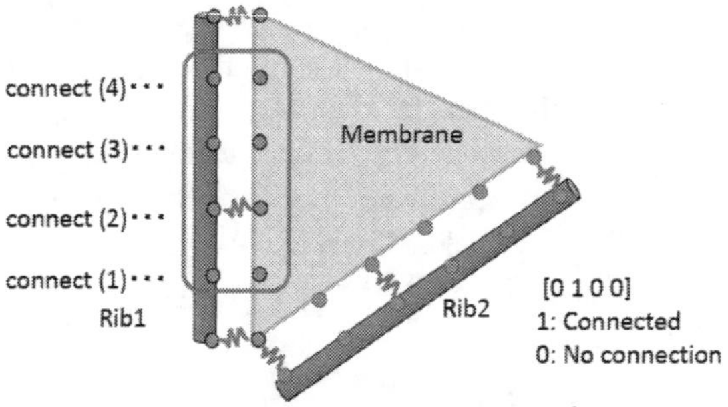

Figure 4 How to express connection

3. PROPOSAL OF A BASIC STRUCTURE FOR LARGER LIGHT WEIGHT SPACE STRUCTURE

3.1. Effect of non-Uniform boom

In this section, non-uniform boom is focused on because this may be a potent member to control stiffness at appropriate point. To evaluate the efficiency of non-uniform boom and how it affect the structural characteristics, reaction force of a boom, wrinkle intensity on membrane and 1st mode are examined. Following topics are also argued in terms of these evaluation items as well.

As for wrinkle evaluation which is the most interest when talking about sail structure. We use this as an evaluation of wrinkle and we call that as wrinkle intensity which is expressed in Eq.(1)

$$\phi_w = \frac{1}{2}\boldsymbol{\varepsilon}_w^T \mathbf{C} \boldsymbol{\varepsilon}_w \tag{1}$$

Where, $\boldsymbol{\varepsilon}_w$ is wrinkle mode deformation, **C** is elasticity matrix.

$$\phi = \phi_e + \phi_w \tag{2}$$

This wrinkle intensity is derived from the Eq. (2) which shows the total strain energy can be efficiently decomposed into elastic energy part and apparent energy part (=zero energy), then the latter term is defined as the wrinkle intensity. Wrinkle intensity means the strain energy density function due to geometrical contraction corresponded to wrinkle, which is based on the tension-field theory.(Akita, Nakashino, Natori et al.,2007, Akita and Natori, 2008)

Here, these evaluations are obtained by using 1/4 model since we do not need to consider asymmetric mode because one of the evaluation is out-plane bending mode. Besides, booms are assumed to move out-of-plane direction only. Also it is assumed that the satellite on the center is large and heavy enough, the satellite can be treated as fixed to the coordinate.

Regarding this model and as a comparative experiment, stiffness of a blue colored part of a boom shown as Table 1 which is 1/3 part from the root is set as equal, two times, three times of the nominal thickness respectively. Then we optimized the connecting points between boom and membrane according to the previous research (Torisaka, Satoh, Natori, 2013) which uses satisficing multi objective optimization method (Kamenoi T., Sugimoto H., Nakayama H.,1992,).

Table 1 shows the structural characteristics including wrinkle intensity contour and principal stress on membrane. As regarding the former, blue colored region shows that it is tensioned and as going to red color, wrinkle is increasing. The wrinkle intensity value in this table means the summation of each point on the membrane. Then the red point in the principal stress shows positive principle stress, blue point (however it is not appeared in these figure) is minus principal stress and green point is positive principle stress at the Gauss point which is judged as wrinkle.

Table 2 Use of non-uniform boom and its effect to structural characteristics

	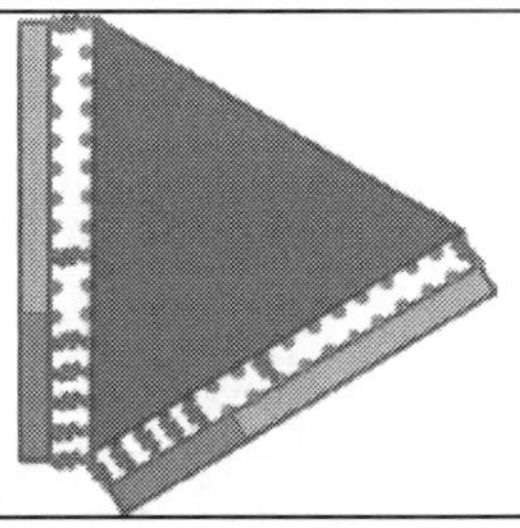	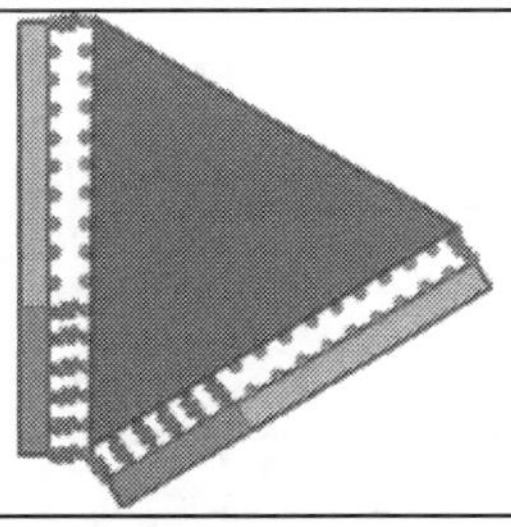	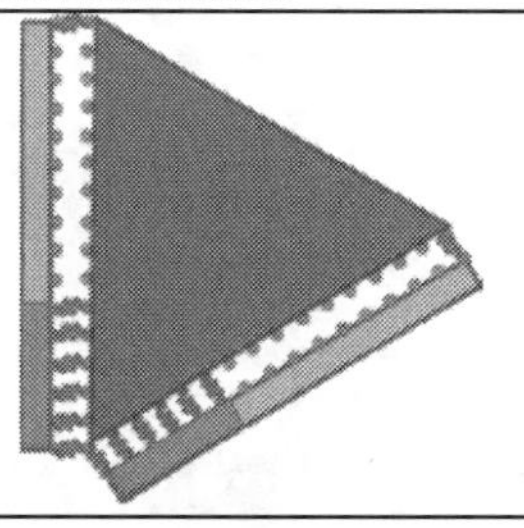
Thickness	x 1	x 2	x 3
Connection point	11110010000000	11111000000000	11111000000000
Num. of Connection	5	5	5
Reaction Force[N]	1.444E+08	1.378E+08	1.378E+08
Wrinkle Intensity [Pa]	3.574E+08	2.591E+08	2.591E+08
1st mode [Hz]	41.67	41.68	43.13
Wrinkle Intensity contour	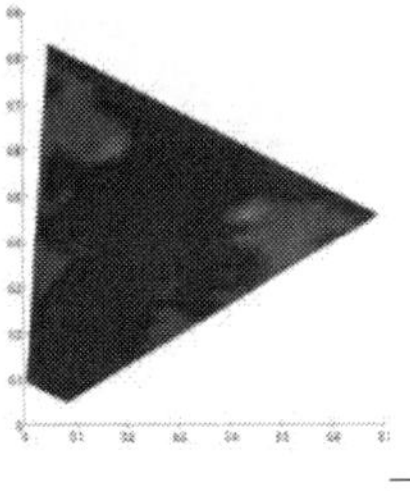	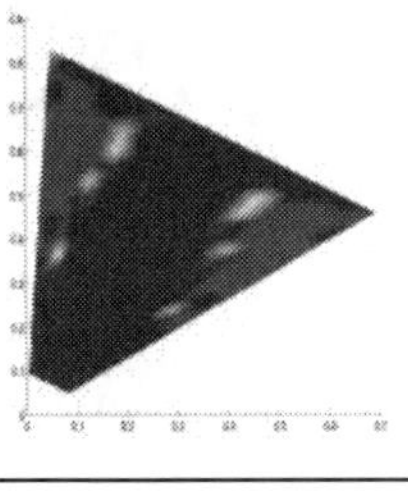	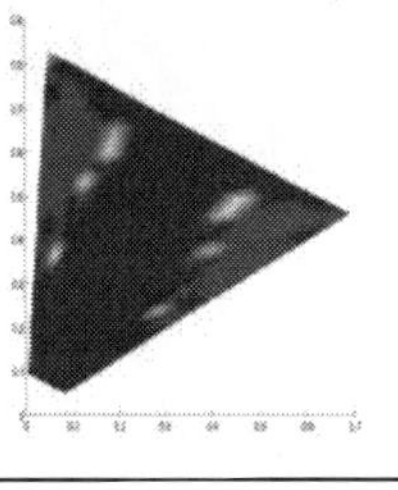
Principal Stress	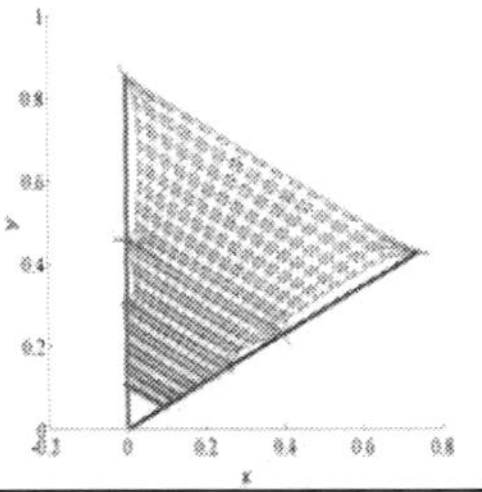	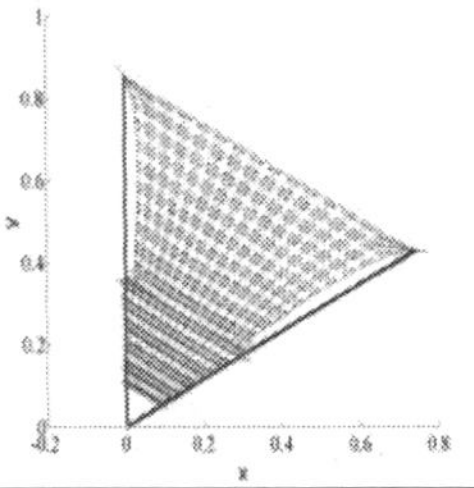	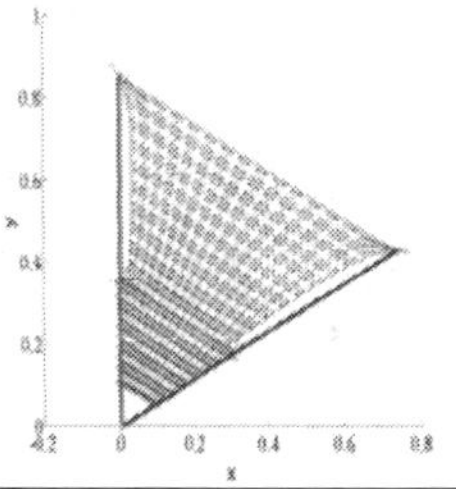

From this table, x2 and x3 model is the best in terms of the wrinkle intensity and reaction force, x1 model is the best in stiffness. The model difference except for boom thickness is connection points between boom and membrane, such as one of the connection is moved to the center side in the case of x1 model. So we focused on why this solution is derived.

The mechanism of x1 model keeps its stiffness is thought to be that the center sided connection point stretches the membrane and keeps the membrane stiffness however it increases the wrinkle. From our previous trial & error calculation, improving stiffness is most difficult, so the priority of keeping stiffness of wrinkle is selected while it sacrifice stiffness. Further, we had found the fact that the wrinkle concentration along a boom reduces the structural stiffness (Torisaka A., Satoh Y., Natori M.C. et al., 2012), so in terms of this point, x1 model effectively maintain the stiffness owing to the stiffness of a membrane by stretching it. Which is to say, the mechanism of keeping stiffness is to stretch the membrane and improve the membrane stiffness.

As for other two models, on the other hand, the mechanism of keeping stiffness is different. These models do not give load to membrane because wrinkle intensity and reaction force is decreased than x1 model. Additionally, the tendency of increasing stiffness is obtained with the increase of boom thickness at 1/3 part. The reason is thought to be that the main mechanism of keeping stiffness is boom stiffness, in other words, non-uniform boom is able to substitute the center sided connection point for membrane stretching.

From above, using non-uniform use is thought to be important. Besides, both the non-uniform boom and the connection between boom and membrane should be took into designing procedure together.

3.2. Cable connection on membrane

(a) Root or Tip extension

Here at first, we want to show which is better to assign tension on the assumption of actual final phase to adjust its shape. Then comparative study is carried out about the root extension model and tip extension model with the square model. **Figure 5** (a). is tip extension in which the forced displacement is given on the tip of a boom, and the root is fixed. The corner of the membrane is also forcibly moved with the boom tip. **Figure 5** (b). is root extension in which the forced displacement is given at the root of a boom. This line of a membrane is globally moved.

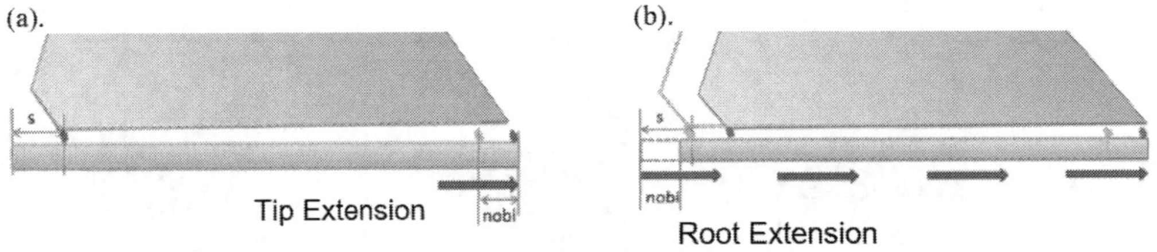

Figure 5 Tip extension and Root extension

Figure 6 shows the result in terms of wrinkle intensity and eigen frequency. This data shows tip extension improves stiffness effectively rather than root extension. It also efficiently reduces wrinkle intensity until 0.5% deformation, while the phenomena as it is increasing after that point is thought to be excessive tension. On the other hand in terms of eigen frequency, increasing rate of stiffness becomes larger after 0.5% deformation. So the mechanism of stiffness improve and reduction of wrinkle before and after this 0.5% point is thought to be different. The reason of this phenomena after this point is due to tension on entire membrane. As that evidence, **Figure 7** shows the tensioned region about tip extension model is larger, and the wrinkle intensity along the booms becomes smaller that is thought to be the mechanism of improving stiffness.

Then the reason of wrinkle reduction before 0.5% deformation is thought to be stress concentration which is caused locally on the membrane. When focusing on the center body side of the membrane edge, blue points which mean there is compression are appeared much more about root extension, so the wrinkle is accumulated around this region.

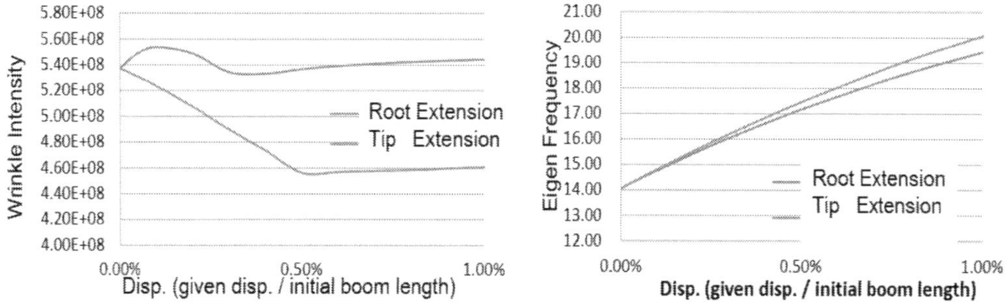

Figure 6 Wrinkle intensity and eigen frequency by root extension and tip extension

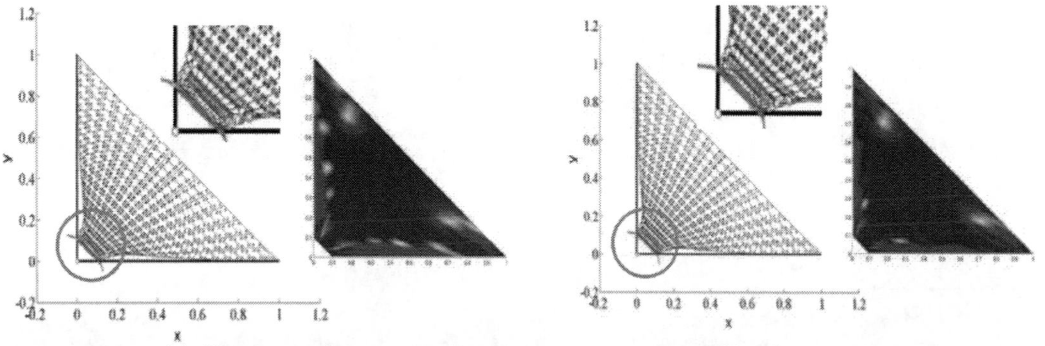

(a) Root extension (b) Tip extension

Figure 7 Principle stress and wrinkle intensity

From above, we think tip extension has advantage in structural characteristics and applied to the following arguments.

(b) - 1 Cable connection on membrane (one cable per membrane)

Already we examined about the efficiency of cable use which is in our previous study, but it has just two connection point on tip of the boom. Here we introduce the connection on the membrane and investigate how it affect the structure. Four kinds of cable attachment are examined as no cable model, one cable and no connection, one cable and one connection, and one cable and two connection. Then these four types are applied to three basic models as [00000000000000] model, [11111000000000] model and [11111111111111] model.

Figure 8 shows the results. The obtained effect is reduction of wrinkle and stiffness improvement when we put connection point between cable and membrane. Compared with Table 3, one cable connection on membrane decreases wrinkle along the boom, and one more connection point works much more. Another significant phenomena is that the wrinkle is gathered towards the oblique line of the membrane that means the model reduces the risk of membrane breaking, and tensioned region which helps to improve stiffness are broaden by this way. So we consider introducing cable connection on membrane is effective and worth to do.

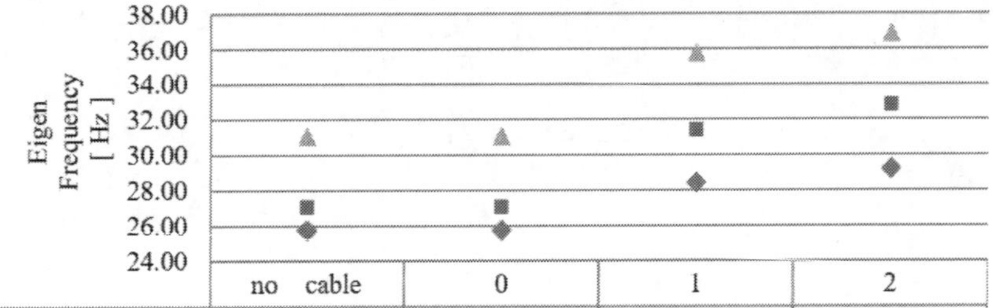

	no cable	0	1	2
◆00000000000000	25.75	25.75	28.43	29.19
■11111000000000	27.07	27.07	31.40	32.80
▲11111111111111	31.04	31.04	35.76	36.87

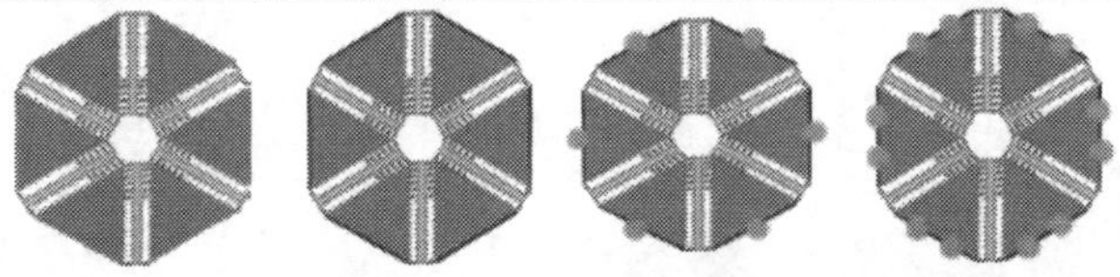

Figure 8 Eigen frequency by cable connection on membrane (one cable per membrane)

Table 3 wrinkle intensity by cable connection on membrane (one cable per membrane)

	No cable	0	1	2
00000000000000	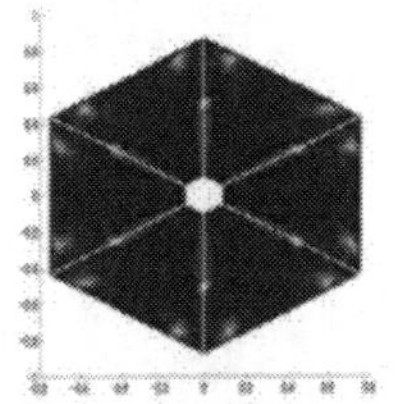	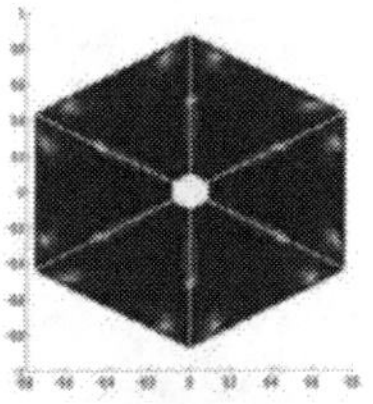	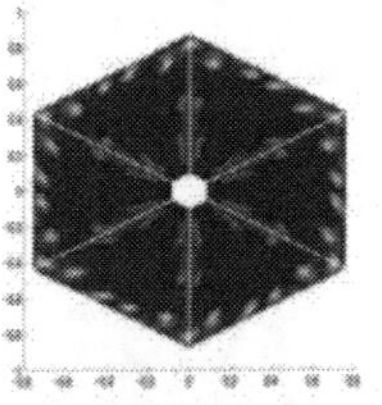	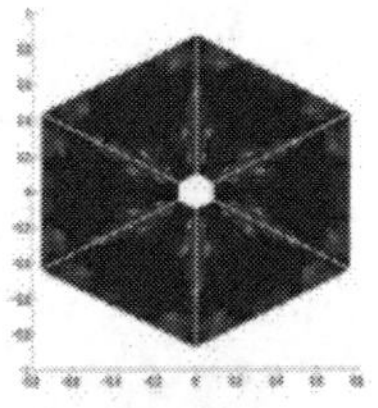
11111000000000	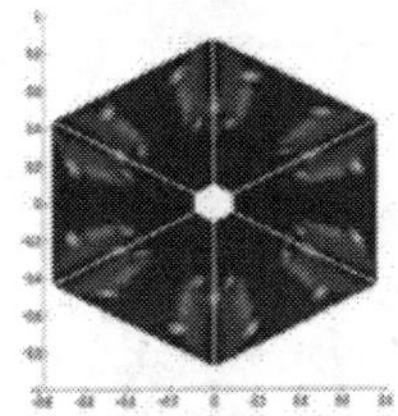	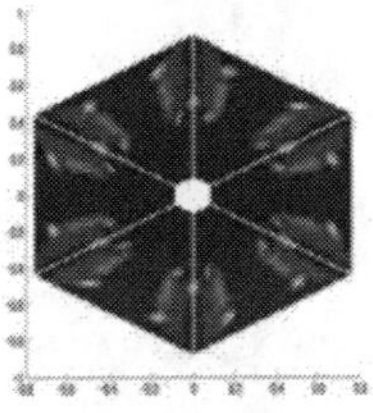	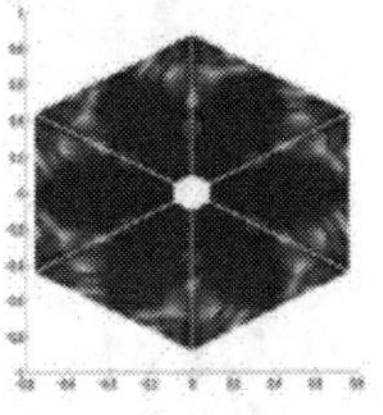	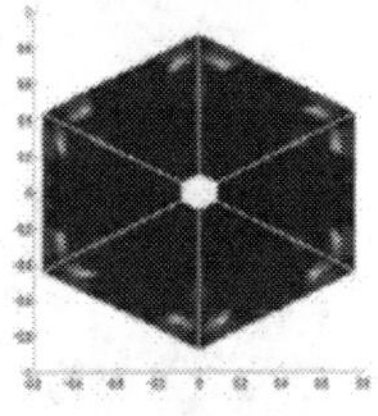
11111111111111	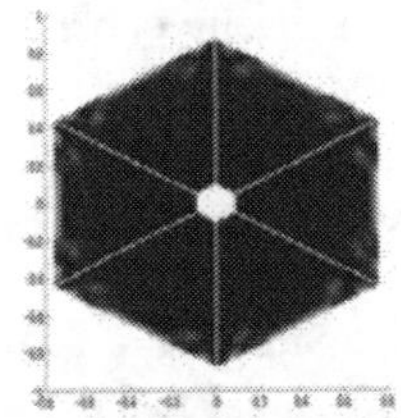	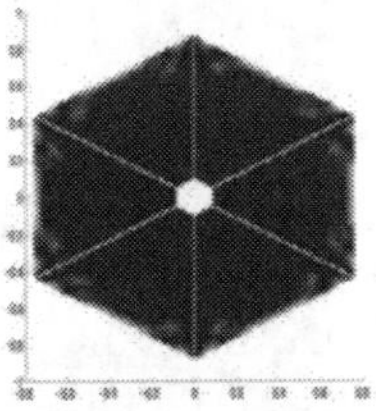	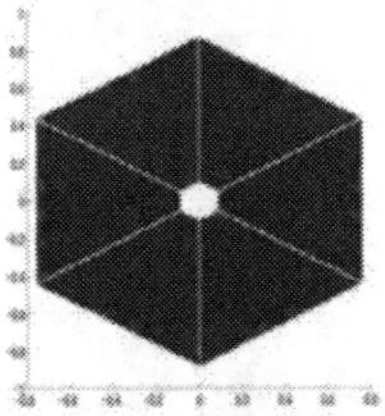	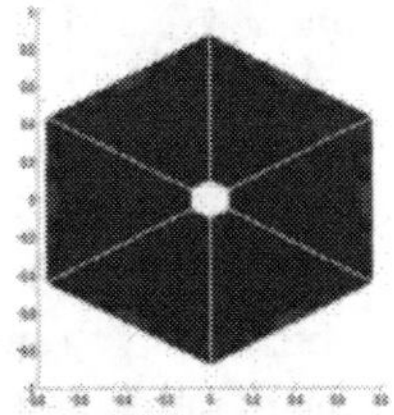

(b) – 2 Cable connection on membrane (two cable per membrane)

Same kinds of examination is carried out on two cable per membrane model, and same results are obtained with two cable model. **Figure 9** is for the purpose of reference.

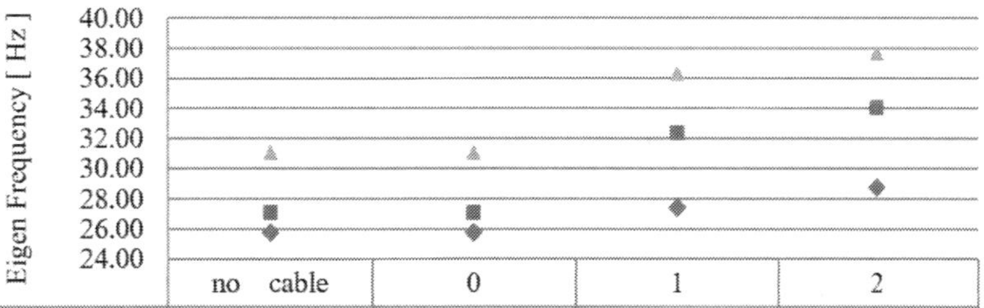

	no cable	0	1	2
◆00000000000000	25.75	25.75	27.41	28.75
■11111000000000	27.07	27.07	32.36	34.04
▲1111111111111	31.04	31.04	36.28	37.63

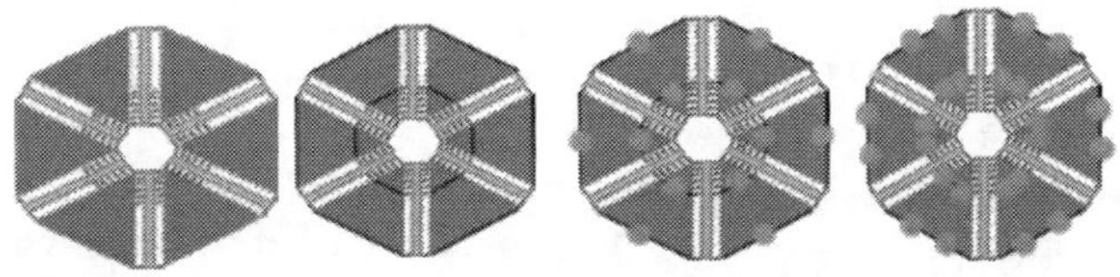

Figure 9 Eigen frequency by cable connection on membrane (two cable per membrane)

3.3. Membrane with steric support of a beam

In this topic, the boom angle Θ is introduced as **Figure 10** to make the steric membrane structure and examine how it affect to structural characteristics. The parameters of the structure is shown in this table. Then cable connection to a membrane is fixed as two points like **Figure 10**.

Table 4 Parameter of a hexagonal model

Boom length L[m]	0.8774
Initial distance between boom and membrane d [m]	0.005
Extension of a boom [%]	0.237
Adjustment angle θ [deg]	45
Connection between boom and membrane	11111000000000
Number of a cable	1
Number of cable connection point on a membrane	2
Boom angle Θ [deg]	0～10

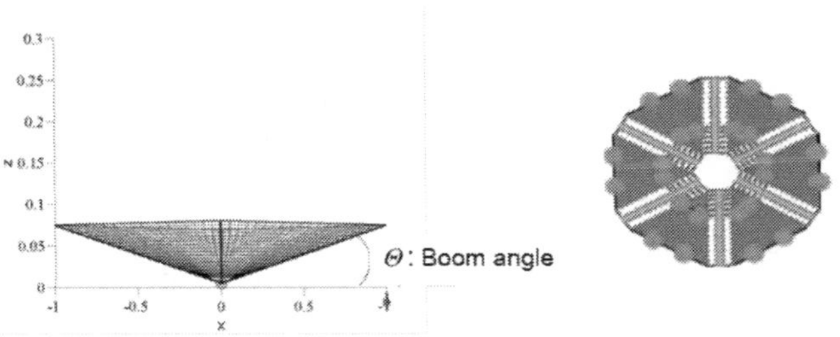

Figure 10 Steric boom- membrane structure

Figure 11 shows the results with dual axis of eigen frequency and wrinkle intensity. Stiffness improves as the slope angle increase. The efficient improve appears on 1st eigne frequency as blue solid line as 126% at 4degree, 160% at 7 degree. So from only the stiffness view point, larger slope angle seems good. While in terms of wrinkle intensity, there is optimum point at around four degree and wrinkle intensity increases steeply after this degree, so the design will end up with around this design region.

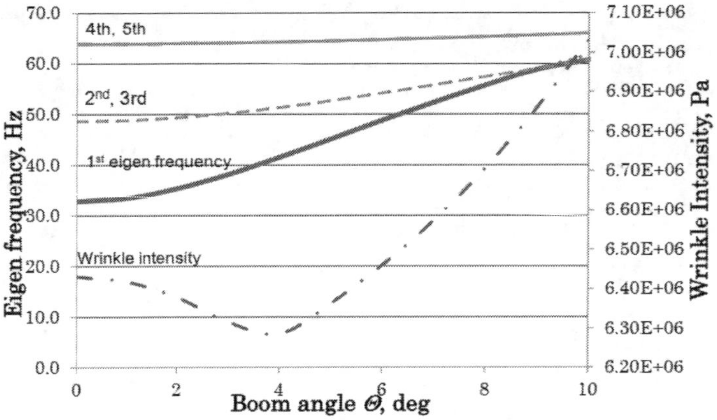

Figure 11 Effect of boom angle on eigen frequency

While the reason of this mechanism have not yet clarified, one possibility may be explained as resulting from the geometric stiffness change of a membrane due to boom buckling which is inferred from the mode shape shown in **Figure 12**. It's also clear that boom deformation depends on the balance of reaction force, tension on membrane and cable tension. So this point should be considered into the designing and it would be our future work along with finding out the mechanism of improving stiffness by the boom angle.

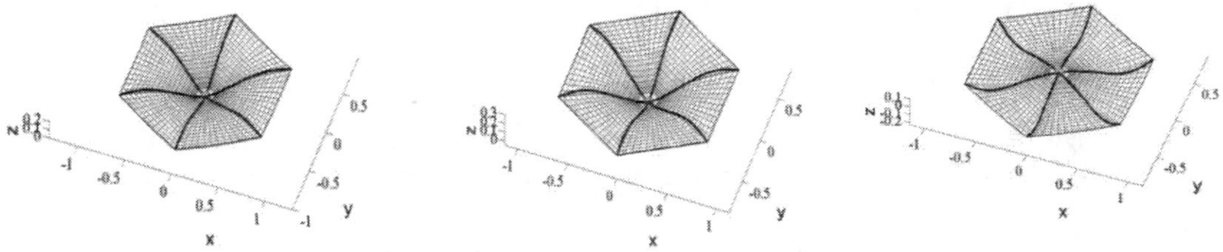

(a) Θ= 0deg (32.79 Hz) Θ= 4 deg (41.54 Hz) Θ= 7 deg (52.32 Hz)

Figure 12 1st mode of steric supported boom-membrane

4. CONCLUSIOS

For proposing a structural concept of boom- membrane structure for larger light weight space structure, structural characteristic by influence of boom design, extension, cable use are examined with both square and hexagonal model. Their effects are summarized as below (including the results of previous study which is not colored here.)

Condition		Expected effect
Connection between boom and membrane	Around main body	Reduction of reaction force on a boom
		Reduction of maximum wrinkle intensity
	Around the tip of a boom	Stiffness improvement
		Prevent stress concentration on the oblique line of membrane (1/5 point from the tip of a boom with Square model only)
Non-Uniform boom use		Stiffness improvement of in-plane mode
		Reduction of connection number (between boom and membrane)

Cable use	Stiffness improvement of symmetric out-plane mode
	Displacement reduction in out-plane
Cable connection on membrane	Reduction of wrinkle
	Stiffness improvement
Sterically supported boom-membrane structure	Noticeable stiffness improvement (out-plane 1st model)

ACKNOWLEDGMENTS

This study has been supported by Japan Society for the Promotion of Science (JSPS), Grant-in-Aid for Young Scientists (B), Grant Number 25820413.

REFERENCES

Akita, T., Nakashino, K., Natori, M.C. et al. (2007) A simple computer implementation of membrane wrinkle behavior via a projection technique, *International Journal for Numerical Methods in Engineering*, 72(10), pp.1231-1259.

Akita T. and Natori M. C (2008) Sensitivity Analysis Method for Membrane Wrinkling Based on the Tension-Field Theory, *AIAA Journal*, Vol. 46, pp.1516-1527

Furuya H. and Masuoka T. (2004) Concept of Rotationally Skew Fold Membrane for Spinning Solar Sail *55th International Astronautical Congress*, Vancouver, Canada, Oct 2004, IAC-04-I.1.05

Furuya H., Inoue Y. and Masuoka T. (2005) Deployment Characteristics of Rotationally Skew Fold Membrane for Spinning Solar Sail *46th AIAA/ASME/ASCE/AHS/ASC Structures, Structural Dynamics, and Materials Conference*, Honolulu, Hawaii , Apr. 2005,AIAA 2005-2045

Greschik G. and Mikulas M. M. (2002) Design Study of a Square Solar Sail Architecture, *Journal of spacecraft and rockets*., Vol. 39, No. 5 September-October 2002

Natori M. C., Watanabe H., Kishimoto N., Higuchi K. (2007) Folding Patterns of Deployable Membrane Space Structures Considering Their Thickness Effects *18th Int'l. Conf. Adaptive Structures and Technologies*, Ottawa, Ontario, Canada, Oct. 2007.

Sakamoto H., Furuya H., Satou Y. et al. (2013) ORIGAMI-Based Membrane Storage and Deployment Techonology for De-orbiting Satellites *64th International Astronautical Congress,* Beijing, China, 23-27 Sept 2013, IAC-13-B4.6A.4

Sakamoto H. and Park K.C. (2005) Dynamic Wrinkle Reduction Strategies for Cable-Suspended Membrane Structures *Journal of spacecraft and rockets*, Vol. 42, No. 5,September - October 2005

Sawada H, Mori O and Okuizumi N et al. (2011) Mission Report on The Solar Power Sail Deployment Demonstration of IKAROS, *52nd AIAA / ASME / ASCE / AHS / ASC Structures, Structural Dynamics and Materials Conference,* Denver, USA, 4-7 April 2013, AIAA 2011-1887

Stohlman O.R., Fernandez J. M., Lappas V. J. et al. (2013) Testing of the Deorbitsail Drag Sail

Subsystem, *54th AIAA / ASME / ASCE / AHS / ASC Structures, Structural Dynamics and Materials Conference,* Boston, USA, 8-11 April 2013, AIAA 2013-1807.

Tamaru H., Akita T., Natori M.C. et al. (2012) Stiffness Characteristics of Membrane Space Structures Supported by Booms, *ASME 2012 International Mechanical Engineering Congress & Exposition,* Texas, USA, November 2012, IMECE2012-88254

Torisaka A., Satoh Y., Natori M.C. et al. (2012) Structural Characteristics of Membrane Space Structures with Booms and Cables. *Proceeding of the JSAS/JSME/JAXA Structural Conference,* Hamamatsu, Japan, Aug.2012

Torisaka A., Satoh Y., Natori M.C. et al. (2013) Study on optimum connecting points of boom-membrane structure as a basic model for larger lightweight space structure. *International Conference on Adaptive Structures and Technologies (ICAST2013)*, Aruba, Netherland, Oct.2013

ICAST2015 #39

A novel method of Peak Detecting for Self-Powered Synchronized Switch Harvesting on Inductance Using Reed Switch

Ya Shan Shih[1,2*], Dejan Vasic[1], and Wen Jong Wu[2]

[1] ENS Cachan, SATIE LAB, 61 Avenue du Président Wilson, 94235 Cachan, France.
[2] Department of Engineering Science and Ocean Engineering, National Taiwan University, Taipei, Taiwan.

Abstract

Synchronized switch harvesting on inductance (SSHI) has long been studied so as to enhance the power gain of piezoelectric energy harvesting. The idea of the SSHI is to launch an inductance to the circuit on the peak of the waveform so that the inductance resonates with the clamped capacitance of the piezoelectric patch. The resonance causes a rapid inversion of the voltage, which then the switch is turned off, halting the voltage on its inverted state. This technique enhances the power output when appropriate parameters are met. It is notable that detecting the optimal switching time is inevitable in this technique. In spite of the several simple ways already in use, such as voltage peak detection using bipolar transistor or zero velocity crossing detection, these methods requires a considerable amount of energy in respect to the harvested energy. On account of the harvesting efficiency, which the total harvested amount is from the scale of nano-watts to micro-watts, it is mostly impractical to introduce techniques or elements that requires external power sources or extra power consumptions when it comes to a high portion of energy loss considering the limited amount of energy harvested. Thereby, researchers have endeavored to realize the peak detecting in various mechanical and electrical methods so that little or few energies are required. In this work, we propose a new hybrid peak detecting technique utilizing reed switch, which is a non-contact mechanical method that requires little energy consumption. Using a pair of magnets as the proof mass, not only the resonance frequency can be easily tuned, it could also serve as the non-contact control of the reed switch. However, mechanical chattering, a phenomena which exists in all the mechanical switches, is yet to be solved. The phenomena of chattering is the contact bounce when the two metal parts of the switch strike together. In attribution to the elasticity and momentum, the two parts bounces off and on in the transient state for a short moment until settling down to stable. Experiments were conducted to find the losses due to the chatter, and to also find the performance of the new structure. Results showed a promising outcome of 88.9% in comparison to the external powered SSHI. The following sections are divided in to 5 parts: introduction to SSHI and some switching techniques, experiment setup, experiment results, discussion, and conclusion along with some plans of future work.

1. Introduction

Synchronized switching on inductance (SSHI) is a technique to improve the harvested power of the piezoelectric energy harvester, proposed by Lefeuvre et al during 2004 [1-3] originally introduced as synchronized switch damping (SSD) for vibration control [4]. With low electro-mechanical coupling systems, it was discovered that this approach could increase the harvested power by 400% in comparison to the standard circuit of a rectifier consisting four diodes. By applying a non-linear processing electronic

* ysshih@ntumems.net

circuit of a switch and an inductor to the standard rectifier, the switch was turned on and off on the maximum and the minimum of the mechanical displacement. On the triggering moment, the LC resonance of the inductance and the parasitic capacitance of the piezoelectric patch, was launched. This quick resonance was switched off after half the period, causing an inversion of the piezoelectric voltage. Since then, self-powered switching on inductance has long been studied on synchronized switching techniques. Some other techniques of synchronized switch harvesting (SSH) were afterwards proposed after the SSHI parallel and SSHI series (figure 1): DSSH[5], SSDCI [6], SECE [7], ESSH[8]… etc. [9]. However, using a complex circuit also remains and create new issues within the design, such as power efficiency, threshold voltage, frequency selectivity and also parasitic bandwidth filters.

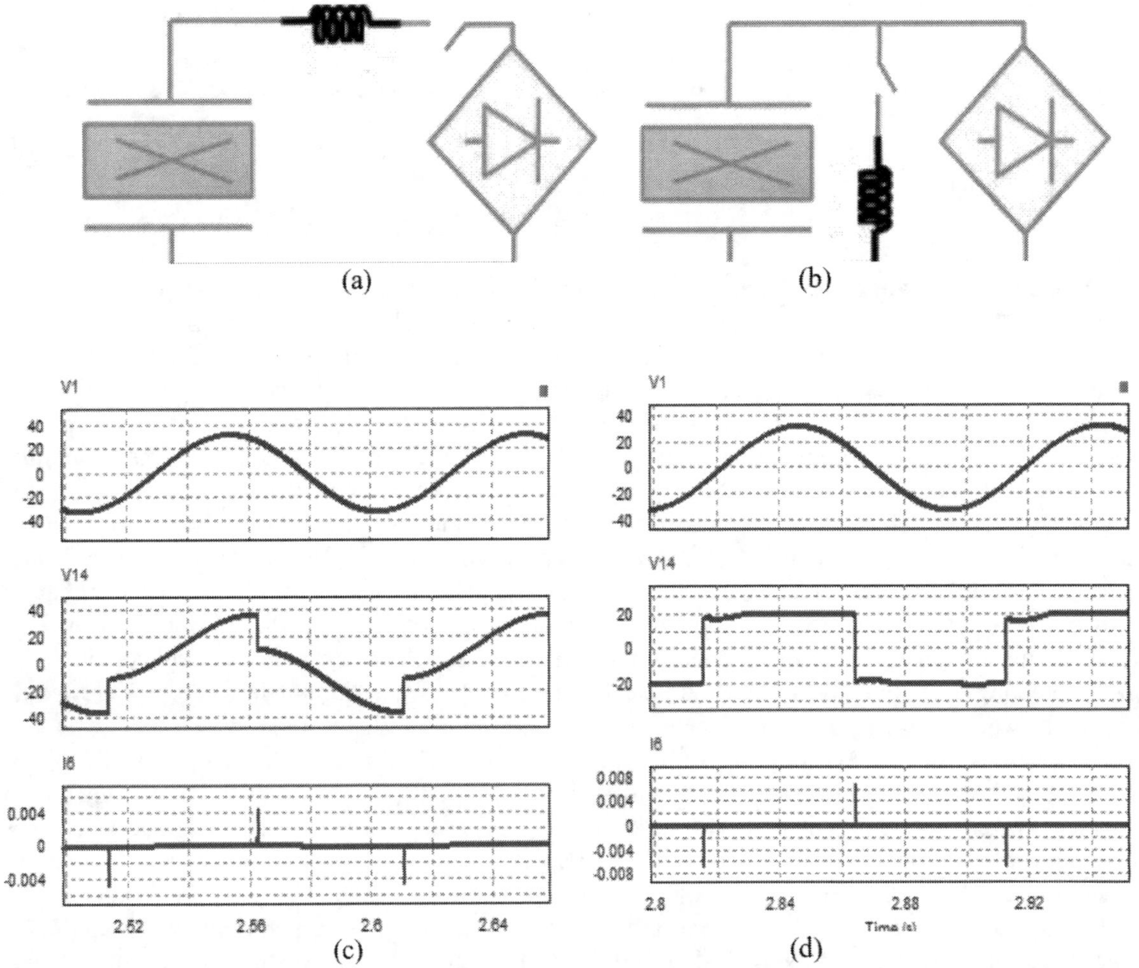

Figure 1. SSHI circuit (a) in the series form (b) in the parallel form (c) waveforms showing from top to low: neutral piezoelectric output, rectified voltage over the load of 2MΩ, and the correspondent current of a series SSHI connection (d) waveforms showing from top to low: neutral piezoelectric output, rectified voltage over the load of 2MΩ, and the correspondent current of a parallel SSHI connection.

One of the critical issues is to drive the switch on the correct instant. It would make no sense to use externally powered switches, while we aim to harvest power from the vibration. Therefore, a self-powered “smart switch” was developed for this task [2, 10]. The stand-alone smart switch utilizes a comparator and a voltage follower, acting as the energy source in the same time (figure 2.a) [11, 12]. The smart switch should be designed that the exerted energy itself is as little energy as possible. Other approaches of smart switches such as velocity controlled techniques were also introduced by Chen et al [13]. With the velocity control technique, the piezoelectric patch was divided in to three patches: a main patch, and two sub patches (figure 2.b). The main patch is for the main energy harvesting. As for the sub patches, one is for the velocity sensing and the other for supplying a comparator. The velocity sensor was composed of a current sensing resistor, and a low pass filter to filter the high frequency noise. This sensor outputs the current trace as the form of voltage. Thence, with the comparator connected to the ground, the zero crossing point of the current, which is identical to the waveform maximum or minimum, can be defined. The comparator were provided its power supply of $+V_{cc}$ and $-V_{cc}$ by the other sub patch. By this mean, the maximum and minimum of the voltage signal could be detected and switched.

Comparing the two methods, we find that the smart switch requires one single patch, which is easier to fabricate and apply. Moreover, the transistors and the MOSFETs that were applied in the smart switch may require less power than the comparator. Yet the velocity control has a finer switching instant, attributing to that the threshold voltage of the transistors which are used for voltage comparisons. The differences between the voltage buffer and the piezoelectric voltage has to reach the threshold to trigger the switch, therefore there exists a small time lag in the self-powered switch.

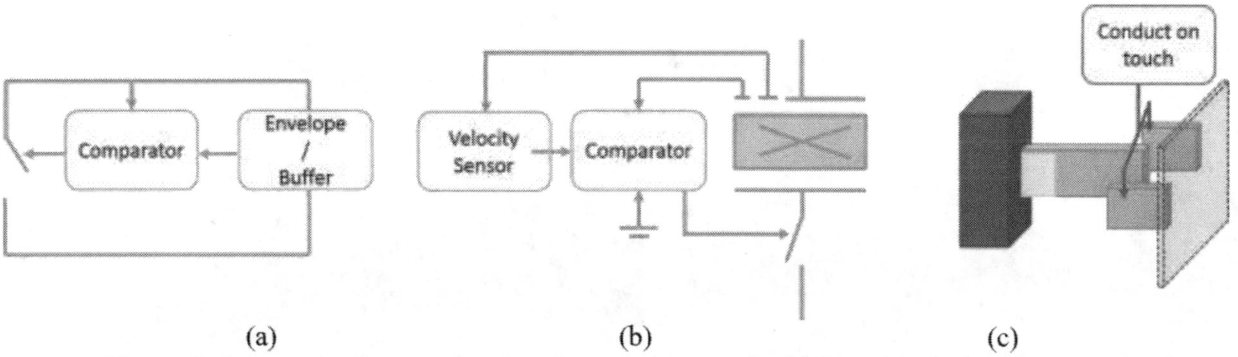

Figure 2. Schematic diagram showing the switching method of (a) electrical self-powered switch (b) electrical velocity sensing switch and (c) mechanical stopper switch.

Yet another method was the mechanical method. Wu et al had integrated stoppers, a method to increase bandwidth, with the mechanical switch [14] (figure 2.c). In their work, introducing the optimized synchronous electric charge extraction (OSECE) [15], as an improved version of SECE proposed by Lefeuvre et al [7]. By taking advantage of the stopper, the switch was switched on as the vibrating part impacts with the stopper. In this way, the components which may establish loss are merely two diodes and no extra energy was required. The outcome was superb, with a wide bandwidth, and enhanced power. Still, the impact of the beam may damage the piezoelectric patch itself, speeding up the fatigue time, and reducing the lifespan of the device.

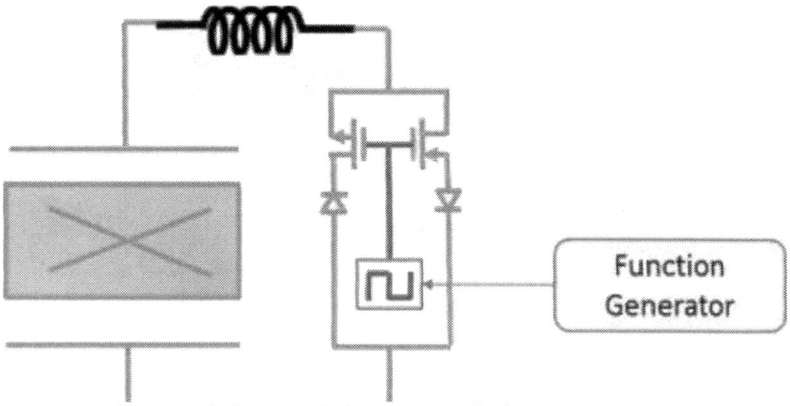

Figure 3. External driven switch demonstration.

Although externally powered switching is not practical in application, it is herein used for a standard, representing a loss-less switching. The method utilizes an extra set of signal from the function generator, to provide an in-phase square wave for the driving of the NMOS and PMOS set (figure 3). The NMOS and PMOS were each connected to one diode to limit the current flow in single direction. Thereby, the two sets of switches forms a bidirectional switch. The driving power of the NMOS and PMOS are provided by the function generator, so as to consider zero power consumption. It is, however noted, that the cross voltage of the MOSFETs and the inductor resistive loss still do take up a small portion of energy.

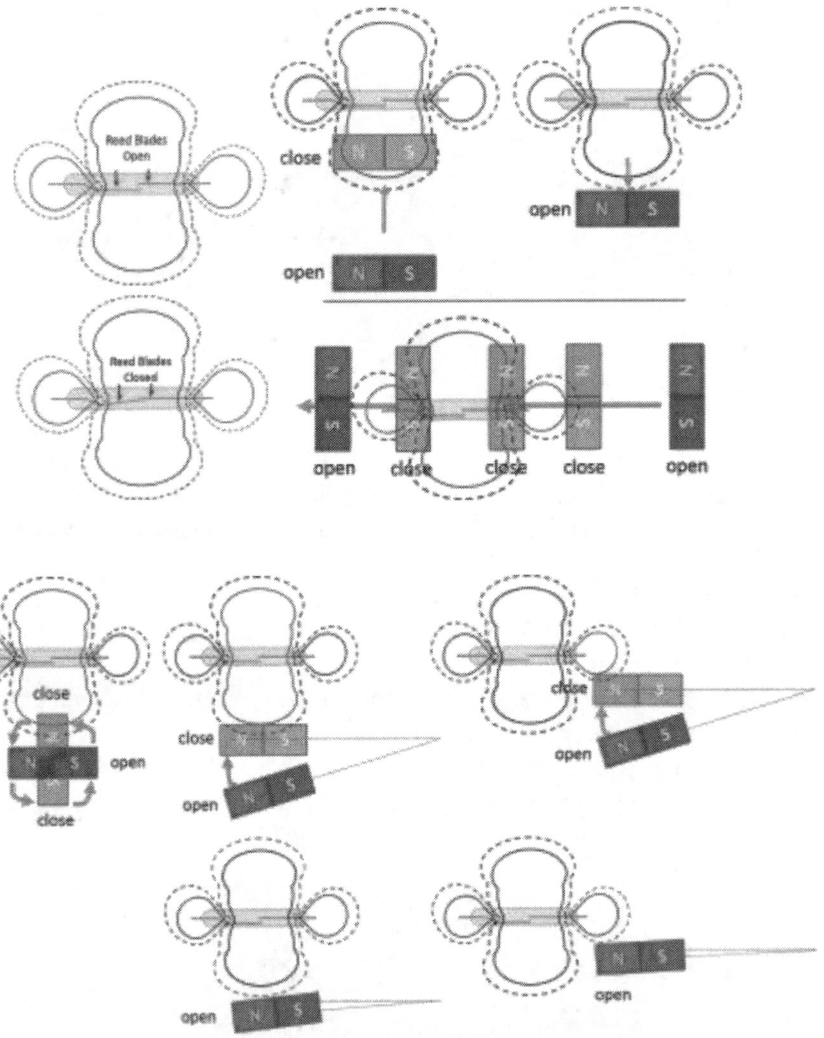

Figure 4. Various ways of triggering the reed switch via permanent magnets

2. Design Concepts and Setup

Reed switches are electrical switches, which can be driven and operated with magnetic fields. With the tiny reed blades enclosed within the glass casing, a normally open switch can be switched on when a magnetic field approaches the reacting area (depicted as the solid line) (figure 4). The size of this area can be determined by the quantity of the magnetic field. As the magnetic field is removed from the hysteresis area that is a bit larger than the active area (as the dashed line in the plot), the switch would be re-opened. For the normally closed switch, it works vice versa. Opening when applied with magnetic field; reclosing when magnetic fields removed. The altering of the magnetic fields can be performed by different movements of magnets. The magnets can generate field changes with orthogonal, parallel, or rotary movements. Here in the case of cantilever beam, a pivoted motion is applied, using the wings of the active area. It has to be noticed that the fields are actually in three dimensional, therefore, choosing a smaller area, such as the wing, can reduce trouble of overlapping fields. Hence, by installing two reed switches in a solid place nearby the upper and lower tip of the beam, they can be used as mechanical detectors, detecting whether the magnet has reached the location.

Not only driving the reed switch, the magnets used in this work can also serve as the proof mass to tune the resonance frequency. The design concept is rather simple, that is, to use reed switch as a mechanical switch. As the beam wavers to the top, the switch is turned on and LC resonance is launched (figure 5). Unlike electrical switches, mechanical switches do not require extra power, nor does it need a threshold voltage. Therefore an ultra-low voltage operating SSHI can be realized. However, all mechanical switches are subjected to contact bouncing, cutting off the resonance and causing unwanted losses. Therefore an N-channel MOSFET was introduced as the circuit switch, making the reed switch to serve as the peak detector. With the parasitic capacitance of the MOSFET, the ringing was eliminated but nevertheless not sufficient. Thereby, a de-bouncer, such as a low pass filter, is inevitable to filter the chattering signals. Capacitances are required in low pass filters, and its quantity in our case, is decided by two rules: first, the capacitance should be able to provide the required power to drive the N-MOS switch; second, it should also be small enough to eliminate the chattering frequency of the mechanical switch. Hence, it is independent of the driving frequency, and can be used in different low resonance frequency occasions.

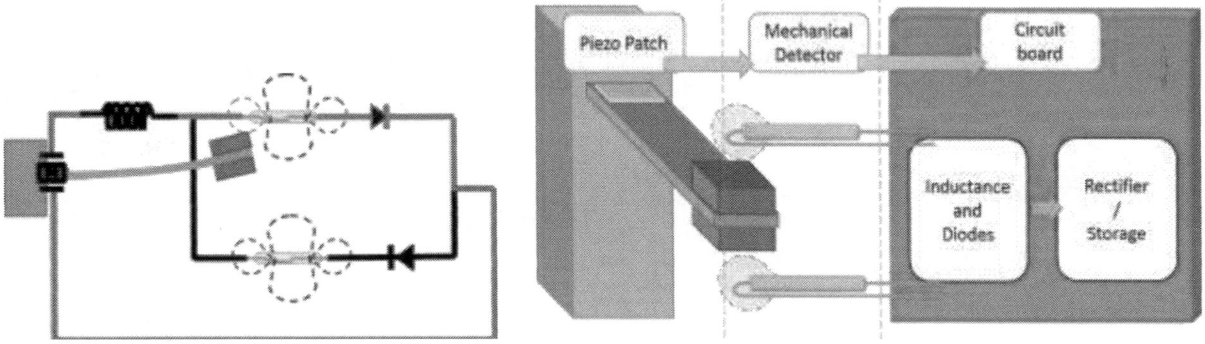

Figure 5. The design of the reed switching technique (a) circuit plot, brown lines indicating the active loop (b) system schematic diagram

3. Experimental Results

In the experiment, a National Instrument Data Acquisition Card© (NI-DAQ card) was used as the analogue-digital interface for driving and obtaining the signals (figure 6). To support with, a LabVIEW© program was created to operate the DAQ card. Attributing to the limited power of the DAQ card output, a power amplifier was set in between the shaker and the DAQ card, so that the shaker can be driven with enough power. The output signal of the beam was connected to the parallel SSHI with the reed switches (figure 5.a), whilst the overall result could be sent back to the oscillator and, in the same time, the DAQ card while the shaker is at work. It is then the DAQ card which converts the signal, and offers the created interface program the collected signal so that we could perform analysis and information storage.

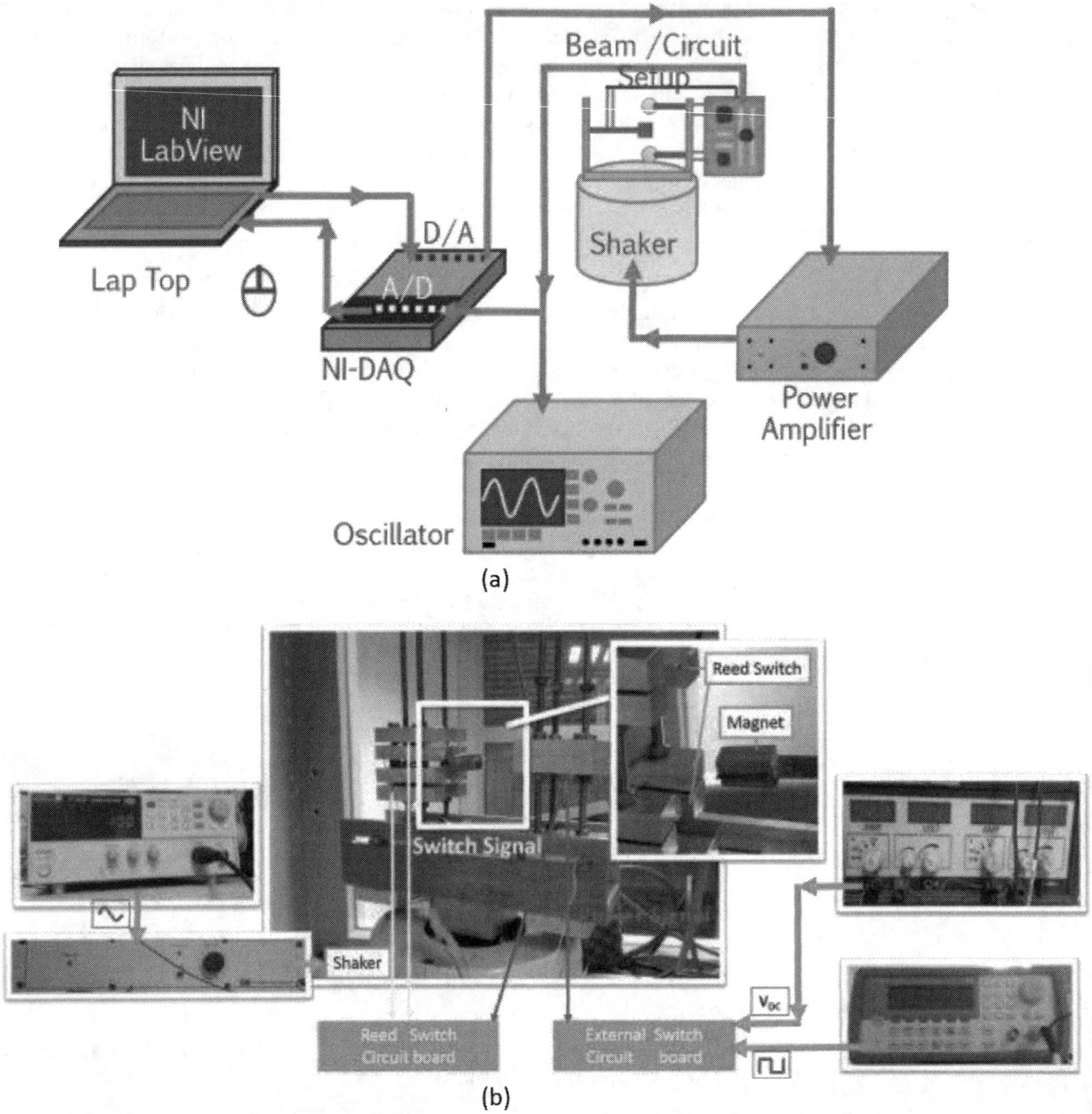

Figure 6. (a) Schematic diagram of experiment setup (b) realizing the experiment.

The experimental piezoelectric output voltage with reed switches is shown as waveform 1 in the figure 7(a). This voltage is preliminarily correct, with some time differences before the optimal switching time, indicated as wave form 3. However, there are considerable oscillations, which are also happening in the switching instance. This phenomenon, called bouncing or chattering, decreases the inversion voltage in comparison to externally powered SSHI, as shown in the figure 7.b. Waveform 3 in both plots are the driving signal of the externally powered SSHI from a function generator, as a reference of the optimal

SSHI, externally powered SSHI does not use any power from the piezoelectric beam to drive the switches. In the following section, we will look closer into the inversion moment, and inspect the chattering which occurs within.

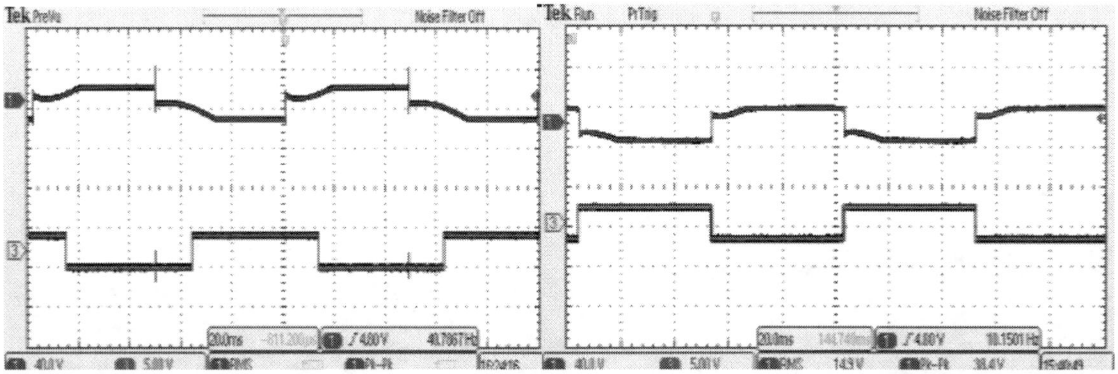

Figure 7. SSHI waveforms using (a) reed switch (b) externally powered switch. Waveform 1 indicating the piezoelectric voltage (rectified), and 2 indicating the function generator signal of optimal switching time.

Chattering

The chattering, or bouncing, as mentioned before, is a parasitic phenomenon of continuous on-off switching due to the elastic contact movement of the mechanical switch. This phenomenon, however, will cause unwanted LC oscillations and excessive load resistances within the inversion period, which lasts approximately one to a few milliseconds. There exist a one to two perfectly detachments, and several wavering contacts with lower contact resistances. The on and off switching and also the change between high and low resistances results in the redundant peaking oscillations, as shown in the figure 8.a. Theses oscillations which are attributed to the redundant closures causes power losses, exceeding currents and short drags in the inversion. The parasitic capacitance strikes on a LC resonance of a natural period in the scale of nano-seconds. This phenomenon occurs twice to three times each inversion, as shown in the figure 8.b.

Experiment was performed to observe how the reed switch works and also to examine the loss chattering causes. Parallel and series SSHI were introduced as the two standards to test the efficiency of the switch. Figure 9b shows the switching time of the external SSHI (on peak). Figure 9.c, shows the switching instant of the reed switch in comparison to the optimal time of the external SSHI. It is worth to note that, the reed switch, due to the mechanical hinder, can only be pre-switched in time rather than belated. It could be tuned in a more ideal location, however, it takes precision and the displacement is altered as the switching is launched. In figure 9.a, there were two overlapping curves both identified as classical rectifier, they were performed to ensure that the parallel and series SSHI were conducted in identical driving forces, with frequency of 10 Hz.

Result turn out that in series SSHI, the reed driven series SSHI has an efficiency of 88.9%, considering the external powered SSHI to be 100% as the reference. The externally powered enhancing effect of series SSHI in comparison to the classic was 2.42 folds. On the other side, the parallel connections show similar output, where the efficiency was nearly 100%. This result can be related to the topology of the circuit: for the parallel circuit, the piezoelectric patch was normally on. On closure, is the only moment that the switch loss take place. For this reason, the switch loss could have been less dominant in parallel SSHIs. One reason for the equally output can be that the reed switch possesses nearly zero voltage drop, however the MOSFETs which were used as the electrical switches still requires small cross voltages. On the other hand, the circuit of series SSHI is normally open, and the current flows through the circuit only on the instant of switch closure. The bouncing would have lowered the transmitted current, which also pulls down the power output. Also all the current would pass by the switch, which had wavering loads. Hence with larger currents, the loss over the reed switch is more severe in comparison to the constant cross-voltage of the MOSFET switch.

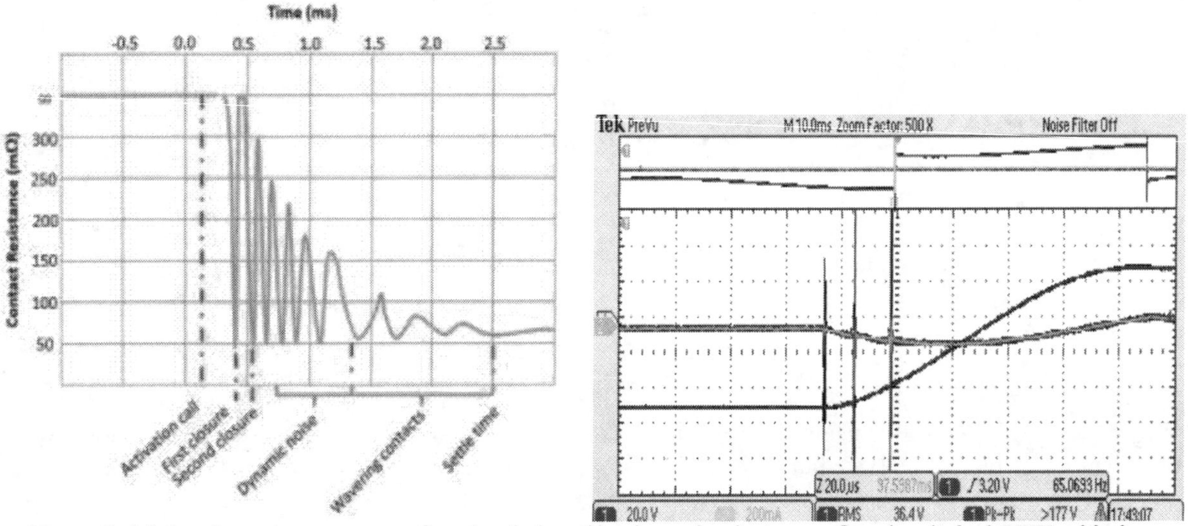

Figure 8. (a) the chattering response of reed switches (b) a inversing instance of reed switched SSHI with the presence of chattering

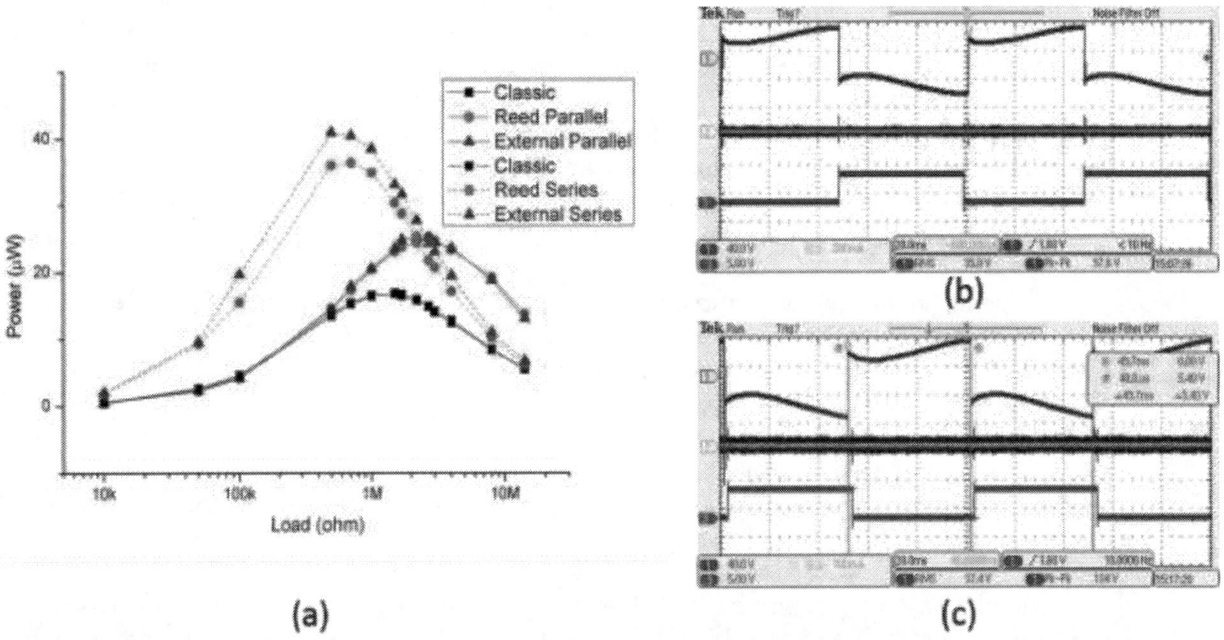

Figure 9. Experiment results showing open loop SSHI (a) comparing series and parallel SSHI with classical circuit, externally powered switch, and reed switches. (b) Switching instant of the externally powered SSHI, waveform 3, which is on the peak of the piezoelectric output waveform 1, (c) Switching instant of the reed switch (1), in comparison to the waveform 3, the switching signal of externally powered switch. Waveform 2 are the overall current flow, flowing out from the piezoelectric patch.

Another experiment was conducted to inspect how the distance between the tip of the beam and the reed switch will affect the power output. That is, the relationship between the power enhancement and switching instant of the switch. The circuit was connected to parallel SSHI. Experiment results show that, after the ideal switching point, the enhancing effect is greatly reduced. Moreover, switching too early even creates losses. However, this result shows a broader bandwidth (figure 10.a), with a smoother output within the bandwidth. The figure 10.b shows a depiction of ideal and non-ideal positions. When the tip displacement is too large, the switch meets its closure before the maximum point of the waveform. During this period, a portion of the rectified signal showed to be zero, at this moment, no power was harvested, but only energy consumed.

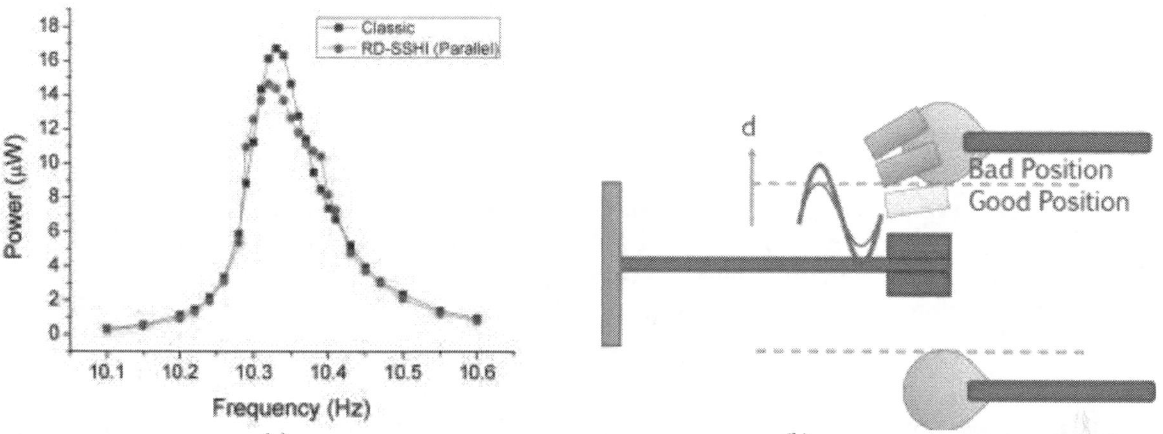

(a) (b)

Figure 10. (a) Frequency scan of the reed switched parallel SSHI (b) demonstration of ideal position and non-ideal positions of the beam displacement

Discussions

There are several merits and cons with reed switches. First of all, in comparison to electrical methods of self-powered SSHI, the reed switch can provide zero threshold driving conditions. The chattering, however, drags down some of the harvested power. A solving method can be found in embedded systems, de-bouncers, which were often applied for mechanical switches to avoid multiple signals caused by the chattering. Since experimental results show a fair efficiency of 88.9% of series SSHI in comparison to the externally powered series SSHI, we should carefully design that the power consumption of the de-bouncer is less than the lost power from chattering. Another problem in the reed method is the displacement of the beam is restricted to enable optimal output. A stopper, can be applied to eliminate the over-deflection, and a bi-stable magnet, can be used to keep the displacement large enough. The optimal distance between the beam tip and the reed is however a tradeoff. Reducing the distance may be able to broaden the bandwidth, but in the same time lowering the optimal output. It could be predicted that with distance high enough would narrow the bandwidth, however enlarging the optimal power.

Conclusion and Future work

We have presented a new mechanical structure as a zero-voltage drop switch and a peak detector for synchronized switching on inductance techniques. The structure can be self-powered, with only two diodes required. Despite the chattering of the reed switch, a well-tuned situation of the structure showed a good efficiency in parallel SSHIs, and fair efficiency of 88.9% in series SSHIs in comparison to externally powered cases, which we consider as 100%. The loss of the series SSHI was due to the only path and the large current flow which has to go through the switch. In the spec of bandwidth, by tuning the distance a broader bandwidth response with smooth power output can also be obtained, however, there is a trade-off between the optimal power and the bandwidth. This design is simple to set up, and requires few components. It is also frequency independent, since reed switches can work up till kilo hertz. However it is restricted to the displacement of the beam.

De-bouncers are to be designed in order to reduce the loss of the switch. It is noted that the loss of the de-bouncer should also be carefully taken care of. Structural designs to restrict the displacement such as stoppers and bi-stable techniques can be applied. Bi-stable techniques can be trivial since that the set up itself contains magnets in before hands.

Acknowledgement

We give special thanks to Campus France and the NSC (National Science Council) of Taiwan who fund this project of interchange. For the NSC project number of 103-2917-I-002 -157. And the project Eifel of boursiers excellence via Campus France.

References

[1] E. Lefeuvre, A. Badel, C. Richard, and D. Guyomar, "High-performance piezoelectric vibration energy reclamation," 2004, pp. 379-387.

[2] D. Guyomar, C. Richard, E. Lefeuvre, and L. Petit, "Piezoelectric Non-linear Systems for Standalone Vibration Con-trol and Energy Reclamation," 2004.

[3] E. Lefeuvre, A. Badel, C. Richard, L. Petit, and D. Guyomar, "A comparison between several vibration-powered piezoelectric generators for standalone systems," *Sensors and Actuators A: Physical,* vol. 126, pp. 405-416, 2/14/ 2006.

[4] C. Richard, D. Guyomar, D. Audigier, and G. Ching, "Semi-passive damping using continuous switching of a piezoelectric device," in *1999 Symposium on Smart Structures and Materials*, 1999, pp. 104-111.

[5] M. Lallart, L. Garbuio, L. Petit, C. Richard, and D. Guyomar, "Double synchronized switch harvesting (DSSH): A new energy harvesting scheme for efficient energy extraction," *Ultrasonics, Ferroelectrics, and Frequency Control, IEEE Transactions on,* vol. 55, pp. 2119-2130, 2008.

[6] W. Wu, A. Wickenheiser, T. Reissman, and E. Garcia, "Modeling and experimental verification of synchronized discharging techniques for boosting power harvesting from piezoelectric transducers," *Smart Materials and Structures,* vol. 18, p. 055012, 2009.

[7] E. Lefeuvre, A. Badel, C. Richard, and D. Guyomar, "Piezoelectric energy harvesting device optimization by synchronous electric charge extraction," *Journal of Intelligent Material Systems and Structures,* vol. 16, pp. 865-876, 2005.

[8] H. Shen, J. Qiu, H. Ji, K. Zhu, and M. Balsi, "Enhanced synchronized switch harvesting: a new energy harvesting scheme for efficient energy extraction," *Smart materials and Structures,* vol. 19, p. 115017, 2010.

[9] D. Guyomar and M. Lallart, "Recent progress in piezoelectric conversion and energy harvesting using nonlinear electronic interfaces and issues in small scale implementation," *Micromachines,* vol. 2, pp. 274-294, 2011.

[10] C. Richard, D. Guyomar, and E. Lefeuvre, "Self-powered electronic breaker with automatic switching by detecting maxima or minima of potential difference between its power electrodes," *FR2005/003000, publication number: WO/2007/063194,* 2007.

[11] L. Mickaël and G. Daniel, "An optimized self-powered switching circuit for non-linear energy harvesting with low voltage output," *Smart Materials and Structures,* vol. 17, p. 035030, 2008.

[12] L. Junrui and L. Wei-Hsin, "Improved Design and Analysis of Self-Powered Synchronized Switch Interface Circuit for Piezoelectric Energy Harvesting Systems," *Industrial Electronics, IEEE Transactions on,* vol. 59, pp. 1950-1960, 2012.

[13] C. Yu-Yin, D. Vasic, F. Costa, W. Wen-Jong, and C. K. Lee, "Self-powered piezoelectric energy harvesting device using velocity control synchronized switching technique," in *IECON 2010 - 36th Annual Conference on IEEE Industrial Electronics Society*, 2010, pp. 1785-1790.

[14] Y. Wu, A. Badel, F. Formosa, W. Liu, and A. Agbossou, "Nonlinear vibration energy harvesting device integrating mechanical stoppers used as synchronous mechanical switches," *Journal of Intelligent Material Systems and Structures,* p. 1045389X14533437, 2014.

[15] Y. Wu, A. Badel, F. Formosa, W. Liu, and A. E. Agbossou, "Piezoelectric vibration energy harvesting by optimized synchronous electric charge extraction," *Journal of Intelligent Material Systems and Structures,* vol. 24, pp. 1445-1458, 2013.

ICAST2015 #036

Equivalent Circuit Modeling and Experimental Validation of a Piezoelectric Energy Harvester Attached on a Thin Plate with AC-DC Conversion

Amirreza Aghakhani[1], Ipek Basdogan[1*], and Alper Erturk[2]

[1] Department of Mechanical Engineering, Koc University, Istanbul 34450, Turkey
[2] G. W. Woodruff School of Mechanical Engineering, Georgia Institute of Technology, Atlanta, GA 30332-0405, USA

Abstract

Plate-like structures are widely used in numerous automotive, marine and aerospace applications. Power output investigations of patch-based piezoelectric energy harvesters integrated to these structures require accurate models for energy harvesting performance evaluation and optimization. The equivalent circuit modeling of the cantilever-based vibration energy harvesters for estimation of electrical response has been proposed in the recent years for predicting the electrical outputs of the harvesters. However, equivalent circuit modeling of piezoelectric patch harvesters integrated to plate-like structures including nonlinear circuits has not been studied in the existing literature. Considering these needs, a multi-mode equivalent circuit model of a piezoelectric energy harvester integrated to a thin plate is developed and verified experimentally in the present study. Equivalent circuit parameters are obtained from analytical distributed-parameter model of the plate and harvester which governs the electromechanical coupling behavior of piezoelectric patch and vibration of the host plate. The multi-mode circuit representation of the harvester is built via electronic circuit simulation software SPICE. Using the SPICE software, electrical outputs of the piezoelectric energy harvester are computed for the standard AC input-AC output and AC input-DC output problems. In the AC-AC case, only a resistive load is connected to the harvester, whereas for the AC-DC case, a full-wave rectifier and a smoothing capacitor are connected to the circuit before the resistive load to convert the AC voltage to stable DC voltage. In the AC-AC problem, voltage FRFs are calculated for various resistive loads and validated by the published analytical closed-form solution. In the AC-DC problem, simulation results of the DC voltage and power outputs are computed for a wide range of load resistance values and validated with comparisons against the analytical single-mode representation of the harvester. Finally, experimental measurements of DC voltage FRFs are conducted for a case study and verified against the numerical model.

1. INTRODUCTION

The application of self-powered wireless sensors and health monitoring systems have been growing increasingly where the external power requirement has decreased to micro-watt range in the past few decades [1, 2]. Meanwhile, development of vibration-based energy harvesters has become promising in terms of enabling constant power supply from ambient vibration sources for the low power

* ibasdogan@ku.edu.tr

electronics [3]. There exist several transduction mechanisms in which vibrational energy can be converted to electrical energy through electrostatic [4, 5], electromagnetic [6, 7], magnetostrictive [8, 9], electroactive and electrostrictive, polymers [10, 11] and piezoelectric [12, 13] transducers. Among these conversion methods, piezoelectric energy harvesters have drawn more attention, since they have high power density, and they are commonly used in microsystems and macroscale devices [14].

In the literature of the piezoelectric energy harvesting, most researches focus on cantilever piezoelectric beams subjected to base excitations [13]. Electromechanical coupling behavior of these beam harvesters have been modeled using analytical and numerical methods with experimental verifications by several researchers [15-19].

Piezoeceramic patches also can be bonded to thin plates to convert vibrational energy of their host plate to electrical energy. This implementation can be even more convenient and practical for use on the surfaces of aerospace, automotive, and marine applications enabling more compact and broadband energy harvesting. However, piezoelectric energy harvesting on two-dimensional lightweight and flexible structures has been rarely reported in the existing literature. Some of the following studies are worthwhile to mention here regarding the harvesters attached to plate-like structures. De Marqui et al. [20], presented an electromechanically coupled finite element model of a plate with embedded piezoceramic layers. Rupp et al. [21], developed a finite element model of a harvesting structure to conduct topology and shape optimizations for distribution of piezoelectric materials with objective of maximizing the harvested power output. Harne et al. [22], built an analytical model for representing electroelastic dynamics of a vibrating panel with corrugated piezoelectric spring. More recently, Aridogan et al. [23] have presented analytical distributed-parameter electroelastic model of patch-based piezoelectric energy harvesters structurally integrated to a fully clamped plate with experimental verification.

In modeling of piezoelectric energy harvesters using cantilever beams and/or plate-like structures, researchers mainly focused on predicting the electrical outputs for the simplified circuit consisting of only resistive loads [19, 24]. In reality, energy harvesting circuits are more sophisticated. For instance, researchers employed AC-DC converters [24, 25] to rectify the AC voltage and later, proposed synchronized switch harvesting on inductor (SSHI) technique [26-28] to boost up power output of the energy harvesters. All the aforementioned circuits include nonlinear electrical parts. However, analytical analysis of such harvesters becomes very difficult when multi-vibration modes of the structure are considered. Instead, researchers utilized Lumped-parameter modeling of energy harvesters with nonlinear electrical components to predict the electrical output strictly near a single vibration mode. Even though this approach gives good prediction around the resonance modes, it fails when the vibration modes are closely distributed and also neglects effect of the other modes on the electrical response. Using the equivalent circuit representations of electromechanical transducers [29, 30], Elvin and Elvin [31] built an equivalent circuit model (ECM) for a piezoelectric energy harvester which could capture the dynamics of a harvester for multiple vibration modes. Parameters of the equivalent circuit were obtained from the Rayleigh-Ritz method. Thus, the lumped parameters of the mechanical domain were modeled as a RLC (resistance, inductor and capacitor) circuit and the electromechanical coupling term was represented as an ideal transformer. They first validated the results of the equivalent circuit model for the AC configuration with comparisons against the existing analytical solutions of a cantilever piezoelectric beam. Later, they used the lumped parameters as well as full–wave rectifier in SPICE software for simulation of the capacitor charging scenario. As an alternative, Elvin and Elvin [32] and Yang and Tang [33] suggested obtaining the system parameters from the finite element model of the system for accurate modeling of complex structures and then using them in SPICE software for simulating the nonlinear circuits. Yang and Tang [33] also presented the equivalent circuit modeling of piezoelectric beam harvesters using the analytical distributed-parameter model introduced by Erturk and Inman [18]. They obtained the equivalent circuit parameters through analogizing of second-order circuit equation with electromechanical coupled equations of the beam harvester.

In the existing literature of energy harvesting, modeling of piezoelectric patch harvesters integrated to plate-like structures including nonlinear circuits has not been studied. Considering these needs in the harvesting methodologies, this study takes the expertise in the field of piezoelectric harvesters integrated to thin plates to the next level and develop more generalized models for such applications.

First, system parameters are identified from the analytical distributed-parameter model developed by Aridogan et al. [23] and results from SPICE simulation are validated for the AC configuration. Then, rectified voltage and power outputs for each single vibration mode obtained through the SPICE software is computed and verified against the Lumped-parameter model presented by Shu and Lien [25]. Comparison of experimental results against the SPICE simulation are also presented for a case study exhibiting similar trend in both results.

2. ANALYTICAL DISTRIBUTED-PARAMETER MODEL

Figure 1 presents the piezoeceramic patch structurally integrated to a thin plate, which is excited by a transverse point force at position. A perfectly bonded isotropic piezoceramic patch with a length of l_p, width of w_p and thickness of covers a rectangular region. The length, width and thickness of the host plate are a, b and h_s respectively. Kirchhoff's plate theory is used for the distributed-parameter model since thickness of the plate is comparably smaller than other dimensions.

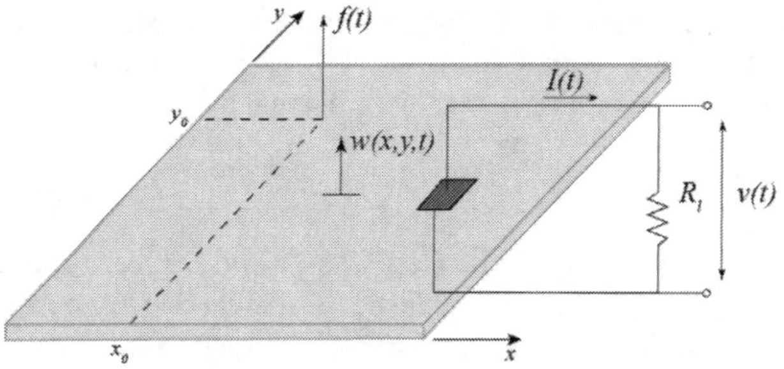

Figure 1. Patch-based piezoelectric energy harvester connected to a resistive load

The partial differential equation of governing the forced vibration of a piezoceramic patch attached to a thin plate is (details are explained in [23]).

$$D\left(\frac{\partial^4 w(x,y,t)}{\partial x^4}+2\frac{\partial^4 w(x,y,t)}{\partial x^2 \partial y^2}+\frac{\partial^4 w(x,y,t)}{\partial y^4}\right)+c\frac{\partial w(x,y,t)}{\partial t}+\rho_s h_s \frac{\partial^2 w(x,y,t)}{\partial t^2}$$
$$-\theta v(t)\left\{\left[\frac{\mathrm{d}\delta(x-x_1)}{\mathrm{d}x}-\frac{\mathrm{d}\delta(x-x_2)}{\mathrm{d}x}\right]\left[H(y-y_1)-H(y-y_2)\right]\right.$$
$$\left.+\left[\frac{\mathrm{d}\delta(y-y_1)}{\mathrm{d}y}-\frac{\mathrm{d}\delta(y-y_2)}{\mathrm{d}y}\right]\left[H(x-x_1)-H(x-x_2)\right]\right\}$$
$$=f(t)\delta(x-x_0)\delta(y-y_0) \tag{1}$$

Where $v(t)$ is the voltage across the external resistive load and the electromechanical term θ is defined as $\theta=\bar{e}_{31}h_{pc}$, which $\bar{e}_{31}$ is the effective piezoelectric constant and $h_{pc}=(h_s+h_p)/2$ is reference distance. The density of the plate is ρ_s, the flexural rigidity of the plate is $D=Y_s h_s^3/(12-12v_s^2)$, where Young's modulus is Y_s and Poisson's ratio is v_s. The viscous air damping coefficient is c_a. The Dirac delta functions are $\delta(x)$ and $\delta(y)$, and the Heaviside functions are $H(x)$ and $H(y)$ along x and y axes, respectively. It should be noted that the volume of the piezoceramic patch is assumed to be significantly smaller than the host plate, and the piezoceramic patch's inertial and stiffness effects are neglected. The governing differential equation of the electrical circuit can be derived by substituting the electrical displacement from piezoelectric constitutive equation into the Gauss law as:

$$C_p \frac{\mathrm{d}v(t)}{\mathrm{d}t} + \frac{v(t)}{R_l} + \theta \left\{ \int_{y=y_1}^{y_2} \int_{x=x_1}^{x_2} \left[\frac{\partial^3 w(x,y,t)}{\partial x^2 \partial t} + \frac{\partial^3 w(x,y,t)}{\partial y^2 \partial t} \right] \mathrm{d}x \, \mathrm{d}y \right\} = 0 \tag{2}$$

where the capacitance of the piezoceramic is defined as $C_p = \left(\bar{\varepsilon}_{33}^S w_p l_p\right) / h_p$.

By following the modal analysis procedure for the equations (1) and (2), electromechanically coupled ordinary differential equations of the plate in modal coordinates can be written as:

$$\frac{\mathrm{d}^2 \eta_{mn}(t)}{\mathrm{d}t^2} + 2\zeta_{mn}\omega_{mn} \frac{\mathrm{d}\eta_{mn}(t)}{\mathrm{d}t} + \omega_{mn}^2 \eta_{mn}(t) - \widetilde{\theta}_{mn} v(t) = f_{mn}(t) \tag{3}$$

$$C_p \frac{\mathrm{d}v(t)}{\mathrm{d}t} + \frac{v(t)}{R_l} + \sum_{n=1}^{\infty} \sum_{m=1}^{\infty} \widetilde{\theta}_{mn} \frac{\mathrm{d}\eta_{mn}(t)}{\mathrm{d}t} = 0 \tag{4}$$

where the plate modes are designated by m and n. The modal forcing f_{mn}, and the electromechanical coupling term $\widetilde{\theta}_{mn}$ are expressed in the following form:

$$f_{mn} = \int_0^b \int_0^a f(t)\delta(x - x_0)\delta(y - y_0)\phi_{mn}(x,y)\, \mathrm{d}x\, \mathrm{d}y = f(t)\phi_{mn}(x_0, y_0) \tag{5}$$

$$\widetilde{\theta}_{mn} = \theta \left[\int_{y_1}^{y_2} \left. \frac{\partial \phi_{mn}(x,y)}{\partial x} \right|_{x_1}^{x_2} \mathrm{d}y + \int_{x_1}^{x_2} \left. \frac{\partial \phi_{mn}(x,y)}{\partial y} \right|_{y_1}^{y_2} \mathrm{d}x \right]. \tag{6}$$

3. EQUIVALENT CIRCUIT MODELING

The above analytical distributed-parameter model of the plate can capture any vibration modes and has been verified experimentally. Although this model is capable of predicting the power output of the system accurately, its validity is limited with characteristics of the electrical circuit elements connected to the harvester. If only a resistive load is considered, then the distributed parameter model is very efficient and accurate. However, if there is an additional non-linear circuit element, then another approach is required to predict the system behavior accurately.

Considering these needs, this paper focuses on equivalent circuit modeling and analysis of a patch-based harvester integrated to a thin plate while connected to a full-wave rectifier. These models will be based on the equivalent circuit modeling technique which is obtained by superposition of infinite number of lumped-parameter models of each vibration mode. Figure 2 shows the equivalent circuit model of the piezoelectric energy harvester integrated to a thin plate. Each vibrational mode of the harvester is represented as a second-order circuit which is connected to a piezoelectric patch (C_P) and a resistive load (R_l) through an ideal transformer. By applying Kirchhoff's voltage law to the multi-vibration mode circuit and analogizing with equation (3), system parameters of the equivalent circuit model can be determined. Table 1 gives a summary of analogy between the analytical distributed-parameter expression and the equivalent circuit model.

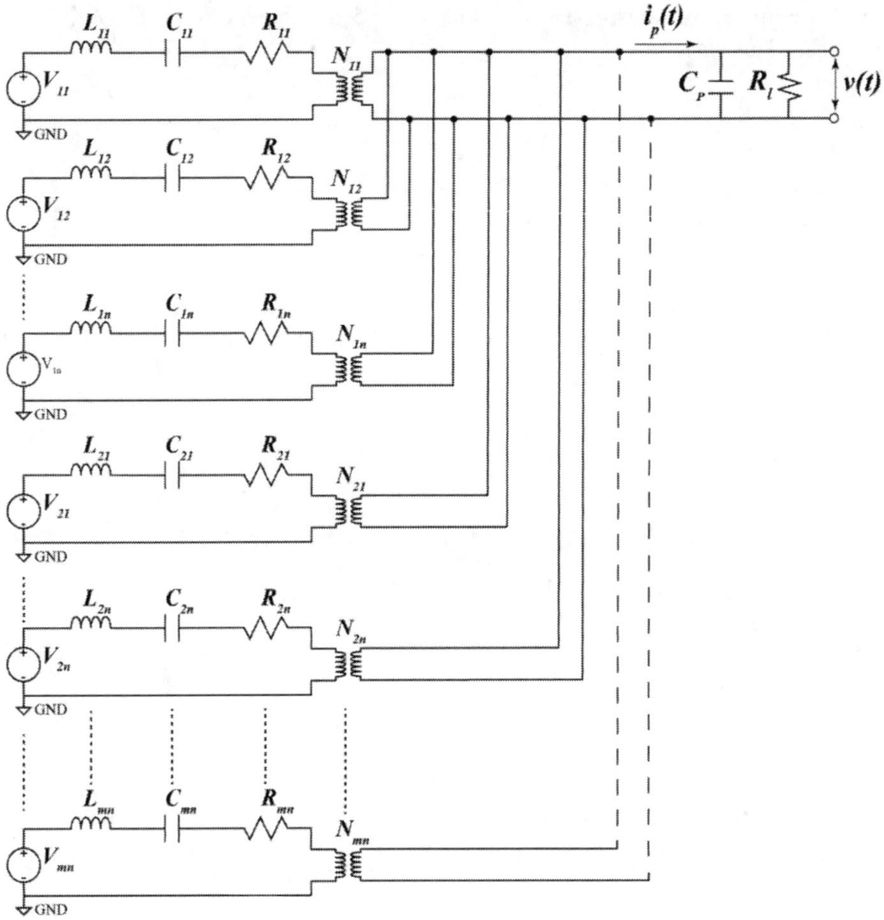

Figure 2. Multi-vibration mode equivalent circuit representation of the piezoelectric patch harvester connected to resistive load

Table 1. Analogy between electrical and mechanical domains of piezoelectric patch-based harvester integrated to a thin plate

Equivalent circuit parameters at mn-th mode	Mechanical counterparts
Voltage source: $V_{mn}(t)$	Modal point force: $f_{mn}(t)$
Inductance: $L_{mn}(t)$	1
Resistance: $R_{mn}(t)$	$2\zeta_{mn}\omega_{mn}$
Capacitance: $C_{mn}(t)$	$1/\omega_{mn}$
Ideal Transformer ratio: N_{mn}	$-\tilde{\theta}_{mn}$
Electrical charge: $q_{mn}(t)$	Modal time response: $\eta_{mn}(t)$
Electrical current: $i_{mn}(t)$	$d\eta_{mn}(t)/dt$

Figure 3 demonstrates simple electrical circuit of the harvester patch as a dependent current source i_p in parallel with its internal capacitance C_p. Applying Kirchhoff's current law and analogizing with equation (4), the current flowing to piezoceramic capacitance and the velocity proportional current source can be written as:

$$i_c = C_p \frac{\mathrm{d}v(t)}{\mathrm{d}t}, \; i_p = -\sum_{n=1}^{\infty}\sum_{m=1}^{\infty}\widetilde{\theta}_{mn}\frac{\mathrm{d}\eta_{mn}(t)}{\mathrm{d}t} \tag{7}$$

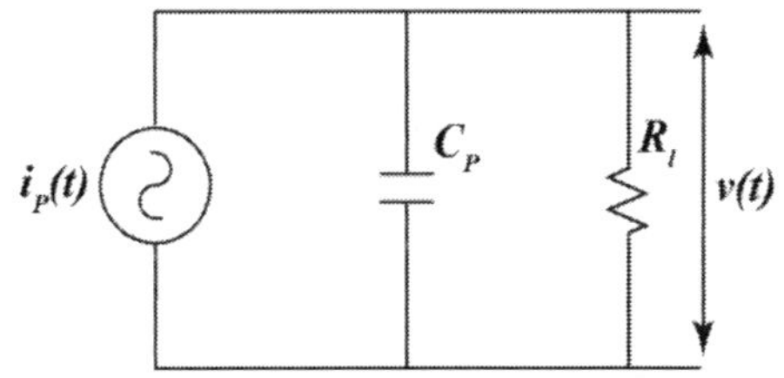

Figure 3. Electric circuit for connection of piezoelectric patch to resistive load

4. MODEL VALIDATION

In this section, equivalent circuit model of a fully clamped plate with a perfectly bonded piezoceramic patch is built in SPICE software and simulation results are validated for the standard AC and DC problems. AC and DC configurations are validated using the distributed parameter model developed by Aridogan et al. and Lumped Parameter model introduced by Shu and Lien, respectively. The material, geometric, dielectric and electroelastic properties of the aluminum plate and piezoceramic patch are given in Table 2.

Table 2. Geometric, material and electroelastic properties

Property	Aluminum	Piezoceramic
Length (mm)	580	72.4
Width (mm)	540	72.4
Thickness (mm)	1.96	0.267
Young's modulus (GPa)	70	66
Mass density (kg m^{-3})	2700	7800
Piezoelectric constant d_{31} (pm/V)	-	-190
Permittivity constant $\bar{\varepsilon}_{33}^{S}$ (nF/m)	-	10.38

4.1 AC input - AC output problem

This section focuses on validation of the equivalent circuit model by comparing the voltage frequency response functions by comparing them against the results of the analytical distributed-parameter model. Voltage response of the harvester under the effect of a harmonic force is shown in Figure 4 for short-circuit and open-circuit cases ($100\,\Omega$ and $1\,\mathrm{M}\Omega$). Results of distributed analytical model and SPICE simulation for equal number of vibrational modes are compared and it can be observed that the equivalent circuit model accurately predicts the voltage response .

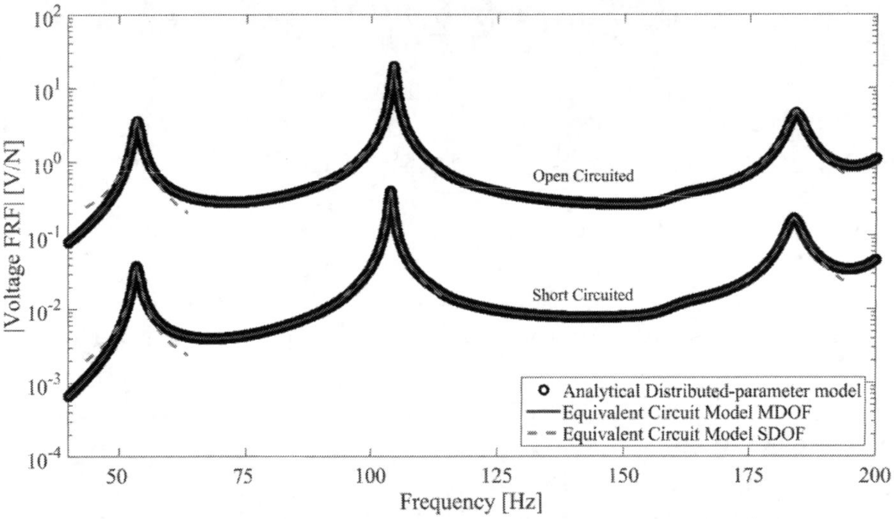

Figure 4. Comparison of analytical distributed-parameter results versus multi-degree-of-freedom and single-degree-of-freedom equivalent circuit simulations

4.2 AC input - DC output problem

The Lumped Parameter model introduced by Shu and Lien considers only single-mode expression of the harvester. For that reason, single-mode results of the equivalent circuit model are compared with the analytical multi-mode solution in the vicinity of the first two modes as shown in Figure 5. It can be observed that single-mode results of the equivalent circuit model are in good agreement with the analytical multi-mode results.

Once the equivalent circuit model is validated for single (around the resonance neighborhood) and multi-mode cases in the AC-AC configuration, the voltage and power output of the equivalent circuit model of the harvester must be also validated for the AC-DC configuration. For that purpose, using the lumped-parameters for each vibrational mode, the harvesting circuit is simulated in SPICE software and validated with the analytical model proposed by Shu and Lien . In their study, the authors developed an analytical formulation of the rectified DC voltage output for the piezoelectric energy harvesters using a lumped-parameter approach. The steady-state voltage (V_c) for the piezoelectric patch-based harvester can be written as [25]:

$$V_c = \frac{L_{mn}\omega R_l}{R_l C_p \omega + \pi/2} u_0 \tag{8}$$

where the displacement amplitude u_0 is derived in terms of the point forcing amplitude and the system parameters as:

$$u_0 = \frac{F_0 \phi_{mn}(x_0, y_0)}{\left\{ \left(R_{mn}\omega + \frac{2\omega N_{mn}^2 R_l}{(C_p \omega R_l + \pi/2)^2} \right)^2 + \left(1/C_{mn} - \omega^2 L_{mn} + \frac{\omega N_{mn}^2 R_l}{C_p \omega R_l + \pi/2} \right)^2 \right\}^{\frac{1}{2}}} \tag{9}$$

For detailed derivation information, reader is referred to [25] .

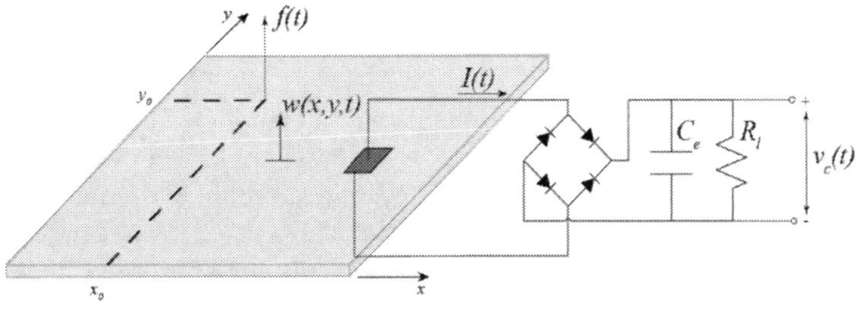

Figure 5. Patch-based piezoelectric energy harvester connected to a full-wave rectifier, a reservoir capacitance and a resistive load

A patch-based piezoelectric energy harvesting circuit is configured for the validation purpose. Figure 6 illustrates the AC-DC patch-based piezoelectric energy harvesting circuit consisting of a full-bridge diode, a smoothing capacitor and a resistive load. To generate a stable DC voltage v_c, it can be assumed that the capacitor C_e is large enough to achieve constant DC voltage output . It should be noted that, ideal transformer and diodes are used in the simulations such that the results can be compared to Shu and Lien's analytical model. Analysis of the harvesting circuit is performed for each single vibration mode in SPICE software. By conducting transient analysis for the harvester circuit, steady-state voltage values are extracted for various values of resistive loads. Figure 7 and 8 present the SPICE simulation results for the peak DC voltage and power outputs and compare the results with Shu and Lien's analytical model for a wide range of resistive loads. Again, the equivalent circuit model accurately predicts the DC voltage and power outputs. It can be also seen from the graph that there exists an optimum resistance in which the DC power output becomes maximum.

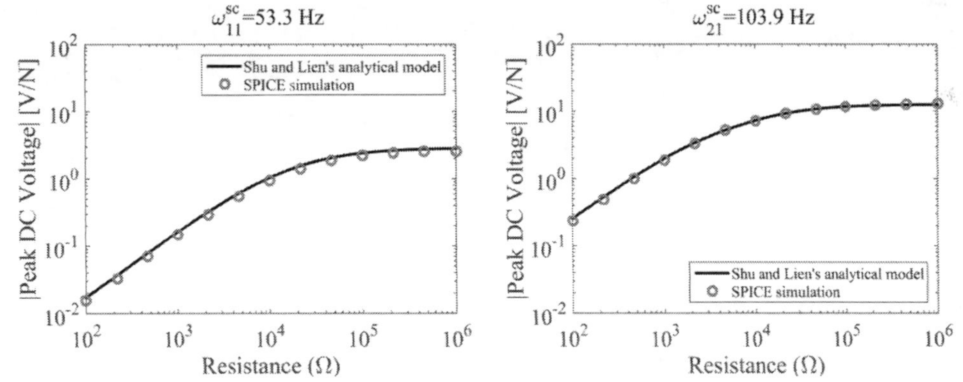

Figure 6. Peak DC voltage output vs resistance for mode 11 (left) and mode 21 (right)

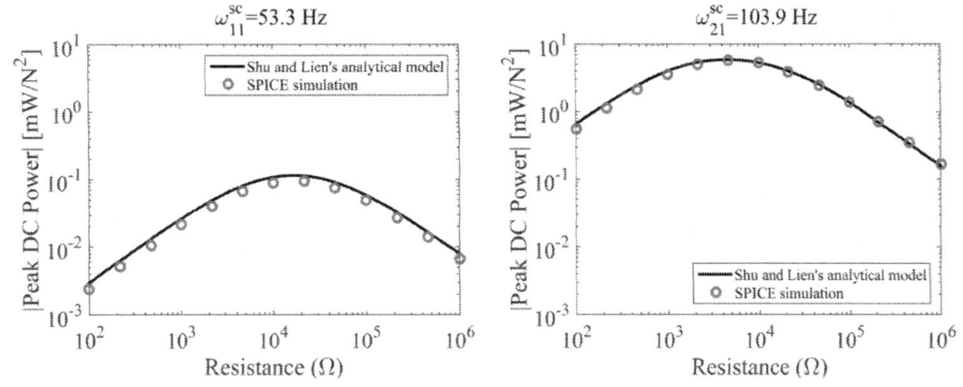

Figure 7. Peak DC power output vs resistance for mode 11 (left) and mode 21 (right)

4.3 Experimental verification for AC input – DC output case

In the experiment section, rectangular aluminum plate is employed as a thin host plate which is fully clamped. The transversely isotropic piezoceramic patch (T105-A4E-602 manufactured by PiezoSystems, Inc.) is used as the harvester patch and bonded on the aluminum plate. The geometric, material and electroelastic properties of the plate and the patch-based harvesters are as given in Table 2. The center of the piezoceramic patch is located at 0.175 and 0.272 m away from the bottom left of the plate. A full-wave bridge rectifier consisting of four diodes, a smoothing capacitor and resistive load are used in the electrical circuit to convert the AC voltage output of piezo patch to the corresponding DC voltage. The top and bottom electrodes of the piezo patch are connected to input alternating voltage nodes of the rectifier as it is shown in Figure 5. The positive and negative voltage output nodes of the rectifier are then connected to a smoothing capacitor in parallel with a resistive load. The host plate is excited via modal shaker in this study where the shaker's rod is attached to 0.105 and 0.1 m away from the lower right corner and the point force excitation acting on the plate is measured by a force transducer (PCB 208C02). Experimental measurements are conducted with quasi-dynamic excitation method, where the plate is excited in a specified frequency intervals, dwell time and constant step size. The reason for this kind of testing is to allow the smoothing capacitor to be charged at one frequency and produce low-ripple steady DC voltage over the resistor.

Figure 8 shows the experimental and numerical frequency response functions of DC voltage output per unit force excitation for the first two vibrational mode of the harvester with the resistive load of R_l = 1 kΩ.

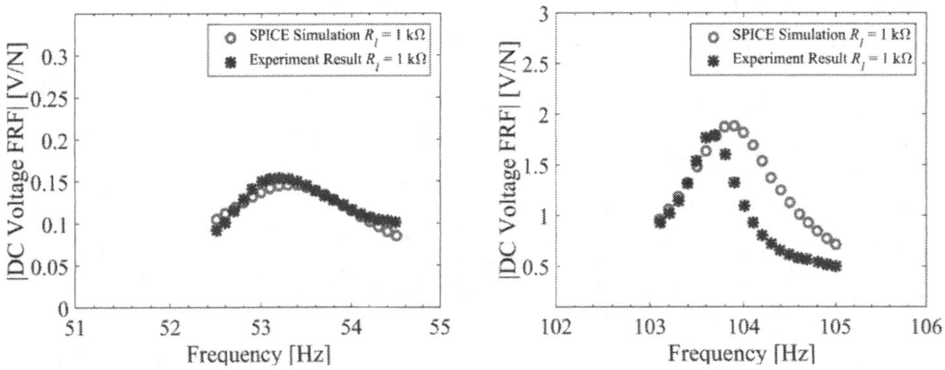

Figure 8. Comparison of Experimental results with SPICE simulations for the DC voltage FRFs in Mode 11 (left) and Mode 21 (right)

5. CONCLUSION

In this paper, an equivalent circuit model of a piezoelectric patch-base harvester attached on a thin plate was developed and validated for the AC input-AC output and AC input-DC output configurations. The equivalent circuit parameters considering the multi-vibration modes of the harvester were identified from the analogy between analytical distributed-parameter model and the second-order circuitry equations. Then, using these system parameters, simulations were conducted in SPICE software.

The simulation results of voltage FRFs were presented for the short-circuit and open-circuit conditions and verified against the analytical model in the AC-AC configuration. It was shown that multi-vibration-mode and single-vibration-mode equivalent circuit models accurately predicts the coupled electromechanical behavior. Then, single-mode equivalent circuit model was connected to full-wave bridge rectifier, a smoothing capacitor and a resistive load. Steady-state electrical responses were investigated for the first two modes of the harvester. Peak DC voltage and power outputs across different resistive loads were numerically computed in the SPICE software and validated with the existing analytical single-degree-of-freedom model proposed by Shu and Lien [25]. Finally, quasi-dynamic experimental measurements were conducted for a case study to verify the numerical model. The multi-mode equivalent circuit model developed in this study enables exploiting the harvesting capabilities of the piezoelectric patch-based harvesters attached to thin plates with any linear/ nonlinear electrical components. With these analytical and numerical models of patch-based harvesters, design and optimization studies can be easily performed to maximize the power output of such systems.

REFERENCES

1. Sazonov, E., Li, H., Curry, D., and Pillay, P., "Self-powered sensors for monitoring of highway bridges," *Sensors Journal, IEEE*, 9(11), 2009, pp. 1422-1429.
2. Hu, Y., Zhang, Y., Xu, C., Lin, L., Snyder, R. L., and Wang, Z. L., "Self-powered system with wireless data transmission," *Nano Letters*, 11(6), 2011, pp. 2572-2577.
3. Roundy, S., Wright, P. K., and Rabaey, J., "A study of low level vibrations as a power source for wireless sensor nodes," *Computer communications*, 26(11), 2003, pp. 1131-1144.
4. Naruse, Y., Matsubara, N., Mabuchi, K., Izumi, M., and Suzuki, S., "Electrostatic micro power generation from low-frequency vibration such as human motion," *Journal of Micromechanics and Microengineering*, 19(9), 2009.
5. Sidek, O., Khalid, M. A., Ishak, M. Z., and Miskam, M. A., "Design and simulation of SOI-MEMS electrostatic vibration energy harvester for micro power generation," Proc. *Electrical, Control and Computer Engineering (INECCE), 2011 International Conference on*, IEEE, pp. 207-212.
6. Chye, W. C., Dahari, Z., Sidek, O., and Miskam, M. A., "Electromagnetic micro power generator—A comprehensive survey," Proc. *Industrial Electronics & Applications (ISIEA), 2010 IEEE Symposium on*, IEEE, pp. 376-382.
7. Beeby, S. P., Torah, R. N., Tudor, M. J., Glynne-Jones, P., O'Donnell, T., Saha, C. R., and Roy, S., "A micro electromagnetic generator for vibration energy harvesting," *Journal of Micromechanics and Microengineering*, 17(7), 2007, pp. 1257-1265.
8. Bayrashev, A., Robbins, W. P., and Ziaie, B., "Low frequency wireless powering of microsystems using piezoelectric–magnetostrictive laminate composites," *Sensors and Actuators A: Physical*, 114(2), 2004, pp. 244-249.
9. Zhang, H., "Power generation transducer from magnetostrictive materials," *Applied Physics Letters*, 98(23), 2011, p. 2505.
10. Liu, Y., Ren, K. L., Hofmann, H. F., and Zhang, Q., "Investigation of electrostrictive polymers for energy harvesting," *Ultrasonics, Ferroelectrics, and Frequency Control, IEEE Transactions on*, 52(12), 2005, pp. 2411-2417.
11. Chiba, S., Waki, M., Kornbluh, R., and Pelrine, R., "Innovative power generators for energy harvesting using electroactive polymer artificial muscles," Proc. *The 15th International Symposium on: Smart Structures and Materials & Nondestructive Evaluation and Health Monitoring*, International Society for Optics and Photonics, pp. 692715-692715-692719.
12. Saadon, S., and Sidek, O., "A review of vibration-based MEMS piezoelectric energy harvesters," *energy conversion and management*, 52(1), 2011, pp. 500-504.
13. Anton, S. R., and Sodano, H. A., "A review of power harvesting using piezoelectric materials (2003-2006)," *Smart Materials and Structures*, 16(3), 2007, pp. R1-R21.
14. Cook-Chennault, K., Thambi, N., and Sastry, A., "Powering MEMS portable devices—a review of non-regenerative and regenerative power supply systems with special emphasis on piezoelectric energy harvesting systems," *Smart Materials and Structures*, 17(4), 2008, p. 043001.
15. Sodano, H. A., Park, G., and Inman, D., "Estimation of electric charge output for piezoelectric energy harvesting," *Strain*, 40(2), 2004, pp. 49-58.

16. Liao, Y., and Sodano, H. A., "Model of a single mode energy harvester and properties for optimal power generation," *Smart Materials and Structures*, 17(6), 2008, p. 065026.
17. Erturk, A., "Assumed-modes modeling of piezoelectric energy harvesters: Euler–Bernoulli, Rayleigh, and Timoshenko models with axial deformations," *Computers & Structures*, 106, 2012, pp. 214-227.
18. Erturk, A., and Inman, D. J., "A distributed parameter electromechanical model for cantilevered piezoelectric energy harvesters," *Journal of Vibration and Acoustics*, 130(4), 2008, p. 041002.
19. Erturk, A., and Inman, D. J., "An experimentally validated bimorph cantilever model for piezoelectric energy harvesting from base excitations," *Smart Materials and Structures*, 18(2), 2009, p. 025009.
20. Junior, C. D. M., Erturk, A., and Inman, D. J., "An electromechanical finite element model for piezoelectric energy harvester plates," *Journal of Sound and Vibration*, 327(1), 2009, pp. 9-25.
21. Rupp, C. J., Evgrafov, A., Maute, K., and Dunn, M. L., "Design of piezoelectric energy harvesting systems: a topology optimization approach based on multilayer plates and shells," *Journal of Intelligent Material Systems and Structures*, 20(16), 2009, pp. 1923-1939.
22. Harne, R. L., "Concurrent attenuation of, and energy harvesting from, surface vibrations: experimental verification and model validation," *Smart Materials and Structures*, 21(3), 2012, p. 035016.
23. Aridogan, U., Basdogan, I., and Erturk, A., "Analytical modeling and experimental validation of a structurally integrated piezoelectric energy harvester on a thin plate," *Smart Materials and Structures*, 23(4), 2014, p. 045039.
24. Ottman, G. K., Hofmann, H. F., Bhatt, A. C., and Lesieutre, G. A., "Adaptive piezoelectric energy harvesting circuit for wireless remote power supply," *Power Electronics, IEEE Transactions on*, 17(5), 2002, pp. 669-676.
25. Shu, Y., and Lien, I., "Analysis of power output for piezoelectric energy harvesting systems," *Smart materials and structures*, 15(6), 2006, p. 1499.
26. Guyomar, D., Badel, A., Lefeuvre, E., and Richard, C., "Toward energy harvesting using active materials and conversion improvement by nonlinear processing," *Ultrasonics, Ferroelectrics, and Frequency Control, IEEE Transactions on*, 52(4), 2005, pp. 584-595.
27. Badel, A., Guyomar, D., Lefeuvre, E., and Richard, C., "Piezoelectric energy harvesting using a synchronized switch technique," *Journal of Intelligent Material Systems and Structures*, 17(8-9), 2006, pp. 831-839.
28. Shu, Y., Lien, I., and Wu, W., "An improved analysis of the SSHI interface in piezoelectric energy harvesting," *Smart Materials and Structures*, 16(6), 2007, p. 2253.
29. Tilmans, H. A., "Equivalent circuit representation of electromechanical transducers: I. Lumped-parameter systems," *Journal of Micromechanics and Microengineering*, 6(1), 1996, p. 157.
30. Tilmans, H. A., "Equivalent circuit representation of electromechanical transducers: II. Distributed-parameter systems," *Journal of Micromechanics and Microengineering*, 7(4), 1997, p. 285.
31. Elvin, N. G., and Elvin, A. A., "A general equivalent circuit model for piezoelectric generators," *Journal of Intelligent Material Systems and Structures*, 20(1), 2009, pp. 3-9.
32. Elvin, N. G., and Elvin, A. A., "A coupled finite element—circuit simulation model for analyzing piezoelectric energy generators," *Journal of Intelligent Material Systems and Structures*, 20(5), 2009, pp. 587-595.
33. Yang, Y., and Tang, L., "Equivalent circuit modeling of piezoelectric energy harvesters," *Journal of Intelligent Material Systems and Structures*, 2009.

ICAST2015 #022

Harvesting Strategy to Increase Electrical Energy for Piezoelectric Active Harvesting

Kenji Yoshimizu[1*], Yuta Yamamoto[1], and Kanjuro Makihara[1]

[1] Department of Aerospace Engineering, Tohoku University, Sendai, Japan

Abstract

In this paper, we present control strategies to increase electrical energy for active harvesting. Vibrational energy harvesting extracts electrical energy from structural vibration. To increase the harvested energy, synchronized switch harvesting on inductor (SSHI) technique was proposed. With the switching action, SSHI effectively converts mechanical energy into electrical energy. We experimentally confirm that the vibrational amplitude and the harvested energy are decreased with SSHI. To solve this critical problem, we devise control strategies that are SCVS (Switching Considering Vibration Suppression) and ASCVS (Adaptive SCVS) to increase the harvested energy. Original SSHI conducts switching action at every peak of vibrational displacement. Our control strategies are designed to temporarily stop the switching action. The vibrating structure is excited to vibrate by a vibration shaker. While switching action is stopped, vibrational amplitude will be recovered from the suppressed amplitude by excitation force. The proposed control strategies can maintain the vibrational amplitude as high as possible. Accordingly, the harvested energy increases. We experimentally demonstrate that the proposed strategies increase the harvested energy. Experimental results show that the stored energy with our techniques is up to 10.5 times greater than that with the original SSHI technique. Moreover, ASCVS (Adaptive SCVS) is shown to be more flexible to increase harvested energy in various vibration conditions.

1. INTRODUCTION

Recently, the technology of energy-harvesting has been attracting a lot of attention. Energy-harvesting technology generates electrical energy from natural energy sources such as wind energy, solar energy, tidal energy, and vibrational energy. It is expected to become a substitute of primary batteries that are used in low-power electrical devices. The energy-harvesting devices can prevent from battery exchanging. This study focuses on to structural vibration (vibrational energy-harvesting) such as bridges, tunnels, trains, airplanes and so on. Vibrational energy-harvesting converts vibrational energy into electrical energy by utilizing electromagnetic materials [1], piezoelectric materials including lead zirconate titanate (PZT) [2]. In this study, we focus on the PZT as a converter.

Vibrational energy-harvesting is expected to be used for a structural health-monitoring system that monitors health diagnostics of structures such as trains, tunnels, airplanes, bridges. Especially, because most existing bridges were built during the period of rapid economic growth, bridges that get older have been a big problem. As the technique of a structural health-monitoring system uses a lot of sensors and wireless communication devices, the system can monitor structures in real time. Because structures

* yoshimizu@ssl.mech.tohoku.ac.jp

generate vibrational energy at every moment, the system prevents from battery exchanging and becomes completely independent system from external power supply.

A lot of experimental and theoretical studies have been conducted with vibrational energy harvesters that exploit a piezoelectric transducer. A standard harvester is a conventional style for vibrational energy-harvesting [3]. The circuit that consists of a rectifier and a storage capacitor is connected to a piezoelectric transducer. The generated electrical energy is rectified and stored into the storage capacitor. To increase harvested energy, Ottman et al. [4] connected DC/DC converter with a standard harvester. Kong et al. [5] proposed a method that can increase harvested energy efficiently to use impedance matching between structural and electrical impedance.

Guyomar et al. [6] expanded SSD (Synchronized Switching Dumping) technique proposed by Richard et al. [7] to energy-harvesting. They proposed the technique of Synchronized Switch Harvesting on Inductor (SSHI). With a harvesting circuit including an inductor and a switch, SSHI controls electrical current by switching action. SSHI inverts piezoelectric voltage in sync with structural vibration. Then, the generated voltage of a piezoelectric transducer can be amplified and the harvested energy can be increased effectively.

Great number of strategies are proposed to increase the harvested energy. SSHI needs the device that measures the peak of structural displacement and conducts the switching action. Qiu et al. [8] compared four different harvester circuit including SSHI theoretically and experimentally. They made the peak detector and the switching circuit by analog devices. Makihara et al. [9] and Yamamoto et al. [10] proposed and fabricated a self-powered harvester. With a digital processor, the harvester can change some parameters easily, and calculate complex problem. Shen et al. [11] proposed the technique that improved SSHI. The harvested power was up to 3 times greater than a standard harvester. Guyomar and Lallart [12] compared various harvesters that are SSHI and several different type of improved SSHI. They compared it with constant vibrational amplitude or constant external force.

2. OBJECTIVE STATEMENT

Many of harvesting strategies that increase electrical energy have been proposed. Because many of these strategies assumed that harvested electrical energy is much less than mechanical energy of structural vibration. The amplitude of vibrational displacement was assumed to be constant. However, actually vibrational energy harvesting converts the mechanical energy of structural vibration into electrical energy and mechanical energy is consequently reduced [13]. Accordingly, it makes the amplitude of structural vibration suppressed. Because the voltage generated by piezoelectric transducer depends on the amplitude of structural vibration, the decrease in the amplitude leads to the reduction of the generated voltage. As a result, the harvested energy is decreased. There have not been harvesting strategies that take account of the suppression of the structural vibration.

In this paper, we devise two control strategies that are SCVS (Switching Considering Vibration Suppression) and ASCVS (Adaptive SCVS) to increase the harvested energy. Our harvesting strategies are to control the timing of switching action based on SSHI. Original SSHI conducts switching action at every peak of vibrational displacement. Because the switching action causes the suppression of vibrational amplitude, our control strategies are designed to temporarily stop the switching action. The vibrating structure is excited to vibrate by a vibration shaker. While switching action is stopped, vibrational amplitude will be recovered from the suppressed amplitude by excitation force. The proposed control strategies can maintain the vibrational amplitude as high as possible. Accordingly, the harvested energy will increase. We intend to increase the stored electrical energy only by changing the switching strategy, not by exchanging harvesting elements. Our group has already proposed a new control to increase harvested energy [14]. However, the previous paper did not explain the detailed mechanism. This paper provides the detailed mechanism of the new control by using analytical approach.

3. PROPOSED HARVESTING STRATEGY TO INCREASE ELECTRIC ENERGY

3.1 Equation of motion of electromechanical model

We consider the equation of the electromechanical model that is shown in Fig.1. The structure mass was excited by external force. The equation of a piezoelectric transducer can be expressed by

$$f_\mathrm{p} = k_\mathrm{p}u - b_\mathrm{p}Q, V_\mathrm{p} = -b_\mathrm{p}u + Q/C_\mathrm{p}^\mathrm{s}, \tag{1}$$

where, f_p is tensile force that arises from a piezoelectric transducer. u is extension of a piezoelectric transducer. V_p is the generated voltage of a piezoelectric transducer. Q is the electrical charge of a piezoelectric transducer. b_p is piezoelectric coefficient. k_p is stiffness of a piezoelectric transducer. C_p^s is the clamped capacitance of a piezoelectric transducer. The equation of the electromechanical model can be expressed by

$$m\ddot{x} + D\dot{x} + kx = f_\mathrm{ext} - f_\mathrm{p}, \tag{2}$$

where, m is the structure mass, D is the structure damping coefficient. k is stiffness of the structure, f_ext is external force and x is structural displacement. We assume that the extension of the piezoelectric transducer is same as the displacement of the structure. Then, Eq. (2) can be written as

$$m\ddot{x} + D\dot{x} + \left(k + k_\mathrm{p}\right)x = f_\mathrm{ext} + b_\mathrm{p}Q. \tag{3}$$

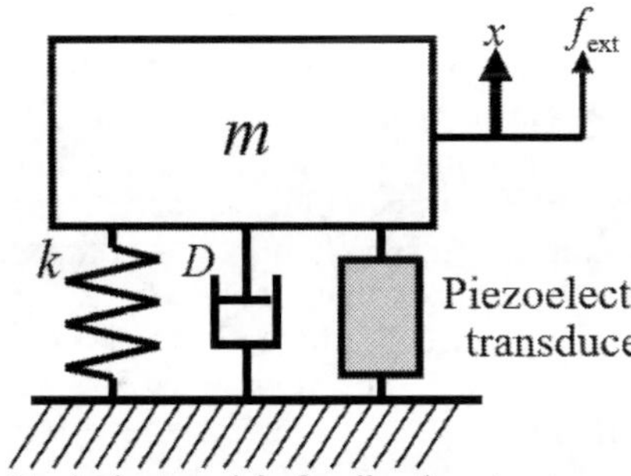

Fig. 1 Equivalent model of a vibrating structure.

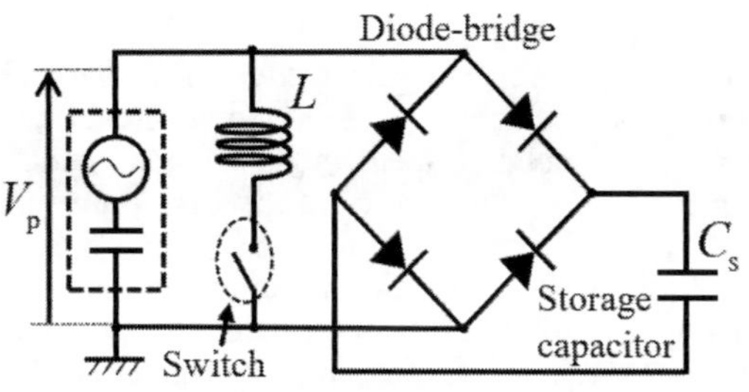

Fig. 2 Harvesting circuit for SSHI technique.

3.2 Method of SSHI

In this section, we explain the technique of SSHI that can increase electrical energy effectively. SSHI increases piezoelectric voltage effectively with the switching action. The control strategy of SSHI [6] is as follows.

[Switching control of original SSHI] At every peak of structural vibration, the switch is turned on during a half period of electrical vibration.

The mechanism of increasing piezoelectric voltage is explained here. Figure 2 shows the harvester circuit for SSHI. It consists of a piezoelectric transducer, an inductor, a switch, a diode-bridge and a storage capacitor. We consider the equation of the circuit of SSHI. If there is no current into the storage capacitor, the equation of the circuit has two cases depending on whether the switch is on or off. The equations can be written as

$$L\ddot{Q} + R\dot{Q} + Q/C_\mathrm{p}^\mathrm{s} = b_\mathrm{p}x, \tag{4}$$

$$\dot{Q} = 0. \tag{5}$$

Equation (4) describes the electrical dynamics of the circuit when the switch is on. Equation (5) describes the electrical dynamics of the circuit when the switch is off. Here, L is the inductance of an inductor, R is the resistance of a circuit. We assumed that the change of the electrical charge is much faster than the change of the displacement of structural vibration. Therefore, when the switch is on, structural displacement is constant. When the switch is turned on, we set the time t to 0. Equation (4) can be expressed as

$$Q(t) = \left(Q_\mathrm{before} - b_\mathrm{p}C_\mathrm{p}^\mathrm{s}x_0\right)\left\{\cos\left(\omega_\mathrm{e\_d}t\right) + \frac{\xi_\mathrm{e}}{\sqrt{1-\xi_\mathrm{e}^2}}\sin\left(\omega_\mathrm{e\_d}t\right)\right\}\exp\left(-\xi_\mathrm{e}\omega_\mathrm{e}t\right) + b_\mathrm{p}C_\mathrm{p}^\mathrm{s}x_0, \tag{6}$$

$$\omega_e \equiv 1/\sqrt{LC_p^S}\,,\ \xi_e \equiv R/2L\omega_e\,,\ \omega_{e\_d} \equiv \sqrt{1-\xi_e^{\,2}}\,\omega_e\,. \tag{7}$$

Here, $Q_{\text{before}}(<0)$ and $x_0(>0)$ are defined as electric charge and structural displacement, respectively, just before the switch is turned on. When $\xi_e \ll 1$, the Eq. (7) can be transformed into

$$Q(t) = \left(Q_{\text{before}} - b_p C_p^S x_0\right)\exp\left(-\xi_e\omega_e t\right)\cos\left(\omega_e t\right) + b_p C_p^S x_0\,, \tag{8}$$

where, ω_e is natural angular frequency of the LC_p^S resonant circuit. During a half period of electrical vibration between $t=0$ and $t=\pi/\omega_e$. the switch is turned on. Then, electric charge after $t=\pi/\omega_e$, Q_{after} becomes

$$Q_{\text{after}} = -\gamma Q_{\text{before}} + \left(1+\gamma\right) b_p C_p^S x_0\,. \tag{9}$$

The voltages before and after the switching can be written as

$$V_{\text{after}} = -\gamma V_{\text{before}}\,, \tag{10}$$

where, V_{after} is the voltage of piezoelectric transducer after the switch is turned off. V_{before} is the voltage of piezoelectric transducer just before the switch is turned on. We define the voltage inversion coefficient γ as $\exp\left(-\xi_e\pi\right)$. If we neglect the resistance of the circuit ($R=0$ and $\gamma=1$), Eq. (9) and Eq. (10) can be written as

$$Q_{\text{after}} = -Q_{\text{before}} + 2b_p C_p^S x_0\,,\ V_{\text{after}} = -V_{\text{before}}\,. \tag{11}$$

According to Eq. (11), when Q_{before} and x_0 have opposite signs, the switching action reverses the electric charge and increases the electric charge by $2C_p^S b_p x_0$. The increase of charge depends on structural displacement x_0. If the vibration is sinusoidal wave, the point of turning on the switch is the peak of vibrational displacement. The switching action reverses the voltage of piezoelectric transducer.

Figure 3 shows the history of increasing piezoelectric voltage with switching action. Piezoelectric voltage V_p is a sum of $-b_p x$ and Q/C_p^S. When the displacement becomes the peak of structural vibration, the switching action is conducted. Because the change of electric charge is much faster than the change of structural displacement, the switching action is finished immediately as Fig. 3 shows. When the switch is off, electric charge is constant. To conduct the switching action at every peak of vibrational displacement, SSHI can effectively increase the voltage of a piezoelectric transducer.

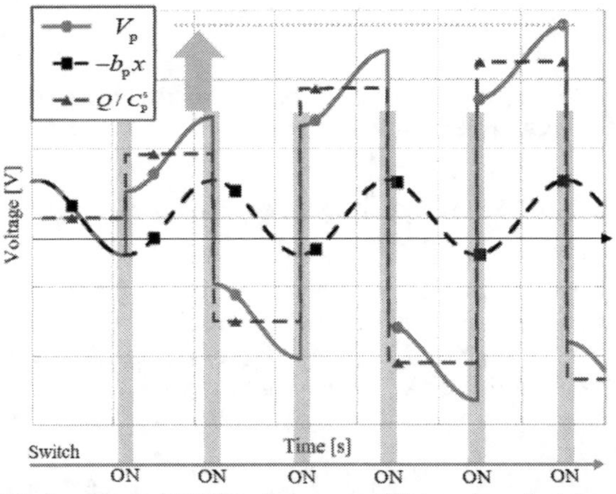

Fig. 3 Ideal concept of SSHI technique amplifying piezoelectric voltage.

3.3 Effect of vibration suppression by SSHI

As shown in the previous section, SSHI is expected to increase piezoelectric voltage effectively. However, SSHI causes vibration suppression. In this section, we consider the influence on mechanical system by SSHI. We focus on the sign of electric charge with SSHI. The sign of electric charge is opposite to the sign of vibrational velocity. From the view point of vibration control, when we consider electric charge as control input, the polarity of the electric charge is same as the polarity of control input of D (derivative) control. Therefore, SSHI not only amplifies piezoelectric voltage but also suppresses the amplitude of structural vibration.

In Eq. (1), a piezoelectric transducer generates the voltage that depends on vibrational displacement. Therefore, the decrease in vibrational amplitude leads to the reduction of the generated voltage. Because electric charge after the switching action can be written as Eq. (9), the amount of increase in electric charge with the switching action can be expressed by

$$\Delta Q \equiv |Q_{\text{after}}| - |Q_{\text{before}}| = (1-\gamma)Q_{\text{before}} + (1+\gamma)b_{\text{p}}C_{\text{p}}^{\text{s}}x_0 . \qquad (12)$$

When the following equation is satisfied,

$$(1+\gamma)b_{\text{p}}C_{\text{p}}^{\text{s}}x_0 < -(1-\gamma)Q_{\text{before}} , \qquad (13)$$

the electric charge after the switching action decreases. Therefore, piezoelectric voltage decreases as Fig. 4 shows. Accordingly, SSHI suppresses the vibrational amplitude that depends on the mechanical system. Then, piezoelectric voltage decreases. To amplify the piezoelectric voltage, a new technique that maintains vibrational amplitude as high as possible is required.

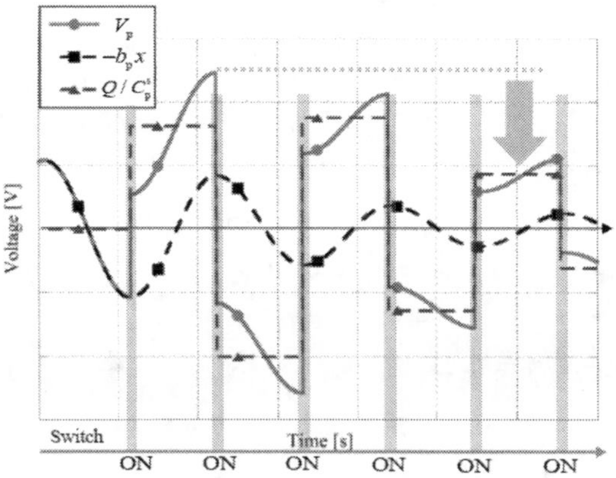

Fig. 4 Problem of SSHI technique that suppresses displacement of structural vibration and reduces voltage of piezoelectric transducer.

3.4 New harvesting strategies to increase electric energy

As shown in the previous section, SSHI suppresses vibrational amplitude, which leads to decrease in piezoelectric voltage. In this section, we propose new harvesting strategies to increase electrical energy. In section 3.4.1, we propose the switching control of SCVS. Our group has already proposed the technique of SCVS. However, the mechanism has not been explained in detail. In section 3.4.1, we reveal the detailed mechanism of SCVS by using analytical approach. Then, in section 3.4.2, we propose an adaptive switching control of SCVS (Adaptive SCVS).

Our proposed strategies are as follow.

[Switching control of SCVS] When the skip number of the peaks of structural vibration become 2*N* times (*N* = 0, 1, 2, 3…), the switch is turned on, during a half period of electrical vibration. (In section 3.4.1).

[Switching control of ASCVS] When Eqs. (23) and (24) are satisfied, the switch is turned on, during a half period of electrical vibration. (In section 3.4.2).

3.4.1 Switching Considering Vibration Suppression (SCVS)

Original SSHI conducts switching action at every peak of vibrational displacement. Because the switching action with SSHI causes the suppression of the amplitude, our control strategy is designed to temporary skip the switching action as Fig. 5 (N = 0, 1, 2, 3…) shows. We refer to the strategy as SCVS. As N = 0 shows the original SSHI, SCVS includes the original SSHI.

Here, we explain the advantage of our strategy. We define external force as

$$f_{\text{ext}} = f_0 \sin \omega_{\text{n}} t , \qquad (14)$$

where, ω_n is natural angular frequency of a structure, f_0 is the amplitude of external force. In Eq. (11) the timing of switching action is the peak of structural displacement when Q_{before} and x_0 have opposite signs. We set an initial condition as $x_0 (< 0)$ that is the peak of structural displacement $(t = 0)$. When the switch is turned on, we set the time t to 0. We assume that the switching action is much faster than the change of structural displacement. We set the electric charge as

$$Q(t) = Q_{after} \left(t > 0 \right), \tag{15}$$

$$Q(t) = 0 \left(t < 0 \right). \tag{16}$$

With Eqs. (14) and (15), Eq. (3) can be expressed as

$$m\ddot{x} + D\dot{x} + k_{all} x = f_0 \sin \omega_n t + b_p Q_{after}, k_{all} \equiv k + k_p. \tag{17}$$

We change variables as

$$y \equiv x - \frac{b_p}{k_{all}} Q_{after}. \tag{18}$$

Equation (17) can lead to

$$m\ddot{y} + D\dot{y} + k_{all} y = f_0 \sin \omega_n t. \tag{19}$$

Equation (19) can be expressed as

$$y(t) = \exp(-\xi \omega_n t) \left(y_0 + \frac{f_0}{D\omega_n} \right) \left(\frac{\xi}{\sqrt{1-\xi^2}} \sin \omega_d t + \cos \omega_d t \right) - \frac{f_0}{D\omega_n} \cos \omega_n t, \tag{20}$$

where, $y_0 \equiv x_0 - b_p Q_{after} / k_{all}$, $\omega_n \equiv \sqrt{k_{all}/m}$, $\xi \equiv D / 2\sqrt{m k_{all}}$, and $\omega_d \equiv \sqrt{1-\xi^2}\, \omega_n$. When $\xi \ll 1$, Eq. (20) can be changed to

$$y(t) = -\left[\frac{f_0}{D\omega_n} - \exp(-\xi \omega_n t) \left(y_0 + \frac{f_0}{D\omega_n} \right) \right] \cos \omega_n t. \tag{21}$$

If electric charge is zero, the amplitude of displacement is $f_0 / D\omega_n$ at the steady state. We set the initial condition of x_0 as $-f_0 / D\omega_n$. Finally, Eq. (21) can be changed to

$$x(t) = \frac{b_p}{k_{all}} Q_{after} - \left[\frac{f_0}{D\omega_n} + \frac{b_p}{k_{all}} Q_{after} \exp(-\xi \omega_n t) \right] \cos \omega_n t. \tag{22}$$

In Eq. (11), the sign of Q_{after} is negative. Equation (22) shows that the center of oscillation is shifted as $b_p Q_{after} / k_{all}$ and the amplitude of structural vibration is suppressed as $b_p Q_{after} / k_{all}$. When k_{all} is small or b_p is large, the effect of suppression of vibrational amplitude is increased. Here, the third term of the right side of Eq. (22) decreases with time goes by. Then, while the switching action is not conducted, the suppressed amplitude of structural vibration will be recovered. Because the voltage generated by a piezoelectric transducer depends on the amplitude of structural vibration, recovered amplitude of structural vibration increases the generated voltage. When the relation between electric charge and structural displacement does not satisfy as Eq. (13), electric charge increases with the switching action. Because, in Eq. (9), electric charge after the switching action depends on structural displacement, recovered amplitude of structural vibration increases the generated voltage by the switching action. Accordingly, piezoelectric voltage increases by skipping the switching action. From the above, our new control strategy is to skip the switching actions for N times (N= 1, 2, 3...).

3.4.2 Adaptive Switching Considering Vibration Suppression (ASCVS)

By skipping the switching action temporally, SCVS reduces the suppression effect and amplifies piezoelectric voltage. However, SCVS has to determine the skip number of switching actions in advance. Because the appropriate number of skipped switching actions depends on the mechanical system, to change the skip number of switching actions with the state of structural vibration is a more adaptive switching strategy than SCVS. In this section, we propose an adaptive switching strategy.

As mentioned in section 3.3, when Eq. (13) is satisfied, electric charge after the switching action decreases. In Eq. (1), the voltage of a piezoelectric transducer can be written as the sum of vibrational displacement and electric charge. Because the decrease of electric charge causes the decrease of piezoelectric voltage directly, our control strategy is to adaptively amplify electric charge as high as possible.

As mentioned in section 3.4.1, as the switching action is skipped, the suppressed amplitude of structural vibration is recovered. Because, in Eq. (9), electric charge after the switching action depends on structural displacement, electric charge after the switching action increases with the skip number of the switching actions. It has the maximum value. Electric charge after the switching action can be estimated in Eq. (9).

Thus, our proposed strategy is as follows. When the estimated charge after the switching action is maximum value, the switch is turned on at the peak of structural vibration. We refer to the strategy as ASCVS (Adaptive SCVS).

Here, we explain the mechanism of our switching strategy. To estimate the electric charge after the switching action in Eq. (9), γ is required. With the value of piezoelectric voltage and structural displacement, electric charge can be calculated using Eq. (1). Because the voltage inversion coefficient γ depends on only electrical circuit, it is constant value for a certain system. Using Eq. (10) we can measure the voltage inversion coefficient γ experimentally.

Figure 5 shows the history of structural vibration and the estimated charge after the switching action. As Fig. 5 shows, the estimated charge after the switching action increases with the skip number of switching actions. The estimated charge has the maximum value. Therefore, to find the maximum value of the estimated charge, we define the condition that can be written as

$$\Delta Q_N^{\text{after}} \equiv \left|Q_N^{\text{after}}\right| - \left|Q_{N\text{-}1}^{\text{after}}\right| \leq 0. \tag{23}$$

Here, Q_N^{after} is the estimated charge when the skip number of switching actions is N times. $Q_{N\text{-}1}^{\text{after}}$ is the estimated charge when the skip number of switching actions is N-1 times. In Eq. (9) the timing of calculation is the peak of structural vibration when Q_{before} and x_0 have opposite signs. The estimated charge rises with the skipped switching action. Equation (23) shows that we can find the maximum value by comparing current estimated charge after the switching action $\left(\left|Q_N^{\text{after}}\right|\right)$ and one step before $\left(\left|Q_{N\text{-}1}^{\text{after}}\right|\right)$.

However, even if Eq. (23) is satisfied, electric charge after the switching action decreases when Eq. (13) is satisfied. To solve this problem, we add an additional condition as

$$\Delta Q \equiv \left|Q_{\text{after}}\right| - \left|Q_{\text{before}}\right| > 0. \tag{24}$$

Therefore, with the conditions in Eqs. (23) and (24), the switching action can be conducted at the appropriate timing. ASCVS does not depend on the mechanical system such as spring, damping and mass. ASCVS does not have to determine the skip number of switching actions in advance. Thus, the switching strategy is more adaptive than SCVS.

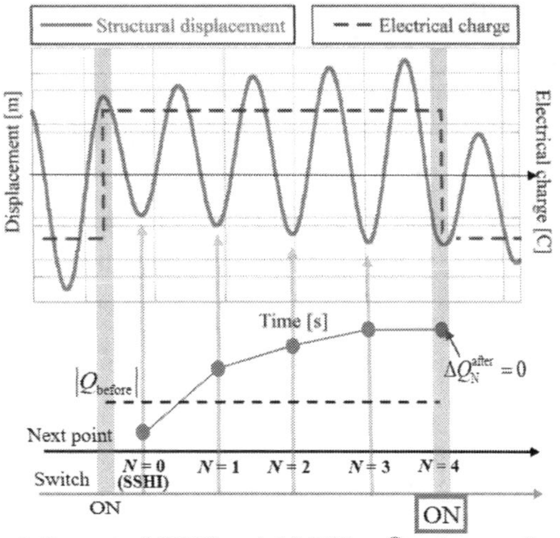

Fig. 5 Concept of SCVS and ASCVS. -●- shows estimated electric charge after switching action ($\left|Q_N^{\text{after}}\right|$) for ASCVS.

4. EXPERIMENTAL RESULTS AND DISCUSSION

4.1 Experimental set up

Figure 6 shows the experimental system. It consists of a piezoelectric transducer, an excitation shaker, a pantograph type spring and structural mass. Here, $m = 23.3$ [kg]. $k_{all} = 5.3\times10^5$ [N/m], $b_p = 2.8\times10^5$ [V/m]. Natural frequency of the structure is 23.8 Hz. A laser displacement sensor (KK-G35/LK-G30, Keyence Co.) is used for measuring structural displacement. The experimental structure is forced to oscillate by a vibration shaker at natural frequency. The amplitude of external force is 1.5 N. A personal computer (PC) is used for controlling the switching action. At first, a laser displacement sensor measures displacement. Second, the PC reads the data from a laser displacement sensor and calculates the peak of structural displacement. Third, when the condition of SCVS (When the skip number of the peaks of structural vibration become $2N$ times ($N = 0, 1, 2, 3\ldots$)) or ASCVS (When Eqs. (23) and (24) are satisfied) is satisfied the switching action is conducted.

4.2 Experimental results with amplifying piezoelectric voltage

In this section, we compare piezoelectric voltage with the original SSHI and the proposed strategies. We intended to change the skip number of switching actions N from 1 to 25 and compare the maximum voltages of a piezoelectric transducer.

Figure 7 shows the history of structural displacement and piezoelectric voltage with the original SSHI. With the switching action at every peak of structural vibraion, the amplitude of structural vibration is gradually decreased. When Eq. (13) is satisfied with the suppresion effect, electric charge decreases with the swtiching action. As a result, piezoelectric voltage decreased.

Fig. 6 Experimental setup for harvesting.

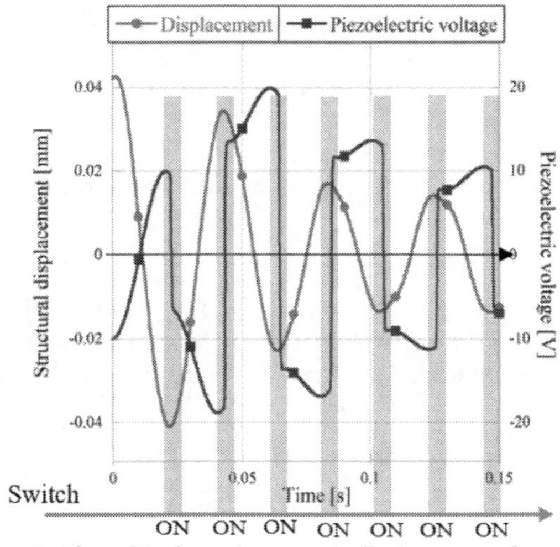

Fig. 7 Piezoelectric voltage and displacement for the original SSHI method.

Figure 8 shows the history of structural displacement with the original SSHI and SCVS ($N = 2$). Figure 8 shows that after the switching action is conducted the amplitude with the proposed strategy is suppressed. Due to the skip the switching actions, the amplitude recovers and much higher than that of the original SSHI. Figure 9 shows the history of piezoelectric voltage with the original SSHI and SCVS ($N = 2$). As shown in Fig. 9, the amplitude of piezoelectric voltage with SCVS is much higher than that with the original SSHI.

Figure 10 shows the history of piezoelectric voltage and the estimated charge after the switching action with ASCVS. As shown in Fig. 10, when Eq. (23) is satisfied to compare current the estimated electric charge $\left(|Q_N^{after}|\right)$ and one step before $\left(|Q_{N-1}^{after}|\right)$, the switching action is conducted. From 0.1 s to 0.6 s, the estimated charge rises with the structural displacement. The switching action is conducted at the maximum value of the estimated charge. After the switching actions, piezoelectric voltage rises.

With SCVS, we intended to change the skip number of switching actions N from 1 to 25. Figure 11 compares the maximum piezoelectric voltages, respectively, where, the skip number of switching actions is N. Here, when $N = 0$, the switching strategy is the original SSHI. The dotted line shows

ASCVS. Table 1 shows the maximum piezoelectric voltage and the RMS value of structural displacement, respectively, when Original SSHI, SCVS ($N = 4$), SCVS ($N = 25$) and ASCVS.

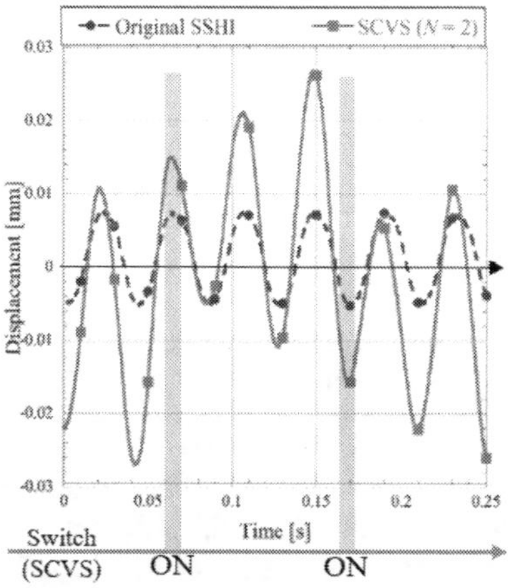

Fig. 8 Displacements of SCVS ($N = 2$) and the original SSHI.

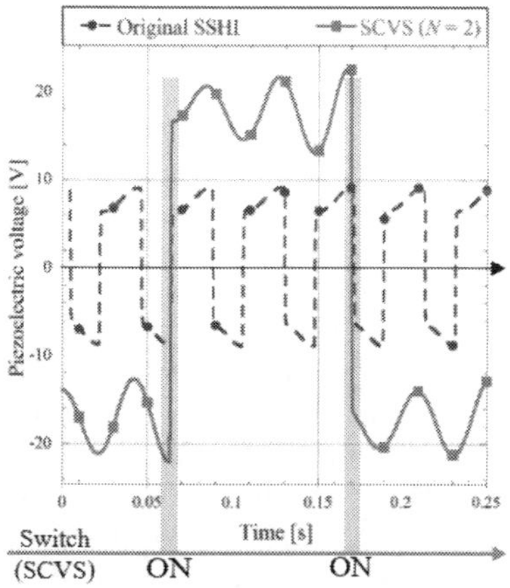

Fig. 9 Piezoelectric voltages of SCVS ($N = 2$) and the original SSHI.

In Fig. 11, piezoelectric voltage rises with the skip number of switching actions. In Eq. (22), the amplitude of vibrational displacement recovers with the skipped switching actions. As the skip number of switching actions N increases, piezoelectric voltage increases. Table 1 shows that as the skip number of switching actions N increase, the RMS value of structural displacement increases. Without control, the RMS value of structural displacement is 43.1μm. In Table1, SCVS ($N = 25$) provides 23 % reduction of the vibrational amplitude. ASCVS provides 29 % reduction. While, the original SSHI provides 87 % reduction. The experimental results show our control strategies reduce the effect of vibration suppression.

In Table1, the voltage with SCVS ($N = 25$) is up to 4.97 times greater than that with the original SSHI. The voltage with ASCVS is 4.76 times greater than that with the original SSHI. In Fig. 2, a piezoelectric transducer is connected to a storage capacitor in pararell. Therefore, the maximum piezoelectric voltage and the voltage of a storage capacitor are the same if we wait for a long time. The stored energy of a storage capacitor can be expressed as the square of the voltage. Thus, if we compose the harvesting circuit in Fig. 2, the harvested energy with SCVS ($N = 25$) can be up to 24.7 times greater than that with the original SSHI. the harvested energy with ASCVS can be 22.7 times greater than that with the original SSHI.

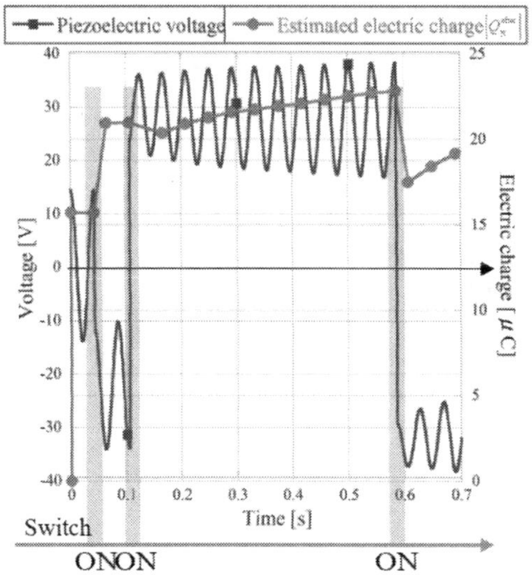

Fig. 10 Piezoelectric voltage and estimated charge for ASCVS.

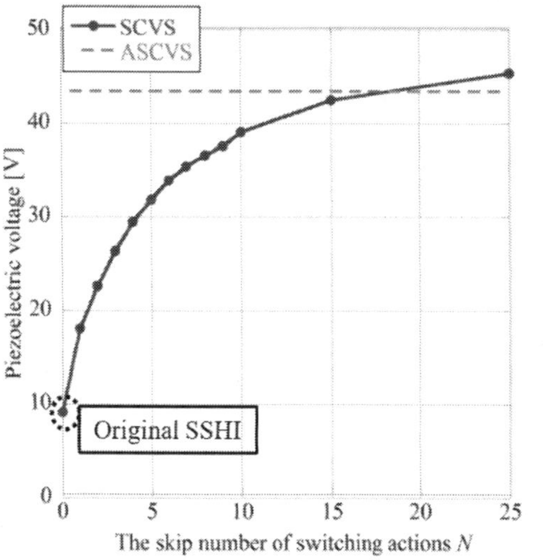

Fig. 11 Piezoelectric voltages for various switching conditions.

Table 1 Displacement and piezoelectric voltage.

	Original SSHI	SCVS (N = 5)	SCVS (N = 25)	ASCVS
Displacement RMS [μm]	5.3	20.0	33.1	30.5
Piezoelectric voltage [V]	9.1	31.9	45.4	43.5

4.3 Experimental results with resistor consumption

In this section, we conduct harvesting experiments by measuring resistor consumption. We compose the harvester circuit in Fig. 3 with a resistor connected to a smoothing capacitor parallel. We compare the original SSHI, SCVS (N = 1, N = 2, and N = 3), and ASCVS. Capacitance of a smoothing capacitor is 47 μF. Resistances of resistors are 1 kΩ to 1 MΩ. Figure12 shows the voltage value of resistors for various value of resistance. Figure 13 shows the electrical power of each strategies for various value of resistance. In Fig. 11, as resistance increases, the voltage increases with all strategies. Because the decrease of electric charge of a piezoelectric transducer depends on the value of resistance, as resistance increases, the reduction of electric charge decreases. Below 10 kΩ, voltage with all strategies are almost the same. Above 10 kΩ, the voltages of the proposed strategies are larger than the original SSHI. In Eq. (22), the suppression of the amplitude of vibrational displacement depends on electric charge. By skipping the switching actions, the proposed strategies can increase the amplitude. Thus, the proposed strategies can amplify the voltage with large resistance.

With SCVS, as resistance increases, the large skip number of switching actions can amplify piezoelectric voltage. As the skip number of switching actions increases, the amplitude of structural displacement increases. However, electric charge decreases. As a result, it has the optimal point of the best skip number of the switching actions with resistance respectively. In Fig. 13, when resistance is bigger than 10 kΩ, the proposed strategies can increase electrical consumption than the original SSHI.

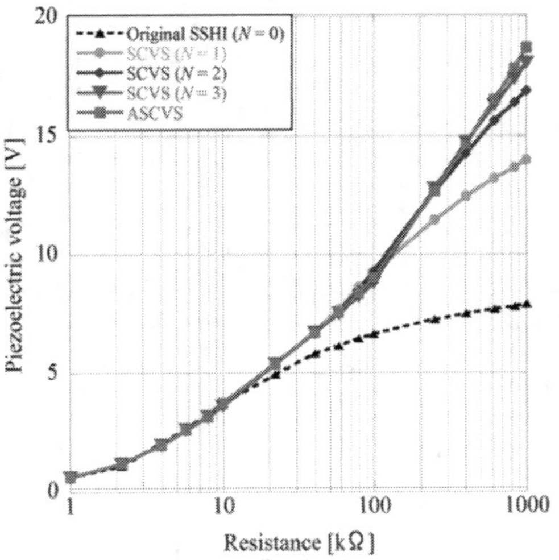

Fig. 12 Measured voltage at load resistors.

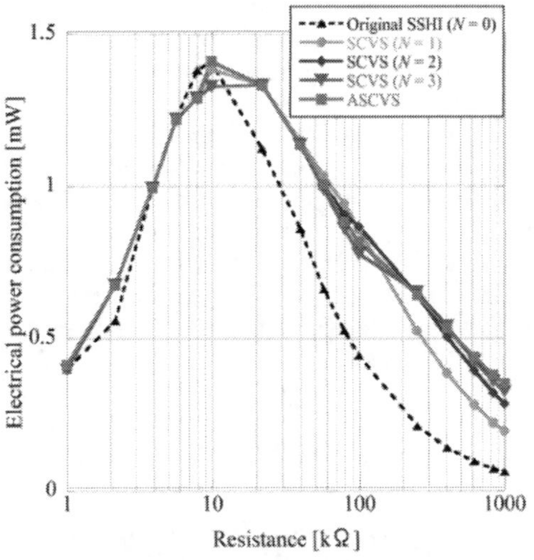

Fig. 13 Electrical power consumption on load resistors.

4.4 Experimental results with charging storage capacitor

To drive electric devices such as wireless communication system, generated power from energy harvesting is charged to a storage capacitor during some period of time. Then, electric devices such as a wireless communication system drive periodically. In this section, we intend to charge a storage capacitor during some period of time with switching strategies and compare the stored energy, respectively.

We compose the harvesting circuit as shown in Fig. 2. Capacitance of a storage capacitor is 47 μF. Figures 14 shows the stored energy during 60 s with some switching strategies, respectively. Figure 15 shows the maximum stored energy during 30 s, 60 s, and, 90 s with the original SSHI, SCVS, and ASCVS, respectively. Table 2 shows the maximum stored energy with some switching strategies. In Fig.15 and Table 2, the stored energy with the original SSHI reach steady state before 30s. However, SCVS and ASCVS can generate more energy than the original SSHI. In section 4.2, with SCVS, piezoelectric voltage rises with the skip number of switching actions. However, in Fig.14, the stored energy has a peak number of skipped switching actions. Because the switching action has the effect of amplifying piezoelectric voltage, in short period of time (30 s), the high frequency of the switching actions can rise piezoelectric voltage quickly. On the other hand, in long period of time in Table 2, the peak of skipped switching actions shift to large number of skipped switching actions.

ASCVS can generate larger energy than the other switching strategies during 30s, 60s and 90s. ASCVS can conduct the switching actions adaptively. The results show that ASCVS can charge electric energy effectively. In Table 2, the stored energy with SCVS and ASCVS is up to 10.5 times greater than that with the original SSHI. SCVS has to determine the skip number of switching actions in advance. However, ASCVS does not have to pre-decide. Therefore, the experimental results show that ASCVS is more adaptive than SCVS.

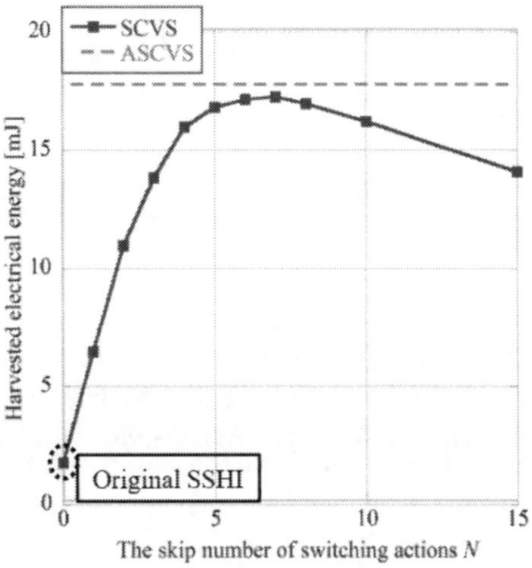

Fig. 14 Stored energy during 60 s.

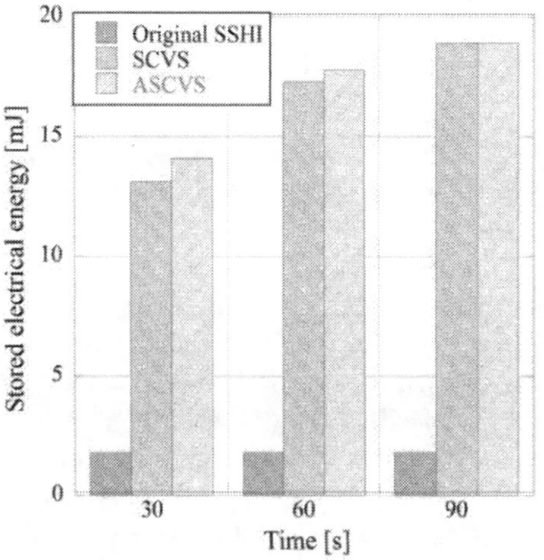

Fig. 15 Maximum stored energy during 30 s, 60 s, and 90 s with the original SSHI, SCVS, and ASCVS.

Table 2 Maximum stored energy with switching strategies.

	Original SSHI	SCVS	ASCVS
Stored energy [mJ] (30 s)	1.8	13.1 (N = 4)	14.1
Stored energy [mJ] (60 s)	1.8	17.3 (N = 7)	17.8
Stored energy [mJ] (90 s)	1.8	18.9 (N = 8)	18.9

5. Conclusion

In this study, we devised two control strategies that were SCVS and ASCVS to increase the harvested energy. Our harvesting strategies were to modulate the timing of switching action based on SSHI. Original SSHI conducted switching action at every peak of vibrational displacement. Because the switching action caused the suppression of vibrational amplitude, our control strategies were designed to temporarily skip the switching action. SCVS was designed to skip the switching actions for several times. ASCVS was designed to conduct the switching when the estimated charge is maximum value. We provided the detail mechanism of the devised new controls. Through analytical approach, we experimentally confirmed the effect of vibration suppression by the switching action and confirmed the decrease in piezoelectric voltage. Experimental results showed that the generated voltage with our strategies that are SCVS (N = 25) and ASCVS were up to 4.97 and 4.76 times greater than that with the original SSHI. Above 10 kΩ, the proposed strategies increased electrical power. ASCVS generated larger electrical energy than other switching strategies. The stored energy with SCVS and ASCVS was up to 10.5 times greater than that with the original SSHI. The experimental results showed that ASCVS was more adaptive switching strategy than SCVS.

ACKNOWLEDGMENTS

This work was supported in part by New Energy and Industrial Technology Development Organization (NEDO) under the Ministry of Economy Trade and Industry (METI) of Japan.

This work was partly supported by the JSPS Core-to-Core Program, A. Advanced Research Networks, "International research core on smart layered materials and structures for energy saving".

REFERENCES

1. Jeon, Y. B., Sood, R., Jeong, J. H., and Kim, S. G., "MEMS Power Generator with Transverse Mode Thin Film PZT," *Sensors and Actuators A*, Vol. 122, 2005, pp. 16-22.
2. Jones, P. G., Tudor, M. J., Beedy, S. P., and White, N. M., "An Electromagnetic, Vibration-powered Generator for Intelligent Sensor Systems," *Sensors and Actuators A*, Vol. 110, 2004, pp. 344-349.
3. Ottman, G. K., Hofmann, H. F., Bhatt, A. C., and Lesicutre, G. A., "Adaptive Piezoelectric Energy Harvesting Circuit for Wireless Remote Power Supply," *IEEE Transactions on Power Electronics*, Vol. 17, No. 5, 2002, pp. 669-676.
4. Ottman, G. K., Hofmann, H. F., and Lesicutre, G. A., "Optimized Piezoelectric Energy Harvesting Circuit Using Step-Down Converter in Discontinuous Conduction Mode," *IEEE Transactions on Power Electronics*, Vol. 18, No. 2, 2003, pp. 696-703.
5. Kong, N., Ha, D. S., Erturk, A., and Inman, D. J., "Resistive Impedance Matching Circuit for Piezoelectric Energy Harvesting," *Journal of Intelligent Material Systems and Structures*, Vol. 21, 2010, pp. 1293-1302.
6. Guyomar, D., Badel, A., Lefeuvre, E., and Richard, C., "Toward Energy Harvesting Using Active Materials and Conversion Improvement by Nonlinear Processing," *IEEE Transactions on Ferroelectrics and Frequency Control*, Vol. 52, No. 4, 2005, pp. 584-595
7. Richard, C., Guyomar, D., Audigier, D., and Ching, G., "Semi-Passive Damping Using Continuous Switching of a Piezoelectric Device," *Smart Structures and Materials*, Vol. 3672, 1999, pp. 104-111.
8. Qiu, J., Jiang, H., Ji, H., and Zhu, K., "Comparison between Four Piezoelectric Energy Harvesting Circuits," *Frontiers of Mechanical Engineering in China*, Vol. 4, 2009, pp. 153-159.
9. Makihara, K., Takeuchi, S., Shimose, S., and Onoda, J., "Innovative Digital Self-Powered Autonomous System for Multimodal Vibration Suppression," *AIAA Journal*, Vol. 50, No. 9, 2012, pp. 2004-2011.
10. Yamamoto, Y., Makihara, K., and Yoshimizu, K., "Synthetic Assessment of Self-Powered Energy-Harvesting Including Robustness Evaluation," *Mechanical Engineering Journal*, in press, 2015.
11. Shen, H., Qiu, J., Ji, H., Zhu, K., and Balsi, M., "Enhanced Synchronized Switch Harvesting: a New Energy Harvesting Scheme for Efficient Energy Extraction," *Smart Materials and Structures*, Vol. 19, No. 11, 2010, 115017.
12. Guyomar, D., and Lallart, M., "Recent Progress in Piezoelectric Conversion and Energy Harvesting Using Nonlinear Electronic Interfaces and Issues in Small Scale Implementation," *Micromachines*, Vol. 2, 2011, pp. 274-294.
13. Lesieutre, G. A., Ottman, G. K., and Hofmann, H. F., "Damping as a Result of Piezoelectric Energy Harvesting," *Journal of Sound and Vibration*, Vol. 269, 2004, pp. 991-1001.
14. Makihara, K., Yamamoto, Y., Yoshimizu, K., Horiguchi, C., Sakaguchi, H., and Fujimoto, K., "A Novel Controller to Increase Harvested Energy from Negating Vibration-Suppression Effect," *Smart Materials and Structures*, Vol. 24, 2015, Article No. 037005.

ICAST2015 #043

Fabrication of Commercial Prototype of Active Harvester

Kanjuro Makihara[1*] and Yuta Yamamoto[1]

[1] Department of Aerospace Engineering, Tohoku University, Sendai, Japan

Abstract

In this paper, the fabrication of a commercial prototype of self-powered active harvester is presented. The active harvester consists of a digital processor and a piezoelectric sensor, which allows the application of a technical method to improve the energy-conversion efficiency. The harvesting is implemented by measuring the vibration displacements, processing the data digitally, and adequately regulating electric switches. These operations are managed by a built-in digital processor. The driving power for the digital processor is provided with a part of the energy harvested from structural vibrations. Thus, the harvester operates flexibly with the digital processor to enhance electrical energy generation, and requires neither batteries nor an external power supply. We refer to the device as a self-powered energy harvester. An advantage of digital processing is that observations by a Kalman filter can be used to estimate modal information. As well as to reduce sensor noise, the processor is also able to manage the process of wireless communication for realizing non-powered health-monitoring. The self-powered harvester is shown to be versatile and promising, opening the way for a future form of energy harvesters.

1. INTRODUCTION

Recently, much attention has been paid to energy harvesting from natural energy sources such as solar power, wind power, radio wave, and mechanical vibration. The technology has turned to be a clean power source that can arrest global warming by reducing the use of fossil fuels. Energy harvesting is also expected to be used for portable power sources. It can replace primary batteries and eliminate battery exchange for low-power electric devices. This study addresses the harvesting of energy from structural vibrations (*i.e.*, vibrational energy harvesting), which is implemented by utilizing electromagnetic materials [1], piezoelectric materials including lead zirconate titanate (PZT), macro fiber composites (MFCs), polyvinylidene difluoride (PVDF) [2, 3], and magnetostrictive materials [4]. Of these substances, here we explore the use of piezoelectric materials for energy-harvesting devices.

The conventional method of energy harvesting, which uses only a diode bridge, is a simple AC-to-DC conversion scheme [5, 6]. It is commonly known as a passive harvester or a standard rectifier. Numerous experimental and theoretical studies have been conducted with vibration-based harvesters that exploit piezoelectric materials. Kauffman and Lesieutre [7] presented a low-order model that predicted the mechanical excitation near the resonance frequency, for piezoelectric harvesting devices.

* makihara@ssl.mech.tohoku.ac.jp

The model showed good agreement with the experimentally obtained energy harvesting performance, and this success led to an optimization scheme and alternative configuration that increased the power output. Scruggs [8] determined the maximum power generation for a vibratory harvester that was excited by a stationary broadband stochastic disturbance. He described the relationship between classical impedance matching techniques and optimal stochastic control technologies, and found that control techniques based on impedance matching could not be used to optimize power generation because the dynamic controller is always anti-causal. Karthikeyan and Amrutur [9] proposed a linear low-drop-diode equivalent circuit, which replaced four diodes in a full-wave bridge rectifier, and found that the circuit could directly extract power from the input signal without any additional power supply. Their simulations and experiments achieved an increase in output power of up to 169% compared with the conventional circuit. Takezawa *et al.* [10] proposed a geometrical optimization method for piezoelectric harvesters that thinly encapsulate mechanical devices and exploit resonances from higher-order vibration modes. By combining nonlinear equations of states and the classical solid isotropic material with penalization scheme, they developed an optimization method for designing both the material layout and polarization direction of the piezoelectric material and demonstrated its validity with an advanced simulation. Caliò *et al.* [11], in a thorough review of state-of-the-art piezoelectric harvesting, emphasized material operating models and device configurations by comparing them in terms of power density and bandwidth. They noted that both non-resonance harvesting and frequency-tuning methods are powerful tools for advancing the growth of vibration harvesting techniques.

In order to increase the output power relative to standard energy-harvesting circuits, Guyomar *et al.* [12] proposed the technique of synchronized switch harvesting on an inductor (SSHI), in which a nonlinear electric element is appended to the conventional harvesting circuit. Specifically, their SSHI technique utilized a circuit composed of on-off switches to provide two electric states: one in which electric current flows and another in which it does not. Many proposals have been advanced and studies conducted to improve the SSHI scheme [13-19]. Badel *et al.* [13] achieved a four-fold increase in harvested energy over the standard circuit by using the SSHI circuit. Makihara *et al.* [14] used a selector switch to make the SSHI circuit more compact and more sophisticated, and enhanced harvesting performance by reducing unnecessary energy consumption. Qiu *et al.* [15] compared four different SSHI conditioning circuits and their output power equations. They proposed a simple source-less circuit to implement SSHI scheme based on the trigger circuitry design. Kan *et al.* [16] found that the energy conversion efficiency of a piezoelectric generator is denoted by its effective electromechanical coupling coefficient, which depends only on the geometry parameters. The performance parameters with the structural parameters can be regarded as variables in the case of the simply supported and clamped boundary conditions. Shen *et al.* [17] proposed a new harvesting technique using piezoelectric micro generators called enhanced synchronized switch harvesting. The new technique dramatically increases the harvested power by almost 300%, compared with the standard harvesting. Dicken *et al.* [18] presented a systematic analysis of some power extraction circuits that allow the damping force to be increased under both ideal and realistic constraints. They also compared several principal types of harvesting methods including the SSHI scheme and advanced techniques, and showed that all investigated types could extract more output power than with a simple bridge rectifier by inverting and charging the piezoelectric capacitor through the inductor. Zhu *et al.* [19] investigated the underlying mechanisms of the SSHI and synchronous charge extraction techniques, and established theoretical models to evaluate their harvesting efficiencies. They showed that these switching circuits could fulfill switch harvesting requirements with different harvesting mechanisms. Guyomar *et al.* [20] showed 4-mW power for a base acceleration of 1 G by using the SSHI scheme, and emphasized a 50%

improvement in output power compared to a standard harvester. They demonstrated great harvesting with only a 1.5-cm^3 harvesting device, and compared their theoretical predictions of output power with experimental results. Lallart *et al.* [21] investigated the effect of control delay on the harvesting performance of nonlinear SSHI techniques, and predicted the output power that could be extracted under real circumstances. They showed that harvesting performance was different for positive and negative control delays and negligible as long as the delay was small. Their study revealed that whereas the parallel SSHI approach exhibits non-symmetric behavior as a function of the delay, the serial SSHI scheme displays symmetric behavior.

This paper presents the detailed concept of our self-powered harvester and the fabrication of the harvester towards a commercial prototype. First, the concept of self-powered harvester is explained by clarifying the feature of the harvester. Next, the fabricated prototype of the self-powered harvesting is presented. Then, the harvesting performance of the self-powered harvester is evaluated and compared with that of the passive harvester.

2. CONCCEPT OF SELF-POWERED HARVESTER

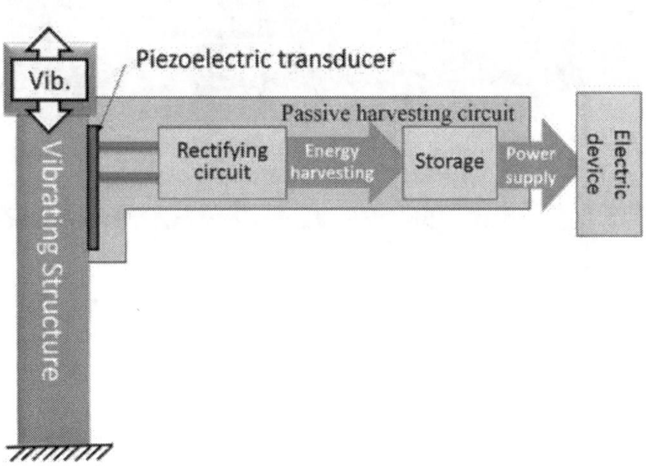

Figure 1. Diagram of passive energy harvester

Figure 1 shows a circuit diagram of a passive harvester. In the harvesting circuit, the piezoelectric transducer, which is embedded in the structure, generates alternating voltage. The circuit consists of a rectifying circuit (diode-bridge) and a storage device. The harvesting performance of the passive harvester is generally low for the harvesting requirement. Thus, a more powerful harvesting method has been required to develop. We utilized the SSHI technique, which is an effective way of improving the efficiency of vibration-based harvesting. Equipped with a circuit including an inductor and switches, SSHI inverts piezoelectric charges synchronized with structural vibrations. The voltage generated at the piezoelectric transducer is then amplified. SSHI is a promising way to improve the efficiency of harvesting. This technique increases the voltage of the piezoelectric transducer; however, it requires a control authority to manipulate the switches. We use a digital processor to address this problem.

Figure 2 shows the circuit diagram of our self-powered harvester. The harvester contains the passive harvester. In the controller circuit, a digital processor is the essential element. The electrical energy acquired by the harvesting circuit is used to drive the processor. A piezoelectric sensor attached to the surface of the structure generates a voltage that is synchronized with the displacement of the vibrating structure. On the basis of a Kalman-filter process, the processor estimates the modal displacements and velocities from the structural displacement information. The processor then sends signal outputs depending on the switching principle that will be explained in sections 3.2-3.4. The proposed harvester achieves active harvesting without any power supply or external control management whatsoever. Consequently, this energy-harvesting system is considered a self-powered device.

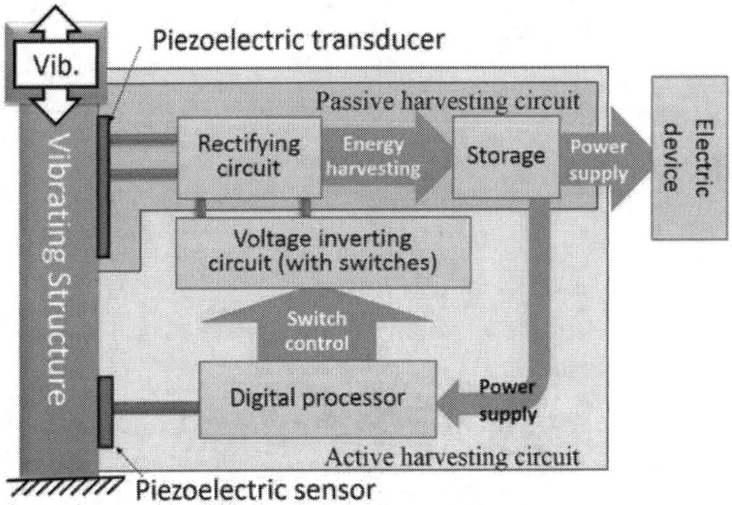

Figure 2. Diagram of self-powered energy harvester that includes a digital processor

3. FABRICATION OF PROTOTYPE HARVESTER

3.1 Prototype harvester towards commercial usage

Figure 3 illustrates a prototype of the developed harvester for commercial usage. The digital self-powered harvester was designed and fabricated by the mixed team of academic and business groups. The fabricated prototype contains a digital processor, a DC/DC converter, and harvesting switches. It has a length of 100 mm and a width of 130 mm. The circuit is to be more densely packed to provide a concise commercial harvester. The processor is located at the center. Some digital devices are integrated on the board. There are setting jumpers to configure the harvesting settings.

Here, we describe the development of software for our self-powered harvester, which includes a digital processor that controls the harvesting devices. The advantage of using a digital control device is that it allows the application of advanced control logic. The self-powered harvester realizes switch control for the SSHI technique by adopting the switch control principle based on linear quadratic regulator (LQR) control. This principle is derived from the modal displacement and modal velocity of the vibrating structure. A Kalman filter, a well-known state observer, is employed to estimate the modal values. Because the filter has the ability of modal decomposition, it can handle multimodal vibrations. Furthermore, because the Kalman filter also has the ability to reduce noise, it is conducive to the construction of a robust harvesting system. The self-powered harvester achieves high-performance digital control using a digital processor, with which the LQR control and Kalman filter are programmed.

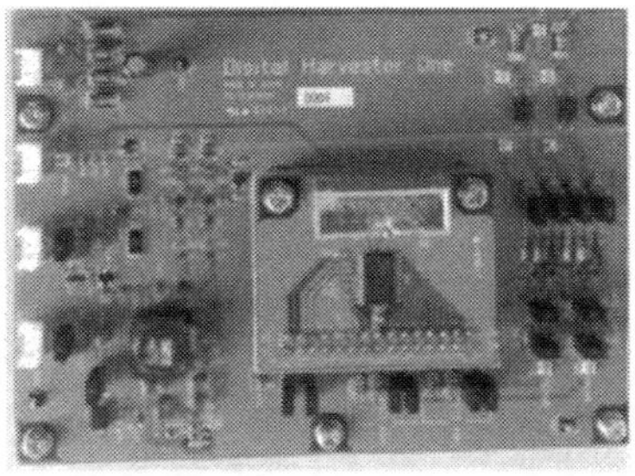

Figure 3. One of the prototype harvesters integrated with digital devices

3.2 Switch control based on LQR

We consider a spring-mass-damper system with N degree of freedom (N-DOF). A piezoelectric transducer is installed between the first mass (upper mass in experimental apparatus) and the base. The guiding principle of the switch control for multimodal vibration is based on our insight into semi-active vibration suppression [22]. The essential part of the circuit for semi-active vibration suppression is equivalent to the SSHI harvesting circuit. In the previous study, we proposed a switch control for semi-active vibration suppression that is based on the LQR control. The proposed semi-active method works well in the experiments of suppressing multimodal vibrations. The semi-active method suppresses vibrations effectively by converting vibration energy to electrical energy. The purpose of the vibration suppression is rapidly decreasing the vibration energy. Focusing on the increase in electrical energy, the switch control of semi-active vibration suppression can be thought as effective energy harvesting. Thus, we apply the switch control of vibration suppression to the harvesting from multimodal vibration. The force produced by a piezoelectric transducer is expressed from the fundamental equation of piezoelectric materials as

$$f_{\mathrm{p}} = k_{\mathrm{p}} u - b_{\mathrm{p}} Q. \tag{1}$$

Here, b_{p} is piezoelectric coefficient. It is derived from piezoelectric constant to represent the relationship between electrical charge and force at the piezoelectric transducer. Then, the equation of motion for the N-DOF structure including the piezoelectric transducer can be written as

$$\mathbf{M}\ddot{\vec{x}} + \mathbf{D}\dot{\vec{x}} + \mathbf{K}_{\mathrm{s}}\vec{x} = -\mathbf{K}_{\mathrm{p}} u + \mathbf{B}_{\mathrm{p}} Q + \vec{w}. \tag{2}$$

The elongation of the transducer is related to the displacement of the structure with a constant vector $\vec{\alpha}$ as

$$u = [\alpha \quad 0 \quad \cdots \quad 0]\vec{x} = \vec{\alpha}^{\mathrm{T}}\vec{x}, \tag{3}$$

where α is the proportionality coefficient of elongation and displacement. The equation of motion for the structure that includes the piezoelectric transducer can be written as

$$\mathbf{M}\ddot{\vec{x}} + \mathbf{D}\dot{\vec{x}} + \mathbf{K}\vec{x} = \mathbf{B}_{\mathrm{p}} Q + \vec{w}, \tag{4}$$

where the synthetic stiffness matrix $\mathbf{K}$ is defined as $\mathbf{K} \equiv \mathbf{K}_s + \mathbf{K}_p \vec{\alpha}^T$. In the modal coordinates, the equation of motion is expressed as

$$\ddot{\vec{\xi}} + \boldsymbol{\Xi}\dot{\vec{\xi}} + \boldsymbol{\Omega}\vec{\xi} = \boldsymbol{\Phi}^T \mathbf{B}_p Q + \boldsymbol{\Phi}^T \vec{w}, \tag{5}$$

where

$$\vec{x} \equiv \boldsymbol{\Phi}\vec{\xi}, \tag{6}$$

$$\boldsymbol{\Phi} \equiv \left[\vec{\phi}_1, \vec{\phi}_2, \cdots, \vec{\phi}_N\right],\ \boldsymbol{\Omega} \equiv \mathrm{diag}\left[\omega_1^{\ 2}, \omega_2^{\ 2}, \cdots \omega_N^{\ 2}\right],\ \boldsymbol{\Xi} \equiv \mathrm{diag}\left[2\zeta_1\omega_1, 2\zeta_2\omega_2, \cdots 2\zeta_N\omega_N\right]. \tag{7}$$

Here, the damping matrix $\boldsymbol{\Xi}$ is assumed as Rayleigh's damping. The transformation matrix $\boldsymbol{\Phi}$ is decided so as to satisfy the equation $\boldsymbol{\Phi}^T \mathbf{M} \boldsymbol{\Phi} = \mathbf{I}$. Equation (5) can then be transformed to

$$\dot{\vec{z}} = \mathbf{A}\vec{z} + \mathbf{B}Q + \mathbf{E}\vec{w}, \tag{8}$$

where

$$\vec{z} \equiv [\vec{\xi}^T, \dot{\vec{\xi}}^T]^T,\ \mathbf{A} \equiv \begin{bmatrix} \mathbf{0} & \mathbf{I} \\ -\Omega & -\Xi \end{bmatrix},\ \mathbf{B} \equiv \begin{bmatrix} \mathbf{0} \\ \Phi^T \mathbf{B}_p \end{bmatrix},\ \mathbf{E} \equiv \begin{bmatrix} \mathbf{0} \\ \Phi^T \end{bmatrix}. \tag{9}$$

If Q in Eq. (8) is regarded as an active control input, LQR control theory specifies that the performance index

$$J \equiv \int_0^\infty (\vec{z}^T \mathbf{W}_1 \vec{z} + Q W_2 Q)\, \mathrm{d}t \tag{10}$$

is minimized by

$$Q = Q_T \equiv -\mathbf{F}\vec{z}, \tag{11}$$

where

$$\mathbf{F} \equiv W_2^{-1} \mathbf{B}^T \mathbf{P}_1. \tag{12}$$

Here, $\mathbf{P}_1$ is the positive solution of

$$\mathbf{P}_1 \mathbf{B} W_2^{-1} \mathbf{B}^T \mathbf{P}_1 - \mathbf{A}^T \mathbf{P}_1 - \mathbf{P}_1 \mathbf{A} - \mathbf{W}_1 = \mathbf{0}. \tag{13}$$

The performance index J comprises two competing factors: the first and second terms in Eq. (10). Our method only refers to the polarity of Q_T, as will be discussed. Thus, the meanings of $\mathbf{W}_1$ and W_2 in the switching control are quite different from those in the standard control guidelines. We attempt to semi-actively control the structural vibration by switching the circuit connection point instead of actively supplying a control charge or voltage to the actuators. In our semi-active approach, Q_T is used as the switching criterion. The significant meanings of weighting matrices for switching controls is elaborately explained in ref. [23].

3.3 State observation with Kalman filter

We implement the modal estimation of a multi-DOF system as a modal truncation to facilitate an effective estimation. To execute a multimodal estimation, the Kalman filter for the proposed system is derived, and harvesting is achieved using the estimated values of the modal velocities and displacements. The sensor equation is written as

$$V_{\mathrm{sen}} = \mathbf{C}z. \tag{14}$$

Because the system has one transducer and one sensor, the values Q and V_{sen} are scalars. The Kalman filter for estimating z is

$$\dot{\hat{z}} = \mathbf{A}\hat{z} + \mathbf{B}Q + \Gamma(V_{\mathrm{sen}} - \mathbf{C}\hat{z}), \tag{15}$$

where the observer gain matrix Γ is defined as

$$\Gamma \equiv \mathbf{P}_2\mathbf{C}^{\mathrm{T}}\mathbf{S}_2^{-1}. \tag{16}$$

Here, $\mathbf{P}_2$ is a positive definite solution of

$$\mathbf{AP}_2 + \mathbf{P}_2\mathbf{A}^{\mathrm{T}} - \mathbf{P}_2\mathbf{C}^{\mathrm{T}}\mathbf{S}_2^{-1}\mathbf{CP}_2 + \mathbf{S}_1 = \mathbf{0}. \tag{17}$$

The noise covariance matrices for Eqs. (8) and (14) are $\mathbf{S}_1$ and $\mathbf{S}_2$, respectively.

3.4 Switching action based on switching criterion

The Kalman-filter estimates the state vector z. The estimated value $\hat{z}$ is then used to calculate the switching criterion as

$$Q_{\mathrm{T}} \equiv -\mathbf{F}\hat{z}. \tag{18}$$

Because our harvester has no electrical energy source for active input, the optimal input cannot be achieved as desired. We regulate switches to make the piezoelectric charge Q as close to the optimal input (criterion) Q_{T} as possible. That is, we regulate the switches so that the polarity of the piezoelectric charge Q is equal to the criterion Q_{T}. The switch allows electric current to flow from the positive terminal of the piezoelectric transducer to the negative terminal and *vice versa*, respectively.

4. EXPERIMENTAL SETUP

Experimental setup for evaluating the harvesting performance is presented. A harvesting experiment was performed under 2DOF vibration excitation. The setup (Fig. 4) consists of two masses, two springs, and a piezoelectric transducer. The pantograph structure between the upper mass and the platform worked as a spring. The mass-spring vibrating system consisted of two masses (upper and lower) and two springs (pantograph spring and coil spring). For converting vibration energy to electrical energy, a piezoelectric transducer was attached to the pantograph-shaped spring. It was inserted in the pantograph structure and fixed. In addition to the role of a spring, the pantograph-structure had a role of expanding the elongation of the transducer (in horizontal axis) into the displacement (in vertical axis) because the allowable elongation of piezoelectric transducers is small. The elongation of the transducer was proportional to the structural displacement. A piezoelectric sheet was used as a displacement sensor. This piezoelectric sheet was attached on the pin-connected part of the structure. The bending of the sheet generates the sensor voltage being in proportion to the structural vibration. The electrodes of the transducer were connected to the harvesting circuit, and the circuit harvested the vibrational energy in the 2DOF structure. The vibration apparatus was excited by a vibration shaker.

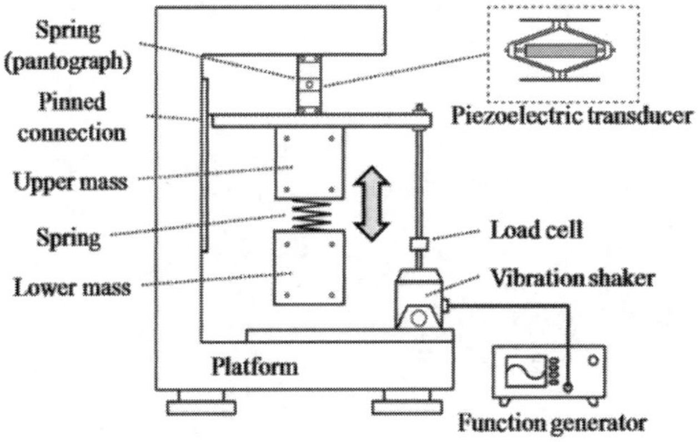

Figure 4. Experimental setup for vibration system

5. HARVESTING PERFORMANCE OF THE DEVELOPED HARVESTER

First, a self-powered harvesting experiment was conducted with the developed harvester (Fig. 3). The upper mass was excited to vibrate with the vibration shaker. The vibration has reached the steady state. Figure 5 shows histories of the displacement of the upper mass and piezoelectric voltage. The harvested voltage (*i.e.*, stored voltage) is depicted in blue line, and the harvested voltage is 20 V. When the absolute value of piezoelectric voltage exceeds the storage voltage, the voltage generated at the piezoelectric transducer flows into the storage capacitor, after which the piezoelectric voltage remains

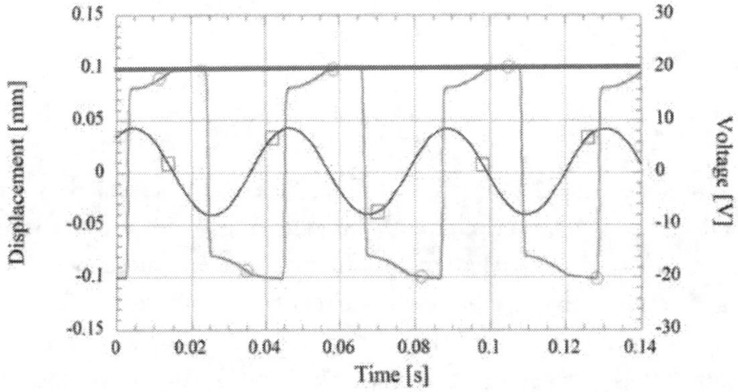

Figure 5. Histories of displacement (black line), piezoelectric voltage (red line), and harvested voltage (blue line) with self-powered harvesting

nearly constant, like plateaus at the tops of the voltage waveform. The gap between these plateaus and capacitor voltages is due to the forward voltage of the diodes in the rectifier. The wave characteristics of voltage is exactly the same as what is expected. Thus, control action is quite accurate, and active harvesting is carried out as expected. The harvester acquired a harvested voltage of as much as 20 V.

Next, for a comparison, a passive harvesting experiment was conducted. The vibration excitation was the same as the previous experiment. The harvested energy in blue line is only 10 V. The harvested energy with the passive harvesting is only 1/4 times of the harvested one with the self-powered harvesting. We conclude that the developed harvester can generate high-level energy, even though some amount of energy is consumed in the digital control devices.

In these harvesting results, there is a serious matter to comment on. The above experiments were conducted under the same excitation condition. The magnitude of vibration in Fig. 5 is 0.04 mm, while that in Fig. 6 is 0.12 mm. Thus, we can see that the energy harvesting with the self-powered device causes the vibration suppression, which has been investigated, for example, in [24]. The effect of vibration suppression may sometimes be useful for engineering purposes. However, from the view of energy harvesting, the decrease in the vibration magnitude is harmful, because it results in the decrease in harvested energy. In order to solve this critical problem, a new control method is proposed in ref. [25]. The new control is designed to pause harvesting action occasionally, and to increase the final amount of harvested energy. The aim of the new control method is to recover the amplitude of the vibration to its original level during the pauses. This recovery of mechanical energy is expected to boost the energy-harvesting effect and to negate the vibration-suppression effect. Subsequently, we expect that the proposed controller could successfully increase the resultant amount of harvested energy.

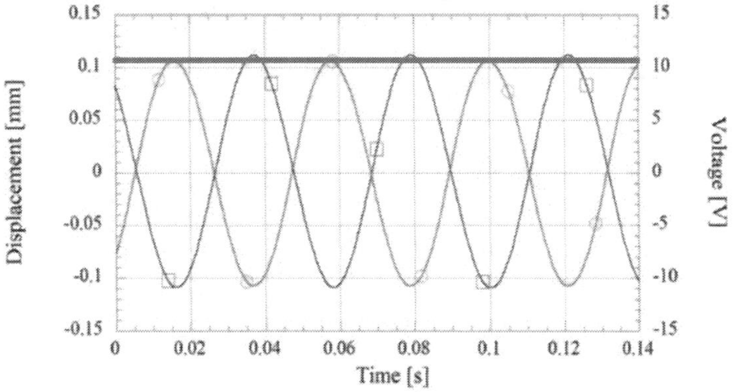

Figure 6. Histories of displacement (black line), piezoelectric voltage (red line), and harvested voltage (blue line) with passive harvesting

6. CONCLUSIONS

A prototype of a self-powered harvester was presented. The harvester included a built-in digital processor driven by harvested energy, which led to self-powered digital control for effective harvesting. Our proposed harvester and principle of switching control can be applied to energy harvesting from multi-modal vibrations, which is more difficult than that from single-modal vibrations. The self-powered harvester is shown to generate 4 times larger harvested energy than the passive harvester. This is quite striking number for the vibration-based harvester.

ACKNOWLEDGMENTS

This work was supported in part by the New Energy and Industrial Technology Development Organization (NEDO) under the Ministry of Economy Trade and Industry (METI) of Japan. This work was partly supported by the JSPS Core-to-Core Program (A). Advanced Research Networks, "International research core on smart layered materials and structures for energy saving".

REFERENCES

1. Glynne-Jones, P., Tudor, M. J., Beeby, S. P., and White, N. M., "An Electromagnetic Vibration-Powered Generator for Intelligent Sensor Systems," *Sensors and Actuators A: Physical*, Vol. 110, 2004, pp. 344–349.
2. Stephen, N. G., "On Energy Harvesting from Ambient Vibration," *Journal of Sound and Vibration*, Vol. 293, 2006, pp. 409–425.
3. Shu, Y. C., and Lien, I. C., "Analysis of Power Output for Piezoelectric Energy Harvesting Systems," *Smart Materials and Structures*, Vol. 15, 2006, pp. 1499–1512.
4. Staley, M. E., and Flatau, A. B., "Characterization of Energy Harvesting Potential of Terfenol-D and Galfenol," *Proceedings of SPIE*, Vol. 5764, 2005, pp. 630–640.
5. Kazmierski, T. J., and Beeby, S., *Energy Harvesting Systems: Principles Modeling and Applications*, London, Springer, 2010.
6. Beeby, S., and White, N., *Energy Harvesting for Autonomous Systems (Smart Materials Structures and Systems,* London, Artech House, 2010.
7. Kauffman, J. L., and Lesieutre, G. A., "A Low-Order Model for the Design of Piezoelectric Energy Harvesting Devices," *Journal of Intelligent Material Systems and Structures*, Vol. 20, 2009, pp. 495–504.
8. Scruggs, J. T., "On the Causal Power Generation Limit for a Vibratory Energy Harvester in Broadband Stochastic Response," *Journal of Intelligent Material Systems and Structures*, Vol. 21, 2010, pp. 1249–1262.
9. Karthikeyan, L., and Amrutur, B., "Signal-Powered Low-Drop-Diode Equivalent Circuit for Full-Wave Bridge Rectifier," *IEEE, Transactions on Power Electronics*, Vol. 27, 2012, pp. 4192–4201.
10. Takezawa, A., Kitamura, M., Watanabe, S. L., and Silva, E. C. N., "Design Methodology of Piezoelectric Energy-Harvesting Skin using Topology Optimization," *Structural and Multidisciplinary Optimization*, Vol. 49, 2014, pp. 281–297.

11. Caliò, R., Rongala, U. B., Camboni, D., Milazzo, M., Stefanini, C., de Petris, G., and Oddo, C. M., "Piezoelectric Energy Harvesting Solutions," *Sensors,* Vol. 14, No. 3, 2014, pp. 4755–4790.
12. Guyomar, D., Badel, A., Lefeuvre, E., and Richard, C., "Toward Energy Harvesting using Active Materials and Conversion Improvement by Nonlinear Processing," *IEEE, Transactions on Ultrasonics Ferroelectrics and Frequency Control*, Vol. 52, 2005, pp. 584–595.
13. Badel, A., Guyomar, D., Lefeuvre, E., and Richard, C., "Piezoelectric Energy Harvesting using a Synchronized Switch Technique," *Journal of Intelligent Material Systems and Structures*, Vol. 17, 2006, pp. 831–839.
14. Makihara, K., Miyakawa, T., and Onoda, J., "Low Energy Dissipation Electric Circuit for Energy Harvester," *Smart Materials and Structures*, Vol. 15, No. 5, 2006, pp. 1493–1498.
15. Qiu, J., Jiang, H., Hongli, J., and Kongjun, Z., "Comparison between Four Piezoelectric Energy Harvesting Circuits," *Frontiers of Mechanical Engineering in China*, Vol. 4, 2009, pp. 153–159.
16. Kan, J., Qiu, J., Tang, K., Zhu, K., and Shao, C., "Modeling and Simulation of Piezoelectric Composite Diaphragms for Energy Harvesting," *International Journal of Applied Electromagnetics and Mechanics*, Vol. 30, 2009, pp. 95–106.
17. Shen, H., Qiu, J., Ji, H., Zhu, K., and Balsi, M., "Enhanced Synchronized Switch Harvesting: a New Energy Harvesting Scheme for Efficient Energy Extraction," *Smart Materials and Structures*, Vol. 19, No. 11, 2010, Article No. 115017.
18. Dicken, J., Mitcheson, P. D., Stoianov, I., and Yeatman, E. M., "Power-Extraction Circuits for Piezoelectric Energy Harvesters in Miniature and Low-Power Applications," *IEEE, Transactions on Power Electronics*, Vol. 27, 2012, pp. 4514–4529.
19. Zhu, L., Chen, R., and Liu, X., "Theoretical Analyses of the Electronic Breaker Switching Method for Nonlinear Energy Harvesting Interfaces," *Journal of Intelligent Material Systems and Structures*, Vol. 23, 2012, pp. 441–451.
20. Guyomar, D., Sebald, G., and Kuwano, H., "Energy Harvester of 1.5 cm^3 Giving Output Power of 2.6 mw with Only 1 G Acceleration," *Journal of Intelligent Material Systems and Structures,* Vol. 22, 2011, pp. 415–420.
21. Lallart, M., Wu, Y.-C., and Guyomar, D., "Switching Delay Effects on Nonlinear Piezoelectric Energy Harvesting Techniques," *IEEE, Transactions on Industrial Electronics*, Vol. 59, 2012, pp. 464–472.
22. Makihara, K., Takeuchi, S., Shimose, S., and Onoda, J., "Innovative Digital Self-Powered Autonomous System for Multimodal Vibration Suppression," *AIAA Journal*, Vol. 50, No. 9, 2012, pp. 2004–2011.
23. Makihara, K., Onoda, J., and Minesugi, K., "Comprehensive Assessment of Semi-Active Vibration Suppression Including Energy Analysis," *Transactions of ASME, Journal of Vibration and Acoustics*, Vol. 129, No. 1, 2007, pp. 84–93.
24. Lesieutre, G. A., Ottman, G. K., and Hofmann, H. F., "Damping as a Result of Piezoelectric Energy Harvesting," *Journal of Sound and Vibration*, Vol. 269, 2004, pp. 991–1001.
25. Makihara, K., Yamamoto, Y., Yoshimizu, K., Horiguchi, C., Sakaguchi, H., and Fujimoto, K., "A Novel Controller to Increase Harvested Energy from Negating Vibration-Suppression Effect," *Smart Materials and Structures*, Vol. 24, 2015, Article No. 037005.

NOMENCLATURE

k_p	:	Constant-charge stiffness of a piezoelectric transducer
u	:	Elongation of a piezoelectric transducer
b_p	:	Piezoelectric coefficient
Q	:	Electrical charge in piezoelectric transducer
$\mathbf{M}$	:	Mass matrix
$\mathbf{D}$	:	Damping matrix
$\mathbf{K}_s$	:	Stiffness matrix of structure
$\mathbf{K}_p$	:	Stiffness matrix of piezoelectric transducer
$\mathbf{B}_p$	:	Piezoelectric coefficient matrix
$\vec{x}$	:	Displacement vector
$\vec{w}$	:	External force vector
$\mathbf{K}$	:	Synthetic stiffness matrix
$\vec{\xi}$	:	Modal displacement vector
$\vec{\phi}_N$	:	Eigenvectors synthetic stiffness matrix
ω_N	:	Natural angular frequencies
ζ_N	:	Modal damping coefficients
$\vec{z}$	:	State vector
J	:	Evaluation function for LQR
$\mathbf{W}_1, W_2$	:	Parameters for controller design
Q_T	:	Target electrical charge as criterion
$\mathbf{F}$	:	Positive feedback gain
V_{sen}	:	Voltage of piezoelectric sensor
$\mathbf{C}$	:	Coefficient matrix of sensor voltage and modal displacements
$\mathbf{\Gamma}$	:	Observer gain matrix
$\mathbf{S}_1, \mathbf{S}_2$	:	Parameters for Kalman filter design
$\hat{\bullet}$	:	Estimated value by Kalman filter

ICAST2015 #013

Polymer based harvester as power source for wireless sensors in pipes

Sherif Keddis, Norbert Schwesinger

Microstructured Mechatronical Systems
Technical University of Munich
Munich, Germany

Abstract

Autarkic sensors need most a lower amount of energy than data transceivers. Both are commonly supplied via wires or batteries with electricity. In this paper, we present an investigation on wireless sensors for fluid control inside a pipe. With common technologies, this request leads to high requirements on tightness since poor sealing could easily lead to failures. Replacement of batteries inside pipes is complicated and almost accompanied by a flow interruption. In this case, the application of energy harvesters as power supply is a good alternative. We studied a flexible piezoelectric energy harvester of PVDF (Poly-Vinylidene–Di-Fluoride). The harvesting unit consists of piezoelectric PVDF-foils as active layers and Aluminum-foils as electrodes. The layers are stacked alternating on each other and wound to a spool. A LDPE-film wraps the spool and prevents the inflow of liquids. The device has following parameters:

- No. of windings: 4
- width 15mm; Ø 20mm
  - PVDF: 25µm;
  - Aluminum: 6µm;
  - LDPE: 25µm

A ring shaped bluff body inside the pipe induces turbulence in the fluid stream. As the harvester is arranged downstream of the bluff body, it is forced to oscillate. By bending the active layers, a separation of electrical charges occurs. First experiments were carried out in a wind channel as dry environment to ensure reliability. At a wind velocity of about 7m/s, the spool oscillates with a frequency of about 15Hz. A peak-peak-Voltage of about 4V can be measured. In case of power adjustment, power values of about 0.54µW can be obtained. This power is stored in a capacitor. Assuming a data transmission unit consumes about 0.2 mWs during one operational period of 1 second, the duty cycle time can be calculated to about 6.2 minutes. Since power densities in liquids are greater than in gases, this result justifies the application of polymer based energy harvesters as power supply for wireless sensors inside pipes.

1 INTRODUCTION

Energy harvesting became an increasing importance during the last years. Originally situated in the MEM's area, the engineering branch spreads out to larger dimensions and volumes. Several SMA's (Small and Medium Enterprises) have been founded to develop, produce and sell appropriate energy harvesting devices. Until today, one can find new foundations in this field. Unfortunately, many of these foundations are gone meanwhile. Although closing of enterprises is not an extraordinary event, the number of company closures at that early stage of development is quite astonishing. Did the products of these companies not meet the expectations of the customers? Was the maturity of development of the products for a market launch inadequate? Were the companies not able to supply the market with a sufficient number of devices? Could it be possible that the products found no regarding acceptance on the market? Were the characteristics of the direct competitive products - batteries – ignored? On the other hand, countless research groups worldwide made investigations in this field, many of them with support from governmental institutions or industrial funding organizations. The result of these developments is evident in confusing large numbers of scientific papers in a permanently growing number of engineering and physical journals oral as well as corresponding poster presentations. Some books on the topic complete the scope of scientific work [PrIn, Tan, KaBe]. Often it seems that energy harvesting deals with itself and has the look lost on why it is needed. Energy harvesters (EH) should serve as electrical energy supply for small consumers. They usually convert the energy losses of mainly technical processes into electrical energy. Since for conversion often small physical effects are used, the conversion efficiency is usually very low. In addition, since the cost of energy harvesters significantly exceeds that of batteries, the use of EHs must be technically and economically viable. In order to assess this properly, we have identified various criteria. If a concrete application fulfills one of the criteria, then a sensible use of EH is given. In case of no agreement, it is better to supply consumers with electrical energy by batteries or the power grid. The identified criteria are listed below.

1. No replacements during lifetime (of the consumer)
2. Ability to deliver a certain amount of energy
3. Acting as EH and Sensor simultaneously
4. Simultaneous use of a single conditioning circuit by numerous harvesting units
5. Ban on the use of power grids or batteries

2 ENERGY HARVESTING APPLICATIONS

An example of an economically justifiable application field for EHs is flow monitoring systems. Sensors to monitor flow conditions and fluid parameters are ideally placed inside the distribution pipes. To supply these sensors with power one can commonly use the electric power grid or batteries. Assuming a supply by the electrical power grid, pipes need to be opened for wiring. Unfortunately, this is one of the main causes of leakage due to sealing complications. On the other hand, replacing depleted batteries is either complicated or totally impossible when pipes are unreachable (e.g., buried). Furthermore, it could be accompanied by an economically unfavorable flow interruption. Due to inaccessibility, these conditions completely fulfill the first criterion.

Placing an EH inside the pipe that generates sufficient energy to power the sensor and enables wireless transmission is therefore a reasonable solution.

This paper reports the results of several investigations conducted to determine the optimal configuration of piezoelectric EH in a pipe with the goal of generating enough energy to power a wireless sensor autonomously. First results show, that the use of PVDF-foils is a very promising approach.

3 ENERGY HARVESTING IN FLOWING MEDIA

Some EHs have been previously introduced that generate electrical energy from flowing media. Typical concepts to achieve this are the piezoelectric conversion used by [Tay, Pob, Wang, Kwon, Ye, Liu] as well as the electromagnetic conversion presented by [Tor, Hoff]. Electromagnetic harvesters require a relative movement of a magnet to a coil, or vice versa, are difficult to scale down due to frictional losses (e.g., friction of rotational components) [Akay] and loose efficiency with shrinking dimensions. This makes these EHs unsuitable for applications inside standard water pipes. Piezoelectric harvesters on the other hand show a more promising scalability. Research shows that at least two piezoelectric materials are being used: PZT (lead-zirconate-titanate) [Kwo, Liu] and PVDF [Tay, Wang]. The latter has much lower piezoelectric coefficients than the former, which leads to small-scale harvesters generating relatively low power. Although this makes PZT more appealing as the active material, its brittleness and fatigue can have a negative impact on the operational life span. Additionally, an application in water distribution systems could be problematic due to the porosity of the material and possible water absorption. A new harvesting device is needed that could overcome the problems described.

4 DESIGN AND IMPLEMENTATION

The developed harvester bases on former investigations by [Tay, All, Pob]. Basic idea is the use of flow-induced turbulence in the wake of a bluff body to achieve mechanical stress in piezoelectric materials. PVDF has excellent mechanical properties, is chemical resistant and shows no water absorption. Therefore, it is well suited for in-pipe applications with no concern about aging, fatigue or other health issues. In order to overcome the low power output traced back to the lower piezoelectric coefficient, PVDF is shaped like a spool. Hereby, a PVDF-film is sandwiched between two electrode layers of aluminum and wound to a spool. This procedure maximizes the active area. Unfortunately, this leads to a direct contact between both electrode-films resulting in a short circuit (see Fig. 1).

Therefore, a second PVDF-film is added to resolve this problem. Keeping in mind the right

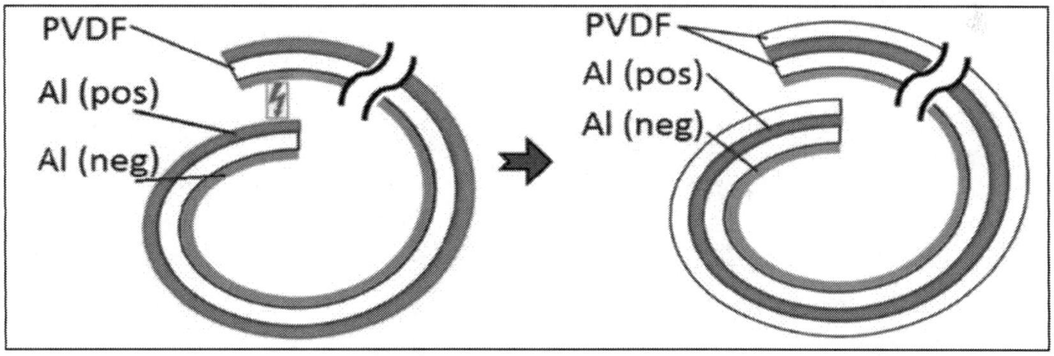

Figure 1: Short circuit prevention using two PVDF-foils (right) instead of one (left)

orientation of the polarized PVDF-films, the active piezoelectric area is now doubled. With increasing number of windings, the active area increases linear. The PVDF-film used in this configuration has a width and thickness of 12 mm and 25 μm, respectively. The width of the aluminum-film is just 8 mm, while the thickness is about 6.0 μm. Both electrode films are centrally layered alternating with the PVDF-foils and wound to form a spool. The length of the layers is determined by the diameter of the mandrel and the number of windings. It can be approximated by using the volume V of the foil in an unrolled form (1) and equating it with the volume of the wound foil on the mandrel (2):

$$V=W\cdot L\cdot T \tag{1}$$
$$V=W\cdot\pi\cdot(r_2^{\,2}-r_1^{\,2}) \tag{2}$$

where W denotes the width of the PVDF-film, L its length and T its thickness. The radius of the mandrel is given by r_1 while r_2 denotes the radius of the most outer winding, which is dependent on n, the number of windings:

$$r_2=n\cdot T+r_1 \tag{3}$$

Inserting Equation (3) in (2) and Equation (2) in (1) and solving Equation (1) for **L** leads to:

$$L=\frac{\pi\,((n\cdot T+r_1)^2-r_1^2)}{T} \tag{4}$$

For later investigations taking place in water as well as to seal this device against liquid penetration and in view of comparable results for both liquid and gaseous fluids, two layers of LDPE (low-density polyethylene) of 25.0μm thickness are added as the inner and outer winding of the spool. These two layers welded together form a sealed ring-shaped spool as energy harvester. LDPE has a low young's modulus, which allows for an efficient energy transfer. In order to convert the kinetic energy of the flowing media into mechanical strain in the piezoelectric films, thus generating electric charges, a ring-shaped bluff body is positioned upstream of the EH to induce vortices. Alternating vortices cause the harvester to oscillate and undergo mechanical strain.

5 EXPERIMENTAL INVESTIGATIONS

To ensure system reliability and safeness, first tests to find the optimal harvester configuration inside the pipe were carried out with air as flowing media. The EH was tested in a 1"-pipe using a closed-loop wind channel from "ELV" (Uniwind). Different EH configurations were tested depending on the wind speed of 7 m/s. The efficiency of the conversion of kinetic energy to mechanical stress inside the active layers was indicated by measuring the open-circuit voltage.

5.1 Harvester Configuration

Akaydin et al. [Akay] calculated that "most of the losses occur during the conversion of flow's kinetic energy into mechanical/vibrational energy". Therefore, finding a suitable form and fixation method inside the pipe is crucial. Based on a mandrel diameter of about 27mm and by means of live measurements in air, three separately tested fixation methods showed promising output signals. The first method leads to a shape similar to a flag or "eel"[Tay, All]. The spool is fixed to the pipe bottom with just one screw and the free end can oscillate in the flow. In other configurations, the spool is fixed with several screws in order to achieve a corresponding mechanical pre-stress.

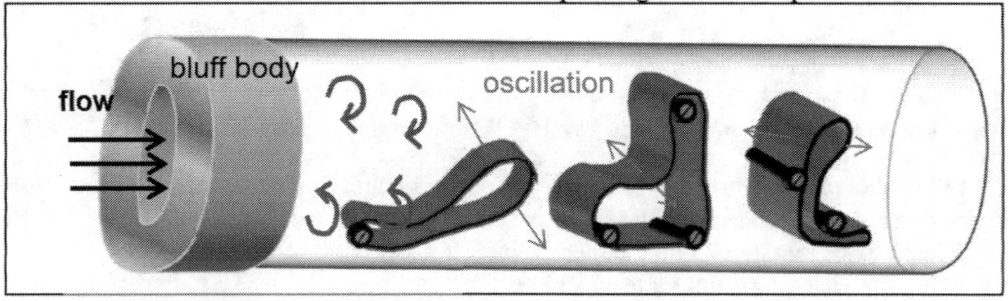

Figure 2: Initial different fixation shapes investigated: "eel" (left), boot (center) and L-shape (right)

Figure 2 illustrates all tested different alternatives. One shape (middle) looks like a boot fixed with three screws. Vortices induced by the bluff body cause one side of the harvester to bend inwards, while the pre-stress inside the materials try to return the layers to the starting position. The third fixation method looks like an "L". This form has a free horizontal oscillating end. Pre-stress inside the layers cause the loop to hold up against the flow, while the screw fixed upstream of the harvester limits the backward movement.

To avoid a contact of the free end of the spool with the wall of the pipe, the diameter of the mandrel was reduced to 20mm. In order to maximize the active area without exceeding a certain stiffness that could restrict the harvester's movement, the ideal number of windings had to be identified. Tests were made with two to six windings. Since 4 windings delivered best results in view of Voltage output to material inset, the number of windings was kept to 4 windings. The shortened spool is mechanically stiffer and allows only for an "eel"- fixation inside the pipe. Appropriate pressure fluctuations due to vortices and the spool-stiffness cause a stable oscillation of the EH. Due to promising result, only the short spool and the corresponding fixation method are used throughout the remaining experiments to determine the optimal position of the harvester behind the bluff body. In contradiction to first investigations, the harvester was examined in the "reversed-eel" mode (see Fig. 3) at different positions.

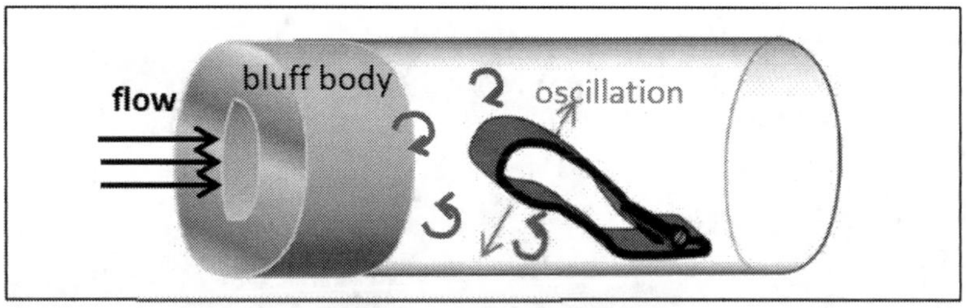

Figure 3: EH-spool in "reversed-eel" fixation mode

For different distances between the bluff body and the fixation screw, the RMS-Voltage was measured. Regarding results are shown in Table 1.

Table 1: RMS-voltage of the reversed "eel" at different positions from the bluff body

EH position	2 cm	2.5 cm	3.5 cm	5.5 cm
RMS-voltage	1.2 V	1.4 V	1.1 V	0.8 V

From Table 1, an optimal distance of about 2,5cm can be obtained. At shorter or larger distances, the strain in the loop obviously drops down leading consistently to a drop of the RMS-Voltage, too.

5.2 Impedance Matching

After determining the optimal EH position and fixation mode inside the pipe that maximizes the mechanical strain, the harvesting output is characterized by means of the delivered energy. Main goal is the energy supply for an application specific sum of consumers to enable their operation for the entire life span under certain conditions. For example, if the load consists of a sensor and a data transfer unit, then the EH must provide both components with a sufficient amount of energy. Because parameters are often measured cyclically and data transmitted cyclically too, the energy is required even during cyclical operating times. One can define therefore a duty cycle for each specific application. If a duty cycle requires for instance 200µWs, the EH must be able to generate this amount of energy in a limited time frame. In our investigations we assumed a duty cycle of about 20ms and a data measurement and transmission each 10 minutes. The energy supplied by the EH, is measured across a load. The measured value was extrapolated to a period of 10 minutes to fit the assumed application requirements. To deliver the maximum energy, which is exactly 50 % of the total energy generated, the impedance of the generator must be adapted to the electrical load. To do so, the voltage drop across a load resistance was measured depending on the value of the resistance. Power and energy were finally calculated by means of the Ohm's law.

6. OUTPUT EVALUATION

The measured voltage during the investigations indicates the efficiency of the conversion from kinetic flow energy to mechanical strain inside the piezoelectric material. Evaluating the voltage courses of the initial EH fixation modes as shown in Figure 2 reveals that the output is almost identical. Figure 4 shows a comparison of the voltage output of different spool fixation modes. Expected higher outputs using spools with mechanical pre-stressed materials could not be confirmed. Although L- and boot-shaped harvesters let expect higher mechanical strains in comparison to the "eel", the efficiency decreases. Reason for that could be the smaller space of free oscillations.

The last fixation mode investigated is the reversed "eel". This configuration showed an increase of the output voltage by a factor of five. Due to the reversed orientation, the harvester undergoes greater

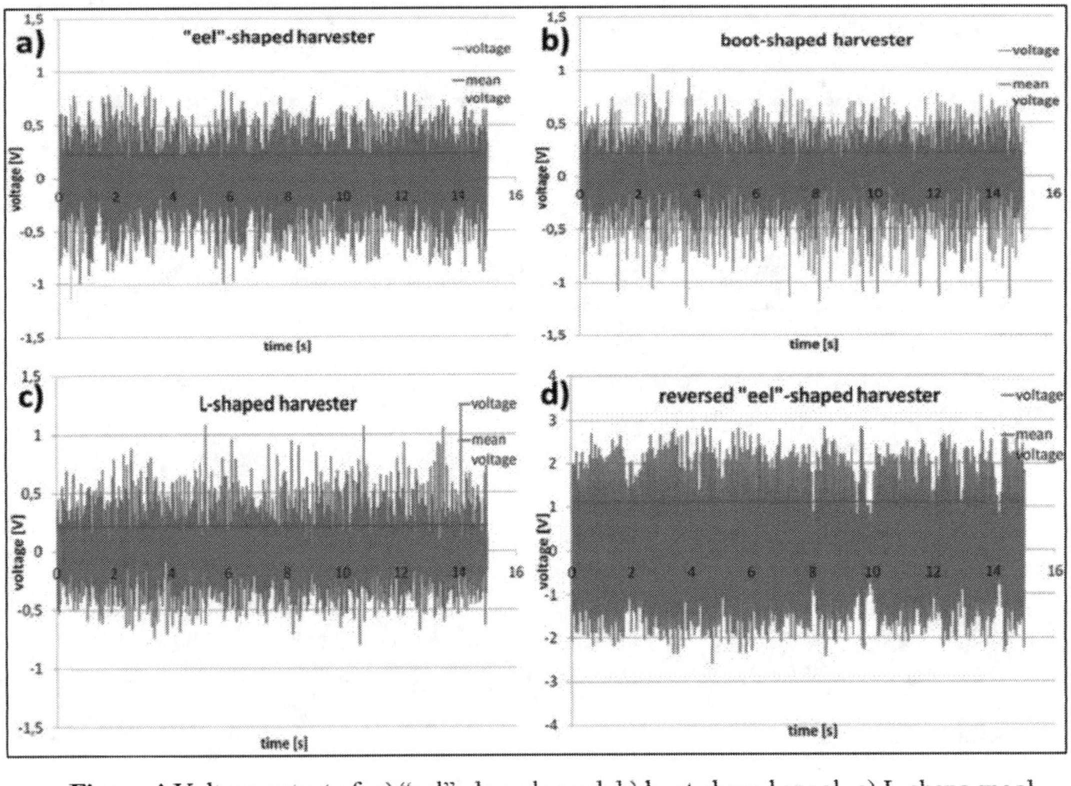

Figure 4: Voltage output of: a) "eel"-shaped spool, b) boot-shaped spool, c) L-shape spool and d) reversed "eel"-shaped spool

forces, resulting in an oscillation with larger amplitudes and therefore increased mechanical stress.
Achieving such a stable oscillation is only possible using shorter spools because of the increased stiffness that prevents the device from collapsing with the flow. The stiffness is controlled by the number of windings. While harvesters with more than four windings are too stiff to oscillate, harvesters with less than four windings are too flexible. In both cases, the Voltage decreases due to unstable oscillation or a collapse in the worst case. The voltage course of the reversed "eel" shown in Figure 4d is measured with four windings at the optimal position.
As this configuration proved to achieve the most efficient conversion of kinetic energy to mechanical stress, this harvester is still under investigation regarding its energy output. Figure 5 shows the voltage output and the resulting power at different loads for the "reversed eel"-mode. At the optimal load of 3MΩ a power output of 0.54μW could be determined. This power corresponds to energy of about 0.32mWs for a period of 10 minutes. This amount exceeds the assumed requirement of 0.2mWs. The piezoelectric PVDF-spool-EH has the capability to supply an assumed sensor-transmitter-system with a sufficient amount of energy for defined operation conditions.

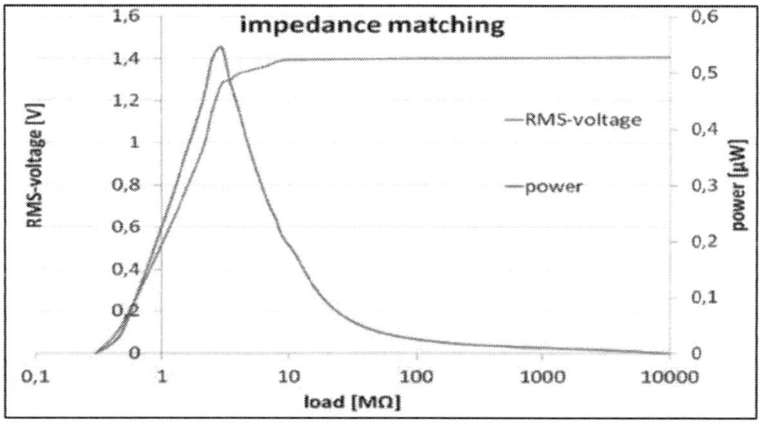

Figure 5: Impedance matching for the reversed "eel"- shaped spool

7 CONCLUSION

An improved flow EH has been developed, that tackles the disadvantages of similar systems. The piezoelectric effect of PVDF is used to convert energy from flowing media inside standard pipes into electrical energy. The flexibility and mechanical robustness of this material together with its bio-compatibility (lead-free) make it a promising alternative to common PZT used in other energy harvesters. To overcome the low piezoelectric coefficients of PVDF, the thin films are wound to a spool to increase the active area and allow for a noticeable charge generation.

The conversion efficiency of the harvester was tested in a wind channel. Tests have shown, the fixation mode of the spool inside the pipe is crucial for a high energy output. Different modes were examined and promising results were achieved using a "reversed-eel"-fixation mode, that is facing the flow and oscillating due to induced vortices. The load resistance was adapted to the internal load of the generator in order to extract the maximum power. At a flow velocity of about 7m/s and an adapted load of 3MΩ the harvester continuously delivers a power of 0.54µW. Depending on the desired application, this output power corresponds to an energy of 324µWs in 10 minutes, which is sufficient to supply a sensor-transmitter-system efficiently with energy during the duty cycle.

Presently the EH is examined in water. It is expected that the higher kinetic energy of flowing water, due to the higher density of the fluid, will lead to a higher power output at lower flow velocities for the reversed-eel-EH. A safe water leak tightness of the whole system is another part of the ongoing work. The design of a final packaging that includes all components of the system is shown in Figure 6. EH-applications like this will avoid the usual sealing problems accompanied by penetrating pipes in order to connect the components and allow for a maintenance-free deployment of sensors at inaccessible locations.

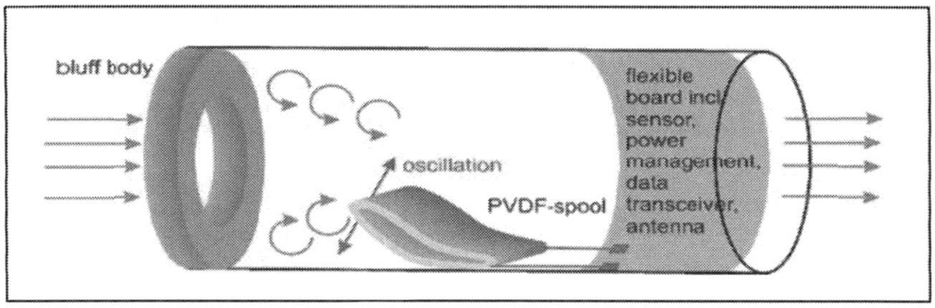

Figure 6: Concept of an autarkic in-pipe sensor-transmitter-system

ACKNOWLEDGMENTS

This research was supported by the German Federal Ministry for Economic Affairs and Energy under the AiF sponsorship with the contract-no.:KF 3101002UW3

REFERENCES

[Akay] Akaydin, H. D., et. al. (2010): "Energy harvesting from highly unsteady fluid flows using piezoelectric materials," Journal of Intelligent Material Systems and Structures, 21, 1263–1278

[All] Allen, J. and A. Smits (2001): "Energy harvesting eel," Journal of fluids and structures, 15, 629–640

[Hoff] Hoffmann, D., A. et. al. (2013): "Energy harvesting from fluid flow in water pipelines for smart meter- ing applications," in Journal of Physics: Conference Series, volume 476, IOP Publishing, volume 476, 012104

[Kwo] Kwon, S.-D. (2010): "A t-shaped piezoelectric cantilever for fluid energy harvest- ing," Applied Physics Letters, 97, 164102

[KaBe] T.J. Kazmierski, S. Beeby; Energy harvesting systems, Springer, 2011

[Liu] Liu, H., et. al. (2012): "Development of piezoelectric microcantilever flow sensor with wind-driven energy harvesting capability," Applied Physics Letters, 100, 223905

[Pob] Pobering, S. and N. Schwesinger (2003): "A New Micro Hydro Power Device; Nanotech 2003 February 23-27 2003, San Francisco, Proceedings Vol. 1, S. 372-375

[PrIn] S. Priya, D.J. Inman; Energy Harvesting Technologies, Springer, 2009

[Tay] Taylor, G. W. et.al. (2001): "The energy harvesting eel: a small subsurface ocean/river power generator," Oceanic Engineering, IEEE Journal of, 26, 539–547

[Tan] Y. K. Tan; Sustainable Energy Harvesting Technologies - Past, Present and Future, InTech, 2012

[Tor] Torah, et. al.: "Self-powered autonomous wireless sensor node using vibration energy harvesting," Measurement science and technology, 19, 125202

[Wang] Wang, D.-A. and H.-H. Ko (2010): "Piezoelectric energy harvesting from flow-induced vibration," Journal of Micromechanics and Microengineering, 20, 025019

[Ye] Ye, G. and K. Soga (2011): "Energy harvesting from water distribution systems," Journal of Energy Engineering.

ICAST2015 #044

A Self-regenerative Electromagnetic Damper for Stayed Cables

Maziar Jamshidi and CC Chang

Department of Civil and Environmental Engineering, Hong Kong University of Science and Technology, Clear Water Bay, Kowloon, Hong Kong

Abstract

In the field of health monitoring and control of large-scale civil engineering structures, maintaining a long term energy supply to various sensors and electronic components is an important factor to their successful implementation. This requirement becomes more crucial as the information and data transmission for current health monitoring and control systems relies more and more on wireless communication. As vibration energy could be prevalent in these structures, one possible option to address this issue is to harvest and reuse this type of energy whenever possible. Among various energy scavenging techniques, devices that adopt electromagnetic (EM) mechanism appear to have a great potential for such an application. In this research, the possibility of scavenging vibration energy from a stayed cable using a tunable EM damper is explored. The EM damper is equipped with a standard energy harvesting circuit, made up of a buck-boost converter which facilitates adjustability and energy harvesting. The proposed EM damper operates in a semi-active way, where the duty-cycle of the buck-boost converter is tuned in such a way that the device can generate the equivalent damping coefficient calculated from the control algorithm. Here LQG control is applied as an output feedback controller which utilizes the velocity and acceleration measurements. It is demonstrated that for the proposed electrical topology there is an applicable range that beyond that, EM damper can no longer emulate the optimal damping coefficient obtained from the control law. The maximum allowable damping as a function of circuit parameters and input voltage is given, which is important for design of such devices. Next, a numerical model for this self-regenerative EM damper – stayed cable system is established and the effectiveness of damper in suppressing the cable vibration and the harvested power is studied under different levels of external white-noise excitations. The damping performance of the self-regenerative EM damper for vibration mitigate is also compared with that of optimal viscous dampers to show its advantage. Numerical results suggest that the proposed self-regenerative EM damper is capable of suppressing the excessive vibration of cables similar to conventional semi-active techniques, if it is designed carefully. Meanwhile it is able to harvest sufficient amount of energy which can be utilized to power up sensors or other electronic components.

1. INTRODUCTION

Suppressing the structural vibrations induced by surrounding excitation in flexible structures, such as stayed-cables, has been a great concern of engineers for a long time. In this regard various controlling strategies have been proposed to tackle this problem, including employing passive dampers, active actuators, and semi-active devices. Among them active schemes outperform others since they utilize the feedback of sensors to calculate the optimal input force, which is not necessarily always dissipative. However their main drawback is a permanent demand for an external power supply, which may not be available under sever circumstances. On the other hand semi-active devices are tunable dampers that both benefit from sensors' feedback and similar to passive dampers they always perform in dissipative mode. Therefore the energy demand for such semi-active devices is comparatively much less.

For many years electromechanical devices, like Piezoelectric and Electromagnetic (EM) instruments, have been used as actuator, sensor and vibrational energy harvester, owing to the fact that they are capable to convert the mechanical energy into electrical and vice versa. They have been intensively studied as an actuator or energy harvester for micro and macro structures. The major advantage of these devices is that the electrical energy obtained from conversion of vibrational energy can be stored in a super-capacitor or rechargeable battery, instead of being dissipated as heat. It is shown, however, for large scale structures with lower frequency EM harvester has larger power [1]. Many researchers have studied the application of EM dampers and harvesters for civil structures [1-7]. Among them some researchers integrated energy harvesting into EM damper in a passive way, otherwise, EM devices are merely considered either as a controlling device with no harvesting capability or the other way around.

In this study, however, EM damper primarily is used as a damper that absorbs the vibrational kinetic energy of stayed-cable and at the same time can harvest it through a harvesting circuit. Such device has already been studied in passive configuration, in which the properties of the device are constant [5]. In contrary, here EM damper is meant to operate in a semi-active setting, where the device properties can be tuned based on the feedback of the system to counteract the external disturbance in a more effective way. In particular, this device will be installed on stayed-cables as an alternative to current passive viscous dampers. It is suggested by [8] that semi-active control is more successful in increasing the damping and, hence, reducing the vibration of the cable compared to passive optimum damper attached at same position. Therefore more energy can be harvested from the system. In this paper the components and mechanism of an EM semi-active damper/harvester is first described. The limitations dominate the operation of such technique is discussed and defined. And finally numerical computer simulation is carried out to demonstrate the device capacity in reducing the vibration of a typical cable, under white-noise disturbance, and also the average power which is stored in the harvesting circuit.

2. DEVICE MECHANISM AND CHARACTRISTICS

As mentioned, the primary function of this device is to mitigate the vibrations induced in cable due to external disturbance. In this way it can be replaced with conventional passive viscous dampers which are currently used for stayed-cables. However, in contrast to viscous dampers, EM dampers are capable of harvesting the vibrational energy in form of electrical energy in some storage device instead of dissipating it as heat. To suit this function some sort of harvesting circuit should be connected to the device. Another thing about EM dampers is that, its damping properties can readily be adjusted which makes them a good candidate for semi-active control.

To achieve and incorporate both desired function,s the architecture which is depicted schematically in Figure 1 is proposed. As can be seen, the EM transducer is attached to the harvesting/control circuit, which consists of a full bridge rectifier, a buck-boost converter that facilitates energy harvesting and also provides adjustability to the device, and a processor which is the control unit of the device. Based on the feedback of the collocated sensors and designated control algorithm it tunes the circuit parameters in order to minimize the cable responses in a semi-active manner.

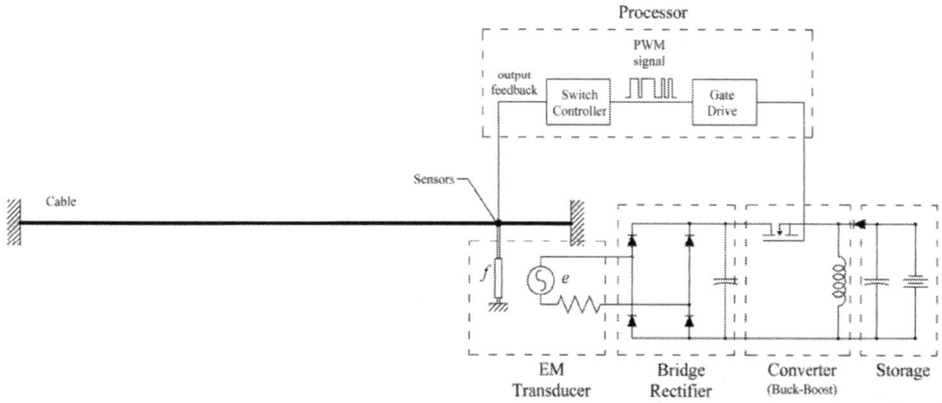

Figure 1. Architecture and components of EM damper/harvester system

To describe the mechanism of the proposed device, first by employing Faraday's law of induction and Lorentz law, it can be shown the following coupling relationship between the mechanical and electrical domains exists.

$$\begin{aligned} F_d &= ki \\ e &= k\dot{x}_d \end{aligned} \tag{1}$$

where F_d and $\dot{x}_d$ are damper's force and velocity, and e and i are AC induced voltage and the current flowing inside the circuit, and k is the coupling factor of the EM transducer. In this way the control design problem is to generating the desired current (i) for a given input voltage (e). The purpose of buck-boost converter, which is shown in Figure 2, in fact is to emulate a linear relationship between the input voltage and current, or in the other word to introduce a resistance load in series to the circuit, with the difference that the consumed power in this resistor instead of being dissipated as heat, like a regular resister, will to be stored in the rechargeable battery. Buck-boost convert is very popular for vibrational harvesting applications, and has been used by many researchers [9, 10]. Figure 2 illustrates the topography of a typical buck-boost converter. The key component of the converter is a MOSFET switch which turns on and off with a high frequency with a given duty cycle. It can be proven that for a buck boost converter the following linear relation holds between the rectifier voltage and the current flowing through the circuit [9].

$$R_{in} = \frac{2Lf_{sw}}{D^2} \tag{2}$$

where L is the inductance, f_{sw} is switching frequency and D is the switch duty cycle, which is the time duration when the switch is on over the duration of one cycle.

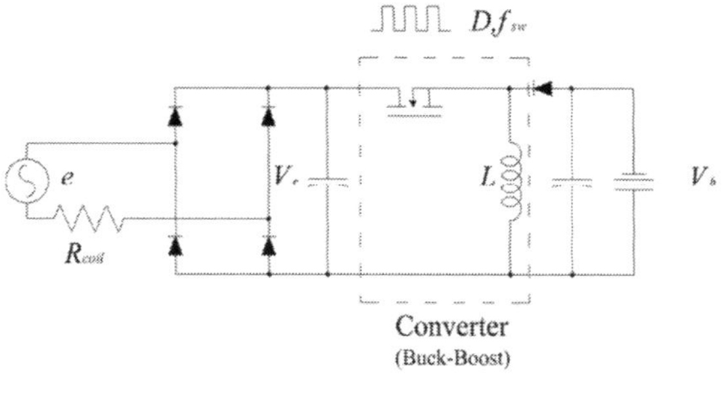

Figure 2. Circuit with Buck-boost converter

The underlying assumption beneath Eq. (2) is that the converter should always work in discontinuous conduction mode (DCM), which means, the current across the inductor L should drop to zero before staring the next switching cycle. The condition that assures such constraint is give below [9].

$$V_r < \frac{1-D}{D} V_b \tag{3}$$

where V_r and V_b are voltage across the rectifier and battery respectively. Since rechargeable battery can sustain a constant voltage, V_b is assumed to be constant. However rectifier output voltage depends on the AC voltage e which in turn depends on the velocity of the device. Generally as long as the convertor satisfies Eq. (3) it is operating in DCM. Subsequently Eq. (2) is valid and R_{in} can be adjusted to generate the desired i based on the given e. Thus the device can be commanded to produce the favorable force based on the given velocity at the point which it is attached to the cable. Considering Eq. (2) it is clear that by changing the switch duty cycle (D), value of R_{in} can be adjusted easily, therefore only by feeding a signal with variable duty cycle into the MOSFET switch the equivalent R_{in} can be emulated. Accordingly the converter can be replaced by an equivalent resistance; so the device force-velocity relationship can be expressed as below,

$$F_d = \frac{k^2}{R_{coil} + R_{in}} \dot{x}_d \tag{4}$$

where R_{coil} is coil inherent resistance. By analogy EM damper/harvester can be treated as a semi-active viscous damper with variable damping coefficient. Therefore similar to that case to find the suitable value for equivalent damping coefficient it is necessary to find the ratio of required control input force to the device velocity, provided that force and velocity are in dissipative regime. Form Eq. (4) and Eq. (2) the equivalent damping coefficient of semi-active device can be written as below,

$$C_{eq} = \frac{k^2}{R_{coil} + \frac{2Lf_{SW}}{D^2}} \tag{5}$$

By rearranging the above equation, the duty cycle, which is the only variable in the equation, can be obtained as a function of equivalent damping coefficient,

$$D^2 = \frac{2Lf_{SW}.C_{eq}}{k^2 - R_{coil}.C_{eq}} \tag{6}$$

The above equation points out some aspect of the EM damper/harvester which differentiates it from the ideal semi-active viscous damper. To describe these distinctions Eq. (6) is plotted in Figure 3(a) with respect to C_{eq}. The first thing that should be noticed is that basically duty cycle cannot be greater than one, so the maximum physically realizable value for the equivalent damping ratio will be,

$$C_{eq}^{\max} = \frac{k^2}{2Lf_{SW} + R_{coil}} \tag{7}$$

Thus larger values for R_{coil}, L and f_{sw} will squeeze the feasible region that EM damper/harvester can be applicable. Beside this intrinsic feasibility condition, DCM assumption which governs the equations imposes an upper bound on the maximum value for the equivalent damping coefficient and basically dictates the applicable range of the device. This coefficient can be calculated as below,

$$C_{eq}^{DCM} = \frac{k^2}{2Lf_{SW}(1+{}^{V_r}\!/_{V_b})^2 + R_{coil}} \tag{8}$$

In the contrary to $C_{eq}^{\max}$ which is only a function of constant parameters of the circuit, C_{eq}^{DCM} is related to the input voltage and subsequently cable velocity response. Therefore for designing a semi-active EM damper/harvester the upper bound C_{eq}^{DCM} should be set large enough in order to guarantee that the device is always within the applicable range. Beyond C_{eq}^{DCM} this device may still provide some sort of damping, but it is no longer equal to optimal damping derived from the control force. It is illustrated in Figure 3(b) as input voltage increases the maximum capable damping coefficient decreases rapidly. In fact an EM damper/harvester may jump out of the applicable range either by experiencing greater excitation than it is designed for, or by imposing more restrict control force and so a need for larger damping coefficient. Therefore for design of such devices the prediction of the input excitation and required equivalent damping plays an essential role.

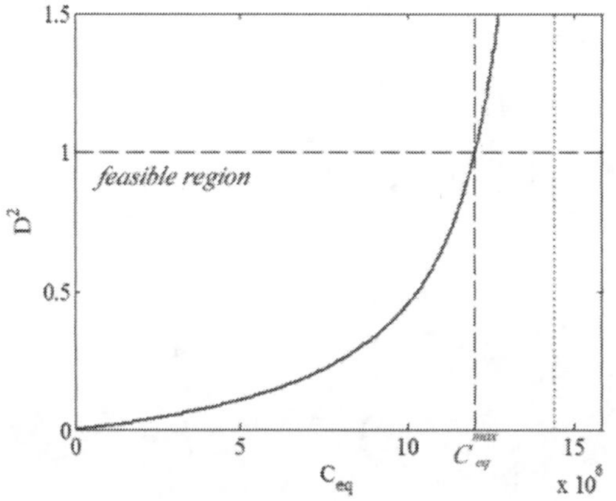

(a). duty-cycle vs. damping coeff. feasible reogion

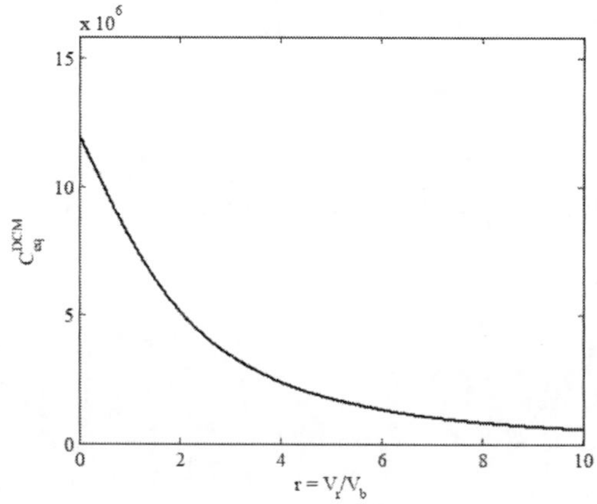

(b). Upper bound for damping coeff. for a given R_{coil}

Figure 3. EM damper applicable range

3. CABLE DYNAMICS AND VIBRATION CONTROL

As it mentioned earlier, herein EM damper/harvester is meant to mitigate the vibration of cables in cable-stayed bridge. The governing equation for vibration of a cable with no sag can be described as below,

$$m(x)\frac{\partial^2 y}{\partial t^2}+c(x)\frac{\partial y}{\partial t}-T\frac{\partial^2 y}{\partial x^2}=w(x,t)+u(t).\delta(x-a) \tag{9}$$

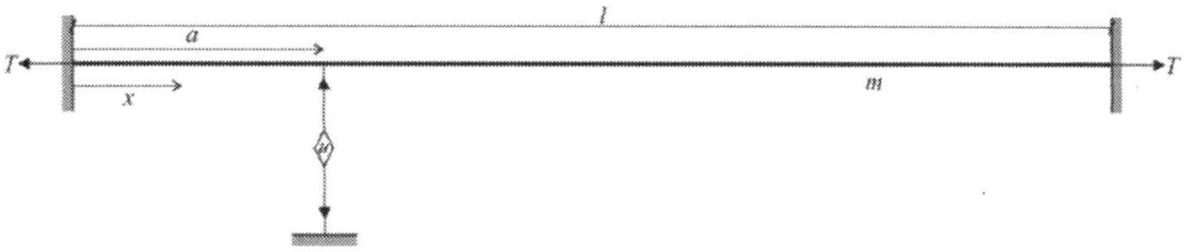

Figure 4. Cable equipped with a EM damper/harvester

In the above equation $m(x)$ and $c(x)$ are mass and damping per unit length of the cable. T is the axial constant tension force along the cable, $w(x,t)$ is distributed lateral dynamic load, u is actuator force imposed at distance a from the support . It is easier to transform Eq. (9) in modal coordinate.

$$\mathbf{M}^*\ddot{\mathbf{q}}+\mathbf{C}^*\dot{\mathbf{q}}+\mathbf{K}^*\mathbf{q}=\mathbf{w}^*+\mathbf{u}^* \tag{10}$$

where $\mathbf{q}$ is modal displacement vector and $\mathbf{M}^*$, $\mathbf{C}^*$ and $\mathbf{K}^*$ are modal mass, damping and stiffness matrices and are defined as follows,

$$\mathbf{M}^* = \frac{m\ell}{2}\mathbf{I};\quad \mathbf{C}^* = \frac{c\ell}{2}\mathbf{I};\quad \mathbf{K}^* = \frac{T\pi^2}{2\ell}diag\langle 1 \quad 4 \quad \cdots \quad n^2 \rangle \tag{11}$$

and $\mathbf{w}^*$ and $\mathbf{u}^*$ are modal external force and control input vectors which can be stated as below,

$$\mathbf{w}^* = \left[\cdots \int_0^\ell w(x,t)\varphi_n(x)dx \cdots\right]^{\mathrm{T}};\quad \mathbf{u}^* = \left[\cdots \varphi_n(a) \cdots\right]^{\mathrm{T}} u(t) \tag{12}$$

The modal shapes and natural frequencies of a vibrating string are,

$$\varphi(x) = \sin\left(\frac{n\pi}{\ell}x\right);\quad \omega_n = \frac{n\pi}{\ell}\sqrt{\frac{T}{m}} \tag{13}$$

For the sake of vibration control it would be more convenient to alternatively represent the equations in state-space form.

$$\dot{\mathbf{z}} = \mathbf{A}\mathbf{z} + \mathbf{B}u + \mathbf{H}\mathbf{w}^* \tag{14}$$

$$\mathbf{z} = \langle \dot{\mathbf{q}} \quad \mathbf{q} \rangle^{\mathrm{T}} \tag{15}$$

$$\mathbf{A} = \begin{pmatrix} -2\mathbf{\Omega}\mathbf{Z} & -\mathbf{\Omega}^2 \\ \mathbf{I} & \mathbf{0} \end{pmatrix};\quad \mathbf{B} = \begin{pmatrix} \mathbf{M}^{*-1}\boldsymbol{\varphi}^{\mathrm{T}}(a) \\ \mathbf{0} \end{pmatrix};\quad \mathbf{H} = \begin{pmatrix} \mathbf{M}^{*-1} \\ \mathbf{0} \end{pmatrix} \tag{16}$$

where $\mathbf{\Omega}$ and $\mathbf{Z}$ are diagonal matrices of natural frequencies and damping ratios. It is assumed that the cable velocity and acceleration at the cable location are measured. Thus the output matrices of the system would be,

$$\mathbf{y} = \mathbf{C}\mathbf{z} + \mathbf{D}\langle u \quad \mathbf{w}^* \rangle^{\mathrm{T}} \tag{17}$$

$$\mathbf{C} = \begin{bmatrix} \boldsymbol{\varphi}(a) & \mathbf{0} \\ -2\boldsymbol{\varphi}(a)\mathbf{\Omega}\mathbf{Z} & -\boldsymbol{\varphi}(a)\mathbf{\Omega}^2 \end{bmatrix};\quad \mathbf{D} = \begin{bmatrix} \mathbf{0} & \mathbf{0} \\ \boldsymbol{\varphi}(a)\mathbf{m}^{*-1}\boldsymbol{\varphi}(a)^{\mathrm{T}} & \boldsymbol{\varphi}(a)\mathbf{m}^{*-1} \end{bmatrix} \tag{18}$$

where $\boldsymbol{\varphi}(a)$ is vector of mode shapes at the location of damper.

$$\boldsymbol{\varphi}(a) = \left\langle \sin\left(\frac{j\pi}{\ell}a\right) \right\rangle^{\mathrm{T}} \quad j = 1,2,\ldots,n \tag{19}$$

To find the control input force (u) linear quadratic Gaussian (LQG) control is employed, as an output optimal controller that reconstructs system's states based on the velocity and acceleration measurements at the location of the damper via Kalman filter. Controller state matrix $\mathbf{Q}$ is defined in such a way that the state term in the objective function corresponds to summation of elastic and kinetic energy of the cable.

$$\mathbf{Q} = \frac{1}{2}\begin{pmatrix} \mathbf{m}^* & \mathbf{0} \\ \mathbf{0} & \mathbf{k}^* \end{pmatrix} \tag{20}$$

The input control weight, R is determined based on the first mode damping ratio of the closed-loop system, which will be explained in the next section. Additionally observer weighting matrices are selected as below,

$$\mathbf{Q}_n = E\left[\mathbf{w}^*\mathbf{w}^{*T}\right] = \alpha\mathbf{r}\mathbf{r}^T; \quad \mathbf{R}_n = E\left[\mathbf{v}\mathbf{v}^T\right] = 10^{-3}\mathbf{I} \tag{21}$$

In the above equations $\mathbf{r}$ is the external disturbance influence vector. Also since the performance of the observer is not the main concern, the covariance of measurements noise is negligible. The structural control block diagram of the system is demonstrated in Figure 5.

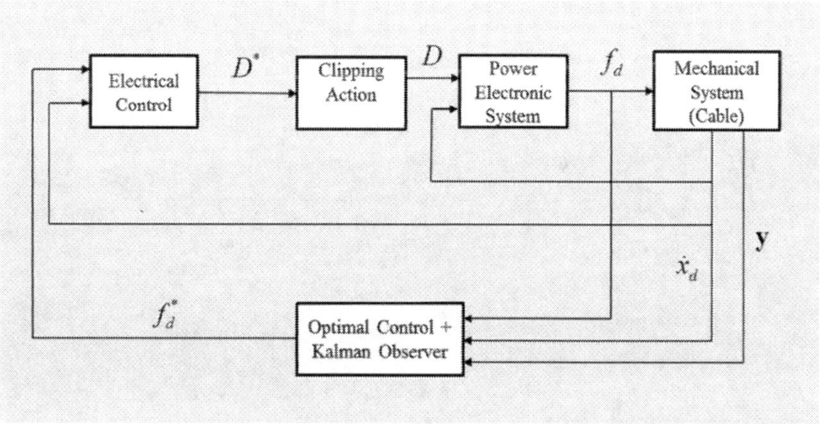

Figure 5. Structural control block diagram

4. NUMERICAL SIMULATION

To illustrate and examine the performance of the semi-active EM damper with harvesting feature, a typical cable from stonecutters cable stayed bridge in Hong Kong is taken which its properties are given in Table 1 [5].

Table 1. Properties of the cable used in numerical simulation

Length (m)	Mass per length (kg/m)	Tension Force (kN)	Fundamental natural frequency, f (Hz)	First mode damping ratio, ξ (%)	Damper location (a/L)
306.69	98.6	5529.6	2.59	0.15	0.03

Simulation is done in MATLAB Simulink (Figure 6). First twenty modes of cable are considered in analysis which are sufficient to capture the response of the cable accurately. The simulation time step is $\Delta t = 5\times10^{-5}$. To find the input control force weight in the LQG objective function, the first mode damping ratio of cable and controller closed-loop system is plotted against different values of control input force weight R, by calculating the poles of the closed-loop system (Figure 7). It can be observed that significant change in the damping ratio occurs between 10^{-8} to 10^{-3}. Larger weights have almost no effect on increasing the damping ratio of the system and for smaller values damping ratio will be saturated up to 0.18 and therefore further decreasing in the weight does not increase the damping of the system. Accordingly the value of $R = 5\times10^{-8}$ is chosen which corresponds to damping ratio of 0.174 that is shown by a red line on Figure 7.

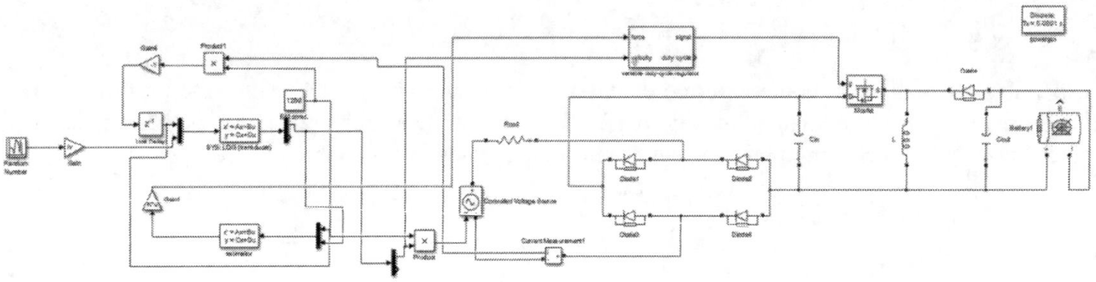

Figure 6. Cable damper model in Simulink

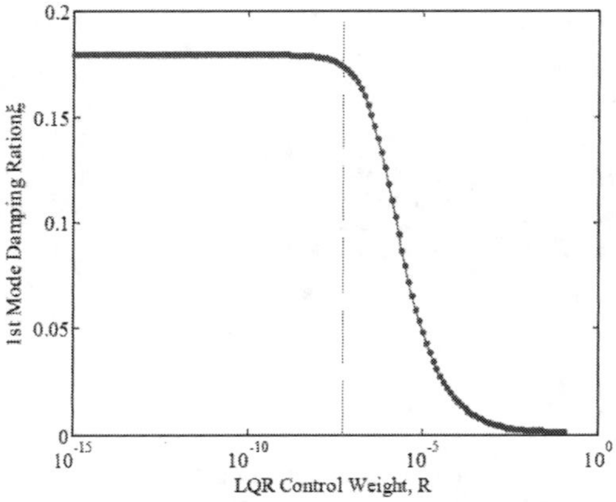

Figure 7. First mode damping ratio of the closed-loop system

The properties of harvester circuit are given in Table 2 (different components of harvester are illustrated in Figure 2). For the sake of numerical simulation runtime, switching frequency is chosen to be 1 kHz, however in real case it would be of the order of 10-20 kHz. A rechargeable NiMH battery with nominal voltage of 4.8 V is selected with 20% initial state of charge in order to maintain a constant voltage.

Table 2. Circuit properties

Inductance, L (μH)	Switching frequency, f_{sw} (Hz)	Coil Resistance, R_{coil} (Ohm)	Battery Voltage, V_b (V)	EM motor coefficient, k (N/A)
10	1000	0.1, 0.5, 1	4.8	1200

Three values for coil resistance are selected to investigate the influence of device applicable range on the controlling and harvesting performance. For the given parameters of the circuit, if the design damping coefficient is selected to be $C_{design} = 2000\ \text{kN-s/m}$, then according to Eq. (8), for different values of $R_{coil} = 0.1, 0.5, 1.0\ \Omega$, maximum voltage ratio that can be imposed to the EM transducer to emulate the required resistance are 4.6 and 2.31 for the first two resistance and for $R_{coil} = 1.0\ \Omega$, it will lay outside the applicable range for all values of input voltage and therefore it is expected not to be able to follow LQG control input force.

Based on these parameters for the cable-damper system, computer simulation is performed. The cable is excited with a white noise Gaussian disturbance with a uniform distribution along the cable. Figure 8 shows the displacement of the mid-span of the cable under excitation with RMS of 400 N/m for 100 seconds of simulation. Red line shows the response of EM damper and the blue line displays the response of semi-active ideal damper with no force limitation as a benchmark. It can be seen that the proposed EM damper/harvester design and ideal semi-active control are in accordance with each other; therefore the EM damper is able to generate the required force under this condition. Battery input power, or in other words, harvested power is also depicted in Figure 9. The spiky shape of the power time history basically is due to the impulsive nature of velocity at the very close location which the device is attached to the cable.

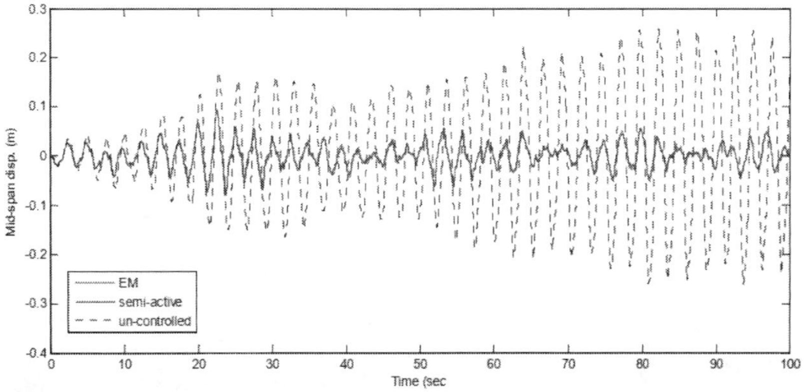

Figure 8. Comparison of cable mid-span displacement time history – *disturbance RMS = 400 N/m*

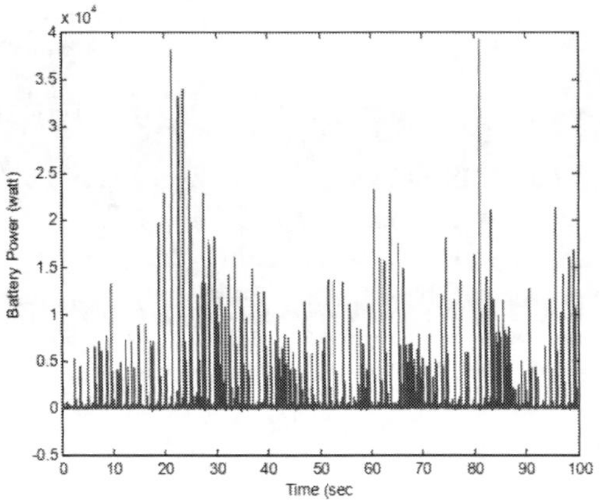

Figure 9. Harvested power – *disturbance RMS = 400 N/M*

Next the response of cable with optimal passive damper and ideal semi-active device is compared with EM damper/harvester (with three different coil resistances) for various levels of external excitations. The optimal value for the damping coefficient in the passive design is determined via the method proposed by [11]. For the cable under consideration $C_{opt} = 245$ kN-s/m and the maximum damping ratio that passive damper can add to the system is around 1.6%.

RMS and maximum displacement of the mid-point of the cable are considered as the performance criteria for different control strategies. In Figure 10(a) RMS of mid-span displacement of the cable is plotted against RMS of the external excitation. Solid lines are results of semi-active ideal damper (blue), optimal passive viscous damper (red) and cable with no supplemental damping (black). On the other hand the dotted lines are response of proposed semi-active EM damper/harvester. As it is expected for the lowest value of coil resistance ($R_{coil} = 0.1$) EM damper exhibits the similar response with the ideal semi-active damper. However for $R_{coil} = 0.5$ it starts to deviate from the ideal case which according to the earlier findings, it means that the input voltage has higher value compared to the estimated ratio, for longer portion of the excitation. However its performance is still better than the optimal passive damper and in this case EM damper/harvester outperforms the optimal passive design even it is not always in DCM condition. Lastly as coil resistance increases to 1 Ohm it approaches to the response of a cable with no damper. This means that for this level of excitation and control force the current circuit parameters should be redesigned to increase the upper bound of the equivalent damping coefficient. However it can be observed even beyond DCM mode EM damper still provides some level of damping which is an indication of its robustness. In other word even beyond the design limits it still can reduce vibration and harvest energy, and it does not completely fail under extreme unpredicted condition.

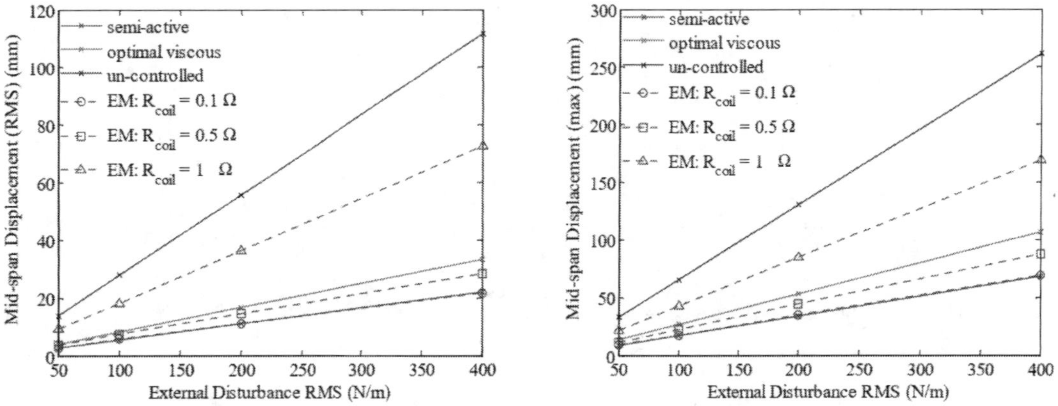

(a) (b)

Figure 10. Response comparison for different control methods

The mean stored power versus the RMS of external disturbance is plotted in Figure 11. As it can be anticipated the level of harvested power increases by the input vibration. Figure 11 also displays the power for other coil resistances. It is obvious by adding the resistance the harvested power will decrease because firstly more power will be dissipated in the circuit instead of being harvester and the next reason is the EM transducer receives less power from the cable, since it is not capable to generate higher damping. The average power for 200 seconds excitation ranges from 280 mWatt for 50 N/m to 18.5 Watt for 200 N/m, for $R_{coil} = 0.1$. The typical power consumption of a wireless smart sensor is from tens to hundreds of milliwats [12]. Therefore the results suggest that the harvested power can be employed to power up wireless sensors. However the level of wind drag force which stayed-cables experience under realistic conditions should be investigated.

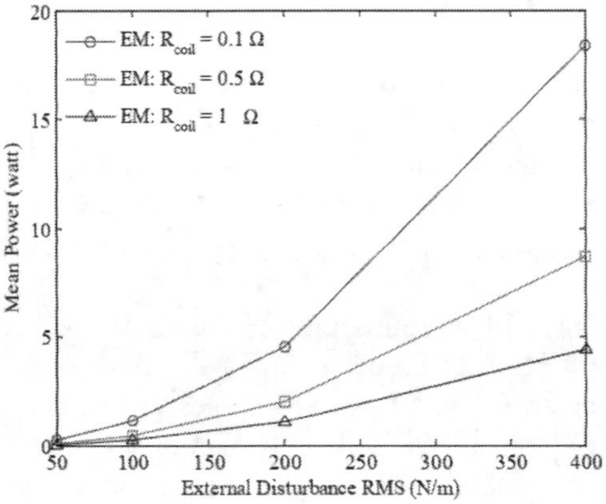

Figure 11. Mean harvested power versus excitation RMS

5. CONCLUSION

In this paper a new smart EM damper/harvester is developed by utilizing a buck-boost converter with variable duty cycle. The duty cycle is tuned according to the required force calculated by LQG control law. However it is demonstrated that the rang of damping coefficients that such device can provide has an upper bound which is defined by circuit parameters and also the level of input voltage. It is illustrated that for design purpose this maximum value should be selected large enough in such a way that the demanded damping always falls in the applicable range of the device. This in turn requires knowledge of damper velocity range to predict the possible input voltage to the EM transducer.

The main function of device is to mitigate the vibration by adding supplemental damping; therefore it can be replaced with current dampers installed on the stayed cables. At the same time through the harvesting circuit the vibrational energy can be stored instead of being dissipated, which gives an additional advantage to the typical semi-active dampers. To illustrate its performance, a numerical simulation for a typical stayed-cable under white-noise excitation is carried out. It is shown that as long as the applicable range is large enough its performance matches with the ideal semi-active case. However by increasing the coil resistance and thus compressing the applicable range device's performance deteriorates. The harvested power is also suggests that the amount of power that can be harvested is sufficient to power up wireless sensors or other electronic components. However as the next step the typical drag force that a cable experiences under the action of wind should be investigated to have a practical range of external disturbance. Moreover to have more realistic results electrical power loss in different electrical components namely, capacitors, inductor, switch, diodes, etc., can be taken into account. In general in this study is shown that smart EM damper/harvester is a viable replacement for passive dampers.

6. REFERENCES

1. Cassidy, I.L., J.T. Scruggs, and S. Behrens. *Design of electromagnetic energy harvesters for large-scale structural vibration applications*. 2011.

2. Palomera-Arias, R., J. Connor, and J. Ochsendorf, *Feasibility Study of Passive Electromagnetic Damping Systems*. Journal of Structural Engineering, 2008. **134**(1): p. 164-170.

3. Zhu, S., W.-a. Shen, and Y.-l. Xu, *Linear electromagnetic devices for vibration damping and energy harvesting: Modeling and testing*. Engineering Structures, 2012. **34**: p. 198-212.

4. Zuo, L. and W. Cui, *Dual-Functional Energy-Harvesting and Vibration Control: Electromagnetic Resonant Shunt Series Tuned Mass Dampers*. Journal of Vibration and Acoustics, 2013. **135**(5): p. 051018-051018.

5. Shen, W. and S. Zhu, *Harvesting energy via electromagnetic damper: Application to bridge stay cables*. Journal of Intelligent Material Systems and Structures, 2015. **26**(1): p. 3-19.

6. Scruggs, J.T., I.L. Cassidy, and S. Behrens, *Multi-objective optimal control of vibratory energy harvesting systems*. Journal of Intelligent Material Systems and Structures, 2012. **23**(18): p. 2077-2093.

7. Zhang, C. and J. Ou, *Control Structure Interaction of Electromagnetic Mass Damper System for Structural Vibration Control*. Journal of Engineering Mechanics, 2008. **134**(5): p. 428-437.

8. Johnson, E., et al., *Semiactive Damping of Stay Cables*. Journal of Engineering Mechanics, 2007. **133**(1): p. 1-11.

9. Lefeuvre, E., et al., *Buck-Boost Converter for Sensorless Power Optimization of Piezoelectric Energy Harvester.* Power Electronics, IEEE Transactions on, 2007. **22**(5): p. 2018-2025.

10. Ottman, G.K., H.F. Hofmann, and G.A. Lesieutre, *Optimized piezoelectric energy harvesting circuit using step-down converter in discontinuous conduction mode.* Power Electronics, IEEE Transactions on, 2003. **18**(2): p. 696-703.

11. Pacheco, B., Y. Fujino, and A. Sulekh, *Estimation Curve for Modal Damping in Stay Cables with Viscous Damper.* Journal of Structural Engineering, 1993. **119**(6): p. 1961-1979.

12. Jerome, P.L., et al., *Performance monitoring of the Geumdang Bridge using a dense network of high-resolution wireless sensors.* Smart Materials and Structures, 2006. **15**(6): p. 1561.

ICAST2015 #037

Developing a Piezo-Electret Composite for Smart Structure Applications Using an Electrospinning Process

Hsin-Jung Chu[1], Po-Cheng Lai[1], Shih-Lun Tai[1], and Chih-Kung Lee[1,2,3,*]

[1] Institute of Applied Mechanics, National Taiwan University, Taipei, Taiwan
[2] Engineering Science & Ocean Engineering, National Taiwan University, Taipei, Taiwan
[3] Research Center for Applied Sciences, Academic Sinica, Nankang, Taipei, Taiwan

Abstract

An electret material is a viable smart material as it possesses a quasi-permanent electrical charge after charging. However, its drawback is that it loses its electrical charge at high humidity. In this paper, we describe a newly developed piezo-electret composite fabricated using an electrospinning process that allows us to circumvent the humidity limitation. P(VDF-TrFE) (75/25) was chosen as the base material after the electrospinning process. We coated cyclic olefin copolymer (COC) onto the base material to form a P(VDF-TrFE)/COC composite nanofibrous membrane. Our results show that the piezo-electret nanofibrous membrane can enhance charge storage stability at high humidity after the COC coating.

1. INTRODUCTION

With the development of flexible devices, flexible smart materials for applications such as for flexible loudspeakers, wearable medical devices, smart phones and flexible audio displays,[1-4] have been extensively investigated. With such a wide variety of potential applications, the field of flexible smart materials and its applications has generated much interest. The traditional type electret actuators have a typical structure where an electret-based diaphragm is placed beside the perforated electrodes and separated by a set of spacers[5]. An electret generates an internal and an external electric field in the configuration. This is similar to electrets and dielectric layers used in an electrostatic loudspeaker. The difference between an electret loudspeaker and an electrostatic loudspeaker is that electrets do not need to have an externally applied high voltage to keep its surface voltage. It is thus very clear that surface potential affects an electret loudspeaker directly. Materials, based on an electret, have been previously used as viable smart materials since electrets possess a quasi-permanent electrical charge after charging[6]. Many different polymers have been investigated for their suitability as a base electret material. Those polymers considered include polymers such as polypropylene (PP), polyethyleneterephthalate (PET), cyclic olefin copolymer (COC), polytetrafluoroethylene (PTFE) and fluoroethylenepropylene (FEP)[6-8]. However, a typical electret faces a drawback in that it loses its electrical charge at high humidity. This humidity limitation has thus restricted the use of electrets for applications which are not under high temperature or high humidity conditions. From recent research work, we have found that a PTFE/COC composite can enhance charge storage stability at high humidity[5] while possessing a better thermal resistance. However, PTFE has some disadvantages when other protection layers are bonded to it. Even by coating the electrode material with aluminum, for example, there is no improvement. However, in our research, we have identified that by adopting an

[*] cklee@ntumems.net

ethylene-ethyl acrylate copolymer (EEA) as the interfacial layer, it can promote the further evaporation of an aluminum thin film electrode. Based on our above-mentioned methodologies, we propose a new piezo-electret composite fabricated using an electrospinning process which can circumvent the humidity restriction. Both cyclic olefin copolymer (COC) and P(VDF-TrFE) were chosen as the base material. We know that COC is an amorphous, non-polar material with high dimensional stability which can improve a composite's elastic modulus. In addition, it is characterized by stable charge storage at high humidity. P(VDF-TrFE) is known for its good chemical stability, low moisture absorption, high sensitivity and good polarity. Additionally, P(VDF-TrFE) is a piezoelectric material which possesses strong piezoelectric properties and an easy fabrication process which makes it a natural as a smart material.

The electrospinning method we adopted was developed more than 70 years ago. The electrospinning process was initiated as part of the fabrication process due to the following reasons: (1) it is light-weight, (2) is easy to process, (3) is of low cost, and (4) allows us to easily fine-tune the processing parameters[9]. Using an electrospinning process, we can easily control the diameter and thickness of the final output. It provides the platform to create various fiber morphologies in order to optimize the final composite polymer. Furthermore, it enables us to create a single layer P(VDF-TrFE)/COC composite nanofibrous membrane which possesses proper porosity and flexibility to meet for various application requirements.

Porosity in ferroelectric polymers has been found to be useful for several special applications. For example, one previous research work incorporated a piezoelectric material into clothing to create the power source for a wearable electronic device[10,11]. An electro-spun piezoelectric nanofibrous membrane was also used as a smart air filter due to its excellent air and water vapor permeability and good flexibility when compared against other solid piezoelectric films. After adopting an electrospinning process, we used a multiple pin array corona charging process to polarize the P(VDF-TrFE)/COC composite nanofibrous membrane. The corona charging occurred when a sufficiently high voltage was exerted between the sharp pin and the flat plate electrode which then generated an electric field and an ion flow without a major breakdown[12]. This process was found to improve the ferroelectric properties of P(VDF-TrFE), which in turn also improved the actuator performance.

In this study, we took a P(VDF-TrFE) nanofibrous membrane as the base material and then coated with COC to form a P(VDF-TrFE)/COC composite nanofibrous membrane. The surface of the P(VDF-TrFE) and P(VDF-TrFE)/COC membrane was observed using a scanning electron microscope (SEM). The surface potential was measured using a non-contact electrostatic voltmeter. This composite nanofibrous membrane has the potential to be used for applications such as electret loudspeakers, wearable devices, etc.

2. EXPERIMENTAL SET-UP

2.1 Materials and Preparation

The P(VDF/TrFE) (75/25) copolymer powder was purchased from Kureha. The copolymer (COC) was purchased from TOPAS Advanced Polymers. The P(VDF-TrFE) (75/25) powder with 20 wt% concentration was dissolved in a dimethylacetamide/methyl ethyl ketone (DMaC/MEK) solvent. The COC powder with 7.5 wt% concentration was dissolved in a toluene solvent.

The P(VDF-TrFE) nanofibrous membrane was fabricated using an Electrospinning process as shown in Figure 1. The apparatus contained a flat plate collector, a syringe pump, a high voltage power supply, a syringe pump, and a syringe with a 23G needle (0.33 mm inner diameter and 0.63 outer diameter). The gap between the needle and the plate collector was 10 cm. The syringe pump rate was 0.6 ml/h and a high voltage of 15 kV was applied to the needle[13,14]. At the above conditions, an uniform and smooth P(VDF-TrFE) fiber for the membrane construction was developed. A spin-coater (2000 rpm @ 40s) was used to coat the COC to the P(VDF-TrFE) to form a P(VDF-TrFE)/COC composite nanofibrous membrane. We prepared two types of samples: Type A was electro-spun,

Type B was electro-spun and annealed at a temperature of 135 deg C for two hours and at 70 deg C for one hour to enhance crystallization[15]. Then, Type A and Type B underwent multiple pin corona charging (Figure 2) in order to polarize the polymer uniformly.

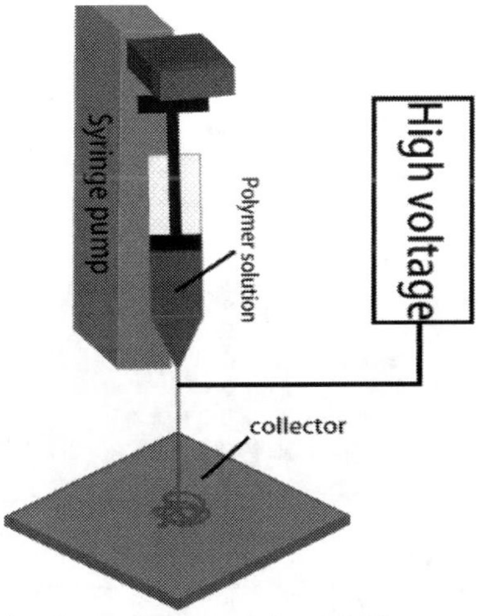

Figure 1. Schematic diagram of the adopted Electrospinning set-up

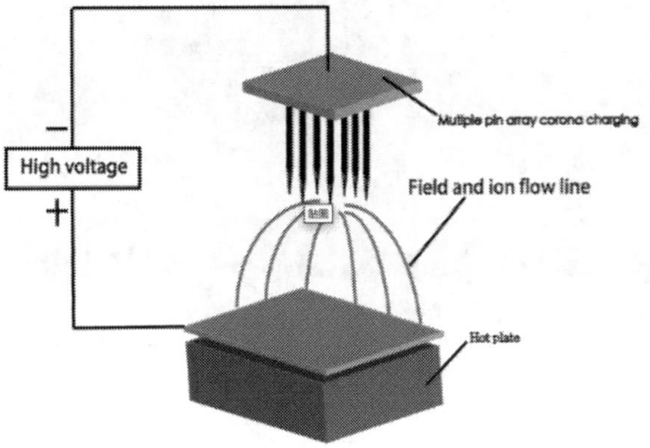

Figure 2. Schematic diagram of the constructed corona charging set-up

2.2 Measurements

After corona charging, the samples were placed in a damp-proof cabinet (30 deg C @ 60% RH). The surface potential of the samples were then measured using a non-contact electrostatic voltmeter. The surface of Type A (the electro-spun type) and Type B (the electro-spun and annealed type) were measured and examined using a scanning electron microscope (SEM) at 10 kV. The composite membrane was then metallized with a gold layer using an ion beam sputtering device for easier SEM observation.

3. RESULTS AND DISCUSSION

For investigating the surface between the P(VDF-TrFE) and P(VDF-TrFE)/COC composite, we used the scanning electron microscope to measure the sample surface. In Figure 3(a), we found that the P(VDF-TrFE) membrane is a non-woven structure with many pores which resulted in high air and water vapor permeability. We found that in our newly fabricated electro-spun sample, the fiber surface was smooth and free from bumps and voids (Figure 3(b)). From Figures 4(a) and 4(b) it appears that the COC penetrated the cavities of the electro-spun P(VDF-TrFE) and filled the spaces and gaps between the fibers. Results show that the final porosity of the P(VDF-TrFE)/COC composite was less than that of the P(VDF-TrFE) membrane.

(a) (b)

Figure 3. SEM image of the P(VDF-TrFE) membrane

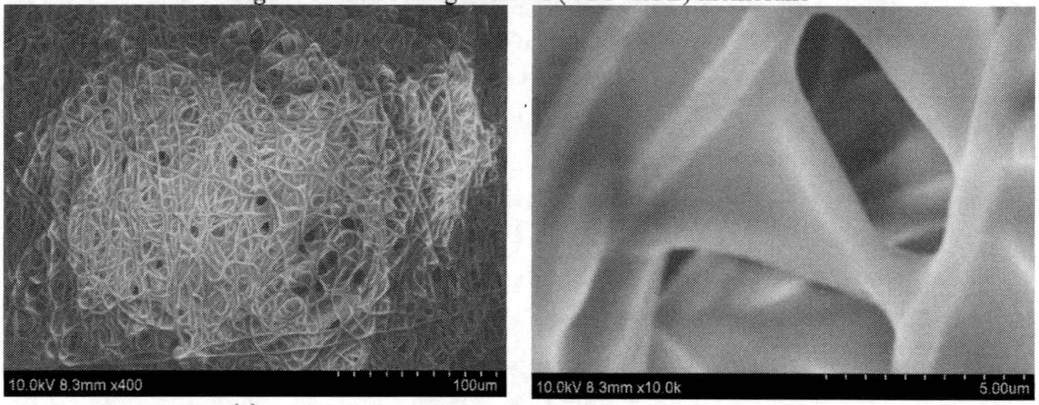

(a) (b)

Figure 4. SEM image of the P(VDF-TrFE)/COC composite membrane

Different processes and structures were compared in Figures 5, 6, and 7. Figure 5 shows that the P(VDF-TrFE) fabricated using an annealing process possessed the best surface potential. For the Type A sample (the electro-spun type), it was eventually stable at around 190V. For the Type B sample (the electro-spun and annealed type), it was stable at 20 V (e.g. 30 deg C @ 60% RH). These different phenomena can be attributed to the charge storage ability associated with good crystallization in the P(VDF-TrFE).

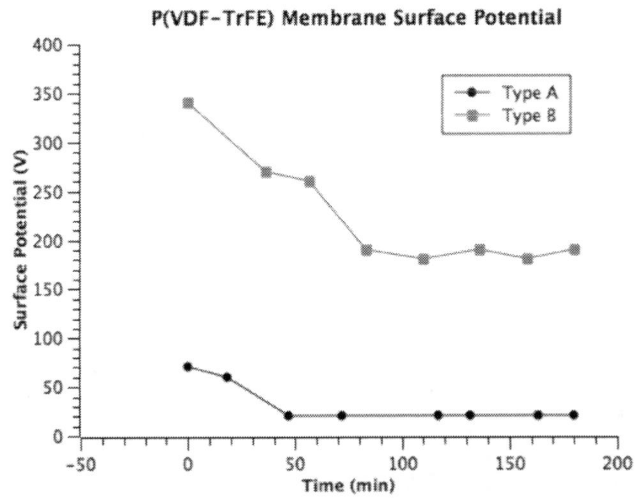

Figure 5. P(VDF-TrFE) membrane surface potential

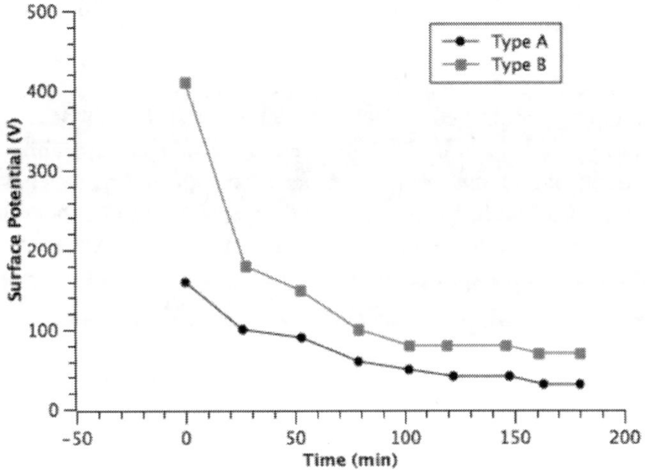

Figure 6. P(VDF-TrFE)/COC* membrane surface potential

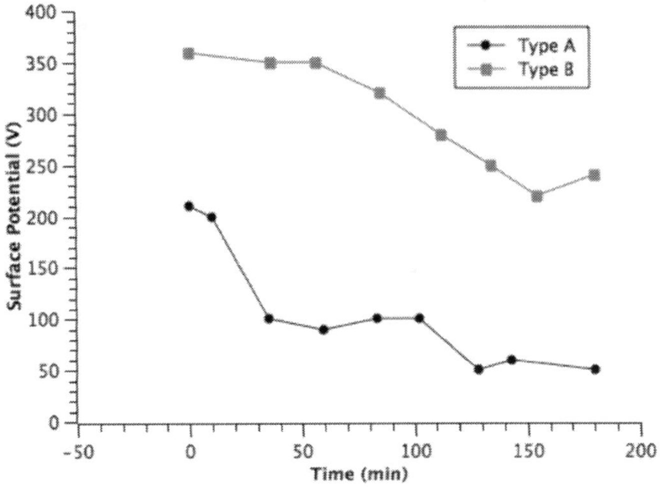

Figure 7. P(VDF-TrFE)/COC** membrane surface potential

In Figures 6 and 7, COC* represents the COC coating before annealing while COC** represents the COC coating after annealing and corona charging. In Figure 6, we can see that the Type

A sample performed better only when compared to the PVDF structure alone without any treatment. Nevertheless, the Type B sample performance was worse than the P(VDF-TrFE) structure alone without any treatment. We theorize that the COC has characteristics of an electret, which can improve the charge storage ability and thus lead to a better surface potential after completion of fabrication. But, as COC possesses a large elastic modulus, it can limit the deformation of the P(VDF-TrFE) and result in an overall lower performance for smart sensor development. To circumvent this limitation, we have developed a new technique which can be adopted. In our work, we coat COC onto the P(VDF-TrFE) to form a P(VDF-TrFE)/COC membrane and which undergoes annealing and corona charging. The result is a new P(VDF-TrFE)/COC** membrane. This material is stable at 60 V for material processed under a Type A condition and 240 V for a Type B condition. We show that under the processing technique, we can obtain a P(VDF-TrFE)/COC material with improved performance.

4. Conclusions

A composite material processing technique for a piezo-electret composite is presented in this paper. We confirm that a charge storage ability depends on good crystallization of P(VDF-TrFE) and COC which can reduce the decay of a surface potential. In other words, the use of COC can improve the humidity resistance of a composite. However, COC can affect the crystallization and polarization of P(VDF-TrFE), which may lead to a lower performance for smart material applications. To circumvent this limitation, an improved processing technique has been developed. In our revised process, we coat the P(VDF-TrFE) with COC and then undertake an annealing and corona charging procedure. Results show that in our newly developed composite material, we can obtain better crystallization and have improved charge storage stability even at high temperature and high humidity (e.g. 30 deg C @ 60% RH). Therefore, the new composite nanofibrous material can result in a new improved piezo-electret material for application in a wide variety of smart material applications.

ACKNOWLEDGMENTS

This study was sponsored by Taiwan's Ministry of Science & Technology under Grant No. NSC 102-2221-E-002-188-MY3.

REFERENCES

1. Xiao, L., Chen, Z., Feng, C., Liu, L., Bai, Z.-Q., Wang, Y., Qian, L., Zhang, Y., Li, Q., Jiang, K., and Fan, S., "Flexible, stretchable, transparent carbon nanotube thin film loudspeakers," *Nano letters*, Vol. 8, No. 12, 2008, pp. 4539-4545.
2. Sugimoto, T., Ono, K., Ando, A., Kurozumi, K., Hara, A., Morita, Y., and Miura, A., "PVDF-driven flexible and transparent loudspeaker," *Applied Acoustics*, Vol. 70, No. 8, 2009, pp. 1021-1028.
3. Heydt, R., Kornbluh, R., Eckerle, J., and Pelrine, R., "Sound radiation properties of dielectric elastomer electroactive polymer loudspeakers," in *Conference on SPIE Smart Structures and Materials: Electroactive Polymer Actuators and Devices,* Vol. 6168, 2006
4. Medley, A. P., Billson, D. R., Hutchins, D. A., and Davis, L. A. J., "Properties of an electrostatic transducer," *The Journal of the Acoustical Society of America*, Vol. 120, No. 5, 2006, pp. 2658-2667.
5. Ko, W.-C., Tseng, C.-K, Wu, W.-J., and Lee. C.-K., "Charge storage and mechanical properties of porous PTFE and composite PTFE/COC electrets," *e-Polymers*, Vol. 10, No. 1, 2010, pp. 1-10

6. Sessler, G. M., and Broadhurst, M. G., *Electrets*, Vol. 33, *Springer Verlag*, Berlin, Germany, 1980.
7. Yang, G. M., Sessler, G. M., and Hatke, W., "Electret properties of cycloolefin copolymer," in *Conference on electrical insulation and dielectric phenomena*, Vol.2, 1997, pp. 467-470.
8. Sessler, G. M., "Electrets: recent developments," *Journal of Electrostatics*, Vol.51-52, 2001, pp. 137-145.
9. Huang, Z.-M., Zhang, Y.-Z., Kotaki, M., and Ramakrishna, S., "A review on polymer nanofibers by electrospinning and their applications in nanocomposites," *Composites science and technology,* Vol. 63, No.15, 2003, pp. 2223-2253.
10. Patel, I., Siores, E., and Shah, T., "Utilisation of smart polymers and ceramic based piezoelectric materials for scavenging wasted energy," *Sensors and Actuator A-physical*, Vol. 159, No.2, 2010, pp. 213-218.
11. Swallow, L. M., Luo, J. K., Siores, E., Patel, I., and Dodds, D. "A piezoelectric fibre composite based energy harvesting device for potential wearable applications," *Smart Materials and Structures*, Vol. 7, No. 2, 2008, pp. 1-7.
12. Giacometti, J. A., Fedosov, S., and Costa, M. M., "Corona charging of polymers: recent advances on constant current charging," *Brazilian Journal of Physics*, Vol. 29, No. 2, 1999, pp. 269-279.
13. Noyori, M., Neo, Y., and Mimura, H., "Single-crystalline poly (vinylidene fluoride–trifluoroethylene) nanofiber webs fabricated by electrospinning," *Japanese Journal of Applied Physics*, Vol. 54, No. 2, 2015, pp. 1-4.
14. He, F., Sarkar, M., Lau, S., Fan, J., and Chan, L. H., "Preparation and characterization of porous poly (vinylidene fluoride-trifluoroethylene) copolymer membranes via electrospinning and further hot pressing," *Polymer Testing*, Vol. 30, No. 4, 2011, pp. 436-441.
15. Chang, W.-C., Wang, A.-B., Lee, C.-K., Ko, W.-C., Chen, H.-L., and Lin, C.-T. "Development of a photoconductive piezoelectronic material from composite of P (VDF-TrFE) and TiOPc," *In Applications of Ferroelectrics held jointly with 2012 European Conference on the Applications of Polar Dielectrics and 2012 International Symp Piezoresponse Force Microscopy and Nanoscale Phenomena in Polar Materials (ISAF/ECAPD/PFM),* 2012, pp. 1-4.

ICAST2015#009

Developing Self-sensing Sensors and Actuators by Using a Novel Piezo-Electret Material

Po-Cheng Lai[1] and Chih-Kung Lee[1,2,3,*]

[1]Institute of Applied Mechanics, National Taiwan University, Taipei, Taiwan
[2] Department of Engineering Science and Ocean Engineering, National Taiwan University, Taipei, Taiwan
[3]Research Center for Applied Sciences, Academic Sinica, Nankang, Taipei, Taiwan

Abstract

To form the simplest type of smart structure, typically a discrete point sensor and an actuator pair are bonded together. However the drawback of this type of structure is that the cross coupling signal initiated from the actuator to the sensor can impose a serious system limitation such as a reduction in bandwidth, etc. To reduce the cross coupling, a piezoelectric sensor and actuator made of two layers of PVF_2 with a polarization profile facing each other and containing a ground electrode facing the outside has been developed to bypass the limitations. In past research, a self-sensing sensor and actuator pair was previously developed with an attempt to minimize the system complexity of smart structures by having a single piezoelectric lamina provide both the sensing and actuating functions. Despite such efforts, an internal coupling prevents a clean and successful separation of the piezoelectric sensor signal riding on top of the corresponding actuating signal which is typically 6 orders of a magnitude larger. This can limit the practical application of traditional self-sensing sensors/actuators.

In this paper, a new configuration of a self-sensing sensor/actuator based on a newly developed piezo-electret material was designed, constructed, and tested. The methodology for developing this new hybrid material will be presented first. We also show the advantages of having a material that is both piezoelectric and an electret. More specifically, the electret component, driven by applying high voltage through a non-contacting electrode, was used to serve as both the actuator and the piezoelectric function of the sensor. It served as the interface of the vibration-induced signal through an electrode evaporated on top of the piezo-electret layer. The piezoelectric constant d_{33} of this hybrid material was found to be 45 pC/N, which is actually larger than the piezoelectric constant of pure PVF_2. In addition, the hysteresis loop (also called P-E curve) of this piezo-electret material was measured. At 2500 V and under a 0.2 Hz sampling rate test condition using a thin film analyzer of an electro-ceramic material, the remanent polarization (P_r) was found to be around 3 $\mu C/cm^2$ and the coercive electric field (E_c) to be around 5 MV/m.

By using this novel piezo-electret material and adopting the concept of using a modal sensor/actuator pair, we examined the modal effect on a one-dimensional cantilever plate by using an experimental set-up which consisted of a dynamic signal analyzer, a high speed bipolar amplifier, and a current amplifier. The transfer function and the phase delay of this self-sensing system are also shown. Potential applications of this newly developed technology include critical activedamping vibration control, structural health monitoring, and active noise control.

cklee@ntumems.net

1. INTRODUCTION

In this paper, a novel piezo-electret material was developed. Electrets were put forth in previous arguments by G. M. Sessler and *et al.* It is know that an electret material possesses dielectric properties and also contains a quasi-permanent electrical charge [1]. With these characteristics, an electret material can be driven by controlling an externally applied electric field. By integrating poly(vinylidenefluoride-co-trifluoroethylene) (i.e. P(VDF-TrFE), also called VF_2/VF_3) and expanding polytetrafluoroethylene (i.e. e-PTFE) film, self-sensing sensors and actuators were developed and are presented herein.

2. ELECTRETS

An electret is a piece of dielectric material exhibiting a quasi-permanent electrical charge. For the charges stored in electrets, four common types were identified (see Fig. 1). [1]

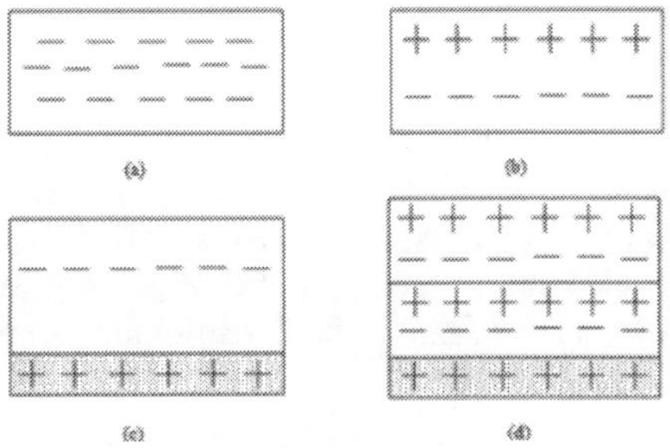

Figure 1. Four common charges storage types of electrets. (Note: the shaded area represents the electrode).
(a) Only negative (or positive) charges,
(b) Charges,
(c) Negative (or positive) charges with single electrode, and
(d) Charges with single electrode.

From a material viewpoint, many materials can be made into electrets. For example, polytetrafluoroethene (PTFE), polypropylene (PP), polystyrene (PS) are such possibilities, although in general PTFE and PP have been commonly used. However, test results show that PTFE exerts better thermal stability under high temperaturesand after corona charging than that of PP film.

3. FABRICATION AND CHARACTERISTICS

P(VDF-TrFE) 70/30 powders (Piezotech Co., Saint Louis, France) was dissolved in dimethyl sulfoxide (DMSO; ECHO Chemical, Co., Ltd.). The mass concentration used was 15wt%. The solution was stirred for at least 9 hours. The temperature of the fabrication was set at 80 deg Cto increase the solubility of the P(VDF-TrFE) 70/30 solution.

A 2mm thick Teflon substrate was made in a size of 10cm by 10cm. The e-PTFE (EF-Materials Industries Inc.) was tiled on the Teflon substrate. The average porosity of the e-PTFE was 83% containing a 0.2 to 0.3μmpore size. The thickness of the e-PTFE was 25μm. The sample was then soaked in acetone (ECHO Chemical, Co., Ltd) and the 15wt% P(VDF-TrFE) solution was then spin-coated onto the infiltrated e-PTFE film. The spin-coating processing parameters were set at 500rpm for 5 seconds and then at 1500rpm for 15secondsusing a spin coater (SC-80R, YOTEC Instruments Co., Ltd). The sample was then baked at 70 deg C for one hour to remove the organic solvent. To transform the P(VDF-TrFE) to a β phase and to achieve better piezoelectricity, the temperature of the sample was increased to 135 deg C for three hours, and then decreased to 65 deg C for 2 hours. After the sample was cooled to room temperature, the above steps to change the P(VDF-TrFE) to a β phase was repeated again.

The corona charging process was performed by externally applying -20kV using a high voltage power supply (Matsusada Precision, Inc.). One needle was made with brass which was used to pole and to provide the surface charge needed to be stored within the electret. The sample was put on a grounded flat metal hot plate with temperature set at 90 deg C. After 2 hours, the sample was gradually cooled to room temperature. The purpose of this step was to prepare the sample for the poling process to be performed later.

Cyclic Olefin Copolymer 6013-S04 (Topas® COC 6013-S04; Polyplastics Co., Ltd.) was used to protect the charges from escaping out of the sample. Out of the various characteristics and models of COC, we chose COC 6013-S04 due to its good ability for charge storage, its good mechanical properties and for its high glass transition temperature. The COC pellets were dissolved in toluene (ECHO Chemical, Co., Ltd.) with a 5wt% and at 80 deg Cfor at least 9 hours. Then, the COC solution was spin-coated onto the sample by setting the spin-coating processing condition to 500rpm for 5 seconds and 1500rpm for 15 seconds. To completely remove the solvent, the sample was then heated at 80 deg C for 30minutes.

Poly(ethylene-co-ethyl acrylate) (EEA; Sigma-Aldrich Co.) particles were dissolved in toluene (ECHO Chemical, Co., Ltd.) with processing parameters at 0.5wt% and at 80 deg C overnight. The solution was heated at 80 deg C as the solubility increased at this temperature. The EEA solution was then spin-coated onto the sample at 500rpm for 5 seconds and at 1500rpm for 15 seconds. The EEA layer was adopted to achieve good adhesion between the COC and the metal electrodes. In fact, this layer served the same function as that of a TetraPak carton which facilitated the bonding of the metal layer ontothe plastic. To release the solvent again, the sample was heated at 80 deg C for 30minutes. The other side of the sample was also spin-coated with the COC layer and the EEA layer for the same reason. As the final step, the corona charging method was applied onto the sample for 2 hours again for poling and charging.

(a) (b)

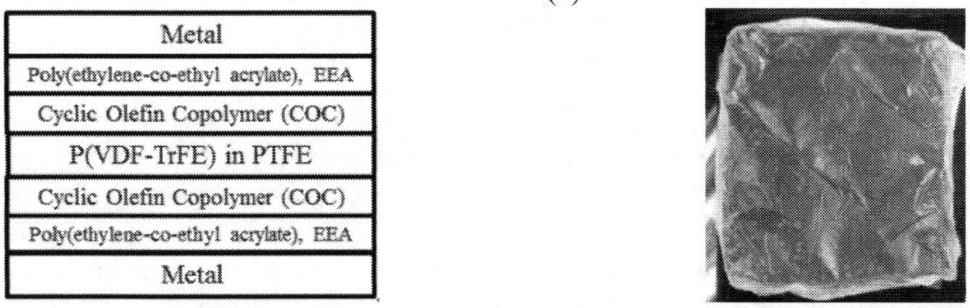

Figure 2.(a) Schematic layout of the laminate and
(b) Real sample of the piezo-electret material

A Ferroelectric Measurement System (TF Analyzer 2000, AixACCT Co.) was used to measure the P-E curve (also called hysteresis loop) of the piezo-electret material. The sample was put into silicon oil to prevent an arcing phenomenon. Under the same previous test conditions (e.g. 2500 V externally applied voltage and a 0.2 Hz sampling rate), the remanent polarization (Pr) was identified to be around 3μC/cm^2 and the coercive electric field (Ec) was about 5 MV/m. The P-E curve of the piezo-electret material is shown in Figure 3.

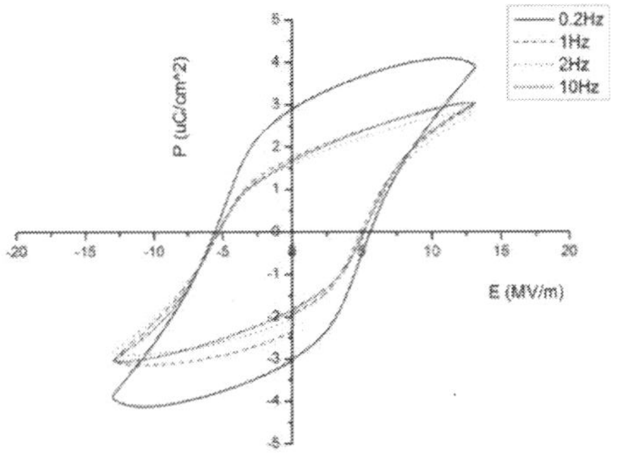

Figure 3.P-E curve (hysteresis loop) of the piezo-electret material under 2500V and at different sampling rates (0.2Hz, 1Hz, 2Hz and 10Hz)

Table 1. Measured d_{33}value of different electret or piezoelectric material

Material	*d_{33} (unit: p C/N)*
Piezo- Electret Material	45
Pure PVDF	17
Pure P(VDF-TrFE) 70/30	22
EMFi	110

The measurement was done in an instrument called a "d_{33} meter". The calibration was operated using the following steps. The meter was turn on for 15 minutes to warm up the instrument. The read out force of the meter was $250*10^{-3}$N. Calibration was then executed using a standard sample which was made from the d_{33} meter manufacturer. The value of standard sample was recorded and then the sample was turned to the opposite direction. An average the two values were taken and then the zero point value was adjusted to make the absolute d_{33} values of the standard sample to be the same in the normal and in the opposite direction. Then, the skew of the d_{33} meter was adjusted to make sure the value was identical to the one marked on the sample by the manufacturer. After the steps of nulling the d_{33} meter, the sample was put at the center of the clamp with suitable force by adjusting the clamping knob [2]. The measured d_{33}results are shown in Table 1. Our newly developed piezo-electret material was found to have a higher d_{33}value than that of pure PVDF or pure P(VDF-TrFE). The piezoelectric constant of the electro mechanical film (EMFi), which is one type of electret, also possessed a higher d_{33}value than that of the piezo-electret material. However, the disadvantage is that the EMFi must be made by performing a bi-axial extension process which makes for a higher, more expensive fabrication cost. Fortunately, P(VDF-TrFE) does not require a more involved bi-axial extension process which makes it easier to use and lower in cost to adopt as our piezo-electret material.

In this study, we found that the d_{33} of our piezo-electret material was higher than pure P(VDF-TrFE) 70/30. The reasons leading to a higher, more improved d_{33} coefficient are as follows: 1) a higher total charge, 2) air voids increased by the strain as a characteristic of the piezo, and 3) increased volume between the piezo material and electret material. Therefore these three characteristics combined allow our newly developed piezo-electret material to possess an overall higher d_{33} coefficient.

In this paper, a piezo-electret material with good longitudinal piezoelectric strain coefficients $d_{33}$is shown. The longitudinal piezoelectric strain coefficient is defined as follows. Equation (1) shows the displacement using the IEEE Standards ANSI/IEEE 176-1987 along the 3-direction, where D is the electric displacement, *T*is stress, *E* the electric field, and ε_{33}^T the permittivity measured at a constant stress.[2]

$$D_3 = d_{33}T_3 + \varepsilon_{33}^T E_3 + s_a T_3^2 + 2d_2 T_3 E_3 + \varepsilon_2 E_3^2 \quad (1)$$

We can simplify the relationship of d_{33} in Equation (1) to Equation (2):

$$d_{nj} = \frac{\partial D_n}{\partial T_j}, \text{ where } n=1, 2 \text{ or } 3 \text{ and } j = 1, 2, 3, 4, 5 \text{ or } 6 \tag{2}$$

Such that for the case where n=3 and j=3, d_{33} can be written as:

$$d_{33} = \frac{\partial D_3}{\partial T_3} = \frac{\frac{Q}{A}}{\frac{F}{A}} = \frac{Q}{F} \tag{3}$$

In Equation (3), $Q$is the charge, $F$is the applied force, and A the area. It is clear from Equation (3) that $d_{33}$is proportional to the charges.

In addition, the charge storage phenomenon (surface potential) was verified by using an electrostatic voltmeter (Model 542, Trek Inc.). Pure e-PTFE and P(VDF-TrFE) in both e-PTFE and COC/P(VDF-TrFE) in e-PTFE/COC were compared. Under general atmosphericconditions, the surface potential first decayed exponentially as reported before. It can be shown that a steady state surface potential will be reached quickly, which matches other previously reported electret work.

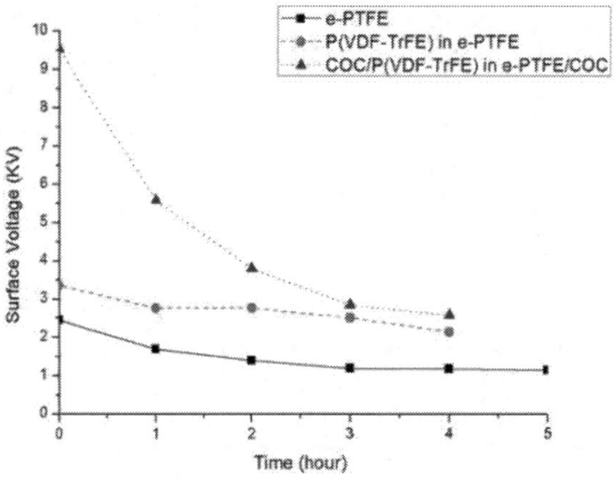

Figure 4. Charge storage phenomenon in pure e-PTFE and P(VDF-TrFE) in PTFE and with COC outside the P(VDF-TrFE) in PTFE.

3.Collocated & Self-sensing Sensor/Actuator

Self-sensing techniques have been discussed for more than 25 years. This technique has typically used a single piezoelectric layer to serve as both an actuator and a sensor simultaneously. Some have previously reported that the advantages for this type of configuration have included easier fabrication, lower overall costs, etc. The first collocated system can be seen in Figure 5.

Dosch *et al*used electrical bridge circuits in a closed loop to achieve the self-sensing behavior [3]. This type of circuit can be used for strain rate and position sensing (see Figures 6(a) and 6(b)). However, variations on temperature, humidity and other external environmental factors can lead to electrical properties of PZT changes that can seriously impact the performance of self-sensing sensors/actuators.

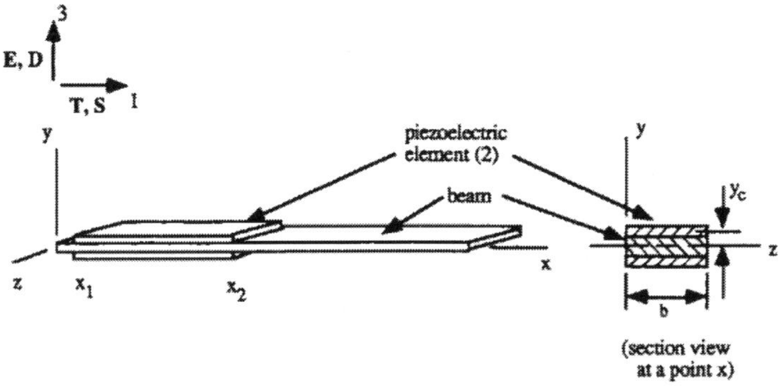

Figure 5. Collocated sensor/actuator

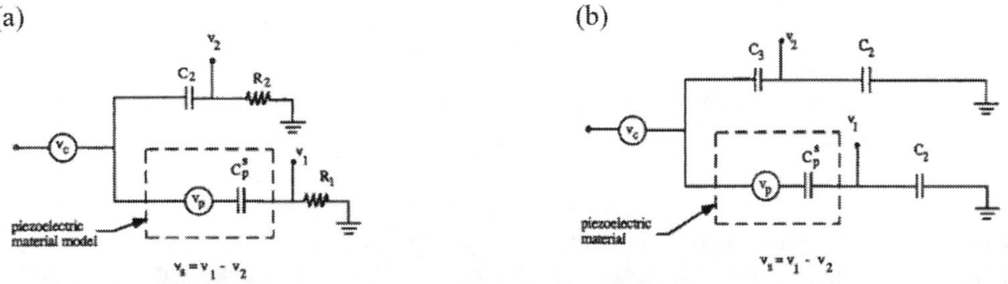

Figure 6. Bridge circuit for (a) stain rate sensing and (b) position sensing/strain sensing [3]

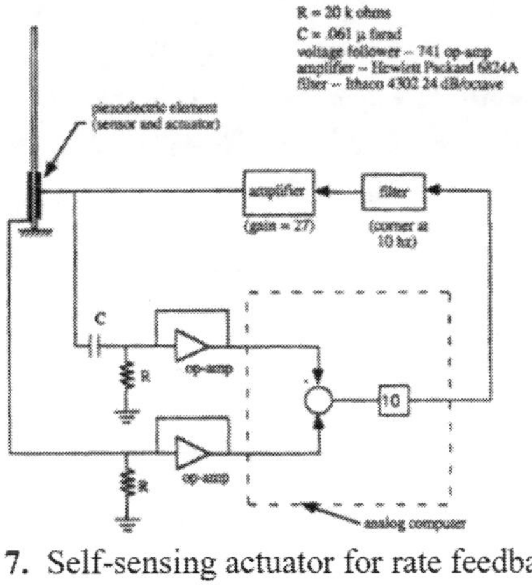

Figure 7. Self-sensing actuator for rate feedback [3]

A time sharing method was another technique which we adopted for solving this problem (see Figures 8(a) and 8(b)) [4]. However, the signal could not be captured with certainty in many cases. For example, when the signal appeared at the time when the device was switched from a sensor to an actuator or vice versa, the sensing signal was missed or destroyed. This drawback can seriously limit the performance of this time-sharing self-sensing configuration.

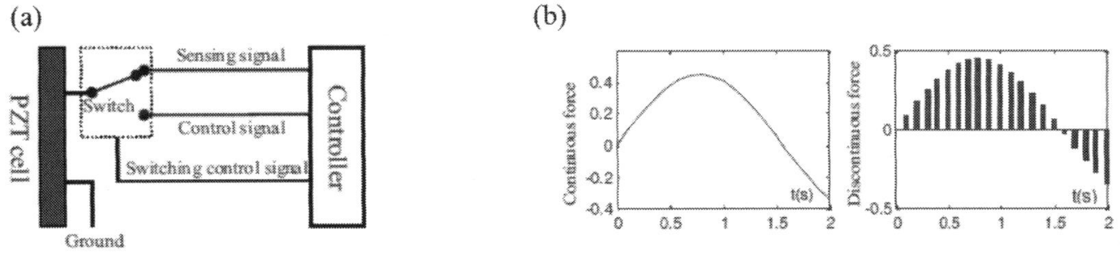

Figure 8. (a) Time sharing actuator and (b) Smooth continuous force and chopped force

Lee *et al* grounded the stainless steel shim between the piezoelectric sensor and actuator to solve the EMI (Electro-Magnetic Interference) issue induced by the close proximity of the two piezoelectric laminae [5, 6]. Their test system configuration and feedback control loop are shown in Figures 9(a) and 9(b).

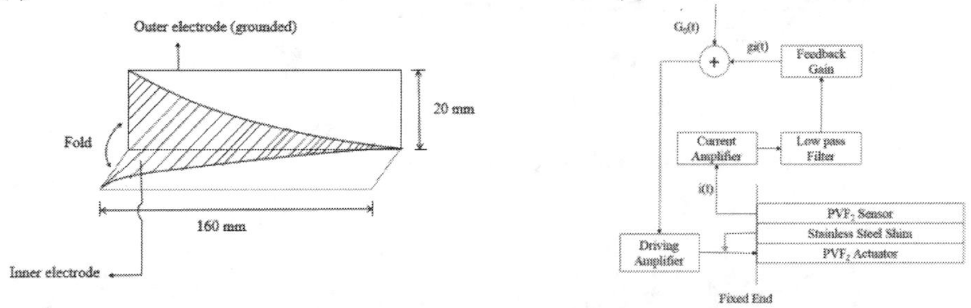

Figure 9. (a) Construction of PVDF sensor and (b) Configuration of the feedback control loop [5, 6]

In this paper, we demonstrate a newly developed piezo-electret material which possesses inherent good electric isolation characteristics. The piezo-electret material is an excellent material which possesses the advantageous characteristics of both an electret and a piezo material simultaneously. Electrets can be driven using an externally applied electric field through a non-contact electrode. Piezoelectric material can produce charge signals which can be retrieved using an effective surface electrode which is directly evaporated onto its surface and is detectable by using a proper interfacing circuit.

To obtain better isolation, the structure also contains an air gap (i.e. an air void). Equation (4) and (5) show the governing equations of a capacitor and its impedance.

$$\mathrm{C} = \varepsilon_0 \varepsilon_r \frac{A}{d} \tag{4}$$

$$Z = \frac{1}{j\omega C} = \frac{1}{j\omega\left(\varepsilon_0 \varepsilon_r \dfrac{A}{d}\right)} \tag{5}$$

The relative permittivity (ε_r) of air approached 1. It is the smallest relative permittivity in the real world. By using an air gap, high impedance was obtained. Therefore, the leakage current was reduced to a minimum.

Figure 10. Schematic diagram for piezo-electret material (the red arrow side)

2. 1D cantilever Beam vs 1D cantilever Plate

In this paper, the testing structure fabricated was a one-dimensional (1D) cantilever plate. However, the difference between a one-dimensional cantilever plate and a one-dimensional cantilever beam is nothing but plane stress approximation for a 1D cantilever plate and plane strain approximation for a 1D cantilever beam. The difference and the resultant computation adjustment factor in bending stiffness are shown in traditional elasticity to be as follows: [7]

$$D_{11} = \left(\frac{1}{1-\upsilon^2}\right)EI \tag{6}$$

In Equation (6), υ is the Poisson's ratio of the material, E the Young's modulus, and I the moment of inertia of the structure. Undera 1D cantilever plate and a 1D cantilever beam, we found E and I to be identical. Therefore, the difference seems to be only on the Poisson's ratio. For example, if υ is 0.33, we obtained 1.122, i.e.a 12% error. On the other hand, if υ is 0.3, we obtained 1.099 which is a 11% difference. In the above discussions, the difference of a 1D cantilever plate and a 1D cantilever beam is a 12% difference or so. Therefore, the theory of a 1D cantilever beam can actually be applied in this paper. The only thing worth noting is that we need to adjust the 12% difference when comparing the theoretical and the experimental results.

In the free vibration of the Euler-Bernoulli beam theory, the governing equation is shown as in Equation (6). In Equation (7), $y$is the distance of surface from neutral surface of the 1D cantilever beam,$E$is the Young's modulus of the 1D cantilever beam, and$I$ the moment of inertia of the 1D cantilever beam. In addition, $\rho$is the density of 1-D cantilever beam,b the width of the 1D cantilever beam,and t the thickness of the 1D cantilever beam.

$$\frac{\partial^2}{\partial x^2}\left(EI\frac{\partial^2 y}{\partial x^2}\right) + \rho bt\frac{\partial^2 y}{\partial t^2} = 0 \tag{7}$$

By using a numerical method to solve Equation 7, we can identify the first three eigenvalues as shown in Equation (8).

$$\begin{aligned} \alpha_1 L &= 1.875104 \\ \alpha_2 L &= 4.694091 \\ \alpha_3 L &= 7.854757 \\ \alpha_4 L &= 10.995541 \end{aligned} \tag{8}$$

After some calculations, the natural frequency (f_n) can be seen in Equation (9).

$$\omega_n = (\alpha_n L)^2\sqrt{\frac{EI}{\rho btL^4}} = 2\pi f_n\,, \quad n = 1,2,3,\ldots\ldots \tag{9}$$

In this paper, a 1D cantilever plate was made using a piezo-electret material, a stainless steel shim and a metal electrode. For calculating the natural frequencies, we adopted the following parameters in MKS units: 200 GPa for the Young's modulus (E) of the stainless steel shim, $1.47*10^{-13}$ for themoment of inertia (I), 7900 kg/m³ for the density (ρ) of stainless steel shim,0.006 m for the width of the cantilever plate (b), and $6.65*10^{-4}$m for the thickness of the cantilever plate(t).

Using Equations (8) and (9), we can theoretically predict the first, second and third natural frequencies as follows:

$$\begin{aligned} f_1 &\approx 59.89\text{Hz} \\ f_2 &\approx 375.3\text{Hz} \\ f_3 &\approx 1050.92\text{Hz} \\ f_4 &\approx 2059.56\text{Hz} \end{aligned} \tag{10}$$

The piezo-electret material was measured using a LCR meter (LCR-819, GW Instek Co.). A resistor and a capacitor in a parallel configuration were identified to be the best circuit model. The resistor was measured to be kilo ohms and the measured capacitor was around a few nano Farads. Looking at the above values, the newly developed piezo-electret material in this configuration was neither an ideal current source nor an ideal voltage source. In fact, the equivalent circuit can be treated as a less than ideal voltage source. Therefore, a non-inverting voltage amplifier with proper negative feedback was used in order to interface the voltage signal generated within the piezo-electret. The schematic of the non-inverting voltage amplifier is seen inFigure 11, and its own transfer function was calculated in Equations (11-13). In the experiments performed, the resistors R_1 and R_2 were set to be 1kΩ.

$$V_{out} = V_{in} + i \cdot R_2 \tag{11}$$

$$= V_{in} + \left(\frac{V_{in}}{R_1} \right) \cdot R_2 \qquad (12)$$

$$= V_{in} \left(1 + \frac{R_2}{R_1} \right) \qquad (13)$$

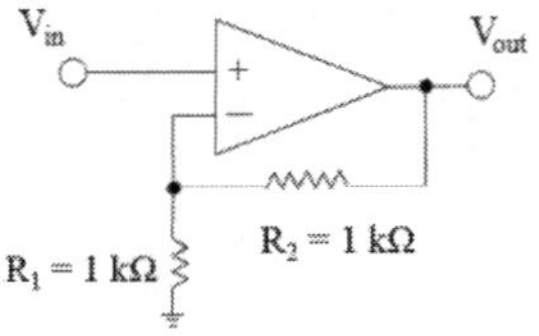

Figure 11. Non-inverting voltage amplifier with negative feedback

The schematic of the experimental set-up is shown in Figure 12. The dynamic signal analyzer was used to deliver the actuating signal. To induce a voltage signal high enough to drive the newly developed piezo-electret laminate, the actuating signal was further amplified using a high-speed bipolar amplifier before injecting into the actuating piezo-electret layer. The sensing signal induced by the same piezo-electret sensing layer was interfaced by a non-inverting amplifier as mentioned above. This configuration completed the self-sensing system. The high isolation inherent within the self-sensing piezo-electret provided the opportunity to achieve a robust self-sensing action. The effect of this self-sensing piezo-electret lamina was demonstrated using a dynamic analyzer to analyze the transfer function between the output sensing signal and the input actuating signal.

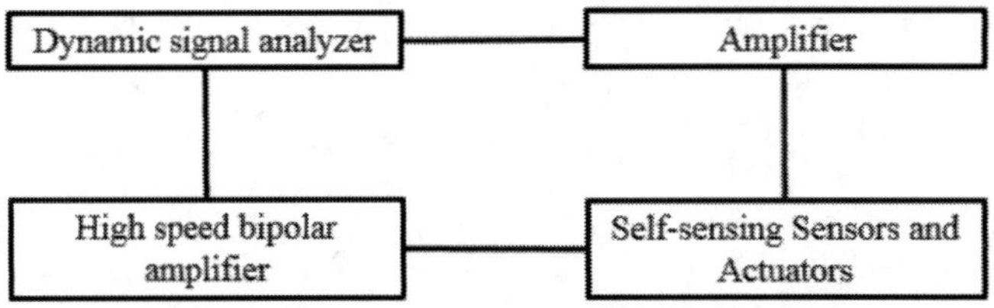

Figure 12. Schematic of the measurement system

(a) (b)

Figure 13. (a) SMA/IPIX line and (b)BNC/SMA connector

To avoid signal coupling, a BNC/SMA/IPIX was used as shown in Figures 13 (a) and (b). The wire was shielded by the grounding jacket. The electromagnetic interference (EMI) was improved by using a co-axial cable. A BNC/SMA connector was used for connecting the signal to the instruments.

The transfer function of the power amplifier (HAS 4052, NF Corporation) and the operational amplifier (AM 501, Trek Inc.) were measured using a dynamic signal analyzer (SR 780, Stanford Research Systems). In Figure 5, the transfer functions are shown to be fairly flat which did not interfere with the structural vibration modal signal tested.

(a) (b)

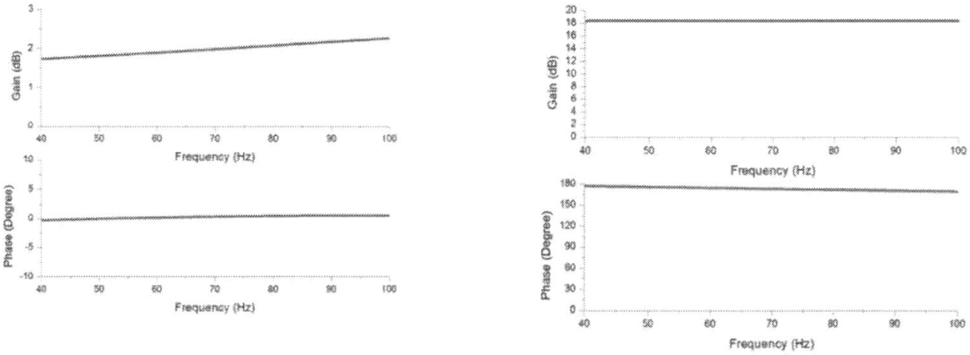

Figure 14. Transfer functions of (a) a high-speed bipolar amplifier and (b) a non-inverting operational amplifier

The schematic shown in Figure 15 was fabricated to verify the sensing and actuating performance provided by the newly developed piezo-electret material. The actuator electrode was the Mode 1 modal electrode and the sensor electrode took the uniform shape. The design chosen was due to the fact that we wanted to excite the Mode 1 vibration only. The red dots pattern ( )worked as a sensor and the blue chevron pattern ( ) worked as an actuator. The leakage current decreased significantly due to the low capacitance associated with the air gap. The measured transfer function of the two-layer piezo-electrets sensor and actuator can be seen in Figure 16.

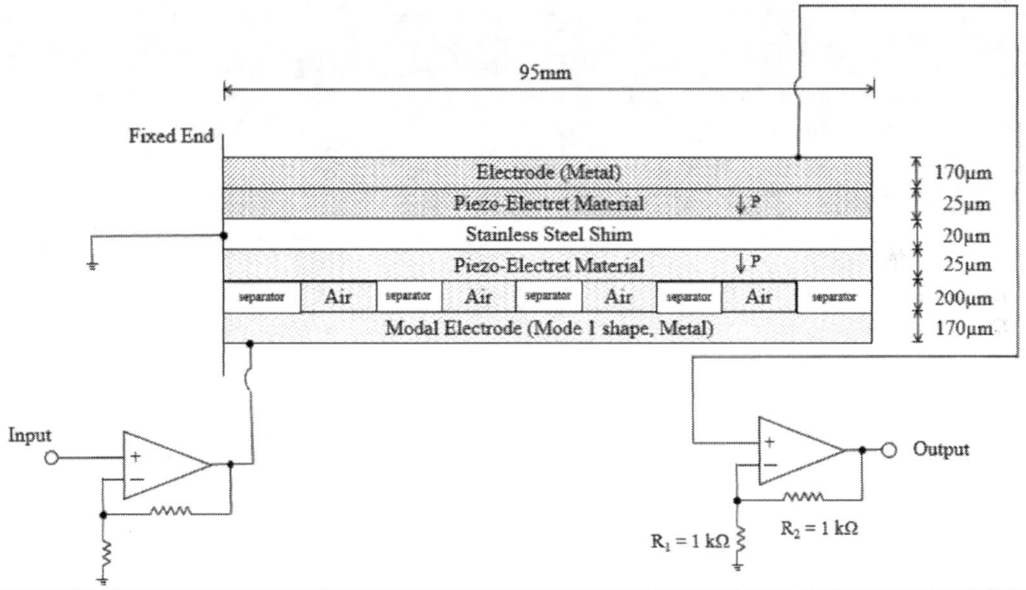

Figure 15. Schematic of a two-layer piezo-electret sensor and actuator

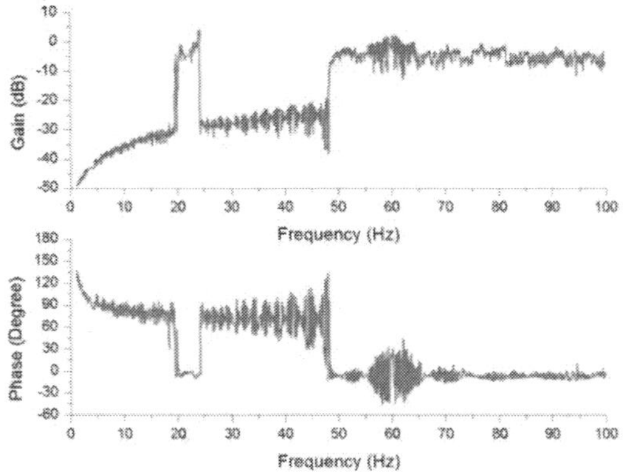

Figure 16. Transfer function of a self-sensing piezo-electret

The schematic of the self-sensing sensor and actuator fabricated by using the newly developed piezo-electret material is shown in Figure 17. Both the actuator electrode and sensor electrode took on a uniform shape. In Figure 17, the red dots pattern ( )worked not only as an actuator, but also as a sensor. The signal obtained, shown in Figure 18, possessed more noise than in the experimental data as seen in Figure 18. This difference shows that better shielding and isolation must be developed to further advance the practicality of a single piece self-sensing sensor/actuator system.

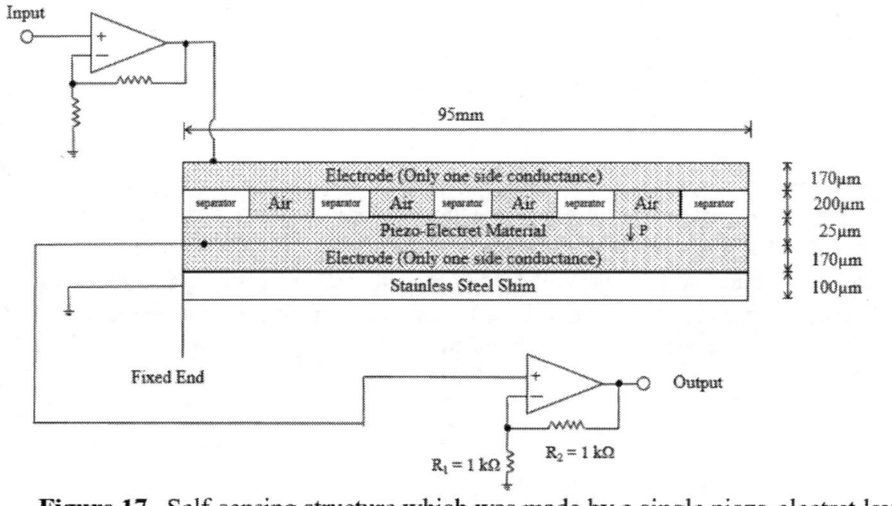

Figure 17. Self-sensing structure which was made by a single piezo-electret layer

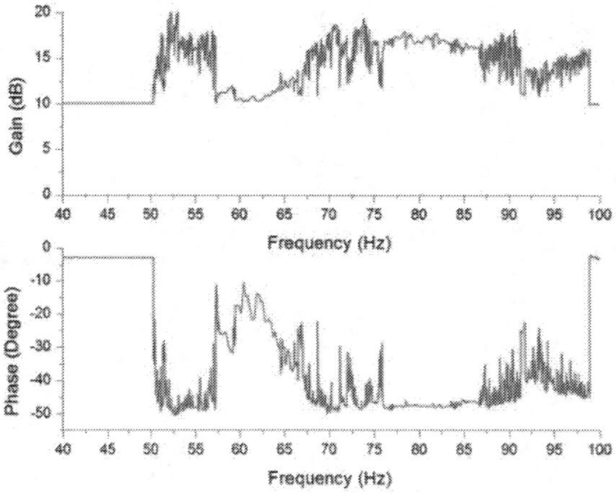

Figure 18. Transfer function of self-sensing structure made of a single piezo-electret layer

4. CONCLUSIONS

A piezo-electret material was developed and shown to possess both piezoelectric and electret characteristics in this study. In addition, the two different physical properties were demonstrated to provide unexpected advantages. We found that the longitudinal piezoelectric strain coefficients (d_{33}) value of piezoelectricity also improved when the two materials were combined to form the newly developed piezo-electret material. Additionally, the remanent polarization and coercive electric field obtained demonstrate that this piezo-electret material can be used as both a sensor and an actuator. Finally, our preliminary test results show the possibility of using this newly developed piezo-electret material to construct a self-sensing sensor and actuator. The measured values of the P-E loop and the remanent polarization (Pr) obtained was about 3μC/cm2 and the coercive electric field (Ec) was about 5 MV/m.

In our newly developed self-sensing system, we also presented the schematic suitable for using the newly developed piezo-electret material. With our new piezo-electret material, we can take advantage of the characteristic properties of both an electret and a piezoelectric material. The combination of the two characteristics available simultaneously, can offer us with new opportunities and possibilities for wider smart structure applications. Electrets aredriven by externally applied electric fields and a P(VDF-TrFE) generatedcharge signal is a response to an induced vibration. Applications can use either the electret advantage or the piezoelectric advantage of the piezo-electret material to use as an actuator (or sensor) and then as sensor (or actuator). With the two completely different physical properties present simultaneously, this piezo-electret material can offer a better isolation for applications that require effective noise reduction.

In summary, our newly developed piezo-electret self-sensing sensor/actuator has the potential for many news applications for smart structures, structure health monitoring (SHM), aerospace engineering, bio-chips, microscopy, and acoustical systems.

ACKNOWLEDGMENTS

This study was sponsored by Taiwan's Ministry of Science & Technology under Grant No. NSC 102-2221-E-002-188-MY3.

REFERENCES

[1] G. M. Sessler and M. G. Broadhurst, *Electrets* vol. 33: Springer Verlag, Berlin, Germany, 1980..

[2] S. Anton, K. Farinholt, and A. Erturk, "Piezoelectret foam–based vibration energy harvesting," *Journal of Intelligent Material Systems and Structures,* p. 1045389X14541501, 2014.

[3] J. J. Dosch, D. J. Inman, and E. Garcia, "A self-sensing piezoelectric actuator for collocated control," *Journal of Intelligent Material Systems and Structures,* vol. 3, pp. 166-185, 1992.

[4] B. R. Wang, "Piezoelectric self-sensing actuator for vibration suppression based on time-sharing method," in *Control and Automation (ICCA), 2010 8th IEEE International Conference on*, 2010, pp. 1403-1408.

[5] C. K. Lee, W. W. Chiang, and T. O'Sullivan, "Piezoelectric modal sensor/actuator pairs for critical active damping vibration control," *The Journal of the Acoustical Society of America,* vol. 90, pp. 374-384, 1991.

[6] C. K. Lee, and F. C. Moon, "Modal sensors/actuators," *Journal of Applied Mechanics,* vol. 57, pp. 434-441, 1990.

[7] C. K. Lee, "Piezoelectric Laminates: Theory and Experiments for Distributed Sensors and Actuators," *Intelligent Structural Systems*, ed. H. S. Tzou and G. L. Anderson, Kluwer Academic Publishers, pp. 75-167, Netherlands Dordrecht, 1992.

ICAST2015 064

Ni-Mn-In type magnetic shape memory alloy free-standing film for magnetically actuated devices

Hiroyuki Miki[1*], Koki Tsuchiya[2], Eijiro Abe[2], Makoto Ohtsuka[3], Marcel Gueltig[4], Hinnerk Ossmer[4], Manfred Kohl[4] and Toshiyuki Takagi[5]

[1] Frontier Research Institute for Interdisciolinary Sciences, Tohoku University, Sendai, Japan.
[2] School of Engineering, Tohoku University, Sendai, Japan.
[3] Institute of Multidisciplinary Research for Advanced Materials, Tohoku University, Sendai, Japan.
[4] IMT, Karlsruhe Institute of Technology, Karlsruhe, Germany
[5] Institute of Fluid Science, Tohoku University, Sendai, Japan.

Abstract

Shape memory behavior on the Ni-Co-Mn-In alloy film was investigated for the practical application as a magnetically driven actuator device. The Ni-Co-Mn-In alloy films of 5 μm thickness were fabricated by the dual magnetron sputtering. This film was heat-treated and shows the shape memory behavior of one-way shape memory effect by heating and applying magnetic field.

1. INTRODUCTION

Ferromagnetic shape memory alloys (FSMAs) show the martensitic transformation (MT) induced not only by the temperature and the applied stress but also the magnetic field. Therefore, FSMAs have been expected to be applied to a magnetic driven actuator which may enable quick response and remote control [1].

Recently, Ni-Mn-X (X = In, Sn) type FSMAs have been found. For the Ni-Mn-In bulk samples, their MT has been observed from ferromagnetic austenitic (A) phase to antiferromagnetic or paramagnetic martensitic phase [2] and the MT temperature decreases by a magnetic field [3]. This fact suggests the magnetic induced reserve MT (meta-magnetic phase transition). These alloys are expected to exhibit the shape memory effect (SME) associated with the magnetic field induced reverse MT [4].

Additionally, it was reported for Ni-Mn-In bulk samples contained Co that the MT temperature (M_s) decreases while Curie temperature (T_C) increases with increasing Co concentration [2].

In this study, Ni-Mn-In alloy films containing cobalt was deposited on a poly-vinyl alcohol (PVA) substrate or an alumina substrate using a dual magnetron sputtering apparatus with $Ni_{45}Mn_{40}In_{15}$ and Co targets. Magnetic properties and shape memory behavior on the Ni-Co-Mn-In alloy films are investigated for the practical application for the magnetically actuated devices.

* miki@fris.tohoku.ac.jp

2. EXPERIMENTAL DETAILS

2.1 Film preparation

The Ni-Mn-In alloy film with Co addition was made to a film thickness of approximately 5 μm on a PVA substrate using dual source magnetron sputtering equipment (Fig. 1). In the process, substrate temperature was kept constantly in 323 K by water cooling, and a film was deposited by applying the high radio frequency (W_{RF}) power at 200W for $Ni_{45}Mn_{40}In_{15}$ target and the constant direct current (W_{DC}) of the Co target at 8W at the same time.

After having delaminated from the PVA substrate, the sputtering film was firstly homogenization heat-treated in a vacuum atmosphere of 2×10^{-4} Pa and furnace-cooled to room temperature. In this study, first step heat-treatment just after delamination was applied for 3.6 ks in 1023 K. After this process, heat-treated film was flatten out between alumina plates and reheat-treated for 3.6 ks in 1173 K.

2.2 Property evaluation

The composition analysis of the alloy film was determined using Inductively-coupled Plasma (ICP) Atomic Emission Spectrometry analyzer (SPECTRO ARCOS) made in AMETEK Co. The composition of the sputtering film was $Ni_{50.4}Co_{3.7}Mn_{32.8}In_{13.1}$.

A four-terminal method of electrical resistance measuring equipment was used for the measurement of the martensitic transition temperature, and SQUID magnetization measuring equipment (MPMS XL) of Quantum Design Co. was used for the magnetization measurement.

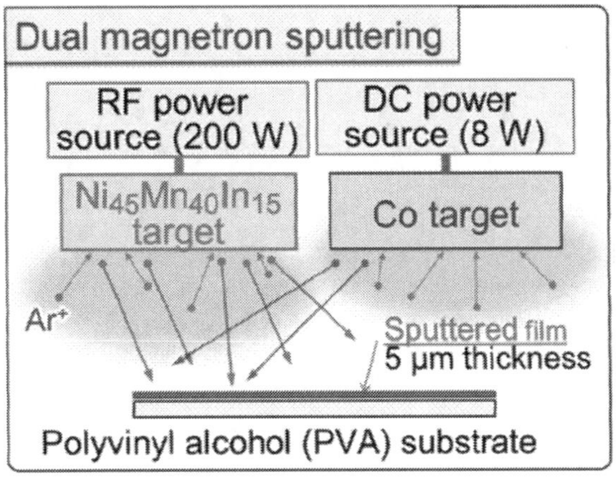

Figure 1. Deposition procedure of the Ni-Co-Mn-In alloy film using dual magnetron sputtering equipment.

The helium-free superconducting magnet which could make a magnetic field of up to 5 T made in Sumitomo Heavy Industries, Ltd. was used to investigate the magnetic field induced shape-recovery effect. The shape variation under the different applying magnetic field were observed in the constant-temperature unit

A mechanical property was investigated by a tensile test. The tensile test were carried out at the temperature range of 295~375.5 K and with applied magnetic field in the range of 0~1 T. The examination performed with cyclic load-unload testing under the constant strain rate of 2×10^{-5} s^{-1} and increasing strain step of 0.2 %.

3. RESULTS AND DISCUSSIONS

3.1 Phase transition of the Ni-Co-Mn-In alloy film

Martensitic phase transition temperature and the Curie temperature obtained from the temperature dependence of the electrical resistance measurement for Ni-Co-Mn-In alloy film. If we suppose temperature hysteresis to be T_{hys} = $(A_s\text{-}M_f)/2$, it is 8.4 K for Ni-Co-Mn-In alloy film which is equivalent with the temperature hysteresis of a practical shape-memory alloy actuator material ($Ti\text{-}_{37.0}Ni\text{-}_{9.5}Cu$ [5]).

The magnetization curves on the heat-treated Ni-Co-Mn-In alloy film under the different applied magnetic field were shown in Fig. 2. Drastically change of the magnetization due to the martensitic phase transition was observed. The variation of the magnetization induced by martensitic phase transition was 26 emu/g in the applied magnetic field at 0.05 T that is in the same range to $Ni_{45}Co_5Mn_{36.5}In_{13.5}$ alloy bulk materials [4].

In addition, with the increase of the magnetic field, it was confirmed that differences between M phase and magnetization of the A phase increase, and, in martensitic phase transition temperature M_f

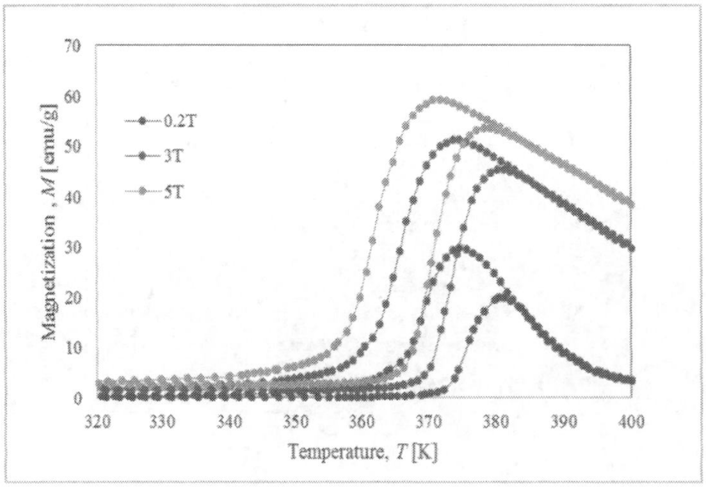

Figure 2. Thermomagnetization curves of the $Ni_{50.4}Co_{3.7}Mn_{32.8}In_{13.1}$ film in the magnetic field of H = 0.2, 3 and 5 T.

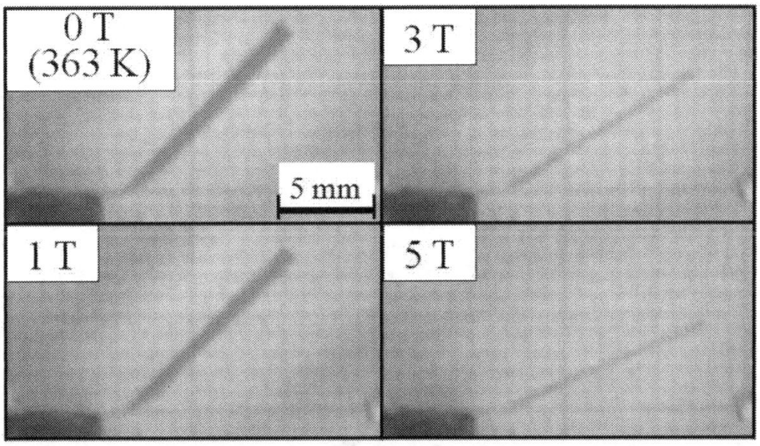

Figure 3. The shape memory effect by magnetic field for the heat-treated $Ni_{50.4}Co_{3.7}Mn_{32.8}In_{13.1}$ film.

decrease 8.5 K with applying a magnetic field of 5 T. The amount of change is larger than the previously reported ΔM_f (-1 K/T) of Ni-Co-Mn- In [6].

3.2 Shape-memory property of the Ni-Co-Mn-In alloy film

Shape recovery in the heat treated alloy film which was bent to 65° at room temperature was seen in the process to increase temperature. After shape recovering to 45° near 363 K, temperature in the constant-temperature unit kept at 363 K and magnetic field was applied after temperature became stable. The shape recovery procedure under the different applied magnetic field of the heat-treated Ni. Co.Mn.In alloy film is shown in Fig. 3.

Shape recovery started at the same time of the field applied, and remarkable shape recovery was confirmed at over 2 T. It was confirmed that the obtained alloy film showed the magnetic field induced reverse martensitic transition which stabilized A phase by applied magnetic field.

Accordingly as a result of obtained each stress strain curve, it is dependence on temperature of the mechanical property in the M phase. If the plastic deformation of the M phase is caused by a rearrangement of twin variants, thermal energy contributes as a driving force. Consequently, in the temperature domain of near to a reverse transition point, possibility to induce plastic deformation with smaller stress is shown.

3.3 Mechanical property of the Ni-Co-Mn-In alloy film

The stress strain curves which the tensile test of the heat-treated Ni-Co-Mn-In alloy film are shown in Fig. 4. In a stress strain curve of Fig. 4, both an elastic deformation and a plastic deformation domain were unclear.

The result of the room temperature testing under zero magnetic field is shown in Fig. 4(a). Because the alloy film is single M phase at the room temperature, the transformation occurs in M phase.

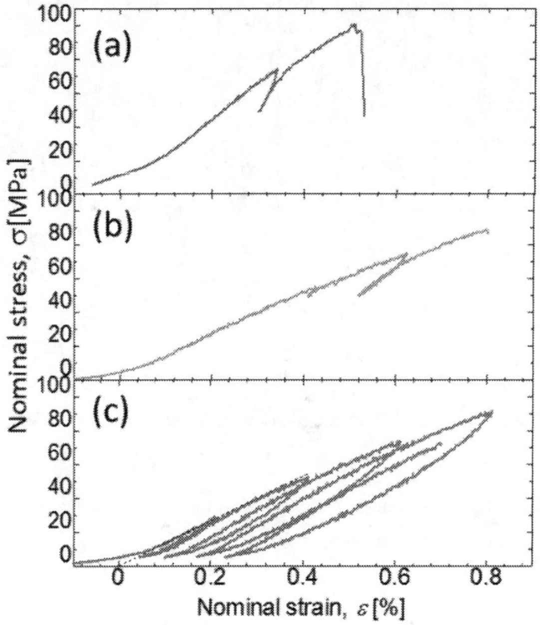

Figure 4. Stress-Strain curves of the heat-treated $Ni_{50.4}Co_{3.7}Mn_{32.8}$ $In_{13.1}$ film at (a)RT, (b)T =372.5 K and (c)T = 375.5 K applied magnetic field of 1 T.

The examination condition of Fig. 4 (b) at $(A_s+A_f)/2$ (372.5K), zero magnetic field. At an initial (no load) state, the alloy film shows a two phase coexistence state of M phase and A phase. However, in the examination temperature range of Ms <T <Af of (b), transformation which originate from stress induced martensitic phase transition occurs with the growth of the M phase.

At the examination temperature of Fig. 4(c), A phase is stable in a no load state. A phase is stabilized further by an applying magnetic field. However, at the examination condition of Fig. 4(c), the stress induced martensitic phase transition occurs around martensitic phase transition temperature.

4. CONCLUSION

Ni-Mn-In alloy film containing Co was fabricated by a sputtering. The MT and one-way SME induced by applying magnetic field were observed, and the MT temperatures decreased with increasing magnetic field. They show the possibility of metamagnetic SME and the possibility for magnetically actuated devices.

ACKNOWLEDGMENTS

This work was partly supported by the Amada Foundation in Japan, the Hosokawa Powder Technology Foundation in Japan, and the JSPS Core-to-Core Program, A. Advanced Research Networks, "International research core on smart layered materials and structures for energy saving".

Part of the work was carried out under the Collaborative Research Project of the Institute of Fluid Science and the International Collaborative Research Project of Frontier Research Institute for Interdisciplinary Sciences, Tohoku University

REFERENCES

1. Matsumoto, M. et al., "Progress of ferromagnetic shape memory alloy research", *Journal of JSAEM*, Vol. 8, No. 4, 2000, pp. 445-453.
2. Y. Sutou et al., "Magnetic and martensitic transformations of NiMnX (X=In,Sn,Sb) ferromagnetic shape memory alloys", *Applied Physics Letters*, Vol. 85, No. 11, 2004, 4358.
3. Oikawa, K. et al., "Effect of magnetic field on martensitic transition of Ni46Mn41In13 Heusler alloy", *Applied Physics Letters*, Vol. 88, No. 12, 2006, 122507.
4. Ito, W. et al., "Martensitic and Magnetic Transformation Behaviors in Heusler-Type NiMnIn and NiCoMnIn Metamagnetic Shape Memory Alloys", *Metallurgical and Materials Transactions A*, Vol. 38A, 2007, pp. 759-766.
5. Ohtsuka, M. et al.," Magnetic-Field Induced Phase Transformation and Metamagnetic Shape Memory Effect of Ni-Mn Base Alloys", *Annual Reports of High Field Laboratory for Superconducting Materials, Institute for Materials Research, Tohoku University*, 2009, 4-7.
6. Miyazaki, S. et al., "High Power and Large Actuation Micro-Actuator Material", *Materia Japan*, Vol. 38, No. 4, 1999, pp. 279-280.

ICAST2015 #077

Application of Shape Memory Materials into Actuators

Ryosuke Matsui[1*], Kohei Takeda[1] and Hisaaki Tobushi[1]

[1] Department of Mechanical Engineering, Aichi Institute of Technology, Toyota, Japan

Abstract

In the intelligent materials, the development of shape memory alloy (SMA) has attracted high attention because the unique properties of the shape memory effect (SME) and superelasticity (SE) appear. If we use the SME and SE in practical applications, not only large recovery strain but also high recovery stress can be obtained. If we apply the recovery stress to the driving force, we can develop the novel SMA actuators. In the shape memory materials, shape memory polymer (SMP) has also been developed. In SMP, large recovery strain of more than 100% can be obtained. Although elastic modulus and yield stress are large at high temperatures and small at low temperatures in SMAs, they are large at low temperatures and small at high temperatures in SMPs. If the composite materials with SMA and SMP having different properties are developed, the new actuation properties of the material can be obtained. In the present study, the development of the novel actuators using SMA and SMP will be investigated.

1. INTRODUCTION

In the intelligent materials, the development of shape memory alloy (SMA) has attracted high attention because the unique properties of the shape memory effect (SME) and superelasticity (SE) appear [1-4]. If we use the SME and SE in practical applications, not only large recovery strain but also high recovery stress can be obtained. If we apply the recovery stress to the driving force, we can develop the novel SMA actuators. The main features of the SME and SE are induced due to the martensitic transformation (MT). In the shape memory materials, shape memory polymer (SMP) has also been developed [4-6]. In SMP, large recovery strain of more than 100 % can be obtained. The main features of SMP appear due to the glass transition. We can use not only the shape recovery and shape fixity but also recovery stress. We can therefore develop the novel SMP actuators by using the recovery stress.

Although elastic modulus and yield stress are large at high temperatures and small at low temperatures in SMAs, they are large at low temperatures and small at high temperature in SMPs. The dependence of rigidity and strength on temperature is therefore quite opposite between SMA and SMP elements. If the composite materials with SMA and SMP having different properties are developed, the new actuation properties of the material can be obtained [7-9].

If the above-mentioned shape memory actuators are developed, the shape memory material works by itself for measuring temperature and driving actuation. That is, sensors measuring temperature and motors or mechanical systems driving actuators are not necessary, and the simple mechanism can therefore be achieved [10].

In the present paper, the development of the novel actuators using SMA and SMP will be investigated. First, based on the two-way SMA actuation principle, a solar-powered active blind model and a space deployable structure model will be proposed as one of applications of SMA. Following these SMA

* r_matsui@aitech.ac.jp

actuators, based on the mechanical properties of SMA and SMP, the three-way motion will be discussed in the shape memory composite (SMC) actuator.

2. SHAPE MEMORY ALLOY ACTUATOR

2.1 Two-way SMA actuation

In order to understand the basic deformation properties of SMA, the force-deflection curves of the SMA helical spring obtained by the tension test at various temperatures T are shown in Figure 1. The diameter of a TiNi SMA wire, the mean spring diameter, the number of active coils and the initial length of the coil were 0.7 mm, 8.7 mm, 6 and 4.2 mm, respectively. The loading rate was 4 mm/s and the maximum deflection was 100 mm. The reverse transformation finish temperature A_f obtained by the differential scanning calorimetry test was 360 K. As can be seen, the deformation resistance increases in proportion to temperature T. In the case of temperatures below A_f, the residual deflection appears after unloading. The residual deflection disappears by heating under no-load, showing the SME. In the case of temperatures above A_f, the force-deflection curve draws a large hysteresis loop and the deflection disappears during unloading, showing the SE.

In order to understand the condition to use the driving force in the SMA actuators, the recovery force of the SMA spring which appears due to the reverse transformation by heating is shown on the force-deflection curves in Figure 2. In Figure 2, the various deformation behaviors (1)-(6) of the spring during heating from the maximum loading point A and the unloading finish point B at temperatures below the reverse transformation start temperature A_s are shown. The shape recovery under no-load (1) corresponds to the SME. The shape recovery under a constant force appears in the path (4). In the cases of perfect restriction of the deflection (3) and (6), the higher recovery forces appear than those in the cases (2) and (5) accompanying the shape recovery, respectively.

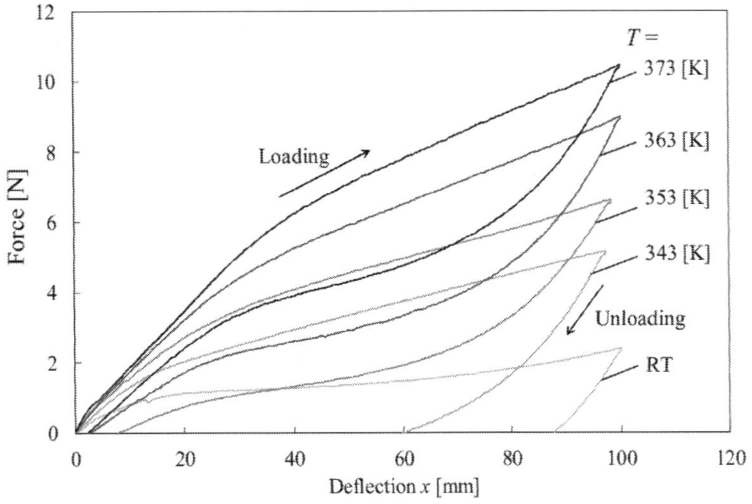

Figure 1. Load-deflection curves of the SMA spring at various temperatures T.

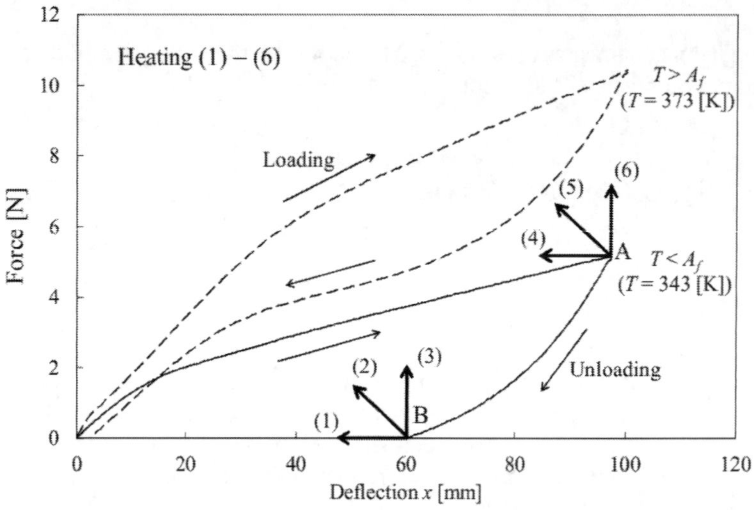

Figure 2. Recovery forces of the SMA spring by heating subjected to various conditions (1) – (6).

The typical two-way motion of the SMA spring is explained in Figure 3. A dead weight W is hung on the spring at temperature $T < A_s$ and then the spring lengthens. If the spring is heated up to $T > A_f$ under the load W, the spring shrinks by a length h. The shrinking behavior corresponds to the shape recovery under a constant force in the path (4) shown in Figure 2. Following the heating process, if the spring is cooled down to $T < A_s$, the spring lengthens by a deflection h. The two-way motion of the distance h is achieved during heating and cooling. The effective work Wh is obtained during the thermal cycling and it can be applied to the SMA heat engine as the driving source [11, 12].

The two-way actuation element used in practical applications is shown in Figure 4 [1, 2]. The SMA coil and a bias coil are arranged in series. The rigidity (spring constant) of the SMA coil is higher than that of the bias coil at $T > A_f$ and lower at $T < A_s$. The shaft in the element moves to the right during heating and to the left during cooling. The two-way motion x is therefore achieved during heating and cooling. The behavior of the SMA coil corresponds to the path (5) shown in Figure 2.

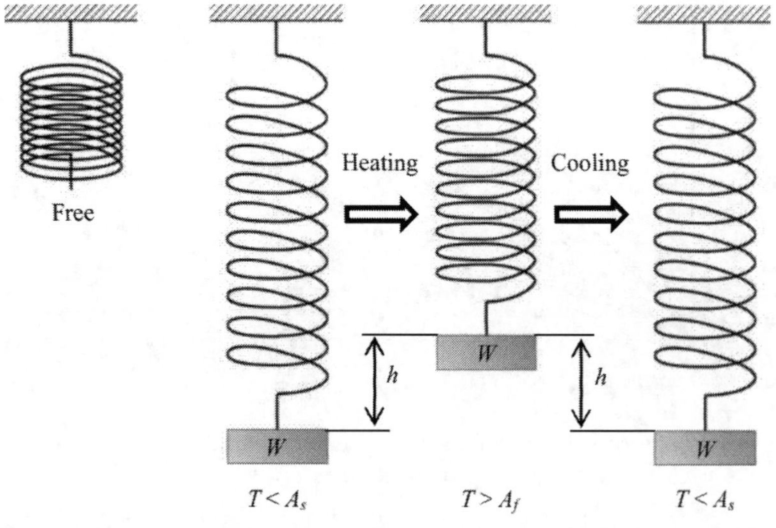

Figure 3. Two-way motion of SMA coil.

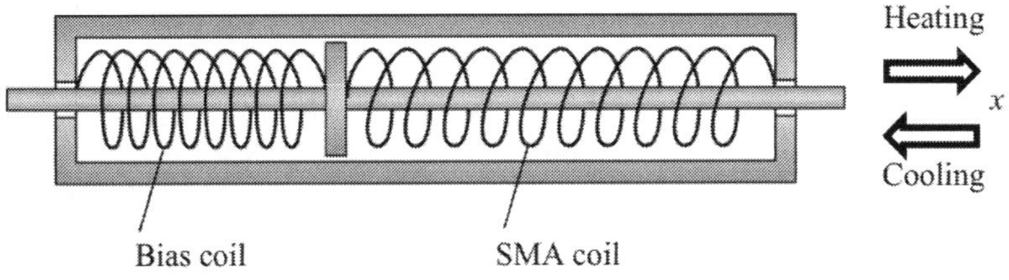

Figure 4. Two-way actuation element using SMA coil and bias coil.

2.2 Solar-powered active blind model

The structure and the photographs of two-way motion of the solar-powered active blind model are shown in Figures 5 and 6, respectively. In the initial state, the superelastic alloy (SEA) tape was mounted to be in a flat plane and the SMA tape was mounted at the total angle of twist $\phi = \pi / 2$. The SMA tape was heated by sunlight through a Fresnel lens. Since recovery torque appears in SMA tape by heating, the blind is closed through a crank and a lever. When the sunlight is shut out, the SMA tape is cooled and the SEA tape recovers a flat plane, resulting in opening the blind. The fatigue life of the SMA tape in the proposed models must be close to the fatigue life obtained by the pulsating fatigue test which is longer than 3×10^6 cycles at the maximum angle of twist $\phi_m = \pi / 2$ with the maximum angle of twist per unit length $\theta_m = 39.3$ rad/m. Therefore, if the proposed models are twisted 100 times a day, the service life will be longer than 80 years.

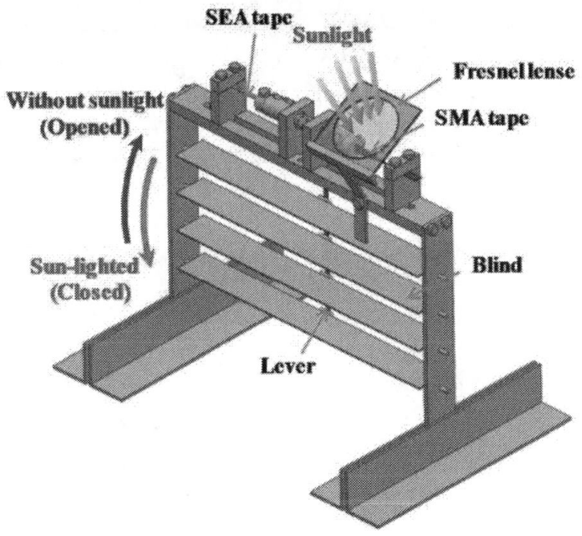

Figure 5. Structure of the solar-powered active blind model driven by twisting of SMA and SEA tapes

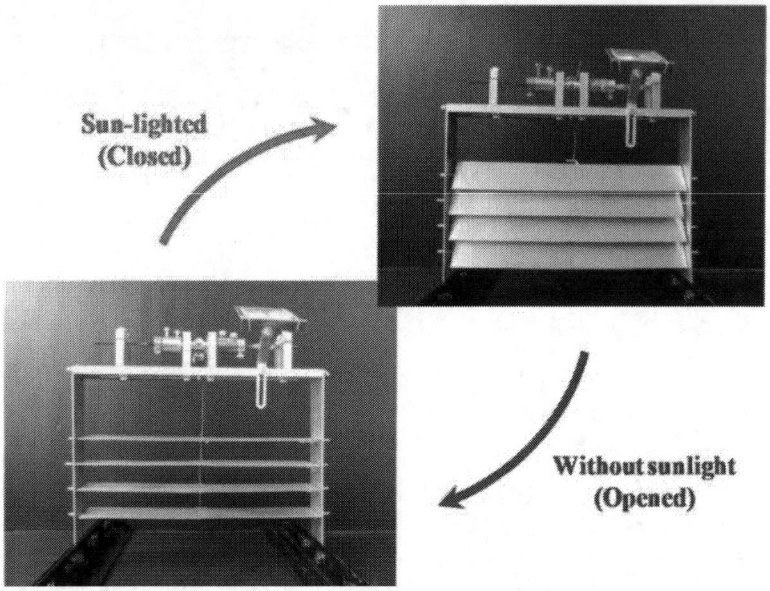

Figure 6. Photographs of two-way motion for opening and closing a blind driven by sunlight through SMA and SEA tapes.

2.3 Space deployable structure model

Recently, there are many literatures about deployable structures using SMA [13, 14]. We fabricated a space deployable structure model characterized by a simple structure and light weight by using SMA wires and an aluminum-evaporated polyimide film, and then conducted the deployment test in micro-gravity (micro-G) environment. Figure 7(a) shows the structure, power circuit and measurement system in the deployment test and Figure 7(b) is a photograph of the model after deployment. The model consists of the circular polyimide film with a diameter with 500 mm and the SMA wire (straight line shape-memorized) with a diameter of 1.0 mm. During deployment the wire was electrically-heated by a constant voltage generator. The folding and deploying procedures of the model are shown in Figure 8.

Figure 9 shows the typical deployment property of the model obtained by the test. This figure indicates a relationship between average deployment ratio of the deployable object and heating time of the SMA wire in the micro-G environment. The deployment ratio is calculated from deployment degree of the objects detected by an upper camera, hence 0% means storage state and 100% means complete deployment. The model deployed drastically at around 6 seconds and deployed completely in 18.3 seconds.

We found that the deployable structure model with a SMA wire characterized by a simple structure and light weight performs adequate ability to deploy in space.

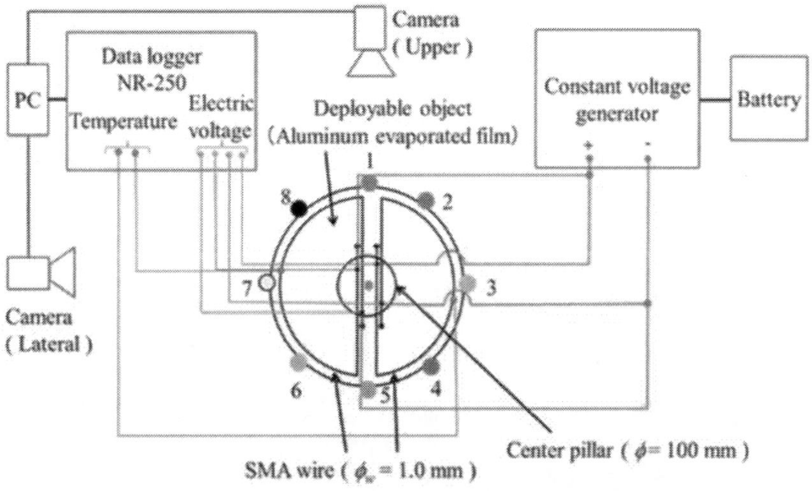

(a) Structure, power circuit and measurement system in the deployment test.

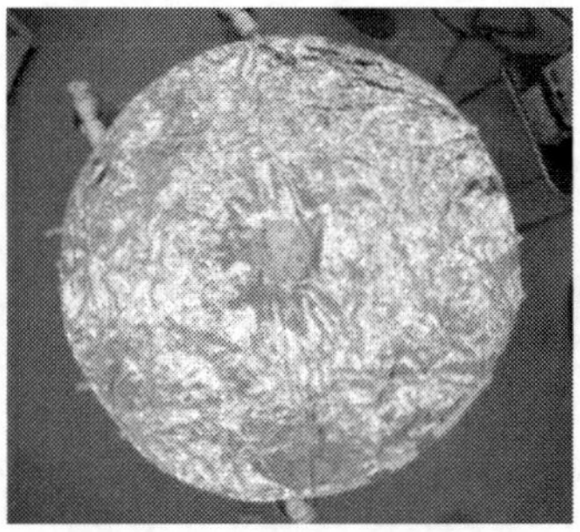

(b) Photograph of the model after deployment.

Figure 7. (a) Structure, power circuit and measurement system in the deployment test and **(b)** Photograph of the model after deployment.

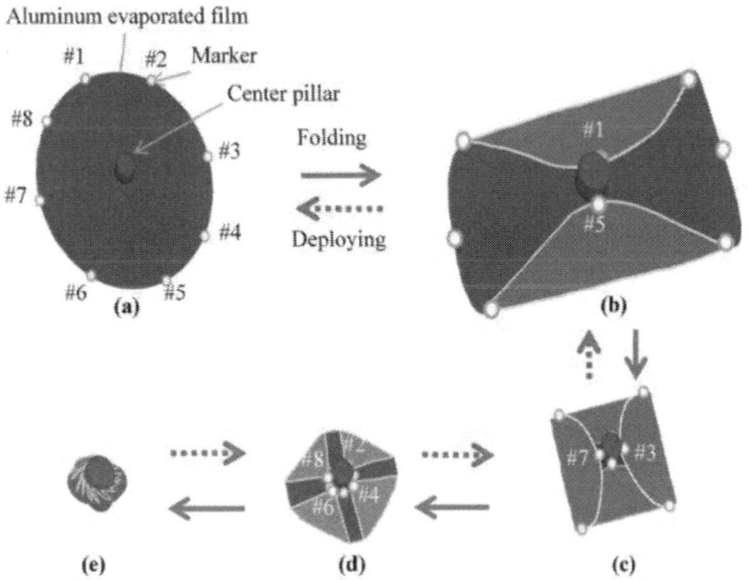

Figure 8. Folding and deploying procedures of the model.

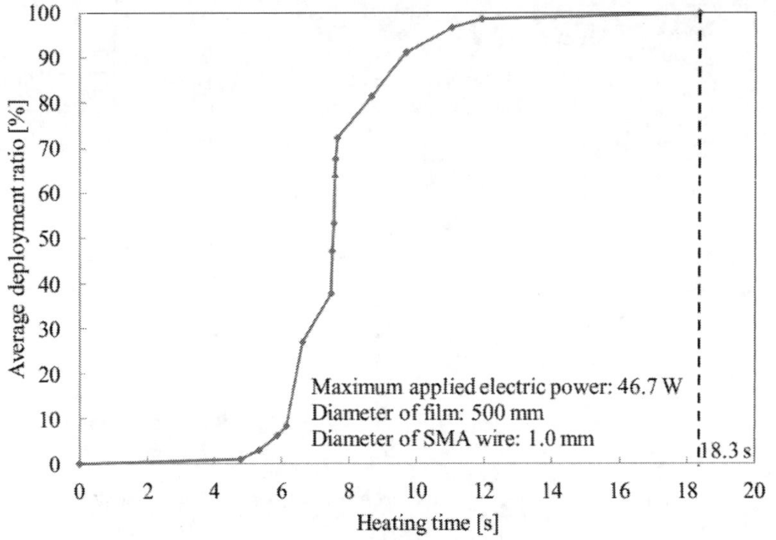

Figure 9. Relationship between average development ratio and heating time

3. SHAPE MEMORY COMPOSITE ACTUATOR

3.1 Characteristics of shape memory composite with SMA and SMP

In order to discuss the characteristics of SMC with SMA and SMP, the dependence of elastic modulus of SMA, SMP and steel on temperature and that of yield stress are shown in Figures 10 and 11, respectively. The symbols σ_M and σ_A represent the MT stress and reverse transformation stress, respectively. Elastic modulus and σ_M are small at temperatures below A_s and large above A_f in SMAs. The stress σ_A appears above A_f. Both σ_M and σ_A increase in proportion to temperature [1-4]. If steel is used as a bias element in combination with SMA element in the temperature region above and below A_f, the two-way shape memory effect (TWSME) can be achieved by heating and cooling.

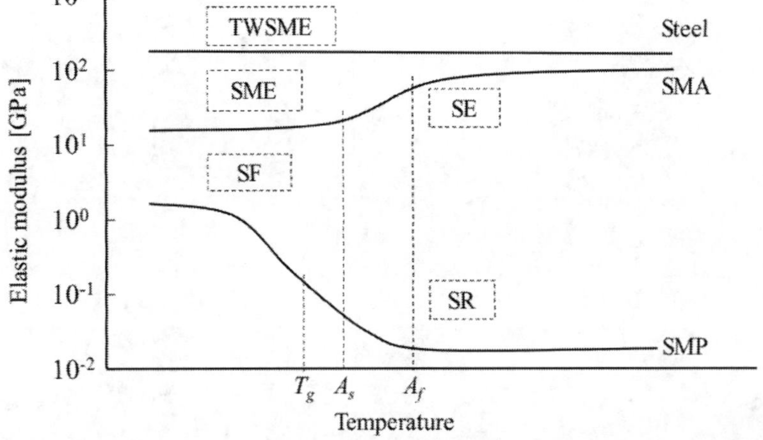

Figure 10. Dependence of elastic modulus on temperature for SMA, SMP and steel.

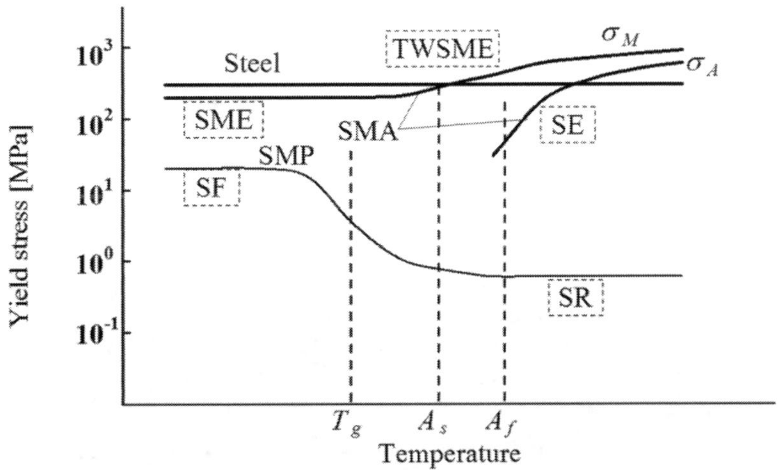

Figure 11. Dependence of yield stress on temperature for SMA, SMP and steel.

On the other hand, elastic modulus and yield stress are large at temperatures below T_g and small above T_g in SMPs. The dependence of rigidity and strength on temperature is therefore quite opposite between SMA and SMP elements. If the composite materials with SMA and SMP having different properties are developed, the new actuation properties of the material can be developed.

3.2 Three-way shape memory composite actuation

The SMC belt was fabricated by using two kinds of shape memory alloy tapes and three SMP tapes [15]. The glass transition temperature of the SMP tape was between the phase transformation temperatures of the SMA and SEA tapes. The SMA and SEA tapes were arranged facing in the opposite directions for the memorized round shape as shown in Figure 12. The SMP tape passed through the SMA and SEA tapes were sandwiched between two SMP tapes from upper and lower sides. The laminated material was set in the mold for heat-treating of the SMC belt.

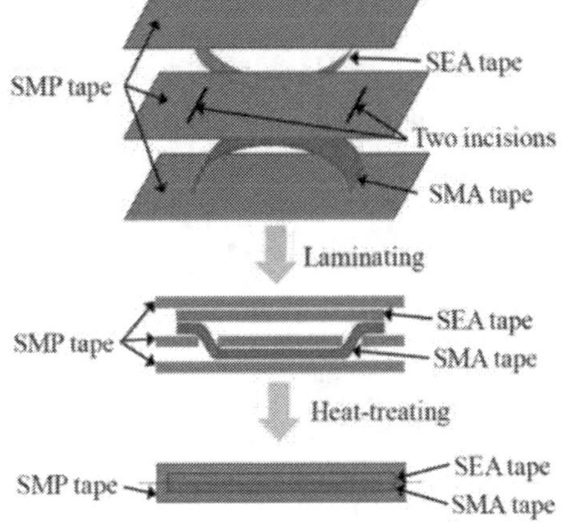

Figure 12. Arrangement of SMA, SEA and SMP tapes for laminating and heat-treating of SMC belt.

The principle of the three-way (reciprocating) bending actuation in the SMC belt during heating and cooling is shown in Figure 13. The SMC belt bends to convex downwards (in the direction of the memorized round shape of the SEA tape) by the recovery force of the SEA tape during heating (1) – (2). It bends to convex upwards (in the direction of the memorized round shape of the SMA tape) by the higher recovery force of the SMA tape at higher temperature (2) – (3). It regains its original shape during cooling (3) – (4).

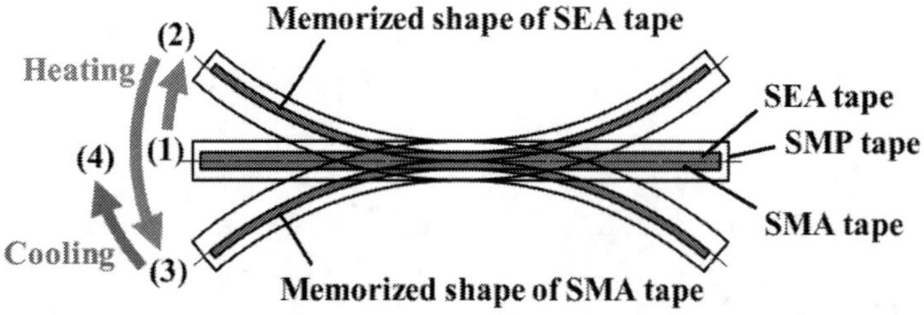

Figure 13. Principle of three-way bending behavior in SMC belt during heating and cooling.

The photographs of the bending motion of the fabricated SMC belt during heating and cooling are shown in Figure 14. As can be seen, the three-way (reciprocating) bending actuation can be obtained by a simple SMC structure.

In the three-point holding cyclic test with heating and cooling for the recovery force, the fabricated SMC belt was put in the supports of the three-point holding device in the chamber of the shape-memory characteristics testing machine. The span was 35 mm. In setting the SMC belt, the center of the SMC belt was put in the punch and the position was held constant. Keeping the positions of the three points (the center and two supports) constant, the SMC belt was heated and cooled. The average heating and cooling rates were ± 0.02 K/s.

The relationship between the recovery force at the center of the SMC belt and temperature obtained by the three-point holding cyclic test is shown in Figure 15. The symbols $A_{f,SEA}$, $A_{f,SMA}$ and T_g shown in Figure 15 represent the reverse-transformation finish temperatures of the SEA and the SMA tape and the glass transition temperature of the SMP tape, respectively. The behavior of the recovery force is different between the heating and cooling process, and the curve describes a large hysteresis loop. In the heating process at the number of cycles $N = 1$, since temperatures around the A_f of the SEA tape are lower than T_g of the SMP tape, the elastic modulus of the SMP is high and therefore the rigidity of the SMP matrix is high. The recovery force of the SMA tape does not appear at this temperature, and therefore the recovery force of the SMC is small. Although the elastic modulus of the SMP decreases at temperatures around T_g of the SMP, the position of the holding part of the punch is kept constant, and therefore the recovery force of the SMC belt decreases. The recovery force starts to increase during keeping temperature at 365 K above A_f. Since the recovery force of the SMA tape becomes stronger than that of the SEA tape at temperatures above $A_{f,SMA}$, the recovery force of the SMC belt increases. During keeping temperature above A_f, the recovery force of the SMA tape becomes stronger than that of the SEA tape, and therefore the recovery force of the SMC belt increases.

In the cooling process, at temperatures around A_f of the SMA tape, since the recovery force of the SMA tape is higher than that of the SEA tape, the recovery force of the SMC belt is almost constant. At temperatures around T_g of the SMP, since the recovery force of the SMA tape is higher than that of the SEA tape, the recovery force of the SMC belt is still high. At temperatures below A_s of the SMA tape, the recovery force of the SMC belt is still high.

With respect to the recovery force under cyclic loading after $N = 2$, negative force does not appear and recovery force is higher than that in $N = 1$ at the same temperature. At temperatures between rhombohedral transformation starting temperature $M_s'_{SMA}$ and finishing temperature $M_f'_{SMA}$, the recovery force of the SMA tape decreases rapidly and the recovery force of the SMC belt decreases correspondingly. It is found that the behavior of we found that the recovery force is almost constant during by cyclic heating and cooling after $N = 2$.

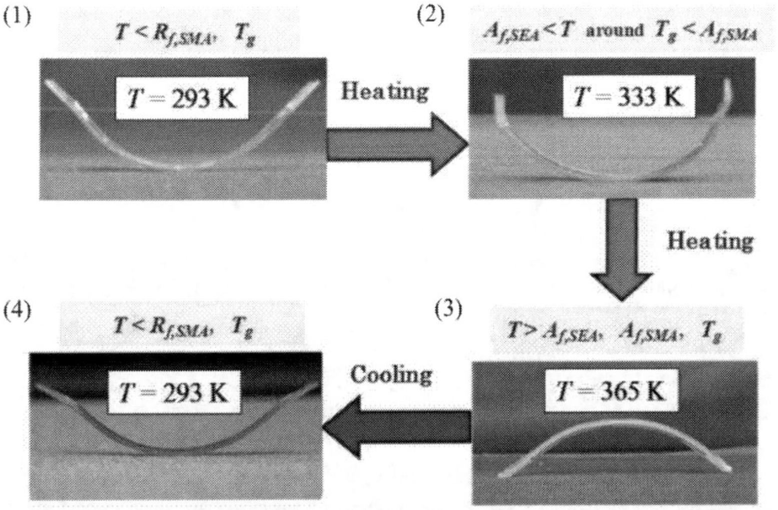

Figure 14. Photographs of three-way bending motion of SMC belt during heating and cooling.

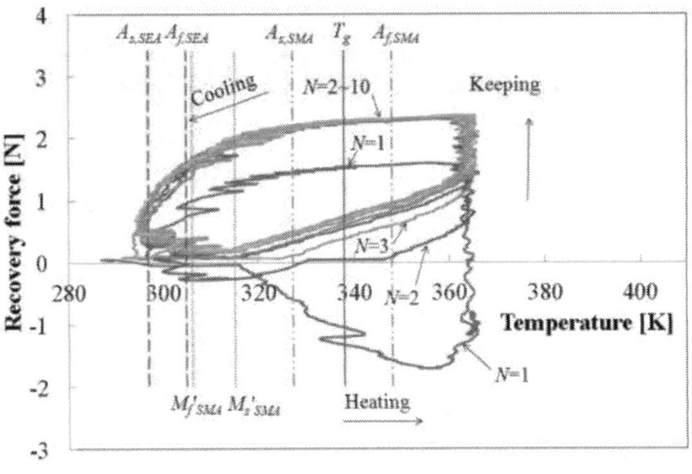

Figure 15. Relationship between recovery force and temperature during heating and cooling obtained by the three-point holding cyclic test for the SMC belt.

4. SUBJECTS FOR DEVELOPMENT OF SHAPE MEMORY ACTUATORS

In order to develop the proposed shape memory actuators, the main subjects to be solved are as follows.

(1) Cyclic deformation and fatigue properties

Actuators work repeatedly in practical applications. The constant deformation properties are requested under cycling. The conditions to use the stable deformation properties are therefore necessary to be clarified. With respect to actuators used for a long time, the fatigue property is very important in all actuators. The SMA and SMP with high-fatigue strength are necessary to be developed.

(2) Interface properties

In SMC actuators, there exist interfaces among SMA and SMP with different properties. In general, the micro-structure of the interface between two materials is different from each element and the stress concentration therefore occurs under deformation, resulting in reduction of strength. It is important to develop the fabrication method to increase strength of the interface between various elements in each actuator.

(3) Multi-axial deformation properties

The deformation properties of shape memory materials are complex since the functional properties appear due to the phase transformation and they depend on the thermomechanical hysteresis. In particular, the deformation properties under multi-axial stress are very complex [16, 17]. Although they are complex, we can use them to novel actuators. One possible application is as follows. By using one SMA shaft, both axial and rotational motions can be achieved simultaneously during heating and cooling. That is, we can develop actuators with very simple structure. It is necessary to clarify the deformation properties of SMA and SMP under multi-axial stress for the development of these actuators.

5. CONCLUSIONS

If the shape memory alloy and shape memory polymer are applied into actuators, the novel intelligent shape memory actuators can be developed. In the present paper, the development of shape memory alloy actuators and a shape memory composite actuator was discussed. The main points confirmed are as follows.

(1) With respect to the shape memory alloy actuators, based on the two-way shape memory effect, the temperature-dependent actuations can be obtained by using various shape memory alloys with different phase transformation temperatures.

(2) With respect to the shape memory composite actuators, by the combination of various kinds of shape-memory alloy and polymer, the three-way actuation can be obtained.

(3) The subjects for development of the shape memory actuators are to develop shape memory alloy and polymer with high-fatigue strength and to find the fabrication method to increase strength of the interface among shape memory alloy, polymer and other materials.

ACKNOWLEDGMENTS

The experimental work for this study was carried out with the assistance of students in Aichi Institute of Technology, to whom the authors wish to express their gratitude. The authors also wish to extend thanks to the administration of Scientific Research (C) in Grant-in-Aid for Scientific Research by the Japan Society for Promotion of Science for financial support.

REFERENCES

1. Funakubo, H. ed., "Shape Memory Alloys", *Gordon and Breach Science Pub.*, 1987, pp. 1-227.
2. Duerig, T.W., Melton, K.N., Stockel, D. and Wayman, C.M., "Engineering Aspects of Shape Memory Alloys", *Butterworth-Heinemann*, 1990, pp. 3-413.
3. Otsuka, K. and Wayman, C.M. eds. "Shape Memory Materials", *Cambridge University Press*, 1998, pp. 1-266.
4. Tobushi, H., Matsui, R., Takeda, K. and Pieczyska E.A., "Mechanical Properties of Shape Memory Materials", *Nova Science Pub.*, 2013, pp. 3-234.
5. Hayashi, S., "Properties and Applications of Polyurethane Series Shape Memory Polymer", *Inter. Progress in Urethanes*, Vol. 6, 1993, pp. 90-115.

6. Huang, W.M., Yang, B. and Fu, Y.Q., "Polyurethane Shape Memory Polymers", *CRC Press*, 2012, pp. 1-50.
7. Wei, H., Liu, L., Zhang, Z., Du, H., Liu, Y. and Leng, J., "Design and analysis of smart release devices based on shape memory polymer composites", *Composite Structure*, Vol. 133, 2015, pp. 642-651.
8. Murasawa, G., Tohgo, K. and Ishi, H., "Deformation Behavior of NiTi/Polymer Shape Memory Alloy Composites – Experimental Verification", *J. Composite Mater.* Vo. 38, 2004, pp. 399-416.
9. Murasawa, G., Tohgo, K. and Ishi, H., "The Effect of Fiber Volume Fraction and Aspect Ratio on the Creation of Internal Stress in the Matrix and Deformation for Short-fiber Shape Memory Alloy Composite", *Smart Mater. Struct.*, Vol. 15, 2006, pp. 33-40.
10. Benafan, O., Brown, J., Calkins, F.T., Kumar, P., Stebner, A.P., Turner, T.L., Vaidyanathan, R., Webster, J. and Young, M.L., *Int. J. Mech. Mater. Des.*, Vol. 10, 2014, pp. 1-42.
11. Sakuma, T. and Iwata, U., "Working Characteristics of a Reciprocating-Type Heat Engine Using Shape Memory Alloys", *JSME Inter. J., Ser. B*, Vol. 41, 1998, pp. 344-350.
12. Tobushi, H., Date, K. and Miyamoto, K., "Characteristics and Development of Shape-Memory Alloy Heat Engine", *J. Solid Mech. Mater. Eng.*, Vol. 4, 2010, pp. 1094-1102.
13. Jeong, J.W., Yoo, Y.I., Shin, D.K., Lim, J.H., Kim, K.W. and Lee, J.J., "A novel tape spring hinge mechanism for quasi-static deployment of a satellite deployable using shape memory alloy", *Review of Scientific Instruments*, Vol. 85, 2014, 025001-1-10.
14. Fragnito, M. and Vetrella, S., "A space release/deployment system actuated by shape memory wires", *Acta Astronautica*, Vol. 51, 2002, pp. 761-770.
15. Tobushi, H., Hayashi, S., Pieczyska, E.A., Date, K. and Nishimura, Y., "Three-Way Actuation of Shape Memory Composite", *Arch. Mech.*, Vol. 63, 2011, pp. 443-457.
16. Tokuda, M., Ye, M., Takakura, M. and Sittner, S., "Calculation of Mechanical Behaviors of Shape Memory Alloy under Multi-axial Loading Conditions", *Int. J. Mech. Sci.*, Vol. 40, 1998, pp. 227-235.
17. Nishimura, F., Watanabe, N. and Tanaka, K., "Evolution of Martensite Start Condition in General Thermomechanical Loads of Fe-based Shape Memory Alloy", *Int. J. Mech. Sci.*, 2000, Vol. 42, pp. 347-365.

ICAST 2015
Optimum System Design Lab
School of System Informatics
Kobe University
1-1, Rokkodai, Nada
Kobe 657-8501
Japan

ISBN 978-1-5108-2189-7

CURRAN ASSOCIATES INC.
proceedings.com